A
B
C
D
E

The Mathematical Theory of Elasticity

Second Edition

The Mathematical Theory of Elasticity

Second Edition

Richard B. Hetnarski
Józef Ignaczak

CRC Press
Taylor & Francis Group
Boca Raton London New York

CRC Press is an imprint of the
Taylor & Francis Group, an **informa** business

CRC Press
Taylor & Francis Group
6000 Broken Sound Parkway NW, Suite 300
Boca Raton, FL 33487-2742

First issued in paperback 2018

ISBN-13: 978-1-4398-2888-5 (hbk)
ISBN-13: 978-1-138-37435-5 (pbk)

Library of Congress Cataloging-in-Publication Data

Hetnarski, Richard B.
The mathematical theory of elasticity / Richard B. Hetnarski and Józef Ignaczak. -- 2nd ed.
p. cm.
Includes bibliographical references and index.
ISBN 978-1-4398-2888-5 (alk. paper)
1. Elasticity. I. Ignaczak, Józef. II. Title.

QA931.H57 2010
531'.382--dc22 2010013692

Visit the Taylor & Francis Web site at
http://www.taylorandfrancis.com

and the CRC Press Web site at
http://www.crcpress.com

Contents

Preface

The second edition of *The Mathematical Theory of Elasticity* is an updated, improved, and revised version of the original edition of the book published in 2004. The purpose of the book has remained the same: to present mathematical theory of elasticity and its applications in a form suitable for a wide range of readers, including graduate students, those preparing PhD theses, and those conducting research in continuum mechanics. Therefore, the book is not only a graduate textbook, but also serves as a complementary text to existing books on elasticity in that it provides classical results on elasticity as well as results obtained in recent years by various researchers, including the authors and their collaborators. Also, the book provides a bridge between the mathematical theory of elasticity and applied elasticity through specific applications given in examples and problems. It covers in one volume the areas of elastostatics, thermoelastostatics, elastodynamics, and thermoelastodynamics. Special emphasis is placed on the problems of elastodynamics and thermoelastodynamics, as most books on elasticity deal mainly with elastostatics and thermoelastostatics.

The book consists of 13 chapters, and is divided into two parts. Part I, Historical Note, Theory, Examples, Problems, comprises Chapters 1 through 8. Part II, Applications and Problems, comprises Chapters 9 through 13.

In Part I, Chapter 1 tells about some of the creators of the theory of elasticity; Chapter 2 provides the mathematical preliminaries; and Chapters 3 through 8, constituting the main portion of Part I, cover the fundamentals of linear elasticity and applications. Chapters 2 through 8 contain worked examples that illustrate the theory involved, and at the end of each chapter, except Chapter 1, a number of problems are included.

While making a selection of the material, it was the authors' intention to provide the reader with both typical and new results of classical type, and to outline new areas of research. Therefore, Chapters 3 through 8 cover kinematics, motion and equilibrium, constitutive relations, formulation of problems, and variational principles. New topics, such as the convolutional variational principles of elastodynamics due to M. E. Gurtin and the pure stress formulations of classical elastodynamics that have emerged recently are also discussed in detail. A new three-dimensional compatibility-related variational principle of elastostatics corresponding to a two-dimensional result due to L. S. Leibenson (*Theory of Elasticity*, 2nd edn., Gostekhizdat, Moscow, 1947) is formulated in Chapter 5, while a number of unsolved PhD-level problems on incompatible elastodynamics are suggested at the end of Chapter 6. The problems at the end of Chapter 7 on the complete solutions of elasticity include a Galerkin-type tensor solution as well as a Lamé-type tensor solution, both related to the pure stress treatment of elastodynamics.

Part II, comprising Chapters 9 through 13, deals with applications only. Chapters 9 and 10 treat, respectively, the solutions to particular three- and two-dimensional problems of elastostatics, and include detailed derivations of classical solutions, such as Boussinesq's and Cerruti's solutions of three-dimensional elastostatics for a semispace subject to a concentrated boundary force, the solutions to two-dimensional counterparts of Boussinesq's and Cerruti's problems, and the solutions to stationary three- and two-dimensional thermoelastic problems for a semispace involving thermal singularities and inclusions (consult the book by Witold Nowacki, *Thermoelasticity*, Addison-Wesley, Reading, MA, 1962). Kirsch's problem of an infinite sheet with a circular hole subject to uniform tension at infinity is revisited in Chapter 10; what is new is the analysis of the associated displacements that is ignored in most of the books on elasticity. Also, the concept of a displacement concentration factor, as opposed to a well-known stress concentration factor, is presented. Chapters 11 and 12 deal, respectively, with particular solutions of three- and two-dimensional problems of elastodynamics and thermoelastodynamics. Apart from the classical results, such as singular solutions of three- and two-dimensional elastodynamics and thermoelastodynamics for an infinite body, these chapters also include (i) a tensorial classification of elastic waves, (ii) solutions describing stress waves due to the initial stress and stress-rate fields in an infinite body, (iii) the dynamical thermal stresses produced by an instantaneous spherical temperature inclusion in an infinite body, and (iv) Saint-Venant's principle of elastodynamics in terms of stresses only.

The authors obtained results (ii)–(iv) some years ago and these results were published for the first time in the first edition of the book, and have been retained in the second edition. The same applies to the closed-form solution that describes dynamical thermal stresses produced by an instantaneous source of heat in an infinite elastic sheet, which is published in Chapter 12. Finally, Chapter 13 covers a number of closed-form solutions to the one-dimensional initial–boundary value problems of isothermal and non-isothermal elastodynamics, including Green's functions for the infinite and semi-infinite solids, as well as the solution that describes the response of a semispace to short laser pulses.

Throughout the book, a direct (i.e., vectorial and tensorial) notation as well as Cartesian coordinates have been used. The associated terminology and general scheme of the notation follow that of M. E. Gurtin (*The Linear Theory of Elasticity*, Encyclopedia of Physics, chief editor: S. Flügge, vol. VIa/2, editor: C. Truesdell, Springer, Berlin, Germany, 1972). However, the authors have taken care to make the material easily readable to the beginners and to those who have already gained an insight into the subject. The theory is developed in the style of Gurtin's treatise, but in addition to the presentation of basic concepts and theorems, Chapters 2 through 8 include 155 examples that illustrate the theory and make the book more comprehensible. The book contains also 118 end-of-chapter problems for which an extensive (over 300 pages long) Solutions Manual is available from the publisher.

Specific applications taken up in the book are developed by using the integral representations of an external thermomechanical load to solve typical boundary value problems of elastostatics and thermoelastostatics, and by taking advantage of the Laplace transform with respect to time when solving the initial–boundary value problems of elastodynamics and thermoelastodynamics.

The authors are greatly indebted to Jonathan Plant (senior editor—Mechanical, Aerospace, Nuclear & Energy Engineering, Taylor & Francis/CRC Press) for his initiative, leading to the publication of the second edition of the book. The authors took this opportunity to make numerous improvements, and the text has been complemented by the inclusion of some recent research results. Also, a three-part table of Laplace transforms has been added, a new section and additional examples have been incorporated, and an appendix with the exposition of recent developments in thermoelasticity is now a part of the book. Moreover, the authors made an effort to eliminate all mistakes and imperfections found in the first edition.

Richard B. Hetnarski
Józef Ignaczak

Authors

Dr. Richard B. Hetnarski, PE, is professor emeritus in the Department of Mechanical Engineering at Rochester Institute of Technology, Rochester, New York. He received his MS degree in mechanical engineering from Gdańsk University of Technology, Gdańsk, Poland, in 1952; his MS degree in mathematics from the University of Warsaw, Warsaw, Poland, in 1960; and his doctor of technical sciences degree from the Institute of Fundamental Technological Research of the Polish Academy of Sciences in Warsaw in 1964. In the period 1964–1965, he held postdoctorate fellowships at Columbia University in New York and at Northwestern University in Evanston, Illinois. In Poland, he worked in the industry, including at the Institute of Aviation in Warsaw (1955–1959) and at the Institute of Fundamental Technological Research of the Polish Academy of Sciences (1959–1969).

In 1969, Dr. Hetnarski and his family emigrated from Poland to the United States, where he became visiting associate professor in the Department of Theoretical and Applied Mechanics at Cornell University in Ithaca, New York, for the period 1969–1970. From 1970 to 1998, he held positions in the Department of Mechanical Engineering at Rochester Institute of Technology, first as New York State Science and Technology Foundation Distinguished Visiting Professor (1970–1971), then as professor (1971–1992), and finally as James E. Gleason Professor (1992–1998).

In 1979, Dr. Hetnarski held a summer faculty fellowship at the NASA Lewis Research Center in Cleveland, Ohio. Subsequently, he become a NASA employee, and as an aeronautical engineer worked at NASA until the end of January 1980. Later, he spent five months as visiting professor at the University of Paderborn in Germany. He also taught a UNESCO-sponsored course on thermoelasticity at the International Centre for Mechanical Sciences in Udine, Italy.

Dr. Hetnarski is the founder and president of the International Congresses on Thermal Stresses (ICTS). These Congresses are held every two years, consecutively on three continents.

Dr. Hetnarski has been active in research, publishing, translating, and editing. He is the author of many papers on mechanics and mathematics. In 1978, he founded the *Journal of Thermal Stresses*, and has been its editor in chief ever since. He served for 13 years (1988–2001) as an associate editor of *Applied Mechanics Reviews*. He is also the author of three books, including two textbooks: *Thermal Stresses* (by N. Noda, R. B. Hetnarski, and Y. Tanigawa), 2nd edn., Taylor & Francis, Boca Raton, FL, 2003; and *Thermal Stresses—Advanced Theory and Applications* (by R. B. Hetnarski and M. R. Eslami), Springer, Berlin, Germany, 2009. He has translated three books from Russian, from German, and from English into Polish, and one (with M. Piterman) from Russian into English. He has also edited seven books on mechanics, including the five-volume *Thermal Stresses* handbook, which was published by Elsevier in Amsterdam and by Lastran Corp. in Rochester during the period 1986–1999.

During his tenure at Rochester Institute of Technology, Dr. Hetnarski served as a consultant to the U.S. Air Force at Wright-Patterson Air Force Base in Dayton, Ohio, and to a number of industrial companies in the Rochester area. He also has served as an expert witness in numerous legal product liability cases.

Professor Józef Ignaczak received his MS degree in mathematics from Warsaw University, Warsaw, Poland, in 1957; his DSc degree in mechanics from the Polish Academy of Sciences, Warsaw, Poland, in 1960; and a habilitated docent degree in mechanics from the Polish Academy of Sciences in 1963. From 1957 to 1964 he worked as an assistant, senior assistant, and adjunct at the Polish Academy of Sciences. In 1964, he served as an assistant professor, an associate professor in 1972, and since 1983 as a professor at the Polish Academy of Sciences.

In addition to the professional positions that he held in Poland, Professor Ignaczak was a research associate at Brown University in Providence, Rhode Island, from 1961 to 1962; a senior lecturer at Monash University in Victoria, Australia, from 1965 to 1969; a reader at Ibadan University in Nigeria, Africa, from 1974 to 1975; a visiting research professor at Naples University in Naples, Italy, in 1985 and 1986; a visiting research fellow at Flinders University of South Australia, Adelaide, South Australia, in 1990; and a visiting research professor at the Rochester Institute of Technology in Rochester, New York, in 1992, 1993, 1994–1995, and 1998.

Before retiring from the Polish Academy of Sciences in 2005, he was a member of the American Mathematical Society, the Planetary Society, the Acoustical Society of America, and the New York Academy of Sciences. He also served as a consulting editor of the *Contemporary Who's Who* published by the American Biographical Institute (ABI).

Professor Ignaczak has been a member of the editorial board of the *Journal of Thermal Stresses* since 1978, and has played an active role in supervising the progress of eight postgraduate and PhD students. The major part of his scientific effort has been directed to the development of linear elastodynamics and dynamic coupled classical and nonclassical thermoelasticity. He is the author of 3 books and over 100 research papers published between 1957 and 2010. Among his most important papers are "A completeness problem for stress equations of motion in the linear elasticity" (without which pure stress variational principles of elastodynamics would not have been formulated), "Rayleigh waves in a nonhomogeneous isotropic elastic semi-space," "Domain of influence theorem in thermoelasticity with one relaxation time," "A tensorial classification of elastic waves," and "Soliton-like waves in a low-temperature nonlinear rigid heat conductor" (coauthor Richard B. Hetnarski). His 850 page book, *Mathematical Theory of Elasticity* (coauthor Richard B. Hetnarski), published by Taylor & Francis, New York, in 2004, also deserves to be mentioned.

In honor of his outstanding work and achievements in his field, Professor Ignaczak was awarded the Golden Cross of Merit and the Polonia Restituta Cross (KKOOP) from the Polish State Council in 1974 and 1983, respectively. He was also awarded the Twentieth Century Achievement Award, the Presidential Seal of Honor, and the Platinum Record for Exceptional Performance from the ABI in 1996; the Honorary Achievement Award from the United States Symposium TS1997 on thermal stresses in 1997; the 2000 Millennium

Medal of Honor from ABI in 1998; and a 50 year Anniversary Medal from the Institute of Fundamental Technological Research of the Polish Academy of Sciences in 2002.

Professor Ignaczak has also served as a member of the International Committee for Thermal Stress Symposia in Hamamatsu (1995), Rochester (1997), Osaka (2001), Blacksburg (2003), Vienna (2005), and Urbana-Champaign (2009).

Notations

Throughout the book, direct notation as well as Cartesian coordinates are used. The terminology and general scheme of the notation follow that of M. E. Gurtin (*The Linear Theory of Elasticity*, Encyclopedia of Physics, chief editor: S. Flügge, vol. VIa/2, editor: C. Truesdell, Springer, Berlin, Germany, 1972). In particular, scalars appear as italic light face letters, vectors are written as lower case letters in boldface, second-order tensors as upper case letters in boldface, and fourth-order tensors as upper case sans-serif letters in boldface. Also, within a section the same letters may be used for quantities other than those listed below.

LIST OF SYMBOLS

Symbol	Name
$\mathbf{A}$	Second-order tensor, Beltrami solution, thermal expansion tensor
$\mathbf{A}(\mathbf{m})$	Acoustic tensor for a direction **m**
B	Body
$\mathbf{B}$	Second-order tensor
$\boldsymbol{\mathsf{C}}$	Elasticity tensor
D	Torsional rigidity, flexural rigidity
$\mathbf{D}$	Finite strain tensor
E	Young's modulus
E^3	Three-dimensional Euclidean space
E^2	Two-dimensional Euclidean space
$\mathbf{E}$	Infinitesimal strain tensor
$\mathbf{E}^{\perp}$	Normal part of **E** with respect to a plane
$\mathbf{E}^{\parallel}$	Tangential part of **E** with respect to a plane
$\mathcal{E}_E(\mathbf{E})$	Strain energy density of a progressive wave
$\mathcal{E}_E(\mathbf{E}^{\perp})$	Normal strain energy density of a progressive wave
$\mathcal{E}_S(\mathbf{S})$	Stress energy density of a progressive wave
$\mathcal{E}_S(\mathbf{S}^{\perp})$	Normal stress energy density of a progressive wave
$\mathcal{E}_S(\mathbf{S}^{\parallel})$	Tangential stress energy density of a progressive wave
F	Airy stress function, force
$\mathbf{F}$	Deformation gradient
$F\{\cdot\}$	Functional
G	Shear modulus, Green's function
$H(\cdot)$	Heaviside function
$\mathbf{H}$	Harmonic second-order tensor field, compatibility related fourth-order tensor
I	Moment of inertia of a cross section
J	Polar moment of inertia of a cross section
K	Stress concentration factor
K_r	Displacement concentration factor
$K(t)$	Kinetic energy
$\boldsymbol{\mathsf{K}}$	Compliance tensor
L	Length
$L\{f(t)\}$	Laplace transform of $f(t)$

(continued)

Symbol	Name
J	Polar moment of inertia of a cross section
K	Stress concentration factor
K_r	Displacement concentration factor
$K(t)$	Kinetic energy
$\mathbf{K}$	Compliance tensor
L	Length
$L\{f(t)\}$	Laplace transform of $f(t)$
$L^{-1}\{\overline{f}(p)\}$	Inverse Laplace transform of $\overline{f}(p)$
M	Bending moment
$\mathbf{M}$	Stress–temperature tensor
$\mathcal{M}_{\mathbf{x},ct}(f)$	Mean value of a function f over the surface of ball with its center at $\mathbf{x}$ and of radius ct
$\mathcal{N}_{\mathbf{x},ct}(f)$	Two-dimensional counterpart of $\mathcal{M}_{\mathbf{x},ct}(f)$
$\mathbf{0}$	Origin, zero vector, zero tensor
P	Part of B, concentrated force
$P(t)$	Stress power
Q	Heat supply field, shear force
$\mathbf{Q}$	Orthogonal tensor
R	Region in E^3, distance between two points
$\mathbf{S}$	Stress tensor
$\widehat{\mathbf{S}}(B)$	Mean stress
$\mathbf{S}^{\perp}$	Normal part of $\mathbf{S}$ with respect to a plane
$\mathbf{S}^{\parallel}$	Tangential part of $\mathbf{S}$ with respect to a plane
T	Temperature change, time interval
$U_C\{\mathbf{E}\}$	Strain energy
$\mathcal{U}(t)$	Total energy of B at time t
$U(l,t)$	Stress energy of a semi-infinite cylinder $B(l)$ stored over time interval $[0,t]$
$\mathcal{V}$	Vector space associated with E^3
$\mathbf{W}$	Rotation tensor
$W(\mathbf{E})$	Stored energy function
$W'(\mathbf{S})$	Complementary strain energy
$\widehat{W}(\mathbf{S})$	Stress energy density
$W_t\{\cdot\}$	Functional involving convolutions
$\mathbf{a}$	Direction of motion of a progressive wave
$\mathbf{b}$	Body force
c	Velocity of propagation, specific heat
c_1	Irrotational velocity
c_2	Isochoric velocity
$c(\mathbf{S}^0)$	Velocity of a stress progressive wave
$\mathbf{e}$	Unit vector along the axis of symmetry of a transversely isotropic body
$\mathbf{e}_i$	Orthonormal basis
$\mathbf{f}$	Pseudo body force field
$\mathbf{g}$	Galerkin vector field
$\mathbf{g}(P)$	Linear momentum of P
$\mathbf{h}(P)$	Angular momentum of P
i	$\sqrt{-1}$, function with the values $i(t)=t$
k	Bulk modulus, polar radius of gyration of a cross section, spring stiffness, heat conductivity coefficient
$\mathbf{k}$	Unit vector along x_3 axis
$\boldsymbol{\ell}$	Concentrated force
$\mathbf{m}$	Direction of propagation
m	The constant $\frac{1+\nu}{1-\nu}\alpha$, mass of P
$\mathbf{n}$	Outward unit normal on ∂B

Symbol	Name
p	Pressure, admissible process, elastic process, thermoelastic process
$\mathbf{s}$	Surface traction
$\widehat{\mathbf{s}}$	Prescribed surface traction
s	Admissible state, elastic state, thermoelastic state, solution
t	Time
$\mathbf{u}$	Displacement
$\widehat{\mathbf{u}}$	Prescribed displacement on boundary
$\mathbf{u}_0$	Initial displacement
$\dot{\mathbf{u}}_0$	Initial velocity
$v(B)$	Volume of B
$\mathbf{w}$	Rigid displacement
$\mathbf{x}$, $\mathbf{y}$	Points in space
x_i	Cartesian components of $\mathbf{x}$
α	Coefficient of thermal expansion, angle of twist, scalar field in a tensorial solution of elastodynamics
$\boldsymbol{\beta}$	Vector field in a tensorial solution of elastodynamics
γ	The constant $(3\lambda + 2\mu)\alpha$
$\delta(\cdot)$	Dirac delta function
δ_{ij}	Kronecker's delta
$\delta v(B)$	Volume change
ϵ_{ijk}	Three-dimensional alternator
$\epsilon_{\alpha\beta}$	Two-dimensional alternator
θ	Absolute temperature
θ_0	Reference temperature
κ	Thermal diffusivity
λ	Lamé modulus, wavelength
μ	Shear modulus
ν	Poisson's ratio
ρ	Density
σ	Normal component of a stress vector
τ	Dimensionless time, shear stress
φ	Scalar field in Boussinesq–Papkovitch–Neuber solution, scalar field in Green–Lamé solution
ϕ	Thermoelastic displacement potential, Prandtl's stress function, biharmonic function
χ	Biharmonic scalar field in Love's solution
$\boldsymbol{\chi}$	Second-order tensor field of Galerkin type in elastodynamics
ψ	Warping function
$\boldsymbol{\psi}$	Vector field in Boussinesq–Papkovitch–Neuber solution, vector field in Green–Lamé solution
$\boldsymbol{\omega}$	Rotation vector, vector field in a tensorial solution of elastodynamics
$\mathbf{1}$	Unit tensor
sym	Symmetric part of a tensor
skw	Skew part of a tensor
tr	Trace of a tensor
$\otimes$	Tensor product of two vectors
∇	Gradient
$\widehat{\nabla}$	Symmetric gradient
curl	Curl
div	Divergence
$\Delta = \nabla^2$	Laplacian
$\Box_1^2$, $\Box_2^2$	Wave operators
$*$	Convolution

(continued)

Symbol	Name
$[\![\cdot]\!]$	Jump in a function
da	Element of area
dv	Element of volume
$(\dot{})$	Time derivative
$(\,)^T$	Transpose of a tensor
$\emptyset$	Empty set

Note: In this book, the meaning of the expressions half-space, semi-space, and semi-infinite body is the same. These expressions are interchangeable.

Some Quantities in SI Units

Name	Symbol	Dimensions
Angular momentum of P	$\mathbf{h}\ (P)$	$(\mathrm{kg}\cdot \mathrm{m}^2)/\mathrm{s}$
Area moment of inertia	I	m^4
Body force vector per unit of volume, in components	b_i	$\mathrm{N/m}^3$
Bulk modulus	k	$\mathrm{Pa} = \mathrm{N/m}^2$
Coefficient of linear thermal expansion	α	$1/\mathrm{K}$
Density	ρ	$\mathrm{kg/m}^3$
Displacement vector components	u_i	m
Force	$F,\ P$	$\mathrm{N} = (\mathrm{kg}\cdot\mathrm{m})/\mathrm{s}^2$
Heat supply field	Q	K/s
Internal heat generated per unit of volume and unit of time	W	$\mathrm{J}/(\mathrm{m}^3\cdot\mathrm{s}) = \mathrm{W/m}^3$
Lamé constants	λ and μ	$\mathrm{Pa} = \mathrm{N/m}^2$
Linear momentum of P	$\mathbf{g}\ (P)$	$(\mathrm{kg}\cdot\mathrm{m})/\mathrm{s}$
Polar moment of inertia	J	m^4
Pressure	p	$\mathrm{Pa} = \mathrm{N/m}^2$
Shear force	Q	N
Specific heat	c	$\mathrm{J}/(\mathrm{kg}\cdot\mathrm{K})$
Stored energy per unit of volume	W	$\mathrm{J/m}^3$
Stress tensor components	S_{ij}	$\mathrm{Pa} = \mathrm{N/m}^2$
Temperature	$\theta,\ T$	K
Thermal conductivity	k	$\mathrm{W}/(\mathrm{m}\cdot\mathrm{K})$
Thermal diffusivity	κ	m^2/s
Thermoelastic displacement potential	ϕ	m^2
Time	t	s
Velocity of propagation	c	m/s
Young's modulus	E	$\mathrm{Pa} = \mathrm{N/m}^2$
γ	$\gamma = (3\lambda + 2\mu)\alpha$	$\mathrm{N}/(\mathrm{m}^2\cdot\mathrm{K})$

Notes:

Values of the specific heat, c, the thermal conductivity, k, the density, ρ, and the thermal diffusivity, κ, for common materials may be found in M. N. Ozisik, *Heat Conduction*, John Wiley, New York, 1980, pp. 2–7.

Values of the coefficient of linear thermal expansion α and Young's modulus, E, for common materials may be found in N. Noda, R. B. Hetnarski, and Y. Tanigawa, *Thermal Stresses*, 2nd edn., Taylor & Francis, New York, 2003, p. 7.

Relations between Elasticity Constants

	Lamé Constants		Young's Modulus	Poisson's Ratio	Bulk Modulus
	λ	Shear Modulus μ	E	ν	k
λ, μ	λ	μ	$\frac{\mu(3\lambda+2\mu)}{\lambda+\mu}$	$\frac{\lambda}{2(\lambda+\mu)}$	$\frac{3\lambda+2\mu}{3}$
λ, E	λ	$\frac{E-3\lambda}{4}+\frac{\sqrt{E^2+9\lambda^2+2E\lambda}}{4}$	E	$-\frac{E+\lambda}{4\lambda}+\frac{\sqrt{E^2+9\lambda^2+2E\lambda}}{4\lambda}$	$\frac{E+3\lambda}{6}+\frac{\sqrt{E^2+9\lambda^2+2E\lambda}}{6}$
λ, ν	λ	$\frac{\lambda(1-2\nu)}{2\nu}$	$\frac{\lambda(1+\nu)(1-2\nu)}{\nu}$	ν	$\frac{\lambda(1+\nu)}{3\nu}$
λ, k	λ	$\frac{3(k-\lambda)}{2}$	$\frac{9k(k-\lambda)}{3k-\lambda}$	$\frac{\lambda}{3k-\lambda}$	k
μ, E	$\frac{(2\mu-E)\mu}{E-3\mu}$	μ	E	$\frac{E-2\mu}{2\mu}$	$\frac{\mu E}{3(3\mu-E)}$
μ, ν	$\frac{2\mu\nu}{1-2\nu}$	μ	$2\mu(1+\nu)$	ν	$\frac{2\mu(1+\nu)}{3(1-2\nu)}$
μ, k	$\frac{3k-2\mu}{3}$	μ	$\frac{9k\mu}{3k+\mu}$	$\frac{1}{2}\left[\frac{3k-2\mu}{3k+\mu}\right]$	k
E, ν	$\frac{\nu E}{(1+\nu)(1-2\nu)}$	$\frac{E}{2(1+\nu)}$	E	ν	$\frac{E}{3(1-2\nu)}$
E, k	$\frac{3k(3k-E)}{9k-E}$	$\frac{3Ek}{9k-E}$	E	$\frac{1}{2}\left[\frac{3k-E}{3k}\right]$	k
ν, k	$\frac{3k\nu}{1+\nu}$	$\frac{3k(1-2\nu)}{2(1+\nu)}$	$3k(1-2\nu)$	ν	k

Laplace Transforms

The presented collection of Laplace transforms consists of three parts:

Part A: Properties of Laplace Transforms
Part B: Laplace Transforms Used in This Book
Part C: Some Laplace Transforms of Exponential Form

Those in need of inverse Laplace transforms should consult large collections, such as

V. A. Ditkin and A. P. Prudnikov, *Reference Book on Operational Calculus,* Fizmatgis, Moscow, 1965 (in Russian).
A. Erdélyi (Ed.), *Bateman Manuscript Project, Tables of Integral Transforms*, vol. 1, McGraw-Hill, New York, 1954.

Also, one can find tables of substantial length in some other books, for example, in

H. S. Carslaw and J. C. Jaeger, *Operational Methods in Applied Mathematics*, Dover Publications, New York, 1963 (contains 53 Laplace transform formulas). The book was originally published by Oxford University Press in 1941.
G. Doetsch, *Anleitung zum praktischen Gebrauch der Laplace–Transformation*, 2nd edn., R. Oldenbourg, München, 1961 (contains 256 Laplace transform formulas).*
F. Oberhettinger and L. Badii, *Tables of Laplace Transforms*, Springer, New York, 1973.
M. R. Spiegel, *Mathematical Handbook of Formulas and Tables*, Schaum's Outline Series, McGraw-Hill, New York, 1968 (contains 168 Laplace transform formulas).

The formulas in Part A of the table are related to the discussion on Laplace transforms in Section 2.3.4. These are transforms of a class of functions, and not just transforms of specific functions. The list cannot be exhausted. In fact, some entries in Part B of the table are of the same class, *viz.*, formulas 7, 8, and 16, and they could be included in Part A.

The Laplace transforms used in the book are collected in Part B for easy reference. In the left column, after each consecutive number of a formula, the section number where the formula is used for the first time is given.

Note that the quantity R, which appears in formulas 13, 14, 15, 16, and 18, although in the text as a nondimentional variable, may be treated as a nonnegative parameter as far as this table is considered.

Note that formula 14./11.2.1 is equivalent to formula 29./13.2.1, and formula 19./12.1.4 is equivalent to formula 23./12.2.2. We retain the additional formulas for clarity.

Transforms in Part C of the table are useful in thermoelasticity but they are not presented in any general tables of Laplace transforms. A possible reason for this is that derivation of these inverse transforms is a cumbersome and time-consuming process. But a method for easy calculation of these transforms has been found, and it is described in Section 2.3.4.

* The book *Anleitung zum praktischen Gebrauch der Laplace-Transformation* by G. Doetsch was translated from German into Polish by the first author (RBH) and published under the title *Praktyka Przekształcenia Laplace'a*, PWN—Polish Scientific Publishers, Warsaw, 1964, p. 307. While writing this section, the authors consulted the Polish edition of this book.

PART A: PROPERTIES OF LAPLACE TRANSFORMS

No.	$L^{-1}\{\bar{f}(p)\} = f(t)$
1.	$L^{-1}\{a\bar{f}(p) + b\bar{g}(p)\} = aL^{-1}\{\bar{f}(p)\} + bL^{-1}\{\bar{g}(p)\} = af(t) + bg(t)$ (a and b are constants)
2.	$L^{-1}\{e^{-ap}\bar{f}(p)\} = f(t-a)H(t-a), \quad a > 0$
3.	$L^{-1}\{\bar{f}(p+a)\} = e^{-at}f(t), \quad a > 0$
4.	$L^{-1}\left\{\frac{1}{a}\bar{f}\left(\frac{p}{a}\right)\right\} = f(at), \quad a > 0$
5.	$L^{-1}\{\bar{f}(ap)\} = \frac{1}{a}f\left(\frac{t}{a}\right), \quad a > 0$
6.	$L^{-1}\{p\bar{f}(p) - f(0)\} = f'(t)$
7.	$L^{-1}\{\bar{f}'(p)\} = -tf(t)$
8.	$L^{-1}\{p^2\bar{f}(p) - pf(0) - f'(0)\} = f''(t)$
9.	$L^{-1}\{\bar{f}''(p)\} = t^2 f(t)$
10.	$L^{-1}\{p^n\bar{f}(p) - p^{n-1}f(0) - p^{n-2}f'(0) - \cdots - f^{(n-1)}(0)\} = f^{(n)}(t)$
11.	$L^{-1}\{\bar{f}^{(n)}(p)\} = (-1)^n t^n f(t)$
12.	$L^{-1}\left\{\frac{\bar{f}(p)}{p}\right\} = \int_0^t f(\tau)\,d\tau$
13.	$L^{-1}\left\{\int_p^\infty \bar{f}(\tau)\,d\tau\right\} = \frac{f(t)}{t}$
14.	$L^{-1}\{\bar{f}(p)\,\bar{g}(p)\} = \int_0^t f(t-\tau)\,g(\tau)\,d\tau = f(t) * g(t)$ (convolution)

PART B: LAPLACE TRANSFORMS USED IN THIS BOOK

No./Section	$L^{-1}\{\bar{f}(p)\} = f(t)$
1./2.3.4	$L^{-1}\left\{\frac{1}{p}\right\} = 1$
2./2.3.4	$L^{-1}\left\{\frac{1}{p+a}\right\} = e^{-at}$
3./2.3.4	$L^{-1}\left\{\frac{1}{p^2}\right\} = t$
4./6.2.1	$L^{-1}\left\{\frac{1}{p^2 + 1/H^2}\right\} = H \sin\frac{t}{H}$, H is a positive constant
5./6.2.1	$L^{-1}\left\{\frac{p}{p^2 + 1/H^2}\right\} = \cos\frac{t}{H}$, H is a positive constant
6./11.1.2	$L^{-1}\left\{\frac{e^{-\alpha p}}{p^2}\right\} = H(t-\alpha)(t-\alpha),\ \alpha > 0$
7./11.1.4	$L^{-1}\left\{-\frac{1}{c^2}\left(\nabla^2 - \frac{p^2}{c^2}\right)^{-1} g(\mathbf{x})\right\} = \frac{1}{4\pi c^2}\int_\Omega \frac{g(\boldsymbol{\xi})}{\lvert\mathbf{x}-\boldsymbol{\xi}\rvert}\delta\left(t - \frac{\lvert\mathbf{x}-\boldsymbol{\xi}\rvert}{c}\right) dv(\boldsymbol{\xi}),\quad \mathbf{x},\ \xi \in E^3$
8./11.1.4	$L^{-1}\left\{-\frac{1}{c^2}\left(\nabla^2 - \frac{p^2}{c^2}\right)^{-1} g(\mathbf{x})\right\} = \frac{t}{4\pi}\int_0^{2\pi}\int_0^{\pi} g(\mathbf{x}+\mathbf{n}ct)\sin\theta\, d\theta\, d\varphi$ where $\mathbf{n} = [\sin\theta\cos\varphi,\ \sin\theta\sin\varphi,\ \cos\theta]$, $\mathbf{x} \in E^3$
9./11.1.4	$L^{-1}\left\{\left[\left(\nabla^2 - \frac{p^2}{c_1^2}\right)\left(\nabla^2 - \frac{p^2}{c_2^2}\right)\right]^{-1} h(\mathbf{x})\right\} = \frac{c_1^2 c_2^2}{c_1^2 - c_2^2}\int_0^t (t-\tau)\tau\left\{\frac{c_1^2}{4\pi}\int_0^{2\pi}\int_0^{\pi} h(\mathbf{x}+\mathbf{n}c_1\tau)\sin\theta\, d\theta\, d\varphi - \frac{c_2^2}{4\pi}\int_0^{2\pi}\int_0^{\pi} h(\mathbf{x}+\mathbf{n}c_2\tau)\sin\theta\, d\theta\, d\varphi\right\} d\tau$ where $\mathbf{n} = [\sin\theta\cos\varphi,\ \sin\theta\sin\varphi,\ \cos\theta]$, $\mathbf{x} \in E^3$
10./11.1.4	$L^{-1}\left\{\left[\left(\nabla^2 - \frac{p^2}{c_1^2}\right)\left(\nabla^2 - \frac{p^2}{c_2^2}\right)\right]^{-1} p^2 h(\mathbf{x})\right\} = \frac{c_1^2 c_2^2}{c_1^2 - c_2^2} t\left[\frac{c_1^2}{4\pi}\int_0^{2\pi}\int_0^{\pi} h(\mathbf{x}+\mathbf{n}c_1 t)\sin\theta\, d\theta\, d\varphi - \frac{c_2^2}{4\pi}\int_0^{2\pi}\int_0^{\pi} h(\mathbf{x}+\mathbf{n}c_2 t)\sin\theta\, d\theta\, d\varphi\right]$ where $\mathbf{n} = [\sin\theta\cos\varphi,\ \sin\theta\sin\varphi,\ \cos\theta]$, $\mathbf{x} \in E^3$

(continued)

No./Section	$L^{-1}\{\bar{f}(p)\} = f(t)$
11./11.1.5	$L^{-1}\left\{\dfrac{1}{(p+\alpha)^2+\beta^2}\right\} = e^{-\alpha t}\dfrac{\sin\beta t}{\beta}$
12./11.1.5	$L^{-1}\left\{\dfrac{e^{-\frac{p}{c_1}(r-a)}}{p}\right\} = H\left(t-\dfrac{r-a}{c_1}\right)$
13./11.2.1	$L^{-1}\left\{e^{-R\sqrt{p}}\right\} = \dfrac{R}{\sqrt{4\pi t^3}}e^{-\frac{R^2}{4t}}$
14./11.2.1	$L^{-1}\left\{\dfrac{e^{-R\sqrt{p}}}{p-1}\right\} = \dfrac{1}{2}\left[e^{t-R}\operatorname{erfc}\left(\dfrac{R}{2\sqrt{t}}-\sqrt{t}\right)+e^{t+R}\operatorname{erfc}\left(\dfrac{R}{2\sqrt{t}}+\sqrt{t}\right)\right]$
15./11.2.1	$L^{-1}\left\{\dfrac{e^{-R\sqrt{p}}}{p}\right\} = \operatorname{erfc}\left(\dfrac{R}{2\sqrt{t}}\right)$
16./11.2.1	$L^{-1}\left\{\bar{f}(R,p)e^{-Rp}\right\} = f(R,t-R)H(t-R)$
17./11.2.1	$L^{-1}\left\{\dfrac{1}{p-1}\right\} = e^{t}$
18./11.2.1	$L^{-1}\left\{e^{-Rp}\right\} = \delta(t-R)$
19./12.1.4	$L^{-1}\left\{K_0\left(\dfrac{p}{c}\rho\right)\right\} = H\left(t-\dfrac{\rho}{c}\right)\left(t^2-\dfrac{\rho^2}{c^2}\right)^{-1/2}$
20./12.1.4	$L^{-1}\left\{-\dfrac{1}{c^2}\left(\nabla^2-\dfrac{p^2}{c^2}\right)^{-1}f(\mathbf{x})\right\} = t\,\dfrac{1}{2\pi ct}\displaystyle\int_{\Sigma(\mathbf{x},ct)}\dfrac{f(\xi_1,\xi_2)\,d\xi_1\,d\xi_2}{\sqrt{c^2t^2-(x_1-\xi_1)^2-(x_2-\xi_2)^2}}$ where $\Sigma(\mathbf{x},ct) = \{\xi : \lvert\xi-\mathbf{x}\rvert < ct\}$, $\mathbf{x},\ \xi \in E^2$
21./12.2.2	$L^{-1}\left\{\left[\left(\nabla^2-\dfrac{p^2}{c_1^2}\right)\left(\nabla^2-\dfrac{p^2}{c_2^2}\right)\right]^{-1}p^2h(\mathbf{x})\right\} = \dfrac{c_1^2c_2^2}{c_1^2-c_2^2}\left[c_1^2t\mathcal{N}_{\mathbf{x},c_1t}(h)-c_2^2t\mathcal{N}_{\mathbf{x},c_2t}(h)\right]$ where $\mathcal{N}_{\mathbf{x},c_it}(h) = \dfrac{1}{2\pi c_it}\displaystyle\int_{\Sigma(\mathbf{x},c_it)}\dfrac{h(\xi_1,\xi_2)\,d\xi_1\,d\xi_2}{\sqrt{c_i^2t^2-(x_1-\xi_1)^2-(x_2-\xi_2)^2}}$, $(i=1,2)$ and $\Sigma(\mathbf{x},c_it) = \{\xi : \lvert\xi-\mathbf{x}\rvert < c_it\}$, $\mathbf{x},\ \xi \in E^2$
22./12.2.2	$L^{-1}\{K_0(a\sqrt{p})\} = \dfrac{1}{2t}e^{-a^2/4t}, \quad a>0$
23./12.2.2	$L^{-1}\{K_0(bp)\} = H(t-b)(t^2-b^2)^{-1/2}, \quad b>0$
24./12.2.2	$L^{-1}\left\{\dfrac{1}{\sqrt{p}}K_1(a\sqrt{p})\right\} = \dfrac{1}{a}e^{-a^2/4t}, \quad a>0$
25./12.2.2	$L^{-1}\{K_1(bp)\} = \dfrac{t}{b}\dfrac{H(t-b)}{\sqrt{t^2-b^2}}, \quad b>0$
26./12.2.2	$L^{-1}\left\{K_1'(a\sqrt{p})\right\} = -\left(\dfrac{1}{a^2}+\dfrac{1}{2t}\right)e^{-a^2/4t}$
27./12.2.2	$L^{-1}\left\{K_1'(bp)\right\} = -\dfrac{t^2}{b^2}\dfrac{H(t-b)}{\sqrt{t^2-b^2}}$

No./Section	$L^{-1}\{\bar{f}(p)\} = f(t)$
28./13.2.1	$L^{-1}\left\{\dfrac{e^{-\lvert x\rvert p}}{p-1}\right\} = e^{t-\lvert x\rvert}H(t-\lvert x\rvert)$
29./13.2.1	$L^{-1}\left\{\dfrac{e^{-\lvert x\rvert\sqrt{p}}}{p-1}\right\} = U(\lvert x\rvert, t)$
	where $U(\lvert x\rvert, t) = \dfrac{1}{2}e^{t}\left[e^{-\lvert x\rvert}\,\mathrm{erfc}\left(\dfrac{\lvert x\rvert}{2\sqrt{t}} - \sqrt{t}\right) + e^{\lvert x\rvert}\mathrm{erfc}\left(\dfrac{\lvert x\rvert}{2\sqrt{t}} + \sqrt{t}\right)\right]$

PART C: LAPLACE TRANSFORMS OF EXPONENTIAL FORM

Explanation of this table is located in Section 2.3.4. Functions $S(t)$ and $T(t)$ are denoted by S and T, respectively, in the table. The form of $S(t)$ and $T(t)$ is as follows:

$$S(t) = \frac{e^{at}}{2}\left[\exp\left(-c\sqrt{a+b}\right)\operatorname{erfc}\left(\frac{c}{2\sqrt{t}} - \sqrt{(a+b)t}\right) + \exp\left(c\sqrt{a+b}\right)\operatorname{erfc}\left(\frac{c}{2\sqrt{t}} + \sqrt{(a+b)t}\right)\right]$$

$$T(t) = \frac{e^{at}}{2\sqrt{a+b}}\left[\exp\left(-c\sqrt{a+b}\right)\operatorname{erfc}\left(\frac{c}{2\sqrt{t}} - \sqrt{(a+b)t}\right) - \exp\left(c\sqrt{a+b}\right)\operatorname{erfc}\left(\frac{c}{2\sqrt{t}} + \sqrt{(a+b)t}\right)\right]$$

No.	$\bar{f}(p) = L\{f(t)\}$	$f(t) = L^{-1}\{\bar{f}(p)\}$
1.	$\dfrac{\exp(-c\sqrt{p+b})}{p-a}$	S
2.	$\dfrac{\exp(-c\sqrt{p+b})}{(p-a)^2}$	$tS - \dfrac{c}{2}T$
3.	$\dfrac{\exp(-c\sqrt{p+b})}{(p-a)^3}$	$\dfrac{1}{2}\left\{\left[\dfrac{c^2}{4(a+b)} + t^2\right]S + \left[\dfrac{1}{4(a+b)} - t\right]cT - \dfrac{c}{2(a+b)}\sqrt{t/\pi}\exp\left(-\dfrac{c^2}{4t} - bt\right)\right\}$
4.	$\dfrac{\exp(-c\sqrt{p+b})}{(p-a)^4}$	$\dfrac{1}{6}\left\{\left[-\dfrac{3c^2}{8(a+b)^2} + \dfrac{3c^2t}{4(a+b)} + t^3\right]S + \left[-\dfrac{3c}{8(a+b)^2} - \dfrac{c^3}{8(a+b)} + \dfrac{3ct}{4(a+b)} - \dfrac{3ct^2}{2}\right]T + \left[\dfrac{3c}{4(a+b)^2} - \dfrac{ct}{a+b}\right]\sqrt{t/\pi}\exp\left(\dfrac{c^2}{4t} - bt\right)\right\}$
5.	$\dfrac{\sqrt{p+b}\exp(-c\sqrt{p+b})}{p-a}$	$(a+b)T + \dfrac{1}{\sqrt{\pi t}}\exp\left(-\dfrac{c^2}{4t} - bt\right)$
6.	$\dfrac{\sqrt{p+b}\exp(-c\sqrt{p+b})}{(p-a)^2}$	$-\dfrac{c}{2}S + \left[\dfrac{1}{2} + (a+b)t\right]T + \sqrt{t/\pi}\exp\left(-\dfrac{c^2}{4t} - bt\right)$
7.	$\dfrac{\sqrt{p+b}\exp(-c\sqrt{p+b})}{(p-a)^3}$	$\dfrac{1}{2}\left\{-c\left[\dfrac{1}{4(a+b)} + t\right]S + \left[-\dfrac{1}{4(a+b)} + \dfrac{c^2}{4} + t + (a+b)t^2\right]T + \left[\dfrac{1}{2(a+b)} + t\right]\sqrt{t/\pi}\exp\left(-\dfrac{c^2}{4t} - bt\right)\right\}$

No.	$\bar{f}(p) = L\{f(t)\}$	$f(t) = L^{-1}\{\bar{f}(p)\}$
8.	$\dfrac{\sqrt{p+b}\exp(-c\sqrt{p+b})}{(p-a)^4}$	$\begin{aligned} &\frac{1}{6}\left\{\left[\frac{3c}{8(a+b)^2}-\frac{c^3}{8(a+b)}-\frac{3ct}{4(a+b)}-\frac{3}{2}ct^2\right]S \right.\\ &+\left[\frac{3}{8(a+b)^2}-\left(\frac{3}{4(a+b)}-\frac{3}{4}c^2\right)t+\frac{3}{2}t^2+(a+b)t^3\right]T \\ &+\left[-\frac{3}{4(a+b)^2}+\frac{c^2}{4(a+b)}+\frac{t}{a+b}+t^2\right] \\ &\left.\times\sqrt{t/\pi}\exp\left(-\frac{c^2}{4t}-bt\right)\right\}\end{aligned}$
9.	$\dfrac{(p+b)\exp(-c\sqrt{p+b})}{(p-a)}$	$(a+b)S+\dfrac{c}{2t^2}\sqrt{t/\pi}\exp\left(-\dfrac{c^2}{4t}-bt\right)$
10.	$\dfrac{(p+b)\exp(-c\sqrt{p+b})}{(p-a)^2}$	$[1+(a+b)t]S-\dfrac{(a+b)c}{2}T$
11.	$\dfrac{(p+b)\exp(-c\sqrt{p+b})}{(p-a)^3}$	$\begin{aligned} &\frac{1}{2}\left\{\left[\frac{c^2}{4}+2t+(a+b)t^2\right]S-\left[\frac{3}{4}c+(a+b)ct\right]T\right.\\ &\left.-\frac{c}{2}\sqrt{t/\pi}\exp\left(-\frac{c^2}{4t}-bt\right)\right\}\end{aligned}$
12.	$\dfrac{(p+b)\exp(-c\sqrt{p+b})}{(p-a)^4}$	$\begin{aligned} &\frac{1}{6}\left\{\left[\frac{3c^2}{8(a+b)}+\frac{3c^2t}{4}+3t^2+(a+b)t^3\right]S\right.\\ &+\left[\frac{3c}{8(a+b)}-\frac{c^3}{8}-\frac{9ct}{4}-\frac{3(a+b)ct^2}{2}\right]T \\ &\left.-\left[\frac{3c}{4(a+b)}+ct\right]\sqrt{t/\pi}\exp\left(-\frac{c^2}{4t}-bt\right)\right\}\end{aligned}$

Part I
(Chapters 1–8)

Historical Note, Theory, Examples, and Problems

1 Creators of the Theory of Elasticity

This chapter constitutes a historical note on the creators of the theory of elasticity, and discusses some of their achievements that contributed to the progress of this science. We start with Galileo Galilei who, in the sixteenth and seventeenth centuries, was the first to work on strength and fracture of beams and, especially, on cantilever beams. Robert Hooke in the seventeenth century was the one who discovered the law of elasticity that now carries his name. We end with notes on the six scientists of the twentieth century: M. T. Huber, S. P. Timoshenko, W. Nowacki, G. Fichera, B. A. Boley, and M. E. Gurtin.

1.1 HISTORICAL NOTE: CREATORS OF THE THEORY OF ELASTICITY

A body is called elastic if it returns to its original shape upon the removal of applied forces. All bodies exhibit elastic behavior under sufficiently small loads. The mathematical analysis of elastic behavior of a solid body is called the theory of elasticity. The interdependence between loads applied to the body and the stresses and deformations that these loads produce have been of importance in various practical applications for a very long time. It is therefore not surprising that development of elasticity theory began many centuries ago.

The first name to be linked with the history of the theory of elasticity is Galileo Galilei (1564–1642), called the father of science. He was a physicist, astronomer, mathematician, and philosopher.

In his second dialogue in the *Discoursi e Dimostrazioni matematiche* (Leiden, 1638),* Galileo† gives 17 propositions that set the direction of the analysis of strength and fracture of beams for future development. In particular, the problem of a cantilever beam loaded at its free end has been associated with Galileo's name.‡ Besides dealing with beams of constant cross section, Galileo analyzed the shape of the cross section in order to create beams of constant strength. A shortcoming of Galileo's analysis was the assumption that the longitudinal fibers in a strained beam are inextensible.§

The relation between applied forces and the extension of a body is credited to Robert Hooke (1635–1702). In his book *De potentia restitutiva* (London, 1678), he writes that 18 years earlier (in 1660), he discovered the theory of springs, but had not revealed it as

---

* Readers are advised to acquaint themselves with the English translation of [1].

† Galileo Galilei is usually referred to by his first name, rather than his surname.

‡ The figure on the front page of this book shows a cantilever beam with a weight at its free end and the figure on the last page of the book shows a weight hanging on the rope. The figures are reproduced from Galileo's book of which the title page is shown in Figure 1.1. The book, under the title *Dialogues Concerning Two New Sciences*, was published by Dover Publications, New York, 1954. The title page shown in Figure 1.1 was also reproduced from that book.

§ A delightful book on Galileo was written by Sobel [2]. The book is based on the extensive collection of letters that Virginia Galilei, who became Sister Maria Celeste of the convent of St. Matthew at Arcetri near Florence, wrote to her father. Galileo's letters to his daughter did not survive.

he had hoped to obtain a patent on its application. The theory was concisely presented in a short sentence *"Ut tensio sic vis*; that is, The Power of any spring is in the same proportion with the Tension thereof." He published this discovery at the end of his earlier *Book of the Description of Helioscopes* as an anagram, stated in this book in Section 3.3.1. By a "spring" Hooke did not mean a helical wire but any extensible body that returns to its original shape when the forces were removed.

Although Hooke's theory was one-dimensional, the present-day six relations between components of the stress tensor and the strain tensor are referred to as the *generalized Hooke's law*.

Apparently, the first to apply Hooke's law to Galileo's problem was Edmé Mariotte (1620–1684). Mariotte lived most of his life in Dijon, France. In 1666, he became a member of the French Academy of Sciences. He is credited with the introduction of experimental methods in France. As a result of his experiments with air, the Boyle–Mariotte law was established. Mariotte discovered Hooke's law independently in 1680. In his work [3], he pointed out the fact that in a loaded beam, some fibers extended while other contracted. Unfortunately, he went somewhat too far in stating that half of the fibers were in each category. His contribution to the theory of elasticity came as a result of his work on the design of water pipelines for the Palace de Versailles. His experiments on wood and glass rods showed that Galileo's theory gave values of a breaking force too large, so he developed his own theory, which included elastic properties of material. He analyzed not only cantilever beams but also beams on two supports and beams built in at both ends.*

The results of Mariotte's experiments brought a few others to the field, especially Gottfried Wilhelm Leibniz (1646–1716) [5] and, somewhat later, Pierre Varignon (1654–1722) [6], a French mathematician born in Caen in Normandy, a friend of Leibniz, Newton, and the members of the Bernoulli family. He contributed to the spread of differential calculus and devoted himself to its various applications. Both Leibniz and Varignon tried to formalize the theories of Galileo and Mariotte. Varignon also occupied himself with the determination of the position of a neutral surface, or zero-stress plane.

Leibniz occupies an equally grand place in both the history of philosophy and the history of mathematics. He invented infinitesimal calculus independently of Newton, and his notation is the one in general use since then. He also invented the binary system, the foundation of virtually all modern computer architectures. In philosophy, he is mostly remembered for optimism, that is, his conclusion that our universe is, in a restricted sense, the best possible one God could have made. He was, along with René Descartes and Baruch Spinoza, one of the three greatest seventeenth-century rationalists, but his philosophy also looks back to the scholastic tradition and anticipates modern logic and analysis. Leibniz also made major contributions to physics and technology, and anticipated notions that surfaced much later in biology, medicine, geology, probability theory, psychology, linguistics, and information science. He also wrote on politics, law, ethics, theology, history, philosophy, and philology, even occasional verse. His contributions to this vast array of subjects are scattered in journals and in tens of thousands of letters and unpublished manuscripts. As of 2009, there is no complete edition of Leibniz's writings.

Extensive contributions to the subject were made by members of the Bernoulli family. The family settled in Basel, Switzerland, at the end of the sixteenth century, after religious persecution forced its departure from Antwerp. Members of this family became famous

---

* See the description of Mariotte's experiments in [4].

DISCORSI
E
DIMOSTRAZIONI
MATEMATICHE,
intorno à due nuoue ſcienze
Attenenti alla
MECANICA & i MOVIMENTI LOCALI;
del Signor
GALILEO GALILEI LINCEO,
Filoſofo e Matematico primario del Sereniſſimo
Grand Duca di Toſcana.
Con vna Appendice del centro di grauità d'alcuni Solidi.

IN LEIDA,
Appreſſo gli Elſevirii. M. D. C. XXXVIII.

FIGURE 1.1 Title page of Galileo's book *Discorsi e Dimostrazioni Matematiche, intorno a due nuove scienze* (Leiden, 1638).

mathematicians. Brothers Jacob and John Bernoulli became foreign members of the French Academy of Sciences in 1699. In the course of the next 90 years, that is, until 1790, at least one Bernoulli was always among its members. Jacob and John contributed substantially to the development of differential calculus, which was started by Leibniz and, independently, in England by Isaac Newton (1642–1727).

It was Jacob Bernoulli (1654–1705) who first considered a mathematical form of a bent lamina, or "elastica" [7]. Thus, he extended the works of Galileo and Mariotte, who were both interested in the problem of beam strength, but not in beam deflections. Jacob

Bernoulli's results on the shape of bent beams were widely accepted by others, but they were fundamentally incorrect. He considered only one equation of equilibrium, namely the one of moments, while for the plane problem we need also consider two equations of forces.

John Bernoulli (1667–1748) is credited with the formulation of the virtual displacement principle. It may be mentioned that his lectures prompted Marquis Guillaume François Antoine de l'Hospital to write the first book on calculus in 1696.

Daniel Bernoulli (1700–1782), a son of John, also contributed to the analysis of bent beams, in addition to writing his superb book *Hydrodynamica*. In his letter to Leonhard Euler, he suggested the derivation of elastica by applying variational calculus. It was also Daniel Bernoulli who derived the differential equation of the lateral vibration of bars and conducted experiments to determine various vibration modes.

Leonhard Euler (1707–1783) was born near Basel and was educated at the University of Basel, which, due to the fame of John Bernoulli, became a celebrated mathematics research center. Euler earned his master's degree at the age of 16, and published his first scientific paper at 20. In 1727, Euler moved to St. Petersburg, and in 1730, at the age of 23, became a member of the Russian Academy. He lived an enormously productive life in St. Petersburg until 1741, then at the Prussian Academy in Berlin from 1741 to 1766, and from 1766, again in St. Petersburg.

Euler's most important results on elastica are included in his book [8], which was the first devoted to variational calculus. Pages 245–310 constitute an appendix, *Additamentum I. De Curvis Elasticis*, where problems of elastica were treated extensively. His work considers the effects of an angle between the direction of a force applied and the tangent at the point where the force is applied. For a very small angle, the problem becomes that of the stability of a column due to an axial compressive force. The problem was easily solved and, thus, an expression for the critical compressive load for a column was established. This constituted the birth of the elastic stability theory. While staying in Berlin, Euler extended his research on the buckling of columns.

Joseph Louis Lagrange (1736–1813) was born in Turin, Italy, where at the age of 19 he became a professor of mathematics at the Royal Artillery School. He soon published a number of papers devoted to variational calculus and, as a consequence, came into correspondence with Euler. This, in turn, led to Lagrange's election in 1759 as a foreign member of the Prussian Academy. In 1766, when Euler was returning to St. Petersburg, Lagrange replaced him at the Prussian Academy. In Berlin, Lagrange published a large number of papers and also wrote his famous book *Méchanique analytique*, published after a long delay in Paris in 1788.* By that time, Lagrange had settled in Paris, since the conditions in Berlin deteriorated after the death of Frederick the Great. He was fortunate to avoid the guillotine of the French revolution—some famous scientists were not so lucky—and soon started to teach at a newly established École Polytechnique, which was founded in 1794.

In his paper on columns [9], Lagrange makes progress on the earlier work of Euler, while extending his own past results. While considering the buckling of a column with hinges, he not only derived the expression for the critical force, but also showed that there exist an infinite number of buckling curves. Among other problems, he considered columns

* The first volume of the second edition appeared in 1811, and the second volume in 1815, after the death of its author.

of varying cross section. In a number of papers, Lagrange also analyzed the deflection of various springs.

École Polytechnique was conceived as a new type of school, open to boys of all backgrounds, with competitive entrance exams. Unlike older schools, learning theoretical subjects was a requirement for all students. The first two years were devoted exclusively to basic sciences, while engineering subjects were taught in the third year. Soon, the engineering subjects were entirely eliminated, and École Polytechnique became a school of basic sciences only, as it remains today. Many great contributors to the theory of elasticity were educated at this school. Schools modeled on École Polytechnique were later established in cities throughout Europe.

Charles Augustin Coulomb (1736–1806) considered the bending theory of beams. In his memoir [10], he places the neutral axis in a bent beam at the middle of the cross section of what looks like a rectangular shape. Then, he presented a correct calculation of the moments of forces over the cross section. The theory was the most accurate of those based on Hooke's law and on the assumption that stress in a beam is a result of the extension and contraction of longitudinal fibers. Coulomb is also credited with the first study of the resistance of thin wires to torsion. Because of these achievements, and because of his many experiments, he is considered to be the most important contributor to the mechanics of elastic solids in the eighteenth century.

Thomas Young (1773–1829) was a child prodigy. He started his scientific work early in his life. By 1802, he was already a member of England's Royal Society. He was the first to introduce the modulus of elasticity in relation to his research on tension and compression. Young was the first to introduce shear as a form of elastic strain, and observed that resistance of a body to shear is different than to extension or compression. He was a physician by training, but his contributions were, among many, to mechanics, physics, and to reading the Egyptian hieroglyphs by contributing to deciphering of the Rosetta Stone. One of his achievements was the explanation of how the eye accommodates, and he was the one to discover astigmatism. It was his, the three-color theory of how the eye's retina reacts to light. His scientific curiosity and the high level of mastering of a multitude of subjects was such that the title of his biography was *The Last Man Who Knew Everything.**

The scientist credited with establishing what is now known as the modern theory of elasticity is Claus Louis Marie Henrie Navier (1785–1836). Born in Dijon, France, he was one of the great alumni of École Polytechnique in Paris, where he studied from 1802 to 1804. He was also educated as an engineer. In his early research, Navier also erroneously placed the position of the neutral axis. However, in the first printed edition of his lectures in 1826, he stated, with a proof given, that for elastic materials the neutral axis passes through the centroid of the cross section. In a memoir† read to the French Academy on May 14, 1821, he presented for the first time the general differential equations of equilibrium and motion that must be satisfied at each point in a body, and the conditions at points on its surface (the boundary conditions). These equations for the components of displacement were derived from the consideration of molecular interaction with forces acting along the lines connecting the particles and proportional to their relative distances. Because of a simplified model, however, Navier's original equations contained only one material constant.

* The book is by Andrew Robinson, published by Plume, a member of Penguin Group, London, U.K., 2006.
† Published in [11].

Augustin Cauchy (1789–1857), another alumnus of École Polytechnique, where he later taught, became a member of the French Academy of Sciences in 1816. Being both a mathematician and an engineer, Cauchy's initial interest in the theory of elasticity arose from a study of Navier's paper, and his subsequent contribution to the subject were very substantial. Unlike Navier, who considered intermolecular forces, Cauchy considered the notion of pressure on a plane, known to him from hydrodynamics. He observed that this pressure in an elastic body does not act normal to the plane. He thus introduced for the first time the notion of stress. It was Cauchy who wrote three equations of equilibrium for a tetrahedron, and showed that shear stresses were symmetric, and that stress on any plane can be described by three normal and three shear stress components. It was Cauchy who derived three differential equations of equilibrium for an elemental rectangular parallelepiped. He also observed that for small deformations, the unit elongations and the change of the right angle between any two initially perpendicular directions may be expressed by six relations, now called strain–displacement relations. Cauchy also observed that for small deformations, the stresses are linear functions of strains, thus providing an important generalization of Hooke's law, using two material constants. With the elimination of stresses, the equations reduce to three differential equations, similar to Navier's but containing two constants, not one. In 1828, Cauchy published two papers in which, on the basis of a law of molecular interaction, he generalized his earlier results to anisotropic media, resulting in equations with 15 material constants. It was shown later by the English mathematician George Green (1793–1841) that there should be 21 constants in such general cases. Green's great contribution was the introduction of the principle of conservation of elastic energy.

It is worth mentioning that discussions on whether 15 or 21 constants are needed for anisotropy continued for many years.* Those influenced by Navier and Cauchy accepted one constant for an isotropic body and 15 for an anisotropic body, while others, following Green's reasoning, accepted two and 21 constants, respectively.

Among researchers who considered the problem of how many constants are needed was G. Wertheim (1815–1861). Born in Vienna and educated in Paris, he conducted extensive experiments that consisted not only of tensile tests, but also of longitudinal and lateral vibrations. Assuming initially the one-constant theory, he found the tensile modulus for a number of materials. He found that processes that increase the density, like hammering or rolling, result in an increase of the modulus. He received higher values of the modulus from the vibration experiments than from static tests, and he concluded that the difference of these values may establish the ratio of specific heat at constant stress and constant volume. Wertheim experimented also with the effect of changing temperature and the influence of electric current in a wire on the values of the tensile modulus. From his research on optical properties of elastic materials, which constitutes the birth of photoelasticity, he established a list of colors that could be associated with stresses in an extended rod made of a transparent material. From his experiment on torsion, he found that the ratio of the lateral to longitudinal displacements (the present-day Poisson's ratio) was closer to 1/3, and did not agree with the theoretical value of 1/4 deduced by Siméon Denise Poisson (1781–1840). Wertheim contributed substantially to the knowledge of elastic behavior, but he did not manage to answer the question regarding the number of necessary constants.

* For more information, see [12].

Although Poisson was born in a poor family, because of his unusual abilities he managed to be educated at the École Polytechnique, and later he taught there. He became a member of the French Academy of Sciences in 1812. His research contributed substantially not only to elastostatics but also elastodynamics. His most important results are contained in two seminal publications [13,14]. First, after he obtained equations of equilibrium, similar to those of Navier and Cauchy, Poisson proved that they are not only necessary but also sufficient for a body, or any portion thereof, to remain in a state of equilibrium. Second, by integrating the equations of motion, he showed that a local disturbance produces two kinds of waves: one, faster moving, in which particles move perpendicular to the wave front, with volume changes occurring; and another, in which particles move tangent to the wave front, with only distortion, not volume change, occurring. As for "Poisson's ratio," Poisson utilized the general equations for an isotropic body of Mikhail Vasilevich Ostrogradsky (1801–1861), a Russian mathematician who later lived in Paris. Poisson found theoretically that for a prismatic bar in tension this ratio has a value of 1/4. As for other topics, Poisson worked on finding stresses in a hollow sphere subjected to external or internal pressure, as well as on radial vibrations of a sphere.

Deflection of elastic surfaces was earlier investigated by Euler, but it was Jacques Bernoulli (1759–1789), a nephew of Daniel, who obtained a not quite correct equation for the deflection of plates in the form $D(w_{,1111} + w_{,2222}) = q$. Later, Poisson was the first to derive the equation $D(w_{,1111} + 2w_{,1122} + w_{,2222}) = q$, wherein the flexural rigidity D included the value of Poisson's ratio of 1/4. In these equations, q stands for the loading intensity, or force per area, and w is the deflection. Bernoulli also established a set of three boundary conditions for the edge along which forces were distributed. This set was later reduced to two by Gustav Robert Kirchhoff (1824–1887), who discovered the correct boundary conditions for a free edge of a plate.

Kirchhoff was born and educated in Königsberg in Eastern Prussia (the city is now called Kaliningrad and belonging to Russia). After working in Berlin and Breslau, he settled in Heidelberg for 20 years, until 1875, and then returned to Berlin. He was known as both an excellent theoretical physicist and an able experimenter. Kirchhoff created an elegant plate theory on two assumptions, now generally accepted: (1) lines normal to the middle plane before deformation remain normal to the surface after deformation, and (2) elements in the middle surface do not change length during deformation. Then, after writing an equation for the potential energy of a deflected plate, he used the principle of virtual work to obtain the differential equation of bending. By combining these two equations, he obtained the fourth-order differential equation of bending of plates, earlier derived in a different way by Poisson. Besides reducing Poisson's three boundary conditions to two, as was already mentioned, he analyzed vibrations of a circular plate with a free edge and calculated various modes of vibration. He extended his analysis to plate deflections that are not very small. He managed to perform exact integration of the general equation for lateral vibration of bars of variable cross section in some cases (e.g., a thin wedge or a long cone). Among other achievements, he contributed to the theory of thin bars by making an observation, now called *Kirchhoff's dynamic analogy*, that allows the application of already known solutions in the dynamics of a rigid body to the deformation of thin bars. Kirchhoff's excellent book on Mechanics appeared as the first volume of his lectures [15].

Alfred Clebsch (1833–1872) was, like Kirchhoff, born in Königsberg. After his studies at Königsberg University, he lectured there, later moving to the University of Berlin and, still later, to the Karlsruhe Polytechnicum. It was there, in 1861, that he wrote his famous book [16], in which his main contribution to the subject was presented. Only one book on the theory of elasticity existed at that time, the one by Lamé. Compared to Lamé's book, which was geared more toward acoustics and optics, Clebsch's book was more directed to the needs of engineers, even though the mathematical methods were more strongly addressed than the engineering applications would require. The more important results contained in the book are related to (a) torsion, (b) plane stress, and (c) thin bars and thin plates. With respect to (a) torsion, Saint-Venant's problem (see later) was treated not by the semi-inverse method, but by considering a mathematical problem of which a solution allowed axial tension, torsion, and bending to be jointly treated. With respect to (b) plane stress, working on a cylindrical bar, Clebsch considered a case opposite to that of Saint-Venant's, that is, he assumed that stresses and forces appearing in Saint-Venant's problem vanish, while those vanishing in Saint-Venant's problem are nonzero. Thus, a thin slice between two cross sections represents a plate with forces at the cylindrical boundary and in the plane of the cross section. In this way, the results are applied to a general solution of plane stress for a circular plate. With respect to (c) thin bars and thin plates, Clebsch modified Kirchhoff's theory of thin bars and wrote equations for bending of plates for nonsmall deflections. Also, he considered a circular plate with clamped edge subject to a concentrated load at any point of its surface.

Lord Kelvin (William Thomson) (1824–1907) was born in Belfast, studied at Glasgow, and later graduated from Cambridge. In 1846, he became a professor at Glasgow and taught there for 53 years. Kelvin, jointly with Peter Guthrie Tait, a professor in Edinburgh, wrote the *Treatise on Natural Philosophy*, of which the first volume was published in 1867. It is in this book, which treats mechanics of rigid bodies, elastic bodies, and fluids, that most of Kelvin's contributions to the theory of elasticity are recorded. Kelvin's systematic presentation of the theory of elasticity in the *Treatise* was the first in the English language and substantially influenced its development in England. He gave a complete account of Kirchhoff's dynamic analogy, explained the sufficiency of Kirchhoff's theory of plates for small deflections, and gave a physical interpretation of the reduction in the number of boundary conditions proposed by Kirchhoff. Earlier, Kelvin conducted a series of tensile tests on bars and studied temperature changes during such tests. He observed that the value of the modulus of elasticity depends on whether the tests are conducted suddenly (over a short time) or slowly (over a long time). In the case of the former, there was an adiabatic extension that increased the value of the modulus, since there was no time for the heat generated as a consequence of deformation to disperse, as compared to the latter case, where there is an isothermal extension due to the distribution of heat. These experiments led to a proof of the existence of a strain energy function that depends on strain relative to some fixed state, but not on the manner in which strain was produced.

Barré de Saint-Venant (1797–1886) studied at École Polytechnique and graduated from École de Ponts et Chaussées, the top French engineering school of that time, where he later lèctured. He became a member of the French Academy of Sciences in 1868. In 1853, Saint-Venant presented his famous memoir on torsion to the academy [17]. In it, he presented the state of the art of elasticity theory, including his own contributions. The main thrust of the memoir was torsion. Saint-Venant observed that the direct method of finding displacements, strains, and stresses by using the differential equilibrium equations, the strain–displacement relations, and Hooke's law, subject to boundary conditions, was not a practical method for engineers. The reason was that general methods of integration of equations of elasticity were not known. Therefore, Saint-Venant proposed a semi-inverse method in which only some information on displacements and stresses is needed. From the equations of elasticity theory, the remaining information is obtained. Engineers could get approximate results from the strength of materials methods first, and then obtain exact solutions by applying the proposed method.

Saint-Venant presented solutions for torsion and bending of prismatic bars. In the case of torsion, he assumed approximate displacements in the plane of the cross section and augmented them by axial displacement, a yet unknown warping function. The strains and then the stresses were calculated, and subsequently the equation of the warping function was established.

Since the displacements are the same in each cross section, the stresses are also the same and, therefore, the forces producing the torsion moment at the ends of the bar would have to be applied in a very specific manner for the results to be exact. But Saint-Venant formulated the proposition, now called the Saint-Venant principle, that requires only the resultant force and resultant moment to remain unchanged, since at some distance from the ends of the rod the distribution of displacements and stresses is sufficiently accurate in typical applications.

Saint-Venant did not write a book, but he translated Clebsch's book into French and also edited Navier's book. His editorial notes in these books were much more extensive than the original texts.

Gabriel Lamé (1795–1870) and Benoit-Paul-Émile Clapeyron (1799–1864) graduated together from École Polytechnique and worked for about 10 years at the Institute of Ways of Communication in St. Petersburg, a Russian engineering school established in 1809. They both taught mathematics and physics and contributed to Russian engineering projects. They returned to Paris in 1831, and Lamé soon became a professor at École Polytechnique. In 1852 his book appeared, the first treatise on the theory of elasticity, *Leçons sur la Théorie Mathématique de l'Élasticité des Corps Solides*. The book contained earlier results of Lamé and Clapeyron presented to the French Academy, but equations in the book were changed, as Lamé came to the conclusion that two, and not one, material constants were needed for the description of an isotropic body. We now refer to these constants as the Lamé constants. The book also presented Clapeyron's theorem, that the sum of products of forces acting on a body and the displacements along the lines of action of these forces is equal to twice the strain energy stored in the body. We recall that Euler earlier used the expression for the strain energy of bending bars for the derivation of the differential equation of the elastic curve.

Luigi Federico Ménabréa, an Italian military engineer, introduced the principle of least work. Considering trusses with redundant elements, he observed that the forces acting in them must make the strain energy a minimum, that is, the derivatives of the strain energy with respect to these forces must be zero. The proof of this principle was later provided by an Italian engineer Alberto Castigliano (1847–1884).

Joseph Vallentin Boussinesq (1842–1929), a student of Saint-Venant, contributed not only to the theory of elasticity, but also to optics, hydrodynamics, and thermodynamics. His book on *Application des potentials à l'étude de l'équilibre et du mouvement des solides élastiques...* was considered one of the most important books on the theory of elasticity since Saint-Venant's memoir on torsion and bending. Using potential functions, Boussinesq found solutions to a number of problems, one of which, now known as Boussinesq's problem, deals with the finding of stresses and deformations in a semi-infinite body under a force acting perpendicular to its boundary. Using this solution and the principle of superposition,

Boussinesq solved the problem of a semi-infinite body acted upon by a distributed load on a part of the boundary. He dealt with problems in which displacements, and not forces, were specified. For instance, he discussed the problem of a rigid cylinder pressed into a semi-infinite body on the bounding plane. He showed the validity of Saint-Venant's principle, that displacements and stresses depend only locally on the manner in which the forces are applied.

Albert-Aimé Flamant, a collaborator of Saint-Venant, applied the solution of Boussinesq's problem to finding the stress distribution in a semi-infinite plate of unit thickness acted upon by a force perpendicular to its boundary. A number of other important two-dimensional problems were solved by J. H. Michell (1863–1940), in part on the basis of the work done by Boussinesq and Flamant. Among them, he solved the problem of a force applied inside an infinite plate, found a solution for a semi-infinite plate with a force inclined to the straight boundary, and for a wedge clamped at the base and loaded at the peak.

Jean-Marie-Constant Duhamel (1797–1872) was born in Saint-Malo, France, and educated at École Polytechnique. Among his teachers were Fourier and Poisson. His main contribution to the theory of elasticity is his *Mémoir sur le calcul des actions moléculaires développées par les changements du température dans les corps solides* [18]. There, he generalized Navier's equations of equilibrium by including temperature changes, and showed that thermal stresses can be found in a manner similar to stresses caused by body forces and forces applied on the surfaces. He was the first to observe that uniform heating does not lead to stresses. He was probably the first to apply the superposition principle to stress theory. Later, Duhamel made a contribution to the theory of vibrations in elasticity and presented a method for the analysis of forced vibrations for elastic bodies.

Franz Neumann (1798–1895) was educated in Berlin and taught at the University of Königsberg from 1826. Navier, Cauchy, and Poisson were still alive when Neumann started publishing papers on the theory of elasticity, most of which were applications to optics and related to the number of necessary material constants for isotropic bodies and crystals of various levels of symmetry, which was still disputed. Neumann was the first to derive a formula for the modulus in tension for prismatic crystals of any orientation. The results of his experiments allowed him to establish the correct number of necessary material constants and reject the assumptions made earlier by Navier and Poisson on this subject. Experiments carried out by him and his pupils established a foundation for photoelastic stress analysis. Neumann contributed to the theory of thermal stresses. Namely, he analyzed thermal stresses in a plate with the temperature varying in the plane of the plate. After deriving the governing equations, he applied them to a circular plate and a circular ring. It is likely that Neumann was the first to study residual stresses.

One of Neumann's students was Woldemar Voigt (1850–1919). Born in Leipzig, he studied in Königsberg, later becoming a professor at the University of Göttingen. It was Voigt who, at last, settled the matter that an elastic isotropic body is governed by two constants, and not by one, as was assumed by Navier, Poisson, and Saint Venant, and that a general anisotropic body is governed by 21 constants, and not by 15, as was assumed by Poisson.

Lord Rayleigh (John William Strutt) (1842–1919) was educated in Cambridge where he later (in 1879) received a chair of experimental physics. His famous book, *The Theory of Sound*, was published in 1877, and it is in this book that his principal contributions to the theory of elasticity are to be found. The problems discussed are vibration of strings, bars, membranes, plates, and shells. He showed how the application of generalized forces and generalized coordinates combined with the Betti–Rayleigh reciprocal principle simplified the treatment of redundant structures. Rayleigh was a master in approximate methods, and his idea, taken up later by Walter Ritz, led to the well-known Rayleigh–Ritz method. Rayleigh's one contribution that was not contained in his treatise was the theory of elastic surface waves, which has since then been associated with his name. As anticipated by the author, the theory of surface waves has been used subsequently in the analysis of earthquakes.

We should mention the works of Augustine Edward Hough Love (1863–1940), an English geophysicist who was also educated at Cambridge. His contributions are to the theory of thin shells in which he expanded on the earlier results of Rayleigh. In *Some Problems in Geodynamics* (1911) Love presented the idea of "Love waves," which are Rayleigh waves transmitted through the surface of an elastic body, and he was the first to observe them. He also presented a rigorous theory of plates. No doubt, the most important influence that he exerted on twentieth century researchers was his *Treatise on the Mathematical Theory of Elasticity*, published initially in two volumes in 1892–1893.* The book is the principal source on what was achieved in the theory of elasticity until the end of the nineteenth century.

Piotr Fiodorovich Papkovich (1887–1946) was a Russian researcher in elasticity and in shipbuilding mechanics. He lectured at the St. Petersburg Polytechnical Institute, the Naval Academy, and the Leningrad Shipbuilding Institute. His works were devoted to the general theorems of the stability of elastic systems and of small elastic oscillations. His best-known contribution to the theory of elasticity is the representation of the displacement solutions in terms of a harmonic vector and a harmonic scalar [19],† employed often in treating three-dimensional problems of elastostatics.

Maksymilian Tytus Huber (1872–1950), an eminent Polish educator and researcher, received his doctoral degree at Lwów Polytechnical School in 1904. In the same year, a Polish journal Czasopismo Techniczne, published his paper‡ in which he presented the maximum distortion energy criterion.

Huber became rector (president) of the Lwów Polytechnic in 1914, but the First World War disrupted his tenure of office; he became a prisoner of war in Russia until 1918, when he returned to his chair at the Lwów Polytechnic. He moved to the Warsaw Polytechnic in 1928 where he was a professor until the outbreak of the Second World War, and lived in Warsaw until the destruction of the city by Germans in 1944. In 1945, Huber became a professor at Gdańsk University of Technology.§

Huber's contributions to mechanics in general, and to the theory of elasticity in particular, were significant. His development of the theory of orthotropic plates allowed application of this theory to various engineering projects. He contributed to the vibration theory and its application in the industry. Over his 50 years of active work, he taught mechanics courses

---

* The book's fourth edition, published in 1944, is available from Dover Publications, New York.

† The method was developed independently, and in a different way, by Heinz Neuber in 1934, and is known as Papkovich–Neuber method.

‡ The original title in Polish: *Właściwa praca odkształcenia jako miara wytężenia materiału. The Maximum Distortion Energy Criterion* was later independently proposed by Richard von Misses (in 1913) and Heinrich Hencky (in 1924).

§ One of the authors (RBH), as the first-year student at Gdańsk University of Technology, took a course in *Strength of Materials* from M. T. Huber in 1946–1947.

to generations of students and researchers. He wrote basic books on the mechanics of the rigid body, on strength of materials, and a two-volume extensive treatise on the theory of elasticity. Because of his achievements in mechanics and his influence on the directions of research in Poland, some called him the Polish Timoshenko.*

Stephen P. Timoshenko (1878–1972) is considered the man who introduced mechanics to the United States. Born in Ukraine, he started his career as a teacher and researcher. He was an instructor at the Institute of Engineers of Ways of Communications (1902–1903), an assistant professor at the Polytechnic Institute in St. Petersburg (1903–1906), professor at the University of Kiev (1906–1911), and professor at the Electrotechnic Institute and Institute of Engineers of Ways of Communications (1912–1917). During the Russian revolution he left Russia and was a professor at the Polytechnic Institute of Zagreb, Croatia (1920–1922). He emigrated to the United States in 1922 and, until 1927, worked at Westinghouse Electric and Manufacturing Company in East Pittsburgh, Pennsylvania, as a research engineer. His university life in the United States started in 1927, when he became a professor at the University of Michigan. From 1936 until his retirement in 1944, he was a professor at Stanford.

Through his lectures, papers, and books, Timoshenko influenced generations of students and engineers in the United States and around the world. He had a practical, yet mathematically sound, approach to engineering problems. His books on the theory of elasticity, strength of materials, theory of elastic stability, theory of vibrations, and theory of structures became standard textbooks in many countries. He had a visionary and fresh approach to new problems, so that even short publications on new subjects, like what we call now the *Timoshenko beams*, created a lasting following by numerous researchers. Because of his enthusiasm and hard work, the Applied Mechanics Division of the ASME was created, and the *Journal of Applied Mechanics* was established.

Nikolai Ivanovich Muskhelishvili (1891–1976) was a professor at both the university and the technical university in Tbilisi since 1922. His research concentrated on the theory of elasticity, analytical geometry, and integral equations. His famous book on *Some Basic Problems of the Mathematical Theory of Elasticity* was translated into English.† The book contains an advanced method of solving problems of elasticity using complex variables.

Witold Nowacki (1911–1986) was a young civil engineer when, as an officer in the Polish army fighting the invading German army in 1939, he became a prisoner of war. He spent over five years in a prisoner-of-war camp. While in the camp, he organized courses for his colleagues, and it was there that he discovered his true vocation: research in mechanics. After the war, he joined M. T. Huber at Gdańsk University of Technology, and soon became a professor in the Department of Civil Engineering. When the Polish Academy of Sciences was founded in Warsaw in 1952, Witold Nowacki became the head of its Technology Sciences Division. From that time until his retirement in 1981, he held positions of increasing importance at the Polish Academy of Sciences, eventually becoming its president in 1977. At the same time he held the chair of structural mechanics at Warsaw University of Technology and, from 1955 until his retirement, he was a professor at the University of Warsaw. Despite his huge administrative duties, Witold Nowacki managed to publish over 200 original research papers in structural mechanics, theory of

* See the richly illustrated biography in [20].

† The book was published by Noordhoff, Groningen, 1953.

plates and shells, theory of vibrations, theory of elasticity, thermoelasticity, viscoelasticity, magnetothermoelasticity, and thermodiffusion in solid bodies. He is the author of 17 textbooks and monographs, and an autobiography. Eight of his books had two editions. There are 14 translations of his books from Polish into eight languages: Chinese, Czech, English, French, German, Romanian, Russian, and Serbo-Croatian. Altogether, counting first editions, second editions, and translations, there exist 40 books by Witold Nowacki. One of his books is an extensive monograph on the theory of elasticity [21]. Moreover, his organizational efforts were enormous. He significantly contributed to the creation of the Institute of Fundamental Technological Research in Warsaw* and to the creation of the International Centre for Mechanical Sciences in Udine, Italy. He initiated the Polish journal on mechanics, *Archiwum Mechaniki Stosowanej*, now *Archive of Mechanics*. He created the Polish Society of Theoretical and Applied Mechanics [22,23].

Gaetano Fichera (1922–1996) was born near Catania in Sicily. He studied mathematics at the University of Rome and graduated in 1941. In 1943 he was conscripted into the Italian army. After Italy signed a separate armistice with the Allies in September 1943, the Germans treated soldiers who did not join Mussolini's fascist army as traitors. As Gaetano Fichera was one of them, he was taken prisoner by the Germans and, subsequently, was condemned to death. Of his escape attempts, it was the third that was successful and saved his life. After that, he joined the partisans and fought against Germans. In 1945, he returned to the University of Rome. In 1949, he became a professor at the University of Trieste, and in 1956 he received a chair at the University of Rome, La Sapienza, where he stayed until his death. Besides his extensive research and teaching, he was very active as a member of the Academia Nazionale dei Lincei.

Gaetano Fichera is the author of more than 200 papers and 18 books. Although he contributed to a number of branches of mathematics, his first love seems to be the mathematical theory of elasticity. In this area, his work concentrated on wave propagation, the theory of boundary value problems, and existence and uniqueness theorems. Extensive publications on the latter two topics should be noted [24,25].

Professor Fichera was a man of great charm and personal culture. He had many students and collaborators around the globe. His hospitality was world-renown.†

We should mention that Gaetano Fichera's last paper [26], *Rend. Mat. Acc. Lincei*, Serie IX, 8 (1997), 197–227, published after his death, was on an extension of two papers published by the present authors [27,28].

Bruno Adrian Boley was born in Trieste, Italy, in 1924. He emigrated to the United States and received his engineering education at the College of the City of New York and at the Polytechnic Institute of Brooklyn, where he served as an assistant professor from 1943 to 1948. For the next two years he worked in industry, but in 1950 he returned to the university life for good, first as an associate professor at the Ohio State University, then as professor at Columbia University and, from 1968, as professor at Cornell University where he was also

* It may be worth mentioning that it was at this institute that both the authors started their life in research in mechanics. While being students in the Department of Mathematics at the University of Warsaw, they attended a course on *Theory of Elasticity* given by Witold Nowacki and, subsequently, were offered by him research positions at the Institute of Fundamental Technological Research. They both were Witold Nowacki's doctoral students there, although not at the same time.

† It happened that one of the authors (RBH) was the last foreign professor invited by Gaetano Fichera for a visit and to deliver a special lecture at the University of Rome (in May–June 1996). It was during his stay in Rome that Professor Fichera suddenly fell ill (on May 30) and died after a surgery two days later, on June 1, 1996.

chairman of the Department of Theoretical and Applied Mechanics.* In 1973, Bruno Boley became dean at Northwestern University. In 1987, he returned to Columbia University.

Of Bruno Boley's books, *Theory of Thermal Stresses*, coauthored by J. H. Weiner, became the principal source of knowledge on the subject for both students and engineers. He has written over 100 papers on structural dynamics, elastic stability, applied mathematics, heat conduction in solids, and, of course, on thermal stresses. In addition to his teaching, writing, editing, and research, Bruno Boley has contributed his efforts to the organization of International Union of Theoretical and Applied Mechanics (IUTAM) and the International Association of Structural Mechanics in Reactor Technology (IASMiRT).

We cannot close this short list of creators of the theory of elasticity without including in it Morton E. Gurtin. He was born in Jersey City in 1934 of parents who emigrated with their parents from Ukraine. He received his mechanical engineering education at Rensselaer Polytechnic Institute. His most important contributions to the theory of elasticity took place at Brown University in Providence, Rhode Island, where he was first a research associate, then an assistant professor, and then an associate professor. In 1966 he became a professor at Carnegie Mellon University in Pittsburgh.

Morton Gurtin's monograph on the theory of elasticity [29], an elegant, modern, and thorough treatise, has provided guidance and inspiration in writing this book. We recognize his influence on the present generation of researchers, not only in the linear theory of elasticity but also in nonlinear theories.

* * *

The theory of elasticity, like the whole science of solid mechanics, has not stagnated. In fact, the second half of the twentieth century witnessed great progress on many fronts. The new areas that have been developed, for example, the theory of fracture and the mechanics of composite materials, to mention only two. The development of the finite element method and other numerical methods, combined with the proliferation of computers, has allowed many problems to be solved that were not solvable before. But writing on the recent history as well as trying to predict the future direction of research do not belong within the scope of this book.

Looking at this chapter after it was written, the authors feel saddened that they wrote so little about so few. There were many who, justifiably, should be called creators of the theory of elasticity, and their histories are often fascinating. Readers may consult two books that the authors also used:

S. P. Timoshenko, *History of Strength of Materials*, McGraw-Hill, New York, 1953, 452 pp.

I. Todhunter and K. Pearson, *A History of the Theory of Elasticity and the Strength of Materials*, vol. I, 936 pp.; vol. II, Part I, 762; vol. II, Part II, 546 pages; Dover Publications, New York, 1960.

The authors should also mention the historical sketches in two earlier books on the *Mathematical Theory of Elasticity*:

* One the authors (RBH) was Professor Boley's postdoctoral student at Columbia University in 1964–1965. Upon his emigration from Poland in 1969, he spent a year (1969–1970) on Professor Boley's invitation as a visiting associate professor in the Department of Theoretical and Applied Mechanics at Cornell University.

A. E. H. Love, *A Treatise on the Mathematical Theory of Elasticity,* 4th edn., Dover Publications, New York, 1944.

I. S. Sokolnikoff, *Mathematical Theory of Elasticity,* 2nd edn., McGraw-Hill, New York, 1956.

Readers interested in the history of Thermoelasticity and Thermal Stress Analysis, may acquaint themselves with Historical Note in R. B. Hetnarski and M. R. Eslami, *Thermal Stresses – Advanced Theory and Applications*, Springer, 2009, pp. XVII–XXXII.

REFERENCES

1. *Dialogues Concerning Two New Sciences*, Dover Publications, New York, 1954.
2. D. Sobel, *Galileo's Daughter*, Penguin Books, New York, 1999.
3. *Traité du mouvement des eaux*, Paris, 1686, Partie V. Disc. 2, pp. 370–400.
4. S. P. Timoshenko, *History of Strength of Materials*, McGraw-Hill, New York, 1953, pp. 21–24.
5. *Demonstrationes novae de Residentia solidorum*, Acta Eruditorum Lipsiae, July 1684.
6. *De la Résistence des Solides en général pour tout ce qu'on peut...*, Mémoires de l'Académie, Paris, 1702.
7. *Curvatura Laminae Elasticae, Acta Eruditorum Lipsiae*, June 1694, p. 262, and an addendum in the same journal in Dec. 1695, p. 537.
8. *Methodus inveniendi lineas curvas maximi minimive sive motus scientia analytice exposita*, Bousquet, Lausanne, 1744.
9. *Sur la figure des colonnes*, *Miscellanea Taurinesia*, vol. V, Royal Society of Turin, 1770–1773, p. 123.
10. *Essai sur un application des règles de Maximis et Minimis à quelques Problems de Statique, relatifs à l'Architecture*, Mémoires... par divers Savants, Paris, 1776, pp. 343–382.
11. *Mémoirs de l'Académie de Sciences*, vol. VII, Paris, 1827.
12. I. Todhunter and K. Pearson, *A History of the Theory of Elasticity and the Strength of Materials*, vol. I, Dover Publications, New York, 1960, pp. 496–505.
13. *Mémoire sur l'équilibre de le mouvement des corps élastiques*, *Mém Acad.*, 8, 1829.
14. *Mémoire sur les équations générales de l'équilibre et du mouvement des corps élastiques et des fluids*, *J. École Polytech.*, 20, 1831.
15. Vorlesungen über matematische Physik, Mechanik, Leipzig, 1876.
16. *Theorie der Elasticität fester Körper*, B. G. Teubner, 1862.
17. *Mém. acad. sci. savants etrangers*, XIV, 1855, 233–560.
18. *Mém. acad. sci. savants étrangers*, 5, 1838, 440–498.
19. P. F. Papkovich, An expression for a general integral of the equations of the theory of elasticity in terms of harmonic functions, *Izvest. Akad. Nauk SSSR Ser. Phys. Math.*, 1932, 1425–1435.
20. Z. S. Olesiak and Z. W. Engel, *Maksymilian Tytus Huber*, PIB, Radom, 2006, 249pp (in Polish).
21. *Theory of Elasticity* (in Polish), PWN—Polish Scientific Publishers, Warsaw, 1970, 769pp.
22. Zbigniew, S. Olesiak, *Tribute to Witold Nowacki (1911–1986), J. Thermal Stresses*, 12, 1989, 289–292.
23. W. Nowacki, *Autobiographical Notes* (in Polish), PWN—Polish Scientific Publishers, Warsaw, 1985, 295pp.
24. G. Fichera, *Existence Theorems in Elasticity*, Encyclopedia of Physics, chief editor: S. Flügge, Vol. VIa/2, editor C. Truesdell, Springer, Berlin, Germany, 1972.
25. G. Fichera, *Boundary Value Problems in Elasticity with Unilateral Constraints*, Encyclopedia of Physics, chief editor: S. Flügge, Vol. VIa/2, editor C. Truesdell, Springer, Berlin, Germany, 1972.

26. G. Fichera, A boundary value problem connected with response of semi-space to a short laser pulse, *Rend. Mat. Acc. Lincei*, 8, 1997, 197–227.
27. R. B. Hetnarski and J. Ignaczak, Generalized thermoelasticity: Closed-form solutions, *J. Thermal Stresses*, 16, 1993, 473–498.
28. R. B. Hetnarski and J. Ignaczak, Generalized thermoelasticity: Response of semi-space to a short laser pulse, *J. Thermal Stresses*, 17, 1994, 377–396.
29. M. E. Gurtin, *The Linear Theory of Elasticity*, Encyclopedia of Physics, chief editor: S. Flügge, vol. VIa/2, editor: C. Truesdell, Springer, Berlin, Germany, 1972, pp. 1–295.

2 Mathematical Preliminaries

In this chapter, the basic results from the tensor calculus are recalled, using both direct and index notations. In particular, a summary of results in vector algebra, the permutation symbol ϵ_{ijk}, the properties of tensors, invariants of a second-order symmetric tensor, eigenvalues and eigenvectors of a second-order symmetric tensor, and a summary of results in the differential and integral calculus of tensor functions, including the integral theorems such as the divergence and Stokes' theorems, are presented. The Laplace transformation is introduced, and the Dirac delta function and the Heaviside function are explained. The chapter contains a number of worked examples, and also end-of-chapter problems with solutions provided in the Solutions Manual.

2.1 VECTORS AND TENSORS

A *vector* will be understood as an element of a *vector space* $\mathcal{V}$. In $\mathcal{V}$, an operation of addition of two elements and multiplication of an element by a scalar is defined in a usual way, that is, for any two vectors $\mathbf{u}$ and $\mathbf{v}$

$$\mathbf{u} + \mathbf{v} \in \mathcal{V} \tag{2.1.1}$$

and

$$\lambda \mathbf{u} \in \mathcal{V} \tag{2.1.2}$$

where λ is a scalar.

The *inner product* of $\mathbf{u}$ and $\mathbf{v}$ will be denoted by $\mathbf{u} \cdot \mathbf{v}$. If Cartesian coordinates are introduced in such a way that the set of vectors $\{\mathbf{e}_i\} = \{\mathbf{e}_1, \mathbf{e}_2, \mathbf{e}_3\}$ with an origin $\mathbf{0}$ stands for an *orthonormal basis*, and if $\mathbf{u}$ is a vector and $\mathbf{x}$ is a point of a three-dimensional Euclidean space E^3, then Cartesian coordinates of $\mathbf{u}$ and $\mathbf{x}$ are given by

$$u_i = \mathbf{u} \cdot \mathbf{e}_i, \quad x_i = \mathbf{x} \cdot \mathbf{e}_i \tag{2.1.3}$$

Apart from the direct vector notation, we will use indicial notation in which subscripts range is from 1 to 3 and a summation convention over repeated subscripts is observed. For example,

$$\mathbf{u} \cdot \mathbf{v} \equiv \sum_{i=1}^{3} u_i v_i = u_i v_i \tag{2.1.4}$$

From the definition of an orthonormal basis $\{\mathbf{e}_i\}$, it follows that

$$\mathbf{e}_i \cdot \mathbf{e}_j = \delta_{ij} \quad (i,j = 1,2,3) \tag{2.1.5}$$

where δ_{ij} is called the *Kronecker symbol* defined by

$$\delta_{ij} = \begin{cases} 1 & \text{if } i = j \\ 0 & \text{if } i \neq j \end{cases} \tag{2.1.6}$$

We introduce the *permutation symbol* ϵ_{ijk}, also called the *alternating symbol* or the *alternator*, defined by

$$\epsilon_{ijk} = \begin{cases} 1 & \text{if } (ijk) \text{ is an even permutation of } (123), \\ -1 & \text{if } (ijk) \text{ is an odd permutation of } (123), \\ 0 & \text{otherwise, that is, if two subscripts are repeated.} \end{cases}$$

From this definition it follows that

$$\epsilon_{ijk} = -\epsilon_{ikj} = \epsilon_{kij} = -\epsilon_{kji} \tag{2.1.7}$$

and

$$\epsilon_{113} = \epsilon_{221} = \epsilon_{233} = \epsilon_{122} = 0 \tag{2.1.8}$$

The permutation symbol will be used for the definition of the *vector product* $\mathbf{u} \times \mathbf{v}$ *of two vectors* $\mathbf{u}$ and $\mathbf{v}$

$$(\mathbf{u} \times \mathbf{v})_i = \epsilon_{ijk} u_j v_k \tag{2.1.9}$$

We may observe that the following identity holds true:

$$\epsilon_{mis}\epsilon_{jks} = \delta_{mj}\delta_{ik} - \delta_{mk}\delta_{ij} \tag{2.1.10}$$

This identity, which is called $\epsilon - \delta$ identity, will be useful to prove a number of vector relations involving the inner and vector products.

An alternative definition of the permutation symbol, given in terms of the orthonormal basis vectors $\mathbf{e}_i$, is

$$\epsilon_{ijk} = \mathbf{e}_i \cdot (\mathbf{e}_j \times \mathbf{e}_k) \tag{2.1.11}$$

Using this definition a generalized form of Equation 2.1.10 is obtained:

$$\epsilon_{ijk}\epsilon_{pqr} = \begin{vmatrix} \delta_{ip} & \delta_{iq} & \delta_{ir} \\ \delta_{jp} & \delta_{jq} & \delta_{jr} \\ \delta_{kp} & \delta_{kq} & \delta_{kr} \end{vmatrix} \tag{2.1.12}$$

Letting $k = r$ in this identity, we obtain Equation 2.1.10.

Example 2.1.1

Show that a 3×3 determinant can be expressed by using the alternating symbol ϵ_{ijk}:

$$\begin{vmatrix} a_1 & a_2 & a_3 \\ b_1 & b_2 & b_3 \\ c_1 & c_2 & c_3 \end{vmatrix} = \epsilon_{ijk} a_i b_j c_k \tag{a}$$

Solution

By using the definitions of a determinant and ϵ_{ijk}, we obtain

$$\begin{aligned} \text{LHS} &= a_1 \begin{vmatrix} b_2 & b_3 \\ c_2 & c_3 \end{vmatrix} - a_2 \begin{vmatrix} b_1 & b_3 \\ c_1 & c_3 \end{vmatrix} + a_3 \begin{vmatrix} b_1 & b_2 \\ c_1 & c_2 \end{vmatrix} \\ &= a_1(b_2c_3 - b_3c_2) - a_2(b_1c_3 - c_1b_3) + a_3(b_1c_2 - b_2c_1). \end{aligned} \tag{b}$$

$$\begin{aligned} \text{RHS} &= \epsilon_{1jk} a_1 b_j c_k + \epsilon_{2jk} a_2 b_j c_k + \epsilon_{3jk} a_3 b_j c_k \\ &= \epsilon_{12k} a_1 b_2 c_k + \epsilon_{21k} a_2 b_1 c_k + \epsilon_{31k} a_3 b_1 c_k \\ &\quad + \epsilon_{13k} a_1 b_3 c_k + \epsilon_{23k} a_2 b_3 c_k + \epsilon_{32k} a_3 b_2 c_k \\ &= a_1 b_2 c_3 - a_2 b_1 c_3 + a_3 b_1 c_2 \\ &\quad - a_1 b_3 c_2 + a_2 b_3 c_1 - a_3 b_2 c_1 = \text{LHS}. \end{aligned} \tag{c}$$

Hence, LHS = RHS. □

A *second-order tensor* is defined as a linear transformation from $\mathcal{V}$ to $\mathcal{V}$, that is, a tensor $\mathbf{T}$ is a linear mapping that associates with each vector $\mathbf{v}$ a vector $\mathbf{u}$ by

$$\mathbf{u} = \mathbf{T}\mathbf{v} \tag{2.1.13}$$

The components of $\mathbf{T}$ are denoted by T_{ij}

$$T_{ij} = \mathbf{e}_i \cdot \mathbf{T}\mathbf{e}_j \tag{2.1.14}$$

so the relation 2.1.13 in index notation takes the form

$$u_i = T_{ij} v_j \tag{2.1.15}$$

The Kronecker symbol δ_{ij} represents an *identity tensor* that in direct notation is written as $\mathbf{1}$.

A *product of two tensors* $\mathbf{A}$ and $\mathbf{B}$ is defined by

$$(\mathbf{AB})\mathbf{v} = \mathbf{A}(\mathbf{B}\mathbf{v}) \tag{2.1.16}$$

for every vector $\mathbf{v}$. Thus, in Cartesian coordinates

$$(\mathbf{AB})_{ij} = A_{ik} B_{kj} \tag{2.1.17}$$

This follows from Equation 2.1.16:

$$(\mathbf{AB})_{ij} v_j = A_{ip} B_{pj} v_j \tag{2.1.18}$$

or

$$[(\mathbf{AB})_{ij} - A_{ip} B_{pj}] v_j = 0 \tag{2.1.19}$$

Since v_j is arbitrary, the expression in brackets is zero and Equation 2.1.17 is received. The *transpose* of $\mathbf{T}$, denoted by $\mathbf{T}^T$, is defined as a unique tensor satisfying the property

$$\mathbf{Tu} \cdot \mathbf{v} = \mathbf{u} \cdot \mathbf{T}^T \mathbf{v} \tag{2.1.20}$$

for every $\mathbf{u}$ and $\mathbf{v}$. From this definition it follows that

$$T_{ik} u_k v_i = u_k T^T_{ki} v_i \tag{2.1.21}$$

$$\left(T_{ik} - T^T_{ki}\right) u_k v_i = 0 \tag{2.1.22}$$

or

$$T_{ik} = T^T_{ki} \tag{2.1.23}$$

because u_k and v_i are arbitrary vectors.

If $\mathbf{T} = \mathbf{T}^T$, then the tensor $\mathbf{T}$ is *symmetric*. Also, if $\mathbf{T} = -\mathbf{T}^T$ then the tensor $\mathbf{T}$ is *skew* or *antisymmetric*. Therefore, $\mathbf{T}$ is symmetric if $T_{ij} = T_{ji}$, and skew if $T_{ij} = -T_{ji}$.

Every tensor $\mathbf{T}$ can be expressed by a sum of a symmetric tensor sym $\mathbf{T}$ and skew tensor skw $\mathbf{T}$, that is,

$$\mathbf{T} = \operatorname{sym} \mathbf{T} + \operatorname{skw} \mathbf{T} \tag{2.1.24}$$

where

$$\operatorname{sym} \mathbf{T} = \frac{1}{2} \left(\mathbf{T} + \mathbf{T}^T\right) \tag{2.1.25}$$

and

$$\operatorname{skw} \mathbf{T} = \frac{1}{2} \left(\mathbf{T} - \mathbf{T}^T\right) \tag{2.1.26}$$

In index notation

$$T_{ij} = T_{(ij)} + T_{[ij]} \tag{2.1.27}$$

where

$$T_{(ij)} = \frac{1}{2} \left(T_{ij} + T_{ji}\right) \tag{2.1.28}$$

$$T_{[ij]} = \frac{1}{2} \left(T_{ij} - T_{ji}\right) \tag{2.1.29}$$

Thus, we may write

$$T_{(ij)} = (\mathrm{sym}\,\mathbf{T})_{ij} \tag{2.1.30}$$

$$T_{[ij]} = (\mathrm{skw}\,\mathbf{T})_{ij} \tag{2.1.31}$$

If a tensor $\mathbf{P}$ is skew, then it contains only three independent components and there exists a one-to-one correspondence between $\mathbf{P}$ and a vector $\boldsymbol{\omega}$ such that

$$\mathbf{Pu} = \boldsymbol{\omega} \times \mathbf{u} \tag{2.1.32}$$

for every vector $\mathbf{u}$.

From this relation, it follows that

$$\omega_i = -\frac{1}{2}\,\epsilon_{ijk}P_{jk} \tag{2.1.33}$$

$$P_{ij} = -\epsilon_{ijk}\omega_k \tag{2.1.34}$$

The vector $\boldsymbol{\omega}$ is called the *axial vector* corresponding to $\mathbf{P}$. Indeed, writing Equation 2.1.32 in index notation,

$$P_{ij}u_j = \epsilon_{ijk}\omega_j u_k \tag{2.1.35}$$

and changing subscripts on the right side, $j \to k$ and $k \to j$, we find

$$[P_{ij} - \epsilon_{ikj}\omega_k]u_j = 0 \tag{2.1.36}$$

or, since u_j is an arbitrary vector and $\epsilon_{ikj} = -\epsilon_{ijk}$,

$$P_{ij} = -\epsilon_{ijk}\omega_k \tag{2.1.37}$$

which is Equation 2.1.34. By multiplying Equation 2.1.37 by $-\frac{1}{2}\epsilon_{pij}$ we find

$$-\frac{1}{2}\,\epsilon_{pij}P_{ij} = \frac{1}{2}\,\epsilon_{ijk}\epsilon_{pij}\omega_k = -\frac{1}{2}\,\epsilon_{ijk}\epsilon_{ipj}\omega_k \tag{2.1.38}$$

By the $\epsilon - \delta$ identity the RHS equals

$$-\frac{1}{2}\,(\delta_{jp}\delta_{kj} - \delta_{jj}\delta_{pk})\omega_k = -\frac{1}{2}\,(\delta_{pk} - 3\delta_{pk})\omega_k = \delta_{pk}\omega_k = \omega_p \tag{2.1.39}$$

From this follows Equation 2.1.33.

For any tensor $\mathbf{T}$ the trace of $\mathbf{T}$ is denoted by $\mathrm{tr}(\mathbf{T})$. In index notation,

$$\mathrm{tr}(\mathbf{T}) = T_{ii} \tag{2.1.40}$$

The determinant of $\mathbf{T}$ is denoted by $\det(\mathbf{T})$ and it is

$$\det(\mathbf{T}) = \begin{vmatrix} T_{11} & T_{12} & T_{13} \\ T_{21} & T_{22} & T_{23} \\ T_{31} & T_{32} & T_{33} \end{vmatrix} \tag{2.1.41}$$

It can be proved that for any two tensors $\mathbf{A}$ and $\mathbf{B}$

$$\mathrm{tr}(\mathbf{AB}) = \mathrm{tr}(\mathbf{BA}) \tag{2.1.42}$$

and

$$\det(\mathbf{AB}) = \det(\mathbf{A}) \det(\mathbf{B}) \tag{2.1.43}$$

We will now give alternative definitions of a vector and a tensor using certain rules of coordinate transformations described by an *orthogonal tensor* $\mathbf{Q}$. We say that $\mathbf{Q}$ is orthogonal if and only if

$$\mathbf{Q}^T\mathbf{Q} = \mathbf{Q}\mathbf{Q}^T = \mathbf{1} \tag{2.1.44}$$

For an orthonormal basis $\mathbf{e}_i$ and orthogonal $\mathbf{Q}$, the vectors

$$\mathbf{e}'_i = \mathbf{Q}\mathbf{e}_i \tag{2.1.45}$$

form an orthonormal basis. Also, for two orthonormal bases $\mathbf{e}_i$ and $\mathbf{e}'_i$, there exists a unique orthogonal tensor $\mathbf{Q}$ such that $\mathbf{e}'_i = \mathbf{Q}\mathbf{e}_i$. The proof will be given in Example 2.1.7.

For a vector $\mathbf{w}$, of which components in the basis $\mathbf{e}_i$ are w_i and in the basis $\mathbf{e}'_i$ are w'_i, the following rules of transformation hold:

$$w'_i = Q_{ji} w_j \tag{2.1.46}$$

and

$$w_j = Q_{ji} w'_i \tag{2.1.47}$$

Similarly, for a tensor $\mathbf{T}$ we get

$$T'_{ij} = Q_{ki} Q_{lj} T_{kl} \tag{2.1.48}$$

$$T_{kl} = Q_{ki} Q_{lj} T'_{ij} \tag{2.1.49}$$

In these transformation rules, the components of the orthogonal tensor $\mathbf{Q}$ may be interpreted as the directional cosines between the primed and unprimed coordinate axes (Figure 2.1)

$$Q_{ji} = \mathbf{e}'_i \cdot \mathbf{e}_j = \cos\left(\mathbf{e}'_i, \mathbf{e}_j\right) \tag{2.1.50}$$

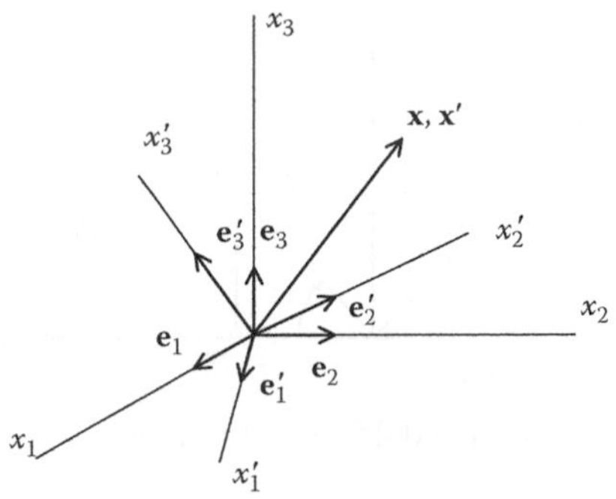

FIGURE 2.1 The primed and unprimed coordinate axes with the origin at $\mathbf{x} = 0$.

The formulas 2.1.46 and 2.1.47, and 2.1.48 and 2.1.49 provide alternative definitions of a vector and a tensor in terms of the rules of the coordinate transformations. The set of three quantities w_j, or nine quantities T_{kl}, referred to the basis $\mathbf{e}_i$, which transform to another set w_i', or T_{ij}', in the basis $\mathbf{e}_i'$ according to Equations 2.1.46 and 2.1.47, or Equations 2.1.48 and 2.1.49, is defined as a vector, or a tensor, respectively.

Example 2.1.2

In the x_i system, a vector $\mathbf{w}$ and a tensor $\mathbf{T}$ are given by

$$\mathbf{w} = [-1, 1, 1]^{\mathrm{T}} \quad \text{(a)}$$

and

$$\mathbf{T} = \begin{bmatrix} 0 & 1 & 0 \\ -1 & 0 & 1 \\ 0 & -1 & 0 \end{bmatrix} \quad \text{(b)}$$

respectively. Show that in the x_i' system, which is obtained by rotating the x_i system about the x_3 axis through an angle of 45°

$$\mathbf{w}' = \left[0, \sqrt{2}, 1\right]^{\mathrm{T}} \quad \text{(c)}$$

and

$$\mathbf{T}' = \begin{bmatrix} 0 & 1 & 1/\sqrt{2} \\ -1 & 0 & 1/\sqrt{2} \\ -1/\sqrt{2} & -1/\sqrt{2} & 0 \end{bmatrix} \quad \text{(d)}$$

Solution

It follows from Equations 2.1.46 and 2.1.48 that

$$\mathbf{w}' = \mathbf{Q}^{\mathrm{T}}\mathbf{w} \quad \text{(e)}$$

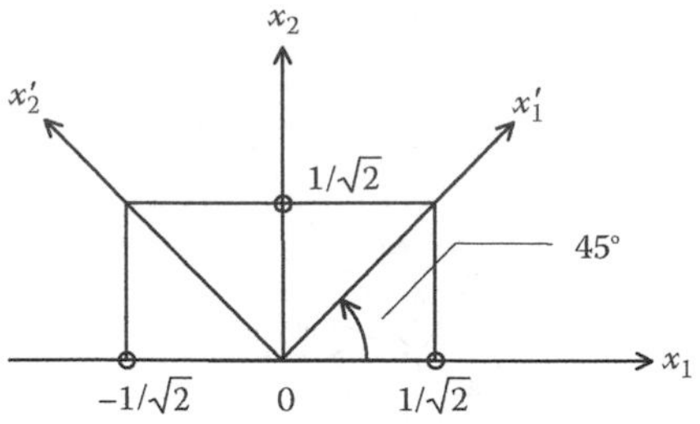

FIGURE 2.2 The rectangular axes rotated 45° about x_3 axis.

and

$$\mathbf{T}' = \mathbf{Q}^{\mathrm{T}}\mathbf{T}\mathbf{Q} \tag{f}$$

where (see Equation 2.1.50)

$$Q_{ij}^{\mathrm{T}} = \mathbf{e}_i' \cdot \mathbf{e}_j \tag{g}$$

The primed axes x_i' and the unprimed axes x_i are shown in Figure 2.2

The x_3 and x_3' axes coincide, and are orthogonal to the (x_1, x_2) plane. The unit vectors $\mathbf{e}_i$ and $\mathbf{e}_i'$ are therefore given by

$$\mathbf{e}_1 = (1, 0, 0), \quad \mathbf{e}_2 = (0, 1, 0), \quad \mathbf{e}_3 = (0, 0, 1) \tag{h}$$

and

$$\mathbf{e}_1' = \left(1/\sqrt{2}, 1/\sqrt{2}, 0\right), \quad \mathbf{e}_2' = \left(-1/\sqrt{2}, 1/\sqrt{2}, 0\right), \quad \mathbf{e}_3' = (0, 0, 1) \tag{i}$$

Substituting the vectors $\mathbf{e}_i$ and $\mathbf{e}_i'$ from (h) and (i) into (g), we obtain

$$\mathbf{Q}^{\mathrm{T}} = \begin{bmatrix} 1/\sqrt{2} & 1/\sqrt{2} & 0 \\ -1/\sqrt{2} & 1/\sqrt{2} & 0 \\ 0 & 0 & 1 \end{bmatrix} \tag{j}$$

Hence, using Equation (e), we obtain

$$\begin{bmatrix} w_1' \\ w_2' \\ w_3' \end{bmatrix} = \begin{bmatrix} 1/\sqrt{2} & 1/\sqrt{2} & 0 \\ -1/\sqrt{2} & 1/\sqrt{2} & 0 \\ 0 & 0 & 1 \end{bmatrix} \begin{bmatrix} -1 \\ 1 \\ 1 \end{bmatrix} = \begin{bmatrix} 0 \\ \sqrt{2} \\ 1 \end{bmatrix} \tag{k}$$

Similarly, by substituting $\mathbf{Q}^{\mathrm{T}}$ from Equation (j) into Equation (f), we obtain

$$\mathbf{T}' = \begin{bmatrix} 1/\sqrt{2} & 1/\sqrt{2} & 0 \\ -1/\sqrt{2} & 1/\sqrt{2} & 0 \\ 0 & 0 & 1 \end{bmatrix} \begin{bmatrix} 0 & 1 & 0 \\ -1 & 0 & 1 \\ 0 & -1 & 0 \end{bmatrix} \begin{bmatrix} 1/\sqrt{2} & -1/\sqrt{2} & 0 \\ 1/\sqrt{2} & 1/\sqrt{2} & 0 \\ 0 & 0 & 1 \end{bmatrix}$$

$$= \begin{bmatrix} 0 & 1 & 1/\sqrt{2} \\ -1 & 0 & 1/\sqrt{2} \\ -1/\sqrt{2} & -1/\sqrt{2} & 0 \end{bmatrix} \tag{l}$$

This completes the solution. □

In what follows, a concept of the *tensor product of two vectors* **a** and **b** denoted by $\mathbf{a} \otimes \mathbf{b}$ will be useful. It is defined by

$$(\mathbf{a} \otimes \mathbf{b})\mathbf{u} = (\mathbf{b} \cdot \mathbf{u})\mathbf{a} \tag{2.1.51}$$

for any vector **u**. In components

$$(\mathbf{a} \otimes \mathbf{b})_{ij} = a_i b_j \tag{2.1.52}$$

Indeed, from Equation 2.1.51 follows

$$(\mathbf{a} \otimes \mathbf{b})_{ij} u_j = b_k u_k a_i \tag{2.1.53}$$

which by changing $k \to j$ yields

$$[(\mathbf{a} \otimes \mathbf{b})_{ij} - b_j a_i] u_j = 0 \tag{2.1.54}$$

and Equation 2.1.52 follows. Clearly, $\mathbf{a} \otimes \mathbf{b}$ is a tensor, and

$$\mathrm{tr}(\mathbf{a} \otimes \mathbf{b}) = \mathbf{a} \cdot \mathbf{b} \tag{2.1.55}$$

Similarly to the inner product of two vectors, we introduce the *inner product of two tensors* **A** and **B** by

$$\mathbf{A} \cdot \mathbf{B} = \mathrm{tr}(\mathbf{A}^T \mathbf{B}) = A_{ij} B_{ij} \tag{2.1.56}$$

The *magnitude* of **A** is defined by

$$|\mathbf{A}| = (\mathbf{A} \cdot \mathbf{A})^{\frac{1}{2}} \tag{2.1.57}$$

With the stated definitions it is easy to prove that the nine tensors $\mathbf{e}_i \otimes \mathbf{e}_j$ are orthonormal, therefore, for any tensor **T** the following relation holds:

$$\mathbf{T} = T_{ij} \mathbf{e}_i \otimes \mathbf{e}_j \tag{2.1.58}$$

The proof is based on the relation

$$\mathbf{a} \cdot \mathbf{A}\mathbf{b} = \mathbf{A} \cdot (\mathbf{a} \otimes \mathbf{b}) \tag{2.1.59}$$

where
- **A** is an arbitrary tensor
- **a** and **b** are arbitrary vectors

In components Equation 2.1.59 takes the form

$$a_k A_{ki} b_i = A_{ki} a_k b_i \tag{2.1.60}$$

Substituting in Equation 2.1.59

$$\mathbf{a} = \mathbf{e}_k, \quad \mathbf{b} = \mathbf{e}_l, \quad \mathbf{A} = \mathbf{e}_i \otimes \mathbf{e}_j \tag{2.1.61}$$

we obtain

$$\mathbf{e}_k \cdot [(\mathbf{e}_i \otimes \mathbf{e}_j)\mathbf{e}_l] = (\mathbf{e}_i \otimes \mathbf{e}_j) \cdot (\mathbf{e}_k \otimes \mathbf{e}_l) \tag{2.1.62}$$

We note that for any four vectors **a**, **b**, **c**, **d** the following identity is satisfied:

$$\mathbf{a} \cdot [(\mathbf{b} \otimes \mathbf{c})\mathbf{d}] = a_k b_k c_j d_j = (\mathbf{a} \cdot \mathbf{b})(\mathbf{c} \cdot \mathbf{d}) \tag{2.1.63}$$

Taking here

$$\mathbf{a} = \mathbf{e}_k, \quad \mathbf{b} = \mathbf{e}_i, \quad \mathbf{c} = \mathbf{e}_j, \quad \mathbf{d} = \mathbf{e}_l \tag{2.1.64}$$

we get

$$\mathbf{e}_k \cdot [(\mathbf{e}_i \otimes \mathbf{e}_j)\mathbf{e}_l] = (\mathbf{e}_k \cdot \mathbf{e}_i)(\mathbf{e}_j \cdot \mathbf{e}_l) \tag{2.1.65}$$

Since

$$\mathbf{e}_k \cdot \mathbf{e}_i = \delta_{ki} \quad \text{and} \quad \mathbf{e}_j \cdot \mathbf{e}_l = \delta_{jl} \tag{2.1.66}$$

we have from Equations 2.1.62 and 2.1.65

$$(\mathbf{e}_i \otimes \mathbf{e}_j) \cdot (\mathbf{e}_k \otimes \mathbf{e}_l) = \delta_{ki}\delta_{jl} \tag{2.1.67}$$

which means that nine tensors $\mathbf{e}_i \otimes \mathbf{e}_j$ are orthonormal.

To prove Equation 2.1.58, we use the vector identity

$$(\mathbf{a} \otimes \mathbf{b})\mathbf{c} = \mathbf{a}(\mathbf{b} \cdot \mathbf{c}) \tag{2.1.68}$$

or, in components,

$$a_i b_j c_j = a_i b_j c_j \tag{2.1.69}$$

Putting in this identity

$$\mathbf{a} = T_{ij}\mathbf{e}_i, \quad \mathbf{b} = \mathbf{e}_j, \quad \mathbf{c} = \mathbf{v} \tag{2.1.70}$$

where **v** is an arbitrary vector, we obtain

$$(T_{ij}\mathbf{e}_i \otimes \mathbf{e}_j)\mathbf{v} = T_{ij}\mathbf{e}_i(\mathbf{e}_j \cdot \mathbf{v}) \tag{2.1.71}$$

Since $\mathbf{e}_j \cdot \mathbf{v} = v_j$ then

$$\text{RHS} = \mathbf{e}_i T_{ij} v_j = \mathbf{T}\mathbf{v} \tag{2.1.72}$$

This relation follows from the fact that

$$(\mathbf{a})_p = \mathbf{a} \cdot \mathbf{e}_p \tag{2.1.73}$$

and

$$(\mathbf{e}_i T_{ij} v_j)_p = \mathbf{e}_i T_{ij} v_j \cdot \mathbf{e}_p = T_{ij} v_j (\mathbf{e}_i \cdot \mathbf{e}_p) = T_{ij} v_j \delta_{ip} = T_{pj} v_j \tag{2.1.74}$$

Hence Equation 2.1.71 can be presented as

$$[(T_{ij}\mathbf{e}_i \otimes \mathbf{e}_j) - \mathbf{T}]\mathbf{v} = \mathbf{0} \tag{2.1.75}$$

Since $\mathbf{v}$ is arbitrary, we obtain Equation 2.1.58, which constitutes the *decomposition formula for a tensor* $\mathbf{T}$ in terms of the nine orthonormal tensors $\mathbf{e}_i \otimes \mathbf{e}_j$.

The formula 2.1.58 is an extension of the vector decomposition formula

$$\mathbf{a} = a_i \mathbf{e}_i \tag{2.1.76}$$

where $\mathbf{a}$ is an arbitrary vector.

Example 2.1.3

Show the following three identities

$$(\mathbf{a} \otimes \mathbf{b} - \mathbf{b} \otimes \mathbf{a})\mathbf{u} = \mathbf{u} \times (\mathbf{a} \times \mathbf{b}) \tag{a}$$

$$\mathbf{T}(\mathbf{a} \otimes \mathbf{b}) = (\mathbf{T}\mathbf{a}) \otimes \mathbf{b} \tag{b}$$

$$(\mathbf{a} \otimes \mathbf{b})(\mathbf{u} \otimes \mathbf{v}) = (\mathbf{b} \cdot \mathbf{u})(\mathbf{a} \otimes \mathbf{v}) \tag{c}$$

Solution

(a) Writing in components, we have

$$(\mathrm{LHS})_i = [(\mathbf{b} \cdot \mathbf{u})\mathbf{a} - (\mathbf{a} \cdot \mathbf{u})\mathbf{b}]_i = b_k u_k a_i - a_k u_k b_i \tag{d}$$

$$\begin{aligned}(\mathrm{RHS})_i &= \epsilon_{ijk} u_j \epsilon_{kpq} a_p b_q = \epsilon_{ijk} \epsilon_{pqk} u_j a_p b_q \\ &= (\delta_{ip}\delta_{jq} - \delta_{iq}\delta_{jp}) u_j a_p b_q = u_q b_q a_i - b_i a_j u_j \end{aligned} \tag{e}$$

Thus, LHS = RHS.

(b) By the definition of the product of two tensors, see Equation 2.1.17 where $\mathbf{A} \equiv \mathbf{T}$, $\mathbf{B} = \mathbf{a} \otimes \mathbf{b}$

$$(\mathrm{LHS})_{ij} = T_{ik} a_k b_j \tag{f}$$

$$(\mathrm{RHS})_{ij} = T_{ik} a_k b_j \tag{g}$$

Thus, LHS = RHS.

(c)

$$(\mathrm{LHS})_{ij} = a_i b_k u_k v_j \tag{h}$$

$$(\mathrm{RHS})_{ij} = b_k u_k a_i v_j \tag{i}$$

Thus, LHS = RHS. □

Example 2.1.4

Show that if **1** is a unit tensor and **T** is an arbitrary tensor then the following identities are satisfied:

$$\mathbf{1T} = \mathbf{T1} = \mathbf{T} \tag{a}$$

$$\mathbf{1} \cdot \mathbf{T} = \mathbf{T} \cdot \mathbf{1} = \operatorname{tr} \mathbf{T} \tag{b}$$

$$\mathbf{1} \cdot (\mathbf{a} \otimes \mathbf{b}) = (\mathbf{a} \otimes \mathbf{b}) \cdot \mathbf{1} = \operatorname{tr}(\mathbf{a} \otimes \mathbf{b}) = \mathbf{a} \cdot \mathbf{b} \tag{c}$$

$$\mathbf{e}_k \otimes \mathbf{e}_k = \mathbf{1} \ \ \text{(summation convention holds)} \tag{d}$$

Solution

Writing in components, we obtain

(a)

$$(\mathbf{1T})_{ij} = \delta_{ik} T_{kj} = T_{ij} \tag{e}$$

$$(\mathbf{T1})_{ij} = T_{ik} \delta_{kj} = T_{ij} \tag{f}$$

(b) From the definition 2.1.56 where $\mathbf{A} = \mathbf{1}$ and $\mathbf{B} = \mathbf{T}$

$$\mathbf{1} \cdot \mathbf{T} = \delta_{ij} T_{ij} = T_{ii} = \operatorname{tr} \mathbf{T} \tag{g}$$

$$\mathbf{T} \cdot \mathbf{1} = T_{ij} \delta_{ij} = T_{ii} = \operatorname{tr} \mathbf{T} \tag{h}$$

(c)

$$\delta_{ij} a_i b_j = a_i b_j \delta_{ij} = a_i b_i = \mathbf{a} \cdot \mathbf{b} = \operatorname{tr}(\mathbf{a} \otimes \mathbf{b}) \tag{i}$$

(d)

$$(\text{LHS})_{ij} = (\mathbf{e}_k)_i (\mathbf{e}_k)_j = (\mathbf{e}_k \cdot \mathbf{e}_i)(\mathbf{e}_k \cdot \mathbf{e}_j) = \delta_{ki} \delta_{kj} = \delta_{ij} \tag{j}$$

$$(\text{RHS})_{ij} = \delta_{ij}$$

Thus LHS = RHS. □

Example 2.1.5

Show that if **A** is a symmetric tensor and **B** is a skew tensor then

$$\mathbf{A} \cdot \mathbf{B} = 0 \tag{a}$$

Solution

By the definition of the inner product of two tensors,

$$\text{LHS} = A_{ij} B_{ij} \tag{b}$$

Since $A_{ij} = A_{ji}$ and $B_{ij} = -B_{ji}$ then

$$\text{LHS} = -A_{ij}B_{ji} \tag{c}$$

By symmetry of **A**

$$\text{LHS} = -A_{ji}B_{ji} = -\mathbf{A} \cdot \mathbf{B} = -\text{LHS} \tag{d}$$

Hence LHS = 0. □

Example 2.1.6

Show that for an orthonormal basis $\mathbf{e}_i$ and an orthogonal tensor $\mathbf{Q}$ the vectors $\mathbf{e}'_i = \mathbf{Q}\mathbf{e}_i$ form an orthonormal basis.

Solution

Using the representation 2.1.58 in which **T** is replaced by **Q** we obtain

$$\mathbf{Q} = Q_{ij}\mathbf{e}_i \otimes \mathbf{e}_j = Q_{kp}\mathbf{e}_k \otimes \mathbf{e}_p \tag{a}$$

Hence

$$\mathbf{Q}\mathbf{e}_i = Q_{kp}(\mathbf{e}_k \otimes \mathbf{e}_p)\mathbf{e}_i \tag{b}$$

and

$$\mathbf{e}'_i \cdot \mathbf{e}'_j = \mathbf{e}'_i \cdot \mathbf{Q}\mathbf{e}_j = Q_{kp}\mathbf{e}'_i \cdot [(\mathbf{e}_k \otimes \mathbf{e}_p)\mathbf{e}_j] \tag{c}$$

Using 2.1.63 we get

$$\mathbf{e}'_i \cdot \mathbf{e}'_j = Q_{kp}(\mathbf{e}'_i \cdot \mathbf{e}_k)(\mathbf{e}_p \cdot \mathbf{e}_j) = Q_{kp}(\mathbf{e}'_i \cdot \mathbf{e}_k)\delta_{pj} = Q_{kj}(\mathbf{e}'_i \cdot \mathbf{e}_k) \tag{d}$$

On the other hand,

$$\begin{aligned}
\mathbf{e}'_i \cdot \mathbf{e}_k &= (\mathbf{Q}\mathbf{e}_i) \cdot \mathbf{e}_k = [(Q_{pq}\mathbf{e}_p \otimes \mathbf{e}_q)\mathbf{e}_i] \cdot \mathbf{e}_k = Q_{pq}\mathbf{e}_k \cdot [(\mathbf{e}_p \otimes \mathbf{e}_q)\mathbf{e}_i] \\
&= Q_{pq}(\mathbf{e}_k \cdot \mathbf{e}_p)(\mathbf{e}_q \cdot \mathbf{e}_i) \quad \text{see Equation 2.1.65} \\
&= Q_{pq}\delta_{kp}\delta_{qi} = Q_{ki} \qquad \text{(e)}
\end{aligned}$$

Therefore,

$$\mathbf{e}'_i \cdot \mathbf{e}'_j = Q_{kj}Q_{ki} = \delta_{ij} \tag{f}$$

since **Q** is orthogonal. To check the last result we use the definition of orthogonality, see Equation 2.1.44,

$$(\mathbf{Q}\mathbf{Q}^T)_{ij} = Q_{ik}Q^T_{kj} = Q_{ik}Q_{jk} = \delta_{ij} \tag{g}$$

$$(\mathbf{Q}^T\mathbf{Q})_{ij} = Q^T_{ik}Q_{kj} = Q_{ki}Q_{kj} = \delta_{ij} \tag{h}$$

This completes the solution. □

Example 2.1.7

Show that for two orthonormal bases $\mathbf{e}_i$ and $\mathbf{e}'_i$, there exists a unique orthogonal tensor $\mathbf{Q}$ such that $\mathbf{e}'_i = \mathbf{Q}\mathbf{e}_i$.

Solution

We introduce a tensor $\mathbf{Q}$ by the formula

$$Q_{ki} = \mathbf{e}'_i \cdot \mathbf{e}_k \tag{a}$$

For an arbitrary position vector $\mathbf{x}$, see Figure 2.1,

$$\mathbf{x} = x_p \mathbf{e}_p \tag{b}$$

Also

$$\mathbf{x} = x'_p \mathbf{e}'_p \tag{c}$$

Taking the inner product with $\mathbf{e}'_i$ on both sides of (b) we obtain

$$\mathbf{e}'_i \cdot \mathbf{x} = \mathbf{e}'_i \cdot x_p \mathbf{e}_p = x_p Q_{pi} \tag{d}$$

Using (c),

$$\mathbf{e}'_i \cdot \mathbf{x} = \mathbf{e}'_i \cdot x'_p \mathbf{e}'_p = x'_p \delta_{ip} = x'_i \tag{e}$$

Comparing, we get

$$x'_i = x_p Q_{pi} \tag{f}$$

Similarly, by (c)

$$x_i = \mathbf{x} \cdot \mathbf{e}_i = x'_p \mathbf{e}'_p \cdot \mathbf{e}_i = x'_p Q_{ip} \tag{g}$$

Substituting this into (f) after changing $p \to k$ and $i \to p$, we obtain

$$x'_i = x'_k Q_{pk} Q_{pi} \tag{h}$$

Now, since $x'_i = \delta_{ik} x'_k$ we may write

$$(\delta_{ik} - Q_{pk} Q_{pi}) x'_i = 0 \tag{i}$$

Since x'_i is an arbitrary vector,

$$Q_{pk} Q_{pi} = \delta_{ik} \tag{j}$$

which is one of the two orthogonality conditions. Similarly, substitution x'_i given by (f) into (g) yields

$$x_i = x_k Q_{kp} Q_{ip} \tag{k}$$

or

$$(\delta_{ik} - Q_{kp}Q_{ip})x_k = 0 \tag{l}$$

or

$$Q_{kp}Q_{ip} = \delta_{ik} \tag{m}$$

which constitutes the other orthogonality condition. □

In the tensor algebra an important notion is that of an *eigenvector* and an associated *eigenvalue* of a tensor **T**. We will call λ an eigenvalue corresponding to an eigenvector **u** of a tensor **T** if

$$\mathbf{Tu} = \lambda\mathbf{u} \tag{2.1.77}$$

In addition to the name *eigenvalue* the names a *characteristic value* or a *principal value* are used. Similarly, *eigenvector* is called a *characteristic vector* or a *principal vector*, or a principal direction.

A subspace $\mathcal{W}$ of the vector space $\mathcal{V}$ consisting of all vectors **u** that satisfy $\mathbf{Tu} = \lambda\mathbf{u}$ is called a *characteristic space for* **T** *corresponding to* λ. For any symmetric tensor **T** the following theorem called the *spectral theorem* holds true [1]:

For a symmetric tensor **T** there exists an orthonormal basis $\mathbf{n}_1$, $\mathbf{n}_2$, $\mathbf{n}_3$ and three principal values λ_1, λ_2, λ_3 of **T** such that

$$\mathbf{Tn}_1 = \lambda_1\mathbf{n}_1, \quad \mathbf{Tn}_2 = \lambda_2\mathbf{n}_2, \quad \mathbf{Tn}_3 = \lambda_3\mathbf{n}_3 \tag{2.1.78}$$

and

$$\mathbf{T} = \lambda_1\mathbf{n}_1 \otimes \mathbf{n}_1 + \lambda_2\mathbf{n}_2 \otimes \mathbf{n}_2 + \lambda_3\mathbf{n}_3 \otimes \mathbf{n}_3 \tag{2.1.79}$$

Conversely, if **T** is of the form 2.1.79 with $\{\mathbf{n}_i\}$ orthonormal, then the relations 2.1.78 are satisfied. The formula 2.1.79 is called the spectral decomposition of **T**.

Special cases of the spectral theorem:

(a) Principal values λ_1, λ_2, λ_3 are all distinct. In this case the characteristic spaces $\mathcal{W}_1$, $\mathcal{W}_2$, and $\mathcal{W}_3$ for **T** are the lines spanned by $\mathbf{n}_1$, $\mathbf{n}_2$, and $\mathbf{n}_3$, respectively.

(b) Two principal values are equal, say $\lambda_2 = \lambda_3$. Then Equation 2.1.79 takes the form

$$\mathbf{T} = \lambda_1\mathbf{n}_1 \otimes \mathbf{n}_1 + \lambda_2(\mathbf{1} - \mathbf{n}_1 \otimes \mathbf{n}_1) \tag{2.1.80}$$

Conversely, if Equation 2.1.80 holds with $\lambda_1 \neq \lambda_2$ then λ_1 and λ_2 are the only principal values of **T**, and **T** has two distinct characteristic spaces $\mathcal{W}_1$ and $\mathcal{W}_2$: the line spanned by $\mathbf{n}_1$ and the plane perpendicular to $\mathbf{n}_1$.

(c) All three principal values are equal: $\lambda_1 = \lambda_2 = \lambda_3 = \lambda$. Then Equation 2.1.79 takes the form

$$\mathbf{T} = \lambda\mathbf{1} \tag{2.1.81}$$

and the vector space $\mathcal{V}$ is the only characteristic space for **T**.

The proof of the spectral theorem in its entirety will not be given. We present, however, the proof of the relation 2.1.79.

First, we note that **T** can be represented by the formula, see Equation 2.1.58,

$$\mathbf{T} = T_{ij}\mathbf{e}_i \otimes \mathbf{e}_j \tag{2.1.82}$$

where $\mathbf{e}_i$ is an arbitrary orthonormal basis. We identify $\mathbf{e}_i$ with $\mathbf{n}_i$, where $\mathbf{n}_i$ are the unit eigenvectors of **T** associated with eigenvalues λ_i. This means, that for any fixed index k_0 from the set $(1, 2, 3)$

$$\mathbf{T}\mathbf{n}_{k_0} = \lambda_{k_0}\mathbf{n}_{k_0} \quad \text{(no summation over } k_0) \tag{2.1.83}$$

This relation, together with Equation 2.1.82, takes the form

$$\lambda_{k_0}\mathbf{n}_{k_0} = T_{ij}(\mathbf{n}_i \otimes \mathbf{n}_j)\mathbf{n}_{k_0} \tag{2.1.84}$$

Now,

$$(\mathbf{n}_i \otimes \mathbf{n}_j)\mathbf{n}_{k_0} = (\mathbf{n}_j \cdot \mathbf{n}_{k_0})\mathbf{n}_i \tag{2.1.85}$$

Since

$$\mathbf{n}_j \cdot \mathbf{n}_{k_0} = \delta_{jk_0} \tag{2.1.86}$$

therefore, from Equation 2.1.84 we get

$$\lambda_{k_0}\mathbf{n}_{k_0} = T_{ik_0}\mathbf{n}_i \tag{2.1.87}$$

Taking tensor product of $\mathbf{n}_{k_0}$ with Equation 2.1.87 we obtain

$$\lambda_{k_0}\mathbf{n}_{k_0} \otimes \mathbf{n}_{k_0} = T_{ik_0}\mathbf{n}_i \otimes \mathbf{n}_{k_0} \tag{2.1.88}$$

Now, by taking the sum of Equation 2.1.88 over k_0 from 1 to 3, and using Equation 2.1.82 we arrive at

$$\sum_{i=1}^{3} \lambda_i \mathbf{n}_i \otimes \mathbf{n}_i = T_{ij}\mathbf{n}_i \otimes \mathbf{n}_j = \mathbf{T} \tag{2.1.89}$$

This completes the proof of relation 2.1.79.

The formula 2.1.79 is consistent with the symmetry of tensor **T**. Some discussion of relations between formulas 2.1.79 through 2.1.81, and spectral decomposition for particular tensors is given in the examples.

Example 2.1.8

Show that if $\lambda_2 = \lambda_3$ then Equation 2.1.79 implies Equation 2.1.80.

Solution

Taking $\lambda_2 = \lambda_3$ in Equation 2.1.79 we get

$$\mathbf{T} = \lambda_1 \mathbf{n}_1 \otimes \mathbf{n}_1 + \lambda_2 [\mathbf{n}_2 \otimes \mathbf{n}_2 + \mathbf{n}_3 \otimes \mathbf{n}_3] \tag{a}$$

Setting in Example 2.1.4 part (d) $\mathbf{e}_k = \mathbf{n}_k$, we have

$$\mathbf{n}_2 \otimes \mathbf{n}_2 + \mathbf{n}_3 \otimes \mathbf{n}_3 = \mathbf{1} - \mathbf{n}_1 \otimes \mathbf{n}_1 \tag{b}$$

From here and from (a) the result 2.1.80 follows. □

Example 2.1.9

Show that if $\lambda_1 = \lambda_2 = \lambda_3 = \lambda$ then

$$\mathbf{T} = \lambda \mathbf{1} \tag{a}$$

Solution

The solution follows directly from Equation 2.1.79 and (b) in Example 2.1.8. □

Example 2.1.10

Show that the eigenvalues λ of a tensor $\mathbf{T}$ satisfy the characteristic equation

$$\lambda^3 - I_1 \lambda^2 + I_2 \lambda - I_3 = 0 \tag{a}$$

where the coefficients I_1, I_2, I_3 are expressed in terms of $\mathbf{T}$ by

$$I_1 = T_{ii} = \operatorname{tr} \mathbf{T} \tag{b}$$

$$I_2 = \frac{1}{2} (T_{ii} T_{jj} - T_{ij} T_{ji}) = \frac{1}{2} [(\operatorname{tr} \mathbf{T})^2 - \operatorname{tr} (\mathbf{T}^2)] \tag{c}$$

$$I_3 = \epsilon_{ijk} T_{1i} T_{2j} T_{3k} = \det \mathbf{T} \tag{d}$$

Here I_1, I_2, I_3 are called the first, the second, and the third invariant, respectively, of the tensor $\mathbf{T}$.

Solution

Let λ be an eigenvalue of $\mathbf{T}$ corresponding to an eigenvector $\mathbf{n}$. Then

$$\mathbf{T}\mathbf{n} = \lambda \mathbf{n} \tag{e}$$

or in components

$$(T_{ij} - \lambda\delta_{ij})n_j = 0 \tag{f}$$

Since $|\mathbf{n}| = 1$, (f) is satisfied if and only if

$$\det(\mathbf{T} - \lambda\mathbf{1}) = 0 \tag{g}$$

Hence, by virtue of (a) in Example 2.1.1, we get

$$\epsilon_{ijk}(T_{1i} - \lambda\delta_{1i})(T_{2j} - \lambda\delta_{2j})(T_{3k} - \lambda\delta_{3k}) = 0 \tag{h}$$

or

$$\begin{aligned}\epsilon_{ijk}\big[T_{1i}T_{2j}T_{3k} &- \lambda\left(T_{1i}T_{2j}\delta_{3k} + T_{1i}T_{3k}\delta_{2j} + T_{2j}T_{3k}\delta_{1i}\right)\\ &+\lambda^2\left(T_{1i}\delta_{2j}\delta_{3k} + T_{2j}\delta_{1i}\delta_{3k} + T_{3k}\delta_{1i}\delta_{2j}\right) - \lambda^3\delta_{1i}\delta_{2j}\delta_{3k}\big] = 0\end{aligned} \tag{i}$$

or equivalently

$$\begin{aligned}\epsilon_{ijk}T_{1i}T_{2j}T_{3k} &- \lambda\left(\epsilon_{ij3}T_{1i}T_{2j} + \epsilon_{i2k}T_{1i}T_{3k} + \epsilon_{1jk}T_{2j}T_{3k}\right)\\ &+ \lambda^2\left(\epsilon_{i23}T_{1i} + \epsilon_{1j3}T_{2j} + \epsilon_{12k}T_{3k}\right) - \lambda^3 = 0\end{aligned} \tag{j}$$

Now, by (a) in Example 2.1.1

$$\epsilon_{ijk}T_{1i}T_{2j}T_{3k} = \det(\mathbf{T}) \tag{k}$$

Also, by direct calculations, we get

$$\epsilon_{ij3}T_{1i}T_{2j} = T_{11}T_{22} - T_{12}T_{21} \tag{l}$$

$$\epsilon_{i2k}T_{1i}T_{3k} = T_{11}T_{33} - T_{13}T_{31} \tag{m}$$

$$\epsilon_{1jk}T_{2j}T_{3k} = T_{22}T_{33} - T_{23}T_{32} \tag{n}$$

and

$$\epsilon_{ij3}T_{1i}T_{2j} + \epsilon_{i2k}T_{1i}T_{3k} + \epsilon_{1jk}T_{2j}T_{3k} = \frac{1}{2}\left[(T_{kk})^2 - T_{ik}T_{ki}\right] \tag{o}$$

So, Equation (j) takes the form

$$\det(\mathbf{T}) - \frac{1}{2}\lambda[(\operatorname{tr}\mathbf{T})^2 - (\operatorname{tr}\mathbf{T}^2)] + \lambda^2(\operatorname{tr}\mathbf{T}) - \lambda^3 = 0 \tag{p}$$

and introducing the first, second, and third invariant of **T** by

$$I_1 = \operatorname{tr}\mathbf{T} \tag{q}$$

$$I_2 = \frac{1}{2}\,[(\operatorname{tr}\mathbf{T})^2 - (\operatorname{tr}\mathbf{T}^2)] \tag{r}$$

$$I_3 = \det(\mathbf{T}) \tag{s}$$

we arrive at (a). This completes the solution. □

The next two examples show the prescription for actual calculations.

Example 2.1.11

Find the eigenvalues and eigenvectors of the tensor **T** whose matrix is

$$\mathbf{T} = \begin{bmatrix} 1 & 1 & 0 \\ 1 & 2 & 0 \\ 0 & 0 & 1 \end{bmatrix} \quad \text{(a)}$$

Solution

It follows from Example 2.1.10 that

$$(\mathbf{T} - \lambda \mathbf{1})\mathbf{n} = \mathbf{0} \quad \text{(b)}$$

where λ is an eigenvalue corresponding to an eigenvector $\mathbf{n}$. Since $|\mathbf{n}| \neq 0$, therefore

$$\det(\mathbf{T} - \lambda \mathbf{1}) = 0 \quad \text{(c)}$$

or

$$\begin{vmatrix} 1-\lambda & 1 & 0 \\ 1 & 2-\lambda & 0 \\ 0 & 0 & 1-\lambda \end{vmatrix} = 0 \quad \text{(d)}$$

or

$$(1-\lambda)(\lambda^2 - 3\lambda + 1) = 0 \quad \text{(e)}$$

Hence

$$\lambda_1 = \frac{3-\sqrt{5}}{2}, \quad \lambda_2 = 1, \quad \lambda_3 = \frac{3+\sqrt{5}}{2} \quad \text{(f)}$$

are the eigenvalues such that

$$0 < \lambda_1 < \lambda_2 < \lambda_3 \quad \text{(g)}$$

Let $\mathbf{n}^{(i)}$ denote a unit eigenvector corresponding to λ_i $(i = 1,2,3)$. To find $\mathbf{n}^{(1)}$ we substitute λ_1 into Equation (b) and obtain

$$\begin{bmatrix} \frac{\sqrt{5}-1}{2} & 1 & 0 \\ 1 & \frac{\sqrt{5}+1}{2} & 0 \\ 0 & 0 & \frac{\sqrt{5}-1}{2} \end{bmatrix} \begin{bmatrix} n_1^{(1)} \\ n_2^{(1)} \\ n_3^{(1)} \end{bmatrix} = \begin{bmatrix} 0 \\ 0 \\ 0 \end{bmatrix} \quad \text{(h)}$$

Hence

$$\frac{\sqrt{5}-1}{2}n_1^{(1)} + n_2^{(1)} = 0$$

$$n_1^{(1)} + \frac{\sqrt{5}+1}{2}n_2^{(1)} = 0 \tag{i}$$

$$n_3^{(1)} = 0$$

Since

$$\left[n_1^{(1)}\right]^2 + \left[n_2^{(1)}\right]^2 = 1 \tag{j}$$

a solution to Equations (i) and (j) takes the form

$$n_1^{(1)} = \pm\sqrt{\frac{2}{5-\sqrt{5}}}, \quad n_2^{(1)} = \mp\frac{\sqrt{5}-1}{2}\sqrt{\frac{2}{5-\sqrt{5}}}, \quad n_3^{(1)} = 0 \tag{k}$$

Similarly, by substituting $\lambda_2 = 1$ into Equation (b) we obtain

$$\begin{bmatrix} 0 & 1 & 0 \\ 1 & 1 & 0 \\ 0 & 0 & 0 \end{bmatrix}\begin{bmatrix} n_1^{(2)} \\ n_2^{(2)} \\ n_3^{(2)} \end{bmatrix} = \begin{bmatrix} 0 \\ 0 \\ 0 \end{bmatrix} \tag{l}$$

Hence, and from the condition

$$\left[n_1^{(2)}\right]^2 + \left[n_2^{(2)}\right]^2 + \left[n_3^{(2)}\right]^2 = 1 \tag{m}$$

we obtain

$$n_1^{(2)} = 0, \quad n_2^{(2)} = 0, \quad n_3^{(2)} = \pm 1 \tag{n}$$

Finally, substituting λ_3 into Equation (b) we get

$$\begin{bmatrix} -\frac{1+\sqrt{5}}{2} & 1 & 0 \\ 1 & \frac{1-\sqrt{5}}{2} & 0 \\ 0 & 0 & -\frac{1+\sqrt{5}}{2} \end{bmatrix}\begin{bmatrix} n_1^{(3)} \\ n_2^{(3)} \\ n_3^{(3)} \end{bmatrix} = \begin{bmatrix} 0 \\ 0 \\ 0 \end{bmatrix} \tag{o}$$

or

$$-\frac{1+\sqrt{5}}{2}n_1^{(3)} + n_2^{(3)} = 0$$

$$n_1^{(3)} + \frac{1-\sqrt{5}}{2}n_2^{(3)} = 0 \tag{p}$$

$$n_3^{(3)} = 0$$

Since

$$\left[n_1^{(3)}\right]^2 + \left[n_2^{(3)}\right]^2 = 1 \tag{q}$$

therefore a solution of Equations (p) and (q) takes the form

$$n_1^{(3)} = \pm\sqrt{\frac{2}{5+\sqrt{5}}}, \quad n_2^{(3)} = \pm\frac{1+\sqrt{5}}{2}\sqrt{\frac{2}{5+\sqrt{5}}}, \quad n_3^{(3)} = 0 \tag{r}$$

This completes the solution. □

Example 2.1.12

Let $\{\mathbf{e}_i^*\}$ be an orthonormal basis formed from the eigenvectors of a tensor **T** and let $\mathbf{T}^*$ be a tensor referred to the basis $\{\mathbf{e}_i^*\}$. Let $\{\mathbf{e}_i\}$ be an orthonormal basis for **T**. Define the tensor **A** in terms of components by

$$a_{ij} = \mathbf{e}_i^* \cdot \mathbf{e}_j \tag{a}$$

Show that

$$\mathbf{T}^* = \mathbf{A}\mathbf{T}\mathbf{A}^T \tag{b}$$

is a tensor represented by a diagonal matrix.

Solution

From the definition of the tensor **A**, it follows that the components of **A** are the directional cosines between coordinate systems associated with $\mathbf{e}_i^*$ and $\mathbf{e}_j$. This means that **A** is an orthogonal tensor, and (b) is obtained from the definition of a transformation formula for tensors.

From the spectral theorem, formula 2.1.79, it follows that

$$\mathbf{T}^* = \lambda_1\left(\mathbf{e}_1^* \otimes \mathbf{e}_1^*\right) + \lambda_2\left(\mathbf{e}_2^* \otimes \mathbf{e}_2^*\right) + \lambda_3\left(\mathbf{e}_3^* \otimes \mathbf{e}_3^*\right) \tag{c}$$

which means that $\mathbf{T}^*$ is a tensor represented by a diagonal matrix.

Note that in components (b) reads

$$T_{ij}^* = A_{ik}T_{kp}A_{pj}^T = A_{ik}T_{kp}A_{jp} \tag{d}$$

so, if we let $A_{ik} = Q_{ki}$, we get

$$T_{ij}^* = Q_{ki}Q_{pj}T_{kp} \tag{e}$$

which is the transformation formula from $\{\mathbf{e}_i\}$ to $\{\mathbf{e}_i^*\}$ given by Equation 2.1.48 if $\{\mathbf{e}_i^*\} = \{\mathbf{e}_i'\}$. □

Example 2.1.13

Let $\mathbf{T}$ be the tensor from Example 2.1.11, and let $\{\mathbf{e}_i^*\}$ be an orthonormal basis formed from the eigenvectors of $\mathbf{T}$. Define a tensor $\mathbf{Q}^T$ in components by

$$Q_{ij}^T = \mathbf{e}_i^* \cdot \mathbf{e}_j \tag{a}$$

Show that

$$\mathbf{T}^* = \mathbf{Q}^T\mathbf{T}\mathbf{Q} \tag{b}$$

is a tensor represented by a diagonal matrix.

Solution

Using Equations (k), (n), and (r) from Example 2.1.11, and the upper set of the ± signs, we find, respectively,

$$\begin{aligned}
\mathbf{e}_1^* &= \left(\sqrt{\frac{2}{5-\sqrt{5}}}, -\frac{\sqrt{5}-1}{2}\sqrt{\frac{2}{5-\sqrt{5}}}, 0\right)\\
\mathbf{e}_2^* &= (0,0,1)\\
\mathbf{e}_3^* &= \left(\sqrt{\frac{2}{5+\sqrt{5}}}, \frac{1+\sqrt{5}}{2}\sqrt{\frac{2}{5+\sqrt{5}}}, 0\right)
\end{aligned} \tag{c}$$

Since

$$\mathbf{e}_1 = (1,0,0), \quad \mathbf{e}_2 = (0,1,0), \quad \mathbf{e}_3 = (0,0,1) \tag{d}$$

therefore, it follows from Equation (a) that

$$\mathbf{Q}^T = \begin{bmatrix} \sqrt{\frac{2}{5-\sqrt{5}}} & -\frac{\sqrt{5}-1}{2}\sqrt{\frac{2}{5-\sqrt{5}}} & 0 \\ 0 & 0 & 1 \\ \sqrt{\frac{2}{5+\sqrt{5}}} & \frac{\sqrt{5}+1}{2}\sqrt{\frac{2}{5+\sqrt{5}}} & 0 \end{bmatrix} \tag{e}$$

and

$$\mathbf{Q} = \begin{bmatrix} \sqrt{\frac{2}{5-\sqrt{5}}} & 0 & \sqrt{\frac{2}{5+\sqrt{5}}} \\ -\frac{\sqrt{5}-1}{2}\sqrt{\frac{2}{5-\sqrt{5}}} & 0 & \frac{\sqrt{5}+1}{2}\sqrt{\frac{2}{5+\sqrt{5}}} \\ 0 & 1 & 0 \end{bmatrix} \tag{f}$$

Hence, using Equation (a) from Example 2.1.11 and Equations (e) and (f), we obtain

$$\mathbf{T}^* = \mathbf{Q}^T\mathbf{T}\mathbf{Q} = \begin{bmatrix} \dfrac{10-4\sqrt{5}}{5-\sqrt{5}} & 0 & 0 \\ 0 & 1 & 0 \\ 0 & 0 & \dfrac{10+4\sqrt{5}}{5+\sqrt{5}} \end{bmatrix} \tag{g}$$

It is easy to show that

$$\mathbf{Q}^T\mathbf{Q} = \mathbf{1} \tag{h}$$

and

$$\text{tr}\,\mathbf{T} = \text{tr}\,\mathbf{T}^* = 4 \tag{i}$$

Therefore, $\mathbf{T}^*$ is a diagonal tensor obtained from a tensor transformation formula 2.1.48 in which $\{\mathbf{e}'_i\} = \{\mathbf{e}^*_i\}$.
This completes the solution. □

To discuss other properties of principal values and principal directions of $\mathbf{T}$, we introduce the inverse of $\mathbf{T}$ denoted by $\mathbf{T}^{-1}$ and defined by the relations

$$\mathbf{T}\mathbf{T}^{-1} = \mathbf{T}^{-1}\mathbf{T} = \mathbf{1} \tag{2.1.90}$$

If $\mathbf{T}^{-1}$ exists, we call tensor $\mathbf{T}$ an *invertible tensor.*

The concept of $\mathbf{T}^{-1}$, which is the inverse of $\mathbf{T}$, is closely related to that of an orthogonal tensor. A tensor $\mathbf{A}$ is said to be an orthogonal tensor if $\mathbf{A}$ is invertible and $\mathbf{A}^{-1} = \mathbf{A}^T$. Thus $\mathbf{A}$ is an orthogonal tensor if and only if

$$\mathbf{A}\mathbf{A}^T = \mathbf{A}^T\mathbf{A} = \mathbf{1} \tag{2.1.91}$$

This definition of an orthogonal tensor is the same as given by Equation 2.1.44 where $\mathbf{Q}$ is to be replaced by $\mathbf{A}$. The definition 2.1.44 was introduced in connection with a particular tensor $\mathbf{Q}$ defined by the directional cosines of two coordinate systems.

It is easy to prove that for any invertible tensors $\mathbf{A}$ and $\mathbf{B}$

$$(\mathbf{A}\mathbf{B})^{-1} = \mathbf{B}^{-1}\mathbf{A}^{-1} \tag{2.1.92}$$

$$(\mathbf{A}^{-1})^{-1} = \mathbf{A} \tag{2.1.93}$$

$$(\mathbf{A}^T)^{-1} = (\mathbf{A}^{-1})^T \tag{2.1.94}$$

$$\mathbf{A}^{-1}(\mathbf{A}\mathbf{u}) = \mathbf{A}(\mathbf{A}^{-1}\mathbf{u}) = \mathbf{u} \tag{2.1.95}$$

for every vector $\mathbf{u}$. Also, it can be proved that a tensor $\mathbf{A}$ is invertible if and only if its matrix $[\mathbf{A}]$ is invertible with $[\mathbf{A}]^{-1} = [\mathbf{A}^{-1}]$.

With these definitions, returning to the eigenvalue problems, we present the following examples.

Example 2.1.14

Show that the principal values of $\mathbf{T}^{-1}$ are reciprocal of principal values of $\mathbf{T}$.

Solution

We multiply the equation

$$\mathbf{Tu} = \lambda\mathbf{u} \tag{a}$$

from the left by $\mathbf{T}^{-1}$ and obtain

$$\mathbf{T}^{-1}(\mathbf{Tu}) = \mathbf{T}^{-1}\lambda\mathbf{u} \tag{b}$$

But

$$\mathbf{T}^{-1}(\mathbf{Tu}) = (\mathbf{T}^{-1}\mathbf{T})\mathbf{u} = \mathbf{1u} = \mathbf{u} \tag{c}$$

Therefore, from (b)

$$\frac{\mathbf{u}}{\lambda} = \mathbf{T}^{-1}\mathbf{u} \tag{d}$$

which means that the principal value of $\mathbf{T}^{-1}$ equals λ^{-1}. This completes the solution. □

Example 2.1.15

Show that if tensors $\mathbf{T}$ and $\mathbf{Q}$ are symmetric and $\mathbf{TQ}$ is symmetric then the tensors $\mathbf{TQ}$ and $\mathbf{QT}$ have the same principal values.

Solution

Let

$$(\mathbf{TQ} - \lambda\mathbf{1})\mathbf{u} = \mathbf{0} \tag{a}$$

Since $\mathbf{TQ}$ is symmetric then

$$\mathbf{TQ} = (\mathbf{TQ})^T = \mathbf{Q}^T\mathbf{T}^T \tag{b}$$

But $\mathbf{Q}$ and $\mathbf{T}$ are symmetric. Thus

$$\mathbf{TQ} = \mathbf{QT} \tag{c}$$

and, therefore,

$$(\mathbf{QT} - \lambda\mathbf{1})\mathbf{u} = \mathbf{0} \tag{d}$$

which ends the proof. □

Similarly to the concept of the second-order tensor **T**, defined as a linear transformation from a vector space $\mathcal{V}$ into $\mathcal{V}$, we introduce a notion of a fourth-order tensor C as a linear transformation that assigns to a second-order tensor **U** another second-order tensor **T**

$$\mathbf{T} = \mathsf{C}[\mathbf{U}] \tag{2.1.96}$$

or, in components,

$$T_{ij} = C_{ijkl}U_{kl} \tag{2.1.97}$$

The components of $\mathsf{C}[\cdot]$ are defined in terms of the basis $\{\mathbf{e}_i\}$ by

$$C_{ijkl} = \mathbf{e}_i \cdot \mathsf{C}[\mathbf{e}_k \otimes \mathbf{e}_l]\mathbf{e}_j \tag{2.1.98}$$

or, alternatively,

$$C_{ijkl} = (\mathbf{e}_i \otimes \mathbf{e}_j) \cdot \mathsf{C}[\mathbf{e}_k \otimes \mathbf{e}_l] \tag{2.1.99}$$

Observe that Equation 2.1.99 is a generalization of Equation 2.1.14, that is,

$$T_{ij} = \mathbf{e}_i \cdot \mathbf{T}\mathbf{e}_j$$

In this equation components of the second-order tensor **T** are expressed in terms of the basis vectors $\mathbf{e}_i$ $(i = 1, 2, 3)$ while in Equation 2.1.99 components of the fourth-order tensor C are expressed in terms of *basis tensors* $\mathbf{e}_i \otimes \mathbf{e}_j$ $(i, j = 1, 2, 3)$.

The equivalence of Equations 2.1.98 and 2.1.99 corresponds to Equation 2.1.62. If C is a fourth-order identity tensor, the equivalence of Equations 2.1.98 and 2.1.99 reduces to Equation 2.1.62.

Let $\{\mathbf{e}'_i\}$ be another orthonormal basis, such that

$$\mathbf{e}'_i = \mathbf{Q}\mathbf{e}_i \tag{2.1.100}$$

and let C'_{ijkl} be components of C with respect to $\{\mathbf{e}'_i\}$. The set of 81 quantities C_{mnpq} that transform to C'_{ijkl} according to the formula

$$C'_{ijkl} = Q_{mi}Q_{nj}Q_{pk}Q_{ql}C_{mnpq} \tag{2.1.101}$$

may be also defined as a fourth-order tensor. This definition is an alternative to that in which a fourth-order tensor is a linear transformation from a tensor space $\mathcal{W}$ into $\mathcal{W}$.

Similarly to a transpose of a second-order tensor the *transpose* C^T *of the fourth-order tensor* C is defined as a unique fourth-order tensor that satisfies the relation

$$\mathbf{A} \cdot \mathsf{C}[\mathbf{B}] = \mathsf{C}^T[\mathbf{A}] \cdot \mathbf{B} \tag{2.1.102}$$

for all second-order tensors **A** and **B**.

In components

$$C^T_{ijkl} = C_{klij} \tag{2.1.103}$$

Also

$$|\mathsf{C}| = \sup_{|\mathbf{A}|=1}\{|\mathsf{C}[\mathbf{A}]|\} \tag{2.1.104}$$

is defined as the magnitude $|\mathsf{C}|$ of C. From this definition it follows that

$$|\mathsf{C}[\mathbf{A}]| \leq |\mathsf{C}||\mathbf{A}| \qquad \text{for every tensor } \mathbf{A} \tag{2.1.105}$$

2.2 SCALAR, VECTOR, AND TENSOR FIELDS

In this section, we are going to discuss scalar, vector, and tensor fields defined on an open region R of a three-dimensional Euclidean space E^3.

By a *scalar field* on R we mean a function f that assigns to each point $\mathbf{x} \in R$, a scalar $f(\mathbf{x})$. A similar definition holds for vector and tensor fields. For a differentiable scalar function $f(\mathbf{x})$, we define the gradient of f at $\mathbf{x}$ by

$$\mathbf{u} = \nabla f \tag{2.2.1}$$

or in components

$$u_i = \frac{\partial f}{\partial x_i} \equiv f_{,i} = \mathbf{e}_i \cdot \nabla f = (\nabla f)_i \tag{2.2.2}$$

The symbol ∇ is called the *del operator* and it can be expressed in terms of the basis $\{\mathbf{e}_k\}$ by

$$\nabla = \mathbf{e}_k \frac{\partial}{\partial x_k} \tag{2.2.3}$$

For a differentiable vector function $\mathbf{v}(\mathbf{x})$ we define the gradient of $\mathbf{v}$ at point $\mathbf{x}$ by a second-order tensor $\mathbf{V}$

$$\mathbf{V} = \nabla \mathbf{v} \tag{2.2.4}$$

or in components

$$V_{ij} = (\nabla \mathbf{v})_{ij} = v_{i,j} \equiv \frac{\partial v_i}{\partial x_j} = \mathbf{e}_i \cdot \nabla \mathbf{v}(\mathbf{x})\mathbf{e}_j \tag{2.2.5}$$

This definition may be extended to include the gradient of a second-order tensor field $\mathbf{T} = \mathbf{T}(\mathbf{x})$. Such a gradient is represented by a tensor of the third-order R, which in components is given by

$$R_{ijk} = T_{ij,k} = \frac{\partial T_{ij}}{\partial x_k} = (\mathbf{e}_i \otimes \mathbf{e}_j) \cdot \nabla \mathbf{T}\, \mathbf{e}_k = \mathbf{e}_i \cdot [\nabla \mathbf{T}(\mathbf{x})\mathbf{e}_k]\mathbf{e}_j \tag{2.2.6}$$

In this formula R stands for a linear operator that transforms a vector space into a second-order tensor space, that is

$$\mathsf{R}\mathbf{v} = \mathbf{U} \tag{2.2.7}$$

or in components

$$R_{ijk}v_k = U_{ij} \tag{2.2.8}$$

The notion of a gradient of a vector field $\mathbf{v}$ serves for the definition of the *divergence of* $\mathbf{v}$, div $\mathbf{v}$, and for the definition of the curl $\mathbf{v}$

$$\operatorname{div} \mathbf{v}(\mathbf{x}) = \operatorname{tr} \nabla\mathbf{v}(\mathbf{x}) = v_{i,i}(\mathbf{x}) \tag{2.2.9}$$

$$[\operatorname{curl} \mathbf{v}(\mathbf{x})] \times \mathbf{a} = [\nabla\mathbf{v}(\mathbf{x}) - \nabla\mathbf{v}^T(\mathbf{x})]\mathbf{a} \quad \text{for every vector } \mathbf{a} \tag{2.2.10}$$

An expression for curl $\mathbf{v}(\mathbf{x})$ in components reads

$$[\operatorname{curl} \mathbf{v}(\mathbf{x})]_i = \epsilon_{ijk} v_{k,j}(\mathbf{x}) \tag{2.2.11}$$

Note that div $\mathbf{v}$ is a scalar field and curl $\mathbf{v}$ represents a vector field.

In the following a notion of the *symmetric gradient of* $\mathbf{v}$, denoted by $\widehat{\nabla}\mathbf{v}(\mathbf{x})$, will be useful. The symmetric gradient of $\mathbf{v}$ is defined by

$$\widehat{\nabla}\mathbf{v}(\mathbf{x}) = \operatorname{sym} \nabla\mathbf{v}(\mathbf{x}) = \frac{1}{2}\,[\nabla\mathbf{v}(\mathbf{x}) + \nabla\mathbf{v}^T(\mathbf{x})] \tag{2.2.12}$$

Remark 2.2.1: From the definition of the del operator, it follows that

$$\operatorname{div} \mathbf{v} = \nabla \cdot \mathbf{v} \tag{2.2.13}$$

and

$$\operatorname{curl} \mathbf{v} = \nabla \times \mathbf{v} \tag{2.2.14}$$

The notions of the divergence and curl of a vector field may be extended to the divergence and curl of a tensor field. Let $\mathbf{T}$ be a differentiable tensor field on R. The *divergence of* $\mathbf{T}$ at x, denoted by div $\mathbf{T}(\mathbf{x})$, is the unique vector such that

$$[\operatorname{div} \mathbf{T}(\mathbf{x})] \cdot \mathbf{a} \overset{\text{def}}{=} \operatorname{div}\,[\mathbf{T}^T(\mathbf{x})\,\mathbf{a}] \quad \text{for any constant vector } \mathbf{a} \tag{2.2.15}$$

Similarly, the *curl of* $\mathbf{T}(\mathbf{x})$, denoted by curl $\mathbf{T}(\mathbf{x})$, is the unique tensor such that

$$[\operatorname{curl} \mathbf{T}(\mathbf{x})]\mathbf{a} \overset{\text{def}}{=} \operatorname{curl}\,[\mathbf{T}^T(\mathbf{x})\,\mathbf{a}] \quad \text{for any constant vector } \mathbf{a} \tag{2.2.16}$$

In components,

$$[\operatorname{div} \mathbf{T}(\mathbf{x})]_i = T_{ij,j}(\mathbf{x}) \tag{2.2.17}$$

and

$$[\text{curl}\,\mathbf{T}(\mathbf{x})]_{ij} = \epsilon_{ipq} T_{jq,p}(\mathbf{x}) \tag{2.2.18}$$

We also define the *Laplacian of a scalar field* f, of a *vector field* $\mathbf{v}$, and *of a tensor field* $\mathbf{T}$, by

$$\Delta f(\mathbf{x}) = \text{div}\,\nabla f \tag{2.2.19}$$

$$\Delta \mathbf{v}(\mathbf{x}) = \text{div}\,\nabla \mathbf{v} \tag{2.2.20}$$

$$[\Delta \mathbf{T}(\mathbf{x})]\mathbf{a} = \Delta[\mathbf{T}(\mathbf{x})\ \mathbf{a}] \quad \text{for any constant vector } \mathbf{a} \tag{2.2.21}$$

In components

$$\Delta f = f_{,kk} \tag{2.2.22}$$

$$(\Delta \mathbf{v})_i = v_{i,kk} \tag{2.2.23}$$

$$(\Delta \mathbf{T})_{ij} = T_{ij,kk} \tag{2.2.24}$$

Equation 2.2.22 in terms of the del operator takes the form

$$\Delta f = (\nabla \cdot \nabla) f = \nabla^2 f \tag{2.2.25}$$

This provides motivation for using ∇^2 instead of Δ in some applications. Note that ∇ is a vector operator, while ∇^2 is a scalar operator.

Example 2.2.1

Show that for any two vector fields $\mathbf{u}$ and $\mathbf{v}$ the following identities hold:

(1) div $(\mathbf{u} \times \mathbf{v}) = \mathbf{v} \cdot \text{curl}\,\mathbf{u} - \mathbf{u} \cdot \text{curl}\,\mathbf{v}$
(2) curl $(\mathbf{u} \times \mathbf{v}) = (\text{div}\,\mathbf{v})\mathbf{u} - (\text{div}\,\mathbf{u})\mathbf{v} + (\mathbf{v} \cdot \nabla)\mathbf{u} - (\mathbf{u} \cdot \nabla)\mathbf{v}$

where

$$\mathbf{u} \cdot \nabla = u_j \frac{\partial}{\partial x_j}$$

that is,

$$(\mathbf{u} \cdot \nabla) f = u_i f_{,i} = \mathbf{u} \cdot \nabla f \quad \text{for a scalar field } f$$

and

$$(\mathbf{u} \cdot \nabla)\mathbf{v} = (\mathbf{u} \cdot \nabla) v_i \mathbf{e}_i = u_k v_{i,k} \mathbf{e}_i \quad \text{for a vector field } \mathbf{v}.$$

Solution

(1) Observe that both the RHS and the LHS are scalars. Therefore,

$$\text{LHS} = (\varepsilon_{ijk}u_j v_k)_{,i} = \epsilon_{ijk}(u_{j,i}v_k + u_j v_{k,i})$$
$$= v_k\epsilon_{ijk}u_{j,i} + u_j\varepsilon_{ijk}v_{k,i} \tag{a}$$

Since

$$(\text{curl}\,\mathbf{u})_k = \epsilon_{kpq}u_{q,p} \tag{b}$$

and

$$\epsilon_{ijk} = \epsilon_{kij} = -\epsilon_{jik} \tag{c}$$

$$\text{RHS} = \mathbf{v}\cdot(\text{curl}\,\mathbf{u}) - \mathbf{u}\cdot(\text{curl}\,\mathbf{v}) \tag{d}$$

Thus, LHS = RHS.

(2) Observe that the LHS (RHS) is a vector field. Therefore, taking the *ith* component of the LHS we obtain

$$(\text{LHS})_i = \epsilon_{ijk}(\epsilon_{kqr}u_q v_r)_{,j} = \epsilon_{ijk}\epsilon_{kqr}(u_{q,j}v_r + u_q v_{r,j}) \tag{e}$$

Since

$$\epsilon_{ijk}\epsilon_{kqr} = \varepsilon_{ijk}\varepsilon_{qrk} = \delta_{iq}\delta_{jr} - \delta_{ir}\delta_{qj} \tag{f}$$

therefore,

$$(\text{LHS})_i = u_{i,r}v_r - u_{j,j}v_i + u_i v_{j,j} - u_j v_{i,j} \tag{g}$$
$$(\text{RHS})_i = v_{j,j}u_i - u_{j,j}v_i + v_k u_{i,k} - u_k v_{i,k} \tag{h}$$

Hence, LHS = RHS. □

Example 2.2.2

Let **a** and **b** be constant vectors. For any vector field **u** and any symmetric tensor field **S** show that

$$\text{div}\,[(\mathbf{u}\cdot\mathbf{a})\mathbf{S}\mathbf{b}] = \mathbf{a}\cdot[(\nabla\mathbf{u})\mathbf{S} + \mathbf{u}\otimes\text{div}\,\mathbf{S}]\mathbf{b}$$

Solution

We observe that the LHS (RHS) is a scalar field. Computing the LHS we write

$$\text{LHS} = (u_k a_k S_{ij} b_j)_{,i} = a_k b_j(u_{k,i}S_{ij} + u_k S_{ij,i})$$
$$= a_k[u_{k,i}S_{ij}b_j + u_k S_{ji,i}b_j] = \mathbf{a}\cdot[(\nabla\mathbf{u})\mathbf{S} + \mathbf{u}\otimes\text{div}\,\mathbf{S}]\mathbf{b}$$

Therefore, LHS = RHS. □

Example 2.2.3

Let $\mathbf{v}$ be a vector field, $\mathbf{V} = \text{sym}\,\nabla\mathbf{v}$, $\mathbf{W} = \text{skw}\,\nabla\mathbf{v}$, and $\boldsymbol{\omega}$ be the axial vector of $\mathbf{W}$.
Prove that

(1) $\boldsymbol{\omega} = \frac{1}{2}\,\text{curl}\,\mathbf{v}$
(2) $|\nabla\mathbf{v}|^2 = |\mathbf{V}|^2 + |\mathbf{W}|^2 = |\mathbf{V}|^2 + \frac{1}{2}|\,\text{curl}\,\mathbf{v}|^2 = \text{div}\,[(\nabla\mathbf{v}^T)\mathbf{v}] - \mathbf{v}\cdot\nabla^2\mathbf{v}$
(3) $\nabla\mathbf{v}\cdot(\nabla\mathbf{v})^T = |\mathbf{V}|^2 - |\mathbf{W}|^2 = \text{div}\,\{[(\mathbf{v}\cdot\nabla)\mathbf{v}] - (\text{div}\,\mathbf{v})\mathbf{v}\} + (\text{div}\,\mathbf{v})^2$
(4) $\mathbf{VW} + \mathbf{WV} = \text{skw}\,(\nabla\mathbf{v})^2$.

Solution

(1) Using the definition of the axial vector, see Equation 2.1.33 with $P_{jk} = W_{jk}$, we obtain

$$\omega_i = -\frac{1}{2}\,\epsilon_{ijk}W_{jk} \tag{a}$$

Since

$$W_{jk} = \frac{1}{2}\,(v_{j,k} - v_{k,j}) \tag{b}$$

therefore,

$$\begin{aligned}\omega_i &= -\frac{1}{4}\,\epsilon_{ijk}(v_{j,k} - v_{k,j}) = \frac{1}{4}\,\epsilon_{ijk}v_{k,j} - \frac{1}{4}\,\epsilon_{ijk}v_{j,k}\\ &= \frac{1}{4}\,\epsilon_{ijk}v_{k,j} + \frac{1}{4}\,\epsilon_{ijk}v_{j,k} = \frac{1}{2}\,\epsilon_{ijk}v_{k,j} = \frac{1}{2}\,(\text{curl}\,\mathbf{v})_i\end{aligned} \tag{c}$$

This proves (1).

(2) From the definition of the axial vector $\boldsymbol{\omega}$

$$W_{ij} = -\epsilon_{ijk}\omega_k \tag{d}$$

Hence

$$W_{ij}W_{ij} = \epsilon_{ijk}\omega_k\epsilon_{ijs}\omega_s \tag{e}$$

By $\epsilon - \delta$ identity we get

$$W_{ij}W_{ij} = (\delta_{jj}\delta_{ks} - \delta_{js}\delta_{jk})\omega_k\omega_s = 3\omega_k\omega_k - \omega_k\omega_k = 2\omega_k\omega_k \tag{f}$$

or, in direct notation,

$$|\mathbf{W}|^2 = 2\boldsymbol{\omega}^2 \tag{g}$$

Now, since by (1)

$$\boldsymbol{\omega} = \frac{1}{2}\,\text{curl}\,\mathbf{v} \tag{h}$$

therefore,

$$|\mathbf{W}|^2 = 2\left(\frac{1}{2}\ \text{curl}\,\mathbf{v}\right)^2 = \frac{1}{2}\ (\text{curl}\,\mathbf{v})^2 \tag{i}$$

Next, since

$$\nabla\mathbf{v} = \mathbf{V} + \mathbf{W} \text{ and } \mathbf{V}\cdot\mathbf{W} = 0 \tag{j}$$

therefore,

$$|\nabla\mathbf{v}|^2 = |\mathbf{V}|^2 + |\mathbf{W}|^2 = |\mathbf{V}|^2 + \frac{1}{2}\ (\text{curl}\,\mathbf{v})^2 \tag{k}$$

On the other hand

$$\begin{aligned}|\nabla\mathbf{v}|^2 &= v_{i,j}v_{i,j} = (v_iv_{i,j})_{,j} - v_iv_{i,jj} = (v^T_{j,i}v_i)_{,j} - v_iv_{i,jj}\\ &= \text{div}\,[(\nabla\mathbf{v})^T\mathbf{v}] - \mathbf{v}\cdot\nabla^2\mathbf{v}\end{aligned} \tag{l}$$

This proves (2).

(3) Since

$$\nabla\mathbf{v} = \mathbf{V} + \mathbf{W} \ \text{ and } \ (\nabla\mathbf{v})^T = \mathbf{V} - \mathbf{W} \tag{m}$$

hence

$$\nabla\mathbf{v}\cdot(\nabla\mathbf{v})^T = |\mathbf{V}|^2 - |\mathbf{W}|^2 \tag{n}$$

On the other hand

$$\begin{aligned}\nabla\mathbf{v}\cdot(\nabla\mathbf{v})^T &= v_{i,j}v_{j,i} = (v_{i,j}v_j - v_{k,k}v_i)_{,i} + (v_{k,k})^2\\ &= \text{div}\,[(\nabla\mathbf{v})\mathbf{v} - \mathbf{v}\,\text{div}\,\mathbf{v}] + [\text{div}\,\mathbf{v}]^2\end{aligned} \tag{o}$$

This proves (3).

(4)

$$\begin{aligned}(\nabla\mathbf{v})^2 &= (\nabla\mathbf{v})(\nabla\mathbf{v}) = (\mathbf{V} + \mathbf{W})(\mathbf{V} + \mathbf{W})\\ &= \mathbf{V}^2 + \mathbf{W}^2 + \mathbf{VW} + \mathbf{WV}\end{aligned} \tag{p}$$

Since for any two tensors **A** and **B**

$$(\mathbf{AB})^T = \mathbf{B}^T\mathbf{A}^T \tag{q}$$

and

$$(\nabla\mathbf{v})^T = \mathbf{V} - \mathbf{W} \tag{r}$$

therefore,

$$[(\nabla \mathbf{v})^2]^T = [(\nabla \mathbf{v})^T][(\nabla \mathbf{v})^T] = (\mathbf{V} - \mathbf{W})(\mathbf{V} - \mathbf{W})$$
$$= (\mathbf{V})^2 + (\mathbf{W})^2 - (\mathbf{VW}) - (\mathbf{WV}) \tag{s}$$

Hence, from (p) and (s) we obtain

$$\frac{(\nabla \mathbf{v})^2 - [(\nabla \mathbf{v})^2]^T}{2} = \text{skw } (\nabla \mathbf{v})^2 = \mathbf{VW} + \mathbf{WV} \tag{t}$$

□

2.3 INTEGRAL THEOREMS

2.3.1 Divergence Theorem

Let f be a scalar field, $\mathbf{u}$ a vector field, and $\mathbf{T}$ a tensor field on a region $R \subset E^3$. Let $\mathbf{n}$ be a unit outer normal vector to ∂R where ∂R stands for the boundary of R. Then

$$\int_{\partial R} f\mathbf{n}\, da = \int_R \nabla f\, dv \tag{2.3.1}$$

$$\int_{\partial R} \mathbf{u} \otimes \mathbf{n}\, da = \int_R \nabla \mathbf{u}\, dv \tag{2.3.2}$$

$$\int_{\partial R} \mathbf{u} \cdot \mathbf{n}\, da = \int_R \operatorname{div} \mathbf{u}\, dv \tag{2.3.3}$$

$$\int_{\partial R} \mathbf{n} \times \mathbf{u}\, da = \int_R \operatorname{curl} \mathbf{u}\, dv \tag{2.3.4}$$

$$\int_{\partial R} \mathbf{Tn}\, da = \int_R \operatorname{div} \mathbf{T}\, dv \tag{2.3.5}$$

This theorem is called the *Divergence Theorem*. The proof will not be provided here.

Example 2.3.1 (Green's Theorem)

By using the divergence theorem prove that for any two scalar fields f and g on R the following identities, called Green's Theorem, are satisfied:

$$\int_R [(\nabla f) \cdot (\nabla g) + f\, \nabla^2 g]\, dv = \int_{\partial R} f(\nabla g) \cdot \mathbf{n}\, da = \int_{\partial R} f \frac{\partial g}{\partial n}\, da \tag{a}$$

and

$$\int_R (f\, \nabla^2 g - g \nabla^2 f)\, dv = \int_{\partial R} \left(f \frac{\partial g}{\partial n} - g \frac{\partial f}{\partial n} \right) da \tag{b}$$

Here $\partial/\partial n$ stands for the normal derivative, that is, for any function h on R

$$\frac{\partial h}{\partial n} = \mathbf{n} \cdot (\nabla h) \tag{c}$$

Solution

(a) The LHS written in index notation takes the form

$$\text{LHS} = \int_R (f_{,i}g_{,i} + fg_{,ii})\, dv = \int_R (fg_{,i})_{,i}\, dv \tag{d}$$

Using Equation 2.3.3 in which $\mathbf{u} = f\nabla g$ we get

$$\text{LHS} = \int_{\partial R} fg_{,i}n_i\, da \tag{e}$$

and by (c), in which $h = g$, we obtain

$$\text{LHS} = \int_{\partial R} f\, \frac{\partial g}{\partial n}\, da \tag{f}$$

Therefore, (a) is proved.

(b) Replacing in (a) f by g and g by f we obtain

$$\int_R [(\nabla g) \cdot (\nabla f) + g\nabla^2 f] dv = \int_{\partial R} g\, \frac{\partial f}{\partial n}\, da \tag{g}$$

Subtracting (a) from (g) we obtain (b). Thus, (b) is proved. □

2.3.2 Stokes' Theorem

Let **u** and **T** denote a vector and tensor fields, respectively, on R, and let C be a closed curve in R. Then

$$\oint_C \mathbf{u} \cdot \mathbf{s}\, dt = \int_S (\text{curl}\, \mathbf{u}) \cdot \mathbf{n}\, da \tag{2.3.6}$$

$$\oint_C \mathbf{Ts}\, dt = \int_S (\text{curl}\, \mathbf{T})^T \mathbf{n}\, da \tag{2.3.7}$$

where

S is any surface contained in R, bounded by C

n is the unit vector normal to S

s is a unit vector tangent to C

The directions of **n** and **s** are shown in Figure 2.3. This theorem is called Stokes' Theorem.

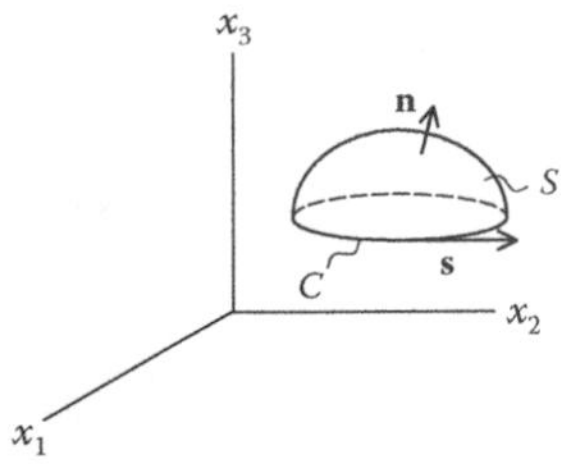

FIGURE 2.3 The directions of **n**, **s**, and other symbols used in Stokes' Theorem.

In these formulas the curve C can be described by a vector equation $\mathbf{y} = \mathbf{y}(t)$, $t_0 \le t \le t_1$, where $\mathbf{y}(t_0) = \mathbf{y}(t_1)$ and $\mathbf{s} = d\mathbf{y}/dt$. If t is a length of the arc then $|\mathbf{s}| = 1$. If C is shrunk to a point, S reduces to a closed surface, and

$$\int_S (\text{curl}\,\mathbf{u}) \cdot \mathbf{n}\, da = 0 \tag{2.3.8}$$

$$\int_S (\text{curl}\,\mathbf{T})^T \mathbf{n}\, da = \mathbf{0} \tag{2.3.9}$$

Proofs are omitted.

Example 2.3.2

Show that for a scalar field f defined on C and S

$$\oint_C f\mathbf{s}\, dt = \int_S \mathbf{n} \times \nabla f\, da \qquad \text{(a)}$$

where S and C have the same meaning as in Stokes' Theorem.

Solution

Let k be one of the three numbers 1, 2, 3. Introduce a vector field **u** by the formula $u_i = f\delta_{ik}$. Substituting u_i into Equation 2.3.6 we obtain

$$\text{LHS} = \oint_C u_i s_i\, dt = \oint_C f\delta_{ik} s_i\, dt = \oint_C f s_k\, dt \qquad \text{(b)}$$

$$\text{RHS} = \int_S n_i \epsilon_{ipq} u_{q,p}\, da = \int_S n_i \epsilon_{ipq} (f\delta_{qk})_{,p}\, da$$

$$= \int_S n_i \epsilon_{ipk} f_{,p}\, da = \int_S \epsilon_{kip} n_i f_{,p}\, da \qquad \text{(c)}$$

Since by Equation 2.3.6 LHS = RHS, then from (b) and (c) we have (a). □

Example 2.3.3

Show that for a vector field **u** defined on C and S

$$\oint_C (\mathbf{u} \times \mathbf{s})\, dt = \int_S [(\text{div}\,\mathbf{u})\mathbf{n} - (\nabla\mathbf{u})^T \mathbf{n}]\, da \qquad \text{(a)}$$

Solution

By Example 2.3.2

$$\oint_C \varphi s_q\, dt = \int_S \epsilon_{qab} n_a \varphi_{,b}\, da \qquad \text{(b)}$$

where φ is an arbitrary scalar field. Letting $\varphi = \epsilon_{ipq}u_p$, where i and q are fixed, we find from (b)

$$\oint_C \epsilon_{ipq}u_p s_q\, dt = \int_S \epsilon_{qab}n_a(\epsilon_{ipq}u_p)_{,b}\, da = \int_S \epsilon_{qab}\epsilon_{qip}n_a u_{p,b}\, da$$

$$= \int_S (\delta_{ai}\delta_{bp} - \delta_{ap}\delta_{bi})n_a u_{p,b}\, da = \int_S (n_i u_{p,p} - n_p u_{p,i})\, da$$

Thus, we arrive at (a). □

Example 2.3.4

Show that

$$\oint_C (\text{curl}\,\mathbf{u}) \cdot \mathbf{s}\, dt = \int_S \left[\frac{\partial}{\partial n}(\text{div}\,\mathbf{u}) - \mathbf{n} \cdot \nabla^2 \mathbf{u}\right] da$$

Solution

By (b) in Example 2.3.3 in which

$$\varphi = \epsilon_{qij}u_{j,i} \quad \text{with } q \text{ fixed,} \tag{a}$$

we find

$$\oint_C \epsilon_{qij}u_{j,i}s_q\, dt = \int_S \epsilon_{qab}n_a(\epsilon_{qij}u_{j,i})_{,b}\, da = \int_S \epsilon_{qab}\epsilon_{qij}n_a u_{j,ib}\, da$$

$$= \int_S (\delta_{ai}\delta_{bj} - \delta_{aj}\delta_{ib})n_a u_{j,ib}\, da$$

$$= \int_S (n_i u_{b,bi} - n_j u_{j,bb})\, da \equiv \text{LHS} \tag{b}$$

Since for any scalar f

$$\mathbf{n} \cdot \nabla f = \frac{\partial}{\partial n} f = n_i f_{,i}$$

from (b) we arrive at

$$\text{LHS} = \int_S \left[\frac{\partial}{\partial n}(\text{div}\,\mathbf{u}) - \mathbf{n} \cdot (\nabla^2 \mathbf{u})\right] da$$

This completes the solution. □

2.3.3 Theorems on Irrotational and Solenoidal Fields

Before formulating the theorems, we give definitions of an irrotational field and a solenoidal field.

A vector field **u** on R is said to be *irrotational* in R if

$$\operatorname{curl} \mathbf{u} = \mathbf{0} \quad \text{in } R \tag{2.3.10}$$

A vector field **u** on R is said to be *solenoidal* in R if

$$\int_S \mathbf{u} \cdot \mathbf{n}\, da = 0 \qquad \text{for every closed regular surface } S \text{ in } R. \tag{2.3.11}$$

The surface integral on the LHS of Equation 2.3.11 is called the *outward flux of* **u** *across S*.
With these definitions we have the following

2.3.3.1 Theorem on Irrotational Fields

Let R be a simply connected region of E^3.

(a) If **u** is a vector field on R and

$$\operatorname{curl} \mathbf{u} = \mathbf{0} \tag{2.3.12}$$

then there exists a scalar f such that

$$\mathbf{u} = \nabla f \tag{2.3.13}$$

(b) If **T** is a tensor field on R such that

$$\operatorname{curl} \mathbf{T} = \mathbf{0} \tag{2.3.14}$$

then there exists a vector field **v** such that

$$\mathbf{T} = \nabla \mathbf{v} \tag{2.3.15}$$

(c) If

$$\operatorname{curl} \mathbf{T} = \mathbf{0} \quad \text{and} \quad \operatorname{tr} \mathbf{T} = 0 \tag{2.3.16}$$

then there exists a skew tensor field **W** such that

$$\mathbf{T} = \operatorname{curl} \mathbf{W} \tag{2.3.17}$$

Proofs are omitted.

2.3.3.2 Theorem on Solenoidal Fields

(a) If **u** is a vector field on R, and

$$\int_S \mathbf{u} \cdot \mathbf{n}\, da = 0 \qquad \text{for every closed regular surface } S \subset R \tag{2.3.18}$$

then there exists a vector field $\mathbf{w}$ such that

$$\mathbf{u} = \text{curl}\,\mathbf{w} \tag{2.3.19}$$

(b) If $\mathbf{T}$ is a tensor field on R such that

$$\int_S \mathbf{T}^T\mathbf{n}\,da = \mathbf{0} \quad \text{for every closed regular surface } S \subset R \tag{2.3.20}$$

then there exists a tensor field $\mathbf{W}$ with the property that

$$\mathbf{T} = \text{curl}\,\mathbf{W} \tag{2.3.21}$$

Proofs are omitted.

Remark 2.3.1: Note that in both theorems, the relations 2.3.12 and 2.3.13, 2.3.14 and 2.3.15, ..., 2.3.20 and 2.3.21 are compatible in the following sense. Equation 2.3.13 $\Rightarrow$ 2.3.12, ..., 2.3.21 $\Rightarrow$ 2.3.20. For example, let us write Equation 2.3.21 in index notation

$$T_{ij} = \epsilon_{iab} W_{jb,a} \tag{2.3.22}$$

Substituting this into LHS of 2.3.20 we get

$$(\text{LHS})_i = \int_S T_{ij}^T n_j\,da = \int_S \epsilon_{jab} W_{ib,a} n_j\,da \tag{2.3.23}$$

On the basis of the divergence theorem

$$(\text{LHS})_i = \int_{R^*} \epsilon_{jab} W_{ib,aj}\,dv \tag{2.3.24}$$

where R^* is the region bounded by the surface S. This expression vanishes because under the integral we have the inner product of a symmetric tensor and a skew tensor, see Example 2.1.5. Therefore, Equation 2.3.21 $\Rightarrow$ Equation 2.3.20.

Remark 2.3.2: The irrotational and solenoidal fields are the main ingredients of the following decomposition theorem, called *Helmholtz's Theorem.*

Helmholtz's Theorem. If $\mathbf{u}$ is a vector field on R then there exist a scalar field f and a vector field $\mathbf{v}$ such that

$$\mathbf{u} = \nabla f + \text{curl}\,\mathbf{v} \tag{2.3.25}$$

$$\text{div}\,\mathbf{v} = 0 \tag{2.3.26}$$

The proof is omitted.

2.3.4 The Laplace Transformation, the Dirac Delta Function, and the Heaviside Function

The Laplace transformation, denoted by $L\{f(t)\}$, is exceptionally useful method in applied mathematics, as it allows to solve differential equations which, with the aid of this method, are first reduced to algebraic equations and next to a contour integration of a solution to these algebraic equations. As differential equations are utilized in a multitude of problems in physics, engineering, control theory, and almost everywhere else, this method has become a standard method in attacking and solving many kinds of problems. The first who occupied himself with integrals that are of the class of Laplace transforms was Leonhard Euler (1707–1783), who worked on such integrals as early as 1744, that is, before Laplace was even born. But it was Pierre-Simon Laplace (1749–1827) who, starting in 1782, used such integrals to solving problems, and in 1785 introduced this method in a way that now is so useful, and it was Laplace that fruitfully worked on expansion of applications of this method. The Laplace transform is a linear operator of the form

$$\bar{f}(p) = L\{f(t)\} = \int_0^\infty e^{-pt} f(t)\, dt \tag{2.3.27}$$

That is, a function $f(t)$, $(t \geq 0)$, is transformed into a function $\bar{f}(p)$, where p, is called the *parameter of the Laplace transform*, and is a complex number. The reason that the Laplace transform acts on a function that usually is of variable t comes from the frequent application of these transforms to functions of the time. Eventually, after solving the problem in the transform domain, we receive a solution in a form of the function of p, which then must be returned to the real domain of t, that is, it must be acted upon by an inverse Laplace transform, L^{-1}. The complex inversion formula has the form [2]

$$f(t) = L^{-1}\{\bar{f}(p)\} = \frac{1}{2\pi i} \lim_{T\to\infty} \int_{c-iT}^{c+iT} e^{pt} \bar{f}(p)\, dp \tag{2.3.28}$$

and c, a real number, is chosen so that the line $\text{Re}\{p\} = c$ in the complex p plane lies to the right of all the singular points of $\bar{f}(p)$. This means that $c \geq \text{Re}(p_k)$ for each singularity p_k of $\bar{f}(p)$ as well as $i^2 = -1$.

From the definition of the Laplace transform, Equation 2.3.27, it follows that for two functions $f(t)$ and $g(t)$

$$L\{f(t) + g(t)\} = L\{f(t)\} + L\{g(t)\} \tag{2.3.29}$$

and

$$L\{cf(t)\} = c\,L\{f(t)\} \tag{2.3.30}$$

where c is a constant.

We should also note that because of the properties of the Laplace transformation, some information on the solutions may already be obtained from the Laplace transform domain. In achieving this we are helped by two theorems:

1. Initial value theorem

$$f(+0) = \lim_{p\to\infty} p\bar{f}(p) \tag{2.3.31}$$

and

2. Final value theorem

$$f(\infty) = \lim_{p\to 0} p\bar{f}(p) \tag{2.3.32}$$

subject to the condition that all poles of $p\bar{f}(p)$ lie on the left side of the plane. The latter theorem allows to find out the behavior of the solution for large values of variable t without further calculations.

The use of Equation 2.3.28 for finding the inverse Laplace transform is not the only method for finding such inverse transforms. First of all, the Laplace transform and the inverse Laplace transform correspond one-to-one to each other, that is, if we know the Laplace transform of a function as a result of using Equation 2.3.27, then we also know the inverse transform for the transformed function. There are numerous other methods for finding inverse Laplace transforms, and some of these methods are designed for just one specific function or a family of functions. An example of a method for a family of inverse Laplace transforms of the exponential type is presented in detail later in this section. The known inverse Laplace transforms are provided in the front matter. If a needed inverse Laplace transform cannot be found in any ready to use tables and cannot be calculated by Equation 2.3.28, and cannot be discovered in the literature, than there are methods of numerical inversion of Laplace transforms.

Example 2.3.5

Calculate the Laplace transform of the derivative of a function if the Laplace transform of the function itself is known.

Solution

The Laplace transform of a function $f(t)$ is given by Equation 2.3.27. Integrating the RHS by parts, we find

$$\int_0^{+\infty} e^{-pt} f(t)\,dt$$

$$= \left[\frac{f(t)e^{-pt}}{-p}\right]_0^{+\infty} - \int_0^{+\infty} \frac{e^{-pt}}{-p} f'(t)\,dt$$

$$= \left[-\frac{f(0)}{-p}\right] + \frac{1}{p} L\{f'(t)\} \tag{a}$$

Thus,

$$L\left\{\frac{df(t)}{dt}\right\} = p\,L\{f(t)\} - f(0) \tag{b}$$

and the result has been obtained. □

It is appropriate to ask a question on what are sufficient conditions for a function $f(t)$ to possess a Laplace transform $L\{f(t)\} \equiv \bar{f}(p)$ and, in addition, that the inverse Laplace transform $L^{-1}\bar{f}(p)$ can be obtained from $L\{f(t)\}$.* The answer to this question is as follows.

A function $f(t)$ is admissible if both the following conditions hold true for $t \geq 0$:

(1) The function $f(t)$ is *sectionally continuous* on every finite interval for $t \geq 0$. The term *sectionally continuous* means that any closed interval $[t_a, t_b]$ can be divided into a finite number of subintervals such that the function $f(t)$ is continuous inside each subinterval and has finite limits when t tends from inside to the endpoints of such subintervals.

(2) The function $f(t)$ is of *exponential order*. This means that there exist constants α, M, and T such that

$$|f(t)|e^{-\alpha t} < M \quad \text{for all } t > T \tag{2.3.33}$$

Under conditions (1) and (2) the integral 2.3.27 converges for $p > \alpha$. The convergence is uniform if $p \geq \alpha_0 > \alpha$, where α_0 is fixed.

The applications of the technique are based on general properties of Laplace transforms (Part A) provided in the front matter. Some of the formulas in that table may be derived by direct use of the definition of the Laplace transform, that is, by Equation 2.3.27.

So far, we have dealt with the application of Laplace transforms to a single function or a combination of functions. However, the usefulness of the method lies in the application of Laplace transforms to differential equations. The result of such application is the transformation of a differential equation from the real, or t, domain, into an algebraic equation in the transformed, or p, domain. We will show how it works on a simple example.

Example 2.3.6

Given a differential equation of the first order

$$f'(t) + af(t) = g(t), \quad \text{where } a = \text{const} \tag{a}$$

and the function $g(t)$ is the *disturbing function* or the *forcing function*. Solve this equation by the Laplace transform technique.

Solution

Applying the Laplace transform to this equation gives

$$L\{f'(t)\} + aL\{f(t)\} = L\{g(t)\} \tag{b}$$

Using Equation (b) from Example 2.3.5 for the first term on the LHS of this equation, and using the notation $L\{f(t)\} = \bar{f}(p)$ and $L\{g(t)\} = \bar{g}(p)$ we get

$$p\bar{f}(p) - f(0+) + a\bar{f}(p) = \bar{g}(p) \tag{c}$$

We see that Equation (c) is an algebraic equation in terms of $\bar{f}(p)$. Moreover, there appears the term $f(0+)$. For selection of a specific solution among infinite number of

* Theoretical foundations and basic assumptions, as well as restrictions for the use of the Laplace transform technique, are given in specialized texts. An elegant and concise exposition is presented in [3].

solutions, we need to assume a value of the solution at some point. In many applications it is the value at the initial point, that is, for $t = 0+$, and we call it the *initial value*. Thus, in this method, the initial value has been automatically used in the transformed equation. The solution of this transformed equation in the p domain is, therefore,

$$\bar{f}(p) = \bar{g}(p)\frac{1}{p+a} + f(0+)\frac{1}{p+a} \tag{d}$$

To find the solution in the real domain, that is, in t domain, we need to apply the inverse Laplace transform to Equation (d). The first term on the RHS is the product of two function in p domain, and the second term on the RHS is the ratio of the initial value $f(0+)$ and the function in the p domain, that is, $(p+a)$. Using (Part A, formula 14, and Table 4, Part B, formula 2) Laplace transforms provided in the front matter, we have

$$f(t) = e^{-at}\int_0^t g(\tau)e^{a\tau}\,d\tau + f(0+)\,e^{-at} \tag{e}$$

and this is the solution with the given initial value $f(0+)$. □

This example has demonstrated the standard scheme of the procedure:

1. Differential equation and initial condition(s).

⇓

2. Laplace transform.

⇓

3. Algebraic equation in p domain.

⇓

4. Solution of the algebraic equation in p domain.

⇓

5. Inverse Laplace transform.

⇓

6. Solution in the real, or t, domain.

2.3.4.1 On Some Laplace Transforms of the Exponential Form

In problems of mechanics, including thermoelasticity, we encounter Laplace transforms of the form [4]

$$\bar{f}(p) = \frac{\sqrt{(p+b)^m}\,\exp\left(-c\sqrt{p+b}\right)}{(p-a)^n} \tag{2.3.34}$$

where

m is a nonnegative integer
n is a natural number
a, b, and c are constants

These transforms are not provided in the published tables known to the authors, except for two simple cases [5].* We are to present an effective and general method for inversion of these functions [7].

* A special case of this transform, namely for $b = 0$, is also given in [6].

It is known that

$$L^{-1}\left\{\overline{S}(p)\right\} = S(t) \tag{2.3.35}$$

where

$$\overline{S}(p) = \frac{\exp\left(-c\sqrt{p+b}\right)}{(p-a)^n} \tag{2.3.36}$$

and

$$S(t) = \frac{e^{at}}{2}\left[\exp\left(-c\sqrt{a+b}\right)\operatorname{erfc}\left(\frac{c}{2\sqrt{t}} - \sqrt{(a+b)t}\right) + \exp\left(c\sqrt{a+b}\right)\operatorname{erfc}\left(\frac{c}{2\sqrt{t}} + \sqrt{(a+b)t}\right)\right] \tag{2.3.37}$$

Here erfc$[f(t)]$ denotes the complementary error function.

We may show that by differentiation of $\overline{S}(p)$ with respect to a and c, inverse transforms of the form given by Equation 2.3.34 can be calculated. Indeed, if $\overline{S}(p)$ is differentiated $n-1$ times with respect to a and m times with respect to c, we find

$$\begin{aligned}\frac{\partial^m}{\partial c^m}\frac{\partial^{n-1}}{\partial a^{n-1}}\overline{S}(p) &= \frac{(-1)^m(n-1)!\sqrt{(p+b)^m}\exp\left(-c\sqrt{p+b}\right)}{(p-a)^n}\\ &= (-1)^m(n-1)!\overline{f}(p)\end{aligned} \tag{2.3.38}$$

We note that

$$L^{-1}\left\{\frac{\partial^m}{\partial c^m}\frac{\partial^{n-1}}{\partial a^{n-1}}\overline{S}(p)\right\} = \frac{\partial^m}{\partial c^m}\frac{\partial^{n-1}}{\partial a^{n-1}}L^{-1}\left\{\overline{S}(p)\right\} \tag{2.3.39}$$

and

$$\begin{aligned}&L^{-1}\left\{\frac{(-1)^m(n-1)!\sqrt{(p+b)^m}\exp\left(-c\sqrt{p+b}\right)}{(p-a)^n}\right\}\\ &= (-1)^m(n-1)!\,L^{-1}\left\{\frac{\sqrt{(p+b)^m}\exp\left(-c\sqrt{p+b}\right)}{(p-a)^n}\right\}\\ &= (-1)^m(n-1)!\,L^{-1}\left\{\overline{f}(p)\right\}\\ &= (-1)^m(n-1)!f(t)\end{aligned} \tag{2.3.40}$$

Therefore, from Equations 2.3.38 through 2.3.40 it follows that

$$f(t) = L^{-1}\left\{\frac{\sqrt{(p+b)^m}\ \exp\left(-c\sqrt{p+b}\right)}{(p-a)^n}\right\}$$

$$= \frac{(-1)^m}{(n-1)!}\frac{\partial^m}{\partial c^m}\frac{\partial^{n-1}}{\partial a^{n-1}}S(t) \tag{2.3.41}$$

The method of finding the inverse Laplace transforms for function $\bar{f}(p)$ of Equation 2.3.34 has, therefore, been found. The difficulty in using the method lies in the fact that the required differentiation leads to lengthy and time-consuming calculations. We are now to present an algorithm that simplifies this task.

Let us introduce a function $T(t)$ in the form

$$T(t) = \frac{e^{at}}{2\sqrt{a+b}}\left[\exp\left(-c\sqrt{a+b}\right)\ \text{erfc}\left(\frac{c}{2\sqrt{t}} - \sqrt{(a+b)t}\right)\right.$$

$$\left. - \exp\left(c\sqrt{a+b}\right)\ \text{erfc}\left(\frac{c}{2\sqrt{t}} + \sqrt{(a+b)t}\right)\right] \tag{2.3.42}$$

Now, by differentiation of $S(t)$ and $T(t)$ with respect both to a and to c the following recurrence relations are established:

$$\frac{\partial S(t)}{\partial a} = tS(t) - \frac{c}{2}T(t) \tag{2.3.43}$$

$$\frac{\partial S(t)}{\partial c} = -(a+b)T(t) - \frac{1}{\sqrt{\pi t}}\exp\left(-\frac{c^2}{4t} - bt\right) \tag{2.3.44}$$

$$\frac{\partial T(t)}{\partial a} = -\frac{c}{2(a+b)}S(t) + \left(t - \frac{1}{2(a+b)}\right)T(t)$$

$$+ \frac{1}{a+b}\sqrt{t/\pi}\ \exp\left(-\frac{c^2}{4t} - bt\right) \tag{2.3.45}$$

$$\frac{\partial T(t)}{\partial c} = -S(t) \tag{2.3.46}$$

We have arrived at the conclusion that the inverse transforms of the form presented by Equation 2.3.34 for any natural m and n can be obtained on the basis of the calculated first derivatives of $S(t)$ and $T(t)$ shown in Equations 2.3.43 through 2.3.46. No additional calculation of derivatives of $S(t)$ and $T(t)$ is needed. Relations 2.3.43 through 2.3.46 jointly with 2.3.41 represent the algorithm for obtaining inverse transforms for consecutive natural numbers m and n.

Example 2.3.7

Calculate the inverse Laplace transform

$$f(t) = L^{-1}\left\{\frac{\sqrt{p+b}\,\exp\left(-c\sqrt{p+b}\right)}{(p-a)^4}\right\} \tag{a}$$

Solution

From Equation 2.3.41, with values of $m = 1$ and $n = 4$, we write

$$f(t) = -\frac{1}{6}\frac{\partial^4}{\partial c \partial a^3} S(t) \tag{b}$$

Applying formulas 2.3.43 through 2.3.46, we get consecutively

$$\frac{\partial S}{\partial c} = -(a+b)T - \frac{1}{\sqrt{\pi t}} \exp\left(-\frac{c^2}{4t} - bt\right) \tag{c}$$

$$\frac{\partial^2 S}{\partial c \partial a} = \frac{c}{2}S - \left[\frac{1}{2} + (a+b)t\right] T - \sqrt{t/\pi}\,\exp\left(-\frac{c^2}{4t} - bt\right) \tag{d}$$

$$\begin{aligned}\frac{\partial^3 S}{\partial c \partial a^2} &= \left[\frac{1}{4(a+b)} + t\right] cS + \left[\frac{1}{4(a+b)} - \frac{c^2}{4} - t - (a+b)t^2\right] T \\ &\quad - \left[\frac{1}{2(a+b)} + t\right]\sqrt{t/\pi}\,\exp\left(-\frac{c^2}{4t} - bt\right)\end{aligned} \tag{e}$$

$$\begin{aligned}\frac{\partial^4 S}{\partial c \partial a^3} &= \left[-\frac{3c}{8(a+b)^2} + \frac{c^3}{8(a+b)} + \frac{3ct}{4(a+b)} + \frac{3}{2}ct^2\right] S \\ &\quad + \left[-\frac{3}{8(a+b)^2} + \left(\frac{3}{4(a+b)} - \frac{3}{4}c^2\right) t - \frac{3}{2}t^2 - (a+b)t^3\right] T \\ &\quad + \left[\frac{3}{4(a+b)^2} - \frac{c^2}{4(a+b)} - \frac{t}{a+b} - t^2\right]\sqrt{t/\pi}\,\exp\left(-\frac{c^2}{4t} - bt\right)\end{aligned} \tag{f}$$

The last expression after multiplying it by $-\frac{1}{6}$ is the transform we have looked for. It is a function of both $S(t)$ and $T(t)$. □

2.3.4.2 The Dirac Delta Function and the Heaviside Function

We often witness sudden events of high "intensity" and of very short duration, for example, a blow of a hammer at a mechanical structure, or a lightning strike at an electric circuit. We will discuss now a method to describe mathematically and analyze such events. To this end, we introduce a function $\delta_a = \delta_a(t)$ that is zero for $t < 0$ and $t > a$, and is equal $1/a$ in the interval $0 < t < a$, with a being a small positive constant. The area under the graphical representation of $\delta_a(t)$ is equal 1. For $a \to 0$ this function we denote $\delta(t)$, that is,

$$\lim_{a \to 0} \delta_a(t) = \delta(t) \tag{2.3.47}$$

and call it the *Dirac delta function* or the *unit impulse function*. This translates in the language of integration into

$$\int_{-\infty}^{+\infty} \delta(t)\,dt = 1 \tag{2.3.48}$$

We may also show [8] that the Laplace transform of $\delta_a(t)$ is

$$L\{\delta_a(t)\} = \int_0^a \frac{1}{a} e^{-pt}\,dt = \frac{1}{pa}(1 - e^{-pa}) \tag{2.3.49}$$

Introducing the variable $pa = u$ and using l'Hospital's rule, we have

$$\lim_{a\to 0} L\{\delta_a(t)\} = \lim_{a\to 0} \frac{1 - e^{-u}}{u} = 1 \tag{2.3.50}$$

and thus, we state that

$$L\{\delta(t)\} = 1 \tag{2.3.51}$$

From these considerations it is clear that by acting with $\delta(t)$ on a function $f(t)$ we have

$$\int_{-\infty}^{\infty} f(t)\,\delta(t)\,dt = f(0) \tag{2.3.52}$$

This property is called the *filtering property* of the Dirac delta function.

We need to explain that, in reality, $\delta(t)$ is not a function, as it does not satisfy conditions of how we define a function; $\delta(t)$ is a *generalized function*, or a *distribution*. This topic, with examples, is discussed in [8], and in works treating the theory of distributions.

The *Heaviside function* or the *unit step function* is defined by

$$H(t) = \begin{cases} 1 & t > 0 \\ 0 & t < 0 \end{cases} \tag{2.3.53}$$

By comparing the definitions for $H(t)$ and $\delta(t)$ it becomes obvious that

$$\frac{dH(t)}{dt} = \delta(t) \tag{2.3.54}$$

We see that the Heaviside function $H(t)$ is equal 0 for negative values of t and is equal 1 for positive values of t, and has a jump that is equal 1 at $t = 0$, but it is not defined at $t = 0$. However, there are situations in some applications, like the one in Problem 10.5, that it is convenient to specify the value of $H(t)$ at $t = 0$, for example, stating that $H(0) = \frac{1}{2}$. Such a "complemented" Heaviside function satisfies the relation 2.3.54 everywhere except at $t = 0$.

We observe that by multiplying a function $f(t)$ by $H(t - t_0)$, we obtain

$$f(t)\,H(t - t_0) = \begin{cases} f(t) & t > t_0 \\ 0 & t < t_0 \end{cases} \tag{2.3.55}$$

where $t_0 > 0$. The formula 2.3.54 in which $H(t)$ is given by Equation 2.3.53 can be used to obtain an alternative definition of the Dirac delta function in which $\delta(t)$ is an even function of $t\ :\ |t| < \infty$. To this end, we let $H(0) = 1/2$ and note that

$$H(t) + H(-t) = 1 \quad \text{for } |t| < \infty \tag{2.3.56}$$

By differentiating Equation 2.3.56 with respect to t and using Equation 2.3.54 we obtain

$$\delta(t) - \delta(-t) = 0 \quad \text{for } |t| < \infty \tag{2.3.57}$$

or

$$\delta(t) = \delta(-t) \quad \text{for } |t| < \infty \tag{2.3.58}$$

This shows that $\delta(t)$ can be defined as

$$\lim_{a \to 0} \delta_a^*(t) = \delta(t) \quad \text{for } |t| < \infty \tag{2.3.59}$$

where

$$\delta_a^*(t) = \begin{cases} \frac{1}{2a} & |t| < a \\ 0 & |t| > a \end{cases} \tag{2.3.60}$$

With such an alternative definition of $\delta(t)$, the filtering property 2.3.52 holds true provided the function $f(t)$ is continuous at $t = 0$. The even Dirac delta function in which the independent variable t is replaced by a space variable is applicable to problems involving concentrated mechanical forces of elastostatics, while the one-sided Dirac delta function 2.3.47 is useful in a study of suddenly produced processes of elastodynamics.

In the following chapters, we will encounter numerous applications of both the Heaviside function and the Dirac delta function.

2.3.5 Time-Dependent Fields

Since we will consider in the next chapters not only static problems but also dynamic problems, this section deals with functions of position $\mathbf{x} \in R$ and time $t \geq 0$. Similarly to previous sections, in stating various theorems in which the regularity assumptions were omitted, we will also here omit the regularity assumptions imposed on space–time-dependent fields. One of the main concepts of time-dependent problems is the concept of a convolution of two functions.

Let f and g be scalar fields on $R \times T$ where R is a region of E^3, and $T = [0, \infty)$ is the time interval. The convolution $f * g$ of f and g is the function defined by

$$[f * g](\mathbf{x}, t) = \int_0^t f(\mathbf{x}, t - \tau) g(\mathbf{x}, \tau) d\tau \tag{2.3.61}$$

on $R \times T$.

The following properties of convolution are useful in applications:

Let f, g, and h be scalar fields on $R \times T$ that are continuous in time. Then

(i) $f * g = g * f$
(ii) $(f * g) * h = f * (g * h) = f * g * h$
(iii) $f * (g + h) = f * g + f * h$
(iv) $f * g = 0 \Rightarrow f = 0$ or $g = 0$
(v) $\overline{(f * g)}^{\boldsymbol{\cdot}} = \dot{f} * g + f(\mathbf{x}, 0)\dot{g}$ where the overdot means the derivative with respect to time, and we assume that $\dot{f}$ exists.
(vi) $L\{f * g\} = L\{f\}L\{g\}$ where L is the Laplace transform operator with respect to t, that is, for any function $h(\mathbf{x}, t)$

$$L\{h\} = \int_0^{\infty} e^{-pt} h(\mathbf{x}, t)dt \tag{2.3.62}$$

Proofs of these properties are omitted.

The *star product* of two scalar functions defined in Equation 2.3.61 can be extended to mixed fields. For example, if **u** is a vector field and f is a scalar field then

$$[f * \mathbf{u}](\mathbf{x}, t) = \int_0^t f(\mathbf{x}, t-\tau)\mathbf{u}(\mathbf{x}, \tau)d\tau \tag{2.3.63}$$

If **A** and **B** are time-dependent tensor fields then

$$[\mathbf{A} * \mathbf{B}](\mathbf{x}, t) = \int_0^t \mathbf{A}(\mathbf{x}, t-\tau) \cdot \mathbf{B}(\mathbf{x}, \tau)d\tau \tag{2.3.64}$$

If **A** is a tensor and **u** is a vector, both time-dependent, then

$$[\mathbf{A} * \mathbf{u}](\mathbf{x}, t) = \int_0^t \mathbf{A}(\mathbf{x}, t-\tau)\mathbf{u}(\mathbf{x}, \tau)d\tau \tag{2.3.65}$$

In components

$$[f * \mathbf{u}]_i = [f * u_i] \tag{2.3.66}$$

$$[\mathbf{A} * \mathbf{B}] = A_{ij} * B_{ij} \quad \text{(summation holds)} \tag{2.3.67}$$

$$[\mathbf{A} * \mathbf{u}]_i = A_{ij} * u_j \tag{2.3.68}$$

Example 2.3.8

Let **S** be a time-dependent symmetric second-order tensor field on $E^3 \times T$, let ρ be a positive scalar, and let $\mathbf{K}[\cdot]$ be a symmetric invertible fourth-order tensor field on E^3. Let **S** satisfy the differential equation

$$\widehat{\nabla}(\operatorname{div}\mathbf{S}) - \rho\mathbf{K}[\ddot{\mathbf{S}}] = \mathbf{0} \quad \text{on } E^3 \times T \tag{a}$$

subject to the conditions:

$$\left.\begin{aligned} \mathbf{S}(\mathbf{x}, 0) &= \mathbf{S}_0(\mathbf{x}) \\ \dot{\mathbf{S}}(\mathbf{x}, 0) &= \dot{\mathbf{S}}_0(\mathbf{x}) \end{aligned}\right\} \quad \mathbf{x} \in E^3 \tag{b}$$

Show that **S** satisfies (a) and (b) if and only if

$$\widehat{\nabla}(t * \operatorname{div}\mathbf{S}) - \rho\mathbf{K}[\mathbf{S}] = -\rho\mathbf{A} \quad \text{on } E^3 \times T \tag{c}$$

where

$$\mathbf{A} = \mathbf{K}[\mathbf{S}_0] + t\mathbf{K}[\dot{\mathbf{S}}_0] \tag{d}$$

Solution

The solution is given in two steps:

(A) We prove that (a) and (b) $\Rightarrow$ (c)
(B) We prove that (c) $\Rightarrow$ (a) and (b)

To show (A) we apply the Laplace transform technique. If $f = f(t)$ is a function on T and

$$L\{f\} \equiv \bar{f} = \int_0^\infty e^{-pt} f dt \tag{e}$$

is the Laplace transform of f, and there exist $f(0) \equiv f_0$ and $\dot{f}(0) \equiv \dot{f}_0$, then

$$L\{\ddot{f}\} = p^2\bar{f} - pf_0 - \dot{f}_0 \tag{f}$$

Therefore, applying the Laplace transform to both sides of (a) and taking into account (b) we obtain

$$\widehat{\nabla}(\operatorname{div}\bar{\mathbf{S}}) - \rho\mathbf{K}\left[p^2\bar{\mathbf{S}} - p\mathbf{S}_0 - \dot{\mathbf{S}}_0\right] = \mathbf{0} \tag{g}$$

or

$$\widehat{\nabla}\left(\frac{1}{p^2}\operatorname{div}\bar{\mathbf{S}}\right) - \rho\mathbf{K}[\bar{\mathbf{S}}] = -\rho\bar{\mathbf{A}} \tag{h}$$

where

$$\bar{\mathbf{A}} = \frac{1}{p}\mathbf{K}[\mathbf{S}_0] + \frac{1}{p^2}\mathbf{K}[\dot{\mathbf{S}}_0] \tag{i}$$

Applying the convolution property (vi) in the form

$$L^{-1}\{\bar{f}\,\bar{g}\} = f * g \tag{j}$$

to both sides of (h) and (i) and using the relations

$$L^{-1}\left\{\frac{1}{p}\right\} = 1, \quad L^{-1}\left\{\frac{1}{p^2}\right\} = t \tag{k}$$

we receive

$$\widehat{\nabla}(t * \operatorname{div}\mathbf{S}) - \rho\mathbf{K}[\mathbf{S}] = -\rho\mathbf{A} \quad \text{on } E^3 \times T \tag{l}$$

where

$$\mathsf{A} = \mathsf{K}[\mathbf{S}_0] + t\mathsf{K}[\dot{\mathbf{S}}_0] \qquad \text{(m)}$$

This ends the proof of (A).

To show (B) we note that differentiating (c) twice with respect to time we arrive at (a). To see this, we take advantage of the formula

$$\frac{d^2}{dt^2}(t * a) = a \qquad \text{(n)}$$

for any function $a = a(t)$ defined on T. This means that (c) $\Rightarrow$ (a). To show that (c) and (d) $\Rightarrow$ (b) we substitute $t = 0$ to both sides of (c) and obtain

$$\mathsf{K}[\mathbf{S}(\mathbf{x}, 0)] = \mathsf{K}[\mathbf{S}_0(\mathbf{x})] \qquad \text{(o)}$$

Also, by differentiating (c) with respect to t and taking $t = 0$ we get

$$\mathsf{K}[\dot{\mathbf{S}}(\mathbf{x}, 0)] = \mathsf{K}[\dot{\mathbf{S}}_0(\mathbf{x})] \qquad \text{(p)}$$

Since K is invertible, applying K^{-1} to (o) and (p) we arrive at (b). This completes the proof of (B). □

We will now define a *Cauchy Problem* or an *Initial Value Problem* as a problem of finding a solution to an ordinary or partial differential equation subject to assumed initial conditions. For example, the Cauchy problem for a wave equation consists of finding a function $\mathbf{u} = \mathbf{u}(\mathbf{x},t)$ on $E^3 \times [0, \infty)$ that satisfies the equation

$$\nabla^2\mathbf{u} - \frac{1}{c^2}\frac{\partial^2}{\partial t^2}\mathbf{u} = \mathbf{0} \quad \text{on } E^3 \times [0, \infty) \qquad (2.3.69)$$

subject to the conditions

$$\left.\begin{aligned} \mathbf{u}(\mathbf{x}, 0) &= \mathbf{u}_0(\mathbf{x}) \\ \frac{\partial}{\partial t}\mathbf{u}(\mathbf{x}, 0) &= \dot{\mathbf{u}}_0(\mathbf{x}) \end{aligned}\right\} \quad \mathbf{x} \in E^3 \qquad (2.3.70)$$

In Equation 2.3.69, c has the dimension $[c]$ = m/s. Functions $\mathbf{u}^0(x)$ and $\dot{\mathbf{u}}_0(\mathbf{x})$ are prescribed on E^3.*

In Example 2.3.8, the problem of finding $\mathbf{S}$ that satisfies (a) and (b) is a Cauchy problem of linear elastodynamics in terms of stresses $\mathbf{S}$ in which K is a compliance tensor field, and $\mathbf{S}_0$ and $\dot{\mathbf{S}}_0$ are prescribed initial stress and initial stress-rate fields [10].

* For the definition of the Cauchy problem involving a general partial differential equations see, for example, [9].

PROBLEMS

2.1 Use the properties of the alternator ϵ_{ijk} introduced in Section 2.1 to show that

$$(\mathbf{a} \times \mathbf{b}) \times \mathbf{c} = (\mathbf{a} \cdot \mathbf{c})\mathbf{b} - (\mathbf{b} \cdot \mathbf{c})\mathbf{a} \tag{a}$$

$$(\mathbf{a} \times \mathbf{b}) \times (\mathbf{c} \times \mathbf{d}) = [\mathbf{a} \cdot (\mathbf{c} \times \mathbf{d})]\mathbf{b} - [\mathbf{b} \cdot (\mathbf{c} \times \mathbf{d})]\mathbf{a} \tag{b}$$

where **a**, **b**, **c**, and **d** are arbitrary vectors.

2.2 Show that for any vector **u** and a unit vector **n** the following decomposition formula holds true

$$\mathbf{u} = \mathbf{u}^{\perp} + \mathbf{u}^{\parallel} \tag{a}$$

where

$$\mathbf{u}^{\perp} = (\mathbf{u} \cdot \mathbf{n})\mathbf{n} \quad \text{and} \quad \mathbf{u}^{\parallel} = \mathbf{n} \times (\mathbf{u} \times \mathbf{n}) \tag{b}$$

Also, show that

$$\mathbf{u}^{\perp} \cdot \mathbf{u}^{\parallel} = 0, \quad \mathbf{u} \cdot \mathbf{n} = \mathbf{u}^{\perp} \cdot \mathbf{n}, \quad \mathbf{u}^{\parallel} \cdot \mathbf{n} = 0 \tag{c}$$

Note: If $\mathbf{u} = \mathbf{u}(\mathbf{x})$ is a vector field defined on a surface S in E^3, $\mathbf{n} = \mathbf{n}(\mathbf{x})$ is a unit outward normal vector field on S, and P is a plane tangent to S at **x**, then $\mathbf{u}^{\perp}$ and $\mathbf{u}^{\parallel}$ represent the normal and tangent parts of **u**, respectively, with regard to P.

2.3 Show that an alternative form of Equations (a) and (b) in Problem 2.2 reads

$$\mathbf{u} = \mathbf{u}^{\perp} + \mathbf{u}^{\parallel} \tag{a}$$

where

$$\mathbf{u}^{\perp} = (\mathbf{n} \otimes \mathbf{u})\mathbf{n} \quad \text{and} \quad \mathbf{u}^{\parallel} = (\mathbf{1} - \mathbf{n} \otimes \mathbf{n})\mathbf{u} \tag{b}$$

In Equations (b) the symbol $\otimes$ represents the tensor product of two vectors, and **1** is a unit second-order tensor (see Equation 2.1.52).

2.4 Let $\mathbf{T} = \mathbf{T}(\mathbf{x})$ be a symmetric tensor field defined on a surface S in E^3, $\mathbf{n} = \mathbf{n}(\mathbf{x})$ a unit outward normal vector field on S, and P a plane tangent to S at **x**. Show that

$$\mathbf{T} = \mathbf{T}^{\perp} + \mathbf{T}^{\parallel} \tag{a}$$

where

$$\mathbf{T}^{\perp} = 2\,\mathrm{sym}\,(\mathbf{n} \otimes \mathbf{T}\mathbf{n}) - (\mathbf{n} \cdot \mathbf{T}\mathbf{n})\mathbf{n} \otimes \mathbf{n} \tag{b}$$

and

$$\mathbf{T}^{\parallel} = (\mathbf{1} - \mathbf{n} \otimes \mathbf{n})\mathbf{T}(\mathbf{1} - \mathbf{n} \otimes \mathbf{n}) \tag{c}$$

Also, show that

$$\mathbf{T}^{\perp} \cdot \mathbf{T}^{\parallel} = 0, \quad \mathbf{Tn} = \mathbf{T}^{\perp}\mathbf{n}, \quad \mathbf{T}^{\parallel}\mathbf{n} = \mathbf{0} \tag{d}$$

Note: The tensors $\mathbf{T}^{\perp}$ and $\mathbf{T}^{\parallel}$ represent the normal and tangential parts of $\mathbf{T}$, respectively, with regard to the plane P.

2.5 Show that if S is a plane $x_3 = 0$ with the unit outward normal vector $\mathbf{n} = (0, 0, -1)$, then the decomposition formula (a) in Problem 2.4 reads

$$\begin{bmatrix} T_{11} & T_{12} & T_{13} \\ T_{21} & T_{22} & T_{23} \\ T_{31} & T_{32} & T_{33} \end{bmatrix} = \begin{bmatrix} 0 & 0 & T_{13} \\ 0 & 0 & T_{23} \\ T_{31} & T_{32} & T_{33} \end{bmatrix} + \begin{bmatrix} T_{11} & T_{12} & 0 \\ T_{21} & T_{22} & 0 \\ 0 & 0 & 0 \end{bmatrix}$$

2.6 Let $\mathbf{T}$ be a second-order tensor with components T_{ij}, and let $\det \mathbf{T} \neq 0$. Show that

$$\det \mathbf{T} = \epsilon_{ijk} T_{i1} T_{j2} T_{k3} \tag{a}$$

$$\epsilon_{pqr}(\det \mathbf{T}) = \epsilon_{ijk} T_{ip} T_{jq} T_{kr} \tag{b}$$

$$\epsilon_{ijk}\epsilon_{pqr}(\det \mathbf{T}) = \begin{vmatrix} T_{ip} & T_{iq} & T_{ir} \\ T_{jp} & T_{jq} & T_{jr} \\ T_{kp} & T_{kq} & T_{kr} \end{vmatrix} \tag{c}$$

2.7 Let $\mathbf{T}$ be a second-order tensor with components T_{ij} such that $\det \mathbf{T} \neq 0$, and let $\widehat{\mathbf{T}}$ be the tensor with components

$$\widehat{T}_{ij} = \frac{1}{2}\epsilon_{ipq}\epsilon_{jrs} T_{pr} T_{qs} \tag{a}$$

Show that

$$\mathbf{T}\widehat{\mathbf{T}}^T = \widehat{\mathbf{T}}^T\mathbf{T} = (\det \mathbf{T})\mathbf{1} \tag{b}$$

$$\mathbf{T}^{-1} = (\det \mathbf{T})^{-1}\widehat{\mathbf{T}}^T \tag{c}$$

Note: The matrix $[\hat{T}_{ij}]$ is called the *cofactor* of the matrix $[T_{ij}]$, while $[\widehat{T}_{ij}^T]$ is called the *adjoint* of $[T_{ij}]$.

2.8 The x_i' system is obtained by rotating the x_i system about the x_3 axis through an angle $0 < \theta < \pi/2$ as shown in Figure P2.8.

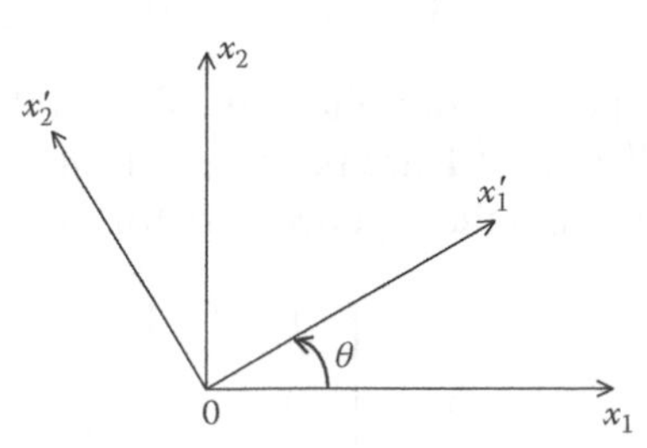

FIGURE P2.8

Let $\mathbf{T}$ be a symmetric second-order tensor referred to the x_i system. Show that

$$\mathbf{T}' = (\mathbf{T}')^T \tag{a}$$

$$\begin{aligned} T'_{11} &= T_{11}\cos^2\theta + T_{12}\sin 2\theta + T_{22}\sin^2\theta \\ T'_{12} &= \frac{1}{2}(T_{22} - T_{11})\sin 2\theta + T_{12}\cos 2\theta \\ T'_{22} &= T_{11}\sin^2\theta - T_{12}\sin 2\theta + T_{22}\cos^2\theta \end{aligned} \tag{b}$$

and

$$\begin{aligned} T'_{13} &= T_{13}\cos\theta + T_{23}\sin\theta \\ T'_{23} &= -T_{13}\sin\theta + T_{23}\cos\theta \\ T'_{33} &= T_{33} \end{aligned} \tag{c}$$

Also, show that an alternative form of the transformation formulas (b) and (c) reads

$$\begin{aligned} T'_{11} + T'_{22} &= T_{11} + T_{22} \\ T'_{22} - T'_{11} + 2iT_{12} &= \exp(2i\theta)(T_{22} - T_{11} + 2iT_{12}) \\ T'_{13} - iT'_{23} &= \exp(i\theta)(T_{13} - iT_{23}) \\ T'_{33} &= T_{33} \end{aligned} \tag{d}$$

where $i = \sqrt{-1}$.

Hence, if the coordinates $\left(x'_1, x'_2, x'_3\right)$ are identified with the cylindrical coordinates (r, θ, x_3), we find

$$\begin{aligned} T_{rr} + T_{\theta\theta} &= T_{11} + T_{22} \\ T_{\theta\theta} - T_{rr} + 2iT_{r\theta} &= \exp(2i\theta)(T_{22} - T_{11} + 2iT_{12}) \\ T_{r3} - iT_{\theta 3} &= \exp(i\theta)(T_{13} - iT_{23}) \\ T_{33} &= T_{33} \end{aligned} \tag{e}$$

[Do not sum over r and θ in Equations (e).]
Hint: Use the formula $\mathbf{T}' = \mathbf{Q}^T\mathbf{T}\mathbf{Q}$ where $\mathbf{Q}^T$ is the matrix

$$\mathbf{Q}^T = \begin{bmatrix} \cos\theta & \sin\theta & 0 \\ -\sin\theta & \cos\theta & 0 \\ 0 & 0 & 1 \end{bmatrix}$$

2.9 A tensor $\mathbf{T}$ is said to be positive definite if $\mathbf{u}\cdot\mathbf{T}\mathbf{u} > 0$ for every $\mathbf{u} \neq \mathbf{0}$. Show that if $\mathbf{T}$ is invertible, then $\mathbf{T}\mathbf{T}^T$ and $\mathbf{T}^T\mathbf{T}$ are positive definite.

2.10 Show that the eigenvalues and the eigenvectors for the matrix

$$\mathbf{T} = \begin{bmatrix} 1 & 0 & 1 \\ 0 & 2 & 0 \\ 1 & 0 & 3 \end{bmatrix} \tag{a}$$

are given by

$$\lambda_1 = 2 - \sqrt{2}, \quad \lambda_2 = 2, \quad \lambda_3 = 2 + \sqrt{2} \tag{b}$$

and

$$n_1^{(1)} = \pm\frac{1}{\sqrt{2}}\frac{1}{1-\sqrt{2}}\frac{1}{\sqrt{2+\sqrt{2}}}$$

$$n_2^{(1)} = 0 \tag{c}$$

$$n_3^{(1)} = \pm\frac{1}{\sqrt{2}}\frac{1}{\sqrt{2+\sqrt{2}}}$$

$$n_1^{(2)} = 0, \quad n_2^{(2)} = \pm 1, \quad n_3^{(2)} = 0 \tag{d}$$

$$n_1^{(3)} = \pm\frac{1}{\sqrt{2}}\frac{1}{1+\sqrt{2}}\frac{1}{\sqrt{2-\sqrt{2}}}$$

$$n_2^{(3)} = 0 \tag{e}$$

$$n_3^{(3)} = \pm\frac{1}{\sqrt{2}}\frac{1}{\sqrt{2-\sqrt{2}}}$$

In Equations (b) through (e) λ_i is an eigenvalue corresponding to the eigenvector $\mathbf{n}^{(i)}$ $(i = 1, 2, 3)$.

2.11 Let $\mathbf{T}$ be the tensor represented by the matrix (a) in Problem 2.10, and let $\{e_i^*\}$ be the orthonormal basis obtained from Equations (c) through (e) in Problem 2.10 in which the upper signs are postulated. Define the tensor $\mathbf{Q}^T$ in terms of components by

$$Q_{ij}^T = \mathbf{e}_i^* \cdot \mathbf{e}_j \tag{a}$$

Show that

$$\mathbf{T}^* = \mathbf{Q}^T\mathbf{T}\mathbf{Q} \tag{b}$$

is a tensor represented by a diagonal matrix. Also, compute the components T_{11}^*, T_{22}^*, and T_{33}^*, and show that

$$\operatorname{tr}\mathbf{T}^* = \operatorname{tr}\mathbf{T} = 6 \tag{c}$$

2.12 Prove the following identities in which ϕ is a scalar field, $\mathbf{u}$ is a vector field, and $\mathbf{S}$ is a tensor field on a region $R \subset E^3$. You may need to use the $\epsilon - \delta$ relation.

$$\text{curl } \nabla\phi = \mathbf{0} \tag{a}$$

$$\text{div curl } \mathbf{u} = 0 \tag{b}$$

$$\text{curl curl } \mathbf{u} = \nabla \text{div } \mathbf{u} - \nabla^2 \mathbf{u} \tag{c}$$

$$\text{curl } \nabla \mathbf{u} = \mathbf{0} \tag{d}$$

$$\text{curl}(\nabla \mathbf{u}^T) = \nabla \text{curl } \mathbf{u} \tag{e}$$

$$\text{If } \nabla \mathbf{u} = -\nabla \mathbf{u}^T \quad \text{then } \nabla \nabla \mathbf{u} = \mathbf{0} \tag{f}$$

$$\text{div curl } \mathbf{S} = \text{curl div } \mathbf{S}^T \tag{g}$$

$$\text{div } (\text{curl } \mathbf{S})^T = 0 \tag{h}$$

$$(\text{curl curl } \mathbf{S})^T = \text{curl } (\text{curl } \mathbf{S}^T) \tag{i}$$

$$\text{curl } (\phi \mathbf{1}) = -[\text{curl } (\phi \mathbf{1})]^T \tag{j}$$

$$\text{div } (\mathbf{S}^T \mathbf{u}) = \mathbf{u} \cdot \text{div } \mathbf{S} + \mathbf{S} \cdot \nabla \mathbf{u} \tag{k}$$

If **S** is symmetric then

$$\text{tr } (\text{curl } \mathbf{S}) = 0 \tag{l}$$

If **S** is symmetric then

$$\text{curl curl } \mathbf{S} = -\nabla^2 \mathbf{S} + 2\widehat{\nabla}(\text{div } \mathbf{S}) - \nabla \nabla (\text{tr } \mathbf{S}) + \mathbf{1}[\nabla^2 (\text{tr } \mathbf{S}) - \text{div div } \mathbf{S}] \tag{m}$$

If **S** is symmetric and $\mathbf{S} = \mathbf{G} - \mathbf{1}(\text{tr } \mathbf{G})$ then

$$\text{curl curl } \mathbf{S} = -\nabla^2 \mathbf{G} + 2\widehat{\nabla}(\text{div } \mathbf{G}) - \mathbf{1} \text{ div div } \mathbf{G} \tag{n}$$

If **S** is skew and $\boldsymbol{\omega}$ is its axial vector then

$$\text{curl } \mathbf{S} = \mathbf{1}(\text{div } \boldsymbol{\omega}) - \nabla \boldsymbol{\omega} \tag{o}$$

2.13 Let f be a scalar field, **u** a vector field, and **T** a tensor field on a region $R \subset E^3$. Let **n** be a unit outer normal vector to ∂R, where ∂R stands for the boundary of R. Show that

$$\int_R (\nabla f)\, dv = \int_{\partial R} f\mathbf{n}\, da \tag{a}$$

$$\int_R (\mathrm{curl}\,\mathbf{u})\, dv = \int_{\partial R} (\mathbf{n} \times \mathbf{u})\, da \tag{b}$$

$$\int_R (\nabla \mathbf{u})\, dv = \int_{\partial R} \mathbf{u} \otimes \mathbf{n}\, da \tag{c}$$

$$\int_R [\mathbf{u} \otimes \mathrm{div}\,\mathbf{T} + (\nabla \mathbf{u})\mathbf{T}^T]\, dv = \int_{\partial R} \mathbf{u} \otimes \mathbf{T}\mathbf{n}\, da \tag{d}$$

2.14 Let $\mathbf{u}$ be a vector field on $R \subset E^3$ subject to one of the conditions:

$$\mathbf{u} = \mathbf{0} \quad \text{on } \partial R \tag{a}$$

or

$$\mathbf{n} \times \mathrm{curl}\,\mathbf{u} = \mathbf{0} \quad \text{on } \partial R \tag{b}$$

Show that

$$\int_R [\mathbf{u} \cdot (\mathrm{curl}\,\mathrm{curl}\,\mathbf{u})] dv = \int_R (\mathrm{curl}\,\mathbf{u})^2 dv \tag{c}$$

2.15 Let $\mathbf{u} = \mathbf{u}(\mathbf{x}, t)$ and $\mathbf{S} = \mathbf{S}(\mathbf{x}, t)$ denote a time-dependent vector field on $E^3 \times [0, \infty)$ and a time-dependent tensor field on $E^3 \times [0, \infty)$, respectively. Let $\rho = \rho(\mathbf{x})$ be a positive scalar field on E^3, and let the pair $[\mathbf{u}, \mathbf{S}]$ satisfy the differential equation

$$\mathrm{div}\,\mathbf{S} - \rho\ddot{\mathbf{u}} = \mathbf{0} \quad \text{on } E^3 \times [0, \infty) \tag{a}$$

subject to the conditions

$$\mathbf{u}(\mathbf{x}, 0) = \mathbf{u}_0(\mathbf{x}),\ \dot{\mathbf{u}}(\mathbf{x}, 0) = \dot{\mathbf{u}}_0(\mathbf{x}) \quad \text{for } \mathbf{x} \in E^3 \tag{b}$$

where $\mathbf{u}_0$ and $\dot{\mathbf{u}}_0$ are prescribed vector fields on E^3. Show that $[\mathbf{u}, \mathbf{S}]$ satisfies Equations (a) and (b) if and only if

$$\mathbf{u} = \rho^{-1} t * (\mathrm{div}\,\mathbf{S}) + \mathbf{u}_0 + t\dot{\mathbf{u}}_0 \quad \text{on } E^3 \times [0, \infty) \tag{c}$$

Here $*$ stands for the convolution product on the t axis.

Note: If $\mathbf{S}$ is identified with a stress field of classical elastodynamics, the formula (c) provides an alternative form of the displacement field of the theory. [See Theorem 2 of Section 4.2.4 with $\mathbf{b} = \mathbf{0}$.]

REFERENCES

1. P. Halmos, *Finite-Dimensional Vector Spaces*, Van Nostrand, New York, 1958.
2. N. Noda, R. B. Hetnarski, and Y. Tanigawa, *Thermal Stresses*, 2nd edn., Taylor & Francis, New York, 2003, Appendix A, Inverse Laplace transform and contour integrals, pp. 469–472.
3. C. R. Wyle and L. C. Barrett, *Advanced Engineering Mathematics*, 5th edn., McGraw-Hill, New York, 1961, pp. 402–408.
4. R. B. Hetnarski, Solution of the coupled problem of thermoelasticity in the form of series of functions, *Archiwum Mechaniki Stosowanej*, 16, 1964, 919–941.
5. V. A. Ditkin and A. P. Prudnikov, *Reference Book on Operational Calculus*, Fizmatgiz, Moscow, 1965, formula 23.124, p. 254 (in Russian).
6. A. Erdélyi (Ed.), *Bateman Manuscript Project, Tables of Integral Transforms*, vol. 1, McGraw-Hill, New York, 1954, formula No. 10, p. 246.
7. R. B. Hetnarski, An algorithm for generating some inverse Laplace transforms of exponential form, *J. Appl. Math. Phys.*, 26, 1975, 249–253.
8. I. S. Sokolnikoff and R. M. Redheffer, *Mathematics of Physics and Modern Engineering*, 2nd edn., McGraw-Hill, New York, 1966, pp. 220–223.
9. I. Stakgold, *Boundary Value Problems of Mathematical Physics*, vol. 2, Macmillan, New York, and Collier–Macmillan, London, U.K., 1968, p. 73.
10. M. E. Gurtin, *The Linear Theory of Elasticity*, Encyclopedia of Physics, chief editor: S. Flügge, vol. VIa/2, editor: C. Truesdell, Springer, Berlin, Germany, 1972, pp. 220–221.

3 Fundamentals of Linear Elasticity

In this chapter, a number of concepts of the solid mechanics are introduced to describe a linear elastic body in terms of the partial differential equations. In particular, the displacement vector, strain tensor, and stress tensor fields are introduced to define a linear elastic body which satisfies the strain–displacement relations, the equations of motion, and the constitutive relations. Emphasis is placed on the compatibility relations, the general solutions of elastostatics, and on an alternative definition of the displacement field of elastodynamics. A discussion of the constitutive relations includes the orthotropic, transversely isotropic, and isotropic materials. The stored energy of an elastic body, the positive definiteness and strong ellipticity of the elasticity fourth-order tensor, and the stress–strain–temperature relations for a thermoelastic body are also discussed. The chapter contains a number of worked examples, and also end-of-chapter problems with solutions provided in the Solutions Manual.

3.1 KINEMATICS

3.1.1 DEFORMATION

We define a material body B as a set of elements $\mathbf{x}$, called particles, for which there is a one-to-one correspondence with the points of a region $\kappa(B)$ of a physical space.

A deformation of B is a map $\boldsymbol{\kappa}$ of B onto a region $\kappa(B)$ in E^3 with $\det(\nabla\boldsymbol{\kappa}) > 0$.

The point $\boldsymbol{\kappa}(\mathbf{x})$ is the place occupied by the particle $\mathbf{x}$ in the deformation $\boldsymbol{\kappa}$, and

$$\mathbf{u}(\mathbf{x}) = \boldsymbol{\kappa}(\mathbf{x}) - \mathbf{x} \tag{3.1.1}$$

is the displacement of $\mathbf{x}$. The condition that $\det(\nabla\boldsymbol{\kappa}) > 0$ means that the mapping $\boldsymbol{\kappa}$ is uniquely invertible, that is, there is the mapping $\boldsymbol{\kappa}^{-1}(\mathbf{x}) = \mathbf{x}$.

We will refer to body B as a reference configuration and to body $\kappa(B)$ as a current configuration, see Figure 3.1.

If the mapping $\boldsymbol{\kappa}$ depends also on time t ($t \in T$) such a mapping defines a motion of body B, and the displacement of $\mathbf{x}$ at time t is

$$\mathbf{u}(\mathbf{x}, t) = \boldsymbol{\kappa}(\mathbf{x}, t) - \mathbf{x} \tag{3.1.2}$$

In this case, the configuration that the body occupies at $t = 0$ is a reference or initial configuration, while the configuration that the body occupies at the current time t is the current configuration.

In the following, we introduce a number of definitions describing a time-independent deformation of B and later we will extend these to a time-dependent case.

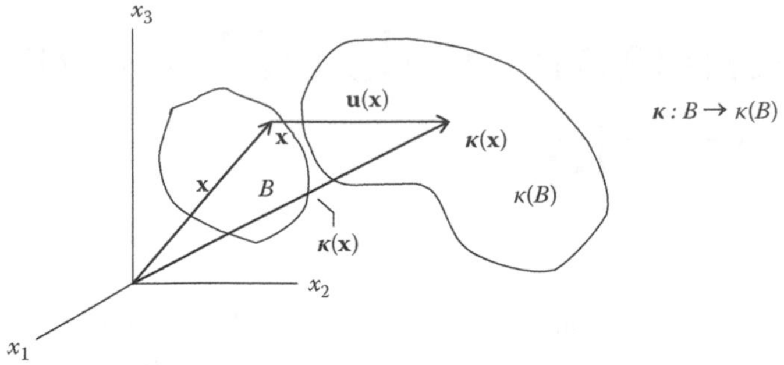

FIGURE 3.1 The deformation of a material body B.

By the *deformation gradient* and the *displacement gradient* we mean the tensor fields $\mathbf{F} = \nabla\boldsymbol{\kappa}$ and $\nabla\mathbf{u}$, respectively. By Equation 3.1.1, we obtain

$$\nabla\mathbf{u} = \mathbf{F} - \mathbf{1} \tag{3.1.3}$$

We now introduce the notion of strain. A *finite strain* tensor $\mathbf{D}$ is defined by

$$\mathbf{D} = \frac{1}{2}\left(\mathbf{F}^T\mathbf{F} - \mathbf{1}\right) \tag{3.1.4}$$

Using Equations 3.1.3 and 3.1.4, we express $\mathbf{D}$ in an alternative form

$$\mathbf{D} = \mathbf{E} + \frac{1}{2}\nabla\mathbf{u}^T\nabla\mathbf{u} \tag{3.1.5}$$

where

$$\mathbf{E} = \frac{1}{2}\left(\nabla\mathbf{u} + \nabla\mathbf{u}^T\right) = \widehat{\nabla}\mathbf{u} \tag{3.1.6}$$

The tensor $\mathbf{E}$ is called the *infinitesimal strain tensor* of linear elasticity. If the field $\epsilon = |\nabla\mathbf{u}|$ is small in the sense that ϵ^2 can be ignored in comparison with ϵ then the finite strain tensor $\mathbf{D}$ and the infinitesimal strain tensor $\mathbf{E}$ coincide.

Example 3.1.1

Express in index notation tensors $\mathbf{D}$ and $\mathbf{E}$.

Solution

By Equations 3.1.3 and 3.1.4 we obtain

$$\begin{aligned} D_{ij} &= \frac{1}{2}\left(F^T_{ik}F_{kj} - \delta_{ij}\right) = \frac{1}{2}\left(F_{ki}F_{kj} - \delta_{ij}\right) \\ &= \frac{1}{2}\left[\left(\delta_{ki} + u_{k,i}\right)\left(\delta_{kj} + u_{k,j}\right) - \delta_{ij}\right] = \frac{1}{2}\left(u_{i,j} + u_{j,i} + u_{k,i}u_{k,j}\right) \\ &= \frac{1}{2}\left(u_{i,j} + u_{j,i} + u^T_{i,k}u_{k,j}\right) \end{aligned} \tag{a}$$

Also, for E_{ij} we obtain

$$E_{ij} = \frac{1}{2}(u_{i,j} + u_{j,i}) = \frac{1}{2}\left(u_{i,j} + u^T_{i,j}\right) \tag{b}$$

□

Example 3.1.2

Show that if

$$\boldsymbol{\kappa}(\mathbf{x}) = \mathbf{y}_0 + \mathbf{Q}(\mathbf{x} - \mathbf{x}_0) \tag{a}$$

where $\mathbf{x}_0$ and $\mathbf{y}_0$ are fixed points of E^3 and $\mathbf{Q}$ is an orthogonal constant tensor then

$$\mathbf{D} = \mathbf{0} \tag{b}$$

and

$$\mathbf{E} = \frac{1}{2}(\mathbf{Q} + \mathbf{Q}^T) - \mathbf{1} \tag{c}$$

Solution

Since $\mathbf{F} = \nabla\boldsymbol{\kappa}$, by (a) we get

$$\mathbf{F} = \mathbf{Q} \tag{d}$$

In index notation

$$\kappa_i = (\mathbf{y}_0)_i + Q_{ip}[x_p - (\mathbf{x}_0)_p] \tag{e}$$

Hence

$$F_{ij} = \kappa_{i,j} = Q_{ip}\delta_{pj} = Q_{ij} \tag{f}$$

which means (d). Also, by Equations 3.1.3 and (d)

$$\nabla\mathbf{u} = \mathbf{Q} - \mathbf{1} \tag{g}$$

Therefore, by Equations 3.1.1, (g) and (a)

$$\mathbf{u} = \mathbf{y}_0 - \mathbf{x}_0 + \nabla\mathbf{u}(\mathbf{x} - \mathbf{x}_0) \tag{h}$$

and by Equation 3.1.4

$$\mathbf{D} = \frac{1}{2}(\mathbf{Q}^T\mathbf{Q} - \mathbf{1}) = \mathbf{0} \tag{i}$$

since $\mathbf{Q}$ is orthogonal. This ends the proof of (b).

To prove (c) we compute **E**

$$\mathbf{E} = \frac{1}{2}\,(\nabla\mathbf{u} + \nabla\mathbf{u}^T) = \frac{1}{2}\,[(\mathbf{Q} - \mathbf{1}) + (\mathbf{Q}^T - \mathbf{1})] = \frac{1}{2}\,(\mathbf{Q} + \mathbf{Q}^T) - \mathbf{1} \tag{j}$$

This ends the proof of (c). □

Remark 3.1.1: In Example 3.1.2, the deformation κ represents a finite rigid deformation because the finite strain tensor **D** for this deformation vanishes.

Example 3.1.3

Show that if κ is the deformation given by (a) in Example 3.1.2 and **E** is an infinitesimal strain tensor then

$$\mathbf{E} = -\frac{1}{2}\nabla\mathbf{u}^T\nabla\mathbf{u} \tag{a}$$

Solution

By Equation 3.1.5

$$\mathbf{D} = \mathbf{E} + \frac{1}{2}\nabla\mathbf{u}^T\nabla\mathbf{u} \tag{b}$$

On the other hand, for the deformation κ defined by (a), as shown in Example 3.1.2 (b), $\mathbf{D} = \mathbf{0}$. Therefore, Equation (b) with LHS $= \mathbf{0}$ implies (a). □

From Examples 3.1.2 and 3.1.3 and from the definition of κ, it follows that in a rigid deformation the infinitesimal strain tensor vanishes if $\epsilon^2 = |\nabla\mathbf{u}|^2$ is ignored as being small in comparison to $\epsilon = |\nabla\mathbf{u}|$. Writing Equation (a) in terms of the symbol $0(\epsilon^2)$, where $0(\epsilon^2)/\epsilon^2$ is bounded as $\epsilon \to 0$, we get

$$\nabla\mathbf{u} + \nabla\mathbf{u}^T = 0(\epsilon^2) \tag{3.1.7}$$

that is, to within an error $0(\epsilon^2)$, the tensor $\nabla\mathbf{u}$, corresponding to the deformation κ given by (a) in Example 3.1.3, is skew. For this reason, we define an *infinitesimal rigid displacement* as a field of the form

$$\mathbf{u}(\mathbf{x}) = \mathbf{u}_0 + \mathbf{W}(\mathbf{x} - \mathbf{x}_0) \tag{3.1.8}$$

where $\mathbf{u}_0$, $\mathbf{x}_0$ are constant vectors and **W** is a skew constant tensor.

In a general deformation κ, the volume change δV is defined by

$$\delta V = V_D - V_R \tag{3.1.9}$$

where

$$V_D = \int_{\kappa(B)} dv \quad \text{and} \quad V_R = \int_B dv \tag{3.1.10}$$

which means that V_R represents the volume of the reference configuration and V_D the volume of the deformed configuration. Since

$$V_D = \int_{\kappa(B)} dv = \int_B \det(\nabla\boldsymbol{\kappa})\, dv \tag{3.1.11}$$

and

$$\nabla\boldsymbol{\kappa} = \mathbf{F} \tag{3.1.12}$$

hence

$$\delta V = \int_B [(\det \mathbf{F}) - 1]\, dv \tag{3.1.13}$$

Example 3.1.4

Show that

(a) $\det \mathbf{F} = \det(\mathbf{1} + \nabla\mathbf{u}) = 1 + \operatorname{tr} \nabla\mathbf{u} + 0(\epsilon^2)$
(b) $\det \mathbf{F} - 1 = \operatorname{div} \mathbf{u} + 0(\epsilon^2)$.

Solution

Using Equation 3.1.3 we obtain

$$\det \mathbf{F} = \det(\mathbf{1} + \nabla\mathbf{u}) = \begin{vmatrix} 1 + u_{1,1} & u_{1,2} & u_{1,3} \\ u_{2,1} & 1 + u_{2,2} & u_{2,3} \\ u_{3,1} & u_{3,2} & 1 + u_{3,3} \end{vmatrix} \tag{a}$$

Expanding this determinant we obtain

$$\det \mathbf{F} = 1 + u_{k,k} + 0(\epsilon^2) \tag{b}$$

which proves (a). Also this proves (b). □

This example and Equation 3.1.13 show that the volume change per unit volume is equal to the divergence of **u** if the term $0(\epsilon^2)$ is neglected. For this reason we call δV given by

$$\delta V = \int_B \operatorname{div} \mathbf{u}\, dv \tag{3.1.14}$$

the *infinitesimal volume change*.

3.1.2 Infinitesimal Strain Field

In the previous section, we defined the infinitesimal strain field **E** by the equation, see Equation 3.1.6,

$$\mathbf{E} = \frac{1}{2}\left(\nabla\mathbf{u} + \nabla\mathbf{u}^T\right) \tag{3.1.15}$$

This equation relating **E** to **u** is also called the *strain–displacement relation*. Note that

$$\nabla\mathbf{u} = \mathbf{E} + \mathbf{W} \tag{3.1.16}$$

where

$$\mathbf{W} = \frac{1}{2}\left(\nabla\mathbf{u} - \nabla\mathbf{u}^T\right) \tag{3.1.17}$$

The tensor **W** is the skew part of $\nabla\mathbf{u}$ and is called the *infinitesimal rotation tensor*. With the tensor field **W** is associated the *infinitesimal rotation vector* $\boldsymbol{\omega}$, the so called *axial vector of* **W**, defined by

$$\boldsymbol{\omega} = \frac{1}{2}\,\mathrm{curl}\,\mathbf{u} \tag{3.1.18}$$

Example 3.1.5

Show that for any vector **a**

$$\mathbf{Wa} = \boldsymbol{\omega} \times \mathbf{a} \tag{a}$$

Solution

By Equations 3.1.17 and 3.1.18 written in components we get

$$W_{ij} = \frac{1}{2}\left(u_{i,j} - u_{j,i}\right) \tag{b}$$

$$\omega_i = \frac{1}{2}\,\epsilon_{ijk}u_{k,j} \tag{c}$$

Equation (a) written in components takes the form

$$W_{ij}a_j = \epsilon_{ikj}\omega_k a_j \tag{d}$$

Computing LHS and RHS of (d) by using (b) and (c) we receive

$$(\mathrm{LHS})_i = W_{ij}a_j = \frac{1}{2}\left(u_{i,j} - u_{j,i}\right)a_j \tag{e}$$

$$\begin{aligned}(\mathrm{RHS})_i &= \frac{1}{2}\,\epsilon_{ikj}\epsilon_{kpq}u_{q,p}a_j = -\frac{1}{2}\,\epsilon_{kij}\epsilon_{kpq}u_{q,p}a_j \\ &= -\frac{1}{2}\left(\delta_{ip}\delta_{jq} - \delta_{iq}\delta_{pj}\right)u_{q,p}a_j = \frac{1}{2}\left(u_{i,j} - u_{j,i}\right)a_j\end{aligned} \tag{f}$$

Hence $(\mathrm{LHS})_i = (\mathrm{RHS})_i$ and the relation (a) has been proved. □

While discussing the volume change in an infinitesimal deformation $\boldsymbol{\kappa}$, we have introduced the notion of the infinitesimal volume change, and we have proved that the volume change per unit volume is equal to div **u**. The field

$$\operatorname{div}\mathbf{u} = \operatorname{tr}\mathbf{E} \tag{3.1.19}$$

is called the *dilatation*. By Equation 3.1.14 the infinitesimal volume change $\delta v(P)$ of a part P of B is defined in terms of **u** by

$$\delta v(P) = \int_P \operatorname{div}\mathbf{u}\, dv = \int_{\partial P} \mathbf{u}\cdot\mathbf{n}\, da \tag{3.1.20}$$

If the deformation is not accompanied by a change of volume, that is, if $\delta v(P) = 0$ for every P, the displacement **u** is called *isochoric*. Since

$$\delta v(P) = \int_P \operatorname{div}\mathbf{u}\, dv = \int_P \operatorname{tr}\mathbf{E}\, dv \tag{3.1.21}$$

we see then that $\delta v(P) = 0$ if and only if $\operatorname{tr}\mathbf{E} = 0$ as P is an arbitrary part of B.

In Section 3.1.1, we introduced the definition of an infinitesimal rigid displacement field **u** by the formula 3.1.8

$$\mathbf{u}(\mathbf{x}) = \mathbf{u}_0 + \mathbf{W}(\mathbf{x} - \mathbf{x}_0) \tag{3.1.22}$$

where

$\mathbf{u}_0$ is a vector
$\mathbf{x}_0$ is a position vector
$\mathbf{W}$ is a skew constant tensor

In the following we give an example describing properties of a rigid displacement.

Example 3.1.6

Show that a displacement **u** is a rigid displacement field if and only if the strain field **E** corresponding to **u** vanishes on B.

Solution

First we prove that if **u** is a rigid displacement then $\mathbf{E} = \mathbf{0}$.

If $\mathbf{u}(\mathbf{x})$ is a rigid displacement then by Equation 3.1.22 for any two position vectors **x** and **y**

$$\mathbf{u}(\mathbf{x}) - \mathbf{u}(\mathbf{y}) = \mathbf{W}(\mathbf{x} - \mathbf{y}) \tag{a}$$

Multiplying this by $(\mathbf{x} - \mathbf{y})$ in the sense of inner product we obtain

$$(\mathbf{x} - \mathbf{y}) \cdot [\mathbf{u}(\mathbf{x}) - \mathbf{u}(\mathbf{y})] = (\mathbf{x} - \mathbf{y}) \cdot [\mathbf{W}(\mathbf{x} - \mathbf{y})] = \mathbf{W} \cdot [(\mathbf{x} - \mathbf{y}) \otimes (\mathbf{x} - \mathbf{y})] = 0 \tag{b}$$

since $(\mathbf{x}-\mathbf{y})\otimes(\mathbf{x}-\mathbf{y})$ is a symmetric tensor and $\mathbf{W}$ is a skew tensor (see Example 2.1.5). Differentiating (b) with respect to $\mathbf{x}$ we obtain

$$(\nabla\mathbf{u}^T)(\mathbf{x}-\mathbf{y})+\mathbf{u}(\mathbf{x})-\mathbf{u}(\mathbf{y})=\mathbf{0} \tag{c}$$

To get (c) from (b) we can also write (b) and (c) in components. Thus, (b) in components reads

$$(x_i-y_i)[u_i(\mathbf{x})-u_i(\mathbf{y})]=0 \tag{d}$$

Differentiating this with respect to x_k we obtain

$$\delta_{ik}[u_i(\mathbf{x})-u_i(\mathbf{y})]+(x_i-y_i)u_{i,k}(\mathbf{x})=0 \tag{e}$$

or

$$u_k(\mathbf{x})-u_k(\mathbf{y})+u^T_{k,i}(x_i-y_i)=0 \tag{f}$$

This result is precisely (c) written in components. Finally, differentiating (c) with respect to $\mathbf{y}$ and then letting $\mathbf{y}=\mathbf{x}$, we get

$$-\nabla\mathbf{u}^T(\mathbf{x})-\nabla\mathbf{u}(\mathbf{x})=\mathbf{0} \tag{g}$$

which means that $\mathbf{E}=\mathbf{0}$. This proves the first part of the statement in the example.

To prove that if $\mathbf{E}=\mathbf{0}$ then $\mathbf{u}$ is a rigid displacement field, we observe that the condition $\mathbf{E}=\mathbf{0}$ is equivalent to

$$\nabla\mathbf{u}=-\nabla\mathbf{u}^T \tag{h}$$

In components

$$u_{i,j}=-u_{j,i} \tag{i}$$

From this it follows that

$$u_{i,jk}=u_{i,kj}=-u_{k,ij}=-u_{k,ji}=u_{j,ki}=u_{j,ik}=-u_{i,jk} \tag{j}$$

Thus

$$u_{i,jk}=0 \tag{k}$$

Solving this equation we find that

$$u_i=u_i^0+A^*_{ik}x_k \tag{l}$$

where u_i^0 and A^*_{ik} are a constant vector and a constant tensor, respectively. Differentiating (l) with respect to x_j we obtain

$$u_{i,j}=A^*_{ik}\delta_{kj}=A^*_{ij} \tag{m}$$

so (l) can be written in the form

$$u_i = u_i^0 + u_{i,j}x_j \tag{n}$$

or, in direct notation

$$\mathbf{u} = \mathbf{u}^0 + (\nabla\mathbf{u})\mathbf{x} \tag{o}$$

Since by (h) $\nabla\mathbf{u}$ is a skew tensor, there exists an axial vector $\boldsymbol{\omega}$ such that

$$(\nabla\mathbf{u})\mathbf{a} = \boldsymbol{\omega} \times \mathbf{a} \tag{p}$$

for every vector $\mathbf{a}$ [see formula (a) in Example 3.1.5]. Using (p) we can write (o) in the form

$$\mathbf{u} = \mathbf{u}^0 + \boldsymbol{\omega} \times \mathbf{x} \tag{q}$$

Since $\mathbf{u}^0$ represents a rigid translation and $\boldsymbol{\omega} \times \mathbf{x}$ represents a rigid rotation about the axis of $\boldsymbol{\omega}$, the displacement $\mathbf{u}$ is a rigid displacement field.

This ends the second part of the solution of Example 3.1.6. □

From this example follows the theorem called Kirchhoff's Theorem.

Kirchhoff's Theorem. If two displacement fields $\mathbf{u}_1$ and $\mathbf{u}_2$ correspond to the same strain field $\mathbf{E}$ then

$$\mathbf{u}_1 - \mathbf{u}_2 = \mathbf{w} \tag{3.1.23}$$

where $\mathbf{w}$ is a rigid displacement field. The proof follows from the fact that $\mathbf{u}_1 - \mathbf{u}_2$ corresponds to the zero infinitesimal strain. Thus, from Example 3.1.6, it follows that $\mathbf{u}_1 - \mathbf{u}_2$ is a rigid displacement. □

If $\mathbf{A}$ is an arbitrary constant tensor and $\mathbf{u}_0$ and $\mathbf{x}_0$ are a constant vector and a position vector, respectively, then a displacement field of the form

$$\mathbf{u}(\mathbf{x}) = \mathbf{u}_0 + \mathbf{A}(\mathbf{x} - \mathbf{x}_0) \tag{3.1.24}$$

is called a *homogeneous displacement field*. Clearly, if $\mathbf{A}$ is skew, 3.1.24 represents a rigid displacement. For an arbitrary $\mathbf{A}$ an interpretation of 3.1.24 is illustrated by the following example.

Example 3.1.7

Show that for an arbitrary constant tensor $\mathbf{A}$ the displacement $\mathbf{u}$ given by 3.1.24 takes the form

$$\mathbf{u}(\mathbf{x}) = \mathbf{u}_1(\mathbf{x}) + \mathbf{u}_2(\mathbf{x}) \tag{a}$$

where

$\mathbf{u}_1(\mathbf{x})$ is a rigid displacement field

$\mathbf{u}_2(\mathbf{x})$ is a displacement field corresponding to the strain field $\mathbf{E} = \text{sym}\,\mathbf{A}$

Solution

We express **A** as the sum of a skew part and a symmetric part

$$\mathbf{A} = \text{skw}\,\mathbf{A} + \text{sym}\,\mathbf{A} \tag{b}$$

or, in index notation

$$A_{ij} = A_{[ij]} + A_{(ij)} \tag{c}$$

where

$$A_{[ij]} = \frac{1}{2}\,(A_{ij} - A_{ji}) \tag{d}$$

and

$$A_{(ij)} = \frac{1}{2}\,(A_{ij} + A_{ji}) \tag{e}$$

So the formula 3.1.24 can be written in the index notation as

$$u_i(\mathbf{x}) = u_i^{(1)}(\mathbf{x}) + u_i^{(2)}(\mathbf{x}) \tag{f}$$

where

$$u_i^{(1)}(\mathbf{x}) = (\mathbf{u}_0)_i + A_{[ij]}[x_j - (\mathbf{x}_0)_j] \tag{g}$$

and

$$u_i^{(2)}(\mathbf{x}) = A_{(ij)}[x_j - (\mathbf{x}_0)_j] \tag{h}$$

By differentiating (f) with respect to x_k we obtain

$$u_{i,k} = A_{[ij]}\delta_{jk} + A_{(ij)}\delta_{jk} = A_{ik} \tag{i}$$

Taking the symmetric part of both sides of (i) we write

$$u_{(i,k)} = A_{(i,k)} \tag{j}$$

By the definition, a displacement field **u** corresponding to the strain field **E** satisfies the strain–displacement relation

$$u_{(i,k)} = E_{ik} \tag{k}$$

So by (h) we get

$$u_i^{(2)}(\mathbf{x}) = E_{ij}[x_j - (\mathbf{x}_0)_j] \tag{l}$$

Now, from (f) it follows that if we put

$$\mathbf{u}^{(1)}(\mathbf{x}) = \mathbf{u}_1(\mathbf{x}) \quad \text{and} \quad u^{(2)}(\mathbf{x}) = \mathbf{u}_2(\mathbf{x}) \tag{m}$$

we obtain (a) since by (g) $\mathbf{u}^{(1)}(\mathbf{x})$ represents a rigid displacement field, and by (*l*) $\mathbf{u}^{(2)}(\mathbf{x})$ represents a displacement field corresponding to $\mathbf{E} = \text{sym}\,\mathbf{A}$. □

The displacement field $\mathbf{u}_2(\mathbf{x})$ in Example 3.1.7 is called a *pure strain* $\mathbf{E}$ from $\mathbf{x}_0$. So formula (a) asserts that an arbitrary homogeneous displacement field is a sum of a rigid displacement field and a displacement field corresponding to pure strain from a position vector $\mathbf{x}_0$. We note that a pure strain $\mathbf{E}$ is a 3×3 constant matrix that can be presented as a sum of simpler matrices that have a direct physical interpretation. This will be explained by the introduction of the following definitions. Let

$$\mathbf{p}_0(\mathbf{x}) = \mathbf{x} - \mathbf{x}_0 \tag{3.1.25}$$

A *simple extension* of magnitude e in the direction of a unit vector $\mathbf{n}$ is represented by the formulas

$$\mathbf{u} = e\,(\mathbf{n} \cdot \mathbf{p}_0)\mathbf{n} \tag{3.1.26}$$

$$\mathbf{E} = e\,\mathbf{n} \otimes \mathbf{n} \tag{3.1.27}$$

A uniform *dilatation* of magnitude e is given by the formulas

$$\mathbf{u} = e\,\mathbf{p}_0 \tag{3.1.28}$$

$$\mathbf{E} = e\,\mathbf{1} \tag{3.1.29}$$

In this case, see Equation 3.1.19,

$$\text{tr}\,\mathbf{E} = 3e \tag{3.1.30}$$

A *simple shear* of magnitude g with respect to perpendicular unit vectors $\mathbf{m}$ and $\mathbf{n}$ is represented by

$$\mathbf{u} = g[(\mathbf{m} \cdot \mathbf{p}_0)\mathbf{n} + (\mathbf{n} \cdot \mathbf{p}_0)\mathbf{m}] \tag{3.1.31}$$

$$\mathbf{E} = g[\mathbf{m} \otimes \mathbf{n} + \mathbf{n} \otimes \mathbf{m}] \tag{3.1.32}$$

Note that in these definitions $\mathbf{E}$ is a strain field corresponding to a displacement field $\mathbf{u}$, that is,

$$\mathbf{E} = \text{sym}\,\nabla\mathbf{u} \tag{3.1.33}$$

In the case when $\mathbf{n} = (1, 0, 0)$ for a simple extension the tensor $\mathbf{E}$ referred to the orthonormal basis $\{\mathbf{n}, \mathbf{e}_2, \mathbf{e}_3\}$ is given by the matrix

$$\mathbf{E} = \begin{bmatrix} e & 0 & 0 \\ 0 & 0 & 0 \\ 0 & 0 & 0 \end{bmatrix} \tag{3.1.34}$$

while the corresponding displacement $\mathbf{u}$ given in components is

$$u_1 = e\left(x_1 - x_1^0\right)$$
$$u_2 = u_3 = 0, \quad \left[x_1^0 = (\mathbf{x}_0)_1\right] \tag{3.1.35}$$

Thus, $u_1 = u_1(x_1)$ represents a linear function of x_1 that increases (decreases) with increasing (decreasing) difference $x_1 - x_1^0$.

Similarly, for the uniform dilatation the strain tensor $\mathbf{E}$ is represented by a diagonal matrix with nonvanishing components e while the associated displacement $\mathbf{u}$ is proportional to the difference $\mathbf{x} - \mathbf{x}_0$. Therefore, $\mathbf{u}$ increases with a distance from $\mathbf{x}_0$. In this case

$$\mathbf{E} = \begin{bmatrix} e & 0 & 0 \\ 0 & e & 0 \\ 0 & 0 & e \end{bmatrix} \tag{3.1.36}$$

Finally, in the simple shear the matrix $\mathbf{E}$ referred to the orthonormal basis $\{\mathbf{m}, \mathbf{n}, \mathbf{e}_3\}$ takes the form

$$\mathbf{E} = \begin{bmatrix} 0 & g & 0 \\ g & 0 & 0 \\ 0 & 0 & 0 \end{bmatrix} \tag{3.1.37}$$

$(\mathbf{m} = \mathbf{e}_1, \ \mathbf{n} = \mathbf{e}_2)$.

To show this we write Equation 3.1.32 in components

$$E_{ij} = g(m_i n_j + n_i m_j) \tag{3.1.38}$$

and obtain

$$\begin{aligned} E_{11} &= g(m_1 n_1 + n_1 m_1) = g(1 \cdot 0 + 0 \cdot 1) = 0 \\ E_{12} &= g(m_1 n_2 + n_1 m_2) = g(1 \cdot 1 + 0 \cdot 0) = g \\ E_{13} &= g(m_1 n_3 + n_1 m_3) = g(1 \cdot 0 + 0 \cdot 0) = 0 \\ E_{21} &= g(m_2 n_1 + n_2 m_1) = g(0 \cdot 0 + 1 \cdot 1) = g \\ E_{23} &= E_{31} = E_{32} = E_{33} = 0 \end{aligned} \tag{3.1.39}$$

The components of displacement $\mathbf{u}$ associated with $\mathbf{E}$ are given by

$$\begin{aligned} u_1 &= g[(\mathbf{p}_0)_1 \cdot 0 + (\mathbf{p}_0)_2 \cdot 1] = g\left(x_2 - x_2^0\right) \\ u_2 &= g[(\mathbf{p}_0)_1 \cdot 1 + (\mathbf{p}_0)_2 \cdot 0] = g\left(x_1 - x_1^0\right) \\ u_3 &= g[(\mathbf{p}_0)_1 \cdot 0 + (\mathbf{p}_0)_2 \cdot 0] = 0 \end{aligned} \tag{3.1.40}$$

We may observe that u_1 and u_2 increase with the increase of $x_2 - x_2^0$ and $x_1 - x_1^0$, respectively, and the displacement $\mathbf{u}$ vanishes when $\mathbf{x} = \mathbf{x}_0$.

Example 3.1.8

Show that for a simple extension, a uniform dilatation, and a simple shear, the relative volume change $\delta v(B)/v(B)$ of a body B is, respectively, e, $3e$, and 0.

Solution

Since $\delta v(B)/v(B)$ is equal to tr **E**, by Equations 3.1.34, 3.1.36, and 3.1.37

(a) $\text{tr}\,\mathbf{E} = e$ for the simple extension
(b) $\text{tr}\,\mathbf{E} = 3e$ for the uniform dilatation
(c) $\text{tr}\,\mathbf{E} = 0$ for the simple shear

then the solution is received. □

The relative volume change $\delta v(B)/v(B)$ in Example 3.1.8 is computed for a constant strain field **E** associated with a homogeneous displacement field. In the case when **E** is a function of **x** the relative volume change is

$$\frac{\delta v(B)}{v(B)} = \text{tr}\,\widehat{\mathbf{E}}(B) \tag{3.1.41}$$

where

$$\widehat{\mathbf{E}}(B) = \frac{1}{v(B)} \int_B \mathbf{E}\, dv \tag{3.1.42}$$

is called the *mean strain*.

Example 3.1.9

A homogeneous displacement field **u** is given by (see Equation 3.1.24)

$$\mathbf{u} = \mathbf{A}\mathbf{x} \tag{a}$$

where **A** is a second-order constant tensor. Show that

(i) **u** corresponds to a pure deformation if and only if

$$\text{skw}\,\mathbf{A} = \mathbf{0} \tag{b}$$

(ii) **u** corresponds to a pure rotation if and only if

$$\text{sym}\,\mathbf{A} = \mathbf{0} \tag{c}$$

Also, find the form of the rotation vector $\boldsymbol{\omega}$.

Solution

From (a) follows

$$\nabla\mathbf{u} = \mathbf{A} \tag{d}$$

Hence

$$\mathbf{E} = \frac{1}{2}\,(\nabla\mathbf{u} + \nabla\mathbf{u}^T) = \text{sym}\,\mathbf{A} \tag{e}$$

and

$$\mathbf{W} = \frac{1}{2} (\nabla\mathbf{u} - \nabla\mathbf{u}^T) = \text{skw}\,\mathbf{A} \tag{f}$$

(i) We need to prove that (aa) if $\mathbf{u}$ corresponds to a pure deformation then skw $\mathbf{A} = \mathbf{0}$ and (bb) if skw $\mathbf{A} = \mathbf{0}$ then $\mathbf{u}$ corresponds to a pure deformation.

In (aa) $\mathbf{u}$ corresponds to the symmetric part of $\nabla\mathbf{u}$. Hence, by (d) skw $\mathbf{A} = \mathbf{0}$.

In (bb) skw $\mathbf{A} = \mathbf{0}$, thus, by (d) $\mathbf{u}$ is such that sym $\nabla\mathbf{u} =$ sym $\mathbf{A} = \mathbf{E}$ represents a pure deformation.

(ii) The proof follows along similar lines to the proof of (i). To find the rotation vector $\boldsymbol{\omega}$ we compute curl $\mathbf{u}$. In components, since sym $\mathbf{A} = \mathbf{0}$, by (a)

$$u_i = A_{[ij]} x_j \tag{g}$$

Hence,

$$\begin{aligned}(\text{curl}\,\mathbf{u})_i &= \epsilon_{ipq} u_{q,p} = \epsilon_{ipq} A_{[qj]} x_{j,p} \\ &= \epsilon_{ipq} A_{[qj]} \delta_{jp} = \epsilon_{ipq} A_{[qp]}\end{aligned} \tag{h}$$

and

$$\omega_i = \frac{1}{2} (\text{curl}\,\mathbf{u})_i = \frac{1}{2} \epsilon_{ipq} A_{[qp]} \tag{i}$$

or

$$\boldsymbol{\omega} = \omega_i \mathbf{e}_i \tag{j}$$

where

$$\begin{aligned}\omega_1 &= \frac{1}{2} (A_{32} - A_{23}) = A_{[32]} \\ \omega_2 &= \frac{1}{2} (A_{13} - A_{31}) = A_{[13]} \\ \omega_3 &= \frac{1}{2} (A_{21} - A_{12}) = A_{[21]}\end{aligned} \tag{k}$$

□

Example 3.1.10

Let **E** and **W** denote the infinitesimal strain field and the rotation tensor field, respectively, corresponding to a displacement field **u** on B. Show that

(i)
$$(\mathbf{1} + \mathbf{E})^{-1} \cong \mathbf{1} - \mathbf{E}$$

(ii)
$$(\mathbf{1} + \mathbf{W})^{-1} \cong \mathbf{1} - \mathbf{W}$$

Solution

(i) We expand LHS of (i) into a power series and obtain

$$(\mathbf{1} + \mathbf{E})^{-1} = \sum_{n=0}^{\infty} (-1)^n \mathbf{E}^n \quad \text{if } |\mathbf{E}| < 1 \tag{a}$$

For infinitesimal strains $\mathbf{E}$ the terms $0(\epsilon^n)$, where $\epsilon = |\nabla \mathbf{u}|$ and $n \geq 2$, are small in comparison to the term $1 + 0(\epsilon)$. Therefore, retaining the first two terms on the RHS of this series, we arrive at (i).

(ii) This part is proved in a similar way to that of (i), since the infinitesimal tensor fields $\mathbf{E}$ and $\mathbf{W}$ are of the same order of magnitude.

□

Example 3.1.11

Given a small displacement

$$\mathbf{u} = L^{-1}\left[x_1^2\mathbf{e}_1 + x_2^2\mathbf{e}_2 + x_3^2\mathbf{e}_3\right] \tag{a}$$

where L has the dimension of length. Calculate the strain tensor $\mathbf{E}$, the rotation tensor $\mathbf{W}$, and the rotation vector $\boldsymbol{\omega}$.

Solution

We rewrite (a) in the form

$$\mathbf{u} = L^{-1}x_k^2\mathbf{e}_k \tag{b}$$

Hence, multiplying (b) by $\mathbf{e}_i$ in the dot sense, we get

$$u_i = L^{-1}x_i^2 \tag{c}$$

Differentiating (c) we obtain

$$\begin{aligned} u_{1,1} &= 2L^{-1}x_1 \\ u_{2,2} &= 2L^{-1}x_2 \\ u_{3,3} &= 2L^{-1}x_3 \end{aligned} \tag{d}$$

Therefore,

$$\nabla \mathbf{u} = 2L^{-1}\begin{bmatrix} x_1 & 0 & 0 \\ 0 & x_2 & 0 \\ 0 & 0 & x_3 \end{bmatrix} \tag{e}$$

Note that $\nabla \mathbf{u}$ is a diagonal tensor, thus

$$\mathbf{E} = \nabla \mathbf{u} \tag{f}$$

and

$$\mathbf{W} = \frac{1}{2}\left(\nabla\mathbf{u} - \nabla\mathbf{u}^T\right) = \mathbf{0} \tag{g}$$

To compute $\boldsymbol{\omega}$ we use the formula

$$\boldsymbol{\omega} = \frac{1}{2}\,\mathrm{curl}\,\mathbf{u} \tag{h}$$

In components,

$$\omega_i = \frac{1}{2}\,\epsilon_{ijk}u_{k,j} \tag{i}$$

or

$$\begin{aligned}
\omega_1 &= \frac{1}{2}\left(u_{3,2} - u_{2,3}\right) = 0\\
\omega_2 &= \frac{1}{2}\left(u_{1,3} - u_{3,1}\right) = 0\\
\omega_3 &= \frac{1}{2}\left(u_{2,1} - u_{1,2}\right) = 0
\end{aligned} \tag{j}$$

Therefore, the displacement (a) corresponds to a deformation that is free from rotation. □

Any infinitesimal strain tensor $\mathbf{E}$ can be expressed as a sum of a *spherical tensor* $\mathbf{E}^{(s)}$ and a *deviatoric tensor* $\mathbf{E}^{(d)}$, called the *deviator*, in the form

$$\mathbf{E} = \mathbf{E}^{(s)} + \mathbf{E}^{(d)} \tag{3.1.43}$$

where

$$\mathbf{E}^{(s)} = \frac{1}{3}\,\mathbf{1}(\mathrm{tr}\,\mathbf{E}) \tag{3.1.44}$$

From this definition it follows that

$$\mathrm{tr}\,\mathbf{E}^{(d)} = 0 \tag{3.1.45}$$

Referring to a simple shear, see Equation 3.1.37 and Example 3.1.8 part (c), we note that a simple shear of magnitude g is represented by a deviatoric strain tensor given by Equation 3.1.37.

Also, a uniform dilatation, discussed in Example 3.1.8 part (b), is represented by a spherical tensor given by Equation 3.1.36.

Example 3.1.12

Let I_1, I_2, and I_3 denote the first, the second, and the third invariant, respectively, of the infinitesimal strain tensor **E**. Thus, [see Example 2.1.10 (b), (c), and (d)]

$$I_1 = \operatorname{tr}\mathbf{E} \tag{a}$$

$$I_2 = \frac{1}{2}\,[(\operatorname{tr}\mathbf{E})^2 - \operatorname{tr}(\mathbf{E}^2)] \tag{b}$$

$$I_3 = \det\mathbf{E} \tag{c}$$

Let $I_1^{(d)}$, $I_2^{(d)}$, $I_3^{(d)}$ be the first, the second, and the third invariant of $\mathbf{E}^{(d)}$, respectively.
Show that

(i) $I_1^{(d)} = 0$
(ii) $I_2^{(d)} = -\frac{1}{2}\operatorname{tr}(\mathbf{E}^{(d)})^2$
(iii) $I_3^{(d)} = \frac{1}{3}\operatorname{tr}(\mathbf{E}^{(d)})^3$

Solution

(i) The result follows directly from the definition of $\mathbf{E}^{(d)}$, since $\operatorname{tr}\mathbf{E}^{(d)} = 0$.
(ii) $I_2^{(d)} = \frac{1}{2}\,[(\operatorname{tr}\mathbf{E}^{(d)})^2 - \operatorname{tr}(\mathbf{E}^{(d)})^2]$. Hence, by (i) we obtain (ii).
(iii) First, we show that the determinant of any tensor **A** can be expressed as

$$\begin{aligned}\det\mathbf{A} &= \frac{1}{6}\,\epsilon_{ijk}\epsilon_{pqr}A_{ip}A_{jq}A_{kr}\\ &= \frac{1}{6}\,[(\operatorname{tr}\mathbf{A})^3 + 2\operatorname{tr}\mathbf{A}^3 - 3(\operatorname{tr}\mathbf{A}^2)\operatorname{tr}\mathbf{A}]\end{aligned} \tag{d}$$

To show (d), we note that by Example 2.1.1, we get

$$\begin{vmatrix} a_1 & a_2 & a_3\\ b_1 & b_2 & b_3\\ c_1 & c_2 & c_3\end{vmatrix} = \epsilon_{pqr}a_pb_qc_r \tag{e}$$

Hence, letting $a_p = A_{ip}$, $b_q = A_{jq}$, $c_r = A_{kr}$ [i, j, k fixed indices in (e)], we obtain

$$\begin{vmatrix} A_{i1} & A_{i2} & A_{i3}\\ A_{j1} & A_{j2} & A_{j3}\\ A_{k1} & A_{k2} & A_{k3}\end{vmatrix} = \epsilon_{pqr}A_{ip}A_{jq}A_{kr} \tag{f}$$

Expansion of the determinant on the LHS of (f) yields

$$\begin{aligned}\epsilon_{pqr}A_{ip}A_{jq}A_{kr} &= A_{i1}(A_{j2}A_{k3} - A_{k2}A_{j3}) + A_{i2}(A_{j3}A_{k1} - A_{j1}A_{k3})\\ &\quad + A_{i3}(A_{j1}A_{k2} - A_{k1}A_{j2})\end{aligned} \tag{g}$$

Now

$$\begin{aligned}\epsilon_{ijk}A_{i1}(A_{j2}A_{k3} - A_{k2}A_{j3}) &= \epsilon_{ijk}A_{i1}A_{j2}A_{k3} + \epsilon_{ikj}A_{i1}A_{k2}A_{j3}\\ &= 2\epsilon_{ijk}A_{i1}A_{j2}A_{k3};\end{aligned} \tag{h}$$

$$\begin{aligned}\epsilon_{ijk}A_{i2}(A_{j3}A_{k1} - A_{j1}A_{k3}) &= \epsilon_{kij}A_{k1}A_{i2}A_{j3} + \epsilon_{jik}A_{j1}A_{i2}A_{k3}\\ &= 2\epsilon_{ijk}A_{i1}A_{j2}A_{k3};\end{aligned} \tag{i}$$

and

$$\begin{aligned} \epsilon_{ijk}A_{i3}(A_{j1}A_{k2} - A_{k1}A_{j2}) &= \epsilon_{jki}A_{j1}A_{k2}A_{i3} + \epsilon_{kji}A_{k1}A_{j2}A_{i3} \\ &= 2\epsilon_{ijk}A_{i1}A_{j2}A_{k3} \end{aligned} \tag{j}$$

Hence, multiplying (f) by ϵ_{ijk} we obtain

$$\epsilon_{ijk}\epsilon_{pqr}A_{ip}A_{jq}A_{kr} = 6\epsilon_{ijk}A_{i1}A_{j2}A_{k3} = 6 \det \mathbf{A} \tag{k}$$

where

$$\mathbf{A} = \begin{bmatrix} A_{11} & A_{12} & A_{13} \\ A_{21} & A_{22} & A_{23} \\ A_{31} & A_{32} & A_{33} \end{bmatrix} \tag{l}$$

The relation

$$\epsilon_{ijk}A_{i1}A_{j2}A_{k3} = \det \mathbf{A} \tag{m}$$

follows from (e) in which

$$a_i = A_{i1} \qquad b_i = A_{i2} \qquad c_i = A_{i3} \tag{n}$$

This ends the proof of the first equality in (d).

To prove the second equality of (d), we note from (k) that by Equation 2.1.12

$$\det \mathbf{A} = \frac{1}{6} \begin{vmatrix} \delta_{ip} & \delta_{iq} & \delta_{ir} \\ \delta_{jp} & \delta_{jq} & \delta_{jr} \\ \delta_{kp} & \delta_{kq} & \delta_{kr} \end{vmatrix} A_{ip}A_{jq}A_{kr} \tag{o}$$

By expanding, this equals to

$$\begin{aligned} &\frac{1}{6}\{\delta_{ip}(\delta_{jq}\delta_{kr} - \delta_{jr}\delta_{kq}) + \delta_{iq}(\delta_{jr}\delta_{kp} - \delta_{jp}\delta_{kr}) + \delta_{ir}(\delta_{jp}\delta_{kq} - \delta_{jq}\delta_{kp})\}A_{ip}A_{jq}A_{kr} \\ &\quad = \frac{1}{6}\{A_{pp}A_{qq}A_{kk} - A_{pp}A_{rq}A_{qr} + A_{qp}A_{rq}A_{pr} - A_{qp}A_{pq}A_{rr} \\ &\qquad + A_{rp}A_{pq}A_{qr} - A_{rp}A_{qq}A_{pr}\} \\ &\quad = \frac{1}{6}\{(\operatorname{tr}\mathbf{A})^3 + 2\operatorname{tr}(\mathbf{A}^3) - 3(\operatorname{tr}\mathbf{A})(\operatorname{tr}\mathbf{A}^2)\} \end{aligned} \tag{p}$$

This ends the proof of the second equality of (d). Now, the proof of (iii) is obtained by letting $\mathbf{A} = \mathbf{E}^{(d)}$ in (d) and by observing that $\operatorname{tr}\mathbf{E}^{(d)} = 0$. □

3.1.3 Compatibility

The strain–displacement relation, see Equation 3.1.15,

$$\mathbf{E} = \frac{1}{2}(\nabla\mathbf{u} + \nabla\mathbf{u}^T) \quad \text{on } B \tag{3.1.46}$$

can be viewed as a system of linear first-order partial differential equations for the displacement **u** if **E** is a prescribed symmetric tensor field on B. Since Equation 3.1.46 represents six scalar equations for the three components of **u**, the system 3.1.46 is to have a solution **u** if some restrictions are imposed on **E**. These restrictions, called the compatibility conditions on **E**, are necessary for the existence of a displacement field **u**. To obtain these conditions, we apply the operator curl curl to both sides of Equation 3.1.46 and arrive at

$$\text{curl curl}\,\mathbf{E} = \mathbf{0} \quad \text{on } B \tag{3.1.47}$$

Proof of relation 3.1.47 in components goes as follows. By the definition of curl of a second-order symmetric tensor **E**, we have

$$A_{ij} \equiv (\text{curl}\,\mathbf{E})_{ij} = \epsilon_{iab} E_{jb,a} \tag{3.1.48}$$

Hence, by Equation 3.1.46 and since

$$\epsilon_{iab} u_{j,ba} = 0 \tag{3.1.49}$$

because ϵ_{iab} is skew and $u_{j,ba}$ is symmetric with respect to a and b, we have

$$A_{ij} = \frac{1}{2}\,\epsilon_{iab}(u_{j,ba} + u_{b,ja}) = \frac{1}{2}\,\epsilon_{iab} u_{b,ja} \tag{3.1.50}$$

Next, applying the operator curl to **A**, because of Equation 3.1.50 we obtain

$$(\text{curl}\,\mathbf{A})_{ij} = \epsilon_{imn} A_{jn,m} = \frac{1}{2}\,\epsilon_{imn}\epsilon_{jab} u_{b,anm} = 0 \tag{3.1.51}$$

since

$$\epsilon_{imn} u_{b,anm} = 0 \tag{3.1.52}$$

This completes the proof of Equation 3.1.47.

We state the following theorem:

Theorem: If **u** is a displacement field corresponding to a strain field **E**, that is, if the strain–displacement relation 3.1.46 is satisfied, then **E** satisfies the equations of compatibility 3.1.47. Conversely, let B be simply connected and let **E** be a symmetric tensor field that satisfies the equation of compatibility 3.1.47. Then there exists a displacement field **u** on B such that **E** and **u** satisfy the strain–displacement relation 3.1.46. ■

Proof of the first part of the theorem has already been given (see the implication Equation 3.1.46 ⇒ Equation 3.1.47). To prove the existence of **u** corresponding to **E** subject to the compatibility conditions 3.1.47, it is sufficient to show that

(i) The vector field **u** given by the line integral (this formula is based on Ref. [1]; see also discussion in Ref. [2])

$$\mathbf{u}(\mathbf{x}) = \int_{\mathbf{x}_0}^{\mathbf{x}} \mathbf{U}(\mathbf{y}, \mathbf{x})\, d\mathbf{y} \tag{3.1.53}$$

where

$$U_{ij}(\mathbf{y}, \mathbf{x}) = E_{ij}(\mathbf{y}) + (x_k - y_k)[E_{ij,k}(\mathbf{y}) - E_{kj,i}(\mathbf{y})] \tag{3.1.54}$$

satisfies the strain–displacement relation 3.1.46

(ii) The line integral 3.1.53 is independent of the path in B from $\mathbf{x}_0$ to $\mathbf{x}$.

To show (i) we rewrite Equation 3.1.53 in components and obtain

$$u_i(\mathbf{x}) = \int_{\mathbf{x}_0}^{\mathbf{x}} U_{ip}(\mathbf{y}, \mathbf{x})\, dy_p \tag{3.1.55}$$

Hence, we get

$$\begin{aligned} u_{i,j}(\mathbf{x}) &= E_{ij}(\mathbf{x}) + \int_{\mathbf{x}_0}^{\mathbf{x}} \delta_{kj}[E_{ip,k}(\mathbf{y}) - E_{kp,i}(\mathbf{y})]\, dy_p \\ &= E_{ij}(\mathbf{x}) + \int_{\mathbf{x}_0}^{\mathbf{x}} [E_{ip,j}(\mathbf{y}) - E_{jp,i}(\mathbf{y})]\, dy_p \end{aligned} \tag{3.1.56}$$

Taking the symmetric part of both sides of this equation, in view of the identity

$$E_{(ip,j)} - E_{(jp,i)} = \frac{1}{2}\,(E_{ip,j} + E_{jp,i} - E_{jp,i} - E_{ip,j}) = 0 \tag{3.1.57}$$

we arrive at

$$u_{(i,j)} = E_{ij} \quad \text{on } B \tag{3.1.58}$$

To prove (ii) it is sufficient to show that

$$\mathbf{U}d\mathbf{y} = d\mathbf{V}(\mathbf{y}) \tag{3.1.59}$$

or

$$U_{ij}\, dy_i = V_{i,j}\, dy_j \tag{3.1.60}$$

which means that the expression under the integral sign in Equation 3.1.53 is a total differential. Since

$$V_{i,jk} = V_{i,kj} \tag{3.1.61}$$

then U_{ij} satisfies the relation

$$U_{ij,k} = U_{ik,j} \tag{3.1.62}$$

Changing indices, we write

$$U_{ij,l} = U_{il,j} \tag{3.1.63}$$

Using the definition of U_{ij}, see Equation 3.1.54, from Equation 3.1.63, we have

$$[E_{ij} + (x_k - y_k)(E_{ij,k} - E_{kj,i})]_{,l} = [E_{il} + (x_k - y_k)(E_{il,k} - E_{kl,i})]_{,j} \tag{3.1.64}$$

This relation is equivalent to

$$\begin{aligned} E_{ij,l} - \delta_{kl}(E_{ij,k} - E_{kj,i}) + (x_k - y_k)(E_{ij,kl} - E_{kj,il}) \\ = E_{il,j} - \delta_{kj}(E_{il,k} - E_{kl,i}) + (x_k - y_k)(E_{il,kj} - E_{kl,ij}) \end{aligned} \tag{3.1.65}$$

Hence

$$(x_k - y_k)(E_{ij,kl} + E_{kl,ij} - E_{il,kj} - E_{kj,il}) = 0 \tag{3.1.66}$$

Since the compatibility condition

$$\text{curl}\,\text{curl}\,\mathbf{E} = \mathbf{0} \tag{3.1.67}$$

is equivalent to

$$E_{ij,kl} + E_{kl,ij} - E_{il,kj} - E_{kj,il} = 0 \tag{3.1.68}$$

we find that the line integral 3.1.53 is independent of the path.

The proof of equivalence of Equations 3.1.67 and 3.1.68 is given in Example 3.1.13. □

Example 3.1.13

Show that six compatibility equations

$$\text{curl}\,\text{curl}\,\mathbf{E} = \mathbf{0} \quad \text{or} \quad \epsilon_{imn}\epsilon_{jab}E_{nb,am} = 0 \tag{a}$$

are equivalent to 81 equations of the form

$$E_{ij,kl} + E_{kl,ij} - E_{il,jk} - E_{jk,il} = 0 \quad (i, j, k, l, m, n = 1, 2, 3) \tag{b}$$

Solution

The solution is based upon selection of particular sets of indices in (a) and (b), and on using the symmetry of tensor **E**, as well as on equality of mixed derivatives. For example, letting $i = j = 1$ in (a) we have

$$\begin{aligned}
\epsilon_{1mn}\epsilon_{1ab}E_{nb,am} &= \epsilon_{12n}\epsilon_{1ab}E_{nb,a2} + \epsilon_{13n}\epsilon_{1ab}E_{nb,a3} \\
&= \epsilon_{1ab}E_{3b,a2} - \epsilon_{1ab}E_{2b,a3} = \epsilon_{12b}E_{3b,22} + \epsilon_{13b}E_{3b,32} \\
&\quad - \epsilon_{12b}E_{2b,23} - \epsilon_{13b}E_{2b,33} \\
&= E_{33,22} - E_{32,32} - E_{23,23} + E_{22,33} = 0
\end{aligned} \tag{c}$$

or

$$E_{33,22} + E_{22,33} = 2E_{32,32} \tag{d}$$

Also, letting $i = j = 2$ and $k = l = 3$ in (b) we arrive at (d).

In this manner we show that for a choice of indices 1, 2, 3 in (b) we obtain an equation contained in the set of six equations (a). Note that the complete set of six equations takes the form:

$$\begin{aligned}
E_{11,22} + E_{22,11} &= 2E_{12,12} \\
E_{22,33} + E_{33,22} &= 2E_{23,23} \\
E_{33,11} + E_{11,33} &= 2E_{31,31} \\
E_{11,23} + E_{23,11} &= E_{12,13} + E_{13,12} \\
E_{22,31} + E_{31,22} &= E_{23,21} + E_{21,23} \\
E_{33,12} + E_{12,33} &= E_{31,32} + E_{32,31}
\end{aligned} \tag{e}$$

□

In the case when the strain tensor components are independent of x_3 and, in addition, E_{13}, E_{23}, and E_{33}, are constants, the six compatibility conditions reduce to the single equation

$$E_{11,22} + E_{22,11} = 2E_{12,12} \tag{3.1.69}$$

which corresponds to a plane strain in a body.

Example 3.1.14

We want strains in a body to be described by a tensor field

$$\mathbf{E} = \begin{bmatrix} a\left(x_1^2 + x_2^2\right) & bx_1x_2 & 0 \\ bx_1x_2 & cx_1x_2 & 0 \\ 0 & 0 & 0 \end{bmatrix} \tag{a}$$

where a, b, and c are constants.

Find the relations between a, b, and c for which **E** is an admissible strain.

Solution

The only compatibility condition to be satisfied takes the form

$$E_{11,22} + E_{22,11} = 2E_{12,12} \tag{b}$$

In our case we obtain

$$2a = 2b \quad \text{or} \quad a = b \tag{c}$$

We see then that tensor $\mathbf{E}$ represents a possible strain as long as $a = b$, and c is an arbitrary constant. □

Example 3.1.15

Show that the compatibility condition 3.1.47 can be written in the equivalent form

$$\nabla^2\mathbf{E} + \nabla\nabla(\text{tr}\,\mathbf{E}) - 2\widehat{\nabla}(\text{div}\,\mathbf{E}) = \mathbf{0} \tag{a}$$

Solution

We use the identity [see Problem 2.12 (m)]

$$\text{curl curl}\,\mathbf{E} = -\nabla^2\mathbf{E} - \nabla\nabla(\text{tr}\,\mathbf{E}) + 2\widehat{\nabla}(\text{div}\,\mathbf{E}) + [\nabla^2(\text{tr}\,\mathbf{E}) - \text{div div}\,\mathbf{E}]\mathbf{1} \tag{b}$$

Also, observe that

$$\text{tr}\,[\nabla^2\mathbf{E} + \nabla\nabla(\text{tr}\,\mathbf{E}) - 2\widehat{\nabla}\,\text{div}\,\mathbf{E}] = 2[\nabla^2(\text{tr}\,\mathbf{E}) - \text{div div}\,\mathbf{E}] \tag{c}$$

Note that Equation 3.1.47, by virtue of (b), takes the form

$$\nabla^2\mathbf{E} + \nabla\nabla(\text{tr}\,\mathbf{E}) - 2\widehat{\nabla}(\text{div}\,\mathbf{E}) = [\nabla^2(\text{tr}\,\mathbf{E}) - \text{div div}\,\mathbf{E}]\mathbf{1} \tag{d}$$

Taking trace of (d) and using (c) yields

$$2[\nabla^2(\text{tr}\,\mathbf{E}) - \text{div div}\,\mathbf{E}] = 3[\nabla^2(\text{tr}\,\mathbf{E}) - \text{div div}\,\mathbf{E}] \tag{e}$$

Hence,

$$\nabla^2(\text{tr}\,\mathbf{E}) - \text{div div}\,\mathbf{E} = 0 \tag{f}$$

This together with (d) implies that the compatibility condition 3.1.47 is equivalent to (a). □

3.2 MOTION AND EQUILIBRIUM

3.2.1 Balance of Momentum: Stress Tensor

Let us introduce a number of definitions that will be needed to describe motion and equilibrium of an elastic body B. We define the *mass* of any part P of B by

$$m = \int_P \rho\, dv \tag{3.2.1}$$

where ρ is the *density* of the body, $\rho = \rho(\mathbf{x}) > 0$.

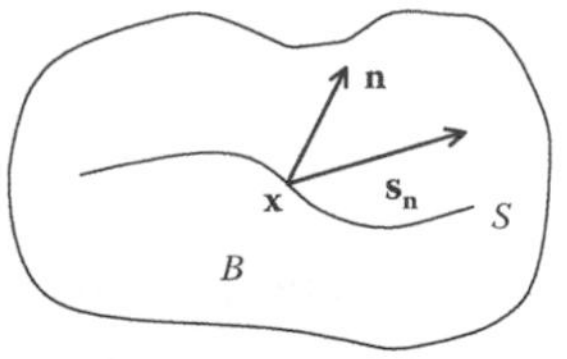

FIGURE 3.2 The stress vector $\mathbf{s_n}$ at a point $\mathbf{x}$ of the surface S.

A motion of the body is described by a vector field $\mathbf{u} = \mathbf{u}(\mathbf{x}, t)$ that represents the displacement of $\mathbf{x}$ at time t. With $\mathbf{u}$ we associate the velocity $\dot{\mathbf{u}}$, acceleration $\ddot{\mathbf{u}}$, strain $\mathbf{E}$, and strain rate $\dot{\mathbf{E}}$.

By a *linear momentum* of P we mean the integral

$$\mathbf{g}(P) = \int_P \dot{\mathbf{u}}\rho \, dv \tag{3.2.2}$$

By a *moment of momentum*, or *angular momentum*, of P about the origin $\mathbf{x} = \mathbf{0}$ we mean the integral

$$\mathbf{h}(P) = \int_P \mathbf{x} \times \dot{\mathbf{u}}\rho \, dv \tag{3.2.3}$$

We assume that two types of forces, both depending on $\mathbf{x}$ and t, act on the body: *body forces* and *surface forces*. The body force $\mathbf{b} = \mathbf{b}(\mathbf{x}, t)$ is the force per unit volume exerted on $\mathbf{x}$ by external agents for any $t \geq 0$. To define the surface force we introduce a surface S in B with unit normal $\mathbf{n}$, Figure 3.2.

Let $\mathbf{s_n} = \mathbf{s_n}(\mathbf{x}, t)$ denote a force per unit area at $\mathbf{x}$ and for $t \geq 0$ exerted by a portion of B on the side of S toward which $\mathbf{n}$ points on a portion of B on the other side of S. The force $\mathbf{s_n}$ is called the *stress vector* at $(\mathbf{x}, t)$. The total force across S is then defined as the integral

$$\int_S \mathbf{s_n} \, da \tag{3.2.4}$$

and the total moment of force across S is defined as

$$\int_S \mathbf{x} \times \mathbf{s_n} \, da \tag{3.2.5}$$

If $\mathbf{x}$ is a point on the boundary ∂B of body B and $\mathbf{n} = \mathbf{n}(\mathbf{x})$ is the outward unit normal to ∂B then $\mathbf{s_n}$ is called the *surface traction.*

The total force acting on P is defined by

$$\mathbf{f}(P) = \int_{\partial P} \mathbf{s_n} \, da + \int_P \mathbf{b} \, dv \tag{3.2.6}$$

and the total moment of force acting on P is defined by

$$\mathbf{m}(P) = \int_{\partial P} \mathbf{x} \times \mathbf{s_n} \, da + \int_P \mathbf{x} \times \mathbf{b} \, dv \tag{3.2.7}$$

We recall that P is an arbitrary portion of B, and ∂P is the bounding surface of P.

We now are in a position to define a *dynamic process* associated with the motion of B as an ordered array of functions $[\mathbf{u}, \mathbf{s_n}, \mathbf{b}]$ that satisfies the conditions

$$\mathbf{f}(P) = \dot{\mathbf{g}}(P) \tag{3.2.8}$$

$$\mathbf{m}(P) = \dot{\mathbf{h}}(P) \tag{3.2.9}$$

for every part P of B and every $t \geq 0$.

Equation 3.2.8 represents the law of balance of linear momentum and Equation 3.2.9 represents the balance of moment of momentum. Clearly, Equation 3.2.8 asserts that the linear momentum rate is the total force acting on P, and Equation 3.2.9 means that the rate of moment of momentum of P is equal to the total moment of force acting on P. From Equations 3.2.2 and 3.2.3, and Equations 3.2.6 and 3.2.7, as well as Equations 3.2.8 and 3.2.9, we obtain

$$\int_{\partial P} \mathbf{s_n}\, da + \int_P \mathbf{b}\, dv = \int_P \ddot{\mathbf{u}}\rho\, dv \tag{3.2.10}$$

$$\int_{\partial P} \mathbf{x} \times \mathbf{s_n}\, da + \int_P \mathbf{x} \times \mathbf{b}\, dv = \int_P \mathbf{x} \times \ddot{\mathbf{u}}\rho\, dv \tag{3.2.11}$$

It can be shown that the vector $\mathbf{s_n}$ in Equations 3.2.10 and 3.2.11 is generated by a second-order tensor field $\mathbf{S}$ through the formula

$$\mathbf{s_n}(\mathbf{x}, t) = \mathbf{S}(\mathbf{x}, t)\mathbf{n}(\mathbf{x}) \tag{3.2.12}$$

This second-order tensor $\mathbf{S}$ is called the *stress tensor*. To prove this we consider a small tetrahedron P_h with height h, Figure 3.3, cut out from body B. The origin of the coordinate system (x_1, x_2, x_3) is at $\mathbf{x}$. Vector $\mathbf{n}$ is the unit outward vector perpendicular to the plane 123. To find the stress vector $\mathbf{s_n}$ on the plane 123 we write down the equilibrium equations for the forces acting on the tetrahedron. Denote the area of the triangle 123 by A_h. Then the areas of the triangles $\mathbf{x}12$, $\mathbf{x}23$, and $\mathbf{x}31$ are, respectively, equal to $A_3 = A_h n_3$, $A_1 = A_h n_1$,

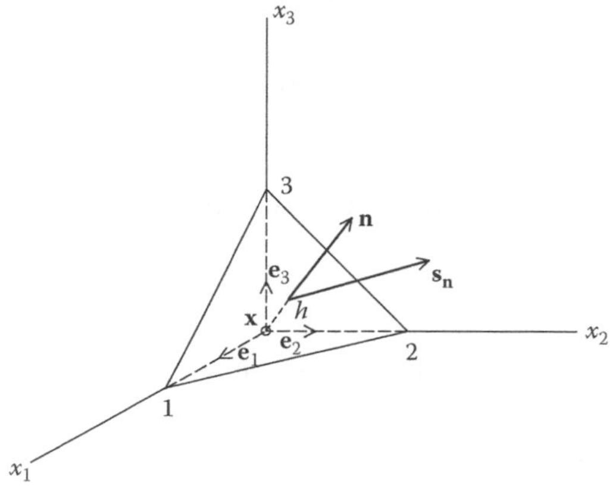

FIGURE 3.3 The tetrahedron subject to the stress vector $\mathbf{s_n}$.

and $A_2 = A_h n_2$. The resultant force on ∂P_h is to vanish as $h \to 0$, which is equivalent to the condition that

$$\frac{1}{A_h} \int_{\partial P_h} \widehat{\mathbf{s}}_{\mathbf{n}} \, da \to \mathbf{0}, \quad \text{as } h \to 0 \tag{3.2.13}$$

where $\widehat{\mathbf{s}}_{\mathbf{n}}$ is the stress vector on ∂P_h. This is equivalent to

$$\frac{1}{A_h} \left[\int_{A_h} \mathbf{s}_{\mathbf{n}} \, da + \int_{A_1} \widehat{\mathbf{s}}_1 \, da + \int_{A_2} \widehat{\mathbf{s}}_2 \, da + \int_{A_3} \widehat{\mathbf{s}}_3 \, da \right] \to 0 \quad \text{as } h \to 0 \tag{3.2.14}$$

where $\widehat{\mathbf{s}}_1$, $\widehat{\mathbf{s}}_2$, and $\widehat{\mathbf{s}}_3$ are the stress vectors on A_1, A_2, and A_3. Now,

$$\frac{1}{A_h} \int_{A_h} \mathbf{s}_{\mathbf{n}} \, da \to \mathbf{s}_{\mathbf{n}}(\mathbf{x}) \quad \text{as } h \to 0 \tag{3.2.15}$$

Similarly,

$$\begin{aligned} \frac{1}{A_h} \int_{A_1} \widehat{\mathbf{s}}_1 \, da &\to -(\mathbf{n} \cdot \mathbf{e}_1) \, \mathbf{s}(\mathbf{e}_1) \quad \text{as } h \to 0 \\ \frac{1}{A_h} \int_{A_2} \widehat{\mathbf{s}}_2 \, da &\to -(\mathbf{n} \cdot \mathbf{e}_2) \, \mathbf{s}(\mathbf{e}_2) \quad \text{as } h \to 0 \\ \frac{1}{A_h} \int_{A_3} \widehat{\mathbf{s}}_3 \, da &\to -(\mathbf{n} \cdot \mathbf{e}_3) \, \mathbf{s}(\mathbf{e}_3) \quad \text{as } h \to 0 \end{aligned} \tag{3.2.16}$$

where $\mathbf{s}_{\mathbf{n}} \equiv \mathbf{s}(\mathbf{n})$ represents the stress vector on a surface with normal $\mathbf{n}$. In particular, $\mathbf{s}(\mathbf{e}_1)$ is the stress vector on the surface with normal $\mathbf{e}_1$. Hence, from Equations 3.2.14 through 3.2.16 as $h \to 0$, we find

$$\mathbf{s}_{\mathbf{n}} = (\mathbf{n} \cdot \mathbf{e}_i) \mathbf{s}(\mathbf{e}_i) \tag{3.2.17}$$

where summation over i is observed. Now, since for any three vectors $\mathbf{a}$, $\mathbf{b}$, and $\mathbf{c}$ we have the identity

$$(\mathbf{a} \cdot \mathbf{b})\mathbf{c} = (\mathbf{c} \otimes \mathbf{b})\mathbf{a} \tag{3.2.18}$$

therefore, letting

$$\mathbf{a} = \mathbf{n}, \quad \mathbf{b} = \mathbf{e}_i, \quad \mathbf{c} = \mathbf{s}(\mathbf{e}_i) \tag{3.2.19}$$

we reduce 3.2.17 to the form

$$\mathbf{s}_{\mathbf{n}} = \mathbf{S}\mathbf{n} \tag{3.2.20}$$

where $\mathbf{S}$ is the second-order tensor defined by

$$\mathbf{S} = \mathbf{s}(\mathbf{e}_i) \otimes \mathbf{e}_i \tag{3.2.21}$$

This completes the proof of the formula 3.2.12. Note that in this proof we let $\mathbf{b} = \mathbf{0}$. If $\mathbf{b}$ is not zero a similar proof leading to Equation 3.2.20 may be given, as the terms involving $\mathbf{b}$ in the equilibrium equations tend to zero when h goes to zero.

Example 3.2.1

Given the stress tensor $\mathbf{S}$ at a given point A

$$\mathbf{S} = \begin{bmatrix} 3 & 1 & 3 \\ 1 & 2 & 1 \\ 3 & 1 & 0 \end{bmatrix} \tag{a}$$

and a plane π described by the equation

$$x_1 + x_2 - x_3 = 1$$

Find

(i) The stress vector $\mathbf{s}$ on a plane through A parallel to the plane
(ii) The magnitude of $\mathbf{s}$
(iii) The angle α between $\mathbf{s}$ and the normal to the plane
(iv) The normal and tangent components of the stress vector $\mathbf{s}$

Solution

(i) The normal unit vector $\mathbf{n}$ to plane $x_1 + x_2 - x_3 = 1$ is $\mathbf{n} = \frac{1}{\sqrt{3}}(\mathbf{e}_1 + \mathbf{e}_2 - \mathbf{e}_3)$. Using the relation

$$\mathbf{s} = \mathbf{S}\mathbf{n} \tag{b}$$

we obtain

$$\begin{Bmatrix} s_1 \\ s_2 \\ s_3 \end{Bmatrix} = \begin{bmatrix} 3 & 1 & 3 \\ 1 & 2 & 1 \\ 3 & 1 & 0 \end{bmatrix} \begin{Bmatrix} \frac{1}{\sqrt{3}} \\ \frac{1}{\sqrt{3}} \\ -\frac{1}{\sqrt{3}} \end{Bmatrix} \tag{c}$$

Hence,

$$s_1 = \frac{1}{\sqrt{3}}, \quad s_2 = \frac{2}{\sqrt{3}}, \quad s_3 = \frac{4}{\sqrt{3}} \tag{d}$$

Therefore,

$$\mathbf{s} = \frac{1}{\sqrt{3}}(\mathbf{e}_1 + 2\mathbf{e}_2 + 4\mathbf{e}_3) \tag{e}$$

(ii) The magnitude of **s** is

$$|\mathbf{s}| = \sqrt{\frac{1+4+16}{3}} = \sqrt{7} \tag{f}$$

(iii) To calculate the angle between **s** and **n** we compute $\mathbf{s}\cdot\mathbf{n}$ and write

$$\mathbf{s}\cdot\mathbf{n} = -\frac{1}{3} \tag{g}$$

Hence the angle θ between **s** and **n** is given by

$$\cos\theta = \frac{\mathbf{s}\cdot\mathbf{n}}{|\mathbf{s}||\mathbf{n}|} = -\frac{1}{3\sqrt{7}} \tag{h}$$

and $\theta = 97.24°$. This means that the component of **s** along **n** has the opposite sense to that of **n**; this fact also follows from (g).

(iv) The normal component $\mathbf{s_n}$ of **s** along **n** is

$$\mathbf{s_n} = (\mathbf{s}\cdot\mathbf{n})\mathbf{n} = -\frac{1}{3}\,\mathbf{n} \tag{i}$$

with

$$(\mathbf{s_n})_1 = -\frac{1}{3\sqrt{3}}, \quad (\mathbf{s_n})_2 = -\frac{1}{3\sqrt{3}}, \quad (\mathbf{s_n})_3 = \frac{1}{3\sqrt{3}} \tag{j}$$

The tangent component of **s** is obtained from the formula

$$\boldsymbol{\tau} = \mathbf{n}\times(\mathbf{s}\times\mathbf{n}) \tag{k}$$

Hence we obtain

$$\boldsymbol{\tau} = \frac{1}{3\sqrt{3}}(4\mathbf{e}_1 + 7\mathbf{e}_2 + 11\mathbf{e}_3) \tag{l}$$

This completes the solution (Figure 3.4). □

Equation 3.2.20 plays a significant role in obtaining a local form of the dynamic equilibrium equation. To find this equation we substitute Equation 3.2.20 into Equation 3.2.10 and write

$$\int_{\partial P} \mathbf{Sn}\,da + \int_P \mathbf{b}\,dv = \int_P \ddot{\mathbf{u}}\rho\,dv \tag{3.2.22}$$

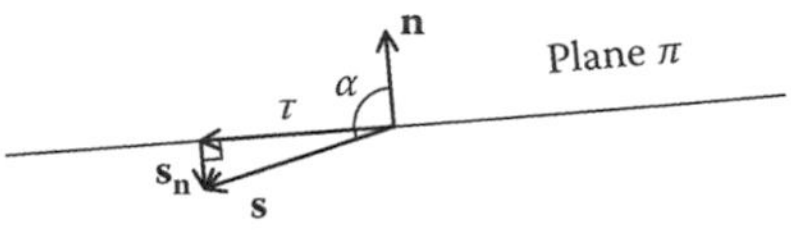

FIGURE 3.4 Decomposition of the vector **s**.

hence, using the divergence theorem, we find

$$\int_P \operatorname{div}\mathbf{S}\, dv + \int_P \mathbf{b}\, dv = \int_P \ddot{\mathbf{u}}\rho\, dv \tag{3.2.23}$$

Since P is an arbitrary portion of B, this implies the local form of the equation of motion

$$\operatorname{div}\mathbf{S} + \mathbf{b} = \rho\ddot{\mathbf{u}} \tag{3.2.24}$$

In the following we prove that the stress tensor $\mathbf{S}$ is symmetric. To this end we write Equation 3.2.11 in the form

$$\int_{\partial P} \mathbf{x}\times\mathbf{Sn}\, da + \int_P \mathbf{x}\times\mathbf{b}\, dv = \int_P \mathbf{x}\times\ddot{\mathbf{u}}\rho\, dv \tag{3.2.25}$$

Also, we note that

$$\int_{\partial P} \mathbf{x}\times\mathbf{Sn}\, da = \int_P \mathbf{x}\times\operatorname{div}\mathbf{S}\, dv + 2\int_P \boldsymbol{\sigma}\, dv \tag{3.2.26}$$

where $\boldsymbol{\sigma}$ is the axial vector corresponding to the skew part of $\mathbf{S}$. In components

$$\sigma_i = \frac{1}{2}\,\epsilon_{ijk}S_{kj} = -\frac{1}{2}\,\epsilon_{ikj}S_{kj} \tag{3.2.27}$$

Hence, from Equations 3.2.25 and 3.2.26 we obtain

$$2\int_P \boldsymbol{\sigma}\, dv + \int_P \mathbf{x}\times(\operatorname{div}\mathbf{S} + \mathbf{b} - \rho\ddot{\mathbf{u}})\, dv = \mathbf{0} \tag{3.2.28}$$

From Equations 3.2.24 and 3.2.28 follows

$$\int_P \boldsymbol{\sigma}\, dv = \mathbf{0} \tag{3.2.29}$$

Since P is an arbitrary portion of B,

$$\boldsymbol{\sigma} = \mathbf{0} \tag{3.2.30}$$

This relation together with the definition of $\boldsymbol{\sigma}$ implies that

$$\mathbf{S} = \mathbf{S}^T \tag{3.2.31}$$

In particular, writing condition 3.2.30 in components for $i = 1$ we have

$$\epsilon_{1jk}S_{kj} = \epsilon_{12k}S_{k2} + \epsilon_{13k}S_{k3} = S_{32} - S_{23} = 0 \tag{3.2.32}$$

In a similar way we find

$$S_{21} - S_{12} = 0 \quad \text{and} \quad S_{13} - S_{31} = 0 \tag{3.2.33}$$

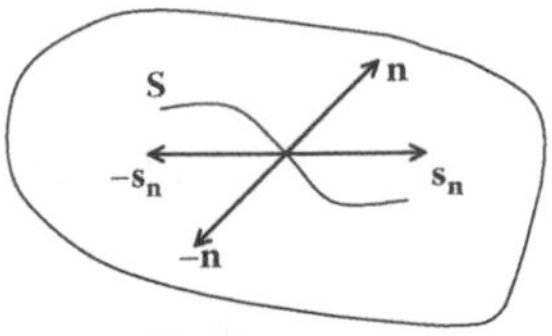

FIGURE 3.5 The graphic representation of the Cauchy reciprocal relation.

We add to the listed properties of the stress tensor (Equations 3.2.12, 3.2.24, and 3.2.31) the Cauchy reciprocal relation

$$\mathbf{s_n} = -\mathbf{s_{-n}} \tag{3.2.34}$$

in which the meanings of $\mathbf{s_n}$ and $\mathbf{s_{-n}}$ follow from the definition of the stress vector given previously. Relation 3.2.34 is interpreted graphically in Figure 3.5.

Proof of the Cauchy reciprocal relation follows from Newton's third law.

Returning to the concept of the dynamic process defined as the array of functions $[\mathbf{u}, \mathbf{s_n}, \mathbf{b}]$ and taking into account the properties of the stress tensor $\mathbf{S}$ we can define a *dynamic process* as an ordered array $[\mathbf{u}, \mathbf{S}, \mathbf{b}]$ in which the functions $\mathbf{u}, \mathbf{S}, \mathbf{b}$ are sufficiently smooth on $\bar{B} \times [0, \infty)$; $\mathbf{S}$ is a symmetric tensor, $\mathbf{S} = \mathbf{S}^T$; and $\mathbf{u}$, $\mathbf{S}$, and $\mathbf{b}$ satisfy the equation of motion

$$\operatorname{div} \mathbf{S} + \mathbf{b} = \rho \ddot{\mathbf{u}} \tag{3.2.35}$$

If the functions $\mathbf{u}$, $\mathbf{S}$, and $\mathbf{b}$ are independent of time, a dynamic process reduces to a static admissible state of the body B, also defined as the array of functions $[\mathbf{u}, \mathbf{S}, \mathbf{b}]$, for which the equation of equilibrium

$$\operatorname{div} \mathbf{S} + \mathbf{b} = \mathbf{0} \tag{3.2.36}$$

is satisfied. Clearly, for a symmetric stress tensor $\mathbf{S}$ a number of properties that hold true for any symmetric tensor can be listed. In particular, if $\mathbf{S}$ satisfies

$$\mathbf{Sn} = s\mathbf{n} \tag{3.2.37}$$

where s is a scalar, then s is called a principal stress and the unit vector $\mathbf{n}$ is called a principal direction of stress. Since $\mathbf{S}$ is symmetric, there exist three mutually perpendicular directions, $\mathbf{n}_1$, $\mathbf{n}_2$, and $\mathbf{n}_3$, and three corresponding principal stresses, $\mathbf{S}_1$, $\mathbf{S}_2$, and $\mathbf{S}_3$.

In the case when $\mathbf{b} = \mathbf{0}$ the equilibrium equation 3.2.36 is satisfied by any constant tensor $\mathbf{S} = \mathbf{S}_0$. Typical examples of constant stress tensor are (i) pure tension or compression, (ii) uniform pressure, and (iii) pure shear.

(i) If $\mathbf{S}$ takes the form

$$\mathbf{S} = \sigma \mathbf{n} \otimes \mathbf{n} \tag{3.2.38}$$

where σ is a constant, $\mathbf{S}$ represents pure tension or compression. For $\sigma > 0$ ($\sigma < 0$) we obtain tension (compression) of the body. For $\mathbf{n} = (1, 0, 0)$ we obtain

$$\mathbf{S} = \begin{bmatrix} \sigma & 0 & 0 \\ 0 & 0 & 0 \\ 0 & 0 & 0 \end{bmatrix} \tag{3.2.39}$$

which corresponds to pure tension along the axis x_1, Figure 3.6.

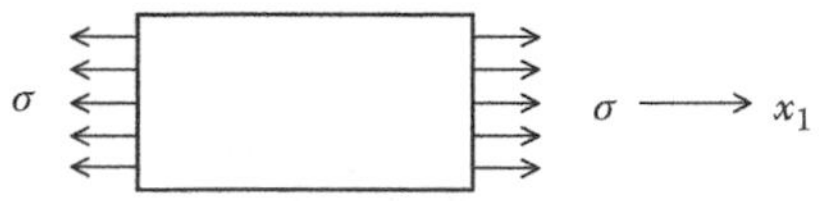

FIGURE 3.6 Pure tension along axis x_1.

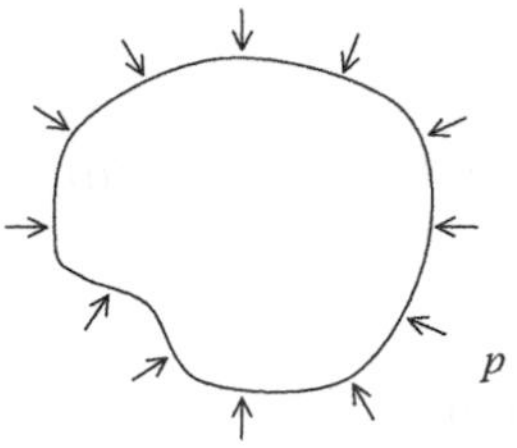

FIGURE 3.7 Body under uniform pressure p.

(ii) If $\mathbf{S}$ takes the form

$$\mathbf{S} = -p\mathbf{1} \tag{3.2.40}$$

where p is a positive constant, $\mathbf{S}$ represents uniform pressure. In this case

$$\mathbf{S} = \begin{bmatrix} -p & 0 & 0 \\ 0 & -p & 0 \\ 0 & 0 & -p \end{bmatrix} \tag{3.2.41}$$

This situation is shown in Figure 3.7.

(iii) If $\mathbf{S}$ takes the form

$$\mathbf{S} = 2\tau \text{ sym } (\mathbf{m} \otimes \mathbf{n}) \equiv \tau(\mathbf{m} \otimes \mathbf{n} + \mathbf{n} \otimes \mathbf{m}) \tag{3.2.42}$$

where $\mathbf{m}$ and $\mathbf{n}$ are orthogonal unit vectors and τ is a constant, $\mathbf{S}$ represents pure shear with shear stress τ relative to the direction pair $\mathbf{m}$, $\mathbf{n}$. For $\mathbf{m} = (1, 0, 0)$ and $\mathbf{n} = (0, 1, 0)$, see Figure 3.8, we obtain

$$\mathbf{S} = \begin{bmatrix} 0 & \tau & 0 \\ \tau & 0 & 0 \\ 0 & 0 & 0 \end{bmatrix} \tag{3.2.43}$$

Example 3.2.2

Let $\mathbf{a}_N$ and $\mathbf{a}_T$ denote the normal and tangent components, respectively, of a vector $\mathbf{a}$ with respect to the plane

$$Ax_1 + Bx_2 + Cx_3 + D = 0 \tag{a}$$

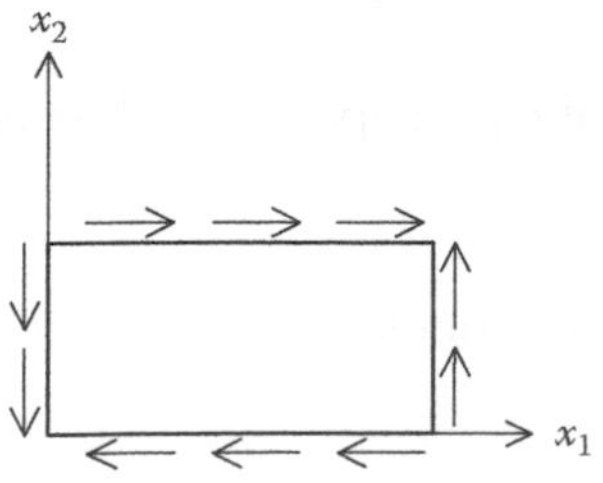

FIGURE 3.8 Pure shear τ in the x_1x_2 plane.

Show that

$$\mathbf{a}_N = \left(\mathbf{a} \cdot \frac{\mathbf{N}}{|\mathbf{N}|}\right) \frac{\mathbf{N}}{|\mathbf{N}|} \tag{b}$$

$$\mathbf{a}_T = \mathbf{a} - \mathbf{a}_N \tag{c}$$

where

$$\mathbf{N} = A\mathbf{e}_1 + B\mathbf{e}_2 + C\mathbf{e}_3 \tag{d}$$

Solution

The vector **N** given by (d) is normal to the plane described by (a). Hence

$$|\mathbf{a}_N| = \mathbf{a} \cdot \frac{\mathbf{N}}{|\mathbf{N}|} \tag{e}$$

and

$$\mathbf{a}_N = |\mathbf{a}_N| \frac{\mathbf{N}}{|\mathbf{N}|} = \left(\mathbf{a} \cdot \frac{\mathbf{N}}{|\mathbf{N}|}\right) \frac{\mathbf{N}}{|\mathbf{N}|} \tag{f}$$

This completes the proof of (b). The result (c) is equivalent to the decomposition formula

$$\mathbf{a} = \mathbf{a}_N + \mathbf{a}_T \tag{g}$$

and this completes the solution. □

Remark 3.2.1: Results of Example 3.2.2 can be used to obtain formula (l) of Example 3.2.1 by an alternative method.

Example 3.2.3

The stress tensor **S** at a point A is given by

$$\mathbf{S} = \begin{bmatrix} 1 & 0 & 2 \\ 0 & 3 & 0 \\ 2 & 0 & 5 \end{bmatrix} \tag{a}$$

Find the principal stresses and the principal stress directions at A.

Solution

The principal stresses and the principal stress directions obey the equation, see Equation 3.2.37,

$$\mathbf{Sn} - s\mathbf{n} = \mathbf{0} \tag{b}$$

in which s is the principal stress and $\mathbf{n}$ is the corresponding principal direction. Equation (b) in components takes the form

$$(S_{ij} - s\delta_{ij})n_j = 0 \tag{c}$$

The nonzero principal stresses are obtained, therefore, from the equation

$$\begin{vmatrix} 1-s & 0 & 2 \\ 0 & 3-s & 0 \\ 2 & 0 & 5-s \end{vmatrix} = 0 \tag{d}$$

Upon expansion of the determinant along the first row we arrive at

$$(1-s)(3-s)(5-s) - 2(2)(3-s) = 0 \quad \text{or} \quad (3-s)(s^2 - 6s + 1) = 0 \tag{e}$$

The principal stresses are then

$$s_1 = 3 - 2\sqrt{2}, \quad s_2 = 3, \quad s_3 = 3 + 2\sqrt{2} \tag{f}$$

To determine the principal directions we write (c) in the expanded form:

$$\begin{aligned} (S_{11} - s)n_1 + S_{12}n_2 + S_{13}n_3 &= 0 \\ S_{21}n_1 + (S_{22} - s)n_2 + S_{23}n_3 &= 0 \\ S_{31}n_1 + S_{32}n_2 + (S_{33} - s)n_3 &= 0 \end{aligned} \tag{g}$$

In addition to (g) we recall that $n_i n_i = 1$ or

$$n_1^2 + n_2^2 + n_3^2 = 1 \tag{h}$$

To find the principal direction $\mathbf{n}^{(1)} = \left(n_1^{(1)}, n_2^{(1)}, n_3^{(1)}\right)$ corresponding to $s = s_1 = 3 - 2\sqrt{2}$ we use (g) and (h) and have

$$\begin{aligned} (1 - s_1)n_1^{(1)} + 2n_3^{(1)} &= 0 \\ (3 - s_1)n_2^{(1)} &= 0 \\ 2n_1^{(1)} + (5 - s_1)n_3^{(1)} &= 0 \\ n_1^{(1)2} + n_2^{(1)2} + n_3^{(1)2} &= 1 \end{aligned} \tag{i}$$

Hence

$$n_2^{(1)} = 0, \quad n_3^{(1)} = -\frac{1 - s_1}{2} n_1^{(1)}, \quad n_1^{(1)} = \pm\left[1 + \left(\frac{1 - s_1}{2}\right)^2\right]^{-\frac{1}{2}} \tag{j}$$

so,

$$\begin{aligned} n_1^{(1)} &= \pm\left[1 + \left(\frac{1 - s_1}{2}\right)^2\right]^{-\frac{1}{2}}, \quad n_2^{(1)} = 0 \\ n_3^{(1)} &= \mp\left(\frac{1 - s_1}{2}\right)\left[1 + \left(\frac{1 - s_1}{2}\right)^2\right]^{-\frac{1}{2}} \end{aligned} \tag{k}$$

Next, for $s = s_2 = 3$, (g) and (h) result in

$$-2n_1^{(2)} + 2n_3^{(2)} = 0$$
$$2n_1^{(2)} + 2n_3^{(2)} = 0 \qquad \text{(l)}$$
$$n_1^{(2)2} + n_2^{(2)2} + n_3^{(2)2} = 1$$

Hence, for $s = s_2 = 3$

$$n_1^{(1)} = 0, \quad n_2^{(2)} = \pm 1, \quad n_3^{(2)} = 0 \qquad \text{(m)}$$

Finally, for $s = s_3 = 3 + 2\sqrt{2}$ we obtain

$$n_1^{(3)} = \pm \left[1 + \left(\frac{1 - s_3}{2}\right)^2\right]^{-\frac{1}{2}}$$
$$n_2^{(3)} = 0 \qquad \text{(n)}$$
$$n_3^{(3)} = \mp \left(\frac{1 - s_3}{2}\right)\left[1 + \left(\frac{1 - s_3}{2}\right)^2\right]^{-\frac{1}{2}}$$

It may be checked that $\mathbf{n}^{(1)}$, $\mathbf{n}^{(2)}$, and $\mathbf{n}^{(3)}$ are mutually orthogonal. This is not an accident. Namely, the orthogonality property is contained in the *spectral theorem* for an arbitrary second-order symmetric tensor. We recall a part of this theorem.

Theorem: If s_1, s_2, and s_3 are all different then $\mathbf{n}^{(1)}$, $\mathbf{n}^{(2)}$, and $\mathbf{n}^{(3)}$ are mutually orthogonal. ■

Proof: First, we show that vectors $\mathbf{n}^{(1)}$ and $\mathbf{n}^{(2)}$, which correspond to s_1 and s_2, are perpendicular. To this end we write the equations satisfied by $\mathbf{n}^{(1)}$ and $\mathbf{n}^{(2)}$

$$\mathbf{S}\mathbf{n}^{(1)} - s_1\mathbf{n}^{(1)} = \mathbf{0} \qquad (3.2.44)$$

$$\mathbf{S}\mathbf{n}^{(2)} - s_2\mathbf{n}^{(2)} = \mathbf{0} \qquad (3.2.45)$$

Taking the scalar product of $\mathbf{n}^{(2)}$ and Equation 3.2.44, and of $\mathbf{n}^{(1)}$ and Equation 3.2.45, and subtracting the results we obtain

$$\mathbf{n}^{(2)} \cdot \mathbf{S}\mathbf{n}^{(1)} - \mathbf{n}^{(1)} \cdot \mathbf{S}\mathbf{n}^{(2)} = (s_1 - s_2)\mathbf{n}^{(1)} \cdot \mathbf{n}^{(2)} \qquad (3.2.46)$$

Since $\mathbf{S}$ is symmetric, the left-hand side of Equation 3.2.46 vanishes, thus, from $s_1 \neq s_2$ it follows

$$\mathbf{n}^{(1)} \cdot \mathbf{n}^{(2)} = 0 \qquad (3.2.47)$$

This proves orthogonality of $\mathbf{n}^{(1)}$ and $\mathbf{n}^{(2)}$. In a similar way other orthogonality conditions can be proved. □

Example 3.2.4

Given a unit vector $\boldsymbol{\ell}$ find the stress vector $\mathbf{s}_\ell$ for

(i) A pure tension in the direction $\mathbf{n}$
(ii) A uniform pressure
(iii) A pure shear

Solution

(i) Stress tensor $\mathbf{S}$ takes the form

$$\mathbf{S} = \sigma \mathbf{n} \otimes \mathbf{n} \tag{a}$$

where σ is a constant. Hence, for the stress vector in the $\boldsymbol{\ell}$ direction we get

$$\mathbf{s}_\ell \equiv \mathbf{S}\boldsymbol{\ell} = \sigma(\mathbf{n} \otimes \mathbf{n})\boldsymbol{\ell} = \sigma(\boldsymbol{\ell} \cdot \mathbf{n})\mathbf{n} \tag{b}$$

The last equality in (b) comes from the fact that

$$n_i n_j \ell_j = [\mathbf{n}(\boldsymbol{\ell} \cdot \mathbf{n})]_i = [(\mathbf{n} \otimes \mathbf{n})\boldsymbol{\ell}]_i \tag{c}$$

If $\boldsymbol{\ell}$ and $\mathbf{n}$ are orthogonal, then $\mathbf{s}_\ell = \mathbf{0}$. If $\boldsymbol{\ell} = \mathbf{n}$ or $\boldsymbol{\ell} = -\mathbf{n}$ then from (b) $\mathbf{s}_\ell = \sigma\mathbf{n}$ or $\mathbf{s}_\ell = -\sigma\mathbf{n}$, respectively.

(ii) Stress tensor $\mathbf{S}$ is of the form

$$\mathbf{S} = -p\mathbf{1} \tag{d}$$

where p is a constant. So, the stress vector

$$\mathbf{s}_\ell = -p\boldsymbol{\ell} \tag{e}$$

(iii) Stress tensor $\mathbf{S}$ is of the form

$$\mathbf{S} = \tau(\mathbf{m} \otimes \mathbf{n} + \mathbf{n} \otimes \mathbf{m}) \tag{f}$$

where τ is a constant and $\mathbf{m}$ and $\mathbf{n}$ are orthonormal vectors. Hence,

$$\mathbf{s}_\ell = \tau[(\mathbf{m} \otimes \mathbf{n})\boldsymbol{\ell} + (\mathbf{n} \otimes \mathbf{m})\boldsymbol{\ell}] = \tau[(\boldsymbol{\ell} \cdot \mathbf{n})\mathbf{m} + (\boldsymbol{\ell} \cdot \mathbf{m})\mathbf{n}] \tag{g}$$

If $\boldsymbol{\ell}$ lies in the plane spanned by the vectors $\mathbf{m}$ and $\mathbf{n}$, and $\boldsymbol{\ell}$ is perpendicular to $\mathbf{m}$, then

$$\boldsymbol{\ell} = \pm\mathbf{n} \tag{h}$$

and

$$\mathbf{s}_\ell = \pm\tau\mathbf{m} \tag{i}$$

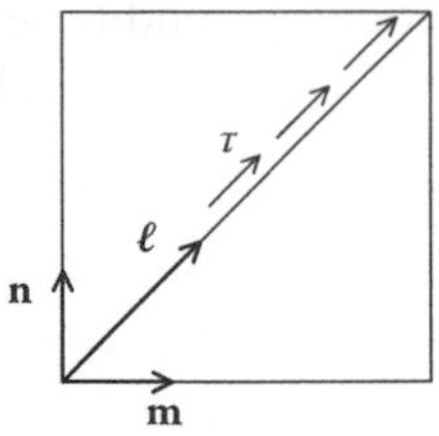

FIGURE 3.9 Shear stress vector on the diagonal of a square.

If $\boldsymbol{\ell}$ lies in the plane spanned by the vectors $\mathbf{m} = (1, 0, 0)$ and $\mathbf{n} = (0, 1, 0)$ and $\boldsymbol{\ell} = (\mathbf{m}+\mathbf{n})/\sqrt{2}$ then $\boldsymbol{\ell}\cdot\mathbf{m} = 1/\sqrt{2}$, $\boldsymbol{\ell}\cdot\mathbf{n} = 1/\sqrt{2}$ and

$$\mathbf{s}_\ell = \frac{\tau}{\sqrt{2}}\,[(1,0,0) + (0,1,0)] = \frac{\tau}{\sqrt{2}}\,(1,1,0) \tag{j}$$

and

$$|\mathbf{s}_\ell| = \tau \tag{k}$$

This is the case of the shear stress vector acting along the diagonal of a square formed by **m** and **n** (see Figure 3.9). □

Example 3.2.5

For a cantilever beam of rectangular cross section $b \times h$ loaded at the free end by force Q, Figure 3.10, the stress components are given by

$$S_{11} = \frac{Mx_2}{I} = -\frac{Q(\ell - x_1)x_2}{I}, \qquad S_{12} = S_{21} = \frac{Q}{2I}\left(\frac{h^2}{4} - x_2^2\right)$$
$$S_{13} = 0, \quad S_{22} = 0, \quad S_{23} = 0, \quad S_{33} = 0 \tag{a}$$

where M is the bending moment at x_1 produced by the shearing force Q, and $I = bh^3/12$ is the second moment of the cross section. Show that the equilibrium equations with zero body forces are satisfied at each point of the beam.

Solution

Equations of equilibrium for this two-dimensional case are

$$\begin{aligned} S_{11,1} + S_{12,2} &= 0 \\ S_{21,1} + S_{22,2} &= 0 \end{aligned} \tag{b}$$

The substitution of Equation (a) into (b) reveals that (b) are satisfied. □

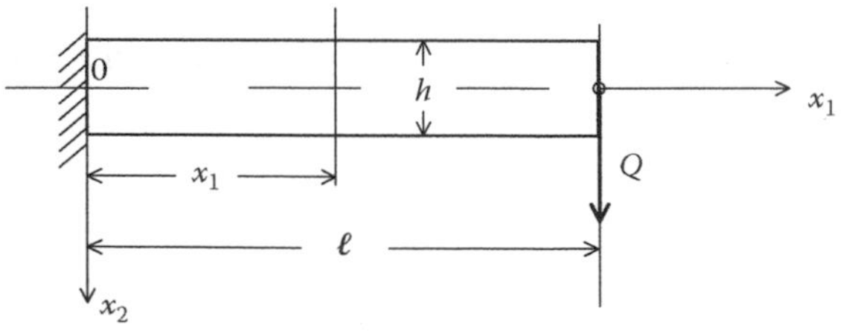

FIGURE 3.10 Cantilever beam loaded at the free end.

Example 3.2.6

The stress tensor **S** at a point **x** is

$$\mathbf{S} = S_0 \begin{bmatrix} 1 & 1 & 1 \\ 1 & 1 & 1 \\ 1 & 1 & 1 \end{bmatrix} \tag{a}$$

where S_0 is a constant. Find the normal and shear stress on the octahedral plane at the point **x**. The octahedral plane is a plane inclined by the same angle to each of the planes of the coordinate system.

Solution

The unit vector **n** perpendicular to the octahedral plane has components

$$n_1 = n_2 = n_3 = \frac{1}{\sqrt{3}} \tag{b}$$

The stress vector is obtained from

$$\mathbf{s_n} = \mathbf{Sn} \tag{c}$$

or

$$\begin{Bmatrix} s_1 \\ s_2 \\ s_3 \end{Bmatrix} = \frac{S_0}{\sqrt{3}} \begin{bmatrix} 1 & 1 & 1 \\ 1 & 1 & 1 \\ 1 & 1 & 1 \end{bmatrix} \begin{Bmatrix} 1 \\ 1 \\ 1 \end{Bmatrix} \tag{d}$$

Hence,

$$s_1 = s_2 = s_3 = S_0\sqrt{3} \tag{e}$$

The magnitude of $\mathbf{s_n}$ is

$$|\mathbf{s_n}| = 3|S_0| \tag{f}$$

The normal stress vector is

$$\boldsymbol{\sigma} = (\mathbf{n} \cdot \mathbf{s_n})\mathbf{n} = 3S_0\mathbf{n} \tag{g}$$

The tangent stress vector is

$$\boldsymbol{\tau} = (\mathbf{s_n} \cdot \mathbf{t})\mathbf{t} \tag{h}$$

where **t** is a unit vector perpendicular to **n**, lying on the octahedral plane. Since

$$\mathbf{s_n} = \boldsymbol{\sigma} + \boldsymbol{\tau} = 3S_0\mathbf{n} + (\mathbf{s_n} \cdot \mathbf{t})\mathbf{t} \tag{i}$$

therefore,

$$|\mathbf{s_n}|^2 = 9S_0^2 + (\mathbf{s_n} \cdot \mathbf{t})^2 \tag{j}$$

and by (f)

$$\mathbf{s_n} \cdot \mathbf{t} = 0 \tag{k}$$

So, there is no shear stress acting on the octahedral plane. □

Example 3.2.7

Given a stress tensor $\mathbf{S}$ in the basis $\{\mathbf{e}_i\}$. Let $\{\mathbf{e}_i^*\}$ be an orthonormal basis formed from the eigenvectors of $\mathbf{S}$, and let $\mathbf{S}^*$ be a tensor obtained from $\mathbf{S}$ by the transformation formula from $\{\mathbf{e}_i\}$ to $\{\mathbf{e}_i^*\}$. The principal stresses of $\mathbf{S}$ corresponding to $\mathbf{e}_1^*$, $\mathbf{e}_2^*$, and $\mathbf{e}_3^*$ are denoted by λ_1, λ_2, and λ_3, respectively. In the coordinate system defined by the basis $\{\mathbf{e}_i^*\}$ we choose an arbitrarily inclined plane passing through a given point $\mathbf{P}^*$ with a unit normal vector $\mathbf{n}^*$.

Show that a locus of the stress vectors $\mathbf{S}^*\mathbf{n}^*$ on the plane is represented by an ellipsoid of the form

$$\left(\frac{x_1^*}{\lambda_1}\right)^2 + \left(\frac{x_2^*}{\lambda_2}\right)^2 + \left(\frac{x_3^*}{\lambda_3}\right)^2 = 1 \tag{a}$$

where

$$x_i^* = \mathbf{e}_i^* \cdot (\mathbf{S}^*\mathbf{n}^*) \tag{b}$$

Solution

Since $\mathbf{S}^*$ is referred to the orthonormal basis $\{\mathbf{e}_i^*\}$, then by Equation (c) in Example 2.1.12, in which $\mathbf{T}^*$ is replaced by $\mathbf{S}^*$,

$$\mathbf{S}^* = \lambda_1(\mathbf{e}_1^* \otimes \mathbf{e}_1^*) + \lambda_2(\mathbf{e}_2^* \otimes \mathbf{e}_2^*) + \lambda_3(\mathbf{e}_3^* \otimes \mathbf{e}_3^*) \tag{c}$$

Hence, we obtain

$$\mathbf{S}^*\mathbf{n}^* = \lambda_1(\mathbf{e}_1^* \otimes \mathbf{e}_1^*)\,\mathbf{n}^* + \lambda_2(\mathbf{e}_2^* \otimes \mathbf{e}_2^*)\,\mathbf{n}^* + \lambda_3(\mathbf{e}_3^* \otimes \mathbf{e}_3^*)\,\mathbf{n}^* \tag{d}$$

Multiplying Equation (d) by $\mathbf{e}_i^*$ in the dot sense and using Equation (b) we obtain

$$x_i^* = \lambda_1\left[(\mathbf{e}_1^* \otimes \mathbf{e}_1^*)\,\mathbf{n}^*\right] \cdot \mathbf{e}_i^* + \lambda_2\left[(\mathbf{e}_2^* \otimes \mathbf{e}_2^*)\,\mathbf{n}^*\right] \cdot \mathbf{e}_i^* + \lambda_3\left[(\mathbf{e}_3^* \otimes \mathbf{e}_3^*)\,\mathbf{n}^*\right] \cdot \mathbf{e}_i^* \tag{e}$$

Since for arbitrary vectors $\mathbf{a}$, $\mathbf{b}$, and $\mathbf{c}$

$$(\mathbf{a} \otimes \mathbf{b})\mathbf{c} = (\mathbf{b} \cdot \mathbf{c})\mathbf{a} \tag{f}$$

then Equation (e) reduces to

$$x_i^* = \lambda_1\left[\mathbf{e}_1^*\left(\mathbf{e}_1^* \cdot \mathbf{n}^*\right)\right] \cdot \mathbf{e}_i^* + \lambda_2\left[\mathbf{e}_2^*\left(\mathbf{e}_2^* \cdot \mathbf{n}^*\right)\right] \cdot \mathbf{e}_i^* + \lambda_3\left[\mathbf{e}_3^*\left(\mathbf{e}_3^* \cdot \mathbf{n}^*\right)\right] \cdot \mathbf{e}_i^* \tag{g}$$

This can be simplified to the form

$$x_i^* = \lambda_1\delta_{i1}n_1^* + \lambda_2\delta_{i2}n_2^* + \lambda_3\delta_{i3}n_3^* \tag{h}$$

Letting $i = 1, 2$, and 3 in Equation (h) we obtain

$$x_1^* = \lambda_1 n_1^*, \quad x_2^* = \lambda_2 n_2^*, \quad x_3^* = \lambda_3 n_3^* \tag{i}$$

These equations together with the condition $|\mathbf{n}^*| = 1$ yield the desired result (a). This completes the solution. □

Note: Equation (a) describes an ellipsoid called the *Lamé ellipsoid.*

Example 3.2.8

Show that as the normal unit vector **n** changes its orientation, the normal stress on a surface element assumes an extreme value when the element is a principal plane of stress and that this extreme value is a principal stress. A principal plane is a plane with the normal vector along the principal direction.

Solution

The normal component of the stress vector is given by

$$\sigma = \mathbf{s_n} \cdot \mathbf{n} = S_{ij} n_i n_j \tag{a}$$

Here, σ is to be treated as a function of **n**, and its extremum is to be found under the condition that $|\mathbf{n}| = 1$. To obtain extreme values of σ under this condition we use a Lagrange multiplier λ, and look for the extremum of the function

$$F = \sigma - \lambda(|\mathbf{n}|^2 - 1) \tag{b}$$

The necessary conditions for the extremum of F using (a) may be written as follows:

$$\frac{\partial F}{\partial n_i} = \frac{\partial}{\partial n_i} [S_{ab} n_a n_b - \lambda(n_a n_a - 1)] = 0 \quad (a, b, i = 1, 2, 3) \tag{c}$$

This is equivalent to

$$2(S_{ia} n_a - \lambda n_i) = 0 \tag{d}$$

or

$$(S_{ij} - \lambda \delta_{ij}) n_j = 0 \tag{e}$$

This condition is satisfied if and only if **n** is a principal direction of **S**. It follows then that σ takes an extreme value on a principal stress plane. By multiplying (e) by n_i and taking into account (a), we obtain

$$\sigma = \lambda \tag{f}$$

Thus, from (e) we conclude that λ is a principal stress. As a result, the principal stress λ is an extreme value of σ. □

3.2.2 Solutions of Elastostatics

In Section 3.2.1, we have shown that the equilibrium equations of elastostatics are of the form

$$\operatorname{div}\mathbf{S} + \mathbf{b} = \mathbf{0} \tag{3.2.48}$$

$$\mathbf{S} = \mathbf{S}^T \tag{3.2.49}$$

where **b** is the body force vector. Equation 3.2.48 expresses the balance of forces, and Equation 3.2.49 expresses the balance of moments.

In this section we assume $\mathbf{b} = \mathbf{0}$, and look for the so-called *stress function*, represented by a tensor field **A**, with the following properties: there is a differential operator $\widehat{L}$, such that

(i)
$$\mathbf{S} = \widehat{L}\mathbf{A} \tag{3.2.50}$$

(ii)
$$\operatorname{div}\,(\widehat{L}\mathbf{A}) = \mathbf{0} \tag{3.2.51}$$

One of the possible representations of **S** was proposed by E. Beltrami (1892) in the form

$$\mathbf{S} = \operatorname{curl}\operatorname{curl}\mathbf{A} \tag{3.2.52}$$

where **A** is a symmetric tensor field. To show this, we prove that **S** given by Equation 3.2.52 satisfies the equations

$$\operatorname{div}\mathbf{S} = \mathbf{0} \tag{3.2.53}$$

$$\mathbf{S} = \mathbf{S}^T \tag{3.2.54}$$

Writing Equations 3.2.52 through 3.2.54 in components, we get

$$S_{ij} = \epsilon_{ipq}\epsilon_{jrs}A_{qs,pr} \tag{3.2.55}$$

$$S_{ij,j} = 0 \tag{3.2.56}$$

$$S_{ij} = S_{ji} \tag{3.2.57}$$

From Equation 3.2.55 we find

$$S_{ij,j} = \epsilon_{ipq}\epsilon_{jrs}A_{qs,prj} \tag{3.2.58}$$

Since

$$\epsilon_{jrs}T_{rj} = 0 \tag{3.2.59}$$

for any symmetric tensor **T**, therefore, letting

$$T_{rj} = A_{qs,prj} \tag{3.2.60}$$

from Equations 3.2.58 and 3.2.59 we obtain Equation 3.2.56. To prove 3.2.57 we need to show that

$$\epsilon_{ipq}\epsilon_{jrs}A_{qs,pr} = \epsilon_{jpq}\epsilon_{irs}A_{qs,pr} \tag{3.2.61}$$

The LHS of Equation 3.2.61, can be written as

$$\epsilon_{jrs}\epsilon_{ipq}A_{qs,pr} \tag{3.2.62}$$

and by changing the indexes:

$$r \to p, \quad s \to q, \quad p \to r, \quad q \to s$$

takes the form

$$\epsilon_{jpq}\epsilon_{irs}A_{sq,rp} \tag{3.2.63}$$

Since $\mathbf{A}$ is symmetric, $\mathbf{A} = \mathbf{A}^T$, and this expression is identical to the RHS of Equation 3.2.61.

The solution in the form of Equation 3.2.52 of the equilibrium equations contains, as special cases, the following:

(i) The Airy representation
(ii) The Maxwell representation
(iii) The Morera representation

We will discuss these representations in detail.

(i) The Airy representation applies to a two-dimensional stress state with

$$\mathbf{A} = \begin{bmatrix} 0 & 0 & 0 \\ 0 & 0 & 0 \\ 0 & 0 & F \end{bmatrix} \tag{3.2.64}$$

where $F = F(x_1, x_2)$. In components, from Equation 3.2.55, we obtain

$$S_{11} = F_{,22}, \quad S_{12} = -F_{,12}, \quad S_{22} = F_{,11}, \quad S_{13} = S_{32} = S_{33} = 0 \tag{3.2.65}$$

For example, since $F = A_{33}$, and $A_{ij} = 0$ for $i, j \neq 3$,

$$\begin{aligned} S_{11} &= \epsilon_{1pq}\epsilon_{1rs}A_{qs,pr} = \epsilon_{1p3}\epsilon_{1r3}A_{33,pr} \\ &= \epsilon_{1r3}A_{33,2r} = A_{33,22} = F_{,22} \end{aligned} \tag{3.2.66}$$

Example 3.2.9

Show that the Airy representation can be written in a compact form

$$S_{\alpha\beta} = \epsilon_{\alpha\delta}\epsilon_{\beta\gamma}F_{,\delta\gamma} \tag{a}$$

where $\epsilon_{\alpha\delta}$ is the two-dimensional alternator and $\alpha, \beta, \gamma, \delta = 1, 2$.
The two-dimensional alternator is defined by

$$\epsilon_{\alpha\delta} = \begin{cases} +1 & \text{for } \alpha = 1, \delta = 2 \\ -1 & \text{for } \alpha = 2, \delta = 1 \\ 0 & \text{otherwise} \end{cases} \tag{b}$$

Solution

From (a) and (b) it follows that (a) is equivalent to Equation 3.2.65. Also, from (a) we obtain

$$S_{\alpha\beta,\beta} = \epsilon_{\alpha\delta}\epsilon_{\beta\gamma}F_{,\delta\gamma\beta}$$

This vanishes because the RHS is the inner product of a symmetric tensor $F_{,\delta\gamma\beta}$ and a skew tensor $\epsilon_{\beta\gamma}$ with respect to indexes β, γ. Finally, it is easy to check that $S_{\alpha\beta} = S_{\beta\alpha}$ □

(ii) The Maxwell representation generates three-dimensional solutions and is obtained by taking **A** in the form

$$\mathbf{A} = \begin{bmatrix} A_1 & 0 & 0 \\ 0 & A_2 & 0 \\ 0 & 0 & A_3 \end{bmatrix} \tag{3.2.67}$$

By substituting Equation 3.2.67 into Equation 3.2.55 we receive

$$\begin{aligned} &S_{11} = A_{2,33} + A_{3,22}, \quad S_{12} = -A_{3,12}, \quad S_{13} = -A_{2,31}, \\ &S_{22} = A_{3,11} + A_{1,33}, \quad S_{23} = -A_{1,23}, \quad S_{33} = A_{1,22} + A_{2,11} \end{aligned} \tag{3.2.68}$$

Obviously, if $A_1 = A_2 = 0$, $A_3 = F$, representation 3.2.68 reduces to the Airy representation, case (i).

(iii) The Morera representation is obtained from Equation 3.2.55 by taking **A** in the form

$$\mathbf{A} = \begin{bmatrix} 0 & B_3 & B_2 \\ B_3 & 0 & B_1 \\ B_2 & B_1 & 0 \end{bmatrix} \tag{3.2.69}$$

and we obtain the stress components

$$\begin{aligned} &S_{11} = -2B_{1,23}, && S_{12} = (B_{1,1} + B_{2,2} - B_{3,3})_{,3} \\ &S_{13} = (B_{3,3} + B_{1,1} - B_{2,2})_{,2}, && S_{22} = -2B_{2,31}, \\ &S_{23} = (B_{2,2} + B_{3,3} - B_{1,1})_{,1}, && S_{33} = -2B_{3,12}. \end{aligned} \tag{3.2.70}$$

An important special case of the Morera representation is the case in which $B_2 = B_3 = 0$, and B_1 is a function of (x_1, x_2) only. In this case the stress components are

$$S_{11} = S_{22} = S_{33} = S_{12} = 0, \quad S_{23} = -\mu\alpha\ [\Psi(x_1, x_2)]_{,1},$$
$$S_{13} = \mu\alpha\ [\Psi(x_1, x_2)]_{,2} \tag{3.2.71}$$

where $\Psi(x_1, x_2) = (B_{1,1})/\mu\alpha$, $\mu > 0$ is the shear modulus, and α is a nonzero constant. This case covers a torsion problem of elastostatics.

An alternative form of the Beltrami solution 3.2.52 is obtained if we define a new tensor field $\mathbf{G}$ by

$$\mathbf{G} = \mathbf{A} - \frac{1}{2}\ (\mathrm{tr}\,\mathbf{A})\mathbf{1} \tag{3.2.72}$$

which is equivalent to

$$\mathbf{A} = \mathbf{G} - (\mathrm{tr}\,\mathbf{G})\mathbf{1} \tag{3.2.73}$$

To show the equivalence of Equations 3.2.72 and 3.2.73 we write Equation 3.2.72 in components

$$G_{ij} = A_{ij} - \frac{1}{2}\ A_{kk}\delta_{ij} \tag{3.2.74}$$

Hence

$$G_{kk} = A_{kk} - \frac{3}{2}\ A_{kk} = -\frac{1}{2}\ A_{kk} \tag{3.2.75}$$

Thus, since $A_{kk} = \mathrm{tr}\,\mathbf{A}$, substituting $\mathrm{tr}\,\mathbf{A}$ in Equation 3.2.72 we arrive at Equation 3.2.73.

To find the alternative form we recall the identity

$$\mathrm{curl}\,\mathrm{curl}\,\mathbf{A} = -\nabla^2\mathbf{A} + 2\widehat{\nabla}(\mathrm{div}\,\mathbf{A}) - \nabla\nabla(\mathrm{tr}\,\mathbf{A}) + [\nabla^2(\mathrm{tr}\,\mathbf{A}) - \mathrm{div}\,\mathrm{div}\,\mathbf{A}]\mathbf{1} \tag{3.2.76}$$

By substituting Equation 3.2.73 into Equation 3.2.76, and using Equation 3.2.52, we obtain

$$\mathbf{S} = -\nabla^2\mathbf{G} + 2\widehat{\nabla}(\mathrm{div}\,\mathbf{G}) - (\mathrm{div}\,\mathrm{div}\,\mathbf{G})\mathbf{1} \tag{3.2.77}$$

This is an alternative form of the Beltrami solution in which the curl curl $\mathbf{A}$ term does not appear. By using form 3.2.77 it seems to be an easier task to obtain the stress components $\mathbf{S}$ than by using Equation 3.2.52. In particular, let us calculate the components S_{12} in the Maxwell representation. From Equations 3.2.67 and 3.2.72 the tensor $\mathbf{G}$ is

$$\mathbf{G} = \frac{1}{2}\begin{bmatrix} A_1 - A_2 - A_3 & 0 & 0 \\ 0 & A_2 - A_3 - A_1 & 0 \\ 0 & 0 & A_3 - A_1 - A_2 \end{bmatrix} \tag{3.2.78}$$

so, Equation 3.2.77 yields

$$S_{12} = G_{1k,k2} + G_{2k,k1}$$
$$= \frac{1}{2}\left[(A_1 - A_2 - A_3)_{,12} + (A_2 - A_3 - A_1)_{,21}\right] = -A_{3,12} \tag{3.2.79}$$

and this is the result shown in Equation 3.2.68.

We note that to establish the Maxwell representation we compute **G** from the matrix 3.2.78. In general case, **S** given by Equation 3.2.77 satisfies the equilibrium equations for any symmetric tensor field **G**. To show this we observe that in view of Equation 3.2.77 and in components, we get

$$S_{ij,j} = -G_{ij,kkj} + G_{ik,kjj} + G_{jk,kij} - G_{kj,kji} = 0 \tag{3.2.80}$$

3.2.3 Properties of Solutions of Elastostatics

We introduce the definition of a so-called *self-equilibrated stress field*. If **S** is a symmetric tensor field on B and

$$\int_S \mathbf{Sn}\, da = \mathbf{0} \tag{3.2.81}$$

$$\int_S \mathbf{x} \times (\mathbf{Sn})\, da = \mathbf{0} \tag{3.2.82}$$

for every closed surface S in B, then **S** is called a self-equilibrated stress field. Equations 3.2.81 and 3.2.82 mean that the resultant force and the resultant moment vanish on S.

This definition allows us to replace the equations of equilibrium in a simply connected body by the condition that **S** is a self-equilibrated stress field. To show this we note that due to the arbitrariness of the surface S and by using the divergence theorem, Equations 3.2.81 and 3.2.82 imply that

$$\operatorname{div} \mathbf{S} = \mathbf{0} \quad \text{and} \quad \mathbf{S} = \mathbf{S}^T \tag{3.2.83}$$

Conversely, if Equations 3.2.83 are satisfied then integrating the first of Equations 3.2.83 over S and using the divergence theorem we arrive at Equation 3.2.81. Also, taking the vector product of **x** and the first of Equations 3.2.83, using the second of Equations 3.2.83, integrating over S and applying the divergence theorem we obtain Equation 3.2.82. The latter procedure is based on the identity

$$\epsilon_{ijk} x_j S_{kp,p} \equiv \epsilon_{ijk}[(x_j S_{kp})_{,p} - x_{j,p} S_{kp}] = 0 \tag{3.2.84}$$

In the replacement procedure it is important that B be a simply connected body. If B is not simply connected there exist stress fields that satisfy the equilibrium equations but are not self-equilibrated.

Example 3.2.10

Let B be a hollow sphere defined by

$$R_1 < |\mathbf{x}| < R_2, \quad R_2 > R_1 > 0 \tag{a}$$

Clearly, the boundary ∂B of B consists of two concentric spherical surfaces

$$|\mathbf{x}| = R_1, \quad |\mathbf{x}| = R_2 \tag{b}$$

Let $\mathbf{S} = \mathbf{S}(\mathbf{x})$ be defined by

$$\mathbf{S}(\mathbf{x}) = F\frac{x_3}{|\mathbf{x}|^5}\,\mathbf{x}\otimes\mathbf{x} \quad \text{on } B \tag{c}$$

where F is a constant force. Show that

$$\operatorname{div}\mathbf{S} = \mathbf{0}, \quad \mathbf{S} = \mathbf{S}^T \quad \text{on } B \tag{d}$$

and

$$\int_{\partial P} S_{3k}n_k\, da = \frac{4}{3}\pi F \tag{e}$$

where ∂P is the surface of a sphere with radius R_0 such that $R_1 < R_0 < R_2$.

Solution

We write $\mathbf{S}$ in components

$$S_{ij} = F x_3 x_i x_j R^{-5} \tag{f}$$

where $R = |\mathbf{x}|$, and compute

$$\begin{aligned} S_{ij,j} &= F[(\delta_{3j}x_i x_j + x_3\delta_{ij}x_j + x_3x_i\delta_{jj})R^{-5} + x_3x_ix_j(-5)R^{-6}x_jR^{-1}] \\ &= F[(x_ix_3 + x_3x_i + 3x_3x_i)R^{-5} - 5x_3x_ix_jx_jR^{-7}] \\ &= F(5x_3x_iR^{-5} - 5x_3x_iR^{-5}) = 0 \end{aligned} \tag{g}$$

Thus, the first equation of (d) is satisfied.

The second equation of (d) comes from (f).

To show (e) we use the spherical coordinates (R, φ, θ) that are related to (x_1, x_2, x_3) by

$$\begin{aligned} x_1 &= R\cos\varphi\sin\theta \\ x_2 &= R\sin\varphi\sin\theta \\ x_3 &= R\cos\theta \end{aligned} \tag{h}$$

with $0 < R < \infty$, $0 \le \varphi \le 2\pi$, and $0 \le \theta \le \pi$ and write down

$$I = \int_{\partial P} S_{3k}n_k\, da = F\int_{\partial P} x_3^2 x_k n_k R_0^{-5}\, da \tag{i}$$

Now,

$$da = R_0^2 \sin\theta d\theta d\varphi \quad \text{and} \quad n_k = \frac{x_k}{R_0} \tag{j}$$

so,

$$\begin{aligned} I &= FR_0^{-3} \int_0^{2\pi}\int_0^{\pi} \left(R_0^2 \cos^2\theta\right) R_0 \sin\theta \, d\theta d\varphi \\ &= 2\pi F \int_0^{\pi} \sin\theta \cos^2\theta \, d\theta = \frac{4}{3}\pi F \end{aligned} \tag{k}$$

This completes Example 3.2.10. □

The example shows that if ∂B does not consist of a single closed surface then a symmetric divergence-free stress field **S** does not have to be self-equilibrated.

Returning to the Beltrami solution, we now show that the Beltrami solution is a self-equilibrated stress field. To prove this we need to check that **S** given by

$$\mathbf{S} = \text{curl}\,\text{curl}\,\mathbf{A} \tag{3.2.85}$$

satisfies the conditions

$$\int_S \mathbf{Sn}\, da = \mathbf{0} \tag{3.2.86}$$

$$\int_S \mathbf{x} \times (\mathbf{Sn})\, da = \mathbf{0} \tag{3.2.87}$$

where S is a closed surface in B.

Introduce a tensor **W** by

$$\mathbf{W} = \text{curl}\,\mathbf{A} \tag{3.2.88}$$

Then Equation 3.2.85 takes the form

$$\mathbf{S} = \text{curl}\,\mathbf{W} \tag{3.2.89}$$

In components this equation reads

$$S_{ij} = \epsilon_{ipq} W_{jq,p} \tag{3.2.90}$$

The LHS of Equation 3.2.86 in components is

$$\int_S S_{ij} n_j \, da = \int_S \epsilon_{ipq} W_{jq,p} n_j \, da \tag{3.2.91}$$

This is equal by the divergence theorem to

$$\int_P \epsilon_{ipq} W_{jq,pj}\, dv \tag{3.2.92}$$

where P is the volume bounded by S. Now, from Equation 3.2.88

$$W_{jq} = \epsilon_{jrs} A_{qs,r} \tag{3.2.93}$$

Substituting Equation 3.2.93 into Equation 3.2.92 we get

$$\int_S S_{ij} n_j\, da = \int_P \epsilon_{ipq} \epsilon_{jrs} A_{qs,rpj}\, dv \tag{3.2.94}$$

This is equal to zero because under the integral ϵ_{jrs} is skew with respect to r and j, and $A_{qs,rpj}$ is symmetric with respect to r and j.

In this way we have proved that Equation 3.2.86 is satisfied. To prove Equation 3.2.87, we write Equation 3.2.87 in components as

$$m_i \equiv \int_S \epsilon_{ipq} x_p s_q\, da = 0 \tag{3.2.95}$$

where

$$s_q = S_{qa} n_a \tag{3.2.96}$$

and

$$S_{aq} = S_{qa} = \epsilon_{qmn} \epsilon_{ars} A_{ns,mr} \tag{3.2.97}$$

Hence, by the divergence theorem

$$m_i = \int_P \epsilon_{ipq} (x_p \epsilon_{qmn} \epsilon_{ars} A_{ns,mr})_{,a}\, dv \tag{3.2.98}$$

which is equivalent to

$$m_i = \int_P \epsilon_{ipq} (\epsilon_{qmn} \epsilon_{prs} A_{ns,mr} + x_p \epsilon_{qmn} \epsilon_{ars} A_{ns,mra})\, dv \tag{3.2.99}$$

or

$$m_i = \int_P \epsilon_{ipq} S_{qp}\, dv + \int_P \epsilon_{ipq} x_p \epsilon_{qmn} \epsilon_{ars} A_{ns,mra}\, dv \tag{3.2.100}$$

Since S_{qp} is symmetric with respect to p and q and ϵ_{ipq} is skew with respect to p and q then the first integral in Equation 3.2.100 is zero. Similarly, the second integral vanishes because the integrand involves a product of ϵ_{ars} and $A_{ns,mra}$ in which ϵ_{ars} is skew with respect to a and r and $A_{ns,mra}$ is symmetric with respect to a and r. As a result $m_i = 0$, which means that the condition 3.2.87 is satisfied. In this way we have proved that the Beltrami solution is a self-equilibrated stress field. We conclude, therefore, that stress fields that are not self-equilibrated cannot be represented by Beltrami solutions. Nevertheless, the Beltrami solutions that are restricted to self-equilibrated stress fields are complete in the following sense:

For any self-equilibrated stress field **S**, there is a symmetric tensor **A** such that

$$\mathbf{S} = \operatorname{curl}\operatorname{curl}\mathbf{A} \tag{3.2.101}$$

We omit the proof.*

We note another property of the Beltrami solution 3.2.101. If **A** is replaced by

$$\mathbf{H} = \mathbf{A} + \mathbf{E} \tag{3.2.102}$$

where **E** is the strain field corresponding to a displacement field **u**, that is

$$\mathbf{E} = \widehat{\nabla}\mathbf{u} = \frac{1}{2}\left(\nabla\mathbf{u} + \nabla\mathbf{u}^T\right) \tag{3.2.103}$$

then

$$\mathbf{S} = \operatorname{curl}\operatorname{curl}\mathbf{A} = \operatorname{curl}\operatorname{curl}\mathbf{H} \tag{3.2.104}$$

This follows from the compatibility relation 3.1.47. Hence, any two stress functions of the Beltrami solution lead to the same stress field **S** if they differ by a strain field.

Looking for a more general solution of the equilibrium equations than that given by the Beltrami solution, we arrive at the Beltrami–Schaefer solution

$$\mathbf{S} = \operatorname{curl}\operatorname{curl}\mathbf{A} + 2\widehat{\nabla}\mathbf{h} - (\operatorname{div}\mathbf{h})\mathbf{1} \tag{3.2.105}$$

where

A is a symmetric tensor field

h is a harmonic vector field on B

To show that **S** given by Equation 3.2.105 satisfies the equilibrium equations, we introduce

$$\mathbf{S}_e = \operatorname{curl}\operatorname{curl}\mathbf{A} \tag{3.2.106}$$

and

$$\mathbf{S}_h = 2\widehat{\nabla}\mathbf{h} - (\operatorname{div}\mathbf{h})\mathbf{1} \tag{3.2.107}$$

* The proof of this statement, based on Stokes' Theorem, is given in Ref. [3].

Clearly, $\mathbf{S}_e$ represents a self-equilibrated stress field, that is

$$\operatorname{div}\mathbf{S}_e = \mathbf{0}, \quad \mathbf{S}_e = \mathbf{S}_e^T \tag{3.2.108}$$

Also, it follows from Equation 3.2.107 that $\mathbf{S}_h$ in components has the form

$$(\mathbf{S}_h)_{ij} = h_{i,j} + h_{j,i} - \delta_{ij}h_{k,k} \tag{3.2.109}$$

Hence

$$(\mathbf{S}_h)_{ij,j} = h_{i,jj} + h_{j,ij} - h_{k,ki} \tag{3.2.110}$$

and, since h_i is harmonic on B

$$h_{i,jj} = 0 \tag{3.2.111}$$

Thus, by Equation 3.2.110, it follows

$$(\mathbf{S}_h)_{ij,j} = 0 \tag{3.2.112}$$

Therefore, from Equations 3.2.109 and 3.2.112 we receive

$$\operatorname{div}\mathbf{S}_h = \mathbf{0}, \quad \mathbf{S}_h = \mathbf{S}_h^T \tag{3.2.113}$$

Now, since by Equations 3.2.105 through 3.2.107

$$\mathbf{S} = \mathbf{S}_e + \mathbf{S}_h \tag{3.2.114}$$

then Equations 3.2.108, 3.2.113, and 3.2.114 imply that the stress field given by Equation 3.2.105 satisfies the equilibrium equations, and the Beltrami–Schaefer solution represents a general solution of elastostatics.

So far in this section we assumed that the body forces vanish. Hence, the Beltrami–Schaefer solution applies to the homogeneous equilibrium equations

$$\operatorname{div}\mathbf{S} = \mathbf{0}, \quad \mathbf{S} = \mathbf{S}^T \tag{3.2.115}$$

If the body force vector field $\mathbf{b}$ does not vanish on B the equilibrium equations take the form

$$\operatorname{div}\mathbf{S} + \mathbf{b} = \mathbf{0}, \quad \mathbf{S} = \mathbf{S}^T \tag{3.2.116}$$

and as a particular solution $\mathbf{S}_p$ to Equation 3.2.116 we may take

$$\mathbf{S}_p = 2\widehat{\nabla\widehat{\mathbf{h}}} - (\operatorname{div}\widehat{\mathbf{h}})\mathbf{1} \tag{3.2.117}$$

where $\widehat{\mathbf{h}}$ satisfies the equation

$$\nabla^2\,\widehat{\mathbf{h}} = -\mathbf{b} \tag{3.2.118}$$

To show that it is so, we write Equation 3.2.117 in components and find

$$(\mathbf{S}_p)_{ij} = \widehat{h}_{i,j} + \widehat{h}_{j,i} - \widehat{h}_{k,k}\delta_{ij} \tag{3.2.119}$$

Hence

$$(\mathbf{S}_p)_{ij,j} = \widehat{h}_{i,jj} \tag{3.2.120}$$

and, by Equation 3.2.118, we arrive at

$$(\mathbf{S}_p)_{ij,j} + b_i = 0 \tag{3.2.121}$$

This equation together with Equation 3.2.117 implies that $\mathbf{S}_p$ satisfies the nonhomogeneous equilibrium equations

$$\text{div}\,\mathbf{S}_p + \mathbf{b} = \mathbf{0}, \quad \mathbf{S}_p = \mathbf{S}_p^T \tag{3.2.122}$$

Example 3.2.11

Show that if $\mathbf{b} = -\nabla\chi$, where χ is a scalar field on B, then a particular solution to the nonhomogeneous equilibrium equations 3.2.122 may be taken in the form

$$\mathbf{S}_p = \chi\mathbf{1} \tag{a}$$

Solution

Equation (a) in components reads

$$(\mathbf{S}_p)_{ij} = \chi\delta_{ij} \tag{b}$$

Hence, we get

$$(\mathbf{S}_p)_{ij} = (\mathbf{S}_p)_{ji} \tag{c}$$

and

$$(\mathbf{S}_p)_{ij,j} = \chi_{,i} \tag{d}$$

This shows that $\mathbf{S}_p$ given by (a) satisfies Equations 3.2.122 with $\mathbf{b} = -\nabla\chi$. □

Example 3.2.12

Show that in a cylinder of arbitrary cross section suspended from the upper end, Figure 3.11, and subject to its own weight ρg, the body force vector $\mathbf{b}$ takes the form

$$\mathbf{b} = [0, 0, -\rho g] \tag{a}$$

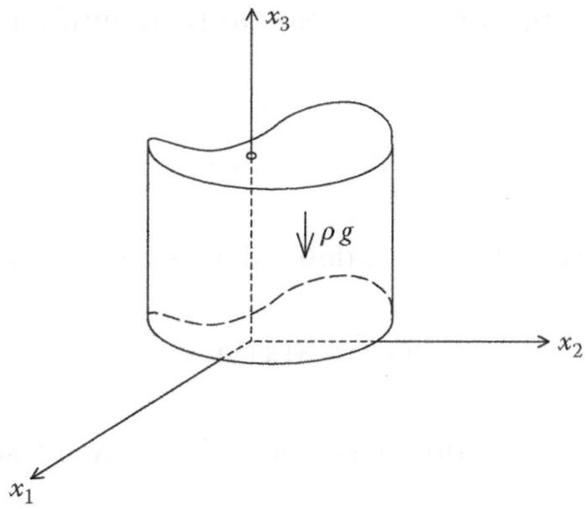

FIGURE 3.11 Cylinder suspended from the upper end subject to its own weight.

and a particular solution $\mathbf{S}_p$ to the equilibrium equations 3.2.122 may be taken as

$$\mathbf{S}_p = \begin{bmatrix} 0 & 0 & 0 \\ 0 & 0 & 0 \\ 0 & 0 & \rho g x_3 \end{bmatrix} \tag{b}$$

Here, ρ is the density of the cylinder and g is the gravity acceleration. The dimensions of ρ and g are

$$[\rho] = [\mathrm{ML}^{-3}] \tag{c}$$

$$[g] = [\mathrm{LT}^{-2}] \tag{d}$$

Solution

It follows from (b) that

$$(\operatorname{div} \mathbf{S}_p)_1 = (\operatorname{div} \mathbf{S}_p)_2 = 0 \quad \text{and} \quad (\operatorname{div} \mathbf{S}_p)_3 = \rho g \tag{e}$$

so Equations 3.2.122 are satisfied. □

We note that this solution can be obtained from Example 3.2.11 by letting $b_1 = b_2 = 0$ and $b_3 = -\chi_{,3}$, where $\chi = \rho g x_3$.

3.2.4 Discussion on the Equations of Equilibrium

In this section, we discuss a number of global results related to solutions of the equilibrium equations

$$\operatorname{div} \mathbf{S} + \mathbf{b} = \mathbf{0}, \quad \mathbf{S} = \mathbf{S}^T \tag{3.2.123}$$

To this end, we introduce the definitions of an *admissible displacement field* **u** and of an *admissible stress field* **S**. By an admissible displacement field **u** we mean a differentiable field on $\bar{B}$, while an admissible stress field **S** is to be identified with a smooth symmetric second-order tensor field on $\bar{B}$, where

$$\bar{B} = B \cup \partial B \tag{3.2.124}$$

The associated *surface force field* is then defined as the stress vector field

$$\mathbf{s}(\mathbf{x}) = \mathbf{S}(\mathbf{x})\,\mathbf{n}(\mathbf{x}) \tag{3.2.125}$$

where $\mathbf{n}(\mathbf{x})$ is the outward unit normal vector to ∂B at $\mathbf{x}$. First, we prove the following theorem:

Theorem: If **S** is an admissible stress field and **u** is an admissible displacement field, then

$$\int_{\partial B} \mathbf{s}\cdot\mathbf{u}\, da = \int_{B} \mathbf{u}\cdot \operatorname{div}\mathbf{S}\, dv + \int_{B} \mathbf{S}\cdot\widehat{\nabla}\mathbf{u}\, dv \tag{3.2.126}$$

■

Proof: Using the divergence theorem and Equation 3.2.125 we obtain

$$\int_{\partial B} \mathbf{s}\cdot\mathbf{u}\, da = \int_{\partial B} S_{ik} n_k u_i\, da = \int_{B} (S_{ik}u_i)_{,k}\, dv \tag{3.2.127}$$

Since

$$(S_{ik}u_i)_{,k} = S_{ik,k}u_i + S_{ik}u_{i,k} \tag{3.2.128}$$

and, by symmetry of **S**,

$$S_{ik}u_{i,k} = S_{ik}u_{(i,k)} \tag{3.2.129}$$

where

$$u_{(i,k)} = (\widehat{\nabla}\mathbf{u})_{ik} \tag{3.2.130}$$

therefore, integrating Equation 3.2.128 over B, and using Equation 3.2.129 we obtain Equation 3.2.126. In Equation 3.2.130

$$u_{(i,k)} = \frac{1}{2}\,(u_{i,k} + u_{k,i}) \tag{3.2.131}$$

which is the notation introduced in Equation 2.1.28.

A direct consequence of the global relation 3.2.126 is the next theorem.

Theorem of Work Expended. If $\mathbf{u}$ is an admissible displacement field, and $\mathbf{S}$ satisfies the equations of equilibrium then

$$\int_{\partial B} \mathbf{s} \cdot \mathbf{u}\, da + \int_{B} \mathbf{b} \cdot \mathbf{u}\, dv = \int_{B} \mathbf{S} \cdot \mathbf{E}\, dv \tag{3.2.132}$$

where $\mathbf{E}$ is the strain field corresponding to the displacement field $\mathbf{u}$.

The relation 3.2.132 is obtained from Equation 3.2.126 by replacing div $\mathbf{S}$ by $-\mathbf{b}$ using Equation 3.2.123. □

From this theorem it follows that the work done by the surface and body forces over the displacement $\mathbf{u}$ is equal to the work done by the stress field over the strain field corresponding to $\mathbf{u}$.

Example 3.2.13

Show that the work expended over a rigid displacement field is equal to zero.

Solution

If $\mathbf{u}$ is a rigid displacement then

$$\mathbf{E} = \widehat{\nabla}\mathbf{u} = \mathbf{0} \tag{a}$$

Compare Example 3.1.6. Hence, by Equation 3.2.132

$$\int_{\partial B} \mathbf{s} \cdot \mathbf{u}\, da + \int_{B} \mathbf{b} \cdot \mathbf{u}\, dv = 0 \tag{b}$$

This relation holds true for any surface force field $\mathbf{s}$ and for any body force $\mathbf{b}$. □

Example 3.2.14

Show that if the relation (b) in Example 3.2.13 is satisfied for any rigid displacement $\mathbf{v}$ then the following balance laws for forces and moments hold true

$$\int_{\partial B} \mathbf{s}\, da + \int_{B} \mathbf{b}\, dv = \mathbf{0} \tag{a}$$

$$\int_{\partial B} \mathbf{x} \times \mathbf{s}\, da + \int_{B} \mathbf{x} \times \mathbf{b}\, dv = \mathbf{0} \tag{b}$$

Solution

An arbitrary rigid displacement $\mathbf{v}$ is represented by

$$\mathbf{v} = \mathbf{a} + \mathbf{W}_0\mathbf{x} \tag{c}$$

where **a** is a vector describing a rigid translation, and $\mathbf{W}_0\mathbf{x}$ describes a rigid rotation, compare Example 3.1.6.

Consider the functional

$$F(\mathbf{v}) = \int_{\partial B} \mathbf{s} \cdot \mathbf{v}\, da + \int_{B} \mathbf{b} \cdot \mathbf{v}\, dv \tag{d}$$

Substituting (c) into (d) we obtain

$$F(\mathbf{v}) = \int_{\partial B} [\mathbf{s} \cdot \mathbf{a} + \mathbf{s} \cdot (\mathbf{W}_0\, \mathbf{x})]\, da + \int_{B} [\mathbf{b} \cdot \mathbf{a} + \mathbf{b} \cdot (\mathbf{W}_0\, \mathbf{x})]\, dv \tag{e}$$

Now, since for any vectors **a** and **c**

$$\mathbf{a} \cdot (\mathbf{W}_0\, \mathbf{c}) = \mathbf{W}_0 \cdot (\mathbf{a} \otimes \mathbf{c}) \tag{f}$$

then (e) can be written as

$$F(\mathbf{v}) = \mathbf{a} \cdot \left[\int_{\partial B} \mathbf{s}\, da + \int_{B} \mathbf{b}\, dv \right] + \mathbf{W}_0 \cdot \left[\int_{\partial B} \mathbf{s} \otimes \mathbf{x}\, da + \int_{B} \mathbf{b} \otimes \mathbf{x}\, dv \right] \tag{g}$$

From (b) in Example 3.2.13 in which $\mathbf{u} = \mathbf{v}$ and (d) in this example follows

$$F(\mathbf{v}) = 0 \tag{h}$$

Since **a** is an arbitrary vector and $\mathbf{W}_0$ is an arbitrary skew tensor, then from (g) and (h) follows that

$$\int_{\partial B} \mathbf{s}\, da + \int_{B} \mathbf{b}\, dv = \mathbf{0} \tag{i}$$

and

$$\text{skw} \left[\int_{\partial B} \mathbf{s} \otimes \mathbf{x}\, da + \int_{B} \mathbf{b} \otimes \mathbf{x}\, dv \right] = \mathbf{0} \tag{j}$$

Equation (i) is identical to (a). As for (j), we write it in components

$$\int_{\partial B} s_{[i} x_{j]}\, da + \int_{B} b_{[i} x_{j]}\, dv = 0 \tag{k}$$

Here we used the notation, see Equation 2.1.29,

$$A_{[ij]} = \frac{1}{2} (A_{ij} - A_{ji}) \tag{l}$$

where **A** is an arbitrary second-order tensor. Multiplying (k) by ϵ_{aij}, and using the identities

$$\epsilon_{aij} s_{[i} x_{j]} = \epsilon_{aij} s_i x_j \tag{m}$$

$$\epsilon_{aij} b_{[i} x_{j]} = \epsilon_{aij} b_i x_j \tag{n}$$

we arrive at

$$\epsilon_{aij} \left[\int_{\partial B} s_i x_j \, da + \int_B b_i x_j \, dv \right] = 0 \tag{o}$$

We note that this is relation (b) written in components, so the solution is complete. □

From this solution it follows that, conversely, if **s** and **b** satisfy the balance laws (a) and (b) then $F(\mathbf{v}) = 0$. This fact together with Examples 3.2.13 and 3.2.14 show that the global balance laws of forces and moments in elastostatics may be replaced by the condition that the work done by **s** and **b** over any rigid displacement vanishes.

We close the discussion on the equations of equilibrium by giving a formula for the mean stress over body B. The *mean stress* $\mathbf{S}_{mean}$ is defined by

$$\mathbf{S}_{mean} = \frac{\int_B \mathbf{S} \, dv}{\int_B dv} \tag{3.2.133}$$

Example 3.2.15

Show that the mean stress corresponding to an admissible stress field **S** that satisfies equations of equilibrium is given by the formula

$$\mathbf{S}_{mean} = \frac{1}{\int_B dv} \operatorname{sym} \left[\int_{\partial B} \mathbf{s} \otimes \mathbf{x} \, da + \int_B \mathbf{b} \otimes \mathbf{x} \, dv \right] \tag{a}$$

Solution

Let **w** be an arbitrary vector field. Introduce the integral

$$\int_{\partial B} \mathbf{w} \otimes \mathbf{s} \, da = \mathbf{T} \tag{b}$$

Since $\mathbf{s} = \mathbf{Sn}$ we obtain in components

$$T_{ij} = \int_{\partial B} w_i s_j \, da = \int_{\partial B} w_i S_{jk} n_k \, da \tag{c}$$

Using the divergence theorem,

$$T_{ij} = \int_B (w_i S_{jk})_{,k} \, dv = \int_B w_{i,k} S_{jk} \, dv + \int_B w_i S_{jk,k} \, dv \tag{d}$$

Since

$$S_{jk,k} = -b_j \tag{e}$$

and $\mathbf{S}$ is symmetric, we arrive at

$$\int_{\partial B} \mathbf{w} \otimes \mathbf{s}\, da + \int_B \mathbf{w} \otimes \mathbf{b}\, dv = \int_B (\nabla \mathbf{w})\mathbf{S}\, dv \tag{f}$$

Taking

$$\mathbf{w} = \mathbf{x} \tag{g}$$

we obtain

$$(\nabla \mathbf{w})\mathbf{S} = \mathbf{S} \tag{h}$$

and (f) reduces to

$$\int_{\partial B} \mathbf{x} \otimes \mathbf{s}\, da + \int_B \mathbf{x} \otimes \mathbf{b}\, dv = \int_B \mathbf{S}\, dv \tag{i}$$

Equation (h) represents a product of two second-order tensors. In components,

$$[(\nabla \mathbf{w})\mathbf{S}]_{ij} = (\nabla \mathbf{w})_{ik} S_{kj} = \delta_{ik} S_{kj} = S_{ij} \tag{j}$$

Finally, dividing (i) by volume of B, using the symmetry of $\mathbf{S}$, and taking the symmetric part of both sides of (i) we arrive at (a). □

It follows from this example that the mean stress $\mathbf{S}_{mean}$ depends only on the associated surface forces and body force fields.

3.2.5 Discussion of the Equations of Motion

In Section 3.2.1, we introduced the concept of the dynamic process $[\mathbf{u}, \mathbf{S}, \mathbf{b}]$ in which $\mathbf{S}$ is a symmetric tensor field and $\mathbf{u}$, $\mathbf{S}$, and $\mathbf{b}$ satisfy the equations of motion

$$\text{div}\, \mathbf{S} + \mathbf{b} = \rho \ddot{\mathbf{u}} \tag{3.2.134}$$

To discuss some global relations for a dynamic process, we now introduce the concept of kinetic energy and of the stress power associated with the process.

We define the *kinetic energy* as

$$K(t) = \frac{1}{2} \int_B \rho \dot{\mathbf{u}}^2\, dv \tag{3.2.135}$$

and the *stress power* as

$$P(t) = \int_B \mathbf{S} \cdot \dot{\mathbf{E}}\, dv \tag{3.2.136}$$

where $\mathbf{E}$ is the strain field associated with $\mathbf{u}$.

With these concepts we formulate the following theorem:

Theorem of Power Expended. If $[\mathbf{u}, \mathbf{S}, \mathbf{b}]$ is a dynamic process, then

$$\int_{\partial B} \mathbf{s} \cdot \dot{\mathbf{u}}\, da + \int_{B} \mathbf{b} \cdot \dot{\mathbf{u}}\, dv = P(t) + \dot{K}(t) \tag{3.2.137}$$

where $\mathbf{s}$ is the stress vector corresponding to $\mathbf{S}$, and $K(t)$ and $P(t)$ are defined by Equations 3.2.135 and 3.2.136. □

Proof: To prove the theorem we note that in components

$$\begin{aligned}\int_{\partial B} \dot{u}_i S_{ij} n_j\, da &= \int_{B} (\dot{u}_i S_{ij})_{,j}\, dv \\ &= \int_{B} \dot{u}_{(i,j)} S_{ij}\, dv + \int_{B} \dot{u}_i S_{ij,j}\, dv \end{aligned} \tag{3.2.138}$$

Since

$$S_{ij,j} = \rho \ddot{u}_i - b_i \tag{3.2.139}$$

and

$$\dot{u}_{(i,j)} = \dot{E}_{ij} \tag{3.2.140}$$

Equation 3.2.138 together with Equations 3.2.135 and 3.2.136 imply Equation 3.2.137. This completes the proof. □

The theorem of power expended asserts that the sum of the rate of working of surface forces and the rate of working of body forces is equal to the sum of the stress power and the rate of change of kinetic energy.

Example 3.2.16

Show that if the stress field $\mathbf{S}$ and strain field $\mathbf{E}$ are independent of position, then the stress power density is given by the formula

$$\mathbf{S} \cdot \dot{\mathbf{E}} = \frac{\int_{\partial B} \mathbf{s} \cdot \dot{\mathbf{u}}\, da}{\int_{B} dv} \tag{a}$$

Here, the stress power density is defined as the integrand of Equation 3.2.136.

Solution

Since $\mathbf{S}$ is time dependent only, the equation of motion reduces to

$$\mathbf{b} = \rho \ddot{\mathbf{u}} \tag{b}$$

Using Equation 3.2.137, and Equations 3.2.135 and 3.2.136, and observing that in Equation 3.2.136 **S** and **E** are time dependent only, we obtain

$$\int_{\partial B} \mathbf{s} \cdot \dot{\mathbf{u}}\, da = (\mathbf{S} \cdot \dot{\mathbf{E}}) \int_{B} dv \tag{c}$$

This result is equivalent to (a). □

This example shows that for a dynamic process for which **S** and **E** are time dependent only, the stress power density is equal to the rate of working of surface forces per unit volume. In this case, a displacement **u** associated with the process is such that $\widehat{\nabla}\mathbf{u} = \mathbf{E}$ is time dependent only and **b** complies with (b).

The next topic in this section deals with finding a displacement field **u** associated with a dynamic process in terms of the stress field **S**, body force vector **b**, and initial conditions imposed on the displacement field **u**. To this end we introduce a so-called pseudo-body force vector **f** by

$$\mathbf{f}(\mathbf{x}, t) = i * \mathbf{b}(\mathbf{x}, t) + \rho(\mathbf{x})[\mathbf{u}_0(\mathbf{x}) + t\dot{\mathbf{u}}_0(\mathbf{x})] \tag{3.2.141}$$

for every $(\mathbf{x}, t) \in \bar{B} \times [0, \infty)$ where

$$i = i(t) = t \quad t \geq 0 \tag{3.2.142}$$

and $\mathbf{u}_0$ and $\dot{\mathbf{u}}_0$ are prescribed fields on B.

The formula for **u** in terms of **S**, **b**, and the initial conditions is contained in the following theorem:

Theorem: An array of functions $[\mathbf{u}, \mathbf{S}, \mathbf{b}]$ is a dynamic process consistent with the initial conditions

$$\mathbf{u}(\mathbf{x}, 0) = \mathbf{u}_0(\mathbf{x}), \quad \dot{\mathbf{u}}(\mathbf{x}, 0) = \dot{\mathbf{u}}_0(\mathbf{x}) \quad \text{for } x \in \bar{B} \tag{3.2.143}$$

if and only if

$$i * \operatorname{div} \mathbf{S} + \mathbf{f} = \rho \mathbf{u} \quad \text{on } \bar{B} \times [0, \infty) \tag{3.2.144}$$

■

Proof: The proof consists of two parts.

Part 1. We assume that $[\mathbf{u}, \mathbf{S}, \mathbf{b}]$ is a dynamic process consistent with conditions 3.2.143 and prove that Equation 3.2.144 is satisfied.

Part 2. We assume that Equation 3.2.144 is satisfied and we show that $[\mathbf{u}, \mathbf{S}, \mathbf{b}]$ is a dynamic process consistent with conditions 3.2.143.

Proof of Part 1. Since $[\mathbf{u}, \mathbf{S}, \mathbf{b}]$ is a dynamic process, $\mathbf{u}$, $\mathbf{S}$, and $\mathbf{b}$ are related by the equation of motion

$$\operatorname{div} \mathbf{S} + \mathbf{b} = \rho \ddot{\mathbf{u}} \tag{3.2.145}$$

By applying the Laplace transform to Equation 3.2.145 and using the initial conditions 3.2.143 we find

$$\operatorname{div} \bar{\mathbf{S}} + \bar{\mathbf{b}} = \rho (p^2 \bar{\mathbf{u}} - p\mathbf{u}_0 - \dot{\mathbf{u}}_0) \tag{3.2.146}$$

where the overbar means the Laplace transform. The Laplace transform is defined by

$$\bar{f}(\mathbf{x}, p) = \int_0^{\infty} e^{-pt} f(\mathbf{x}, t)\, dt \tag{3.2.147}$$

where $f(\mathbf{x}, t)$ is a function on $\bar{B} \times [0, \infty)$.

Now, by applying the Laplace transform to Equation 3.2.141 and using the convolution theorem we arrive at

$$\bar{f}(\mathbf{x}, p) = \frac{1}{p^2}\, \bar{\mathbf{b}}(\mathbf{x}, p) + \rho(\mathbf{x}) \left[\frac{\mathbf{u}_0(\mathbf{x})}{p} + \frac{\dot{\mathbf{u}}_0(\mathbf{x})}{p^2} \right] \tag{3.2.148}$$

Using this equation we rewrite Equation 3.2.146, after dividing it by p^2, in the form

$$\frac{1}{p^2} \operatorname{div} \bar{\mathbf{S}} + \bar{\mathbf{f}} = \rho \bar{\mathbf{u}} \tag{3.2.149}$$

Taking the inverse Laplace transform of Equation 3.2.149 we arrive at Equation 3.2.144. This completes the proof of Part 1.

Proof of Part 2. To show that $[\mathbf{u}, \mathbf{S}, \mathbf{b}]$ is a dynamic process provided that Equation 3.2.144 is satisfied, we differentiate Equation 3.2.144 twice with respect to time and taking into account the definition of pseudo-body force $\mathbf{f}$ as well as the relation that for any function $g = g(\mathbf{x}, t)$

$$\frac{\partial^2}{\partial t^2} (i * g) = g \tag{3.2.150}$$

we obtain the equation of motion

$$\operatorname{div} \mathbf{S} + \mathbf{b} = \rho \ddot{\mathbf{u}} \tag{3.2.151}$$

Hence, $[\mathbf{u}, \mathbf{S}, \mathbf{b}]$ is a dynamic process.

The proof of Equation 3.2.150 is immediate. Taking the Laplace transform of Equation 3.2.150 leads to

$$\overline{(i * g)} p^2 - (i * g)|_{t=0}\, p - \overbrace{(i * g)}^{\cdot}|_{t=0} = \bar{g} \tag{3.2.152}$$

Now, since

$$\overline{i * g} = \bar{i}\,\bar{g}, \quad \bar{i} = p^{-2}, \quad (i * g)|_{t=0} = 0, \quad \text{and} \quad \overset{\cdot}{\overbrace{(i * g)}}|_{t=0} = 0 \tag{3.2.153}$$

we arrive at the identity

$$\bar{g} = \bar{g} \tag{3.2.154}$$

Going back from Equation 3.2.154 to Equation 3.2.152 and applying the inverse Laplace transform to Equation 3.2.152 we obtain Equation 3.2.150.

To show that $[\mathbf{u}, \mathbf{S}, \mathbf{b}]$ is consistent with Equation 3.2.143 we note that Equation 3.2.144 is satisfied on $\bar{B} \times [0, \infty)$. Hence, taking Equation 3.2.144 at $t = 0$ and using the definition 3.2.141 of $\mathbf{f}$ we obtain

$$\mathbf{u}(\mathbf{x}, 0) = \mathbf{u}_0(\mathbf{x}) \tag{3.2.155}$$

Next, by differentiating Equation 3.2.144 with respect to time and taking the result at $t = 0$ we receive

$$\dot{\mathbf{u}}(\mathbf{x}, 0) = \dot{\mathbf{u}}_0(\mathbf{x}) \tag{3.2.156}$$

Therefore, $[\mathbf{u}, \mathbf{S}, \mathbf{b}]$ is consistent with Equation 3.2.143. In arriving at Equation 3.2.155 and Equation 3.2.156, we used the obvious identities that

$$(i * g)|_{t=0} = 0 \quad \text{and} \quad (1 * g)|_{t=0} = 0 \tag{3.2.157}$$

for any function $g = g(\mathbf{x}, t)$. Also, we note that

$$\frac{\partial}{\partial t}(i * g) = 1 * g \tag{3.2.158}$$

This completes the proof of Part 2. □

We make two observations regarding the theorem.

First, the theorem asserts that the equation of motion subject to initial condition 3.2.143 can be reduced to an alternative integro-differential equation 3.2.144 in which the initial data are incorporated.

Second, the theorem provides a representation for the displacement vector $\mathbf{u}$ in terms of the stress field $\mathbf{S}$ (see Equation 3.2.144). If $\mathbf{S}$ is a function of strain tensor $\mathbf{E}$, the formula 3.2.144 gives a representation of the displacement field $\mathbf{u}$ in terms of $\mathbf{E}$ that is an alternative to the displacement line integral representation of elastostatics from Cesàro [1], see Equation 3.1.53. The equivalence of these two representations leads to a dynamic form of compatibility conditions that results in a pure stress equation of motion of elastodynamics to be discussed in Section 4.2.1.

In the following a dynamic process $[\mathbf{u}, \mathbf{S}, \mathbf{b}]$ that appears in the theorem involving the representation of $\mathbf{u}$ in terms of $\mathbf{S}$ is to be called a *dynamic process corresponding to the pseudo-body force field* $\mathbf{f}$. With this definition we present the following theorem:

Theorem: If $[\mathbf{u}, \mathbf{S}, \mathbf{b}]$ is a dynamic process corresponding to pseudo-body force field $\mathbf{f}$, then

$$i * \int_{\partial B} \mathbf{s} * \mathbf{u}\, da + \int_B \mathbf{f} * \mathbf{u}\, dv = i * \int_B \mathbf{S} * \mathbf{E}\, dv + \int_B \rho \mathbf{u} * \mathbf{u}\, dv \tag{3.2.159}$$

■

Proof: First, we show that for any vector field $\mathbf{w} = \mathbf{w}(\mathbf{x}, t)$ and a symmetric tensor field $\mathbf{S} = \mathbf{S}(\mathbf{x}, t)$ the integral relation

$$\int_{\partial B} \mathbf{s} * \mathbf{w}\, da = \int_B \mathbf{w} * \operatorname{div} \mathbf{S}\, dv + \int_B \mathbf{S} * \widehat{\nabla} \mathbf{w}\, dv \tag{3.2.160}$$

where

$$\mathbf{s} = \mathbf{s}(\mathbf{x}, t) = \mathbf{S}(\mathbf{x}, t)\mathbf{n}(\mathbf{x}) \tag{3.2.161}$$

is satisfied. To this end we write Equation 3.2.160 in components

$$\int_{\partial B} s_i * w_i\, da = \int_B w_i * S_{ik,k}\, dv + \int_B S_{ik} * w_{(i,k)}\, dv \tag{3.2.162}$$

and note that by Equation 3.2.161 and the divergence theorem

$$\begin{aligned} \int_{\partial B} s_i * w_i\, da &= \int_{\partial B} S_{ik} n_k * w_i\, da = \int_B (S_{ik} * w_i)_{,k}\, dv \\ &= \int_B w_i * S_{ik,k}\, dv + \int_B S_{ik} * w_{(i,k)}\, dv \end{aligned} \tag{3.2.163}$$

This shows that the integral relation 3.2.160 holds true.

Second, we apply to Equation 3.2.160 the operator $i*$, put $\mathbf{w} = \mathbf{u}$ and write

$$i * \int_{\partial B} \mathbf{s} * \mathbf{u}\, da = i * \int_B \mathbf{u} * \operatorname{div} \mathbf{S}\, dv + i * \int_B \mathbf{S} * \widehat{\nabla} \mathbf{u}\, dv \tag{3.2.164}$$

Since $[\mathbf{u}, \mathbf{S}, \mathbf{b}]$ is a dynamic process corresponding to $\mathbf{f}$, by Equation 3.2.144

$$i * \operatorname{div} \mathbf{S} = \rho \mathbf{u} - \mathbf{f} \tag{3.2.165}$$

Substituting Equation 3.2.165 into Equation 3.2.164 we obtain Equation 3.2.159. Note that in passing from Equation 3.2.164 to Equation 3.2.159 we observed that

$$\mathbf{E} = \widehat{\nabla}\mathbf{u} \tag{3.2.166}$$

This completes the proof of the theorem. □

The relation 3.2.159 is an analogue to the relation 3.2.132 that appears in the theorem of work expended of elastostatics.

Finally, we present the following theorem.

Reciprocal Theorem. If $[\mathbf{u}, \mathbf{S}, \mathbf{b}]$ and $[\widetilde{\mathbf{u}}, \widetilde{\mathbf{S}}, \widetilde{\mathbf{b}}]$ are dynamic processes corresponding to the pseudo-body forces $\mathbf{f}$ and $\widetilde{\mathbf{f}}$, respectively, then the following reciprocal relations are satisfied:

$$i * \int_{\partial B} \mathbf{s} * \widetilde{\mathbf{u}}\, da + \int_B \mathbf{f} * \widetilde{\mathbf{u}}\, dv - i * \int_B \mathbf{S} * \widetilde{\mathbf{E}}\, dv$$
$$= i * \int_{\partial B} \widetilde{\mathbf{s}} * \mathbf{u}\, da + \int_B \widetilde{\mathbf{f}} * \mathbf{u}\, dv - i * \int_B \widetilde{\mathbf{S}} * \mathbf{E}\, dv \tag{3.2.167}$$

and

$$\int_{\partial B} \mathbf{s} * \widetilde{\mathbf{u}}\, da + \int_B \mathbf{b} * \widetilde{\mathbf{u}}\, dv - \int_B \mathbf{S} * \widetilde{\mathbf{E}}\, dv + \int_B \rho(\mathbf{u}_0 \cdot \dot{\widetilde{\mathbf{u}}} + \dot{\mathbf{u}}_0 \cdot \widetilde{\mathbf{u}})\, dv$$
$$= \int_{\partial B} \widetilde{\mathbf{s}} * \mathbf{u}\, da + \int_B \widetilde{\mathbf{b}} * \mathbf{u}\, dv - \int_B \widetilde{\mathbf{S}} * \mathbf{E}\, dv + \int_B \rho(\widetilde{\mathbf{u}}_0 \cdot \dot{\mathbf{u}} + \dot{\widetilde{\mathbf{u}}}_0 \cdot \mathbf{u})\, dv \tag{3.2.168}$$

Proof: Applying the operator $i*$ to Equation 3.2.160 we write

$$i * \int_{\partial B} \mathbf{s} * \mathbf{w}\, da = i * \int_B \mathbf{w} * \operatorname{div}\mathbf{S}\, dv + i * \int_B \mathbf{S} * \widehat{\nabla}\mathbf{w}\, dv \tag{3.2.169}$$

Since $[\mathbf{u}, \mathbf{S}, \mathbf{b}]$ is a dynamic process corresponding to $\mathbf{f}$, then

$$i * \operatorname{div}\mathbf{S} = \rho\mathbf{u} - \mathbf{f} \tag{3.2.170}$$

Hence, Equation 3.2.169 reduces to

$$i * \int_{\partial B} \mathbf{s} * \mathbf{w}\, da + \int_B \mathbf{f} * \mathbf{w}\, dv = i * \int_B \mathbf{S} * \widehat{\nabla}\mathbf{w}\, dv + \int_B \rho\mathbf{u} * \mathbf{w}\, dv \tag{3.2.171}$$

Letting $\mathbf{w} = \widetilde{\mathbf{u}}$ in this equation we obtain

$$i * \int_{\partial B} \mathbf{s} * \widetilde{\mathbf{u}}\, da + \int_B \mathbf{f} * \widetilde{\mathbf{u}}\, dv - i * \int_B \mathbf{S} * \widetilde{\mathbf{E}}\, dv = \int_B \rho\mathbf{u} * \widetilde{\mathbf{u}}\, dv \tag{3.2.172}$$

Since $[\widetilde{\mathbf{u}}, \widetilde{\mathbf{S}}, \widetilde{\mathbf{b}}]$ is a dynamic process corresponding to $\widetilde{\mathbf{f}}$, Equation 3.2.171 holds true if $\mathbf{u}, \mathbf{s}, \mathbf{f}$, and $\mathbf{S}$ are replaced by $\widetilde{\mathbf{u}}, \widetilde{\mathbf{s}}, \widetilde{\mathbf{f}}$, and $\widetilde{\mathbf{S}}$, respectively. We then find

$$i * \int_{\partial B} \widetilde{\mathbf{s}} * \mathbf{w}\, da + \int_{B} \widetilde{\mathbf{f}} * \mathbf{w}\, dv - i * \int_{B} \widetilde{\mathbf{S}} * \widehat{\nabla} \mathbf{w}\, dv = \int_{B} \rho \widetilde{\mathbf{u}} * \mathbf{w}\, dv \tag{3.2.173}$$

Putting $\mathbf{w} = \mathbf{u}$ in Equation 3.2.173 we obtain

$$i * \int_{\partial B} \widetilde{\mathbf{s}} * \mathbf{u}\, da + \int_{B} \widetilde{\mathbf{f}} * \mathbf{u}\, dv - i * \int_{B} \widetilde{\mathbf{S}} * \mathbf{E}\, dv = \int_{B} \rho \widetilde{\mathbf{u}} * \mathbf{u}\, dv \tag{3.2.174}$$

Since the RHS of Equations 3.2.174 and 3.2.172 are the same, we arrive at Equation 3.2.167.

To show that Equation 3.2.168 is satisfied we differentiate twice with respect to time both sides of Equation 3.2.167, use the definition of the pseudo-body force $\mathbf{f}$ and $\widetilde{\mathbf{f}}$, and arrive at Equation 3.2.168. This completes the proof of Equation 3.2.168.

For the homogeneous initial conditions the relation 3.2.168 reduces to

$$\begin{aligned} &\int_{\partial B} \mathbf{s} * \widetilde{\mathbf{u}}\, da + \int_{B} \mathbf{b} * \widetilde{\mathbf{u}}\, dv - \int_{B} \mathbf{S} * \widetilde{\mathbf{E}}\, dv \\ &\quad = \int_{\partial B} \widetilde{\mathbf{s}} * \mathbf{u}\, da + \int_{B} \widetilde{\mathbf{b}} * \mathbf{u}\, dv - \int_{B} \widetilde{\mathbf{S}} * \mathbf{E}\, dv \end{aligned} \tag{3.2.175}$$

Equation 3.2.175 will be utilized in the proof of the Graffi reciprocal theorem which is a dynamic counterpart of the Betti reciprocal theorem for elastostatics.

3.3 CONSTITUTIVE RELATIONS

3.3.1 General Anisotropic Body

In Sections 3.1 and 3.2, we discussed the strain–displacement relations and the laws of momentum balance for a body of which physical properties have not been specified. Although in these laws a notion of a stress field appears, we have not considered the relations that connect stress $\mathbf{S}$ and strain $\mathbf{E}$.

In 1660, Robert Hooke discovered a law that now bears his name. He published it in the form of an anagram *ceiiinosssttuv* in 1676. We do not know if anybody solved the anagram. Hooke published the law in the words *Ut tensio sic vis* in *De Potentia restitutiva*, in London, in 1678. The close translation is *As tension, so force,* or, in an extended form, *the extension of a body is proportional to the force applied to it.* Obviously, Hooke's law applies to elastic bodies and, in its original formulation, refers to a one-dimensional extension. The law has been generalized to a three-dimensional case. The generalization leads to the following definition of a linearly elastic body:

A body B is said to be *linearly elastic* if for every point $\mathbf{x} \in B$ there is a linear transformation C from the space of all symmetric tensors $\mathbf{E}$ into the space of all symmetric tensors $\mathbf{S}$, or

$$\mathbf{S} = \mathsf{C}[\mathbf{E}] \tag{3.3.1}$$

In components this relation takes the form

$$S_{ij} = C_{ijkl}E_{kl} \tag{3.3.2}$$

The tensor $\mathsf{C} = \mathsf{C}(\mathbf{x})$ is called the *elasticity tensor field* on B. We recall, see Equations 2.1.98 and 2.1.99, the formulas for components of C in terms of the basis vectors $\{\mathbf{e}_i\}$

$$C_{ijkl} = \mathbf{e}_i \cdot \mathsf{C}[\mathbf{e}_k \otimes \mathbf{e}_l]\mathbf{e}_j \tag{3.3.3}$$

or, alternatively,

$$C_{ijkl} = (\mathbf{e}_i \otimes \mathbf{e}_j) \cdot \mathsf{C}[\mathbf{e}_k \otimes \mathbf{e}_l] \tag{3.3.4}$$

From the symmetry of **S** and **E** it follows that

$$C_{ijkl} = C_{jikl} = C_{ijlk} \tag{3.3.5}$$

Because of the symmetry of Equation 3.3.5, from among 81 components C_{ijkl} only 36 are different.

It is assumed that the elasticity tensor field C in Equation 3.3.1 is invertible. This means that there is a fourth-order tensor field $\mathsf{K} = \mathsf{K}(\mathbf{x})$ such that

$$\mathsf{K} = \mathsf{C}^{-1} \tag{3.3.6}$$

Therefore, an equivalent form of Equation 3.3.1 is

$$\mathbf{E} = \mathsf{K}[\mathbf{S}] \tag{3.3.7}$$

The tensor K is called the *compliance tensor field.* A body is called *homogeneous* if its density ρ and its elasticity tensor are independent of $\mathbf{x}$. A body that is not homogeneous is called *inhomogeneous*.

Apart from the symmetry relations given by Equation 3.3.5, we introduce the following definition of symmetry of the fourth-order tensor C: the tensor C is symmetric if and only if

$$\mathbf{A} \cdot \mathsf{C}[\mathbf{B}] = \mathbf{B} \cdot \mathsf{C}[\mathbf{A}] \tag{3.3.8}$$

for every $\mathbf{A} = \mathbf{A}^T$ and $\mathbf{B} = \mathbf{B}^T$.
In components,

$$A_{ij}C_{ijkl}B_{kl} = B_{ij}C_{ijkl}A_{kl} \tag{3.3.9}$$

From this relation, replacing on the RHS of Equation 3.3.9 i by k and j by l, we obtain

$$(C_{ijkl} - C_{klij})A_{ij}B_{kl} = 0 \tag{3.3.10}$$

Hence, due to the arbitrariness of **A** and **B** the definition of symmetry of **C** can be replaced by the relation

$$C_{ijkl} = C_{klij} \tag{3.3.11}$$

If the elasticity tensor **C** satisfies the symmetry relation 3.3.11, there is a further reduction of the number of different components C_{ijkl} from 36 to 21. As will be shown in Section 3.3.4, the symmetry of **C** is equivalent to the existence of a stored energy function for a body.

Another useful definition associated with tensor **C** is the definition of a positive semidefinite **C** and a positive definite **C**. We say that **C** is

(a) Positive semidefinite if

$$\mathbf{A} \cdot \mathsf{C}[\mathbf{A}] \geq 0 \quad \text{for every } \mathbf{A} = \mathbf{A}^T \neq \mathbf{0} \tag{3.3.12}$$

(b) Positive definite if

$$\mathbf{A} \cdot \mathsf{C}[\mathbf{A}] > 0 \quad \text{for every } \mathbf{A} = \mathbf{A}^T \neq \mathbf{0} \tag{3.3.13}$$

In general, engineering materials in elastic range of deformation may be described by an elasticity tensor satisfying the properties 3.3.5, 3.3.11, 3.3.12, and 3.3.13.

It may be shown that the positive definiteness of tensor **C** described by Equation 3.3.13 means invertibility of **C**. So, if **C** satisfies the relations 3.3.5 through 3.3.13, the tensor $\mathsf{K} = \mathsf{C}^{-1}$ (see Equation 3.3.6) satisfies similar symmetry relations

$$K_{ijkl} = K_{jikl} = K_{ijlk} \tag{3.3.14}$$

$$K_{ijkl} = K_{klij} \tag{3.3.15}$$

and

$$A_{ij}K_{ijkl}A_{kl} > 0 \quad \text{for every } A_{ij} = A_{ji} \neq 0 \tag{3.3.16}$$

Also, Equations 3.3.14 through 3.3.16 imply that in general there are 21 different components K_{ijkl}.

By an *anisotropic elastic body* we mean a body for which the tensor **C** possesses in general 21 different components. In this case the general constitutive relation 3.3.1 written in a matrix form is

$$\begin{Bmatrix} S_{11} \\ S_{22} \\ S_{33} \\ S_{23} \\ S_{31} \\ S_{12} \end{Bmatrix} = \begin{bmatrix} C_{1111} & C_{1122} & C_{1133} & C_{1123} & C_{1131} & C_{1112} \\ & C_{2222} & C_{2233} & C_{2223} & C_{2231} & C_{2212} \\ & & C_{3333} & C_{3323} & C_{3331} & C_{3312} \\ & \text{sym} & & C_{2323} & C_{2331} & C_{2312} \\ & & & & C_{3131} & C_{3112} \\ & & & & & C_{1212} \end{bmatrix} \begin{Bmatrix} E_{11} \\ E_{22} \\ E_{33} \\ 2E_{23} \\ 2E_{31} \\ 2E_{12} \end{Bmatrix} \tag{3.3.17}$$

Equation 3.3.17 may be written in a compact form

$$[S] = [C^*][E] \tag{3.3.18}$$

where the meaning of the column matrices $[S]$ and $[E]$ and of the symmetric 6×6 matrix $[C^*]$ is clear from Equation 3.3.17.

Equation 3.3.18 describes an anisotropic elastic body that behaves differently in different directions. Depending on the material symmetry, we obtain a number of particular cases that are important in physics and engineering. We will now discuss three such cases.

1. *Orthotropic material*. This is a material that possesses three mutually orthogonal planes of symmetry at each point. For such a material the matrix $[C^*]$ contains nine independent components and is given by

$$[C^*] = \begin{bmatrix} C_{1111} & C_{1122} & C_{1133} & 0 & 0 & 0 \\ & C_{2222} & C_{2233} & 0 & 0 & 0 \\ & & C_{3333} & 0 & 0 & 0 \\ & \text{sym} & & C_{2323} & 0 & 0 \\ & & & & C_{1313} & 0 \\ & & & & & C_{1212} \end{bmatrix} \tag{3.3.19}$$

   An example of an orthotropic material is wood with different properties along the grains and across the grains. Another example of an orthotropic material is a sheet of rolled steel.
2. A special case of an orthotropic material is a *transversely isotropic material*. This is a material for which there exists one plane where elastic properties are the same in any direction. In this case the matrix $[C^*]$ has five different components

$$[C^*] = \begin{bmatrix} C_{1111} & C_{1122} & C_{1133} & 0 & 0 & 0 \\ & C_{1111} & C_{1133} & 0 & 0 & 0 \\ & & C_{3333} & 0 & 0 & 0 \\ & \text{sym} & & C_{1313} & 0 & 0 \\ & & & & C_{1313} & 0 \\ & & & & & G \end{bmatrix} \tag{3.3.20}$$

   where

$$G = \frac{1}{2}\left(C_{1111} - C_{1122}\right) \tag{3.3.21}$$

   An example of a transversely isotropic material is a fibrous composite with elastic properties in the direction of the fibers different than those in planes perpendicular to the fibers.
3. *Isotropic material*. In this case the material properties are independent of a direction. The matrix $[C^*]$ for this material can be obtained from Equation 3.3.20. It takes the form

$$[C^*] = \begin{bmatrix} C_{1111} & C_{1122} & C_{1122} & 0 & 0 & 0 \\ & C_{1111} & C_{1122} & 0 & 0 & 0 \\ & & C_{1111} & 0 & 0 & 0 \\ & \text{sym} & & G & 0 & 0 \\ & & & & G & 0 \\ & & & & & G \end{bmatrix} \tag{3.3.22}$$

where, again

$$G = \frac{1}{2}\left(C_{1111} - C_{1122}\right) \tag{3.3.23}$$

Hence, in the isotropic material there are only two different components of matrix $[C^*]$. Many engineering materials, such as metals, are considered isotropic.

Since matrix $[C^*]$ is invertible, we may write an equivalent relation to Equation 3.3.18 in the form, see Equation 3.3.7,

$$[E] = [K^*][S] \tag{3.3.24}$$

which in a general anisotropic case is equivalent to

$$\begin{Bmatrix} E_{11} \\ E_{22} \\ E_{33} \\ 2E_{23} \\ 2E_{31} \\ 2E_{12} \end{Bmatrix} = \begin{bmatrix} K_{1111} & K_{1122} & K_{1133} & 2K_{1123} & 2K_{1131} & 2K_{1112} \\ & K_{2222} & K_{2233} & 2K_{2223} & 2K_{2231} & 2K_{2212} \\ & & K_{3333} & 2K_{3323} & 2K_{3331} & 2K_{3312} \\ & \text{sym} & & 4K_{2323} & 4K_{2331} & 4K_{2312} \\ & & & & 4K_{3131} & 4K_{3112} \\ & & & & & 4K_{1212} \end{bmatrix} \begin{Bmatrix} S_{11} \\ S_{22} \\ S_{33} \\ S_{23} \\ S_{31} \\ S_{12} \end{Bmatrix} \tag{3.3.25}$$

where K_{ijkl} are components of the compliance tensor, see Equation 3.3.6.

In a particular case of an *orthotropic material* the matrix $[K^*]$ reduces to

$$[K^*] = \begin{bmatrix} K_{1111} & K_{1122} & K_{1133} & 0 & 0 & 0 \\ & K_{2222} & K_{2233} & 0 & 0 & 0 \\ & & K_{3333} & 0 & 0 & 0 \\ & \text{sym} & & 4K_{2323} & 0 & 0 \\ & & & & 4K_{3131} & 0 \\ & & & & & 4K_{1212} \end{bmatrix} \tag{3.3.26}$$

with nine different components.

For a transversely isotropic body we obtain

$$[K^*] = \begin{bmatrix} K_{1111} & K_{1122} & K_{1122} & 0 & 0 & 0 \\ & K_{2222} & K_{2233} & 0 & 0 & 0 \\ & & K_{2222} & 0 & 0 & 0 \\ & \text{sym} & & 4K_{2323} & 0 & 0 \\ & & & & 4K_{3131} & 0 \\ & & & & & 4K_{3131} \end{bmatrix} \tag{3.3.27}$$

with five different components.

For an isotropic body we obtain

$$[K^*] = \begin{bmatrix} K_{1111} & K_{1122} & K_{1122} & 0 & 0 & 0 \\ & K_{1111} & K_{1122} & 0 & 0 & 0 \\ & & K_{1111} & 0 & 0 & 0 \\ & \text{sym} & & 4K_{2323} & 0 & 0 \\ & & & & 4K_{2323} & 0 \\ & & & & & 4K_{2323} \end{bmatrix} \tag{3.3.28}$$

with two different components, because K_{2323} may be expressed by K_{1111} and K_{1122}.

3.3.2 Engineering Material Constants

In particular cases of orthotropic, transversely isotropic, and isotropic material, the matrix $[K^*]$ given by Equations 3.3.26, 3.3.27, and 3.3.28, respectively, may be expressed in terms of the so called *engineering constants*, which give a physical interpretation to the components of the compliance tensor **K**.

For the orthotropic case the matrix $[K^*]$ is

$$[K^*] = \begin{bmatrix} \frac{1}{E_1} & \frac{-\nu_{21}}{E_2} & \frac{-\nu_{31}}{E_3} & 0 & 0 & 0 \\ & \frac{1}{E_2} & \frac{-\nu_{32}}{E_3} & 0 & 0 & 0 \\ & & \frac{1}{E_3} & 0 & 0 & 0 \\ & \text{sym} & & \frac{1}{G_{23}} & 0 & 0 \\ & & & & \frac{1}{G_{13}} & 0 \\ & & & & & \frac{1}{G_{12}} \end{bmatrix} \tag{3.3.29}$$

For the transversely isotropic case the matrix $[K^*]$ is

$$[K^*] = \begin{bmatrix} \frac{1}{E_1} & \frac{-\nu_{12}}{E_1} & \frac{-\nu_{12}}{E_1} & 0 & 0 & 0 \\ & \frac{1}{E_2} & \frac{-\nu_{23}}{E_2} & 0 & 0 & 0 \\ & & \frac{1}{E_2} & 0 & 0 & 0 \\ & \text{sym} & & \frac{1}{G_{23}} & 0 & 0 \\ & & & & \frac{1}{G_{12}} & 0 \\ & & & & & \frac{1}{G_{12}} \end{bmatrix} \tag{3.3.30}$$

where

$$\frac{1}{G_{23}} = \frac{2(1+\nu_{23})}{E_2} \tag{3.3.31}$$

Finally, for the isotropic case the matrix $[K^*]$ is

$$[K^*] = \begin{bmatrix} \frac{1}{E} & \frac{-\nu}{E} & \frac{-\nu}{E} & 0 & 0 & 0 \\ & \frac{1}{E} & \frac{-\nu}{E} & 0 & 0 & 0 \\ & & \frac{1}{E} & 0 & 0 & 0 \\ & \text{sym} & & \frac{1}{G} & 0 & 0 \\ & & & & \frac{1}{G} & 0 \\ & & & & & \frac{1}{G} \end{bmatrix} \tag{3.3.32}$$

where

$$\frac{1}{G} = \frac{2(1+\nu)}{E} \tag{3.3.33}$$

To give the physical interpretation of the constants in the matrices 3.3.29, 3.3.30, and 3.3.32, we start with the isotropic case represented by Equation 3.3.32. As follows from Equations 3.3.24 and 3.3.32, Hooke's law reads

$$\begin{aligned} E_{11} &= \frac{1}{E}\left[S_{11} - \nu(S_{22}+S_{33})\right] \\ E_{22} &= \frac{1}{E}\left[S_{22} - \nu(S_{33}+S_{11})\right] \\ E_{33} &= \frac{1}{E}\left[S_{33} - \nu(S_{11}+S_{22})\right] \\ E_{12} &= \frac{1}{2G}S_{12}, \quad E_{23} = \frac{1}{2G}S_{23}, \quad E_{31} = \frac{1}{2G}S_{31} \end{aligned} \tag{3.3.34}$$

To interpret the two constants E and ν we consider simple tension or compression, and pure shear.

If a body is under simple tension or compression with tensile stress σ in the x_1 direction then the stress tensor **S** is given by

$$\mathbf{S} = \begin{bmatrix} \sigma & 0 & 0 \\ 0 & 0 & 0 \\ 0 & 0 & 0 \end{bmatrix} \tag{3.3.35}$$

and from 3.3.34 we obtain

$$\mathbf{E} = \begin{bmatrix} E_{11} & 0 & 0 \\ 0 & E_{22} & 0 \\ 0 & 0 & E_{33} \end{bmatrix} \quad (3.3.36)$$

where

$$\begin{aligned} E_{11} &= \frac{1}{E} S_{11} \\ E_{22} &= -\frac{\nu}{E} S_{11} = -\nu E_{11} \\ E_{33} &= E_{22} = -\nu E_{11} \\ E_{12} &= E_{13} = E_{31} = 0 \end{aligned} \quad (3.3.37)$$

Therefore, from the first equation of 3.3.37, the constant E is equal to the ratio of S_{11} and E_{11} and it is called *Young's modulus* or the *modulus of elasticity*. Clearly, from

$$S_{11} = EE_{11} \quad (3.3.38)$$

Young's modulus E is a coefficient of proportionality between the strain and the stress in simple tension or compression. As a bar is extended when acted by a tensile stress and shortened when acted by a compressive stress, Equation 3.3.38 suggests that E must be positive. Since E_{11} is dimensionless, E has the dimension of S_{11}

$$[S_{11}] = [\text{Force} \times \text{Length}^{-2}] \quad (3.3.39)$$

As for ν in the second equation of 3.3.37, we note that

$$\nu = -\frac{E_{22}}{E_{11}} = -\frac{E_{33}}{E_{11}} \quad (3.3.40)$$

Hence, the constant ν is the ratio of the lateral contraction to the longitudinal strain of a bar under pure tension. By the contraction we mean the negative strain equal to $-E_{22}$ in the direction perpendicular to tension. The dimensionless constant ν is called *Poisson's ratio*. As a tensile stress should produce a contraction in the direction perpendicular to it, Equation 3.3.40 implies that ν must be positive.

To interpret the constant G in the last three equations of Equations 3.3.34, we consider the case of *pure shear* in which the stress tensor $\mathbf{S}$ is given by

$$\mathbf{S} = \begin{bmatrix} 0 & \tau & 0 \\ \tau & 0 & 0 \\ 0 & 0 & 0 \end{bmatrix} \quad (3.3.41)$$

where τ is the magnitude of shear. Then, from Equations 3.3.34 we obtain

$$\mathbf{E} = \begin{bmatrix} 0 & E_{12} & 0 \\ E_{21} & 0 & 0 \\ 0 & 0 & 0 \end{bmatrix} \tag{3.3.42}$$

where

$$E_{12} = E_{21} = \frac{\tau}{2G} \tag{3.3.43}$$

We note that the strain tensor $\mathbf{E}$ represents a pure shear of the amount $\tau/(2G)$ with respect to the x_1 and x_2 directions, and since

$$S_{12} = \tau = 2GE_{12} \tag{3.3.44}$$

the constant $2G$ is a coefficient of proportionality between S_{12} and E_{12}. The constant G is called the *shear modulus*. Since a positive shear stress produces a positive shear strain, Equation 3.3.44 suggests that G must be positive. The dimension of G is the same as of stress

$$[G] = [\text{Force} \times \text{Length}^{-2}] \tag{3.3.45}$$

Apart from the engineering constants E, ν, and G, another engineering constant called *bulk modulus* will be introduced. To this end, we consider a case of uniform pressure of magnitude $p > 0$ acting on the body. Then, the tensor $\mathbf{S}$ has the form

$$\mathbf{S} = \begin{bmatrix} -p & 0 & 0 \\ 0 & -p & 0 \\ 0 & 0 & -p \end{bmatrix} \tag{3.3.46}$$

By Equation 3.3.34 we obtain

$$\mathbf{E} = \begin{bmatrix} E_{11} & 0 & 0 \\ 0 & E_{11} & 0 \\ 0 & 0 & E_{11} \end{bmatrix} \tag{3.3.47}$$

where

$$E_{11} = -\frac{p(1-2\nu)}{E} = -\frac{p}{3k} \tag{3.3.48}$$

with

$$k = \frac{E}{3(1-2\nu)} \tag{3.3.49}$$

The constant k is called the bulk modulus. This name is motivated by the relation

$$p = -k(\operatorname{tr}\mathbf{E}) \tag{3.3.50}$$

in which $\operatorname{tr}\mathbf{E}$ represents the change of volume, see Equation 3.1.19. Since $\operatorname{tr}\mathbf{E}$ represents a decrease of volume when a positive pressure acts on a body, it follows from Equation 3.3.50 that k must be positive.

Example 3.3.1

Show that in case of uniform pressure of magnitude $p > 0$ acting on an isotropic elastic body, the displacement vector $\mathbf{u}$ may be given in the form

$$\mathbf{u} = -\frac{p}{3k}\mathbf{x} \tag{a}$$

where k is the bulk modulus.

Solution

In this case the stress tensor $\mathbf{S}$ and the strain tensor $\mathbf{E}$ are given by Equations 3.3.46 and 3.3.47, respectively. Therefore, both $\mathbf{E}$ and $\mathbf{S}$ are represented by diagonal matrices,

$$E_{11} = E_{22} = E_{33} = -\frac{p}{3k}, \quad E_{12} = E_{23} = E_{31} = 0 \tag{b}$$

$$S_{11} = S_{22} = S_{33} = -p, \quad S_{12} = S_{23} = S_{31} = 0 \tag{c}$$

To obtain $\mathbf{u}$ we use the formula, see Equation 3.1.46,

$$E_{ij} = \frac{1}{2}(u_{i,j} + u_{j,i}) \tag{d}$$

Hence, we are to integrate the equations

$$\begin{gathered}\frac{\partial u_1}{\partial x_1} = \frac{\partial u_2}{\partial x_2} = \frac{\partial u_3}{\partial x_3} = -\frac{p}{3k} \\ \frac{\partial u_1}{\partial x_2} + \frac{\partial u_2}{\partial x_1} = \frac{\partial u_2}{\partial x_3} + \frac{\partial u_3}{\partial x_2} = \frac{\partial u_3}{\partial x_1} + \frac{\partial u_1}{\partial x_3} = 0\end{gathered} \tag{e}$$

As a result of the integration we obtain

$$u_i = -\frac{p}{3k}x_i + u_i^0 \tag{f}$$

where u_i^0 represents a rigid body motion given by, see Equation 3.1.8,

$$u_i^0 = a_i + W_{[ij]}x_j \tag{g}$$

Here a_i is a constant vector, and $W_{[ij]}$ is the skew part of a constant tensor W_{ij}. The strain field $\mathbf{E}^0$ corresponding to $\mathbf{u}^0$ vanishes since

$$u_{i,j}^0 = W_{[ij]} \tag{h}$$

so, taking the symmetric part of this, we obtain

$$u^0_{(i,j)} = E^0_{ij} = 0 \tag{i}$$

If we assume that the displacement $\mathbf{u}$ vanishes at $\mathbf{x} = \mathbf{0}$ we obtain the required result. □

Passing now to a transversely isotropic case, with matrix $[K^*]$ given by Equations 3.3.30 and 3.3.31, we interpret the constants E_1 and E_2 as Young's moduli in the directions x_1 and x_2, respectively. The interpretation of Poisson's ratios ν_{12} and ν_{23}, and the shear moduli G_{23} and G_{12} follows from the discussion of the isotropic case and from the form of matrix $[K^*]$, Equation 3.3.30.

The discussion of the engineering constants for the orthotropic case, represented by the matrix $[K^*]$ given by Equation 3.3.29, may be carried out in a similar way.

3.3.3 Alternative Forms of the Constitutive Relations for an Isotropic Body

In Sections 3.3.1 and 3.3.2, we described an isotropic elastic body as a body in which both the elasticity tensor field C and the compliance tensor field K have only two different components. In this case, the relation 3.3.1, which is equivalent to Equation 3.3.18 with $[C^*]$ given by Equation 3.3.22, may be also postulated in the form

$$\mathbf{S} = \mathsf{C}[\mathbf{E}] = 2\mu\mathbf{E} + \lambda(\operatorname{tr}\mathbf{E})\mathbf{1} \tag{3.3.51}$$

where the scalars $\mu = \mu(\mathbf{x})$ and $\lambda = \lambda(\mathbf{x})$ are called the *Lamé moduli*. These moduli are related to the shear modulus $G = G(\mathbf{x})$ and the bulk modulus $k = k(\mathbf{x})$ introduced in Equations 3.3.33 and 3.3.49, respectively, by

$$\mu(\mathbf{x}) = G(\mathbf{x}), \quad \lambda(\mathbf{x}) = k(\mathbf{x}) - \frac{2}{3}\,\mu(\mathbf{x}) \tag{3.3.52}$$

In components, the relation 3.3.51 reads

$$S_{ij} = 2\mu E_{ij} + \lambda E_{kk}\delta_{ij} \tag{3.3.53}$$

By taking the trace of Equation 3.3.53 we get

$$S_{kk} = (3\lambda + 2\mu)E_{kk} \tag{3.3.54}$$

So, if $\mu \neq 0$ and $3\lambda + 2\mu \neq 0$, the stress–strain relation 3.3.53 is invertible and by virtue of Equation 3.3.54, we obtain

$$E_{ij} = \frac{1}{2\mu}\left[S_{ij} - \frac{\lambda}{3\lambda + 2\mu}\,S_{kk}\delta_{ij}\right] \tag{3.3.55}$$

or, in direct notation,

$$\mathbf{E} = \mathsf{K}[\mathbf{S}] = \frac{1}{2\mu}\left[\mathbf{S} - \frac{\lambda}{3\lambda + 2\mu}\,(\operatorname{tr}\mathbf{S})\mathbf{1}\right] \tag{3.3.56}$$

Note that this equation was already discussed when it was written in terms of engineering constants, see Equations 3.3.34. Writing Equations 3.3.34 as a single equation we have

$$\mathbf{E} = \frac{1}{E}\,[(1+\nu)\mathbf{S} - \nu(\mathrm{tr}\,\mathbf{S})\mathbf{1}] \tag{3.3.57}$$

where, comparing with Equation 3.3.33,

$$\frac{1}{E}\,(1+\nu) = \frac{1}{2G} \tag{3.3.58}$$

Since Equations 3.3.56 and 3.3.57 are equivalent, the coefficients in these equations must be identical. Therefore,

$$\frac{1}{2\mu} = \frac{1+\nu}{E} \quad \text{and} \quad \frac{\lambda}{(3\lambda+2\mu)2\mu} = \frac{\nu}{E} \tag{3.3.59}$$

From Equations 3.3.52, 3.3.58, and 3.3.59 follow

$$\begin{aligned} \lambda &= \frac{\nu E}{(1+\nu)(1-2\nu)} \\ \mu &= \frac{E}{2(1+\nu)} \\ k &= \frac{1}{3}\,(3\lambda+2\mu) \end{aligned} \tag{3.3.60}$$

The table of elasticity constants provided in the front matter contains all the relations between E, ν, k, λ, and μ.

In components the relation 3.3.57 takes the form

$$E_{ij} = \frac{1}{E}\,[(1+\nu)S_{ij} - \nu S_{kk}\delta_{ij}] \tag{3.3.61}$$

An equivalent form of Equation 3.3.51 is obtained if the deviatoric parts of **S** and **E** are introduced, see Equations 3.1.43 and 3.1.44,

$$\mathbf{S}^{(d)} = \mathbf{S} - \frac{1}{3}\,(\mathrm{tr}\,\mathbf{S})\mathbf{1} \tag{3.3.62}$$

$$\mathbf{E}^{(d)} = \mathbf{E} - \frac{1}{3}\,(\mathrm{tr}\,\mathbf{E})\mathbf{1} \tag{3.3.63}$$

With these notations, Equation 3.3.51 is equivalent to the following pair of relations:

$$\begin{aligned} \mathbf{S}^{(d)} &= 2\mu\mathbf{E}^{(d)} \\ \mathrm{tr}\,\mathbf{S} &= 3k\,\mathrm{tr}\,\mathbf{E} \end{aligned} \tag{3.3.64}$$

The first relation of Equations 3.3.64 is obtained by taking the deviatoric part of Equation 3.3.51 and using the definitions 3.3.62 and 3.3.63, as well as the fact that

$$\mathbf{1}^{(d)} = \mathbf{1} - \frac{1}{3}(\mathrm{tr}\,\mathbf{1})\mathbf{1} = \mathbf{0} \tag{3.3.65}$$

since $\mathrm{tr}\,\mathbf{1} = 3$. The second relation of 3.3.64 is obtained by taking trace of Equation 3.3.51 and using the relation for k in terms of λ and μ, see also Equation 3.3.54. Therefore, Equation 3.3.51 implies Equation 3.3.64 and vice versa.

For an *incompressible isotropic elastic body*, $\mathrm{tr}\,\mathbf{E} = 0$, and from Equation 3.3.50 it follows that in this case the bulk modulus k must be infinite to make p finite. By Equation 3.3.49 it also means that $k \to \infty$ when $\nu \to \frac{1}{2} - 0$.

Example 3.3.2

Show that for an incompressible isotropic elastic body the constitutive relations are

$$\mathbf{S} = 2\mu\mathbf{E} - p\mathbf{1} \tag{a}$$

where

$$p = -\frac{1}{3}\,\mathrm{tr}\,\mathbf{S} \tag{b}$$

Solution

For an incompressible body $\nu = 1/2$ and $\mu = E/3$ and from Equation 3.3.57 we obtain

$$\mathbf{E} = \frac{1}{2\mu}\mathbf{S} - \frac{1}{6\mu}(\mathrm{tr}\,\mathbf{S})\mathbf{1} \tag{c}$$

Hence,

$$\mathbf{S} = 2\mu\mathbf{E} - p\mathbf{1} \tag{d}$$

where p is given by (b).

From this example it follows that for an isotropic and incompressible material there is only one elastic constant, the shear modulus μ. □

Example 3.3.3

Show that if

$$\mathbf{S} = \mathbf{C}[\mathbf{E}] \tag{a}$$

and if $\mathbf{C}$ in components has the form

$$C_{ijkl} = \lambda\delta_{ij}\delta_{kl} + \mu(\delta_{ik}\delta_{jl} + \delta_{il}\delta_{jk}) \tag{b}$$

then

$$S_{ij} = C_{ijkl}E_{kl} = 2\mu E_{ij} + \lambda E_{kk}\delta_{ij} \tag{c}$$

Solution

Substituting C_{ijkl} given by (b) into the relation

$$S_{ij} = C_{ijkl}E_{kl} \tag{d}$$

we obtain

$$S_{ij} = [\lambda\delta_{ij}\delta_{kl} + \mu(\delta_{ik}\delta_{jl} + \delta_{il}\delta_{jk})]E_{kl} \tag{e}$$

Using now the property of the Kronecker symbol we have

$$\delta_{kl}E_{kl} = E_{kk}, \quad \delta_{ik}\delta_{jl}E_{kl} = \delta_{il}\delta_{jk}E_{kl} = E_{ij} \tag{f}$$

Hence, (d) reduces to (c). This completes the solution. □

Example 3.3.4

Show that for an isotropic elastic solid, the principal directions of the stress tensor **S** coincide with the principal directions of the strain tensor **E**.

Solution

Let **n** be a unit vector along the principal direction of the strain tensor **E** and let h denote the associated principal value. Then,

$$\mathbf{E}\mathbf{n} = h\mathbf{n} \tag{a}$$

Since the body is isotropic,

$$\mathbf{S} = 2\mu\mathbf{E} + \lambda(\mathrm{tr}\,\mathbf{E})\mathbf{1} \tag{b}$$

Now, acting on **n** with **S** given by (b), and using (a), we obtain

$$\mathbf{S}\mathbf{n} = 2\mu h\mathbf{n} + \lambda(\mathrm{tr}\,\mathbf{E})\mathbf{n} \tag{c}$$

or

$$\mathbf{S}\mathbf{n} = [2\mu h + \lambda(\mathrm{tr}\,\mathbf{E})]\mathbf{n} \tag{d}$$

This relation shows that **n** is parallel to a principal vector of the tensor **S**. Therefore, **n** is a principal vector for both **E** and **S**. Since **n** is any of the three principal vectors of **E**, then all principal directions of **E** coincide with appropriate principal directions of **S**.

This completes the solution. □

Example 3.3.5

Show that the inverted form of Hooke's law given by Equation 3.3.57 is

$$\mathbf{S} = \frac{E}{1+\nu}\left[\mathbf{E} + \frac{\nu}{1-2\nu}(\mathrm{tr}\,\mathbf{E})\mathbf{1}\right] \tag{a}$$

Solution

We start with Equation 3.3.51

$$\mathbf{S} = 2\mu\mathbf{E} + \lambda(\mathrm{tr}\,\mathbf{E})\mathbf{1} \tag{b}$$

Using relations between elasticity constants provided in the front matter, we note that

$$2\mu = \frac{E}{(1+\nu)}, \quad \lambda = \frac{\nu E}{(1+\nu)(1-2\nu)} \tag{c}$$

Substituting relations (c) into (b) we obtain (a). □

3.3.4 Stored Energy of an Elastic Body

The concept of a stored energy of an elastic body is closely related to the concept of the fourth-order elasticity tensor **C** introduced in Section 3.3.1.

A *stored energy function* W is a scalar valued function defined on a space of all symmetric tensors $\mathbf{E}$ such that $W(\mathbf{0}) = 0$ and

$$\mathbf{C}[\mathbf{E}] = \nabla_{\mathbf{E}} W(\mathbf{E}) \tag{3.3.66}$$

Here **C** is the elasticity tensor, and the RHS of Equation 3.3.66 is the gradient of W defined by

$$\nabla_{\mathbf{E}} W(\mathbf{E}) \cdot \mathbf{A} = \frac{d}{ds} W(\mathbf{E} + s\mathbf{A})\bigg|_{s=0} \tag{3.3.67}$$

for every $\mathbf{A} = \mathbf{A}^T$ and for an arbitrary scalar s. The need for such a definition comes from the fact that we look for the gradient of a tensor-dependent scalar function W. The definition is analogous to that of the gradient of a scalar field on B, see Section 2.2, and it implies that $\nabla_{\mathbf{E}} W(\mathbf{E})$ is a symmetric tensor with components

$$[\nabla_{\mathbf{E}} W(\mathbf{E})]_{ij} = \frac{\partial W(\mathbf{E})}{\partial E_{ij}} \tag{3.3.68}$$

To show that Equation 3.3.67 implies Equation 3.3.68, we write Equation 3.3.67 in the form

$$\left[(\nabla_{\mathbf{E}} W)_{ij} - \frac{\partial W(\mathbf{E})}{\partial E_{ij}}\right] A_{ij} = 0 \tag{3.3.69}$$

and from arbitrariness of $A_{ij} = A_{ji}$ we conclude that Equation 3.3.68 holds true.

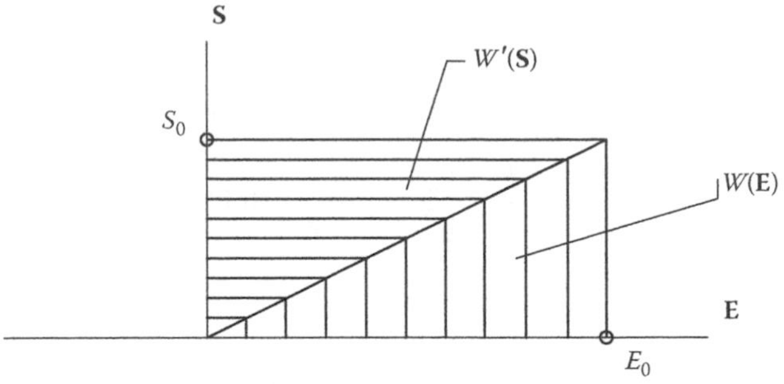

FIGURE 3.12 Stored energy in a one-dimensional case.

In the following we prove that the symmetry of the elasticity tensor **C** is equivalent to the existence of a stored energy function W of the form

$$W(\mathbf{E}) = \frac{1}{2}\,\mathbf{E}\cdot\mathsf{C}[\mathbf{E}] \tag{3.3.70}$$

The energy function W in the case of a one-dimensional stress state is shown in Figure 3.12. It is equal to

$$W(\mathbf{E}) = \frac{1}{2}\int_0^{S_0} E_{11}dS_{11} \tag{3.3.71}$$

In Figure 3.12, $W'(\mathbf{S})$ denotes the *complementary strain energy* density that is equal to

$$W'(\mathbf{S}) = \frac{1}{2}\int_0^{E_0} S_{11}dE_{11} \tag{3.3.72}$$

In the case of a linear stress–strain relation of isothermal and isotropic elasticity

$$W(\mathbf{E}) = W'(\mathbf{S}) = \frac{1}{2}\,S_0E_0 = \frac{1}{2}\,EE_0^2 = \frac{1}{2E}\,S_0^2 \tag{3.3.73}$$

where E is Young's modulus.

First we show that the symmetry of **C** implies existence of W in the form of Equation 3.3.70. To do so we define W by Equation 3.3.70 and show that it is in agreement with the definition 3.3.66. We compute the gradient of W by using the formula 3.3.67 and write

$$[\nabla_{\mathbf{E}}W(\mathbf{E})]\cdot\mathbf{A} = \frac{d}{ds}\,W(\mathbf{E}+s\mathbf{A})\bigg|_{s=0} = \frac{1}{2}\frac{d}{ds}\,\{(\mathbf{E}+s\mathbf{A})\cdot\mathsf{C}[\mathbf{E}+s\mathbf{A}]\}\bigg|_{s=0}$$
$$= \frac{1}{2}\frac{d}{ds}\,\{(\mathbf{E}\cdot\mathsf{C}[\mathbf{E}] + s\mathbf{E}\cdot\mathsf{C}[\mathbf{A}] + s\mathbf{A}\cdot\mathsf{C}[\mathbf{E}]$$
$$+ s^2\mathbf{A}\cdot\mathsf{C}[\mathbf{A}]\}\bigg|_{s=0} = \frac{1}{2}\,\{(\mathbf{E}\cdot\mathsf{C}[\mathbf{A}] + \mathbf{A}\cdot\mathsf{C}[\mathbf{E}]\} \tag{3.3.74}$$

Since $\mathbf{C}$ is symmetric, see Equation 3.3.8, then for every $\mathbf{A} = \mathbf{A}^T$ and $\mathbf{B} = \mathbf{B}^T$

$$\mathbf{A} \cdot \mathbf{C}[\mathbf{B}] = \mathbf{B} \cdot \mathbf{C}[\mathbf{A}] \tag{3.3.75}$$

and Equation 3.3.74 takes the form

$$[\nabla_{\mathbf{E}} W(\mathbf{E})] \cdot \mathbf{A} = \mathbf{A} \cdot \mathbf{C}[\mathbf{E}] \tag{3.3.76}$$

As a symmetric tensor $\mathbf{A}$ is otherwise an arbitrary tensor, Equation 3.3.76 implies Equation 3.3.66, which means that W is a stored energy function.

What is left to be shown is that Equation 3.3.70 implies the symmetry of $\mathbf{C}$. To this end we introduce the concept of a *strain line* $L_{\mathbf{E}}$ (which may, for example, be parameterized by time) connecting two different strains $\mathbf{E}_0$ and $\mathbf{E}_1$ at the same point $\mathbf{x}$. By the strain line we mean a function $\mathbf{E} = \mathbf{E}(\alpha)$ where $\alpha_0 \leq \alpha \leq \alpha_1$, such that $\mathbf{E}(\alpha_0) = \mathbf{E}_0$ and $\mathbf{E}(\alpha_1) = \mathbf{E}_1$. We say that the line $L_{\mathbf{E}}$ is closed if $\mathbf{E}_0 = \mathbf{E}_1$. We associate $L_{\mathbf{E}}$ through Hooke's law with the stress tensor $\mathbf{S} = \mathbf{S}(\alpha)$,

$$\mathbf{S}(\alpha) = \mathbf{C}[\mathbf{E}(\alpha)] \tag{3.3.77}$$

and define the work performed along $L_{\mathbf{E}}$ by

$$w(L_{\mathbf{E}}) = \int_{\alpha_0}^{\alpha_1} \mathbf{S}(\alpha) \cdot \frac{d\mathbf{E}(\alpha)}{d\alpha}\, d\alpha \tag{3.3.78}$$

From this definition and from the definition of the stored energy, Equation 3.3.66, it follows that

$$w(L_{\mathbf{E}}) = W(\mathbf{E}_1) - W(\mathbf{E}_0) \tag{3.3.79}$$

Now we will prove that from the assumption that the work done along any closed $L_{\mathbf{E}}$ is nonnegative follows that this work is zero. We introduce the function

$$\widetilde{\mathbf{E}}(\alpha) = \mathbf{E}(\alpha_0 + \alpha_1 - \alpha), \quad \alpha_0 \leq \alpha \leq \alpha_1 \tag{3.3.80}$$

Thus,

$$\widetilde{\mathbf{E}}(\alpha_0) = \mathbf{E}(\alpha_1) \quad \text{and} \quad \widetilde{\mathbf{E}}(\alpha_1) = \mathbf{E}(\alpha_0) \tag{3.3.81}$$

and from the definition 3.3.78 we find

$$w(L_{\mathbf{E}}) = -w(L_{\widetilde{\mathbf{E}}}) \tag{3.3.82}$$

since $L_{\widetilde{\mathbf{E}}}$ has a direction opposite from $L_{\mathbf{E}}$. As both $w(L_{\widetilde{\mathbf{E}}})$ and $w(L_{\mathbf{E}})$ are nonnegative, we conclude from Equation 3.3.82 that

$$w(L_{\mathbf{E}}) = w(L_{\widetilde{\mathbf{E}}}) = 0 \tag{3.3.83}$$

This completes the proof of the statement.

Next we will prove that from the fact that the work is zero on a closed strain line $L_{\mathbf{E}}$ it follows that **C** is symmetric. We define $L_{\mathbf{E}}$ as a strain line associated with strain

$$\mathbf{E}(\alpha) = (\cos\alpha - 1)\mathbf{A} + (\sin\alpha)\mathbf{B} \tag{3.3.84}$$

where $0 \leq \alpha \leq 2\pi$ and $\mathbf{A}$ and $\mathbf{B}$ are symmetric tensors. We note that $L_{\mathbf{E}}$ is a closed strain line since

$$\mathbf{E}(0) = \mathbf{E}(2\pi) = 0 \tag{3.3.85}$$

The work performed along $L_{\mathbf{E}}$ is given by Equation 3.3.78. Computing the integrand in Equation 3.3.78 we obtain

$$\begin{aligned}\mathbf{C}[\mathbf{E}(\alpha)] \cdot \frac{d\mathbf{E}(\alpha)}{d\alpha} &\\ &= \{(\cos\alpha - 1)\mathbf{C}[\mathbf{A}] + (\sin\alpha)\mathbf{C}[\mathbf{B}]\} \cdot [(-\sin\alpha)\mathbf{A} + (\cos\alpha)\mathbf{B}] \\ &= (\sin\alpha - \cos\alpha\sin\alpha)\mathbf{A}\cdot\mathbf{C}[\mathbf{A}] + (\cos^2\alpha - \cos\alpha)\mathbf{B}\cdot\mathbf{C}[\mathbf{A}] \\ &\quad -(\sin^2\alpha)\mathbf{A}\cdot\mathbf{C}[\mathbf{B}] + (\sin\alpha\cos\alpha)\mathbf{B}\cdot\mathbf{C}[\mathbf{B}]\end{aligned} \tag{3.3.86}$$

Hence,

$$w(L_{\mathbf{E}}) = \int_0^{2\pi} \mathbf{C}[\mathbf{E}(\alpha)] \cdot \frac{d\mathbf{E}(\alpha)}{d\alpha}\, d\alpha = \pi\{\mathbf{B}\cdot\mathbf{C}[\mathbf{A}] - \mathbf{A}\cdot\mathbf{C}[\mathbf{B}]\} \tag{3.3.87}$$

because

$$\begin{aligned}&\int_0^{2\pi} (\sin\alpha - \cos\alpha\sin\alpha)\, d\alpha = 0 \\ &\int_0^{2\pi} \sin\alpha\cos\alpha\, d\alpha = 0 \\ &\int_0^{2\pi} (\cos^2\alpha - \cos\alpha)\, d\alpha = \pi \\ &\int_0^{2\pi} \sin^2\alpha\, d\alpha = \pi\end{aligned} \tag{3.3.88}$$

Since the assumption that $w(L_{\mathbf{E}}) = 0$ implies that the RHS of Equation 3.3.87 vanishes, we obtain

$$\mathbf{B}\cdot\mathbf{C}[\mathbf{A}] = \mathbf{A}\cdot\mathbf{C}[\mathbf{B}] \tag{3.3.89}$$

This relation means that C is symmetric, see Equation 3.3.8. Since the symmetry of C implies the existence of W in the form of Equation 3.3.70, from the relation 3.3.79 it follows that the work along a closed strain line $L_{\mathbf{E}}$ is nonnegative.

This completes the proof that Equation 3.3.70 implies the symmetry of C. This also completes the proof that the symmetry of C means that there is a stored energy function $W(\mathbf{E})$ of the form, see Equation 3.3.70,

$$W(\mathbf{E}) = \frac{1}{2}\,\mathbf{E} \cdot \mathsf{C}[\mathbf{E}] \tag{3.3.90}$$

The property of symmetry of C was also discussed in Section 3.3.1 on general constitutive equations for an anisotropic body, where the symmetry of C caused a reduction of the number of components of C from 36 to 21. Hence, the equivalence of the symmetry of C with the existence of W motivates our discussion on the stored energy of an elastic body. Note that from Equation 3.3.90 it follows that the dimension of a stored energy W is

$$[W] = [\text{Stress}] = [\text{Work} \times \text{L}^{-3}] \tag{3.3.91}$$

and

$$[\text{Work}] = [\text{Force} \times \text{L}] \tag{3.3.92}$$

From this it is clear that W represents a work per unit volume and can be treated as an energy density function defined for each point $\mathbf{x} \in B$. Hence, the total *strain energy of body* B may be presented by

$$U_C\{\mathbf{E}\} = \int_B W\,dv = \frac{1}{2}\int_B \mathbf{E} \cdot \mathsf{C}[\mathbf{E}]\,dv \tag{3.3.93}$$

By using the constitutive relation

$$\mathbf{S} = \mathsf{C}[\mathbf{E}] \tag{3.3.94}$$

and introducing the function

$$\widehat{W}(\mathbf{S}) = \frac{1}{2}\,\mathbf{S} \cdot \mathsf{K}[\mathbf{S}] \tag{3.3.95}$$

we may also define the *stress energy of* B by

$$U_K\{\mathbf{S}\} = \int_B \widehat{W}(\mathbf{S})\,dv = \frac{1}{2}\int_B \mathbf{S} \cdot \mathsf{K}[\mathbf{S}]\,dv \tag{3.3.96}$$

Obviously, from Equations 3.3.93 through 3.3.96 it follows that

$$U_C\{\mathbf{E}\} = U_K\{\mathbf{S}\} \tag{3.3.97}$$

Example 3.3.6

Show that if $\mathbf{C}$ is symmetric, then the constitutive relation

$$\mathbf{S} = \mathbf{C}[\mathbf{E}] \tag{a}$$

may be written in the form

$$\mathbf{S} = \nabla_{\mathbf{E}} W(\mathbf{E}) \tag{b}$$

where $W = W(\mathbf{E})$ is a stored energy function.

Solution

Since $\mathbf{C}$ is symmetric, there is a stored energy function $W = W(\mathbf{E})$ such that Equation 3.3.66 is satisfied. So, Equations 3.3.66 and (a) imply (b). □

This example shows that a stress tensor $\mathbf{S}$ for an anisotropic elastic body may be obtained by taking the gradient of a stored energy for such a body.

Example 3.3.7

Let $\widehat{W} = \widehat{W}(\mathbf{S})$ be a scalar-valued function defined on a space of symmetric tensors $\mathbf{S}$ by the relation

$$\widehat{W}(\mathbf{S}) = \frac{1}{2}\, \mathbf{S} \cdot \mathbf{K}[\mathbf{S}] \tag{a}$$

where $\mathbf{K}$ is the compliance tensor. Show that (A) the constitutive equation (a) from Example 3.3.6 may be written in the form

$$\mathbf{E} = \mathbf{K}[\mathbf{S}] = \nabla_{\mathbf{S}} \widehat{W}(\mathbf{S}) \tag{b}$$

and (B)

$$\widehat{W}(\mathbf{S}) = W(\mathbf{E}) \tag{c}$$

Solution

Since $\mathbf{K}$ is the compliance tensor, (a) from Example 3.3.6 is equivalent to

$$\mathbf{E} = \mathbf{K}[\mathbf{S}] \tag{d}$$

Next, as $\mathbf{K}$ is symmetric (since $\mathbf{C}$ is symmetric), we have by (a)

$$\mathbf{K}[\mathbf{S}] = \nabla_{\mathbf{S}} \widehat{W}(\mathbf{S}) \tag{e}$$

Hence, (d) and (e) imply (b), and this proves (A). To show (B), we use (a) from Example 3.3.6, and obtain

$$\widehat{W}(\mathbf{S}) = \frac{1}{2}\, \mathbf{C}[\mathbf{E}] \cdot \mathbf{K}[\mathbf{C}[\mathbf{E}]] \tag{f}$$

Since $\mathbf{K}$ is the inverse of $\mathbf{C}$, therefore

$$\mathbf{K}[\mathbf{C}[\mathbf{E}]] = \mathbf{E} \tag{g}$$

Hence, by (f), (g), and Equation 3.3.90, we arrive at (c). This proves (B). □

Example 3.3.8

Show that for an isotropic elastic body the functions $W = W(\mathbf{E})$ and $\widehat{W} = \widehat{W}(\mathbf{S})$, given by Equations 3.3.90 and 3.3.95, respectively, take the forms

$$W(\mathbf{E}) = \mu|\mathbf{E}|^2 + \frac{\lambda}{2}\,(\mathrm{tr}\,\mathbf{E})^2 \tag{a}$$

and

$$\widehat{W}(\mathbf{S}) = \frac{1}{4\mu}\left[|\mathbf{S}|^2 - \frac{\lambda}{3\lambda + 2\mu}\,(\mathrm{tr}\,\mathbf{S})^2\right] \tag{b}$$

Solution

For an isotropic body, see Equation 3.3.51,

$$\mathbf{S} = \mathbf{C}[\mathbf{E}] = 2\mu\mathbf{E} + \lambda(\mathrm{tr}\,\mathbf{E})\mathbf{1} \tag{c}$$

Recalling Equation 3.3.90,

$$W(\mathbf{E}) = \frac{1}{2}\,\mathbf{E}\cdot\mathbf{C}[\mathbf{E}] \tag{d}$$

and substituting (c) into this equation, we obtain (a).

To show (b) we use Equation 3.3.56 together with the definition of $\widehat{W}$ given by Equation 3.3.95 and we arrive at (b). □

Example 3.3.9

Show that if the material is isotropic the following five conditions are equivalent:

(i) The elasticity tensor $\mathbf{C}$ is positive definite.
(ii) Young's modulus E is positive and Poisson's ratio ν satisfies the inequality

$$-1 < \nu < 1/2 \tag{a}$$

(iii) The shear modulus μ is positive and Poisson's ratio satisfies relation (a).
(iv) The shear modulus μ is positive and the bulk modulus k is positive.
(v) The shear modulus μ is positive and

$$3\lambda + 2\mu > 0 \tag{b}$$

Solution

We are to show that (i) $\Leftrightarrow$ (ii) $\Leftrightarrow$ (iii) $\Leftrightarrow$ (iv) $\Leftrightarrow$ (v).

The equivalence of (i) and (ii) means that (i) implies (ii) and (ii) implies (i). To show that (i) $\Rightarrow$ (ii) we use the definition of the positive definiteness of the tensor $\mathbf{C}$, see Equation 3.3.13, in the form

$$\mathbf{A} \cdot \mathbf{C}[\mathbf{A}] > 0 \quad \text{for every } \mathbf{A} = \mathbf{A}^T \neq \mathbf{0} \tag{c}$$

In the isotropic case this is equivalent to

$$\frac{E}{1+\nu}\left[|\mathbf{A}|^2 + \frac{\nu}{(1-2\nu)}(\operatorname{tr}\mathbf{A})^2\right] > 0 \tag{d}$$

since, see (a) in Example 3.3.5,

$$\mathbf{C}[\mathbf{A}] = \frac{E}{1+\nu}\left[\mathbf{A} + \frac{\nu}{1-2\nu}(\operatorname{tr}\mathbf{A})\mathbf{1}\right] \tag{e}$$

Replacing $\mathbf{A}$ by $\mathbf{E}$ in relation (d) we obtain

$$\frac{E}{1+\nu}\left[|\mathbf{E}|^2 + \frac{\nu}{(1-2\nu)}(\operatorname{tr}\mathbf{E})^2\right] > 0 \tag{f}$$

Next, we note that $\mathbf{E}$ can be represented as

$$\mathbf{E} = \mathbf{E}^{(d)} + \frac{1}{3}(\operatorname{tr}\mathbf{E})\mathbf{1} \tag{g}$$

where

$$\mathbf{E}^{(d)} = \mathbf{E} - \frac{1}{3}(\operatorname{tr}\mathbf{E})\mathbf{1} \tag{h}$$

We note that

$$\operatorname{tr}\mathbf{E}^{(d)} = 0 \tag{i}$$

and

$$|\mathbf{E}|^2 = |\mathbf{E}^{(d)}|^2 + \frac{1}{3}(\operatorname{tr}\mathbf{E})^2 \tag{j}$$

Hence, an equivalent form of the inequality (f), obtained by substituting (j) into (f), reads

$$E\left[\frac{|\mathbf{E}^{(d)}|^2}{1+\nu} + \frac{1}{3}\frac{(\operatorname{tr}\mathbf{E})^2}{1-2\nu}\right] > 0 \tag{k}$$

Due to the arbitrariness of $\mathbf{E}$ we may take first $\mathbf{E}$ such that $\operatorname{tr}\mathbf{E} = 0$, then (k) implies that $E > 0$ and $1+\nu > 0$ or $E < 0$ and $1+\nu < 0$. Next, by selecting $\mathbf{E}$ such that $|\mathbf{E}^{(d)}| = 0$ we obtain from (k)

$$E > 0 \text{ and } 1-2\nu > 0 \quad \text{or} \quad E < 0 \text{ and } 1-2\nu < 0 \tag{l}$$

Since from the one-dimensional Hooke's law Young's modulus is positive, from the inequalities

$$1 + \nu > 0, \quad 1 - 2\nu > 0 \tag{m}$$

follows

$$-1 < \nu < \frac{1}{2} \tag{n}$$

This ends the proof that (i) ⇒ (ii). A proof that (ii) ⇒ (i) can be carried out by proceeding in the opposite direction starting from (ii).

To show that (ii) ⇒ (iii) we use the table of relations between elasticity constants provided in the front matter from which we write

$$\mu = \frac{E}{2(1 + \nu)} \tag{o}$$

Since in (ii) E is positive and (a) is satisfied, μ is positive and Poisson's ratio ν satisfies (a). The proof that (iii) implies (ii) is obvious.

To show that (iii) and (iv) are equivalent we again use the table of relations between elasticity constants provided in the front matter. In particular, we use the formula

$$k = \frac{2\mu(1 + \nu)}{3(1 - 2\nu)} \tag{p}$$

Finally, to show that (iv) and (v) are equivalent, we note that

$$k = \frac{1}{3}\,(3\lambda + 2\mu) \tag{q}$$

This completes the proofs. □

This example shows that for an isotropic body the positiveness of the stored energy function implies Poisson's ratio may take negative values, so that $\nu > -1$. From the previous discussion, see Equation 3.3.40, we observed that ν must be positive. The restriction of Poisson's ratio ν to positive values only, which comes naturally from engineering tests, is consistent with the positive definiteness of the stored energy.

Positive definiteness of the elasticity tensor **C** allows us to prove the uniqueness of solutions in problems of elasticity. A more general hypothesis than the positive definiteness of **C**, which also guarantees the uniqueness, is the concept of a strong ellipticity of **C**.

Tensor **C** is said to be strongly elliptic if

$$\mathbf{A} \cdot \mathbf{C}[\mathbf{A}] > 0 \tag{3.3.98}$$

for every second-order tensor **A** of the form

$$\mathbf{A} = \mathbf{a} \otimes \mathbf{b}, \tag{3.3.99}$$

where **a** and **b** are arbitrary nonzero vectors. In components, Equation 3.3.98 reads

$$C_{ijkl}a_ia_kb_jb_l > 0 \tag{3.3.100}$$

Due to the symmetry of **C** with respect to the indexes i, j and k, l, we have

$$A_{ij}C_{ijkl}A_{kl} = A_{(ij)}C_{ijkl}A_{(kl)} \tag{3.3.101}$$

Hence, the positive definiteness of **C** implies strong ellipticity of **C**. This becomes clear if we substitute into Equation 3.3.101

$$A_{ij} = a_ib_j \tag{3.3.102}$$

and use the definition of positive definiteness of **C**. The converse statement, that strong ellipticity implies the positive definiteness of **C** is not true. This may be shown by proving that for an isotropic solid the tensor **C** is strongly elliptic if and only if

$$\mu > 0 \quad \text{and} \quad \lambda + 2\mu > 0 \tag{3.3.103}$$

To prove this we note that for an isotropic body, the tensor **C** in components is, see (b) in Example 3.3.3,

$$C_{ijkl} = \lambda\delta_{ij}\delta_{kl} + \mu(\delta_{ik}\delta_{jl} + \delta_{il}\delta_{jk}) \tag{3.3.104}$$

Therefore,

$$\begin{aligned} C_{ijkl}a_ia_kb_jb_l &= [\lambda\delta_{ij}\delta_{kl} + \mu(\delta_{ik}\delta_{jl} + \delta_{il}\delta_{jk})]a_ia_kb_jb_l \\ &= \lambda a_jb_ja_lb_l + \mu(a_ka_kb_lb_l + a_lb_la_jb_j) \\ &= \mu[|\mathbf{a}|^2|\mathbf{b}|^2 - (\mathbf{a}\cdot\mathbf{b})^2] + (\lambda + 2\mu)(\mathbf{a}\cdot\mathbf{b})^2 \end{aligned} \tag{3.3.105}$$

To prove that the inequalities 3.3.103 imply the strong ellipticity of **C** we recall that

$$(\mathbf{a}\cdot\mathbf{b})^2 \leq |\mathbf{a}|^2|\mathbf{b}|^2 \tag{3.3.106}$$

and observe that the RHS of Equation 3.3.105 is positive if the inequalities 3.3.103 are satisfied.

To prove the converse statement, we use Equation 3.3.105 and write the assumption on the strong ellipticity of **C** in the form

$$\mu[|\mathbf{a}|^2|\mathbf{b}|^2 - (\mathbf{a}\cdot\mathbf{b})^2] + (\lambda + 2\mu)(\mathbf{a}\cdot\mathbf{b})^2 > 0 \tag{3.3.107}$$

Since inequality 3.3.107 is to be satisfied for any two nonzero vectors **a** and **b**, we first let in the inequality 3.3.107 $\mathbf{a} = \mathbf{b}$, and find

$$\lambda + 2\mu > 0 \tag{3.3.108}$$

Next, substituting in the inequality 3.3.107 any two vectors **a** and **b** that are orthogonal, we obtain

$$\mu > 0 \tag{3.3.109}$$

This completes the proof of the converse statement.

Example 3.3.10

Show that for an isotropic body, the strong ellipticity of **C** means that

$$\mu > 0 \quad \text{and} \quad \nu \notin \left[\frac{1}{2}, 1\right] \tag{a}$$

Solution

We note from the table of relations between elasticity constants provided in the front matter that

$$\lambda = 2\mu \frac{\nu}{1 - 2\nu} \tag{b}$$

thus

$$\lambda + 2\mu = \frac{2\mu(1 - \nu)}{1 - 2\nu} \tag{c}$$

Using now the equivalence of the strong ellipticity of **C** and the inequalities 3.3.103 we find that the strong ellipticity of **C** is equivalent to the relations (a). This follows from the fact that

$$\frac{1 - \nu}{1 - 2\nu} \tag{d}$$

is positive if and only if ν does not belong to the interval $\left[\frac{1}{2}, 1\right]$.

This completes the solution. □

The ranges of Poisson's ratio ν for a positive definite tensor **C** and for a strongly elliptic tensor **C** are shown in Figure 3.13.

From Example 3.3.10 and from Figure 3.13, in which the ranges of ν both for positive definiteness of **C** and for a strong ellipticity of **C** are shown, it follows that

(a) In general, strong ellipticity does not imply positive definiteness.
(b) Positive definiteness of **C** coincides with the strong ellipticity of **C** for the range of ν from -1 to $\frac{1}{2}$.
(c) From an engineering point of view it is clear that both the positive definiteness of **C** and the strong ellipticity of **C** are acceptable concepts for ν between 0 and $\frac{1}{2}$.

The need for the introduction of the concept of strong ellipticity of **C** comes from the fact that not only a uniqueness result of elasticity can be established but also this property of **C**

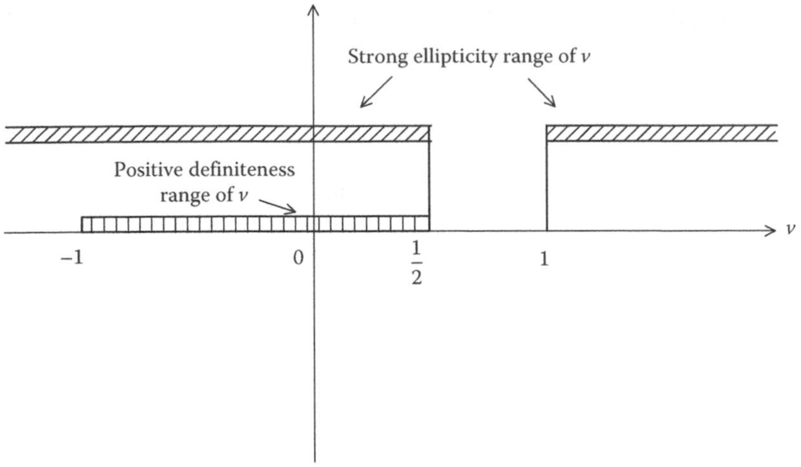

FIGURE 3.13 Ranges of ν for positive definiteness of **C** and for strong ellipticity of **C** in the case of an isotropic body.

is used in an analysis of wave propagation in elastic solids in which the inequalities $\mu > 0$ and $\lambda + 2\mu > 0$ imply that the velocity of a shear wave and the velocity of a longitudinal wave are real.

Example 3.3.11

A particular anisotropic elastic material for which in addition to 3.3.5 and 3.3.11 the following symmetry relations are satisfied

$$C_{ijkl} = C_{ikjl} \tag{a}$$

is called an *anisotropic elastic material subject to the Cauchy relations* [4]. A theory of anisotropic elasticity obeying the Cauchy relations is carrying only 15 independent elasticity constants. Show that for the special case of isotropy the Cauchy relations (a) reduce to the single relation $\lambda = \mu$, where λ and μ are Lamé constants, and hence the elasticity tensor depends only on one modulus as predicted by Cauchy in his days.

Solution

We note that for an isotropic elastic material the elasticity tensor C_{ijkl} takes the form

$$C_{ijkl} = \lambda\, \delta_{ij}\, \delta_{kl} + \mu(\delta_{ik}\, \delta_{jl} + \delta_{il}\, \delta_{jk}) \tag{b}$$

where δ_{ij} is the Kronecker symbol. Then, it follows from (a) and (b) that the Cauchy relations for the isotropic solid take the form

$$C_{i[jk]l} = \lambda\, \delta_{i[j}\, \delta_{k]l} + \mu\, \delta_{[jl}\, \delta_{ik]} = 0 \tag{c}$$

or, equivalently,

$$(\lambda - \mu)\, \delta_{[ik}\, \delta_{lj]} = 0 \tag{d}$$

In Equations (c) and (d) brackets over two indices indicate antisymmetrization. Now, multiplying (d) by the sixth-order tensor

$$A_{abijkl} = \epsilon_{aij}\,\epsilon_{bkl} \tag{e}$$

where ϵ_{abc} is the alternating symbol, and using the identity

$$\delta_{[ik}\,\delta_{lj]}\,\epsilon_{aij}\,\epsilon_{bkl} = 2\delta_{ab} \tag{f}$$

we obtain

$$2\,(\lambda - \mu)\,\delta_{ab} = 0 \tag{g}$$

Equation (g) implies that for an isotropic elastic material of the Cauchy type we have $\lambda = \mu$.

Note that there is the following relation between the pair (λ, μ) and Poisson's ratio ν

$$\lambda = \frac{2\,\mu\,\nu}{1-2\,\nu} \tag{h}$$

Hence, by letting $\lambda = \mu$ in (h) we find that for an isotropic elastic material of the Cauchy type Poisson's ratio $\nu = 1/4$. This completes the solution. □

3.3.5 Stress–Strain–Temperature Relations for a Thermoelastic Body

The constitutive relations considered in the previous sections were postulated under the condition that a body remained at a constant temperature. If a body is heated or cooled unevenly and it is allowed to expand freely, thermally induced strains and associated stresses will be produced. Hence, for an anisotropic body subject to uneven heating the constitutive law 3.3.1 and its equivalent given by Equation 3.3.7 are modified to the form

$$\mathbf{S} = \mathsf{C}[\mathbf{E}] + T\mathbf{M} \tag{3.3.110}$$

and

$$\mathbf{E} = \mathsf{K}[\mathbf{S}] + T\mathbf{A} \tag{3.3.111}$$

where

$$T = \theta - \theta_0, \quad \theta_0 > 0 \tag{3.3.112}$$

is a temperature change.

In Equation 3.3.110, C denotes the elasticity tensor evaluated at a constant temperature $\theta = \theta_0$. The tensor $\mathbf{M}$ is called the *stress–temperature tensor*. We note that $\mathbf{M}$ is symmetric, $\mathbf{M} = \mathbf{M}^T$.

In Equation 3.3.111, K is the compliance tensor evaluated at $\theta = \theta_0$. The tensor $\mathbf{A}$ is called the *thermal expansion tensor*, and it is symmetric.

In Equation 3.3.112, θ is the absolute temperature ($\theta > 0$), and θ_0 is a reference temperature, at which the stress and strain vanish throughout the body.

Since relations 3.3.110 and 3.3.111 are equivalent,

$$\mathsf{K} = \mathsf{C}^{-1} \quad \text{and} \quad \mathbf{A} = -\mathsf{K}[\mathbf{M}] \tag{3.3.113}$$

To prove this it is sufficient to apply the operator K to both sides of Equation 3.3.110. This yields

$$\mathsf{K}[\mathbf{S}] = \mathsf{K}[\mathsf{C}[\mathbf{E}]] + T\mathsf{K}[\mathbf{M}] \tag{3.3.114}$$

Hence, we write

$$\mathbf{E} = \mathsf{K}[\mathbf{S}] - T\mathsf{K}[\mathbf{M}] \tag{3.3.115}$$

Thus, Equation 3.3.115 is identical to Equation 3.3.111 if

$$\mathbf{A} = -\mathsf{K}[\mathbf{M}] \tag{3.3.116}$$

By virtue of Equation 3.3.110, the stress–temperature tensor $\mathbf{M}$ multiplied by T gives the stress resulting from the temperature change T when strain $\mathbf{E}$ vanishes:

$$\mathbf{S} = T\mathbf{M} \quad \text{if } \mathbf{E} = \mathbf{0} \tag{3.3.117}$$

Also, from Equation 3.3.111 it follows that the thermal expansion tensor $\mathbf{A}$ multiplied by T gives the strain $\mathbf{E}$ produced by T when stress $\mathbf{S}$ vanishes:

$$\mathbf{E} = T\mathbf{A} \quad \text{if } \mathbf{S} = \mathbf{0} \tag{3.3.118}$$

Obviously, the first and second terms on the RHS of Equation 3.3.110 represent the mechanical and thermal part, respectively, of the stress tensor for a thermoelastic body. Similarly, the first and second terms on the RHS of Equation 3.3.111 represent the mechanical and the thermal part, respectively, of the strain tensor.

For a nonhomogeneous thermoelastic body the *material tensors* C, K, $\mathbf{M}$, and $\mathbf{A}$ depend on the position $\mathbf{x} \in \bar{B}$.

In components, the constitutive relations 3.3.110 and 3.3.111 for an anisotropic thermoelastic solid take the form

$$S_{ij} = C_{ijkl}E_{kl} + TM_{ij} \tag{3.3.119}$$

and

$$E_{ij} = K_{ijkl}S_{kl} + TA_{ij} \tag{3.3.120}$$

It is assumed that the tensors C and K satisfy the symmetry properties stated in Equations 3.3.5 and 3.3.11 as well as in Equations 3.3.14 and 3.3.15:

$$C_{ijkl} = C_{jikl} = C_{ijlk} = C_{klij} \tag{3.3.121}$$

$$K_{ijkl} = K_{jikl} = K_{ijlk} = K_{klij} \tag{3.3.122}$$

Also, **C** and **K** are assumed to be positive definite, which means

$$B_{ij}C_{ijkl}B_{kl} > 0$$
$$B_{ij}K_{ijkl}B_{kl} > 0 \tag{3.3.123}$$

for every $B_{ij} = B_{ij} \neq 0$.

What regards the tensors **M** and **A**, they satisfy the symmetry relations

$$M_{ij} = M_{ji} \quad \text{and} \quad A_{ij} = A_{ji} \tag{3.3.124}$$

but we do not require that **M** and **A** be positive definite.

Now we pass to an isotropic thermoelastic body. For such a body, Equations 3.3.110 and 3.3.111, written in terms of Lamé moduli λ and μ, and the *coefficient of the thermal linear expansion* α, take the form

$$\mathbf{S} = 2\mu\mathbf{E} + \lambda(\operatorname{tr}\mathbf{E})\mathbf{1} - (3\lambda + 2\mu)\alpha T\mathbf{1} \tag{3.3.125}$$

and

$$\mathbf{E} = \frac{1}{2\mu}\left[\mathbf{S} - \frac{\lambda}{3\lambda + 2\mu}(\operatorname{tr}\mathbf{S})\mathbf{1}\right] + \alpha T\mathbf{1} \tag{3.3.126}$$

This means that for an isotropic body

$$\mathbf{M} = -(3\lambda + 2\mu)\alpha\mathbf{1} \tag{3.3.127}$$

and

$$\mathbf{A} = \alpha\mathbf{1} \tag{3.3.128}$$

Taking the trace of Equation 3.3.126 we receive the dilatation, see Equation 3.1.19,

$$\operatorname{tr}\mathbf{E} = \frac{\operatorname{tr}\mathbf{S}}{3\lambda + 2\mu} + 3\alpha T \tag{3.3.129}$$

If the body is subject to temperature change T only and is not constrained, then in Equation 3.3.129 $\mathbf{S} = \mathbf{0}$ and we obtain

$$\operatorname{tr}\mathbf{E} = 3\alpha T \tag{3.3.130}$$

This represents the change of volume per unit initial volume of an element of the body that undergoes a small deformation caused by a temperature change T. Therefore, for a stress free body subject to T the strain tensor **E** is given by the diagonal matrix with

$$E_{11} = E_{22} = E_{33} = \alpha T \tag{3.3.131}$$

and the meaning of α is clear from the relation

$$\alpha = \frac{E_{11}}{T} = \frac{E_{22}}{T} = \frac{E_{33}}{T}, \qquad T \neq 0 \tag{3.3.132}$$

Since E_{ij} is nondimensional, the dimension of α is

$$[\alpha] = [K^{-1}] \tag{3.3.133}$$

The constitutive equations 3.3.125 and 3.3.126 written in terms of engineering constants are

$$\mathbf{S} = \frac{E}{1+\nu}\left[\mathbf{E} + \frac{\nu}{1-2\nu}\,(\mathrm{tr}\,\mathbf{E})\mathbf{1}\right] - \frac{E}{1-2\nu}\,\alpha T\mathbf{1} \tag{3.3.134}$$

$$\mathbf{E} = \frac{1+\nu}{E}\left[\mathbf{S} - \frac{\nu}{1+\nu}\,(\mathrm{tr}\,\mathbf{S})\mathbf{1}\right] + \alpha T\mathbf{1} \tag{3.3.135}$$

Example 3.3.12

Show that Equation 3.3.135 can be inverted to produce Equation 3.3.134.

Solution

Assume that **E** is given by Equation 3.3.135. We take the trace of Equation 3.3.135 and obtain

$$\left(\frac{1+\nu}{E} - \frac{3\nu}{E}\right)(\mathrm{tr}\,\mathbf{S}) + 3\alpha T = \mathrm{tr}\,\mathbf{E} \tag{a}$$

Hence,

$$\mathrm{tr}\,\mathbf{S} = \frac{E}{1-2\nu}\,(\mathrm{tr}\,\mathbf{E} - 3\alpha T) \tag{b}$$

Since, by Equation 3.3.135,

$$\mathbf{S} = \frac{\nu}{1+\nu}\,(\mathrm{tr}\,\mathbf{S})\mathbf{1} + \frac{E}{1+\nu}\,(\mathbf{E} - \alpha T\mathbf{1}) \tag{c}$$

then substituting (b) into (c) we arrive at Equation 3.3.134.

This completes the solution. □

Example 3.3.13

Let an isotropic homogeneous thermoelastic body kept at a constant temperature $\theta_0 > 0$ be subject to a uniform pressure of magnitude $p > 0$. Show that it is possible to decrease the temperature of the body to a constant level $\theta < \theta_0$ in such a way as to make the body stress free.

Solution

For an isotropic body kept at the temperature $\theta_0 > 0$ and subject to uniform pressure $p > 0$, the stress tensor $\mathbf{S}$ is given by

$$\mathbf{S}^{(1)} = -p\mathbf{1} \tag{a}$$

If the temperature of the body is decreased to θ level, the total stress is

$$\mathbf{S} = \mathbf{S}^{(1)} + \mathbf{S}^{(2)} \tag{b}$$

where

$$\mathbf{S}^{(2)} = -(3\lambda + 2\mu)\alpha(\theta - \theta_0)\mathbf{1} \tag{c}$$

Hence, the condition $\mathbf{S} = \mathbf{0}$, leads to

$$-p - (3\lambda + 2\mu)(\theta - \theta_0)\alpha = 0 \tag{d}$$

So, the body becomes stress free if

$$\theta = \theta_0 - \frac{p}{(3\lambda + 2\mu)\alpha} \tag{e}$$

This completes the solution. □

For numerical calculations we take

$\theta_0 = 20°\text{C}$ or 293.16 K (room temperature);
$p = 100$ MPa (pressure);
Material: structural steel with
$E = 204$ GPa (modulus of elasticity)
$\nu = 0.29$ (Poisson's ratio)
$\alpha = 12 \times 10^{-6}/\text{K}$ (coefficient of linear thermal expansion).

Since

$$3\lambda + 2\mu = \frac{E}{1 - 2\nu} = \frac{204{,}000 \text{ MPa}}{1 - 2 \times 0.29} = 485{,}714 \text{ MPa} \tag{3.3.136}$$

we obtain from (e)

$$\begin{aligned} \theta &= 293.16\,\text{K} - \frac{100 \text{ MPa} \times 10^6}{485{,}714 \text{ MPa} \times 12}\,\text{K} \\ &= (293.16 - 17.16)\text{ K} = 276 \text{ K} \text{ or } 2.84°\text{C}. \end{aligned} \tag{3.3.137}$$

PROBLEMS

3.1 Show that if $\mathbf{u}$ is a pure strain from $\mathbf{x}_0$, then $\mathbf{u}$ admits the decomposition

$$\mathbf{u} = \mathbf{u}_1 + \mathbf{u}_2 + \mathbf{u}_3 \tag{a}$$

where $\mathbf{u}_1$, $\mathbf{u}_2$, and $\mathbf{u}_3$ are simple extensions in mutually perpendicular directions from $\mathbf{x}_0$.

3.2 Show that $\mathbf{u}$ in Problem 3.1 admits an alternative representation

$$\mathbf{u} = \mathbf{u}_d + \mathbf{u}_c \tag{a}$$

where $\mathbf{u}_d$ is a uniform dilatation from $\mathbf{x}_0$, while $\mathbf{u}_c$ is an isochoric pure strain from $\mathbf{x}_0$.

3.3 Show that if $\mathbf{u}$ is a simple shear of amount γ with respect to the pair $(\mathbf{m}, \mathbf{n})$, where $\mathbf{m}$ and $\mathbf{n}$ are perpendicular unit vectors, then $\mathbf{u}$ admits the decomposition

$$\mathbf{u} = \mathbf{u}_1 + \mathbf{u}_2 \tag{a}$$

where

$\mathbf{u}_1$ is a simple extension of amount γ in the direction $\frac{1}{\sqrt{2}}(\mathbf{m} + \mathbf{n})$

$\mathbf{u}_2$ is a simple extension of amount $-\gamma$ in the direction $\frac{1}{\sqrt{2}}(\mathbf{m} - \mathbf{n})$

3.4 Let $\mathbf{u}$ and $\mathbf{E}$ denote a displacement vector field and the corresponding strain tensor field defined on $\overline{B}$. Show that the mean strain $\widehat{\mathbf{E}}(B)$ is represented by the surface integral

$$\widehat{\mathbf{E}}(B) = \frac{1}{v(B)} \int_{\partial B} \text{sym}\,(\mathbf{u} \otimes \mathbf{n})da \tag{a}$$

where $v(B)$ is the volume of B.

3.5 Show that if $\mathbf{u} = \mathbf{0}$ on ∂B then

$$\int_B (\nabla\mathbf{u})^2 dv \leq 2 \int_B |\mathbf{E}|^2 dv \tag{a}$$

where $\mathbf{E}$ is the strain tensor field corresponding to a displacement vector field on $\overline{B}$.

3.6 (i) Let $\mathbf{E}$ be a strain tensor field on E^3 defined by the matrix

$$\mathbf{E} = \begin{bmatrix} \dfrac{N}{E} & 0 & 0 \\ 0 & -\nu\dfrac{N}{E} & 0 \\ 0 & 0 & -\nu\dfrac{N}{E} \end{bmatrix} \tag{a}$$

where E, N, and ν are positive constants. Show that a unique solution $\mathbf{u}$ to the equation $\mathbf{E} = \widehat{\nabla}\mathbf{u}$ on E^3 subject to the condition $\mathbf{u}(\mathbf{0}) = \mathbf{0}$ takes the form

$$\mathbf{u} = \left[\frac{N}{E}x_1, \ -\nu\frac{N}{E}x_2, \ -\nu\frac{N}{E}x_3\right]^T \tag{b}$$

(ii) Let $\mathbf{E}$ be a strain tensor field on E^3 defined by the matrix

$$\mathbf{E} = \begin{bmatrix} \nu\frac{M}{EI}x_1 & 0 & 0 \\ 0 & \nu\frac{M}{EI}x_1 & 0 \\ 0 & 0 & -\frac{M}{EI}x_1 \end{bmatrix} \tag{c}$$

where M, E, I, and ν are positive constants. Show that a unique solution $\mathbf{u} = [u_1, u_2, u_3]^T$ to the equation $\mathbf{E} = \widehat{\nabla}\,\mathbf{u}$ on E^3 subject to the condition $\mathbf{u}(\mathbf{0}) = \mathbf{0}$ takes the form

$$\mathbf{u} = \frac{M}{EI}\left[\frac{1}{2}\left(x_3^2 + \nu x_1^2 - \nu x_2^2\right), \ \nu x_1 x_2, \ -x_1 x_3\right]^T \tag{d}$$

3.7 Given a stress tensor $\mathbf{S}$ at a point A find (i) the stress vector $\mathbf{s}$ on a plane through A parallel to the plane $\mathbf{n}\cdot\mathbf{x} - vt = 0$ ($|\mathbf{n}| = 1, \ v > 0, \ t \geq 0$), (ii) the magnitude of $\mathbf{s}$, (iii) the angle between $\mathbf{s}$ and the normal to the plane, and (iv) the normal and tangential components of the stress vector $\mathbf{s}$.

Answers:
(i) $\mathbf{s} = \mathbf{Sn}$; (ii) $|\mathbf{s}| = |\mathbf{Sn}|$; (iii) $\cos\theta = \mathbf{s}\cdot\mathbf{n}/|\mathbf{s}|$; (iv) $\mathbf{s} = \mathbf{s}_n + \mathbf{s}_\tau$, where $\mathbf{s}_n = (\mathbf{n}\cdot\mathbf{s})\mathbf{n}$ and $\mathbf{s}_\tau = \mathbf{n}\times(\mathbf{s}\times\mathbf{n})$.

3.8 Let $\{\mathbf{e}_i\}$ be an orthonormal basis for a stress tensor $\mathbf{S}$, and let $\{\mathbf{e}_i^*\}$ be an orthonormal basis formed by the eigenvectors of $\mathbf{S}$. Then a tensor $\mathbf{S}^*$ obtained from $\mathbf{S}$ by the transformation formula from $\{\mathbf{e}_i\}$ to $\{\mathbf{e}_i^*\}$ takes the form (see Example 2.1.12)

$$\mathbf{S}^* = \lambda_1\mathbf{e}_1^*\otimes\mathbf{e}_1^* + \lambda_2\mathbf{e}_2^*\otimes\mathbf{e}_2^* + \lambda_3\mathbf{e}_3^*\otimes\mathbf{e}_3^* \tag{a}$$

where λ_i is an eigenvalue (principal stress) of $\mathbf{S}$ corresponding to the eigenvector $\mathbf{e}_i^*$. Show that the function

$$g(\mathbf{n}^*) = \left|\mathbf{s}_\tau^*\right| = |\mathbf{n}^*\times(\mathbf{S}^*\mathbf{n}^*\times\mathbf{n}^*)| \tag{b}$$

representing the tangent stress vector magnitude with regard to a plane with a normal $\mathbf{n}^*$ in the $\{\mathbf{e}_i^*\}$ basis, assumes the extreme values:

$$\left|\mathbf{s}_\tau^*\right|_1 = \frac{1}{2}|\lambda_2 - \lambda_3| \tag{c}$$

$$\left|\mathbf{s}_\tau^*\right|_2 = \frac{1}{2}|\lambda_3 - \lambda_1| \tag{d}$$

and

$$|\mathbf{s}_\tau^*|_3 = \frac{1}{2}|\lambda_1 - \lambda_2| \tag{e}$$

at

$$\mathbf{n}_1^* = [0,\ \pm 1/\sqrt{2},\ \pm 1/\sqrt{2}]^T \tag{f}$$

$$\mathbf{n}_2^* = [\pm 1/\sqrt{2},\ 0,\ \pm 1/\sqrt{2}]^T \tag{g}$$

and

$$\mathbf{n}_3^* = [\pm 1/\sqrt{2},\ \pm 1/\sqrt{2},\ 0]^T \tag{h}$$

respectively. Hence, if $\lambda_1 > \lambda_2 > \lambda_3$ then the largest tangential stress vector magnitude is

$$|\mathbf{s}_\tau^*|_2 = \frac{1}{2}|\lambda_3 - \lambda_1| \tag{i}$$

and this extreme vector acts on the plane that bisects the angle between $\mathbf{e}_1^*$ and $\mathbf{e}_3^*$.

3.9 Let $D = \{\mathbf{x} : x_1 \geq 0,\ x_1 \tan\theta \geq x_2 \geq 0\}$ be a two-dimensional wedge region shown in Figure P3.9, and let $S_{\alpha\beta} = S_{\alpha\beta}(\mathbf{x}),\ [\mathbf{x} = (x_1, x_2);\ \alpha, \beta = 1, 2]$ be a symmetric tensor field on D defined by

$$S_{11} = dx_2 + ex_1 - \rho g x_1, \quad S_{22} = -\gamma x_1, \quad S_{12} = S_{21} = -ex_2 \tag{a}$$

where d, e, g, ρ, and γ are constants ($g > 0,\ \rho > 0,\ \gamma > 0$).

(i) Show that

$$\operatorname{div} \mathbf{S} + \mathbf{b} = \mathbf{0} \quad \text{on } D \tag{b}$$

where

$$\mathbf{b} = [\rho g,\ 0]^T \quad \text{on } D \tag{c}$$

(ii) Using the transformation formula from the x_α system to the system x'_α [see Equation (b) in Problem 2.8] find the components $S'_{\alpha\beta}$ in terms of $S_{\alpha\beta}$, and show that

$$S'_{12} = 0 \quad \text{and} \quad S'_{22} = 0 \quad \text{for } x_2 = x_1 \tan\theta \tag{d}$$

provided

$$e = \frac{\gamma}{\tan^2\theta} \quad \text{and} \quad d = \frac{\rho g}{\tan\theta} - \frac{2\gamma}{\tan^3\theta} \tag{e}$$

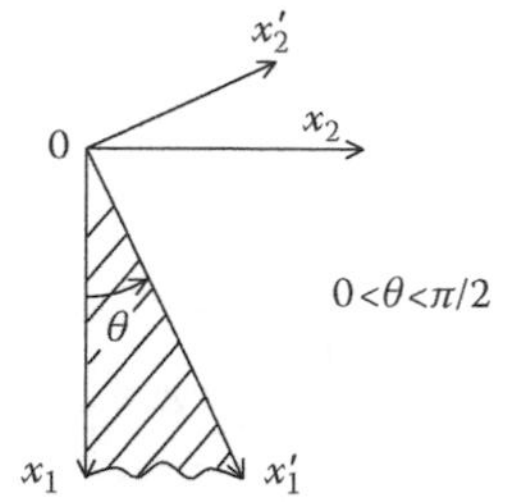

FIGURE P3.9

(iii) Give diagrams of S_{11} and S_{12} over a horizontal section $x_1 = x_1^0 = \text{const.}$
(iv) Give a diagram of S_{22} over the vertical section $x_2 = 0$.

3.10 Let B denote a cylinder of length ℓ and of arbitrary cross section, suspended from the upper end and subject to its own weight ρg (see Example 3.2.12 and Figure P3.10). Then the stress tensor $\mathbf{S} = \mathbf{S}(\mathbf{x})$ on B takes the form

$$\mathbf{S} = \begin{bmatrix} 0 & 0 & 0 \\ 0 & 0 & 0 \\ 0 & 0 & \rho g x_3 \end{bmatrix} \tag{a}$$

since, in this case, the body force vector field is given by $\mathbf{b} = [0, 0, -\rho g]^T$, and $\operatorname{div}\mathbf{S} + \mathbf{b} = \mathbf{0}$ on B. The stress vector $\mathbf{s}$ associated with $\mathbf{S}$ on ∂B has the following properties: $\mathbf{s} = [0, 0, \rho g \ell]^T$ on the end plane $x_3 = \ell$; and $\mathbf{s} = \mathbf{0}$ on the plane $x_3 = 0$ and on the lateral surface of the cylinder as $\mathbf{n} = [n_1, n_2, 0]^T$ on the surface. Assuming that the cylinder is made of a homogeneous isotropic elastic material, the associated strain tensor field $\mathbf{E}$ takes the form (see Equations 3.3.57)

$$\mathbf{E} = \frac{\rho g x_3}{E}\begin{bmatrix} -\nu & 0 & 0 \\ 0 & -\nu & 0 \\ 0 & 0 & 1 \end{bmatrix} \tag{b}$$

where E and ν are Young's modulus and Poisson's ratio, respectively.

(i) Show that a solution $\mathbf{u}$ of the equation

$$\mathbf{E} = \widehat{\nabla}\mathbf{u} \quad \text{on } B \tag{c}$$

subject to the condition

$$\mathbf{u}(0, 0, \ell) = \mathbf{0} \tag{d}$$

takes the form

$$\mathbf{u} = \frac{\rho g}{E}\left[-\nu x_1 x_3, -\nu x_2 x_3, \frac{\nu}{2}\left(x_1^2 + x_2^2\right) + \frac{1}{2}\left(x_3^2 - \ell^2\right)\right]^T \tag{e}$$

(ii) Plot $u_3 = u_3(0, 0, x_3)$ over the range $0 \le x_3 \le \ell$.

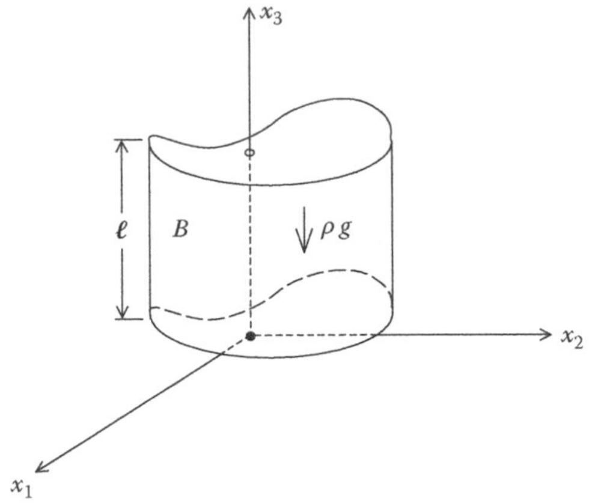

FIGURE P3.10

3.11 For a transversely isotropic elastic body each material point of the body possesses an axis of rotational symmetry, which means that the elastic properties are the same in any direction on any plane perpendicular to the axis, but they are different than those in the direction of the axis. If the x_3 axis coincides with the axis of symmetry, then the stress–strain relation for such a body takes the form

$$\begin{bmatrix} S_{11} \\ S_{22} \\ S_{33} \\ S_{32} \\ S_{31} \\ S_{12} \end{bmatrix} = \begin{bmatrix} c_{11} & c_{12} & c_{13} & 0 & 0 & 0 \\ c_{12} & c_{11} & c_{13} & 0 & 0 & 0 \\ c_{13} & c_{13} & c_{33} & 0 & 0 & 0 \\ 0 & 0 & 0 & c_{44} & 0 & 0 \\ 0 & 0 & 0 & 0 & c_{44} & 0 \\ 0 & 0 & 0 & 0 & 0 & (c_{11}-c_{12})/2 \end{bmatrix} \begin{bmatrix} E_{11} \\ E_{22} \\ E_{33} \\ 2E_{23} \\ 2E_{31} \\ 2E_{12} \end{bmatrix} \tag{a}$$

where $\mathbf{S}$ and $\mathbf{E}$ are the stress and strain tensors, respectively, and five numerically independent moduli $c_{11}, c_{33}, c_{12}, c_{13}$, and c_{44} are related to the components C_{ijkl} of the fourth-order elasticity tensor C by (see Equation 3.3.20)

$$\begin{aligned} c_{11} &= C_{1111}, \quad c_{12} = C_{1122}, \quad c_{13} = C_{1133}, \\ c_{33} &= C_{3333}, \quad c_{44} = C_{1313} \end{aligned} \tag{b}$$

Show that if the axis of symmetry of a transversely isotropic body coincides with the direction of an arbitrary unit vector $\mathbf{e}$, then the stress–strain relation takes the form

$$\begin{aligned} \mathbf{S} = \mathsf{C}[\mathbf{E}] &= (c_{11}-c_{12})\mathbf{E} + \{c_{12}(\operatorname{tr}\mathbf{E}) - (c_{12}-c_{13})\mathbf{e}\cdot(\mathbf{E}\mathbf{e})\}\mathbf{1} \\ &\quad - (c_{11}-c_{12}-2c_{44})\{\mathbf{e}\otimes(\mathbf{E}\mathbf{e}) + (\mathbf{E}\mathbf{e})\otimes\mathbf{e}\} \\ &\quad - \{(c_{12}-c_{13})(\operatorname{tr}\mathbf{E}) - (c_{11}+c_{33}-2c_{13}-4c_{44})\mathbf{e}\cdot(\mathbf{E}\mathbf{e})\}\mathbf{e}\otimes\mathbf{e} \end{aligned} \tag{c}$$

3.12 Show that the stress–strain relation (c) in Problem 3.11 is invertible provided

$$c \equiv (c_{11}+c_{12})c_{33} - 2c_{13}^2 > 0, \quad c_{11} > |c_{12}|, \; c_{44} > 0 \tag{a}$$

and that the strain–stress relation reads

$$\mathbf{E} = \mathbf{K}[\mathbf{S}] = (c_{11} - c_{12})^{-1}\mathbf{S} + \frac{1}{2}\Big[\{c^{-1}c_{33} - (c_{11} - c_{12})^{-1}\} (\operatorname{tr}\mathbf{S})$$

$$- \{c^{-1}(c_{33} + 2c_{13}) - (c_{11} - c_{12})^{-1}\} \mathbf{e} \cdot (\mathbf{Se}) \Big] \mathbf{1}$$

$$- \left\{ (c_{11} - c_{12})^{-1} - \frac{1}{2}c_{44}^{-1} \right\} \{\mathbf{e} \otimes (\mathbf{Se}) + (\mathbf{Se}) \otimes \mathbf{e}\}$$

$$- \frac{1}{2}\Big[\left\{ c^{-1}(c_{33} + 2c_{13}) - (c_{11} - c_{12})^{-1} \right\} (\operatorname{tr}\mathbf{S})$$

$$- \{c^{-1}(2c_{11} + c_{33} + 2c_{12} + 4c_{13}) + (c_{11} - c_{12})^{-1} - 2c_{44}^{-1}\} \mathbf{e} \cdot (\mathbf{Se}) \Big] \mathbf{e} \otimes \mathbf{e} \tag{b}$$

3.13 Prove that the inequalities (a) in Problem 3.12 are necessary and sufficient conditions for the elasticity tensor C (compliance tensor K) to be positive definite. This means that the strain energy density (stress energy density) of a transversely isotropic body is positive definite if and only if the inequalities (a) in Problem 3.12 hold true.

3.14 Consider a plane $\mathbf{x} \cdot \mathbf{n} - vt = 0$, where $\mathbf{n}$ is a unit vector and v and t are constants $(v > 0, t \geq 0)$. Let $\mathbf{S}$ be the stress tensor obtained from Equation (c) of Problem 3.11 in which $0 < \mathbf{e} \cdot \mathbf{n} < 1$, and the strain tensor is defined by

$$\mathbf{E} = \operatorname{sym}(\mathbf{n} \otimes \mathbf{a}) \tag{a}$$

where $\mathbf{a}$ is an arbitrary vector orthogonal to $\mathbf{n}$. Let $\mathbf{S}^{\perp}$ and $\mathbf{S}^{\parallel}$ represent the normal and tangential parts of $\mathbf{S}$ with respect to the plane (see Problem 2.4 in which $\mathbf{T}$ is replaced by $\mathbf{S}$). Show that

$$\mathbf{S} = (c_{11} - c_{12})\operatorname{sym}(\mathbf{n} \otimes \mathbf{a}) - (c_{11} - c_{12} - 2c_{44}) \cos\theta \operatorname{sym}(\mathbf{e} \otimes \mathbf{a}) \tag{b}$$

$$\mathbf{S}^{\perp} = [(c_{11} - c_{12}) \sin^2\theta + 2c_{44} \cos^2\theta] \operatorname{sym}(\mathbf{n} \otimes \mathbf{a}) \tag{c}$$

$$\mathbf{S}^{\parallel} = (c_{11} - c_{12} - 2c_{44}) \cos\theta \operatorname{sym}[(\mathbf{n}\cos\theta - \mathbf{e}) \otimes \mathbf{a}] \tag{d}$$

where

$$\cos\theta = \mathbf{e} \cdot \mathbf{n}, \quad 0 < \theta < \pi/2 \tag{e}$$

3.15 Show that for a transversely isotropic elastic body the stress energy density

$$\widehat{W}(\mathbf{S}) = \frac{1}{2}\mathbf{S} \cdot \mathsf{K}[\mathbf{S}] \tag{a}$$

corresponding to the stress tensor given by Equation (b) of Problem 3.14 takes the form

$$\widehat{W}(\mathbf{S}) = \frac{1}{2}\mathbf{a}^2\left[\frac{1}{2}(c_{11} - c_{12})\sin^2\theta + c_{44}\cos^2\theta\right] \tag{b}$$

Also, show that

$$\widehat{W}(\mathbf{S}) = \widehat{W}(\mathbf{S}^{\perp}) - \widehat{W}(\mathbf{S}^{\parallel}) \tag{c}$$

where $\widehat{W}(\mathbf{S}^{\perp})$ and $\widehat{W}(\mathbf{S}^{\parallel})$ represent the "normal" and "tangential" stress energies, respectively, given by

$$\begin{aligned}\widehat{W}(\mathbf{S}^{\perp}) &\equiv \frac{1}{2}\mathbf{S}^{\perp}\cdot\mathsf{K}[\mathbf{S}^{\perp}]\\ &= \widehat{W}(\mathbf{S})\left[1 + \frac{1}{8}c_{44}^{-1}(c_{11} - c_{12})^{-1}(c_{11} - c_{12} - 2c_{44})^2\sin^2 2\theta\right]\end{aligned} \tag{d}$$

and

$$\begin{aligned}\widehat{W}(\mathbf{S}^{\parallel}) &\equiv \frac{1}{2}\mathbf{S}^{\parallel}\cdot\mathsf{K}[\mathbf{S}^{\parallel}]\\ &= \widehat{W}(\mathbf{S})\,\frac{1}{8}\,c_{44}^{-1}(c_{11} - c_{12})^{-1}(c_{11} - c_{12} - 2c_{44})^2\sin^2 2\theta\end{aligned} \tag{e}$$

Here $\mathbf{S}^{\perp}$ and $\mathbf{S}^{\parallel}$ are given by Equations (c) and (d), respectively, of Problem 3.14.

3.16 Let $\varphi(\theta) = \widehat{W}(\mathbf{S}^{\parallel})/\widehat{W}(\mathbf{S})^{\perp}$, where $\widehat{W}(\mathbf{S}^{\perp})$ and $\widehat{W}(\mathbf{S}^{\parallel})$ denote the normal and tangential stress energy densities of Problem 3.15. Show that

$$\max_{0\leq\theta\leq\pi/2}[\varphi(\theta)] = \varphi(\pi/4) = \frac{A^2}{1 + A^2} \tag{a}$$

where

$$A = \frac{1}{2\sqrt{2}}\frac{|c_{11} - c_{12} - 2c_{44}|}{(c_{11} - c_{12})^{1/2}c_{44}^{1/2}} \tag{b}$$

Note: When the body is isotropic we have

$$c_{11} = c_{33} = \lambda + 2\mu, \quad c_{12} = c_{13} = \lambda, \quad c_{44} = \mu \tag{c}$$

where λ and μ are the Lamé material constants. In this case Equation (b) reduces to $A = 0$, which means that for an isotropic body the tangential stress energy corresponding to the stress (d) of Problem 3.14 vanishes.

3.17 Let **u**, **E**, and **S** denote the displacement vector, strain tensor, and stress tensor fields, respectively, corresponding to a body force **b** and a temperature change T. Suppose that the fields **u**, **E**, and **S** satisfy the equations

$$\mathbf{E} = \widehat{\nabla}\mathbf{u} \quad \text{on } B \tag{a}$$

$$\operatorname{div}\mathbf{S} + \mathbf{b} = \mathbf{0} \quad \text{on } B \tag{b}$$

$$\mathbf{S} = \mathsf{C}[\mathbf{E}] + T\mathbf{M} \quad \text{on } B \tag{c}$$

where B is a bounded domain in E^3; and C and $\mathbf{M}$ denote the elasticity and stress–temperature tensors, respectively, independent of $\mathbf{x} \in \overline{B}$. Also, suppose that an alternative equation to Equation (c) reads

$$\mathbf{E} = \mathsf{K}[\mathbf{S}] + T\mathbf{A} \quad \text{on } B \tag{d}$$

where K and $\mathbf{A}$ represent the compliance and thermal expansion tensors, respectively. Let $\widehat{f} = \widehat{f}(B)$ denote the mean value of a function $f = f(\mathbf{x})$ on $\overline{B}$

$$\widehat{f}(B) = \frac{1}{v(B)} \int_B f(\mathbf{x})\, dv \tag{e}$$

where $v(B)$ stands for the volume of B. Show that

$$\widehat{\mathbf{E}}(B) = \frac{1}{v(B)} \int_{\partial B} \operatorname{sym}(\mathbf{u} \otimes \mathbf{n})\, da \tag{f}$$

and

$$\widehat{\mathbf{S}}(B) = \frac{1}{v(B)} \left[\int_{\partial B} \operatorname{sym}(\mathbf{x} \otimes \mathbf{S}\mathbf{n})\, da + \int_B \operatorname{sym}(\mathbf{x} \otimes \mathbf{b})\, dv \right] \tag{g}$$

where **n** is the outward unit normal on ∂B. Also, show that

$$\widehat{\mathbf{E}}(B) = \mathsf{K}[\widehat{\mathbf{S}}(B)] + \widehat{T}(B)\mathbf{A} \tag{h}$$

and

$$\widehat{\mathbf{S}}(B) = \mathsf{C}[\widehat{\mathbf{E}}(B)] + \widehat{T}(B)\mathbf{M} \tag{i}$$

3.18 The volume change $\delta v(B)$ associated with the fields **u**, **E**, and **S** in Problem 3.17 is defined by (see Equation 3.1.41)

$$\delta v(B) = v(B)\operatorname{tr}\widehat{\mathbf{E}}(B) \tag{a}$$

Show that

(i)

$$\delta v(B) = 0, \quad \widehat{\mathbf{S}}(B) = \widehat{T}(B)\mathbf{M}$$
$$\text{if } \mathbf{u} = 0 \text{ on } \partial B \tag{b}$$

and

(ii)

$$\widehat{\mathbf{S}}(B) = \mathbf{0}, \quad \widehat{\mathbf{E}}(B) = \widehat{T}(B)\mathbf{A}, \quad \delta v(B) = v(B)\widehat{T}(B)\text{tr}\,\mathbf{A}$$
$$\text{if } \mathbf{Sn} = \mathbf{0} \text{ on } \partial B \text{ and } \mathbf{b} = \mathbf{0} \text{ on } \overline{B} \tag{c}$$

Note: Equations (c) imply that the volume change $\delta v(B)$ of a homogeneous isotropic thermoelastic body with zero stress vector on ∂B and zero body force vector on $\overline{B}$ subject to a temperature change T on B is given by

$$\delta v(B) = 3\alpha \widehat{T}(B) v(B) \tag{d}$$

where α is the coefficient of linear thermal expansion of the body.

REFERENCES

1. E. Cesàro, Sulle formole del Volterra, fondomentali nella teoria delle distorsioni elastiche, *Rend. Acad. Sci. Fis. Mat. Nap.*, 12, 1906, 311–321.
2. Y. C. Fung, *Foundations of Solid Mechanics*, Prentice-Hall, Englewood Cliffs, NJ, 1965, pp. 101–103.
3. M. E. Gurtin, *The Linear Theory of Elasticity*, Encyclopedia of Physics, chief editor: S. Flügge, vol. VI a/2, editor: C. Truesdell, Springer, Berlin, Germany, 1972, pp. 56–57.
4. F. W. Hehl and Y. Itin, The Cauchy relations in linear elasticity theory, *J. Elasticity*, 66, 2002, 185–192.

4 Formulation of Problems of Elasticity

In Chapter 3, the field equations describing a linear elastic solid were discussed. In this chapter, these field equations are used to formulate the boundary value problems of elastostatics and the initial boundary value problems of elastodynamics; in particular, the mixed boundary value problems of isothermal and nonisothermal elastostatics, as well as the pure displacement and the pure stress problems of classical elastodynamics are discussed. The Betti reciprocal theorem of elastostatics and Graffi's reciprocal theorem of elastodynamics together with the uniqueness theorems are also presented. An emphasis is made on a pure stress initial boundary value problem of incompatible elastodynamics in which a body possesses initially distributed defects. The chapter contains both solved examples and end-of-chapter problems with solutions provided in the Solutions Manual; problems in which the stress reciprocity relation of incompatible elastodynamics are discussed deserve special attention.

4.1 BOUNDARY VALUE PROBLEMS OF ELASTOSTATICS

4.1.1 Field Equations of Elastostatics

In Chapter 3, we developed kinematics, equations of equilibrium, and constitutive equations for a linear elastic body. We start by recalling the fundamental system of field equations.

The strain–displacement relation

$$\mathbf{E} = \widehat{\nabla}\mathbf{u} = \frac{1}{2}\left(\nabla\mathbf{u} + \nabla\mathbf{u}^T\right) \tag{4.1.1}$$

The equations of equilibrium

$$\operatorname{div}\mathbf{S} + \mathbf{b} = \mathbf{0}, \quad \mathbf{S} = \mathbf{S}^T \tag{4.1.2}$$

The stress–strain relation

$$\mathbf{S} = \mathsf{C}[\mathbf{E}] \tag{4.1.3}$$

In these equations, $\mathbf{b}$ is a prescribed body force field, and the functions $\mathbf{u}, \mathbf{E}, \mathbf{S}$ are to be found at each point of the body. Now we will reduce Equations 4.1.1 through 4.1.3 to the displacement equations of elastostatics. To this end, we substitute Equation 4.1.1 into Equation 4.1.3 to obtain the stress–displacement equation

$$\mathbf{S} = \mathsf{C}[\widehat{\nabla}\mathbf{u}] \tag{4.1.4}$$

Next, substituting Equation 4.1.4 into Equation 4.1.2, we obtain the displacement equation of equilibrium

$$\operatorname{div} \mathbf{C}[\nabla \mathbf{u}] + \mathbf{b} = \mathbf{0} \tag{4.1.5}$$

Obviously, if $\mathbf{u}$ is a solution of Equation 4.1.5, then the stress tensor $\mathbf{S}$ is obtained from Equation 4.1.4 and the strain tensor $\mathbf{E}$ is obtained from Equation 4.1.1. This implies that finding a solution $[\mathbf{u}, \mathbf{E}, \mathbf{S}]$ to Equations 4.1.1 through 4.1.3 may be reduced to finding a displacement $\mathbf{u}$ that satisfies Equation 4.1.5.

In components, Equation 4.1.5 reads

$$(C_{ijkl} u_{k,l})_{,j} + b_i = 0 \tag{4.1.6}$$

We note that for an anisotropic nonhomogeneous body, the task of finding an analytical solution to Equation 4.1.6 is hopeless except for simple particular cases. In the following, we discuss an isotropic elastic body. For a nonhomogeneous isotropic body, the stress–strain relation, written in terms of Lamé coefficients, has the form

$$\mathbf{S}(\mathbf{x}) = 2\mu(\mathbf{x})\mathbf{E}(\mathbf{x}) + \lambda(\mathbf{x})[\operatorname{tr} \mathbf{E}(\mathbf{x})]\,\mathbf{1} \tag{4.1.7}$$

Substituting Equation 4.1.1 into this relation, we obtain

$$\mathbf{S}(\mathbf{x}) = \mu(\mathbf{x})(\nabla \mathbf{u} + \nabla \mathbf{u}^T) + \lambda(\mathbf{x})(\operatorname{div} \mathbf{u})\,\mathbf{1} \tag{4.1.8}$$

In components, this reads

$$S_{ij} = \mu(u_{i,j} + u_{j,i}) + \lambda u_{k,k}\delta_{ij} \tag{4.1.9}$$

Computing $S_{ij,j}$ and taking into account the fact that λ and μ depend on $\mathbf{x}$, we get

$$S_{ij,j} = \mu_{,j}(u_{i,j} + u_{j,i}) + \mu(u_{i,jj} + u_{j,ij}) + \lambda_{,i}u_{k,k} + \lambda u_{k,ki} \tag{4.1.10}$$

Thus, the displacement equation of equilibrium 4.1.6 for a nonhomogeneous isotropic body written in components is

$$\mu u_{i,jj} + (\lambda + \mu)u_{k,ki} + (u_{i,j} + u_{j,i})\mu_{,j} + u_{k,k}\lambda_{,i} + b_i = 0 \tag{4.1.11}$$

This equation presented in direct notation, see Section 2.2, takes the form

$$\mu \nabla^2 \mathbf{u} + (\lambda + \mu)\nabla(\operatorname{div} \mathbf{u}) + 2(\widehat{\nabla \mathbf{u}})\nabla\mu + (\operatorname{div} \mathbf{u})\nabla\lambda + \mathbf{b} = \mathbf{0} \tag{4.1.12}$$

Equation 4.1.11 is associated with C. L. M. H. Navier.

Example 4.1.1

Show that for a homogeneous isotropic body the displacement equation of equilibrium 4.1.12 reduces to

$$\mu \nabla^2 \mathbf{u} + (\lambda + \mu)\nabla(\operatorname{div} \mathbf{u}) + \mathbf{b} = \mathbf{0} \tag{a}$$

or

$$\nabla^2 \mathbf{u} + \frac{1}{1-2\nu} \nabla(\operatorname{div} \mathbf{u}) + \frac{\mathbf{b}}{\mu} = \mathbf{0} \tag{b}$$

Solution

Since the body is homogeneous,

$$\nabla \mu = \mathbf{0} \quad \text{and} \quad \nabla \lambda = \mathbf{0} \tag{c}$$

and Equation 4.1.12 reduces to (a). Equation (b) is obtained from (a) by observing that

$$\frac{\lambda + \mu}{\mu} = \frac{1}{1-2\nu} \tag{d}$$

This completes the solution. □

Example 4.1.2

Show that the displacement equation of equilibrium for a homogeneous isotropic body may be written in the form

$$(\lambda + 2\mu)\nabla(\operatorname{div} \mathbf{u}) - \mu \operatorname{curl} \operatorname{curl} \mathbf{u} + \mathbf{b} = \mathbf{0} \tag{a}$$

Solution

Using (a) from Example 4.1.1 as well as the identity

$$\nabla^2 \mathbf{u} = \nabla \operatorname{div} \mathbf{u} - \operatorname{curl} \operatorname{curl} \mathbf{u} \tag{b}$$

we reduce (a) of Example 4.1.1 to the form of (a) in this example. □

Example 4.1.3

Use the result (a) from Example 4.1.1 to show that if

$$\operatorname{div} \mathbf{b} = 0 \quad \text{and} \quad \operatorname{curl} \mathbf{b} = \mathbf{0} \tag{a}$$

then

$$\nabla^2 \nabla^2 \mathbf{u} = \mathbf{0} \tag{b}$$

Solution

Applying the operator div to both sides of (a) of Example 4.1.1 and writing in components, we find

$$\mu u_{i,kki} + (\lambda + \mu) u_{k,kii} + b_{i,i} = 0 \tag{c}$$

Hence,

$$(\lambda + 2\mu)u_{i,ikk} + b_{i,i} = 0 \tag{d}$$

Since $b_{i,i}$ is assumed to be zero and

$$\lambda + 2\mu > 0 \tag{e}$$

then

$$u_{i,ikk} = 0 \tag{f}$$

Next, by applying the operator curl to (a) of Example 4.1.1 we get

$$\mu\nabla^2(\text{curl}\,\mathbf{u}) = \mathbf{0} \tag{g}$$

because

$$\text{curl}\,[\nabla(\text{div}\,\mathbf{u})] = \mathbf{0} \tag{h}$$

and, by second equation of (a), curl $\mathbf{b} = \mathbf{0}$. Hence, since $\mu > 0$ Equations (f) and (g) take the form

$$\text{div}(\nabla^2\mathbf{u}) = 0 \quad \text{and} \quad \text{curl}(\nabla^2\mathbf{u}) = \mathbf{0} \tag{i}$$

Recalling (b) of Example 4.1.2, which is valid for any vector field $\mathbf{u}$, and replacing in that equation $\mathbf{u}$ by $\nabla^2\mathbf{u}$ and using (i) we arrive at the equation

$$\nabla^2\nabla^2\mathbf{u} = \mathbf{0} \tag{j}$$

This completes the solution. □

Example 4.1.4

Show that the stress vector $\mathbf{s}$ on the boundary ∂B of an isotropic body B may be represented by

$$\mathbf{s} = 2\mu\,\frac{\partial\mathbf{u}}{\partial\mathbf{n}} + \mu\mathbf{n} \times \text{curl}\,\mathbf{u} + \lambda(\text{div}\,\mathbf{u})\mathbf{n} \tag{a}$$

where $\mathbf{u}$ is a displacement vector, $\mathbf{n}$ is an outward vector normal to ∂B, and

$$\frac{\partial\mathbf{u}}{\partial\mathbf{n}} = (\nabla\mathbf{u})\mathbf{n} \tag{b}$$

Solution

The stress vector $\mathbf{s}$ is calculated from Equation 3.2.12

$$\mathbf{s} = \mathbf{S}\mathbf{n} \tag{c}$$

where, see Equation 4.1.8,

$$\mathbf{S} = \mu(\nabla\mathbf{u} + \nabla\mathbf{u}^T) + \lambda(\operatorname{div}\mathbf{u})\mathbf{1} \tag{d}$$

Substituting (d) into (c) and writing in components, we obtain

$$s_i = S_{ij}n_j = \mu(u_{i,j}n_j + u_{j,i}n_j) + \lambda u_{k,k}n_i \tag{e}$$

or

$$s_i = 2\mu u_{i,j}n_j + \mu(u_{j,i} - u_{i,j})n_j + \lambda u_{k,k}n_i \tag{f}$$

To show that this equation is the same as (a) we need to prove that

$$(u_{j,i} - u_{i,j})n_j = \epsilon_{ipq}n_p\epsilon_{qab}u_{b,a} \tag{g}$$

Now, if we use the $\epsilon - \delta$ relation

$$\epsilon_{ipq}\epsilon_{abq} = \delta_{ia}\delta_{pb} - \delta_{ib}\delta_{ap} \tag{h}$$

we arrive at

$$(u_{j,i} - u_{i,j})n_j = (\delta_{ia}\delta_{pb} - \delta_{ib}\delta_{ap})n_pu_{b,a} = (u_{p,i} - u_{i,p})n_p \tag{i}$$

This ends the solution.

This example shows that if $\operatorname{div}\mathbf{u} = 0$ on ∂B and $\operatorname{curl}\mathbf{u} = \mathbf{0}$ on ∂B then the stress vector $\mathbf{s}$ is given by a simple formula

$$\mathbf{s} = 2\mu\frac{\partial\mathbf{u}}{\partial\mathbf{n}} \quad \text{on } \partial B \tag{j}$$

□

We now derive the stress equations of elastostatics. These are obtained by eliminating $\mathbf{u}$ and $\mathbf{E}$ from Equations 4.1.1 through 4.1.3. To this end, first, we write the constitutive relation 4.1.3 in its inverted equivalent form

$$\mathbf{E} = \mathsf{K}[\mathbf{S}] \tag{4.1.13}$$

Next, eliminating $\mathbf{u}$ and $\mathbf{E}$ from Equations 4.1.1 through 4.1.3 we arrive at the following form of the stress equations of elastostatics for a nonhomogeneous anisotropic body:

$$\operatorname{div}\mathbf{S} + \mathbf{b} = \mathbf{0}, \quad \mathbf{S} = \mathbf{S}^T \tag{4.1.14}$$

$$\operatorname{curl}\operatorname{curl}\mathsf{K}[\mathbf{S}] = \mathbf{0} \tag{4.1.15}$$

Equations 4.1.14 are identical to Equations 4.1.2, and Equation 4.1.15 is obtained as follows: We combine Equation 4.1.1 with Equation 4.1.13 and have

$$\mathbf{E} = \widehat{\nabla}\mathbf{u} = \mathsf{K}[\mathbf{S}] \tag{4.1.16}$$

Applying now curl curl operation to Equation 4.1.16 we arrive at Equation 4.1.15, which is called the *stress compatibility equation* as opposed to the *strain compatibility equation* given in Equation 3.1.47, which we rewrite here:

$$\text{curl}\,\text{curl}\,\mathbf{E} = \mathbf{0} \tag{4.1.17}$$

For a simply connected body B the stress equations of elastostatics (nine scalar partial differential equations), Equations 4.1.14 and 4.1.15, imply the fundamental system of field equations 4.1.1 through 4.1.3 in the following sense. If $\mathbf{S}$ is a solution of Equations 4.1.14 and 4.1.15, then $\mathbf{S}$ satisfies Equations 4.1.2. Also, if we define the strain tensor $\mathbf{E}$ in terms of $\mathbf{S}$ by Equation 4.1.13, we find by Equation 4.1.15 that $\mathbf{E}$ satisfies the compatibility equation 4.1.17. Hence, by the theorem given in Section 3.1.3 there exists a displacement field $\mathbf{u}$ on B such that $\mathbf{E}$ and $\mathbf{u}$ satisfy the strain–displacement equation 4.1.1. In this way, for a solution $\mathbf{S}$ to Equations 4.1.14 and 4.1.15 there are a vector field $\mathbf{u}$ and a strain tensor $\mathbf{E}$ such that a triple $[\mathbf{u}, \mathbf{E}, \mathbf{S}]$ satisfies Equations 4.1.1 through 4.1.3.

In case of a multiply connected body the stress equations of elastostatics 4.1.14 and 4.1.15 are complemented by additional conditions to secure the existence of a triple $[\mathbf{u}, \mathbf{E}, \mathbf{S}]$ that satisfies Equations 4.1.1 through 4.1.3 [1,2].

For an isotropic homogeneous elastic body the strain–stress relations 4.1.13 take the form, see Equation 3.3.56,

$$\mathbf{E} = \frac{1}{2\mu}\left[\mathbf{S} - \frac{\lambda}{3\lambda + 2\mu}\,(\text{tr}\,\mathbf{S})\mathbf{1}\right] \tag{4.1.18}$$

or, by using the table of relations between elasticity constants provided in the front matter,

$$\mathbf{E} = \frac{1}{2\mu}\left[\mathbf{S} - \frac{\nu}{1+\nu}\,(\text{tr}\,\mathbf{S})\mathbf{1}\right] \tag{4.1.19}$$

and the stress compatibility equation 4.1.15 takes much simpler form than that of the anisotropic case. To obtain the stress equations of elastostatics for an isotropic body, we first note that the strain compatibility equation 4.1.17 is equivalent to, see (a) in Example 3.1.15,

$$\nabla^2\mathbf{E} + \nabla\nabla(\text{tr}\,\mathbf{E}) - 2\widehat{\nabla}(\text{div}\,\mathbf{E}) = \mathbf{0} \tag{4.1.20}$$

In components this equation reads

$$E_{ij,kk} + E_{kk,ij} - (E_{ik,kj} + E_{jk,ki}) = 0 \tag{4.1.21}$$

Also, note that Equation 4.1.19 written in components is

$$E_{ij} = \frac{1}{2\mu}\left[S_{ij} - \frac{\nu}{1+\nu}\,S_{aa}\delta_{ij}\right] \tag{4.1.22}$$

Putting $i = j = k$ in Equation 4.1.22, we get

$$E_{kk} = \frac{1}{2\mu}\,\frac{1-2\nu}{1+\nu}\,S_{aa} \tag{4.1.23}$$

Also, calculating div **E** in components, in view of Equation 4.1.22, we write

$$E_{ik,k} = \frac{1}{2\mu}\left(S_{ik,k} - \frac{\nu}{1+\nu}\, S_{aa,i}\right) \tag{4.1.24}$$

This relation, with the help of the equilibrium equation 4.1.14, which is given here as

$$S_{ik,k} + b_i = 0 \tag{4.1.25}$$

reduces to

$$E_{ik,k} = \frac{1}{2\mu}\left(-\, b_i - \frac{\nu}{1+\nu}\, S_{aa,i}\right) \tag{4.1.26}$$

Hence, we obtain

$$E_{ik,kj} = \frac{1}{2\mu}\left(-\, b_{i,j} - \frac{\nu}{1+\nu}\, S_{aa,ij}\right) \tag{4.1.27}$$

Similarly,

$$E_{jk,ki} = \frac{1}{2\mu}\left(-\, b_{j,i} - \frac{\nu}{1+\nu}\, S_{aa,ij}\right) \tag{4.1.28}$$

Next, substituting Equation 4.1.22 into Equation 4.1.21 and using Equations 4.1.23, 4.1.27, and 4.1.28 and multiplying by 2μ, we get

$$\begin{aligned} S_{ij,kk} - \frac{\nu}{1+\nu}\, S_{aa,kk}\delta_{ij} + \frac{1-2\nu}{1+\nu}\, S_{aa,ij} \\ + \frac{2\nu}{1+\nu}\, S_{aa,ij} + b_{i,j} + b_{j,i} = 0 \end{aligned} \tag{4.1.29}$$

This equation after simplification becomes

$$S_{ij,kk} + \frac{1}{1+\nu}\, S_{aa,ij} - \frac{\nu}{1+\nu}\, S_{aa,kk}\delta_{ij} + b_{i,j} + b_{j,i} = 0 \tag{4.1.30}$$

Taking the trace of this equation, thus letting $i = j = a$ in Equation 4.1.30, we arrive at

$$\left(1 + \frac{1}{1+\nu} - \frac{3\nu}{1+\nu}\right) S_{aa,kk} + 2b_{a,a} = 0 \tag{4.1.31}$$

or

$$\frac{1-\nu}{1+\nu}\, S_{aa,kk} + b_{a,a} = 0 \tag{4.1.32}$$

Substituting this equation into Equation 4.1.30 we obtain the final form of the stress compatibility equation of elastostatics for a homogeneous isotropic body

$$S_{ij,kk} + \frac{1}{1+\nu} S_{aa,ij} + \frac{\nu}{1-\nu} b_{a,a}\delta_{ij} + b_{i,j} + b_{j,i} = 0 \tag{4.1.33}$$

This equation is known as the *Beltrami–Michell stress compatibility equation*.

We note that a complete set of stress equations of elastostatics for a homogeneous isotropic simply connected body consists of

The equilibrium equations

$$S_{ik,k} + b_i = 0, \quad S_{ik} = S_{ki} \tag{4.1.34}$$

The stress compatibility equation

$$S_{ij,kk} + \frac{1}{1+\nu} S_{aa,ij} + \frac{\nu}{1-\nu} b_{a,a}\delta_{ij} + b_{i,j} + b_{j,i} = 0 \tag{4.1.35}$$

In direct notation these equations read, respectively,

$$\operatorname{div} \mathbf{S} + \mathbf{b} = \mathbf{0}, \quad \mathbf{S} = \mathbf{S}^T \tag{4.1.36}$$

$$\nabla^2\mathbf{S} + \frac{1}{1+\nu} \nabla\nabla(\operatorname{tr}\mathbf{S}) + \frac{\nu}{1-\nu} (\operatorname{div}\mathbf{b})\mathbf{1} + 2\widehat{\nabla}\mathbf{b} = \mathbf{0} \tag{4.1.37}$$

Example 4.1.5

Show that for an isotropic body if $\mathbf{b} = \mathbf{0}$ then

$$\nabla^2\nabla^2\mathbf{S} = \mathbf{0} \tag{a}$$

Solution

For a homogeneous isotropic body with body forces equal to zero

$$\nabla^2\mathbf{S} + \frac{1}{1+\nu} \nabla\nabla(\operatorname{tr}\mathbf{S}) = \mathbf{0} \tag{b}$$

In components

$$S_{ij,kk} + \frac{1}{1+\nu} S_{kk,ij} = 0 \tag{c}$$

Taking the trace of this equation, we obtain, after putting $i = j = a$,

$$S_{aa,kk} = 0 \tag{d}$$

Applying now the Laplace operator to (c), we write

$$S_{ij,kkaa} + \frac{1}{1+\nu} S_{kk,ijaa} = 0 \tag{e}$$

Becuase of (d) the second term in (e) vanishes and (e) reduces to

$$S_{ij,kkaa} = 0 \tag{f}$$

which is equivalent to (a). This completes the solution. □

The result of this example may also be obtained by using (b) of Example 4.1.3. To this end, (b) of Example 4.1.3 should be combined with the stress–displacement relation, Equation 4.1.9.

Example 4.1.6

Show that if the components S_{i3} $(i = 1, 2, 3)$ of a stress tensor field **S** corresponding to zero body forces are given by

$$S_{i3} = -\frac{x_3}{2(1+\nu)} S_{aa,i} + \Psi_i \tag{a}$$

where Ψ_i are harmonic, then S_{i3} satisfies the equation

$$S_{i3,kk} + \frac{1}{1+\nu} S_{kk,i3} = 0 \tag{b}$$

Solution

From (a) we obtain

$$S_{i3,k} = -\frac{1}{2(1+\nu)} (\delta_{3k} S_{aa,i} + x_3 S_{aa,ik}) + \Psi_{i,k} \tag{c}$$

$$S_{i3,kk} = -\frac{1}{2(1+\nu)} (2\delta_{3k} S_{aa,ik} + x_3 S_{aa,ikk}) + \Psi_{i,kk} \tag{d}$$

The last term in (d) is equal to zero because Ψ_i is harmonic, and (d) reduces to

$$S_{i3,kk} + \frac{1}{1+\nu} S_{aa,i3} + \frac{1}{2(1+\nu)} x_3 S_{aa,ikk} = 0 \tag{e}$$

Since **S** is a stress tensor corresponding to zero body forces, its trace is harmonic, see (d) in Example 4.1.5. This means that the third term in (e) vanishes, and (e) reduces to (b).

This completes the solution. □

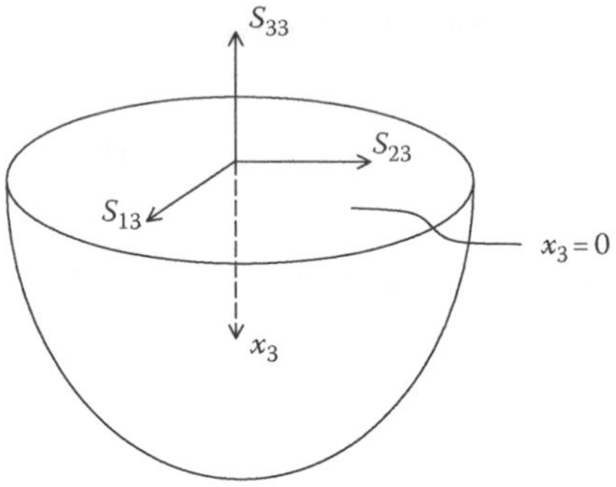

FIGURE 4.1 Components of a stress vector S_{i3} on the boundary of a semispace.

Note: In Example 4.1.6 the representation of stress components S_{i3} in the form of (a) allows us to express the stress vector on the boundary plane $x_3 = 0$ of a semispace $x_3 \geq 0$, see Figure 4.1, in terms of harmonic functions because

$$S_{i3}\big|_{x_3=0} = \Psi_i\big|_{x_3=0} \tag{4.1.38}$$

Since S_{aa} is harmonic, S_{i3} is expressed in terms of four harmonic functions S_{aa}, Ψ_1, Ψ_2, Ψ_3. These functions are restricted by the equilibrium equation

$$S_{i3,i} = 0 \tag{4.1.39}$$

which, by (c), is equivalent to the equation

$$\Psi_{i,i} - \frac{1}{2(1+\nu)} S_{aa,3} = 0 \tag{4.1.40}$$

Example 4.1.7

Show that a stress tensor **S** of homogeneous isotropic elastostatics corresponding to zero body forces may be represented by the formula

$$\mathbf{S} = -\frac{1}{2(1+\nu)} \text{ sym } [\mathbf{x} \otimes \nabla(\text{tr}\,\mathbf{S})] + \mathbf{H} \tag{a}$$

where **H** is a harmonic symmetric second-order tensor field, that is

$$\nabla^2\mathbf{H} = \mathbf{0}, \quad \mathbf{H} = \mathbf{H}^T \tag{b}$$

Solution

Tensor **S** satisfies the stress compatibility equation

$$\nabla^2\mathbf{S} + \frac{1}{1+\nu} \nabla\nabla(\text{tr}\,\mathbf{S}) = 0 \tag{c}$$

This equation implies that

$$\nabla^2(\mathrm{tr}\,\mathbf{S}) = 0 \tag{d}$$

To show that **S** as given by (a) satisfies (c) we write (a), (c), and (d) in components:

$$S_{ij} = -\frac{1}{4(1+\nu)}\,(x_i S_{aa,j} + x_j S_{aa,i}) + H_{ij} \tag{e}$$

$$S_{ij,kk} + \frac{1}{1+\nu}\,S_{aa,ij} = 0 \tag{f}$$

$$S_{aa,kk} = 0 \tag{g}$$

Next, computing the gradient of (e) we obtain

$$S_{ij,k} = -\frac{1}{4(1+\nu)}\,(\delta_{ik} S_{aa,j} + x_i S_{aa,jk} + \delta_{jk} S_{aa,i} + x_j S_{aa,ik}) + H_{ij,k} \tag{h}$$

This leads to the Laplacian of **S** in the form

$$S_{ij,kk} = -\frac{1}{4(1+\nu)}\,(2\delta_{ik} S_{aa,jk} + 2\delta_{jk} S_{aa,ik} + x_i S_{aa,jkk} + x_j S_{aa,ikk}) + H_{ij,kk} \tag{i}$$

Hence, from (b) written in components, and (g), we see that the stress tensor **S** given by (e) satisfies (f).

This completes the solution. □

Notes:

(1) The representation (a) in Example 4.1.7 is a generalization of the representation (a) in Example 4.1.6. As opposed to (a) in Example 4.1.6 it contains six harmonic scalar functions $H_{ij} = H_{ji}$ $(i, j = 1, 2, 3)$ and one harmonic scalar function S_{aa}. This means that **S** is expressed in terms of seven harmonic functions. Since **S** must also satisfy the equilibrium equation in the form

$$S_{ij,j} = 0 \tag{4.1.41}$$

the harmonic functions are subject to the constraints, see (h) in Example 4.1.7 with $k = j$

$$-\frac{1}{4(1+\nu)}\,(S_{aa,i} + x_i S_{aa,kk} + 3S_{aa,i} + x_k S_{aa,ik}) + H_{ij,j} = 0 \tag{4.1.42}$$

Since by (g) from Example 4.1.7 the second term in 4.1.42 vanishes, this equation reduces to

$$-\frac{1}{4(1+\nu)}\,(4S_{aa,i} + x_k S_{aa,ik}) + H_{ij,j} = 0 \tag{4.1.43}$$

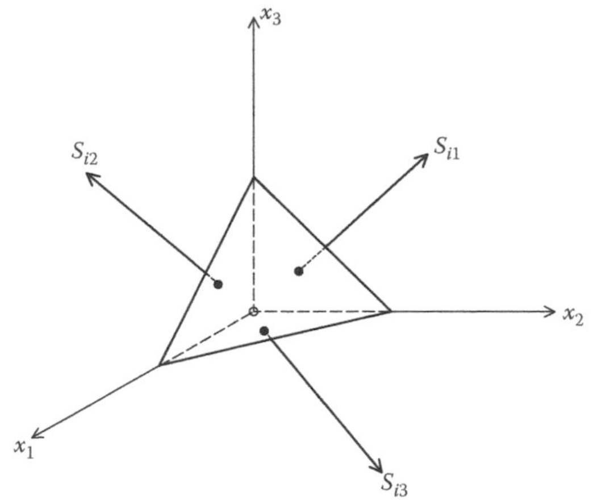

FIGURE 4.2 Vectors S_{i1}, S_{i2}, and S_{i3} on the boundary of a pyramid.

(2) The representation (a) in Example 4.1.7 may be useful in solving a stress boundary value problem for a homogeneous isotropic elastic pyramid shown in Figure 4.2, where S_{i1}, S_{i2}, and S_{i3} are prescribed surface loads on the planes $x_1 = 0$, $x_2 = 0$, and $x_3 = 0$, respectively. For the surface load on the plane $x_1 = 0$ we have

$$S_{i1} = -\frac{1}{4(1+\nu)}\,(x_i S_{aa,1} + x_1 S_{aa,i}) + H_{i1} \tag{4.1.44}$$

Thus

$$\begin{aligned} S_{11}\big|_{x_1=0} &= H_{11}\big|_{x_1=0} \\ S_{21}\big|_{x_1=0} &= -\frac{1}{4(1+\nu)}\, x_2 S_{aa,1}\big|_{x_1=0} + H_{21}\big|_{x_1=0} \\ S_{31}\big|_{x_1=0} &= -\frac{1}{4(1+\nu)}\, x_3 S_{aa,1}\big|_{x_1=0} + H_{31}\big|_{x_1=0} \end{aligned} \tag{4.1.45}$$

We observe, therefore, that the normal load on the plane $x_1 = 0$ is expressed by the harmonic function H_{11} only, while shear loads on the plane $x_1 = 0$ are expressed both by the harmonic components of tensor $\mathbf{H}$ and by the harmonic function $S_{aa,1}$.

Similar representations of the boundary conditions on the planes $x_2 = 0$ and $x_3 = 0$ are obtained by using (e) of Example 4.1.7.

Example 4.1.8

Given a stress field $S_{ij} = S_{ij}(\mathbf{x})$ of the form

$$S_{ij}(\mathbf{x}) = A_{ija} x_a + S^0_{ij}, \quad i, j, a = 1, 2, 3 \tag{a}$$

where A_{ija} are constants, $A_{ija} = A_{jia}$, $S^0_{ij} = S^0_{ji}$ = constant, and $A_{iaa} = 0$. Show that for a homogeneous isotropic simply connected body B subject to zero body forces, the stress field $S_{ij}(\mathbf{x})$ given by (a) satisfies the equilibrium equation

$$S_{ij,j} = 0 \text{ on } B \tag{b}$$

and a displacement vector $u_i = u_i(\mathbf{x})$ corresponding to $S_{ij} = S_{ij}(\mathbf{x})$ and satisfying the condition $u_i = u_i(\mathbf{0}) = 0$ takes the form

$$u_i(\mathbf{x}) = E^0_{ia}x_a + \frac{1}{2\mu}\left[\left(A_{iab} - \frac{1}{2}A_{abi}\right)x_a x_b - \frac{\nu}{1+\nu}\left(A_{ppa}x_a x_i - \frac{1}{2}A_{ppi}x_a x_a\right)\right] \tag{c}$$

where

$$E^0_{ij} = \frac{1}{2\mu}\left(S^0_{ij} - \frac{\nu}{1+\nu}S^0_{kk}\delta_{ij}\right) \tag{d}$$

Solution

Applying the div operator to (a) and using the condition $A_{iaa} = 0$ we find that (b) holds true. To obtain (c) we note that the strain field E_{ij} corresponding to S_{ij} takes the form

$$\begin{aligned} E_{ij} &= \frac{1}{2\mu}\left(S_{ij} - \frac{\nu}{1+\nu}S_{kk}\delta_{ij}\right) \\ &= \frac{1}{2\mu}\left(S^0_{ij} - \frac{\nu}{1+\nu}S^0_{kk}\delta_{ij}\right) + \frac{1}{2\mu}\left(A_{ija}x_a - \frac{\nu}{1+\nu}A_{ppa}x_a\delta_{ij}\right) \end{aligned} \tag{e}$$

It follows from (d) and (e) that

$$\text{curl curl } \mathbf{E} = \mathbf{0} \quad \text{or} \quad \epsilon_{ipq}\epsilon_{jrs}E_{qs,pr} = 0 \tag{f}$$

Hence, $\mathbf{E}$ given by (e) satisfies the compatibility relation (f) and, because of Equations 3.1.53 and 3.1.54, we obtain

$$u_i(\mathbf{x}) = \int_{x_0}^{x} U_{ij}(\mathbf{y}, \mathbf{x})\, dy_j \tag{g}$$

where

$$U_{ij}(\mathbf{y}, \mathbf{x}) = E_{ij}(\mathbf{y}) + (x_k - y_k)[E_{ij,k}(\mathbf{y}) - E_{kj,i}(\mathbf{y})] \tag{h}$$

and the line integral in (g) is path independent. Substitution of (e) into (g) yields

$$\begin{aligned} u_i(\mathbf{x}) &= \int_{x_0}^{x} E^0_{ij} dy_j + \frac{1}{2\mu}\int_{x_0}^{x}\left(A_{ija}y_a - \frac{\nu}{1+\nu}A_{ppa}y_a\delta_{ij}\right)dy_j \\ &+ \frac{1}{2\mu}\int_{x_0}^{x}(x_k - y_k)B_{ijk}dy_j \end{aligned} \tag{i}$$

where

$$B_{ijk} = A_{ijk} - A_{kji} - \frac{\nu}{1+\nu}(A_{ppk}\delta_{ij} - A_{ppi}\delta_{kj}) \tag{j}$$

Next, we select a point $\mathbf{x}_0 = \mathbf{0}$, and chose for the integration path, a straight line joining points $\mathbf{0}$ and $\mathbf{x}$. Then

$$\mathbf{x} = \mathbf{a}t, \quad 0 \le t \le t_0 \tag{k}$$

where $\mathbf{a}$ is a constant vector parallel to the straight line of the integration path, and t_0 is selected in such a way that $\mathbf{x} \in \overline{B}$. Thus, we obtain

$$y_j = a_j \tau, \quad 0 \le \tau \le t \tag{l}$$

and

$$dy_j = a_j d\tau \tag{m}$$

and the line integrals in (i) may be computed using the formulas

$$\int_{x_0}^{x} dy_j = \int_0^t a_j d\tau = a_j t = x_j$$

$$\int_{x_0}^{x} y_a dy_j = \int_0^t a_a \tau a_j d\tau = a_a a_j \int_0^t \tau \, d\tau$$

$$= \frac{1}{2}(a_a t)(a_j t) = \frac{1}{2} x_a x_j \tag{n}$$

and

$$\int_{x_0}^{x} (x_k - y_k) dy_j = \int_0^t (x_k - a_k \tau) a_j d\tau = x_k x_j - \frac{1}{2} x_k x_j = \frac{1}{2} x_k x_j \tag{o}$$

Substituting (n) and (o) into (i) we arrive at (c), where E_{ij}^0 is given by (d). Also, we check that $u_i = u_i(\mathbf{0}) = 0$.

This completes the solution. □

Note: It is easy to show that by taking the symmetric gradient of (c) we have $(u_{i,j} + u_{j,i})/2 = E_{ij}$, where E_{ij} is given by (e). This proves that (c) is a displacement field corresponding to the stress field of the form (a). Also, we note that any other displacement corresponding to the stress (a) differs from (c) by the rigid rotation of the form $u_i^*(\mathbf{x}) = W_{ik} x_k$, where $W_{ik} = -W_{ki}$ is a constant skew tensor.

4.1.2 Field Equations of Thermoelastostatics with Prescribed Temperature

In this section, we develop the displacement–temperature and the stress–temperature field equations of thermoelastostatics for an isotropic body. These equations are generalizations of Equation (a) of Example 4.1.1 and Equations 4.1.36 and 4.1.37, and will be derived assuming that the temperature change T is a prescribed field on B.

To derive the displacement–temperature equations of equilibrium for this case we recall, see Equations 4.1.1, 4.1.2, and 3.3.125, the strain–displacement relation

$$\mathbf{E} = \widehat{\nabla} \mathbf{u} \tag{4.1.46}$$

the equations of equilibrium

$$\operatorname{div}\mathbf{S} + \mathbf{b} = \mathbf{0}, \quad \mathbf{S} = \mathbf{S}^T \tag{4.1.47}$$

and the stress–strain–temperature relation

$$\mathbf{S} = 2\mu\mathbf{E} + \lambda(\operatorname{tr}\mathbf{E})\mathbf{1} - (3\lambda + 2\mu)\alpha T\mathbf{1} \tag{4.1.48}$$

By eliminating **E** and **S** from these equations, that is, substituting Equation 4.1.46 into Equation 4.1.48 and Equation 4.1.48 into Equation 4.1.47, we arrive at the following *displacement–temperature field equation*

$$\mu\nabla^2\mathbf{u} + (\lambda + \mu)\nabla(\operatorname{div}\mathbf{u}) - (3\lambda + 2\mu)\alpha\nabla T + \mathbf{b} = \mathbf{0} \tag{4.1.49}$$

To obtain the stress–temperature compatibility equations of thermoelastostatics, we proceed in the following way.

First, we write Equation 4.1.49 in components,

$$\mu u_{i,kk} + (\lambda + \mu)u_{k,ki} = \gamma T_{,i} - b_i \tag{4.1.50}$$

where

$$\gamma = (3\lambda + 2\mu)\alpha \tag{4.1.51}$$

Using the relation 4.1.46 written in components

$$2E_{ij} = u_{i,j} + u_{j,i} \tag{4.1.52}$$

the Equation 4.1.50, after differentiation with respect to x_j and taking the symmetric part of both sides with respect to indexes i and j, is transformed to

$$\mu E_{ij,kk} + (\lambda + \mu)E_{kk,ij} = \gamma T_{,ij} - b_{(i,j)} \tag{4.1.53}$$

Now, let us recall the strain–stress–temperature relation, see Equation 3.3.126,

$$2\mu E_{ij} = S_{ij} - \frac{\lambda}{3\lambda + 2\mu} S_{aa}\delta_{ij} + 2\mu\alpha T\delta_{ij} \tag{4.1.54}$$

Taking the trace of this equation, we obtain

$$2\mu E_{kk} = \frac{2\mu}{3\lambda + 2\mu} S_{kk} + 6\mu\alpha T \tag{4.1.55}$$

Next, we substitute E_{ij} from Equation 4.1.54 and E_{kk} from Equation 4.1.55 into Equation 4.1.53, and arrive at

$$S_{ij,kk} + \frac{2(\lambda + \mu)}{3\lambda + 2\mu} S_{aa,ij} - \frac{\lambda\delta_{ij}}{3\lambda + 2\mu} S_{aa,kk} + 2\mu\alpha(T_{,ij} + T_{,kk}\delta_{ij}) + 2b_{(i,j)} = 0 \tag{4.1.56}$$

Taking the trace of Equation 4.1.56, we obtain

$$S_{aa,kk} + \frac{4\mu(3\lambda + 2\mu)\alpha}{\lambda + 2\mu} T_{,kk} + \frac{3\lambda + 2\mu}{\lambda + 2\mu} b_{k,k} = 0 \tag{4.1.57}$$

Substituting $S_{aa,kk}$ from Equation 4.1.57 into Equation 4.1.56 yields

$$\begin{aligned} S_{ij,kk} &+ \frac{2(\lambda + \mu)}{3\lambda + 2\mu} S_{aa,ij} + 2\mu\alpha \left(T_{,ij} + \frac{3\lambda + 2\mu}{\lambda + 2\mu} T_{,kk}\delta_{ij} \right) \\ &+ \frac{\lambda}{\lambda + 2\mu} b_{k,k}\delta_{ij} + 2b_{(i,j)} = 0 \end{aligned} \tag{4.1.58}$$

Using the table of relations between elasticity constants provided in the front matter, we note the equalities

$$\begin{aligned} \frac{2(\lambda + \mu)}{3\lambda + 2\mu} &= \frac{1}{1 + \nu}, \quad \frac{3\lambda + 2\mu}{\lambda + 2\mu} = \frac{1 + \nu}{1 - \nu}, \\ \frac{\lambda}{\lambda + 2\mu} &= \frac{\nu}{1 - \nu}, \quad 2\mu = \frac{E}{1 + \nu} \end{aligned} \tag{4.1.59}$$

Finally, substitution of relations 4.1.59 into Equation 4.1.58 provides the *stress–temperature compatibility equations of thermoelastostatics* for an isotropic body

$$\begin{aligned} S_{ij,kk} &+ \frac{1}{1 + \nu} S_{aa,ij} + \frac{E\alpha}{1 + \nu} \left(T_{,ij} + \frac{1 + \nu}{1 - \nu} T_{,kk}\delta_{ij} \right) \\ &+ \frac{\nu}{1 - \nu} b_{k,k}\delta_{ij} + 2b_{(i,j)} = 0 \end{aligned} \tag{4.1.60}$$

Equations 4.1.47 and 4.1.60, written in direct notation, yield the *stress–temperature field equations* for an isotropic body

$$\operatorname{div} \mathbf{S} + \mathbf{b} = \mathbf{0}, \quad \mathbf{S} = \mathbf{S}^T \tag{4.1.61}$$

$$\begin{aligned} \nabla^2 \mathbf{S} &+ \frac{1}{1 + \nu} \nabla\nabla(\operatorname{tr} \mathbf{S}) + \frac{E\alpha}{1 + \nu} \left(\nabla\nabla T + \frac{1 + \nu}{1 - \nu} \nabla^2 T \mathbf{1} \right) \\ &+ \frac{\nu}{1 - \nu} (\operatorname{div} \mathbf{b})\mathbf{1} + 2\widehat{\nabla \mathbf{b}} = \mathbf{0} \end{aligned} \tag{4.1.62}$$

We note that the displacement–temperature equation 4.1.49 and the stress–temperature equations 4.1.61 and 4.1.62 are complete in the following sense:

(a) Completeness of the displacement–temperature equation 4.1.49

Let $\mathbf{u}$ be a solution to Equation 4.1.49, and define $\mathbf{E}$ in terms of $\mathbf{u}$ by Equation 4.1.46, and $\mathbf{S}$ by Equation 4.1.48. Then the triple $[\mathbf{u}, \mathbf{E}, \mathbf{S}]$ satisfies Equations 4.1.46 through 4.1.48.

(b) Completeness of the stress–temperature equations 4.1.61 and 4.1.62

If $\mathbf{S}$ satisfies Equations 4.1.61 and 4.1.62, then the corresponding strain $\mathbf{E}$ defined by Equation 4.1.54 satisfies the equation of compatibility. So, if B is simply connected, there is a displacement field $\mathbf{u}$ that satisfies the strain–displacement relation 4.1.46. As a result, the triple $[\mathbf{u}, \mathbf{E}, \mathbf{S}]$ satisfies Equations 4.1.46 through 4.1.48.

Also, it is easy to see that if $T = \text{const}$, Equations 4.1.49 and 4.1.62 reduce to the displacement and stress field equations, respectively, of isothermal isotropic elastostatics, see Equation (a) in Example 4.1.1 and Equation 4.1.37.

Example 4.1.9

Show that if $\mathbf{b} = \mathbf{0}$ then a particular solution of thermoelastostatics for a homogeneous isotropic body takes the form

$$\mathbf{u} = \nabla\phi \tag{a}$$

$$\mathbf{E} = \nabla\nabla\phi \tag{b}$$

$$\mathbf{S} = 2\mu(\nabla\nabla\phi - \nabla^2\phi\mathbf{1}) \tag{c}$$

where ϕ is a scalar field that satisfies Poisson's equation

$$\nabla^2\phi = mT \tag{d}$$

with

$$m = \frac{1+\nu}{1-\nu}\alpha \tag{e}$$

Solution

The displacement–temperature field equation of thermoelastostatics with zero body force takes the form, see Equation 4.1.49,

$$\mu\nabla^2\mathbf{u} + (\lambda + \mu)\nabla(\operatorname{div}\mathbf{u}) - (3\lambda + 2\mu)\alpha\nabla T = \mathbf{0} \tag{f}$$

Substituting (a) into this equation we arrive at

$$\nabla[(\lambda + 2\mu)\nabla^2\phi - (3\lambda + 2\mu)\alpha T] = \mathbf{0} \tag{g}$$

Dividing this equation by $\lambda + 2\mu$ and considering that (see Equation 4.1.59)

$$\frac{3\lambda + 2\mu}{\lambda + 2\mu} = \frac{1+\nu}{1-\nu} \tag{h}$$

we find that if ϕ is a solution to (d) then $\mathbf{u}$ given by (a) satisfies the displacement–temperature equation (f). To obtain (b) we substitute (a) into Equation 4.1.46. Finally, to obtain **S** in the form of (c), we use Equation 4.1.48, in which **E** is given by (b), and obtain

$$\mathbf{S} = 2\mu\nabla\nabla\phi + \lambda\nabla^2\phi\mathbf{1} - (3\lambda + 2\mu)\alpha T\mathbf{1} \tag{i}$$

Since by (d)

$$(\lambda + 2\mu)\nabla^2\phi = (3\lambda + 2\mu)\alpha T \tag{j}$$

therefore Equations (i) and (j) imply that **S** is given by (c).

This completes the solution. □

Notes:

(1) The particular solution given in this example plays an important role in solving boundary value problems of thermoelastostatics for an isotropic body. The function ϕ is called the *thermoelastostatic displacement potential* [3].*

(2) Poisson's equation (d) possesses a solution in the form of a Newtonian potential in which the density of the potential is determined by a prescribed temperature change T

$$\phi(\mathbf{x}) = -\frac{m}{4\pi}\int_B \frac{T(\boldsymbol{\xi})}{R(\mathbf{x}, \boldsymbol{\xi})}\, dv(\boldsymbol{\xi}) \tag{4.1.63}$$

where

$$R(\mathbf{x}, \boldsymbol{\xi}) = |\mathbf{x} - \boldsymbol{\xi}| \quad \mathbf{x} \in B \tag{4.1.64}$$

In the case of a *nucleus of thermoelastic strain*, which is defined by

$$\mathbf{E}^* = \alpha T_0\, \delta(\mathbf{x} - \boldsymbol{\xi}^*) \tag{4.1.65}$$

where T_0 is a constant temperature change, and $\delta(\mathbf{x} - \boldsymbol{\xi}^*)$ is a three-dimensional Dirac delta function,† with $\boldsymbol{\xi}^*$ being a fixed point belonging to B, the temperature field T is given by $T = T_0\delta(\mathbf{x} - \boldsymbol{\xi}^*)$ and the formula 4.1.63 reduces to

$$\phi(\mathbf{x}, \boldsymbol{\xi}^*) = -\frac{mT_0}{4\pi}\frac{1}{R(\mathbf{x}, \boldsymbol{\xi}^*)} \tag{4.1.66}$$

This equation allows us to calculate in closed forms the displacement, strain, and stress corresponding to a nucleus of thermoelastic strain in an infinite solid.

---

* See the application of Goodier's thermoelastic displacement potential in solutions of various problems in [2].

† See the description of the Dirac delta function in Section 2.3.4.

Example 4.1.10

Assume that a temperature change T in an infinite isotropic elastic space is given by the formula

$$T = T(r) = T_0 H(a - r) \tag{a}$$

where
T_0 is a positive constant
$H = H(x)$ is the Heaviside function defined by Equation 2.3.53

Moreover, in (a) r is a radial coordinate in a spherical coordinate system (r, θ, φ), and a is a positive constant.

Equation (a) means that the temperature change T is constant inside a solid sphere of radius $r = a$, and it vanishes outside of the sphere.

Show that a solution $\phi(r)$ to Poisson's equation (d) in Example 4.1.9 that satisfies the conditions

(i) $|\phi(0)| < \infty$
(ii) $\phi(r)$ and $\phi'(r)$ are continuous functions across $r = a$
(iii) $\phi(\infty) = 0$

is given by

$$\phi(r) = -\frac{ma^2 T_0}{6}\left[\left(3 - \frac{r^2}{a^2}\right)H(a - r) + \frac{2a}{r}\,H(r - a)\right] \tag{b}$$

with m given in Example 4.1.9.
Hint. The Laplace operator for a spherically symmetric problem is of the form

$$\nabla^2 = \frac{\partial^2}{\partial r^2} + \frac{2}{r}\frac{\partial}{\partial r} \tag{c}$$

Solution

First, we prove that $\phi(r)$ given by (b) satisfies Poisson's equation (d) of Example 4.1.9 in the form

$$\nabla^2 \phi = mT_0 H(a - r) \tag{d}$$

We check first that this equation is satisfied inside the sphere, $0 \leq r < a$. To this end, we apply the operator ∇^2 given by (c) (in this example) to $\phi(r)$ given by (b) for $r < a$, and obtain

$$\nabla^2 \phi = -\frac{ma^2 T_0}{6}\left[-\frac{2}{a^2} + \frac{2}{r}\left(-\frac{2r}{a^2}\right)\right] = mT_0 \tag{e}$$

Hence, function ϕ given by (b) satisfies Poisson's equation (d) of Example 4.1.9 for $r < a$.

To see what equation is satisfied by ϕ outside of the sphere we note that from (b)

$$\phi(r) = -\frac{ma^2 T_0}{6}\frac{2a}{r} \quad \text{for } r > a \tag{f}$$

and from this equation we obtain

$$\nabla^2 \phi = -\frac{1}{3}\, m a^3 T_0 \left[\frac{2}{r^3} + \frac{2}{r} \left(-\frac{1}{r^2} \right) \right] = 0 \tag{g}$$

Therefore, we have proved that ϕ given by (b) satisfies (d) of Example 4.1.9 for $r > a$.

To show that ϕ satisfies conditions (i)–(iii) we note that condition (i) is satisfied because from (b) follows

$$\phi(0) = -\frac{m a^2 T_0}{2} < 0 \tag{h}$$

and condition (iii) is satisfied because in view of (b) we have

$$\phi(r) = -\frac{m a^3 T_0}{3} \frac{1}{r} \quad \text{for } r > a \tag{i}$$

Hence,

$$\phi(\infty) = 0 \tag{j}$$

As for condition (ii), we compute the first derivative of $\phi(r)$ and obtain

$$\phi'(r) = -\frac{m a^2 T_0}{6} \left[-\frac{2r}{a^2} H(a - r) - \frac{2a}{r^2} H(r - a) \right] \tag{k}$$

From (b) follows

$$\phi(a - 0) = -\frac{m a^2 T_0}{3}, \quad \phi(a + 0) = -\frac{m a^2 T_0}{3} \tag{l}$$

Hence,

$$\phi(a - 0) = \phi(a + 0) = \phi(a) \tag{m}$$

which means that the function ϕ is continuous across $r = a$.

To prove the continuity of $\phi'(r)$ across $r = a$, we use (k) and obtain

$$\phi'(a - 0) = \frac{m a T_0}{3}, \quad \phi'(a + 0) = \frac{m a T_0}{3} \tag{n}$$

Hence, ϕ' is a continuous function across $r = a$.

This completes the solution. □

Notes:

(1) A constant temperature change of the sphere does not produce an infinite value of a solution to Poisson's equation at the center $\mathbf{x} = \mathbf{0}$. This explains condition (i).

(2) A constant temperature change on a sphere of finite radius $r = a$ does produce a vanishing solution at infinity. This explains condition (iii).

(3) Condition (ii) ensures that a radial displacement corresponding to ϕ is a continuous function across $r = a$.

Example 4.1.11

Find the radial displacement u_r and thermal stresses S_{rr}, $S_{\theta\theta}$, and $S_{\varphi\varphi}$ corresponding to the distribution of the temperature change T given in the previous example.

Solution

Equations (a) and (c) in Example 4.1.9 transformed to a spherical symmetry are

$$u_r = \frac{\partial \phi}{\partial r} \tag{a}$$

$$S_{rr} = 2\mu\left(\frac{\partial^2 \phi}{\partial r^2} - mT\right) \tag{b}$$

$$S_{\theta\theta} = S_{\varphi\varphi} = 2\mu\left(\frac{1}{r}\frac{\partial \phi}{\partial r} - mT\right) \tag{c}$$

Hence, by using ϕ from (b) in Example 4.1.10, we arrive at

$$u_r = \frac{mT_0 r}{3}\left[H(a-r) + \frac{a^3}{r^3} H(r-a)\right] \tag{d}$$

$$S_{rr} = -\frac{4}{3}\mu m T_0\left[H(a-r) + \frac{a^3}{r^3} H(r-a)\right] \tag{e}$$

$$S_{\theta\theta} = S_{\varphi\varphi} = -\frac{4}{3}\mu m T_0\left[H(a-r) - \frac{a^3}{r^3} H(r-a)\right] \tag{f}$$

This completes the solution. □

Note: Equations (d) and (e) imply that

$$u_r(a+0) - u_r(a-0) = 0 \tag{4.1.67}$$

$$S_{rr}(a+0) - S_{rr}(a-0) = 0 \tag{4.1.68}$$

while (f) leads to the discontinuity formula

$$[S_{\theta\theta}](a) \stackrel{\text{def}}{=} S_{\theta\theta}(a+0) - S_{\theta\theta}(a-0) = \frac{8}{3}\mu m T_0 \tag{4.1.69}$$

This means that radial displacement and radial stress are continuous across $r = a$, while hoop stresses $S_{\theta\theta}$ and $S_{\varphi\varphi}$ reveal a discontinuity at $r = a$ of the magnitude $(8/3)\mu m T_0$.

This simple example shows that a discontinuity of a temperature field may produce a thermal stress discontinuity in an elastic body.

Example 4.1.12

Show that if $\mathbf{S}$ is a stress field of the homogeneous isotropic thermoelastostatics corresponding to zero body forces, then $\mathbf{S}$ satisfies the tensor equation

$$\nabla^2[\nabla^2\mathbf{S} - 2\mu m(\nabla\nabla T - \nabla^2 T\mathbf{1})] = \mathbf{0} \tag{a}$$

where m is defined by (e) in Example 4.1.9.

Solution

Letting $b_i = 0$ in Equation 4.1.58 we get

$$S_{ij,kk} + \frac{2(\lambda+\mu)}{3\lambda+2\mu} S_{aa,ij} + 2\mu\alpha\left(T_{,ij} + \frac{3\lambda+2\mu}{\lambda+2\mu} T_{,kk}\delta_{ij}\right) = 0 \tag{b}$$

Taking the trace of this equation, we obtain (see also Equation 4.1.57, with $b_i = 0$)

$$S_{aa,kk} = -\frac{4\mu(3\lambda+2\mu)}{\lambda+2\mu} \alpha T_{,kk} \tag{c}$$

Now, if the Laplace operator is applied to (b), and (c) is used, we arrive at

$$S_{ij,kkaa} - \frac{2(\lambda+\mu)}{3\lambda+2\mu}\frac{4\mu(3\lambda+2\mu)}{\lambda+2\mu}\alpha T_{,kkij}$$

$$+ 2\mu\alpha\left(T_{,ijaa} + \frac{3\lambda+2\mu}{\lambda+2\mu} T_{,kkaa}\delta_{ij}\right) = 0 \tag{d}$$

Since

$$2\mu - 8\mu\frac{(\lambda+\mu)}{\lambda+2\mu} = -2\mu\frac{3\lambda+2\mu}{\lambda+2\mu} \tag{e}$$

then (d) reduces to the form

$$[S_{ij,kk} - 2\mu m(T_{,ij} - T_{,kk}\delta_{ij})]_{,aa} = 0 \tag{f}$$

where m is defined by (e) of Example 4.1.9. And this relation, when written in direct notation, is identical to (a). This completes the solution. □

Note: From (a) it follows that if the temperature change T is harmonic, that is

$$\nabla^2 T = 0 \quad \text{on } B \tag{4.1.70}$$

then $\mathbf{S}$ is a biharmonic tensor field, that is

$$\nabla^2\nabla^2\mathbf{S} = \mathbf{0} \tag{4.1.71}$$

Consider a more general case, when T satisfies Poisson's equation

$$\nabla^2 T = -\frac{Q}{\kappa} \quad \text{where} \quad \frac{Q}{\kappa} = \frac{W}{k} \tag{4.1.72}$$

Here W is the internal heat generated per unit volume per unit time, and Q is called the prescribed heat supply field, while k denotes the thermal conductivity, and κ means the thermal diffusivity. Then (a) reduces to

$$\nabla^2\nabla^2\mathbf{S} = -2\mu\frac{m}{\kappa}(\nabla\nabla Q - \nabla^2 Q\,\mathbf{1}) \tag{4.1.73}$$

For an infinite thermoelastic body with prescribed Q, integration of this equation leads directly to the determination of the stress tensor $\mathbf{S}$.

4.1.3 Concept of an Elastic State: Energy and Reciprocal Theorems

To make further development of the theory more comprehensive and consistent, and maybe more elegant and easier to grasp, we introduce the concept of an *elastic state*. This concept is based on the notion of an *admissible state*, which is defined as an ordered array of functions $s = [\mathbf{u}, \mathbf{E}, \mathbf{S}]$ such that $\mathbf{u}$ is an admissible displacement, $\mathbf{E}$ is a symmetric second-order strain tensor field, and $\mathbf{S}$ is an admissible stress field, see Section 3.2.4, and the fields $\mathbf{u}$, $\mathbf{E}$, and $\mathbf{S}$ do not have to be related.

If we define addition of two admissible states s and $\widetilde{s}$ by

$$s + \widetilde{s} = [\mathbf{u} + \widetilde{\mathbf{u}}, \mathbf{E} + \widetilde{\mathbf{E}}, \mathbf{S} + \widetilde{\mathbf{S}}] \tag{4.1.74}$$

and scalar multiplication of s by a scalar η as

$$\eta s = [\eta\mathbf{u}, \eta\mathbf{E}, \eta\mathbf{S}] \tag{4.1.75}$$

then the set of all admissible states may be identified with a vector space. Thus, elements of this vector space are identified as $s, \widetilde{s}, \approx{s}$, and so on.

An ordered array of functions $s = [\mathbf{u}, \mathbf{E}, \mathbf{S}]$ is called *an elastic state* corresponding to the body force $\mathbf{b}$ if s is an admissible state and the functions $\mathbf{u}$, $\mathbf{E}$, and $\mathbf{S}$ satisfy the system of fundamental field equations of elastostatics:

$$\mathbf{E} = \frac{1}{2}(\nabla\mathbf{u} + \nabla\mathbf{u}^T) \tag{4.1.76}$$

$$\operatorname{div}\mathbf{S} + \mathbf{b} = \mathbf{0} \tag{4.1.77}$$

$$\mathbf{S} = \mathsf{C}[\mathbf{E}] \tag{4.1.78}$$

We will also use a concept of an *external force system* for s. This is defined as a pair $[\mathbf{b}, \mathbf{s}]$ where

$$\mathbf{s} = \mathbf{Sn} \tag{4.1.79}$$

with $\mathbf{n}$ being an outward unit vector normal to ∂B.

With these definitions we now formulate the *principle of superposition* of elastic states. If $s = [\mathbf{u}, \mathbf{E}, \mathbf{S}]$ and $\widetilde{s} = [\widetilde{\mathbf{u}}, \widetilde{\mathbf{E}}, \widetilde{\mathbf{S}}]$ are elastic states corresponding to the external force systems $[\mathbf{b}, \mathbf{s}]$ and $[\widetilde{\mathbf{b}}, \widetilde{\mathbf{s}}]$, respectively, and a and b are scalars, then $a[\mathbf{u}, \mathbf{E}, \mathbf{S}] + b[\widetilde{\mathbf{u}}, \widetilde{\mathbf{E}}, \widetilde{\mathbf{S}}]$ is an elastic state corresponding to the external force system $a[\mathbf{b}, \mathbf{s}] + b[\widetilde{\mathbf{b}}, \widetilde{\mathbf{s}}]$, where

$$a[\mathbf{b}, \mathbf{s}] + b[\widetilde{\mathbf{b}}, \widetilde{\mathbf{s}}] = [a\mathbf{b} + b\widetilde{\mathbf{b}}, a\mathbf{s} + b\widetilde{\mathbf{s}}] \tag{4.1.80}$$

The proof of the principle of superposition is based on the definition of an elastic state associated with linear partial differential equations 4.1.76 and 4.1.78.

With the definitions introduced earlier in this section, Theorem of work expended presented in Section 3.2.4 leads to the following

Theorem of Work and Energy. If $s = [\mathbf{u}, \mathbf{E}, \mathbf{S}]$ is an elastic state corresponding to the external force system $[\mathbf{b}, \mathbf{s}]$ then

$$\int_{\partial B} \mathbf{s} \cdot \mathbf{u}\, da + \int_{B} \mathbf{b} \cdot \mathbf{u}\, dv = 2U_C\{\mathbf{E}\} \tag{4.1.81}$$

where $U_C\{\mathbf{E}\}$ is the total strain energy of body B, see Equation 3.3.93,

$$U_C\{\mathbf{E}\} = \frac{1}{2}\int_B \mathbf{E}\cdot\mathbf{C}[\mathbf{E}]\,dv \tag{4.1.82}$$

Proof: Since s is an elastic state, then Equations 4.1.76 and 4.1.77 are satisfied, and this implies that the assumptions of the theorem of work expended, see Section 3.2.4, hold true. Hence, Equation 3.2.132 is satisfied, or

$$\int_{\partial B} \mathbf{s}\cdot\mathbf{u}\,da + \int_B \mathbf{b}\cdot\mathbf{u}\,dv = \int_B \mathbf{S}\cdot\mathbf{E}\,dv \tag{4.1.83}$$

Recalling again that s is an elastic state, which means, in particular, that Equation 4.1.78 is satisfied, Equation 4.1.83 yields

$$\int_{\partial B} \mathbf{s}\cdot\mathbf{u}\,da + \int_B \mathbf{b}\cdot\mathbf{u}\,dv = \int_B \mathbf{E}\cdot\mathbf{C}[\mathbf{E}]\,dv \tag{4.1.84}$$

This equation, jointly with Equation 4.1.82 implies Equation 4.1.81. This completes the proof. □

The theorem asserts that the work done by an external force system is equal to double the total strain energy.

Example 4.1.13

Show that if $\mathbf{C}$ is positive definite, then for any elastic state s the work done by external forces is nonnegative and vanishes only if the displacement field is a rigid body motion.

Solution

Since $\mathbf{C}$ is positive definite, $\mathbf{E}\cdot\mathbf{C}[\mathbf{E}] > 0$, and by Equation 4.1.82 $U_C\{\mathbf{E}\} \geq 0$, and this implies that the work done by external forces is nonnegative. Now we are to show that if $U_C\{\mathbf{E}\} = 0$, then $\mathbf{u}$ associated with $\mathbf{E}$ is a rigid body motion. To this end we note that the positive definiteness of $\mathbf{C}$, together with the condition that $U_C\{\mathbf{E}\} = 0$, implies that $\mathbf{E} = \mathbf{0}$ on B. The latter condition, together with Equation 4.1.76, leads to the conclusion that $\mathbf{u}$ is a rigid body motion, see Example 3.1.6. This completes the solution. □

Example 4.1.14

Let s be an elastic state with a positive definite tensor $\mathbf{C}$, and corresponding to zero body forces. Assume also that $\mathbf{u}\cdot\mathbf{s} = 0$ on ∂B. Show that s takes the form

$$s = [\mathbf{u}, \mathbf{0}, \mathbf{0}] \tag{a}$$

where $\mathbf{u}$ is a rigid motion.

Solution

By Equation 4.1.84 and from the assumptions that $\mathbf{b} = \mathbf{0}$ and $\mathbf{u} \cdot \mathbf{s} = 0$ on ∂B it follows that

$$\int_B \mathbf{E} \cdot \mathbf{C}[\mathbf{E}]\, dv = 0 \tag{b}$$

This together with the positive definiteness of $\mathbf{C}$ and the definition of an elastic state implies that

$$\mathbf{E} = \mathbf{S} = \mathbf{0} \tag{c}$$

Finally, using the results of the previous example, we note that equation $\mathbf{E} = \mathbf{0}$ in (c) implies that $\mathbf{u}$ is a rigid body motion. This completes the solution. □

The concept of an elastic state corresponding to an external force system allows us to formulate another important theorem that is called the Betti reciprocal theorem.

The Betti Reciprocal Theorem. Let the elasticity tensor $\mathbf{C}$ be symmetric, and let

$$s = [\mathbf{u}, \mathbf{E}, \mathbf{S}] \quad \text{and} \quad \widetilde{s} = [\widetilde{\mathbf{u}}, \widetilde{\mathbf{E}}, \widetilde{\mathbf{S}}] \tag{4.1.85}$$

be elastic states corresponding to external force systems $[\mathbf{b}, \mathbf{s}]$ and $[\widetilde{\mathbf{b}}, \widetilde{\mathbf{s}}]$, respectively. Then the following reciprocity relation holds

$$\begin{aligned} \int_{\partial B} \mathbf{s} \cdot \widetilde{\mathbf{u}}\, da + \int_B \mathbf{b} \cdot \widetilde{\mathbf{u}}\, dv &= \int_{\partial B} \widetilde{\mathbf{s}} \cdot \mathbf{u}\, da + \int_B \widetilde{\mathbf{b}} \cdot \mathbf{u}\, dv \\ &= \int_B \mathbf{S} \cdot \widetilde{\mathbf{E}}\, dv = \int_B \widetilde{\mathbf{S}} \cdot \mathbf{E}\, dv \end{aligned} \tag{4.1.86}$$

Proof: Since $\mathbf{C}$ is symmetric, and s and $\widetilde{s}$ are elastic states, then

$$\mathbf{S} \cdot \widetilde{\mathbf{E}} = \widetilde{\mathbf{E}} \cdot \mathbf{C}[\mathbf{E}] = \mathbf{E} \cdot \mathbf{C}[\widetilde{\mathbf{E}}] = \mathbf{E} \cdot \widetilde{\mathbf{S}} \tag{4.1.87}$$

Hence, by integrating Equation 4.1.87 over B we arrive at the last equality in Equation 4.1.86.

Next, by the theorem of work expended, of Section 3.2.4, where $\mathbf{u}$ is replaced by $\widetilde{\mathbf{u}}$ and $\mathbf{E}$ is replaced by $\widetilde{\mathbf{E}}$ we have

$$\int_{\partial B} \mathbf{s} \cdot \widetilde{\mathbf{u}}\, da + \int_B \mathbf{b} \cdot \widetilde{\mathbf{u}}\, dv = \int_B \mathbf{S} \cdot \widetilde{\mathbf{E}}\, dv \tag{4.1.88}$$

In a similar way, by the same theorem,

$$\int_{\partial B} \widetilde{\mathbf{s}} \cdot \mathbf{u}\, da + \int_B \widetilde{\mathbf{b}} \cdot \mathbf{u}\, dv = \int_B \widetilde{\mathbf{S}} \cdot \mathbf{E}\, dv \tag{4.1.89}$$

Since the RHS of Equations 4.1.88 and 4.1.89 are equal by Equation 4.1.87, Equations 4.1.88 and 4.1.89 imply the first equality in Equation 4.1.86. This completes the proof. □

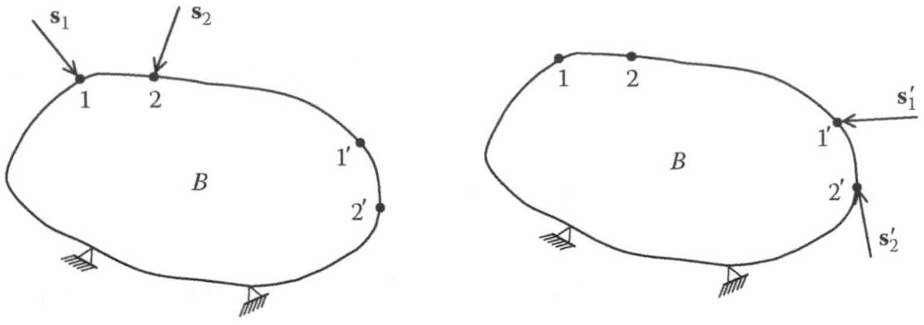

FIGURE 4.3 Two systems of concentrated boundary forces acting on B.

This theorem may be specialized for the case of a body B acted upon by two systems of concentrated forces, each applied at different points on its boundary. Let the forces of the first system be denoted by $\{\mathbf{s}_i\}$ and the forces of the second system by $\{\mathbf{s}_i^{'}\}$, $i = 1, 2, \ldots, N$. Next, let $\mathbf{u}_i$, $\mathbf{E}_i$, and $\mathbf{S}_i$ be, respectively, the displacement, strain, and stress produced by $\mathbf{s}_i$ at any point $\mathbf{x} \in \bar{B}$. Similarly, let $\mathbf{u}_i^{'}$, $\mathbf{E}_i^{'}$, and $\mathbf{S}_i^{'}$ denote, respectively, the displacement, strain, and stress produced by $\mathbf{s}_i^{'}$ at $\mathbf{x} \in \bar{B}$. Since there are no body forces acting on B, from Equation 4.1.86 follows

$$\begin{aligned} \sum_{i=1}^{N} \int_{\partial B} \mathbf{s}_i \cdot \mathbf{u}_i^{'} \, da &= \sum_{i=1}^{N} \int_{\partial B} \mathbf{s}_i^{'} \cdot \mathbf{u}_i \, da \\ &= \sum_{i=1}^{N} \int_{B} \mathbf{S}_i \cdot \mathbf{E}_i^{'} \, dv = \sum_{i=1}^{N} \int_{B} \mathbf{S}_i^{'} \cdot \mathbf{E}_i \, dv \end{aligned} \tag{4.1.90}$$

The concentrated forces $\mathbf{s}_i$ and $\mathbf{s}_i^{'}$, as well as points of application of these forces, are shown in Figure 4.3. The displacements $\mathbf{u}_i$ and $\mathbf{u}_i^{'}$ in Equation 4.1.90 are the boundary displacements produced by the forces $\mathbf{s}_i$ and $\mathbf{s}_i^{'}$, respectively.

Since $\mathbf{s}_i$ and $\mathbf{s}_i^{'}$ are concentrated forces acting on ∂B

$$\int_{\partial B} \mathbf{s}_i \cdot \mathbf{u}_i^{'} \, da = \mathbf{s}_i(i) \cdot \mathbf{u}_i(i^{'}) \tag{4.1.91}$$

and

$$\int_{\partial B} \mathbf{s}_i^{'} \cdot \mathbf{u}_i \, da = \mathbf{s}_i(i^{'}) \cdot \mathbf{u}_i(i) \tag{4.1.92}$$

Therefore, the first part of Equation 4.1.90 reduces to

$$\sum_{i=1}^{N} \mathbf{s}_i(i) \cdot \mathbf{u}_i(i^{'}) = \sum_{i=1}^{N} \mathbf{s}_i(i^{'}) \cdot \mathbf{u}_i(i) \tag{4.1.93}$$

In the case of $N = 2$, shown in Figure 4.3, we obtain

$$\mathbf{s}_1(1) \cdot \mathbf{u}_1(1^{'}) + \mathbf{s}_2(2) \cdot \mathbf{u}_2(2^{'}) = \mathbf{s}_1(1^{'}) \cdot \mathbf{u}_1(1) + \mathbf{s}_2(2^{'}) \cdot \mathbf{u}_2(2) \tag{4.1.94}$$

In this equation $\mathbf{u}_1(1)$ is the boundary displacement at point 1 due to the force $\mathbf{s}_1(1)$, and $\mathbf{u}_2(2)$ is the boundary displacement at point 2 due to $\mathbf{s}_2(2)$. Similar interpretation is given to primed quantities in Equations 4.1.91 through 4.1.94.

If $N = 1$, Equation 4.1.93 asserts that work done by $\mathbf{s}_1(1)$ on the displacement $\mathbf{u}_1(1')$ is equal to work done by $\mathbf{s}_1(1')$ on $\mathbf{u}_1(1)$.

Example 4.1.15

Given an anisotropic infinite elastic body. Let $\widetilde{\mathbf{b}}^{(1)}$ denote a concentrated force that acts at a point $\mathbf{x} = \boldsymbol{\xi}$ in the x_1 direction and is represented in components by

$$\widetilde{b}_i^{(1)} = \delta(\mathbf{x} - \boldsymbol{\xi})\delta_{i1} \quad (i = 1, 2, 3) \tag{a}$$

where $\delta = \delta(\mathbf{x})$ is the Dirac delta function in E^3. Such a force will produce a displacement field that depends on both $\mathbf{x}$ and $\boldsymbol{\xi}$, and this displacement will be denoted by

$$\widetilde{u}_i^{(1)} = \widetilde{u}_i^{(1)}(\mathbf{x}, \boldsymbol{\xi}) \tag{b}$$

Let $u_i = u_i(\mathbf{x})$ be a displacement field produced by a body force $\mathbf{b} = \mathbf{b}(\mathbf{x})$ where $\mathbf{x} \in \Omega \subset E^3$ so $\mathbf{b}$ is distributed on bounded region Ω of the three-dimensional space. Assume also that both the elastic states corresponding to the body force $\mathbf{b}$ and the elastic state $\widetilde{s}$ corresponding to $\widetilde{\mathbf{b}}^{(1)}$ vanish at infinity.

By using Equation 4.1.86 show that

$$u_1(\mathbf{x}) = \int_{\Omega} \widetilde{u}_i^{(1)}(\mathbf{x}, \boldsymbol{\xi}) b_i(\boldsymbol{\xi})\, dv(\boldsymbol{\xi}) \tag{c}$$

where

$$dv(\boldsymbol{\xi}) = d\xi_1\, d\xi_2\, d\xi_3 \tag{d}$$

Solution

Equation 4.1.86 in which B is the whole space E^3, and the elastic states s and $\widetilde{s}^{(1)} \equiv \widetilde{s}$ vanish at infinity, reduces to

$$\int_{B=E^3} \mathbf{b} \cdot \widetilde{\mathbf{u}}^{(1)}\, dv(\boldsymbol{\xi}) = \int_{B=E^3} \widetilde{\mathbf{b}}^{(1)} \cdot \mathbf{u}\, dv(\boldsymbol{\xi}) \tag{e}$$

Substituting Equations (a) and (b) into Equation (e), and recalling that $\mathbf{b}$ is distributed over region Ω only, and writing in components, we find

$$\int_{\Omega} b_i(\boldsymbol{\xi}) \widetilde{u}_i^{(1)}(\mathbf{x}, \boldsymbol{\xi})\, dv(\boldsymbol{\xi}) = \int_{E^3} \delta(\mathbf{x} - \boldsymbol{\xi})\delta_{i1} u_i(\boldsymbol{\xi})\, dv(\boldsymbol{\xi}) \tag{f}$$

Hence, by using the filtering property of the Dirac delta function

$$\int_{-\infty}^{\infty} \delta(\mathbf{x} - \boldsymbol{\xi}) f(\boldsymbol{\xi})\, d\boldsymbol{\xi} = f(\mathbf{x}) \tag{g}$$

where f is an arbitrary function, we arrive at formula (c). This completes the solution. □

Note: A direct consequence of this example is the formula

$$u_k(\mathbf{x}) = \int_{\Omega} \widetilde{u}_i^{(k)}(\mathbf{x}, \boldsymbol{\xi})\, b_i(\boldsymbol{\xi})\, dv(\boldsymbol{\xi}) \tag{4.1.95}$$

where

$$\widetilde{u}_i^{(k)} = \widetilde{u}_i^{(k)}(\mathbf{x}, \boldsymbol{\xi}) \tag{4.1.96}$$

is the displacement field produced at point $\mathbf{x}$ of an infinite anisotropic elastic body by a concentrated force $\widetilde{\mathbf{b}}^{(k)}$, which acts at $\mathbf{x} = \boldsymbol{\xi}$ along the x_k direction.

Example 4.1.16

Let B denote a bounded region of E^3 occupied by an anisotropic nonhomogeneous elastic solid, and let $s = [\mathbf{u}, \mathbf{E}, \mathbf{S}]$ denote an elastic state on B corresponding to an external force system $[\mathbf{b}, \mathbf{s}]$. Let $\widetilde{s} = [\widehat{\mathbf{u}}^{(k)}, \widehat{\mathbf{E}}^{(k)}, \widehat{\mathbf{S}}^{(k)}]$, $(k = 1, 2, 3)$, denote an elastic state on B corresponding to an external force system $[\delta(\mathbf{x} - \boldsymbol{\xi})\mathbf{e}_k, \mathbf{0}]$, where $\delta = \delta(\mathbf{x} - \boldsymbol{\xi})$ is the three-dimensional Dirac delta function, $\mathbf{e}_k$ is a unit vector along x_k axis, and $\mathbf{x}$, $\boldsymbol{\xi} \in \overline{B}$. Use the reciprocal relation 4.1.86 to show that

$$\mathbf{u}_k(\mathbf{x}) = \int_{\partial B} \mathbf{s}(\boldsymbol{\xi}) \cdot \widehat{\mathbf{u}}^{(k)}(\mathbf{x}, \boldsymbol{\xi}) da(\boldsymbol{\xi}) + \int_{B} \mathbf{b}(\boldsymbol{\xi}) \cdot \widehat{\mathbf{u}}^{(k)}(\mathbf{x}, \boldsymbol{\xi}) dv(\boldsymbol{\xi}) \quad \text{for } \mathbf{x} \in B \tag{a}$$

Solution

Note that

$$\widetilde{\mathbf{s}} = \widehat{\mathbf{S}}^{(k)}\mathbf{n} = \mathbf{0} \quad \text{for } \mathbf{x} \in \partial B \quad \text{and } k = 1, 2, 3 \tag{b}$$

Hence, substituting the force system $[\widetilde{\mathbf{b}}, \widetilde{\mathbf{s}}] \equiv [\delta(\mathbf{x} - \boldsymbol{\xi})\mathbf{e}_k, \mathbf{0}]$ into the relation 4.1.86, and taking into account the fact that $\widetilde{s}$ depends on $\mathbf{x}$ and $\boldsymbol{\xi}$, which means that $\widetilde{s} = \widetilde{s}(\mathbf{x}, \boldsymbol{\xi})$, we obtain

$$\int_{B} \delta(\mathbf{x} - \boldsymbol{\xi})\mathbf{e}_k \cdot \mathbf{u}(\boldsymbol{\xi}) dv(\boldsymbol{\xi}) = \int_{\partial B} \mathbf{s}(\boldsymbol{\xi}) \cdot \widehat{\mathbf{u}}^{(k)}(\mathbf{x}, \boldsymbol{\xi}) da(\boldsymbol{\xi}) + \int_{B} \mathbf{b}(\boldsymbol{\xi}) \cdot \widehat{\mathbf{u}}^{(k)}(\mathbf{x}, \boldsymbol{\xi}) dv(\boldsymbol{\xi}) \tag{c}$$

Finally, using the filtering property of the Dirac delta function we obtain Equation (a). This completes the solution. □

Note: The formula (a) provides the displacement field $\mathbf{u}$ on B if the fields $\mathbf{b}$ and $\mathbf{s}$ are prescribed on B and ∂B, respectively. Since for an arbitrary nonhomogeneous anisotropic elastic body, the task of finding an analytical form of $\widehat{\mathbf{u}}^{(k)} = \widehat{\mathbf{u}}^{(k)}(\mathbf{x}, \boldsymbol{\xi})$ is rather hopeless, therefore Equation (a) has a theoretical value only.

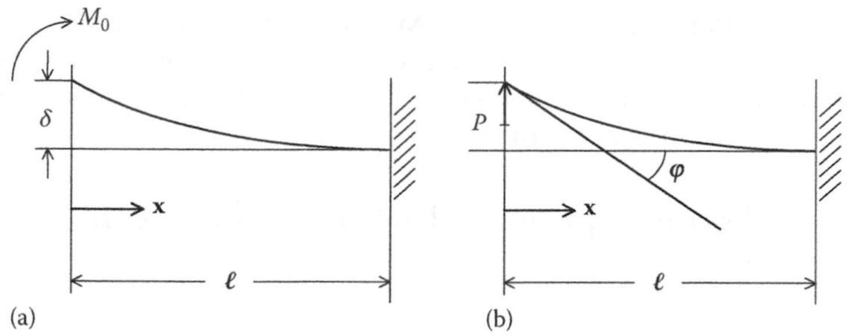

FIGURE 4.4 A cantilever beam subject to a moment M_0 at the free end (a) and to a force P at the free end (b).

Example 4.1.17

Given a cantilever beam of length ℓ with the moment of inertia of the cross section I and Young's modulus of the material E, subject to a moment M_0 at the free end, see Figure 4.4. From the strength of materials theory we know that the deflection of the free end of this beam is

$$\delta = \frac{M_0 \ell^2}{2EI} \tag{a}$$

Using the Betti reciprocal theorem find the rotation angle φ of the free end of the beam when a force P is applied at its end.

Solution

From the Betti reciprocal theorem follows

$$P\delta = M_0 \varphi \tag{b}$$

Therefore,

$$\varphi = \frac{P\delta}{M_0} = \frac{P\ell^2}{2EI} \tag{c}$$

This completes the solution. □

Example 4.1.18

An ordered array of functions $s = [\mathbf{u}, \mathbf{E}, \mathbf{S}]$ is called a *thermoelastic state corresponding to an external force–temperature system* $[\mathbf{b}, \mathbf{s}, T]$ if s is an admissible state and the functions $\mathbf{u}$, $\mathbf{E}$, and $\mathbf{S}$ satisfy the field equations of thermoelastostatics (see Equations 4.1.76 through 4.1.78 in which Equation 4.1.78 is replaced by $\mathbf{S} = \mathbf{C}[\mathbf{E}] + T\mathbf{M}$)

$$\begin{aligned} \mathbf{E} &= \frac{1}{2}\left(\nabla \mathbf{u} + \nabla \mathbf{u}^T\right) \\ \operatorname{div} \mathbf{S} + \mathbf{b} &= \mathbf{0} \\ \mathbf{S} &= \mathbf{C}[\mathbf{E}] + T\mathbf{M} \end{aligned} \tag{a}$$

Here, T is a temperature change field, and $\mathbf{M}$ represents the stress–temperature tensor field (see Equation 3.3.110). Prove the following generalization of the isothermal theorem of work and energy. If $s = [\mathbf{u}, \mathbf{E}, \mathbf{s}]$ is a thermoelastic state corresponding to the external force–temperature system $[\mathbf{b}, \mathbf{s}, T]$ then

$$\int_{\partial B} \mathbf{s} \cdot \mathbf{u}\, da + \int_{B} \mathbf{b} \cdot \mathbf{u}\, dv - \int_{B} T\mathbf{M} \cdot \mathbf{E}\, dv = \int_{B} \mathbf{E} \cdot \mathbf{C}[\mathbf{E}]\, dv \tag{b}$$

Solution

Multiplication of Equations $(a)_2$ and $(a)_3$ by $\mathbf{u}$ and $\mathbf{E}$, respectively, in the dot sense, and using Equation $(a)_1$, leads to

$$\mathbf{u} \cdot \operatorname{div} \mathbf{S} + \mathbf{u} \cdot \mathbf{b} = 0 \tag{c}$$

and

$$(\widehat{\nabla \mathbf{u}}) \cdot \mathbf{S} = \mathbf{E} \cdot \mathbf{C}[\mathbf{E}] + T\mathbf{M} \cdot \mathbf{E} \tag{d}$$

Next, by integrating Equations (c) and (d) over B, and using the divergence theorem we obtain, respectively,

$$-\int_{B} (\widehat{\nabla \mathbf{u}}) \cdot \mathbf{S}\, dv + \int_{\partial B} \mathbf{u} \cdot \mathbf{s}\, da + \int_{B} \mathbf{u} \cdot \mathbf{b}\, dv = 0 \tag{e}$$

and

$$\int_{B} (\widehat{\nabla \mathbf{u}}) \cdot \mathbf{S}\, dv = \int_{B} \mathbf{E} \cdot \mathbf{C}[\mathbf{E}]\, dv + \int_{B} T\mathbf{M} \cdot \mathbf{E}\, dv \tag{f}$$

Finally, by eliminating the integral $\int_B (\widehat{\nabla \mathbf{u}}) \cdot \mathbf{S}\, dv$ from Equations (e) and (f), we obtain (b). This completes the solution. □

Example 4.1.19

Prove the following thermoelastic reciprocal theorem. Let $s = [\mathbf{u}, \mathbf{E}, \mathbf{S}]$ and $\tilde{s} = [\tilde{\mathbf{u}}, \tilde{\mathbf{E}}, \tilde{\mathbf{S}}]$ be thermoelastic states corresponding to the external force–temperature systems $[\mathbf{b}, \mathbf{s}, T]$ and $[\tilde{\mathbf{b}}, \tilde{\mathbf{s}}, \tilde{T}]$, respectively. Then

$$\begin{aligned} &\int_{\partial B} \mathbf{s} \cdot \tilde{\mathbf{u}}\, da + \int_{B} \mathbf{b} \cdot \tilde{\mathbf{u}}\, dv - \int_{B} T\mathbf{M} \cdot \tilde{\mathbf{E}}\, dv \\ &= \int_{\partial B} \tilde{\mathbf{s}} \cdot \mathbf{u}\, da + \int_{B} \tilde{\mathbf{b}} \cdot \mathbf{u}\, dv - \int_{B} \tilde{T}\mathbf{M} \cdot \mathbf{E}\, dv \end{aligned} \tag{a}$$

Solution

Since s and $\tilde{s}$ are thermoelastic states corresponding to the external force–temperature systems $[\mathbf{b}, \mathbf{s}, T]$ and $[\tilde{\mathbf{b}}, \tilde{\mathbf{s}}, \tilde{T}]$, respectively, therefore, using the definition of s and $\tilde{s}$ from

Example 4.1.18, and proceeding similarly to the proof of isothermal Betti reciprocal theorem (see Equation 4.1.86), we obtain

$$\int_B \tilde{\mathbf{u}} \cdot \mathbf{b}\, dv + \int_{\partial B} \tilde{\mathbf{u}} \cdot \mathbf{s}\, da - \int_B TM \cdot \tilde{\mathbf{E}}\, dv = \int_B \tilde{\mathbf{E}} \cdot \mathbf{C}[\mathbf{E}]\, dv \qquad \text{(b)}$$

and

$$\int_B \mathbf{u} \cdot \tilde{\mathbf{b}}\, dv + \int_{\partial B} \mathbf{u} \cdot \tilde{\mathbf{s}}\, da - \int_B \tilde{T}M \cdot \mathbf{E}\, dv = \int_B \mathbf{E} \cdot \mathbf{C}[\tilde{\mathbf{E}}]\, dv \qquad \text{(c)}$$

Hence, using the symmetry of the elasticity tensor $\mathbf{C}$, and eliminating the integral $\int_B \mathbf{E} \cdot \mathbf{C}[\tilde{\mathbf{E}}]\, dv$ from Equations (b) and (c), we arrive at Equation (a). This completes the solution. □

4.1.4 Formulation of Boundary Value Problems

The most general boundary value problem of elastostatics is a *mixed problem*. By a mixed boundary value problem of elastostatics, we mean the problem of finding an elastic state $s = [\mathbf{u}, \mathbf{E}, \mathbf{S}]$ corresponding to a body force $\mathbf{b}$ and satisfying the following boundary conditions: the displacement condition

$$\mathbf{u} = \hat{\mathbf{u}} \quad \text{on } \partial B_1 \qquad (4.1.97)$$

and the load condition, also called the traction condition

$$\mathbf{s} = \mathbf{S}\mathbf{n} = \hat{\mathbf{s}} \quad \text{on } \partial B_2 \qquad (4.1.98)$$

where

$$\partial B_1 \cup \partial B_2 = \partial B \qquad (4.1.99)$$

and

$$\partial B_1 \cap \partial B_2 = \emptyset \qquad (4.1.100)$$

with ∂B denoting the boundary of body B. In Equations 4.1.97 and 4.1.98 vectors $\hat{\mathbf{u}}$ and $\hat{\mathbf{s}}$ are prescribed functions.

An elastic state s that satisfies the boundary conditions 4.1.97 and 4.1.98 is called a solution to the mixed problem.

If $\partial B_2 = \emptyset$, the mixed problem becomes a *displacement boundary value problem*.

If $\partial B_1 = \emptyset$, the mixed problem becomes a *traction boundary value problem*.

The three boundary value problems we have just introduced will be called, in short, the mixed, displacement, and traction problems, respectively. The following definitions will also be useful.

(1) *A displacement field corresponding to a solution to a mixed problem* is a vector field $\mathbf{u}$ with the property that there are symmetric tensor fields $\mathbf{E}$ and $\mathbf{S}$ such that $s = [\mathbf{u}, \mathbf{E}, \mathbf{S}]$ is a solution to the mixed problem.

(2) *A stress field corresponding to a solution to a mixed problem* is a tensor field $\mathbf{S}$ with the property that there are $\mathbf{u}$ and $\mathbf{E}$ such that $s = [\mathbf{u}, \mathbf{E}, \mathbf{S}]$ is a solution to the mixed problem.
(3) An analogous definition may be stated for *a strain field corresponding to a solution to a mixed problem.*

The importance of definitions (1) and (2) is seen when the mixed problem is characterized in terms of displacements and stresses, respectively. Definition (3) is of lesser importance in applications.

The displacement equation of equilibrium, see Equation 4.1.5, and the stress equations of elastostatics, see Equations 4.1.14 and 4.1.15, will now be used to characterize a mixed problem in terms of displacements and stresses, respectively.

Theorem 1 (Mixed Problem in Terms of Displacements)

A vector field $\mathbf{u}$ corresponds to a solution to the mixed problem if and only if

$$\operatorname{div} \mathsf{C}[\,\nabla\mathbf{u}\,] + \mathbf{b} = \mathbf{0} \quad \text{on } B \tag{4.1.101}$$

$$\mathbf{u} = \widehat{\mathbf{u}} \quad \text{on } \partial B_1 \tag{4.1.102}$$

$$\mathsf{C}[\,\nabla\mathbf{u}\,]\mathbf{n} = \widehat{\mathbf{s}} \quad \text{on } \partial B_2 \tag{4.1.103}$$

■

Proof: We are to show that

(a) If $\mathbf{u}$ is a displacement field corresponding to a solution to the mixed problem, $\mathbf{u}$ satisfies Equations 4.1.101 through 4.1.103
(b) If $\mathbf{u}$ satisfies Equations 4.1.101 through 4.1.103, then $\mathbf{u}$ corresponds to a solution to the mixed problem

Proof of part (a). Since $\mathbf{u}$ corresponds to a solution to the mixed problem, there exist $\mathbf{E} = \widehat{\nabla}\mathbf{u}$ and $\mathbf{S}$ given by $\mathbf{S} = \mathsf{C}[\nabla\mathbf{u}]$ such that $s = [\mathbf{u}, \mathbf{E}, \mathbf{S}]$ is an elastic state subject to the conditions 4.1.97 and 4.1.98. By the definition of an elastic state, s satisfies Equations 4.1.76 through 4.1.78. Thus, eliminating $\mathbf{S}$ and $\mathbf{E}$ from Equations 4.1.76 through 4.1.78 and from the boundary condition 4.1.98 we arrive at Equations 4.1.101 through 4.1.103. This completes the proof of part (a).

Proof of part (b). If $\mathbf{u}$ is a solution to Equations 4.1.101 through 4.1.103, then if we define $\mathbf{E}$ by $\mathbf{E} = \widehat{\nabla}\mathbf{u}$ and $\mathbf{S}$ by $\mathbf{S} = \mathsf{C}[\nabla\mathbf{u}]$, we find from Equation 4.1.101 that the equilibrium equation is satisfied in the form

$$\operatorname{div} \mathbf{S} + \mathbf{b} = \mathbf{0} \quad \text{on } B \tag{4.1.104}$$

and from Equation 4.1.103 we observe that this equation reduces to

$$\mathbf{S}\mathbf{n} = \widehat{\mathbf{s}} \quad \text{on } \partial B_2 \tag{4.1.105}$$

Hence, $s = [\mathbf{u}, \mathbf{E}, \mathbf{S}]$ is a solution to the mixed problem. Thus, **u** corresponds to a solution to the mixed problem. This completes the proof of part (b), as well as the proof of Theorem 1. □

In other words, Theorem 1 implies that solving Equations 4.1.101 through 4.1.103 is equivalent to finding a solution to the mixed problem.

Theorem 2 (Mixed Problem in Terms of Stresses)

Let body B be simply connected. A tensor field **S** corresponds to a solution to the mixed problem if and only if

$$\operatorname{div}\mathbf{S} + \mathbf{b} = \mathbf{0} \quad \text{on } B \tag{4.1.106}$$

$$\operatorname{curl}\operatorname{curl}\mathsf{K}[\mathbf{S}] = \mathbf{0} \quad \text{on } B \tag{4.1.107}$$

$$\mathbf{u}[\mathbf{S}] = \widehat{\mathbf{u}} \quad \text{on } \partial B_1 \tag{4.1.108}$$

$$\mathbf{Sn} = \widehat{\mathbf{s}} \quad \text{on } \partial B_2 \tag{4.1.109}$$

where $\mathbf{u}[\mathbf{S}]$ is a stress-dependent vector field on ∂B_1 defined by the line integral

$$\mathbf{u}[\mathbf{S}] = \int_{\mathbf{x}_0}^{\mathbf{x}} \mathbf{U}(\mathbf{y}, \mathbf{x})\, d\mathbf{y} \tag{4.1.110}$$

in which $\mathbf{x}_0$ is an arbitrary point on ∂B_1 and the integrand **U** is given in components by

$$U_{ij}(\mathbf{y}, \mathbf{x}) = E_{ij}(\mathbf{y}) + (x_k - y_k)[E_{ij,k}(\mathbf{y}) - E_{kj,i}(\mathbf{y})] \tag{4.1.111}$$

and

$$E_{ij}(\mathbf{y}) = K_{ijkl} S_{kl}(\mathbf{y}) \tag{4.1.112}$$

with K_{ijkl} being components of K.

The proof of this theorem is based on two observations.

(1) The stress equations of elastostatics for a simply connected body B have the form 4.1.106 and 4.1.107, see Equations 4.1.14 and 4.1.15.

(2) In the boundary condition given by Equation 4.1.108, we equate a stress-dependent displacement that is a solution to the equation

$$\widehat{\nabla}\mathbf{u} = \mathsf{K}[\mathbf{S}] \quad \text{on } \bar{B} \tag{4.1.113}$$

to a field $\widehat{\mathbf{u}}$ prescribed on ∂B_1. The function **u** that satisfies Equation 4.1.113 and is restricted to ∂B_1 is given by Equations 4.1.110 through 4.1.112 obtained in Section 3.1.3, see Equations 3.1.53 and 3.1.54 in which E_{ij} is replaced by Equation 4.1.112.

The proof may be completed by using observations (1) and (2) and proceeding in a way similar to that of the proof of Theorem 1. ■

Notes: For a homogeneous isotropic body, a description of the mixed problem reads as follows:

(1) *In terms of displacements*, a vector field $\mathbf{u}$ corresponds to a solution of the mixed problem if and only if

$$\mu\nabla^2\mathbf{u} + (\lambda + \mu)\nabla(\operatorname{div}\mathbf{u}) + \mathbf{b} = \mathbf{0} \quad \text{on } B \tag{4.1.114}$$

$$\mathbf{u} = \widehat{\mathbf{u}} \quad \text{on } \partial B_1 \tag{4.1.115}$$

$$\mu(\nabla\mathbf{u} + \nabla\mathbf{u}^T)\mathbf{n} + \lambda(\operatorname{div}\mathbf{u})\mathbf{n} = \widehat{\mathbf{s}} \quad \text{on } \partial B_2 \tag{4.1.116}$$

(2) *In terms of stresses*, a tensor field $\mathbf{S}$ corresponds to a solution of the mixed problem if and only if

$$\operatorname{div}\mathbf{S} + \mathbf{b} = \mathbf{0} \quad \text{on } B \tag{4.1.117}$$

$$\nabla^2\mathbf{S} + \frac{1}{1+\nu}\nabla\nabla(\operatorname{tr}\mathbf{S}) + \frac{1}{1-\nu}(\operatorname{div}\mathbf{b})\mathbf{1} + 2\widehat{\nabla\mathbf{b}} = \mathbf{0} \quad \text{on } B \tag{4.1.118}$$

$$\mathbf{u}[\mathbf{S}] = \widehat{\mathbf{u}} \quad \text{on } \partial B_1 \tag{4.1.119}$$

$$\mathbf{S}\mathbf{n} = \widehat{\mathbf{s}} \quad \text{on } \partial B_2 \tag{4.1.120}$$

Here $\mathbf{u}[\mathbf{S}]$ is defined by Equations 4.1.110 through 4.1.112 in which

$$K_{ijkl} = \frac{1+\nu}{2E}(\delta_{ik}\delta_{jl} + \delta_{il}\delta_{jk}) - \frac{\nu}{E}\delta_{ij}\delta_{kl} \tag{4.1.121}$$

This formula for K_{ijkl} comes from the strain–stress relation for an isotropic body, see Equation 3.3.61, that is, the relation

$$E_{ij} = K_{ijkl}S_{kl} \tag{4.1.122}$$

If Equation 4.1.121 is substituted into Equation 4.1.122 we obtain Equation 3.3.61. □

We note that Theorems 1 and 2 lead to the following natural formulation of the displacement problem and the traction problem, respectively.

(1) *The displacement problem*: A vector field $\mathbf{u}$ corresponds to a solution to the displacement problem if and only if

$$\operatorname{div}\mathsf{C}[\nabla\mathbf{u}] + \mathbf{b} = \mathbf{0} \quad \text{on } B \tag{4.1.123}$$

$$\mathbf{u} = \widehat{\mathbf{u}} \quad \text{on } \partial B \tag{4.1.124}$$

(2) *The traction problem*: A tensor field $\mathbf{S}$ corresponds to a solution to the traction problem if and only if

$$\operatorname{div}\mathbf{S} + \mathbf{b} = \mathbf{0} \quad \text{on } B \tag{4.1.125}$$

$$\operatorname{curl}\operatorname{curl}\mathsf{K}[\mathbf{S}] = \mathbf{0} \quad \text{on } B \tag{4.1.126}$$

$$\mathbf{S}\mathbf{n} = \widehat{\mathbf{s}} \quad \text{on } \partial B \tag{4.1.127}$$

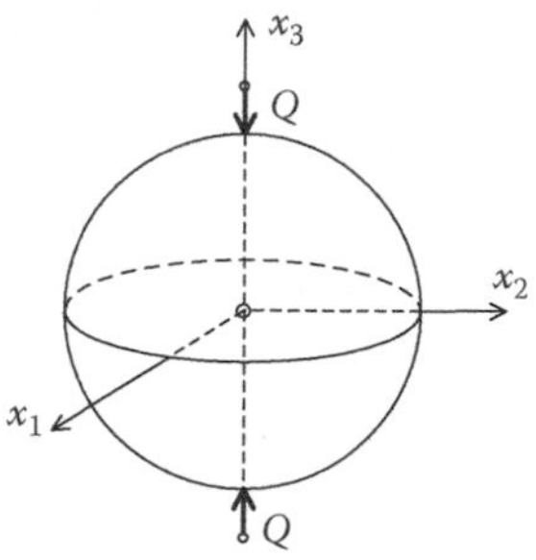

FIGURE 4.5 A sphere acted upon by two concentrated forces.

For an isotropic homogeneous body, the natural formulations of the displacement and traction problems are obtained by replacing in Equation 4.1.123 and in Equation 4.1.126 tensors **C** and **K** by their isotropic representations. As a result, Equations 4.1.123 and 4.1.126 take the form of Equations 4.1.114 and 4.1.118, respectively.

An example of a traction problem is the problem of an elastic solid sphere under two concentrated loads, shown in Figure 4.5.

In this problem, the boundary load **s** is axially symmetric, and the concentrated forces are received as limits of continuously distributed normal loads acting on the upper and lower halves of the spherical surface [4].

4.1.5 Uniqueness

The boundary value problems formulated in Section 4.1.4 are well posed in the following sense. First, they possess a solution in a class of functions. Second, a solution to any of these boundary value problems is unique in a class of functions. Third, solutions are continuously dependent on data.

The existence of solutions and the continuous dependence on data are not discussed in this book.* By data in the boundary value problems, we mean the prescribed tensor of elasticity **C** and the prescribed array of functions $\{\mathbf{b}, \widehat{\mathbf{u}}, \widehat{\mathbf{s}}\}$.

In the following, we will prove the uniqueness of solutions of the boundary value problems. We formulate the uniqueness theorem for the mixed problem.

Uniqueness Theorem for the Mixed Problem. If the elasticity tensor **C** is positive definite, then any two solutions of the mixed problem are equal to within a rigid displacement. If $\partial B_1 \neq \emptyset$ then a rigid displacement vanishes.

Proof: Let $s = [\mathbf{u}, \mathbf{E}, \mathbf{S}]$ and $\widetilde{s} = [\widetilde{\mathbf{u}}, \widetilde{\mathbf{E}}, \widetilde{\mathbf{S}}]$ be two solutions of the mixed problem, and

$$s^* = s - \widetilde{s} = [\mathbf{u}^*, \mathbf{E}^*, \mathbf{S}^*] = [\mathbf{u} - \widetilde{\mathbf{u}}, \mathbf{E} - \widetilde{\mathbf{E}}, \mathbf{S} - \widetilde{\mathbf{S}}] \tag{4.1.128}$$

By the principle of superposition, see Equation 4.1.80, s^* is an elastic state corresponding to zero body forces and zero boundary data, that is,

$$\mathbf{u}^* = \mathbf{0} \quad \text{on } \partial B_1 \tag{4.1.129}$$

$$\mathbf{s}^* \equiv \mathbf{S}^*\mathbf{n} = \mathbf{0} \quad \text{on } \partial B_2 \tag{4.1.130}$$

From these boundary conditions it follows that

$$\mathbf{u}^* \cdot \mathbf{s}^* = 0 \quad \text{on } \partial B \tag{4.1.131}$$

Using the solution presented in Example 4.1.14 in which s is replaced by s^* we conclude that s^* takes the form

* The existence and continuous dependence on data in theory of linear elasticity is to be found in [5].

$$s^* = [\mathbf{u}^{(0)}, \mathbf{0}, \mathbf{0}] \tag{4.1.132}$$

where $\mathbf{u}^{(0)}$ is a rigid body motion. This proves the first part of the theorem.

To prove the second part, we take three noncolinear points on ∂B_1. Since at each of these three points the condition $\mathbf{u}^* = \mathbf{0}$ is satisfied, and by Equation 4.1.132 $\mathbf{u}^* = \mathbf{u}^{(0)}$, then $\mathbf{u}^{(0)} \equiv \mathbf{0}$.

Notes:

(1) From the definition of a displacement problem and from the uniqueness theorem, it follows that a solution to the displacement problem is unique. Thus, there exists only one elastic state $s = [\mathbf{u}, \mathbf{E}, \mathbf{S}]$ corresponding to the body force $\mathbf{b}$ and the boundary displacement $\widehat{\mathbf{u}}$.

(2) A solution to the traction problem is unique to within a rigid body motion, which means that this solution is to be identified with an elastic state $s = [\mathbf{u}, \mathbf{E}, \mathbf{S}]$, corresponding to $\mathbf{b}$ and $\widehat{\mathbf{s}}$, in which $\mathbf{u}$ is determined to within a rigid body motion while $\mathbf{E}$ and $\mathbf{S}$ are determined in a unique way.

4.2 INITIAL–BOUNDARY VALUE PROBLEMS OF ELASTODYNAMICS

4.2.1 Field Equations

In this section we discuss the geometric relations, the equation of motion, and the constitutive relation for a linear elastic body in which the displacement vector field $\mathbf{u}$, strain tensor field $\mathbf{E}$, and stress tensor field $\mathbf{S}$ depend on the position vector $\mathbf{x} \in B$ and on the time t.

The geometric relation is of the same form as in elastostatics, see Equation 4.1.1,

$$\mathbf{E} = \widehat{\nabla}\mathbf{u} = \frac{1}{2}(\nabla\mathbf{u} + \nabla\mathbf{u}^T) \tag{4.2.1}$$

The equations of motion take the form, see Equations 3.2.24 and 3.2.31,

$$\operatorname{div}\mathbf{S} + \mathbf{b} = \rho\ddot{\mathbf{u}} \quad \mathbf{S} = \mathbf{S}^T \tag{4.2.2}$$

The stress–strain relation is, see Equation 4.1.3,

$$\mathbf{S} = \mathsf{C}[\mathbf{E}] \tag{4.2.3}$$

The system of field equations 4.2.1 through 4.2.3 may be viewed as a generalization and parameterization of that of elastostatics given by Equations 4.1.1 through 4.1.3 in the following sense:

(1) The inertia term $\rho\ddot{\mathbf{u}}$ is included in Equation 4.2.2.

(2) The functions $\mathbf{u}$, $\mathbf{E}$, $\mathbf{S}$ and the body force $\mathbf{b}$ are now functions of time t.

With regard to the density ρ and the elasticity tensor C, we assume that they depend on $\mathbf{x}$ only. Therefore, the relation 4.2.3 describes an instantaneous response of the body to an applied load.

Finally, Equations 4.2.1 through 4.2.3 are to be satisfied on the Cartesian product $\bar{B} \times [0, \infty)$.

By eliminating **E** and **S** from Equations 4.2.1 through 4.2.3 we arrive at the *displacement equation of motion*

$$\text{div}\,\mathsf{C}[\nabla\mathbf{u}] + \mathbf{b} = \rho\ddot{\mathbf{u}} \tag{4.2.4}$$

In components Equation 4.2.4 reads

$$(C_{ijkl}u_{k,l})_{,j} + b_i = \rho\ddot{u}_i \tag{4.2.5}$$

Equation 4.2.4 is complete in the following sense: If **u** is a solution of Equation 4.2.4, then the strain tensor **E** follows from Equation 4.2.1 and the stress tensor **S** follows from the equation

$$\mathbf{S} = \mathsf{C}[\nabla\mathbf{u}] \tag{4.2.6}$$

So, finding a solution [**u**, **E**, **S**] to Equations 4.2.1 through 4.2.3 may be reduced to finding a field **u** that satisfies Equation 4.2.4.

Next, dividing the first of Equations 4.2.2 by ρ and operating on the resulting equation with the symmetric gradient $\widehat{\nabla}$ and using the strain–displacement relation 4.2.1, we obtain

$$\widehat{\nabla}\left[\rho^{-1}\,(\text{div}\,\mathbf{S} + \mathbf{b})\right] = \widehat{\nabla}\ddot{\mathbf{u}} = \ddot{\mathbf{E}} \tag{4.2.7}$$

If we assume that the tensor C is invertible, so that the relation 4.2.6 may be written as

$$\mathbf{E} = \mathsf{K}[\mathbf{S}] \tag{4.2.8}$$

where K is the compliance tensor field, then from 4.2.7 we arrive at the *stress equation of motion* [6]

$$\widehat{\nabla}\left[\rho^{-1}\,(\text{div}\,\mathbf{S})\right] - \mathsf{K}[\ddot{\mathbf{S}}] = -\mathbf{B} \tag{4.2.9}$$

where

$$\mathbf{B} = \widehat{\nabla}\left(\rho^{-1}\,\mathbf{b}\right) \tag{4.2.10}$$

Obviously, **B** is a prescribed second-order symmetric tensor field if **b** and ρ are prescribed.

In components Equations 4.2.9 and 4.2.10 read, respectively,

$$(\rho^{-1}S_{(ik,k)})_{,j)} - K_{ijkl}\ddot{S}_{kl} = -B_{ij} \tag{4.2.11}$$

$$B_{ij} = (\rho^{-1}b_{(i})_{,j)} \tag{4.2.12}$$

Notes:

(1) The stress equation of motion 4.2.9 is obtained by elimination of displacement **u** and strain tensor **E** from Equations 4.2.1 through 4.2.3.

(2) In Equations 4.2.11 and 4.2.12 the parentheses on the indexes i and j indicate the symmetric part of a tensor with indexes i and j, see Equation 2.1.28.

A completeness problem for Equation 4.2.9 will be discussed later. At present, we give the displacement equation of motion and the stress equation of motion for an isotropic homogeneous body.

If the elasticity tensor **C** and the compliance tensor **K** describe a homogeneous isotropic body, then from Equations 4.2.4 and 4.2.9 we obtain, respectively,

Displacement equation of motion

$$\mu\nabla^2\mathbf{u} + (\lambda + \mu)\nabla(\operatorname{div}\mathbf{u}) + \mathbf{b} = \rho\ddot{\mathbf{u}} \tag{4.2.13}$$

Stress equation of motion

$$\widehat{\nabla}(\operatorname{div}\mathbf{S}) - \frac{\rho}{2\mu}\left[\ddot{\mathbf{S}} - \frac{\lambda}{3\lambda + 2\mu}(\operatorname{tr}\ddot{\mathbf{S}})\mathbf{1}\right] = -\widehat{\nabla}\mathbf{b} \tag{4.2.14}$$

In components, Equations 4.2.13 and 4.2.14 take the form, respectively,

$$\mu u_{i,kk} + (\lambda + \mu)u_{k,ki} + b_i = \rho\ddot{u}_i \tag{4.2.15}$$

and

$$S_{(ik,kj)} - \frac{\rho}{2\mu}\left(\ddot{S}_{ij} - \frac{\lambda}{3\lambda + 2\mu}\ddot{S}_{kk}\delta_{ij}\right) = -b_{(i,j)} \tag{4.2.16}$$

The displacement equation of motion 4.2.13 is called *Navier's equation of motion.* It is widely used in the analysis of elastic wave propagation. While Navier's equation of motion was published approximately 175 years ago, the stress equation of motion 4.2.9 was proposed only about 45 years ago.

If the inertia term $\rho\ddot{\mathbf{u}}$ in Navier's equation 4.2.13 vanishes, the equation reduces to the displacement equation of equilibrium of elastostatics, Equation 4.1.114. Also, for a general anisotropic body the displacement equation of motion, Equation 4.2.4, reduces to the displacement equation of equilibrium, Equation 4.1.123, if the inertia term $\rho\ddot{\mathbf{u}}$ in Equation 4.2.4 is ignored. There is no such correspondence between the stress equation of motion 4.2.9 and the stress equations of elastostatics, Equations 4.1.125 and 4.1.126. The reason is that in the elastostatics and for a simply connected body, the stress field equations consist of the three scalar equations of equilibrium Equation 4.1.125, and six scalar stress compatibility conditions, Equation 4.1.126, while the stress equation of elastodynamics in the form of a single tensor equation 4.2.9 consists of six scalar equations only. A theorem that a stress tensor of elastodynamics is characterized by this single equation 4.2.9 will be stated later when we come to the formulations of initial–boundary value problems of elastodynamics.

Returning to Navier's equation 4.2.13, which is valid for an isotropic body, we derive a number of properties of a solution to this equation.

First, we discuss the so-called irrotational and isochoric motions. By using the vector identity, see (b) in Example 4.1.2,

$$\nabla^2\mathbf{u} = \nabla\operatorname{div}\mathbf{u} - \operatorname{curl}\operatorname{curl}\mathbf{u} \tag{4.2.17}$$

we note that Equation 4.2.13 may be reduced to the form

$$(\lambda + 2\mu)\nabla \operatorname{div} \mathbf{u} - \mu \operatorname{curl} \operatorname{curl} \mathbf{u} + \mathbf{b} = \rho \ddot{\mathbf{u}} \tag{4.2.18}$$

Now, if we introduce the so-called *irrotational velocity* c_1 and *isochoric velocity* c_2 by means of the formulas

$$c_1 = \sqrt{\frac{\lambda + 2\mu}{\rho}} \quad \text{and} \quad c_2 = \sqrt{\frac{\mu}{\rho}} \tag{4.2.19}$$

Equation 4.2.18 reduces to

$$c_1^2 \nabla \operatorname{div} \mathbf{u} - c_2^2 \operatorname{curl} \operatorname{curl} \mathbf{u} + \frac{\mathbf{b}}{\rho} = \ddot{\mathbf{u}} \tag{4.2.20}$$

If $\mu > 0$ and $\lambda + 2\mu > 0$, see Equation 3.3.103, then c_1 and c_2 are real valued numbers. Using the velocities c_1 and c_2 we define the two wave operators

$$\Box_\alpha^2 = \nabla^2 - \frac{1}{c_\alpha^2} \frac{\partial^2}{\partial t^2} \qquad \alpha = 1, 2 \tag{4.2.21}$$

Since

$$\frac{\lambda + \mu}{\mu} = \frac{\lambda + 2\mu}{\mu} - 1 = \left(\frac{c_1}{c_2}\right)^2 - 1 \tag{4.2.22}$$

Navier's equation 4.2.13 may also be written as

$$\Box_2^2 \mathbf{u} + \left[\left(\frac{c_1}{c_2}\right)^2 - 1\right] \nabla(\operatorname{div} \mathbf{u}) + \frac{\mathbf{b}}{\mu} = \mathbf{0} \tag{4.2.23}$$

To explain the terms *irrotational velocity* and *isochoric velocity* we take for now that $\mathbf{b} = \mathbf{0}$, and first assume that a motion $\mathbf{u}$ is subject to the constraint

$$\operatorname{curl} \mathbf{u} = \mathbf{0} \tag{4.2.24}$$

Since, by Equation 4.2.17

$$\nabla(\operatorname{div} \mathbf{u}) = \nabla^2 \mathbf{u} \tag{4.2.25}$$

so, from Equation 4.2.18, with $\mathbf{b} = \mathbf{0}$, we get

$$\Box_1^2 \mathbf{u} = \mathbf{0} \tag{4.2.26}$$

Hence, a curl-free field $\mathbf{u}$ corresponding to zero body force satisfies the wave equation with velocity c_1, and this explains why c_1 is called irrotational velocity.

To show that c_2 is associated with an isochoric motion, that is, with a motion without change of volume, we assume that $\mathbf{u}$ satisfies the condition, see 3.1.19 through 3.1.21,

$$\operatorname{div} \mathbf{u} = 0 \tag{4.2.27}$$

Then from Equation 4.2.23, with $\mathbf{b} = \mathbf{0}$, follows

$$\Box_2^2\, \mathbf{u} = \mathbf{0} \tag{4.2.28}$$

which shows that an isochoric displacement field $\mathbf{u}$ propagates with velocity c_2. Therefore, for an isotropic body a curl-free motion and a divergence-free motion comply with the wave equations with irrotational and isochoric velocities, respectively.

Next, we show that if $\mathbf{b} = \mathbf{0}$, then

$$\Box_1^2 \operatorname{div} \mathbf{u} = 0 \tag{4.2.29}$$

$$\Box_2^2 \operatorname{curl} \mathbf{u} = \mathbf{0} \tag{4.2.30}$$

Equation 4.2.29 is obtained from Equation 4.2.18, with $\mathbf{b} = \mathbf{0}$, after we apply to it the operator div and use the identities:

$$\operatorname{div} \nabla(\operatorname{div} \mathbf{u}) = \nabla^2(\operatorname{div} \mathbf{u}) \tag{4.2.31}$$

and

$$\operatorname{div} \operatorname{curl} \operatorname{curl} \mathbf{u} = 0 \tag{4.2.32}$$

In components, Equations 4.2.31 and 4.2.32 read

$$u_{k,kii} = (u_{k,k})_{,ii} \tag{4.2.33}$$

$$\epsilon_{iab}\epsilon_{bjk}u_{k,jai} = 0 \tag{4.2.34}$$

On the LHS of 4.2.34 we multiply a skew tensor with indexes i and a by a symmetric tensor with indexes i and a, and such a product vanishes.

Equation 4.2.30 is obtained by applying the operator curl to Equation 4.2.23, with $\mathbf{b} = \mathbf{0}$, and using the identity $\operatorname{curl} \nabla(\operatorname{div} \mathbf{u}) = \mathbf{0}$.

Since $\operatorname{div} \mathbf{u}$ is a dilatation, Equation 4.2.29 means that the dilatation satisfies a wave equation with velocity c_1.

Also, Equation 4.2.30 implies that the vector field

$$\boldsymbol{\omega} = \frac{1}{2} \operatorname{curl} \mathbf{u} \tag{4.2.35}$$

which represents a rotation field, satisfies a wave equation with velocity c_2.

Now we show that for a homogeneous isotropic body a stress tensor field $\mathbf{S}$, in addition to Equation 4.2.14, satisfies also the following

Dynamic stress compatibility equation of Beltrami–Michell type

$$\Box_2^2\,\mathbf{S} + \frac{2(\lambda+\mu)}{3\lambda+2\mu}\,\nabla\nabla(\operatorname{tr}\mathbf{S}) + \left(\frac{1}{c_2^2} - \frac{1}{c_1^2}\right)\frac{\lambda}{3\lambda+2\mu}\,(\operatorname{tr}\ddot{\mathbf{S}})\mathbf{1} + \mathbf{F} = \mathbf{0} \tag{4.2.36}$$

where

$$\mathbf{F} = 2\widehat{\nabla\mathbf{b}} + \frac{\lambda}{\lambda+2\mu}\,(\operatorname{div}\mathbf{b})\mathbf{1} \tag{4.2.37}$$

In components Equations 4.2.36 and 4.2.37 read, respectively,

$$\Box_2^2\,S_{ij} + \frac{2(\lambda+\mu)}{3\lambda+2\mu}\,S_{kk,ij} + \left(\frac{1}{c_2^2} - \frac{1}{c_1^2}\right)\frac{\lambda}{3\lambda+2\mu}\,\ddot{S}_{kk}\delta_{ij} + F_{ij} = 0 \tag{4.2.38}$$

and

$$F_{ij} = b_{i,j} + b_{j,i} + \frac{\lambda}{\lambda+2\mu}\,b_{k,k}\delta_{ij} \tag{4.2.39}$$

To obtain Equation 4.2.38 we differentiate Navier's equation 4.2.15 with respect to x_j and take the symmetric part of the result with respect to the indexes i and j and arrive at

$$\mu u_{(i,j)kk} + (\lambda+\mu)u_{k,kij} + b_{(i,j)} = \rho\ddot{u}_{(i,j)} \tag{4.2.40}$$

Now, since

$$u_{(i,j)} = E_{ij} \tag{4.2.41}$$

and

$$E_{ij} = \frac{1}{2\mu}\left[S_{ij} - \frac{\lambda}{3\lambda+2\mu}\,S_{aa}\delta_{ij}\right] \tag{4.2.42}$$

from Equation 4.2.40 we obtain

$$\begin{aligned}&\left[S_{ij} - \frac{\lambda}{3\lambda+2\mu}\,S_{aa}\delta_{ij}\right]_{,kk} + \frac{2(\lambda+\mu)}{3\lambda+2\mu}\,S_{aa,ij}\\ &-\frac{\rho}{\mu}\left[\ddot{S}_{ij} - \frac{\lambda}{3\lambda+2\mu}\,\ddot{S}_{aa}\delta_{ij}\right] + 2b_{(i,j)} = 0\end{aligned} \tag{4.2.43}$$

Taking the trace of this equation yields

$$\Box_1^2\,S_{aa} + \frac{3\lambda+2\mu}{\lambda+2\mu}\,b_{a,a} = 0 \tag{4.2.44}$$

or

$$S_{aa,kk} = \frac{1}{c_1^2}\,\ddot{S}_{aa} - \frac{3\lambda+2\mu}{\lambda+2\mu}\,b_{a,a} \tag{4.2.45}$$

Substitution of Equation 4.2.45 into Equation 4.2.43 results in Equation 4.2.38. This completes the derivation of Equation 4.2.36.

If **S** is independent of time, then Equation 4.2.36 reduces to the Beltrami–Michell stress compatibility equation of elastostatics, that is, Equation 4.1.37, in which ν is to be substituted by $\lambda/[2(\lambda + \mu)]$. See also Equation 4.1.62 in which we let $T = 0$.

Applying the operator $\square_1^2$ to Equation 4.2.38 we arrive at the nonhomogeneous biwave equation for S_{ij}

$$\square_1^2\, \square_2^2 S_{ij} = M_{ij} \tag{4.2.46}$$

where

$$M_{ij} = \frac{2(\lambda + \mu)}{\lambda + 2\mu}\, b_{r,rij} - \square_1^2\, (b_{i,j} + b_{j,i}) - \frac{\lambda}{\lambda + 2\mu}\, \delta_{ij} \square_2^2\, b_{r,r} \tag{4.2.47}$$

while receiving Equation 4.2.46 we used the identity

$$\square_1^2 = \square_2^2 + \left(\frac{1}{c_1^2} - \frac{1}{c_2^2}\right) \frac{\partial^2}{\partial t^2} \tag{4.2.48}$$

Equation 4.2.46 written in direct notation reads

$$\square_1^2 \square_2^2\, \mathbf{S} = \mathbf{M} \tag{4.2.49}$$

where

$$\mathbf{M} = \frac{2(\lambda + \mu)}{\lambda + 2\mu}\, \nabla\nabla(\operatorname{div} \mathbf{b}) - 2\square_1^2\, (\widehat{\nabla \mathbf{b}}) - \frac{\lambda}{\lambda + 2\mu}\, \square_2^2\, (\operatorname{div} \mathbf{b})\mathbf{1} \tag{4.2.50}$$

If body force **b** is divergence free and its symmetric gradient vanishes, then it follows from Equation 4.2.47 or Equation 4.2.50 that the stress field **S** satisfies the homogeneous biwave equation

$$\square_1^2 \square_2^2\, \mathbf{S} = \mathbf{0} \tag{4.2.51}$$

Also, in this case, by Equation 4.2.45 we obtain

$$\square_1^2\, (\operatorname{tr} \mathbf{S}) = 0 \tag{4.2.52}$$

Equations 4.2.36 and 4.2.49 are necessary field equations to be satisfied by a stress field **S** corresponding to a prescribed body force field **b**. In general, they are not sufficient to find a stress field **S** corresponding to a solution to an initial–boundary value problem of elastodynamics for an isotropic body. These equations may be useful in solving a problem for an infinite elastic body or in finding a class of admissible stress fields of isotropic elastodynamics.

Example 4.2.1

Let $[\mathbf{u}, \mathbf{E}, \mathbf{S}]$ be a triple of functions satisfying field equations 4.2.1, 4.2.2, and 4.2.3 with $\mathbf{b} = \mathbf{0}$. We define the total energy per unit volume e and the energy flux vector $\mathbf{q}$, respectively, by

$$e = \frac{1}{2}\,\mathbf{E} \cdot \mathbf{C}[\mathbf{E}] + \frac{1}{2}\,\rho \dot{\mathbf{u}}^2 \tag{a}$$

$$\mathbf{q} = -\mathbf{S}\dot{\mathbf{u}} \tag{b}$$

Show that the following local form of balance of energy holds true:

$$\dot{e} = -\operatorname{div} \mathbf{q} \tag{c}$$

We note that the first term on the RHS of (a) represents the strain elastic energy per unit volume, see Equation 3.3.90, while the second term stands for the kinetic energy per unit volume.

Solution

We differentiate (a) with respect to time and use the symmetry of tensor $\mathbf{C}$, to obtain

$$\begin{aligned} \dot{e} &= \frac{1}{2}\,(\dot{\mathbf{E}} \cdot \mathbf{C}[\mathbf{E}] + \mathbf{E} \cdot \mathbf{C}[\dot{\mathbf{E}}]) + \rho \dot{\mathbf{u}} \cdot \ddot{\mathbf{u}} \\ &= \dot{\mathbf{E}} \cdot \mathbf{C}[\mathbf{E}] + \rho \dot{\mathbf{u}} \cdot \ddot{\mathbf{u}} \end{aligned} \tag{d}$$

Since the constitutive equation 4.2.3 is to be satisfied, which means that

$$\mathbf{S} = \mathbf{C}[\mathbf{E}] \tag{e}$$

we reduce (d) to the form

$$\dot{e} = \dot{\mathbf{E}} \cdot \mathbf{S} + \rho \dot{\mathbf{u}} \cdot \ddot{\mathbf{u}} \tag{f}$$

Next, by applying the divergence operator to (b), we obtain

$$\operatorname{div} \mathbf{q} = -\operatorname{div}(\mathbf{S}\dot{\mathbf{u}}) \tag{g}$$

or in components

$$q_{i,i} = -(S_{ik}\dot{u}_k)_{,i} = -(S_{ik,i}\dot{u}_k + S_{ik}\dot{u}_{k,i}) \tag{h}$$

Since $\mathbf{S}$ satisfies the equations of motion 4.2.2, with $\mathbf{b} = \mathbf{0}$, and the geometric relation 4.2.1 is satisfied, (h) may be written in direct notation as

$$-\operatorname{div} \mathbf{q} = \rho \ddot{\mathbf{u}} \cdot \dot{\mathbf{u}} + \mathbf{S} \cdot \dot{\mathbf{E}} \tag{i}$$

By comparing (f) and (i) we arrive at the resulting equation (c). This completes the solution. □

Note: Equation (c) has the same form as that of the first law of thermodynamics for a rigid heat conductor with zero heat supply if the function e and the vector field $\mathbf{q}$ are identified with the conductor's internal energy and the heat flux vector, respectively [7].

Example 4.2.2

Show that if **u** satisfies Navier's equation 4.2.13, then the corresponding stress tensor **S** also satisfies the dynamic stress compatibility equation of Beltrami–Michell type, Equation 4.2.36.

Solution

We assume that **S** is given by the stress–strain relation

$$S_{ij} = 2\mu E_{ij} + \lambda E_{kk}\delta_{ij} \tag{a}$$

where

$$E_{ij} = u_{(i,j)} \tag{b}$$

and u_i satisfies Navier's equation

$$\mu u_{i,kk} + (\lambda + \mu)u_{k,ki} + b_i = \rho \ddot{u}_i \tag{c}$$

To show that S_{ij} satisfies Equation 4.2.38, which is equivalent to Equation 4.2.36, we write Equation 4.2.38 in the form of Equation 4.2.43. Equation 4.2.43, by virtue of Equations 4.2.42 and 4.2.41 which are equivalent to (a) and (b), respectively, takes the form of Equation 4.2.40, which can be rewritten as

$$v_{(i,j)} = 0 \tag{d}$$

where

$$v_i = \mu u_{i,kk} + (\lambda + \mu)u_{k,ki} + b_i - \rho \ddot{u}_i \tag{e}$$

Now, by virtue of (c) $v_i = 0$, (d) is satisfied. Hence, S_{ij} satisfies the dynamic stress compatibility equation of Beltrami–Michell type, Equation 4.2.38, provided S_{ij} complies with (a) through (c). This completes the solution. □

Example 4.2.3

Let $b_i = 0$, and let S_{ij} be a solution to the dynamic stress compatibility equation of Beltrami–Michell type, Equation 4.2.38 with $F_{ij} = 0$,

$$\Box_2^2 S_{ij} + \frac{2(\lambda + \mu)}{3\lambda + 2\mu} S_{kk,ij} + \left(\frac{1}{c_2^2} - \frac{1}{c_1^2}\right) \frac{\lambda}{3\lambda + 2\mu} \ddot{S}_{kk}\delta_{ij} = 0 \tag{a}$$

subject to the condition

$$S_{ik,ki} - \frac{1}{c_3^2} \ddot{S}_{kk} = 0 \tag{b}$$

where

$$\frac{1}{c_3^2} = \frac{\rho}{3\lambda + 2\mu} \tag{c}$$

Also, suppose that S_{ij} satisfies the homogeneous initial conditions

$$S_{ij}(\mathbf{x}, 0) = 0, \quad \dot{S}_{ij}(\mathbf{x}, 0) = 0 \quad \mathbf{x} \in B \tag{d}$$

and define a vector field u_i by the formula

$$u_i = \frac{1}{\rho} \int_0^t (t - \tau) S_{ij,j} \, d\tau \tag{e}$$

Show that u_i satisfies Navier's equation (c) of Example 4.2.2 with $b_i = 0$.

Solution

The differentiation of (a) with respect to x_j yields

$$\Box_2^2 \, S_{ij,j} + \frac{2(\lambda + \mu)}{3\lambda + 2\mu} \, S_{kk,ijj} + \left(\frac{1}{c_2^2} - \frac{1}{c_1^2} \right) \frac{\lambda}{3\lambda + 2\mu} \, \ddot{S}_{kk,i} = 0 \tag{f}$$

We note also that (a) with $i = j \equiv a$ reduces to Equation 4.2.44 with $b_a = 0$, that is,

$$\Box_1^2 \, S_{aa} = 0 \tag{g}$$

From this equation, after differentiating it with respect to x_i, we obtain

$$S_{kk,jji} = \frac{1}{c_1^2} \, \ddot{S}_{aa,i} \tag{h}$$

Next, by differentiating (b) with respect to x_i and after a suitable change of dummy indexes, we obtain

$$S_{ab,bai} = \frac{1}{c_3^2} \, \ddot{S}_{kk,i} \tag{i}$$

Equation (h) may be reduced by virtue of (i) to the form

$$S_{kk,jji} = \left(\frac{c_3}{c_1} \right)^2 S_{ab,bai} \tag{j}$$

So, if we substitute $S_{kk,jji}$ from (j) and $\ddot{S}_{kk,i}$ from (i) into (f), we obtain

$$\Box_2^2 \, S_{ij,j} + \frac{\lambda + \mu}{\mu} \, S_{ab,bai} = 0 \tag{k}$$

Now, let us rewrite the vector equation (e) in a compact form

$$u_i = \rho^{-1} t * S_{ij,j} \tag{l}$$

where * stands for the convolution product on the time axis, see Equation 2.3.27. It follows from (l) that

$$u_{a,a} = \rho^{-1} t * S_{ab,ba} \tag{m}$$

and

$$u_{a,ai} = \rho^{-1} t * S_{ab,bai} \tag{n}$$

Also, let us rewrite (k) in an expanded form

$$S_{ij,jaa} - \frac{1}{c_2^2} \ddot{S}_{ij,j} + \frac{\lambda + \mu}{\mu} S_{ab,bai} = 0 \tag{o}$$

and note that by the homogeneous initial conditions (d)

$$\rho^{-1} t * \ddot{S}_{ij,j} = \rho^{-1} \overline{\overline{t * S_{ij,j}}}^{\,\cdot\cdot} = \ddot{u}_i \tag{p}$$

To show that this equation holds true we observe that for an arbitrary function $f(t)$ such that $f(0) = \dot{f}(0) = 0$, we have

$$t * \ddot{f} = \ddot{\overline{t * f}} \tag{q}$$

Applying the operator $\rho^{-1} t*$ to (o), and using (n) and (p), we arrive at Navier's equation, see Equation 4.2.23 with $\mathbf{b} = \mathbf{0}$,

$$\Box_2^2 u_i + \frac{(\lambda + \mu)}{\mu} u_{a,ai} = 0 \tag{r}$$

This completes the solution. □

Notes:

(1) This example shows that a suitable restricted solution S_{ij} to the dynamic stress compatibility equation of Beltrami–Michell type, see (a), generates a vector field u_i, by (e), that satisfies Navier's equation. It is easy to see, by taking the trace of the stress equation of motion, see Equation 4.2.16 with $b_i = 0$, that the restriction imposed on S_{ij} in the example, see (b), is equivalent to the trace of the stress equation of motion in the form of Equation 4.2.16 with $b_i = 0$.

(2) It is to be shown in the discussion of initial–boundary value problems of elastodynamics that (e), extended to nonhomogeneous initial displacement and velocity, generates a displacement field provided a stress tensor of elastodynamics is known.

4.2.2 Field Equations of Dynamic Theory of Thermal Stresses

For a homogeneous isotropic body, a dynamic theory of thermal stresses is described by the following system of field equations:

The strain–displacement relation

$$\mathbf{E} = \widehat{\nabla}\mathbf{u} \tag{4.2.53}$$

the equation of motion

$$\operatorname{div}\mathbf{S} + \mathbf{b} = \rho\ddot{\mathbf{u}}, \quad \mathbf{S} = \mathbf{S}^T \tag{4.2.54}$$

the stress–strain–temperature relation

$$\mathbf{S} = 2\mu\mathbf{E} + \lambda(\operatorname{tr}\mathbf{E})\mathbf{1} - (3\lambda + 2\mu)\alpha T\mathbf{1} \tag{4.2.55}$$

Here **u**, **E**, and **S** represent the displacement, strain, and stress fields, respectively, and they depend on both **x** and t with $\mathbf{x} \in B$ and $t \geq 0$. The vector field **b** represents the body force, which also depends on **x** and t, ρ is a constant density, and μ and λ are Lamé constants. Finally, T is a temperature change that satisfies the parabolic heat conduction equation

$$\nabla^2 T - \frac{1}{\kappa}\dot{T} = -\frac{Q}{\kappa} \tag{4.2.56}$$

where $Q = Q(\mathbf{x}, t)$ denotes a heat supply field, and κ is the thermal diffusivity,

$$\kappa = \frac{k}{\rho c} \tag{4.2.57}$$

with k representing heat conductivity, ρ meaning density, and c being specific heat at constant volume. Note that $Q/\kappa = W/k$, where W stands for the internal heat generation per unit volume per unit time. We note the dimensions: $[Q] = \mathrm{K/s}$ and $[W] = \mathrm{N/(m^2\,s)}$.

Equations 4.2.53 through 4.2.55 reduce to the field equations 4.1.16 through 4.1.18 of thermoelastostatics if the fields **u**, **E**, and **S**, as well as **b** and T, are time independent.

Similarly as in the case of thermoelastostatics, T is a solution to the heat conduction equation 4.2.56. Once such a solution is found, the term $(3\lambda + 2\mu)\alpha T\mathbf{1}$ in Equation 4.2.55 is known.

In the following, we present the displacement–temperature and stress–temperature fields equations of the dynamic theory. To derive the first type of these equations we eliminate **E** and **S** from Equations 4.2.53 through 4.2.55 and arrive at the dynamic displacement–temperature field equation

$$\mu\nabla^2\mathbf{u} + (\lambda + \mu)\nabla(\operatorname{div}\mathbf{u}) - (3\lambda + 2\mu)\alpha\nabla T + \mathbf{b} = \rho\ddot{\mathbf{u}} \tag{4.2.58}$$

The stress–temperature field equation of the dynamic theory is found by eliminating **u** and **E** from Equations 4.2.53 through 4.2.55. To this end, we rewrite Equation 4.2.55 in an equivalent form, see Equation 3.3.126,

$$\mathbf{E} = \frac{1}{2\mu}\left[\mathbf{S} - \frac{\lambda}{3\lambda + 2\mu}(\operatorname{tr}\mathbf{S})\mathbf{1}\right] + \alpha T\mathbf{1} \tag{4.2.59}$$

Next, operating on Equation 4.2.54 with the symmetric gradient $\widehat{\nabla}$ and using the strain—displacement relation 4.2.53, we obtain

$$\widehat{\nabla}(\operatorname{div}\mathbf{S}+\mathbf{b}) = \rho\widehat{\nabla}\ddot{\mathbf{u}} = \rho\ddot{\mathbf{E}} \tag{4.2.60}$$

This equation together with Equation 4.2.59 lead to the stress–temperature field equation of the dynamic theory. Namely, substitution of $\mathbf{E}$ from Equation 4.2.59 into Equation 4.2.60 results in

$$\widehat{\nabla}(\operatorname{div}\mathbf{S}) - \frac{\rho}{2\mu}\left[\ddot{\mathbf{S}} - \frac{\lambda}{3\lambda+2\mu}(\operatorname{tr}\ddot{\mathbf{S}})\mathbf{1}\right] = -\widetilde{\mathbf{B}} \tag{4.2.61}$$

where

$$\widetilde{\mathbf{B}} = \widehat{\nabla}\mathbf{b} + \rho\ddot{T}\mathbf{1} \tag{4.2.62}$$

We may observe that Equations 4.2.58 and 4.2.61 reduce to the displacement and stress equations of isothermal elastodynamics, Equations 4.2.13 and 4.2.14, respectively, if $T=0$. Also, it follows from Equations 4.2.58 and 4.2.61 that the appearance of time-dependent temperature change T accounts for an additional term of body force type in these equations.

Proceeding along the lines appropriate to isothermal elastodynamics, see Section 4.2.1, we may show that a stress tensor field $\mathbf{S}$ of the dynamic theory, in addition to Equation 4.2.61, also satisfies the following stress compatibility equation of Beltrami–Michell type

$$\Box_2^2\,\mathbf{S} + \frac{2(\lambda+\mu)}{3\lambda+2\mu}\nabla\nabla(\operatorname{tr}\mathbf{S}) + \left(\frac{1}{c_2^2}-\frac{1}{c_1^2}\right)\frac{\lambda}{3\lambda+2\mu}(\operatorname{tr}\ddot{\mathbf{S}})\mathbf{1} + \widetilde{\mathbf{F}} = \mathbf{0} \tag{4.2.63}$$

where

$$\begin{aligned}\widetilde{\mathbf{F}} = {} & 2\widehat{\nabla}\mathbf{b} + \frac{\lambda}{\lambda+2\mu}(\operatorname{div}\mathbf{b})\mathbf{1} \\ & + 2\mu\alpha\left(\nabla\nabla T + \frac{3\lambda+2\mu}{\lambda+2\mu}\nabla^2 T\mathbf{1}\right) - \frac{5\lambda+4\mu}{\lambda+2\mu}\alpha\rho\ddot{T}\mathbf{1}\end{aligned} \tag{4.2.64}$$

By applying to Equation 4.2.63 the wave operator $\Box_1^2$ it may also be shown that

$$\Box_2^2\left\{\Box_1^2\,\mathbf{S} - m\left[2\mu(\nabla\nabla T - \nabla^2 T\mathbf{1}) + \rho\ddot{T}\mathbf{1}\right]\right\} = \mathbf{M} \tag{4.2.65}$$

where $\mathbf{M}$ is defined by Equation 4.2.50, and m is given in Example 4.1.8,

$$m = \frac{3\lambda+2\mu}{\lambda+2\mu}\alpha \tag{4.2.66}$$

A derivation of Equations 4.2.63 and 4.2.65 is similar to that of Equations 4.2.36 and 4.2.49 of isothermal elastodynamics, and will not be given.

Notes:

(1) Equations 4.2.63 and 4.2.65 reduce to Equations 4.2.36 and 4.2.49, respectively, if $T = 0$.
(2) If $\mathbf{b} = \mathbf{0}$, Equations 4.2.63 and 4.2.65 reduce to those of a dynamic theory of thermal stresses without body forces [8].

Example 4.2.4

Show that if $\mathbf{b} = \mathbf{0}$ then a particular solution to the dynamic theory of thermal stresses for a homogeneous isotropic body takes the form

$$\mathbf{u} = \nabla\phi \tag{a}$$

$$\mathbf{E} = \nabla\nabla\phi \tag{b}$$

$$\mathbf{S} = 2\mu(\nabla\nabla\phi - \nabla^2\phi\mathbf{1}) + \rho\ddot{\phi}\mathbf{1} \tag{c}$$

where ϕ is a time-dependent scalar field that satisfies the nonhomogeneous wave equation

$$\square_1^2\, \phi = mT \tag{d}$$

where

$$m = \frac{3\lambda + 2\mu}{\lambda + 2\mu}\,\alpha \tag{e}$$

Solution

We start from the displacement–temperature field equation, see Equation 4.2.58 with $\mathbf{b} = \mathbf{0}$,

$$\mu\nabla^2\mathbf{u} + (\lambda + \mu)\nabla(\operatorname{div}\mathbf{u}) - (3\lambda + 2\mu)\alpha\nabla T = \rho\ddot{\mathbf{u}} \tag{f}$$

Substituting into this equation the displacement $\mathbf{u}$ from (a), we write

$$\nabla\,[(\lambda + 2\mu)\nabla^2\phi - \rho\ddot{\phi}] = (3\lambda + 2\mu)\alpha\nabla T \tag{g}$$

Since the function ϕ satisfies (d), (g) is also satisfied. Therefore, $\mathbf{u}$ given by (a) satisfies the displacement–temperature equation (f) provided (d) is satisfied.

Substitution of (a) into the strain–displacement equation 4.2.53 yields (b). Also, (b) together with Equation 4.2.55 and (d) imply that $\mathbf{S}$ is given by (c). To show this, we substitute (b) into Equation 4.2.55 and obtain

$$\mathbf{S} = 2\mu\nabla\nabla\phi + \lambda\nabla^2\phi\mathbf{1} - (3\lambda + 2\mu)\alpha T\mathbf{1} \tag{h}$$

Now, from (d), which we write in the form

$$(\lambda + 2\mu)\nabla^2\phi - \rho\ddot{\phi} = (3\lambda + 2\mu)\alpha T \tag{i}$$

we find

$$\lambda \nabla^2 \phi = \rho \ddot{\phi} - 2\mu \nabla^2 \phi + (3\lambda + 2\mu)\alpha T \tag{j}$$

Substituting (j) into (h) results in (c). This completes the solution. □

Notes:

(1) The function ϕ in this example is often called the *dynamic thermoelastic displacement potential.* The particular solution, (a) through (c), plays an important role in reducing a solution of a problem of the dynamic theory of thermal stresses to a solution of an isotropic isothermal elastodynamics.

(2) If $\mathbf{b} = \mathbf{0}$ then $\mathbf{M} = \mathbf{0}$, see Equation 4.2.65, and Equation 4.2.65 reduces to

$$\Box_2^2 \left\{ \Box_1^2 \mathbf{S} - m \left[2\mu (\nabla\nabla T - \nabla^2 T \mathbf{1}) + \rho \ddot{T} \mathbf{1} \right] \right\} = \mathbf{0} \tag{4.2.67}$$

By substituting $\mathbf{S}$ from (c) into 4.2.67 we find that this equation is satisfied provided ϕ satisfies (d).

(3) If T satisfies the parabolic heat conduction equation 4.2.56 then ϕ is a solution of the equation

$$\Box_1^2 D_T^2 \phi = -\frac{m}{\kappa} Q \tag{4.2.68}$$

where $Q/\kappa = W/k$, and D_T^2 is a parabolic operator with diffusivity κ defined by

$$D_T^2 = \nabla^2 - \frac{1}{\kappa} \frac{\partial}{\partial t} \tag{4.2.69}$$

Equation 4.2.68 is obtained by applying to (d) the operator D_T^2 and using the fact that T satisfies Equation 4.2.56.

(4) For an infinite isotropic body subject to a temperature field T that satisfies the heat conduction equation 4.2.56, a stress field $\mathbf{S}$ may be obtained directly by the integration of the stress equation

$$\Box_1^2 D_T^2 \mathbf{S} = -\frac{m}{\kappa} \left[2\mu (\nabla\nabla Q - \nabla^2 Q \mathbf{1}) + \rho \ddot{Q} \mathbf{1} \right] \tag{4.2.70}$$

in which Q is a prescribed function. If $\mathbf{S}$ and Q are time independent, then Equation 4.2.70 reduces to Equation 4.1.73 of a steady-state thermoelasticity.

(5) The wave equation (d) has a solution in the form of a so-called *retarded potential*

$$\phi(\mathbf{x}, t) = -\frac{m}{4\pi} \int\limits_{R \le c_1 t} \frac{T(\boldsymbol{\xi}, t - R/c_1)}{R(\mathbf{x}, \boldsymbol{\xi})} \, dv(\boldsymbol{\xi}) \tag{4.2.71}$$

In this formula, R is the distance between the points $\mathbf{x}$ and $\boldsymbol{\xi}$, and the integration is over a ball of radius $R = c_1 t$ with its center at $\mathbf{x}$. We may show that the function ϕ given by Equation 4.2.71 satisfies the homogeneous initial conditions

$$\phi(\mathbf{x}, 0) = \dot{\phi}(\mathbf{x}, 0) = 0 \tag{4.2.72}$$

4.2.3 Concept of an Elastic Process: Energy and Reciprocal Theorems

A concept of an elastic process is based on that of an admissible process. By an admissible process we mean an array of functions $p = [\mathbf{u}, \mathbf{E}, \mathbf{S}]$ in which $\mathbf{u}$ is an admissible motion, $\mathbf{E}$ is a symmetric tensor field, and $\mathbf{S}$ is an admissible stress field, see Section 3.2.4.

Similarly as in the case of an admissible state of elastostatics, see Section 4.1.3, a set of all admissible processes may be identified with a vector space if we define addition and scalar multiplication in the way we did for admissible states. In particular, if p and $\widetilde{p}$ are two different admissible processes, and ω is a scalar, then

$$p^* = p + \widetilde{p} = [\mathbf{u} + \widetilde{\mathbf{u}}, \mathbf{E} + \widetilde{\mathbf{E}}, \mathbf{S} + \widetilde{\mathbf{S}}] \tag{4.2.73}$$

and

$$p_1 = \omega p = [\omega \mathbf{u}, \omega \mathbf{E}, \omega \mathbf{S}] \tag{4.2.74}$$

are admissible processes.

An *elastic process* corresponding to a body force $\mathbf{b}$ is defined as an admissible process $p = [\mathbf{u}, \mathbf{E}, \mathbf{S}]$ that complies with the fundamental system of field equations of elastodynamics:

$$\mathbf{E} = \widehat{\nabla} \mathbf{u} = \frac{1}{2} (\nabla \mathbf{u} + \nabla \mathbf{u}^T) \tag{4.2.75}$$

$$\text{div}\, \mathbf{S} + \mathbf{b} = \rho \ddot{\mathbf{u}}, \quad \mathbf{S} = \mathbf{S}^T \tag{4.2.76}$$

$$\mathbf{S} = \mathsf{C}[\mathbf{E}] \tag{4.2.77}$$

Similarly as in elastostatics, the surface load field $\mathbf{s}$ is defined by

$$\mathbf{s}(\mathbf{x}, t) = \mathbf{S}(\mathbf{x}, t)\mathbf{n}(\mathbf{x}) \tag{4.2.78}$$

and a pair $[\mathbf{s}, \mathbf{b}]$ is called an external force system for p. We observe that in linear elastodynamics the normal vector to ∂B is time independent, $\mathbf{n} = \mathbf{n}(\mathbf{x})$. Associated with an elastic process p is the kinetic energy of body B, see Equation 3.2.135,

$$K(t) = \frac{1}{2} \int_B \rho \dot{\mathbf{u}}^2 \, dv \tag{4.2.79}$$

and the strain energy of body B, see Equation 3.3.93,

$$U(t) = U_C\{\mathbf{E}\} = \frac{1}{2} \int_B \mathbf{E} \cdot \mathsf{C}[\mathbf{E}] \, dv \tag{4.2.80}$$

A function

$$\mathcal{U}(t) = K(t) + U(t) \tag{4.2.81}$$

is called the *total energy of B at time t*.

By differentiation of Equation 4.2.81 with respect to time and using Equations 4.2.79 and 4.2.80 in which $\mathbf{C}$ is assumed to be symmetric, we obtain

$$\begin{aligned}\dot{\mathcal{U}} &= \dot{K} + \frac{1}{2}\int_B (\dot{\mathbf{E}} \cdot \mathbf{C}[\mathbf{E}] + \mathbf{E} \cdot \mathbf{C}[\dot{\mathbf{E}}])\, dv \\ &= \dot{K} + \int_B \dot{\mathbf{E}} \cdot \mathbf{C}[\mathbf{E}]\, dv = \dot{K} + \int_B \dot{\mathbf{E}} \cdot \mathbf{S}\, dv \end{aligned} \tag{4.2.82}$$

The second term in the latter formula is called the *stress power*, see Equation 3.2.136,

$$P(t) = \int_B \mathbf{S} \cdot \dot{\mathbf{E}}\, dv \tag{4.2.83}$$

Hence, the formula 4.2.82 means that the rate of change of the total energy equals the sum of the rate of the kinetic energy and the stress power,

$$\dot{\mathcal{U}} = \dot{K} + P \tag{4.2.84}$$

By the theorem of power expended, see Equation 3.2.137,

$$\dot{\mathcal{U}} = \int_{\partial B} \mathbf{s} \cdot \dot{\mathbf{u}}\, da + \int_B \mathbf{b} \cdot \dot{\mathbf{u}}\, dv \tag{4.2.85}$$

Therefore, the rate of change of the total energy equals the rate at which work is done by the surface and body forces. A direct consequence of Equation 4.2.85 is the following.

Theorem: Let $\mathbf{C}$ be symmetric and let $p = [\mathbf{u}, \mathbf{E}, \mathbf{S}]$ be an elastic process corresponding to an external force system $[\mathbf{s}, \mathbf{b}]$. Also, it is assumed that

$$\mathbf{b} = \mathbf{0} \text{ on } B \quad \text{and} \quad \mathbf{s} \cdot \dot{\mathbf{u}} = 0 \quad \text{on } \partial B \tag{4.2.86}$$

Then the total energy is constant

$$\mathcal{U}(t) = \mathcal{U}(0) \quad t \geq 0 \tag{4.2.87}$$

■

Proof: From Equation 4.2.85 and the assumptions 4.2.86 we get

$$\dot{\mathcal{U}}(t) = 0 \tag{4.2.88}$$

Integrating Equation 4.2.88 we arrive at Equation 4.2.87. This completes the proof. □

We close this section with *Graffi's reciprocal theorem,* which is a counterpart of Betti's reciprocal theorem of elastostatics.

Graffi's Reciprocal Theorem. Let $[\mathbf{u}, \mathbf{E}, \mathbf{S}]$ be an elastic process corresponding to the external force system $[\mathbf{s}, \mathbf{b}]$ and to the initial data $[\mathbf{u}_0, \dot{\mathbf{u}}_0]$. Let $[\widetilde{\mathbf{u}}, \widetilde{\mathbf{E}}, \widetilde{\mathbf{S}}]$ be another elastic process corresponding to $[\widetilde{\mathbf{s}}, \widetilde{\mathbf{b}}]$ and to $[\widetilde{\mathbf{u}}_0, \dot{\widetilde{\mathbf{u}}}_0]$. Then the following integral relations hold true:

$$i * \int_{\partial B} \mathbf{s} * \widetilde{\mathbf{u}}\, da + \int_B \mathbf{f} * \widetilde{\mathbf{u}}\, dv = i * \int_{\partial B} \widetilde{\mathbf{s}} * \mathbf{u}\, da + \int_B \widetilde{\mathbf{f}} * \mathbf{u}\, dv \tag{4.2.89}$$

$$\int_B \mathbf{S} * \widetilde{\mathbf{E}}\, dv = \int_B \widetilde{\mathbf{S}} * \mathbf{E}\, dv \tag{4.2.90}$$

$$\begin{aligned} &\int_{\partial B} \mathbf{s} * \widetilde{\mathbf{u}}\, da + \int_B \mathbf{b} * \widetilde{\mathbf{u}}\, dv + \int_B \rho(\mathbf{u}_0 \cdot \dot{\widetilde{\mathbf{u}}} + \dot{\mathbf{u}}_0 \cdot \widetilde{\mathbf{u}})\, dv \\ &\quad = \int_{\partial B} \widetilde{\mathbf{s}} * \mathbf{u}\, da + \int_B \widetilde{\mathbf{b}} * \mathbf{u}\, dv + \int_B \rho(\widetilde{\mathbf{u}}_0 \cdot \dot{\mathbf{u}} + \dot{\widetilde{\mathbf{u}}}_0 \cdot \mathbf{u})\, dv \end{aligned} \tag{4.2.91}$$

Here,

$$i = i(t) = t \quad t \geq 0 \tag{4.2.92}$$

and $\mathbf{f}$ and $\widetilde{\mathbf{f}}$ are pseudo-body forces corresponding to $[\mathbf{u}, \mathbf{E}, \mathbf{S}]$ and to $[\widetilde{\mathbf{u}}, \widetilde{\mathbf{E}}, \widetilde{\mathbf{S}}]$, respectively, defined by

$$\mathbf{f}(\mathbf{x}, t) = i * \mathbf{b}(\mathbf{x}, t) + \rho\ [\mathbf{u}_0(\mathbf{x}) + t\dot{\mathbf{u}}_0(\mathbf{x})] \tag{4.2.93}$$

$$\widetilde{\mathbf{f}}(\mathbf{x}, t) = i * \widetilde{\mathbf{b}}(\mathbf{x}, t) + \rho\ [\widetilde{\mathbf{u}}_0(\mathbf{x}) + t\dot{\widetilde{\mathbf{u}}}_0(\mathbf{x})] \tag{4.2.94}$$

Proof: By the reciprocal theorem given in Section 3.2.5, Equations 3.2.167 and 3.2.168 hold true. Therefore, to prove Equations 4.2.89 through 4.2.91 it is sufficient to prove that Equation 4.2.90 holds true. To this end, we note that by symmetry of $\mathbf{C}$ and by the definition of an elastic process, we have

$$\mathbf{S} * \widetilde{\mathbf{E}} = \widetilde{\mathbf{E}} * \mathbf{S} = \widetilde{\mathbf{E}} * \mathbf{C}[\mathbf{E}] = \mathbf{E} * \mathbf{C}[\widetilde{\mathbf{E}}] = \mathbf{E} * \widetilde{\mathbf{S}} \tag{4.2.95}$$

Note that by symmetry of $\mathbf{C}$, see Equation 3.3.8,

$$\begin{aligned} \widetilde{\mathbf{E}} * \mathbf{C}[\mathbf{E}] &\equiv \int_0^t \widetilde{\mathbf{E}}(\mathbf{x}, t-\tau) \cdot \mathbf{C}[\mathbf{E}(\mathbf{x}, \tau)]\, d\tau \\ &= \int_0^t \mathbf{E}(\mathbf{x}, \tau) \cdot \mathbf{C}[\widetilde{\mathbf{E}}(\mathbf{x}, t-\tau)]\, d\tau \\ &= \int_0^t \mathbf{E}(\mathbf{x}, t-\tau) \cdot \mathbf{C}[\widetilde{\mathbf{E}}(x, \tau)]\, d\tau = \mathbf{E} * \mathbf{C}[\widetilde{\mathbf{E}}] \end{aligned} \tag{4.2.96}$$

Hence, by integrating Equation 4.2.95 over B we arrive at Equation 4.2.90. This completes the proof. □

Note: If the initial data are homogeneous for both processes, then Equation 4.2.91 reduces to

$$\int_{\partial B} \mathbf{s} * \tilde{\mathbf{u}}\, da + \int_{B} \mathbf{b} * \tilde{\mathbf{u}}\, dv = \int_{\partial B} \tilde{\mathbf{s}} * \mathbf{u}\, da + \int_{B} \tilde{\mathbf{b}} * \mathbf{u}\, dv \tag{4.2.97}$$

Example 4.2.5

In this example we show an application of Equation 4.2.97 when $\tilde{s}$ corresponds to an instantaneous concentrated force acting in an infinite elastic body, and s corresponds to a time-dependent body force $\mathbf{b}$ distributed over a bounded region Ω of an infinite body. Let

$$\tilde{s} = [\tilde{\mathbf{u}}^{(k)}, \tilde{\mathbf{E}}^{(k)}, \tilde{\mathbf{S}}^{(k)}] \quad (k = 1, 2, 3) \tag{a}$$

be an elastic process corresponding to an instantaneous concentrated force $\tilde{\mathbf{b}}^{(k)}$ acting at a point $\boldsymbol{\xi}$ and at time τ in an infinite elastic body E^3, that is,

$$\tilde{\mathbf{b}}^{(k)} = \delta(\mathbf{x} - \boldsymbol{\xi})\delta(t - \tau)\mathbf{e}_k \tag{b}$$

where

$\mathbf{e}_k$ is a unit vector along the x_k axis

$\delta(\mathbf{x}-\boldsymbol{\xi})$ and $\delta(t-\tau)$ are three-dimensional and one-dimensional Dirac delta functions, respectively

Then, the functions $\tilde{\mathbf{u}}^{(k)}$, $\tilde{\mathbf{E}}^{(k)}$, and $\tilde{\mathbf{S}}^{(k)}$ depend on $\mathbf{x}$, $\boldsymbol{\xi}$, t and τ, in particular

$$\tilde{\mathbf{u}}^{(k)} = \tilde{\mathbf{u}}^{(k)}(\mathbf{x} - \boldsymbol{\xi}; t - \tau) \tag{c}$$

Show that a displacement field $\mathbf{u} = \mathbf{u}(\mathbf{x}, t)$ associated with the elastic process $s = [\mathbf{u}, \mathbf{E}, \mathbf{S}]$ and corresponding to body forces $\mathbf{b} = \mathbf{b}(\mathbf{x}, t)$ distributed over Ω is in components given by

$$u_k(\mathbf{x}, t) = \int_{\Omega} \int_{0}^{t} \mathbf{b}(\boldsymbol{\xi}, \tau) \cdot \tilde{\mathbf{u}}^{(k)}(\mathbf{x} - \boldsymbol{\xi}; t - \tau)\, d\tau\, dv(\boldsymbol{\xi}) \tag{d}$$

Solution

We apply the reciprocity relation 4.2.97 when $B = E^3$. Since the concentrated force $\tilde{\mathbf{b}}^{(k)}$ and body force vector field $\mathbf{b}$ distributed over Ω should not produce stresses at infinity, surface integrals in Equation 4.2.97 restricted to E^3 vanish, and the equation reduces to

$$\int_{E^3} \tilde{\mathbf{b}}^{(k)} * \mathbf{u}\, dv(\boldsymbol{\xi}) = \int_{E^3} \mathbf{b} * \tilde{\mathbf{u}}^{(k)}\, dv(\boldsymbol{\xi}) \tag{e}$$

Writing this equation in an expanded form, and using (b) and (c), we get

$$\int_{E^3}\int_0^t \mathbf{u}(\boldsymbol{\xi},\tau)\cdot\delta(\mathbf{x}-\boldsymbol{\xi})\delta(t-\tau)\mathbf{e}_k\, d\tau dv(\boldsymbol{\xi})$$

$$= \int_{\Omega}\int_0^t \mathbf{b}(\boldsymbol{\xi},\tau)\cdot\widetilde{\mathbf{u}}^{(k)}(\mathbf{x}-\boldsymbol{\xi};t-\tau)\, d\tau dv(\boldsymbol{\xi}) \tag{f}$$

Hence, by using the filtering property of the Dirac delta function

$$\int_{E^3} \delta(\mathbf{x}-\boldsymbol{\xi})f(\boldsymbol{\xi})\, dv(\boldsymbol{\xi}) = f(\mathbf{x}) \tag{g}$$

and

$$\int_0^t \delta(t-\tau)g(\tau)\, d\tau = g(t) \tag{h}$$

we write from (f)

$$\mathbf{e}_k\cdot\mathbf{u}(\mathbf{x},t) = \int_{\Omega}\int_0^t \mathbf{b}(\boldsymbol{\xi},\tau)\cdot\widetilde{\mathbf{u}}^{(k)}(\mathbf{x}-\boldsymbol{\xi};t-\tau)\, d\tau dv(\boldsymbol{\xi}) \tag{i}$$

This is identical to the formula (d), and the solution is completed. □

4.2.4 Formulation of Initial–Boundary Value Problems

In an initial–boundary value problem, the data consist not only of a boundary load and body forces but also of initial data. In a classical formulation the initial data are imposed on the displacement vector and the velocity vector, and in most textbooks on wave propagation in elastic solids such initial data are postulated. Recently, other formulations were proposed, in which the initial data are imposed on stress and stress rate tensor fields. This amounts to a study of elastic waves in a body with initial tensorial incompatibilities in the sense that compatibility equations are not satisfied initially by strain and strain-rate tensor fields. In this section the problems covering both types of initial conditions will be discussed.

To begin with a classical type of an initial–boundary value problem, we assume that the displacement field $\mathbf{u} = \mathbf{u}(\mathbf{x}, t)$ associated with an elastic process corresponding to a body force $\mathbf{b}$, see Section 4.2.3, satisfies the initial conditions

$$\mathbf{u}(\mathbf{x},0) = \mathbf{u}_0(\mathbf{x}), \quad \dot{\mathbf{u}}(\mathbf{x},0) = \dot{\mathbf{u}}_0(\mathbf{x}), \quad \mathbf{x}\in B \tag{4.2.98}$$

where $\mathbf{u}_0(\mathbf{x})$ and $\dot{\mathbf{u}}_0(\mathbf{x})$ are prescribed fields. Also, we assume the following boundary conditions associated with the process:
the displacement condition

$$\mathbf{u} = \widehat{\mathbf{u}} \quad \text{on } \partial B_1 \times [0,\infty) \tag{4.2.99}$$

the traction condition

$$\mathbf{s} = \mathbf{S}\mathbf{n} = \widehat{\mathbf{s}} \quad \text{on } \partial B_2 \times [0, \infty) \tag{4.2.100}$$

where

$$\partial B_1 \cup \partial B_2 = \partial B \tag{4.2.101}$$

and

$$\partial B_1 \cap \partial B_2 = \emptyset \tag{4.2.102}$$

and $\widehat{\mathbf{u}}$ and $\widehat{\mathbf{s}}$ are prescribed vector fields.

A mixed initial–boundary value problem is a problem in which we are to find an elastic process that complies with the initial conditions given by Equation 4.2.98 and the mixed boundary conditions given by Equations 4.2.99 and 4.2.100.

By a solution to the mixed boundary value problem, we mean an elastic process that satisfies Equations 4.2.98 through 4.2.100.

If $\partial B_2 = \emptyset$, the mixed initial–boundary value problem becomes a *displacement initial–boundary value problem*, and if $\partial B_1 = \emptyset$ it becomes a *traction initial–boundary value problem*. In short, these problems will be called *the mixed, displacement, and traction problems of elastodynamics*. Since all these problems may be formulated either in terms of displacements or in terms of stresses, we introduce the definition of a displacement field corresponding to a solution to a mixed problem, and the definition of a stress field corresponding to a solution to a mixed problem, in a way analogous to the definitions (1) and (2) in Section 4.1.4.

With these definitions we formulate three theorems of which the first characterizes a mixed problem in terms of displacements, the second describes a traction problem in terms of stresses, and the third deals with a mixed problem in terms of stresses.

Theorem 1 (Mixed Problem in Terms of Displacements)

A vector field $\mathbf{u}$ corresponds to a solution to a mixed problem of elastodynamics if and only if

$$\operatorname{div} \mathsf{C}\,[\nabla \mathbf{u}] + \mathbf{b} = \rho \ddot{\mathbf{u}} \quad \text{on } B \times [0, \infty) \tag{4.2.103}$$

$$\mathbf{u}(\mathbf{x}, 0) = \mathbf{u}_0(\mathbf{x}), \quad \dot{\mathbf{u}}(\mathbf{x}, 0) = \dot{\mathbf{u}}_0(\mathbf{x}) \quad \mathbf{x} \in B \tag{4.2.104}$$

$$\mathbf{u} = \widehat{\mathbf{u}} \quad \text{on } \partial B_1 \times [0, \infty) \tag{4.2.105}$$

$$\mathsf{C}[\nabla \mathbf{u}]\mathbf{n} = \widehat{\mathbf{s}} \quad \text{on } \partial B_2 \times [0, \infty) \tag{4.2.106}$$

The proof of this theorem is analogous to that of Theorem 1 of Section 4.1.4 and is based on recovering of a strain tensor $\mathbf{E}$ and the stress tensor $\mathbf{S}$ from a displacement $\mathbf{u}$ that satisfies Equations 4.2.103 through 4.2.106 by employing the geometric and constitutive relations, respectively. ■

Theorem 2 (Traction Problem in Terms of Stresses)

A tensor **S** corresponds to a solution to a traction problem of elastodynamics if and only if

$$\widehat{\nabla}\left[\rho^{-1}\,(\operatorname{div}\mathbf{S})\right] - \mathsf{K}[\ddot{\mathbf{S}}] = -\mathbf{B} \quad \text{on } B \times [0, \infty) \tag{4.2.107}$$

$$\mathbf{S}(\mathbf{x}, 0) = \mathsf{C}[\nabla\mathbf{u}_0], \quad \dot{\mathbf{S}}(\mathbf{x}, 0) = \mathsf{C}[\nabla\dot{\mathbf{u}}_0] \quad \text{on } B \tag{4.2.108}$$

$$\mathbf{S}\mathbf{n} = \widehat{\mathbf{s}} \quad \text{on } \partial B \times [0, \infty) \tag{4.2.109}$$

In Equation 4.2.107,

$$\mathbf{B} = \widehat{\nabla}\left(\rho^{-1}\,\mathbf{b}\right) \tag{4.2.110}$$

■

Proof: The proof consists of two parts:

(a) We assume that $s = [\mathbf{u}, \mathbf{E}, \mathbf{S}]$ is a solution to a traction initial–boundary value problem and show that **S** satisfies Equations 4.2.107 through 4.2.109.
(b) We assume that **S** satisfies Equations 4.2.107 through 4.2.109 and show that there is a displacement vector **u** and an associated strain tensor **E**, such that the triple $s = [\mathbf{u}, \mathbf{E}, \mathbf{S}]$ is an elastic process corresponding to the traction boundary condition, Equation 4.2.109, and the displacement initial data, Equation 4.2.98.

Proof of part (a). Since s is an elastic process corresponding to a body force **b** and displacement initial conditions 4.2.98, and it complies with the boundary condition 4.2.109, by eliminating **u** and **E** from Equations 4.2.75 through 4.2.77 we arrive at Equation 4.2.107 with **B** given by Equation 4.2.110, see also Equations 4.2.9 through 4.2.10. To complete the proof of part (a) it is, therefore, sufficient to show that **S** satisfies the initial conditions 4.2.108. With this in mind, we apply the gradient operator to initial conditions 4.2.98, and obtain

$$\nabla\mathbf{u}(\mathbf{x}, 0) = \nabla\mathbf{u}_0(\mathbf{x}), \quad \nabla\dot{\mathbf{u}}(\mathbf{x}, 0) = \nabla\dot{\mathbf{u}}_0(\mathbf{x}) \tag{4.2.111}$$

Since the constitutive equation in the form

$$\mathbf{S}(\mathbf{x}, t) = \mathsf{C}\,[\nabla\mathbf{u}(\mathbf{x}, t)] \tag{4.2.112}$$

is to hold also at $t = 0$ then

$$\mathbf{S}(\mathbf{x}, 0) = \mathsf{C}\,[\nabla\mathbf{u}(\mathbf{x}, 0)] \tag{4.2.113}$$

Differentiation of Equation 4.2.112 with respect to t and letting $t = 0$, yields

$$\dot{\mathbf{S}}(\mathbf{x}, 0) = \mathsf{C}\,[\nabla\dot{\mathbf{u}}(\mathbf{x}, 0)] \tag{4.2.114}$$

Hence, from Equations 4.2.111 through 4.2.114 we arrive at initial data for $\mathbf{S}$ and $\dot{\mathbf{S}}$:

$$\mathbf{S}(\mathbf{x},0) = \mathsf{C}\,[\nabla \mathbf{u}_0(\mathbf{x})], \quad \dot{\mathbf{S}}(\mathbf{x},0) = \mathsf{C}\,[\nabla \dot{\mathbf{u}}_0(\mathbf{x})] \tag{4.2.115}$$

This completes the proof of part (a).

Proof of part (b). Let $\mathbf{S}$ satisfy Equations 4.2.107 through 4.2.109 and define a displacement $\mathbf{u}$ and the tensor $\mathbf{E}$ by the relations, see Equations 3.2.141 and 3.2.144,

$$\mathbf{u} = \rho^{-1}\,(i * \operatorname{div}\mathbf{S} + \mathbf{f}) \tag{4.2.116}$$

$$\mathbf{E} = \mathsf{K}[\mathbf{S}] \tag{4.2.117}$$

In Equation 4.2.116, $\mathbf{f}$ is the pseudo-body force field

$$\mathbf{f}(\mathbf{x},t) = i * \mathbf{b}(\mathbf{x},t) + \rho(\mathbf{x})[\mathbf{u}_0(\mathbf{x}) + t\dot{\mathbf{u}}_0(\mathbf{x})] \tag{4.2.118}$$

and

$$i = i(t) = t \tag{4.2.119}$$

Since Equation 4.2.116 is to be satisfied on $\bar{B} \times [0,\infty)$, by Theorem of Section 3.2.5 $\mathbf{u}$, $\mathbf{S}$, and $\mathbf{b}$ satisfy the equations of motion, Equation 4.2.76, subject to the displacement initial condition 4.2.98. To complete the proof of part (b) we need to show that the strain–displacement relations, Equation 4.2.75, are satisfied. By virtue of Equations 4.2.107, 4.2.110, and 4.2.117, we obtain

$$\widehat{\nabla}\,\left[\rho^{-1}\,(\operatorname{div}\mathbf{S} + \mathbf{b})\right] = \ddot{\mathbf{E}} \tag{4.2.120}$$

Also, differentiating Equation 4.2.116 twice with respect to t we write

$$\ddot{\mathbf{u}} = \rho^{-1}\,(\operatorname{div}\mathbf{S} + \mathbf{b}) \tag{4.2.121}$$

Hence, substituting Equation 4.2.121 into Equation 4.2.120 we get

$$\widehat{\nabla}\ddot{\mathbf{u}} = \ddot{\mathbf{E}} \tag{4.2.122}$$

Since $\mathbf{u}$ satisfies the initial conditions 4.2.98, it follows from Equation 4.2.117 and its time derivative, and from the initial conditions satisfied by $\mathbf{S}$, see Equation 4.2.108, that

$$\mathbf{E}(\mathbf{x},0) = \widehat{\nabla}\mathbf{u}(\mathbf{x},0), \quad \dot{\mathbf{E}}(\mathbf{x},0) = \widehat{\nabla}\dot{\mathbf{u}}(\mathbf{x},0) \tag{4.2.123}$$

Therefore, integrating Equation 4.2.122 twice with respect to t and using the initial conditions 4.2.123 we arrive at the strain–displacement equation

$$\mathbf{E} = \widehat{\nabla}\,\mathbf{u} \quad \text{on } \bar{B} \times [0,\infty) \tag{4.2.124}$$

This completes the proof of part (b) of the theorem. □

Notes:

(1) The tensorial initial data in Theorem 2 are generated by the initial displacement and initial velocity fields through the constitutive relation of elastodynamics. They imply that the initial strain $\mathbf{E}_0$ and initial strain rate $\dot{\mathbf{E}}_0$ both satisfy the compatibility condition, see Equation 3.1.47,

$$\text{curl curl}\,\mathbf{E}_0 = \mathbf{0}, \quad \text{curl curl}\,\dot{\mathbf{E}}_0 = \mathbf{0} \tag{4.2.125}$$

since

$$\mathbf{E}_0 = \widehat{\nabla}\mathbf{u}_0, \quad \dot{\mathbf{E}}_0 = \widehat{\nabla}\dot{\mathbf{u}}_0 \tag{4.2.126}$$

(2) A generalization of Theorem 2 may be formulated so that the initial stress and initial stress rate are arbitrarily prescribed symmetric tensor fields. The associated initial strain and initial strain rate fields following from the constitutive relations, in general, do not satisfy the compatibility conditions. Such a generalized theorem describes then an *elastic process with incompatible initial data*, that is, an elastic process of elastodynamics of a body with continuously distributed defects. Later we will discuss well-posedness of the generalized problem.

Example 4.2.6

Suppose that a homogeneous isotropic infinite elastic body is subject to the initial stress $\mathbf{S}^{(0)}(\mathbf{x})$ of a compact support

$$S_{ij}^{(0)} = [2\mu\delta_{(i1}\delta_{1j)} + \lambda\delta_{ij}]\varphi(x_1)$$
$$\text{for } |x_1| \le a, \; |x_2| < \infty, \; |x_3| < \infty \tag{a}$$

where λ and μ are the Lamé moduli, a is a positive constant, and $\varphi = \varphi(x_1)$ is the function defined by

$$\varphi(x_1) = \begin{cases} 0 & \text{for } -\infty < x_1 \le -a \\ 1 + \dfrac{x_1}{a} & \text{for } -a \le x_1 \le 0 \\ 1 - \dfrac{x_1}{a} & \text{for } 0 \le x_1 \le a \\ 0 & \text{for } a \le x_1 < \infty \end{cases} \tag{b}$$

The associated strain field $\mathbf{E}^{(0)}(\mathbf{x})$ is then given by

$$E_{ij}^{(0)}(\mathbf{x}) = \delta_{(i1}\delta_{1j)}\varphi(x_1)$$
$$\text{for } |x_1| \le a, \; |x_2| < \infty, \; |x_3| < \infty \tag{c}$$

Show that $\mathbf{S}^{(0)}(\mathbf{x})$ satisfies the equilibrium equation

$$\text{div}\,\mathbf{S}^{(0)} + \mathbf{b} = \mathbf{0} \tag{d}$$

on E^3 except for the planes $x_1 = -a$, $x_1 = 0$, and $x_1 = +a$, where

$$\mathbf{b} = [b_1,\ 0,\ 0]^T \tag{e}$$

and

$$b_1(x_1) = \frac{\lambda + 2\mu}{a} \begin{cases} 0 & \text{for } -\infty < x_1 < -a \\ -1 & \text{for } -a < x_1 < 0 \\ +1 & \text{for } 0 < x_1 < a \\ 0 & \text{for } a < x_1 < \infty \end{cases} \tag{f}$$

Also, show that the body possesses "defects" on the planes $x_1 = -a$, $x_1 = 0$ and $x_1 = +a$, and a defect on the plane $x_1 = 0$ is represented by the jump

$$u_1(0-0) - u_1(0+0) = a \tag{g}$$

where $u_1 = u_1(x_1)$ is a displacement in the x_1 direction corresponding to the strain $\mathbf{E}^{(0)}(\mathbf{x})$.

Solution

To show that $\mathbf{S}^{(0)}$ satisfies Equation (d) we substitute $\mathbf{S}^{(0)}$ and $\mathbf{b}$ from Equations (a) and (e), respectively, into LHS of Equation (d) and check that Equation (d) is satisfied provided $x_1 \neq -a$, $x_1 \neq 0$, and $x_1 \neq a$. Next, we note that $\varphi = \varphi(x_1)$ is continuous everywhere on the x_1 axis, but $\varphi'(x_1)$ does not exist on the planes $x_1 = -a$, $x_1 = 0$, and $x_1 = a$. This implies that $\mathbf{E}^{(0)}(\mathbf{x})$ does not satisfy the compatibility condition

$$\text{curl curl}\, \mathbf{E}^{(0)}(\mathbf{x}) = \mathbf{0}$$
$$\text{for } x_1 = -a,\ x_1 = 0,\ x_1 = +a \tag{h}$$

Therefore, the body possesses "defects" on these planes. To describe a defect on the plane $x_1 = 0$ we use the strain–displacement relation

$$E_{11}^{(0)}(x_1) = \frac{\partial u_1}{\partial x_1} = \varphi(x_1), \quad |x_1| < \infty \tag{i}$$

Assuming that

$$u_1(-a) = u_1(+a) = 0 \tag{j}$$

and integrating Equation (i) we find

$$u_1(x_1) = \begin{cases} 0 & \text{for } -\infty < x_1 \leq -a \\ \dfrac{x_1^2}{2a} + x_1 + \dfrac{a}{2} & \text{for } -a \leq x_1 \leq 0 \\ -\dfrac{x_1^2}{2a} + x_1 - \dfrac{a}{2} & \text{for } 0 \leq x_1 \leq a \\ 0 & \text{for } a \leq x_1 < \infty \end{cases} \tag{k}$$

Hence, we obtain

$$u_1(0-0) - u_1(0+0) = a \tag{l}$$

This completes the solution. □

Theorem 3 (Mixed Problem in Terms of Stresses)

A tensor field $\mathbf{S}$ corresponds to a solution to a mixed problem of elastodynamics if and only if

$$\widehat{\nabla}\left[\rho^{-1}\left(i * \operatorname{div}\mathbf{S} + \mathbf{f}\right)\right] - \mathsf{K}[\mathbf{S}] = \mathbf{0} \quad \text{on } \bar{B} \times [0,\infty) \tag{4.2.127}$$

$$\rho^{-1}\left(i * \operatorname{div}\mathbf{S} + \mathbf{f}\right) = \widehat{\mathbf{u}} \quad \text{on } \partial B_1 \times [0,\infty) \tag{4.2.128}$$

$$\mathbf{Sn} = \widehat{\mathbf{s}} \quad \text{on } \partial B_2 \times [0,\infty) \tag{4.2.129}$$

■

Proof: We use the following equivalent definition of a solution to the mixed problem of elastodynamics: $p = [\mathbf{u}, \mathbf{E}, \mathbf{S}]$ is a solution to the mixed problem if and only if

$$\mathbf{E} = \widehat{\nabla}\mathbf{u} \quad \text{on } \bar{B} \times [0,\infty) \tag{4.2.130}$$

$$\rho^{-1}\left(i * \operatorname{div}\mathbf{S} + \mathbf{f}\right) = \mathbf{u} \quad \text{on } \bar{B} \times [0,\infty) \tag{4.2.131}$$

$$\mathbf{S} = \mathsf{C}[\mathbf{E}] \quad \text{on } \bar{B} \times [0,\infty) \tag{4.2.132}$$

and

$$\mathbf{u} = \widehat{\mathbf{u}} \quad \text{on } \partial B_1 \times [0,\infty) \tag{4.2.133}$$

$$\mathbf{Sn} = \widehat{\mathbf{s}} \quad \text{on } \partial B_2 \times [0,\infty) \tag{4.2.134}$$

To show that this definition is equivalent to that given at the beginning of this section, it is sufficient to observe that by Theorem of Section 3.2.5 the integro-differential equation 4.2.131 is satisfied if and only if $\mathbf{u}$, $\mathbf{S}$, and $\mathbf{b}$ satisfy the equations of motion, Equations 4.2.76, and the initial conditions 4.2.98. Obviously, the definition given by Equations 4.2.130 through 4.2.134 means that a solution to the mixed problem of elastodynamics satisfies a system of integro-differential equations 4.2.130 through 4.2.132 subject to the boundary conditions 4.2.133 and 4.2.134 only, that is, the initial data have been incorporated into the field equations.

To prove Theorem 3 itself, using the new definition, we need to show that:

(a) If $p = [\mathbf{u}, \mathbf{E}, \mathbf{S}]$ is a solution of the mixed problem of elastodynamics, then $\mathbf{S}$ satisfies Equations 4.2.127 through 4.2.129.
(b) If $\mathbf{S}$ is a solution to Equations 4.2.127 through 4.2.129, then there is a vector field $\mathbf{u}$ and a tensor field $\mathbf{E}$ such that the triple $[\mathbf{u}, \mathbf{E}, \mathbf{S}]$ complies with Equations 4.2.130 through 4.2.134.

Proof of (a). Since $p = [\mathbf{u}, \mathbf{E}, \mathbf{S}]$ is a solution to the mixed problem, p satisfies Equations 4.2.130 through 4.2.134. Writing down Equation 4.2.132 in the equivalent form

$$\mathbf{E} = \mathsf{K}[\mathbf{S}] \tag{4.2.135}$$

and eliminating **u** and **E** from Equations 4.2.130, 4.2.131, and 4.2.135, we arrive at Equation 4.2.127. Hence, to complete proof of part (a) we need to show that **S** satisfies the boundary condition given by Equation 4.2.128. To this end we note that Equation 4.2.131 restricted to $\partial B_1 \times [0, \infty)$ together with the displacement boundary condition 4.2.133 imply Equation 4.2.128. This completes proof of part (a) of Theorem 3.

Proof of (b). Let **S** be a solution to Equations 4.2.127 through 4.2.129, and define **u** and **E** by

$$\mathbf{u} = \rho^{-1}\ (i * \operatorname{div}\mathbf{S} + \mathbf{f}) \quad \text{on } B \times [0, \infty) \tag{4.2.136}$$

and

$$\mathbf{E} = \mathsf{K}[\mathbf{S}] \quad \text{on } B \times [0, \infty) \tag{4.2.137}$$

Then, from Equations 4.2.127, 4.2.136, and 4.2.137, it follows that Equations 4.2.130 through 4.2.132 are satisfied. Also, since Equation 4.2.136 restricted to $\partial B_1 \times [0, \infty)$ together with the boundary condition 4.2.128 imply the boundary condition 4.2.133, the triple $[\mathbf{u}, \mathbf{E}, \mathbf{S}]$ satisfies Equations 4.2.130 through 4.2.134. This completes the proof of (b). □

Notes:

(1) The equivalent definition of a solution to the mixed problem, see Equations 4.2.130 through 4.2.134, and the characterization of the solution in terms of stresses (Theorem 3) play an important role in formulation of the variational principles of elastodynamics to be discussed in Chapter 6.

(2) Theorems 1 through 3 hold true for the elastic processes that are sufficiently smooth on $\bar{B} \times [0, \infty)$, for example in Theorem 2 the tensor field **S** is to be of class $C^{2,2}$ on $\bar{B} \times [0, \infty)$. The symbol $C^{2,2}$ stands for a class of functions $f = f(\mathbf{x}, t)$ that possess the partial derivatives with respect **x** and t up to the second order, and these derivatives are continuous.

Example 4.2.7

Let **S** be a solution to the initial value problem. Find a symmetric second-order tensor field $\mathbf{S} = \mathbf{S}(\mathbf{x}, t)$ on $E^3 \times [0, \infty)$ that satisfies the field equation

$$\widehat{\nabla}\left[\rho^{-1}\,(\operatorname{div}\mathbf{S})\right] - \mathsf{K}[\ddot{\mathbf{S}}] = -\mathbf{A} \quad \text{on } E^3 \times [0, \infty) \tag{a}$$

subject to the initial conditions

$$\mathbf{S}(\mathbf{x}, 0) = \mathbf{S}_0(\mathbf{x}), \quad \dot{\mathbf{S}}(\mathbf{x}, 0) = \dot{\mathbf{S}}_0(\mathbf{x}) \quad \mathbf{x} \in E^3 \tag{b}$$

Here

$\mathbf{S}_0$ and $\dot{\mathbf{S}}_0$ are arbitrary symmetric tensor fields on E^3
A is a prescribed symmetric tensor field on $E^3 \times [0, \infty)$

Show that $\mathbf{S}$ is a solution to the problem described by (a) and (b) if and only if $\mathbf{S}$ is a solution to the integro-differential equation

$$\widehat{\nabla}\left[\rho^{-1}(i * \operatorname{div}\mathbf{S})\right] - \mathbf{K}[\mathbf{S}] = -\mathbf{B}^* \quad \text{on } E^3 \times [0,\infty) \tag{c}$$

where

$$\mathbf{B}^* = i * \mathbf{A} + \mathbf{K}[\mathbf{S}_0 + t\dot{\mathbf{S}}_0] \tag{d}$$

Solution

We apply the Laplace transform to (a) and using (b) we obtain

$$\widehat{\nabla}\left[\rho^{-1}(\operatorname{div}\bar{\mathbf{S}})\right] - \mathbf{K}[p^2\bar{\mathbf{S}} - p\mathbf{S}_0 - \dot{\mathbf{S}}_0] = -\mathbf{A} \tag{e}$$

By dividing (e) by p^2 and taking the inverse Laplace transform we arrive at (c) since

$$L^{-1}\left\{\frac{1}{p^2}\right\} = t \quad \text{and} \quad L^{-1}\left\{\frac{1}{p}\right\} = 1 \tag{f}$$

This completes the solution. □

Notes:

(1) The initial value problem in this example is a generalization of the initial–boundary value problem of Theorem 2, see Equations 4.2.107 through 4.2.110, in the following sense. If $\mathbf{A}$ is of the form of $\mathbf{B}$ given by Equation 4.2.110 but extended to $E^3 \times [0,\infty)$ and the initial tensor fields $\mathbf{S}_0$ and $\dot{\mathbf{S}}_0$ in (b) are restricted to those given by Equation 4.2.108 but extended to E^3, then the initial value problem described by (a) and (b) reduces to that of Theorem 2 formulated for E^3.

(2) In this example, the initial value problem described by (a) and (b) has been replaced by an equivalent problem of finding a solution to an integro-differential equation (c) in which the arbitrary initial tensor field data are incorporated. Equation (c) may be dealt with by using an associated variational principle to be discussed later.

(3) The initial value problem stated by (a) and (b) with arbitrary symmetric tensor $\mathbf{A}$ on $E^3 \times [0,\infty)$ and arbitrary symmetric tensors $\mathbf{S}_0$ and $\dot{\mathbf{S}}_0$ on E^3 describes stress waves in an infinite elastic body with continuously distributed defects; see Note (2) given after Theorem 2.

Example 4.2.8

A pure traction problem of nonhomogeneous anisotropic dynamic theory of thermal stresses is formulated in the following way. Find a symmetric second-order tensor field $\mathbf{S} = \mathbf{S}(\mathbf{x}, t)$ on $\overline{B} \times [0,\infty)$ that satisfies the field equation

$$\widehat{\nabla}\left[\rho^{-1}(\operatorname{div}\mathbf{S} + \mathbf{b})\right] - \mathbf{K}[\ddot{\mathbf{S}} - \ddot{\mathbf{S}}_A] = \mathbf{0} \quad \text{on } B \times [0,\infty) \tag{a}$$

subject to the initial conditions

$$\mathbf{S}(\mathbf{x},0) = \dot{\mathbf{S}}(\mathbf{x},0) = 0 \quad \text{on } B \tag{b}$$

$$\mathbf{S}_A(\mathbf{x},0) = \dot{\mathbf{S}}_A(\mathbf{x},0) = 0 \quad \text{on } B \tag{c}$$

and the traction boundary condition

$$\mathbf{Sn} = \widehat{\mathbf{s}} \quad \text{on } \partial B \times [0,\infty) \tag{d}$$

In Equation (a) $\mathbf{S}_A = \mathbf{S}_A(\mathbf{x},t)$ is an actuation stress tensor field that is defined in terms of a time-dependent temperature field $\theta = \theta(\mathbf{x},t)$ and the stress–temperature tensor field $\mathbf{M} = \mathbf{M}(\mathbf{x})$ (see Section 3.3.5) by

$$\mathbf{S}_A(\mathbf{x},t) = \theta\,\mathbf{M} \quad \text{on } \overline{B} \times [0,\infty) \tag{e}$$

and in Equation (d) $\widehat{\mathbf{s}}$ is a prescribed vector field. In addition to Equations (a) through (e), let us define an auxiliary isothermal problem as follows. Find a symmetric second-order tensor field $\mathbf{S}_f = \mathbf{S}_f(\mathbf{x},t)$ on $\overline{B} \times [0,\infty)$ that satisfies the equations

$$\widehat{\nabla}\left[\rho^{-1}\,(\operatorname{div}\mathbf{S}_f + \mathbf{b}_f)\right] - \mathbf{K}[\ddot{\mathbf{S}}_f] = \mathbf{0} \quad \text{on } B \times [0,\infty) \tag{f}$$

$$\mathbf{S}_f\,(\mathbf{x},\,0) = \dot{\mathbf{S}}_f(\mathbf{x},0) = 0 \quad \text{on } B \tag{g}$$

and

$$\mathbf{S}_f\mathbf{n} = \widehat{\mathbf{s}_f} \quad \text{on } \partial B \times [0,\infty) \tag{h}$$

where

$$\widehat{\mathbf{s}_f} = \widehat{\mathbf{s}} \quad \text{on } \partial B \times [0,\infty) \tag{i}$$

and

$$\widehat{\nabla}\left(\rho^{-1}\mathbf{b}_f\right) = \widehat{\nabla}\left(\rho^{-1}\mathbf{b}\right) + \mathbf{K}[\ddot{\mathbf{S}}_A] \quad \text{on } \overline{B} \times [0,\infty) \tag{j}$$

Show that the following body force analogy for the transient thermal stresses holds true [9]. A unique solution $\mathbf{S} = \mathbf{S}(\mathbf{x},t)$ to the pure stress problem of dynamic theory of thermal stresses given by Equations (a) through (e) can be obtained by finding a unique solution $\mathbf{S}_f = \mathbf{S}_f(\mathbf{x},t)$ to the isothermal pure stress problem given by Equations (f) through (j).

Solution

By subtracting Equation (f) from Equation (a) and using Equation (j) we write

$$\widehat{\nabla}\left[\rho^{-1}(\operatorname{div}\overline{\mathbf{S}})\right] - \mathbf{K}[\ddot{\overline{\mathbf{S}}}] = \mathbf{0} \quad \text{on } B \times [0,\infty) \tag{k}$$

where

$$\overline{\mathbf{S}}(\mathbf{x},t) = \mathbf{S}(\mathbf{x},t) - \mathbf{S}_f(\mathbf{x},t) \quad \text{on } \overline{B} \times [0,\infty) \tag{l}$$

Also, by using Equations (b), (d), (h), and (i), we obtain

$$\bar{\mathbf{S}}(\mathbf{x}, 0) = \dot{\bar{\mathbf{S}}}(\mathbf{x}, 0) = 0 \quad \text{on } B \tag{m}$$

and

$$\bar{\mathbf{S}}\mathbf{n} = \mathbf{0} \quad \text{on } \partial B \times [0, \infty) \tag{n}$$

Next, it follows from Equations (k) through (n) that $\bar{\mathbf{S}}(\mathbf{x}, t) = 0$ on $\overline{B} \times [0, \infty)$ (see Theorem 2 of Section 4.2.5). Therefore, $\mathbf{S}(\mathbf{x}, t) = \mathbf{S}_f(\mathbf{x}, t)$ on $\overline{B} \times [0, \infty)$. This completes the solution. □

Notes:

(1) The body force analogy for the traction problem can be extended to include a mixed initial–boundary value problem of the uncoupled dynamic thermoelasticity characterized in terms of stresses. Such an extension is given in [9], where it is shown that for the mixed problem the nonisothermal transient stresses can be identified with a solution to an isothermal transient stress problem, while the associated displacements are different.

(2) The body force analogy in the uncoupled dynamic thermoelasticity may be used to control dynamic shape of structures by thermal actuation (see, for example, [9]).

4.2.5 UNIQUENESS

We consider two uniqueness theorems: (1) for a classical mixed initial boundary value problem, and (2) for a nonconventional initial–boundary value problem that describes an elastic body with continuously distributed defects.

Theorem 1: If the density field $\rho(\mathbf{x})$ is positive, and the elasticity tensor $\mathbf{C}$ is symmetric and positive definite, then the mixed problem of elastodynamics has at most one solution. ■

Proof: Let $p = [\mathbf{u}, \mathbf{E}, \mathbf{S}]$ denote a difference between two solutions, then p corresponds to vanishing body forces, and the following initial and boundary conditions

$$\mathbf{u}(\mathbf{x}, 0) = \mathbf{0}, \quad \dot{\mathbf{u}}(\mathbf{x}, 0) = \mathbf{0} \quad \mathbf{x} \in B \tag{4.2.138}$$

$$\mathbf{u} = \mathbf{0} \quad \text{on } \partial B_1 \times [0, \infty) \tag{4.2.139}$$

$$\mathbf{S}\mathbf{n} = \mathbf{0} \quad \text{on } \partial B_2 \times [0, \infty) \tag{4.2.140}$$

The first of initial conditions 4.2.138 implies that

$$\mathbf{E}(\mathbf{x}, 0) = \mathbf{0} \tag{4.2.141}$$

Also, it follows from the boundary conditions 4.2.139 and 4.2.140 that

$$\dot{\mathbf{u}} \cdot \mathbf{s} = 0 \quad \text{on } \partial B \times [0, \infty) \tag{4.2.142}$$

where $\mathbf{s} = \mathbf{Sn}$. Therefore, using the Theorem of Section 4.2.3 in which $\mathbf{b} = \mathbf{0}$ on $B \times [0, \infty)$ and the condition 4.2.142 is satisfied, we arrive at the conclusion, see Equation 4.2.87,

$$\mathcal{U}(t) = \mathcal{U}(0) \quad t \geq 0 \tag{4.2.143}$$

where

$$\mathcal{U}(t) = \frac{1}{2} \int_B [\rho \dot{\mathbf{u}}^2 + \mathbf{E} \cdot \mathbf{C}[\mathbf{E}]] \, dv \tag{4.2.144}$$

Now, by virtue of the second of initial conditions 4.2.138 and from Equation 4.2.141, we write

$$\mathcal{U}(0) = 0 \tag{4.2.145}$$

Hence, from Equation 4.2.143

$$\mathcal{U}(t) = 0 \quad t \geq 0 \tag{4.2.146}$$

Since $\rho > 0$ and $\mathbf{C}$ is positive definite, Equation 4.2.146 implies that

$$\dot{\mathbf{u}} = \mathbf{0} \quad \text{and} \quad \mathbf{E} = \mathbf{0} \quad \text{on } B \times [0, \infty) \tag{4.2.147}$$

Integrating with respect to t the first of these equations and using the first of initial conditions 4.2.138, we obtain

$$\mathbf{u} = \mathbf{0} \quad \text{on } B \times [0, \infty) \tag{4.2.148}$$

This result together with the strain–displacement relation implies that the second of Equations 4.2.147 is also satisfied. This completes the proof. □

Note: We recall that in elastostatics a traction boundary value problem has a solution that is unique to within a rigid body motion. On the other hand, Theorem 1 implies that in elastodynamics a displacement initial–boundary value problem as well as a traction initial–boundary value problem possess at most one solution. The uniqueness of the traction problem of elastodynamics follows from prescribed initial data and smoothness of the solution in the space–time domain.

The following theorem covers a nonconventional traction initial–boundary value problem in terms of stresses. In such a problem the data are determined by arbitrary symmetric tensor fields and by a prescribed traction field on the whole boundary of body B.

Theorem 2: Let $\mathbf{S}$ be a solution to the following initial–boundary value problem: Find a symmetric second-order tensor field $\mathbf{S}$ on $\bar{B} \times [0, \infty)$ that satisfies the field equation

$$\widehat{\nabla} \left(\rho^{-1} \operatorname{div} \mathbf{S}\right) - \mathsf{K}[\ddot{\mathbf{S}}] = -\mathbf{A} \quad \text{on } B \times [0, \infty) \tag{4.2.149}$$

the initial conditions

$$\mathbf{S}(\mathbf{x},0)=\mathbf{S}_0(\mathbf{x}),\quad \dot{\mathbf{S}}(x,0)=\dot{\mathbf{S}}_0(\mathbf{x})\quad x\in B \tag{4.2.150}$$

and the boundary condition

$$\mathbf{Sn}=\widehat{\mathbf{s}}\quad \text{on } \partial B\times[0,\infty) \tag{4.2.151}$$

Here

A is an arbitrary symmetric second-order tensor field prescribed on $B\times[0,\infty)$
$\mathbf{S}_0$ and $\dot{\mathbf{S}}_0$ are prescribed symmetric tensor fields on B

$\widehat{\mathbf{s}}$ is a prescribed vector field on $\partial B\times[0,\infty)$. Also, ρ and K stand for the density and the compliance tensor field on B, respectively. Then the problem described by Equations 4.2.149 through 4.2.151 has at most one solution. ■

Proof: Due to linearity of Equations 4.2.149 through 4.2.151 it is sufficient to show that a solution to the problem corresponding to the homogeneous data vanishes. In other words, we are to show that the field equation

$$\widehat{\nabla}\left(\rho^{-1}\ \operatorname{div}\mathbf{S}\right)-\mathsf{K}[\ddot{\mathbf{S}}]=\mathbf{0}\quad \text{on } B\times[0,\infty) \tag{4.2.152}$$

subject to the conditions

$$\mathbf{S}(\mathbf{x},0)=\dot{\mathbf{S}}(\mathbf{x},0)=\mathbf{0}\quad \text{on } B \tag{4.2.153}$$

$$\mathbf{Sn}=\mathbf{0}\quad \text{on } \partial B\times[0,\infty) \tag{4.2.154}$$

implies that

$$\mathbf{S}=\mathbf{0}\quad \text{on } \bar{B}\times[0,\infty) \tag{4.2.155}$$

Proof. Writing Equation 4.2.152 in components yields

$$(\rho^{-1}S_{(ik,k}),_{j)}-K_{ijkl}\ddot{S}_{kl}=0 \tag{4.2.156}$$

Multiplying this relation by $\dot{S}_{ij}$ we get

$$\dot{S}_{ij}(\rho^{-1}S_{ik,k})_{,j}-K_{ijkl}\dot{S}_{ij}\ddot{S}_{kl}=0 \tag{4.2.157}$$

In Equation 4.2.157, parentheses on indexes i and j are not present since **S** is symmetric, $S_{ij}=S_{ji}$, and for any second-order tensor T_{ij}

$$S_{ij}T_{(ij)}=S_{ij}T_{ij} \tag{4.2.158}$$

Next, we reduce Equation 4.2.157 to the form

$$(\dot{S}_{ij}\rho^{-1}S_{ik,k})_{,j}-\dot{S}_{ij,j}\rho^{-1}S_{ik,k}-K_{ijkl}\dot{S}_{ij}\ddot{S}_{kl}=0 \tag{4.2.159}$$

The boundary condition 4.2.154, written in components, is

$$S_{ij}n_j = 0 \quad \text{on } \partial B \times [0, \infty) \tag{4.2.160}$$

and this, due to the fact that n_j is time independent, implies that

$$\dot{S}_{ij}n_j = 0 \quad \text{on } \partial B \times [0, \infty) \tag{4.2.161}$$

Now, integrating Equation 4.2.159 over B, and using the divergence theorem, we obtain

$$\int_{\partial B} \dot{S}_{ij}n_j\, \rho^{-1}\, S_{ik,k}\, da - \frac{1}{2}\frac{d}{dt}\int_B \rho^{-1}\, S_{ij,j}S_{ik,k}\, dv - \int_B K_{ijkl}\dot{S}_{ij}\ddot{S}_{kl}\, dv = 0 \tag{4.2.162}$$

Because of the symmetry of the tensor K and by virtue of the homogeneous boundary condition 4.2.161, it follows that Equation 4.2.162 reduces to

$$\frac{1}{2}\frac{d}{dt}\int_B \left(\rho^{-1}\, S_{ij,j}S_{ik,k} + K_{ijkl}\dot{S}_{ij}\dot{S}_{kl}\right) dv = 0 \tag{4.2.163}$$

Integrating this equation with respect to time from 0 to t and using the initial conditions 4.2.153, in particular the conditions

$$S_{ij,j}(\mathbf{x}, 0) = 0, \quad \dot{S}_{ij}(\mathbf{x}, 0) = 0 \tag{4.2.164}$$

we find

$$\int_B \left(\rho^{-1}\, S_{ij,j}S_{ik,k} + K_{ijkl}\dot{S}_{ij}\dot{S}_{kl}\right) dv = 0 \tag{4.2.165}$$

Since $\rho > 0$ and K is positive definite

$$S_{ij,j} = 0, \quad \dot{S}_{ij} = 0 \quad \text{on } \bar{B} \times [0, \infty) \tag{4.2.166}$$

Using again the first of initial conditions 4.2.153, after integrating the second equation of 4.2.166, we arrive at

$$S_{ij} = 0 \quad \text{on } \bar{B} \times [0, \infty) \tag{4.2.167}$$

This equation implies that the first equation of 4.2.166 is also satisfied, and this completes the proof of the theorem. □

Notes:

(1) The initial–boundary value problem of this theorem generalizes that of Theorem 2 of Section 4.2.4 on the traction problem in terms of stresses in the following sense. If $\mathbf{A}$ is identified with tensor $\mathbf{B}$ given by Equation 4.2.110, and the tensors $\mathbf{S}_0$ and $\dot{\mathbf{S}}_0$ are identified by Equations 4.2.108, then this problem indeed reduces to that of Theorem 2.

(2) In the case when $\mathbf{A}$ is arbitrary, the problem stated by Equations 4.2.149 through 4.2.151 describes stress waves in an elastic body with time dependent continuously distributed defects [10]. In this case, by applying curl curl operator to Equation 4.2.149 and using the relation $\mathbf{E} = \mathsf{K}[\mathbf{S}]$, we obtain

$$\text{curl}\,\text{curl}\,\widehat{\nabla}\left(\rho^{-1}\ \text{div}\,\mathbf{S}\right) = \mathbf{0} \tag{4.2.168}$$

and

$$\operatorname{curl}\operatorname{curl}\ddot{\mathbf{E}} = \operatorname{curl}\operatorname{curl}\mathbf{A} \tag{4.2.169}$$

If the RHS of Equation 4.2.169 is not zero, there are time-dependent continuously distributed defects over the body B and for $t \geq 0$.

(3) If $\mathbf{A} = \mathbf{0}$ and $\mathbf{S}_0$ and $\dot{\mathbf{S}}_0$ are arbitrary symmetric tensor fields, the problem stated by Equations 4.2.149 through 4.2.151 describes an elastodynamic problem of a body with initially distributed defects.

Example 4.2.9

Show that if the boundary ∂B of a body B is traction free for every time $t \geq 0$, the initial displacement field $\mathbf{u}_0$ and the initial velocity $\dot{\mathbf{u}}_0$ vanish on B, and if $\nabla(\mathbf{b}/\rho)$ is skew, then $\mathbf{S} = \mathbf{0}$ on $\bar{B} \times [0, \infty)$, and $\mathbf{u} = i * (\mathbf{b}/\rho)$ represents a rigid body motion.

Solution

From Theorem 2 of Section 4.2.4 and from the assumptions in this example it follows that $\mathbf{S}$ satisfies the homogeneous field equation, see Equation 4.2.107 with $\mathbf{B} = \mathbf{0}$, the homogeneous initial conditions 4.2.108, see Equation 4.2.108 with $\mathbf{u}_0 = \mathbf{0}$ and $\dot{\mathbf{u}}_0 = \mathbf{0}$, and the homogeneous boundary condition 4.2.109. Hence, by the uniqueness theorem, Theorem 2, Equations 4.2.152 through 4.2.155, $\mathbf{S} = \mathbf{0}$ on $\bar{B} \times [0, \infty)$. On the other hand, by the formulas 4.2.116 through 4.2.118 $\mathbf{u} = i * (\mathbf{b}/\rho)$. Since by Equation 4.2.117 the strain field $\mathbf{E}$ corresponding to $\mathbf{u}$ vanishes, $\mathbf{u} = i * (\mathbf{b}/\rho)$ must be a rigid body motion. This completes the solution. □

Notes:

(1) The result of this example describes a rigid body motion that is exclusive for elastodynamics.

(2) The solution presented has been obtained apparently for the first time by a pure stress treatment of elastodynamics. The result has not been derived in the literature by the conventional displacement method of elastodynamics.

To conclude this section we note that a mixed initial–boundary problem of classical elastodynamics and that of elastodynamics with continuously distributed defects may possess at most one solution. To prove that these two types of problems are well-posed one needs also to prove the existence of solutions to these problems as well as to show continuous dependence of solutions on data in a suitable class of functions. In this book, we do not consider these two issues.

PROBLEMS

4.1 For a homogeneous isotropic elastic body occupying a region $B \subset E^3$ subject to zero body forces, the displacement equation of equilibrium takes the form [see, Example 4.1.1, Equation (b) with $\mathbf{b} = \mathbf{0}$]

$$\nabla^2 \mathbf{u} + \frac{1}{1-2\nu} \nabla(\operatorname{div} \mathbf{u}) = \mathbf{0} \quad \text{on } B \tag{a}$$

where ν is Poisson's ratio.

Show that if $\mathbf{u} = \mathbf{u}(\mathbf{x})$ is a solution to Equation (a) then $\mathbf{u}$ also satisfies the equation

$$\nabla^2 \left[\mathbf{u} + \frac{\mathbf{x}}{2(1-2\nu)} (\operatorname{div} \mathbf{u}) \right] = \mathbf{0} \quad \text{on } B \tag{b}$$

4.2 An alternative form of Equation (a) in Problem 4.1 reads [see Example 4.1.2, Equation (a) with $\mathbf{b} = \mathbf{0}$]

$$\nabla(\operatorname{div} \mathbf{u}) - \frac{1-2\nu}{2-2\nu} \operatorname{curl} \operatorname{curl} \mathbf{u} = \mathbf{0} \quad \text{on } B \tag{a}$$

Show that if $\mathbf{u} = \mathbf{u}(\mathbf{x})$ is a solution to Equation (a) then

$$\int_B \left[(\operatorname{div} \mathbf{u})^2 + \frac{1-2\nu}{2-2\nu} (\operatorname{curl} \mathbf{u})^2 \right] dv$$

$$= \int_{\partial B} \mathbf{u} \cdot \left[(\operatorname{div} \mathbf{u})\, \mathbf{n} + \frac{1-2\nu}{2-2\nu} (\operatorname{curl} \mathbf{u}) \times \mathbf{n} \right] da \tag{b}$$

where $\mathbf{n}$ is the unit outward normal vector field on ∂B.

Hint: Multiply Equation (a) by $\mathbf{u}$ in the dot product sense, integrate the result over B, and use the divergence theorem.

Note: Since $-1 < \nu < 1/2$ [see Example 3.3.9, Equation (a)], then Equation (b) implies that a displacement boundary value problem of homogeneous isotropic elastostatics may have at most one solution.

4.3 Show that for a homogeneous isotropic infinite elastic body subject to a temperature change $T = T(\mathbf{x})$ its volume change is represented by the formula

$$\operatorname{tr} \mathbf{E}(\mathbf{x}) = \frac{1+\nu}{1-\nu} \alpha\, T(\mathbf{x}) \quad \text{for } \mathbf{x} \in E^3 \tag{a}$$

where ν and α denote Poisson's ratio and coefficient of thermal expansion, respectively.

Hint: Apply the reciprocal relation (a) from Example 4.1.19 to the external force–temperature systems $[\mathbf{b}, \mathbf{s}, T] = [\mathbf{0}, \mathbf{0}, T]$ and $[\widetilde{\mathbf{b}}, \widetilde{\mathbf{s}}, \widetilde{T}] = [\mathbf{0}, \mathbf{0}, \delta(\mathbf{x} - \boldsymbol{\xi})]$ on E^3. Also note that for an isotropic body $\mathbf{M} = -(3\lambda + 2\mu)\alpha \mathbf{1}$ and $\operatorname{tr} \widetilde{\mathbf{E}} = [(3\lambda + 2\mu)/(\lambda + 2\mu)]\, \alpha \widetilde{T}$.

4.4 Let Assume T_0 to be a constant temperature, and let a_i ($i = 1, 2, 3$) be positive constants of the length dimension. Show that for a homogeneous isotropic infinite elastic body subject to the temperature change

$$T(\mathbf{x}) = T_0[H(x_1 + a_1) - H(x_1 - a_1)] \times [H(x_2 + a_2) - H(x_2 - a_2)]$$
$$\times [H(x_3 + a_3) - H(x_3 - a_3)] \tag{a}$$

where $H(x)$ denotes the Heaviside function defined in Section 2.3.4, the stress components S_{ij} are represented by the formula

$$S_{ij}(\mathbf{x}) = A_0\left[\int_{-a_1}^{a_1} d\xi_1 \int_{-a_2}^{a_2} d\xi_2 \int_{-a_3}^{a_3} d\xi_3 \times \frac{\partial^2}{\partial\xi_i\partial\xi_j} \right.$$

$$\times [(x_1-\xi_1)^2 + (x_2-\xi_2)^2 + (x_3-\xi_3)^2]^{-1/2}$$

$$\left. + 4\pi\delta_{ij} \int_{-a_1}^{a_1} d\xi_1 \int_{-a_2}^{a_2} d\xi_2 \int_{-a_3}^{a_3} d\xi_3 \delta(x_1-\xi_1)\delta(x_2-\xi_2)\delta(x_3-\xi_3)\right] \tag{b}$$

where

$$A_0 = -\frac{\mu}{2\pi}\frac{1+\nu}{1-\nu}\alpha T_0 \tag{c}$$

Also, show that the integrals on the RHS of Equation (b) can be calculated in terms of elementary functions, and for the exterior of the parallelepiped

$$|x_1| \le a_1, \quad |x_2| \le a_2, \quad |x_3| \le a_3 \tag{d}$$

we obtain

$$S_{12} = A_0 \ln\left[\frac{(x_3+a_3+r_{+1.+2.-3})}{(x_3-a_3+r_{+1.+2.+3})}\frac{(x_3-a_3+r_{+1.-2.+3})}{(x_3+a_3+r_{+1.-2.-3})}\right.$$

$$\left.\times \frac{(x_3-a_3+r_{-1.+2.+3})}{(x_3+a_3+r_{-1.+2.-3})}\frac{(x_3+a_3+r_{-1.-2.-3})}{(x_3-a_3+r_{-1.-2.+3})}\right] \tag{e}$$

$$S_{23} = A_0 \ln\left[\frac{(x_1+a_1+r_{-1.+2.+3})}{(x_1-a_1+r_{+1.+2.+3})}\frac{(x_1-a_1+r_{+1.+2.-3})}{(x_1+a_1+r_{-1.+2.-3})}\right.$$

$$\left.\times \frac{(x_1-a_1+r_{+1.-2.+3})}{(x_1+a_1+r_{-1.-2.+3})}\frac{(x_1+a_1+r_{-1.-2.-3})}{(x_1-a_1+r_{+1.-2.-3})}\right] \tag{f}$$

$$S_{31} = A_0 \ln\left[\frac{(x_2+a_2+r_{+1.-2.+3})}{(x_2-a_2+r_{+1.+2.+3})}\frac{(x_2-a_2+r_{-1.+2.+3})}{(x_2+a_2+r_{-1.-2.+3})}\right.$$

$$\left.\times \frac{(x_2-a_2+r_{+1.+2.-3})}{(x_2+a_2+r_{+1.-2.-3})}\frac{(x_2+a_2+r_{-1.-2.-3})}{(x_2-a_2+r_{-1.+2.-3})}\right] \tag{g}$$

and

$$S_{11} = A_0\left[\tan^{-1}\left(\frac{x_2+a_2}{x_1-a_1}\frac{x_3+a_3}{r_{+1.-2.-3}}\right) - \tan^{-1}\left(\frac{x_2+a_2}{x_1-a_1}\frac{x_3-a_3}{r_{+1.-2.+3}}\right)\right.$$

$$- \tan^{-1}\left(\frac{x_2-a_2}{x_1-a_1}\frac{x_3+a_3}{r_{+1.+2.-3}}\right) + \tan^{-1}\left(\frac{x_2-a_2}{x_1-a_1}\frac{x_3-a_3}{r_{+1.+2.+3}}\right)$$

$$- \tan^{-1}\left(\frac{x_2+a_2}{x_1+a_1}\frac{x_3+a_3}{r_{-1.-2.-3}}\right) + \tan^{-1}\left(\frac{x_2+a_2}{x_1+a_1}\frac{x_3-a_3}{r_{-1.-2.+3}}\right)$$

$$\left. + \tan^{-1}\left(\frac{x_2-a_2}{x_1+a_1}\frac{x_3+a_3}{r_{-1.+2.-3}}\right) - \tan^{-1}\left(\frac{x_2-a_2}{x_1+a_1}\frac{x_3-a_3}{r_{-1.+2.+3}}\right)\right] \tag{h}$$

$$S_{22} = A_0 \left[\tan^{-1}\left(\frac{x_3 + a_3}{x_2 - a_2}\frac{x_1 + a_1}{r_{-1.+2.-3}}\right) - \tan^{-1}\left(\frac{x_3 + a_3}{x_2 - a_2}\frac{x_1 - a_1}{r_{+1.+2.-3}}\right) \right.$$
$$- \tan^{-1}\left(\frac{x_3 - a_3}{x_2 - a_2}\frac{x_1 + a_1}{r_{-1.+2.+3}}\right) + \tan^{-1}\left(\frac{x_3 - a_3}{x_2 - a_2}\frac{x_1 - a_1}{r_{+1.+2.+3}}\right)$$
$$- \tan^{-1}\left(\frac{x_3 + a_3}{x_2 + a_2}\frac{x_1 + a_1}{r_{-1.-2.-3}}\right) + \tan^{-1}\left(\frac{x_3 + a_3}{x_2 + a_2}\frac{x_1 - a_1}{r_{+1.-2.-3}}\right)$$
$$\left. + \tan^{-1}\left(\frac{x_3 - a_3}{x_2 + a_2}\frac{x_1 + a_1}{r_{-1.-2.+3}}\right) - \tan^{-1}\left(\frac{x_3 - a_3}{x_2 + a_2}\frac{x_1 - a_1}{r_{+1.-2.+3}}\right) \right] \quad \text{(i)}$$

$$S_{33} = A_0 \left[\tan^{-1}\left(\frac{x_1 + a_1}{x_3 - a_3}\frac{x_2 + a_2}{r_{-1.-2.+3}}\right) - \tan^{-1}\left(\frac{x_1 + a_1}{x_3 - a_3}\frac{x_2 - a_2}{r_{-1.+2.+3}}\right) \right.$$
$$- \tan^{-1}\left(\frac{x_1 - a_1}{x_3 - a_3}\frac{x_2 + a_2}{r_{+1.-2.+3}}\right) + \tan^{-1}\left(\frac{x_1 - a_1}{x_3 - a_3}\frac{x_2 - a_2}{r_{+1.+2.+3}}\right)$$
$$- \tan^{-1}\left(\frac{x_1 + a_1}{x_3 + a_3}\frac{x_2 + a_2}{r_{-1.-2.-3}}\right) + \tan^{-1}\left(\frac{x_1 + a_1}{x_3 + a_3}\frac{x_2 - a_2}{r_{-1.+2.-3}}\right)$$
$$\left. + \tan^{-1}\left(\frac{x_1 - a_1}{x_3 + a_3}\frac{x_2 + a_2}{r_{+1.-2.-3}}\right) - \tan^{-1}\left(\frac{x_1 - a_1}{x_3 + a_3}\frac{x_2 - a_2}{r_{+1.+2.-3}}\right) \right] \quad \text{(j)}$$

where

$$r_{\pm 1.\pm 2.\pm 3} = [(x_1 \mp a_1)^2 + (x_2 \mp a_2)^2 + (x_3 \mp a_3)^2]^{1/2} \quad \text{(k)}$$

Note that Equation (f) follows from Equation (e) by the transformation of indices

$$1 \to 2, \quad 2 \to 3, \quad 3 \to 1$$

and Equation (g) follows from Equation (f) by the transformation of indices

$$2 \to 3, \quad 3 \to 1, \quad 1 \to 2$$

Also, Equation (i) follows from Equation (h) by the transformation of indices

$$1 \to 2, \quad 2 \to 3, \quad 3 \to 1$$

and Equation (j) follows from Equation (i) by the transformation of indices

$$2 \to 3, \quad 3 \to 1, \quad 1 \to 2.$$

Hint: To find S_{12} use the formula

$$\int \frac{du}{\sqrt{u^2 + a^2}} = \ln\left(u + \sqrt{u^2 + a^2}\right) \quad \text{(l)}$$

and to calculate S_{11} take advantage of the formulas

$$\int \frac{du}{\left(\sqrt{u^2 + a^2}\right)^3} = \frac{1}{a^2}\frac{u}{\sqrt{u^2 + a^2}} \quad \text{(m)}$$

and

$$\int \frac{du}{(u^2+b^2)\sqrt{u^2+a^2}} = \frac{1}{b\sqrt{a^2-b^2}}\tan^{-1}\left(\frac{u\sqrt{a^2-b^2}}{b\sqrt{u^2+a^2}}\right) \tag{n}$$

where a and b are constants subject to the conditions

$$a \neq 0, \quad b \neq 0, \quad |a| > |b| \tag{o}$$

4.5 Let $\mathbf{u} = \mathbf{u}(\mathbf{x}, t)$ be a solution of the vector wave equation

$$\nabla^2\mathbf{u} - \frac{1}{c^2}\frac{\partial^2\mathbf{u}}{\partial t^2} = -\frac{\mathbf{f}}{c^2} \quad \text{on } B \times (0, \infty) \tag{a}$$

subject to the initial conditions

$$\mathbf{u}(\mathbf{x}, 0) = \mathbf{u}_0(\mathbf{x}), \quad \dot{\mathbf{u}}(\mathbf{x}, 0) = \dot{\mathbf{u}}_0(\mathbf{x}) \quad \text{on } \overline{B} \tag{b}$$

where $\mathbf{f} = \mathbf{f}(\mathbf{x}, t)$ is a prescribed vector field on $\overline{B} \times [0, \infty)$; and $\mathbf{u}_0(\mathbf{x})$ and $\dot{\mathbf{u}}_0(\mathbf{x})$ are prescribed vector fields on $\overline{B}$; and $c > 0$.

Also, let $\widetilde{\mathbf{u}} = \widetilde{\mathbf{u}}(\mathbf{x}, t)$ be a solution of the vector wave equation

$$\nabla^2\widetilde{\mathbf{u}} - \frac{1}{c^2}\frac{\partial^2\widetilde{\mathbf{u}}}{\partial t^2} = -\frac{\widetilde{\mathbf{f}}}{c^2} \quad \text{on } B \times (0, \infty) \tag{c}$$

subject to the conditions

$$\widetilde{\mathbf{u}}(\mathbf{x}, 0) = \widetilde{\mathbf{u}}_0(\mathbf{x}), \quad \dot{\widetilde{\mathbf{u}}}_0(\mathbf{x}, 0) = \dot{\widetilde{\mathbf{u}}}_0(\mathbf{x}) \quad \text{on } \overline{B} \tag{d}$$

where $\widetilde{\mathbf{f}} = \widetilde{\mathbf{f}}(\mathbf{x}, t) \neq \mathbf{f}(\mathbf{x}, t)$, $\widetilde{\mathbf{u}}_0(\mathbf{x}) \neq \mathbf{u}_0(\mathbf{x})$, and $\dot{\widetilde{\mathbf{u}}}_0(\mathbf{x}) \neq \dot{\mathbf{u}}_0(\mathbf{x})$ are prescribed functions on $\overline{B} \times [0, \infty)$, $\overline{B}$, and $\overline{B}$, respectively. Show that the following reciprocal relation holds true

$$\begin{aligned}\frac{1}{c^2}\int_B \left(\mathbf{u} * \widetilde{\mathbf{f}} + \mathbf{u} \cdot \dot{\widetilde{\mathbf{u}}}_0 + \dot{\mathbf{u}} \cdot \widetilde{\mathbf{u}}_0\right) dv + \int_{\partial B} \mathbf{u} * \frac{\partial \widetilde{\mathbf{u}}}{\partial n}\, da \\ = \frac{1}{c^2}\int_B \left(\widetilde{\mathbf{u}} * \mathbf{f} + \widetilde{\mathbf{u}} \cdot \dot{\mathbf{u}}_0 + \dot{\widetilde{\mathbf{u}}} \cdot \mathbf{u}_0\right) dv + \int_{\partial B} \widetilde{\mathbf{u}} * \frac{\partial \mathbf{u}}{\partial n}\, da\end{aligned} \tag{e}$$

where $*$ represents the inner convolutional product, that is, for any two vector fields $\mathbf{a} = \mathbf{a}(\mathbf{x}, t)$ and $\mathbf{b} = \mathbf{b}(\mathbf{x}, t)$ on $\overline{B} \times [0, \infty)$

$$\mathbf{a} * \mathbf{b} = \int_0^t \mathbf{a}(\mathbf{x}, t - \tau) \cdot \mathbf{b}(\mathbf{x}, \tau)\, d\tau \tag{f}$$

4.6 Let $\widetilde{\mathbf{U}} = \widetilde{\mathbf{U}}(\mathbf{x}, t)$ be a symmetric second-order tensor field that satisfies the wave equation

$$\left(\nabla^2 - \frac{1}{c^2}\frac{\partial^2}{\partial t^2}\right)\widetilde{\mathbf{U}} = -\frac{\widetilde{\mathbf{F}}}{c^2} \quad \text{on } B \times (0, \infty) \tag{a}$$

subject to the initial conditions

$$\widetilde{\mathbf{U}}(\mathbf{x}, 0) = \widetilde{\mathbf{U}}_0(\mathbf{x}), \quad \dot{\widetilde{\mathbf{U}}}(\mathbf{x}, 0) = \dot{\widetilde{\mathbf{U}}}_0(\mathbf{x}) \quad \text{on } B \tag{b}$$

where $\widetilde{\mathbf{F}} = \widetilde{\mathbf{F}}(\mathbf{x}, t)$, $\widetilde{\mathbf{U}}_0(\mathbf{x})$, and $\dot{\widetilde{\mathbf{U}}}_0(\mathbf{x})$ are prescribed symmetric second-order tensor fields on $\overline{B} \times [0, \infty)$, $\overline{B}$, and $\overline{B}$, respectively. Let $\mathbf{u} = \mathbf{u}(\mathbf{x}, t)$ be a solution to Equations (a) and (b) of Problem 4.5.
Show that

$$\frac{1}{c^2}\int_B \left(\widetilde{\mathbf{F}} * \mathbf{u} + \dot{\widetilde{\mathbf{U}}}_0\mathbf{u} + \widetilde{\mathbf{U}}_0\dot{\mathbf{u}}\right) dv + \int_{\partial B} \frac{\partial \widetilde{\mathbf{U}}}{\partial n} * \mathbf{u}\, da$$
$$= \frac{1}{c^2}\int_B \left(\widetilde{\mathbf{U}} * \mathbf{f} + \dot{\widetilde{\mathbf{U}}}\mathbf{u}_0 + \widetilde{\mathbf{U}}\dot{\mathbf{u}}_0\right) dv + \int_{\partial B} \widetilde{\mathbf{U}} * \frac{\partial \mathbf{u}}{\partial n}\, da \tag{c}$$

where for any tensor field $\mathbf{T} = \mathbf{T}(\mathbf{x}, t)$ on $\overline{B} \times [0, \infty)$ and for any vector field $\mathbf{v} = \mathbf{v}(\mathbf{x}, t)$ on $\overline{B} \times [0, \infty)$

$$\mathbf{T} * \mathbf{v} = \int_0^t \mathbf{T}(\mathbf{x}, t - \tau)\mathbf{v}(\mathbf{x}, \tau)\, d\tau \tag{d}$$

4.7 Let $\mathbf{G} = \mathbf{G}(\mathbf{x}, \boldsymbol{\xi}; t)$ be a symmetric second-order tensor field that satisfies the wave equation

$$\Box_1^2 \mathbf{G} = -\mathbf{1}\delta(\mathbf{x} - \boldsymbol{\xi})\delta(t) \quad \text{for } \mathbf{x} \in E^3,\ \boldsymbol{\xi} \in E^3,\ t > 0 \tag{a}$$

subject to the homogeneous initial conditions

$$\mathbf{G}(\mathbf{x}, \boldsymbol{\xi}; 0) = \mathbf{0}, \quad \dot{\mathbf{G}}(\mathbf{x}, \boldsymbol{\xi}; 0) = \mathbf{0} \quad \text{for } \mathbf{x} \in E^3,\ \boldsymbol{\xi} \in E^3 \tag{b}$$

where

$$\Box_1^2 = \frac{\partial^2}{\partial x_k \partial x_k} - \frac{1}{c^2}\frac{\partial^2}{\partial t^2} \quad (k = 1, 2, 3) \tag{c}$$

Show that a solution $\mathbf{u}$ to Equations (a) and (b) of Problem 4.5 admits the integral representation

$$\mathbf{u}(\mathbf{x}, t) = \frac{1}{c^2}\int_B \left(\mathbf{G} * \mathbf{f} + \dot{\mathbf{G}}\mathbf{u}_0 + \mathbf{G}\dot{\mathbf{u}}_0\right) dv(\boldsymbol{\xi})$$
$$+ \int_{\partial B} \left(\mathbf{G} * \frac{\partial \mathbf{u}}{\partial n} - \frac{\partial \mathbf{G}}{\partial n} * \mathbf{u}\right) da(\boldsymbol{\xi}) \tag{d}$$

Hint: Apply the reciprocal relation (c) of Problem 4.6 in which $\widetilde{\mathbf{F}}/c^2 = \mathbf{1}\delta(\mathbf{x}-\boldsymbol{\xi})\delta(t)$ and $\widetilde{\mathbf{U}} = \mathbf{G}(\mathbf{x}, \boldsymbol{\xi}; t)$.

4.8 Show that a unique solution to Equations (a) and (b) of Problem 4.7 takes the form

$$\mathbf{G}(\mathbf{x}, \boldsymbol{\xi}; t) = \frac{1}{4\pi\,|\mathbf{x}-\boldsymbol{\xi}|}\,\delta\left(t - \frac{|\mathbf{x}-\boldsymbol{\xi}|}{c}\right)\mathbf{1} \tag{a}$$

and, hence, reduce Equation (d) from Problem 4.7 to the Poisson–Kirchhoff integral representation

$$\begin{aligned}
\mathbf{u}(\mathbf{x}, t) = {} & \frac{1}{4\pi c^2}\int_B \frac{\mathbf{f}(\boldsymbol{\xi}, t - |\mathbf{x}-\boldsymbol{\xi}|/c)}{|\mathbf{x}-\boldsymbol{\xi}|}\, dv(\boldsymbol{\xi}) + \frac{\partial}{\partial t}\left[t\,\mathcal{M}_{\mathbf{x},ct}(\mathbf{u}_0)\right] \\
& + t\,\mathcal{M}_{\mathbf{x},ct}(\dot{\mathbf{u}}_0) + \frac{1}{4\pi}\int_{\partial B}\left\{\frac{1}{|\mathbf{x}-\boldsymbol{\xi}|}\frac{\partial \mathbf{u}}{\partial n}(\boldsymbol{\xi}, t - |\mathbf{x}-\boldsymbol{\xi}|/c)\right. \\
& - \mathbf{u}(\boldsymbol{\xi}, t - |\mathbf{x}-\boldsymbol{\xi}|/c)\,\frac{\partial}{\partial n}\frac{1}{|\mathbf{x}-\boldsymbol{\xi}|} \\
& \left. + \frac{1}{c|\mathbf{x}-\boldsymbol{\xi}|}\left[\frac{\partial}{\partial n}|\mathbf{x}-\boldsymbol{\xi}|\right]\left[\frac{\partial \mathbf{u}}{\partial t}(\boldsymbol{\xi}, t - |\mathbf{x}-\boldsymbol{\xi}|/c)\right]\right\} da(\boldsymbol{\xi})
\end{aligned} \tag{b}$$

where for any vector field $\mathbf{v} = \mathbf{v}(\mathbf{x})$ on $B \subset E^3$ the symbol $\mathcal{M}_{\mathbf{x},ct}(\mathbf{v})$ represents the mean value of $\mathbf{v}$ over the spherical surface with a center at $\mathbf{x}$ and of radius ct, that is,

$$\begin{aligned}
\mathcal{M}_{\mathbf{x},ct}(\mathbf{v}) = {} & \frac{1}{4\pi}\int_0^{2\pi} d\varphi \int_0^{\pi} d\theta\, \sin\theta \\
& \times \mathbf{v}(x_1 + ct\sin\theta\cos\varphi,\; x_2 + ct\sin\theta\sin\varphi,\; x_3 + ct\cos\theta)
\end{aligned} \tag{c}$$

and we adopt the convention that all relevant quantities vanish for negative time arguments.

Note: If $B = E^3$ and $\mathbf{f} = \mathbf{0}$ on $E^3 \times [0, \infty)$ then Equation (b) reduces to the form

$$\mathbf{u}(\mathbf{x}, t) = \frac{\partial}{\partial t}\left[t\,\mathcal{M}_{\mathbf{x},ct}(\mathbf{u}_0)\right] + t\,\mathcal{M}_{\mathbf{x},ct}(\dot{\mathbf{u}}_0) \tag{d}$$

4.9 Let $\mathbf{G}^* = \mathbf{G}^*(\mathbf{x}, \boldsymbol{\xi}; t)$ be a solution to the initial–boundary value problem:

$$\Box_1^2 \mathbf{G}^* = -\mathbf{1}\delta(\mathbf{x}-\boldsymbol{\xi})\delta(t) \quad \text{for } \mathbf{x},\ \boldsymbol{\xi} \in B,\ t > 0 \tag{a}$$

$$\mathbf{G}^*(\mathbf{x}, \boldsymbol{\xi}; 0) = \mathbf{0}, \quad \dot{\mathbf{G}}^*(\mathbf{x}, \boldsymbol{\xi}; 0) = \mathbf{0} \quad \text{for } \mathbf{x},\ \boldsymbol{\xi} \in B \tag{b}$$

and

$$\mathbf{G}^*(\mathbf{x}, \boldsymbol{\xi}; t) = \mathbf{0} \quad \text{for } \mathbf{x} \in \partial B,\ t > 0,\ \boldsymbol{\xi} \in \overline{B} \tag{c}$$

and let $\mathbf{u} = \mathbf{u}(\mathbf{x}, t)$ be a solution to the initial–boundary value problem

$$\square_1^2 \mathbf{u} = -\frac{\mathbf{f}}{c^2} \quad \text{on } B \times (0, \infty) \tag{d}$$

$$\mathbf{u}(\mathbf{x}, 0) = \mathbf{u}_0(\mathbf{x}), \quad \dot{\mathbf{u}}(\mathbf{x}, 0) = \dot{\mathbf{u}}_0(\mathbf{x}) \quad \text{on } B \tag{e}$$

$$\mathbf{u}(\mathbf{x}, t) = \mathbf{g}(\mathbf{x}, t) \quad \text{on } \partial B \times [0, \infty) \tag{f}$$

where the functions $\mathbf{f}$, $\mathbf{u}_0$, $\dot{\mathbf{u}}_0$, and $\mathbf{g}$ are prescribed. Use the representation formula (d) of Problem 4.7 to show that

$$\mathbf{u}(\mathbf{x}, t) = \frac{1}{c^2} \int_B (\mathbf{G}^* * \mathbf{f} + \dot{\mathbf{G}}^* \mathbf{u}_0 + \mathbf{G}^* \dot{\mathbf{u}}_0)\, dv(\boldsymbol{\xi}) - \int_{\partial B} \left(\frac{\partial \mathbf{G}^*}{\partial n} * \mathbf{g} \right) da(\boldsymbol{\xi}) \tag{g}$$

4.10 A tensor field $\mathbf{S}$ corresponds to the solution of a traction problem of classical elastodynamics if and only if

$$\widehat{\nabla}(\rho^{-1} \operatorname{div} \mathbf{S}) - \mathsf{K}[\ddot{\mathbf{S}}] = -\mathbf{B} \quad \text{on } B \times (0, \infty) \tag{a}$$

$$\mathbf{S}(\mathbf{x}, 0) = \mathbf{S}^{(0)}(\mathbf{x}), \quad \dot{\mathbf{S}}(\mathbf{x}, 0) = \dot{\mathbf{S}}^{(0)}(\mathbf{x}) \quad \text{on } B \tag{b}$$

$$\mathbf{S}\mathbf{n} = \hat{\mathbf{s}} \quad \text{on } \partial B \times [0, \infty) \tag{c}$$

[see Theorem 3 of Section 4.2.4 in which $\mathbf{B}$ is expressed in terms of a body force $\mathbf{b}$, and $\mathbf{S}^{(0)}$ and $\dot{\mathbf{S}}^{(0)}$ are defined in terms of two vector fields].

A tensor field $\mathbf{S}$ corresponding to an external load $[\mathbf{B}, \mathbf{S}^{(0)}, \dot{\mathbf{S}}^{(0)}, \widehat{\mathbf{s}}]$ is said to be of σ-type if $\mathbf{S}$ satisfies Equations (a) through (c) with an arbitrary symmetric second-order tensor field $\mathbf{B}$ and arbitrary symmetric initial tensor fields $\mathbf{S}^{(0)}$ and $\dot{\mathbf{S}}^{(0)}$, not necessarily related to the data of classic elastodynamics. Show that if $\mathbf{S}$ and $\widetilde{\mathbf{S}}$ are two different tensorial fields of σ-type corresponding to the external loads $[\mathbf{B}, \mathbf{S}^{(0)}, \dot{\mathbf{S}}^{(0)}, \widehat{\mathbf{s}}]$ and $[\widetilde{\mathbf{B}}, \widetilde{\mathbf{S}}^{(0)}, \dot{\widetilde{\mathbf{S}}}^{(0)}, \widehat{\widetilde{\mathbf{s}}}]$, respectively, then the following reciprocal relation holds true

$$\int_B \left\{ \widetilde{\mathbf{B}} * \mathbf{S} + \widetilde{\mathbf{S}}^{(0)} \cdot \mathsf{K}[\dot{\mathbf{S}}] + \dot{\widetilde{\mathbf{S}}}^{(0)} \cdot \mathsf{K}[\mathbf{S}] \right\} dv + \int_{\partial B} \rho^{-1} (\operatorname{div} \widetilde{\mathbf{S}}) * (\mathbf{S}\,\mathbf{n})\, da$$
$$= \int_B \left\{ \mathbf{B} * \widetilde{\mathbf{S}} + \mathbf{S}^{(0)} \cdot \mathsf{K}[\dot{\widetilde{\mathbf{S}}}] + \dot{\mathbf{S}}^{(0)} \cdot \mathsf{K}[\widetilde{\mathbf{S}}] \right\} dv + \int_{\partial B} \rho^{-1} (\operatorname{div} \mathbf{S}) * (\widetilde{\mathbf{S}}\mathbf{n})\, da \tag{d}$$

4.11 Let $S_{ij}^{(kl)} = S_{ij}^{(kl)}(\mathbf{x}, \boldsymbol{\xi}; t)$ be a solution of the following equation

$$(\rho^{-1} S^{(kl)}_{(is,s}),_{j)} - K_{ijpq} \ddot{S}^{(kl)}_{pq} = 0$$
$$\text{for } \mathbf{x} \in E^3, \ \boldsymbol{\xi} \in E^3, \ t > 0, \ i, j, k, l = 1, 2, 3 \tag{a}$$

subject to the initial conditions

$$S_{ij}^{(kl)}(\mathbf{x}, \boldsymbol{\xi}; 0) = 0, \quad \dot{S}_{ij}^{(kl)}(\mathbf{x}, \boldsymbol{\xi}; 0) = C_{ijkl}\delta(\mathbf{x} - \boldsymbol{\xi})$$
$$\text{for } \mathbf{x} \in E^3, \ \boldsymbol{\xi} \in E^3; \ i, j, k, l = 1, 2, 3 \tag{b}$$

where K_{ijkl} denotes the components of the compliance tensor **K**, and C_{ijkl} stands for the components of elasticity tensor **C**, that is,

$$C_{ijkl}K_{klmn} = \delta_{(im}\delta_{nj)} \tag{c}$$

Let $S_{ij} = S_{ij}(\mathbf{x}, t)$ be a solution of the equation

$$(\rho^{-1}S_{(ik,k)})_{,j)} - K_{ijkl}\ddot{S}_{kl} = 0 \quad \text{for } \mathbf{x} \in B, \ t > 0 \tag{d}$$

subject to the homogeneous initial conditions

$$S_{ij}(\mathbf{x}, 0) = 0, \quad \dot{S}_{ij}(\mathbf{x}, 0) = 0 \quad \text{for } \mathbf{x} \in B \tag{e}$$

and the boundary condition

$$S_{ij}n_j = \widehat{s}_i \quad \text{on } \partial B \times [0, \infty) \tag{f}$$

Use the reciprocal relation (d) of Problem 4.10 to show that

$$S_{kl}(\mathbf{x}, t) = \int_{\partial B} \rho^{-1}\left(S_{im,m} * S_{ij}^{(kl)}n_j - \widehat{s}_i * S_{im,m}^{(kl)}\right) da(\boldsymbol{\xi}) \tag{g}$$

Note: Equation (g) provides a solution to the traction initial–boundary value problem of classical elastodynamics if the field $S_{im,m}$ on $\partial B \times [0, \infty)$ is found from an associated integral equation on $\partial B \times [0, \infty)$. The idea of solving a traction problem of elastodynamics in terms of displacements through an associated boundary integral equation is due to V. D. Kupradze [11].

4.12 Consider the pure stress initial–boundary value problem of linear elastodynamics for a homogeneous isotropic *incompressible* elastic body B (see Equation 4.2.16 with $\mu > 0$ and $\lambda \to \infty$). Find a tensor field $\mathbf{S} = \mathbf{S}(\mathbf{x}, t)$ on $\overline{B} \times [0, \infty)$ that satisfies the equation

$$\widehat{\nabla}(\operatorname{div} \mathbf{S}) - \frac{\rho}{2\mu}\left[\ddot{\mathbf{S}} - \frac{1}{3}(\operatorname{tr}\ddot{\mathbf{S}})\mathbf{1}\right] = -\widehat{\nabla}\,\mathbf{b} \quad \text{on } B \times (0, \infty) \tag{a}$$

subject to the initial conditions

$$\mathbf{S}(\mathbf{x}, 0) = \mathbf{S}_0(\mathbf{x}), \quad \dot{\mathbf{S}}(\mathbf{x}, 0) = \dot{\mathbf{S}}_0(\mathbf{x}) \quad \text{on } B \tag{b}$$

and the traction boundary condition

$$\mathbf{S}\mathbf{n} = \widehat{\mathbf{s}} \quad \text{on } \partial B \times [0, \infty) \tag{c}$$

Here, $\mathbf{b}, \widehat{\mathbf{s}}$, $\mathbf{S}_0$, and $\dot{\mathbf{S}}_0$ are prescribed functions ($\mu > 0, \rho > 0$).
Show that the problem (a) through (c) may have at most one solution [12].

REFERENCES

1. B. A. Boley and J. H. Weiner, *Theory of Thermal Stresses,* Wiley, New York, 1960, pp. 92–95.
2. N. Noda, R. B. Hetnarski, and Y. Tanigawa, *Thermal Stresses,* 2nd edn., Taylor & Francis, New York, 2003, pp. 173–177.
3. J. N. Goodier, On the integration of the thermoelastic equations, *Phil. Mag.*, 23, 1937, 1017.
4. E. Sternberg and F. Rosenthal, The elastic sphere under concentrated loads, *J. Appl. Mech.*, 19(4), 1952, 413.
5. G. Fichera, *Existence Theorems in Elasticity,* Encyclopedia of Physics, chief editor: S. Flügge, vol. VIa/2, editor: C. Truesdell, Springer, Berlin, Germany, 1972, pp. 347–389.
6. J. Ignaczak, A completeness problem for the stress equation of motion in the linear theory of elasticity, *Archiwum Mechaniki Stosowanej*, 15, 1963, 225–234.
7. D. E. Carlson, *Linear Thermoelasticity,* Encyclopedia of Physics, chief editor: S. Flügge, vol. VIa/2, editor: C. Truesdell, Springer, Berlin, Germany, 1972, Eq. (3.6), p. 301.
8. W. Nowacki, *Thermoelasticity,* 2nd edn., PWN—Polish Scientific Publishers, Warsaw, and Pergamon Press, Oxford, U.K., 1986, p. 57.
9. H. Irschik and M. Gusenbauer, Body force analogy for the transient thermal stresses, *J. Thermal Stresses*, 30, 2007, 965–975.
10. J. Ignaczak and C. R. A. Rao, Stress characterization of elastodynamics with continuously distributed defects, *J. Elasticity*, 30, 1993, 219–250.
11. V. D. Kupradze, T. G. Gegelia, M. O. Basheleishvili, and T. V. Burchuladze, *Three-Dimensional Problems of the Mathematical Theory of Elasticity and Thermoelasticity*, North-Holland, Amsterdam, the Netherlands, 1979, pp. 1–929.
12. R. Wojnar, Uniqueness theorem for stress equations of isochoric motions of linear elasticity, *Arch. Mech.*, 26, 1974, 747–750.

5 Variational Formulation of Elastostatics

A solution to a boundary value problem of elastostatics is shown in this chapter to be an array of functions at which a functional attains an extremum. The variational formulation includes the principle of minimum potential energy, the principle of minimum complementary energy, the Hu–Washizu principle, as well as the compatibility-related principle for a traction problem. The chapter contains a number of worked examples in which the Rayleigh–Ritz method is used for finding a minimum of a functional. Included are end-of-chapter problems with solutions provided in the Solutions Manual.

5.1 MINIMUM PRINCIPLES

5.1.1 THE PRINCIPLE OF MINIMUM POTENTIAL ENERGY

For a general anisotropic elastic body, we recall the concept of the strain energy and of the stress energy.

By the *strain energy of a body B*, we mean the integral, see Equation 3.3.93,

$$U_C\{\mathbf{E}\} = \frac{1}{2}\int_B \mathbf{E}\cdot\mathsf{C}[\mathbf{E}]\,dv \tag{5.1.1}$$

and by the *stress energy of a body B* we mean, see Equation 3.3.96,

$$U_K\{\mathbf{S}\} = \frac{1}{2}\int_B \mathbf{S}\cdot\mathsf{K}[\mathbf{S}]\,dv \tag{5.1.2}$$

Here $\mathbf{E}$ and $\mathbf{S}$ are the strain and the stress tensor fields, respectively, and C and K stand for the elasticity and compliance fourth-order tensor fields, respectively. It is assumed that both C and K are symmetric and positive definite on B, and $\mathsf{K} = \mathsf{C}^{-1}$. Since the stress–strain relation is postulated in the form $\mathbf{S} = \mathsf{C}[\mathbf{E}]$, therefore, see Equation 3.3.97,

$$U_K\{\mathbf{S}\} = U_C\{\mathbf{E}\} \tag{5.1.3}$$

Before formulating the principle of minimum potential energy, we introduce a concept of a *kinematically admissible state*.

By a kinematically admissible state, we mean an admissible state $s = [\mathbf{u}, \mathbf{E}, \mathbf{S}]$ that satisfies

(1) The strain–displacement relation

$$\mathbf{E} = \widehat{\nabla}\mathbf{u} \quad \text{on } B \tag{5.1.4}$$

(2) The stress–strain relation

$$\mathbf{S} = \mathsf{C}[\mathbf{E}] \quad \text{on } B \tag{5.1.5}$$

(3) The displacement boundary condition

$$\mathbf{u} = \widehat{\mathbf{u}} \quad \text{on } \partial B_1 \tag{5.1.6}$$

Although in this definition nothing is said of the traction boundary condition on ∂B_2, the minimum principle we are to formulate is related to a mixed boundary value problem.

(I) **The Principle of Minimum Potential Energy.** Let R be the set of all kinematically admissible states. Define a functional* $F = F\{\cdot\}$ on R by

$$F\{s\} = U_C\{\mathbf{E}\} - \int_B \mathbf{b} \cdot \mathbf{u}\, dv - \int_{\partial B_2} \widehat{\mathbf{s}} \cdot \mathbf{u}\, da \tag{5.1.7}$$

for every $s = [\mathbf{u}, \mathbf{E}, \mathbf{S}] \in R$. Let s be a solution to the mixed problem. Then

$$F\{s\} \leq F\{\widetilde{s}\} \quad \text{for every } \widetilde{s} \in R \tag{5.1.8}$$

and the equality holds true if s and $\widetilde{s}$ differ by a rigid displacement.

In Equation 5.1.7 $\mathbf{b}$ is a prescribed body force field on B, and $\widehat{\mathbf{s}}$ stands for a prescribed surface load on ∂B_2, which appears in the definition of a mixed problem, see Equations 4.1.101 through 4.1.103.

Proof: We take two arbitrary elements s and $\widetilde{s}$ belonging to R, and denote their difference by s'

$$s' = \widetilde{s} - s \tag{5.1.9}$$

Since R is a vector space, see Section 4.1.3, then s' is also an admissible state, and in view of Equations 5.1.4 through 5.1.6

$$\mathbf{E}' = \widehat{\nabla}\mathbf{u}' \quad \text{on } B \tag{5.1.10}$$

$$\mathbf{S}' = \mathsf{C}[\mathbf{E}'] \quad \text{on } B \tag{5.1.11}$$

$$\mathbf{u}' = \mathbf{0} \quad \text{on } \partial B_1 \tag{5.1.12}$$

Here,

$$\mathbf{u}' = \widetilde{\mathbf{u}} - \mathbf{u}, \quad \mathbf{E}' = \widetilde{\mathbf{E}} - \mathbf{E}, \quad \mathbf{S}' = \widetilde{\mathbf{S}} - \mathbf{S} \tag{5.1.13}$$

Next, we are to show that

$$U_C\{\widetilde{\mathbf{E}}\} - U_C\{\mathbf{E}\} = U_C\{\mathbf{E}'\} + \int_B \mathbf{S} \cdot \mathbf{E}'\, dv \tag{5.1.14}$$

* A function $F = F\{\cdot\}$ defined on a set of functions R is called a functional.

In order to do this, we use the definition of the strain energy, see Equations 5.1.1 and 5.1.13, to obtain

$$U_C\{\widetilde{\mathbf{E}}\} = \frac{1}{2}\int_B \widetilde{\mathbf{E}} \cdot \mathbf{C}[\widetilde{\mathbf{E}}]\, dv = \frac{1}{2}\int_B (\mathbf{E}' + \mathbf{E}) \cdot \mathbf{C}[\mathbf{E}' + \mathbf{E}]\, dv$$

$$= U_C\{\mathbf{E}'\} + U_C\{\mathbf{E}\} + \int_B \mathbf{E}' \cdot \mathbf{C}[\mathbf{E}]\, dv \tag{5.1.15}$$

Since the stress–strain relation is to be observed in the form

$$\mathbf{S} = \mathbf{C}[\mathbf{E}] \tag{5.1.16}$$

then Equation 5.1.15 implies that

$$U_C\{\widetilde{\mathbf{E}}\} = U_C\{\mathbf{E}'\} + U_C\{\mathbf{E}\} + \int_B \mathbf{E}' \cdot \mathbf{S}\, dv \tag{5.1.17}$$

This completes the proof of Equation 5.1.14.

Since

$$S_{ij}E'_{ij} = S_{ij}u'_{i,j} = (S_{ij}u'_i)_{,j} - S_{ij,j}u'_i \tag{5.1.18}$$

then by integrating Equation 5.1.18 over B and using the divergence theorem we write

$$\int_B \mathbf{E}' \cdot \mathbf{S}\, dv = \int_{\partial B} (\mathbf{Sn}) \cdot \mathbf{u}'\, da - \int_B (\operatorname{div} \mathbf{S}) \cdot \mathbf{u}'\, dv \tag{5.1.19}$$

Hence, Equation 5.1.17 may be reduced to the form

$$U_C\{\widetilde{\mathbf{E}}\} - U_C\{\mathbf{E}\} = U_C\{\mathbf{E}'\} + \int_{\partial B_2} \mathbf{s} \cdot \mathbf{u}'\, da - \int_B (\operatorname{div} \mathbf{S}) \cdot \mathbf{u}'\, dv \tag{5.1.20}$$

In obtaining Equation 5.1.20 we have also used the fact that $\mathbf{u}' = \mathbf{0}$ on ∂B_1.

Using the definition of the functional F given in Equation 5.1.7, we obtain

$$F\{\widetilde{s}\} = U_C\{\widetilde{\mathbf{E}}\} - \int_B \mathbf{b} \cdot \widetilde{\mathbf{u}}\, dv - \int_{\partial B_2} \widehat{\mathbf{s}} \cdot \widetilde{\mathbf{u}}\, da \tag{5.1.21}$$

Subtracting Equation 5.1.7 from Equation 5.1.21, and using Equation 5.1.20, we get

$$F\{\widetilde{s}\} - F\{s\} = U_C\{\mathbf{E}'\} + \int_{\partial B_2} \mathbf{s} \cdot \mathbf{u}'\, da - \int_B (\operatorname{div} \mathbf{S}) \cdot \mathbf{u}'\, dv$$

$$- \int_B \mathbf{b} \cdot \widetilde{\mathbf{u}}\, dv - \int_{\partial B_2} \widehat{\mathbf{s}} \cdot \widetilde{\mathbf{u}}\, da + \int_B \mathbf{b} \cdot \mathbf{u}\, dv + \int_{\partial B_2} \widehat{\mathbf{s}} \cdot \mathbf{u}\, da \tag{5.1.22}$$

Since $\widetilde{\mathbf{u}} - \mathbf{u} = \mathbf{u}'$, Equation 5.1.22 reduces to

$$F\{\widetilde{s}\} - F\{s\} = U_C\{\mathbf{E}'\} - \int_B (\operatorname{div}\mathbf{S} + \mathbf{b}) \cdot \mathbf{u}'\, dv + \int_{\partial B_2} (\mathbf{s} - \widehat{\mathbf{s}}) \cdot \mathbf{u}'\, da \tag{5.1.23}$$

Therefore, since s is a solution to the mixed problem, the second and third terms on the RHS of Equation 5.1.23 vanish, and

$$F\{\widetilde{s}\} - F\{s\} = U_C\{\mathbf{E}'\} \tag{5.1.24}$$

From this relation and from the positive definiteness of $\mathbf{C}$, which also means that the strain energy is nonnegative, it follows that

$$F\{\widetilde{s}\} - F\{s\} \geq 0 \tag{5.1.25}$$

which is the same as Equation 5.1.8. Hence, to complete the proof of the principle we need to show that the equality in Equation 5.1.25 holds true if s and $\widetilde{s}$ differ by a rigid displacement. To this end we note that, by Equation 5.1.24, the equation

$$F\{\widetilde{s}\} = F\{s\} \tag{5.1.26}$$

means that

$$U_C\{\mathbf{E}'\} = 0 \tag{5.1.27}$$

Equation 5.1.27 together with the positive definiteness of $\mathbf{C}$ imply that

$$\mathbf{E}' = \widetilde{\mathbf{E}} - \mathbf{E} = \mathbf{0} \tag{5.1.28}$$

This relation together with the statement given in Example 3.1.6 implies that s and $\widetilde{s}$ differ by a rigid displacement. This completes the proof of the principle. □

Note: The principle of minimum potential energy asserts that the difference between the strain energy and the work done by the body force and surface traction, defined over a set of kinematically admissible states, attains a minimum at a solution to the mixed problem. Hence, a direct method of calculus of variations may be used in conjunction with this principle to obtain an approximate solution to the mixed problem.

Example 5.1.1

Let R_1 denote a set of admissible displacement fields that satisfy the boundary condition 5.1.6, and define a functional $F_1\{\cdot\}$ on R_1 by the formula

$$F_1\{\mathbf{u}\} = \frac{1}{2}\int_B (\nabla\mathbf{u}) \cdot \mathbf{C}[\nabla\mathbf{u}]\, dv - \int_B \mathbf{b} \cdot \mathbf{u}\, dv - \int_{\partial B_2} \widehat{\mathbf{s}} \cdot \mathbf{u}\, da \tag{a}$$

for every $\mathbf{u} \in R_1$.

Show that if $\mathbf{u}$ corresponds to a solution to the mixed problem, then

$$F_1\{\mathbf{u}\} \leq F_1\{\widetilde{\mathbf{u}}\} \quad \text{for every } \widetilde{\mathbf{u}} \in R_1 \tag{b}$$

Solution

Since $\mathbf{u}$ corresponds to a solution to the mixed problem, from the definitions of R and R_1, where R is defined in the principle of minimum potential energy, it follows that $\mathbf{u} \in R_1$ implies that $s = [\mathbf{u}, \mathbf{E}, \mathbf{S}] \in R$, so the conditions 5.1.4 through 5.1.6 are satisfied. Hence, substituting Equation 5.1.4 into Equation 5.1.7, we note that the functional $F\{s\}$ given by Equation 5.1.7 reduces to the functional $F_1\{\mathbf{u}\}$ given by (a). Therefore, it follows from Equation 5.1.8 that the inequality (b) is satisfied. This completes the solution.

□

Note: The functional $F_1 = F_1\{\mathbf{u}\}$ in this example depends on a vector field $\mathbf{u}$ only, as opposed to the functional $F = F\{s\}$ where s stands for a triple of functions $s = [\mathbf{u}, \mathbf{E}, \mathbf{S}]$. This is a reason why in applications most often the functional $F_1\{\cdot\}$ is used.

5.1.2 The Rayleigh–Ritz Method

This method allows to find an approximate solution to the mixed problem by minimizing the functional F_1 introduced by (a) in Example 5.1.1.

We assume that the displacement field $\mathbf{u}$ corresponding to a solution to the mixed problem may be approximated by

$$\mathbf{u} \approx \mathbf{u}^{(N)} = \widehat{\mathbf{u}}^{(N)} + \sum_{n=1}^{N} a_n \mathbf{f}_n \quad \text{on } B \tag{5.1.29}$$

where $\widehat{\mathbf{u}}^{(N)}$ is a function on B such that

$$\widehat{\mathbf{u}}^{(N)} \approx \widehat{\mathbf{u}} \quad \text{on } \partial B_1 \tag{5.1.30}$$

and $\widehat{\mathbf{u}}$ is a prescribed vector field in the mixed problem, and $\mathbf{f}_n$ stands for a set of given functions on B that vanish on ∂B_1, that is,

$$\mathbf{f}_n = \mathbf{0} \quad \text{on } \partial B_1 \tag{5.1.31}$$

In addition, a_n are unknown constants to be determined from the condition that F_1 attains a minimum. Finally, N is a prescribed number of functions $\mathbf{f}_n$ to be used in the approximation.

Substitution of Equation 5.1.29 into (a) of Example 5.1.1 yields

$$\begin{aligned} F_1\{\mathbf{u}^{(N)}\} &= F_1\left\{\widehat{\mathbf{u}}^{(N)} + \sum_{n=1}^{N} a_n \mathbf{f}_n\right\} \\ &= \frac{1}{2}\int_B \nabla\left(\widehat{\mathbf{u}}^{(N)} + \sum_{n=1}^{N} a_n \mathbf{f}_n\right) \cdot \mathbf{C}\left[\nabla\left(\widehat{\mathbf{u}}^{(N)} + \sum_{n=1}^{N} a_n \mathbf{f}_n\right)\right] dv \\ &\quad - \int_B \mathbf{b} \cdot \left(\widehat{\mathbf{u}}^{(N)} + \sum_{n=1}^{N} a_n \mathbf{f}_n\right) dv - \int_{\partial B_2} \widehat{\mathbf{s}} \cdot \left(\widehat{\mathbf{u}}^{(N)} + \sum_{n=1}^{N} a_n \mathbf{f}_n\right) da \end{aligned} \tag{5.1.32}$$

Using the symmetry of **C**, we write Equation 5.1.32 in an expanded form:

$$F_1\{\mathbf{u}^{(N)}\} = F_1\{\widehat{\mathbf{u}}^{(N)}\} + \frac{1}{2}\int_B \sum_{n=1}^{N} a_n(\nabla \mathbf{f}_n)\cdot \mathbf{C}\left[\sum_{k=1}^{N} a_k(\nabla \mathbf{f}_k)\right] dv$$
$$+ \int_B \sum_{n=1}^{N} a_n(\nabla \mathbf{u}^{(N)})\cdot \mathbf{C}[\nabla \mathbf{f}_n]\, dv - \int_B \sum_{n=1}^{N} a_n \mathbf{b}\cdot \mathbf{f}_n\, dv$$
$$- \int_{\partial B_2} \sum_{n=1}^{N} a_n \widehat{\mathbf{s}}\cdot \mathbf{f}_n\, da \tag{5.1.33}$$

Now, introduce the following notations:

$$A_{nk} = \int_B \nabla \mathbf{f}_n \cdot \mathbf{C}[\nabla \mathbf{f}_k]\, dv \tag{5.1.34}$$

$$b_n = \int_B \{(\nabla \widehat{\mathbf{u}}^{(N)})\cdot \mathbf{C}[\nabla f_n] - \mathbf{b}\cdot \mathbf{f}_n\}\, dv - \int_{\partial B_2} \widehat{\mathbf{s}}\cdot \mathbf{f}_n\, da \tag{5.1.35}$$

Then using the fact that

$$\mathbf{C}\left[\sum_{k=1}^{N} a_k(\nabla \mathbf{f}_k)\right] = \sum_{k=1}^{N} a_k \mathbf{C}[\nabla \mathbf{f}_k] \tag{5.1.36}$$

we reduce Equation 5.1.33 to the form

$$F_1\{\mathbf{u}^{(N)}\} = F_1\{\widehat{\mathbf{u}}^{(N)}\} + \frac{1}{2}\sum_{n,k=1}^{N} A_{nk}a_n a_k + \sum_{n=1}^{N} b_n a_n \tag{5.1.37}$$

Note that due to symmetry and positive definiteness of the tensor **C**, the matrix $[A_{nk}]$ is also symmetric and positive definite, and this implies that

$$\frac{\partial}{\partial a_i}\left[\frac{1}{2}\sum_{n,k=1}^{N} A_{nk}a_n a_k\right] = \sum_{k=1}^{N} A_{ki}a_k \tag{5.1.38}$$

where $i = 1, 2, \ldots, N$. Obviously, $F_1\{\mathbf{u}^{(N)}\}$ is a quadratic function of the parameters a_1, $a_2, \ldots, a_N$, so if we introduce the notation

$$F_1\{\mathbf{u}^{(N)}\} \equiv \varphi(a_1, a_2, \ldots, a_N) \tag{5.1.39}$$

then the necessary condition for the function $\varphi = \varphi(a_1, a_2, \ldots, a_N)$ to attain a minimum takes the form

$$\frac{\partial \varphi}{\partial a_i}(a_1, a_2, \ldots, a_N) = 0 \quad i = 1, \ldots, N \tag{5.1.40}$$

Since $F_1\{\widehat{\mathbf{u}}^{(N)}\}$ does not depend on $a_1, \ldots, a_N$, Equation 5.1.40, in view of Equations 5.1.37 through 5.1.39, reads

$$\sum_{k=1}^{N} A_{ki} a_k = -b_i \quad i = 1, 2, \ldots, N \tag{5.1.41}$$

Due to the symmetry and positive definiteness of the matrix $[A_{nk}]$, the system of linear algebraic equations 5.1.41 has a unique solution. Hence, this solution provides a unique approximation of the displacement field corresponding to a solution to the mixed problem in the form of Equation 5.1.29.

It may be shown that this approximation provides the displacement of which the magnitude is generally smaller than that of the displacement corresponding to the exact solution to the mixed problem. If $N \to \infty$ the approximate displacement $\mathbf{u}^{(N)}$ converges to the displacement corresponding to the exact solution.

Example 5.1.2

The deflection $w = w(x)$ of an elastic beam of length ℓ which is simply supported at its ends and subject to a concentrated force P at $x = \xi$ $(0 < \xi < \ell)$ satisfies the differential equation

$$EI \frac{d^4 w}{dx^4} = P\delta(x - \xi) \quad 0 < x < \ell \tag{a}$$

subject to the boundary conditions

$$w(0) = w''(0) = w(\ell) = w''(\ell) = 0 \tag{b}$$

where

E is Young's modulus

I is the moment of inertia of a cross section A of the beam (see Figure 5.1)

The boundary conditions (b) assert that displacements and bending moments at the ends of the beam are zero.

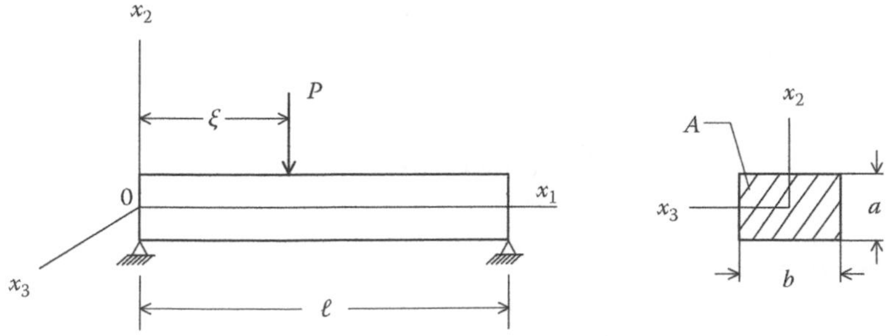

FIGURE 5.1 Simply supported beam subject to a concentrated force.

(i) Prove the following principle of minimum potential energy for the problem (a)–(b). Let R_1 be a set of functions $w = w(x)$ on $[0, \ell]$ that satisfy the boundary conditions (b), and define a functional $\widehat{F}_1\{\cdot\}$ on R_1 by

$$\widehat{F}_1\{w\} = \frac{1}{2} EI \int_0^\ell \left(\frac{d^2 w}{dx^2} \right)^2 dx - P \int_0^\ell \delta(x - \xi) w(x) dx \tag{c}$$

for every $w \in R_1$. Let w be a solution to the problem (a)–(b). Then

$$\widehat{F}_1\{w\} \leq \widehat{F}_1\{\widetilde{w}\} \quad \text{for every } \widetilde{w} \in R_1 \tag{d}$$

(ii) Using the Rayleigh–Ritz method leading to a minimum of the functional $\widehat{F}_1\{\cdot\}$ on R_1 show that

$$w(x) = \sum_{n=1}^{\infty} a_n w_n(x), \quad 0 \leq x \leq \ell \tag{e}$$

where

$$a_n = \frac{2P\ell^3}{EI\pi^4} \frac{\sin(n\pi\xi/\ell)}{n^4}, \quad w_n(x) = \sin(n\pi x/\ell) \quad n = 1, 2, 3, \ldots. \tag{f}$$

Solution

Proof of (i) is similar to that of the minimum principle in Example 5.1.1, and can be safely omitted.

To show (ii) we note that $w_n(x) \in R_1$ for $n = 1, 2, 3, \ldots$, and by substituting (e) into (c) we obtain

$$\varphi(a_1, a_2, a_3, \ldots) \equiv \widehat{F}_1\{w\} = \frac{1}{2} EI \int_0^\ell \left[\sum_{n=1}^{\infty} a_n \left(\frac{n\pi}{\ell} \right)^2 w_n(x) \right]^2 dx \\ - P \sum_{n=1}^{\infty} a_n w_n(\xi) \tag{g}$$

Next, by differentiating (g) with respect to a_k $(k = 1, 2, 3, \ldots)$, we write

$$\frac{\partial \varphi}{\partial a_k} = EI \int_0^\ell \left[\sum_{n=1}^{\infty} a_n \left(\frac{n\pi}{\ell} \right)^2 w_n(x) \right] \left(\frac{k\pi}{\ell} \right)^2 w_k(x)\, dx - P w_k(\xi) \tag{h}$$

Since

$$\int_0^\ell w_n(x) w_k(x)\, dx = \frac{\ell}{2} \delta_{nk} \tag{i}$$

therefore the equation

$$\frac{\partial \varphi}{\partial a_k} = 0, \quad k = 1, 2, 3, \ldots \infty \tag{j}$$

takes the form

$$EIa_k\left(\frac{k\pi}{\ell}\right)^4\frac{\ell}{2}-P\sin\frac{k\pi\xi}{\ell}=0 \tag{k}$$

or

$$a_k=\frac{2P\ell^3}{EI\pi^4}\frac{1}{k^4}\sin\frac{k\pi\xi}{\ell} \tag{l}$$

Hence, a minimum of the functional $\widetilde{F}_1\{\cdot\}$ on R_1 is attained at the function $w = w(x)$ defined by Equations (e) and (f). This completes the solution. □

Note:

Since [1]

$$\sum_{n=1}^{\infty}\frac{1}{n^4}=\frac{\pi^4}{90} \tag{m}$$

therefore the deflection of a beam shown in Figure 5.1 satisfies the inequality

$$|w(x)|\leq 0.0222\frac{P\ell^3}{EI},\quad 0<x<\ell \tag{n}$$

5.1.3 The Principle of Minimum Complementary Energy

This principle deals with a minimum of a functional related to the mixed problem, but expressed in terms of stresses. To formulate this principle we introduce a concept of a *statically admissible stress field*. By such a field we mean a symmetric second-order tensor field $\mathbf{S}$ that satisfies

(1) The equation of equilibrium

$$\operatorname{div}\mathbf{S}+\mathbf{b}=\mathbf{0}\quad\text{on } B \tag{5.1.42}$$

(2) The traction boundary condition

$$\mathbf{s}=\mathbf{Sn}=\widehat{\mathbf{s}}\quad\text{on } \partial B_2 \tag{5.1.43}$$

(II) **The Principle of Minimum Complementary Energy.** Let P denote a set of all statically admissible stress fields, and let $G = G\{\cdot\}$ be a functional on P defined by

$$G\{\mathbf{S}\}=U_K\{\mathbf{S}\}-\int_{\partial B_1}\mathbf{s}\cdot\widehat{\mathbf{u}}\,da \tag{5.1.44}$$

for every $\mathbf{S}\in P$.

If $\mathbf{S}$ is a stress field corresponding to a solution to the mixed problem, then

$$G\{\mathbf{S}\}\leq G\{\widetilde{\mathbf{S}}\}\quad\text{for every } \widetilde{\mathbf{S}}\in P \tag{5.1.45}$$

and the equality holds if $\mathbf{S} = \widetilde{\mathbf{S}}$. Here $U_K\{\mathbf{S}\}$ is the stress energy of B defined by Equation 5.1.2, and $\widehat{\mathbf{u}}$ stands for a prescribed displacement on ∂B_1 in the mixed problem.

Proof: Let $\mathbf{S}$ be a stress field corresponding to a solution to the mixed problem, and let $\widetilde{\mathbf{S}}$ be an arbitrary element of P, and define the difference

$$\mathbf{S}' = \widetilde{\mathbf{S}} - \mathbf{S} \tag{5.1.46}$$

Since $\mathbf{S} \in P$, $\widetilde{\mathbf{S}} \in P$, then $\mathbf{S}' \in P$, and this implies in view of Equations 5.1.42 and 5.1.43 that

$$\operatorname{div} \mathbf{S}' = \mathbf{0} \quad \text{on } B \tag{5.1.47}$$

and

$$\mathbf{s}' = \mathbf{S}'\mathbf{n} = \mathbf{0} \quad \text{on } \partial B_2 \tag{5.1.48}$$

Next, by the definition of $U_K\{\mathbf{S}\}$, see Equation 5.1.2,

$$U_K\{\mathbf{S}\} = \frac{1}{2}\int_B \mathbf{S} \cdot \mathsf{K}[\mathbf{S}]\, dv \tag{5.1.49}$$

Hence, calculating $U_K\{\widetilde{\mathbf{S}}\}$ according to this formula, by virtue of Equation 5.1.46 we obtain

$$\begin{aligned} U_K\{\widetilde{\mathbf{S}}\} &= U_K\{\mathbf{S}' + \mathbf{S}\} = \frac{1}{2}\int_B (\mathbf{S}' + \mathbf{S}) \cdot \mathsf{K}[\mathbf{S}' + \mathbf{S}]\, dv \\ &= U_K\{\mathbf{S}'\} + \int_B \mathbf{S}' \cdot \mathsf{K}[\mathbf{S}]\, dv + U_K\{\mathbf{S}\} \end{aligned} \tag{5.1.50}$$

Since $\mathbf{S}$ is a stress field corresponding to a solution to the mixed problem

$$\mathbf{E} = \mathsf{K}[\mathbf{S}] = \widehat{\nabla} \mathbf{u} \tag{5.1.51}$$

therefore, the second term on the RHS of Equation 5.1.50 can be written as

$$\int_B \mathbf{S}' \cdot \mathsf{K}[\mathbf{S}]\, dv = \int_B \mathbf{S}' \cdot (\nabla \mathbf{u})\, dv = \int_{\partial B} \mathbf{u} \cdot (\mathbf{S}'\mathbf{n})\, da - \int_B (\operatorname{div} \mathbf{S}') \cdot \mathbf{u}\, dv \tag{5.1.52}$$

Equation 5.1.52 has been obtained by using the identity

$$\mathbf{S}' \cdot (\nabla \mathbf{u}) = S'_{ij} u_{i,j} = (S'_{ij} u_i)_{,j} - S'_{ij,j} u_i \tag{5.1.53}$$

and the divergence theorem.

Next, since $\mathbf{S}'$ satisfies Equations 5.1.47 and 5.1.48, Equations 5.1.50 and 5.1.52 imply that

$$U_K\{\widetilde{\mathbf{S}}\} - U_K\{\mathbf{S}\} = U_K\{\mathbf{S}'\} + \int_{\partial B_1} \mathbf{u} \cdot \mathbf{s}' \, da \tag{5.1.54}$$

Now, computing the difference $G\{\widetilde{\mathbf{S}}\} - G\{\mathbf{S}\}$, and using Equation 5.1.54, we get

$$G\{\widetilde{\mathbf{S}}\} - G\{\mathbf{S}\} = U_K\{\mathbf{S}'\} + \int_{\partial B_1} \mathbf{u} \cdot \mathbf{s}' \, da - \int_{\partial B_1} \widetilde{\mathbf{s}} \cdot \widehat{\mathbf{u}} \, da + \int_{\partial B_1} \mathbf{s} \cdot \mathbf{u} \, da \tag{5.1.55}$$

Thus, since $\mathbf{s}' + \mathbf{s} = \widetilde{\mathbf{s}}$, see Equation 5.1.46, and $\mathbf{u}$ is the displacement field corresponding to a solution to the mixed problem, then

$$\mathbf{u} = \widehat{\mathbf{u}} \quad \text{on } \partial B_1 \tag{5.1.56}$$

and the surface integral on the RHS of Equation 5.1.55, which is a sum of three surface integrals over ∂B_1, vanishes, and Equation 5.1.55 reduces to

$$G\{\widetilde{\mathbf{S}}\} - G\{\mathbf{S}\} = U_K\{\mathbf{S}'\} \tag{5.1.57}$$

Hence, since the tensor K is positive definite, it follows from Equation 5.1.57 that

$$G\{\widetilde{\mathbf{S}}\} \geq G\{\mathbf{S}\} \quad \text{for every } \widetilde{\mathbf{S}} \in P \tag{5.1.58}$$

We note that the equality sign in Equation 5.1.58 together with the relation 5.1.57 imply that

$$U_K\{\mathbf{S}'\} = 0 \tag{5.1.59}$$

So, appealing again to the positive definiteness of K, from Equations 5.1.59 and 5.1.46, we write

$$\mathbf{S}' = \widetilde{\mathbf{S}} - \mathbf{S} = \mathbf{0} \tag{5.1.60}$$

This completes the proof of the principle. □

Notes:

(1) The principle restricted to a traction problem ($\partial B_1 = 0$) asserts that the stress energy of a body subject to the boundary condition

$$\mathbf{s} = \widehat{\mathbf{s}} \quad \text{on } \partial B \tag{5.1.61}$$

attains a minimum over the set of all statically admissible stress fields.

(2) The concept of the complementary strain energy was introduced by Equation 3.3.72.

(3) An approximate stress field $\mathbf{S}^{(N)}$ of a traction problem may be obtained by finding a minimum of the stress energy $U_K\{\mathbf{S}^{(N)}\}$, where

$$\mathbf{S}^{(N)} = \widehat{\mathbf{S}}^{(N)} + \sum_{k=1}^{N} a_k \mathbf{S}_k \quad \text{on } B \tag{5.1.62}$$

In this equation the stress field $\widehat{\mathbf{S}}^{(N)}$ is selected in such a way that it satisfies the equilibrium equation

$$\operatorname{div} \widehat{\mathbf{S}}^{(N)} + \mathbf{b} = \mathbf{0} \quad \text{on } B \tag{5.1.63}$$

subject to the condition

$$\widehat{\mathbf{S}}^{(N)}\mathbf{n} = \widehat{\mathbf{s}} \quad \text{on } \partial B \tag{5.1.64}$$

and $\mathbf{S}_k$ is to satisfy the homogeneous equilibrium equation

$$\operatorname{div} \mathbf{S}_k = \mathbf{0} \quad \text{on } \partial B \tag{5.1.65}$$

and the traction free boundary condition

$$\mathbf{S}_k\mathbf{n} = \mathbf{0} \quad \text{on } \partial B \tag{5.1.66}$$

Obviously, $\mathbf{S}^{(N)}$ is statically admissible stress field that contains unknown coefficients a_k $(k = 1, 2, \ldots, N)$. These coefficients are obtained from the condition of a minimum of the function

$$U_K\{\mathbf{S}^{(N)}\} \equiv \psi(a_1, a_2, \ldots, a_N) \tag{5.1.67}$$

in the form

$$\frac{\partial \psi}{\partial a_k} = 0 \quad k = 1, 2, \ldots, N \tag{5.1.68}$$

Similarly as in Section 5.1.2, Equation 5.1.68 represents a system of linear algebraic equations for a_k with a unique solution. Finding all a_k amounts to finding the approximate solution to the traction problem in a form of Equation 5.1.62.

Example 5.1.3

Consider a generalized plane stress traction problem of homogeneous isotropic isothermal elastostatics with zero body forces in a region C_0 of (x_1, x_2) plane (see Section 8.1.1). For such a problem a stress field $\overline{S}_{\alpha\beta}$ does not depend of elasticity constants, and the stress energy to be minimized may be simplified by equating Poisson's ratio to zero [see Equations (a) and (b) in Problem 5.1 taken at $\nu = 0$], that is, we let

$$U_K\{\overline{\mathbf{S}}\} = \frac{1}{4\mu} \int_{C_0} \overline{S}_{\alpha\beta}\overline{S}_{\alpha\beta}\, da \tag{a}$$

where $\bar{S}_{\alpha\beta} = \bar{S}_{\beta\alpha}$ $(\alpha, \beta = 1, 2)$ are the components of the stress tensor $\bar{\mathbf{S}}$ corresponding to a solution $\bar{s} = [\bar{\mathbf{u}}, \bar{\mathbf{E}}, \bar{\mathbf{S}}]$ of the traction problem, and μ is the shear modulus. Let $\widetilde{P}$ be a set of all statically admissible stress fields, that is, $\widetilde{\mathbf{S}} \in \widetilde{P}$ if and only if

$$\operatorname{div} \widetilde{\mathbf{S}} = \mathbf{0} \quad \text{on } C_0, \qquad \widetilde{\mathbf{S}}\mathbf{n} = \widehat{\mathbf{s}} \quad \text{on } \partial C_0 \tag{b}$$

where $\widehat{\mathbf{s}}$ is a boundary load in the traction problem. Then the two-dimensional principle of minimum complementary energy holds true. If $\bar{\mathbf{S}}$ is the stress field corresponding to a solution of the traction problem, then

$$U_K\{\bar{\mathbf{S}}\} \le U_K\{\widetilde{\mathbf{S}}\} \quad \text{for every } \widetilde{\mathbf{S}} \in \widetilde{P} \tag{c}$$

and the equality holds if $\bar{\mathbf{S}} = \widetilde{\mathbf{S}}$.

Solution

The proof of this principle is similar to that of Principle (II) for the three-dimensional case, and can be safely omitted. □

Example 5.1.4

Use the Rayleigh–Ritz method to find the stress field $\widetilde{\mathbf{S}}$ corresponding to an approximate solution of the generalized plane stress traction problem of homogeneous isotropic isothermal elastostatics for a rectangular region subject to a boundary parabolic tension shown in Figure 5.2.

Solution

The region of plate is described by the inequalities

$$|x_1| < a_1, \qquad |x_2| < a_2 \tag{a}$$

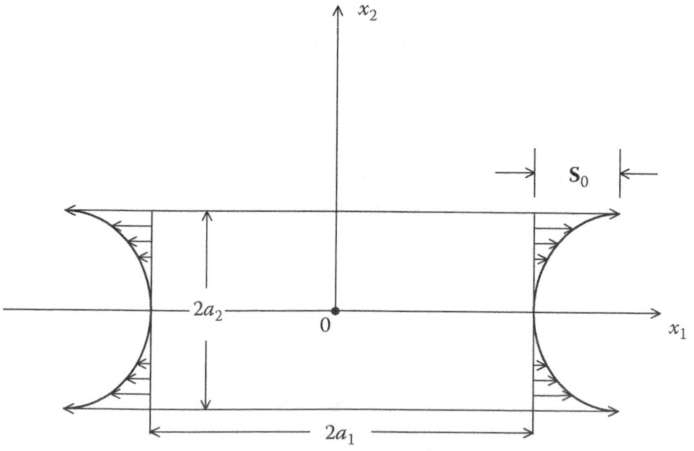

FIGURE 5.2 Rectangular plate under parabolic tension.

and the boundary conditions take the form

$$\overline{S}_{22}(x_1, \pm a_2) = \overline{S}_{12}(x_1, \pm a_2) = 0 \quad \text{for } |x_1| < a_1$$
$$\overline{S}_{11}(\pm a_1, x_2) = S_0 \left(\frac{x_2}{a_2}\right)^2, \quad \overline{S}_{12}(\pm a_1, x_2) = 0 \quad \text{for } |x_2| < a_2 \tag{b}$$

where S_0 is a positive constant of the stress dimension.

To find the stress field $\widetilde{\mathbf{S}}$ corresponding to an approximate solution of the problem we use the two-dimensional principle of minimum complementary energy from Example 5.1.3. The set $\widetilde{P}$ is defined as a set of tensors $\widetilde{\mathbf{S}}$ of the form

$$\widetilde{\mathbf{S}} = \mathbf{S}^{(0)} + \widehat{a}\mathbf{S}^{(1)} \tag{c}$$

where $\widehat{a}$ is a dimensionless parameter, and the tensor fields $\mathbf{S}^{(0)}$ and $\mathbf{S}^{(1)}$ satisfy the conditions, respectively,

$$\operatorname{div} \mathbf{S}^{(0)} = \mathbf{0} \quad \text{for } |x_1| < a_1, |x_2| < a_2 \tag{d}$$

$$\left.\begin{aligned} &S_{22}^{(0)} = S_{12}^{(0)} = 0 \quad \text{for } |x_1| < a_1, x_2 = \pm a_2 \\ &S_{11}^{(0)} = S_0 \left(\frac{x_2}{a_2}\right)^2, \quad S_{12}^{(0)} = 0 \quad \text{for } x_1 = \pm a_1, \ |x_2| < a_2 \end{aligned}\right\} \tag{e}$$

and

$$\operatorname{div} \mathbf{S}^{(1)} = \mathbf{0} \quad \text{for } |x_1| < a_1, |x_2| < a_2 \tag{f}$$

$$\left.\begin{aligned} S_{22}^{(1)} &= S_{12}^{(1)} = 0 \quad \text{for } |x_1| < a_1, x_2 = \pm a_2 \\ S_{11}^{(1)} &= S_{12}^{(1)} = 0 \quad \text{for } x_1 = \pm a_1, |x_2| < a_2 \end{aligned}\right\} \tag{g}$$

With such a choice of $\mathbf{S}^{(0)}$ and $\mathbf{S}^{(1)}$, the tensor $\widetilde{\mathbf{S}}$ is a statically admissible stress field, and $\widetilde{\mathbf{S}}$ represents an approximate stress tensor corresponding to a solution of the traction problem if the function

$$U_K\{\widetilde{\mathbf{S}}\} = \frac{1}{4\mu} \int_{-a_1}^{a_1} \int_{-a_2}^{a_2} \widetilde{\mathbf{S}} \cdot \widetilde{\mathbf{S}} \, dx_1 dx_2 \equiv \psi(\widehat{a}) \tag{h}$$

attains a minimum at a particular value of the parameter $\widehat{a}$.

Substituting $\widetilde{\mathbf{S}}$ from Equation (c) into Equation (h), and differentiating the result with respect to $\widehat{a}$ we find that the only solution of the equation

$$\psi'(\widehat{a}) = 0 \tag{i}$$

takes the form

$$\widehat{a} = -\frac{\int_{-a_1}^{a_1} \int_{-a_2}^{a_2} \mathbf{S}^{(0)} \cdot \mathbf{S}^{(1)} dx_1 dx_2}{\int_{-a_1}^{a_1} \int_{-a_2}^{a_2} \mathbf{S}^{(1)} \cdot \mathbf{S}^{(1)} dx_1 dx_2} \tag{j}$$

To obtain $\mathbf{S}^{(0)}$ and $\mathbf{S}^{(1)}$ we let

$$S^{(0)}_{\alpha\beta} = -F^{(0)}_{,\alpha\beta} + \delta_{\alpha\beta}\nabla^2 F^{(0)} \tag{k}$$

$$S^{(1)}_{\alpha\beta} = -F^{(1)}_{,\alpha\beta} + \delta_{\alpha\beta}\nabla^2 F^{(1)} \tag{l}$$

where $F^{(0)}$ and $F^{(1)}$ are functions of the Airy type (see Example 3.2.9). Then Equations (d) and (f) are satisfied identically, and we need to find $F^{(0)}$ and $F^{(1)}$ in such a way that the boundary conditions (e) and (g) are satisfied. To this end we let

$$F^{(0)} = \frac{S_0}{12a_2^2}x_2^4 \tag{m}$$

and

$$F^{(1)} = \frac{S_0}{a_1^4 a_2^2}\left(x_1^2 - a_1^2\right)^2\left(x_2^2 - a_2^2\right)^2 \tag{n}$$

Substitution of (m) and (n) into (k) and (l), respectively, leads to

$$\begin{aligned} S^{(0)}_{11} &= F^{(0)}_{,22} = S_0\left(\frac{x_2}{a_2}\right)^2 \\ S^{(0)}_{22} &= F^{(0)}_{,11} = 0, \quad S^{(0)}_{12} = -F^{(0)}_{,12} = 0 \end{aligned} \tag{o}$$

and

$$\begin{aligned} S^{(1)}_{11} &= F^{(1)}_{,22} = 4S_0\left(\frac{x_1^2}{a_1^2} - 1\right)^2\left(3\frac{x_2^2}{a_2^2} - 1\right) \\ S^{(1)}_{22} &= F^{(1)}_{,11} = 4S_0\left(\frac{a_2}{a_1}\right)^2\left(3\frac{x_1^2}{a_1^2} - 1\right)\left(\frac{x_2^2}{a_2^2} - 1\right)^2 \\ S^{(1)}_{12} &= -F^{(1)}_{,12} = -16S_0\left(\frac{a_2}{a_1}\right)\left(\frac{x_1}{a_1}\right)\left(\frac{x_2}{a_2}\right)\left(\frac{x_1^2}{a_1^2} - 1\right)\left(\frac{x_2^2}{a_2^2} - 1\right) \end{aligned} \tag{p}$$

and it follows from Equations (o) and (p) that $\mathbf{S}^{(0)}$ and $\mathbf{S}^{(1)}$ satisfy the boundary conditions (e) and (g), respectively. Finally, substituting Equations (o) and (p) into Equation (j), and performing the integrations, we arrive at

$$\widehat{a} = -\left[\frac{64}{7} + \left(\frac{a_2}{a_1}\right)^2\frac{256}{49} + \left(\frac{a_2}{a_1}\right)^4\frac{64}{7}\right]^{-1} \tag{q}$$

For a square plate $a_1 = a_2 = a$, and we receive $\widehat{a} = -0.04253$. In this case the stress components $\widetilde{S}_{11}$, $\widetilde{S}_{22}$, and $\widetilde{S}_{12}$, calculated from Equations (c), (o), and (p), take the forms

$$\begin{aligned} \widetilde{S}_{11} &= S_0\left[\left(\frac{x_2}{a}\right)^2 - 0.1702\left(\frac{x_1^2}{a^2} - 1\right)^2\left(3\frac{x_2^2}{a^2} - 1\right)\right] \\ \widetilde{S}_{22} &= -0.1702\, S_0\left(3\frac{x_1^2}{a^2} - 1\right)\left(\frac{x_2^2}{a^2} - 1\right)^2 \\ \widetilde{S}_{12} &= 0.6805\, S_0\left(\frac{x_1}{a}\right)\left(\frac{x_2}{a}\right)\left(\frac{x_1^2}{a^2} - 1\right)\left(\frac{x_2^2}{a^2} - 1\right) \end{aligned} \tag{r}$$

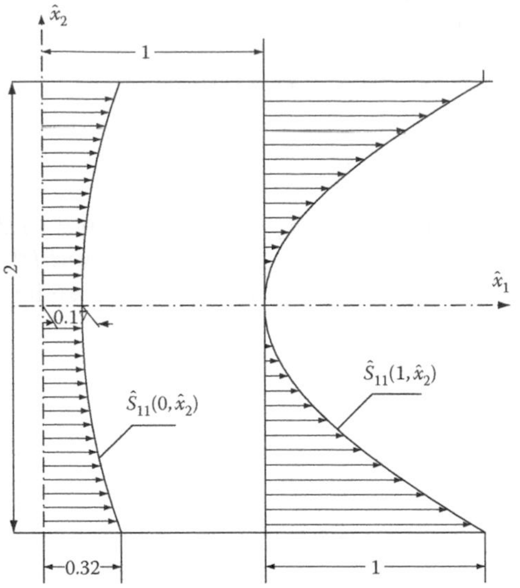

FIGURE 5.3 Plot of $\widehat{S}_{11}(0,\widehat{x}_2)$ and $\widehat{S}_{11}(1,\widehat{x}_2)$ for $|\widehat{x}_2| \leq 1$. Because of symmetry, only the right half is shown.

If we introduce the dimensionless variables

$$\widehat{x}_1 = \frac{x_1}{a}, \qquad \widehat{x}_2 = \frac{x_2}{a}, \qquad \widehat{S}_{\alpha\beta} = \frac{\widetilde{S}_{\alpha\beta}}{S_0}$$

then Equations (r) reduce to the form

$$\begin{aligned} \widehat{S}_{11} &= \widehat{x}_2^2 - 0.1702\left(\widehat{x}_1^2 - 1\right)^2\left(3\widehat{x}_2^2 - 1\right) \\ \widehat{S}_{22} &= -0.1702\left(3\widehat{x}_1^2 - 1\right)\left(\widehat{x}_2^2 - 1\right)^2 \\ \widehat{S}_{12} &= 0.6805\,\widehat{x}_1\widehat{x}_2\left(\widehat{x}_1^2 - 1\right)\left(\widehat{x}_2^2 - 1\right) \end{aligned} \tag{s}$$

In particular, we obtain

$$\widehat{S}_{11}(0,0) = \widehat{S}_{22}(0,0) = 0.1702, \quad \widehat{S}_{12}(0,0) = 0$$

and

$$\widehat{S}_{11}(0,\widehat{x}_2) = 0.1702 + 0.4894\widehat{x}_2^2$$

A graph of the functions $\widehat{S}_{11}(0,\widehat{x}_2)$ and $\widehat{S}_{11}(1,\widehat{x}_2)$ for $|\widehat{x}_2| \leq 1$ is shown in Figure 5.3. □

5.1.4 The Principle of Minimum Complementary Energy for Nonisothermal Elasticity

To extend the result of the previous section to the case of a body subject to both an external mechanical load and prescribed temperature change, we recall the fundamental system of field equations for such a body. Assuming that the body is anisotropic with regard to both mechanical and thermal properties, the system consists of:

(1) The strain–displacement relation

$$\mathbf{E} = \widehat{\nabla}\mathbf{u} \tag{5.1.69}$$

(2) The equation of equilibrium

$$\operatorname{div}\mathbf{S} + \mathbf{b} = \mathbf{0} \tag{5.1.70}$$

(3) the constitutive relation, see Equation 3.3.110,

$$\mathbf{S} = \mathsf{C}[\mathbf{E}] + T\mathbf{M} \tag{5.1.71}$$

Here, $\mathbf{u}, \mathbf{E}, \mathbf{S}, \mathsf{C}, \mathbf{b}$ have the same meaning as in isothermal elasticity, $\mathbf{M}$ is the stress–temperature tensor, and T is the temperature change. If we introduce the notations

$$\mathbf{b}' = \mathbf{b} + \operatorname{div}(T\mathbf{M}) \tag{5.1.72}$$

$$\mathbf{S}' = \mathbf{S} - T\mathbf{M} \tag{5.1.73}$$

$$\mathbf{s}' = \mathbf{s} - T\mathbf{M}\mathbf{n} \tag{5.1.74}$$

then Equations 5.1.69 through 5.1.71 may be written as

$$\mathbf{E} = \widehat{\nabla}\mathbf{u} \tag{5.1.75}$$

$$\operatorname{div}\mathbf{S}' + \mathbf{b}' = \mathbf{0} \tag{5.1.76}$$

$$\mathbf{S}' = \mathsf{C}[\mathbf{E}] \tag{5.1.77}$$

Hence, the system of fundamental field equations of nonisothermal elastostatics from which a triple $[\mathbf{u}, \mathbf{E}, \mathbf{S}]$ subject to suitable boundary conditions is to be found has been reduced to the system 5.1.75 through 5.1.77, which is of the same form as that of an isothermal elastostatics. As a consequence, a number of results of isothermal elastostatics may be used to obtain solutions to problems of nonisothermal elastostatics. In particular, let $s' = [\mathbf{u}, \mathbf{E}, \mathbf{S}']$ be a solution to Equations 5.1.75 through 5.1.77, subject to the traction boundary condition

$$\mathbf{s}' = -T\mathbf{M}\mathbf{n} \quad \text{on } \partial B \tag{5.1.78}$$

Then a solution $s = [\mathbf{u}, \mathbf{E}, \mathbf{S}]$ to the nonisothermal field equations 5.1.69 through 5.1.71 subject to the condition

$$\mathbf{s} = \mathbf{0} \quad \text{on } \partial B \tag{5.1.79}$$

takes the form

$$s = [\mathbf{u}, \mathbf{E}, \mathbf{S}' + T\mathbf{M}] \tag{5.1.80}$$

A body force analogy may be formulated as follows: Let $[\mathbf{u}, \mathbf{E}, \mathbf{S}]$ be a thermoelastic state corresponding to the body force $\mathbf{b}$ and temperature change T. Let $\mathbf{b}'$ and $\mathbf{S}'$ be given by Equations 5.1.72 and 5.1.73, respectively. Then $[\mathbf{u}, \mathbf{E}, \mathbf{S}']$ satisfies Equations 5.1.75 through 5.1.77. Conversely, let $[\mathbf{u}, \mathbf{E}, \mathbf{S}']$ be a solution to Equations 5.1.75 through 5.1.77. If $\mathbf{S}$ and $\mathbf{b}$ are defined by Equations 5.1.73 and 5.1.72, respectively, then $[\mathbf{u}, \mathbf{E}, \mathbf{S}]$ is a thermoelastic state corresponding to the body force $\mathbf{b}$ and the temperature change T.

The first part of this analogy has already been proved. To prove the converse statement we need to reverse the steps, by passing from Equations 5.1.75 through 5.1.77 to Equations 5.1.69 through 5.1.71 using Equations 5.1.72 through 5.1.74.

The body force analogy asserts that $[\mathbf{u}, \mathbf{E}, \mathbf{S}]$ is a thermoelastic state corresponding to the body force $\mathbf{b}$ and the temperature change T if and only if $[\mathbf{u}, \mathbf{E}, \mathbf{S}']$ is an elastic state corresponding to the body force $\mathbf{b}'$.

Using this analogy we may formulate an extension of the theorem on minimum of complementary energy given in Section 5.1.3 to include the temperature change T.

The Extended Principle of Minimum Complementary Energy. Let P denote a set of all statically admissible stress fields, and let $G_T = G_T\{\cdot\}$ be a functional on P defined by

$$G_T\{\mathbf{S}\} = U_K\{\mathbf{S}'\} - \int_{\partial B_1} \mathbf{s}' \cdot \widehat{\mathbf{u}}\, da \tag{5.1.81}$$

for every $\mathbf{S} \in P$.

If $\mathbf{S}$ is a stress field corresponding to a solution to the mixed problem of thermoelasticity, then

$$G_T\{\mathbf{S}\} \leq G_T\{\widetilde{\mathbf{S}}\} \quad \text{for every } \widetilde{\mathbf{S}} \in P \tag{5.1.82}$$

and the equality holds true if $\mathbf{S} = \widetilde{\mathbf{S}}$.

Proof: The proof of the extended principle of minimum complementary energy is similar to that given in Section 5.1.3.

Notes:

(1) We observe that the RHS of Equation 5.1.81 depends on $\mathbf{S}$. Indeed, by virtue of Equations 5.1.72 through 5.1.74

$$U_K\{\mathbf{S}'\} - \int_{\partial B_1} \mathbf{s}' \cdot \widehat{\mathbf{u}}\, da = U_K\{\mathbf{S} - T\mathbf{M}\} - \int_{\partial B_1} (\mathbf{s} - T\mathbf{M}\mathbf{n}) \cdot \widehat{\mathbf{u}}\, da \tag{5.1.83}$$

(2) The RHS of Equation 5.1.83 may be simplified significantly if we note that the temperature change T and the stress–temperature tensor $\mathbf{M}$ are prescribed, and that the tensor K is symmetric. Moreover, we recall that, see Equation 3.3.113,

$$\mathbf{A} = -\mathsf{K}[\mathbf{M}] \tag{5.1.84}$$

where $\mathbf{A}$ is the thermal expansion tensor. To obtain the simplified formula, we first note that by the symmetry of K, we have

$$U_K\{\mathbf{S} - T\mathbf{M}\} = \frac{1}{2}\int_B (\mathbf{S} - T\mathbf{M}) \cdot \mathsf{K}[\mathbf{S} - T\mathbf{M}]\, dv$$

$$= U_K\{\mathbf{S}\} + U_K\{T\mathbf{M}\} + \int_B \mathbf{S} \cdot \mathsf{K}[-T\mathbf{M}]\, dv \tag{5.1.85}$$

Hence, by Equations 5.1.81, 5.1.84, and 5.1.85 we obtain

$$G_T\{\mathbf{S}\} = U_K\{\mathbf{S}\} + \int_B T\mathbf{S} \cdot \mathbf{A}\, dv - \int_{\partial B_1} \mathbf{s} \cdot \widehat{\mathbf{u}}\, da + G_T^{(0)} \tag{5.1.86}$$

where

$$G_T^{(0)} = U_K\{T\mathbf{M}\} + \int_{\partial B_1} T(\mathbf{Mn}) \cdot \widehat{\mathbf{u}}\, da \tag{5.1.87}$$

Since $G_T^{(0)}$ is entirely expressed in terms of the data of the mixed problem, then in the functional G_T that occurs in the Extended principle the term $G_T^{(0)}$ may be ignored, and G_T be replaced by G_T^* where

$$G_T^*\{\mathbf{S}\} = U_K\{\mathbf{S}\} + \int_B T\mathbf{S} \cdot \mathbf{A}\, dv - \int_{\partial B_1} \mathbf{s} \cdot \widehat{\mathbf{u}}\, da \tag{5.1.88}$$

(3) In the case of a traction free boundary, often encountered in problems of thermoelasticity, the surface integral in Equation 5.1.88 vanishes, and the extended principle may be used effectively to obtain an approximate solution to the traction problem by means of the Rayleigh–Ritz method.

Example 5.1.5

Formulate a two-dimensional extended principle of minimum complementary energy for a thin elastic sheet subject to a temperature field $\overline{T} = \overline{T}(x_1, x_2)$ (see Section 8.1.2). Assume that the sheet occupies a region C_0 of (x_1, x_2) plane, its boundary ∂C_0 is traction-free, and the temperature satisfies Poisson's equation

$$\nabla^2 \overline{T} = -\frac{Q}{\kappa} \quad \text{on } C_0 \tag{a}$$

subject to the boundary condition

$$\overline{T} = 0 \quad \text{on } \partial C_0 \tag{b}$$

Note that $Q(x_1, x_2)/\kappa = W(x_1, x_2)/k$, where $Q(x_1, x_2)$ is a prescribed heat supply field, W stands for the internal heat generated per unit volume per unit time, κ denotes the thermal diffusivity, and k means the thermal conductivity.

Solution

A two-dimensional counterpart of the functional 5.1.88 for a homogeneous isotropic thin elastic sheet takes the form

$$G_T^*\{\overline{\mathbf{S}}\} = \frac{1}{4\mu} \int\limits_{C_0} \left[\overline{S}_{\alpha\beta}\overline{S}_{\alpha\beta} - \frac{\nu}{1+\nu} \left(\overline{S}_{\gamma\gamma}\right)^2 \right] da + \alpha \int\limits_{C_0} \overline{T}\, \overline{S}_{\gamma\gamma} da \tag{c}$$

where
 $\overline{\mathbf{S}} = \overline{\mathbf{S}}(x_1, x_2)$ is the stress tensor field on C_0
 μ, ν, and α stand for the shear modulus, Poisson's ratio, and coefficient of thermal expansion, respectively

The functional $G_T^*\{\cdot\}$ is defined on a set $\overline{P}$ of the stress fields that satisfy the conditions

$$\operatorname{div} \overline{\mathbf{S}} = \mathbf{0} \quad \text{on } C_0, \qquad \overline{\mathbf{S}}\mathbf{n} = 0 \quad \text{on } \partial C_0 \tag{d}$$

and the two-dimensional extended principle reads:
Let $\overline{P}$ denote a set of all stress fields that satisfy Equations (d), and let $G_T^* = G_T^*\{\cdot\}$ be a functional on $\overline{P}$ defined by Equation (c) for every $\overline{\mathbf{S}} \in \overline{P}$. If $\overline{\mathbf{S}}$ is the stress field corresponding to a solution of the traction problem of thermoelasticity with the temperature $\overline{T} = \overline{T}(x_1, x_2)$, then

$$G_T^*\{\overline{\mathbf{S}}\} \le G_T^*\{\widetilde{\mathbf{S}}\} \quad \text{for every } \widetilde{\mathbf{S}} \in \overline{P} \tag{e}$$

and the equality holds true if $\overline{\mathbf{S}} = \widetilde{\mathbf{S}}$.
 Proof of this principle is similar to that of the three-dimensional one, and is omitted here. This completes the solution. □

Example 5.1.6

Let $\overline{F} = \overline{F}(x_1, x_2)$ be an Airy stress function that generates the stress $\overline{\mathbf{S}}$ corresponding to a solution of the traction problem of thermoelasticity discussed in Example 5.1.5, that is,

$$\overline{S}_{\alpha\beta} = \delta_{\alpha\beta}\nabla^2\overline{F} - \overline{F}_{,\alpha\beta} \quad \text{on } C_0 \tag{a}$$

and

$$\overline{S}_{\alpha\beta} n_\beta = 0 \quad \text{on } \partial C_0 \tag{b}$$

Let P_F be a set of all functions $\widetilde{F}$ on C_0 that satisfy the boundary condition

$$\left(\delta_{\alpha\beta}\nabla^2\widetilde{F} - \widetilde{F}_{,\alpha\beta}\right) n_\beta = 0 \quad \text{on } \partial C_0 \tag{c}$$

or, alternatively,

$$\widetilde{F} = 0, \quad \frac{\partial \widetilde{F}}{\partial n} = 0 \quad \text{on } \partial C_0 \tag{d}$$

Define a functional $\widehat{G} = \widehat{G}\{\cdot\}$ on P_F by

$$\widehat{G}\{F\} = \frac{1}{2E}\int_{C_0} (\nabla^2 F)^2 da - \frac{\alpha}{\kappa}\int_{C_0} QF da \tag{e}$$

for every $F \in P_F$. Show that if $\overline{F}$ generates $\overline{\mathbf{S}}$ that corresponds to a solution of the traction problem then

$$\widehat{G}\{\overline{F}\} \leq \widehat{G}\{\widetilde{F}\} \quad \text{for every } \widetilde{F} \in P_F \tag{f}$$

and the equality holds true if $\overline{F} = \widetilde{F}$.

Solution

First, we note that if $\widetilde{F} \in P_F$ then $\widetilde{\mathbf{S}}$ generated by $\widetilde{F}$ belongs to the set $\overline{P}$ of Example 5.1.5. Therefore, substituting $\widetilde{\mathbf{S}}$ given by

$$\widetilde{S}_{\alpha\beta} = \delta_{\alpha\beta}\nabla^2\widetilde{F} - \widetilde{F}_{,\alpha\beta} \tag{g}$$

into Equation (c) of Example 5.1.5, we receive

$$G_T^*\{\widetilde{\mathbf{S}}\} = \frac{1}{4\mu(1+\nu)}\int_{C_0} \left[\left(\nabla^2\widetilde{F}\right)^2 - 2(1+\nu)\left(\widetilde{F}_{,11}\widetilde{F}_{,22} - \widetilde{F}^2_{,12}\right)\right] da$$
$$+ \alpha\int_{C_0} \overline{T}\,\nabla^2\widetilde{F} da \tag{h}$$

Next, we show that

$$\int_{C_0} \left(\widetilde{F}_{,11}\widetilde{F}_{,22} - \widetilde{F}^2_{,12}\right) da = 0 \tag{i}$$

and

$$\int_{C_0} \overline{T}\,\nabla^2\widetilde{F} da = -\frac{1}{\kappa}\int_{C_0} Q\widetilde{F} da \tag{j}$$

To show that (i) holds true, we note that

$$\widetilde{F}_{,11}\widetilde{F}_{,22} - \widetilde{F}^2_{,12} = \widetilde{F}_{,11}\widetilde{S}_{11} + \widetilde{F}_{,12}\widetilde{S}_{12} = (\widetilde{F}_{,1}\widetilde{S}_{1\alpha})_{,\alpha} - \widetilde{F}_{,1}\widetilde{S}_{1\alpha,\alpha} \tag{k}$$

Since $\widetilde{S}_{\alpha\beta}$ satisfies the conditions

$$\widetilde{S}_{\alpha\beta,\beta} = 0 \quad \text{on } C_0, \quad \widetilde{S}_{\alpha\beta}n_\beta = 0 \quad \text{on } \partial C_0 \tag{l}$$

therefore, integrating Equation (k) over C_0, using the divergence theorem, and the conditions (l) we obtain the relation (i).

Note that the identity

$$\widetilde{F}_{,11}\widetilde{F}_{,22} - \widetilde{F}^2_{,12} = \widetilde{F}_{,22}\widetilde{S}_{22} + \widetilde{F}_{,12}\widetilde{S}_{12} = (\widetilde{F}_{,2}\widetilde{S}_{2\alpha})_{,\alpha} - \widetilde{F}_{,2}\widetilde{S}_{2\alpha,\alpha} \tag{m}$$

together with the conditions (l) leads also to Equation (i).

To prove that (j) holds true we note that

$$\overline{T}\,\widetilde{F}_{,\gamma\gamma} = (\overline{T}\,\widetilde{F}_{,\gamma})_{,\gamma} - \overline{T}_{,\gamma}\widetilde{F}_{,\gamma} = (\overline{T}\,\widetilde{F}_{,\gamma})_{,\gamma} - (\overline{T}_{,\gamma}\widetilde{F})_{,\gamma} + \overline{T}_{,\gamma\gamma}\widetilde{F} \tag{n}$$

Hence, integrating Equation (n) over C_0, using the divergence theorem, and taking into account Equation (a) from Example 5.1.5 as well as the conditions (d), we receive Equation (j). Finally, if we note that

$$4\mu(1+\nu) = 2E \tag{o}$$

then, by virtue of Equations (i) and (j), the functional (h) on $\overline{P}$ is equivalent to the functional (e) on P_F; and if $\overline{F}$ generates the stress $\overline{\mathbf{S}}$ corresponding to a solution of the traction problem of thermoelasticity then the inequality (f) holds true. This completes the solution. □

Example 5.1.7

Using the Rayleigh–Ritz method, leading to a minimum of the functional $\widehat{G}\{\cdot\}$ of Example 5.1.6, show that the stress $\overline{\mathbf{S}}$ corresponding to an approximate solution of the traction problem of thermoelasticity for a thin rectangular plate, $|x_1| < a_1$, $|x_2| < a_2$, subject to a concentrated heat source of intensity Q_0 at $x_1 = x_2 = 0$, and kept at a zero temperature on its boundary, takes the form (see Figure 5.4)

$$\begin{aligned}
\overline{S}_{11} &= \frac{F_0}{4}\left(\frac{\pi}{a_2}\right)^2 \left[1 - \cos\frac{\pi}{a_1}(x_1 - a_1)\right] \cos\frac{\pi}{a_2}(x_2 - a_2)\\
\overline{S}_{22} &= \frac{F_0}{4}\left(\frac{\pi}{a_1}\right)^2 \cos\frac{\pi}{a_1}(x_1 - a_1)\left[1 - \cos\frac{\pi}{a_2}(x_2 - a_2)\right]\\
\overline{S}_{12} &= -\frac{F_0}{4}\left(\frac{\pi}{a_1}\right)\left(\frac{\pi}{a_2}\right) \sin\frac{\pi}{a_1}(x_1 - a_1)\sin\frac{\pi}{a_2}(x_2 - a_2)
\end{aligned} \tag{a}$$

where

$$\frac{F_0}{4} = \frac{4Q_0\alpha E}{\pi^4\kappa}\cdot\frac{a_1^3a_2^3}{3a_1^4 + 2a_1^2a_2^2 + 3a_2^4} \tag{b}$$

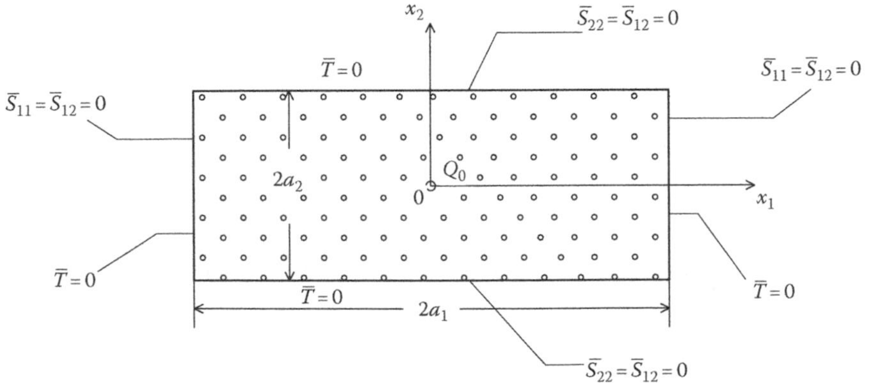

FIGURE 5.4 A rectangular plate subject to a concentrated heat source Q_0 at its center, and kept at a zero boundary temperature.

In particular, for a thin square plate when $a_1 = a_2 = a$ we obtain

$$\begin{aligned}\bar{S}_{11} &= S_0\left[1-\cos\frac{\pi}{a}(x_1-a)\right]\cos\frac{\pi}{a}(x_2-a)\\ \bar{S}_{22} &= S_0\cos\frac{\pi}{a}(x_1-a)\left[1-\cos\frac{\pi}{a}(x_2-a)\right]\\ \bar{S}_{12} &= -S_0\sin\frac{\pi}{a}(x_1-a)\sin\frac{\pi}{a}(x_2-a)\end{aligned} \tag{c}$$

where $S_0 = Q_0\alpha E/2\pi^2\kappa$.

Solution

The functional $\widehat{G}\{\cdot\}$ of Example 5.1.6 takes now the form

$$\widehat{G}\{F\} = \frac{1}{2E}\int_{-a_1}^{a_1}\int_{-a_2}^{a_2}(\nabla^2 F)^2 dx_1 dx_2 - \frac{\alpha}{\kappa}\int_{-a_1}^{a_1}\int_{-a_2}^{a_2} QF dx_1 dx_2 \tag{d}$$

where $F \in P_F$, and $Q = Q_0\delta(x_1)\delta(x_2)$.
An approximate solution to the problem is obtained if we let

$$F(x_1, x_2) = F_0\varphi(x_1)\psi(x_2) \tag{e}$$

where

$$\varphi(x_1) = \left[\sin\frac{\pi}{2a_1}(x_1-a_1)\right]^2 \tag{f}$$

$$\psi(x_2) = \left[\sin\frac{\pi}{2a_2}(x_2-a_2)\right]^2 \tag{g}$$

and F_0 is an unknown parameter to be determined by the Rayleigh–Ritz method. To prove this, we first note that $F \in P_F$ since

$$\begin{aligned}&F(\pm a_1, x_2) = 0, \quad \frac{\partial F}{\partial x_1}(\pm a_1, x_2) = 0, \quad |x_2| < a_2\\ &F(x_1, \pm a_2) = 0, \quad \frac{\partial F}{\partial x_2}(x_1, \pm a_2) = 0, \quad |x_1| < a_1\end{aligned} \tag{h}$$

Next, by computing $\nabla^2 F$, we receive

$$\begin{aligned}\nabla^2 F = \frac{F_0}{4}\Bigg\{&\left(\frac{\pi}{a_1}\right)^2\cos\frac{\pi}{a_1}(x_1-a_1)\left[1-\cos\frac{\pi}{a_2}(x_2-a_2)\right]\\ &+\left(\frac{\pi}{a_2}\right)^2\cos\frac{\pi}{a_2}(x_2-a_2)\left[1-\cos\frac{\pi}{a_1}(x_1-a_1)\right]\Bigg\}\end{aligned} \tag{i}$$

Finally, substituting Equations (e) and (i) into RHS of Equation (d), and taking into account the filtrating property of the Dirac delta function, we obtain

$$\widehat{G}\{F\} = \frac{1}{2E}\frac{F_0^2}{16}\frac{\pi^4}{a_1^3 a_2^3}(3a_1^4 + 2a_1^2 a_2^2 + 3a_2^4) - \frac{\alpha}{\kappa}Q_0F_0 \equiv G(F_0) \tag{j}$$

It follows from Equation (j) that the function $G = G(F_0)$ attains a minimum at the only root of the equation

$$G'(F_0) = 0 \tag{k}$$

and this root is given by Equation (b). Therefore, substituting F from Equation (e) into the relations

$$\overline{S}_{11} = F_{,22}, \quad \overline{S}_{22} = F_{,11}, \quad \overline{S}_{12} = -F_{,12} \tag{l}$$

we arrive at Equations (a) and (b). This completes the solution. □

5.2 VARIATIONAL PRINCIPLES

In Sections 5.1.1 through 5.1.4, we formulated (I) *the Principle of minimum potential energy* and (II) *the Principle of minimum complementary energy* and in each of these principles a functional is defined on a set of admissible states subject to suitable constraints.

In (I), the admissible states are to be kinematically admissible, which means that a state satisfies the strain–displacement relation, the stress–strain relation, and the displacement boundary condition.

In (II), the admissible states are such that the associated stress field is a statically admissible stress field **S**, which means that **S** satisfies the equations of equilibrium and the traction boundary condition.

Both (I) and (II) allow us to obtain a solution to the mixed problem by finding a minimum of the corresponding functionals, so a task of finding a solution may be reduced to that of the calculus of variations.

In applications, it is of importance to have as few constraints on the admissible states as possible. In the following, we formulate two variational principles in which the functionals are defined on a set of admissible states having indeed few constraints. These principles include as particular cases the Principles (I) and (II).

To formulate these variational principles we recall the definition of the first variation of a scalar-valued functional $H = H\{s\}$, where s is an element of a subset of the set of all admissible states.

Let $\mathcal{A}$ denote the subset, and let s and $\widetilde{s}$ be two elements of $\mathcal{A}$. Then, since $\mathcal{A}$ is a vector space, we have

$$s + \omega\widetilde{s} \in \mathcal{A} \quad \text{for every scalar } \omega \tag{5.2.1}$$

and the value $H\{s + \omega\widetilde{s}\}$ is well defined.

By *the first variation of* $H\{s\}$, we mean the number

$$\delta_{\widetilde{s}} H\{s\} = \frac{d}{d\omega} H\{s + \omega\widetilde{s}\}\bigg|_{\omega=0} \tag{5.2.2}$$

provided such number exists, and we say that

$$\delta_{\widetilde{s}} H\{s\} \equiv \delta H\{s\} = 0 \tag{5.2.3}$$

if $\delta_{\widetilde{s}} H\{s\}$ exists and equals zero for any $\widetilde{s}$ consistent with the condition 5.2.1.

In the following, we also use

Fundamental Lemma of the Calculus of Variations. Let a scalar-valued function $f = f(\mathbf{x})$ be continuous on $\bar{B}$ and satisfy the integral relation

$$\int_B f(\mathbf{x})\tilde{g}(\mathbf{x})\, dv = 0 \tag{5.2.4}$$

for every smooth scalar-valued function $\tilde{g} = \tilde{g}(\mathbf{x})$ on $\bar{B}$ that vanishes near ∂B. Then

$$f(\mathbf{x}) = 0 \quad \text{on } \bar{B} \tag{5.2.5}$$

The proof of this lemma may be found in any textbook on the calculus of variations. □

Also, note that the lemma may be extended to the case when f and $\tilde{g}$ are vector or tensor-valued functions; then $f\tilde{g}$ represents the inner product. In addition, the lemma may be extended to the case when f and $\tilde{g}$ are functions of position and time; then the product $f\tilde{g}$ under the integral is a convolution product on the time axis. The extended lemma involving the convolution product is particularly useful in the formulation of variational principles of elastodynamics.

With such preliminaries from the calculus of variations, we first formulate a variational principle of elastostatics in which the admissible states are not required to meet any of the field equations or boundary conditions, that is, the set $\mathcal{A}$ is to be identified with the set of all admissible states.

(H) **Hu–Washizu Principle.** Let $\mathcal{A}$ denote the set of all admissible states of elastostatics, and let $H = H\{\cdot\}$ be the functional on $\mathcal{A}$ defined by

$$\begin{aligned} H\{s\} = U_C\{\mathbf{E}\} - \int_B \mathbf{S}\cdot\mathbf{E}\, dv - \int_B (\operatorname{div}\mathbf{S} + \mathbf{b})\cdot\mathbf{u}\, dv \\ + \int_{\partial B_1} \mathbf{s}\cdot\widehat{\mathbf{u}}\, da + \int_{\partial B_2} (\mathbf{s} - \widehat{\mathbf{s}})\cdot\mathbf{u}\, da \end{aligned} \tag{5.2.6}$$

for every $s = [\mathbf{u}, \mathbf{E}, \mathbf{S}] \in \mathcal{A}$. Then

$$\delta H\{s\} = 0 \tag{5.2.7}$$

if and only if s is a solution to the mixed problem.

Proof: The proof consists of two parts:

(a) We prove that if s is a solution to the mixed problem, then Equation 5.2.7 is satisfied.
(b) We assume that Equation 5.2.7 is satisfied and show that s is a solution to the mixed problem.

Part (a) of the proof. We take $s = [\mathbf{u}, \mathbf{E}, \mathbf{S}]$ and $\tilde{s} = [\tilde{\mathbf{u}}, \tilde{\mathbf{E}}, \tilde{\mathbf{S}}]$ belonging to $\mathcal{A}$. Then $s + \omega\tilde{s} \in \mathcal{A}$ for every scalar ω. Also, note that by the definition of the strain energy $U_C\{\mathbf{E}\}$, see Equation 5.1.1, and the symmetry of $\mathbf{C}$, we obtain

$$U_C\{\mathbf{E} + \omega\widetilde{\mathbf{E}}\} = \frac{1}{2}\int_B (\mathbf{E} + \omega\widetilde{\mathbf{E}}) \cdot \mathbf{C}[\mathbf{E} + \omega\widetilde{\mathbf{E}}]\, dv \tag{5.2.8}$$

and

$$U_C\{\mathbf{E} + \omega\widetilde{\mathbf{E}}\} = U_C\{\mathbf{E}\} + \omega^2 U_C\{\widetilde{\mathbf{E}}\} + \omega\int_B \widetilde{\mathbf{E}} \cdot \mathbf{C}[\mathbf{E}]\, dv \tag{5.2.9}$$

Hence, replacing s by $s + \omega\widetilde{s}$ in Equation 5.2.6, we arrive at the relation

$$\begin{aligned}
H\{s + \omega\widetilde{s}\} &= U_C\{\mathbf{E}\} + \omega^2 U_C\{\widetilde{\mathbf{E}}\} + \omega\int_B \widetilde{\mathbf{E}} \cdot \mathbf{C}[\mathbf{E}]\, dv \\
&\quad - \int_B [\mathbf{S} \cdot \mathbf{E} + \omega^2\widetilde{\mathbf{S}} \cdot \widetilde{\mathbf{E}} + \omega(\mathbf{S} \cdot \widetilde{\mathbf{E}} + \widetilde{\mathbf{S}} \cdot \mathbf{E})]\, dv \\
&\quad - \int_B \{(\operatorname{div}\mathbf{S} + \mathbf{b}) \cdot \mathbf{u} + \omega^2(\operatorname{div}\widetilde{\mathbf{S}}) \cdot \widetilde{\mathbf{u}} \\
&\quad + \omega[(\operatorname{div}\mathbf{S} + \mathbf{b}) \cdot \widetilde{\mathbf{u}} + (\operatorname{div}\widetilde{\mathbf{S}}) \cdot \mathbf{u}]\}dv \\
&\quad + \int_{\partial B_1} (\mathbf{s} + \omega\widetilde{\mathbf{s}}) \cdot \widehat{\mathbf{u}}\, da \\
&\quad + \int_{\partial B_2} \{(\mathbf{s} - \widehat{\mathbf{s}}) \cdot \mathbf{u} + \omega^2\widetilde{\mathbf{s}} + \omega[\widetilde{\mathbf{s}} \cdot \mathbf{u} + (\mathbf{s} - \widehat{\mathbf{s}}) \cdot \widetilde{\mathbf{u}}]\}\, da
\end{aligned} \tag{5.2.10}$$

Differentiating Equation 5.2.10 with respect to ω, taking the result at $\omega = 0$, and using the formula, see Equation 5.2.2,

$$\delta_{\widetilde{s}} H\{s\} = \frac{d}{d\omega} H(s + \omega\widetilde{s})\Big|_{\omega=0} \tag{5.2.11}$$

we obtain

$$\begin{aligned}
\delta_{\widetilde{s}} H\{s\} &= \int_B \{(\mathbf{C}[\mathbf{E}] - \mathbf{S}) \cdot \widetilde{\mathbf{E}} - (\operatorname{div}\mathbf{S} + \mathbf{b}) \cdot \widetilde{\mathbf{u}} - \widetilde{\mathbf{S}} \cdot \mathbf{E} - \mathbf{u} \cdot \operatorname{div}\widetilde{\mathbf{S}}\}\, dv \\
&\quad + \int_{\partial B_1} \widetilde{\mathbf{s}} \cdot \widehat{\mathbf{u}}\, da + \int_{\partial B_2} [\widetilde{\mathbf{s}} \cdot \mathbf{u} + (\mathbf{s} - \widehat{\mathbf{s}}) \cdot \widetilde{\mathbf{u}}]\, da
\end{aligned} \tag{5.2.12}$$

Now, since by the divergence theorem and the symmetry of $\widetilde{\mathbf{S}}$:

$$\int_B \mathbf{u} \cdot (\operatorname{div}\widetilde{\mathbf{S}})\, dv = \int_{\partial B} \mathbf{u} \cdot \widetilde{\mathbf{s}}\, da - \int_B \widetilde{\mathbf{S}} \cdot \widehat{\nabla}\mathbf{u}\, dv \tag{5.2.13}$$

and

$$\int_{\partial B} \mathbf{u} \cdot \widetilde{\mathbf{s}}\, da = \int_{\partial B_1} \mathbf{u} \cdot \widetilde{\mathbf{s}}\, da + \int_{\partial B_2} \mathbf{u} \cdot \widetilde{\mathbf{s}}\, da \tag{5.2.14}$$

therefore, Equation 5.2.12 reduces to

$$\delta_{\tilde{s}} H\{s\} = \int_B (\mathbf{C}[\mathbf{E}] - \mathbf{S}) \cdot \tilde{\mathbf{E}}\, dv - \int_B (\operatorname{div} \mathbf{S} + \mathbf{b}) \cdot \tilde{\mathbf{u}}\, dv$$
$$+ \int_B (\widehat{\nabla} \mathbf{u} - \mathbf{E}) \cdot \tilde{\mathbf{S}}\, dv + \int_{\partial B_1} (\widehat{\mathbf{u}} - \mathbf{u}) \cdot \tilde{\mathbf{s}}\, da + \int_{\partial B_2} (\mathbf{s} - \widehat{\mathbf{s}}) \cdot \tilde{\mathbf{u}}\, da \tag{5.2.15}$$

To give a proof of part (a) of Principle (H) we assume that $s = [\mathbf{u}, \mathbf{E}, \mathbf{S}]$ is a solution to the mixed problem. Then Equation 5.2.15 implies that

$$\delta_{\tilde{s}} H\{s\} = 0 \quad \text{for every } \tilde{s} \in \mathcal{A} \tag{5.2.16}$$

which shows that Equation 5.2.7 is satisfied. This completes the proof of part (a).

Proof of part (b). To prove (b) we assume that Equation 5.2.7 is satisfied, and hence Equation 5.2.16 holds true, and, by using an Extended Lemma of calculus of variations, we show that s is a solution of the mixed problem. To recover the field equations and the boundary conditions that are satisfied by s, we apply different extensions of the Fundamental Lemma of calculus of variations, see Equations 5.2.4 through 5.2.5.

If we choose $\tilde{s} = [\tilde{\mathbf{u}}, \mathbf{0}, \mathbf{0}]$, where $\tilde{\mathbf{u}}$ vanishes near ∂B, then Equations 5.2.15 and 5.2.16 imply that

$$\int_B (\operatorname{div} \mathbf{S} + \mathbf{b}) \cdot \tilde{\mathbf{u}}\, dv = 0 \tag{5.2.17}$$

Since $\tilde{\mathbf{u}}$ is an arbitrary field on $\bar{B}$ that vanishes near ∂B, then by the extension of the fundamental lemma, Equation 5.2.17 implies that

$$\operatorname{div} \mathbf{S} + \mathbf{b} = \mathbf{0} \quad \text{on } B \tag{5.2.18}$$

and this shows that s complies with the equation of equilibrium of elastostatics.

Next, if we select $\tilde{s} = [\tilde{\mathbf{u}}, \mathbf{0}, \mathbf{0}]$ in such a way that $\tilde{\mathbf{u}}$ vanishes near ∂B_1 only, and otherwise is arbitrary on $\bar{B}$, then Equations 5.2.15 and 5.2.16 and Equation 5.2.18 imply that

$$\int_{\partial B_2} (\mathbf{s} - \widehat{\mathbf{s}}) \cdot \tilde{\mathbf{u}}\, da = 0 \tag{5.2.19}$$

and from an extension of the Fundamental Lemma, it follows that

$$\mathbf{s} = \widehat{\mathbf{s}} \quad \text{on } \partial B_2 \tag{5.2.20}$$

Now, let $\tilde{s} = [\mathbf{0}, \tilde{\mathbf{E}}, \mathbf{0}]$ where $\tilde{\mathbf{E}}$ is an arbitrary symmetric tensor field on $\bar{B}$ that vanishes near ∂B. Then, by Equations 5.2.15 and 5.2.16, and Equations 5.2.18 and 5.2.20, we obtain

$$\int_B (\mathbf{C}[\mathbf{E}] - \mathbf{S}) \cdot \tilde{\mathbf{E}}\, dv = 0 \tag{5.2.21}$$

Hence, from an extension of the Fundamental Lemma, it follows that

$$\mathbf{S} = \mathbf{C}[\mathbf{E}] \quad \text{on } B \tag{5.2.22}$$

Similarly, if we select $\widetilde{s}$ in such a way that $\widetilde{s} = [\mathbf{0}, \mathbf{0}, \widetilde{\mathbf{S}}]$ where $\widetilde{\mathbf{S}}$ is an arbitrary symmetric second-order tensor field on $\bar{B}$ that vanishes near ∂B, by virtue of Equations 5.2.15 and 5.2.16, 5.2.18, 5.2.20, and 5.2.22, we arrive at the strain–displacement relation

$$\mathbf{E} = \widehat{\nabla}\mathbf{u} \quad \text{on } B \tag{5.2.23}$$

Finally, if we select $\widetilde{s}$ in the form $\widetilde{s} = [\mathbf{0}, \mathbf{0}, \widetilde{\mathbf{S}}]$, where $\widetilde{\mathbf{S}}$ is an arbitrary symmetric second-order tensor field on $\bar{B}$ such that $\widetilde{\mathbf{s}} = \widetilde{\mathbf{S}}\mathbf{n}$ vanishes near ∂B_2 only, then by Equations 5.2.15 and 5.2.16, 5.2.18, 5.2.20, 5.2.22, and 5.2.23, we obtain

$$\int_{\partial B_1} (\widehat{\mathbf{u}} - \mathbf{u}) \cdot \widetilde{\mathbf{s}} \, da = 0 \tag{5.2.24}$$

and this implies that

$$\mathbf{u} = \widehat{\mathbf{u}} \quad \text{on } \partial B_1 \tag{5.2.25}$$

As a result, $s = [\mathbf{u}, \mathbf{E}, \mathbf{S}]$ is a solution to the mixed problem. This completes the proof of part (b), as well as the proof of (H).

If the set of all admissible states $\mathcal{A}$ is suitably restricted, and the functional $H\{s\}$ in Equation 5.2.6 is modified accordingly, the Hu–Washizu principle reduces to a restricted form, which is also called the Hellinger–Reissner principle.

(R) **Hellinger–Reissner Principle.** Let $\mathcal{A}_1$ denote the set of all admissible states that satisfy the strain–displacement relation, and let $H_1 = H_1\{s\}$ be the functional on $\mathcal{A}_1$ defined by

$$H_1\{s\} = U_K\{\mathbf{S}\} - \int_B \mathbf{S} \cdot \mathbf{E} \, dv + \int_B \mathbf{b} \cdot \mathbf{u} \, dv + \int_{\partial B_1} \mathbf{s} \cdot (\mathbf{u} - \widehat{\mathbf{u}}) \, da + \int_{\partial B_2} \widehat{\mathbf{s}} \cdot \mathbf{u} \, da \tag{5.2.26}$$

for every $s = [\mathbf{u}, \mathbf{E}, \mathbf{S}] \in \mathcal{A}_1$. Then

$$\delta H_1\{s\} = 0 \tag{5.2.27}$$

if and only if s is a solution of the mixed problem.

The proof of (R) is similar to that of (H) and it will not be given here.

In the following examples, we are to show how the principles (H) and (R) may be further restricted to obtain the minimum principles discussed in Sections 5.1.1 through 5.1.4.

Example 5.2.1

Show that if $s = [\mathbf{u}, \mathbf{E}, \mathbf{S}]$ is a kinematically admissible state, which means that s satisfies the strain–displacement and stress–strain relations and the displacement boundary condition, then the functional $H = H\{s\}$, given by Equation 5.2.6, reduces to $F = F\{s\}$, where F is the functional of the principle of minimum potential energy, given by Equation 5.1.7.

Solution

Since s is kinematically admissible

$$\int_B \mathbf{u} \cdot (\operatorname{div} \mathbf{S})\, dv = \int_{\partial B_1} \widehat{\mathbf{u}} \cdot \mathbf{s}\, da + \int_{\partial B_2} \mathbf{u} \cdot \mathbf{s}\, da - \int_B \mathbf{S} \cdot \mathbf{E}\, dv \tag{a}$$

Substituting (a) in Equation 5.2.6, we obtain

$$\begin{aligned} H\{s\} &= U_C\{\mathbf{E}\} + \int_B \mathbf{u} \cdot (\operatorname{div} \mathbf{S})\, dv - \int_{\partial B_1} \widehat{\mathbf{u}} \cdot \mathbf{s}\, da \\ &\quad - \int_{\partial B_2} \mathbf{u} \cdot \mathbf{s}\, da - \int_B (\operatorname{div} \mathbf{S} + \mathbf{b}) \cdot \mathbf{u}\, dv \\ &\quad + \int_{\partial B_1} \mathbf{s} \cdot \widehat{\mathbf{u}}\, da + \int_{\partial B_2} (\mathbf{s} - \widehat{\mathbf{s}}) \cdot \mathbf{u}\, da \\ &= U_C\{\mathbf{E}\} - \int_B \mathbf{b} \cdot \mathbf{u}\, dv - \int_{\partial B_2} \widehat{\mathbf{s}} \cdot \mathbf{u}\, da \end{aligned} \tag{b}$$

Hence, comparing (b) with Equation 5.1.7, we conclude that

$$H\{s\} = F\{s\} \tag{c}$$

where F is the functional of the principle of minimum potential energy. This completes the solution. □

Example 5.2.2

Show that if $s = [\mathbf{u}, \mathbf{E}, \mathbf{S}]$ is a statically admissible state, which means a state for which $\mathbf{S}$ satisfies the equations of equilibrium and the traction boundary condition and, in addition, s satisfies the strain–displacement relation, then the functional $H_1 = H_1\{s\}$, given by Equation 5.2.26, reduces to the functional $G = G\{\mathbf{S}\}$ of the principle of minimum complementary energy, given by Equation 5.1.44.

Solution

$s = [\mathbf{u}, \mathbf{E}, \mathbf{S}]$ satisfies the field equations

$$\mathbf{E} = \widehat{\nabla} \mathbf{u} \quad \text{on } B \tag{a}$$

$$\operatorname{div} \mathbf{S} + \mathbf{b} = \mathbf{0} \quad \text{on } B \tag{b}$$

and the boundary condition

$$\mathbf{s} = \widehat{\mathbf{s}} \quad \text{on } \partial B_2 \tag{c}$$

Hence, using the divergence theorem, we obtain

$$\begin{aligned}\int_B \mathbf{S}\cdot\mathbf{E}\,dv &= \int_B \mathbf{S}\cdot(\widehat{\nabla}\mathbf{u})\,dv = \int_B \mathbf{S}\cdot(\nabla\mathbf{u})\,dv \\ &= \int_{\partial B} \mathbf{s}\cdot\mathbf{u}\,da - \int_B \mathbf{u}\cdot(\operatorname{div}\mathbf{S})\,dv \\ &= \int_{\partial B_1} \mathbf{s}\cdot\mathbf{u}\,da + \int_{\partial B_2} \widehat{\mathbf{s}}\cdot\mathbf{u}\,da + \int_B \mathbf{b}\cdot\mathbf{u}\,dv \end{aligned} \tag{d}$$

and (d) implies that

$$-\int_B \mathbf{S}\cdot\mathbf{E}\,dv + \int_B \mathbf{b}\cdot\mathbf{u}\,dv = -\int_{\partial B_1} \mathbf{s}\cdot\mathbf{u}\,da - \int_{\partial B_2} \widehat{\mathbf{s}}\cdot\mathbf{u}\,da \tag{e}$$

Substituting (e) into the RHS of Equation 5.2.26, we obtain

$$\begin{aligned} H_1\{s\} &= U_K\{\mathbf{S}\} - \int_{\partial B_1} \mathbf{s}\cdot\mathbf{u}\,da - \int_{\partial B_2} \widehat{\mathbf{s}}\cdot\mathbf{u}\,da \\ &\quad + \int_{\partial B_1} \mathbf{s}\cdot(\mathbf{u}-\widehat{\mathbf{u}})\,da + \int_{\partial B_2} \widehat{\mathbf{s}}\cdot\mathbf{u}\,da = U_K\{\mathbf{S}\} - \int_{\partial B_1} \mathbf{s}\cdot\widehat{\mathbf{u}}\,da \end{aligned} \tag{f}$$

Therefore, by comparing the RHS of (f) with the RHS of Equation 5.1.44, we conclude that

$$H_1\{s\} = G\{\mathbf{S}\} \tag{g}$$

where G is the functional of the principle of minimum complementary energy, given by Equation 5.1.44. This completes the solution. □

(C) **Compatibility-Related Principle.** Consider a traction problem for a body B subject to an external load $[\mathbf{b}, \mathbf{s}]$. Let Q denote the set of all admissible states that satisfy the equation of equilibrium, the strain–stress relations, and the traction boundary condition; and let $I\{\cdot\}$ be the functional on Q defined by

$$I\{s\} = U_K\{\mathbf{S}\} = \frac{1}{2}\int_B \mathbf{S}\cdot\mathbf{K}[\mathbf{S}]\,dv \tag{5.2.28}$$

for every $s = [\mathbf{u}, \mathbf{E}, \mathbf{S}] \in Q$. Then

$$\delta I\{s\} = 0 \tag{5.2.29}$$

if and only if s is a solution of the traction problem.

Proof: We take $s = [\mathbf{u}, \mathbf{E}, \mathbf{s}]$ and $\tilde{s} = [\tilde{\mathbf{u}}, \tilde{\mathbf{E}}, \tilde{\mathbf{S}}]$ belonging to Q. Then $s + \omega\tilde{s} \in Q$ for every scalar ω. Also, by the symmetry of K we obtain

$$U_k\{\mathbf{S} + \omega\tilde{\mathbf{S}}\} = \frac{1}{2}\int_B \{\mathbf{S}\cdot\mathsf{K}[\mathbf{S}] + \omega^2\tilde{\mathbf{S}}\cdot\mathsf{K}[\tilde{\mathbf{S}}] + 2\omega\tilde{\mathbf{S}}\cdot\mathsf{K}[\mathbf{S}]\}\,dv \tag{5.2.30}$$

Hence

$$\frac{d}{d\omega}U_K\{\mathbf{S} + \omega\tilde{\mathbf{S}}\}\bigg|_{\omega=0} = \delta I\{s\} = \int_B \tilde{\mathbf{S}}\cdot\mathsf{K}[\mathbf{S}]\,dv \tag{5.2.31}$$

In the following we prove that:

(i) If s is a solution of the traction problem, then Equation 5.2.29 is satisfied, and
(ii) If Equation 5.2.29 holds true, then s is a solution of the traction problem.

To prove (i) note that if s is a solution of the traction problem, then

$$\operatorname{div}\mathbf{S} + \mathbf{b} = \mathbf{0}, \quad \mathbf{E} = \widehat{\nabla}\mathbf{u}, \quad \mathbf{E} = \mathsf{K}[\mathbf{S}] \quad \text{on } B \tag{5.2.32}$$

and

$$\mathbf{Sn} = \mathbf{s} \quad \text{on } \partial B \tag{5.2.33}$$

and since $s + \omega\tilde{s} \in Q$ for every scalar ω, therefore

$$\operatorname{div}\tilde{\mathbf{S}} = \mathbf{0}, \quad \tilde{\mathbf{E}} = \mathsf{K}[\tilde{\mathbf{S}}] \quad \text{on } B \tag{5.2.34}$$

and

$$\tilde{\mathbf{S}}\mathbf{n} = \mathbf{0} \quad \text{on } \partial B \tag{5.2.35}$$

Also, note that $\tilde{\mathbf{S}}$ can be taken in the form

$$\tilde{\mathbf{S}} = \text{curl curl }\tilde{\mathbf{A}} \tag{5.2.36}$$

where $\tilde{\mathbf{A}} = \tilde{\mathbf{A}}^T$ is an arbitrary second-order tensor field on $\tilde{B}$ with the property that $\tilde{\mathbf{A}}$, $\nabla\tilde{\mathbf{A}}$, and $\nabla\,\nabla\tilde{\mathbf{A}}$ vanish near ∂B. Substituting $\tilde{\mathbf{S}}$ from Equation 5.2.36 into Equation 5.2.31 we find

$$\delta I\{s\} = \int_B (\text{curl curl}\tilde{\mathbf{A}})\cdot\mathsf{K}[\mathbf{S}]\,dv \tag{5.2.37}$$

Finally, applying the divergence theorem and taking into account the vanishing properties of $\tilde{\mathbf{A}}$ near ∂B we obtain

$$\delta I\{s\} = \int_B \tilde{\mathbf{A}}\cdot(\text{curl curl }\mathsf{K}[\mathbf{S}])\,dv \tag{5.2.38}$$

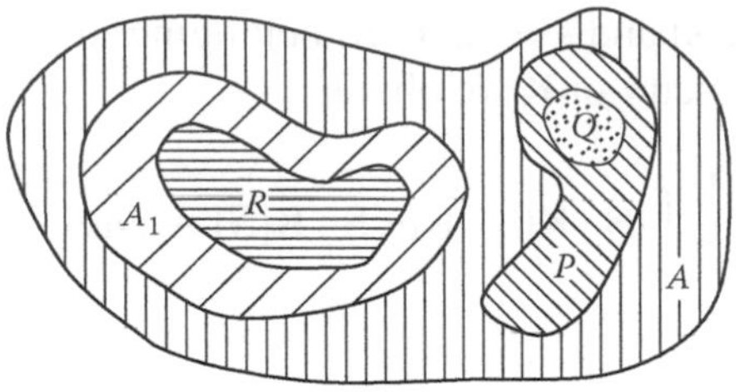

FIGURE 5.5 Domains of functionals $F\{\cdot\}$, $G\{\cdot\}$, $H\{\cdot\}$, $H_1\{\cdot\}$, and $I\{\cdot\}$ defined in the principles (I), (II), (H), (R) and (C), respectively.

Since s is a solution of the traction problem

$$\text{curl curl } \mathsf{K}[\mathbf{S}] = \mathbf{0} \tag{5.2.39}$$

therefore, Equation 5.2.29 is satisfied. This completes the proof of (i).
To prove (ii) we note that due to arbitrariness of $\widetilde{\mathbf{A}}$, Equation 5.2.31, in which $\widetilde{\mathbf{S}}$ is defined by Equation 5.2.36, is equivalent to Equation 5.2.38. As a result, if Equation 5.2.29 holds true, then $\mathbf{S}$ satisfies the compatibility condition 5.2.39. Therefore, there is a vector field $\mathbf{u}$ such that $s = [\mathbf{u}, \mathbf{E}, \mathbf{S}]$ is a solution of the traction problem. This completes the proof of (ii), as well as the proof of (C). □

Figure 5.5 shows the domains $R, P, \mathcal{A}, \mathcal{A}_1$, and Q of the functionals $F\{\cdot\}, G\{\cdot\}, H\{\cdot\}, H_1\{\cdot\}$, and $I\{\cdot\}$, respectively.

In Figure 5.5, $\mathcal{A}$ is the set of all admissible states over which the functional $H = H\{s\}$ of the Hu–Washizu principle is defined, see Equation 5.2.6.

$\mathcal{A}_1$ is the set of all admissible states that satisfy the strain–displacement relation. The functional $H_1 = H_1\{s\}$ of the Hellinger–Prange–Reissner principle is defined over $\mathcal{A}_1$, see Equation 5.2.26.

R is the set of all admissible states that satisfy the strain–displacement relation, the stress–strain relation, and the displacement boundary condition, that is, R is the set of all kinematically admissible states. The functional $F = F\{s\}$ of the principle of minimum potential energy is defined over R, see Equation 5.1.7.

P is the set of all admissible states that satisfy the equation of equilibrium and the traction boundary condition, that is, P is the set of all statically admissible states. The functional $G = G\{\mathbf{S}\}$ of the principle of the minimum complementary energy is defined over P, see Equation 5.1.44.

Q is the set of all admissible states that satisfy the equation of equilibrium, the strain–stress relations, and the traction boundary condition on ∂B. The functional $I = I\{\cdot\}$ of the compatibility-related principle is defined over Q, see Equation 5.2.28.

The correspondences between the five domains and the five functionals are then as follows:

$$\mathcal{A} \to H\{\cdot\} \quad \mathcal{A}_1 \to H_1\{\cdot\}$$

$$R \to F\{\cdot\} \quad P \to G\{\cdot\} \quad Q \to I\{\cdot\}$$

Also note that $\mathcal{A} \supset \mathcal{A}_1 \supset R$, $\mathcal{A} \supset P \supset Q$, and $\mathcal{A}_1 \cap P = \emptyset$, where $\emptyset$ is an empty set.

The Principles (H) and (R) are useful in solving problems of elastostatics for bodies of complex geometries, like plates and shells of various configurations, and also in conjunction with modern methods of computational analysis of engineering structures, such as the finite element method.

PROBLEMS

5.1 Consider a generalized plane stress traction problem of homogeneous isotropic elastostatics for a region C_0 of (x_1, x_2) plane (see Section 8.1.1). For such a problem the stress energy is represented by the integral

$$\overline{U}_K\{\overline{\mathbf{S}}\} = \frac{1}{2}\int_{C_0} \overline{\mathbf{S}} \cdot \mathsf{K}[\overline{\mathbf{S}}]\, da \tag{a}$$

where $\overline{\mathbf{S}}$ is the stress tensor corresponding to a solution $\overline{s} = [\overline{\mathbf{u}}, \overline{\mathbf{E}}, \overline{\mathbf{S}}]$ of the traction problem, and

$$\overline{\mathbf{E}} = \mathsf{K}[\overline{\mathbf{S}}] = \frac{1}{2\mu}\left[\overline{\mathbf{S}} - \frac{\nu}{1+\nu}(\mathrm{tr}\,\overline{\mathbf{S}})\mathbf{1}\right] \quad \text{on } C_0 \tag{b}$$

$$\mathrm{div}\,\overline{\mathbf{S}} + \overline{\mathbf{b}} = \mathbf{0} \quad \text{on } C_0 \tag{c}$$

$$\overline{\mathbf{E}} = \widehat{\nabla}\overline{\mathbf{u}} \quad \text{on } C_0 \tag{d}$$

and

$$\overline{\mathbf{S}}\mathbf{n} = \widehat{\mathbf{s}} \quad \text{on } \partial C_0 \tag{e}$$

Let $\overline{Q}$ denote the set of all admissible states that satisfy Equations (b) through (e) except for Equation (d). Define the functional $\overline{I}\{\cdot\}$ on $\overline{Q}$ by

$$\overline{I}\{\overline{s}\} \equiv \overline{U}_K\{\overline{\mathbf{S}}\} \quad \text{for every } \overline{s} \in \overline{Q} \tag{f}$$

Show that

$$\delta\overline{I}\{\overline{s}\} = 0 \tag{g}$$

if and only if $\overline{s}$ is a solution to the traction problem.

Hint: The proof is similar to that of the compatibility-related principle (C) of Section 5.2. First, we note that if $\overline{s} \in \overline{Q}$ and $\widetilde{\overline{s}} \in \overline{Q}$ then $\overline{s} + \omega\widetilde{\overline{s}} \in \overline{Q}$ for every scalar ω, and

$$\delta\overline{I}\{\overline{s}\} = \int_{C_0} \widetilde{\overline{\mathbf{S}}} \cdot \overline{\mathbf{E}}\, da \tag{h}$$

Next, by letting

$$\widetilde{S}_{\alpha\beta} = \epsilon_{\alpha\gamma 3}\epsilon_{\beta\delta 3}\widetilde{F}_{,\gamma\delta} \tag{i}$$

where $\widetilde{F}$ is an Airy stress function such that $\widetilde{F}$, $\widetilde{F}_{,\alpha}$, and $\widetilde{F}_{,\alpha\beta}$ $(\alpha, \beta = 1, 2)$ vanish near ∂C_0, we find that

$$\delta\bar{I}\{\bar{s}\} = \int_{C_0} \widetilde{F}\epsilon_{\alpha\gamma 3}\epsilon_{\beta\delta 3}\overline{E}_{\alpha\beta,\gamma\delta}\, da \tag{j}$$

The proof then follows from (j).

5.2 Consider an elastic prismatic bar in simple tension shown in Figure P5.2

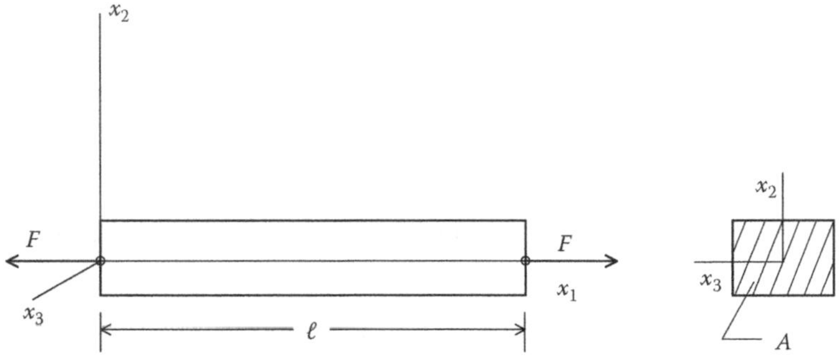

FIGURE P5.2

The stress energy of the bar takes the form

$$U_K\{\mathbf{S}\} = \int_0^{\ell} \left(\int_A \frac{1}{2E} S_{11}^2 da \right) dx_1 = \frac{1}{2E} \int_0^{\ell} \left(\frac{F}{A} \right)^2 A dx = \frac{F^2 \ell}{2EA} \tag{a}$$

where

A is the cross section of the bar

E denotes Young's modulus

The strain energy of the bar is obtained from

$$U_C\{\mathbf{E}\} = U_K\{\mathbf{S}\} = \frac{EAe^2}{2\ell} \tag{b}$$

where e is an elongation of the bar produced by the force $F = AEE_{11} = AEe/\ell$. The elastic state of the bar is then represented by

$$s = [u_1, E_{11}, S_{11}] = [e, e/\ell, F/A] \tag{c}$$

(i) Define a potential energy of the bar as $\widehat{F}\{s\} \equiv \varphi(e)$ and show that the relation

$$\delta\varphi(e) = 0 \tag{d}$$

is equivalent to the condition

$$\frac{\partial U_C}{\partial e} = F \tag{e}$$

(ii) Define a complementary energy of the bar as $\widehat{G}(s) \equiv \psi(F)$ and show that the condition

$$\delta\psi(F) = 0 \tag{f}$$

is equivalent to the equation

$$\frac{\partial U_K}{\partial F} = e \tag{g}$$

Hint: The functions $\varphi = \varphi(e)$ and $\psi = \psi(F)$ are given by

$$\varphi(e) = \frac{EA}{2\ell}e^2 - Fe$$

and

$$\psi(F) = \frac{\ell}{2EA}F^2 - Fe$$

respectively.

Note: Equations (e) and (g) constitute the *Castigliano theorem* [2].

5.3 The complementary energy of a cantilever beam loaded at the end by force P takes the form (see Figure P5.3)

$$\psi(P) = \frac{1}{2E}\int_B S_{11}^2 dv - Pu_2(\ell)$$

$$= \frac{1}{2E}\int_0^\ell \left\{\int_A \frac{M^2(x_1)}{I^2} x_2^2\, dA\right\} dx_1 - Pu_2(\ell) \tag{a}$$

where $M = M(x_1)$ and I stand for the bending moment and the moment of inertia of the area A with respect to the x_3 axis, respectively given by

$$M(x_1) = P(\ell - x_1), \quad I = \int_A x_2^2\, da \tag{b}$$

Use the minimum complementary energy principle for the cantilever beam in the form

$$\delta\psi(P) = 0 \tag{c}$$

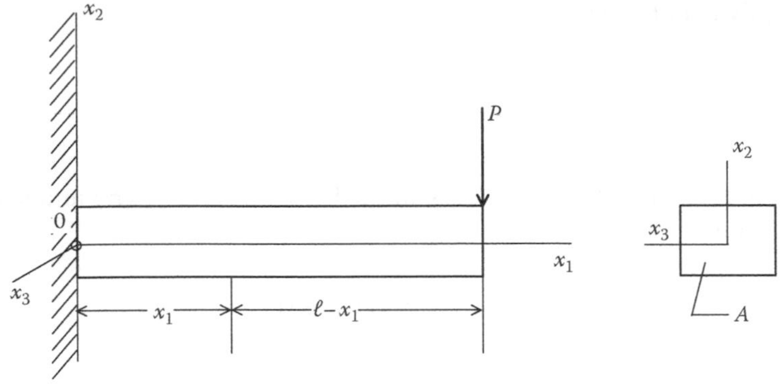

FIGURE P5.3

to show that the magnitude of deflection at the end of the beam is

$$u_2(\ell) = \frac{P\ell^3}{3EI} \tag{d}$$

5.4 An elastic beam which is clamped at one end and simply supported at the other end is loaded at an internal point $x_1 = \xi$ by force P (see Figure P5.4)

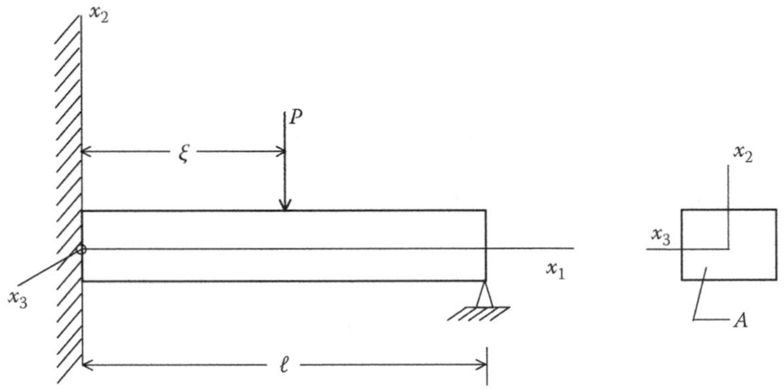

FIGURE P5.4

The potential energy of the beam, treated as a functional depending on a deflection of the beam $u_2 = u_2(x_1)$, takes the form, see [3],

$$\varphi\{u_2\} = \frac{EI}{2} \int_0^{\ell} \left(\frac{d^2 u_2}{dx_1^2} \right)^2 dx_1 - P u_2(\xi) \tag{a}$$

and $u_2 \in \widetilde{P} = \{u_2 = u_2(x_1) : u_2(0) = u_2'(0) = 0; u_2(l) = u_2''(l) = 0\}$. Let $u_2 = u_2(x_1)$ be a solution of the equation

$$EI \frac{d^4 u_2}{dx_1^4} = P\delta(x_1 - \xi) \quad \text{for } 0 < x_1 < \ell \tag{b}$$

subject to the conditions

$$u_2(0) = u_2'(0) = 0, \quad u_2(\ell) = u_2''(\ell) = 0 \tag{c}$$

Show that

$$\delta\varphi\{u_2\} = 0 \tag{d}$$

if and only if u_2 is a solution to the boundary value problem (b)–(c).

5.5 Use the Rayleigh–Ritz method to show that an approximate deflection of the beam of Problem 5.4 takes the form ($x_1 = x$)

$$u_2(x) = -c\,\ell^3 \left(\frac{x}{\ell}\right)^2 \left(1 - \frac{x}{\ell}\right)\left(1 - \frac{2}{3}\frac{x}{\ell}\right) \tag{a}$$

where

$$c = -\frac{5}{4}\frac{P}{EI}\left(\frac{\xi}{\ell}\right)^2 \left(1 - \frac{\xi}{\ell}\right)\left(1 - \frac{2}{3}\frac{\xi}{\ell}\right) \tag{b}$$

Also, show that for $\xi = \ell/2$ we obtain

$$u_2(\ell/2) = 0.0086\,\frac{\ell^3 P}{EI} \tag{c}$$

5.6 The potential energy of a rectangular thin elastic membrane fixed at its boundary and subjected to a vertical load $f = f(x_1, x_2)$ is

$$I\{u\} = \int_{-a_1}^{a_1}\int_{-a_2}^{a_2} \left(\frac{T_0}{2} u_{,\alpha}u_{,\alpha} - fu\right) dx_1 dx_2 \tag{a}$$

where $u \in \widehat{P}$, and

$$\widehat{P} = \{u = u(x_1, x_2) : u(\pm a_1, x_2) = 0 \quad \text{for } |x_2| < a_2;$$

$$u(x_1, \pm a_2) = 0 \quad \text{for } |x_1| < a_1\} \tag{b}$$

Here, $u = u(x_1, x_2)$ is a deflection of the membrane in the x_3 direction, and T_0 is a uniform tension of the membrane (see Figure P5.6).
Let the load function $f = f(x_1, x_2)$ be represented by the series

$$f(x_1, x_2) = \sum_{m,n=1}^{\infty} f_{mn} \sin\frac{m\pi(x_1 - a_1)}{2a_1} \sin\frac{n\pi(x_2 - a_2)}{2a_2} \tag{c}$$

Use the Rayleigh–Ritz method to show that the functional $I\{u\}$ attains a minimum over $\widehat{P}$ at

$$u(x_1, x_2) = \sum_{m,n=1}^{\infty} u_{mn} \sin\frac{m\pi(x_1 - a_1)}{2a_1} \sin\frac{n\pi(x_2 - a_2)}{2a_2} \tag{d}$$

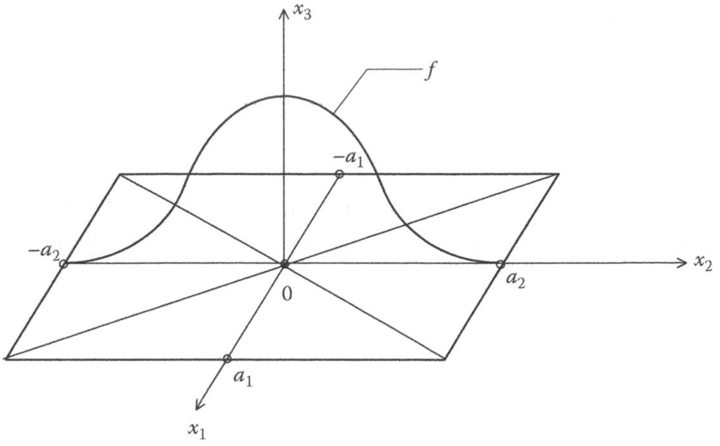

FIGURE P5.6

where

$$u_{mn} = \frac{1}{T_0} \frac{f_{mn}}{\left[(m\pi/2a_1)^2 + (n\pi/2a_2)^2\right]} \quad m, n = 1, 2, 3, \ldots \tag{e}$$

5.7 Use the solution obtained in Problem 5.6 to find the deflection of a square membrane of side a that is held fixed at its boundary and is vertically loaded by a load f of the form

$$f(x_1, x_2) = f_0[H(x_1 + \epsilon) - H(x_1 - \epsilon)][H(x_2 + \epsilon) - H(x_2 - \epsilon)] \tag{a}$$

where $H = H(x)$ is the Heaviside function defined in Equation 2.3.53, and f_0 and ϵ are positive constants $(0 < \epsilon < a)$. Also, compute a deflection of the square membrane at its center when $\epsilon = a/8$.

5.8 The potential energy of a rectangular thin elastic plate that is simply supported along all the edges and is vertically loaded by a force P at a point (ξ_1, ξ_2) takes the form

$$\widehat{I}\{w\} = \frac{D}{2} \int_{-a_1}^{a_1} \int_{-a_2}^{a_2} \left(\nabla^2 w\right)^2 dx_1 dx_2 - Pw(\xi_1, \xi_2) \tag{a}$$

where $w \in \widetilde{P}$, and

$$\begin{aligned} \widetilde{P} = \{w = w(x_1, x_2) : w(\pm a_1, x_2) = 0,& \\ \nabla^2 w(\pm a_1, x_2) = 0 \quad \text{for } |x_2| < a_2;& \\ w(x_1, \pm a_2) = 0, \quad \nabla^2 w(x_1, \pm a_2) = 0 \quad \text{for } |x_1| < a_1\}& \end{aligned} \tag{b}$$

Here $w = w(x_1, x_2)$ is a deflection of the plate, and D is the bending rigidity of the plate (see Figure P5.8).

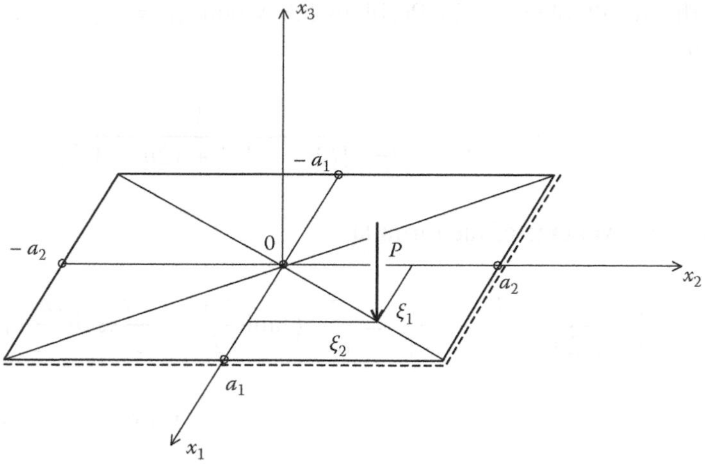

FIGURE P5.8

Show that a minimum of the functional $\widehat{I}\{\cdot\}$ over $\widetilde{P}$ is attained at a function $w = w(x_1, x_2)$ represented by the series

$$w(x_1, x_2) = \sum_{m,n=1}^{\infty} w_{mn} \sin \frac{m\pi}{2a_1}(x_1 - a_1) \sin \frac{n\pi}{2a_2}(x_2 - a_2) \tag{c}$$

where

$$w_{mn} = \frac{P}{Da_1a_2} \frac{\sin \frac{m\pi}{2a_1}(\xi_1 - a_1) \sin \frac{n\pi}{2a_2}(\xi_2 - a_2)}{\left[(m\pi/2a_1)^2 + (n\pi/2a_2)^2\right]^2} \tag{d}$$

Hint: Use the series representation of the concentrated load P

$$\begin{aligned} &P\delta(x_1 - \xi_1)\delta(x_2 - \xi_2) \\ &= \frac{P}{a_1a_2} \sum_{m,n=1}^{\infty} \sin \frac{m\pi}{2a_1}(\xi_1 - a_1) \sin \frac{n\pi}{2a_2}(\xi_2 - a_2) \sin \frac{m\pi}{2a_1}(x_1 - a_1) \\ &\quad \times \sin \frac{n\pi}{2a_2}(x_2 - a_2) \end{aligned}$$

for every $|x_1| < a_1, |x_2| < a_2; |\xi_1| < a_1, |\xi_2| < a_2$ (e)

5.9 Show that the central deflection of a square plate of side a that is simply supported along all the edges, and is loaded by a force P at its center, takes the form

$$w(0, 0) \approx 0.0459 \frac{Pa^2}{D} \tag{a}$$

Hint: Use the result obtained in Problem 5.8 when $\xi_1 = \xi_2 = 0$, $x_1 = x_2 = 0$, $a_1 = a_2 = a$

$$w(0,0) = \frac{16Pa^2}{D\pi^4} \sum_{m,n=1}^{\infty} \frac{1}{[(2m-1)^2 + (2n-1)^2]^2} \tag{b}$$

Also, by taking advantage of the formula

$$\sum_{m=1}^{\infty} \frac{1}{[(2m-1)^2 + x^2]^2} = \frac{\pi}{8x^3}\left(\tanh\frac{\pi x}{2} - \frac{\pi x}{2}\operatorname{sech}^2\frac{\pi x}{2}\right) \quad \text{for every } x > 0 \tag{c}$$

which is obtained by differentiating with respect to x the formula [4]

$$\sum_{k=1}^{\infty} \frac{1}{(2k-1)^2 + x^2} = \frac{\pi}{4x} \tanh\frac{\pi x}{2} \tag{d}$$

we reduce Equation (b) to the simple form

$$w(0,0) = \frac{2Pa^2}{D\pi^3} \sum_{n=1}^{\infty} \frac{1}{(2n-1)^3} \times \left[\tanh\frac{\pi}{2}(2n-1) - \frac{\pi}{2}(2n-1)\operatorname{sech}^2\frac{\pi}{2}(2n-1)\right] \tag{e}$$

The result (a) then follows by truncating the series (e).

REFERENCES

1. I. S. Gradshteyn and I. M. Ryzhik, *Tables of Integrals, Sums, Series, and Products,* 4th edn., G. I. F. M. L., Moscow 1962, p. 53 (in Russian); English translation: Academic Press, London, U.K., 1980.
2. I. H. Shames, *Mechanics of Deformable Solids,* Prentice-Hall, Englewood Cliffs, NJ, 1964, p. 357.
3. T. R. Tauchert, *Energy Principles in Structural Mechanics*, McGraw-Hill, New York, 1974, p. 47.
4. I. S. Gradshteyn and I. M. Ryzhik, *Tables of Integrals, Sums, Series, and Products,* 4th edn., G. I. F. M. L., Moscow 1962, p. 50 (in Russian); English translation: Academic Press, London, U.K., 1980.

6 Variational Principles of Elastodynamics

In this chapter, a variational characterization of a solution to an initial boundary value problem of elastodynamics is presented. The characterization includes the classical Hamilton–Kirchhoff Principle and a number of convolutional variational principles of Gurtin's type that describe completely a solution to an initial–boundary value problem. In particular, the convolutional principles without counterparts in elastostatics are discussed. Also, a pure stress variational principle of incompatible elastodynamics is formulated. The chapter contains a number of worked examples, and also end-of-chapter problems with solutions provided in the Solutions Manual.

6.1 THE HAMILTON–KIRCHHOFF PRINCIPLE

We begin with a variational principle of elastodynamics, which represents a generalization of the principle of minimum potential energy of elastostatics, see Section 5.1.1, to include dynamics effects. In the variational principle, the domain of a functional is a set of admissible processes, as opposed to the domain of the functional of the principle of minimum potential energy consisting of kinematically admissible states. To formulate the dynamic principle, we introduce a notion of *kinematically admissible process*, and by this we mean an admissible process that satisfies the strain–displacement relation, the stress–strain relation, and the displacement boundary condition. Obviously, these three constraints are to be satisfied for any point $\mathbf{x}$ of the body and for any time $t \geq 0$. Such a principle is useful in deriving the equation of motion as well as the traction boundary condition in various applications.

(H–K) **The Hamilton–Kirchhoff Principle.** Let $\mathcal{P}$ denote the set of all kinematically admissible processes $p = [\mathbf{u}, \mathbf{E}, \mathbf{S}]$ on $\bar{B} \times [0, \infty)$ satisfying the conditions

$$\mathbf{u}(\mathbf{x}, t_1) = \mathbf{u}_1(\mathbf{x}), \quad \mathbf{u}(\mathbf{x}, t_2) = \mathbf{u}_2(\mathbf{x}) \quad \text{on } \bar{B} \tag{6.1.1}$$

where t_1 and t_2 are arbitrary points on the t-axis such that $0 \leq t_1 < t_2 < \infty$, and $\mathbf{u}_1(\mathbf{x})$ and $\mathbf{u}_2(\mathbf{x})$ are prescribed fields on $\bar{B}$.

Let $\mathcal{K} = \mathcal{K}\{p\}$ be the functional on $\mathcal{P}$ defined by

$$\mathcal{K}\{p\} = \int_{t_1}^{t_2} [F(t) - K(t)]\, dt \tag{6.1.2}$$

where

$$F(t) = U_C\{\mathbf{E}\} - \int_B \mathbf{b} \cdot \mathbf{u}\, dv - \int_{\partial B_2} \widehat{\mathbf{s}} \cdot \mathbf{u}\, da \tag{6.1.3}$$

and

$$K(t) = \frac{1}{2} \int_B \rho \dot{\mathbf{u}}^2 \, dv \tag{6.1.4}$$

for every $p = [\mathbf{u}, \mathbf{E}, \mathbf{S}] \in \mathcal{P}$. Then

$$\delta \mathcal{K}\{p\} = 0 \tag{6.1.5}$$

if and only if p satisfies the equation of motion and the traction boundary condition.

Proof: We take two elements of $\mathcal{P}$: p and $\widetilde{p}$, where $p = [\mathbf{u}, \mathbf{E}, \mathbf{S}]$ and $\widetilde{p} = [\widetilde{\mathbf{u}}, \widetilde{\mathbf{E}}, \widetilde{\mathbf{S}}]$. Then

$$p + \omega \widetilde{p} \in \mathcal{P} \quad \text{for every scalar } \omega \tag{6.1.6}$$

From Equation 6.1.6 it follows that $\widetilde{p}$ satisfies the strain–displacement relation, the stress–strain relation, the boundary condition

$$\widetilde{\mathbf{u}} = \mathbf{0} \quad \text{on } \partial B_1 \times [0, \infty) \tag{6.1.7}$$

and the conditions

$$\widetilde{\mathbf{u}}(\mathbf{x}, t_1) = \widetilde{\mathbf{u}}(\mathbf{x}, t_2) = \mathbf{0} \quad \text{on } \bar{B} \tag{6.1.8}$$

Condition 6.1.7 results from the fact that $\mathbf{u} + \omega \widetilde{\mathbf{u}}$ is a kinematically admissible motion that implies $\mathbf{u} + \omega \widetilde{\mathbf{u}} = \widehat{\mathbf{u}}$ on $\partial B_1 \times [0, \infty)$ and $\mathbf{u} = \widehat{\mathbf{u}}$ on $\partial B_1 \times [0, \infty)$. When $\omega = 1$ the last two boundary conditions imply Equation 6.1.7. Similar reasonings lead to the conditions 6.1.8.

To compute the first variation of the functional $\mathcal{K}\{\cdot\}$, we recall the formula, see Equation 5.2.2, in which s is replaced by p,

$$\delta_{\widetilde{p}} \mathcal{K}\{p\} = \frac{d}{d\omega} \mathcal{K}\{p + \omega \widetilde{p}\} \bigg|_{\omega=0} \tag{6.1.9}$$

Now, from Equations 6.1.2 through 6.1.4, we write

$$\mathcal{K}\{p + \omega \widetilde{p}\} = \int_{t_1}^{t_2} \bigg[U_C\{\mathbf{E} + \omega \widetilde{\mathbf{E}}\} - \int_B \mathbf{b} \cdot (\mathbf{u} + \omega \widetilde{\mathbf{u}}) \, dv - \int_{\partial B_2} \widehat{\mathbf{s}} \cdot (\mathbf{u} + \omega \widetilde{\mathbf{u}}) \, da - \frac{1}{2} \int_B \rho (\dot{\mathbf{u}} + \omega \dot{\widetilde{\mathbf{u}}})^2 \, dv \bigg] dt \tag{6.1.10}$$

Since by the definition of the strain energy $U_C\{\mathbf{E}\}$ and by the symmetry of $\mathbf{C}$, see Equations 5.2.9,

$$U_C\{\mathbf{E} + \omega \widetilde{\mathbf{E}}\} = U_C\{\mathbf{E}\} + \omega^2 U_C\{\widetilde{\mathbf{E}}\} + \omega \int_B \widetilde{\mathbf{E}} \cdot \mathbf{S} \, dv \tag{6.1.11}$$

then by differentiating Equation 6.1.10 with respect to ω, using Equation 6.1.11, and taking the result at $\omega = 0$, from Equation 6.1.9 we get

$$\delta_{\tilde{p}}\mathcal{K}\{p\} = \int_{t_1}^{t_2} \left[\int_B \mathbf{S} \cdot \widetilde{\mathbf{E}}\, dv - \int_B \mathbf{b} \cdot \widetilde{\mathbf{u}}\, dv - \int_{\partial B_2} \widehat{\mathbf{s}} \cdot \widetilde{\mathbf{u}}\, da - \int_B \rho\dot{\mathbf{u}} \cdot \dot{\widetilde{\mathbf{u}}}\, dv \right] dt \tag{6.1.12}$$

Also, note that by integrating by parts, we obtain

$$\int_{t_1}^{t_2} \rho\dot{\mathbf{u}} \cdot \dot{\widetilde{\mathbf{u}}}\, dt = \rho\dot{\mathbf{u}} \cdot \widetilde{\mathbf{u}} \Big|_{t=t_1}^{t=t_2} - \int_{t_1}^{t_2} \rho\ddot{\mathbf{u}} \cdot \widetilde{\mathbf{u}}\, dt \tag{6.1.13}$$

and using the conditions 6.1.8, we obtain

$$\int_{t_1}^{t_2} \rho\dot{\mathbf{u}} \cdot \dot{\widetilde{\mathbf{u}}}\, dt = -\int_{t_1}^{t_2} \rho\ddot{\mathbf{u}} \cdot \widetilde{\mathbf{u}}\, dt \tag{6.1.14}$$

In addition, since $\widetilde{\mathbf{u}}$ and $\widetilde{\mathbf{E}}$ satisfy the strain–displacement relation, that is,

$$\widetilde{\mathbf{E}} = \widehat{\nabla}\widetilde{\mathbf{u}} \tag{6.1.15}$$

the first term of the integrand on the RHS of Equation 6.1.12 takes the form

$$\int_B \mathbf{S} \cdot \widetilde{\mathbf{E}}\, dv = \int_B \mathbf{S} \cdot \nabla\widetilde{\mathbf{u}}\, dv \tag{6.1.16}$$

and, by the divergence theorem, we arrive at

$$\begin{aligned} \int_B \mathbf{S} \cdot \widetilde{\mathbf{E}}\, dv &= \int_{\partial B} \widetilde{\mathbf{u}} \cdot \mathbf{s}\, da - \int_B \widetilde{\mathbf{u}} \cdot (\operatorname{div}\mathbf{S})\, dv \\ &= \int_{\partial B_1} \mathbf{s} \cdot \widetilde{\mathbf{u}}\, da + \int_{\partial B_2} \mathbf{s} \cdot \widetilde{\mathbf{u}}\, da - \int_B (\operatorname{div}\mathbf{S}) \cdot \widetilde{\mathbf{u}}\, dv \end{aligned} \tag{6.1.17}$$

By virtue of Equation 6.1.7, the first integral on the RHS of Equation 6.1.17 vanishes, hence, using Equations 6.1.14 and 6.1.17 we reduce Equation 6.1.12 to the form

$$\delta_{\tilde{p}}\mathcal{K}\{p\} = -\int_{t_1}^{t_2} \left[\int_B (\operatorname{div}\mathbf{S} + \mathbf{b} - \rho\ddot{\mathbf{u}}) \cdot \widetilde{\mathbf{u}}\, dv - \int_{\partial B_2} (\mathbf{s} - \widehat{\mathbf{s}}) \cdot \widetilde{\mathbf{u}}\, da \right] dt \tag{6.1.18}$$

Proof of the (H–K) principle will be done if we show that

(a) If the equation of motion and the traction boundary condition are satisfied, then Equation 6.1.5 holds true.
(b) If Equation 6.1.5 is satisfied, then the equation of motion and the traction boundary condition are met.

Proof of (a). If

$$\operatorname{div}\mathbf{S}+\mathbf{b}=\rho\ddot{\mathbf{u}} \quad \text{on } \partial B\times[0,\infty) \tag{6.1.19}$$

and

$$\mathbf{s}=\widehat{\mathbf{s}} \quad \text{on } \partial B_2\times[0,\infty) \tag{6.1.20}$$

then it follows from Equation 6.1.18 and from the definition of $\delta\mathcal{K}\{p\}$ that

$$\delta\mathcal{K}\{p\}=0 \tag{6.1.21}$$

and this completes the proof of part (a).

Proof of (b). Since

$$\delta_{\tilde{p}}\mathcal{K}\{p\}=0 \quad \text{for every } \widetilde{p} \text{ such that } p+\omega\widetilde{p}\in\mathcal{P} \tag{6.1.22}$$

then Equation 6.1.22 holds true for every admissible $\widetilde{\mathbf{u}}$ consistent with Equations 6.1.7 and 6.1.8. Selecting $\widetilde{\mathbf{u}}$ as an arbitrary vector field on $\bar{B}\times[0,\infty)$ that vanishes near $\partial B\times[0,\infty)$ and complies with the conditions 6.1.8, using Equations 6.1.18 and 6.1.22 we obtain

$$\int_B (\operatorname{div}\mathbf{S}+\mathbf{b}-\rho\ddot{\mathbf{u}})\cdot\widetilde{\mathbf{u}}\,dv=0 \tag{6.1.23}$$

and by an extension of the Fundamental Lemma of calculus of variations, we arrive at the equation of motion

$$\operatorname{div}\mathbf{S}+\mathbf{b}=\rho\ddot{\mathbf{u}} \quad \text{on } \bar{B}\times[0,\infty) \tag{6.1.24}$$

Next, we select $\widetilde{\mathbf{u}}$ in such a way that $\widetilde{\mathbf{u}}$ vanishes near ∂B_1 only and complies with the conditions 6.1.8. Then, using Equations 6.1.18, 6.1.24, and 6.1.22, we find

$$\int_{\partial B_2} (\mathbf{s}-\widehat{\mathbf{s}})\cdot\widetilde{\mathbf{u}}\,da=0 \tag{6.1.25}$$

This result together with the Fundamental Lemma imply that

$$\mathbf{s}=\widehat{\mathbf{s}} \quad \text{on } \partial B_2\times[0,\infty) \tag{6.1.26}$$

This completes the proof of part (b), and of the (H–K) principle.

Notes:

(1) If the functions $\mathbf{u}$, $\mathbf{b}$, and $\widehat{\mathbf{s}}$ are time independent, then from Equation 6.1.4 $K(t)\equiv 0$ and $F(t)$ given by Equation 6.1.3 reduces to the functional $F=F\{\cdot\}$ of the (I) principle of minimum potential energy of elastostatics, Equation 5.1.7. In this sense the (H–K) principle is a generalization on elastodynamics of (I).

(2) The (H–K) principle is based on the knowledge of the displacement $\mathbf{u} = \mathbf{u}(\mathbf{x}, t)$ at two times, t_1 and t_2, $t_2 > t_1$. Since in an initial–boundary value problem of elastodynamics an initial value of the displacement $\mathbf{u}(\mathbf{x}, 0)$ and initial velocity $\dot{\mathbf{u}}(\mathbf{x}, 0)$ are prescribed over B, then the (H–K) principle cannot be used to describe the initial–boundary value problem. Indeed, if $t_1 = 0$, then $\mathbf{u}(\mathbf{x}, 0)$ is known but $\dot{\mathbf{u}}(\mathbf{x}, 0)$ is not. Also, the value $\mathbf{u}(\mathbf{x}, t_2)$ that is prescribed in the (H–K) principle, see Equation 6.1.1, is not known in the initial–boundary value problem. In the following section, we formulate variational principles that do characterize the initial–boundary value problem of elastodynamics. Before that we give examples of the application of the (H–K) principle.

Example 6.1.1

An elastic bar subject to an axial body force $b = b(x, t)$, and the end displacements $\widehat{u}_0 = \widehat{u}_0(t)$ and $\widehat{u}_L = \widehat{u}_L(t)$, $0 \leq x \leq L$, $t \geq 0$, is shown in Figure 6.1.

Assume that the motion occurs only in a direction along the axis of the bar, that is, the displacement vector $\mathbf{u} = \mathbf{u}(\mathbf{x}, t)$ is described bySuch a motion represents the longitudinal waves in the bar.

Use the (H–K) principle to derive the equation of motion for the bar.

$$\mathbf{u}(\mathbf{x}, t) = [u(x, t), 0, 0] \tag{a}$$

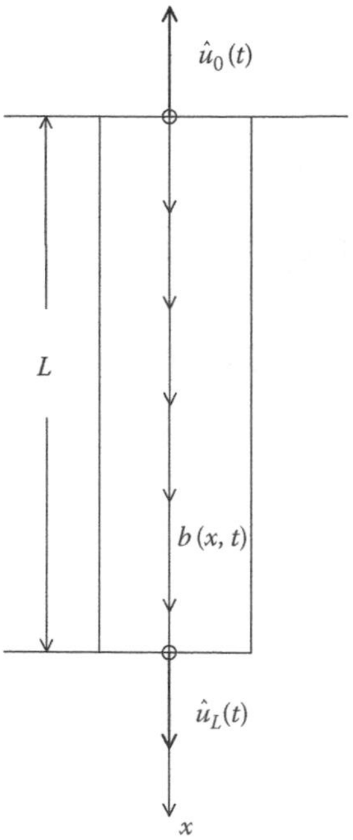

FIGURE 6.1 A bar subject to an axial load and boundary displacements.

Solution

A kinematically admissible process in this example is defined as

$$p = [u, E_{11}, S_{11}] \tag{b}$$

where

$$E_{11} = \frac{\partial u}{\partial x}, \quad S_{11} = E E_{11} \tag{c}$$

$$u(0, t) = \widehat{u}_0(t), \quad u(L, t) = \widehat{u}_L(t) \tag{d}$$

In addition, we assume

$$u(x, t_1) = u_1(x), \quad u(x, t_2) = u_2(x) \tag{e}$$

Let $\mathcal{P}$ denote the set of all kinematically admissible processes p that comply with (a) through (e). As a functional $\mathcal{K}$ defined on $\mathcal{P}$ in this example, we take a one-dimensional counterpart to Equation 6.1.2

$$\mathcal{K}\{p\} = \int_{t_1}^{t_2} [F(t) - K(t)]\, dt \tag{f}$$

where

$$F(t) = \frac{1}{2} \int_B \frac{\partial u}{\partial x} \cdot \left(E \frac{\partial u}{\partial x} \right) dv - \int_B b u \, dv \tag{g}$$

and

$$K(t) = \frac{1}{2}\int_B \rho\left(\frac{\partial u}{\partial t}\right)^2 dv \tag{h}$$

In (g), the term corresponding to the integral over ∂B_2 in Equation 6.1.3 vanishes because ∂B_2 in this one-dimensional problem is an empty set.

Since $dv = A\ dx$, where A denotes the area of the cross section of the bar, and $0 \leq x \leq L$, the functional (f) takes the form

$$\mathcal{K}\{p\} = \mathcal{K}\{u\} = \int_{t_1}^{t_2}\left[\int_0^L \frac{EA}{2}\left(\frac{\partial u}{\partial x}\right)^2 - \frac{\rho A}{2}\left(\frac{\partial u}{\partial t}\right)^2 - Abu\right] dxdt \tag{i}$$

In order to calculate the first variation of $\mathcal{K}\{p\}$, we select $\widetilde{p} \in \mathcal{P}$ in such a way that $(p + \omega\widetilde{p}) \in \mathcal{P}$ for any scalar ω. This means that $\widetilde{p} = [\widetilde{u}, \widetilde{E}_{11}, \widetilde{S}_{11}]$ satisfies the constraints

$$\widetilde{E}_{11} = \frac{\partial \widetilde{u}}{\partial x}, \quad \widetilde{S}_{11} = E\widetilde{E}_{11} \tag{j}$$

$$\widetilde{u}(0, t) = \widetilde{u}(L, t) = 0, \quad \widetilde{u}(x, t_1) = \widetilde{u}(x, t_2) = 0 \tag{k}$$

The first variation of $\mathcal{K}\{p\}$ is calculated from the formula

$$\delta_{\widetilde{u}}\mathcal{K}\{u\} = \frac{d}{d\omega}\,\mathcal{K}\{u + \omega\widetilde{u}\}\Big|_{\omega=0} \tag{l}$$

Now,

$$\mathcal{K}\{u + \omega\widetilde{u}\} = \int_{t_1}^{t_2}\int_0^L\left[\frac{EA}{2}\left(\frac{\partial u}{\partial x} + \omega\,\frac{\partial \widetilde{u}}{\partial x}\right)^2\right.$$

$$\left. - \frac{\rho A}{2}\left(\frac{\partial u}{\partial t} + \omega\,\frac{\partial \widetilde{u}}{\partial t}\right)^2 - Ab(u + \omega\widetilde{u})\right] dxdt \tag{m}$$

Differentiation of this equation with respect to ω, putting $\omega = 0$, and substituting in (l) yields

$$\delta_{\widetilde{u}}\mathcal{K}\{u\} = \int_{t_1}^{t_2}\int_0^L\left(EA\,\frac{\partial u}{\partial x}\,\frac{\partial \widetilde{u}}{\partial x} - \rho A\,\frac{\partial u}{\partial t}\,\frac{\partial \widetilde{u}}{\partial t} - Ab\widetilde{u}\right) dxdt \tag{n}$$

Next, after integrating by parts the product $(\partial u/\partial t)(\partial \widetilde{u}/\partial t)$ we obtain

$$\int_{t_1}^{t_2} \frac{\partial u}{\partial t}\,\frac{\partial \widetilde{u}}{\partial t}\,dt = \frac{\partial u}{\partial t}\,\widetilde{u}\,\Big|_{t_1}^{t_2} - \int_{t_1}^{t_2} \frac{\partial^2 u}{\partial t^2}\,\widetilde{u}\,dt \tag{o}$$

By conditions (k), the first term on the RHS of (o) vanishes, and (o) reduces to

$$\int_{t_1}^{t_2} \frac{\partial u}{\partial t}\,\frac{\partial \widetilde{u}}{\partial t}\,dt = -\int_{t_1}^{t_2} \frac{\partial^2 u}{\partial t^2}\,\widetilde{u}\,dt \tag{p}$$

Also, integrating by parts the product $EA(\partial u/\partial x)(\partial \tilde{u}/dx)$ in (n), we get

$$\int_0^L EA \frac{\partial u}{\partial x}\frac{\partial \tilde{u}}{\partial x}\,dx = EA\frac{\partial u}{\partial x}\tilde{u}\bigg|_{x=0}^{x=L} - \int_0^L \left[\frac{\partial}{\partial x}\left(EA\frac{\partial u}{\partial x}\right)\right]\tilde{u}\,dx \tag{q}$$

Again, by conditions (k), the first term on the RHS of (q) vanishes, and (q) reduces to

$$\int_0^L EA \frac{\partial u}{\partial x}\frac{\partial \tilde{u}}{\partial x}\,dx = -\int_0^L \left[\frac{\partial}{\partial x}\left(EA\frac{\partial u}{\partial x}\right)\right]\tilde{u}\,dx \tag{r}$$

Substituting (p) and (r) into (n) we write the condition

$$\delta\mathcal{K}\{u\} = 0 \tag{s}$$

in the form

$$\int_{t_1}^{t_2}\int_0^L \left[\frac{\partial}{\partial x}\left(EA\frac{\partial u}{\partial x}\right) + Ab - \rho A\frac{\partial^2 u}{\partial t^2}\right]\tilde{u}\,dxdt = 0 \tag{t}$$

Hence, by the Fundamental Lemma of the calculus of variations, due to the arbitrariness of $\tilde{u}$ and conditions (k), we write

$$\frac{\partial}{\partial x}\left(EA\frac{\partial u}{\partial x}\right) + Ab - \rho A\frac{\partial^2 u}{\partial t^2} = 0 \tag{u}$$

This is the desired equation of motion of a bar along which longitudinal waves propagate. This completes the solution. □

Notes:

(1) Equation (u), which is to be satisfied for $0 \le x \le L$ and $t \ge 0$, subject to the initial conditions

$$u(x,0) = u_0(x), \quad \dot{u}(x,0) = \dot{u}_0(x) \tag{6.1.27}$$

and the boundary conditions

$$u(0,t) = \widehat{u}_0(t), \quad u(L,t) = \widehat{u}_L(t) \tag{6.1.28}$$

where $u_0(x)$, $\dot{u}_0(x)$, $\widehat{u}_0(t)$, and $\widehat{u}_L(t)$ are prescribed, constitutes an initial displacement boundary value problem for the bar. In accordance with the discussion on uniqueness, see Section 4.2.5, the problem possesses a unique solution.

(2) Equation (u) was derived under the assumption that E, A, and ρ may be functions of x. However, due to the one-dimensional wave motion in the bar, it is reasonable to assume that the bar be prismatic and, therefore, A be constant. In this case, dividing (u) by A we obtain the familiar equation of motion

$$\frac{\partial}{\partial x}\left(E\frac{\partial u}{\partial x}\right) + b - \rho\frac{\partial^2 u}{\partial t^2} = 0 \tag{6.1.29}$$

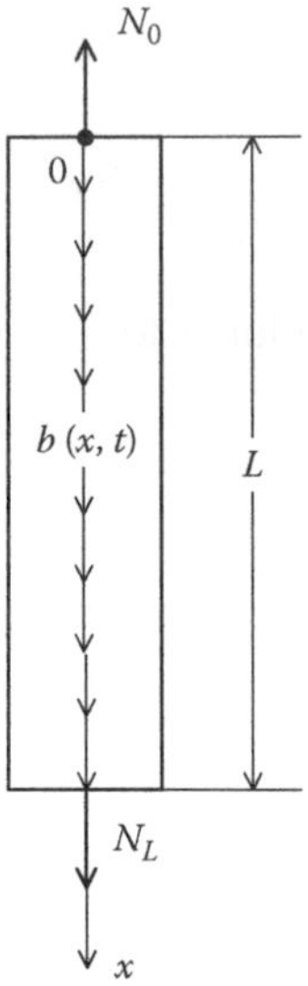

FIGURE 6.2
A bar subject to a body force b and end loads N_0 and N_L.

Example 6.1.2

Consider an elastic bar subject to an axial body force $b = b(x, t)$ and axial end loads $N_0 = N_0(t)$ and $N_L = N_L(t)$, see Figure 6.2.

Such a system of loads produces a motion of the bar along the x-axis, $\mathbf{u}(\mathbf{x}, t) = [u(x, t), 0, 0]$. The situation is similar to that in the previous example. Use a one-dimensional form of the (H–K) principle to derive the equation of motion for the bar and associated traction boundary conditions.

Solution

Since in this example the tractions at the ends of the bar are prescribed, then $\partial B_1 = 0$ and $\partial B_2 = \partial B$, and we select as the domain of a functional the set of all admissible processes $p = [u, E_{11}, S_{11}]$ that satisfy the strain–displacement relation, the stress–strain relation, and the conditions

$$u(x, t_1) = u_1(x), \quad u(x, t_2) = u_2(x) \tag{a}$$

where t_1, t_2, u_1, and u_2 have the same meaning as in the previous example. We denote the domain by $\mathcal{P}_0$. Here, as a functional $\mathcal{K}_0\{u\}$ we take

$$\mathcal{K}_0\{u\} = \int_{t_1}^{t_2} [F_0(t) - K(t)]\, dt \tag{b}$$

where

$$F_0(t) = \frac{1}{2}\int_B E\left(\frac{\partial u}{\partial x}\right)^2 dv - \int_B bu\, dv - [N_L u(L, t) - N_0 u(0, t)] \tag{c}$$

$$K(t) = \frac{1}{2}\int_B \rho\left(\frac{\partial u}{\partial t}\right)^2 dv \tag{d}$$

The term $[N_L u(L, t) - N_0 u(0, t)]$ in (c) corresponds to the last term on the RHS in Equation 6.1.3, that is, to the surface integral over $\partial B_2 = \partial B$.

Since $dv = A\, dx$, where A is the area of the cross section, (b) reduces to

$$\mathcal{K}_0\{u\} = \int_{t_1}^{t_2} \left\{ \int_0^L \left[\frac{EA}{2}\left(\frac{\partial u}{\partial x}\right)^2 - \frac{\rho A}{2}\left(\frac{\partial u}{\partial t}\right)^2 - Abu\right] dx \right.$$
$$\left. - [N_L u(L, t) - N_0 u(0, t)] \right\} dt \tag{e}$$

In this equation, u is a displacement field corresponding to the process $p = [u, E_{11}, S_{11}] \in \mathcal{P}_0$.

Let $\tilde{u}$ be a displacement field corresponding to $\tilde{p} = [\tilde{u}, \tilde{E}_{11}, \tilde{S}_{11}] \in \mathcal{P}_0$, and let $p + \omega\tilde{p} \in \mathcal{P}_0$ for every scalar ω. Then, $\tilde{p}$ satisfies the strain–displacement relation, the stress–strain relation, and the conditions

$$\tilde{u}(x, t_1) = \tilde{u}(x, t_2) = 0 \tag{f}$$

To show that conditions (f) hold true we note that from conditions (a), we get

$$u(x, t_1) = u_1(x), \quad u(x, t_2) = u_2(x) \tag{g}$$

and from the relation

$$u + \omega\widetilde{u} \in \mathcal{P}_0 \tag{h}$$

we write

$$u(x, t_1) + \omega\widetilde{u}(x, t_1) = u_1(x) \tag{i}$$

$$u(x, t_2) + \omega\widetilde{u}(x, t_2) = u_2(x) \tag{j}$$

Taking $\omega = 1$ in (i) and (j) and comparing these equations with conditions (g) we arrive at (f).

To evaluate the first variation of $\mathcal{K}_0\{u\}$ we compute $\mathcal{K}_0\{u + \omega\widetilde{u}\}$ and obtain

$$\begin{aligned}\mathcal{K}_0\{u + \omega\widetilde{u}\} = \int_{t_1}^{t_2} \Bigg\{ \int_0^L \Bigg[\frac{EA}{2} \left(\frac{\partial u}{\partial x} + \omega \frac{\partial \widetilde{u}}{\partial x} \right)^2 \\ - \frac{\rho A}{2} \left(\frac{\partial u}{\partial t} + \omega \frac{\partial \widetilde{u}}{\partial t} \right)^2 - Ab(u + \omega\widetilde{u}) \Bigg] dx \\ - \Big[N_L[u(L, t) + \omega\widetilde{u}(L, t)] - N_0[u(0, t) + \omega\widetilde{u}(0, t)] \Big] \Bigg\} dt\end{aligned} \tag{k}$$

Differentiation of this equation with respect to ω and putting $\omega = 0$ yields

$$\begin{aligned}\delta_{\widetilde{u}}\mathcal{K}_0\{u\} = \int_{t_1}^{t_2} \Bigg\{ \int_0^L \left(EA \frac{\partial u}{\partial x} \frac{\partial \widetilde{u}}{\partial x} - \rho A \frac{\partial u}{\partial t} \frac{\partial \widetilde{u}}{\partial t} - Ab\widetilde{u} \right) dx \\ - \Big[N_L\widetilde{u}(L, t) - N_0\widetilde{u}(0, t) \Big] \Bigg\} dt\end{aligned} \tag{l}$$

Now, by (f)

$$\int_{t_1}^{t_2} \frac{\partial u}{\partial t} \frac{\partial \widetilde{u}}{\partial t}\, dt = - \int_{t_1}^{t_2} \frac{\partial^2 u}{\partial t^2} \widetilde{u}\, dt \tag{m}$$

Also,

$$\int_0^L EA \frac{\partial u}{\partial x} \frac{\partial \widetilde{u}}{\partial x}\, dx = EA \frac{\partial u}{\partial x} \widetilde{u} \Bigg|_{x=0}^{x=L} - \int_0^L \left[\frac{\partial}{\partial x} \left(EA \frac{\partial u}{\partial x} \right) \right] \widetilde{u}\, dx \tag{n}$$

Hence, by (l) through (n) the condition

$$\delta\mathcal{K}_0\{u\} = 0 \tag{o}$$

takes the form

$$\int_{t_1}^{t_2}\left\{\int_0^L\left[\frac{\partial}{\partial x}\left(EA\,\frac{\partial u}{\partial x}\right)+Ab-\rho A\,\frac{\partial^2 u}{\partial t^2}\right]\widetilde{u}\,dx\right.$$
$$+\left[N_L-E(L)A(L)\,\frac{\partial u(L,t)}{\partial x}\right]\widetilde{u}(L,t)$$
$$\left.-\left[N_0-E(0)A(0)\,\frac{\partial u(0,t)}{\partial x}\right]\widetilde{u}(0,t)\right\}dt=0 \tag{p}$$

Now we use the Fundamental Lemma of the calculus of variations in which $\widetilde{u}$ is an arbitrary displacement field defined on $[0, L] \times [0, \infty)$ and such that

$$\widetilde{u}(0,t)=\widetilde{u}(L,t)=0 \tag{q}$$

and then from (p) we find

$$\int_0^L\left[\frac{\partial}{\partial x}\left(EA\,\frac{\partial u}{\partial x}\right)+Ab-\rho A\,\frac{\partial^2 u}{\partial t^2}\right]\widetilde{u}\,dx=0 \tag{r}$$

Due to the arbitrariness of $\widetilde{u}$ and by the Fundamental Lemma, we arrive at the equation of motion,

$$\frac{\partial}{\partial x}\left(EA\,\frac{\partial u}{\partial x}\right)+Ab-\rho A\,\frac{\partial^2 u}{\partial t^2}=0 \tag{s}$$

To obtain the traction boundary condition at $x = 0$, we select as $\widetilde{u}$ an arbitrary field on $[0, L] \times [0, \infty)$ such that $\widetilde{u}(L, t) = 0$, and from (p) and (s) we obtain

$$\int_{t_1}^{t_2}\left[N_0(t)-E(0)A(0)\,\frac{\partial u(0,t)}{\partial x}\right]\widetilde{u}(0,t)\,dt=0 \tag{t}$$

Again referring to the Fundamental Lemma in which $\widetilde{u}(0, t)$ is arbitrary on $[0, \infty)$, from (t) we get

$$N_0(t)=E(0)A(0)\,\frac{\partial u(0,t)}{\partial x} \tag{u}$$

Similarly, by selecting $\widetilde{u}$ in such a way that $\widetilde{u}(L, t)$ is an arbitrary function of t, from (p), (s), and (u) we arrive at the traction boundary condition at $x = L$

$$N_L(t)=E(L)A(L)\,\frac{\partial u(L,t)}{\partial x} \tag{v}$$

This completes the solution. □

Notes:

(1) In this example, the dimensions of E and A are

$$[E] = [\text{Force} \times \text{L}^{-2}], \quad [A] = [\text{L}^2] \tag{6.1.30}$$

so the dimension of N_0 and N_L is that of force,

$$[N_0] = [N_L] = [\text{Force}] \tag{6.1.31}$$

(2) The traction boundary conditions (u) and (v) comply with a general traction boundary condition in which the stress–strain relation is used.
(3) Equations (s) and (u) of Example 6.1.1 are the same, and they describe the longitudinal waves in a bar.

Example 6.1.3

Consider a homogeneous isotropic elastic rod in which the axial longitudinal displacement $u_1 = u_1(x_1, t)$ is coupled with the lateral displacement u_2 and u_3 at a typical point of the cross section with the coordinates x_2 and x_3 (see Figure 6.3).

The coupling is to comply with the following conditions: (i) the stress tensor $\mathbf{S} = \mathbf{S}(\mathbf{x}, t)$ represents a uniaxial stress state in the x_1 direction for every time t, (ii) the displacement u_2 and u_3 vanish at the centroid of the rod for every time t, and (iii) a motion of the rod is produced by a body force of the form $\mathbf{b} = [b(x_1, t), 0, 0]$, the boundary displacements $u_1(0, t) = f(t)$ and $u_1(L, t) = g(t)$, and the initial disturbances $u_1(x_1, 0) = u_1^{(0)}(x_1)$ and $\dot{u}_1(x_1, 0) = \dot{u}_1^{(0)}(x_1)$. Here, $b(x_1, t)$, $f(t)$, $g(t)$, $u_1^{(0)}(x_1)$, and $\dot{u}_1^{(0)}(x_1)$ are prescribed functions.

Use a three-dimensional form of the (H–K) principle to derive the equation of motion for $u_1 = u_1(x_1, t)$.

Solution

To define a functional $\widehat{F}\{p\}$ associated with the (H–K) principle we note that $p = [\mathbf{u}, \mathbf{E}, \mathbf{S}]$ where

$$S_{ij} = \delta_{i1}\delta_{j1}S(x_1, t) \tag{a}$$

$$E_{ij} = \frac{1}{E}[(1 + \nu)S_{ij} - \nu S_{kk}\delta_{ij}] \tag{b}$$

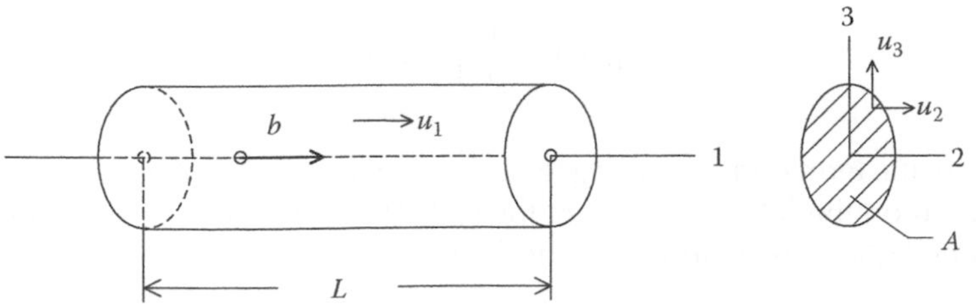

FIGURE 6.3 Rod and cross section A with the origin at the centroid of rod showing lateral displacements.

and

$$E_{ij} = \frac{1}{2}(u_{i,j} + u_{j,i}) \tag{c}$$

Here, $S = S(x_1, t)$ is an unknown stress field in the x_1 direction at time t, and E and ν are Young's modulus and Poisson's ratio, respectively.
Hence, we obtain

$$\begin{aligned} &E_{11} = \frac{S}{E}, && E_{12} = 0, && E_{13} = 0 \\ &E_{21} = 0, && E_{22} = -\frac{\nu}{E}S, && E_{33} = 0 \\ &E_{31} = 0, && E_{32} = 0, && E_{33} = -\frac{\nu}{E}S \end{aligned} \tag{d}$$

and

$$S = E\frac{\partial u_1}{\partial x_1} \tag{e}$$

$$\frac{\partial u_2}{\partial x_2} = -\nu\frac{\partial u_1}{\partial x_1}, \quad \frac{\partial u_3}{\partial x_3} = -\nu\frac{\partial u_1}{\partial x_1} \tag{f}$$

Since $\mathbf{x} = \mathbf{0}$ is a centroid of the rod, therefore $u_2 = 0$ for $x_2 = 0$ and $u_3 = 0$ for $x_3 = 0$. Integration of Equations (f) yields

$$u_2 = u_2(x_1, x_2; t) = -\nu x_2\frac{\partial u_1}{\partial x_1} \tag{g}$$

$$u_3 = u_3(x_1, x_3; t) = -\nu x_3\frac{\partial u_1}{\partial x_1} \tag{h}$$

As a functional $\widehat{F}\{p\}$ we take [see Equations (f) through (h) in Example 6.1.1]

$$\widehat{F}\{p\} = \int_{t_1}^{t_2}\{F_1(t) - K_1(t)\}dt \tag{i}$$

where

$$F_1(t) = \frac{1}{2}\int_B S_{11}E_{11}dv - \int_B bu_1\, dv \tag{j}$$

and

$$K_1(t) = \frac{1}{2}\int_B \rho\dot{\mathbf{u}}^2 dv \tag{k}$$

In Equation (j), the term corresponding to the integral over ∂B_2 in Equation 6.1.3 vanishes because $\mathbf{Sn} = \mathbf{0}$ on the lateral surface of the rod. Therefore, using Equations (d) through (h), we reduce Equations (j) and (k) to the form

$$F_1(t) = \frac{1}{2}EA\int_0^L\left(\frac{\partial u_1}{\partial x_1}\right)^2 dx_1 - A\int_0^L bu_1 dx_1 \tag{l}$$

and

$$K_1(t) = \frac{1}{2}\rho A \int_0^L \left\{ \left(\frac{\partial u_1}{\partial t} \right)^2 + \nu^2 k^2 \left(\frac{\partial^2 u_1}{\partial x_1 \partial t} \right)^2 \right\} dx_1 \tag{m}$$

where

$$Ak^2 = \int_A (x_2^2 + x_3^2)\, da = J \tag{n}$$

is the *polar moment of inertia* of the cross section (k is the *polar radius of gyration* of A). Hence, a functional $\widehat{F}\{\cdot\}$ depending on $u_1 \equiv u \in \widehat{P}$ takes the form

$$\widehat{F}\{u\} = \int_{t_1}^{t_2} \left\{ \int_0^L \left[\frac{EA}{2} \left(\frac{\partial u}{\partial x} \right)^2 - \frac{1}{2}\rho A \left\{ \left(\frac{\partial u}{\partial t} \right)^2 + \nu^2 k^2 \left(\frac{\partial^2 u}{\partial x \partial t} \right)^2 \right\} - Abu \right] \right\} dxdt \tag{o}$$

where $\widehat{P}$ is a domain of $\widehat{F}\{\cdot\}$ defined as a set of functions $u = u(x,t)$ that satisfy the boundary conditions ($x \equiv x_1$)

$$u(0,t) = f(t), \quad u(L,t) = g(t) \quad t > 0 \tag{p}$$

Let $u \in \widehat{P}$ and $\widetilde{u} \in \widehat{P}$. Then $u + \omega\widetilde{u} \in \widehat{P}$ for every scalar ω. Also, suppose that

$$u(x,t_1) = u_1(x), \quad u(x,t_2) = u_2(x) \tag{q}$$

where $u_1(x)$ and $u_2(x)$ are prescribed functions on $[0, L]$. Then

$$\widetilde{u}(x,t_1) = 0, \quad \widetilde{u}(x,t_2) = 0, \tag{r}$$

and

$$\widetilde{u}(0,t) = 0, \quad \widetilde{u}(L,t) = 0, \tag{s}$$

It follows from Equation (o) that the functional $\widehat{F}\{\cdot\}$ at $\nu = 0$ reduces to the functional $\mathcal{K}\{\cdot\}$ of Example 6.1.1 [see Equation (i) in that example].
Also note that

$$\left\{ \frac{\partial}{\partial \omega} \left[\frac{\partial^2}{\partial x \partial t}(u + \omega\widetilde{u}) \right]^2 \right\}_{\omega=0} = 2 \frac{\partial^2 u}{\partial x \partial t} \cdot \frac{\partial^2 \widetilde{u}}{\partial x \partial t} \tag{t}$$

and

$$\begin{aligned} \int_{t_1}^{t_2} dt \int_0^L \frac{\partial^2 u}{\partial x \partial t} \frac{\partial^2 \widetilde{u}}{\partial x \partial t} dx &= \int_{t_1}^{t_2} \left[\frac{\partial^2 u}{\partial x \partial t} \frac{\partial \widetilde{u}}{\partial t} \right]_{x=0}^{x=L} dt - \int_{t_1}^{t_2} \left[\int_0^L \frac{\partial^3 u}{\partial x^2 \partial t} \frac{\partial \widetilde{u}}{\partial t} dx \right] dt \\ &= \int_{t_1}^{t_2} \left[\frac{\partial^2 u}{\partial x \partial t} \frac{\partial \widetilde{u}}{\partial t} \right]_{x=0}^{x=L} dt - \int_0^L \left[\frac{\partial^3 u}{\partial x^2 \partial t} \widetilde{u} \right]_{t=t_1}^{t=t_2} dx + \int_{t_1}^{t_2} \left[\int_0^L \frac{\partial^4 u}{\partial x^2 \partial t^2} \widetilde{u}\, dx \right] dt \end{aligned} \tag{u}$$

Since, by virtue of Equations (r) and (s), the first and second integrals on the RHS of Equation (u) vanish, we receive

$$\int_{t_1}^{t_2}\int_0^L \frac{\partial^2 u}{\partial x\partial t}\frac{\partial^2 \tilde{u}}{\partial x\partial t}\,dxdt = \int_{t_1}^{t_2}\int_0^L \frac{\partial^4 u}{\partial x^2\partial t^2}\,\tilde{u}\,dxdt \tag{v}$$

Therefore, using the definition of $\delta\widehat{F}\{u\}$ and proceeding in a way similar to that of Example 6.1.1, we obtain

$$\delta_{\tilde{u}}\widehat{F}\{u\} = \frac{d}{d\omega}\widehat{F}\{u+\omega\tilde{u}\}\bigg|_{\omega=0} = \delta\widehat{F}\{u\}$$

$$= A\int_{t_1}^{t_2}\int_0^L \left[-E\frac{\partial^2 u}{\partial x^2} + \rho\left(\frac{\partial^2 u}{\partial t^2} - \nu^2 k^2\frac{\partial^4 u}{\partial x^2\partial t^2}\right) - b\right]\tilde{u}\,dxdt \tag{w}$$

and the (H–K) principle reads

$$\delta_{\tilde{u}}\widehat{F}\{u\} = 0 \quad \text{for } \tilde{u}\in\widehat{P} \tag{x}$$

where the LHS of Equation (x) is given by Equation (w).

Hence, by the Fundamental Lemma of the calculus of variations, due to arbitrariness of $\tilde{u}$ and conditions (r) and (s), we arrive at

$$\frac{\partial^2 u}{\partial x^2} + \frac{\nu^2 k^2}{c^2}\frac{\partial^4 u}{\partial x^2\partial t^2} + \frac{b}{E} = \frac{1}{c^2}\frac{\partial^2 u}{\partial t^2} \tag{y}$$

where

$$c = \sqrt{\frac{E}{\rho}} \tag{z}$$

Equation (y) represents the longitudinal wave equation for a rod with effects of lateral inertia. This completes the solution. □

In winding up this section, we note that the (H–K) principle may be used to derive the equation of motion and traction boundary conditions for three-dimensional elastic bodies such as hollow cylinders, thick plates, and open cylindrical and spherical shells. It may also be used to obtain an approximate solution to a particular initial–boundary value problem by an extended Rayleigh–Ritz method [1]. The Rayleigh–Ritz method of elastostatics is discussed in Section 5.1.2.

6.2 GURTIN'S CONVOLUTIONAL VARIATIONAL PRINCIPLES

6.2.1 Dynamic Principles with Counterparts in Elastostatics

In the previous section, we formulated the (H–K) principle in which a displacement vector $\mathbf{u} = \mathbf{u}(\mathbf{x}, t)$ needs to be prescribed at two points t_1 and t_2 of the time axis. If $t_1 = 0$, then $\mathbf{u}(\mathbf{x}, 0)$ may be identified with the initial value of the displacement vector in the formulation of an initial–boundary value problem, however, the value of $\mathbf{u}(\mathbf{x}, t_2)$ is not available in this formulation. This is the reason why the (H–K) principle cannot be used to describe the initial–boundary value problem. A full variational characterization of an initial–boundary value problem of elastodynamics is from Gurtin [2].

In the following, we formulate a number of Gurtin's convolutional principles, some of which have counterparts in elastostatics, and some of which do not. We begin with a variational principle that is a counterpart of the Hu–Washizu principle, or the (H) principle, of elastostatics, see Equations 5.2.6 and 5.2.7. To this end, we recall the definition of the pseudo-body force field $\mathbf{f}$ associated with an initial–boundary value problem. The force field $\mathbf{f}$ is defined by (see Equations 4.2.118 and 4.2.119),

$$\mathbf{f}(\mathbf{x}, t) = i * \mathbf{b}(\mathbf{x}, t) + \rho(x)\,[\mathbf{u}_0(\mathbf{x}) + t\dot{\mathbf{u}}_0] \tag{6.2.1}$$

where

$$i = i(t) = t \tag{6.2.2}$$

(G 1) **Principle of Gurtin.** Let $\mathcal{R}$ denote the set of all admissible processes, and for each $t \geq 0$ define the functional $G_t = G_t\{\cdot\}$ on $\mathcal{R}$ by

$$\begin{aligned} G_t\{p\} = & \int\limits_B \left\{ \frac{1}{2}\, i * \mathbf{E} * \mathsf{C}[\mathbf{E}] + \frac{1}{2}\, \rho\mathbf{u} * \mathbf{u} - i * \mathbf{S} * \mathbf{E} - (i * \operatorname{div}\mathbf{S} + \mathbf{f}) * \mathbf{u} \right\} dv \\ & + \int\limits_{\partial B_1} i * \mathbf{s} * \widehat{\mathbf{u}}\, da + \int\limits_{\partial B_2} i * (\mathbf{s} - \widehat{\mathbf{s}}) * \mathbf{u}\, da \end{aligned} \tag{6.2.3}$$

for every $p = [\mathbf{u}, \mathbf{E}, \mathbf{S}] \in \mathcal{R}$. Then

$$\delta_{\tilde{p}} G_t\{p\} \equiv \delta G_t\{p\} = 0 \quad t \geq 0 \tag{6.2.4}$$

if and only if p is a solution to the mixed initial–boundary value problem.

Proof: We will present the proof in two parts, (a) and (b):

(a) We assume that p is a solution to the mixed problem, and show that Equation 6.2.4 is satisfied.
(b) We assume that Equation 6.2.4 is satisfied for every $\tilde{p} \in \mathcal{R}$, and show that p is a solution to the mixed problem.

Proof of (a). First, we note that if $p = [\mathbf{u}, \mathbf{E}, \mathbf{S}]$ and $\tilde{p} = [\widetilde{\mathbf{u}}, \widetilde{\mathbf{E}}, \widetilde{\mathbf{S}}]$ are two admissible processes belonging to $\mathcal{R}$, then $p + \omega\tilde{p} \in \mathcal{R}$ for every scalar ω, see Section 4.2.3. Next, we note that due to the symmetry of C

$$\widetilde{\mathbf{E}} * \mathsf{C}[\mathbf{E}] = \mathsf{C}[\widetilde{\mathbf{E}}] * \mathbf{E} = \mathbf{E} * \mathsf{C}[\widetilde{\mathbf{E}}] \tag{6.2.5}$$

To show the identity relations 6.2.5, we note that, omitting dependence on **x**,

$$
\begin{aligned}
\widetilde{\mathbf{E}} * \mathsf{C}[\mathbf{E}] \equiv \widetilde{E}_{ij} * C_{ijkl}E_{kl} &= \int_0^t \widetilde{E}_{ij}(t-\tau)C_{ijkl}E_{kl}(\tau)d\tau \\
&= \int_0^t E_{ij}(\tau)C_{ijkl}\widetilde{E}_{kl}(t-\tau)d\tau = \int_0^t E_{ij}(t-\tau)C_{ijkl}\widetilde{E}_{kl}(\tau)d\tau \\
&= \mathbf{E} * \mathsf{C}[\widetilde{\mathbf{E}}] = \mathsf{C}[\widetilde{\mathbf{E}}] * \mathbf{E}
\end{aligned}
\tag{6.2.6}
$$

Hence, if we introduce the notation

$$
U_t\{\mathbf{E}\} = \frac{1}{2}\int_B i * \mathbf{E} * \mathsf{C}[\mathbf{E}]\, dv \tag{6.2.7}
$$

then, using Equation 6.2.5, we obtain

$$
U_t\{\mathbf{E} + \omega\widetilde{\mathbf{E}}\} = U_t\{\mathbf{E}\} + \omega^2 U_t\{\widetilde{\mathbf{E}}\} + \omega\int_B i * \widetilde{\mathbf{E}} * \mathsf{C}[\mathbf{E}]\, dv \tag{6.2.8}
$$

Clearly, since $p + \omega\widetilde{p} \in \mathcal{R}$, then value $G_t\{.\}$ at $(p + \omega\widetilde{p})$ makes sense (is well defined) and by virtue of Equations 6.2.3 and 6.2.8 we obtain

$$
\begin{aligned}
G_t\{p + \omega\widetilde{p}\} = {} & U_t\{\mathbf{E}\} + \omega^2 U_t\{\widetilde{\mathbf{E}}\} + \omega\int_B i * \widetilde{\mathbf{E}} * \mathsf{C}[\mathbf{E}]\, dv \\
& + \int_B \Big\{\frac{1}{2}\,\rho(\mathbf{u} + \omega\widetilde{\mathbf{u}}) * (\mathbf{u} + \omega\widetilde{\mathbf{u}}) - i * (\mathbf{S} + \omega\widetilde{\mathbf{S}}) * (\mathbf{E} + \omega\widetilde{\mathbf{E}}) \\
& - [i * (\operatorname{div}\mathbf{S} + \omega\operatorname{div}\widetilde{\mathbf{S}}) + \mathbf{f}] * (\mathbf{u} + \omega\widetilde{\mathbf{u}})\Big\}\, dv \\
& + \int_{\partial B_1} i * (\mathbf{s} + \omega\widetilde{\mathbf{s}}) * \widehat{\mathbf{u}}\, da \\
& + \int_{\partial B_2} i * (\mathbf{s} + \omega\widetilde{\mathbf{s}} - \widehat{\mathbf{s}}) * (\mathbf{u} + \omega\widetilde{\mathbf{u}})\, da
\end{aligned}
\tag{6.2.9}
$$

Differentiating Equation 6.2.9 with respect to ω and taking $\omega = 0$ we arrive at

$$
\begin{aligned}
\delta_{\widetilde{p}} G_t\{p\} = {} & \int_B \{i * (\mathsf{C}[\mathbf{E}] - \mathbf{S}) * \widetilde{\mathbf{E}} - i * \widetilde{\mathbf{S}} * \mathbf{E} - (i * \operatorname{div}\mathbf{S} + \mathbf{f} - \rho\mathbf{u}) * \widetilde{\mathbf{u}} \\
& - i * \operatorname{div}\widetilde{\mathbf{S}} * \mathbf{u}\}\, dv + \int_{\partial B_1} i * \widetilde{\mathbf{s}} * \widehat{\mathbf{u}}\, da \\
& + \int_{\partial B_2} i * [\widetilde{\mathbf{s}} * \mathbf{u} + (\mathbf{s} - \widehat{\mathbf{s}}) * \widetilde{\mathbf{u}}]\, da
\end{aligned}
\tag{6.2.10}
$$

Now, by using the divergence theorem, we write

$$\begin{aligned} i * \int_B (\operatorname{div} \widetilde{\mathbf{S}}) * \mathbf{u}\, dv &= i * \int_B u_k * \widetilde{S}_{kj,j}\, dv \\ &= i * \int_B [(u_k * \widetilde{S}_{kj})_{,j} - u_{k,j} * \widetilde{S}_{kj}]\, dv \\ &= i * \int_{\partial B} \mathbf{u} * \widetilde{\mathbf{s}}\, da - i * \int_B (\widehat{\nabla} \mathbf{u}) * \widetilde{\mathbf{S}}\, dv \\ &= i * \int_{\partial B_1} \mathbf{u} * \widetilde{\mathbf{s}}\, da + i * \int_{\partial B_2} \mathbf{u} * \widetilde{\mathbf{s}}\, da \\ &\quad - i * \int_B (\widehat{\nabla} \mathbf{u}) * \widetilde{\mathbf{S}}\, dv \end{aligned} \tag{6.2.11}$$

Hence, substitution of Equation 6.2.11 into Equation 6.2.10 yields

$$\begin{aligned} \delta_{\tilde{p}} G_t\{p\} = & \int_B i * (\mathbf{C}[\mathbf{E}] - \mathbf{S}) * \widetilde{\mathbf{E}}\, dv \\ & - \int_B (i * \operatorname{div} \mathbf{S} + \mathbf{f} - \rho \mathbf{u}) * \widetilde{\mathbf{u}}\, dv + \int_B i * (\widehat{\nabla} \mathbf{u} - \mathbf{E}) * \widetilde{\mathbf{S}}\, dv \\ & + \int_{\partial B_1} i * (\widehat{\mathbf{u}} - \mathbf{u}) * \widetilde{\mathbf{s}}\, da + \int_{\partial B_2} i * (\mathbf{s} - \widehat{\mathbf{s}}) * \widetilde{\mathbf{u}}\, da \end{aligned} \tag{6.2.12}$$

Since $p = [\mathbf{u}, \mathbf{E}, \mathbf{S}]$ is a solution to the mixed problem, therefore, by Theorem 3 of Section 4.2.4,

$$\mathbf{S} = \mathbf{C}[\mathbf{E}] \quad \text{on } B \times [0, \infty) \tag{6.2.13}$$

$$i * \operatorname{div} \mathbf{S} + \mathbf{f} = \rho \mathbf{u} \quad \text{on } B \times [0, \infty) \tag{6.2.14}$$

$$\mathbf{E} = \widehat{\nabla} \mathbf{u} \quad \text{on } B \times [0, \infty) \tag{6.2.15}$$

$$\mathbf{u} = \widehat{\mathbf{u}} \quad \text{on } \partial B_1 \times [0, \infty) \tag{6.2.16}$$

$$\mathbf{s} = \widehat{\mathbf{s}} \quad \text{on } \partial B_2 \times [0, \infty) \tag{6.2.17}$$

and the RHS of Equation 6.2.12 vanishes, which implies Equation 6.2.4.

This completes the proof of (a).

Proof of (b). We assume that

$$\delta_{\tilde{p}} G_t\{p\} = 0 \quad \text{for every } \widetilde{p} \in \mathcal{R} \tag{6.2.18}$$

where the LHS of Equation 6.2.18 is given by Equation 6.1.12. We are to prove that p is a solution to the mixed problem, which means that it satisfies Equations 6.2.13 through 6.2.17.

First, we choose $\widetilde{p} \in \mathcal{R}$ in the form $\widetilde{p} = [\mathbf{0}, \widetilde{\mathbf{E}}, \mathbf{0}]$, where $\widetilde{\mathbf{E}}$ is an arbitrary symmetric second-order tensor field on $\bar{B} \times [0, \infty)$ that vanishes near ∂B. Then, from Equation 6.2.18

$$\int_B i * (\mathbf{C}[\mathbf{E}] - \mathbf{S}) * \widetilde{\mathbf{E}}\, dv = 0 \tag{6.2.19}$$

Since $\widetilde{\mathbf{E}}$ is an arbitrary tensor field on $\bar{B} \times [0, \infty)$ that vanishes near ∂B, by an extension of the Fundamental Lemma of the calculus of variations involving the convolution product, Section 5.2, we obtain

$$\mathbf{S} = \mathbf{C}[\mathbf{E}] \quad \text{on } B \times [0, \infty) \tag{6.2.20}$$

which implies Equation 6.2.13.

Next, we choose $\widetilde{p} \in \mathcal{R}$ in the form $\widetilde{p} = [\widetilde{\mathbf{u}}, \mathbf{0}, \mathbf{0}]$, where $\widetilde{\mathbf{u}}$ is an arbitrary vector field on $\bar{B} \times [0, \infty)$ that vanishes near ∂B. Then by Equations 6.2.18 and 6.2.20,

$$\int_B (i * \operatorname{div} \mathbf{S} + \mathbf{f} - \rho \mathbf{u}) * \widetilde{\mathbf{u}}\, dv = 0 \tag{6.2.21}$$

and this equation together with an extension of the Fundamental Lemma of the calculus of variations imply that

$$i * \operatorname{div} \mathbf{S} + \mathbf{f} = \rho \mathbf{u} \quad \text{on } \bar{B} \times [0, \infty) \tag{6.2.22}$$

This shows that Equation 6.2.14 is satisfied.

To show that Equation 6.2.15 holds true, we select $\widetilde{p} \in \mathcal{R}$ in the form $\widetilde{p} = [\mathbf{0}, \mathbf{0}, \widetilde{\mathbf{S}}]$, where $\widetilde{\mathbf{S}}$ is an arbitrary symmetric second-order tensor field on $\bar{B} \times [0, \infty)$ that vanishes near ∂B. Then, from Equations 6.2.18, 6.2.20, and 6.2.22, we get

$$\int_B i * (\widehat{\nabla} \mathbf{u} - \mathbf{E}) * \widetilde{\mathbf{S}}\, dv = 0 \tag{6.2.23}$$

Hence, by an extension of the Fundamental Lemma of the calculus of variations, we obtain

$$\mathbf{E} = \widehat{\nabla} \mathbf{u} \quad \text{on } B \times [0, \infty) \tag{6.2.24}$$

which proves Equation 6.2.15.

Next, if we take $\widetilde{p} \in \mathcal{R}$ in the form $\widetilde{p} = [\widetilde{\mathbf{u}}, \mathbf{0}, \mathbf{0}]$, where $\widetilde{\mathbf{u}}$ is an arbitrary vector field that vanishes near ∂B_1, then from Equations 6.2.18, 6.2.20, and 6.2.24, we find

$$\int_{\partial B_2} i * (\mathbf{s} - \widehat{\mathbf{s}}) * \widetilde{\mathbf{u}}\, da = 0 \tag{6.2.25}$$

and this and an extension of the Fundamental Lemma of the calculus of variations imply that

$$\mathbf{s} = \widehat{\mathbf{s}} \quad \text{on } \partial B_2 \times [0, \infty) \tag{6.2.26}$$

which means that Equation 6.2.17 is satisfied.

Finally, if we select $\widetilde{p} \in \mathcal{R}$ in the form $\widetilde{p} = [\mathbf{0}, \mathbf{0}, \widetilde{\mathbf{S}}]$, where $\widetilde{\mathbf{S}}$ is an arbitrary symmetric second-order tensor field on $\bar{B} \times [0, \infty)$ such that $\widetilde{\mathbf{s}} = \widetilde{\mathbf{S}}\mathbf{n}$ vanishes near ∂B_2, then by Equations 6.2.18, 6.2.20, 6.2.22, 6.2.24, and 6.2.26, we arrive at

$$\int_{\partial B_1} i * (\widehat{\mathbf{u}} - \mathbf{u}) * \widetilde{\mathbf{s}}\, da = 0 \tag{6.2.27}$$

Equation 6.2.27 together with an extension of the Fundamental Lemma of the calculus of variations imply that

$$\mathbf{u} = \widehat{\mathbf{u}} \quad \text{on } \partial B_1 \times [0, \infty) \tag{6.2.28}$$

and this completes the proof of (b). Hence, the proof of the (G 1) principle is complete.

If the set of all admissible processes $\mathcal{R}$ is suitably restricted, and the functional $G_t\{\cdot\}$ in Equation 6.2.3 is modified accordingly, then the (G 1) principle reduces to an analog of the Hellinger–Prange–Reissner principle, see the (R) principle, Section 5.2.

(G 2) **Principle of Gurtin.** Let $\mathcal{R}_1$ denote the set of all admissible processes that satisfy the strain–displacement relation, and for each $t \geq 0$ define the functional $H_t\{.\}$ on $\mathcal{R}_1$ by

$$H_t\{p\} = \int_B \left\{ \frac{1}{2}\, i * \mathbf{S} * \mathsf{K}[\mathbf{S}] - \frac{1}{2}\, \rho \mathbf{u} * \mathbf{u} - i * \mathbf{S} * \mathbf{E} + \mathbf{f} * \mathbf{u} \right\} dv$$
$$+ \int_{\partial B_1} [i * \mathbf{s} * (\mathbf{u} - \widehat{\mathbf{u}})]\, da + \int_{\partial B_2} (i * \widehat{\mathbf{s}} * \mathbf{u})\, da \tag{6.2.29}$$

for every $p = [\mathbf{u}, \mathbf{E}, \mathbf{S}] \in \mathcal{R}_1$. Then

$$\delta H_t\{p\} = 0 \quad t \geq 0 \tag{6.2.30}$$

if and only if p is a solution to the mixed problem.

The proof of this principle is quite similar to that of the (G 1) principle, and it will not be given here.

Notes:

(1) We observe that the (G 1) principle is an analog of the (H) principle of elastostatics, see Section 5.2, in the following sense:

- The domain of the functional $G_t\{\cdot\}$ of the (G 1) principle is the set of all admissible processes, while the functional $H\{\cdot\}$ of the (H) principle is defined on the set of all admissible states. This means that: (a) the admissible processes in the (G 1) principle do not have to satisfy any of the field equations, initial conditions, and the boundary conditions; (b) the admissible states in the (H) principle are not required to meet any of the field equations and boundary conditions.
- The structure of the functionals $G_t\{\cdot\}$ and $H\{\cdot\}$ is similar: if in $G_t\{\cdot\}$ the term containing ρ and the operation $i*$ are omitted, and $\{*\}$ is replaced by $\{\cdot\}$ (the dot product operation), and $\mathbf{f}$ is identified with $\mathbf{b}$, then $G_t\{\cdot\}$ is reduced to $H\{\cdot\}$.

(2) There exists an analogy between the (G 2) principle and the (R) principle, Section 5.2:
- Both the domain of the functional $H_t\{\cdot\}$ of the (G 2) principle and the domain of the functional $H_1\{\cdot\}$ of the (R) principle are restricted by the same strain–displacement relation.
- The structure of the functionals $H_t\{\cdot\}$ and $H_1\{\cdot\}$ is similar to that in Note (1).

Example 6.2.1

Let $p = [\mathbf{u}, \mathbf{E}, \mathbf{S}]$ be a kinematically admissible process, which means that p satisfies the strain–displacement and stress–strain relations, and the displacement boundary condition. For each $t \geq 0$, let $F_t = F_t\{\cdot\}$ be the functional defined on the set of all kinematically admissible processes by

$$F_t\{p\} = \frac{1}{2}\int_B (i * \mathbf{S} * \mathbf{E} + \rho\mathbf{u} * \mathbf{u} - 2\mathbf{f} * \mathbf{u})\, dv - \int_{\partial B_2} i * \widehat{\mathbf{s}} * \mathbf{u}\, da \tag{a}$$

Then,

$$\delta F_t\{p\} = 0 \tag{b}$$

if and only if p is a solution to the mixed problem.

Solution

Let Q denote the set of all kinematically admissible processes $p = [\mathbf{u}, \mathbf{E}, \mathbf{S}]$. Let $\widetilde{p} \in Q$. Then

$$p + \omega\widetilde{p} \in Q \quad \text{for every scalar } \omega \tag{c}$$

which implies

$$\widetilde{\mathbf{u}} = \mathbf{0} \quad \text{on } \partial B_1 \times [0, \infty) \tag{d}$$

Also, since $p \in Q$ and $\widetilde{p} \in Q$, then

$$\int_B i * \mathbf{S} * \widetilde{\mathbf{E}}\, dv = \int_B i * \mathbf{S} * \nabla\widetilde{\mathbf{u}}\, dv \tag{e}$$

and by the divergence theorem we obtain

$$\begin{aligned}\int_B i * \mathbf{S} * \widetilde{\mathbf{E}}\, dv &= \int_{\partial B} i * \mathbf{s} * \widetilde{\mathbf{u}}\, da - \int_B i * (\operatorname{div}\mathbf{S}) * \widetilde{\mathbf{u}}\, dv \\ &= \int_{\partial B_1} i * \mathbf{s} * \widetilde{\mathbf{u}}\, da + \int_{\partial B_2} i * \mathbf{s} * \widetilde{\mathbf{u}}\, da - \int_B i * (\operatorname{div}\mathbf{S}) * \widetilde{\mathbf{u}}\, dv\end{aligned} \tag{f}$$

Now, from the definition of the functional $F_t = F_t\{\cdot\}$ given by (a), we obtain

$$F_t\{p + \omega\tilde{p}\} = \frac{1}{2}\int_B [i * (\mathbf{S} + \omega\tilde{\mathbf{S}}) * (\mathbf{E} + \omega\tilde{\mathbf{E}}) + \rho(\mathbf{u} + \omega\tilde{\mathbf{u}}) * (\mathbf{u} + \omega\tilde{\mathbf{u}}) - 2\mathbf{f} * (\mathbf{u} + \omega\tilde{\mathbf{u}})]\,dv - \int_{\partial B_2} i * \hat{\mathbf{s}} * (\mathbf{u} + \omega\tilde{\mathbf{u}})\,da \qquad \text{(g)}$$

Since

$$(\mathbf{S} + \omega\tilde{\mathbf{S}}) * (\mathbf{E} + \omega\tilde{\mathbf{E}}) = \mathbf{S} * \mathbf{E} + \omega^2\tilde{\mathbf{S}} * \tilde{\mathbf{E}} + \omega(\mathbf{S} * \tilde{\mathbf{E}} + \tilde{\mathbf{S}} * \mathbf{E}) \qquad \text{(h)}$$

and

$$\tilde{\mathbf{S}} * \mathbf{E} = \mathbf{E} * \tilde{\mathbf{S}} = \mathbf{E} * \mathbf{C}[\tilde{\mathbf{E}}] = \tilde{\mathbf{E}} * \mathbf{C}[\mathbf{E}] = \tilde{\mathbf{E}} * \mathbf{S} = \mathbf{S} * \tilde{\mathbf{E}} \qquad \text{(i)}$$

therefore,

$$(\mathbf{S} + \omega\tilde{\mathbf{S}}) * (\mathbf{E} + \omega\tilde{\mathbf{E}}) = \mathbf{S} * \mathbf{E} + \omega^2\tilde{\mathbf{S}} * \tilde{\mathbf{E}} + 2\omega\mathbf{S} * \tilde{\mathbf{E}} \qquad \text{(j)}$$

Differentiation of (g) with respect to ω, using (j) and putting $\omega = 0$, leads to

$$\delta_{\tilde{p}}F_t\{p\} \equiv \frac{d}{d\omega}F_t\{p + \omega\tilde{p}\}\Big|_{\omega=0} = \int_B (i * \mathbf{S} * \tilde{\mathbf{E}} + \rho\mathbf{u} * \tilde{\mathbf{u}} - \mathbf{f} * \tilde{\mathbf{u}})\,dv - \int_{\partial B_2} i * \hat{\mathbf{s}} * \tilde{\mathbf{u}}\,da \qquad \text{(k)}$$

Substituting (f) into (k), we get

$$\delta_{\tilde{p}}F_t\{p\} = \int_B [i * (\operatorname{div}\mathbf{S}) + \mathbf{f} - \rho\mathbf{u}] * \tilde{\mathbf{u}}\,dv + \int_{\partial B_1} i * \mathbf{s} * \tilde{\mathbf{u}}\,da + \int_{\partial B_2} i * (\mathbf{s} - \hat{\mathbf{s}}) * \tilde{\mathbf{u}}\,da \qquad \text{(l)}$$

Now, because of (d), the integral over ∂B_1 vanishes, and we obtain

$$\delta_{\tilde{p}}F_t\{p\} = \int_B [i * (\operatorname{div}\mathbf{S}) + \mathbf{f} - \rho\mathbf{u}] * \tilde{\mathbf{u}}\,dv + \int_{\partial B_2} i * (\mathbf{s} - \hat{\mathbf{s}}) * \tilde{\mathbf{u}}\,da \qquad \text{(m)}$$

Hence, if $p \in Q$ is a solution to the mixed problem, it follows from (m) and from Equations 4.2.131 and 4.2.134, that

$$\delta F_t\{p\} = 0 \qquad \text{(n)}$$

This is equivalent to (b).

Conversely, we assume that

$$\delta_{\tilde{p}}F_t\{p\} = 0 \quad \text{for every } \tilde{p} \in Q \qquad \text{(o)}$$

and show that p is a solution to the mixed problem. To this end, we select $\widetilde{p} \in Q$ in the form $\widetilde{p} = [\widetilde{\mathbf{u}}, \mathbf{0}, \mathbf{0}]$ where $\widetilde{\mathbf{u}}$ is an arbitrary vector field on $\bar{B} \times [0, \infty)$ that vanishes near ∂B. Then, (m) and (o) imply that

$$\int_B [i * (\operatorname{div} \mathbf{S}) + \mathbf{f} - \rho \mathbf{u}] * \widetilde{\mathbf{u}}\, dv = 0 \tag{p}$$

and this together with an extension of the Fundamental Lemma of the calculus of variations leads to

$$i * (\operatorname{div} \mathbf{S}) + \mathbf{f} - \rho \mathbf{u} = 0 \quad \text{on } \bar{B} \times [0, \infty) \tag{q}$$

Finally, if we select $\widetilde{p}$ in the form $\widetilde{p} = [\widetilde{\mathbf{u}}, \mathbf{0}, \mathbf{0}]$ where $\widetilde{\mathbf{u}}$ is arbitrary on $\bar{B} \times [0, \infty)$ but vanishes near ∂B_1, then (m), (o), and (q) yield

$$\int_{\partial B_2} i * (\mathbf{s} - \widehat{\mathbf{s}}) * \widetilde{\mathbf{u}}\, da = 0 \tag{r}$$

This together with the arbitrariness of $\widetilde{\mathbf{u}}$ and an extension of the Fundamental Lemma of the calculus of variations implies that

$$\mathbf{s} = \widehat{\mathbf{s}} \quad \text{on } \partial B_2 \times [0, \infty) \tag{s}$$

This completes the converse part of the principle in this example. As a result, the proof is complete. □

Note: In this example, the principle constitutes an analog of the principle of minimum potential energy of elastostatics, see (I) Principle of Section 5.1.1, in the following sense:

- The domain of the functional $F_t\{\cdot\}$ given by (a) is the set of all kinematically admissible processes Q, while the domain of the functional $F\{\cdot\}$ given by Equation 5.1.7 is the set of all kinematically admissible states R.
- The structure of the functionals $F_t\{\cdot\}$ and $F\{\cdot\}$ is similar, as discussed in Note (1) which follows the (G 2) principle.

Example 6.2.2

Consider the following initial–boundary value problem for the classical wave equation. Find a scalar-valued function $u = u(\mathbf{x}, t)$ on $\bar{B} \times [0, \infty)$ that satisfies the wave equation

$$\nabla^2 u - \frac{1}{c^2} \frac{\partial^2 u}{\partial t^2} = -F \quad \text{on } B \times [0, \infty) \tag{a}$$

the initial conditions

$$u(\mathbf{x}, 0) = u_0(\mathbf{x}), \quad \dot{u}(\mathbf{x}, 0) = \dot{u}_0(\mathbf{x}) \quad \text{on } B \tag{b}$$

and the boundary condition

$$u = \widehat{u} \quad \text{on } \partial B \times [0, \infty) \tag{c}$$

Here $F = F(\mathbf{x}, t)$ is a prescribed function on $\bar{B} \times [0, \infty)$, $u_0(\mathbf{x})$ and $\dot{u}_0(\mathbf{x})$ are prescribed on B, and $\widehat{u}$ is given on $\partial B \times [0, \infty)$. Moreover, in (a) c is a positive constant.

Formulate a variational principle that characterizes the problem stated by (a) through (c).

Solution

We will proceed in two steps:

(A) We replace the initial–boundary value problem by an equivalent boundary value problem in which the initial data are incorporated into a field equation.
(B) We formulate a variational principle for the equivalent problem.

Part (A). We apply the Laplace transform to (a) and using the initial conditions (b) we obtain

$$\nabla^2 \bar{u} - \frac{1}{c^2} (p^2 \bar{u} - p u_0 - \dot{u}_0) = -\bar{F} \tag{d}$$

Dividing this equation by p^2 results in

$$\frac{1}{p^2} \nabla^2 \bar{u} - \frac{1}{c^2} \bar{u} = -\bar{G} \tag{e}$$

where

$$\bar{G} = \frac{\bar{F}}{p^2} + \frac{1}{c^2} \left(\frac{u_0}{p} + \frac{\dot{u}_0}{p^2} \right) \tag{f}$$

Inverting (e) and (f) we get

$$i * \nabla^2 u - \frac{1}{c^2} u = -G \quad \text{on } \bar{B} \times [0, \infty) \tag{g}$$

where

$$i = i(t) = t \tag{h}$$

and

$$G = i * F + \frac{1}{c^2} (u_0 + t \dot{u}_0) \tag{i}$$

Hence, we come to the conclusion that $u = u(\mathbf{x}, t)$ satisfies (a) and the initial conditions (b) if and only if u satisfies the integro-differential equation (g). As a result, the initial–boundary value problem is equivalent to the boundary value problem: Find a function $u = u(\mathbf{x}, t)$ on $\bar{B} \times [0, \infty)$ that satisfies the equation

$$i * \nabla^2 u - \frac{1}{c^2} u = -G \quad \text{on } \bar{B} \times [0, \infty) \tag{j}$$

subject to the boundary condition

$$u = \widehat{u} \quad \text{on } \partial B \times [0, \infty) \tag{k}$$

This completes part (A) of the solution.

Part (B). We need to define a functional in such a way that the vanishing of its first variation over its domain implies the wave equation (a), the initial conditions (b), and the boundary condition (c). More precisely, we are to prove the following variational principle for the wave equation:

Let $\mathcal{U}$ be the set of all functions $u = u(\mathbf{x}, t)$ on $\bar{B} \times [0, \infty)$ that satisfy the boundary condition (c). Define the functional $W_t = W_t\{\cdot\}$ on $\mathcal{U}$ by

$$W_t\{u\} = \frac{1}{2} \int_B \left(i * \nabla u * \nabla u + \frac{1}{c^2}\, u * u - 2G * u \right) dv \tag{l}$$

for every $u \in \mathcal{U}$. Then

$$\delta W_t\{u\} = 0 \tag{m}$$

at a particular $u \in \mathcal{U}$ if and only if u is a solution to the initial–boundary value problem described by (a) through (c).

We will present a proof of the principle in two parts.

(AA) We assume that u is a solution to the initial–boundary value problem (a) through (c), and show that (m) is satisfied.

(BB) We show that the condition

$$\delta_{\tilde{u}} W_t\{u\} \equiv \delta W_t\{u\} = 0 \quad \text{for every } \tilde{u} \in \mathcal{U} \tag{n}$$

implies that u is a solution to the problem.

Proof of (AA). Take two arbitrary elements u and $\tilde{u}$ belonging to $\mathcal{U}$. Then

$$u + \omega\tilde{u} \in \mathcal{U} \quad \text{for every scalar } \omega \tag{o}$$

which implies that

$$\tilde{u} = 0 \quad \text{on } \partial B \times [0, \infty) \tag{p}$$

Next, we compute the value of the functional $W_t\{\cdot\}$ at $u + \omega\tilde{u}$, and obtain

$$\begin{aligned} W_t\{u + \omega\tilde{u}\} = & \frac{1}{2} \int_B \Big[i * (\nabla u + \omega\nabla\tilde{u}) * (\nabla u + \omega\nabla\tilde{u}) \\ & + \frac{1}{c^2} (u + \omega\tilde{u}) * (u + \omega\tilde{u}) - 2G * (u + \omega\tilde{u}) \Big] dv \end{aligned} \tag{q}$$

Now,

$$(\nabla u + \omega\nabla\tilde{u}) * (\nabla u + \omega\nabla\tilde{u}) = (\nabla u) * (\nabla u) + 2\omega(\nabla u) * (\nabla\tilde{u}) + \omega^2(\nabla\tilde{u}) * (\nabla\tilde{u}) \tag{r}$$

Hence, differentiating (q) with respect to ω and putting $\omega = 0$, we obtain

$$\delta W_t\{u\} = \int_B \left[i * \nabla u * \nabla \widetilde{u} + \frac{1}{c^2}\, u * \widetilde{u} - G * \widetilde{u} \right] dv \tag{s}$$

Since,

$$\nabla u * \nabla \widetilde{u} = \text{div}(\nabla u * \widetilde{u}) - (\nabla^2 u) * \widetilde{u} \tag{t}$$

or in components,

$$u_{,i} * \widetilde{u}_{,i} = (u_{,i} * \widetilde{u})_{,i} - u_{,ii} * \widetilde{u} \tag{u}$$

then, multiplying (t) by the $i*$ operator, using the divergence theorem, and (p), we get

$$\begin{aligned} \int_B i * \nabla u * \nabla \widetilde{u}\, dv &= \int_{\partial B} i * \widetilde{u} * \frac{\partial u}{\partial n}\, da - \int_B i * (\nabla^2 u) * \widetilde{u}\, dv \\ &= -\int_B i * (\nabla^2 u) * \widetilde{u}\, dv \end{aligned} \tag{v}$$

Substituting this into the RHS of (s), we obtain

$$\delta W_t\{u\} = -\int_B \left[i * (\nabla^2 u) - \frac{1}{c^2}\, u + G \right] * \widetilde{u}\, dv \tag{w}$$

Since u is a solution to the problem (a) through (c), then u is a solution to the problem (j) and (k), hence, by (j) and (w)

$$\delta W_t\{u\} = 0 \tag{x}$$

This completes the proof of (AA).

Proof of (BB). To prove (BB), we assume that (n) is satisfied for every $\widetilde{u} \in \mathcal{U}$. Selecting $\widetilde{u}$ to be an arbitrary function on $B \times [0, \infty)$ that vanishes near ∂B, it follows from (w) that

$$\int_B \left[i * (\nabla^2 u) - \frac{1}{c^2}\, u + G \right] * \widetilde{u}\, dv = 0 \tag{y}$$

Therefore, by an extension of the Fundamental Lemma of the calculus of variations, we obtain

$$i * (\nabla^2 u) - \frac{1}{c^2}\, u = -G \quad \text{on } B \times [0, \infty) \tag{z}$$

This equation together with the fact that $u \in \mathcal{U}$ implies that (j) and (k) are satisfied, which means that u is a solution to the problem (a) through (c).

This completes the solution. □

Note: The functional $W_t\{\cdot\}$ is an analog of a functional $W\{\cdot\}$ that appears in a variational principle for the Dirichlet problem:

Find $u = u(\mathbf{x})$ that satisfies Poisson's equation

$$\nabla^2 u = -F \quad \text{on } B \tag{6.2.31}$$

and the boundary condition

$$u = \widehat{u} \quad \text{on } \partial B \tag{6.2.32}$$

where F and $\widehat{u}$ are prescribed functions. In this case $W\{\cdot\}$ takes the form

$$W\{u\} = \frac{1}{2} \int_B [(\nabla u) \cdot (\nabla u) - 2Fu] \, dv \tag{6.2.33}$$

and the domain of $W\{\cdot\}$ is a set of functions that satisfy the boundary condition 6.2.32. Thus, both the form and the domain of the functionals $W_t\{\cdot\}$ and $W\{\cdot\}$ are similar.

The functional $W\{\cdot\}$ is of the type used in a variational characterization of the deflection of an elastic membrane [3].

Example 6.2.3

The displacement initial–boundary value problem of longitudinal waves in a thin rod in which the lateral inertial effects are taken into account is described by the differential equation [see Equation (y) in Example 6.1.3 in which $\nu^2 k^2/c^2 = H^2$]

$$\left[\frac{\partial^2}{\partial x^2} \left(1 + H^2 \frac{\partial^2}{\partial t^2} \right) - \frac{1}{c^2} \frac{\partial^2}{\partial t^2} \right] u + \frac{b}{E} = 0 \quad \text{on } [0, L] \times [0, \infty) \tag{a}$$

the initial conditions

$$u(x, 0) = u_0(x), \quad \dot{u}(x, 0) = \dot{u}_0(x) \quad \text{on } [0, L] \tag{b}$$

and the boundary conditions

$$u(0, t) = f(t), \quad u(L, t) = g(t) \quad \text{on } [0, \infty) \tag{c}$$

Here, $u = u(x, t)$ is a displacement of the rod in the x direction; $b = b(x, t)$, $u_0(x)$, $\dot{u}_0(x)$, $f(t)$, and $g(t)$ are prescribed functions; and H, E, and c are positive constants.

Let U be the set of all functions $u = u(x, t)$ on $[0, L] \times [0, \infty)$ that satisfy the boundary conditions (c). Let $\widehat{b} = \widehat{b}(x, t)$ be the function on $[0, L] \times [0, \infty)$ defined by

$$\begin{aligned} \widehat{b}(x, t) = {} & \frac{b}{H} * \sin \frac{t}{H} + \frac{E}{c^2 H^2} \left(u_0 \cos \frac{t}{H} + H \dot{u}_0 \sin \frac{t}{H} \right) \\ & - E \left(u_0'' \cos \frac{t}{H} + H \dot{u}_0'' \sin \frac{t}{H} \right) \end{aligned} \tag{d}$$

where $*$ is the convolution product on the t axis, and $'' = d^2/dx^2$. Define a functional $W_t\{\cdot\}$ on U by

$$W_t\{u\} = \int_0^L \left[\frac{1}{2}\frac{\partial u}{\partial x} * \frac{\partial u}{\partial x} + \frac{1}{2}\frac{1}{c^2H^2}\left(u * u - \frac{1}{H}\sin\frac{t}{H} * u * u \right) - \frac{\widehat{b}}{E} * u \right] dx$$

$$\text{for every } u \in U \tag{e}$$

Show that

$$\delta W_t\{u\} = 0 \tag{f}$$

at a particular $u \in U$ if and only if u is a solution to the problem (a) through (c).

Solution

By applying the Laplace transform to Equation (a) and using the initial conditions (b) we obtain

$$(1 + H^2p^2)\frac{\partial^2 \overline{u}}{\partial x^2} - H^2\frac{\partial^2}{\partial x^2}(pu_0 + \dot{u}_0) - \frac{p^2}{c^2}\overline{u} + \frac{1}{c^2}(pu_0 + \dot{u}_0) + \frac{\overline{b}}{E} = 0 \tag{g}$$

where a bar over a function denotes the Laplace transform of the function with respect to time, and p is the transform parameter. Next, by dividing Equation (g) by $(1 + H^2p^2)$ we arrive at

$$\frac{\partial^2 \overline{u}}{\partial x^2} - \frac{1}{c^2H^2}\left[1 - \frac{1}{H^2(p^2 + 1/H^2)} \right]\overline{u}$$

$$+ \frac{1}{H^2}\left[\frac{\overline{b}}{E(p^2 + 1/H^2)} + \frac{1}{c^2}\frac{pu_0 + \dot{u}_0}{(p^2 + 1/H^2)} - H^2\frac{pu_0'' + \dot{u}_0''}{(p^2 + 1/H^2)} \right] = 0 \tag{h}$$

Since

$$L^{-1}\left\{ \frac{1}{p^2 + 1/H^2} \right\} = H\sin\frac{t}{H}, \quad L^{-1}\left\{ \frac{p}{p^2 + 1/H^2} \right\} = \cos\frac{t}{H} \tag{i}$$

where L^{-1} stands for the inverse Laplace transform, therefore, applying the operator L^{-1} to Equation (h) and using the convolution theorem we find that the problem (a) through (c) is equivalent to the following one: Find a function $u = u(x, t)$ on $[0, L] \times [0, \infty)$ that satisfies the integro-differential equation

$$\frac{\partial^2 u}{\partial x^2} - \frac{1}{c^2H^2}\left(u - \frac{1}{H}\sin\frac{t}{H} * u \right) + \frac{\widehat{b}}{E} = 0 \quad \text{on } [0, L] \times [0, \infty) \tag{j}$$

subject to the boundary conditions

$$u(0, t) = f(t), \quad u(L, t) = g(t) \quad \text{on } [0, \infty) \tag{k}$$

where $\widehat{b}$ is given by Equation (d).

As a result, it is sufficient to show that the condition (f) is equivalent to Equations (j) through (k). To this end we compute the first variation of the functional $W_t\{\cdot\}$ and get

$$\delta W_t\{u\} = \int_0^L \left[\frac{\partial u}{\partial x} * \frac{\partial \widetilde{u}}{\partial x} + \frac{1}{c^2H^2}\left(u * \widetilde{u} - \frac{1}{H}\sin\frac{t}{H} * u * \widetilde{u}\right) - \frac{\widehat{b}}{E} * \widetilde{u}\right] dx \qquad \text{(l)}$$

Since $u \in U$ and $u + \omega\widetilde{u} \in U$ for every scalar ω, then

$$\widetilde{u}(0,t) = \widetilde{u}(L,t) = 0 \quad \text{on } [0,\infty) \qquad \text{(m)}$$

Hence, integrating by parts the first term under the integral (l) and using (m) we arrive at

$$\delta W_t\{u\} = -\int_0^L \left[\frac{\partial^2 u}{\partial x^2} - \frac{1}{c^2H^2}\left(u - \frac{1}{H}\sin\frac{t}{H} * u\right) + \frac{\widehat{b}}{E}\right] * \widetilde{u}\, dx \qquad \text{(n)}$$

Since $\widetilde{u}$ is an arbitrary function belonging to U subject to the conditions (m), by an extension of the Fundamental Lemma of the calculus of variations involving the convolution product, the relation (f), in which the LHS is given by (n), is equivalent to Equations (j) through (k). This completes the solution. □

6.2.2 Dynamical Principles without Counterparts in Elastostatics

It was shown in Section 4.2.4 on formulation of initial–boundary value problems that a mixed problem of elastodynamics may be formulated in terms of stresses only. This implies that a single stress equation subject to boundary conditions in terms of stresses is necessary and sufficient for obtaining a solution to the mixed problem, see Theorem 3 of Section 4.2.4. Since the dynamic stress problem has no counterpart in elastostatics, therefore, the variational principles associated with the dynamic stress problem, which we are to discuss in this Section, also do not have counterparts in elastostatics.

(S 1) **Principle of Gurtin.** Let L denote the set of all admissible stress fields $\mathbf{S}$ on $\bar{B}\times[0,\infty)$.* Let $A_t\{.\}$ be the functional on L defined by

$$A_t\{\mathbf{S}\} = \frac{1}{2}\int_B \{\rho^{-1}\, i * (\operatorname{div}\mathbf{S}) * (\operatorname{div}\mathbf{S}) + \mathbf{S} * \mathsf{K}[\mathbf{S}] - 2\mathbf{S} * \nabla(\rho^{-1}\,\mathbf{f})\}\, dv$$
$$+ \int_{\partial B_1} [(\rho^{-1}\,\mathbf{f} - \widehat{\mathbf{u}}) * \mathbf{s}]\, da + \int_{\partial B_2} [\rho^{-1}\, i * (\widehat{\mathbf{s}} - \mathbf{s}) * (\operatorname{div}\mathbf{S})]\, da \qquad (6.2.34)$$

Then,

$$\delta A_t\{\mathbf{S}\} = 0 \quad \text{for } t \geq 0 \qquad (6.2.35)$$

at $\mathbf{S} \in L$ if and only if $\mathbf{S}$ corresponds to a solution to the mixed problem.

Proof: We take two arbitrary elements $\mathbf{S}$ and $\widetilde{\mathbf{S}}$ belonging to L. Then $\mathbf{S} + \omega\widetilde{\mathbf{S}} \in L$ for every scalar ω. Calculating the value of the functional $A_t\{\cdot\}$ at $\mathbf{S} + \omega\widetilde{\mathbf{S}}$ we obtain

* By the set of all admissible stress fields $\mathbf{S} = \mathbf{S}(\mathbf{x}, t)$ on $\bar{B} \times [0,\infty)$ we mean a set of all smooth symmetric second-order tensor fields.

$$A_t\{\mathbf{S}+\omega\widetilde{\mathbf{S}}\} = \frac{1}{2}\int_B \{\rho^{-1}\, i * (\operatorname{div}\mathbf{S} + \omega \operatorname{div}\widetilde{\mathbf{S}}) * (\operatorname{div}\mathbf{S} + \omega \operatorname{div}\widetilde{\mathbf{S}})$$
$$+ \ (\mathbf{S}+\omega\widetilde{\mathbf{S}}) * \mathsf{K}[\mathbf{S}+\omega\widetilde{\mathbf{S}}] - 2(\mathbf{S}+\omega\widetilde{\mathbf{S}}) * \nabla(\rho^{-1}\,\mathbf{f})\big\}\ dv$$
$$+ \int_{\partial B_1} [(\rho^{-1}\,\mathbf{f} - \widehat{\mathbf{u}}) * (\mathbf{s}+\omega\widetilde{\mathbf{s}})]\ da$$
$$+ \int_{\partial B_2} [\rho^{-1}\, i * (\widehat{\mathbf{s}} - \mathbf{s} - \omega\widetilde{\mathbf{s}}) * (\operatorname{div}\mathbf{S} + \omega \operatorname{div}\widetilde{\mathbf{S}})]\ da \qquad (6.2.36)$$

Differentiation of Equation 6.2.36 with respect to ω and using the symmetry of K yields

$$\delta_{\tilde{\mathbf{S}}} A_t\{\mathbf{S}\} \equiv \frac{d}{d\omega} A_t\{\mathbf{S}+\omega\widetilde{\mathbf{S}}\}\bigg|_{\omega=0} = \int_B \{\rho^{-1}\, i * (\operatorname{div}\mathbf{S}) * (\operatorname{div}\widetilde{\mathbf{S}})$$
$$+ \widetilde{\mathbf{S}} * \mathsf{K}[\mathbf{S}] - \widetilde{\mathbf{S}} * \widehat{\nabla}(\rho^{-1}\,\mathbf{f})\big\}\ dv + \int_{\partial B_1} (\rho^{-1}\,\mathbf{f} - \widehat{\mathbf{u}}) * \widetilde{\mathbf{s}}\ da$$
$$+ \int_{\partial B_2} \rho^{-1}\, i * [-\widetilde{\mathbf{s}} * (\operatorname{div}\mathbf{S}) + (\widehat{\mathbf{s}} - \mathbf{s}) * (\operatorname{div}\widetilde{\mathbf{S}})]\ da \qquad (6.2.37)$$

Now, by the divergence theorem and using index notation we arrive at

$$\int_B \rho^{-1}\, i * (\operatorname{div}\mathbf{S}) * (\operatorname{div}\widetilde{\mathbf{S}})\ dv = \int_B i * (\rho^{-1}\ \operatorname{div}\mathbf{S}) * (\operatorname{div}\widetilde{\mathbf{S}})\ dv$$
$$= \int_B i * \left(\rho^{-1}\ S_{ik,k}\right) * \left(\widetilde{S}_{ij,j}\right)\ dv$$
$$= \int_B i * \left[\left(\rho^{-1} S_{ik,k} * \widetilde{S}_{ij}\right)_{,j} - \left(\rho^{-1} S_{ik,k}\right)_{,j} * \widetilde{S}_{ij}\right]\ dv$$
$$= \int_{\partial B} i * \rho^{-1} S_{ik,k} * \widetilde{s}_i\ da - \int_B i * \left(\rho^{-1} S_{ik,k}\right)_{,j} * \widetilde{S}_{ij}\ dv$$
$$= \int_{\partial B_1} i * \rho^{-1}\ (\operatorname{div}\mathbf{S}) * \widetilde{\mathbf{s}}\ da + \int_{\partial B_2} i * \rho^{-1}\ (\operatorname{div}\mathbf{S}) * \widetilde{\mathbf{s}}\ da$$
$$- \int_B i * \widehat{\nabla}(\rho^{-1}\ \operatorname{div}\mathbf{S}) * \widetilde{\mathbf{S}}\ dv \qquad (6.2.38)$$

Substituting Equation 6.2.38 into Equation 6.2.37 we get

$$\delta_{\tilde{\mathbf{S}}} A_t\{\mathbf{S}\} = -\int_B \big\{\widehat{\nabla}\rho^{-1}[i * (\operatorname{div}\mathbf{S}) + \mathbf{f}] - \mathsf{K}[\mathbf{S}]\big\} * \widetilde{\mathbf{S}}\ dv$$
$$+ \int_{\partial B_1} \rho^{-1}\ [i * (\operatorname{div}\mathbf{S}) - \rho\,\widehat{\mathbf{u}} + \ \mathbf{f}] * \widetilde{\mathbf{s}}\ da$$
$$+ \int_{\partial B_2} \rho^{-1}\ i * (\widehat{\mathbf{s}} - \mathbf{s}) * (\operatorname{div}\widetilde{\mathbf{S}})\ da \qquad (6.2.39)$$

Hence, to complete the proof we need to show that:

(a) If **S** corresponds to a solution to the mixed problem then

$$\delta_{\tilde{\mathbf{S}}} A_t\{\mathbf{S}\} \equiv \delta A_t\{\mathbf{S}\} = 0 \tag{6.2.40}$$

(b) If Equation 6.2.40 holds true for every $\tilde{\mathbf{S}} \in L$ then **S** corresponds to a solution to the mixed problem.

Proof of (a). Since **S** corresponds to a solution to the mixed problem, then by Theorem 3, Equations 4.2.127 through 4.2.129, we obtain Equation 6.2.40. This completes the proof of (a).

Proof of (b). Since Equation 6.2.40 holds true for every $\tilde{\mathbf{S}} \in L$, we select first $\tilde{\mathbf{S}}$ as an arbitrary symmetric second-order tensor field on $\bar{B} \times [0, \infty)$ that vanishes near ∂B. Then, from Equations 6.2.39 and 6.2.40 we obtain

$$\int_B \left\{ \widehat{\nabla} \rho^{-1} [i * (\text{div}\, \mathbf{S}) + \mathbf{f}] - \mathsf{K}[\mathbf{S}] \right\} * \tilde{\mathbf{S}}\, dv = 0 \tag{6.2.41}$$

Hence, by an extension of the Fundamental Lemma of the calculus of variations

$$\widehat{\nabla} \rho^{-1} [i * (\text{div}\, \mathbf{S}) + \mathbf{f}] - \mathsf{K}[\mathbf{S}] = \mathbf{0} \quad \text{on } \bar{B} \times [0, \infty) \tag{6.2.42}$$

Next, if we select $\tilde{\mathbf{S}}$ to be an arbitrary symmetric tensor field on $\bar{B} \times [0, \infty)$ that vanishes near ∂B_2, then in view of Equations 6.2.39, 6.2.40, and 6.2.42 we write

$$\int_{\partial B_1} \left[\rho^{-1}\, i * (\text{div}\, \mathbf{S}) - \widehat{\mathbf{u}} + \rho^{-1}\, \mathbf{f} \right] * \tilde{\mathbf{s}}\, da = 0 \tag{6.2.43}$$

and this, together with an extension of the Fundamental Lemma of the calculus of variations, leads to

$$\rho^{-1} [i * (\text{div}\, \mathbf{S}) + \mathbf{f}] = \widehat{\mathbf{u}} \quad \text{on } \partial B_1 \times [0, \infty) \tag{6.2.44}$$

Finally, if we select $\tilde{\mathbf{S}}$ to be an arbitrary symmetric tensor field on $\bar{B} \times [0, \infty)$ that vanishes near ∂B_1, then from Equations 6.2.39, 6.2.40, 6.2.42, and 6.2.44 we obtain

$$\int_{\partial B_2} \rho^{-1}\, i * (\widehat{\mathbf{s}} - \mathbf{s}) * (\text{div}\, \tilde{\mathbf{S}})\, da = 0 \tag{6.2.45}$$

Hence, by an extension of the Fundamental Lemma of the calculus of variations, we get

$$\mathbf{s} = \widehat{\mathbf{s}} \quad \text{on } \partial B_2 \times [0, \infty) \tag{6.2.46}$$

Therefore, by Theorem 3 of Section 4.2.4, **S** corresponds to a solution to the mixed problem. This completes the proof of (b) and of the (S 1) principle.

The next variational principle will be restricted to the case of a traction boundary value problem. In connection with this, we introduce a concept of a dynamically admissible stress field.

By a *dynamically admissible stress field* we mean a symmetric second-order tensor field $\mathbf{S}$ on $\bar{B} \times [0, \infty)$ that satisfies the traction boundary condition

$$\mathbf{s} = \mathbf{Sn} = \widehat{\mathbf{s}} \quad \text{on } \partial B \times [0, \infty) \tag{6.2.47}$$

(S 2) **Principle of Gurtin.** Let M denote the set of all dynamically admissible stress fields on $\bar{B} \times [0, \infty)$, and let $B_t\{\cdot\}$ be the functional on M defined by

$$B_t\{\mathbf{S}\} = \frac{1}{2} \int_B \left\{\rho^{-1}\, i * (\operatorname{div} \mathbf{S}) * (\operatorname{div} \mathbf{S}) + \mathbf{S} * \mathsf{K}[\mathbf{S}] - 2\mathbf{S} * \widehat{\nabla}(\rho^{-1}\, \mathbf{f})\right\} dv \tag{6.2.48}$$

for every $t \geq 0$. Then,

$$\delta B_t\{\mathbf{S}\} = 0 \tag{6.2.49}$$

at a particular $\mathbf{S} \in M$ if and only if $\mathbf{S}$ corresponds to a solution to the traction problem.

Proof: Let $\mathbf{S} \in M$ and $\widetilde{\mathbf{S}} \in M$. Then

$$\mathbf{S} + \omega\widetilde{\mathbf{S}} \in M \quad \text{for every scalar } \omega \tag{6.2.50}$$

and this implies that

$$\widetilde{\mathbf{s}} = \widetilde{\mathbf{S}}\mathbf{n} = \mathbf{0} \quad \text{on } \partial B \times [0, \infty) \tag{6.2.51}$$

Computing the value of $B_t\{\cdot\}$ at $\mathbf{S} + \omega\widetilde{\mathbf{S}}$ we obtain

$$\begin{aligned} B_t\{\mathbf{S} + \omega\widetilde{\mathbf{S}}\} = \frac{1}{2} \int_B \Big\{\rho^{-1}\, i * \left[(\operatorname{div} \mathbf{S}) + \omega\left(\operatorname{div} \widetilde{\mathbf{S}}\right)\right] * \left[(\operatorname{div} \mathbf{S}) + \omega\left(\operatorname{div} \widetilde{\mathbf{S}}\right)\right] \\ + \left(\mathbf{S} + \omega\widetilde{\mathbf{S}}\right) * \mathsf{K}\left[\mathbf{S} + \omega\widetilde{\mathbf{S}}\right] - 2\left(\mathbf{S} + \omega\widetilde{\mathbf{S}}\right) * \widehat{\nabla}\left(\rho^{-1}\, \mathbf{f}\right)\Big\} dv \end{aligned} \tag{6.2.52}$$

Differentiating Equation 6.2.52 with respect to ω, putting $\omega = 0$, and using the symmetry of K, we obtain

$$\begin{aligned} \delta_{\widetilde{\mathbf{S}}} B_t\{\mathbf{S}\} &\equiv \frac{d}{d\omega}\, B_t\{\mathbf{S} + \omega\widetilde{\mathbf{S}}\}\Big|_{\omega=0} \\ &= \int_B \left\{\rho^{-1}\, i * (\operatorname{div} \mathbf{S}) * \left(\operatorname{div} \widetilde{\mathbf{S}}\right) + \widetilde{\mathbf{S}} * \mathsf{K}[\mathbf{S}] - \widetilde{\mathbf{S}} * \widehat{\nabla}\left(\rho^{-1}\, \mathbf{f}\right)\right\} dv \end{aligned} \tag{6.2.53}$$

This equation, together with Equation 6.2.38 and the boundary condition 6.2.51, implies that

$$\delta_{\widetilde{\mathbf{S}}} B_t\{\mathbf{S}\} = -\int_B \left\{\widehat{\nabla}\rho^{-1}\, [i * (\operatorname{div} \mathbf{S}) + \mathbf{f}] - \mathsf{K}[\mathbf{S}]\right\} * \widetilde{\mathbf{S}}\, dv \tag{6.2.54}$$

Obviously, the condition

$$\delta_{\bar{\mathbf{S}}} B_t\{\mathbf{S}\} = \delta B_t\{\mathbf{S}\} = 0 \tag{6.2.55}$$

is satisfied at a particular $\mathbf{S} \in M$ if and only if $\mathbf{S}$ is the stress field corresponding to a solution to the traction problem. This observation is based on steps similar to those presented in the proof of the (S 1) principle. Therefore, the proof of the (S 2) principle is completed.

Notes:

(1) We observe that the (S 1) principle and the (S 2) principle characterize the stress field corresponding to a solution to an initial–boundary value problem of classical elastodynamics in which initial data are imposed on the displacement and velocity fields.
(2) Since the domain of the (S 1) principle consists of arbitrary symmetric second-order tensor fields that are not required to satisfy any initial or boundary conditions, it may be attractive in the analysis of wave propagation in various engineering structures, such as bars, plates, and shells.

We will now present a variational principle in terms of stresses that covers a traction boundary value problem of a noncompatible elastodynamics, that is, elastodynamics of a body with continuously distributed defects.

Before formulating the principle we formulate the traction problem of noncompatible elastodynamics, see Theorem 2 of Section 4.2.5. Find a symmetric second-order tensor field $\mathbf{S} = \mathbf{S}(\mathbf{x}, t)$ on $\bar{B} \times [0, \infty)$ that satisfies the field equation

$$\widehat{\nabla}\left[\rho^{-1}(\operatorname{div}\mathbf{S})\right] - \mathsf{K}\left[\ddot{\mathbf{S}}\right] = -\mathbf{B} \quad \bar{B} \times [0, \infty) \tag{6.2.56}$$

subject to the initial conditions

$$\mathbf{S}(\mathbf{x}, 0) = \mathbf{S}_0(\mathbf{x}), \quad \dot{\mathbf{S}}(\mathbf{x}, 0) = \dot{\mathbf{S}}_0(\mathbf{x}), \quad \mathbf{x} \in B \tag{6.2.57}$$

and the boundary condition

$$\mathbf{s} = \mathbf{S}\mathbf{n} = \widehat{\mathbf{s}} \quad \text{on } \partial B \times [0, \infty) \tag{6.2.58}$$

Here $\mathbf{S}_0$ and $\dot{\mathbf{S}}_0$ are arbitrary symmetric tensor fields on B, and $\mathbf{B}$ is a prescribed symmetric tensor field on $\bar{B} \times [0, \infty)$. Moreover, ρ, K, and $\widehat{\mathbf{s}}$ have the same meaning as in classical elastodynamics. We note that the problem is equivalent to the following one: Find a symmetric second-order tensor field $\mathbf{S} = \mathbf{S}(\mathbf{x}, t)$ on $\bar{B} \times [0, \infty)$ that satisfies the integro-differential equation

$$\widehat{\nabla}\, \rho^{-1}\, [i * (\operatorname{div}\mathbf{S})] - \mathsf{K}[\mathbf{S}] = -\widehat{\mathbf{B}} \quad \text{on } \bar{B} \times [0, \infty) \tag{6.2.59}$$

subject to the condition

$$\mathbf{s} = \mathbf{S}\mathbf{n} = \widehat{\mathbf{s}} \quad \text{on } \partial B \times [0, \infty) \tag{6.2.60}$$

where

$$\widehat{\mathbf{B}} = i * \mathbf{B} + \mathsf{K}[\mathbf{S}_0 + t\dot{\mathbf{S}}_0] \tag{6.2.61}$$

Now we are in a position to formulate the following variational principle:

(S 3) **Principle of Incompatible Elastodynamics.** Let N denote the set of all symmetric second order stress fields on $\bar{B} \times [0, \infty)$ that satisfy the traction boundary condition 6.2.60. Let $C_t\{.\}$ be the functional on N defined by

$$C_t\{\mathbf{S}\} = \frac{1}{2}\int_B \left\{\rho^{-1}\, i * (\operatorname{div}\mathbf{S}) * (\operatorname{div}\mathbf{S}) + \mathbf{S} * \mathsf{K}[\mathbf{S}] - 2\mathbf{S} * \widehat{\mathbf{B}}\right\}\, dv \tag{6.2.62}$$

Then

$$\delta C_t\{\mathbf{S}\} = 0 \tag{6.2.63}$$

at a particular $\mathbf{S} \in N$, if and only if $\mathbf{S}$ is a solution to the traction problem described by Equations 6.2.56 through 6.2.58.

The proof of the (S 3) principle is similar to that of the (S 2) principle, and will not be given here.

Notes:

(1) If the fields $\mathbf{B}$, $\mathbf{S}_0$, and $\dot{\mathbf{S}}_0$ are suitably restricted, that is, Equations 6.2.56 through 6.2.58 reduce to those of a traction problem of classical elastodynamics, then the (S 3) principle becomes the (S 2) principle.

(2) When the fields $\mathbf{B}$, $\mathbf{S}_0$, and $\dot{\mathbf{S}}_0$ are arbitrarily prescribed, the (S 3) principle may be useful in the study of elastic waves in bodies with various types of defects.

Example 6.2.4

A pure stress initial–boundary value problem of two-dimensional elastodynamics for a homogeneous isotropic body C_0 subject to plane-strain conditions is described by the equation (see Example 8.2.3 with $b_\alpha = 0$ and $\widehat{s}_\alpha = 0$)

$$S_{(\alpha\gamma,\gamma\beta)} - \frac{\rho}{2\mu}\left(\ddot{S}_{\alpha\beta} - \nu \ddot{S}_{\gamma\gamma}\delta_{\alpha\beta}\right) = 0 \quad \text{on } C_0 \times (0, \infty) \tag{a}$$

the initial conditions

$$S_{\alpha\beta}(\mathbf{x}, 0) = S^{(0)}_{\alpha\beta}(\mathbf{x}), \quad \dot{S}_{\alpha\beta}(\mathbf{x}, 0) = \dot{S}^{(0)}_{\alpha\beta}(\mathbf{x}) \quad \text{on } C_0 \tag{b}$$

and the boundary condition

$$S_{\alpha\beta} n_\beta = 0 \quad \text{on } \partial C_0 \times (0, \infty) \tag{c}$$

Here, $S_{\alpha\beta} = S_{\alpha\beta}(\mathbf{x}, t)$ represents a two-dimensional stress wave produced by the initial tensor fields $S^{(0)}_{\alpha\beta}$ and $\dot{S}^{(0)}_{\alpha\beta}$ on C_0; ρ, μ, and ν denote the density, shear modulus, and

Poisson's ratio, respectively; and n_α is a unit outward normal vector field on ∂C_0 $(\alpha, \beta = 1, 2)$. Let $\boldsymbol{\mathcal{S}}$ be the set of all symmetric second-order tensor fields $S_{\alpha\beta}$ on $C_0 \times [0, \infty)$ that satisfy the boundary condition (c). Define a functional $\Sigma_t\{\cdot\}$ on $\boldsymbol{\mathcal{S}}$ by

$$\Sigma_t\{S_{\alpha\beta}\} = \frac{1}{2}\int_{C_0}\left\{\frac{t}{\rho} * S_{\alpha\beta,\beta} * S_{\alpha\gamma,\gamma} + \frac{1}{2\mu}\left(S_{\alpha\beta} - \nu S_{\gamma\gamma}\delta_{\alpha\beta}\right) * S_{\alpha\beta}\right.$$

$$\left. - \frac{1}{\mu}\left[\left(S^{(0)}_{\alpha\beta} + t\dot{S}^{(0)}_{\alpha\beta}\right) - \nu\left(S^{(0)}_{\gamma\gamma} + t\dot{S}^{(0)}_{\gamma\gamma}\right)\delta_{\alpha\beta}\right] * S_{\alpha\beta}\right\} da \tag{d}$$

Show that

$$\delta\Sigma_t\{S_{\alpha\beta}\} = 0 \tag{e}$$

at a particular $S_{\alpha\beta} \in \boldsymbol{\mathcal{S}}$ if and only if $S_{\alpha\beta}$ is a solution to the traction problem (a) through (c).

Solution

First, we replace Equations (a) through (c) by the single integro-differential equation

$$\frac{t}{\rho} * S_{(\alpha\gamma,\gamma\beta)} - \frac{1}{2\mu}\left(S_{\alpha\beta} - \nu S_{\gamma\gamma}\delta_{\alpha\beta}\right) = -\frac{1}{2\mu}\left[\left(S^{(0)}_{\alpha\beta} + t\dot{S}^{(0)}_{\alpha\beta}\right)\right.$$

$$\left. - \nu\left(S^{(0)}_{\gamma\gamma} + t\dot{S}^{(0)}_{\gamma\gamma}\right)\delta_{\alpha\beta}\right] \quad \text{on } C_0 \times [0, \infty) \tag{f}$$

subject to the condition

$$S_{\alpha\beta}n_\beta = 0 \quad \text{on } \partial C_0 \times [0, \infty) \tag{g}$$

Next, we take $S_{\alpha\beta} \in \boldsymbol{\mathcal{S}}$ and $S_{\alpha\beta} + \omega\widetilde{S}_{\alpha\beta} \in \boldsymbol{\mathcal{S}}$ for every scalar ω, and find that

$$\widetilde{S}_{\alpha\beta}n_\beta = 0 \quad \text{on } \partial C_0 \times [0, \infty) \tag{h}$$

Finally, computing the first variation of $\Sigma_t\{\cdot\}$, by virtue of the divergence theorem and (h), we arrive at

$$\delta\Sigma_t\{S_{\alpha\beta}\} = \frac{d}{d\omega}\Sigma_t\{S_{\alpha\beta} + \omega\widetilde{S}_{\alpha\beta}\}\Big|_{\omega=0}$$

$$= \int_{C_0}\left\{-\frac{t}{\rho} * S_{(\alpha\gamma,\gamma\beta)} + \frac{1}{2\mu}(S_{\alpha\beta} - \nu S_{\gamma\gamma}\delta_{\alpha\beta})\right.$$

$$\left. - \frac{1}{2\mu}\left[\left(S^{(0)}_{\alpha\beta} + t\dot{S}^{(0)}_{\alpha\beta}\right) - \nu\left(S^{(0)}_{\gamma\gamma} + t\dot{S}^{(0)}_{\gamma\gamma}\right)\delta_{\alpha\beta}\right]\right\} * \widetilde{S}_{\alpha\beta} da \tag{i}$$

Hence, it follows from (i) that (e) is satisfied if and only if $S_{\alpha\beta}$ is a solution to Equations (f) and (g). This completes the solution. □

In closing this section, we note that the convolutional variational principles (G 1), (G 2), (S 1), (S 2), and (S 3) were obtained under the assumption that the data and corresponding solutions are sufficiently smooth functions of both space and time variables. This means that an analysis of elastic waves in a body subject to discontinuous data, such as a mechanical shock, cannot be covered by these principles. An extension of the principles to include discontinuous loadings, bodies with corners, edges, and cracks may constitute a subject of a separate study.

As far as the (S 3) principle is concerned, we want to emphasize that it may be useful in solving practical problems of wave propagation in elastic bodies with smoothly distributed defects, in particular, smoothly distributed initial stress velocity fields. An infinite plane with an elliptical hole, containing continuously distributed defects, could be a subject of a separate study.

PROBLEMS

6.1 A symmetrical elastic beam of flexural rigidity EI, density ρ, and length L is acted upon by: (i) the transverse force $F = F(x_1, t)$, (ii) the end shear forces V_0 and V_L, and (iii) the end bending moments M_0 and M_L shown in Figure P6.1.

The strain energy of the beam is

$$F(t) = \frac{1}{2}\int_0^L EI\left(u_2''\right)^2 dx_1 \tag{a}$$

the kinetic energy of the beam is

$$K(t) = \frac{1}{2}\int_0^L \rho(\dot{u}_2)^2 dx_1 \tag{b}$$

and the energy of external forces is

$$V(t) = -\int_0^L Fu_2 dx_1 + V_0 u_2(0,t) + M_0 u_2'(0,t) - V_L u_2(L,t) - M_L u_2'(L,t) \tag{c}$$

where the prime denotes differentiation with respect to x_1.

Let U be the set of functions $u_2 = u_2(x_1, t)$ that satisfy the conditions

$$u_2(x_1, t_1) = u(x_1), \quad u_2(x_1, t_2) = v(x_1) \tag{d}$$

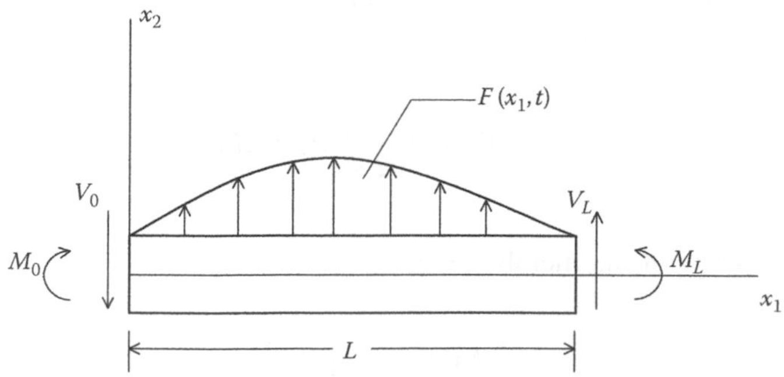

FIGURE P6.1

where t_1 and t_2 are two arbitrary points on the t-axis ($0 \le t_1 < t_2$), and $u(x_1)$ and $v(x_1)$ are prescribed functions on $[0, L]$. Define a functional $\widehat{K}\{\cdot\}$ on U by

$$\widehat{K}\{u_2\} = \int_{t_1}^{t_2} [F(t) + V(t) - K(t)]dt \tag{e}$$

Show that

$$\delta\widehat{K}\{u_2\} = 0 \tag{f}$$

if and only if u_2 satisfies the equation of motion

$$\left(EIu_2''\right)'' + \rho\ddot{u}_2 = F \quad \text{on } [0, L] \times [0, \infty) \tag{g}$$

and the boundary conditions

$$\left[\left(EIu_2''\right)'\right](0, t) = -V_0 \quad \text{on } [0, \infty) \tag{h}$$

$$\left[\left(EIu_2''\right)\right](0, t) = M_0 \quad \text{on } [0, \infty) \tag{i}$$

$$\left[\left(EIu_2''\right)'\right](L, t) = -V_L \quad \text{on } [0, \infty) \tag{j}$$

$$\left[\left(EIu_2''\right)\right](L, t) = M_L \quad \text{on } [0, \infty) \tag{k}$$

The field equation (g) and boundary conditions (h) through (k) describe flexural waves in the beam.

6.2 A thin elastic membrane of uniform area density $\widehat{\rho}$ is stretched to a uniform tension $\widehat{T}$ over a region C_0 of the x_1x_2 plane. The membrane is subject to a vertical load $f = f(\mathbf{x}, t)$ on $C_0 \times [0, \infty)$ and the initial conditions

$$u(\mathbf{x}, 0) = u_0(\mathbf{x}), \quad \dot{u}(\mathbf{x}, 0) = \dot{u}_0(\mathbf{x}) \quad \text{for } \mathbf{x} \in C_0$$

where

$u = u(\mathbf{x}, t)$ is a vertical deflection of the membrane on $\overline{C_0} \times [0, \infty)$

$u_0(\mathbf{x})$ and $\dot{u}_0(\mathbf{x})$ are prescribed functions on C_0

Also, $u = u(\mathbf{x}, t)$ on $\partial C_0 \times [0, \infty)$ is represented by a given function $g = g(\mathbf{x}, t)$. The strain energy of the membrane is

$$F(t) = \frac{\widehat{T}}{2} \int_{C_0} u_{,\alpha} u_{,\alpha} da \tag{a}$$

The kinetic energy of the membrane is

$$K(t) = \frac{\widehat{\rho}}{2} \int_{C_0} (\dot{u})^2 da \tag{b}$$

The external load energy is

$$V(t) = -\int_{C_0} fu\,da \tag{c}$$

Let U be the set of functions $u = u(\mathbf{x}, t)$ on $C_0 \times [0, \infty)$ that satisfy the conditions

$$u(\mathbf{x}, t_1) = a(\mathbf{x}), \quad u(\mathbf{x}, t_2) = b(\mathbf{x}) \quad \mathbf{x} \in C_0 \tag{d}$$

and

$$u(\mathbf{x}, t) = g(\mathbf{x}, t) \quad \text{on } \partial C_0 \times [0, \infty) \tag{e}$$

where t_1 and t_2 have the same meaning as in Problem 6.1, and $a(\mathbf{x})$ and $b(\mathbf{x})$ are prescribed functions on C_0. Define a functional $\widehat{K}\{\cdot\}$ on U by

$$\widehat{K}\{u\} = \int_{t_1}^{t_2} [F(t) + V(t) - K(t)]dt \tag{f}$$

Show that the condition

$$\delta\widehat{K}\{u\} = 0 \quad \text{on } U \tag{g}$$

implies the wave equation

$$\left(\nabla^2 - \frac{1}{c^2}\frac{\partial^2}{\partial t^2}\right)u = -\frac{f}{\widehat{T}} \quad \text{on } C_0 \times [0, \infty) \tag{h}$$

where

$$c = \sqrt{\frac{\widehat{T}}{\widehat{\rho}}} \tag{i}$$

Note that $[\widehat{T}] = [\text{Force} \times \text{L}^{-1}]$, $[\widehat{\rho}] = [\text{Density} \times \text{L}]$, $[c] = [\text{LT}^{-1}]$, where L and T are the length and time units, respectively.

6.3 Transverse waves propagating in a thin elastic membrane are described by the field equation (see Problem 6.2)

$$\left(\nabla^2 - \frac{1}{c^2}\frac{\partial^2}{\partial t^2}\right)u = -\frac{f}{\widehat{T}} \quad \text{on } C_0 \times [0, \infty) \tag{a}$$

the initial conditions

$$u(\mathbf{x}, 0) = u_0(\mathbf{x}), \quad \dot{u}(\mathbf{x}, 0) = \dot{u}_0(\mathbf{x}) \quad \text{on } C_0 \tag{b}$$

and the boundary condition

$$u(\mathbf{x}, t) = g(\mathbf{x}, t) \quad \text{on } \partial C_0 \times [0, \infty) \tag{c}$$

Let $\widehat{U}$ be a set of functions $u = u(\mathbf{x}, t)$ on $C_0 \times [0, \infty)$ that satisfy the boundary condition (c). Define a functional $\mathcal{F}_t\{\cdot\}$ on $\widehat{U}$ in such a way that

$$\delta \mathcal{F}_t\{u\} = 0 \tag{d}$$

if and only if $u = u(\mathbf{x}, t)$ is a solution to the initial–boundary value problem (a) through (c).

6.4 A homogeneous isotropic thin elastic plate defined over a region C_0 of x_1x_2 plane, and clamped on its boundary ∂C_0, is subject to a transverse load $p = p(\mathbf{x}, t)$ on $C_0 \times [0, \infty)$. The strain energy of the plate is

$$F(t) = \frac{D}{2} \int_{C_0} (\nabla^2 w)^2 da \tag{a}$$

The kinetic energy of the plate is

$$K(t) = \frac{1}{2} \widehat{\rho} \int_{C_0} \dot{w}^2 da \tag{b}$$

The external load energy is

$$V(t) = -\int_{C_0} pw \, da \tag{c}$$

Here, $w = w(\mathbf{x}, t)$ is a transverse deflection of the plate on $C_0 \times [0, \infty)$, D is the bending rigidity of the plate ($[D] = [\text{Force} \times \text{Length}]$), and $\widehat{\rho}$ is the area density of the plate ($[\widehat{\rho}] = [\text{Density} \times \text{Length}]$).

Let W be the set of functions $w = w(\mathbf{x}, t)$ on $C_0 \times [0, \infty)$ that satisfy the conditions

$$w(\mathbf{x}, t_1) = a(\mathbf{x}), \quad w(\mathbf{x}, t_2) = b(\mathbf{x}) \tag{d}$$

and

$$w = 0, \quad \frac{\partial w}{\partial n} = 0 \quad \text{on } \partial C_0 \times [0, \infty) \tag{e}$$

where t_1, t_2, $a(\mathbf{x})$, and $b(\mathbf{x})$ have the same meaning as in Problem 6.2 and $\partial/\partial n$ is the normal derivative on ∂C_0. Define a functional $\widehat{K}\{\cdot\}$ on W by

$$\widehat{K}\{w\} = \int_{t_1}^{t_2} \{F(t) + V(t) - K(t)\} dt \tag{f}$$

Show that

$$\delta\widehat{K}\{w\} = 0 \quad \text{on } W \tag{g}$$

if and only if $w = w(\mathbf{x}, t)$ satisfies the differential equation

$$\nabla^2\nabla^2 w + \frac{\widehat{\rho}}{D}\frac{\partial^2 w}{\partial t^2} = \frac{p}{D} \quad \text{on } C_0 \times [0, \infty) \tag{h}$$

and the boundary conditions

$$w = \frac{\partial w}{\partial n} = 0 \quad \text{on } \partial C_0 \times [0, \infty) \tag{i}$$

6.5 Transverse waves propagating in a clamped thin elastic plate are described by the equations (see Problem 6.4)

$$\nabla^2\nabla^2 w + \frac{\widehat{\rho}}{D}\frac{\partial^2 w}{\partial t^2} = \frac{p}{D} \quad \text{on } C_0 \times [0, \infty) \tag{a}$$

$$w(\mathbf{x}, 0) = w_0(\mathbf{x}), \quad \dot{w}(\mathbf{x}, 0) = \dot{w}_0(\mathbf{x}) \quad \text{on } C_0 \tag{b}$$

and

$$w = \frac{\partial w}{\partial n} = 0 \quad \text{on } \partial C_0 \times [0, \infty) \tag{c}$$

where $w_0(\mathbf{x})$ and $\dot{w}_0(\mathbf{x})$ are prescribed functions. Let W^* denote the set of functions $w = w(\mathbf{x}, t)$ that satisfy the homogeneous boundary conditions (c). Find a functional $\widehat{\mathcal{F}}_t\{\cdot\}$ on W^* with the property that

$$\delta\widehat{\mathcal{F}}_t\{w\} = 0 \quad \text{on } W^* \tag{d}$$

if and only if w is a solution to the initial–boundary value problem (a) through (c).

6.6 Free longitudinal vibrations of a bar are defined as solutions of the form

$$u(x, t) = \phi(x)\sin(\omega t + \gamma) \tag{a}$$

to the homogeneous wave equation

$$\frac{\partial}{\partial x}\left(E\,\frac{\partial u}{\partial x}\right) - \rho\,\frac{\partial^2 u}{\partial t^2} = 0 \quad \text{on } [0, L] \times [0, \infty) \tag{b}$$

subject to one of the homogeneous boundary conditions

$$u(0, t) = u(L, t) = 0 \quad \text{on } [0, \infty) \tag{c}$$

or

$$\frac{\partial u}{\partial x}(0,t) = \frac{\partial u}{\partial x}(L,t) = 0 \quad \text{on } [0,\infty) \tag{d}$$

Here

ω is a circular frequency of vibrations
γ is a dimensionless constant
$\phi = \phi(x)$ is an unknown function that complies with Equations (b) and (c), or Equations (b) and (d)

Substituting $u = u(x,t)$ from Equation (a) into Equations (b) through (d) we obtain

$$\frac{d}{dx}\left(E\,\frac{d\phi}{dx}\right) + \lambda\phi = 0 \quad \text{on } [0,L] \tag{e}$$

$$\phi(0) = \phi(L) = 0 \tag{f}$$

or

$$\phi'(0) = \phi'(L) = 0 \tag{g}$$

where the prime stands for derivative with respect to x, and

$$\lambda = \rho\omega^2 \tag{h}$$

Therefore, introduction of (a) into (b) through (d) results in an eigenproblem in which an eigenfunction $\phi = \phi(x)$ corresponding to an eigenvalue λ is to be found. An eigenproblem that covers both boundary conditions (c) and (d) can be written as

$$\frac{d}{dx}\left(E\,\frac{d\phi}{dx}\right) + \lambda\phi = 0 \quad \text{on } [0,L] \tag{i}$$

$$\phi'(0) - \alpha\phi(0) = 0, \quad \phi'(L) + \beta\phi(L) = 0 \tag{j}$$

where $|\alpha| + |\beta| > 0$. Let U be the set of functions $\phi = \phi(x)$ on $[0,L]$ that satisfy the boundary conditions (j). Define a functional $\pi\{\cdot\}$ on U by

$$\pi\{\phi\} = \frac{1}{2}\int_0^L \left[E\left(\frac{d\phi}{dx}\right)^2 - \lambda\phi^2\right] dx$$

$$+ \frac{1}{2}\alpha E(0)[\phi(0)]^2 + \frac{1}{2}\beta E(L)[\phi(L)]^2 \tag{k}$$

Show that

$$\delta\pi\{\phi\} = 0 \quad \text{over } U \tag{l}$$

if and only if $\phi = \phi(x)$ is an eigenfunction corresponding to an eigenvalue λ in the eigenproblem (i) and (j).

6.7 Free lateral vibrations of a bar clamped at the end $x = 0$ and supported by a spring of stiffness k at the end $x = L$ are defined as solutions of the form

$$u(x,t) = \phi(x)\sin(\omega t + \gamma) \tag{a}$$

to the equation [see Problem 6.1, Equation (g) in which $u_2 = u$, and $F = 0$]

$$\frac{\partial^2}{\partial x^2}\left(EI\frac{\partial^2 u}{\partial x^2}\right) + \rho\frac{\partial^2 u}{\partial t^2} = 0 \quad \text{on } [0,L]\times[0,\infty) \tag{b}$$

subject to the boundary conditions

$$u(0,t) = u'(0,t) = 0 \quad \text{on } [0,\infty) \tag{c}$$

$$u''(L,t) = 0, \quad (EIu'')'(L,t) - ku(L,t) = 0 \quad \text{on } [0,\infty) \tag{d}$$

Let $\rho = \text{const}$, and $\lambda = \rho\omega^2$. Then the associated eigenproblem reads

$$(EI\phi'')'' - \lambda\phi = 0 \quad \text{on } [0,L] \tag{e}$$

$$\phi(0) = \phi'(0) = 0 \tag{f}$$

$$\phi''(L) = 0, \quad (EI\phi'')'(L) - k\phi(L) = 0 \tag{g}$$

Let V denote the set of functions $\phi = \phi(x)$ on $[0,L]$ that satisfy the boundary conditions (f) and (g). Define a functional $\pi\{\cdot\}$ on V by

$$\pi\{\phi\} = \frac{1}{2}\int_0^L EI(\phi'')^2 dx + \frac{1}{2}k\,[\phi(L)]^2 - \frac{\lambda}{2}\int_0^L \phi^2 dx \tag{h}$$

Show that

$$\delta\pi\{\phi\} = 0 \quad \text{over } V \tag{i}$$

if and only if (λ, ϕ) is a solution to the eigenproblem (e) through (g).

6.8 Show that the eigenvalues λ_i and the eigenfunctions $\phi_i = \phi_i(x)$ for the longitudinal vibrations of a uniform elastic bar having one end clamped and the other end free are given by the relations

$$\omega_i \equiv \sqrt{\frac{\lambda_i}{\rho}} = \frac{(2i-1)\pi}{2L}\sqrt{\frac{E}{\rho}}$$

$$\phi_i(x) = \sin\frac{(2i-1)\pi x}{2L}, \quad i = 1, 2, \ldots, \; 0 \le x \le L$$

(See Problem 6.6.)

6.9 Show that the eigenvalues λ_i and the eigenfunctions $\phi_i = \phi_i(x)$ for the lateral vibrations of a uniform, simply supported beam are given by

$$\omega_i \equiv \sqrt{\frac{\lambda_i}{\rho}} = \frac{\pi^2 i^2}{L^2}\sqrt{\frac{EI}{\rho}}$$

$$\phi_i(x) = \sin\frac{i\pi x}{L}, \quad i = 1, 2, ..., \; 0 \le x \le L$$

(See Problem 6.1.)

6.10 Show that the eigenvalues λ_{mn} and the eigenfunctions $\phi_{mn}(x_1, x_2)$ for the transversal vibrations of a rectangular membrane: $0 \le x_1 \le a_1$, $0 \le x_2 \le a_2$, that is clamped on its boundary, are given by

$$\omega_{mn} \equiv \sqrt{\frac{\lambda_{mn}}{\widehat{\rho}}} = \pi\sqrt{\frac{\widehat{T}}{\widehat{\rho}}\left(\frac{m^2}{a_1^2} + \frac{n^2}{a_2^2}\right)}$$

$$\phi_{mn}(x_1, x_2) = \sin\frac{m\pi x_1}{a_1}\sin\frac{n\pi x_2}{a_2} \quad m, n = 1, 2, 3, ...$$

$$0 \le x_1 \le a_1, \quad 0 \le x_2 \le a_2$$

(See Problem 6.2.)

6.11 Show that the eigenvalues λ_{mn} and the eigenfunctions $\phi_{mn} = \phi_{mn}(x_1, x_2)$ for the transversal vibrations of a thin elastic rectangular plate
$0 \le x_1 \le a_1$, $0 \le x_2 \le a_2$ that is simply supported on its boundary are given by the relations

$$\omega_{mn} \equiv \sqrt{\frac{\lambda_{mn}}{\widehat{\rho}}} = \pi^2\left(\frac{m^2}{a_1^2} + \frac{n^2}{a_2^2}\right)\sqrt{\frac{D}{\widehat{\rho}}}$$

$$\phi_{mn}(x_1, x_2) = \sin\frac{m\pi x_1}{a_1}\sin\frac{n\pi x_2}{a_2}, \quad m, n = 1, 2, 3, ...$$

$$0 \le x_1 \le a_1, \quad 0 \le x_2 \le a_2$$

(See Problem 6.4.)

REFERENCES

1. T. R. Tauchert, *Energy Principles in Structural Mechanics*, McGraw-Hill, New York, 1974, p. 315.
2. M. E. Gurtin, *The Linear Theory of Elasticity*, Encyclopedia of Physics, chief editor: S. Flügge, vol. VIa/2, editor: C. Truesdell, Springer, Berlin, Germany, 1972, pp. 226–230.
3. I. S. Sokolnikoff, *Mathematical Theory of Elasticity*, McGraw-Hill, New York, 1956, pp. 399–400.

7 Complete Solutions of Elasticity

In this chapter, a number of general solutions of the homogeneous isotropic elastostatics and elastodynamics are presented. The general solutions to the displacement equation of equilibrium include (i) Boussinesq–Papkovitch–Neuber representation in terms of the potentials satisfying Poisson's equations, and (ii) Boussinesq–Somigliana–Galerkin representation in terms of the potential satisfying a biharmonic equation; and it is shown that both solutions (i) and (ii) are complete in the sense that the potentials exist for any displacement that satisfies the displacement equation of equilibrium. For the displacement equation of homogeneous isotropic elastodynamics, a complete Green–Lamé representation in terms of the potentials satisfying the wave equations, and a complete Cauchy–Kovalevski–Somigliana representation in terms of the vectorial potential satisfying a biwave equation, are discussed. The chapter also contains worked examples and end-of-chapter problems, including the one related to the stress equations of homogeneous isotropic elastodynamics for which a complete Galerkin-type tensor solution is to be obtained. As in the previous chapters, the solutions to problems are given in the Solutions Manual.

7.1 COMPLETE SOLUTIONS OF ELASTOSTATICS

In this section, we discuss general solutions of the displacement equation of equilibrium for a homogeneous isotropic elastic body. Such general solutions are useful in obtaining analytical solutions of elastostatic boundary value problems for bodies of simple geometries such as a semi-infinite space or an infinite layer.

We recall the displacement equation of equilibrium for the homogeneous isotropic body subject to a body force $\mathbf{b}$, in the form, see Equation (b) in Example 4.1.1,

$$\nabla^2 \mathbf{u} + \frac{1}{1-2\nu}\nabla(\operatorname{div}\mathbf{u}) + \frac{\mathbf{b}}{\mu} = \mathbf{0} \tag{7.1.1}$$

A vector field $\mathbf{u} = \mathbf{u}(\mathbf{x})$ on $\bar{B}$ that satisfies Equation 7.1.1 on $\bar{B}$ is called an *elastic displacement field corresponding to* $\mathbf{b}$. With this definition we are to show that $\mathbf{u}$ can be expressed in terms of a vector field $\boldsymbol{\psi}$ and a scalar field φ that satisfy vector and scalar Poisson's equations, respectively, i.e., equations of lesser complexity than Equation 7.1.1. More precisely, we are to prove the following theorem:

(T 1) **Boussinesq–Papkovitch–Neuber Solution.** Let

$$\mathbf{u} = \boldsymbol{\psi} - \frac{1}{4(1-\nu)}\nabla(\mathbf{x}\cdot\boldsymbol{\psi} + \varphi) \tag{7.1.2}$$

where φ and $\boldsymbol{\psi}$ are fields on B that satisfy Poisson's equations

$$\nabla^2 \boldsymbol{\psi} = -\frac{1}{\mu}\,\mathbf{b} \tag{7.1.3}$$

and

$$\nabla^2 \varphi = \frac{1}{\mu}\,\mathbf{x}\cdot\mathbf{b} \tag{7.1.4}$$

Then $\mathbf{u}$ is an elastic displacement field corresponding to $\mathbf{b}$.

Proof: We need to show that $\mathbf{u}$ given by Equation 7.1.2 satisfies Equation 7.1.1 provided $\boldsymbol{\psi}$ and φ satisfy Equations 7.1.3 and 7.1.4, respectively. To this end we rewrite Equations 7.1.1 through 7.1.4 in components

$$u_{i,kk} + \frac{1}{1-2\nu}\,u_{k,ki} + \frac{b_i}{\mu} = 0 \tag{7.1.5}$$

$$u_i = \psi_i - \frac{1}{4(1-\nu)}\,\left(x_p\psi_p + \varphi\right)_{,i} \tag{7.1.6}$$

$$\psi_{i,kk} = -\frac{1}{\mu}\,b_i \tag{7.1.7}$$

$$\varphi_{,kk} = \frac{1}{\mu}\,x_p b_p \tag{7.1.8}$$

Next, since

$$\left(x_p\psi_p + \varphi\right)_{,i} = \delta_{pi}\psi_p + x_p\psi_{p,i} + \varphi_{,i} = \psi_i + x_p\psi_{p,i} + \varphi_{,i} \tag{7.1.9}$$

then Equation 7.1.6 may be rewritten as

$$u_i = \frac{1}{4(1-\nu)}\,\left[(3-4\nu)\psi_i - x_p\psi_{p,i} - \varphi_{,i}\right] \tag{7.1.10}$$

Differentiating Equation 7.1.10 with respect to x_k, we get

$$u_{i,k} = \frac{1}{4(1-\nu)}\,\left[(3-4\nu)\psi_{i,k} - \psi_{k,i} - x_p\psi_{p,ik} - \varphi_{,ik}\right] \tag{7.1.11}$$

Differentiating Equation 7.1.11 we respect to x_k, we obtain

$$u_{i,kk} = \frac{1}{4(1-\nu)}\,\left[(3-4\nu)\psi_{i,kk} - 2\psi_{k,ki} - x_p\psi_{p,ikk} - \varphi_{,ikk}\right] \tag{7.1.12}$$

So, by letting $i = k$ in Equation 7.1.11, we receive

$$u_{k,k} = \frac{1}{4(1-\nu)}\,\left[2(1-2\nu)\psi_{k,k} - x_p\psi_{p,kk} - \varphi_{,kk}\right] \tag{7.1.13}$$

Now, because of Equations 7.1.7 and 7.1.8, we write

$$\varphi_{,kk} + x_p \psi_{p,kk} = \frac{1}{\mu} x_p b_p - \frac{1}{\mu} x_i b_i = 0 \tag{7.1.14}$$

and Equation 7.1.13 reduces to

$$u_{k,k} = \frac{1 - 2\nu}{2(1 - \nu)} \psi_{k,k} \tag{7.1.15}$$

Also, by letting $i = p$ in Equation 7.1.7 and differentiating the result with respect to x_i, we get

$$\psi_{p,kki} = -\frac{1}{\mu} b_{p,i} \tag{7.1.16}$$

and, by differentiating Equation 7.1.8 with respect to x_i, we obtain

$$\varphi_{,kki} = \frac{1}{\mu} \left(b_i + x_p b_{p,i}\right) \tag{7.1.17}$$

Hence, multiplying Equation 7.1.16 by x_p and using Equation 7.1.17, we get

$$x_p \psi_{p,kki} + \varphi_{,kki} = \frac{1}{\mu} b_i \tag{7.1.18}$$

Substitution of Equations 7.1.7 and 7.1.18 into Equation 7.1.12 yields

$$\begin{aligned} u_{i,kk} &= \frac{1}{4(1 - \nu)} \left[(3 - 4\nu)\left(-\frac{1}{\mu}\right) b_i - \frac{1}{\mu} b_i - 2\psi_{k,ki}\right] \\ &= \frac{1}{4(1 - \nu)} \left[4(1 - \nu)\left(-\frac{1}{\mu}\right) b_i - 2\psi_{k,ki}\right] \end{aligned} \tag{7.1.19}$$

Now, it follows from Equations 7.1.15 and 7.1.19 that

$$u_{i,kk} + \frac{1}{1 - 2\nu} u_{k,ki} + \frac{b_i}{\mu} = 0 \tag{7.1.20}$$

This completes the proof of the (T 1) Theorem. □

Example 7.1.1

Show that a particular elastic displacement field corresponding to $\mathbf{b}$ is given by Equation 7.1.2 in which $\boldsymbol{\psi}$ and φ are taken as

$$\boldsymbol{\psi}(\mathbf{x}) = \frac{1}{4\pi\mu} \int_B \frac{\mathbf{b}(\mathbf{y})}{|\mathbf{x} - \mathbf{y}|} dv(\mathbf{y}) \tag{a}$$

and

$$\varphi(\mathbf{x}) = -\frac{1}{4\pi\mu} \int_B \frac{\mathbf{y}\cdot\mathbf{b}(\mathbf{y})}{|\mathbf{x}-\mathbf{y}|}\, dv(\mathbf{y}) \tag{b}$$

Solution

We use the fact that a solution of Poisson's equation

$$\nabla^2\phi = -4\pi\rho \quad \text{on } B \tag{c}$$

where $\rho = \rho(\mathbf{x})$ is a known function on B and $\phi = \phi(\mathbf{x})$ is to be found, is given by the Newtonian potential

$$\phi(\mathbf{x}) = \int_B \frac{\rho(\mathbf{y})}{|\mathbf{x}-\mathbf{y}|}\, dv(\mathbf{y}) \tag{d}$$

Hence, if $\boldsymbol{\psi} = \boldsymbol{\psi}(\mathbf{x})$ and $\varphi = \varphi(\mathbf{x})$ are selected in the form of Newtonian potentials (a) and (b), respectively, then Equations 7.1.3 and 7.1.4 are satisfied, and Equation 7.1.2 represents a particular solution to Equation 7.1.1. This completes the solution. □

Example 7.1.2

Show that if $\mathbf{b} = \nabla h$, where h is a known scalar-valued field on B, then a particular solution to Equation 7.1.1 may be taken in the form

$$\mathbf{u} = \nabla\varphi \quad \text{on } B \tag{a}$$

where $\varphi = \varphi(\mathbf{x})$ is a solution to Poisson's equation

$$\nabla^2\varphi = -\frac{1}{\lambda+2\mu}\, h \tag{b}$$

Solution

We rewrite Equation 7.1.1 in the alternative form, see Equation (a) in Example 4.1.1,

$$\nabla^2\mathbf{u} + \frac{\lambda+\mu}{\mu}\nabla(\operatorname{div}\mathbf{u}) + \frac{1}{\mu}\mathbf{b} = \mathbf{0} \tag{c}$$

Substituting (a) into (c), and taking into account the fact that

$$\mathbf{b} = \nabla h \tag{d}$$

we arrive at

$$\nabla\left(\nabla^2\varphi + \frac{1}{\lambda+2\mu}\, h\right) = \mathbf{0} \tag{e}$$

Hence, if $\varphi = \varphi(\mathbf{x})$ is a solution to Poisson's equation (b), $\mathbf{u}$ given by (a) satisfies (c). This completes the solution. □

Example 7.1.3

Show that if the body force $\mathbf{b}$ takes the form

$$\mathbf{b} = \text{curl}\,\mathbf{k} \tag{a}$$

where $\mathbf{k} = \mathbf{k}(\mathbf{x})$ is a known vector field on B, then a vector field $\mathbf{u}$ of the form

$$\mathbf{u} = \text{curl}\,\boldsymbol{\omega} \tag{b}$$

satisfies Equation (c) in Example 7.1.2 provided

$$\nabla^2 \boldsymbol{\omega} = -\frac{1}{\mu}\,\mathbf{k} \tag{c}$$

Solution

Equation (b) written in components takes the form

$$u_i = \epsilon_{ijk}\omega_{k,j} \tag{d}$$

Hence

$$u_{i,i} = \epsilon_{ijk}\omega_{k,ji} = 0 \tag{e}$$

or

$$\text{div}(\text{curl}\,\boldsymbol{\omega}) = 0 \tag{f}$$

Substituting (b) into (c) of Example 7.1.2, and using (a), (c), and (f), we receive

$$\nabla^2\mathbf{u} + \frac{\lambda+\mu}{\mu}\,\nabla(\text{div}\,\mathbf{u}) + \frac{1}{\mu}\,\mathbf{b} = \text{curl}\,\nabla^2\boldsymbol{\omega} + \frac{1}{\mu}\,\text{curl}\,\mathbf{k} = \text{curl}\left(\nabla^2\boldsymbol{\omega} + \frac{1}{\mu}\,\mathbf{k}\right) = \mathbf{0} \tag{g}$$

This completes the solution. □

Note: A combination of the particular solutions (a) of Example 7.1.2 and (b) of Example 7.1.3 leads to the following conclusion:

If the body force field $\mathbf{b}$ is represented by

$$\mathbf{b} = \nabla h + \text{curl}\,\mathbf{k} \tag{7.1.21}$$

where h and $\mathbf{k}$ are prescribed fields on B, then a particular solution $\mathbf{u}$ of the displacement equation of equilibrium (c) in Example 7.1.2 may be taken in the form

$$\mathbf{u} = \nabla\varphi + \text{curl}\,\boldsymbol{\omega} \tag{7.1.22}$$

where φ and $\boldsymbol{\omega}$ are, respectively, particular solutions to Poisson's equations

$$\nabla^2\varphi = -\frac{1}{\lambda + 2\mu}\, h \tag{7.1.23}$$

and

$$\nabla^2\boldsymbol{\omega} = -\frac{1}{\mu}\, \mathbf{k} \tag{7.1.24}$$

Example 7.1.4

Let $b_i^{(1)}$ denote a concentrated body force acting at the origin of the coordinate system in the direction of the x_1 axis in an infinite elastic body. Such a force can be represented by

$$b_i^{(1)}(\mathbf{x}) = \delta_{i1}\,\delta(\mathbf{x}) \tag{a}$$

where

$$\delta(\mathbf{x}) = \delta(x_1)\,\delta(x_2)\,\delta(x_3) \tag{b}$$

and $\delta(x_i)$, $i = 1, 2, 3$, is a one-dimensional Dirac delta function, and δ_{ik}, $i, k = 1, 2, 3$, is the Kronecker symbol.

Find an elastic displacement field corresponding to $b_i^{(1)}$.

Solution

We use the result of Example 7.1.1 to solve this problem. Equations (a) and (b) in Example 7.1.1 written in components and specified to an infinite isotropic body, and to the body force given by (a) and (b) of this example, take the form

$$\psi_i(\mathbf{x}) = \frac{1}{4\pi\mu} \int_{E^3} \frac{\delta_{i1}\,\delta(\mathbf{y})}{|\mathbf{x} - \mathbf{y}|}\, dv(\mathbf{y}) \tag{c}$$

and

$$\varphi(\mathbf{x}) = -\frac{1}{4\pi\mu} \int_{E^3} \frac{\delta_{i1}\, y_i\,\delta(\mathbf{y})}{|\mathbf{x} - \mathbf{y}|}\, dv(\mathbf{y}) \tag{d}$$

Using the filtering properties of the Dirac delta function

$$\int_{-\infty}^{\infty} f(x - y)\,\delta(y)\, dy = f(x) \tag{e}$$

and

$$\int_{-\infty}^{\infty} f(x)\,\delta(x)\, dx = f(0) \tag{f}$$

we find from (c) and (d)

$$\psi_i(\mathbf{x}) = \frac{1}{4\pi\mu}\frac{\delta_{i1}}{|\mathbf{x}|} \quad \mathbf{x} \neq \mathbf{0} \tag{g}$$

$$\varphi(\mathbf{x}) = 0 \tag{h}$$

The latter equation (h) follows from the property (f) applied to the triple integral (d), which then takes the form

$$\varphi(\mathbf{x}) = -\frac{1}{4\pi\mu}\int_{-\infty}^{\infty} y_1\delta(y_1)\left[\int_{-\infty}^{\infty}\int_{-\infty}^{\infty}\frac{\delta(y_2)\delta(y_3)}{|\mathbf{x}-\mathbf{y}|}\,dy_2dy_3\right]dy_1 \tag{i}$$

Since

$$\int_{-\infty}^{\infty}\int_{-\infty}^{\infty}\frac{\delta(y_2)\delta(y_3)}{|\mathbf{x}-\mathbf{y}|}\,dy_2dy_3 \equiv \frac{1}{[(x_1-y_1)^2+x_2^2+x_3^2]^{1/2}} \tag{j}$$

therefore, (i) reduces to

$$\varphi(\mathbf{x}) = -\frac{1}{4\pi\mu}\int_{-\infty}^{\infty}\frac{y_1\delta(y_1)dy_1}{[(x_1-y_1)^2+x_2^2+x_3^2]^{1/2}} \tag{k}$$

and using (f), we arrive at (h).

Hence, if we substitute $\psi_i(\mathbf{x})$ given by (g) and $\varphi(\mathbf{x}) = 0$ into Equation 7.1.6 we arrive at

$$u_i^{(1)} = \psi_i - \frac{1}{4(1-\nu)}(x_p\psi_p)_{,i} = \frac{1}{4\pi\mu}\left[\frac{\delta_{i1}}{|\mathbf{x}|} - \frac{1}{4(1-\nu)}\left(\frac{x_1}{|\mathbf{x}|}\right)_{,i}\right] \tag{l}$$

Now, we introduce the notation

$$r = |\mathbf{x}| \tag{m}$$

and we note that

$$\left(\frac{x_1}{r}\right)_{,i} = \frac{\delta_{i1}}{r} - \frac{x_1x_i}{r^3} \tag{n}$$

and reduce (l) to

$$u_i^{(1)} = \frac{1}{16\pi\mu(1-\nu)}\frac{1}{r}\left[(3-4\nu)\delta_{i1} + \frac{x_1x_i}{r^2}\right] \tag{o}$$

This completes the solution. □

Notes:

(1) If for the fixed k ($k = 1, 2, 3$), an infinite elastic body is loaded by a concentrated body force $\mathbf{b}^{(k)}$ of the form

$$b_i^{(k)} = \delta_{ik}\delta(\mathbf{x}) \tag{7.1.25}$$

then an elastic displacement field $u_i^{(k)}$ corresponding to $b_i^{(k)}$ is given by

$$u_i^{(k)} = \frac{1}{16\pi\mu(1-\nu)}\frac{1}{r}\left[(3-4\nu)\delta_{ik} + \frac{x_i x_k}{r^2}\right] \tag{7.1.26}$$

(2) Equation 7.1.26 represents a symmetric second-order tensor field. If this tensor is denoted by $\mathbf{U}$, then

$$U_{ij} = U_{ji} = u_i^{(j)} = u_j^{(i)} \tag{7.1.27}$$

Moreover, if $\boldsymbol{\ell}$ is a constant concentrated force acting at $\mathbf{x} = \mathbf{0}$ in an arbitrary direction (see Figure 7.1), then an elastic displacement $\mathbf{u}$ corresponding to $\boldsymbol{\ell}$ is given by

$$\mathbf{u} = \mathbf{U}\boldsymbol{\ell} \tag{7.1.28}$$

or, in components,

$$u_i = U_{ij}l_j \tag{7.1.29}$$

Since $[U_{ij}] = [\text{Stress}^{-1} \times \text{Length}^{-1}]$, and $[l_j] = [\text{Force}]$, therefore $[u_i] = [\text{Length}]$.

(3) If $\boldsymbol{\ell}$ is a concentrated force acting at an arbitrary point $\boldsymbol{\xi}$ in an infinite body, then Equation 7.1.28, with arguments shown in explicit form, is

$$\mathbf{u}(\mathbf{x}, \boldsymbol{\xi}) = \mathbf{U}(\mathbf{x}, \boldsymbol{\xi})\boldsymbol{\ell} \tag{7.1.30}$$

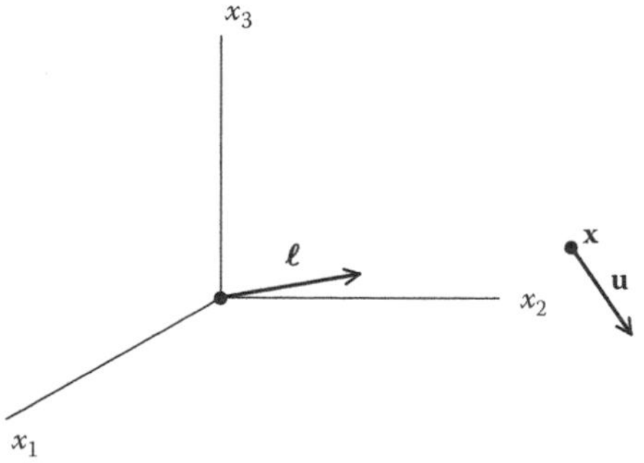

FIGURE 7.1 Displacement $\mathbf{u}$ in an infinite elastic body at point $\mathbf{x}$ due to force $\boldsymbol{\ell}$ acting at the origin.

where

$$\mathbf{U}(\mathbf{x},\boldsymbol{\xi}) = \frac{1}{16\pi\mu(1-\nu)}\frac{1}{R}\left[(3-4\nu)\mathbf{1} + \frac{(\mathbf{x}-\boldsymbol{\xi})\otimes(\mathbf{x}-\boldsymbol{\xi})}{R^2}\right] \tag{7.1.31}$$

with

$$R = |\mathbf{x}-\boldsymbol{\xi}| \tag{7.1.32}$$

and we recall that $\mathbf{1}$ is the unit tensor and $\otimes$ denotes the tensor product of two vectors. The solution 7.1.30 through 7.1.32 is due to Kelvin [1].

Example 7.1.5

Show that the stress tensor S_{ij} corresponding to the displacement u_i given by Equation 7.1.6 takes the form

$$S_{ij} = \frac{2\mu}{4(1-\nu)}\left[(1-2\nu)\left(\psi_{i,j}+\psi_{j,i}\right) + 2\nu\delta_{ij}\psi_{k,k} - x_k\psi_{k,ij} - \varphi_{,ij}\right] \tag{a}$$

Solution

The stress tensor S_{ij} for a homogeneous isotropic body is given by Equation 4.1.9

$$S_{ij} = \mu(u_{i,j}+u_{j,i}) + \lambda u_{k,k}\delta_{ij} \tag{b}$$

Since, see the table of relations between elasticity constants provided in the front matter,

$$\lambda = \frac{2\mu\nu}{1-2\nu} \tag{c}$$

we may rewrite (b) as

$$S_{ij} = 2\mu\left[\frac{1}{2}(u_{i,j}+u_{j,i}) + \frac{\nu}{1-2\nu}u_{k,k}\delta_{ij}\right] \tag{d}$$

Now, from Equation 7.1.11,

$$u_{i,j} = \frac{1}{4(1-\nu)}\left[(3-4\nu)\psi_{i,j} - \psi_{j,i} - x_p\psi_{p,ij} - \varphi_{,ij}\right] \tag{e}$$

and from Equation 7.1.15

$$u_{k,k} = \frac{1-2\nu}{2(1-\nu)}\psi_{k,k} \tag{f}$$

Substitution of (e) and (f) into (d) yields (a). This completes the solution. □

Note: Equation (a) may be used to obtain a representation of the traction vector $s_i = S_{ij}n_j$ on the boundary ∂B of the body B in terms of the functions ψ_i and φ.

We present now a heuristic method of obtaining another general solution to the displacement equation of equilibrium, Equation 7.1.1. This solution is called the *Boussinesq–Somigliana–Galerkin solution*. To this end we note that the displacement equation of equilibrium, Equation 7.1.1, written in components is, see Equation 7.1.5,

$$\nabla^2 u_i + a u_{k,ki} = -\frac{b_i}{\mu} \tag{7.1.33}$$

where

$$a = \frac{1}{1-2\nu} \tag{7.1.34}$$

Equation 7.1.33 may be rewritten in the operator form

$$L_{ij} u_j = -\frac{b_i}{\mu} \tag{7.1.35}$$

where

$$L_{ij} = \delta_{ij}\nabla^2 + a\partial_i\partial_j \tag{7.1.36}$$

with

$$\partial_i = \frac{\partial}{\partial x_i} \tag{7.1.37}$$

We note that the L_{ij} operator is a 3×3 symmetric matrix differential operator, with elements being linear combinations of second order partial derivative operators. Equation 7.1.35 may be treated as a system of linear algebraic equations with u_1, u_2, and u_3 as unknowns. Treating L_{ij} as numbers, and assuming that b_1, b_2, and b_3 are known, by Cramer's rule we obtain

$$u_i = \frac{W_i}{W} \tag{7.1.38}$$

where

$$W = \begin{vmatrix} L_{11} & L_{12} & L_{13} \\ L_{21} & L_{22} & L_{23} \\ L_{31} & L_{32} & L_{33} \end{vmatrix} \tag{7.1.39}$$

and

$$W_1 = \begin{vmatrix} -\dfrac{b_1}{\mu} & L_{12} & L_{13} \\ -\dfrac{b_2}{\mu} & L_{22} & L_{23} \\ -\dfrac{b_3}{\mu} & L_{32} & L_{33} \end{vmatrix} \tag{7.1.40}$$

$$W_2 = \begin{vmatrix} L_{11} & -\frac{b_1}{\mu} & L_{13} \\ L_{21} & -\frac{b_2}{\mu} & L_{23} \\ L_{31} & -\frac{b_3}{\mu} & L_{33} \end{vmatrix} \quad (7.1.41)$$

$$W_3 = \begin{vmatrix} L_{11} & L_{12} & -\frac{b_1}{\mu} \\ L_{21} & L_{22} & -\frac{b_2}{\mu} \\ L_{31} & L_{32} & -\frac{b_3}{\mu} \end{vmatrix} \quad (7.1.42)$$

We rewrite Equation 7.1.39 as

$$W = \begin{vmatrix} \nabla^2 + a\partial_1^2 & a\partial_1\partial_2 & a\partial_1\partial_3 \\ a\partial_1\partial_2 & \nabla^2 + a\partial_2^2 & a\partial_2\partial_3 \\ a\partial_3\partial_1 & a\partial_3\partial_2 & \nabla^2 + a\partial_3^2 \end{vmatrix} \quad (7.1.43)$$

Expanding the determinant W, we obtain

$$W = (1+a)\nabla^2\nabla^2\nabla^2 \quad (7.1.44)$$

Now, we introduce the notation

$$f_i = -\frac{1}{W}\frac{b_i}{\mu} \quad (7.1.45)$$

Then, for u_1 we obtain

$$u_1 = \begin{vmatrix} f_1 & L_{12} & L_{13} \\ f_2 & L_{22} & L_{23} \\ f_3 & L_{23} & L_{33} \end{vmatrix} = \begin{vmatrix} f_1 & a\partial_1\partial_2 & a\partial_1\partial_3 \\ f_2 & \nabla^2 + a\partial_2^2 & a\partial_2\partial_3 \\ f_3 & a\partial_3\partial_2 & \nabla^2 + a\partial_3^2 \end{vmatrix} \quad (7.1.46)$$

Expanding the determinant on the RHS of Equation 7.1.46 we get

$$\begin{aligned} u_1 &= \nabla^2\nabla^2 f_1 + a\nabla^2\partial_2^2 f_1 + a\nabla^2\partial_3^2 f_1 - a\nabla^2\partial_1\partial_2 f_2 - a\nabla^2\partial_3\partial_1 f_3 \\ &= (1+a)\nabla^2\nabla^2 f_1 - a\partial_1\nabla^2(\partial_1 f_1 + \partial_2 f_2 + \partial_3 f_3) \end{aligned} \quad (7.1.47)$$

The latter formula may be presented as

$$u_1 = (1+a)\nabla^2\nabla^2 f_1 - a\partial_1\partial_k(\nabla^2 f_k) \quad (7.1.48)$$

Similarly, we write u_2 and u_3 in the form

$$u_2 = (1+a)\nabla^2\nabla^2 f_2 - a\partial_2\partial_k(\nabla^2 f_k) \quad (7.1.49)$$

$$u_3 = (1+a)\nabla^2\nabla^2 f_3 - a\partial_3\partial_k(\nabla^2 f_k) \quad (7.1.50)$$

In direct notation,

$$\mathbf{u} = (1 + a)\nabla^2(\nabla^2\mathbf{f}) - a\nabla[\operatorname{div}(\nabla^2\mathbf{f})] \tag{7.1.51}$$

Now, if we introduce a new field **g** by the formula

$$(1 + a)\nabla^2\mathbf{f} = \mathbf{g} \tag{7.1.52}$$

then using Equations 7.1.34, 7.1.45, and 7.1.51 we arrive at

$$\mathbf{u} = \nabla^2\mathbf{g} - \frac{1}{2(1-\nu)}\nabla(\operatorname{div}\mathbf{g}) \tag{7.1.53}$$

where

$$\nabla^2\nabla^2\mathbf{g} = -\frac{\mathbf{b}}{\mu} \tag{7.1.54}$$

Note that Equation 7.1.54 is found from Equation 7.1.45 by multiplying Equation 7.1.45 by W and using Equations 7.1.44 and 7.1.52.

The displacement field **u** given by the formulas 7.1.53 and 7.1.54 is called the *Boussinesq–Somigliana–Galerkin solution.*

The heuristic method used in obtaining the solution 7.1.53 and 7.1.54 is formal, as it contains a number of rational functions with arguments being differential operators, see Equation 7.1.45. The result, however, makes sense, and this is proved in the following theorem:

(T 2) **Boussinesq–Somigliana–Galerkin Solution.** Let **u** be a vector field given by, see Equations 7.1.53 and 7.1.54,

$$\mathbf{u} = \nabla^2\mathbf{g} - \frac{1}{2(1-\nu)}\nabla(\operatorname{div}\mathbf{g}) \tag{7.1.55}$$

where

$$\nabla^2\nabla^2\mathbf{g} = -\frac{\mathbf{b}}{\mu} \tag{7.1.56}$$

Then **u** is an elastic displacement field corresponding to **b**.

Proof: Equations 7.1.55 and 7.1.56 in components read, respectively,

$$u_i = g_{i,kk} - \frac{1}{2(1-\nu)} g_{k,ki} \tag{7.1.57}$$

$$g_{i,kkll} = -\frac{b_i}{\mu} \tag{7.1.58}$$

We need to show that u_i given by Equations 7.1.57 and 7.1.58 satisfies the equation, see Equation 7.1.5,

$$u_{i,kk} + \frac{1}{1-2\nu} u_{k,ki} + \frac{b_i}{\mu} = 0 \tag{7.1.59}$$

In order to do so, we substitute Equation 7.1.57 into Equation 7.1.59, use Equation 7.1.58, and obtain

$$g_{i,kkll} - \frac{1}{2(1-\nu)} g_{k,kill} + \frac{1}{1-2\nu}\left[g_{l,lkk} - \frac{1}{2(1-\nu)} g_{k,kll}\right]_{,i} + \frac{b_1}{\mu} = 0 \tag{7.1.60}$$

This completes the proof of the (T 2) theorem. □

Note: The heuristic method that led to the (T 2) theorem is useful in obtaining an analogous solution of the displacement equation of motion. Such a solution will be discussed in Section 7.2.

Example 7.1.6

Show that the stress tensor field S_{ij} related to the Boussinesq–Somigliana–Galerkin solution given by Equations 7.1.57 and 7.1.58, takes the form

$$S_{ij} = \mu\left[g_{i,jkk} + g_{j,ikk} + \frac{1}{1-\nu}(\nu\delta_{ij}g_{k,kll} - g_{k,kij})\right] \tag{a}$$

Solution

From (d) in Example 7.1.5 we write

$$S_{ij} = \mu\left(u_{i,j} + u_{j,i} + \frac{2\nu}{1-2\nu} u_{k,k}\delta_{ij}\right) \tag{b}$$

Calculating $u_{k,k}$ from Equation 7.1.57 we obtain

$$u_{k,k} = \frac{1-2\nu}{2(1-\nu)} g_{k,kii} \tag{c}$$

Substituting Equations 7.1.57 and (c) into (b) we arrive at (a).

This completes the solution. □

The displacement representations given in the (T 1) and (T 2) theorems allow us to solve boundary value problems of elastostatics in terms of auxiliary functions that satisfy the equations of well-known types, such as Poisson's equation or a nonhomogeneous biharmonic equation. A question arises as to what extent the representations are complete. We say that a representation for the displacement **u** expressed in terms of auxiliary functions is complete if these auxiliary functions exist for any **u** that satisfies the displacement equation of equilibrium. In particular, in the (T 1) theorem the representation for **u** is given by Equation 7.1.2, and the auxiliary functions are φ and $\boldsymbol{\psi}$. In the (T 2) theorem, the representation for **u** is given by Equation 7.1.55, and the auxiliary function is **g**.

In the following theorem, we prove that the two representations discussed in the (T 1) and (T 2) theorems are complete in the stated sense.

(T 3) Completeness of the Boussinesq–Papkovich–Neuber and the Boussinesq–Somigliana–Galerkin Representations. Let **u** be a solution to the displacement equation of equilibrium with the body force **b**. Then there exists a field **g** on B that satisfies Equations 7.1.55 and 7.1.56. Also, there exist fields φ and $\boldsymbol{\psi}$ that satisfy Equations 7.1.2 through 7.1.4.

Proof: To prove the first part of this theorem, that the representation 7.1.55 and 7.1.56 is complete, we rewrite Equation 7.1.55 in the form

$$\nabla^2\mathbf{g} + \frac{1}{1-2\widehat{\nu}}\nabla(\operatorname{div}\mathbf{g}) + \frac{\widehat{\mathbf{b}}}{\mu} = \mathbf{0} \tag{7.1.61}$$

where

$$\frac{\widehat{\mathbf{b}}}{\mu} = -\mathbf{u} \tag{7.1.62}$$

and

$$\frac{1}{1-2\widehat{\nu}} = -\frac{1}{2(1-\nu)} \quad \text{or} \quad \widehat{\nu} = \frac{3}{2} - \nu \tag{7.1.63}$$

Hence, there exists a field **g** satisfying Equation 7.1.61 in the form, see Equations 7.1.2 through 7.1.4,

$$\mathbf{g} = \widehat{\boldsymbol{\psi}} - \frac{1}{4(1-\widehat{\nu})}\nabla(\mathbf{x}\cdot\widehat{\boldsymbol{\psi}} + \widehat{\varphi}) \tag{7.1.64}$$

where

$$\nabla^2\widehat{\boldsymbol{\psi}} = -\frac{1}{\mu}\widehat{\mathbf{b}} \tag{7.1.65}$$

$$\nabla^2\widehat{\varphi} = \frac{1}{\mu}\mathbf{x}\cdot\widehat{\mathbf{b}} \tag{7.1.66}$$

The vector field **g** is given in an implicit form in terms of **u**, if $\widehat{\boldsymbol{\psi}}$ and $\widehat{\varphi}$ are expressed in terms of the Newtonian potentials

$$\widehat{\boldsymbol{\psi}}(\mathbf{x}) = \frac{1}{4\pi\mu}\int_B \frac{\widehat{\mathbf{b}}(\mathbf{y})}{|\mathbf{x}-\mathbf{y}|}\,dv(\mathbf{y}) \tag{7.1.67}$$

and

$$\widehat{\varphi}(\mathbf{x}) = -\frac{1}{4\pi\mu}\int_B \frac{\mathbf{y}\cdot\widehat{\mathbf{b}}(\mathbf{y})}{|\mathbf{x}-\mathbf{y}|}\,dv(\mathbf{y}) \tag{7.1.68}$$

To show that Equation 7.1.56 is satisfied we check that **u** given by Equation 7.1.55 satisfies the relation

$$\nabla^2\mathbf{u} + \frac{1}{1-2\nu}\,\nabla(\operatorname{div}\mathbf{u}) = \nabla^2\nabla^2\mathbf{g} \tag{7.1.69}$$

This is easy to do if we write Equation 7.1.69 in components

$$u_{i,kk} + \frac{1}{1-2\nu}\,u_{k,ki} = g_{i,kkll} \tag{7.1.70}$$

Indeed, substituting u_i given by Equation 7.1.57 into the LHS of Equation 7.1.70 we obtain

$$\begin{aligned}&\left[g_{i,kk} - \frac{1}{2(1-\nu)}\,g_{k,ki}\right]_{,ll} + \frac{1}{1-2\nu}\left[g_{l,lkk} - \frac{1}{2(1-\nu)}\,g_{k,kll}\right]_{,i}\\&= g_{i,kkll} - \frac{1}{2(1-\nu)}\,g_{k,klli} + \frac{1}{1-2\nu}\,\frac{(1-2\nu)}{2(1-\nu)}\,g_{k,klli} = g_{i,kkll}\end{aligned} \tag{7.1.71}$$

Now, since **u** is an elastic displacement corresponding to the body force **b**, then the LHS of Equation 7.1.69 is equal to $-\mathbf{b}/\mu$. Hence, Equation 7.1.69 implies that Equation 7.1.56 is satisfied, and this proves that the representation 7.1.55 and 7.1.56 is complete.

To prove the second part of the theorem, namely, to prove that the representation 7.1.2 through 7.1.4 is complete, we need to show the existence of φ and $\boldsymbol{\psi}$, in terms of an elastic displacement **u** corresponding to **b**, such that Equations 7.1.2 through 7.1.4 are satisfied. The proof is based on the existence of a function **g**, which is assured by the first part of the theorem.

Let us define the functions φ and $\boldsymbol{\psi}$ by the formulas

$$\varphi = 2\operatorname{div}\mathbf{g} - \mathbf{x}\cdot\nabla^2\mathbf{g} \tag{7.1.72}$$

$$\boldsymbol{\psi} = \nabla^2\mathbf{g} \tag{7.1.73}$$

where **g** is the field occurring in Equations 7.1.55 and 7.1.56. Since the existence of **g** was established in the proof of the first part of the theorem, then φ and $\boldsymbol{\psi}$ given by Equations 7.1.72 and 7.1.73 also exist. In addition, it follows from Equations 7.1.56 and 7.1.73 that

$$\nabla^2\boldsymbol{\psi} = -\frac{\mathbf{b}}{\mu} \tag{7.1.74}$$

Also, it follows from Equations 7.1.56 and 7.1.72 that

$$\nabla^2\varphi = \frac{1}{\mu}\,\mathbf{x}\cdot\mathbf{b} \tag{7.1.75}$$

To show this, we write Equation 7.1.72 in components:

$$\varphi = 2g_{k,k} - x_p g_{p,ss} \tag{7.1.76}$$

Computing the gradient and the Laplacian of φ we obtain

$$\varphi_{,i} = 2g_{k,ki} - g_{i,ss} - x_p g_{p,ssi} \tag{7.1.77}$$

and

$$\varphi_{,ii} = 2g_{k,kii} - g_{i,iss} - g_{i,iss} - x_p g_{p,ssii} = -x_p g_{p,ssii} \tag{7.1.78}$$

By virtue of Equation 7.1.56, Equation 7.1.78 implies that Equation 7.1.75 is satisfied.

In order to complete the proof we need to show that Equation 7.1.55 reduces to Equation 7.1.2, which means that the Boussinesq–Somigliana–Galerkin solution reduces to the Boussinesq–Papkovich–Neuber solution. With this in mind, we note that Equations 7.1.72 and 7.1.73 can be written as

$$\varphi = 2(\operatorname{div}\mathbf{g}) - \mathbf{x}\cdot\boldsymbol{\psi} \tag{7.1.79}$$

$$\boldsymbol{\psi} = \nabla^2\mathbf{g} \tag{7.1.80}$$

Hence,

$$\operatorname{div}\mathbf{g} = \frac{1}{2}(\varphi + \mathbf{x}\cdot\boldsymbol{\psi}) \tag{7.1.81}$$

Substituting Equations 7.1.80 and 7.1.81 into Equation 7.1.55, we arrive at Equation 7.1.2, which is the Boussinesq–Papkovitch–Neuber solution.

This completes the proof of the theorem. □

Notes:

(1) We observe that a solution given by Equation 7.1.22 of the displacement equation of equilibrium corresponding to the body force **b**, see Equation 7.1.21,

$$\mathbf{b} = \nabla h + \operatorname{curl}\mathbf{k} \tag{7.1.82}$$

is not complete in the following sense. Since, see Equation 7.1.22,

$$\mathbf{u} = \nabla\varphi + \operatorname{curl}\boldsymbol{\omega} \tag{7.1.83}$$

then

$$\operatorname{div}\mathbf{u} = \nabla^2\varphi \tag{7.1.84}$$

and if $h = 0$, then, see Equation 7.1.23,

$$\operatorname{div}\mathbf{u} = \nabla^2\varphi = 0 \tag{7.1.85}$$

Therefore, the representation of **u** given by Equation 7.1.83 describes divergence free displacement only, that is, displacement without change of volume. Thus, the representation does not cover an arbitrary elastic displacement corresponding to a body force **b**. On the other hand, both the Boussinesq–Papkovitch–Neuber solution and the Boussinesq–Somigliana–Galerkin solution are complete solutions.

(2) Comparing the (T 1) Boussinesq–Papkovitch–Neuber solution, in short the (T 1) solution, with the (T 2) Boussinesq–Somigliana–Galerkin solution, in short the (T 2) solution, we make the following observations:

(a) The displacement $\mathbf{u}$ in the (T 1) solution is expressed as a linear combination of the first derivatives of the auxiliary functions φ and $\boldsymbol{\psi}$, while the displacement $\mathbf{u}$ in the (T 2) solution is expressed as a linear combination of the second derivatives of the auxiliary function $\mathbf{g}$. The coefficients in the linear combination in the (T 1) solution depend on the spatial variable $\mathbf{x}$, while those in the (T 2) solution are constants.

(b) In case of zero body forces, the functions φ and $\boldsymbol{\psi}$ in the (T 1) solution are harmonic, while the function $\mathbf{g}$ in the (T 2) solution is biharmonic.

(c) Finding suitable harmonic functions in curvilinear coordinates in case of the (T 1) solution may be less painful than obtaining biharmonic functions in such coordinates.

(d) A large number of published papers devoted to analytical solutions of equations of elastostatics for geometries like semispace, layer, thick cylinder, sphere, etc., indicate that the (T 1) solution and the (T 2) solution do not exclude each other, and both can be used effectively.

(e) We will return to the (T 1) solution and the (T 2) solution while solving particular boundary value problems in Chapter 9.

In closing this section, we present two special cases of the (T 1) and the (T 2) solutions that are extremely useful in solving axisymmetric problems.

(1) **Boussinesq's Solution for Axial Symmetry.** If $\mathbf{b} = \mathbf{0}$ and $\boldsymbol{\psi} = \psi\mathbf{k}$, then the (T 1) solution reduces to

$$\mathbf{u} = \psi\mathbf{k} - \frac{1}{4(1-\nu)}\nabla(\varphi + z\psi) \tag{7.1.86}$$

where

$$z = \mathbf{x} \cdot \mathbf{k} \tag{7.1.87}$$

with $\mathbf{k}$ being a unit vector along the x_3 axis, which is assumed to be the axis of symmetry of the body, and φ and ψ are scalar-valued harmonic functions.

In cylindrical coordinates (r, θ, z) and for an axially symmetric problem,

$$\mathbf{u} = [u_r(r,z), 0, u_z(r,z)] \tag{7.1.88}$$

where

$$u_r = -\frac{1}{4(1-\nu)}\frac{\partial}{\partial r}(\varphi + z\psi) \tag{7.1.89}$$

$$u_z = \psi - \frac{1}{4(1-\nu)}\frac{\partial}{\partial z}(\varphi + z\psi) \tag{7.1.90}$$

and $\varphi = \varphi(r,z)$ and $\psi = \psi(r,z)$ are harmonic functions,

$$\nabla^2\varphi = 0, \quad \nabla^2\psi = 0 \tag{7.1.91}$$

with

$$\nabla^2 = \frac{\partial^2}{\partial r^2} + \frac{1}{r}\frac{\partial}{\partial r} + \frac{\partial^2}{\partial z^2} \tag{7.1.92}$$

Example 7.1.7

Derive components of the stress tensor **S** associated with the displacements 7.1.89 and 7.1.90.

Solution

The stress tensor **S** associated with the displacements 7.1.89 and 7.1.90 is given by the constitutive relations

$$\begin{aligned} S_{rr} &= 2\mu\left[E_{rr} + \frac{\nu}{1-2\nu}(E_{rr} + E_{\theta\theta} + E_{zz})\right] \\ S_{\theta\theta} &= 2\mu\left[E_{\theta\theta} + \frac{\nu}{1-2\nu}(E_{rr} + E_{\theta\theta} + E_{zz})\right] \\ S_{zz} &= 2\mu\left[E_{zz} + \frac{\nu}{1-2\nu}(E_{rr} + E_{\theta\theta} + E_{zz})\right] \\ S_{rz} &= 2\mu E_{rr}, \quad S_{r\theta} = S_{z\theta} = 0 \end{aligned} \tag{a}$$

where

$$\begin{aligned} E_{rr} &= \frac{\partial u_r}{\partial r}, \quad E_{\theta\theta} = \frac{u_r}{r}, \quad E_{zz} = \frac{\partial u_z}{\partial z} \\ E_{rz} &= \frac{1}{2}\left(\frac{\partial u_r}{\partial z} + \frac{\partial u_z}{\partial r}\right), \quad E_{r\theta} = E_{z\theta} = 0 \end{aligned} \tag{b}$$

Introduce the function

$$\phi(r,z) = \varphi(r,z) + z\psi(r,z) \tag{c}$$

Than the displacements 7.1.89 and 7.1.90 can be written in the form

$$u_r = -\frac{1}{4(1-\nu)}\frac{\partial}{\partial r}\phi \tag{d}$$

and

$$u_z = \psi - \frac{1}{4(1-\nu)}\frac{\partial\phi}{\partial z} \tag{e}$$

Using (b), (d), and (e), we obtain

$$E_{rr} + E_{\theta\theta} + E_{zz} = \frac{\partial \psi}{\partial z} - \frac{1}{4(1-\nu)}\nabla^2 \phi \tag{f}$$

where the Laplacian operator ∇^2 is given by Equation 7.1.92. Also, note that, because of (c),

$$\phi_{,k} = \varphi_{,k} + \delta_{3k}\psi + z\psi_{,k} \tag{g}$$

and

$$\nabla^2 \phi = \phi_{,kk} = \varphi_{,kk} + 2\delta_{3k}\psi_{,k} + z\psi_{,kk} \tag{h}$$

Since, due to Equations 7.1.91,

$$\varphi_{,kk} = \nabla^2 \varphi = 0, \quad \psi_{,kk} = \nabla^2 \psi = 0 \tag{i}$$

therefore, from (h), we obtain

$$\nabla^2 \phi = 2\frac{\partial \psi}{\partial z} \tag{j}$$

and (f) reduces to

$$E_{rr} + E_{\theta\theta} + E_{zz} = \frac{1-2\nu}{2(1-\nu)}\frac{\partial \psi}{\partial z} \tag{k}$$

Now, by $(b)_1$ and (d) we get

$$E_{rr} = -\frac{1}{4(1-\nu)}\frac{\partial^2 \phi}{\partial r^2} \tag{l}$$

Hence, substituting (k) and (l) into $(a)_1$ we obtain

$$S_{rr} = -\frac{\mu}{2(1-\nu)}\left(\frac{\partial^2 \phi}{\partial r^2} - 2\nu\frac{\partial \psi}{\partial z}\right) \tag{m}$$

Also, by $(b)_2$ and (d), we get

$$E_{\theta\theta} = -\frac{1}{4(1-\nu)}\frac{1}{r}\frac{\partial \phi}{\partial r} \tag{n}$$

and substituting (k) and (n) into $(a)_2$ we find

$$S_{\theta\theta} = -\frac{\mu}{2(1-\nu)}\left(\frac{1}{r}\frac{\partial \phi}{\partial r} - 2\nu\frac{\partial \psi}{\partial z}\right) \tag{o}$$

Next, by $(b)_3$ and (e), we get

$$E_{zz} = \frac{\partial \psi}{\partial z} - \frac{1}{4(1-\nu)}\frac{\partial^2 \phi}{\partial z^2} \tag{p}$$

Therefore, from (k), (p), and $(a)_2$, we write

$$S_{zz} = -\frac{\mu}{2(1-\nu)}\left[\frac{\partial^2\phi}{\partial z^2} - 2(2-\nu)\frac{\partial\psi}{\partial z}\right] \tag{q}$$

Finally, by $(b)_4$, (d), and (f), we get

$$E_{rz} = -\frac{1}{4(1-\nu)}\left[\frac{\partial^2\phi}{\partial r\partial z} - 2(1-\nu)\frac{\partial\psi}{\partial r}\right] \tag{r}$$

and substituting (r) into $(a)_4$ we arrive at

$$S_{rz} = -\frac{\mu}{2(1-\nu)}\left[\frac{\partial^2\phi}{\partial r\partial z} - 2(1-\nu)\frac{\partial\psi}{\partial r}\right] \tag{s}$$

Hence, the stress tensor $\mathbf{S}$ associated with the displacement vector $\mathbf{u} = [u_r, 0, u_z]$ is given by the formulas

$$\begin{aligned}
S_{rr} &= -\frac{\mu}{2(1-\nu)}\left(\frac{\partial^2\phi}{\partial r^2} - 2\nu\frac{\partial\psi}{\partial z}\right)\\
S_{\theta\theta} &= -\frac{\mu}{2(1-\nu)}\left(\frac{1}{r}\frac{\partial\phi}{\partial r} - 2\nu\frac{\partial\psi}{\partial z}\right)\\
S_{zz} &= -\frac{\mu}{2(1-\nu)}\left[\frac{\partial^2\phi}{\partial z^2} - 2(2-\nu)\frac{\partial\psi}{\partial z}\right]\\
S_{rz} &= -\frac{\mu}{2(1-\nu)}\left[\frac{\partial^2\phi}{\partial r\partial z} - 2(1-\nu)\frac{\partial\psi}{\partial r}\right]\\
S_{r\theta} &= 0, \quad S_{z\theta} = 0
\end{aligned} \tag{t}$$

This completes the solution. □

(2) **Love's Solution.** If $\mathbf{b} = \mathbf{0}$ and $\mathbf{g} = \chi\mathbf{k}$, then the (T 2) solution reduces to Love's solution

$$\mathbf{u} = (\nabla^2\chi)\mathbf{k} - \frac{1}{2(1-\nu)}\nabla(\nabla\chi\cdot\mathbf{k}) \tag{7.1.93}$$

where

$$\nabla^2\nabla^2\chi = 0 \tag{7.1.94}$$

Here, χ is a scalar-valued function depending on r and z only, and $\mathbf{k}$ is a unit vector along the x_3 axis. The function $\chi(r,z)$ is called *Love's function* [2,3].

In cylindrical coordinates (r,θ,z) and for an axially symmetric problem,

$$u_r = -\frac{1}{2(1-\nu)}\frac{\partial^2}{\partial r\partial z}\chi \tag{7.1.95}$$

$$u_\theta = 0 \tag{7.1.96}$$

$$u_z = \frac{1}{2(1-\nu)}\left[2(1-\nu)\nabla^2 - \frac{\partial^2}{\partial z^2}\right]\chi \tag{7.1.97}$$

Example 7.1.8

Show that the stress components corresponding to the displacements u_r and u_z given by Equations 7.1.95 and 7.1.97, respectively, take the form

$$S_{rr} = \frac{\mu}{1-\nu}\frac{\partial}{\partial z}\left(\nu\nabla^2 - \frac{\partial^2}{\partial r^2}\right)\chi \tag{a}$$

$$S_{rz} = \frac{\mu}{1-\nu}\frac{\partial}{\partial r}\left[(1-\nu)\nabla^2 - \frac{\partial^2}{\partial z^2}\right]\chi \tag{b}$$

$$S_{\theta\theta} = \frac{\mu}{1-\nu}\frac{\partial}{\partial z}\left(\nu\nabla^2 - \frac{1}{r}\frac{\partial}{\partial r}\right)\chi \tag{c}$$

$$S_{zz} = \frac{\mu}{1-\nu}\frac{\partial}{\partial z}\left[(2-\nu)\nabla^2 - \frac{\partial^2}{\partial z^2}\right]\chi \tag{d}$$

$$S_{r\theta} = S_{\theta z} = 0 \tag{e}$$

Note that in the cylindrical coordinates (r, θ, z) double subscripts do not mean summation. Thus, S_{rr} is a single component of the stress tensor **S** in the r direction, and should not be confused with the trace of **S**. Similarly, S_{zz} and $S_{\theta\theta}$ stand for the components of **S** in the z and θ directions, respectively.

Solution

The stress tensor **S** corresponding to the displacements u_r and u_z given by Equations 7.1.95 and 7.1.97 is defined by the equations [cf. Equations (a) in Example 7.1.7]

$$\left.\begin{aligned}
S_{rr} &= 2\mu\left[E_{rr} + \frac{\nu}{1-2\nu}(E_{rr} + E_{\theta\theta} + E_{zz})\right]\\
S_{\theta\theta} &= 2\mu\left[E_{\theta\theta} + \frac{\nu}{1-2\nu}(E_{rr} + E_{\theta\theta} + E_{zz})\right]\\
S_{zz} &= 2\mu\left[E_{zz} + \frac{\nu}{1-2\nu}(E_{rr} + E_{\theta\theta} + E_{zz})\right]\\
S_{rz} &= 2\mu E_{rz}, \quad S_{r\theta} = S_{z\theta} = 0
\end{aligned}\right. \tag{f}$$

where

$$\left.\begin{aligned}
&E_{rr} = \frac{\partial u_r}{\partial r}, \quad E_{\theta\theta} = \frac{u_r}{r}, \quad E_{zz} = \frac{\partial u_z}{\partial z}\\
&E_{rz} = \frac{1}{2}\left(\frac{\partial u_r}{\partial z} + \frac{\partial u_z}{\partial r}\right), \quad E_{r\theta} = E_{z\theta} = 0
\end{aligned}\right. \tag{g}$$

Substituting u_r and u_z from Equations 7.1.95 and 7.1.97, respectively, into (g) we obtain

$$E_{rr} = -\frac{1}{2(1-\nu)}\frac{\partial^3\chi}{\partial r^2\partial z} \tag{h}$$

$$E_{\theta\theta} = -\frac{1}{2(1-\nu)}\frac{1}{r}\frac{\partial^2\chi}{\partial r\partial z} \tag{i}$$

$$E_{zz} = \frac{1}{2(1-\nu)}\frac{\partial}{\partial z}\left[2(1-\nu)\nabla^2 - \frac{\partial^2}{\partial z^2}\right]\chi \tag{j}$$

$$E_{rz} = \frac{1}{2(1-\nu)} \frac{\partial}{\partial r} \left[(1-\nu)\nabla^2 - \frac{\partial^2}{\partial z^2} \right] \chi \tag{k}$$

$$E_{\theta r} = E_{z\theta} = 0 \tag{l}$$

Now, Equations (h) through (j) imply that

$$E_{rr} + E_{\theta\theta} + E_{zz} = \frac{1-2\nu}{2(1-\nu)} \frac{\partial}{\partial z} \nabla^2 \chi \tag{m}$$

Therefore, substituting (h) and (m) into $(f)_1$ we arrive at (a). Also, substituting (k) into $(f)_4$ we obtain (b). Next, substituting (i) and (m) into $(f)_2$ we get (c). Finally, if (j) and (m) are inserted into $(f)_3$, we find (d). This completes the solution. □

7.2 COMPLETE SOLUTIONS OF ELASTODYNAMICS

In the previous section, we discussed two complete solutions of the displacement equation of equilibrium for a homogeneous isotropic body. Now we are to present two complete solutions to the displacement equation of motion for such a body.

Let us recall the displacement equation of motion for a homogeneous isotropic elastic body, see Equation 4.2.23,

$$\Box_2^2 \mathbf{u} + \left[\left(\frac{c_1}{c_2} \right)^2 - 1 \right] \nabla(\operatorname{div} \mathbf{u}) + \frac{\mathbf{b}}{\mu} = \mathbf{0} \tag{7.2.1}$$

where

$$\Box_2^2 = \nabla^2 - \frac{1}{c_2^2}\frac{\partial^2}{\partial t^2}, \quad \frac{1}{c_1^2} = \frac{\rho}{\lambda + 2\mu}, \quad \frac{1}{c_2^2} = \frac{\rho}{\mu} \tag{7.2.2}$$

We assume that the body force $\mathbf{b}$ is represented by Helmholtz's decomposition formula, see Helmholtz's Theorem, Section 2.3.3,

$$\mathbf{b} = -\nabla h - \operatorname{curl} \mathbf{k}, \quad \operatorname{div} \mathbf{k} = 0 \tag{7.2.3}$$

where h and $\mathbf{k}$ are prescribed fields on $\bar{B} \times [0, \infty)$.

A solution $\mathbf{u}$ on $\bar{B} \times [0, \infty)$ to Equation 7.2.1 will be called an *elastic motion* corresponding to $\mathbf{b}$. The notion of an elastic motion is a dynamic counterpart to an elastic displacement field of elastostatics.

Before we formulate a theorem on the first dynamic complete solution we also recall an equivalent displacement equation of motion, compare Equation 4.2.20

$$\nabla(\operatorname{div} \mathbf{u}) - \frac{c_2^2}{c_1^2} \operatorname{curl}\operatorname{curl} \mathbf{u} - \frac{1}{c_1^2} \ddot{\mathbf{u}} + \frac{\mathbf{b}}{\lambda + 2\mu} = \mathbf{0} \tag{7.2.4}$$

(U 1) **Green–Lamé Solution.** Let

$$\mathbf{u} = \nabla \varphi + \operatorname{curl} \boldsymbol{\psi} \tag{7.2.5}$$

where φ and $\boldsymbol{\psi}$ satisfy, respectively, the equations

$$\square_1^2 \varphi = \frac{h}{\lambda + 2\mu} \tag{7.2.6}$$

$$\square_2^2 \boldsymbol{\psi} = \frac{\mathbf{k}}{\mu} \tag{7.2.7}$$

Then $\mathbf{u}$ is an elastic motion corresponding to $\mathbf{b}$ given by Equation 7.2.3.
In Equation 7.2.6,

$$\square_1^2 = \nabla^2 - \frac{1}{c_1^2}\frac{\partial^2}{\partial t^2} \tag{7.2.8}$$

Equations 7.2.5 through 7.2.7 define Green–Lamé solution.

Proof: We need to show that $\mathbf{u}$ given by Equations 7.2.5 through 7.2.7 satisfies Equation 7.2.1, which is equivalent to Equation 7.2.4. To this end, we first note that

$$\operatorname{div} \mathbf{u} = \nabla^2 \varphi \tag{7.2.9}$$

Hence, substituting $\mathbf{u}$ from Equation 7.2.5 into the LHS of Equation 7.2.4, which we denote by $\boldsymbol{\ell}$, we get

$$\begin{aligned} \boldsymbol{\ell} &\equiv \nabla(\operatorname{div} \mathbf{u}) - \frac{c_2^2}{c_1^2} \operatorname{curl}\operatorname{curl} \mathbf{u} - \frac{1}{c_1^2}\ddot{\mathbf{u}} + \frac{\mathbf{b}}{\lambda + 2\mu} \\ &= \nabla\nabla^2\varphi - \frac{1}{c_1^2}\nabla\ddot{\varphi} - \frac{1}{c_1^2}\operatorname{curl}\ddot{\boldsymbol{\psi}} - \frac{c_2^2}{c_1^2}\operatorname{curl}\operatorname{curl}\operatorname{curl}\boldsymbol{\psi} \\ &\quad + \frac{1}{\lambda + 2\mu}(-\nabla h - \operatorname{curl} \mathbf{k}) \end{aligned} \tag{7.2.10}$$

Since

$$\operatorname{curl}\operatorname{curl} \boldsymbol{\psi} = \nabla \operatorname{div} \boldsymbol{\psi} - \nabla^2 \boldsymbol{\psi} \tag{7.2.11}$$

then

$$\operatorname{curl}\operatorname{curl}\operatorname{curl} \boldsymbol{\psi} = -\operatorname{curl}(\nabla^2 \boldsymbol{\psi}) \tag{7.2.12}$$

Therefore, and taking into account that

$$\frac{c_2^2}{c_1^2} = \frac{\mu}{\lambda + 2\mu} \tag{7.2.13}$$

we arrive at

$$\boldsymbol{\ell} = \nabla\left(\square_1^2 \varphi - \frac{h}{\lambda + 2\mu}\right) + \frac{\mu}{\lambda + 2\mu}\operatorname{curl}\left(\square_2^2 \boldsymbol{\psi} - \frac{\mathbf{k}}{\mu}\right) \tag{7.2.14}$$

Hence, since φ and $\boldsymbol{\psi}$ satisfy Equations 7.2.6 and 7.2.7, respectively, $\boldsymbol{\ell} = \mathbf{0}$, and this completes the proof of the theorem. □

Example 7.2.1

Let $\square_0^2$ be the wave operator defined by

$$\square_0^2 = \nabla^2 - \frac{1}{c_0^2}\frac{\partial^2}{\partial t^2} \tag{a}$$

where c_0 stands for a positive constant, and let $f = f(t)$ be a function on $[0, \infty)$. Show that a solution of the wave equation

$$\square_0^2 \boldsymbol{\psi} = -\frac{1}{c_0^2} f(t)\delta(\mathbf{x} - \boldsymbol{\xi})\mathbf{e} \quad \text{on } E^3 \times [0, \infty) \tag{b}$$

where $\boldsymbol{\xi}$ is a fixed point, $\delta(\mathbf{x})$ is the three-dimensional Dirac delta function, and $\mathbf{e}$ is a constant unit vector, is given by

$$\boldsymbol{\psi}(\mathbf{x}, \boldsymbol{\xi}; t) = \frac{1}{4\pi c_0^2}\frac{1}{R} f\left(t - \frac{R}{c_0}\right)\mathbf{e} \tag{c}$$

where

$$R = |\mathbf{x} - \boldsymbol{\xi}| \tag{d}$$

Solution

It is sufficient to show that

$$\square_0^2 \psi_0 = -\frac{1}{c_0^2} f(t)\,\delta(\mathbf{x} - \boldsymbol{\xi}) \tag{e}$$

where

$$\psi_0 = \psi_0(\mathbf{x}, \boldsymbol{\xi}; t) = \frac{1}{4\pi c_0^2}\frac{1}{R} f\left(t - \frac{R}{c_0}\right) \tag{f}$$

Indeed, by multiplying Equation (e) by $\mathbf{e}$ we find that $\boldsymbol{\psi} = \psi_0\,\mathbf{e}$, given by Equation (c), satisfies Equation (b).

We compute first, in components, $\psi_{0,k}$ and the result is

$$\psi_{0,k} = \frac{1}{4\pi c_0^2}\left[\left(\frac{1}{R}\right)_{,k} f\left(t - \frac{R}{c_0}\right) + \frac{1}{R}\left(-\frac{1}{c_0}\right)R_{,k}f'\left(t - \frac{R}{c_0}\right)\right] \tag{g}$$

Next, we calculate $\psi_{0,kk}$ and find

$$\begin{aligned}\psi_{0,kk} = \frac{1}{4\pi c_0^2}\Bigg\{&\left(\frac{1}{R}\right)_{,kk} f\left(t - \frac{R}{c_0}\right) - \frac{2}{c_0}\left(\frac{1}{R}\right)_{,k} R_{,k}f'\left(t - \frac{R}{c_0}\right)\\ &- \frac{1}{c_0}\frac{1}{R}\left[R_{,kk}f'\left(t - \frac{R}{c_0}\right) - R_{,k}R_{,k}\frac{1}{c_0}f''\left(t - \frac{R}{c_0}\right)\right]\Bigg\}\end{aligned} \tag{h}$$

Since

$$R_{,k} = \frac{x_k - \xi_k}{R} \tag{i}$$

and

$$\left(\frac{1}{R}\right)_{,k} = -\frac{1}{R^2} R_{,k} \tag{j}$$

hence, from (i) it follows that

$$R_{,k} R_{,k} = 1 \tag{k}$$

and from (i) and (j)

$$\left(\frac{1}{R}\right)_{,k} R_{,k} = -\frac{1}{R^2} \tag{l}$$

Moreover,

$$R_{,kk} = \frac{3}{R} - \frac{(x_k - \xi_k)(x_k - \xi_k)}{R^3} = \frac{2}{R} \tag{m}$$

Substitution of (k) through (m) into (h) yields

$$\psi_{0,kk} = \frac{1}{4\pi c_0^2} \left[\left(\frac{1}{R}\right)_{,kk} f\left(t - \frac{R}{c_0}\right) + \frac{1}{c_0^2 R} f''\left(t - \frac{R}{c_0}\right) \right] \tag{n}$$

Also, differentiating (f) twice with respect to time, we get

$$\ddot{\psi}_0 = \frac{1}{4\pi c_0^2} \frac{1}{R} f''\left(t - \frac{R}{c_0}\right) \tag{o}$$

We compute the LHS of (e)

$$\Box_0^2 \psi_0 = \frac{1}{4\pi c_0^2} \left(\frac{1}{R}\right)_{,kk} f\left(t - \frac{R}{c_0}\right) \tag{p}$$

Now, since

$$\left(\frac{1}{R}\right)_{,kk} = \nabla^2 \left(\frac{1}{R}\right) = -4\pi \delta(\mathbf{x} - \boldsymbol{\xi}) \tag{q}$$

Equation (p) may be reduced to the form

$$\Box_0^2 \psi_0 = -\frac{1}{c_0^2} \delta(\mathbf{x} - \boldsymbol{\xi}) f\left(t - \frac{R}{c_0}\right) \tag{r}$$

Using the relation

$$f\left(t - \frac{R}{c_0}\right)\delta(\mathbf{x} - \boldsymbol{\xi}) = f(t)\delta(\mathbf{x} - \boldsymbol{\xi}) \tag{s}$$

we find that ψ_0 satisfies (e). This completes the solution. □

Note: The result in this example can be obtained in easier way if a spherical coordinate system with the origin at the point $\boldsymbol{\xi}$ is used, and one notes that

$$\nabla^2 = \frac{\partial^2}{\partial R^2} + \frac{2}{R}\frac{\partial}{\partial R} \tag{7.2.15}$$

In this case, the solution ψ_0 depends on the radial spherical coordinate R only.

Example 7.2.2

Show that a solution to the nonhomogeneous wave equation

$$\Box_0^2\, \varphi(\mathbf{x}, t) = -F(\mathbf{x}, t) \tag{a}$$

subject to the homogeneous initial conditions

$$\varphi(\mathbf{x}, 0) = 0, \dot{\varphi}(\mathbf{x}, 0) = 0 \tag{b}$$

is given by the integral

$$\varphi(\mathbf{x}, t) = \frac{1}{4\pi}\int_{B_0} \frac{F(\boldsymbol{\xi}, t - |\mathbf{x} - \boldsymbol{\xi}|/c_0)}{|\mathbf{x} - \boldsymbol{\xi}|}\, dv(\boldsymbol{\xi}) \tag{c}$$

where

$$B_0 = \{\boldsymbol{\xi} : |\mathbf{x} - \boldsymbol{\xi}| \leq c_0 t\} \tag{d}$$

In (a), $F = F(\mathbf{x}, t)$ is a prescribed function on $E^3 \times [0, \infty)$. If $F(\mathbf{x}, t)$ is prescribed over a finite domain B of E^3, (c) describes a wave propagating with velocity c_0 from B to infinity.

Solution

Let $g = g(\mathbf{x}, t; \boldsymbol{\xi}, s)$ denote Green's function for an unbounded domain for the wave operator $\Box_0^2$, that is, $g(\mathbf{x}, t; \boldsymbol{\xi}, s)$ satisfies the equation

$$\Box_0^2\, g = -\delta(\mathbf{x} - \boldsymbol{\xi})\delta(t - s) \quad \text{on } E^3 \times [0, \infty) \tag{e}$$

where

$\boldsymbol{\xi}$ is a fixed point of E^3

s is a fixed point on the t axis

Such a Green function may be derived from the solution given in the previous example if we let

$$f(t) = c_0^2 \delta(t - s) \tag{f}$$

Then from (f) in Example 7.2.1 we write

$$g(\mathbf{x}, t; \boldsymbol{\xi}, s) = \frac{1}{4\pi} \frac{1}{|\mathbf{x} - \boldsymbol{\xi}|} \delta \left[(t - s) - \frac{|\mathbf{x} - \boldsymbol{\xi}|}{c_0} \right] \tag{g}$$

Now, multiplying both sides of (e) by $F(\boldsymbol{\xi}, s)$, assuming $F(\boldsymbol{\xi}, s)$ to vanish outside of the region B, and integrating the result with respect to $\boldsymbol{\xi}$ over B, and with respect to s over the interval $(0, t]$, we obtain

$$\Box_0^2 \varphi = -F(\mathbf{x}, t) \tag{h}$$

where

$$\varphi(\mathbf{x}, t) = \int_0^t \int_B g(\mathbf{x}, t; \boldsymbol{\xi}, s) F(\boldsymbol{\xi}, s)\, dv(\boldsymbol{\xi})\, ds \tag{i}$$

or, using (g),

$$\varphi(\mathbf{x}, t) = \frac{1}{4\pi} \int_0^t \left\{ \int_B \frac{F(\boldsymbol{\xi}, s)}{|\mathbf{x} - \boldsymbol{\xi}|} \delta \left[(t - s) - \frac{|\mathbf{x} - \boldsymbol{\xi}|}{c_0} \right] dv(\boldsymbol{\xi}) \right\} ds \tag{j}$$

From the filtering property of the Dirac delta function

$$\int_0^t h(s) \delta(t_0 - s)\, ds = h(t_0), \quad t_0 \in (0, t) \tag{k}$$

where $h(s)$ is an arbitrary function, and from the fact that the body is initially at rest, in particular, that $F(\mathbf{x}, t) = 0$ for $t \leq 0$, we obtain from (j)

$$\varphi(\mathbf{x}, t) = \frac{1}{4\pi} \int_{B_0} \frac{F(\boldsymbol{\xi}, t - |\mathbf{x} - \boldsymbol{\xi}|/c_0)}{|\mathbf{x} - \boldsymbol{\xi}|} dv(\boldsymbol{\xi}) \tag{l}$$

where B_0 is given by (d). Obviously, B_0 is the interior of a sphere with the center at $\mathbf{x}$ and with radius $c_0 t$.

Finally, if $F = F(\mathbf{x}, t)$ is smooth on $B \times [0, \infty)$ then from (l) we find that

$$\varphi(\mathbf{x}, 0) = 0, \quad \dot{\varphi}(\mathbf{x}, 0) = 0, \quad \mathbf{x} \in E^3 \tag{m}$$

This completes the solution. □

Note: The function $\varphi(\mathbf{x}, t)$ given by (l) is called a *retarded potential* because

$$\{F\}_0 \equiv F\left(\boldsymbol{\xi}, t - \frac{|\mathbf{x} - \boldsymbol{\xi}|}{c_0}\right) \tag{7.2.16}$$

represents a retarded value of the function $F(\boldsymbol{\xi}, t)$ relative to the point $\mathbf{x}$. The physical meaning of $\{F\}_0$ is that a wave propagating from the point $\boldsymbol{\xi}$ reaches the point $\mathbf{x}$ after a finite time $|\mathbf{x} - \boldsymbol{\xi}|/c_0$.

Example 7.2.3

Let the body force $\mathbf{b}$ be prescribed by the formula, see Equation 7.2.3,

$$\mathbf{b} = -\nabla h - \text{curl}\,\mathbf{k}, \quad \text{div}\,\mathbf{k} = 0 \quad \text{on } B \times [0, \infty) \tag{a}$$

Show that a solution to the displacement equation of motion for a homogeneous isotropic elastic body corresponding to $\mathbf{b}$ may be obtained in the form

$$\begin{aligned} \mathbf{u}(\mathbf{x}, t) = &-\frac{1}{4\pi(\lambda + 2\mu)} \int_{B_1} \nabla_x \frac{\{h\}_1}{|\mathbf{x} - \boldsymbol{\xi}|}\, dv(\boldsymbol{\xi}) \\ &- \frac{1}{4\pi\mu} \int_{B_2} \text{curl}_x \frac{\{\mathbf{k}\}_2}{|\mathbf{x} - \boldsymbol{\xi}|}\, dv(\boldsymbol{\xi}) \end{aligned} \tag{b}$$

Here,

$$B_i = \{\boldsymbol{\xi} : |\mathbf{x} - \boldsymbol{\xi}| \le c_i t\}, \quad i = 1, 2 \tag{c}$$

$$\{h\}_1 = h\left(\boldsymbol{\xi}, t - \frac{|\mathbf{x} - \boldsymbol{\xi}|}{c_1}\right) \tag{d}$$

$$\{\mathbf{k}\}_2 = \mathbf{k}\left(\boldsymbol{\xi}, t - \frac{|\mathbf{x} - \boldsymbol{\xi}|}{c_2}\right) \tag{e}$$

Moreover, the symbols ∇_x and curl_x indicate that gradient and curl are taken with respect to x, and c_1 and c_2 are defined by Equation 7.2.2.

Solution

Using the Green–Lamé solution corresponding to the body force $\mathbf{b}$ given by (a), we find that $\mathbf{u}$ is represented by (compare Equations 7.2.5 through 7.2.7)

$$\mathbf{u} = \nabla\varphi + \text{curl}\,\boldsymbol{\psi} \tag{f}$$

where

$$\square_1^2\,\varphi = \frac{h}{\lambda + 2\mu}, \quad \square_2^2\,\boldsymbol{\psi} = \frac{\mathbf{k}}{\mu} \tag{g}$$

It follows from the solution in Example 7.2.2, Equations (c) and (d), that φ and $\boldsymbol{\psi}$ satisfying the first and the second of equations (g), respectively, may be represented by

$$\varphi(\mathbf{x}, t) = -\frac{1}{4\pi(\lambda + 2\mu)} \int_{B_1} \frac{\{h\}_1}{|\mathbf{x} - \boldsymbol{\xi}|}\, dv(\boldsymbol{\xi}) \tag{h}$$

and

$$\boldsymbol{\psi}(\mathbf{x}, t) = -\frac{1}{4\pi\mu} \int_{B_2} \frac{\{\mathbf{k}\}_2}{|\mathbf{x} - \boldsymbol{\xi}|} \, dv(\boldsymbol{\xi}) \tag{i}$$

where B_1 and B_2, and $\{h\}_1$ and $\{\mathbf{k}\}_2$ are defined by (c) through (e). Hence, substituting (h) and (i) into (f), we arrive at (b). This completes the solution. □

In the following, we prove that the Green–Lamé solution, given by Equations 7.2.5 through 7.2.7, is complete in the sense similar to that of the Boussinesq–Papkovitch–Neuber solution of elastostatics, see Equations 7.1.2 through 7.1.4. More precisely, we prove the following theorem:

(U 2) **Completeness of the Green–Lamé Solution.** Let $\mathbf{u}$ be an elastic motion corresponding to $\mathbf{b}$, and let

$$\mathbf{b} = -\nabla h - \operatorname{curl} \mathbf{k}, \quad \operatorname{div} \mathbf{k} = 0 \quad \text{on } B \times [0, \infty) \tag{7.2.17}$$

where h and $\mathbf{k}$ are prescribed on $B \times [0, \infty)$. Then there exist a scalar function $\varphi(\mathbf{x}, t)$ and a vector-valued function $\boldsymbol{\psi}(\mathbf{x}, t)$ such that $\mathbf{u}(\mathbf{x}, t)$ is represented by

$$\mathbf{u} = \nabla\varphi + \operatorname{curl} \boldsymbol{\psi} \tag{7.2.18}$$

with

$$\operatorname{div} \boldsymbol{\psi} = 0 \tag{7.2.19}$$

where $\varphi(\mathbf{x}, t)$ and $\boldsymbol{\psi}(\mathbf{x}, t)$ satisfy the nonhomogeneous wave equations

$$\Box_1^2 \varphi = \frac{h}{\lambda + 2\mu} \tag{7.2.20}$$

$$\Box_2^2 \boldsymbol{\psi} = \frac{\mathbf{b}}{\mu} \tag{7.2.21}$$

Proof: Since $\mathbf{u}$ is an elastic motion corresponding to $\mathbf{b}$, then $\mathbf{u}$ satisfies the displacement equation of motion, see Equation 7.2.4,

$$\ddot{\mathbf{u}} = c_1^2 \nabla(\operatorname{div} \mathbf{u}) - c_2^2 \operatorname{curl} \operatorname{curl} \mathbf{u} + \frac{\mathbf{b}}{\rho} \tag{7.2.22}$$

Integrating this equation twice with respect to t, and using the notation, see Equation 6.2.2,

$$i(t) = t \tag{7.2.23}$$

we obtain

$$\mathbf{u} = c_1^2 \nabla(i * \operatorname{div} \mathbf{u}) - c_2^2 \operatorname{curl}(i * \operatorname{curl} \mathbf{u}) + \frac{1}{\rho} \mathbf{f} \tag{7.2.24}$$

where **f** is the pseudo-body force field defined by, see Equation 6.2.1,

$$\mathbf{f} = i * \mathbf{b} + \rho(\mathbf{u}_0 + t\dot{\mathbf{u}}_0) \tag{7.2.25}$$

$$\mathbf{u}_0 = \mathbf{u}(\mathbf{x}, 0), \quad \dot{\mathbf{u}}_0 = \dot{\mathbf{u}}(\mathbf{x}, 0) \tag{7.2.26}$$

and $*$ stands for the convolution product on the time axis. Note that in view of Equation 7.2.17 the pseudo-body force field **f** may also be written in the form

$$\mathbf{f} = -\nabla(i * h) - \operatorname{curl}(i * \mathbf{k}) + \rho(\mathbf{u}_0 + t\dot{\mathbf{u}}_0) \tag{7.2.27}$$

Since $\mathbf{u}_0$ and $\dot{\mathbf{u}}_0$ are prescribed functions on B, then by Helmholtz's theorem, see Equations 2.3.25 and 2.3.26, there exist fields φ_0, $\boldsymbol{\psi}_0$, $\dot{\varphi}_0$, and $\dot{\boldsymbol{\psi}}_0$ on B, such that

$$\mathbf{u}_0 = \nabla\varphi_0 + \operatorname{curl}\boldsymbol{\psi}_0, \quad \operatorname{div}\boldsymbol{\psi}_0 = 0 \tag{7.2.28}$$

$$\dot{\mathbf{u}}_0 = \nabla\dot{\varphi}_0 + \operatorname{curl}\dot{\boldsymbol{\psi}}_0, \quad \operatorname{div}\dot{\boldsymbol{\psi}}_0 = 0 \tag{7.2.29}$$

Substituting Equations 7.2.28 and 7.2.29 into Equation 7.2.27, we find

$$\mathbf{f} = -\nabla[i * h - \rho(\varphi_0 + t\dot{\varphi}_0)] - \operatorname{curl}[i * \mathbf{k} - \rho(\boldsymbol{\psi}_0 + t\dot{\boldsymbol{\psi}}_0)] \tag{7.2.30}$$

Now, we define the functions $\varphi = \varphi(\mathbf{x}, t)$ and $\boldsymbol{\psi} = \boldsymbol{\psi}(\mathbf{x}, t)$ by

$$\varphi = c_1^2\, i * \operatorname{div}\mathbf{u} - \frac{1}{\rho}\, i * h + \varphi_0 + t\dot{\varphi}_0 \tag{7.2.31}$$

$$\boldsymbol{\psi} = -c_2^2\, i * \operatorname{curl}\mathbf{u} - \frac{1}{\rho}\, i * \mathbf{k} + \boldsymbol{\psi}_0 + t\dot{\boldsymbol{\psi}}_0 \tag{7.2.32}$$

and from Equations 7.2.24, 7.2.30, 7.2.31, and 7.2.32, we arrive at

$$\mathbf{u} = \nabla\varphi + \operatorname{curl}\boldsymbol{\psi} \tag{7.2.33}$$

which is the desired Equation 7.2.18. To obtain Equation 7.2.19, we note that by Equations 7.2.17, 7.2.28, and 7.2.29

$$\operatorname{div}\mathbf{k} = 0, \quad \operatorname{div}\boldsymbol{\psi}_0 = 0, \quad \operatorname{div}\dot{\boldsymbol{\psi}}_0 = 0 \tag{7.2.34}$$

Hence, applying operator div to Equation 7.2.32, and using the fact that

$$\operatorname{div}(\operatorname{curl}\mathbf{u}) = 0 \tag{7.2.35}$$

we arrive at Equation 7.2.19.

To complete the proof of the theorem, it remains to show that φ and $\boldsymbol{\psi}$ satisfy Equations 7.2.20 and 7.2.21, respectively. To this end, we differentiate Equations 7.2.31 and 7.2.32 twice with respect to time, and obtain

$$\ddot{\varphi} = c_1^2 \operatorname{div}\mathbf{u} - \frac{h}{\rho} \tag{7.2.36}$$

$$\ddot{\boldsymbol{\psi}} = -c_2^2 \operatorname{curl}\mathbf{u} - \frac{\mathbf{k}}{\rho} \tag{7.2.37}$$

By applying the div operation to Equation 7.2.33, we get

$$\operatorname{div}\mathbf{u} = \nabla^2\varphi \tag{7.2.38}$$

Substituting Equation 7.2.38 into Equation 7.2.36 we arrive at Equation 7.2.20. Next, we apply the operation curl to Equation 7.2.33 and we find

$$\operatorname{curl}\mathbf{u} = \operatorname{curl}\operatorname{curl}\boldsymbol{\psi} = \nabla(\operatorname{div}\boldsymbol{\psi}) - \nabla^2\boldsymbol{\psi} \tag{7.2.39}$$

Hence, and because of Equations 7.2.19 and 7.2.37, we arrive at Equation 7.2.21. This completes the proof of (U 2) theorem. □

Example 7.2.4

Show that if $\mathbf{u}$ is an elastic motion corresponding to zero body forces, then there are vectors $\mathbf{u}_1$ and $\mathbf{u}_2$ on $\bar{B} \times [0, \infty)$ such that

$$\mathbf{u} = \mathbf{u}_1 + \mathbf{u}_2 \tag{a}$$

where

$$\Box_1^2\, \mathbf{u}_1 = \mathbf{0}, \quad \operatorname{curl}\mathbf{u}_1 = \mathbf{0} \tag{b}$$

$$\Box_2^2\, \mathbf{u}_2 = \mathbf{0}, \quad \operatorname{div}\mathbf{u}_2 = 0 \tag{c}$$

Solution

It follows from (U 2) theorem that there exist functions φ and $\boldsymbol{\psi}$ such that

$$\mathbf{u} = \nabla\varphi + \operatorname{curl}\boldsymbol{\psi}, \quad \operatorname{div}\boldsymbol{\psi} = 0 \tag{d}$$

and

$$\Box_1^2\, \varphi = 0 \tag{e}$$

$$\Box_2^2\, \boldsymbol{\psi} = \mathbf{0} \tag{f}$$

Hence, if we define $\mathbf{u}_1$ and $\mathbf{u}_2$ by

$$\mathbf{u}_1 = \nabla\varphi, \quad \mathbf{u}_2 = \operatorname{curl}\boldsymbol{\psi} \tag{g}$$

then (d) and (g) imply that

$$\mathbf{u} = \mathbf{u}_1 + \mathbf{u}_2 \tag{h}$$

Also, by applying the operations ∇ and curl to (e) and (f), respectively, and using (g), we arrive at

$$\Box_1^2\, \mathbf{u}_1 = \mathbf{0} \tag{i}$$

and

$$\Box_2^2\, \mathbf{u}_2 = \mathbf{0} \tag{j}$$

The second equation of (b) and the second equation of (c) are a direct consequence of (g). This completes the solution. □

Although the Green–Lamé solution of the (U 1) theorem is useful in solving a large number of initial-boundary value problems for a homogeneous isotropic elastic body, such problems may also be approached by using the so called Cauchy–Kovalevski–Somigliana solution. The latter is contained in the following theorem:

(U 3) **Cauchy–Kovalevski–Somigliana Solution.** Let

$$\mathbf{u} = \Box_1^2\, \mathbf{g} + \left(\frac{c_2^2}{c_1^2} - 1\right) \nabla(\operatorname{div} \mathbf{g}) \tag{7.2.40}$$

where $\mathbf{g}$ satisfies the nonhomogeneous biwave equation

$$\Box_1^2 \Box_2^2\, \mathbf{g} = -\frac{\mathbf{b}}{\mu} \tag{7.2.41}$$

Then $\mathbf{u}$ is an elastic motion corresponding to $\mathbf{b}$.

Proof: Since $\mathbf{u}$ is an elastic motion corresponding to $\mathbf{b}$, then $\mathbf{u}$ satisfies the equation, see Equation 7.2.1,

$$\boldsymbol{\ell} \equiv \Box_2^2\, \mathbf{u} + \left(\frac{c_1^2}{c_2^2} - 1\right) \nabla(\operatorname{div} \mathbf{u}) + \frac{\mathbf{b}}{\mu} = \mathbf{0} \tag{7.2.42}$$

Hence, substituting Equation 7.2.40 into Equation 7.2.42, we find

$$\begin{aligned} \boldsymbol{\ell} = \Box_2^2 &\left[\Box_1^2\, \mathbf{g} + \left(\frac{c_2^2}{c_1^2} - 1\right) \nabla(\operatorname{div} \mathbf{g})\right] \\ &+ \left(\frac{c_1^2}{c_2^2} - 1\right) \nabla \left[\Box_1^2\, (\operatorname{div} \mathbf{g}) + \left(\frac{c_2^2}{c_1^2} - 1\right) \nabla^2(\operatorname{div} \mathbf{g})\right] + \frac{\mathbf{b}}{\mu} \end{aligned} \tag{7.2.43}$$

Now,

$$\begin{aligned} \Box_1^2\, (\operatorname{div} \mathbf{g}) + \left(\frac{c_2^2}{c_1^2} - 1\right) \nabla^2(\operatorname{div} \mathbf{g}) &= \left(\frac{c_2^2}{c_1^2}\, \nabla^2 - \frac{1}{c_1^2}\frac{\partial^2}{\partial t^2}\right) \operatorname{div} \mathbf{g} \\ &= \frac{c_2^2}{c_1^2}\, \Box_2^2\, (\operatorname{div} \mathbf{g}) \end{aligned} \tag{7.2.44}$$

Substituting Equation 7.2.44 into Equation 7.2.43, we get

$$\begin{aligned} \boldsymbol{\ell} = \Box_2^2 \Box_1^2\, \mathbf{g} &+ \left(\frac{c_2^2}{c_1^2} - 1\right) \Box_2^2\, \nabla(\operatorname{div} \mathbf{g}) \\ &+ \left(1 - \frac{c_2^2}{c_1^2}\right) \nabla \Box_2^2\, (\operatorname{div} \mathbf{g}) + \frac{\mathbf{b}}{\mu} = \Box_1^2 \Box_2^2\, \mathbf{g} + \frac{\mathbf{b}}{\mu} \end{aligned} \tag{7.2.45}$$

Since **g** satisfies Equation 7.2.41,

$$\boldsymbol{\ell} = \mathbf{0} \tag{7.2.46}$$

therefore, **u** is an elastic motion corresponding to **b**. This completes the proof of the (U 3) theorem. □

Note: If **g** is independent of the time, then Equations 7.2.40 and 7.2.41 reduce to

$$\mathbf{u}(\mathbf{x}) = \nabla^2\mathbf{g} - \frac{1}{2(1-\nu)}\nabla(\operatorname{div}\mathbf{g}) \tag{7.2.47}$$

and

$$\nabla^2\nabla^2\mathbf{g} = -\frac{\mathbf{b}}{\mu} \tag{7.2.48}$$

Equations 7.2.47 and 7.2.48 represent the Boussinesq–Somigliana–Galerkin solution of elastostatics for a homogeneous isotropic elastic body, see Equations 7.1.55 and 7.1.56.

Example 7.2.5

Show that the stress tensor S_{ij} associated with vector g_i, which appears in the (U 3) theorem, is given by

$$S_{ij} = \mu\left[\Box_1^2\,(g_{i,j} + g_{j,i}) - \frac{1}{1-\nu}\,g_{k,kij} + \frac{\nu}{1-\nu}\,\Box_2^2\,g_{k,k}\delta_{ij}\right] \tag{a}$$

Solution

We use Hooke's law in the form, see (b) in Example 7.1.6,

$$S_{ij} = \mu\left(u_{i,j} + u_{j,i} + \frac{2\nu}{1-2\nu}\,u_{k,k}\delta_{ij}\right) \tag{b}$$

Now, we write Equation 7.2.40 in components

$$u_i = \Box_1^2\,g_i - \frac{1}{2(1-\nu)}\,g_{k,ki} \tag{c}$$

and compute $u_{k,k}$ to obtain

$$u_{k,k} = \Box_1^2\,g_{k,k} - \frac{1}{2(1-\nu)}\,g_{l,lkk} = \frac{1-2\nu}{2(1-\nu)}\,g_{l,lkk} - \frac{1}{c_1^2}\,\ddot{g}_{l,l} \tag{d}$$

Next, substituting (c) and (d) into the RHS of (b), we arrive at

$$S_{ij} = \mu\left\{\Box_1^2\,(g_{i,j} + g_{j,i}) - \frac{1}{1-\nu}\,g_{k,kij} + \frac{2\nu}{1-2\nu}\left[\frac{1-2\nu}{2(1-\nu)}\,g_{l,lkk} - \frac{1}{c_1^2}\ddot{g}_{l,l}\right]\delta_{ij}\right\} \tag{e}$$

Finally, if we use the identity

$$\frac{2(1-\nu)}{1-2\nu}\,\frac{1}{c_1^2}=\frac{1}{c_2^2} \tag{f}$$

we reduce (e) to the form

$$S_{ij}=\mu\left[\Box_1^2\,(g_{i,j}+g_{j,i})-\frac{1}{1-\nu}\,g_{k,kij}+\frac{\nu}{1-\nu}\,\Box_2^2\,g_{k,k}\delta_{ij}\right] \tag{g}$$

This completes the solution. □

Note: Equation (a) in direct notation takes the form

$$\mathbf{S}=\mu\left[2\Box_1^2\,\widehat{\nabla}\mathbf{g}-\frac{1}{1-\nu}\,\nabla\nabla(\operatorname{div}\mathbf{g})+\frac{\nu}{1-\nu}\,\mathbf{1}\Box_2^2\,(\operatorname{div}\mathbf{g})\right] \tag{7.2.49}$$

□

If $\mathbf{b}=\mathbf{0}$, then the following completeness theorem for the Cauchy–Kovalevski–Somigliana solution holds true:

(U 4) **Completeness of the Cauchy–Kovalevski–Somigliana Solution.** Let **u** be an elastic motion corresponding to zero body forces. Then there exists a vector field **g** such that

$$\mathbf{u}=\Box_1^2\,\mathbf{g}+\left(\frac{c_2^2}{c_1^2}-1\right)\nabla(\operatorname{div}\mathbf{g}) \tag{7.2.50}$$

where

$$\Box_1^2\Box_2^2\,\mathbf{g}=\mathbf{0} \tag{7.2.51}$$

The proof of this theorem is omitted.

Clearly, to solve an initial-boundary value problem by using Equations 7.2.50 and 7.2.51 we need to find a solution **g** to Equation 7.2.51.

(U 5) **Boggio's Theorem.** Let **g** be a solution to the biwave equation

$$\Box_1^2\Box_2^2\,\mathbf{g}=\mathbf{0} \tag{7.2.52}$$

Then

$$\mathbf{g}=\mathbf{g}_1+\mathbf{g}_2 \tag{7.2.53}$$

where $\mathbf{g}_1$ and $\mathbf{g}_2$ are fields on $\bar{B}\times[0,\infty)$ that satisfy the wave equations

$$\Box_1^2\,\mathbf{g}_1=\mathbf{0} \tag{7.2.54}$$

and

$$\square_2^2\,\mathbf{g}_2 = \mathbf{0} \tag{7.2.55}$$

Proof: It is sufficient to show that there exists a field $\mathbf{g}_1$ on $\bar{B} \times [0, \infty)$ such that

$$\square_1^2\,\mathbf{g}_1 = \mathbf{0}, \quad \square_2^2\,(\mathbf{g} - \mathbf{g}_1) = \mathbf{0} \tag{7.2.56}$$

since if there is $\mathbf{g}_1$ that satisfies Equations 7.2.56 then $\mathbf{g}_2$ is defined by

$$\mathbf{g}_2 = \mathbf{g} - \mathbf{g}_1 \tag{7.2.57}$$

Also, note that the equation

$$\square_1^2\,\mathbf{g}_1 = \mathbf{0} \tag{7.2.58}$$

implies that

$$\square_2^2\,\mathbf{g}_1 = \left(\frac{1}{c_1^2} - \frac{1}{c_2^2}\right)\ddot{\mathbf{g}}_1 \tag{7.2.59}$$

Hence, Equations 7.2.56 are satisfied if

$$\square_1^2\,\mathbf{g}_1 = \mathbf{0}, \quad \ddot{\mathbf{g}}_1 = \mathbf{f} \tag{7.2.60}$$

where

$$\mathbf{f} = \left(\frac{1}{c_1^2} - \frac{1}{c_2^2}\right)^{-1}\square_2^2\,\mathbf{g} \tag{7.2.61}$$

Note that since $\mathbf{g}$ is a solution to Equation 7.2.52, then applying the operator $\square_1^2$ to 7.2.61, we obtain

$$\square_1^2\,\mathbf{f} = \mathbf{0} \tag{7.2.62}$$

Therefore, to complete the proof it suffices to show that there exists a field $\mathbf{g}_1$ that satisfies Equations 7.2.60 subject to Equation 7.2.62. Obviously, a solution of the second equation of 7.2.60 takes the form

$$\mathbf{g}_1(\mathbf{x}, t) = i * \mathbf{f}(\mathbf{x}, t) + \mathbf{g}_1(\mathbf{x}, 0) + t\dot{\mathbf{g}}_1(\mathbf{x}, 0) \quad \text{on } \bar{B} \times [0, \infty) \tag{7.2.63}$$

where, see Equation 7.2.23,

$$i(t) = t \tag{7.2.64}$$

$\mathbf{g}_1(\mathbf{x}, 0)$ and $\dot{\mathbf{g}}_1(\mathbf{x}, b)$ are unknown functions on $\bar{B}$, and $*$ stands for the convolution product on the time axis. Next, we show that $\mathbf{g}_1(\mathbf{x}, 0)$ and $\dot{\mathbf{g}}_1(\mathbf{x}, 0)$ may be chosen in such a way that the first equation of 7.2.60 is satisfied. To this end, we apply the operator $\Box_1^2$ to Equation 7.2.63, and obtain

$$\Box_1^2 \mathbf{g}_1 = i * \nabla^2 \mathbf{f}(\mathbf{x}, t) - \frac{1}{c_1^2} \mathbf{f}(\mathbf{x}, t) + \nabla^2 \mathbf{g}_1(\mathbf{x}, 0) + t\nabla^2 \dot{\mathbf{g}}_1(\mathbf{x}, 0) \tag{7.2.65}$$

Also, note that

$$i * \ddot{\mathbf{f}}(\mathbf{x}, t) = \mathbf{f}(\mathbf{x}, t) - \mathbf{f}(\mathbf{x}, 0) - t\dot{\mathbf{f}}(\mathbf{x}, 0) \tag{7.2.66}$$

Hence, applying the operation $i*$ to Equation 7.2.62 and using Equation 7.2.66, we obtain

$$i * \nabla^2 \mathbf{f}(\mathbf{x}, t) - \frac{1}{c_1^2} [\mathbf{f}(\mathbf{x}, t) - \mathbf{f}(\mathbf{x}, 0) - t\dot{\mathbf{f}}(\mathbf{x}, 0)] = 0 \tag{7.2.67}$$

As a result, Equation 7.2.65 may be written in the form

$$\Box_1^2 \mathbf{g}_1 = \nabla^2 \mathbf{g}_1(\mathbf{x}, 0) - \frac{1}{c_1^2} \mathbf{f}(\mathbf{x}, 0) + t\left[\nabla^2 \dot{\mathbf{g}}_1(\mathbf{x}, 0) - \frac{1}{c_1^2} \dot{\mathbf{f}}(\mathbf{x}, 0)\right] \tag{7.2.68}$$

So, if $\mathbf{g}_1(\mathbf{x}, 0)$ and $\dot{\mathbf{g}}_1(\mathbf{x}, 0)$ are taken in the form of the Newtonian potentials

$$\mathbf{g}_1(\mathbf{x}, 0) = -\frac{1}{4\pi c_1^2} \int_B \frac{\mathbf{f}(\boldsymbol{\xi}, 0)}{|\mathbf{x} - \boldsymbol{\xi}|} dv(\boldsymbol{\xi}) \tag{7.2.69}$$

$$\dot{\mathbf{g}}_1(\mathbf{x}, 0) = -\frac{1}{4\pi c_1^2} \int_B \frac{\dot{\mathbf{f}}(\boldsymbol{\xi}, 0)}{|\mathbf{x} - \boldsymbol{\xi}|} dv(\boldsymbol{\xi}) \tag{7.2.70}$$

then the RHS of Equation 7.2.68 vanishes, and $\mathbf{g}_1$ satisfies the first equation of 7.2.60. This completes the proof of the (U 5) theorem. □

Example 7.2.6

Show that the Cauchy–Kovalevski–Somigliana solution corresponding to zero body forces, see Equations 7.2.50 and 7.2.51, reduces to the Green–Lamé solution, see Equations 7.2.5 through 7.2.7 with $h = 0$, $\mathbf{k} = \mathbf{0}$, provided

$$\mathbf{g} = \mathbf{g}_1 + \mathbf{g}_2 \tag{a}$$

$$\Box_1^2 \mathbf{g}_1 = \mathbf{0}, \quad \Box_2^2 \mathbf{g}_2 = \mathbf{0} \tag{b}$$

and

$$\varphi = \left(\frac{c_2^2}{c_1^2} - 1\right) \operatorname{div} \mathbf{g}_1 \tag{c}$$

$$\boldsymbol{\psi} = \left(\frac{c_2^2}{c_1^2} - 1\right) \operatorname{curl} \mathbf{g}_2 \tag{d}$$

Solution

Let $\mathbf{g}$ be a vector field that occurs in the Cauchy–Kovalevski–Somigliana solution corresponding to $\mathbf{b} = \mathbf{0}$. Then, by Equations 7.2.50 and 7.2.51, we obtain

$$\mathbf{u} = \Box_1^2\,\mathbf{g} + \left(\frac{c_2^2}{c_1^2} - 1\right)\nabla(\operatorname{div}\mathbf{g}) \tag{e}$$

where

$$\Box_1^2\Box_2^2\,\mathbf{g} = \mathbf{0} \tag{f}$$

Also, by (f) and by Boggio's theorem, there exist functions $\mathbf{g}_1$ and $\mathbf{g}_2$ with the properties

$$\mathbf{g} = \mathbf{g}_1 + \mathbf{g}_2 \tag{g}$$

$$\Box_1^2\,\mathbf{g}_1 = \mathbf{0}, \quad \Box_2^2\,\mathbf{g}_2 = \mathbf{0} \tag{h}$$

It follows then from the second equation of (h) that

$$\Box_1^2\,\mathbf{g}_2 = \left(1 - \frac{c_2^2}{c_1^2}\right)\nabla^2\mathbf{g}_2 \tag{i}$$

Thus, substituting (g) into (e), and using (h) and (i), we obtain

$$\mathbf{u} = \left(1 - \frac{c_2^2}{c_1^2}\right)\left[\nabla^2\mathbf{g}_2 - \nabla^2\operatorname{div}(\mathbf{g}_1 + \mathbf{g}_2)\right] \tag{j}$$

Next, since

$$\nabla^2\mathbf{g}_2 = \nabla(\operatorname{div}\mathbf{g}_2) - \operatorname{curl}\operatorname{curl}\mathbf{g}_2 \tag{k}$$

therefore, (j) can be written as

$$\mathbf{u} = \left(\frac{c_2^2}{c_1^2} - 1\right)[\nabla(\operatorname{div}\mathbf{g}_1) + \operatorname{curl}\operatorname{curl}\mathbf{g}_2] \tag{l}$$

Finally, if we define functions φ and $\boldsymbol{\psi}$ by (c) and (d), respectively, we arrive at

$$\mathbf{u} = \nabla\varphi + \operatorname{curl}\boldsymbol{\psi} \tag{m}$$

where, because of (b),

$$\Box_1^2\,\varphi = 0, \quad \Box_2^2\,\boldsymbol{\psi} = \mathbf{0} \tag{n}$$

Equations (m) and (n) are identical to Equations 7.2.5 through 7.2.7 with $h = 0$ and $\mathbf{k} = \mathbf{0}$. This completes the solution. □

PROBLEMS

7.1 The displacement $\mathbf{u} = \mathbf{u}(\mathbf{x}, \boldsymbol{\xi})$ at a point $\mathbf{x}$ due to a concentrated force $\boldsymbol{\ell}$ applied at a point $\boldsymbol{\xi}$ of a homogeneous isotropic infinite elastic body is given by $(\mathbf{x} \neq \boldsymbol{\xi})$

$$\mathbf{u}(\mathbf{x}, \boldsymbol{\xi}) = \mathbf{U}(\mathbf{x}, \boldsymbol{\xi})\boldsymbol{\ell}$$

where (see Equations 7.1.30 through 7.1.32)

$$\mathbf{U}(\mathbf{x}, \boldsymbol{\xi}) = \frac{1}{16\pi\mu(1-\nu)}\frac{1}{R}\left[(3-4\nu)\mathbf{1} + \frac{(\mathbf{x}-\boldsymbol{\xi})\otimes(\mathbf{x}-\boldsymbol{\xi})}{R^2}\right]$$

with

$$R = |\mathbf{x} - \boldsymbol{\xi}|$$

Use the stress–displacement relation to show that the associated stress $\mathbf{S} = \mathbf{S}(\mathbf{x}, \boldsymbol{\xi})$ takes the form

$$\mathbf{S}(\mathbf{x}, \boldsymbol{\xi}) = -\frac{1}{8\pi(1-\nu)}\frac{1}{R^3}\left\{\frac{3}{R^2}[(\mathbf{x}-\boldsymbol{\xi})\cdot\boldsymbol{\ell}](\mathbf{x}-\boldsymbol{\xi})\otimes(\mathbf{x}-\boldsymbol{\xi})\right.$$
$$\left. +(1-2\nu)\left\{(\mathbf{x}-\boldsymbol{\xi})\otimes\boldsymbol{\ell} + \boldsymbol{\ell}\otimes(\mathbf{x}-\boldsymbol{\xi}) - [(\mathbf{x}-\boldsymbol{\xi})\cdot\boldsymbol{\ell}]\mathbf{1}\right\}\right\}$$

7.2 The displacement equation of thermoelastostatics for a homogeneous isotropic body subject to a temperature change $T = T(\mathbf{x})$ takes the form (see Equation 4.1.19 with $\mathbf{b} = \mathbf{0}$)

$$\nabla^2\mathbf{u} + \frac{1}{1-2\nu}\nabla(\operatorname{div}\mathbf{u}) - \frac{2+2\nu}{1-2\nu}\alpha\,\nabla T = \mathbf{0} \quad \text{(a)}$$

Let

$$\mathbf{u} = \boldsymbol{\psi} - \frac{1}{4(1-\nu)}\nabla(\mathbf{x}\cdot\boldsymbol{\psi} + \widehat{\varphi}) \quad \text{(b)}$$

where

$$\nabla^2\boldsymbol{\psi} = \mathbf{0} \quad \text{(c)}$$

and

$$\nabla^2\widehat{\varphi} = -4(1+\nu)\alpha T \quad \text{(d)}$$

Show that $\mathbf{u}$ given by Equations (b) through (d) satisfies Equation (a).

7.3 The temperature change T of a homogeneous isotropic infinite elastic body is represented by

$$T(\mathbf{x}) = \widehat{T}\,\delta(\mathbf{x}) \quad \text{(a)}$$

where

$$\delta(\mathbf{x}) = \delta(x_1)\delta(x_2)\delta(x_3) \tag{b}$$

$\delta(x_i)$, $i = 1, 2, 3$, is a one-dimensional Dirac delta function, and $\widehat{T}$ is a constant with the dimension $[\widehat{T}]$ = [Temperature $\times$ Volume]. Show that an elastic displacement $\mathbf{u}(\mathbf{x})$ and stress $\mathbf{S}(\mathbf{x})$ corresponding to $T(\mathbf{x})$ are given by

$$\mathbf{u}(\mathbf{x}) = -\frac{1}{4\pi}\frac{1+\nu}{1-\nu}\alpha\,\widehat{T}\;\nabla\frac{1}{|\mathbf{x}|} \tag{c}$$

and

$$\mathbf{S}(\mathbf{x}) = -\frac{\mu}{2\pi}\frac{1+\nu}{1-\nu}\alpha\,\widehat{T}\,(\nabla\,\nabla - \mathbf{1}\nabla^2)\,\frac{1}{|\mathbf{x}|} \tag{d}$$

Hint: Use the representation (b) through (d) of Problem 7.2 in which $\boldsymbol{\psi} = \mathbf{0}$ and $T = \widehat{T}\delta(\mathbf{x})$. Also, note that

$$\mathbf{S} = -\frac{\mu}{2(1-\nu)}(\nabla\,\nabla - \mathbf{1}\nabla^2)\widehat{\varphi} \tag{e}$$

7.4 A solution $\varphi = \varphi(\mathbf{x}, t)$ to the nonhomogeneous wave equation

$$\Box_0^2\varphi(\mathbf{x}, t) = -F(\mathbf{x}, t) \quad \text{on } E^3 \times [0, \infty) \tag{a}$$

subject to the homogeneous initial conditions

$$\varphi(\mathbf{x}, 0) = 0, \quad \dot{\varphi}(\mathbf{x}, 0) = 0 \quad \text{on } E^3 \tag{b}$$

takes the form (see Example 7.2.2)

$$\varphi(\mathbf{x}, t) = \frac{1}{4\pi}\int\limits_{|\mathbf{x}-\boldsymbol{\xi}|\leq ct}\frac{F(\boldsymbol{\xi}, t - |\mathbf{x}-\boldsymbol{\xi}|/c)}{|\mathbf{x}-\boldsymbol{\xi}|}\,dv(\boldsymbol{\xi}) \text{ on } E^3 \times [0, \infty) \tag{c}$$

Here

$$\Box_0^2 = \nabla^2 - \frac{1}{c^2}\frac{\partial^2}{\partial t^2} \tag{d}$$

Show that an equivalent form of Equation (c) reads

$$\varphi(\mathbf{x}, t) = -\frac{c^2t^2}{4\pi}\int\limits_{|\boldsymbol{\xi}|\leq 1}\frac{F[\mathbf{x} - ct\boldsymbol{\xi}, (1 - |\boldsymbol{\xi}|)t]}{|\boldsymbol{\xi}|}\,dv(\boldsymbol{\xi}) \text{ on } E^3 \times [0, \infty) \tag{e}$$

7.5 Let $\mathbf{S} = \mathbf{S}(\mathbf{x}, t)$ be a solution to the stress equation of motion of a homogeneous anisotropic elastodynamics [see Equation 4.2.9 in which ρ and K are constants]

$$\widehat{\nabla}(\operatorname{div}\mathbf{S}) - \rho\mathsf{K}[\ddot{\mathbf{S}}] = -\mathbf{B} \quad \text{on } B \times [0, \infty) \tag{a}$$

subject to the initial conditions

$$\mathbf{S}(\mathbf{x},0) = \mathbf{S}_0(\mathbf{x}), \quad \dot{\mathbf{S}}(\mathbf{x},0) = \dot{\mathbf{S}}_0(\mathbf{x}) \quad \text{on } B \tag{b}$$

Here, $\mathbf{B} = \mathbf{B}(\mathbf{x}, t)$, $\mathbf{S}_0 = S_0(\mathbf{x})$, and $\dot{\mathbf{S}}_0 = \dot{\mathbf{S}}_0(\mathbf{x})$ are prescribed functions.

Show that the compatibility condition

$$\text{curl curl}\, \mathsf{K}[\mathbf{S}] = \mathbf{0} \quad \text{on } B \times [0, \infty) \tag{c}$$

is satisfied if and only if there exists a vector field $\mathbf{u} = \mathbf{u}(\mathbf{x}, t)$ on $B \times [0, \infty)$ such that

$$\widehat{\nabla}\ddot{\mathbf{u}} = \rho^{-1}\mathbf{B} \quad \text{on } B \times [0, \infty) \tag{d}$$

and

$$\mathbf{S}_0(\mathbf{x}) = \mathsf{K}^{-1}[\widehat{\nabla}\mathbf{u}(\mathbf{x},0)], \quad \dot{\mathbf{S}}_0(\mathbf{x}) = \mathsf{K}^{-1}[\widehat{\nabla}\dot{\mathbf{u}}(\mathbf{x},0)] \quad \text{on } B \tag{e}$$

Note that $\mathbf{B}$ in Equation (a) represents an arbitrary second-order symmetric tensor field on $\overline{\mathbf{B}} \times [0, \infty)$ while $\mathbf{S}_0$ and $\dot{\mathbf{S}}_0$ in Equation (b) stand for arbitrary second-order symmetric tensor fields on B.

7.6 Consider a homogeneous isotropic elastic body occupying a region $\overline{B}$. Let $\mathbf{S} = \mathbf{S}(\mathbf{x}, t)$ be a tensor field defined by*

$$\mathbf{S}(\mathbf{x},t) = \left[\left(\nabla\nabla - \nu\mathbf{1}\Box_2^2\right)\text{tr}\,\boldsymbol{\chi} - 2(1-\nu)\Box_1^2\boldsymbol{\chi}\right] \quad \text{on } \overline{B} \times [0, \infty) \tag{a}$$

where $\boldsymbol{\chi} = \boldsymbol{\chi}(\mathbf{x}, t)$ is a symmetric second-order tensor field that satisfies the equations

$$\Box_1^2\Box_2^2\boldsymbol{\chi} = \frac{1}{1-\nu}\widehat{\nabla}\mathbf{b} \quad \text{on } B \times [0, \infty) \tag{b}$$

and

$$\nabla^2\boldsymbol{\chi} + \nabla\nabla(\text{tr}\,\boldsymbol{\chi}) - 2\widehat{\nabla}(\text{div}\,\boldsymbol{\chi}) = \mathbf{0} \quad \text{on } B \times [0, \infty) \tag{c}$$

Show that $\mathbf{S}$ satisfies the stress equation of motion (see Equation 4.2.14)

$$\widehat{\nabla}(\text{div}\,\mathbf{S}) - \frac{\rho}{2\mu}\left[\ddot{\mathbf{S}} - \frac{\nu}{1+\nu}(\text{tr}\,\ddot{\mathbf{S}})\mathbf{1}\right] = -\widehat{\nabla}\mathbf{b} \quad \text{on } B \times [0, \infty) \tag{d}$$

In Equations (a) and (b)

$$\Box_1^2 = \nabla^2 - \frac{1}{c_1^2}\frac{\partial^2}{\partial t^2}, \quad \Box_2^2 = \nabla^2 - \frac{1}{c_2^2}\frac{\partial^2}{\partial t^2} \tag{e}$$

$$\frac{1}{c_1^2} = \frac{1}{c_2^2}\frac{1-2\nu}{2-2\nu}, \quad \frac{1}{c_2^2} = \frac{\rho}{\mu} \tag{f}$$

* The stress field $\mathbf{S}$ in the form of Equations (a) through (c) is a tensor solution of homogeneous isotropic elastodynamics of the Galerkin type. To show this we let $\boldsymbol{\chi} = -[\mu/(1-\nu)]\widehat{\nabla}\mathbf{g}$, where $\mathbf{g}$ is the Galerkin vector satisfying Equation 7.2.41. Then Equations (b) and (c) are satisfied, and Equation (a) reduces to Equation 7.2.49.

7.7 Let $\mathbf{S}$ be the tensor solution of homogeneous isotropic elastodynamics of Problem 7.6 corresponding to homogeneous initial conditions. Show that the solution is complete, that is, there exists a second-order symmetric tensor field χ such that Equations (a) through (c) of Problem 7.6 are satisfied [4].

7.8 Consider the stress equation of motion

$$\widehat{\nabla}(\operatorname{div}\mathbf{S}) - \frac{\rho}{2\mu}\left[\ddot{\mathbf{S}} - \frac{\nu}{1+\nu}(\operatorname{tr}\ddot{\mathbf{S}})\mathbf{1}\right] = -\widehat{\nabla}\mathbf{b} \quad \text{on } B \times [0,\infty) \tag{a}$$

subject to the initial conditions

$$\mathbf{S}(\mathbf{x},0) = \mathbf{S}_0(\mathbf{x}), \quad \dot{\mathbf{S}}(\mathbf{x},0) = \dot{\mathbf{S}}_0(\mathbf{x}) \tag{b}$$

where $\mathbf{b}$, $\mathbf{S}_0$, and $\dot{\mathbf{S}}_0$ are prescribed functions. Define a scalar field $\alpha = \alpha(\mathbf{x}, t)$ and a vector field $\boldsymbol{\beta} = \boldsymbol{\beta}(\mathbf{x}, t)$ by

$$\alpha(\mathbf{x}, t) = \frac{1}{4\pi c_1^2}\operatorname{div}\int_B \frac{\boldsymbol{\gamma}(\mathbf{y}, t)}{|\mathbf{x}-\mathbf{y}|}\,dv(\mathbf{y}) \tag{c}$$

and

$$\boldsymbol{\beta}(\mathbf{x}, t) = -\frac{1}{4\pi c_2^2}\operatorname{curl}\int_B \frac{\boldsymbol{\gamma}(\mathbf{y}, t)}{|\mathbf{x}-\mathbf{y}|}\,dv(\mathbf{y}) \tag{d}$$

where

$$\boldsymbol{\gamma}(\mathbf{x}, t) = \mathbf{b}(\mathbf{x}, t) + \operatorname{div}[\mathbf{S}_0(\mathbf{x}) + t\dot{\mathbf{S}}_0(\mathbf{x})] \tag{e}$$

Let ϕ and $\boldsymbol{\omega}$ satisfy the equations

$$\Box_1^2\phi = \alpha \quad \text{on } B \times [0,\infty) \tag{f}$$

and

$$\Box_2^2\boldsymbol{\omega} = \boldsymbol{\beta}, \quad \operatorname{div}\boldsymbol{\omega} = 0 \quad \text{on } B \times [0,\infty) \tag{g}$$

subject to the homogeneous initial conditions

$$\left.\begin{aligned} \phi(\mathbf{x},0) &= \dot{\phi}(\mathbf{x},0) = 0 \\ \boldsymbol{\omega}(\mathbf{x},0) &= \dot{\boldsymbol{\omega}}(\mathbf{x},0) = \mathbf{0} \end{aligned}\right\} \quad \text{on } B \tag{h}$$

Let

$$\mathbf{S}(\mathbf{x}, t) = \mathbf{S}_0(\mathbf{x}) + t\dot{\mathbf{S}}_0(\mathbf{x}) + 2c_2^2[\nabla\nabla\phi + \widehat{\nabla}(\operatorname{curl}\boldsymbol{\omega})] + \left(c_1^2 - 2c_2^2\right)\nabla^2\phi\mathbf{1} \tag{i}$$

Show that $\mathbf{S}$ satisfies Equations (a) and (b).*

* The solution (i), in which ϕ and $\boldsymbol{\omega}$ satisfy Equations (f) through (h), represents a tensor solution of homogeneous isotropic elastodynamics of the Lamé-type (see Section 7.2).

7.9 Let $\mathbf{S}$ be a symmetric second-order tensor field on $\overline{B} \times [0, \infty)$ that satisfies the stress equation of motion

$$\widehat{\nabla}(\operatorname{div} \mathbf{S}) - \frac{\rho}{2\mu}\left[\ddot{\mathbf{S}} - \frac{\nu}{1+\nu}(\operatorname{tr} \ddot{\mathbf{S}})\mathbf{1}\right] = -\widehat{\nabla}\mathbf{b} \quad \text{on } B \times [0, \infty) \tag{a}$$

subject to the initial conditions

$$\mathbf{S}(\mathbf{x}, 0) = \mathbf{S}_0(\mathbf{x}), \quad \dot{\mathbf{S}}(\mathbf{x}, 0) = \dot{\mathbf{S}}_0(\mathbf{x}) \quad \text{on } B \tag{b}$$

Show that there are a scalar field $\phi = \phi(\mathbf{x}, t)$ and a vector field $\boldsymbol{\omega} = \boldsymbol{\omega}(\mathbf{x}, t)$ such that

$$\mathbf{S} = \mathbf{S}_0 + t\dot{\mathbf{S}}_0 + 2c_2^2[\nabla\nabla\phi + \widehat{\nabla}(\operatorname{curl} \boldsymbol{\omega})] + \left(c_1^2 - 2c_2^2\right)(\nabla^2\phi)\mathbf{1} \tag{c}$$

$$\Box_1^2\phi = \alpha, \quad \phi(\mathbf{x}, 0) = \dot{\phi}(\mathbf{x}, 0) = 0 \tag{d}$$

$$\Box_2^2\boldsymbol{\omega} = \boldsymbol{\beta}, \quad \operatorname{div} \boldsymbol{\omega} = 0, \quad \boldsymbol{\omega}(\mathbf{x}, 0) = \dot{\boldsymbol{\omega}}(\mathbf{x}, 0) = \mathbf{0} \tag{e}$$

where the fields α and $\boldsymbol{\beta}$ are given by Equations (c) and (d), respectively, of Problem 7.8.*

7.10 Consider the stress equation of motion in the form

$$\widehat{\nabla}(\operatorname{div} \mathbf{S}) - \rho\mathsf{K}[\ddot{\mathbf{S}}] = -\mathbf{B} \quad \text{on } B \times [0, \infty) \tag{a}$$

subject to the homogeneous initial conditions

$$\mathbf{S}(\mathbf{x}, 0) = \mathbf{0}, \quad \dot{\mathbf{S}}(\mathbf{x}, 0) = \mathbf{0} \quad \text{on } B \tag{b}$$

where

$$\mathsf{K}[\mathbf{S}] = \frac{1}{2\mu}\left[\mathbf{S} - \frac{\nu}{1+\nu}(\operatorname{tr} \mathbf{S})\mathbf{1}\right] \tag{c}$$

and $\mathbf{B} = \mathbf{B}(\mathbf{x}, t)$ is an arbitrary symmetric second-order tensor field on $\overline{B} \times [0, \infty)$. Define a vector field $\mathbf{v} = \mathbf{v}(\mathbf{x}, t)$ by

$$\mathbf{v}(\mathbf{x}, t) = -\frac{c_2^2t^2}{4\pi}\int\limits_{|\boldsymbol{\xi}|\leq 1}\frac{\mathbf{f}[\mathbf{x} - c_2t\boldsymbol{\xi}, (1 - |\boldsymbol{\xi}|)t]}{|\boldsymbol{\xi}|}\,dv(\boldsymbol{\xi}) \tag{d}$$

where

$$\mathbf{f}(\mathbf{x}, t) = \left\{\left(\frac{c_1^2}{c_2^2} - 1\right)\nabla g + \frac{1}{\rho c_2^2}\operatorname{div} \mathsf{K}^{-1}[\mathbf{B}]\right\}(\mathbf{x}, t) \tag{e}$$

$$g(\mathbf{x}, t) = -\frac{c_1^2t^2}{4\pi}\int\limits_{|\boldsymbol{\xi}|\leq 1}\frac{h[\mathbf{x} - c_1t\boldsymbol{\xi}, (1 - |\boldsymbol{\xi}|)t]}{|\boldsymbol{\xi}|}\,dv(\boldsymbol{\xi}) \tag{f}$$

* Solution to Problem 7.9 implies that the tensor solution of Lamé-type [Equations (c) through (e)], is complete.

and

$$h(\mathbf{x}, t) = \frac{1}{\rho c_1^2} \operatorname{div} \operatorname{div} \mathbf{K}^{-1}[\mathbf{B}](\mathbf{x}, t) \tag{g}$$

Let

$$\mathbf{S}(\mathbf{x}, t) = \frac{1}{\rho} \mathbf{K}^{-1} \left[\widehat{\nabla}\mathbf{v} + \mathbf{B}\right] * t \tag{h}$$

Show that **S** satisfies Equations (a) and (b).
Hint: Use the result of Problem 7.4 that the function

$$\varphi(\mathbf{x}, t) = -\frac{c^2 t^2}{4\pi} \int\limits_{|\boldsymbol{\xi}| \le 1} \frac{F[\mathbf{x} - ct\boldsymbol{\xi}, (1 - |\boldsymbol{\xi}|)t]}{|\boldsymbol{\xi}|} dv(\boldsymbol{\xi}) \tag{i}$$

satisfies the nonhomogeneous wave equation

$$\left(\nabla^2 - \frac{1}{c^2}\frac{\partial^2}{\partial t^2}\right)\varphi = -F \tag{j}$$

subject to the homogeneous initial conditions

$$\varphi(\mathbf{x}, 0) = 0, \quad \dot{\varphi}(\mathbf{x}, 0) = 0 \tag{k}$$

REFERENCES

1. W. Thomson (Lord Kelvin), On the equations of equilibrium of an elastic solid, *Cambr. Dubl. Math. J.*, 3, 1848, 87–89.
2. A. E. H. Love, *A Treatise on the Mathematical Theory of Elasticty*, Dover Publications, New York, 1944, p. 276.
3. J. L. Nowiński, *Theory of Thermoelasticity with Applications*, Sijthoff & Noordhoff, Alphen aan den Rijn, 1978, pp. 265–266.
4. G. Iwaniec, A Galerkin-type tensor solution for linear elastodynamics, *Bull. Acad. Polon., Sci. Ser. Tech.*, 24, 1976, 15–23.

8 Formulation of Two-Dimensional Problems

In the theory of elasticity, there is a class of problems in which an elastic state depends on two space variables only. For example, such a situation arises when a homogeneous isotropic infinitely long elastic cylinder is subject to a constant normal pressure on its lateral surface. In this chapter, the two-dimensional elastic states as well as the two-dimensional elastodynamic processes are discussed. In particular, a plane strain state and a generalized plane stress state of homogeneous isotropic elastostatics are defined and the associated field equations are obtained. Also, a plane strain process and a generalized plane stress process of homogeneous isotropic elastodynamics are defined, and the associated field equations in terms of the displacements and stresses are obtained. The governing field equations of a nonisothermal two-dimensional elastodynamics are also discussed. The chapter contains a number of worked examples, as well as end-of-chapter problems that cover the general complete solutions of homogeneous isotropic two-dimensional elastostatics and elastodynamics, including a general solution to the stress equation of two-dimensional elastodynamics. The solutions to problems are provided in the Solutions Manual.

8.1 TWO-DIMENSIONAL PROBLEMS OF ELASTOSTATICS

8.1.1 Two-Dimensional Problems of Isothermal Elastostatics

Let B be a cylinder, not necessarily circular, made of a homogeneous isotropic elastic material, and referred to the Cartesian coordinate system, as shown in Figure 8.1. The x_3-axis is parallel to the axis of the cylinder.

The height of the cylinder is $2h$, and its boundary ∂B consists of the lateral surface S and of the upper and lower end surfaces, C_1 and C_2, respectively, or

$$\partial B = S \cup C_1 \cup C_2 \tag{8.1.1}$$

In addition, we assume that

$$S = S_1 \cup S_2, \quad S_1 \cap S_2 = \emptyset \tag{8.1.2}$$

which means that the lateral surface is a sum of two nonoverlapping parts S_1 and S_2 over which two different types of boundary conditions will be prescribed. Surfaces S_1 and S_2 are received by dividing the lateral surface S with two straight lines on S, both parallel to the axis x_3.

We assume that a body force vector $\mathbf{b}$ on B, a surface displacement $\widehat{\mathbf{u}}$ on S_1, and a surface traction $\widehat{\mathbf{s}}$ on S_2 are independent of x_3 and parallel to the (x_1, x_2) plane. In addition, we assume that the cylinder is kept at a constant temperature θ_0.

By *a plane problem of elastostatics*, we mean a problem of finding an elastic state $s = [\mathbf{u}, \mathbf{E}, \mathbf{S}]$ on B that corresponds to $\mathbf{b}$ and satisfies the mixed boundary conditions:

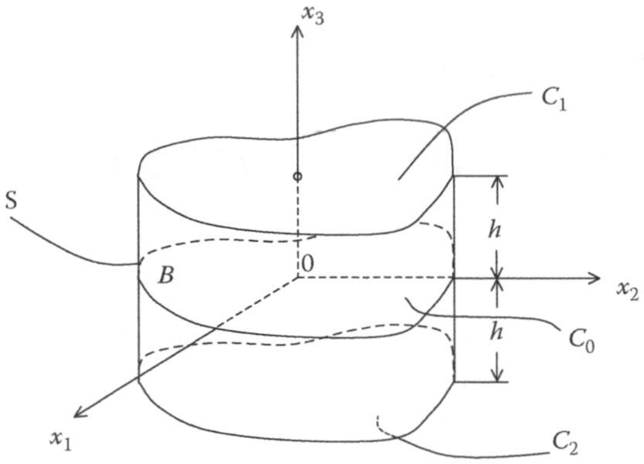

FIGURE 8.1 Cylinder of height $2h$ in the (x_1, x_2, x_3) coordinate system.

$$\mathbf{u} = \widehat{\mathbf{u}} \quad \text{on } S_1 \tag{8.1.3}$$

$$\mathbf{s} = \widehat{\mathbf{s}} \quad \text{on } S_2 \tag{8.1.4}$$

$$\mathbf{s} = \mathbf{0} \quad \text{on } C_1 \text{ and } C_2 \tag{8.1.5}$$

We recall that $s = [\mathbf{u}, \mathbf{E}, \mathbf{S}]$ is an elastic state corresponding to $\mathbf{b}$ if s satisfies the fundamental field equations of elastostatics, see Equations 4.1.76 through 4.1.78.

Since the boundary data, $\widehat{\mathbf{u}}$ and $\widehat{\mathbf{s}}$, and the body force vector $\mathbf{b}$ are all independent of x_3 and parallel to the (x_1, x_2) plane we might expect that the displacement $\mathbf{u}$ is independent of x_3. Taking this into account, we introduce the definition.

A body is said to be in *a state of plane strain* in the (x_1, x_2) plane if the elastic state $s = [\mathbf{u}, \mathbf{E}, \mathbf{S}]$ complies with the conditions

$$u_\alpha = u_\alpha(x_1, x_2), \tag{8.1.6}$$

$$u_3 = 0 \tag{8.1.7}$$

The conditions 8.1.6 and 8.1.7 imply that $\mathbf{E} = \mathbf{E}(x_1, x_2)$ and $\mathbf{S} = \mathbf{S}(x_1, x_2)$. Moreover, Equations 4.1.76 through 4.1.78, restricted to a homogeneous isotropic body, lead to the strain–displacement relations

$$E_{\alpha\beta} = \frac{1}{2}\left(u_{\alpha,\beta} + u_{\beta,\alpha}\right) \tag{8.1.8}$$

$$E_{13} = E_{23} = E_{33} = 0 \tag{8.1.9}$$

the constitutive relations

$$S_{\alpha\beta} = 2\mu E_{\alpha\beta} + \lambda E_{\gamma\gamma}\delta_{\alpha\beta} \tag{8.1.10}$$

$$S_{13} = S_{23} = 0 \tag{8.1.11}$$

$$S_{33} = \lambda E_{\alpha\alpha} \tag{8.1.12}$$

and the equation of equilibrium

$$S_{\alpha\beta,\beta} + b_\alpha = 0 \tag{8.1.13}$$

In Equations 8.1.6 through 8.1.13, and in all plane problems, the Greek subscripts α, β, and γ take values 1 and 2.

We note that in a plane strain problem the equation

$$S_{3i,i} + b_3 = 0 \quad i = 1,2,3 \tag{8.1.14}$$

is satisfied identically since, by virtue of Equations 8.1.11 and 8.1.12, $S_{13} = S_{23} = 0$, and $S_{33,3} = 0$ due to the fact that $E_{\alpha\alpha}$ does not depend of x_3, and $b_3 = 0$. We also note that by taking the trace of Equation 8.1.10 we get

$$S_{\alpha\alpha} = 2(\lambda + \mu)E_{\alpha\alpha} \tag{8.1.15}$$

Hence, from Equation 8.1.12 we obtain, see the table of relations between elasticity constants provided in the front matter,

$$S_{33} = \frac{\lambda}{2(\lambda + \mu)} S_{\alpha\alpha} \equiv \nu S_{\alpha\alpha} \tag{8.1.16}$$

An alternative form of the constitutive relation 8.1.10 is

$$E_{\alpha\beta} = \frac{1}{2\mu}\left(S_{\alpha\beta} - \nu S_{\gamma\gamma}\delta_{\alpha\beta}\right) \tag{8.1.17}$$

Let C_0 be a cross section of the cylinder at $x_3 = 0$, see Figure 8.1, and let ∂C_0 denote its boundary. Then

$$\partial C_0 = \partial C_0^{(1)} \cup \partial C_0^{(2)} \tag{8.1.18}$$

where

$$\partial C_0^{(1)} = S_1 \cap \partial C_0, \quad \partial C_0^{(2)} = S_2 \cap \partial C_0 \tag{8.1.19}$$

Clearly, the closed curve ∂C_0 is a sum of two parts, $\partial C_0^{(1)}$ and $\partial C_0^{(2)}$, over which the displacement and traction vector fields, respectively, are prescribed.

A mixed problem of elastostatics for an isotropic elastic body subject to plane strain conditions is then formulated in the following way:

Find a displacement u_α and stress $S_{\alpha\beta}$ that satisfy the field equations

$$S_{\alpha\beta,\beta} + b_\alpha = 0 \quad \text{on } C_0 \tag{8.1.20}$$

$$S_{\alpha\beta} = \mu(u_{\alpha,\beta} + u_{\beta,\alpha}) + \lambda\delta_{\alpha\beta}u_{\gamma,\gamma} \quad \text{on } C_0 \tag{8.1.21}$$

subject to the mixed boundary conditions

$$u_\alpha = \widehat{u}_\alpha \quad \text{on } \partial C_0^{(1)} \tag{8.1.22}$$

$$S_{\alpha\beta}n_\beta = \widehat{s}_\alpha \quad \text{on } \partial C_0^{(2)} \tag{8.1.23}$$

If a solution $[u_\alpha, S_{\alpha\beta}]$ to the problem stated by Equations 8.1.20 through 8.1.23 is found, then S_{33} is obtained from Equation 8.1.16, and $E_{\alpha\beta}$ from Equation 8.1.17.

Note: Let $s^* = [\mathbf{u}^*, \mathbf{E}^*, \mathbf{S}^*]$ be a three-dimensional elastic state on B that corresponds to zero body forces and an axial load along the x_3-axis applied to the end surfaces C_1 and C_2,

$$\widehat{s_i^*} = \delta_{i3} s_0 \quad \text{on } C_1, \quad \widehat{s_i^*} = -\delta_{i3} s_0 \quad \text{on } C_2 \quad i = 1, 2, 3 \tag{8.1.24}$$

where s_0 is a positive constant. This is equivalent to a simple tension of the cylinder in the x_3 direction of magnitude s_0. Such a three-dimensional elastic state depends on x_1, x_2, x_3, and it is described by

$$\mathbf{u}^* = \left[-\frac{\nu}{E} s_0 x_1, \ -\frac{\nu}{E} s_0 x_2, \ \frac{s_0}{E} x_3 \right] \tag{8.1.25}$$

$$\mathbf{E}^* = \begin{bmatrix} -\frac{\nu}{E} s_0 & 0 & 0 \\ 0 & -\frac{\nu}{E} s_0 & 0 \\ 0 & 0 & \frac{s_0}{E} \end{bmatrix} \tag{8.1.26}$$

and

$$\mathbf{S}^* = \begin{bmatrix} 0 & 0 & 0 \\ 0 & 0 & 0 \\ 0 & 0 & s_0 \end{bmatrix} \tag{8.1.27}$$

where E is Young's modulus, see the table of relations between elasticity constants provided in the front matter,

$$E = \frac{\mu(3\lambda + 2\mu)}{\lambda + \mu} \tag{8.1.28}$$

If we superpose the three-dimensional elastic state s^* on a plane strain elastic state s on the (x_1, x_2) plane, see Equations 8.1.6 through 8.1.13, we arrive at the three-dimensional elastic state s':

$$s' = s + s^* \tag{8.1.29}$$

which is called a *generalized plane strain state*. The latter elastic state occurs in a study of rotating long cylinders [1].

With a two-dimensional problem of elastostatics, *a generalized plane stress problem* may also be associated.

Let an elastic state $s = [\mathbf{u}, \mathbf{E}, \mathbf{S}]$ corresponding to the body force field $\mathbf{b} = (b_1, b_2, 0)$ be subject to the conditions

$$S_{33} = 0 \quad \text{on } B \tag{8.1.30}$$

$$\begin{aligned} u_\alpha(x_1, x_2, x_3) &= u_\alpha(x_1, x_2, -x_3) \\ u_3(x_1, x_2, x_3) &= -u_3(x_1, x_2, -x_3) \end{aligned} \tag{8.1.31}$$

Given a function f on B, let $\bar{f}$ denote its *thickness average*:

$$\bar{f}(x_1,x_2) = \frac{1}{2h}\int_{-h}^{h} f(x_1,x_2,x_2)\,dx_3 \tag{8.1.32}$$

Since the thickness average of an odd function with respect to x_3 is zero, and since the derivative with respect to x_3 of an even function with respect to x_3 is odd, it follows from the strain–displacement and constitutive relations, subject to Equation 8.1.31, that

$$\bar{u}_3 = \bar{E}_{13} = \bar{E}_{23} = \bar{S}_{13} = \bar{S}_{23} = 0 \tag{8.1.33}$$

and

$$\bar{E}_{\alpha\beta} = \frac{1}{2}\left(\bar{u}_{\alpha,\beta} + \bar{u}_{\beta,\alpha}\right) \tag{8.1.34}$$

$$\bar{S}_{\alpha\beta} = 2\mu\bar{E}_{\alpha\beta} + \lambda\delta_{\alpha\beta}\left(\bar{E}_{\gamma\gamma} + \bar{E}_{33}\right) \tag{8.1.35}$$

$$\bar{S}_{33} = 2\mu\bar{E}_{33} + \lambda\left(\bar{E}_{\gamma\gamma} + \bar{E}_{33}\right) \tag{8.1.36}$$

Since, by Equation 8.1.30,

$$\bar{S}_{33} = 0 \tag{8.1.37}$$

hence, it follows from Equation 8.1.36 that

$$\bar{E}_{33} = -\frac{\lambda}{\lambda + 2\mu}\,\bar{E}_{\gamma\gamma} \tag{8.1.38}$$

Substituting Equation 8.1.38 into Equation 8.1.35 we obtain

$$\bar{S}_{\alpha\beta} = \mu\left(\bar{u}_{\alpha,\beta} + \bar{u}_{\beta,\alpha}\right) + \bar{\lambda}\delta_{\alpha\beta}\bar{u}_{\gamma,\gamma} \tag{8.1.39}$$

where

$$\bar{\lambda} = \frac{2\mu\lambda}{\lambda + 2\mu} \tag{8.1.40}$$

Next, applying the thickness average operation to the equilibrium equations, we obtain

$$\bar{S}_{\alpha\beta,\beta} + \bar{b}_\alpha = 0 \tag{8.1.41}$$

A generalized plane stress state $s = [\bar{\mathbf{u}}, \bar{\mathbf{E}}, \bar{\mathbf{S}}]$ corresponding to a body force $\bar{\mathbf{b}} = (\bar{b}_1, \bar{b}_2, 0)$ is defined as an elastic state, which complies with the conditions 8.1.33, 8.1.34, and 8.1.37 through 8.1.41.

As a result, a mixed problem of elastostatics for an isotropic elastic body subject to generalized plane stress conditions may be formulated as follows:

Find a displacement $\bar{u}_\alpha$ and stress $\bar{S}_{\alpha\beta}$ that satisfy the field equations

$$\bar{S}_{\alpha\beta,\beta} + \bar{b}_\alpha = 0 \quad \text{on } C_0 \tag{8.1.42}$$

$$\bar{S}_{\alpha\beta} = \mu\left(\bar{u}_{\alpha,\beta} + \bar{u}_{\beta,\alpha}\right) + \bar{\lambda}\delta_{\alpha\beta}\bar{u}_{\gamma,\gamma} \quad \text{on } C_0 \tag{8.1.43}$$

subject to the mixed boundary conditions

$$\bar{u}_\alpha = \widehat{u}_\alpha \quad \text{on } \partial C_0^{(1)} \tag{8.1.44}$$

$$\bar{S}_{\alpha\beta} n_\beta = \widehat{s}_\alpha \quad \text{on } \partial C_0^{(2)} \tag{8.1.45}$$

where $\bar{b}_\alpha$, $\widehat{u}_\alpha$, and $\widehat{s}_\alpha$ are prescribed functions; and C_0, $\partial C_0^{(1)}$ and $\partial C_0^{(2)}$ have the same meaning as in the case of a plane strain state, see 8.1.18 and 8.1.19.

Once a solution $(\bar{u}_\alpha, \bar{S}_{\alpha\beta})$ to the problem 8.1.42 through 8.1.45 is found, the strain tensor $\bar{E}_{\alpha\beta}$ is computed from the formula

$$\bar{E}_{\alpha\beta} = \frac{1}{2\mu}\left(\bar{S}_{\alpha\beta} - \bar{\nu}\bar{S}_{\gamma\gamma}\delta_{\alpha\beta}\right) \tag{8.1.46}$$

where

$$\bar{\nu} = \frac{\bar{\lambda}}{2\left(\bar{\lambda} + \mu\right)} = \frac{\nu}{1+\nu} \tag{8.1.47}$$

It should be emphasized that the plane strain and generalized plane stress states, discussed in this section, correspond to the boundary conditions and body forces that are independent of x_3 and parallel to the (x_1, x_2) plane. Also note that both the plane strain state and generalized plane stress state are defined on the same representative cross section C_0 of the cylinder lying in the plane $x_3 = 0$ (see Figure 8.1). If there is an elastic state $s = [\mathbf{u}, \mathbf{E}, \mathbf{S}]$ on B corresponding to an external x_3-dependent displacement–force system with $b_3 = 0$ and such that $S_{i3} = 0$ on B $(i = 1, 2, 3)$, then s is called *a plane stress state* on B. In this case s is a three-dimensional elastic state with the property that for any cross section of B perpendicular to the x_3-axis $\mathbf{S}^{\perp} = \mathbf{0}$, where $\mathbf{S} = \mathbf{S}^{\perp} + \mathbf{S}^{\parallel}$, and

$$\mathbf{S}^{\perp} = \begin{bmatrix} 0 & 0 & S_{13} \\ 0 & 0 & S_{23} \\ S_{31} & S_{32} & S_{33} \end{bmatrix}, \quad \mathbf{S}^{\parallel} = \begin{bmatrix} S_{11} & S_{12} & 0 \\ S_{21} & S_{22} & 0 \\ 0 & 0 & 0 \end{bmatrix}$$

The tensors $\mathbf{S}^{\perp}$ and $\mathbf{S}^{\parallel}$ are called, respectively, the normal and tangential components of $\mathbf{S}$ with respect to the $x_3 = 0$ plane.

Example 8.1.1

Formulate the displacement boundary value problem for a body subject to plane strain conditions, in terms of the displacement vector $\mathbf{u} = (u_1, u_2)$.

Solution

In this case, in Equations 8.1.20 through 8.1.23 $\partial C_0 = \partial C_0^{(1)}$, $\partial C_0^{(2)} = \emptyset$. Hence, by eliminating $S_{\alpha\beta}$ from Equations 8.1.20 through 8.1.23 we arrive at the following problem:

Find a displacement field u_α on C_0, that satisfies the field equation

$$\mu u_{\alpha,\gamma\gamma} + (\lambda + \mu) u_{\gamma,\gamma\alpha} + b_\alpha = 0 \quad \text{on } C_0 \tag{a}$$

and the boundary condition

$$u_\alpha = \widehat{u}_\alpha \quad \text{on } \partial C_0 \tag{b}$$

This completes the solution. □

Example 8.1.2

Formulate a displacement boundary value problem for a body subject to generalized plane stress conditions in terms of the displacements.

Solution

We let $\partial C_0^{(2)} = \emptyset$ in Equations 8.1.42 through 8.1.45, eliminate $\bar{S}_{\alpha\beta}$ from Equations 8.1.42 and 8.1.43, and arrive at the following problem:

Find a displacement field $\bar{u}_\alpha$ on C_0, that satisfies the field equation

$$\mu\bar{u}_{\alpha,\gamma\gamma} + (\bar{\lambda} + \mu)\bar{u}_{\gamma,\gamma\alpha} + \bar{b}_\alpha = 0 \quad \text{on } C_0 \tag{a}$$

and the boundary condition

$$\bar{u}_\alpha = \widehat{u}_\alpha \quad \text{on } \partial C_0 \tag{b}$$

This completes the solution. □

Example 8.1.3

(A) Show that in the case of zero body forces, and for a body subject to plane strain conditions the field equations written in terms of stresses take the form

$$S_{\alpha\gamma,\gamma} = 0, \quad S_{\alpha\alpha,\gamma\gamma} = 0 \quad \text{on } C_0 \tag{a}$$

(B) Also, show that in the case of zero body forces and for a body subject to generalized plane stress conditions, the stress tensor $\bar{S}_{\alpha\beta}$ satisfies the field equations

$$\bar{S}_{\alpha\gamma,\gamma} = 0, \quad \bar{S}_{\alpha\alpha,\gamma\gamma} = 0 \quad \text{on } C_0 \tag{b}$$

Solution

(A) The first equation of (a) is obtained from Equation 8.1.13 by letting $b_\alpha = 0$. To obtain the second equation of (a) we note that the strain tensor $E_{\alpha\beta}$ associated with $S_{\alpha\beta}$ satisfies the compatibility relation, see Equation (f) in Example 3.1.13,

$$2E_{12,12} = E_{11,22} + E_{22,11} \tag{c}$$

Now, from Equation 8.1.17 we obtain

$$E_{11} = \frac{1}{2\mu}\left(S_{11} - \nu S_{\gamma\gamma}\right) \tag{d}$$

$$E_{22} = \frac{1}{2\mu}\left(S_{22} - \nu S_{\gamma\gamma}\right) \tag{e}$$

$$E_{12} = \frac{1}{2\mu}S_{12} \tag{f}$$

Substituting (d) through (f) into (c), and multiplying the result by 2μ we obtain

$$2S_{12,12} = S_{11,22} + S_{22,11} - \nu S_{\gamma\gamma,\delta\delta} \tag{g}$$

Writing the equation of equilibrium [the first of (a)] in an extended form yields

$$S_{11,1} + S_{12,2} = 0 \tag{h}$$

$$S_{12,1} + S_{22,2} = 0 \tag{i}$$

Differentiating (h) with respect to x_1 and (i) with respect to x_2 and adding the results, we get

$$2S_{12,12} = -S_{11,11} - S_{22,22} \tag{j}$$

Now, substituting (j) into (g) and rearranging the terms, we obtain

$$S_{11,11} + S_{11,22} + S_{22,11} + S_{22,22} - \nu S_{\gamma\gamma,\alpha\alpha} = 0 \tag{k}$$

or

$$(1 - \nu)S_{\gamma\gamma,\alpha\alpha} = 0 \tag{l}$$

Since $1 - \nu > 0$, then (l) implies the second equation of (a). This completes the solution of part (A).

(B) Due to the similarity between Equations 8.1.13 and 8.1.42, and 8.1.17 and 8.1.46, proceeding as in the case of obtaining (A) we arrive at (B).

This completes the solution of part (B). □

Example 8.1.4

An Airy stress function $F = F(x_1, x_2)$ of a two-dimensional elastostatics is related to $S_{\alpha\beta}$ by, see Example 3.2.9,

$$S_{\alpha\beta} = \epsilon_{\alpha\gamma}\epsilon_{\beta\delta}F_{,\gamma\delta} \tag{a}$$

where $\epsilon_{\alpha\gamma}$ is the two-dimensional alternator defined by

$$\epsilon_{12} = -\epsilon_{21} = 1 \tag{b}$$

$$\epsilon_{11} = \epsilon_{22} = 0 \tag{c}$$

Equation (a) written in components is

$$S_{11} = \epsilon_{1\gamma}\epsilon_{1\delta}F_{,\gamma\delta} = \epsilon_{12}\epsilon_{1\delta}F_{,2\delta} = \epsilon_{12}\epsilon_{12}F_{,22} = F_{,22} \tag{d}$$

Similarly,

$$S_{22} = F_{,11} \tag{e}$$

and

$$S_{12} = -F_{,12} \tag{f}$$

Show that a traction problem for a body subject to zero body forces and plane strain conditions may be reduced to the following boundary value problem for F:

Find $F = F(x_1, x_2)$ on C_0 that satisfies the biharmonic equation

$$\nabla^2\nabla^2 F = 0 \quad \text{on } C_0 \tag{g}$$

and the boundary condition

$$\epsilon_{\alpha\gamma}\epsilon_{\beta\delta}F_{,\gamma\delta}n_\beta = \widehat{s}_\alpha \quad \text{on } \partial C_0 \tag{h}$$

where $\widehat{s}_\alpha$ is a prescribed vector function and n_β is the outward unit vector normal to ∂C_0.

Solution

It follows from (a) and also from (d) and (e) that

$$S_{\alpha\alpha} = F_{,\alpha\alpha} \tag{i}$$

In addition, $S_{\alpha\beta}$ given by (a) satisfies the first equation of part (A) in Example 8.1.3 identically. Hence, substituting (a) of this example into the second equation of (A) of Example 8.1.3, and using (i), we arrive at (g). Obviously, the boundary condition (h) is obtained by inserting (a) into the traction boundary condition

$$S_{\alpha\beta}n_\beta = \widehat{s}_\alpha \quad \text{on } \partial C_0 \tag{j}$$

This completes the solution. □

Notes:

(1) Comparing a mixed boundary value problem for a body under plane strain conditions with that under generalized plane stress conditions, we observe that if a solution to one of these two problems is known, then a solution to the other problem may be found by an appropriate replacement of elastic constants.* In particular, if a plane strain problem with the material constants μ and ν is solved, then a solution to the corresponding generalized plane stress problem is obtained by replacing the pair (μ, ν) by $(\mu, \bar{\nu})$ in the plane strain solution.

(2) From Examples 8.1.3 and 8.2.14, it follows that a stress field $S_{\alpha\beta}$ corresponding to a solution to the traction problem for a body under both plane strain conditions and generalized plane stress conditions does not depend on elastic constants, see (a) and (b) in Example 8.1.3, and (g) and (h) in Example 8.2.14. This fact is utilized in photoelasticity. Namely, thin optically sensitive disks are used in finding two-dimensional stress fields experimentally by photoelastic methods. Such stress fields are the same as in thin plates made of, say, steel, if loading conditions are identical.

* See a detailed discussion in [2].

(3) Since Poisson's ratio ν ranges over the interval, see Example 3.3.9,

$$-1 < \nu < \frac{1}{2} \tag{8.1.48}$$

then, by Equation 8.1.47, "Poisson's ratio" $\bar{\nu}$ for a generalized plane stress state ranges over the interval

$$-\infty < \bar{\nu} < \frac{1}{3} \tag{8.1.49}$$

For engineering materials, the range of ν is restricted to

$$0 < \nu < \frac{1}{2} \tag{8.1.50}$$

Hence, the range of $\bar{\nu}$, corresponding to 8.1.50 is confined to

$$0 < \bar{\nu} < \frac{1}{3} \tag{8.1.51}$$

8.1.2 Two-Dimensional Problems of Nonisothermal Elastostatics

In this section, we discuss two-dimensional field equations for a cylindrical body under plane strain and generalized plane stress conditions, see Section 8.1.1, and subject to a two-dimensional temperature field, zero body forces, and two-dimensional boundary conditions. The cylindrical body is as shown in Figure 8.1.

We begin with a *nonisothermal plane strain state in* the (x_1, x_2) *plane*, which is defined as a thermoelastic state $s = [\mathbf{u}, \mathbf{E}, \mathbf{S}]$ corresponding to a temperature change $T = T(x_1, x_2)$ and satisfying the conditions, see Equations 8.1.6 and 8.1.7,

$$u_\alpha = u_\alpha(x_1, x_2), \quad u_3 = 0 \tag{8.1.52}$$

Clearly, Equation 8.1.52 implies that $\mathbf{E} = \mathbf{E}(x_1, x_2)$ and $\mathbf{S} = \mathbf{S}(x_1, x_2)$. In addition, the field equations of static thermoelasticity, see Equations 4.1.46 through 4.1.48, restricted to the case $\mathbf{b} = \mathbf{0}$, take the form of the strain–displacement relations

$$E_{\alpha\beta} = \frac{1}{2}\,(u_{\alpha,\beta} + u_{\beta,\alpha}) \tag{8.1.53}$$

$$E_{12} = E_{23} = E_{33} = 0 \tag{8.1.54}$$

the equation of equilibrium

$$S_{\alpha\beta,\beta} = 0 \tag{8.1.55}$$

and the constitutive relations

$$S_{\alpha\beta} = 2\mu E_{\alpha\beta} + \lambda E_{\gamma\gamma}\delta_{\alpha\beta} - (3\lambda + 2\mu)\alpha T\delta_{\alpha\beta} \tag{8.1.56}$$

$$S_{13} = S_{23} = 0 \tag{8.1.57}$$

$$S_{33} = \lambda E_{\gamma\gamma} - (3\lambda + 2\mu)\alpha T \tag{8.1.58}$$

By taking the trace of Equation 8.1.56, we obtain

$$S_{\alpha\alpha} = 2(\lambda + \mu)E_{\alpha\alpha} - 2(3\lambda + 2\mu)\alpha T \tag{8.1.59}$$

Hence

$$E_{\alpha\alpha} = \frac{1}{2(\lambda + \mu)} S_{\alpha\alpha} + \frac{3\lambda + 2\mu}{\lambda + \mu} \alpha T \tag{8.1.60}$$

Substituting this in Equation 8.1.58, we get

$$S_{33} = \frac{\lambda}{2(\lambda + \mu)} S_{\alpha\alpha} - \mu \frac{3\lambda + 2\mu}{\lambda + \mu} \alpha T \tag{8.1.61}$$

Also, using Equations 8.1.56 and 8.1.60, we arrive at the inverted form of the constitutive relation

$$E_{\alpha\beta} = \frac{1}{2\mu}\left[S_{\alpha\beta} - \frac{\lambda}{2(\lambda + \mu)} S_{\gamma\gamma}\delta_{\alpha\beta}\right] + \frac{3\lambda + 2\mu}{2(\lambda + \mu)} \alpha T \delta_{\alpha\beta} \tag{8.1.62}$$

Note that Equations 8.1.56 and 8.1.61, respectively, written in terms of μ and ν take the forms

$$S_{\alpha\beta} = 2\mu\left(E_{\alpha\beta} + \frac{\nu}{1 - 2\nu} E_{\gamma\gamma}\delta_{\alpha\beta}\right) - 2\mu \frac{1 + \nu}{1 - 2\nu} \alpha T \delta_{\alpha\beta} \tag{8.1.63}$$

and

$$S_{33} = \nu S_{\alpha\alpha} - 2\mu(1 + \nu)\alpha T, \tag{8.1.64}$$

while Equation 8.1.62 may be written as

$$E_{\alpha\beta} = \frac{1}{2\mu} \left(S_{\alpha\beta} - \nu S_{\gamma\gamma}\delta_{\alpha\beta}\right) + (1 + \nu)\alpha T \delta_{\alpha\beta} \tag{8.1.65}$$

To express Equations 8.1.63 through 8.1.65 in terms of E and ν, we substitute in these equations $\mu = E/2(1 + \nu)$. The presence of the stress components S_{33} given by Equation 8.1.64 complies with the condition that E_{33} vanishes everywhere in the cylinder, see Equation 8.1.54.

A mixed problem of nonisothermal elastostatics for an isotropic elastic body subject to plane strain conditions is now formulated in the following way:

Find a displacement u_α and stress $S_{\alpha\beta}$ that satisfy the field equations

$$S_{\alpha\beta,\beta} = 0 \quad \text{on } C_0 \tag{8.1.66}$$

$$S_{\alpha\beta} = \mu\left(u_{\alpha,\beta} + u_{\beta,\alpha}\right) + \lambda\delta_{\alpha\beta}u_{\gamma,\gamma} - (3\lambda + 2\mu)\,\alpha T \delta_{\alpha\beta} \quad \text{on } C_0 \tag{8.1.67}$$

subject to the mixed boundary conditions

$$u_\alpha = \widehat{u}_\alpha \quad \text{on } \partial C_0^{(1)} \tag{8.1.68}$$

$$S_{\alpha\beta} n_\beta = \widehat{s}_\alpha \quad \text{on } \partial C_0^{(2)} \tag{8.1.69}$$

Here T is a prescribed temperature change on C_0, and $\widehat{u}_\alpha$ and $\widehat{s}_\alpha$ are prescribed functions on $\partial C_0^{(1)}$ and $\partial C_0^{(2)}$, respectively. In a general case, the function $T = T(x_1, x_2)$ is a solution to Poisson's equation (see Equation 4.2.56 with a time independent temperature T and Q),

$$\nabla^2 T = -\frac{Q}{\kappa} = -\frac{W}{k} \quad \text{on } C_0 \tag{8.1.70}$$

subject to suitable boundary conditions on ∂C_0. In Equation 8.1.70 $Q(x_1, x_2) = \kappa W(x_1, x_2)/k$) denotes a prescribed heat supply field, W is the internal heat generated per unit volume per unit time, while k denotes the thermal conductivity, and κ is the thermal diffusivity.

A *nonisothermal generalized plane stress state* in (x_1, x_2) plane corresponding to zero body forces is defined as a thermoelastic state $s = [\bar{\mathbf{u}}, \bar{\mathbf{E}}, \bar{\mathbf{S}}]$ that complies with the two-dimensional field equations:

the strain–displacement relation

$$\bar{E}_{\alpha\beta} = \frac{1}{2}\left(\bar{u}_{\alpha,\beta} + \bar{u}_{\beta,\alpha}\right) \tag{8.1.71}$$

the equilibrium equation

$$\bar{S}_{\alpha\beta,\beta} = 0 \tag{8.1.72}$$

the constitutive relation

$$\bar{S}_{\alpha\beta} = 2\mu \bar{E}_{\alpha\beta} + \lambda \delta_{\alpha\beta}\left(\bar{E}_{\gamma\gamma} + \bar{E}_{33}\right) - (3\lambda + 2\mu)\,\alpha \bar{T} \delta_{\alpha\beta} \tag{8.1.73}$$

where $\bar{E}_{33}$ in Equation 8.1.73 is obtained, in terms of $E_{\gamma\gamma}$ and $\bar{T}$, from the condition

$$\bar{S}_{33} = 2\mu \bar{E}_{33} + \lambda\left(\bar{E}_{\gamma\gamma} + \bar{E}_{33}\right) - (3\lambda + 2\mu)\,\alpha \bar{T} = 0 \tag{8.1.74}$$

In addition, it is assumed that, see Equations 8.1.33,

$$\bar{u}_3 = 0 \tag{8.1.75}$$

and

$$\bar{E}_{13} = \bar{E}_{23} = \bar{S}_{13} = \bar{S}_{23} = 0 \tag{8.1.76}$$

In Equations 8.1.71 through 8.1.76 the bar over a function denotes its thickness average, see Equation 8.1.32,

$$\bar{f}(x_1, x_2) = \frac{1}{2h}\int_{-h}^{h} f(x_1, x_2, x_3)\, dx_3 \tag{8.1.77}$$

By solving Equation 8.1.74 with respect to $\bar{E}_{33}$, we arrive at

$$\bar{E}_{33} = -\frac{\lambda}{\lambda + 2\mu}\,\bar{E}_{\gamma\gamma} + \frac{3\lambda + 2\mu}{\lambda + 2\mu}\,\alpha\bar{T} \tag{8.1.78}$$

Substitution of Equation 8.1.78 into Equation 8.1.73 yields

$$\bar{S}_{\alpha\beta} = 2\mu\bar{E}_{\alpha\beta} + \bar{\lambda}\bar{E}_{\gamma\gamma}\delta_{\alpha\beta} - 2\mu\,\frac{3\lambda + 2\mu}{\lambda + 2\mu}\,\alpha\bar{T}\delta_{\alpha\beta} \tag{8.1.79}$$

where, see Equation 8.1.40,

$$\bar{\lambda} = \frac{2\mu\lambda}{\lambda + 2\mu} \tag{8.1.80}$$

Taking the trace of Equation 8.1.79 we get

$$\bar{S}_{\alpha\alpha} = 2(\bar{\lambda} + \mu)\bar{E}_{\alpha\alpha} - 4\mu\,\frac{3\lambda + 2\mu}{\lambda + 2\mu}\,\alpha\bar{T} \tag{8.1.81}$$

or, using Equation 8.1.80,

$$\bar{S}_{\alpha\alpha} = 2\mu\,\frac{3\lambda + 2\mu}{\lambda + 2\mu}\,\bar{E}_{\alpha\alpha} - 4\mu\frac{3\lambda + 2\mu}{\lambda + 2\mu}\,\alpha\bar{T} \tag{8.1.82}$$

Hence,

$$\bar{E}_{\alpha\alpha} = \frac{\lambda + 2\mu}{2\mu(3\lambda + 2\mu)}\,\bar{S}_{\alpha\alpha} + 2\alpha\bar{T} \tag{8.1.83}$$

Substituting Equation 8.1.83 into Equation 8.1.79 and solving for $\bar{E}_{\alpha\beta}$ we obtain

$$\bar{E}_{\alpha\beta} = \frac{1}{2\mu}\,(\bar{S}_{\alpha\beta} - \bar{\nu}\bar{S}_{\gamma\gamma}\delta_{\alpha\beta}) + \alpha\bar{T}\delta_{\alpha\beta} \tag{8.1.84}$$

where, see Equation 8.1.47,

$$\bar{\nu} = \frac{\nu}{1 + \nu} \tag{8.1.85}$$

Also, substitution of Equation 8.1.83 into Equation 8.1.78 yields

$$\bar{E}_{33} = -\frac{1}{2\mu}\,\frac{\lambda}{3\lambda + 2\mu}\,\bar{S}_{\alpha\alpha} + \alpha\bar{T} \tag{8.1.86}$$

To express the formulas 8.1.79, 8.1.82, 8.1.83, and 8.1.86 in terms of μ and ν we use the relation

$$\lambda = \frac{2\mu\nu}{1 - 2\nu} \tag{8.1.87}$$

and obtain

$$\bar{S}_{\alpha\beta} = 2\mu\left(\bar{E}_{\alpha\beta} + \frac{\nu}{1-\nu}\,\bar{E}_{\gamma\gamma}\delta_{\alpha\beta}\right) - 2\mu\,\frac{1+\nu}{1-\nu}\,\alpha\bar{T}\delta_{\alpha\beta} \tag{8.1.88}$$

$$\bar{S}_{\alpha\alpha} = 2\mu\,\frac{1+\nu}{1-\nu}\,(\bar{E}_{\alpha\alpha} - 2\alpha\bar{T}) \tag{8.1.89}$$

$$\bar{E}_{\alpha\alpha} = \frac{1-\nu}{2\mu(1+\nu)}\,\bar{S}_{\alpha\alpha} + 2\alpha\bar{T} \tag{8.1.90}$$

$$\bar{E}_{33} = -\frac{\nu}{2\mu(1+\nu)}\,\bar{S}_{\alpha\alpha} + \alpha\bar{T} \tag{8.1.91}$$

To express Equations 8.1.88 through 8.1.91 in terms of E and ν we substitute in these equations $\mu = E/2(1+\nu)$.

A mixed problem of nonisothermal elastostatics for an isotropic elastic body subject to generalized plane stress conditions is now formulated in the following way:

Find a displacement $\bar{u}_\alpha$ and stress $\bar{S}_{\alpha\beta}$ that satisfy the field equations

$$\bar{S}_{\alpha\beta,\beta} = 0 \quad \text{on } C_0 \tag{8.1.92}$$

$$\bar{S}_{\alpha\beta} = \mu(\bar{u}_{\alpha,\beta} + \bar{u}_{\beta,\alpha}) + \bar{\lambda}\bar{u}_{\gamma\gamma}\delta_{\alpha\beta} - 2\mu\,\frac{3\lambda+2\mu}{\lambda+2\mu}\,\alpha\bar{T}\delta_{\alpha\beta} \quad \text{on } C_0 \tag{8.1.93}$$

subject to the mixed boundary conditions

$$\bar{u}_\alpha = \widehat{u}_\alpha \quad \text{on } \partial C_0^{(1)} \tag{8.1.94}$$

$$\bar{S}_{\alpha\beta}b_\beta = \widehat{s}_\alpha \quad \text{on } \partial C_0^{(2)} \tag{8.1.95}$$

where $\bar{T}$, $\widehat{u}_\alpha$, and $\widehat{s}_\alpha$ are prescribed functions.

Similarly as in the nonisothermal plane strain case, $\bar{T}$ may be taken as a solution to Poisson's equation

$$\nabla^2\bar{T} = -\frac{Q}{\kappa} \quad \text{on } C_0 \tag{8.1.96}$$

where Q and κ have the same meaning as in Equation 8.1.70.

Notes:

(1) If the temperature change T in a plane strain problem depends also on the time, and if T satisfies the parabolic heat conduction equation

$$\nabla^2 T - \frac{1}{\kappa}\,\dot{T} = -\frac{Q}{\kappa} \quad \text{on } C_0 \times [0,\infty) \tag{8.1.97}$$

subject to the initial condition

$$T(\mathbf{x},0) = T_0(\mathbf{x}), \quad \mathbf{x} \in C_0 \tag{8.1.98}$$

and suitable boundary conditions on ∂C_0, then the two-dimensional thermoelastic state $s = [\mathbf{u}, \mathbf{E}, \mathbf{S}]$ depends on the time, and the problem described by Equations 8.1.66 through 8.1.69 is called a *plane strain problem of quasi-static thermoelasticity*.

(2) If the temperature change $\bar{T}$ in a generalized plane stress problem depends on the time, and if $\bar{T}$ is a solution to the parabolic equation

$$\nabla^2 \bar{T} - \frac{1}{\kappa} \dot{\bar{T}} = -\frac{Q}{\kappa} \quad \text{on } C_0 \times [0, \infty) \tag{8.1.99}$$

with the initial condition

$$\bar{T}(\mathbf{x}, 0) = T_0(\mathbf{x}) \quad \mathbf{x} \in C_0 \tag{8.1.100}$$

and suitable boundary conditions on ∂C_0, then the two-dimensional thermoelastic state $s = [\bar{\mathbf{u}}, \bar{\mathbf{E}}, \bar{\mathbf{S}}]$ depends on the time, and the problem described by Equations 8.1.92 through 8.1.95 is called a *generalized plane stress problem of quasi-static thermoelasticity*.

Example 8.1.5

Formulate a displacement boundary value problem of thermoelastostatics for a body under plane strain conditions subject to zero body forces and the temperature change T, in terms of the displacement vector.

Solution

This is a particular case of the mixed problem given by Equations 8.1.66 through 8.1.69 in which $\partial C_0^{(1)} = \partial C_0$, that is, $\partial C_0^{(2)} = \emptyset$. Hence, by eliminating $S_{\alpha\beta}$ from Equations 8.1.66 and 8.1.67 we arrive at the following problem:

Find a displacement field u_α on C_0 that satisfies the equation

$$\mu u_{\alpha,\gamma\gamma} + (\lambda + \mu) u_{\gamma,\gamma\alpha} - \gamma T_{,\alpha} = 0 \quad \text{on } C_0 \tag{a}$$

and the boundary condition

$$u_\alpha = \widehat{u}_\alpha \quad \text{on } \partial C_0 \tag{b}$$

where γ, as in Equation 4.1.51, means

$$\gamma = (3\lambda + 2\mu)\alpha \tag{c}$$

and $\widehat{u}_\alpha$ and T are prescribed functions. This completes the solution. □

Example 8.1.6

Formulate a displacement boundary value problem of thermoelastostatics for a body under generalized plane stress conditions subject to zero body forces and the temperature change $\bar{T}$, in terms of the displacements.

Solution

Proceeding in a similar way as in the previous example, in particular, by eliminating $\bar{S}_{\alpha\beta}$ from Equations 8.1.92 and 8.1.93 we arrive at the following problem:

Find a displacement field $\bar{u}_\alpha$ on C_0 that satisfies the equation

$$\mu\bar{u}_{\alpha,\gamma\gamma} + (\bar{\lambda} + \mu)\,\bar{u}_{\gamma,\gamma\alpha} - \bar{\gamma}\bar{T}_{,\alpha} = 0 \quad \text{on } C_0 \tag{a}$$

and the boundary condition

$$u_\alpha = \widehat{u}_\alpha \quad \text{on } \partial C_0 \tag{b}$$

where

$$\bar{\gamma} = 2\mu\,\frac{3\lambda + 2\mu}{\lambda + 2\mu}\,\alpha \tag{c}$$

and $\widehat{u}_\alpha$ and $\bar{T}$ are prescribed functions. This completes the solution. □

Example 8.1.7

Show that the stress–temperature compatibility condition of thermoelastostatics for a body under plane strain state and for zero body forces takes the form

$$S_{\gamma\gamma,\alpha\alpha} + \frac{E\alpha}{1-\nu}\,T_{,\gamma\gamma} = 0 \tag{a}$$

Solution

For a plane strain problem of thermoelastostatics and for zero body forces, the equilibrium equation and the strain–stress–temperature relation, see Equations 8.1.55 and 8.1.65, take the form, respectively,

$$S_{\alpha\beta,\beta} = 0 \tag{b}$$

and

$$E_{\alpha\beta} = \frac{1}{2\mu}\,(S_{\alpha\beta} - \nu S_{\gamma\gamma}\delta_{\alpha\beta}) + (1+\nu)\alpha T\delta_{\alpha\beta} \tag{c}$$

In addition, $E_{\alpha\beta}$ satisfies the compatibility relation, see (c) in Example 8.1.3,

$$2E_{12,12} = E_{11,22} + E_{22,11} \tag{d}$$

From (c) we get

$$E_{11} = \frac{1}{2\mu}\,(S_{11} - \nu S_{\gamma\gamma}) + (1+\nu)\alpha T \tag{e}$$

$$E_{22} = \frac{1}{2\mu}\,(S_{22} - \nu S_{\gamma\gamma}) + (1+\nu)\alpha T \tag{f}$$

$$E_{12} = \frac{1}{2\mu}\,S_{12} \tag{g}$$

Substituting (e) through (g) into (d), we obtain

$$2S_{12,12} = S_{11,22} + S_{22,11} - \nu S_{\gamma\gamma,\alpha\alpha} + E\alpha T_{,\gamma\gamma} \tag{h}$$

Now, by equilibrium equations (b), see (j) in Example 8.1.3, we find

$$2S_{12,12} = -S_{11,11} - S_{22,22} \tag{i}$$

Therefore, substituting (i) into (h) and dividing by $1 - \nu$ we arrive at (a). This completes the solution. □

Example 8.1.8

Show that the stress–temperature compatibility condition of thermoelastostatics for a body under generalized plane stress state and for zero body forces takes the form

$$\bar{S}_{\gamma\gamma,\alpha\alpha} + \alpha E\bar{T}_{,\alpha\alpha} = 0 \tag{a}$$

Solution

In this case, the equilibrium equation and the strain–stress–temperature relation, see Equations 8.1.72 and 8.1.84, take the form, respectively,

$$\bar{S}_{\alpha\beta,\beta} = 0 \tag{b}$$

and

$$\bar{E}_{\alpha\beta} = \frac{1}{2\mu}\left(\bar{S}_{\alpha\beta} - \bar{\nu}\bar{S}_{\gamma\gamma}\delta_{\alpha\beta}\right) + \alpha\bar{T}\delta_{\alpha\beta} \tag{c}$$

Equation (c) in components reads

$$\bar{E}_{11} = \frac{1}{2\mu}\left(\bar{S}_{11} - \bar{\nu}\bar{S}_{\gamma\gamma}\right) + \alpha\bar{T} \tag{d}$$

$$\bar{E}_{22} = \frac{1}{2\mu}\left(\bar{S}_{22} - \bar{\nu}\bar{S}_{\gamma\gamma}\right) + \alpha\bar{T} \tag{e}$$

$$\bar{E}_{12} = \frac{1}{2\mu}\bar{S}_{12} \tag{f}$$

The compatibility relation for a generalized plane stress state takes the form, see (d) in Example 8.1.7,

$$2\bar{E}_{12,12} = \bar{E}_{11,22} + \bar{E}_{22,11} \tag{g}$$

Substituting (d) through (f) into (g) and using the relation, see (i) in Example 8.1.7,

$$2\bar{S}_{12,12} = -\bar{S}_{11,11} - \bar{S}_{22,11} \tag{h}$$

we arrive at

$$(1-\bar{\nu})\bar{S}_{\gamma\gamma,\alpha\alpha}+2\mu\alpha\bar{T}_{,\gamma\gamma}=0 \tag{i}$$

If we note that, see Equation 8.1.85,

$$\bar{\nu}=\frac{\nu}{1+\nu},\quad 2\mu=\frac{E}{1+\nu} \tag{j}$$

then dividing (i) by $(1-\bar{\nu})=1/(1+\nu)$ we arrive at (a). This completes the solution. □

Example 8.1.9

Show that a traction problem of thermoelastostatics for a body subject to zero body forces and plane strain conditions may be reduced to the following boundary value problem for an Airy stress function F:

Find $F=F(x_1,x_2)$ on C_0 that satisfies the equation

$$\nabla^2\nabla^2 F+\frac{E\alpha}{1-\nu}\nabla^2 T=0 \quad \text{on } C_0 \tag{a}$$

subject to the condition

$$\left(\nabla^2 F\delta_{\alpha\beta}-F_{,\alpha\beta}\right)n_\beta=\widehat{s}_\alpha \quad \text{on } \partial C_0 \tag{b}$$

where $\widehat{s}_\alpha$ is a prescribed load on ∂C_0.

Solution

A traction problem of thermoelastostatics for a body under plane strain conditions, and subject to zero body forces and temperature change T, is described by the stress–temperature field equations, see (a) and (b) in Example 8.1.7

$$S_{\alpha\beta,\beta}=0 \quad \text{on } C_0 \tag{c}$$

$$S_{\alpha\alpha,\beta\beta}+\frac{E\alpha}{1-\nu}T_{,\beta\beta}=0 \quad \text{on } C_0 \tag{d}$$

and the boundary condition

$$S_{\alpha\beta}n_\beta=\widehat{s}_\alpha \quad \text{on } \partial C_0 \tag{e}$$

Let $F=F(x_1,x_2)$ be an Airy stress function defined by, see (a) in Example 8.2.14,

$$S_{\alpha\beta}=\nabla^2 F\delta_{\alpha\beta}-F_{,\alpha\beta} \tag{f}$$

By substituting (f) into (c) we find that (c) is identically satisfied. Next, by (f), we obtain

$$S_{\alpha\alpha}=\nabla^2 F \tag{g}$$

Hence, substituting (g) and (f) into Equations (d) and (e), respectively, we arrive at (a) and (b). This completes the solution. □

Example 8.1.10

Show that a traction problem of thermoelastostatics for a body corresponding to zero body forces and generalized plane stress conditions may be reduced to the following problem for an Airy stress function $\bar{F}$:

Find $\bar{F} = \bar{F}(x_1, x_2)$ on C_0 that satisfies the equation

$$\nabla^2\nabla^2\bar{F} + E\alpha\nabla^2\bar{T} = 0 \quad \text{on } C_0 \tag{a}$$

subject to the condition

$$\left(\nabla^2\bar{F}\delta_{\alpha\beta} - \bar{F}_{,\alpha\beta}\right)n_\beta = \widehat{s}_\alpha \quad \text{on } \partial C_0 \tag{b}$$

Solution

We proceed in a way similar to that in the previous example. In particular, we express $\bar{S}_{\alpha\beta}$ in terms of $\bar{F}$ by

$$\bar{S}_{\alpha\beta} = \nabla^2\bar{F}\delta_{\alpha\beta} - \bar{F}_{,\alpha\beta} \tag{c}$$

and satisfy identically the equilibrium equation

$$\bar{S}_{\alpha\beta,\beta} = 0 \tag{d}$$

Next, we substitute (c) into (a) of Example 8.1.8 and arrive at (a) of this example. The boundary condition (b) is obtained by inserting (c) into the traction boundary condition

$$\bar{S}_{\alpha\beta}n_\beta = \widehat{s}_\alpha \tag{e}$$

This completes the solution. □

Note: It follows from the last two examples that a stress field $S_{\alpha\beta}$ corresponding to an external load $(\widehat{s}_\alpha, T)$ and to a plane strain state is known if a stress field $\bar{S}_{\alpha\beta}$ corresponding to an external load $(\widehat{s}_\alpha, \bar{T})$ and to a generalized plane stress state is known. Namely, in order to obtain $S_{\alpha\beta}$ from $\bar{S}_{\alpha\beta}$ we replace in the generalized plane stress solution $(\widehat{s}_\alpha, \bar{T})$ by $(\widehat{s}_\alpha, T)$ and $E\alpha$ by $E\alpha/(1-\nu)$.

Example 8.1.11

Show that a traction problem of thermoelastostatics for a body under plane strain conditions (or for a body under generalized plane stress conditions) has a unique solution.

Solution

We will treat here the case of a body under plane strain conditions, and the case of a body under generalized plane stress conditions may be treated in a similar way.

It is sufficient to show that the homogeneous field equations, see Equations 8.1.53 and 8.1.55 and Equation 8.1.65 with $T = 0$,

$$E_{\alpha\beta} = \frac{1}{2}\left(u_{\alpha,\beta} + u_{\beta,\alpha}\right) \quad \text{on } C_0 \tag{a}$$

$$S_{\alpha\beta,\beta} = 0 \quad \text{on } C_0 \tag{b}$$

$$E_{\alpha\beta} = \frac{1}{2\mu}\left(S_{\alpha\beta} - \nu S_{\gamma\gamma}\delta_{\alpha\beta}\right) \quad \text{on } C_0 \tag{c}$$

subject to the homogeneous traction boundary condition

$$S_{\alpha\beta}n_\beta = 0 \quad \text{on } \partial C_0 \tag{d}$$

imply that

$$S_{\alpha\beta} = 0 \quad \text{on } C_0 \tag{e}$$

To prove this we multiply (b) by u_α and obtain

$$u_\alpha S_{\alpha\beta,\beta} = 0 \quad \text{on } C_0 \tag{f}$$

or

$$(u_\alpha S_{\alpha\beta})_{,\beta} - u_{(\alpha,\beta)}S_{\alpha\beta} = 0 \quad \text{on } C_0 \tag{g}$$

Now, integration of (g) over C_0, and using (d) and the divergence theorem, yields

$$\int_{C_0} u_{(\alpha,\beta)}S_{\alpha\beta}\, da = 0 \tag{h}$$

This, by virtue of (a) and (c) is equivalent to

$$\frac{1}{2\mu}\int_{C_0}\left(S_{\alpha\beta}S_{\alpha\beta} - \nu S_{\gamma\gamma}^2\right) da = 0 \tag{i}$$

Next, we write $S_{\alpha\beta}$ in the form

$$S_{\alpha\beta} = S_{\alpha\beta}^{(d)} + \frac{1}{2}S_{\gamma\gamma}\delta_{\alpha\beta} \tag{j}$$

where $S_{\alpha\beta}^{(d)}$ is the deviatoric part of $S_{\alpha\beta}$ given by

$$S_{\alpha\beta}^{(d)} = S_{\alpha\beta} - \frac{1}{2}S_{\gamma\gamma}\delta_{\alpha\beta} \tag{k}$$

Substituting (j) into (i) we get

$$\frac{1}{2\mu}\int_{C_0}\left(S_{\alpha\beta}^{(d)}S_{\alpha\beta}^{(d)} + \frac{1}{2}S_{\gamma\gamma}^2 - \nu S_{\gamma\gamma}^2\right) da = 0 \tag{l}$$

or

$$\int_{C_0} \left(S_{\alpha\beta}^{(d)} S_{\alpha\beta}^{(d)} + \frac{1-2\nu}{2} S_{\gamma\gamma}^2 \right) da = 0 \tag{m}$$

Since $1 - 2\nu > 0$, the integrand in (m) is the sum of two nonnegative terms, hence, by (m) each of the two terms must vanish, that is

$$S_{\alpha\beta}^{(d)} = 0 \quad \text{on } C_0 \tag{n}$$

and

$$S_{\gamma\gamma} = 0 \quad \text{on } C_0 \tag{o}$$

Using the formulas (j), (n), and (o), we arrive at (e).

This completes the proof of the uniqueness theorem stated in the example. □

Notes:

(1) The uniqueness result in the last example settles the question of uniqueness of the solution to a traction problem of two-dimensional elastostatics or thermoelastostatics. In particular, it covers uniqueness for isothermal plane strain and generalized plane stress traction problems.

(2) If heat sources are absent in a two-dimensional problem of thermoelastostatics, body forces are zero, and the boundary is traction free, then from the uniqueness result given in the last example and from Examples 8.1.9 and 8.1.10, it follows that there are no thermal stresses inside a cylinder in plane strain state or a thin disk in a generalized plane stress state. Indeed, in this case, for a plane strain state

$$\nabla^2 T = 0 \quad \text{on } C_0 \tag{8.1.101}$$

and the traction problem described in Example 8.1.9 reduces to the homogeneous equation

$$\nabla^2 \nabla^2 F = 0 \quad \text{on } C_0 \tag{8.1.102}$$

subject to the homogeneous boundary condition

$$\left(\nabla^2 F \delta_{\alpha\beta} - F_{,\alpha\beta} \right) n_\beta = 0 \quad \text{on } \partial C_0 \tag{8.1.103}$$

By the uniqueness theorem of isothermal two-dimensional elastostatics, the problem described by Equations 8.1.102 and 8.1.103 has only zero solution.

In case of a generalized plane stress state, subject to a harmonic temperature change $\bar{T}$, and for a stress free boundary ∂C_0, the situation is similar, that is

$$\bar{S}_{\alpha\beta} = 0 \quad \text{on } C_0 \tag{8.1.104}$$

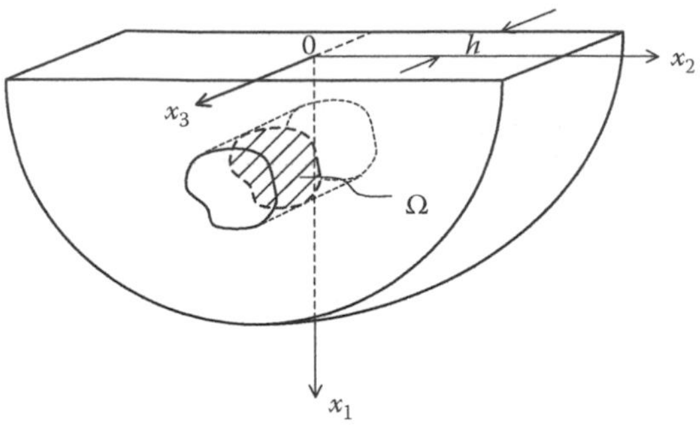

FIGURE 8.2 Semi-infinite disk with a heat supply $Q(x_1, x_2)$ over Ω.

Example 8.1.12

Consider a semi-infinite disk $x_1 \geq 0$, $-\infty < x_2 < \infty$, $-h \leq x_3 \leq h$, in a generalized plane stress state, subject to a heat supply $Q(x_1, x_2)$ over a domain Ω entirely contained inside the semispace, see Figure 8.2.

The two-dimensional domain C_0 over which the generalized plane stress state is defined is described by the inequalities $x_1 \geq 0$ and $-\infty < x_2 < \infty$, and Ω is contained in C_0. The body forces are assumed to be zero on C_0 and the boundary $x_1 = 0$ is traction free. Moreover, stresses $\bar{S}_{\alpha\beta}$ vanish as $x_1^2 + x_2^2 \to \infty$.

Show that $\bar{S}_{\alpha\beta}$ produced by the heat supply Q may be obtained from the formula

$$\bar{S}_{\alpha\beta} = \nabla^2 \bar{F} \delta_{\alpha\beta} - \bar{F}_{,\alpha\beta} \tag{a}$$

where $\bar{F} = \bar{F}(x_1, x_2)$ is a solution to the following boundary value problem:

Find a function $\bar{F} = \bar{F}(x_1, x_2)$ for $x_1 \geq 0$, $-\infty < x_2 < \infty$, that satisfies the field equation

$$\nabla^2 \nabla^2 \bar{F} = \alpha E \frac{Q}{\kappa} \quad \text{on } C_0 \tag{b}$$

subject to the conditions

$$\bar{F} = \bar{F}_{,1} = 0 \quad \text{on } x_1 = 0, \ -\infty < x_2 < \infty \tag{c}$$

and vanishing conditions at infinity. Note that in Equation (b), $Q/\kappa = W/k$.

Solution

In this case, the stress field $\bar{S}_{\alpha\beta}$ corresponds to a temperature change $\bar{T}$ that satisfies Poisson's equation

$$\nabla^2 \bar{T} = -\frac{Q}{\kappa} \quad \text{for } x_1 > 0, \ -\infty < x_2 < \infty \tag{d}$$

Substituting (d) into (a) of Example 8.1.10, and taking into account the formulation of a generalized plane stress problem given in Example 8.1.10, we find that $\bar{S}_{\alpha\beta}$ is related to a stress function $\bar{F}$ by

$$\bar{S}_{\alpha\beta} = \nabla^2 \bar{F} \delta_{\alpha\beta} - \bar{F}_{,\alpha\beta} \tag{e}$$

where

$$\nabla^2\nabla^2\bar{F} = \alpha E\frac{Q}{\kappa} \quad \text{for } x_1 > 0, \ -\infty < x_2 < \infty \tag{f}$$

and

$$\bar{F}_{,22} = 0, \quad \bar{F}_{,12} = 0 \quad \text{for } x_1 = 0, \ -\infty < x_2 < \infty \tag{g}$$

and $\bar{F}$ together with its partial derivatives vanish at infinity. The boundary conditions (g) imply that the boundary of the disk is stress free. Next, we note that if $\bar{F}$ satisfies the boundary conditions

$$\bar{F} = 0, \quad \bar{F}_{,1} = 0 \quad \text{for } x_1 = 0, \ -\infty < x_2 < \infty \tag{h}$$

then differentiating the first boundary condition of (h) twice with respect to x_2, and the second boundary condition of (h) with respect to x_2, we arrive at (g). Hence, a formulation of the problem of finding $\bar{S}_{\alpha\beta}$ corresponding to the heat supply Q in the semi-infinite disk may be reduced to that of solving (b) subject to the conditions (c). This completes the solution. □

Notes:

(1) A boundary value problem for the function $\bar{F}$ is the one in which $\bar{F}$ satisfies the nonhomogeneous field equation (b) subject to the conditions that $\bar{F}$ and its normal derivative $\partial\bar{F}/\partial n = \partial\bar{F}/\partial x_1$ vanish on the boundary of the body. One may show that in case of a general cross section C_0 with a smooth boundary ∂C_0 a plane problem of the thermoelastostatics with a prescribed heat supply Q on $\Omega \subset C_0$ and with traction free boundary ∂C_0 can by formulated as the one in which a function $\bar{F} = \bar{F}(x_1, x_2)$ satisfies the field equation, see (b),

$$\nabla^2\nabla^2\bar{F} = \alpha E\frac{Q}{\kappa} \quad \text{on } C_0 \tag{8.1.105}$$

subject to the boundary conditions

$$\bar{F} = \frac{\partial\bar{F}}{\partial n} = 0 \quad \text{on } \partial C_0 \tag{8.1.106}$$

(2) Let $\bar{F}^* = \bar{F}^*(\mathbf{x}, \boldsymbol{\xi})$ be a stress function that generates a thermoelastic generalized plane stress state in the semispace $x_1 \geq 0$, $|x_2| < \infty$ subject to a zero boundary load, and to the concentrated heat source located at a point $\boldsymbol{\xi} = (\xi_1, \xi_2)$ of the semispace

$$Q^*(\mathbf{x}, \boldsymbol{\xi}) = \delta(x_1 - \xi_1)\delta(x_2 - \xi_2) \tag{8.1.107}$$

where $\delta = \delta(x)$ is the Dirac delta function. Then, the stress function $\bar{F} = \bar{F}(\mathbf{x})$ discussed in Example 8.1.12 may be computed from the formula

$$\bar{F}(\mathbf{x}) = \int_{\Omega} \bar{F}^*(\mathbf{x}, \boldsymbol{\xi})Q(\boldsymbol{\xi})\, da(\boldsymbol{\xi}) \tag{8.1.108}$$

As will be shown in Chapter 9, the function $\bar{F}^* = \bar{F}^*(\mathbf{x}, \boldsymbol{\xi})$ may be obtained in a closed form by using a Fourier integral method.

In modern theory of engineering materials, an important role is played by inhomogenous elastic materials, see definition of such materials in Section 3.3.1. For an inhomogeneous isotropic elastic body, Young's modulus E and Poisson's ratio ν depend on the space variable $\mathbf{x}$, that is, $E = E(\mathbf{x})$ and $\nu = \nu(\mathbf{x})$. In the following example, there is shown a result of two-dimensional isothermal inhomogeneous isotropic elastostatics for determining the so called effective moduli of elastic composites.

Example 8.1.13

Show that a solution $S_{\alpha\beta} = S_{\alpha\beta}(\mathbf{x})$, $\mathbf{x} = (x_1, x_2)$, to a two-dimensional traction problem of isothermal inhomogeneous isotropic elastostatics with zero body forces is invariant under the transformation of material parameters

$$A(x) \to A^*(\mathbf{x}) = mA(\mathbf{x}) + A_0 + A_\alpha x_\alpha \tag{a}$$

$$S(x) \to S^*(\mathbf{x}) = mS(\mathbf{x}) - A_0 - A_\alpha x_\alpha \tag{b}$$

where

$$A(x) = \frac{1 - 2\nu(\mathbf{x})}{\mu(\mathbf{x})}, \quad S(\mathbf{x}) = \frac{1}{\mu(\mathbf{x})} \tag{c}$$

for a body under plane strain conditions, and

$$A(\mathbf{x}) = \frac{1 - \nu(\mathbf{x})}{\mu(\mathbf{x})[1 + \nu(\mathbf{x})]}, \quad S(\mathbf{x}) = \frac{1}{\mu(\mathbf{x})} \tag{d}$$

for a body under generalized plane stress conditions; and $m \neq 0$, A_0, $A_\alpha (\alpha = 1, 2)$ are constants selected in such a way that the bulk and shear compliances A^* and S^*, respectively, be nonnegative.

Solution

First, we note that the strain–stress relations for a two-dimensional isothermal inhomogeneous isotropic elastostatics can be written in the form (see Equations 8.1.17 and 8.1.46)

$$E_{\alpha\beta} = \frac{1}{2} S\, S_{\alpha\beta} + \frac{A - S}{4} S_{\gamma\gamma} \delta_{\alpha\beta} \quad \text{on } C_0 \tag{e}$$

Therefore, the traction boundary value problem in terms of stresses reads: Find a stress field $S_{\alpha\beta} = S_{\alpha\beta}(\mathbf{x})$ that satisfies the equilibrium equation

$$S_{\alpha\beta,\beta} = 0 \quad \text{on } C_0 \tag{f}$$

and the stress compatibility equation

$$\left(\frac{1}{2} S\, S_{11} + \frac{A - S}{4} S_{\gamma\gamma}\right)_{,22} + \left(\frac{1}{2} S\, S_{22} + \frac{A - S}{4} S_{\gamma\gamma}\right)_{,11} - (S\, S_{12})_{,12} = 0 \quad \text{on } C_0 \tag{g}$$

subject to the boundary condition

$$S_{\alpha\beta} n_\beta = \widehat{s}_\alpha \quad \text{on } \partial C_0 \tag{h}$$

where $\widehat{s}_\alpha$ is a prescribed vector field.

Next, we note that an equivalent form of the field equations reads

$$S_{\alpha\beta,\beta} = 0 \quad \text{on } C_0 \tag{i}$$

and

$$\left(\frac{A+S}{2} S_{\gamma\gamma}\right)_{,\alpha\alpha} - S_{,\alpha\beta} S_{\alpha\beta} = 0 \quad \text{on } C_0 \tag{j}$$

Also, it follows from (a) and (b) that

$$\frac{A+S}{2} = \frac{A^* + S^*}{2m} \quad \text{and} \quad S_{,\alpha\beta} = \frac{1}{m} S^*{}_{,\alpha\beta} \quad \text{on } C_0 \tag{k}$$

Therefore, substituting (k) into (j), and multiplying the result by m we obtain

$$\left(\frac{A^* + S^*}{2} S_{\gamma\gamma}\right)_{,\alpha\alpha} - S^*{}_{,\alpha\beta} S_{\alpha\beta} = 0 \quad \text{on } C_0 \tag{l}$$

Hence, from a uniqueness theorem for a two-dimensional traction boundary value problem of inhomogeneous isotropic elastostatics in which the bulk and shear compliances are nonnegative we conclude that

$$S_{\alpha\beta} = S^*_{\alpha\beta} \quad \text{on } \overline{C}_0 = C_0 \cup \partial C_0 \tag{m}$$

where $S^*_{\alpha\beta} = S^*_{\alpha\beta}(\mathbf{x})$ is a solution to the traction problem described by the field equations (i) and (l) subject to the boundary condition (h).

This completes the solution. □

Notes:

(1) The result obtained in Example 8.1.13 is also called a CLM stress invariance theorem [3]. It has been widely used to determine the effective moduli of elastic composites.*

(2) The CLM shift on the compliances can serve as a check for analytical and computational results for both stress fields and effective properties of an elastic composite. Also, it reduces a number of needed experiments or calculations for characterizing materials, and it gives exact relations for the constitutive parameters which are independent of the geometry [6].

(3) The CLM theorem relies heavily on the hypotheses of a two-dimensional inhomogeneous isotropic elastostatics and finding its counterpart to a pure stress two-dimensional inhomogeneous isotropic elastodynamics would be useful.

* See, for example, [4,5].

8.2 TWO-DIMENSIONAL PROBLEMS OF ELASTODYNAMICS

8.2.1 Two-Dimensional Problems of Isothermal Elastodynamics

In Section 8.1.1, we introduced two types of two-dimensional problems: a plane strain problem and a generalized plane stress problem of isothermal elastostatics. In both these problems, an elastic state $s = [\mathbf{u}, \mathbf{E}, \mathbf{S}]$ is generated by solving a two-dimensional boundary value problem in the (x_1, x_2) plane. In case of a plane strain state, the problem is described by Equations 8.1.20 through 8.1.23, and in case of a generalized plane stress state, the problem is described by Equations 8.1.42 through 8.1.45. In both these formulations the body forces, the boundary data, and the solution itself depend on x_1 and x_2 only.

In this section, we generalize the two-dimensional problems of isothermal elastostatics into two-dimensional problems of isothermal elastodynamics by including inertia terms and initial data into consideration. In addition, we assume that the body forces and the boundary data depend not only on x_1 and x_2 but also on the time t.

Proceeding in a manner similar to that of a plane strain problem of elastostatics, see Equations 8.1.20 through 8.1.23, we formulate the following *mixed problem of isothermal elastodynamics for an isotropic elastic body subject to plane strain conditions.*

Find a displacement $u_\alpha = u_\alpha(\mathbf{x}, t)$ and stress $S_{\alpha\beta} = S_{\alpha\beta}(\mathbf{x}, t)$ on $C_0 \times [0, \infty)$ that satisfy the field equations

$$S_{\alpha\beta,\beta} + b_\alpha = \rho\ddot{u}_\alpha \quad \text{on } C_0 \times [0, \infty) \tag{8.2.1}$$

$$S_{\alpha\beta} = \mu(u_{\alpha,\beta} + u_{\beta,\alpha}) + \lambda\delta_{\alpha\beta}u_{\gamma,\gamma} \quad \text{on } C_0 \times [0, \infty) \tag{8.2.2}$$

subject to the initial conditions

$$u_\alpha(\mathbf{x}, 0) = u_\alpha^{(0)}, \quad \dot{u}_\alpha(\mathbf{x}, 0) = \dot{u}_\alpha^{(0)}, \quad \text{on } C_0 \tag{8.2.3}$$

and the mixed boundary conditions

$$u_\alpha = \widehat{u}_\alpha \quad \text{on } \partial C_0^{(1)} \times [0, \infty) \tag{8.2.4}$$

$$S_{\alpha\beta}n_\beta = \widehat{s}_\alpha \quad \text{on } \partial C_0^{(2)} \times [0, \infty) \tag{8.2.5}$$

Here $u_\alpha^{(0)}$ and $\dot{u}_\alpha^{(0)}$ are prescribed functions on C_0, and $\widehat{u}_\alpha$ and $\widehat{s}_\alpha$ are prescribed vector fields on $\partial C_0^{(1)} \times [0, \infty)$ and $\partial C_0^{(2)} \times [0, \infty)$, respectively. The other symbols are the same as in the elastostatics formulation.

Similarly, the following *mixed problem of isothermal elastodynamics for an isotropic elastic body subject to generalized plane stress conditions* may be formulated, see Equations 8.1.42 through 8.1.45.

Find a displacement $\bar{u}_\alpha = \bar{u}_\alpha(\mathbf{x}, t)$ and stress $\bar{S}_{\alpha\beta} = \bar{S}_{\alpha\beta}(\mathbf{x}, t)$ on $C_0 \times [0, \infty)$ that satisfy the field equations

$$\bar{S}_{\alpha\beta,\beta} + \bar{b}_\alpha = \rho\ddot{\bar{u}}_\alpha \quad \text{on } C_0 \times [0, \infty) \tag{8.2.6}$$

$$\bar{S}_{\alpha\beta} = \mu\left(\bar{u}_{\alpha,\beta} + \bar{u}_{\beta,\alpha}\right) + \bar{\lambda}\delta_{\alpha\beta}\bar{u}_{\gamma,\gamma} \quad \text{on } C_0 \times [0, \infty) \tag{8.2.7}$$

the initial conditions

$$\bar{u}_\alpha(\mathbf{x}, 0) = u_\alpha^{(0)}, \quad \dot{\bar{u}}_\alpha(\mathbf{x}, 0) = \dot{u}_\alpha^{(0)} \quad \text{on } C_0 \tag{8.2.8}$$

and the mixed boundary conditions

$$\bar{u}_\alpha = \widehat{u}_\alpha \quad \text{on } \partial C_0^{(1)} \times [0, \infty) \tag{8.2.9}$$

$$\bar{S}_{\alpha\beta} n_\beta = \widehat{s}_\alpha \quad \text{on } \partial C_0^{(2)} \times [0, \infty) \tag{8.2.10}$$

Here $\bar{b}_\alpha$, $u_\alpha^{(0)}$, $\dot{u}_\alpha^{(0)}$, $\widehat{u}_\alpha$, and $\widehat{s}_\alpha$ are prescribed functions.

Observe that the strain tensor $E_{\alpha\beta}$ is related to $S_{\alpha\beta}$ by, see Equation 8.1.17,

$$E_{\alpha\beta} = \frac{1}{2\mu} \left(S_{\alpha\beta} - \nu S_{\gamma\gamma} \delta_{\alpha\beta} \right) \tag{8.2.11}$$

while the strain tensor $\bar{E}_{\alpha\beta}$ is related to $\bar{S}_{\alpha\beta}$ by the formula, see Equation 8.1.46,

$$\bar{E}_{\alpha\beta} = \frac{1}{2\mu} \left(\bar{S}_{\alpha\beta} - \bar{\nu} \bar{S}_{\gamma\gamma} \delta_{\alpha\beta} \right) \tag{8.2.12}$$

where

$$\bar{\nu} = \frac{\nu}{1 + \nu} \tag{8.2.13}$$

Example 8.2.1

Formulate the displacement initial–boundary value problem of isothermal elastodynamics for a body subject to plane strain conditions, in terms of a displacement **u**.

Solution

We restrict the formulation described by Equations 8.2.1 through 8.2.5 to the case $\partial C_0^{(2)} = \emptyset$, eliminate $S_{\alpha\beta}$ from Equations 8.2.1 and 8.2.2, and arrive at the following problem:

Find a displacement field u_α on $C_0 \times [0, \infty)$ that satisfies the field equation

$$\mu u_{\alpha,\gamma\gamma} + (\lambda + \mu) u_{\gamma,\gamma\alpha} + b_\alpha = \rho \ddot{u}_\alpha \quad \text{on } C_0 \times [0, \infty) \tag{a}$$

the initial conditions

$$u_\alpha(\mathbf{x}, 0) = u_\alpha^{(0)}, \quad \dot{u}_\alpha(\mathbf{x}, 0) = \dot{u}_\alpha^{(0)} \quad \text{on } C_0 \tag{b}$$

and the boundary condition

$$u_\alpha = \widehat{u}_\alpha \quad \text{on } \partial C_0 \times [0, \infty) \tag{c}$$

This completes the solution. □

Example 8.2.2

Formulate the displacement initial–boundary value problem of isothermal elastodynamics for a body subject to generalized plane stress conditions in terms of the displacements.

Solution

Similarly as in the previous example, we restrict the formulation, given by Equations 8.2.6 through 8.2.10 to the case $\partial C_0^{(2)} = \emptyset$, eliminate $\bar{S}_{\alpha\beta}$ from Equations 8.2.6 and 8.2.7, and arrive at the formulation:

Find a displacement field $\bar{u}_\alpha$ on $C_0 \times [0, \infty)$ that satisfies the field equation

$$\mu \bar{u}_{\alpha,\gamma\gamma} + (\bar{\lambda} + \mu)\bar{u}_{\gamma,\gamma\alpha} + \bar{b}_\alpha = \rho \ddot{\bar{u}}_\alpha \quad \text{on } C_0 \times [0, \infty) \tag{a}$$

the initial conditions

$$\bar{u}_\alpha(\mathbf{x}, 0) = u_\alpha^{(0)}, \quad \dot{\bar{u}}_\alpha(\mathbf{x}, 0) = \dot{u}_\alpha^{(0)} \quad \text{on } C_0 \tag{b}$$

and the boundary condition

$$\bar{u}_\alpha = \widehat{u}_\alpha \quad \text{on } \partial C_0 \times [0, \infty) \tag{c}$$

This completes the solution. □

Example 8.2.3

Formulate the traction initial–boundary value problem of isothermal elastodynamics for a body subject to plane strain conditions in terms of the stresses.

Solution

First, we write Equations 8.2.1 and 8.2.2 in an alternative form

$$S_{\alpha\gamma,\gamma} + b_\alpha = \rho \ddot{u}_\alpha \quad \text{on } C_0 \times [0, \infty) \tag{a}$$

$$u_{(\alpha,\beta)} = \frac{1}{2\mu}\left(S_{\alpha\beta} - \nu S_{\gamma\gamma}\delta_{\alpha\beta}\right) \quad \text{on } C_0 \times [0, \infty) \tag{b}$$

By differentiating (a) with respect to x_β, we get

$$S_{\alpha\gamma,\gamma\beta} + b_{\alpha,\beta} = \rho \ddot{u}_{\alpha,\beta} \tag{c}$$

and taking the symmetric part of this equation with respect to indices α and β, we obtain

$$S_{(\alpha\gamma,\gamma\beta)} + b_{(\alpha,\beta)} = \rho \ddot{u}_{(\alpha,\beta)} \tag{d}$$

By substituting (b) into (d), we receive the stress equation of motion

$$S_{(\alpha\gamma,\gamma\beta)} - \frac{\rho}{2\mu}\left(\ddot{S}_{\alpha\beta} - \nu \ddot{S}_{\gamma\gamma}\delta_{\alpha\beta}\right) = -b_{(\alpha,\beta)} \tag{e}$$

Next, it follows from Equations 8.2.2 and 8.2.3 that

$$S_{\alpha\beta}(\mathbf{x},0) = S^{(0)}_{\alpha\beta}, \quad \dot{S}_{\alpha\beta}(\mathbf{x},0) = \dot{S}^{(0)}_{\alpha\beta} \quad \text{on } C_0 \tag{f}$$

where

$$S^{(0)}_{\alpha\beta} = \mu\left(u^{(0)}_{\alpha,\beta} + u^{(0)}_{\beta,\alpha}\right) + \lambda\delta_{\alpha\beta}u^{(0)}_{\gamma,\gamma} \tag{g}$$

and

$$\dot{S}^{(0)}_{\alpha\beta} = \mu\left(\dot{u}^{(0)}_{\alpha,\beta} + \dot{u}^{(0)}_{\beta,\alpha}\right) + \lambda\delta_{\alpha\beta}\dot{u}^{(0)}_{\gamma,\gamma} \tag{h}$$

Hence, restricting the mixed problem described by Equations 8.2.1 through 8.2.5 to the case $\partial C^{(1)}_0 = \emptyset$, and using (e) through (h), we arrive at the following initial–boundary value problem in terms of stresses $S_{\alpha\beta}$:

Find a stress field $S_{\alpha\beta}$ on $C_0 \times [0,\infty)$ that satisfies the equation

$$S_{(\alpha\gamma,\gamma\beta)} - \frac{\rho}{2\mu}\left(\ddot{S}_{\alpha\beta} - \nu\ddot{S}_{\gamma\gamma}\delta_{\alpha\beta}\right) = -b_{(\alpha,\beta)} \quad \text{on } C_0 \times [0,\infty) \tag{i}$$

the initial conditions

$$S_{\alpha\beta}(\mathbf{x},0) = S^{(0)}_{\alpha\beta}, \quad \dot{S}_{\alpha\beta}(\mathbf{x},0) = \dot{S}^{(0)}_{\alpha\beta} \quad \text{on } C_0 \tag{j}$$

and the boundary condition

$$S_{\alpha\beta}n_\beta = \widehat{s}_\alpha \quad \text{on } \partial C_0 \times [0,\infty) \tag{k}$$

where $b_{(\alpha,\beta)}$, $S^{(0)}_{\alpha\beta}$, $\dot{S}^{(0)}_{\alpha\beta}$, and $\widehat{s}_\alpha$ are prescribed functions. This completes the solution. □

Example 8.2.4

Formulate the traction initial–boundary value problem of isothermal elastodynamics for a body subject to generalized plane stress conditions in terms of the stresses.

Solution

The process leading to the formulation is similar to that of the previous example, and it is based upon a restriction of the problem described by Equations 8.2.6 through 8.2.10 to the case $\partial C^{(1)}_0 = \emptyset$. As a result we arrive at the formulation:

Find a stress field $\bar{S}_{\alpha\beta}$ on $C_0 \times [0,\infty)$ that satisfies the equation

$$\bar{S}_{(\alpha\gamma,\gamma\beta)} - \frac{\rho}{2\mu}\left(\ddot{\bar{S}}_{\alpha\beta} - \bar{\nu}\ddot{\bar{S}}_{\gamma\gamma}\delta_{\alpha\beta}\right) = -\bar{b}_{(\alpha,\beta)} \quad \text{on } C_0 \times [0,\infty) \tag{a}$$

the initial conditions

$$\bar{S}_{\alpha\beta}(\mathbf{x},0) = \bar{S}^{(0)}_{\alpha\beta}, \quad \dot{\bar{S}}_{\alpha\beta}(\mathbf{x},0) = \dot{\bar{S}}^{(0)}_{\alpha\beta} \quad \text{on } C_0 \tag{b}$$

and the boundary condition

$$\bar{S}_{\alpha\beta}n_\beta = \widehat{s}_\alpha \quad \text{on } \partial C_0 \times [0,\infty) \tag{c}$$

Here, $\bar{S}^{(0)}_{\alpha\beta}$ and $\dot{\bar{S}}^{(0)}_{\alpha\beta}$ are obtained from (g) and (h), respectively, by replacing λ by $\bar{\lambda}$, see Equation 8.2.7. □

Notes:

(1) If $S_{\alpha\beta}$ is a solution to the initial–boundary value problem described by (i) through (k) of Example 8.2.3, then the associated displacement field u_α and strain field $E_{\alpha\beta}$ may be obtained from the formulas

$$u_\alpha(\mathbf{x}, t) = \frac{1}{\rho}(i * S_{\alpha\beta,\beta} + f_\alpha) \tag{8.2.14}$$

$$E_{\alpha\beta}(\mathbf{x}, t) = \frac{1}{2\mu}(S_{\alpha\beta} - \nu S_{\gamma\gamma}\delta_{\alpha\beta}) \tag{8.2.15}$$

where

$$i = i(t) = t \tag{8.2.16}$$

and f_α is the two-dimensional pseudo-body force field, see Equations 6.2.1 and 6.2.2.

$$f_\alpha(\mathbf{x}, t) = i * b_\alpha + \rho\left(u^{(0)}_\alpha + t\dot{u}^{(0)}_\alpha\right) \tag{8.2.17}$$

(2) In the case of a dynamic generalized plane stress state when $\bar{S}_{\alpha\beta}$ is described by (a) through (c) of Example 8.2.4 the appropriate formulas for the displacement $\bar{u}_\alpha$ and strain field $\bar{E}_{\alpha\beta}$ read

$$\bar{u}_\alpha(\mathbf{x}, t) = \frac{1}{\rho}\left(i * \bar{S}_{\alpha\beta,\beta} + \bar{f}_\alpha\right) \tag{8.2.18}$$

$$\bar{E}_{\alpha\beta}(\mathbf{x}, t) = \frac{1}{2\mu}\left(\bar{S}_{\alpha\beta} - \bar{\nu}\bar{S}_{\gamma\gamma}\delta_{\alpha\beta}\right) \tag{8.2.19}$$

where

$$\bar{f}_\alpha(\mathbf{x}, t) = i * \bar{b}_\alpha + \rho\left(u^{(0)}_\alpha + t\dot{u}^{(0)}_\alpha\right) \tag{8.2.20}$$

(3) In the case of zero body forces and for arbitrary initial symmetric tensor fields $S^{(0)}_{\alpha\beta}$ and $\dot{S}^{(0)}_{\alpha\beta}$ in the formulation (i) through (k) of Example 8.2.3, the plane strain problem is the one of a two-dimensional incompatible elastodynamics in which stress waves are produced by the initial incompatibilities continuously distributed throughout the body. Therefore, in this case the problem belongs to a class of problems of incompatible elastodynamics.

As for the stress formulation (a) through (c) of Example 8.2.4 covering a generalized plane stress process, a similar remark applies, that is, if $\bar{S}^{(0)}_{\alpha\beta}$ and $\dot{\bar{S}}^{(0)}_{\alpha\beta}$ are arbitrary symmetric tensor fields, and $\bar{b}_\alpha$ vanishes, then a solution to the problem (a) through (c) of Example 8.2.4 represents stress waves of an incompatible elastodynamics.

8.2.2 Two-Dimensional Problems of Nonisothermal Elastodynamics

The two-dimensional formulations of nonisothermal elastodynamics are obtained by suitable generalization and parametrization of the two-dimensional nonisothermal elastostatics discussed in Section 8.1.2. The generalization is done by including the inertia terms and initial conditions into consideration, and the parameterization means that both a thermoelastic state and the corresponding thermomechanical load depend not only on x_1 and x_2, but also on the time t. As in the two-dimensional thermoelastostatics, we assume that there are no body forces.

As a result of the generalization and parameterization of Equations 8.1.53, 8.1.55, and 8.1.56 we arrive at the following *field equations of nonisothermal elastodynamics for a body subject to plane strain conditions*:

the strain–displacement relations

$$E_{\alpha\beta} = \frac{1}{2}\left(u_{\alpha,\beta} + u_{\beta,\alpha}\right) \tag{8.2.21}$$

the equation of motion

$$S_{\alpha\beta,\beta} = \rho\ddot{u}_{\alpha} \tag{8.2.22}$$

and the constitutive relations

$$S_{\alpha\beta} = 2\mu E_{\alpha\beta} + \lambda E_{\gamma\gamma}\delta_{\alpha\beta} - (3\lambda + 2\mu)\alpha T\delta_{\alpha\beta} \tag{8.2.23}$$

Here u_α, $E_{\alpha\beta}$, and $S_{\alpha\beta}$ are unknown fields on $C_0 \times [0,\infty)$, and T is a solution to the parabolic heat conduction equation

$$\nabla^2 T - \frac{1}{\kappa}\dot{T} = -\frac{Q}{\kappa} \quad \text{on } C_0 \times [0,\infty) \tag{8.2.24}$$

subject to the initial condition

$$T(\mathbf{x},0) = T_0(\mathbf{x}) \quad \text{on } C_0 \tag{8.2.25}$$

and a boundary condition on ∂C_0. We note that in Equation 8.2.24 $Q/\kappa = W/k$.

Similarly, a generalization of Equations 8.1.71, 8.1.72, and 8.1.79 leads to the *field equations of nonisothermal elastodynamics for a body subject to generalized plane stress conditions*:

the strain–displacement relations

$$\bar{E}_{\alpha\beta} = \frac{1}{2}\left(\bar{u}_{\alpha,\beta} + \bar{u}_{\beta,\alpha}\right) \tag{8.2.26}$$

the equation of motion

$$\bar{S}_{\alpha\beta,\beta} = \rho\ddot{\bar{u}}_{\alpha} \tag{8.2.27}$$

and the constitutive relations

$$\bar{S}_{\alpha\beta} = 2\mu\bar{E}_{\alpha\beta} + \bar{\lambda}\bar{E}_{\gamma\gamma}\delta_{\alpha\beta} - 2\mu\,\frac{3\lambda + 2\mu}{\lambda + 2\mu}\,\alpha\bar{T}\delta_{\alpha\beta} \tag{8.2.28}$$

The functions $\bar{u}_\alpha$, $\bar{E}_{\alpha\beta}$, and $\bar{S}_{\alpha\beta}$ are unknown fields on $C_0 \times [0, \infty)$, and $\bar{T}$ is a solution to the parabolic heat conduction equation

$$\nabla^2\bar{T} - \frac{1}{\kappa}\,\dot{\bar{T}} = -\frac{Q}{\kappa} \quad \text{on } C_0 \times [0, \infty) \tag{8.2.29}$$

subject to the initial condition

$$\bar{T}(\mathbf{x}, 0) = \bar{T}_0(\mathbf{x}) \quad \text{on } C_0 \tag{8.2.30}$$

and a boundary condition on ∂C_0.

Note that an equivalent form of the constitutive relations 8.2.23 reads, see Equation 8.1.65,

$$E_{\alpha\beta} = \frac{1}{2\mu}\,(S_{\alpha\beta} - \nu S_{\gamma\gamma}\delta_{\alpha\beta}) + (1 + \nu)\alpha T\delta_{\alpha\beta} \tag{8.2.31}$$

Also, note that an equivalent form of the constitutive relations 8.2.28 is, see Equation 8.1.84,

$$\bar{E}_{\alpha\beta} = \frac{1}{2\mu}\left(\bar{S}_{\alpha\beta} - \bar{\nu}\bar{S}_{\gamma\gamma}\delta_{\alpha\beta}\right) + \alpha\bar{T}\delta_{\alpha\beta} \tag{8.2.32}$$

Now, by eliminating $E_{\alpha\beta}$ from the Equations 8.2.21 through 8.2.23 we arrive at the following formulation of a *mixed problem of nonisothermal elastodynamics for an isotropic elastic body subject to plane strain conditions*:

Find a displacement $u_\alpha = u_\alpha(\mathbf{x}, t)$ and stress $S_{\alpha\beta} = S_{\alpha\beta}(\mathbf{x}, t)$ on $C_0 \times [0, \infty)$ that satisfy the field equations

$$S_{\alpha\beta,\beta} = \rho\ddot{u}_\alpha \quad \text{on } C_0 \times [0, \infty) \tag{8.2.33}$$

$$S_{\alpha\beta} = \mu(u_{\alpha,\beta} + u_{\beta,\alpha}) + \lambda u_{\gamma,\gamma}\delta_{\alpha\beta} - (3\lambda + 2\mu)\alpha T\delta_{\alpha\beta} \quad \text{on } C_0 \times [0, \infty) \tag{8.2.34}$$

subject to the initial conditions

$$u_\alpha(\mathbf{x}, 0) = u_\alpha^{(0)}, \quad \dot{u}_\alpha(\mathbf{x}, 0) = \dot{u}_\alpha^{(0)}, \quad \text{on } C_0 \tag{8.2.35}$$

and the mixed boundary conditions

$$u_\alpha = \widehat{u}_\alpha \quad \text{on } \partial C_0^{(1)} \times [0, \infty) \tag{8.2.36}$$

$$S_{\alpha\beta}n_\beta = \widehat{s}_\alpha \quad \text{on } \partial C_0^{(2)} \times [0, \infty) \tag{8.2.37}$$

Here $u_\alpha^{(0)}$, $\dot{u}_\alpha^{(0)}$, $\widehat{u}_\alpha$, $\widehat{s}_\alpha$, and T are prescribed functions.

Similarly, Equations 8.2.26 through 8.2.28 imply the formulation of *a mixed problem of nonisothermal elastodynamics for a body subject to generalized plane stress conditions*:

Find a displacement $\bar{u}_\alpha = \bar{u}_\alpha(\mathbf{x}, t)$ and stress $\bar{S}_{\alpha\beta} = \bar{S}_{\alpha\beta}(\mathbf{x}, t)$ on $C_0 \times [0, \infty)$ that satisfy the field equations

$$\bar{S}_{\alpha\beta,\beta} = \rho \ddot{\bar{u}}_\alpha \quad \text{on } C_0 \times [0, \infty) \tag{8.2.38}$$

$$\bar{S}_{\alpha\beta} = \mu(\bar{u}_{\alpha,\beta} + \bar{u}_{\beta,\alpha}) + \bar{\lambda}\bar{u}_{\gamma,\gamma}\delta_{\alpha\beta} - 2\mu\frac{3\lambda + 2\mu}{\lambda + 2\mu}\alpha\bar{T}\delta_{\alpha\beta} \quad \text{on } C_0 \times [0, \infty) \tag{8.2.39}$$

subject to the initial conditions

$$\bar{u}_\alpha(\mathbf{x}, 0) = u_\alpha^{(0)}, \quad \dot{\bar{u}}_\alpha(\mathbf{x}, 0) = \dot{u}_\alpha^{(0)}, \quad \text{on } C_0 \tag{8.2.40}$$

and the mixed boundary conditions

$$\bar{u}_\alpha = \widehat{u}_\alpha \quad \text{on } \partial C_0^{(1)} \times [0, \infty) \tag{8.2.41}$$

$$\bar{S}_{\alpha\beta}n_\beta = \widehat{s}_\alpha \quad \text{on } \partial C_0^{(2)} \times [0, \infty) \tag{8.2.42}$$

Here $u_\alpha^{(0)}$, $\dot{u}_\alpha^{(0)}$, $\widehat{u}_\alpha$, $\widehat{s}_\alpha$, and $\bar{T}$ are prescribed functions.

Example 8.2.5

Formulate the displacement initial–boundary value problem of nonisothermal elastodynamics for a body subject to plane strain conditions, in terms of the displacements.

Solution

We restrict the formulation described by Equations 8.2.33 through 8.2.37 to the case $\partial C_0^{(2)} = \emptyset$, eliminate $S_{\alpha\beta}$ from Equations 8.2.33 and 8.2.34, and arrive at the following problem.

Find a displacement field u_α on $C_0 \times [0, \infty)$ that satisfies the field equation

$$\mu u_{\alpha,\gamma\gamma} + (\lambda + \mu)u_{\gamma,\gamma\alpha} - \gamma T_{,\alpha} = \rho\ddot{u}_\alpha \quad \text{on } C_0 \times [0, \infty) \tag{a}$$

the initial conditions

$$u_\alpha(\mathbf{x}, 0) = u_\alpha^{(0)}, \quad \dot{u}_\alpha(\mathbf{x}, 0) = \dot{u}_\alpha^{(0)} \quad \text{on } C_0 \tag{b}$$

and the boundary condition

$$u_\alpha = \widehat{u}_\alpha \quad \text{on } \partial C_0 \times [0, \infty) \tag{c}$$

Here, see (c) in Example 8.1.5,

$$\gamma = (3\lambda + 2\mu)\alpha \tag{d}$$

and $u_\alpha^{(0)}$, $\dot{u}_\alpha^{(0)}$, $\widehat{u}_\alpha$, and T are prescribed functions. This completes the solution. □

Example 8.2.6

Formulate the displacement initial–boundary value problem of nonisothermal elastodynamics for a body subject to generalized plane stress conditions, in terms of the displacements.

Solution

Similarly as in the previous example, a restriction of the formulation 8.2.38 through 8.2.42 to the case $\partial C_0^{(2)} = \emptyset$ and elimination of $\bar{S}_{\alpha\beta}$ from Equations 8.2.38 and 8.2.39 lead to the following problem:

Find a displacement field $\bar{u}_\alpha$ on $C_0 \times [0, \infty)$ that satisfies the field equation

$$\mu \bar{u}_{\alpha,\gamma\gamma} + (\bar{\lambda} + \mu)\bar{u}_{\gamma,\gamma\alpha} - \bar{\gamma}\bar{T}_{,\alpha} = \rho \ddot{\bar{u}}_\alpha \quad \text{on } C_0 \times [0, \infty) \tag{a}$$

the initial conditions

$$\bar{u}_\alpha(\mathbf{x}, 0) = u_\alpha^{(0)}, \quad \dot{\bar{u}}_\alpha(\mathbf{x}, 0) = \dot{u}_\alpha^{(0)} \quad \text{on } C_0 \tag{b}$$

and the boundary condition

$$\bar{u}_\alpha = \widehat{u}_\alpha \quad \text{on } \partial C_0 \times [0, \infty) \tag{c}$$

where

$$\bar{\gamma} = 2\mu \, \frac{3\lambda + 2\mu}{\lambda + 2\mu} \, \alpha \tag{d}$$

and $u_\alpha^{(0)}$, $\dot{u}_\alpha^{(0)}$, $\widehat{u}_\alpha$, and $\bar{T}$ are prescribed functions. This completes the solution. □

Example 8.2.7

Formulate the traction initial–boundary value problem of nonisothermal elastodynamics for a body subject to plane strain conditions in terms of the stresses.

Solution

First we write Equations 8.2.21 through 8.2.23 in an alternative form

$$S_{\alpha\gamma,\gamma} = \rho \ddot{u}_\alpha \tag{a}$$

$$E_{\alpha\beta} = u_{(\alpha,\beta)} = \frac{1}{2\mu} \left(S_{\alpha\beta} - \nu S_{\gamma\gamma}\delta_{\alpha\beta} \right) + (1 + \nu)\alpha T \delta_{\alpha\beta} \tag{b}$$

By differentiating (a) with respect to x_β, we write

$$S_{\alpha\gamma,\gamma\beta} = \rho \ddot{u}_{\alpha,\beta} \tag{c}$$

and taking the symmetric part of this equation with respect to indexes α and β, we find

$$S_{(\alpha\gamma,\gamma\beta)} = \rho \ddot{u}_{(\alpha,\beta)} \tag{d}$$

Now, substituting (b) into (d), we obtain the stress equation of motion of nonisothermal elastodynamics

$$S_{(\alpha\gamma,\gamma\beta)} - \frac{\rho}{2\mu}\left(\ddot{S}_{\alpha\beta} - \nu\ddot{S}_{\gamma\gamma}\delta_{\alpha\beta}\right) = \rho(1+\nu)\alpha\ddot{T}\delta_{\alpha\beta} \tag{e}$$

Next, it follows from Equations 8.2.34 and 8.2.35 that

$$S_{\alpha\beta}(\mathbf{x},0) = S^{(0)}_{\alpha\beta}, \quad \dot{S}_{\alpha\beta}(\mathbf{x},0) = \dot{S}^{(0)}_{\alpha\beta} \tag{f}$$

where

$$S^{(0)}_{\alpha\beta} = \mu\left(u^{(0)}_{\alpha,\beta} + u^{(0)}_{\beta,\alpha}\right) + \lambda u^{(0)}_{\gamma,\gamma}\delta_{\alpha\beta} - (3\lambda + 2\mu)\alpha T^{(0)}\delta_{\alpha\beta} \tag{g}$$

and

$$\dot{S}^{(0)}_{\alpha\beta} = \mu\left(\dot{u}^{(0)}_{\alpha,\beta} + \dot{u}^{(0)}_{\beta,\alpha}\right) + \lambda \dot{u}^{(0)}_{\gamma,\gamma}\delta_{\alpha\beta} - (3\lambda + 2\mu)\alpha \dot{T}^{(0)}\delta_{\alpha\beta} \tag{h}$$

Here $T^{(0)}$ and $\dot{T}^{(0)}$ are the initial distributions of temperature T and temperature rate $\dot{T}$, respectively.

Hence, restricting the mixed problem described by Equations 8.2.33 through 8.2.37 to the case $\partial C_0^{(1)} = \emptyset$, and using (e) through (h), we arrive at the following initial–boundary value problem in terms of stresses $S_{\alpha\beta}$:

Find a stress field $S_{\alpha\beta}$ on $C_0 \times [0,\infty)$ that satisfies the equation

$$S_{(\alpha\gamma,\gamma\beta)} - \frac{\rho}{2\mu}\left(\ddot{S}_{\alpha\beta} - \nu\ddot{S}_{\gamma\gamma}\delta_{\alpha\beta}\right) - \rho(1+\nu)\alpha\ddot{T}\delta_{\alpha\beta} = 0 \quad \text{on } C_0 \times [0,\infty) \tag{i}$$

the initial conditions

$$S_{\alpha\beta}(\mathbf{x},0) = S^{(0)}_{\alpha\beta}, \quad \dot{S}_{\alpha\beta}(\mathbf{x},0) = \dot{S}^{(0)}_{\alpha\beta} \quad \text{on } C_0 \tag{j}$$

and the boundary condition

$$S_{\alpha\beta}n_\beta = \hat{s}_\alpha \quad \text{on } \partial C_0 \times [0,\infty) \tag{k}$$

where T, $S^{(0)}_{\alpha\beta}$, $\dot{S}^{(0)}_{\alpha\beta}$, and $\hat{s}_\alpha$ are prescribed functions. This completes the solution. □

Example 8.2.8

Formulate the traction initial–boundary value problem of nonisothermal elastodynamics for a body subject to generalized plane stress conditions in terms of the stresses.

Solution

The method leading to the formulation is similar to that of the previous example, and it is based upon a restriction of the mixed problem described by Equations 8.2.38 through 8.2.42 to the case $\partial C_0^{(1)} = \emptyset$. As a result we arrive at the formulation:

Find a stress field $\bar{S}_{\alpha\beta}$ on $C_0 \times [0,\infty)$ that satisfies the equation

$$\bar{S}_{(\alpha\gamma,\gamma\beta)} - \frac{\rho}{2\mu}\left(\ddot{\bar{S}}_{\alpha\beta} - \bar{\nu}\ddot{\bar{S}}_{\gamma\gamma}\right) - \rho\alpha\ddot{\bar{T}}\delta_{\alpha\beta} = 0 \quad \text{on } C_0 \times [0,\infty) \tag{a}$$

the initial conditions

$$\bar{S}_{\alpha\beta}(\mathbf{x},0) = \bar{S}^{(0)}_{\alpha\beta}, \quad \dot{\bar{S}}_{\alpha\beta}(\mathbf{x},0) = \dot{\bar{S}}^{(0)}_{\alpha\beta} \quad \text{on } C_0 \tag{b}$$

and the boundary condition

$$\bar{S}_{\alpha\beta} n_\beta = \widehat{\bar{s}}_\alpha \quad \text{on } \partial C_0 \times [0,\infty) \tag{c}$$

Here, $\bar{S}^{(0)}_{\alpha\beta}$ and $\dot{\bar{S}}^{(0)}_{\alpha\beta}$, $\widehat{\bar{s}}_\alpha$, and $\bar{T}$ are prescribed functions. In particular, see Equations 8.2.39 and 8.2.40,

$$\bar{S}^{(0)}_{\alpha\beta} = \mu\left(u^{(0)}_{\alpha,\beta} + u^{(0)}_{\beta,\alpha}\right) + \bar{\lambda} u^{(0)}_{\gamma,\gamma}\delta_{\alpha\beta} - 2\mu\frac{3\lambda+2\mu}{\lambda+2\mu}\,\alpha\bar{T}^{(0)}\delta_{\alpha\beta} \tag{d}$$

and

$$\dot{\bar{S}}^{(0)}_{\alpha\beta} = \mu\left(\dot{u}^{(0)}_{\alpha,\beta} + \dot{u}^{(0)}_{\beta,\alpha}\right) + \bar{\lambda} \dot{u}^{(0)}_{\gamma,\gamma}\delta_{\alpha\beta} - 2\mu\frac{3\lambda+2\mu}{\lambda+2\mu}\,\alpha\dot{\bar{T}}^{(0)}\delta_{\alpha\beta} \tag{e}$$

This completes the solution. □

Notes:

(1) If $S_{\alpha\beta}$ is a solution to the initial–boundary value problem described by (i) through (k) in Example 8.2.7, then the associated displacement field u_α and strain field $E_{\alpha\beta}$ may be generated from the formulas

$$u_\alpha(\mathbf{x},t) = \frac{1}{\rho}\left(i * S_{\alpha\beta,\beta} + \widehat{f}_\alpha\right) \tag{8.2.43}$$

$$E_{\alpha\beta}(\mathbf{x},t) = \frac{1}{2\mu}\left(S_{\alpha\beta} - \nu S_{\gamma\gamma}\delta_{\alpha\beta}\right) + (1+\nu)\alpha T\delta_{\alpha\beta} \tag{8.2.44}$$

where, see Equations 8.2.16 and 8.2.17,

$$i = i(t) = t \tag{8.2.45}$$

and

$$\widehat{f}_\alpha(\mathbf{x},t) = \rho\left(u^{(0)}_\alpha + t\dot{u}^{(0)}_\alpha\right) \tag{8.2.46}$$

(2) In the case of a nonisothermal dynamic generalized plane stress state when $\bar{S}_{\alpha\beta}$ is described by Equations (a) through (c) of Example 8.2.8, the appropriate formulas for the displacement $\bar{u}_\alpha$ and strain field $\bar{E}_{\alpha\beta}$ read

$$\bar{u}_\alpha(\mathbf{x},t) = \frac{1}{\rho}\left(i * \bar{S}_{\alpha\beta,\beta} + \widehat{f}_\alpha\right) \tag{8.2.47}$$

$$\bar{E}_{\alpha\beta}(\mathbf{x},t) = \frac{1}{2\mu}\left(\bar{S}_{\alpha\beta} - \bar{\nu}\bar{S}_{\gamma\gamma}\delta_{\alpha\beta}\right) + \alpha\bar{T}\delta_{\alpha\beta} \tag{8.2.48}$$

(3) If $S_{\alpha\beta}^{(0)}$ and $\dot{S}_{\alpha\beta}^{(0)}$ in the formulation (i) through (k) of Example 8.2.7 are arbitrary symmetric second-order tensor fields, the problem belongs to a class of problems of a nonisothermal incompatible elastodynamics.

Similarly, if $\bar{S}_{\alpha\beta}^{(0)}$ and $\dot{\bar{S}}_{\alpha\beta}^{(0)}$ in the pure stress initial–boundary value problem described by (a) through (c) of Example 8.2.8 are arbitrary symmetric second-order tensor fields, which are not related to $\left(u_\alpha^{(0)}, \bar{T}^{(0)}\right)$ and $\left(\dot{u}_\alpha^{(0)}, \dot{\bar{T}}^{(0)}\right)$ by (d) and (e) in that example, then the problem belongs also to a class of problems of a nonisothermal incompatible elastodynamics.

(4) The initial fields $T^{(0)}$ and $\dot{T}^{(0)}$ that occur in (g) and (h) in Example 8.2.7 are obtained from the formulation of the heat conduction problem 8.2.24 and 8.2.25. From 8.2.25 we obtain

$$T^{(0)} = T_0(\mathbf{x}) \quad \text{on } C_0 \tag{8.2.49}$$

and from 8.2.24 taken at $t = 0$, we get

$$\dot{T}^{(0)} = \kappa \nabla^2 T_0(\mathbf{x}) + Q_0 \quad \text{on } C_0 \tag{8.2.50}$$

where $T_0 = T_0(x)$ is a prescribed temperature on C_0 and $Q_0 = Q(\mathbf{x}, 0)$ is prescribed on C_0.

Similarly, the functions $\bar{T}^{(0)}$ and $\dot{\bar{T}}^{(0)}$ in (d) and (e) of Example 8.2.8 are computed from the formulas, see Equations 8.2.29 and 8.2.30,

$$\bar{T}^{(0)} = \bar{T}_0(\mathbf{x}) \quad \text{on } C_0 \tag{8.2.51}$$

$$\dot{\bar{T}}^{(0)} = \kappa \nabla^2 \bar{T}_0(\mathbf{x}) + Q_0 \quad \text{on } C_0 \tag{8.2.52}$$

Example 8.2.9

Show that a stress tensor $S_{\alpha\beta}$ of nonisothermal elastodynamics for a body subject to plane strain conditions satisfies the stress compatibility equation of the Beltrami–Michell type

$$\Box_2^2 S_{\alpha\beta} + S_{\gamma\gamma,\alpha\beta} + \left(\frac{1}{c_2^2} - \frac{1}{c_1^2}\right) \frac{\lambda \delta_{\alpha\beta}}{2(\lambda + \mu)} \ddot{S}_{\gamma\gamma} + 2\mu m \left(\Box_2^2 T\right) \delta_{\alpha\beta} = 0 \tag{a}$$

where

$$\Box_2^2 = \nabla^2 - \frac{1}{c_2^2} \frac{\partial^2}{\partial t^2} \tag{b}$$

$$\frac{1}{c_1^2} = \frac{\rho}{\lambda + 2\mu}, \quad \frac{1}{c_2^2} = \frac{\rho}{\mu} \tag{c}$$

and, see Example 4.1.8,

$$m = \frac{3\lambda + 2\mu}{\lambda + 2\mu} \alpha \tag{d}$$

Solution

For a body subject to plane strain conditions the displacement equation of motion takes the form, see Example 8.2.5,

$$\mu u_{\alpha,\gamma\gamma} + (\lambda + \mu) u_{\gamma,\gamma\alpha} - \gamma T_{,\alpha} = \rho \ddot{u}_\alpha \tag{e}$$

where

$$\gamma = (3\lambda + 2\mu)\alpha \tag{f}$$

By differentiating (e) with respect to x_β, and taking the symmetric part of the resulting equation with respect to indexes α and β, we obtain

$$\mu u_{(\alpha,\beta)\gamma\gamma} + (\lambda + \mu) u_{\gamma,\gamma\alpha\beta} - \gamma T_{,\alpha\beta} = \rho \ddot{u}_{(\alpha,\beta)} \tag{g}$$

Now, since

$$u_{(\alpha,\beta)} = E_{\alpha\beta} \tag{h}$$

then (g) can be rewritten as

$$2\mu E_{\alpha\beta,\gamma\gamma} + 2(\lambda + \mu) E_{\gamma\gamma,\alpha\beta} - 2\gamma T_{,\alpha\beta} = 2\rho \ddot{E}_{\alpha\beta} \tag{i}$$

Also, recall the strain–stress–temperature relation for a body subject to plane strain conditions, see Equation 8.1.62,

$$E_{\alpha\beta} = \frac{1}{2\mu}\left[S_{\alpha\beta} - \frac{\lambda}{2(\lambda+\mu)} S_{\gamma\gamma}\delta_{\alpha\beta}\right] + \frac{3\lambda + 2\mu}{2(\lambda+\mu)}\,\alpha T \delta_{\alpha\beta} \tag{j}$$

By taking the trace of (j), we get

$$2(\lambda + \mu) E_{\gamma\gamma} = S_{\gamma\gamma} + 2\gamma T \tag{k}$$

Substituting Equations (j) and (k) into Equation (i), we arrive at the tensor equation

$$\begin{aligned} S_{\alpha\beta,\gamma\gamma} &+ S_{\gamma\gamma,\alpha\beta} - \frac{\lambda}{2(\lambda+\mu)} S_{\gamma\gamma,\delta\delta}\delta_{\alpha\beta} + \mu\,\frac{3\lambda+2\mu}{\lambda+\mu}\,\alpha T_{,\gamma\gamma}\delta_{\alpha\beta} \\ &= \frac{1}{c_2^2}\left[\ddot{S}_{\alpha\beta} - \frac{\lambda}{2(\lambda+\mu)}\ddot{S}_{\gamma\gamma}\delta_{\alpha\beta}\right] + \mu\,\frac{3\lambda+2\mu}{\lambda+\mu}\,\alpha\,\frac{1}{c_2^2}\,\ddot{T}\delta_{\alpha\beta} \end{aligned} \tag{l}$$

Next, by taking the trace of Equation (l), we obtain

$$\Box_1^2\, S_{\alpha\alpha} + 2\mu m\, \Box_2^2\, T = 0 \tag{m}$$

where m is given by Equation (d), and

$$\Box_1^2 = \nabla^2 - \frac{1}{c_1^2}\frac{\partial^2}{\partial t^2} \tag{n}$$

An alternative form of Equation (m) reads

$$S_{\gamma\gamma,\delta\delta} = \frac{1}{c_1^2}\ddot{S}_{\gamma\gamma} - 2\mu m\, \Box_2^2\, T \tag{o}$$

Now, if we substitute Equation (o) into Equation (l), we arrive at Equation (a).
This completes the solution. □

Example 8.2.10

Show that a stress tensor $S_{\alpha\beta}$ from the previous example also satisfies the equation

$$\Box_2^2\, \{\Box_1^2\, S_{\alpha\beta} - m\, [2\mu(T_{,\alpha\beta} - T_{,\gamma\gamma}\delta_{\alpha\beta}) + \rho\delta_{\alpha\beta}\ddot{T}]\} = 0 \tag{a}$$

Solution

We apply the operator $\Box_1^2$ to Equation (a) of Example 8.2.9, use the relation (m) from Example 8.2.9, and obtain

$$\Box_1^2\Box_2^2\, S_{\alpha\beta} - 2\mu m\, \Box_2^2\, T_{,\alpha\beta} + \left(\frac{1}{c_2^2} - \frac{1}{c_1^2}\right)\frac{\lambda\delta_{\alpha\beta}}{2(\lambda+\mu)}(-2\mu m)\, \Box_2^2\, \ddot{T}$$

$$+\,2\mu m\, \Box_1^2\Box_2^2\, T\delta_{\alpha\beta} = 0 \tag{b}$$

This may be reduced to the form

$$\Box_2^2\left\{\Box_1^2\, S_{\alpha\beta} + 2\mu m\left(\Box_1^2\, T\delta_{\alpha\beta} - T_{,\alpha\beta} - \frac{\lambda}{2\mu}\frac{1}{c_1^2}\,\ddot{T}\delta_{\alpha\beta}\right)\right\} = 0 \tag{c}$$

which is equivalent to Equation (a).
This completes the solution. □

Example 8.2.11

Show that a stress tensor $\bar{S}_{\alpha\beta}$ of nonisothermal elastodynamics for a body subject to generalized plane stress conditions satisfies the stress compatibility equation of the Beltrami–Michell type:

$$\Box_2^2\, \bar{S}_{\alpha\beta} + \bar{S}_{\gamma\gamma,\alpha\beta} + \left(\frac{1}{c_2^2} - \frac{1}{\bar{c}_1^2}\right)\frac{\lambda}{3\lambda+2\mu}\,\ddot{\bar{S}}_{\gamma\gamma}\delta_{\alpha\beta} + \frac{3\lambda+2\mu}{\lambda+\mu}\,\mu\alpha\Box_2^2\, \bar{T}\delta_{\alpha\beta} = 0 \tag{a}$$

where

$$\frac{1}{\bar{c}_1^2} = \frac{\rho}{\bar{\lambda}+2\mu} = \frac{\rho}{4\mu}\,\frac{\lambda+2\mu}{\lambda+\mu} \tag{b}$$

Solution

We recall the displacement–temperature equation of motion for a body subject to generalized plane stress conditions, see Equation (a) in Example 8.2.6,

$$\mu \bar{u}_{\alpha,\gamma\gamma} + (\bar{\lambda} + \mu)\bar{u}_{\gamma,\gamma\alpha} - \bar{\gamma}\bar{T}_{,\alpha} = \rho \ddot{\bar{u}}_{\alpha} \tag{c}$$

where

$$\bar{\lambda} = \frac{2\mu\lambda}{\lambda + 2\mu}, \quad \bar{\gamma} = 2\mu \frac{3\lambda + 2\mu}{\lambda + 2\mu} \alpha \tag{d}$$

Using the strain–displacement relation

$$\bar{E}_{\alpha\beta} = \bar{u}_{(\alpha,\beta)} \tag{e}$$

we find that Equation (c) implies the equation

$$2\mu \bar{E}_{\alpha\beta,\gamma\gamma} + 2\left(\bar{\lambda} + \mu\right)\bar{E}_{\gamma\gamma,\alpha\beta} - 2\bar{\gamma}\bar{T}_{,\alpha\beta} = 2\rho \ddot{\bar{E}}_{\alpha\beta} \tag{f}$$

Next, we recall the strain–stress–temperature relation in the form, see Equation 8.2.32,

$$\bar{E}_{\alpha\beta} = \frac{1}{2\mu}\left(\bar{S}_{\alpha\beta} - \bar{\nu}\bar{S}_{\gamma\gamma}\delta_{\alpha\beta}\right) + \alpha\bar{T}\delta_{\alpha\beta} \tag{g}$$

where

$$\bar{\nu} = \frac{\nu}{1+\nu} = \frac{\lambda}{3\lambda + 2\mu} \tag{h}$$

Hence, an alternative form of Equation (g) reads

$$\bar{E}_{\alpha\beta} = \frac{1}{2\mu}\left(\bar{S}_{\alpha\beta} - \frac{\lambda}{3\lambda + 2\mu}\bar{S}_{\gamma\gamma}\delta_{\alpha\beta}\right) + \alpha\bar{T}\delta_{\alpha\beta} \tag{i}$$

and it follows from Equation (i) that

$$\bar{E}_{\gamma\gamma} = \frac{1}{2\mu}\frac{\lambda + 2\mu}{3\lambda + 2\mu}\bar{S}_{\gamma\gamma} + 2\alpha\bar{T} \tag{j}$$

Now, substituting Equations (i) and (j) into Equation (f), we arrive at

$$\begin{aligned} &\bar{S}_{\alpha\beta,\gamma\gamma} - \frac{\lambda}{3\lambda + 2\mu}\bar{S}_{\delta\delta,\gamma\gamma}\delta_{\alpha\beta} + \bar{S}_{\gamma\gamma,\alpha\beta} \\ &- \frac{1}{c_2^2}\left(\ddot{\bar{S}}_{\alpha\beta} - \frac{\lambda}{3\lambda + 2\mu}\ddot{\bar{S}}_{\gamma\gamma}\delta_{\alpha\beta}\right) + 2\mu\alpha \, \Box_2^2 \, \bar{T}\delta_{\alpha\beta} = 0 \end{aligned} \tag{k}$$

Taking the trace of this equation, we obtain

$$\bar{\Box}_1^2 \, \bar{S}_{\alpha\alpha} + \frac{3\lambda + 2\mu}{\lambda + \mu}\mu\alpha \, \Box_2^2 \, \bar{T} = 0 \tag{l}$$

where

$$\bar{\square}_1^2 = \nabla^2 - \frac{1}{\bar{c}_1^2}\frac{\partial^2}{\partial t^2} \tag{m}$$

Next, we rewrite Equation (l) in the form

$$\bar{S}_{\delta\delta,\gamma\gamma} = \frac{1}{\bar{c}_1^2}\ddot{\bar{S}}_{\delta\delta} - \frac{3\lambda + 2\mu}{\lambda + \mu}\mu\alpha\,\square_2^2\,\bar{T} \tag{n}$$

Finally, substituting Equation (n) into Equation (k) we arrive at Equation (a).
This completes the solution. □

Example 8.2.12

Show that a stress tensor $\bar{S}_{\alpha\beta}$ from the previous example also satisfies the equation

$$\square_2^2\left\{\bar{\square}_1^2\,\bar{S}_{\alpha\beta} - \bar{m}\left[2\mu(\bar{T}_{,\alpha\beta} - \bar{T}_{,\gamma\gamma}\delta_{\alpha\beta}) + \rho\delta_{\alpha\beta}\ddot{\bar{T}}\right]\right\} = 0 \tag{a}$$

where

$$\bar{m} = \frac{3\lambda + 2\mu}{2(\lambda + \mu)}\alpha \tag{b}$$

Solution

We apply the operator $\bar{\square}_1^2$ to Equation (a) in Example 8.2.11, use the relation (l) in Example 8.2.11, and obtain

$$\bar{\square}_1^2\,\square_2^2\,\bar{S}_{\alpha\beta} - 2\mu\bar{m}\,\square_2^2\,\bar{T}_{,\alpha\beta} + \left(\frac{1}{c_2^2} - \frac{1}{\bar{c}_1^2}\right)\frac{\lambda\delta_{\alpha\beta}}{3\lambda + 2\mu}(-2\mu\bar{m})\,\square_2^2\,\ddot{\bar{T}}$$
$$+ 2\mu\bar{m}\,\bar{\square}_1^2\,\square_2^2\,\bar{T}\delta_{\alpha\beta} = 0 \tag{c}$$

This may be reduced to the form

$$\square_2^2\left\{\bar{\square}_1^2\,\bar{S}_{\alpha\beta} - 2\mu\bar{m}\left[\bar{T}_{,\alpha\beta} - \bar{\square}_1^2\,\bar{T}\delta_{\alpha\beta} + \left(\frac{1}{c_2^2} - \frac{1}{\bar{c}_1^2}\right)\frac{\lambda\delta_{\alpha\beta}}{3\lambda + 2\mu}\ddot{\bar{T}}\right]\right\} = 0 \tag{d}$$

Finally, by taking into account the relation

$$\frac{1}{\bar{c}_1^2} + \left(\frac{1}{c_2^2} - \frac{1}{\bar{c}_1^2}\right)\frac{\lambda}{3\lambda + 2\mu} = \frac{\rho}{2\mu} \tag{e}$$

we reduce Equation (d) to Equation (a).
This completes the solution. □

Example 8.2.13

Show that:

(A) A particular solution to Equation (a) of Example 8.2.10 takes the form

$$S_{\alpha\beta} = 2\mu\left(\phi_{,\alpha\beta} - \phi_{,\gamma\gamma}\delta_{\alpha\beta}\right) + \rho\delta_{\alpha\beta}\ddot{\phi} \tag{a}$$

where ϕ satisfies the equation

$$\Box_1^2\phi = mT,\ \Box_1^2 = \nabla^2 - \frac{1}{c_1^2}\frac{\partial^2}{\partial t^2} \tag{b}$$

(B) A particular solution to Equation (a) of Example 8.2.12 is given by

$$\bar{S}_{\alpha\beta} = 2\mu\left(\bar{\phi}_{,\alpha\beta} - \bar{\phi}_{,\gamma\gamma}\delta_{\alpha\beta}\right) + \rho\delta_{\alpha\beta}\ddot{\bar{\phi}} \tag{c}$$

where $\overline{\phi}$ satisfies the equation

$$\bar{\Box}_1^2\,\bar{\phi} = \bar{m}\bar{T},\quad \bar{\Box}_1^2 = \nabla^2 - \frac{1}{\bar{c}_1^2}\frac{\partial^2}{\partial t^2} \tag{d}$$

Solution

Part (A). Applying the operator $\Box_1^2$ to Equation (a) and using Equation (b), we obtain

$$\Box_1^2 S_{\alpha\beta} = m\left[2\mu\left(T_{,\alpha\beta} - T_{,\gamma\gamma}\delta_{\alpha\beta}\right) + \rho\delta_{\alpha\beta}\ddot{T}\right] \tag{e}$$

Next, applying the operator $\Box_2^2$ to Equation (e) we arrive at Equation (a) of Example 8.2.10.

Part (B). Proof of this part is similar to that of Part A. By applying $\bar{\Box}_1^2$ to (c) and using (d), we get

$$\bar{\Box}_1^2\,\bar{S}_{\alpha\beta} = \bar{m}\left[2\mu\left(\bar{T}_{,\alpha\beta} - \bar{T}_{,\gamma\gamma}\delta_{\alpha\beta}\right) + \rho\delta_{\alpha\beta}\ddot{\bar{T}}\right] \tag{f}$$

Finally, applying $\Box_2^2$ to (f) we find that $\overline{S}_{\alpha\beta}$ satisfies Equation (a) of Example 8.2.12. □

Example 8.2.14

Show that $S_{\alpha\beta}$ and $\bar{S}_{\alpha\beta}$ in the previous example satisfy the stress–temperature equations of motion: Equation (i) of Example 8.2.7 and Equation (a) of Example 8.2.8, respectively.

Solution

We recall Equation (i) of Example 8.2.7 in the form

$$L_{\alpha\beta} \equiv S_{(\alpha\gamma,\gamma\beta)} - \frac{\rho}{2\mu}\left(\ddot{S}_{\alpha\beta} - \nu\ddot{S}_{\gamma\gamma}\delta_{\alpha\beta}\right) - \rho(1+\nu)\alpha\ddot{T}\delta_{\alpha\beta} = 0 \tag{a}$$

Since

$$S_{\alpha\beta} = 2\mu\left(\phi_{,\alpha\beta} - \phi_{,\gamma\gamma}\delta_{\alpha\beta}\right) + \delta_{\alpha\beta}\ddot{\phi} \tag{b}$$

therefore

$$S_{\alpha\gamma,\gamma} = \rho\ddot{\phi}_{,\alpha}, \quad S_{(\alpha\gamma,\gamma\beta)} = \rho\ddot{\phi}_{,\alpha\beta}, \quad S_{\gamma\gamma} = -2\left(\mu\phi_{,\gamma\gamma} - \rho\ddot{\phi}\right) \tag{c}$$

Also note that

$$\nu = \frac{\lambda}{2(\lambda+\mu)} \tag{d}$$

so $L_{\alpha\beta}$ in (a) may be written as

$$L_{\alpha\beta} \equiv S_{(\alpha\gamma,\gamma\beta)} - \frac{\rho}{2\mu}\left[\ddot{S}_{\alpha\beta} - \frac{\lambda}{2(\lambda+\mu)}\ddot{S}_{\gamma\gamma}\delta_{\alpha\beta}\right] - \rho\frac{3\lambda+2\mu}{2(\lambda+\mu)}\alpha\ddot{T}\delta_{\alpha\beta} \tag{e}$$

Substituting (b) and (c) into (e), we obtain

$$\begin{aligned} L_{\alpha\beta} = \rho\ddot{\phi}_{,\alpha\beta} &- \frac{\rho}{2\mu}\left[2\mu(\ddot{\phi}_{,\alpha\beta} - \ddot{\phi}_{,\gamma\gamma}\delta_{\alpha\beta}) + \rho\delta_{\alpha\beta}\phi^{(4)}\right] \\ &- \frac{\rho}{2\mu}\cdot\frac{2\lambda}{2(\lambda+\mu)}\left(\mu\ddot{\phi}_{,\gamma\gamma} - \rho\phi^{(4)}\right)\delta_{\alpha\beta} - \rho\frac{3\lambda+2\mu}{2(\lambda+\mu)}\alpha\ddot{T}\delta_{\alpha\beta} \end{aligned} \tag{f}$$

Hence

$$L_{\alpha\beta} = \rho\delta_{\alpha\beta}\frac{\partial^2}{\partial t^2}\left[\frac{\lambda+2\mu}{2(\lambda+\mu)}\phi_{,\gamma\gamma} - \frac{\rho}{2(\lambda+\mu)}\ddot{\phi} - \frac{3\lambda+2\mu}{2(\lambda+\mu)}\alpha T\right] \tag{g}$$

or

$$L_{\alpha\beta} = \rho\frac{\lambda+2\mu}{2(\lambda+\mu)}\delta_{\alpha\beta}\frac{\partial^2}{\partial t^2}\left(\Box_1^2\phi - mT\right) \tag{h}$$

and by virtue of (b) of the previous example we get

$$L_{\alpha\beta} = 0 \tag{i}$$

which means that $S_{\alpha\beta}$ of the previous example satisfies Equation (i) of Example 8.2.7.

In a similar way we show that $\bar{S}_{\alpha\beta}$ given by Equations (c) and (d) of the previous example satisfies Equation (a) of Example 8.2.8.

This completes the solution. □

Note:

If the two-dimensional domain C_0 of a plane strain problem of the nonisothermal isotropic homogeneous elastodynamics is described by the inequalities

$$0 \le x_1 \le h, \quad |x_2| < \infty \quad (h = \text{const.}) \tag{8.2.53}$$

and the external thermomechanical load is independent of x_2 for every $t \geq 0$, then the problem becomes one-dimensional in which the fields **u**, **E**, and **S** are represented by $(x = x_1)$

$$\mathbf{u} = [u_1(x,t), 0, 0] \tag{8.2.54}$$

$$\mathbf{E} = \begin{bmatrix} E_{11} & 0 & 0 \\ 0 & 0 & 0 \\ 0 & 0 & 0 \end{bmatrix} \tag{8.2.55}$$

and

$$\mathbf{S} = \begin{bmatrix} S_{11} & 0 & 0 \\ 0 & S_{22} & 0 \\ 0 & 0 & S_{33} \end{bmatrix} \tag{8.2.56}$$

where

$$E_{11} = u_{1,1} \tag{8.2.57}$$

$$\left.\begin{aligned} S_{11} &= (\lambda + 2\mu)E_{11} - (3\lambda + 2\mu)\alpha T \\ S_{22} &= S_{23} = \lambda E_{11} - (3\lambda + 2\mu)\alpha T \end{aligned}\right\} \tag{8.2.58}$$

Letting

$$u_1 = u(x,t), \quad S_{11} = S(x,t) \tag{8.2.59}$$

the following one-dimensional initial–boundary value problem may be formulated: Find a pair (u, S) on $[0, h] \times [0, \infty)$ that satisfies the field equations

$$\begin{aligned} &S_{,x} - \rho \ddot{u} = 0 \\ &S = (\lambda + 2\mu)u_{,x} - (3\lambda + 2\mu)\alpha T \quad \text{on } [0, h] \times [0, \infty) \end{aligned} \tag{8.2.60}$$

subject to the initial conditions

$$u(x,0) = u_0(x), \quad \dot{u}(x,0) = \dot{u}_0(x) \quad \text{on } [0, h] \tag{8.2.61}$$

and the boundary conditions

$$S(0,t) = -p(t), \quad S(h,t) = q(t) \quad t \geq 0 \tag{8.2.62}$$

where $T = T(x,t)$, $u_0(x)$, $\dot{u}_0(x)$, $p(t)$, and $q(t)$ are prescribed functions.

The function $T = T(x,t)$ is obtained by solving the one-dimensional parabolic equation

$$T_{,xx} - \frac{1}{\kappa}\dot{T} = -\frac{Q}{\kappa} \quad \text{on } [0, h] \times [0, \infty) \tag{8.2.63}$$

subject to the initial condition

$$T(x,0) = T_0(x) \quad \text{on } [0,h] \tag{8.2.64}$$

and the boundary conditions

$$T(0,t) = f_1(t), \quad T(h,t) = f_2(t) \quad t \geq 0 \tag{8.2.65}$$

Here, $Q = Q(x,t)$, $T_0(x), f_1(t)$, and $f_2(t)$ are prescribed functions.

By eliminating u from the field equations 8.2.60 we arrive at the nonhomogeneous stress equation of motion

$$\left(\frac{\partial^2}{\partial x^2} - \frac{1}{c_1^2}\frac{\partial^2}{\partial t^2}\right) S = \rho m \ddot{T} \quad \text{on } [0,h] \times [0,\infty) \tag{8.2.66}$$

As a result, the following problem may be formulated: Find a stress field $S = S(x,t)$ on $[0,h] \times [0,\infty)$ that satisfies the field equation

$$\left(\frac{\partial^2}{\partial x^2} - \frac{1}{c_1^2}\frac{\partial^2}{\partial t^2}\right) S = \rho m \ddot{T} \quad \text{on } [0,h] \times [0,\infty) \tag{8.2.67}$$

subject to the initial conditions

$$S(x,0) = S_0(x), \quad \dot{S}(x,0) = \dot{S}_0(x) \quad \text{on } [0,h] \tag{8.2.68}$$

and the boundary conditions, (see 8.2.62),

$$S(0,t) = -p(t), \quad S(h,t) = q(t) \quad t \geq 0 \tag{8.2.69}$$

where

$$\begin{aligned} S_0(x) &= (\lambda + 2\mu) u_{0,x} - (3\lambda + 2\mu)\alpha T_0 \\ \dot{S}_0(x) &= (\lambda + 2\mu) \dot{u}_{0,x} - (3\lambda + 2\mu)\alpha \dot{T}_0 \\ \dot{T}_0(x) &= Q(x,0) + \kappa T_{0,xx} \end{aligned} \tag{8.2.70}$$

If the external thermomechanical load of the problem 8.2.67 through 8.2.69 satisfies the conditions

$$\begin{aligned} &T_0(x) = u_0(x) = \dot{u}_0(x) = Q(x,0) = 0, \quad x \in [0,h] \\ &f_1(t) = f_0 H(t), \quad f_2(t) = p(t) = q(t) = 0, \quad t > 0 \end{aligned} \tag{8.2.71}$$

where $H = H(t)$ is the Heaviside function defined in Equation 2.3.53, and f_0 is a constant, then the problem 8.2.67 through 8.2.69 reduces to a thermoelastic shock problem for the layer $0 \leq x \leq h$. For $h \to \infty$, a solution to the problem describes thermal stresses propagating in the semispace $x \geq 0$ due to a sudden heating of its boundary $x = 0$.

PROBLEMS

8.1 The displacement equations of elastostatics for a body subject to plane strain conditions take the form (see Example 8.1.1)

$$\mu u_{\alpha,\gamma\gamma} + (\lambda + \mu)u_{\gamma,\gamma\alpha} + b_\alpha = 0 \tag{a}$$

where u_α is the displacement corresponding to the body force $b_\alpha(\alpha, \gamma = 1, 2)$. Let

$$u_\alpha = \psi_\alpha - \frac{1}{4(1-\nu)}(x_\gamma \psi_\gamma + \varphi)_{,\alpha} \tag{b}$$

where

$$\psi_{\alpha,\gamma\gamma} = -\frac{b_\alpha}{\mu}, \quad \varphi_{,\gamma\gamma} = \frac{(x_\gamma b_\gamma)}{\mu} \tag{c}$$

Show that u_α defined by Equations (b) and (c) satisfies Equation (a).

8.2 The displacement equations of equilibrium for a body subject to generalized plane stress conditions take the form (see Example 8.1.2)

$$\mu \overline{u}_{\alpha,\gamma\gamma} + (\overline{\lambda} + \mu)\overline{u}_{\gamma,\gamma\alpha} + \overline{b}_\alpha = 0 \tag{a}$$

where $\overline{u}_\alpha$ is the displacement corresponding to the body force $\overline{b}_\alpha(\alpha, \gamma = 1, 2)$, and

$$\overline{\lambda} = \frac{2\mu}{\lambda + 2\mu}\lambda \tag{b}$$

Let

$$\overline{u}_\alpha = \overline{\psi}_\alpha - \frac{1}{4(1-\overline{\nu})}(x_\gamma \overline{\psi}_\gamma + \overline{\varphi})_{,\alpha} \tag{c}$$

where

$$\overline{\psi}_{\alpha,\gamma\gamma} = -\frac{\overline{b}_\alpha}{\mu}, \quad \overline{\varphi}_{,\gamma\gamma} = \frac{(x_\gamma \overline{b}_\gamma)}{\mu} \tag{d}$$

and

$$\overline{\nu} = \frac{\nu}{1+\nu} \tag{e}$$

Show that $\overline{u}_\alpha$ defined by Equations (c) through (e) satisfies the equilibrium equation (a).

8.3 The displacement equations of thermoelastostatics for a body under plane strain conditions subject to a temperature change $T = T(\mathbf{x})$ take the form (see Example 8.1.5)

$$u_{\alpha,\gamma\gamma} + \frac{1}{1-2\nu}u_{\gamma,\gamma\alpha} - \frac{\gamma}{\mu}T_{,\alpha} = 0 \tag{a}$$

where

$$\gamma = 2\mu \frac{1+\nu}{1-2\nu}\alpha \tag{b}$$

Let

$$u_\alpha = \nabla^2 g_\alpha - \frac{1}{2(1-\nu)} g_{\gamma,\gamma\alpha} \tag{c}$$

where

$$\nabla^2\nabla^2 g_\alpha = \frac{\gamma}{\mu} T_{,\alpha} \tag{d}$$

Show that u_α given by Equations (c) and (d) satisfies Equation (a).

8.4 The displacement equations of thermoelastostatics for a body under generalized plane stress conditions subject to a temperature change $\overline{T}=\overline{T}(\mathbf{x})$ take the form (see Example 8.1.6)

$$\overline{u}_{\alpha,\gamma\gamma} + \frac{1}{1-2\overline{\nu}}\overline{u}_{\gamma,\gamma\alpha} - \frac{\overline{\gamma}}{\mu}\overline{T}_{,\alpha} = 0 \tag{a}$$

where

$$\overline{\nu} = \frac{\nu}{1+\nu}, \quad \overline{\gamma} = 2\mu\frac{1+\nu}{1-\nu}\alpha \tag{b}$$

Let

$$\overline{u}_\alpha = \nabla^2 \overline{g}_\alpha - \frac{1}{2(1-\overline{\nu})}\overline{g}_{\gamma,\gamma\alpha} \tag{c}$$

where

$$\nabla^2\nabla^2\overline{g}_\alpha = \frac{\overline{\gamma}}{\mu}\overline{T}_{,\alpha} \tag{d}$$

Show that $\overline{u}_\alpha$ given by Equations (c) and (d) satisfies Equation (a).

8.5 The displacement equations of isothermal elastodynamics for a body subject to plane strain conditions take the form (see Example 8.2.14)

$$\mu u_{\alpha,\gamma\gamma} + (\lambda+\mu)u_{\gamma,\gamma\alpha} + b_\alpha = \rho\ddot{u}_\alpha \tag{a}$$

Let

$$b_\alpha = -h_{,\alpha} - \epsilon_{\alpha\beta 3}k_{,\beta} \tag{b}$$

where $h = h(\mathbf{x})$ and $k = k(\mathbf{x})$ are prescribed scalar fields.

Let

$$u_\alpha = \varphi_{,\alpha} + \epsilon_{\alpha\beta 3}\psi_{,\beta} \tag{c}$$

where φ and ψ satisfy, respectively, the equations

$$\Box_1^2\varphi = \frac{h}{\lambda + 2\mu} \tag{d}$$

and

$$\Box_2^2\psi = \frac{k}{\mu} \tag{e}$$

Show that u_α given by Equations (c) through (e) satisfies Equation (a).

8.6 The displacement equations of isothermal elastodynamics for a body subject to generalized plane stress conditions take the form (see Example 8.2.2)

$$\mu\overline{u}_{\alpha,\gamma\gamma} + \left(\overline{\lambda} + \mu\right)\overline{u}_{\gamma,\gamma\alpha} + \overline{b}_\alpha = \rho\ddot{\overline{u}}_\alpha \tag{a}$$

Let

$$\overline{b}_\alpha = -\overline{h}_{,\alpha} - \epsilon_{\alpha\beta 3}\overline{k}_{,\beta} \tag{b}$$

where $\overline{h} = \overline{h}(\mathbf{x})$ and $\overline{k} = \overline{k}(\mathbf{x})$ are prescribed fields.

Let

$$\overline{u}_\alpha = \overline{\varphi}_{,\alpha} + \epsilon_{\alpha\beta 3}\overline{\psi}_{,\beta} \tag{c}$$

where $\overline{\varphi}$ and $\overline{\psi}$ satisfy, respectively, the equations

$$\overline{\Box}_1^2\overline{\varphi} = \frac{\overline{h}}{\overline{\lambda} + 2\mu} \tag{d}$$

and

$$\Box_2^2\,\overline{\psi} = \frac{\overline{k}}{\mu} \tag{e}$$

Here, see Equation 8.1.40,

$$\overline{\Box}_1^2 = \nabla^2 - \frac{1}{\overline{c}_1^2}\frac{\partial^2}{\partial t^2}, \quad \frac{1}{\overline{c}_1^2} = \frac{\rho}{\overline{\lambda} + 2\mu} \tag{f}$$

Show that $\overline{u}_\alpha$ given by Equations (c) through (f) satisfies Equation (a).

8.7 The displacement equations of isothermal elastodynamics for a body subject to plane strain conditions take the form (see Problem 8.5)

$$\mu u_{\alpha,\gamma\gamma} + (\lambda + \mu)u_{\gamma,\gamma\alpha} + b_\alpha = \rho\ddot{u}_\alpha \tag{a}$$

Let u_α be a vector field defined by

$$u_\alpha = \Box_1^2\, g_\alpha + \left(\frac{c_2^2}{c_1^2} - 1\right) g_{\gamma,\gamma\alpha} \tag{b}$$

where

$$\Box_1^2 \Box_2^2\, g_\alpha = -\frac{b_\alpha}{\mu} \tag{c}$$

Show that u_α given by Equations (b) and (c) satisfies Equation (a).

8.8 The stress equations of elastodynamics for a body subject to plane strain conditions take the form (see Example 8.2.3)

$$S_{(\alpha\gamma,\gamma\beta)} - \frac{\rho}{2\mu}\left(\ddot{S}_{\alpha\beta} - \nu \ddot{S}_{\gamma\gamma}\delta_{\alpha\beta}\right) = -b_{(\alpha,\beta)} \tag{a}$$

Let $S_{\alpha\beta}$ be a tensor field defined by

$$S_{\alpha\beta} = 2\mu\left[\Box_1^2 \chi_{\alpha\beta} - \frac{1}{2(1-\nu)}\left(\chi_{\gamma\gamma,\alpha\beta} - \nu\delta_{\alpha\beta}\Box_2^2\chi_{\gamma\gamma}\right)\right] \tag{b}$$

where $\chi_{\alpha\beta}$ is a symmetric second-order tensor field that satisfies the equations

$$\Box_1^2 \Box_2^2\, \chi_{\alpha\beta} = -\frac{b_{(\alpha,\beta)}}{\mu} \tag{c}$$

and

$$\chi_{\alpha\beta,\gamma\gamma} + \chi_{\gamma\gamma,\alpha\beta} - 2\chi_{(\alpha\gamma,\gamma\beta)} = 0 \tag{d}$$

Show that $S_{\alpha\beta}$ given by Equations (b) through (d) satisfies the tensorial equation (a).

Note: If $\chi_{\alpha\beta} = g_{(\alpha,\beta)}$, where g_α is the vector field of Galerkin type from Problem 8.7, then $\chi_{\alpha\beta}$ satisfies Equation (d) identically, and $S_{\alpha\beta}$ is the stress tensor corresponding to the displacement vector u_α of Problem 8.7.

8.9 Let $S_{\alpha\beta}$ be a solution to the stress equations of elastodynamics (see Problem 8.8)

$$S_{(\alpha\gamma,\gamma\beta)} - \frac{\rho}{2\mu}\left(\ddot{S}_{\alpha\beta} - \nu\, \ddot{S}_{\gamma\gamma}\delta_{\alpha\beta}\right) = -b_{(\alpha,\beta)} \tag{a}$$

subject to the homogeneous initial conditions

$$S_{\alpha\beta}(\mathbf{x},0) = 0, \quad \dot{S}_{\alpha\beta}(\mathbf{x},0) = 0 \tag{b}$$

Show that there is a second-order symmetric tensor field $\chi_{\alpha\beta}$ that satisfies Equations (b) through (d) of Problem 8.8, that is, the stress representation of Problem 8.8 is complete.

8.10 A homogeneous isotropic infinite elastic body under plane strain conditions and initially at rest is subject to a temperature change $T = T(\mathbf{x}, t)$ for every $(\mathbf{x}, t) \in E^2 \times [0, \infty)$. Show that the dynamic thermal stresses $S_{\alpha\beta} = S_{\alpha\beta}(\mathbf{x}, t)$ corresponding to the temperature $T = T(\mathbf{x}, t)$ are represented by the formulas (see Examples 8.1.3 and 8.2.14)

$$S_{\alpha\beta} = 2\mu\left(\phi_{,\alpha\beta} - \phi_{,\gamma\gamma}\delta_{\alpha\beta}\right) + \rho\ddot{\phi}\delta_{\alpha\beta} \tag{a}$$

where

$$\phi(\mathbf{x}, t) = -\frac{mc_1}{2\pi}\int_0^t d\tau \int_{|\mathbf{x}-\boldsymbol{\xi}|<c_1\tau} \frac{T(\boldsymbol{\xi}, t-\tau)da(\boldsymbol{\xi})}{\sqrt{c_1^2\tau^2 - |\mathbf{x}-\boldsymbol{\xi}|^2}} \tag{b}$$

and

$$m = \frac{1+\nu}{1-\nu}\alpha, \quad \frac{1}{c_1} = \sqrt{\frac{\rho}{\lambda + 2\mu}} \tag{8.2.72}$$

REFERENCES

1. A. S. Saada, *Elasticity—Theory and Applications,* Pergamon Press, New York, 1974, or Krieger, Malabar, 1987, pp. 229–230.
2. N. Noda, R. B. Hetnarski, and Y. Tanigawa, *Thermal Stresses*, 2nd edn., Chap. 5, Taylor & Francis, New York, 2003.
3. A. Cherkaev, K. Lurie, and G. W. Milton, Invariant properties in the stress in plane elasticity and equivalence classes in composites, *Proc. Royal Soc. Lond.*, A 438, 1992, 519–529.
4. M. F. Thorpe, I. Jasiuk, New results in the theory of elasticity for two-dimensional composites, *Proc. Royal Soc. Lond.*, A 458, 1992, 531–544.
5. M. Ostoja-Starzewski, *Microstructural Randomness and Scaling in Mechanics of Materials*, Chap. #5, Chapman & Hall/CRC Press, Boca Raton, FL, 2008.
6. I. Jasiuk, Stress invariance and exact relations in the mechanics of composite materials: Extensions of the CLM result—A review, *Mech. Mater.*, 41, 2009, 394–404.

Part II
(Chapters 9–13)

Applications and Problems

9 Solutions to Particular Three-Dimensional Boundary Value Problems of Elastostatics

In Chapter 7, a number of general solutions of three-dimensional homogeneous isotropic elastostatics were introduced. In this chapter the Boussinesq–Papkovitch–Neuber representation is used to obtain a closed-form solution to the problem of finding an elastic state in a semispace subject to a concentrated force normal to its boundary. Also, the Boussinesq–Somigliana–Galerkin representation is utilized for obtaining a closed-form solution to the problem of finding an elastic state in a semispace subject to a concentrated force tangent to its boundary. In addition, Love's representation of Chapter 7 is used for obtaining closed-form solutions to three-dimensional boundary value problems of nonisothermal elastostatics: (i) the thermoelastic state in a semi-infinite body corresponding to an internal source of heat, and (ii) the thermoelastic state in a semi-infinite body corresponding to a heat exposure on the boundary plane. An approximate three-dimensional theory of torsion of a homogeneous isotropic elastic prismatic bar is also included in this chapter. The chapter ends up with torsion problems, with the solutions included in the Solutions Manual.

9.1 THREE-DIMENSIONAL SOLUTIONS OF ISOTHERMAL ELASTOSTATICS

In this section a number of closed-form solutions to three-dimensional boundary value problems (BVP) of isothermal elastostatics are presented. In Chapter 7 a closed-form solution to the problem of finding an elastic state $s = [\mathbf{u}, \mathbf{E}, \mathbf{S}]$ corresponding to a concentrated force in a homogeneous isotropic infinite body was obtained (see Equation 7.1.30). In the following we discuss (A) an elastic state in a homogeneous isotropic semispace subjected to a concentrated force normal to its boundary (Boussinesq's solution), and (B) an elastic state in a homogeneous isotropic semispace subject to a concentrated force tangent to its boundary (Cerruti's solution). In both cases the closed-form solutions are obtained by using general solutions of three-dimensional elastostatics introduced in Chapter 7.

9.1.1 An Elastic State in a Homogeneous Isotropic Semispace Subject to a Concentrated Force Normal to Its Boundary: Boussinesq's Solution

Consider a homogeneous isotropic semi-infinite elastic solid subjected to a concentrated normal force on its boundary as shown in Figure 9.1. An elastic state $s = [\mathbf{u}, \mathbf{E}, \mathbf{S}]$ corresponding to the concentrated load is axially symmetric with respect to the $x_3 = z$ axis.

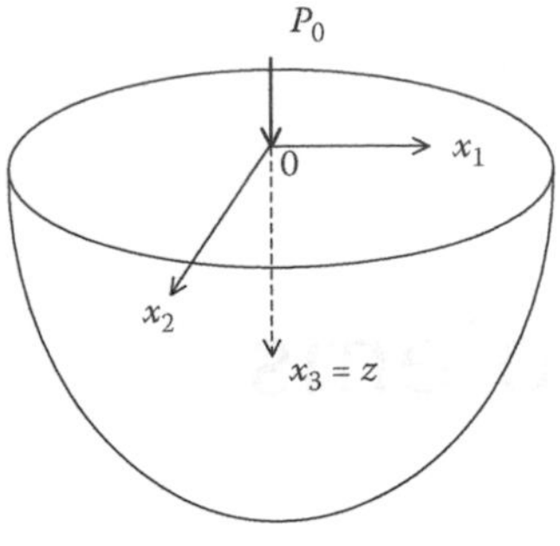

FIGURE 9.1 Normal force P_0 acting on the boundary of a semi-infinite solid.

This means that s is independent of θ of the cylindrical coordinate system (r, θ, z) and $\mathbf{u} = \mathbf{u}(r, z)$, $\mathbf{E} = \mathbf{E}(r, z)$, and $\mathbf{S} = \mathbf{S}(r, z)$. Since action of the force P_0 is not to be felt far away from the load region, s is to satisfy the fundamental system of field equations of three-dimensional elastostatics for a homogeneous isotropic solid in the domain $0 < r < \infty$, $0 \le \theta \le 2\pi$, $0 < z < \infty$ subject to the boundary conditions

$$S_{zz}(r, 0) = -P_0 \frac{\delta(r)}{2\pi r}, \quad S_{rz}(r, 0) = 0 \quad \text{for } r > 0 \tag{9.1.1}$$

and suitable vanishing conditions at infinity. In Equation 9.1.1, $\delta = \delta(r)$ is the Dirac delta function that satisfies the relation

$$\int_0^\infty \delta(r) dr = 1 \tag{9.1.2}$$

Clearly, the boundary conditions 9.1.1 imply that

$$\int_0^\infty \int_0^{2\pi} S_{zz}(r, 0) d\theta \, r \, dr = -P_0 \tag{9.1.3}$$

A problem of finding s corresponding to the situation shown in Figure 9.1 is called Boussinesq's problem. A solution to the problem can be found by using the Boussinesq–Papkovitch–Neuber representation of the displacement $\mathbf{u}$ for an axially symmetric problem (cf. Equations 7.1.88 through 7.1.92)

$$\mathbf{u} = [u_r(r, z), 0, u_z(r, z)] \tag{9.1.4}$$

where

$$u_r = -\frac{1}{4(1-\nu)} \frac{\partial \phi}{\partial r} \tag{9.1.5}$$

$$u_z = \psi - \frac{1}{4(1-\nu)} \frac{\partial \phi}{\partial z} \tag{9.1.6}$$

$$\phi = \varphi + z\psi \tag{9.1.7}$$

and $\varphi = \varphi(r, z)$ and $\psi = \psi(r, z)$ are harmonic functions,

$$\nabla^2 \varphi = 0, \quad \nabla^2 \psi = 0 \tag{9.1.8}$$

with

$$\nabla^2 = \frac{\partial^2}{\partial r^2} + \frac{1}{r}\frac{\partial}{\partial r} + \frac{\partial^2}{\partial z^2} \tag{9.1.9}$$

The associated strain tensor **E** and stress tensor **S** are computed from the formulas [cf. Example 7.1.7, Equations (l), (n), (p), (r), and (t)]

$$E_{rr} = -\frac{1}{4(1-\nu)}\frac{\partial^2 \phi}{\partial r^2} \tag{9.1.10}$$

$$E_{\theta\theta} = -\frac{1}{4(1-\nu)}\frac{1}{r}\frac{\partial \phi}{\partial r} \tag{9.1.11}$$

$$E_{zz} = \frac{\partial \psi}{\partial z} - \frac{1}{4(1-\nu)}\frac{\partial^2 \phi}{\partial z^2} \tag{9.1.12}$$

$$E_{rz} = -\frac{1}{4(1-\nu)}\left[\frac{\partial^2 \phi}{\partial r \partial z} - 2(1-\nu)\frac{\partial \psi}{\partial r}\right] \tag{9.1.13}$$

$$E_{r\theta} = 0, \quad E_{z\theta} = 0 \tag{9.1.14}$$

and

$$S_{rr} = -\frac{\mu}{2(1-\nu)}\left(\frac{\partial^2 \phi}{\partial r^2} - 2\nu\frac{\partial \psi}{\partial z}\right) \tag{9.1.15}$$

$$S_{\theta\theta} = -\frac{\mu}{2(1-\nu)}\left(\frac{1}{r}\frac{\partial \phi}{\partial r} - 2\nu\frac{\partial \psi}{\partial z}\right) \tag{9.1.16}$$

$$S_{zz} = -\frac{\mu}{2(1-\nu)}\left[\frac{\partial^2 \phi}{\partial z^2} - 2(2-\nu)\frac{\partial \psi}{\partial z}\right] \tag{9.1.17}$$

$$S_{rz} = -\frac{\mu}{2(1-\nu)}\left[\frac{\partial^2 \phi}{\partial r \partial z} - 2(1-\nu)\frac{\partial \psi}{\partial r}\right] \tag{9.1.18}$$

$$S_{r\theta} = 0, \quad S_{z\theta} = 0 \tag{9.1.19}$$

To obtain a solution to the problem, we note that the first of the boundary conditions 9.1.1 may be written in the integral form [1]

$$S_{zz}(r, 0) = -\frac{P_0}{2\pi}\int_0^\infty \alpha J_0(\alpha r) d\alpha \tag{9.1.20}$$

where $J_0 = J_0(\alpha r)$ is the Bessel function of order zero and of the first kind that satisfies the differential equation

$$\left(\nabla_r^2 + \alpha^2\right) J_0(\alpha r) = 0 \tag{9.1.21}$$

in which

$$\nabla_r^2 = \frac{\partial^2}{\partial r^2} + \frac{1}{r}\frac{\partial}{\partial r} \tag{9.1.22}$$

Also, note that [2]

$$\int_0^\infty e^{-\alpha z} J_n(\alpha r) d\alpha = r^{-n} \frac{(R-z)^n}{R} \qquad n = 0, 1, 2, ... \tag{9.1.23}$$

where

$$R = \sqrt{r^2 + z^2}, \quad z \geq 0 \tag{9.1.24}$$

and $J_n = J_n(\alpha r)$, $n \geq 0$, is the Bessel function of order n and of the first kind that satisfies the differential equation

$$\left[\nabla_r^2 + \left(\alpha^2 - \frac{n^2}{r^2}\right)\right] J_n(\alpha r) = 0 \tag{9.1.25}$$

Clearly, if $n = 0$, Equation 9.1.25 reduces to Equation 9.1.21, and, by Equation 9.1.21, we obtain

$$\nabla^2[e^{-\alpha z} J_0(\alpha r)] = 0 \tag{9.1.26}$$

and

$$\nabla^2 \nabla^2[\alpha z e^{-\alpha z} J_0(\alpha r)] = 0 \tag{9.1.27}$$

Since, in view of Equations 9.1.7 through 9.1.9,

$$\nabla^2 \phi = 2\frac{\partial \psi}{\partial z}, \quad \nabla^2 \psi = 0, \quad \nabla^2 \nabla^2 \phi = 0 \tag{9.1.28}$$

and S_{zz} is given by Equation 9.1.17, to satisfy the boundary conditions 9.1.1, we postulate ϕ and ψ in the forms

$$\phi(r, z) = \int_0^\infty (A + \alpha z B) e^{-\alpha z} J_0(\alpha r) d\alpha \tag{9.1.29}$$

$$\psi(r, z) = \int_0^\infty C e^{-\alpha z} J_0(\alpha r) d\alpha \tag{9.1.30}$$

where $A = A(\alpha)$, $B = B(\alpha)$, and $C = C(\alpha)$ are unknown functions of α to be determined from the two boundary conditions 9.1.1 and from the first of Equations 9.1.28. Clearly, ϕ and ψ comply with the second and third of Equations 9.1.28, because of Equations 9.1.26 and 9.1.27. Substituting 9.1.29 and 9.1.30 into the first of Equations 9.1.28, we arrive at the relation

$$\int_0^\infty [-2\alpha^2 B e^{-\alpha z} J_0(\alpha r)] d\alpha = 2 \int_0^\infty (-\alpha) C e^{-\alpha z} J_0(\alpha r) d\alpha \tag{9.1.31}$$

This implies that $C = \alpha B$, and Equation 9.1.30 reduces to

$$\psi(r,z) = \int_0^\infty \alpha B e^{-\alpha z} J_0(\alpha r) d\alpha \tag{9.1.32}$$

Therefore, the problem is reduced to finding A and B in such a way that the two boundary conditions 9.1.1 are satisfied.

By substituting Equations 9.1.29 and 9.1.32 into Equation 9.1.18 and letting $z = 0$ we find that the boundary condition

$$S_{rz}(r,0) = 0 \tag{9.1.33}$$

is satisfied provided

$$A = -(1-2\nu)B \tag{9.1.34}$$

As a result, the function ϕ given by Equation 9.1.29 may be written in the form

$$\phi(r,z) = \int_0^\infty B[-(1-2\nu) + \alpha z] e^{-\alpha z} J_0(\alpha r) d\alpha \tag{9.1.35}$$

Finally, by substituting the integrals 9.1.32 and 9.1.35 into Equation 9.1.17, letting $z = 0$, and using the boundary condition 9.1.20, we obtain

$$B\alpha = \frac{P_0}{\mu} \frac{1-\nu}{\pi} \tag{9.1.36}$$

Therefore, using Equations 9.1.32, 9.1.35, and 9.1.36, the following formulas are obtained:

$$\psi = \frac{P_0(1-\nu)}{\mu\pi} \int_0^\infty e^{-\alpha z} J_0(\alpha r) d\alpha \tag{9.1.37}$$

$$\frac{\partial\phi}{\partial r} = \frac{P_0(1-\nu)}{\mu\pi} \int_0^\infty [(1-2\nu) - \alpha z] e^{-\alpha z} J_1(\alpha r) d\alpha \tag{9.1.38}$$

$$\frac{\partial\phi}{\partial z} = \frac{P_0(1-\nu)}{\mu\pi} \int_0^\infty [2(1-\nu) - \alpha z] e^{-\alpha z} J_0(\alpha r) d\alpha \tag{9.1.39}$$

Note that Equation 9.1.38 is obtained from Equation 9.1.35 by using the relation

$$J_1(\alpha r) = -\frac{1}{\alpha} \frac{d}{dr} J_0(\alpha r) \tag{9.1.40}$$

In the following, we show that the integrals 9.1.37 through 9.1.39 can be expressed in terms of elementary functions. To this end we note that, because of Equation 9.1.23 with $n = 0$ and $n = 1$, we get

$$\int_0^\infty e^{-\alpha z} J_0(\alpha r) d\alpha = \frac{1}{R} \tag{9.1.41}$$

and

$$\int_0^\infty e^{-\alpha z} J_1(\alpha r) d\alpha = \frac{1}{r} \frac{R - z}{R} = \frac{r}{R(R + z)} \tag{9.1.42}$$

By differentiating Equations 9.1.41 and 9.1.42 with respect to z, we obtain, respectively

$$\int_0^\infty \alpha e^{-\alpha z} J_0(\alpha r) d\alpha = -\frac{\partial}{\partial z}\left(\frac{1}{R}\right) \tag{9.1.43}$$

and

$$\int_0^\infty \alpha e^{-\alpha z} J_1(\alpha r) d\alpha = -r\frac{\partial}{\partial z}\left[\frac{1}{R(R + z)}\right] \tag{9.1.44}$$

Therefore, using Equations 9.1.41 through 9.1.44, we reduce Equations 9.1.37 through 9.1.39 to the forms

$$\psi = \frac{P_0(1 - \nu)}{\mu\pi} \frac{1}{R} \tag{9.1.45}$$

$$\frac{\partial\phi}{\partial r} = \frac{P_0(1 - \nu)}{\mu\pi}\left\{(1 - 2\nu)\frac{r}{R}\frac{1}{R + z} + zr\frac{\partial}{\partial z}\left[\frac{1}{R(R + z)}\right]\right\} \tag{9.1.46}$$

$$\frac{\partial\phi}{\partial z} = \frac{P_0(1 - \nu)}{\mu\pi}\left[2(1 - \nu)\frac{1}{R} + z\frac{\partial}{\partial z}\left(\frac{1}{R}\right)\right] \tag{9.1.47}$$

Now, since

$$\frac{\partial}{\partial z}\left[\frac{1}{R(R + z)}\right] = -\frac{1}{R^3}, \quad \frac{\partial}{\partial z}\left(\frac{1}{R}\right) = -\frac{z}{R^3} \tag{9.1.48}$$

from Equations 9.1.46 and 9.1.47 we get

$$\frac{\partial\phi}{\partial r} = \frac{P_0(1 - \nu)}{\mu\pi}\left[(1 - 2\nu)\frac{r}{R}\frac{1}{R + z} - \frac{rz}{R^3}\right] \tag{9.1.49}$$

and

$$\frac{\partial\phi}{\partial z} = \frac{P_0(1 - \nu)}{\mu\pi}\left[2(1 - \nu)\frac{1}{R} - \frac{z^2}{R^3}\right] \tag{9.1.50}$$

and substituting Equations 9.1.49 and 9.1.50 into Equations 9.1.5 and 9.1.6, respectively, we arrive at the displacement components

$$u_r = \frac{P_0}{4\mu\pi}\frac{r}{R}\left[\frac{z}{R^2} - (1-2\nu)\frac{1}{R+z}\right] \tag{9.1.51}$$

$$u_z = \frac{P_0}{4\mu\pi}\frac{1}{R}\left[\frac{z^2}{R^2} + 2(1-\nu)\right] \tag{9.1.52}$$

To compute the strain components, we differentiate Equations 9.1.49 and 9.1.50 with respect to r and z, respectively, and obtain

$$\frac{\partial^2\phi}{\partial r^2} = \frac{P_0(1-\nu)}{\mu\pi}\left\{\frac{z}{R^3}\left[(3-2\nu) - 3\frac{z^2}{R^2}\right] - (1-2\nu)\frac{1}{R(R+z)}\right\} \tag{9.1.53}$$

and

$$\frac{\partial^2\phi}{\partial z^2} = \frac{P_0(1-\nu)}{\mu\pi}\left\{\frac{z}{R^3}\left[3\frac{z^2}{R^2} - 2(2-\nu)\right]\right\} \tag{9.1.54}$$

Also, differentiating Equation 9.1.45 with respect to z we write

$$\frac{\partial\psi}{\partial z} = -\frac{P_0(1-\nu)}{\mu\pi}\frac{z}{R^3} \tag{9.1.55}$$

and differentiating Equation 9.1.45 with respect to r we obtain

$$\frac{\partial\psi}{\partial r} = -\frac{P_0(1-\nu)}{\mu\pi}\frac{r}{R^3} \tag{9.1.56}$$

Now, substituting Equation 9.1.53 into Equation 9.1.10, we get

$$E_{rr} = -\frac{P_0}{4\pi\mu}\left\{\frac{z}{R^3}\left[(3-2\nu) - 3\frac{z^2}{R^2}\right] - (1-2\nu)\frac{1}{R(R+z)}\right\} \tag{9.1.57}$$

Next, substituting Equation 9.1.49 into Equation 9.1.11, we obtain

$$E_{\theta\theta} = -\frac{P_0}{4\pi\mu}\left[(1-2\nu)\frac{1}{R(R+z)} - \frac{z}{R^3}\right] \tag{9.1.58}$$

Also, substituting Equations 9.1.54 and 9.1.55 into Equation 9.1.12, we write

$$E_{zz} = -\frac{P_0}{4\pi\mu}\left[\frac{z}{R^3}\left(3\frac{z^2}{R^2} - 2\nu\right)\right] \tag{9.1.59}$$

To obtain E_{rz} note that because of Equations 9.1.45 and 9.1.50 we have

$$\frac{\partial \phi}{\partial z} - 2(1-\nu)\psi = -\frac{P_0(1-\nu)}{\mu\pi}\frac{z^2}{R^3} \tag{9.1.60}$$

Hence, differentiation of Equation 9.1.60 with respect to r and using Equation 9.1.13, we obtain

$$E_{rz} = -\frac{3P_0}{4\pi\mu}\frac{rz^2}{R^5} \tag{9.1.61}$$

It follows from Equations 9.1.57 through 9.1.59 that

$$E_{rr} + E_{\theta\theta} + E_{zz} = -\frac{P_0}{2\pi\mu}(1-2\nu)\frac{z}{R^3} \tag{9.1.62}$$

This result is in compliance with the equation

$$E_{rr} + E_{\theta\theta} + E_{zz} = \frac{1-2\nu}{2(1-\nu)}\frac{\partial \psi}{\partial z} \tag{9.1.63}$$

which is obtained by using Equations 9.1.10 through 9.1.12 and the first of Equations 9.1.28.

To obtain the stress components we proceed in the following way. To get S_{rr} we insert Equations 9.1.53 and 9.1.55 into Equation 9.1.15 and obtain

$$S_{rr} = \frac{P_0}{2\pi R^2}\left[-\frac{3r^2 z}{R^3} + \frac{(1-3\nu)R}{R+z}\right] \tag{9.1.64}$$

Substitution of Equations 9.1.49 and 9.1.55 into Equation 9.1.16 yields

$$S_{\theta\theta} = \frac{(1-2\nu)P_0}{2\pi R^2}\left(\frac{z}{R} - \frac{R}{R+z}\right) \tag{9.1.65}$$

Next, inserting Equations 9.1.54 and 9.1.55 into Equation 9.1.17, we obtain

$$S_{zz} = -\frac{3P_0}{2\pi}\frac{z^3}{R^5} \tag{9.1.66}$$

Finally, by differentiating Equation 9.1.60 with respect to r, and substituting the result into Equation 9.1.18, we get

$$S_{rz} = -\frac{3P_0}{2\pi}\frac{rz^2}{R^5} \tag{9.1.67}$$

It follows from Equations 9.1.64 through 9.1.66 that

$$S_{rr} + S_{\theta\theta} + S_{zz} = -\frac{P_0(1+\nu)}{\pi}\frac{z}{R^3} \tag{9.1.68}$$

Also, Equations 9.1.62 and 9.1.68 imply the well known result (cf. Equation 4.1.23)

$$E_{rr} + E_{\theta\theta} + E_{zz} = \frac{1}{2\mu}\frac{1-2\nu}{1+\nu}(S_{rr} + S_{\theta\theta} + S_{zz}) \tag{9.1.69}$$

In the direct notation this reads

$$\text{tr}\,\mathbf{E} = \frac{1}{2\mu}\frac{1-2\nu}{1+\nu}\,\text{tr}\,\mathbf{S} \tag{9.1.70}$$

9.1.1.1 Analysis of the Solution

The solution is represented by the elastic state $s = [\mathbf{u}, \mathbf{E}, \mathbf{S}]$ in which $\mathbf{u}$ is given by Equations 9.1.4, 9.1.51, and 9.1.52, $\mathbf{E}$ is defined by Equations 9.1.57 through 9.1.59 and 9.1.61, and $\mathbf{S}$ is defined by Equations 9.1.64 through 9.1.67. The state s is singular at $R=0$, and $s \to 0$ as $R \to \infty$. Moreover, s has the following properties: (i) $u_r(0,z)=0$ for $z>0$; $u_r(r,0)<0$ and $u_z(r,0)>0$ for $r>0$; (ii) $\text{tr}\,\mathbf{E}(r,0)=0$ for $r>0$; and the stress components S_{zz} and S_{rz} are independent of elastic constants throughout the semispace $0<r<\infty$, $0<z<\infty$; in addition

$$\int_0^\infty\int_0^{2\pi} S_{zz}(r,z)r\,dr\,d\theta = -P_0 \quad \text{for every} \quad z>0 \tag{9.1.71}$$

To show this we use Equation 9.1.66 and obtain

$$\int_0^\infty\int_0^{2\pi} S_{zz}(r,z)r\,dr\,d\theta = -3P_0 z^3 \int_0^\infty \frac{r\,dr}{R^5} \tag{9.1.72}$$

Since

$$\frac{r}{R^5} = -\frac{1}{3}\frac{\partial}{\partial r}\left(\frac{1}{R^3}\right) \tag{9.1.73}$$

therefore

$$\begin{aligned}\int_0^\infty\int_0^{2\pi} S_{zz}(r,z)r\,dr\,d\theta &= P_0 z^3 \left.\frac{1}{R^3}\right|_{r=0}^{r=\infty} \\ &= P_0 z^3\left(-\frac{1}{z^3}\right) = -P_0\end{aligned} \tag{9.1.74}$$

Similarly, by Equation 9.1.67, we get

$$\int_0^\infty\int_0^{2\pi} S_{rz}(r,z)r\,dr\,d\theta = -P_0 \quad \text{for every } z>0 \tag{9.1.75}$$

To prove Equation 9.1.75 we use Equation 9.1.67 and write

$$\int_0^\infty \int_0^{2\pi} S_{rz}(r,z) r\, dr\, d\theta = -3P_0 z^2 \int_0^\infty \frac{r^2 dr}{R^5} \tag{9.1.76}$$

Since

$$\int_0^\infty \frac{r^2 dr}{R^5} = -\frac{1}{3} \int_0^\infty r \frac{\partial}{\partial r}\left(\frac{1}{R^3}\right) dr \tag{9.1.77}$$

therefore, the integration by parts leads to

$$\int_0^\infty \frac{r^2 dr}{R^5} = -\frac{1}{3}\left[r \frac{1}{R^3}\Bigg|_{r=0}^{r=\infty} - \int_0^\infty \frac{dr}{R^3} \right] \tag{9.1.78}$$

Substituting Equation 9.1.78 into Equation 9.1.76, we get

$$\int_0^\infty \int_0^{2\pi} S_{rz}(r,z) r\, dr\, d\theta = P_0 z \frac{\partial}{\partial z} \int_0^\infty \frac{dr}{R} \tag{9.1.79}$$

or

$$\begin{aligned} \int_0^\infty \int_0^{2\pi} S_{rz}(r,z) r\, dr\, d\theta &= P_0 z \frac{\partial}{\partial z} [\ln(r+R)] \Bigg|_{r=0}^{r=\infty} \\ &= P_0 \frac{z^2}{R(R+r)} \Bigg|_{r=0}^{r=\infty} = -P_0 \end{aligned} \tag{9.1.80}$$

Note that by Equations 9.1.66 and 9.1.67

$$S_{zz}(r,0) = S_{rz}(r,0) \quad \text{for every } r > 0 \tag{9.1.81}$$

In the following, Boussinesq's solution will be used to obtain a solution of three-dimensional elastostatics for a semi-infinite solid subjected to an arbitrary normal load on its boundary. To this end we transform the elastic state $s = s(r,z)$ into Cartesian coordinates (x_1, x_2, x_3) using the formulas

$$\begin{aligned} u_1 &= u_r \cos\theta \\ u_2 &= u_r \sin\theta \\ u_3 &= u_z \end{aligned} \tag{9.1.82}$$

$$\begin{aligned} S_{13} &= S_{rz} \cos\theta \\ S_{23} &= S_{rz} \sin\theta \\ S_{33} &= S_{zz} \end{aligned} \tag{9.1.83}$$

and

$$S_{11} = S_{rr}\cos^2\theta + S_{\theta\theta}\sin^2\theta$$
$$S_{22} = S_{rr}\sin^2\theta + S_{\theta\theta}\cos^2\theta \tag{9.1.84}$$
$$S_{12} = \frac{1}{2}(S_{rr} - S_{\theta\theta})\sin 2\theta$$

where

$$\cos\theta = \frac{x_1}{r}, \quad \sin\theta = \frac{x_2}{r} \tag{9.1.85}$$

Substituting u_r and u_z from Equations 9.1.51 and 9.1.52, and $\cos\theta$ and $\sin\theta$ from Equation 9.1.85 into Equations 9.1.82, we obtain the Cartesian components of the displacement vector

$$u_1 = \frac{P_0}{4\pi\mu}\frac{x_1}{R}\left[\frac{x_3}{R^2} - (1-2\nu)\frac{1}{R+x_3}\right]$$
$$u_2 = \frac{P_0}{4\pi\mu}\frac{x_2}{R}\left[\frac{x_3}{R^2} - (1-2\nu)\frac{1}{R+x_3}\right] \tag{9.1.86}$$
$$u_3 = \frac{P_0}{4\pi\mu}\frac{1}{R}\left[\frac{x_3^2}{R^2} + 2(1-\nu)\right]$$

Similarly, using Equations 9.1.66 and 9.1.67, and Equations 9.1.83, we write

$$S_{13} = -\frac{3P_0}{2\pi}\frac{x_1 x_3^2}{R^5}, \quad S_{23} = -\frac{3P_0}{2\pi}\frac{x_2 x_3^2}{R^5}, \quad S_{33} = -\frac{3P_0}{2\pi}\frac{x_3^3}{R^5} \tag{9.1.87}$$

Finally, substituting S_{rr} and $S_{\theta\theta}$ from Equations 9.1.64 and 9.1.65 into Equations 9.1.84, we get

$$S_{11} = -\frac{P_0}{2\pi R^2}\left\{\frac{3x_1^2 x_3}{R^3} - (1-2\nu)\left[\frac{x_3}{R} - \frac{R}{R+x_3} + \frac{x_1^2(2R+x_3)}{R(R+x_3)^2}\right]\right\}$$
$$S_{22} = -\frac{P_0}{2\pi R^2}\left\{\frac{3x_2^2 x_3}{R^3} - (1-2\nu)\left[\frac{x_3}{R} - \frac{R}{R+x_3} + \frac{x_2^2(2R+x_3)}{R(R+x_3)^2}\right]\right\} \tag{9.1.88}$$
$$S_{12} = -\frac{P_0}{2\pi R^2}x_1 x_2\left[\frac{3x_3}{R^3} - (1-2\nu)\frac{(2R+x_3)}{R(R+x_3)^2}\right]$$

Similar formulas may be obtained for Cartesian components of the strain tensor **E**.

Now, let a homogeneous isotropic elastic semi-infinite solid, described by the inequalities

$$|x_1| < \infty, \quad |x_2| < \infty, \quad 0 \le x_3 < \infty \tag{9.1.89}$$

be subjected to a normal load $p = p(x_1, x_2)$ on a finite domain Ω of the boundary $x_3 = 0$, i.e.,

$$\begin{aligned} S_{33}(x_1, x_2, 0) &= -p(x_1, x_2) \quad \text{for } (x_1, x_2) \in \Omega \\ S_{13}(x_1, x_2, 0) &= S_{23}(x_1, x_2, 0) = 0 \quad \text{for } |x_1| < \infty, |x_2| < \infty \end{aligned} \tag{9.1.90}$$

where $p = p(x_1, x_2)$ is a prescribed function on Ω. Then the displacement $\mathbf{u} = \mathbf{u}(\mathbf{x})$ and stress $\mathbf{S} = \mathbf{S}(\mathbf{x})$ for any point of the solid may be computed from the formulas

$$\begin{aligned} \mathbf{u}(\mathbf{x}) &= \int_\Omega \mathbf{u}^*(x_1 - \xi_1, x_2 - \xi_2; x_3) p(\xi_1, \xi_2) d\xi_1 d\xi_2 \\ \mathbf{S}(\mathbf{x}) &= \int_\Omega \mathbf{S}^*(x_1 - \xi_1, x_2 - \xi_2; x_3) p(\xi_1, \xi_2) d\xi_1 d\xi_2 \end{aligned} \tag{9.1.91}$$

where $\mathbf{u}^*$ and $\mathbf{S}^*$ are obtained from Equations 9.1.86 and 9.1.87 and 9.1.88, respectively; in which we put $P_0 = 1$; and replace x_1 and x_2 by $x_1 - \xi_1$ and $x_2 - \xi_2$, respectively. For example, for S_{33} we obtain

$$S_{33}(\mathbf{x}) = \frac{x_3^2}{2\pi} \frac{\partial}{\partial x_3} \int_\Omega \frac{p(\xi_1, \xi_2) d\xi_1 d\xi_2}{\left[(x_1 - \xi_1)^2 + (x_2 - \xi_2)^2 + x_3^2\right]^{3/2}} \tag{9.1.92}$$

9.1.2 An Elastic State in a Homogeneous Isotropic Semispace Subject to a Concentrated Force Tangent to Its Boundary: Cerruti's Solution

Let us consider a homogeneous isotropic elastic semi-infinite solid on the boundary of which acts a concentrated tangential force P_0 (see Figure 9.2).

A problem of finding an elastic state $s = [\mathbf{u}, \mathbf{E}, \mathbf{S}]$ corresponding to the situation shown in Figure 9.2 is called Cerruti's problem. A solution to the problem is represented by s that satisfies the field equations of three-dimensional homogeneous isotropic elastostatics in the region

$$|x_1| < \infty, \quad |x_2| < \infty, \quad 0 < x_3 < \infty \tag{9.1.93}$$

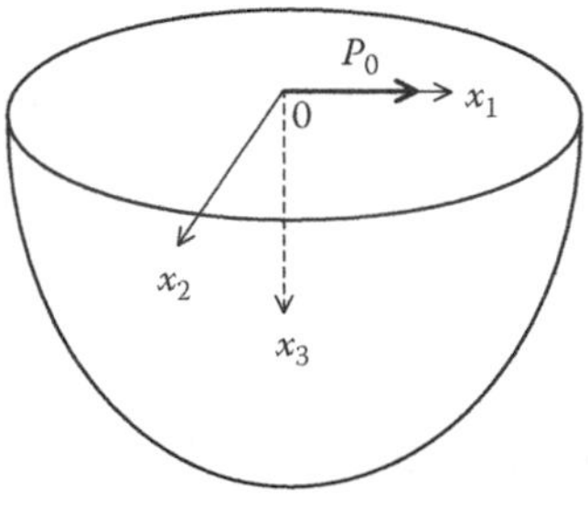

FIGURE 9.2 Tangential force P_0 acting on the boundary of a semi-infinite solid.

subject to the boundary conditions

$$S_{13}(x_1, x_2, 0) = -P_0\, \delta(x_1)\delta(x_2) \tag{9.1.94}$$

$$S_{23}(x_1, x_2, 0) = S_{33}(x_1, x_2, 0) = 0 \tag{9.1.95}$$

and suitable vanishing conditions at $R \to \infty$, where $R = \left(x_1^2 + x_2^2 + x_3^2\right)^{1/2}$. In Equation 9.1.94 $\delta = \delta(x_1)$ and $\delta = \delta(x_2)$

are the Dirac delta functions on the x_1-axis and x_2-axis, respectively. Since

$$\int_{-\infty}^{+\infty} \delta(x_1)dx_1 = 1 \quad \text{and} \quad \int_{-\infty}^{+\infty} \delta(x_2)dx_2 = 1 \tag{9.1.96}$$

from Equation 9.1.94 we get

$$\int_{-\infty}^{+\infty}\int_{-\infty}^{+\infty} S_{13}(x_1, x_2, 0)dx_1dx_2 = -P_0 \tag{9.1.97}$$

To obtain $s = [\mathbf{u}, \mathbf{E}, \mathbf{S}]$ we let [3]

$$s = s^{(1)} + s^{(2)} \tag{9.1.98}$$

where $s^{(1)}$ is generated by a scalar potential ϕ, and $s^{(2)}$ is generated by a Boussinesq–Somigliana–Galerkin solution (cf. Equations 7.1.55 and 7.1.56 with $\mathbf{b} = \mathbf{0}$),

$$s^{(1)} = [\mathbf{u}^{(1)}, \mathbf{E}^{(1)}, \mathbf{S}^{(1)}] \tag{9.1.99}$$

where

$$\begin{aligned} \mathbf{u}^{(1)} &= \nabla\phi \\ \mathbf{E}^{(1)} &= \nabla\nabla\phi \\ \mathbf{S}^{(1)} &= 2\mu\nabla\nabla\phi \end{aligned} \tag{9.1.100}$$

$$\nabla^2\phi = 0 \tag{9.1.101}$$

and

$$s^{(2)} = [\mathbf{u}^{(2)}, \mathbf{E}^{(2)}, \mathbf{S}^{(2)}] \tag{9.1.102}$$

where

$$\mathbf{u}^{(2)} = \nabla^2\mathbf{g} - \frac{1}{2(1-\nu)}\nabla(\text{div}\,\mathbf{g}) \tag{9.1.103}$$

$$\mathbf{E}^{(2)} = \nabla^2(\widehat{\nabla}\mathbf{g}) - \frac{1}{2(1-\nu)}\nabla\,\nabla(\text{div}\,\mathbf{g}) \tag{9.1.104}$$

$$\mathbf{S}^{(2)} = \mu\left\{2\nabla^2(\widehat{\nabla}\,\mathbf{g}) + \frac{1}{1-\nu}\left[\nu\nabla^2(\text{div}\,\mathbf{g})\mathbf{1} - \nabla\nabla(\text{div}\,\mathbf{g})\right]\right\} \tag{9.1.105}$$

$$\nabla^2\nabla^2\mathbf{g} = \mathbf{0} \tag{9.1.106}$$

It follows from Equations 9.1.100 and 9.1.103 through 9.1.105 that in components

$$u_i^{(1)} = \phi_{,i} \quad E_{ij}^{(1)} = \phi_{,ij}, \quad S_{ij}^{(1)} = 2\mu\phi_{,ij} \tag{9.1.107}$$

and

$$u_i^{(2)} = \nabla^2 g_i - \frac{1}{2(1-\nu)} g_{k,ki} \tag{9.1.108}$$

$$E_{ij}^{(2)} = \frac{1}{2}\nabla^2(g_{i,j} + g_{j,i}) - \frac{1}{2(1-\nu)} g_{k,kij} \tag{9.1.109}$$

$$S_{ij}^{(2)} = \mu\left[\nabla^2(g_{i,j} + g_{j,i}) + \frac{1}{1-\nu}(\nu\delta_{ij}g_{k,kll} - g_{k,kij})\right] \tag{9.1.110}$$

In particular, it follows from Equations 9.1.107 through 9.1.110 that the boundary stress components at $x_3 = 0$ take the forms

$$S_{33}^{(1)} = 2\mu\phi_{,33}, \quad S_{32}^{(1)} = 2\mu\phi_{,32}, \quad S_{31}^{(1)} = 2\mu\phi_{,31} \tag{9.1.111}$$

and

$$S_{33}^{(2)} = \mu\left[2\nabla^2 g_{3,3} + \frac{1}{1-\nu}\left(\nu\nabla^2 g_{k,k} - g_{k,k33}\right)\right] \tag{9.1.112}$$

$$S_{32}^{(2)} = \mu\left[\nabla^2(g_{3,2} + g_{2,3}) - \frac{1}{1-\nu} g_{k,k23}\right] \tag{9.1.113}$$

$$S_{31}^{(2)} = \mu\left[\nabla^2(g_{1,3} + g_{3,1}) - \frac{1}{1-\nu} g_{k,k13}\right] \tag{9.1.114}$$

Now, we are to show that s given by Equation 9.1.98 is a solution to Cerruti's problem provided ϕ and g_k are taken in the forms

$$\phi = C\frac{x_1}{R + x_3} \tag{9.1.115}$$

and

$$g_k = AR\delta_{k1} + Bx_1 \ln\left(\frac{R + x_3}{L}\right)\delta_{k3} \tag{9.1.116}$$

where

$$R = \left(x_1^2 + x_2^2 + x_3^2\right)^{1/2} \tag{9.1.117}$$

A, B, and C are suitably selected constants, and L is a positive constant of the length dimension. To this end, we prove that (i) ϕ is harmonic and g_k is biharmonic in the semispace described by the inequalities 9.1.93, (ii) A, B, and C are uniquely determined by the boundary conditions 9.1.94 and 9.1.95, and (iii) $s \to 0$ as $R \to \infty$.

Proof of (i). To show that ϕ is harmonic, we apply the Laplacian operator to Equation 9.1.115 and write

$$\begin{aligned}\nabla^2\phi &= C[x_1(R+x_3)^{-1}]_{,kk}\\ &= C\left[\delta_{1k}(R+x_3)^{-1} - x_1(R+x_3)^{-2}(x_kR^{-1}+\delta_{3k})\right]_{,k}\\ &= C[-2\delta_{1k}(R+x_3)^{-2}(x_kR^{-1}+\delta_{3k})\\ &\quad + 2x_1(R+x_3)^{-3}(x_kR^{-1}+\delta_{3k})(x_kR^{-1}+\delta_{3k})\\ &\quad - x_1(R+x_3)^{-2}(3R^{-1}-x_kx_kR^{-3})]\end{aligned} \tag{9.1.118}$$

Since

$$\delta_{1k}(x_kR^{-1}+\delta_{3k}) = x_1R^{-1} \tag{9.1.119}$$

$$(x_kR^{-1}+\delta_{3k})(x_kR^{-1}+\delta_{3k}) = 2R^{-1}(R+x_3) \tag{9.1.120}$$

therefore, from Equation 9.1.118 follows

$$\begin{aligned}\nabla^2\phi &= C\left[-2x_1R^{-1}(R+x_3)^{-2} + 4x_1R^{-1}(R+x_3)^{-2}\right.\\ &\quad \left.- 2x_1R^{-1}(R+x_3)^{-2}\right] = 0\end{aligned} \tag{9.1.121}$$

This proves that ϕ is harmonic.

To prove that g_k is biharmonic, we take gradient of Equation 9.1.116, and obtain

$$\begin{aligned}g_{k,j} &= Ax_jR^{-1}\delta_{k1} + B\delta_{k3}\left\{\delta_{1j}\left[\ln\left(\frac{R+x_3}{L}\right)\right]\right.\\ &\quad \left.+ x_1(x_jR^{-1}+\delta_{j3})(R+x_3)^{-1}\right\}\end{aligned} \tag{9.1.122}$$

Next, the differentiation of this equation with respect to x_j yields

$$\begin{aligned}g_{k,jj} &= A(3R^{-1} - x_jR^{-2}x_jR^{-1})\delta_{k1} + B\delta_{k3}\left[2\delta_{1j}(x_jR^{-1}+\delta_{j3})(R+x_3)^{-1}\right.\\ &\quad + x_1(3R^{-1} - x_jR^{-2}x_jR^{-1})(R+x_3)^{-1}\\ &\quad \left.- x_1(x_jR^{-1}+\delta_{j3})(x_jR^{-1}+\delta_{j3})(R+x_3)^{-2}\right]\end{aligned}$$

Hence

$$g_{k,jj} = 2A\delta_{k1}R^{-1} + 2B\delta_{k3}x_1R^{-1}(R+x_3)^{-1} \tag{9.1.123}$$

Also note that

$$\nabla^2 R^{-1} = (R^{-1})_{,kk} = -(x_k R^{-3})_{,k}$$
$$= -(3R^{-3} - 3x_k R^{-4} x_k R^{-1}) = 0 \tag{9.1.124}$$

and

$$\nabla^2 \ln\left(\frac{R+x_3}{L}\right) = \left[\ln\left(\frac{R+x_3}{L}\right)\right]_{,kk} = \left[\left(x_k R^{-1} + \delta_{k3}\right)(R+x_3)^{-1}\right]_{,k}$$
$$= \left[(3R^{-1} - x_k R^{-2} x_k R^{-1})(R+x_3)^{-1} - 2R^{-1}(R+x_3)^{-1}\right] = 0 \tag{9.1.125}$$

In addition

$$x_1 R^{-1}(R+x_3)^{-1} = \left[\ln\left(\frac{R+x_3}{L}\right)\right]_{,1} \tag{9.1.126}$$

Therefore, applying ∇^2 to Equation 9.1.123 and using Equations 9.1.124 through 9.1.126, we find that g_k is biharmonic. This completes the proof of (i).

Proof of (ii). To show that A, B, and C are uniquely determined by the boundary conditions 9.1.94 and 9.1.95 we need to compute the stress components S_{33}, S_{13}, and S_{23}. To this end we insert ϕ, given by Equation 9.1.115, into Equations 9.1.111, and obtain

$$S_{33}^{(1)} = 2\mu C\left[x_1(R+x_3)^{-1}\right]_{,33} = 2\mu C x_1 R^{-3} \tag{9.1.127}$$

$$S_{13}^{(1)} = 2\mu C[x_1(R+x_3)^{-1}]_{,13}$$
$$= -2\mu C R^{-1}(R+x_3)^{-1}\left(1 - \frac{x_1^2}{R^2}\,\frac{2R+x_3}{R+x_3}\right) \tag{9.1.128}$$

$$S_{23}^{(1)} = 2\mu C\left[x_1(R+x_3)^{-1}\right]_{,23}$$
$$= 2\mu C x_1 x_2 R^{-3}(R+x_3)^{-1}\left(\frac{2R+x_3}{R+x_3}\right) \tag{9.1.129}$$

Next, from Equation 9.1.122, we get

$$g_{k,k} = (A+B)x_1 R^{-1} \tag{9.1.130}$$

and this implies that

$$g_{k,kp} = (A+B)\left(\delta_{1p}R^{-1} - x_1 x_p R^{-3}\right) \tag{9.1.131}$$

and

$$g_{k,kpq} = -(A+B)R^{-3}\left(\delta_{1p}x_q + \delta_{1q}x_p + x_1\delta_{pq} - 3x_1 x_p x_q R^{-2}\right) \tag{9.1.132}$$

Also, by Equation 9.1.123,

$$\nabla^2 g_i = 2A\delta_{i1}R^{-1} + 2B\delta_{i3}\left[\ln\left(\frac{R+x_3}{L}\right)\right]_{,1} \tag{9.1.133}$$

Hence

$$\nabla^2 g_{i,j} = -2A\delta_{i1}x_jR^{-3} + 2B\delta_{i3}\left[(x_jR^{-1}+\delta_{j3})(R+x_3)^{-1}\right]_{,1} \tag{9.1.134}$$

Since

$$\begin{aligned}\left[(x_jR^{-1}+\delta_{j3})(R+x_3)^{-1}\right]_{,1} = \big[&\left(\delta_{j1}R^{-1} - x_1x_jR^{-3}\right)(R+x_3)^{-1} \\ &- \left(x_jR^{-1}+\delta_{j3}\right)x_1R^{-1}(R+x_3)^{-2}\big]\end{aligned} \tag{9.1.135}$$

therefore

$$\begin{aligned}\nabla^2\left(g_{i,j}+g_{j,i}\right) = -2A\left(\delta_{i1}x_j+\delta_{j1}x_i\right)R^{-3} + 2B\big\{&\delta_{i3}(R+x_3)^{-1} \\ &\left[\left(\delta_{j1}R^{-1} - x_1x_jR^{-3}\right) - \left(x_jR^{-1}+\delta_{j3}\right)x_1R^{-1}(R+x_3)^{-1}\right] \\ &+ \delta_{j3}(R+x_3)^{-1}\big[\left(\delta_{i1}R^{-1} - x_1x_iR^{-3}\right) \\ &- \left(x_iR^{-1}+\delta_{i3}\right)x_1R^{-1}(R+x_3)^{-1}\big]\big\}\end{aligned} \tag{9.1.136}$$

Now, by letting $p = q$ in Equation 9.1.132, we write

$$\nabla^2 g_{k,k} = g_{k,kpp} = -2(A+B)x_1R^{-3} \tag{9.1.137}$$

Next, by letting $p=q=3$ in Equation 9.1.132 and $i=j=3$ in Equation 9.1.136, we obtain, respectively

$$g_{k,k33} = -(A+B)x_1R^{-3}\left(1 - \frac{3x_3^2}{R^2}\right) \tag{9.1.138}$$

and

$$2\nabla^2 g_{3,3} = -4Bx_1R^{-3} \tag{9.1.139}$$

Similarly, by letting $p=2$, $q=3$ in Equation 9.1.132 and $i=2, j=3$ in Equation 9.1.136, we write, respectively

$$g_{k,k23} = 3(A+B)x_1x_2x_3R^{-5} \tag{9.1.140}$$

and

$$\nabla^2(g_{2,3}+g_{3,2}) = -2Bx_1x_2R^{-3}(R+x_3)^{-2}(2R+x_3) \tag{9.1.141}$$

Finally, by letting $p=1$, $q=3$ in Equation 9.1.132 and $i=1$, $j=3$ in Equation 9.1.136, we get, respectively

$$g_{k,k13} = -(A+B)x_3R^{-3}\left(1-\frac{3x_1^2}{R^2}\right) \tag{9.1.142}$$

and

$$\nabla^2(g_{1,3}+g_{3,1}) = -2Ax_3R^{-3} + 2BR^{-1}(R+x_3)^{-1}\left(1-\frac{x_1^2}{R^2}\frac{2R+x_3}{R+x_3}\right) \tag{9.1.143}$$

Now, we substitute $\nabla^2 g_{k,k}$, $g_{k,k33}$, and $2\nabla^2 g_{3,3}$ from Equations 9.1.137, 9.1.138, and 9.1.139, respectively, into Equation 9.1.112, and obtain

$$S_{33}^{(2)} = \mu\left\{-4Bx_1R^{-3} + \frac{1}{1-\nu}\left[-2\nu(A+B)x_1R^{-3} + (A+B)x_1R^{-3}\left(1-\frac{3x_3^2}{R^2}\right)\right]\right\} \tag{9.1.144}$$

Also, the substitution of $g_{k,k23}$ and $\nabla^2(g_{2,3}+g_{3,2})$ from Equations 9.1.140 and 9.1.141, respectively, into Equation 9.1.113, yields

$$S_{32}^{(2)} = \mu\left[-2Bx_1x_2R^{-3}(R+x_3)^{-2}(2R+x_3) - \frac{3}{1-\nu}(A+B)x_1x_2x_3R^{-5}\right] \tag{9.1.145}$$

Finally, by inserting $g_{k,k13}$ and $\nabla^2(g_{1,3}+g_{3,1})$ from Equations 9.1.142 and 9.1.143, respectively, into Equation 9.1.114, we arrive at

$$\begin{aligned} S_{13}^{(2)} = \mu\Bigg[&-2Ax_3R^{-3} + 2BR^{-1}(R+x_3)^{-1}\left(1-\frac{x_1^2}{R^2}\frac{2R+x_3}{R+x_3}\right) \\ &+ \frac{1}{1-\nu}(A+B)x_3R^{-3}\left(1-\frac{3x_1^2}{R^2}\right)\Bigg] \end{aligned} \tag{9.1.146}$$

Therefore, the stress components S_{33}, S_{23}, and S_{13} are given by

$$S_{33} = S_{33}^{(1)} + S_{33}^{(2)} \tag{9.1.147}$$

$$S_{23} = S_{23}^{(1)} + S_{23}^{(2)} \tag{9.1.148}$$

$$S_{13} = S_{13}^{(1)} + S_{13}^{(2)} \tag{9.1.149}$$

where $S_{i3}^{(1)}$ $(i=1,2,3)$ is given by Equations 9.1.127 through 9.1.129 and $S_{i3}^{(2)}$ is given by Equations 9.1.144 through 9.1.146.

To compute the constants A, B, and C, we note that the boundary conditions

$$S_{33}(x_1, x_2, 0) = 0, \quad S_{23}(x_1, x_2, 0) = 0 \tag{9.1.150}$$

are satisfied provided

$$2C - 4B + \frac{1-2\nu}{1-\nu}(A+B) = 0$$

$$-B + C = 0 \tag{9.1.151}$$

Hence, the constants are

$$A = \frac{1}{1-2\nu}B, \quad C = B \tag{9.1.152}$$

To satisfy the boundary condition

$$\int_{-\infty}^{+\infty}\int_{-\infty}^{+\infty} S_{13}(x_1, x_2, 0)dx_1dx_2 = -P_0 \tag{9.1.153}$$

we note that, because of Equations 9.1.149 and 9.1.1 52, we obtain

$$S_{13}(x_1, x_2, x_3) = -\frac{2B\mu}{1-2\nu}\frac{3x_1^2x_3}{R^5} \tag{9.1.154}$$

Also, we note that for $x_3 > 0$

$$I \equiv \int_{-\infty}^{+\infty}\int_{-\infty}^{+\infty} \frac{x_1^2x_3}{R^5}dx_1dx_2 = \frac{2}{3}\pi \tag{9.1.155}$$

To prove 9.1.155 we observe that

$$\int_{-\infty}^{+\infty} \frac{x_1^2}{R^5}dx_1 = 2\int_0^{\infty} \frac{x_1^2}{R^5}dx_1 = -\frac{2}{3}\int_0^{\infty} x_1\frac{\partial}{\partial x_1}\left(\frac{1}{R^3}\right)dx_1$$

$$= -\frac{2}{3}\left(x_1\frac{1}{R^3}\Big|_{x_1=0}^{x_1=\infty} - \int_0^{\infty}\frac{dx_1}{R^3}\right) = \frac{2}{3}\int_0^{\infty}\frac{dx_1}{R^3}$$

therefore

$$I = -\frac{2}{3}\frac{\partial}{\partial x_3}\int_{-\infty}^{+\infty}dx_2\int_0^{\infty}\frac{dx_1}{R} = -\frac{4}{3}\frac{\partial}{\partial x_3}\int_0^{\infty}\int_0^{\infty}\frac{1}{R}dx_1dx_2$$

$$= -\frac{4}{3}\frac{\partial}{\partial x_3}\int_0^{\pi/2}d\theta\int_0^{\infty}\frac{r\,dr}{\sqrt{r^2+x_3^2}} = -\frac{2\pi}{3}\frac{x_3}{\sqrt{r^2+x_3^2}}\Bigg|_{r=0}^{r=\infty} = \frac{2}{3}\pi$$

This completes proof of Equation 9.1.155. Therefore, integrating Equation 9.1.154 over the (x_1, x_2) plane and using Equation 9.1.155, we obtain

$$\int_{-\infty}^{+\infty}\int_{-\infty}^{+\infty} S_{13}(x_1, x_2, x_3)dx_1dx_2 = -\frac{4B\mu\pi}{1-2\nu} \tag{9.1.156}$$

Letting $x_3 = 0$ in Equation 9.1.156 and substituting into Equation 9.1.153, we find that

$$B = \frac{(1-\nu)P_0}{4\pi\mu} \tag{9.1.157}$$

As a result, the boundary conditions 9.1.94 and 9.1.95 are satisfied provided (cf. Equation 9.1.152)

$$A = \frac{P_0}{4\pi\mu}, \quad B = \frac{(1-2\nu)P_0}{4\pi\mu}, \quad C = \frac{(1-2\nu)P_0}{4\pi\mu} \tag{9.1.158}$$

This completes the proof of (ii).

Proof of (iii). To prove that $s \to 0$ as $R \to \infty$, we compute $\mathbf{u}$ and $\mathbf{S}$, and show that $\mathbf{u} \to \mathbf{0}$ and $\mathbf{S} \to \mathbf{0}$ as $R \to \infty$. Then it follows from Hooke's law that $\mathbf{E} \to \mathbf{0}$ as $R \to \infty$. First, we compute the stress components. To find $S_{3i}^{(1)}$ we substitute C from Equation 9.1.158 into Equations 9.1.127 through 9.1.129, and arrive at

$$\begin{aligned}
S_{33}^{(1)} &= \frac{(1-2\nu)P_0}{2\pi}\frac{x_1}{R^3} \\
S_{32}^{(1)} &= \frac{(1-2\nu)P_0}{2\pi}\frac{x_1x_2}{R^3}\frac{2R+x_3}{(R+x_3)^2} \\
S_{31}^{(1)} &= -\frac{(1-2\nu)P_0}{2\pi}\frac{1}{R(R+x_3)}\left(1-\frac{x_1^2}{R^2}\frac{2R+x_3}{R+x_3}\right)
\end{aligned} \tag{9.1.159}$$

Also, by substituting A and B from Equation 9.1.158 into Equations 9.1.144 through 9.1.146, we get

$$\begin{aligned}
S_{33}^{(2)} &= -\frac{(1-2\nu)P_0}{2\pi}\frac{x_1}{R^3} - \frac{3P_0}{2\pi}\frac{x_1x_3^2}{R^5} \\
S_{32}^{(2)} &= -\frac{(1-2\nu)P_0}{2\pi}\frac{x_1x_2}{R^3}\frac{2R+x_3}{(R+x_3)^2} - \frac{3P_0}{2\pi}\frac{x_1x_2x_3}{R^5} \\
S_{31}^{(2)} &= \frac{(1-2\nu)P_0}{2\pi}\frac{1}{R(R+x_3)}\left(1-\frac{x_1^2}{R^2}\frac{2R+x_2}{R+x_3}\right) - \frac{3P_0}{2\pi}\frac{x_1^2x_3}{R^5}
\end{aligned} \tag{9.1.160}$$

To find $S_{i2}^{(1)}$ and $S_{i2}^{(2)}$ we use the formulas (cf. Equations 9.1.107 and 9.1.110)

$$S_{12}^{(1)} = 2\mu C \left(\frac{x_1}{R + x_3} \right)_{,12}$$
$$S_{22}^{(1)} = 2\mu C \left(\frac{x_1}{R + x_3} \right)_{,22} \tag{9.1.161}$$

and

$$S_{12}^{(2)} = \mu \left[\nabla^2 (g_{1,2} + g_{2,1}) - \frac{1}{1 - \nu} g_{k,k12} \right]$$
$$S_{22}^{(2)} = \mu \left[2\nabla^2 g_{2,2} + \frac{1}{1 - \nu} (\nu\, g_{k,kll} - g_{k,k22}) \right] \tag{9.1.162}$$

where (cf. Equation 9.1.116)

$$g_k = AR\delta_{k1} + Bx_1 \ln \left(\frac{R + x_3}{L} \right) \delta_{k3} \tag{9.1.163}$$

Substituting C from Equation 9.1.158 into Equation 9.1.161 we write

$$S_{12}^{(1)} = \frac{(1 - 2\nu)P_0}{2\pi} \frac{x_2}{R^3} \frac{1}{(R + x_3)^2} \left(x_1^2 - R^2 + \frac{2x_1^2 R}{R + x_3} \right)$$
$$S_{22}^{(1)} = \frac{(1 - 2\nu)P_0}{2\pi} \frac{x_1}{R^3} \frac{1}{(R + x_3)^2} \left(x_2^2 - R^2 + \frac{2x_2^2 R}{R + x_3} \right) \tag{9.1.164}$$

Next, substituting A and B from Equation 9.1.158 into Equation 9.1.163, we get

$$g_k = \frac{P_0}{4\pi\mu} \left[R\delta_{k1} + (1 - 2\nu)x_1 \ln \left(\frac{R + x_3}{L} \right) \delta_{k3} \right] \tag{9.1.165}$$

Hence, we obtain (cf. Equations 9.1.130 through 9.1.133 and 9.1.136 and 9.1.137)

$$g_{k,k} = \frac{P_0(1 - \nu)}{2\pi\mu} \frac{x_1}{R} \tag{9.1.166}$$

$$g_{k,kp} = \frac{P_0(1 - \nu)}{2\pi\mu} \left(\delta_{1p} \frac{1}{R} - \frac{x_1 x_p}{R^3} \right) \tag{9.1.167}$$

$$g_{k,kpq} = -\frac{P_0(1 - \nu)}{2\pi\mu} \frac{1}{R^3} \left(\delta_{1p} x_q + \delta_{1q} x_p + x_1 \delta_{pq} - 3x_1 x_p x_q \frac{1}{R^2} \right) \tag{9.1.168}$$

$$g_{i,kk} = \frac{P_0}{2\pi\mu} \left\{ \delta_{i1} \frac{1}{R} + (1 - 2\nu)\delta_{i3} \left[\ln \left(\frac{R + x_3}{L} \right) \right]_{,1} \right\} \tag{9.1.169}$$

$$\nabla^2(g_{i,j}+g_{j,i}) = -\frac{P_0}{2\pi\mu}(\delta_{i1}x_j+\delta_{j1}x_i)\frac{1}{R^3}+\frac{(1-2\nu)P_0}{2\pi\mu}$$
$$\times\left\{\delta_{i3}\frac{1}{(R+x_3)}\left[\left(\delta_{j1}\frac{1}{R}-\frac{x_1x_j}{R^3}\right)-\frac{x_1}{R}\frac{1}{R+x_3}\left(\frac{x_j}{R}+\delta_{j3}\right)\right]\right.$$
$$\left.+\,\delta_{j3}\frac{1}{(R+x_3)}\left[\left(\delta_{i1}\frac{1}{R}-\frac{x_1x_i}{R^3}\right)-\frac{x_1}{R}\frac{1}{R+x_3}\left(\frac{x_i}{R}+\delta_{i3}\right)\right]\right\} \quad (9.1.170)$$

Now, by letting $p = 1, q = 2$ in Equation 9.1.168, and $i = 1, j = 2$ in Equation 9.1.170 and substituting into Equation 9.1.162$_1$, we get

$$S_{12}^{(2)} = -\frac{3P_0}{2\pi}\frac{x_1^2x_2}{R^5} \quad (9.1.171)$$

Next, by letting $p = q = l$ in Equation 9.1.168, we write

$$g_{k,kll} = -2(1-\nu)\frac{P_0}{2\pi\mu}\frac{x_1}{R^3} \quad (9.1.172)$$

Therefore, substituting Equations 9.1.168 and 9.1.172 with $p = q = 2$, and 9.1.70 with $i = j = 2$, into 9.1.162$_2$, we obtain

$$S_{22}^{(2)} = \frac{P_0}{2\pi}\frac{x_1}{R^3}\left[(1-2\nu)-\frac{3x_2^2}{R^2}\right] \quad (9.1.173)$$

To find $S_{i1}^{(1)}$ and $S_{i1}^{(2)}$, we calculate the components (cf. Equations 9.1.107 and 9.1.110)

$$S_{11}^{(1)} = 2\mu C\left(\frac{x_1}{R+x_3}\right)_{,11} \quad (9.1.174)$$

and

$$S_{11}^{(2)} = \mu\left[2\nabla^2 g_{1,1}+\frac{1}{1-\nu}(\nu g_{k,kll}-g_{k,11})\right] \quad (9.1.175)$$

By substituting C from Equation 9.1.158 into Equation 9.1.174, and differentiating, we obtain

$$S_{11}^{(1)} = \frac{(1-2\nu)P_0}{2\pi}\frac{x_1}{R^3}\frac{1}{(R+x_3)^2}\left(x_1^2-3R^2+\frac{2x_1^2R}{R+x_3}\right) \quad (9.1.176)$$

Also, by substituting Equations 9.1.170 with $i=j=1$, 9.1.168 with $p=q=1$, and 9.1.172 into Equation 9.1.175, we get

$$S_{11}^{(2)} = \frac{P_0}{2\pi}\frac{x_1}{R^3}\left[(1-2\nu)-\frac{3x_1^2}{R^2}\right] \quad (9.1.177)$$

It follows from Equations 9.1.159$_1$, 9.1.164$_2$, and 9.1.176, that

$$S_{11}^{(1)} + S_{22}^{(1)} + S_{33}^{(1)} = S_{kk}^{(1)} = \text{tr}\,\mathbf{S}^{(1)} = 0 \tag{9.1.178}$$

as should be expected since, by Equation 9.1.107,

$$S_{kk}^{(1)} = 2\mu\nabla^2\phi = 0 \tag{9.1.179}$$

On the other hand, Equations 9.1.160$_1$, 9.1.173, and 9.1.177 imply that

$$S_{11}^{(2)} + S_{22}^{(2)} + S_{33}^{(2)} = S_{kk}^{(2)} = \text{tr}\,\mathbf{S}^{(2)} = -\frac{P_0(1+\nu)}{\pi}\frac{x_1}{R^3} \tag{9.1.180}$$

Therefore, using Equations 9.1.159, 9.1.160, 9.1.164, 9.1.171, 9.1.173, 9.1.176, 9.1.177, and the relation

$$S_{ij} = S_{ij}^{(1)} + S_{ij}^{(2)} \tag{9.1.181}$$

we arrive at

$$S_{11} = \frac{P_0}{2\pi}\frac{x_1}{R^3}\left\{-\frac{3x_1^2}{R^2} + \frac{(1-2\nu)}{(R+x_3)^2}\left[R^2 - x_2^2\left(1 + \frac{2R}{R+x_3}\right)\right]\right\} \tag{9.1.182}$$

$$S_{12} = \frac{P_0}{2\pi}\frac{x_2}{R^3}\left\{-\frac{3x_1^2}{R^2} - \frac{(1-2\nu)}{(R+x_3)^2}\left[R^2 - x_1^2\left(1 + \frac{2R}{R+x_3}\right)\right]\right\} \tag{9.1.183}$$

$$S_{22} = \frac{P_0}{2\pi}\frac{x_1}{R^3}\left\{-\frac{3x_2^2}{R^2} + \frac{(1-2\nu)}{(R+x_3)^2}\left[3R^2 - x_1^2\left(1 + \frac{2R}{R+x_3}\right)\right]\right\} \tag{9.1.184}$$

$$S_{13} = -\frac{3P_0x_1^2x_3}{2\pi R^5} \tag{9.1.185}$$

$$S_{23} = -\frac{3P_0x_1x_2x_3}{2\pi R^5} \tag{9.1.186}$$

$$S_{33} = -\frac{3P_0x_1x_3^2}{2\pi R^5} \tag{9.1.187}$$

This completes the derivation of **S** in Cerruti's problem. To obtain the associated displacement field **u**, we use the formula

$$u_i = u_i^{(1)} + u_i^{(2)} \tag{9.1.188}$$

where $u_i^{(1)}$ and $u_i^{(2)}$ are given by Equations 9.1.107 and 9.1.108, respectively. Hence, substituting

$$\phi = \frac{(1-2\nu)P_0}{4\pi\mu}\left(\frac{x_1}{R+x_3}\right) \tag{9.1.189}$$

into Equation 9.1.107$_1$, and Equations 9.1.167 with $p = i$ and 9.1.169 into Equation 9.1.108, we obtain

$$u_1 = \frac{P_0}{4\pi\mu R}\left\{1 + \frac{x_1^2}{R^2} + (1-2\nu)\left[\frac{R}{R+x_3} - \frac{x_1^2}{(R+x_3)^2}\right]\right\} \tag{9.1.190}$$

$$u_2 = \frac{P_0}{4\pi\mu}\frac{x_1 x_2}{R^3}\left[1 - (1-2\nu)\frac{R^2}{(R+x_3)^2}\right] \tag{9.1.191}$$

$$u_3 = \frac{P_0}{4\pi\mu}\frac{x_1}{R^2}\left[\frac{x_3}{R} + (1-2\nu)\frac{R}{R+x_3}\right] \tag{9.1.192}$$

It follows from Equations 9.1.182 through 9.1.187 and 9.1.190 through 9.1.192 that

$$\mathbf{S} = \mathbf{S}_0\, 0(R^{-2}) \quad \text{as } R \to \infty \tag{9.1.193}$$

and

$$\mathbf{u} = \mathbf{u}_0\, 0(R^{-1}) \quad \text{as } R \to \infty \tag{9.1.194}$$

where $\mathbf{S}_0$ and $\mathbf{u}_0$ are a given tensor and a given vector, respectively, and the symbol $0(1)$ has the usual meaning: $0[f(R)]/f(R) \to 0(1) < \infty$ as $R \to \infty$, where $f = f(R)$ is the prescribed function. The relations 9.1.193 and 9.1.194 imply that $s = [\mathbf{u}, \mathbf{E}, \mathbf{S}] \to 0$ as $R \to \infty$, and this completes the proof of (iii).

9.2 THREE-DIMENSIONAL SOLUTIONS OF NONISOTHERMAL ELASTOSTATICS

In this section, a number of closed-form solutions to three-dimensional BVP of nonisothermal elastostatics are presented. In Chapter 4, a closed-form solution to the problem of finding a thermoelastic state $s = [\mathbf{u}, \mathbf{E}, \mathbf{S}]$ corresponding to a temperature change T, which is constant inside a sphere of radius $r = a$ and vanishes outside of the sphere in a homogeneous isotropic infinite body, was obtained (see Examples 4.1.10 and 4.1.11). In the following we discuss (A) a thermoelastic state in a homogeneous isotropic infinite body corresponding to a concentrated source of heat, (B) a thermoelastic state in a homogeneous isotropic semi-infinite body corresponding to an internal source of heat, and (C) a thermoelastic state in a homogeneous isotropic semi-infinite body corresponding to a heat exposure on the boundary plane. In all three cases, the closed-form solution is obtained in the form $s = s_1 + s_2$, where s_1 is a particular thermoelastic state of three-dimensional thermoelastostatics, and s_2 is an elastic state of three-dimensional isothermal elastostatics.

9.2.1 A Thermoelastic State due to the Action of a Point Source of Heat in an Infinite Body

Let a homogeneous isotropic infinite elastic body be referred to the Cartesian coordinate system $\{x_i\}$, and let a point source of heat be located at the origin of the system. A temperature field $T = T(\mathbf{x})$ produced by the heat source in the body then satisfies Poisson's equation (see Equation 4.1.72 with $Q = r_0\delta(\mathbf{x})$)

$$\nabla^2 T = -\frac{r_0}{\kappa}\,\delta(\mathbf{x}) \quad \text{for } \mathbf{x} \in E^3 \tag{9.2.1}$$

subject to the vanishing condition at infinity

$$T(\mathbf{x}) \to 0 \quad \text{as } |\mathbf{x}| \to \infty \tag{9.2.2}$$

Here, $\delta(\mathbf{x}) = \delta(x_1)\delta(x_2)\delta(x_3)$, is a three-dimensional Dirac delta function, r_0 represents a constant intensity of the heat source, and κ stands for the diffusivity coefficient [4].

Since

$$\nabla^2 \frac{1}{R} = -4\pi\,\delta(\mathbf{x}) \tag{9.2.3}$$

where

$$R = |\mathbf{x}| \tag{9.2.4}$$

therefore, the only solution to Equation 9.2.1 that satisfies the condition 9.2.2 takes the form

$$T(\mathbf{x}) = \frac{r_0}{4\pi\kappa}\,\frac{1}{R} \tag{9.2.5}$$

Clearly, the temperature field T has a spherical symmetry, $T = T(R)$, where R is the radial component of a point P referred to the spherical coordinates (R, φ, θ) shown in Figure 9.3 and the operator ∇^2 may be written in the form

$$\nabla^2 = \frac{\partial^2}{\partial R^2} + \frac{2}{R}\frac{\partial}{\partial R} \tag{9.2.6}$$

In view of the spherical symmetry, a thermoelastic state $s = [\mathbf{u}, \mathbf{E}, \mathbf{S}]$ corresponding to the temperature $T = T(R)$ is generated by a thermoelastic displacement potential $\phi = \phi(R)$ through the formulas (cf. Example 4.1.9)

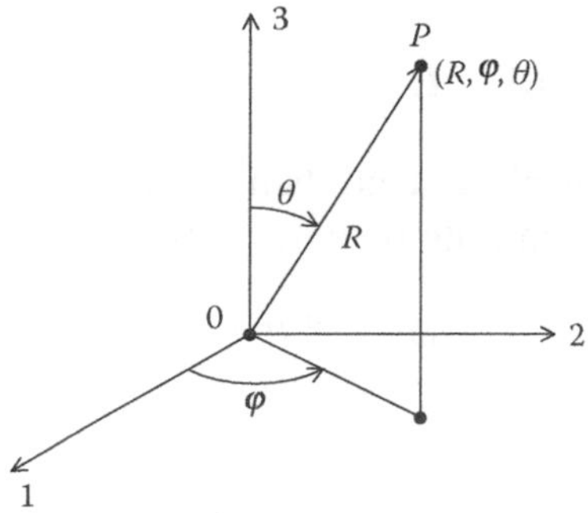

FIGURE 9.3 The spherical coordinates in the heat source problem.

$$\begin{aligned} \mathbf{u} &= \nabla\phi \\ \mathbf{E} &= \nabla\nabla\phi \\ \mathbf{S} &= 2\mu(\nabla\nabla\phi - \nabla^2\phi\mathbf{1}) \end{aligned} \tag{9.2.7}$$

where ϕ satisfies Poisson's equation

$$\nabla^2\phi = mT \tag{9.2.8}$$

with

$$m = \frac{1+\nu}{1-\nu}\alpha \tag{9.2.9}$$

Here, μ, ν, and α denote the shear modulus, Poisson's ratio, and the coefficient of linear thermal expansion, respectively.

In the spherical coordinates Equations 9.2.7 take the form

$$\mathbf{u} = [u_R, 0, 0] = \left[\frac{\partial\phi}{\partial R}, 0, 0\right] \tag{9.2.10}$$

$$\mathbf{E} = \begin{bmatrix} E_{RR} & E_{R\varphi} & E_{R\theta} \\ E_{\varphi R} & E_{\varphi\varphi} & E_{\varphi\theta} \\ E_{\theta R} & E_{\theta\varphi} & E_{\theta\theta} \end{bmatrix} = \begin{bmatrix} \dfrac{\partial^2\phi}{\partial R^2} & 0 & 0 \\ 0 & \dfrac{1}{R}\dfrac{\partial\phi}{\partial R} & 0 \\ 0 & 0 & \dfrac{1}{R}\dfrac{\partial\phi}{\partial R} \end{bmatrix} \tag{9.2.11}$$

$$\mathbf{S} = \begin{bmatrix} S_{RR} & S_{R\varphi} & S_{R\theta} \\ S_{\varphi R} & S_{\varphi\varphi} & S_{\varphi\theta} \\ S_{\theta R} & S_{\theta\varphi} & S_{\theta\theta} \end{bmatrix}$$

$$= \begin{bmatrix} 2\mu\left(\dfrac{\partial^2\phi}{\partial R^2} - \nabla^2\phi\right) & 0 & 0 \\ 0 & 2\mu\left(\dfrac{1}{R}\dfrac{\partial\phi}{\partial R} - \nabla^2\phi\right) & 0 \\ 0 & 0 & 2\mu\left(\dfrac{1}{R}\dfrac{\partial\phi}{\partial R} - \nabla^2\phi\right) \end{bmatrix} \tag{9.2.12}$$

Now, if the following asymptotic conditions are imposed on $s = [\mathbf{u}, \mathbf{E}, \mathbf{S}]$

$$\mathbf{u} \to [u_R^0, 0, 0], \quad \mathbf{E} \to \mathbf{0}, \quad \mathbf{S} \to \mathbf{0} \quad \text{as} \quad R \to \infty \tag{9.2.13}$$

where u_R^0 is a constant radial displacement, then it is easy to show that the only solution to Poisson's equation 9.2.8 for an infinite body takes the form

$$\phi = u_R^0 R + \phi_0 \tag{9.2.14}$$

where

$$u_R^0 = \frac{mr_0}{8\pi\kappa} \tag{9.2.15}$$

and ϕ_0 is an arbitrary constant. By substituting ϕ from Equation 9.2.14 into Equations 9.2.10 through 9.2.12, and using ∇^2 in the form 9.2.6, we arrive at

$$u_R = u_R^0 \tag{9.2.16}$$

$$\mathbf{E} = \frac{u_R^0}{R}\begin{bmatrix} 0 & 0 & 0 \\ 0 & 1 & 0 \\ 0 & 0 & 1 \end{bmatrix} \tag{9.2.17}$$

$$\mathbf{S} = -2\mu\,\frac{u_R^0}{R}\begin{bmatrix} 2 & 0 & 0 \\ 0 & 1 & 0 \\ 0 & 0 & 1 \end{bmatrix} \tag{9.2.18}$$

Therefore, a thermoelastic state $s = [\mathbf{u}, \mathbf{E}, \mathbf{S}]$ corresponding to the action of a point heat source in a homogeneous isotropic infinite elastic body and satisfying the asymptotic conditions 9.2.13 is represented by Equations 9.2.16 through 9.2.18. Note that s is obtained in the form

$$s = s_1 + s_2 \tag{9.2.19}$$

where $s_1 = [\mathbf{u}, \mathbf{E}, \mathbf{S}]$ is a thermoelastic state of three-dimensional theory, while $s_2 = [\mathbf{0}, \mathbf{0}, \mathbf{0}]$ is a zero elastic state of three-dimensional isothermal elastostatics.

Notes:

(1) In the Cartesian coordinates $\{x_i\}$ we have $R = \sqrt{x_i x_i}$, so substitution of ϕ from Equation 9.2.14 into Equations 9.2.7 yields

$$u_i = \phi_{,i} = u_R^0\,\frac{x_i}{R} \tag{9.2.20}$$

$$E_{ij} = \phi_{,ij} = u_R^0\left(\frac{\delta_{ij}}{R} - \frac{x_i x_j}{R^3}\right) \tag{9.2.21}$$

$$S_{ij} = -2\mu u_R^0\left(\frac{\delta_{ij}}{R} + \frac{x_i x_j}{R^3}\right) \tag{9.2.22}$$

(2) In the cylindrical coordinates (r, φ, z), shown in Figure 9.4, we obtain $R = \sqrt{r^2 + z^2}$; so if the potential ϕ is treated as a function of r and z, that is, $\phi = \phi(r, z)$, the formulas 9.2.7 reduce to

$$u_r = \frac{\partial \phi}{\partial r}, \quad u_\varphi = 0, \quad u_z = \frac{\partial \phi}{\partial z} \tag{9.2.23}$$

$$\begin{aligned} E_{rr} &= \frac{\partial^2 \phi}{\partial r^2}, & E_{r\varphi} &= 0, & E_{rz} &= \frac{\partial^2 \phi}{\partial r \partial z} \\ E_{\varphi r} &= 0, & E_{\varphi\varphi} &= \frac{1}{r}\frac{\partial \phi}{\partial r}, & E_{\varphi z} &= 0 \\ E_{zr} &= \frac{\partial^2 \phi}{\partial r \partial z}, & E_{z\varphi} &= 0, & E_{zz} &= \frac{\partial^2 \phi}{\partial z^2} \end{aligned} \tag{9.2.24}$$

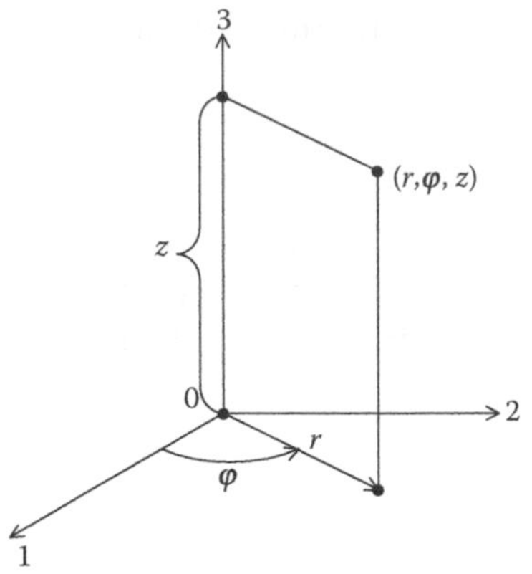

FIGURE 9.4 The cylindrical coordinates in the heat source problem.

$$\begin{aligned}
S_{rr} &= -2\mu\left(\frac{1}{r}\frac{\partial\phi}{\partial r}+\frac{\partial^2\phi}{\partial z^2}\right), \quad S_{r\varphi}=0, \quad S_{rz}=2\mu\frac{\partial^2\phi}{\partial r\partial z} \\
S_{\varphi r} &= 0, \quad S_{\varphi\varphi}=-2\mu\left(\frac{\partial^2\phi}{\partial r^2}+\frac{\partial^2\phi}{\partial z^2}\right), \quad S_{\varphi z}=0 \qquad (9.2.25) \\
S_{zr} &= 2\mu\frac{\partial^2\phi}{\partial r\partial z}, \quad S_{z\varphi}=0, \quad S_{zz}=-2\mu\left(\frac{1}{r}\frac{\partial\phi}{\partial r}+\frac{\partial^2\phi}{\partial r^2}\right)
\end{aligned}$$

Therefore, substituting ϕ from Equation 9.2.14 into Equations 9.2.23 through 9.2.25, we arrive at

$$u_r=u_R^0\,\frac{r}{R}, \quad u_\varphi=0, \quad u_z=u_R^0\,\frac{z}{R} \qquad (9.2.26)$$

$$\begin{aligned}
E_{rr} &= u_R^0\left(\frac{1}{R}-\frac{r^2}{R^3}\right), \quad E_{r\varphi}=0, \quad E_{rz}=-u_R^0\,\frac{rz}{R^3} \\
E_{\varphi r} &= 0, \quad E_{\varphi\varphi}=\frac{u_R^0}{R}, \quad E_{\varphi z}=0 \qquad (9.2.27) \\
E_{zr} &= -u_R^0\,\frac{rz}{R^3}, \quad E_{z\varphi}=0, \quad E_{zz}=u_R^0\left(\frac{1}{R}-\frac{z^2}{R^3}\right)
\end{aligned}$$

$$\begin{aligned}
S_{rr} &= -2\mu u_R^0\,\frac{1}{R}\left(1+\frac{r^2}{R^2}\right), \quad S_{r\varphi}=0, \quad S_{rz}=-2\mu u_R^0\,\frac{zr}{R^3} \\
S_{\varphi r} &= 0, \quad S_{\varphi\varphi}=-2\mu\,\frac{u_R^0}{R}, \quad S_{\varphi z}=0 \qquad (9.2.28) \\
S_{zr} &= -2\mu u_R^0\,\frac{zr}{R^3}, \quad S_{z\varphi}=0, \quad S_{zz}=-2\mu\,\frac{u_R^0}{R}\left(2-\frac{r^2}{R^2}\right)
\end{aligned}$$

The formulas 9.2.28 will be used to obtain a solution to a heat source problem for a semispace.

9.2.2 A Thermoelastic State in a Homogeneous Isotropic Semi-Infinite Body Caused by an Internal Source of Heat

In this section, we are to find a thermoelastic state $s = [\mathbf{u}, \mathbf{E}, \mathbf{S}]$ produced by a concentrated source of heat located at an internal point of a semispace. The source is located at a point $(0, 0, \zeta)$ of the Cartesian system of coordinates (x_1, x_2, x_3) and $\zeta > 0$; the semispace is described by the inequalities

$$|x_1| < \infty, \quad |x_2| < \infty, \quad x_3 \geq 0 \tag{9.2.29}$$

It is assumed that a temperature field $T = T(\mathbf{x})$ produced by a heat source of intensity r_0 satisfies Poisson's equation

$$\nabla^2 T = -\frac{r_0}{\kappa}\,\delta(x_1)\delta(x_2)\delta(x_3 - \zeta) \quad \text{for } |x_1| < \infty, |x_2| < \infty, \;\; x_3 > 0 \tag{9.2.30}$$

subject to the boundary condition

$$T(x_1, x_2, 0) = 0 \quad \text{for } |x_1| < \infty, \quad |x_2| < \infty \tag{9.2.31}$$

and the vanishing condition at infinity

$$T \to 0 \quad \text{as } R = |\mathbf{x}| \to \infty \tag{9.2.32}$$

The heat conduction BVP described by Equations 9.2.30 through 9.2.32 is axially symmetric with respect to the x_3-axis, so it can be replaced by the cylindrical coordinate formulation: Find a temperature $T = T(r, z)$ for $r > 0$, $z > 0$, that satisfies Poisson's equation

$$\nabla^2 T = -\frac{r_0}{\kappa}\,\frac{\delta(r)}{2\pi r}\,\delta(z - \zeta) \quad \text{for } r > 0, \quad z > 0 \tag{9.2.33}$$

subject to the boundary condition

$$T(r, 0) = 0 \quad \text{for } r > 0 \tag{9.2.34}$$

and the asymptotic condition

$$T \to 0 \quad \text{as } R = \sqrt{r^2 + z^2} \to \infty \tag{9.2.35}$$

Here

$$\nabla^2 = \frac{\partial^2}{\partial r^2} + \frac{1}{r}\frac{\partial}{\partial r} + \frac{\partial^2}{\partial z^2} \tag{9.2.36}$$

and

$$\frac{\delta(r)}{2\pi r} = \delta(x_1)\delta(x_2) \tag{9.2.37}$$

In addition, it is assumed that the thermoelastic state $s = [\mathbf{u}, \mathbf{E}, \mathbf{S}]$ produced by the temperature $T = T(r, z)$ satisfies the boundary conditions

$$S_{zz}(r, 0) = S_{rz}(r, 0) = 0 \quad \text{for } r > 0 \tag{9.2.38}$$

and suitable vanishing conditions as $R = \sqrt{r^2 + z^2} \to \infty$.

9.2.2.1 Solution to the Heat Conduction Problem

A solution to Poisson's equation subject to the conditions 9.2.34 and 9.2.35 is represented in the form

$$T = T_1(r, z) + T_2(r, z) \tag{9.2.39}$$

where

$$\nabla^2 T_1 = -\frac{r_0}{\kappa} \frac{\delta(r)}{2\pi r} \delta(z - \zeta) \quad \text{for } r > 0, \quad |z| < \infty \tag{9.2.40}$$

$$T_1 \to 0 \quad \text{as} \quad R \to \infty \tag{9.2.41}$$

and

$$\nabla^2 T_2 = 0 \quad \text{for } r > 0, \quad z > 0 \tag{9.2.42}$$

$$T_2(r, 0) = -T_1(r, 0) \quad \text{for } r > 0 \tag{9.2.43}$$

$$T_2 \to 0 \quad \text{as } R \to \infty \tag{9.2.44}$$

We note that if T_1 satisfies Equations 9.2.40 and 9.2.41 and T_2 meets Equations 9.2.42 through 9.2.44, then T is a solution to the heat conduction problem 9.2.33 through 9.2.35.

Since the problem associated with T_1 is the one for an infinite body, therefore, by virtue of Equations 9.2.1 through 9.2.5 in which x_3 is shifted to $x_3 - \zeta$, we obtain

$$T_1(r, z) = \frac{r_0}{4\pi\kappa} \frac{1}{R_1} \tag{9.2.45}$$

where

$$R_1 = \sqrt{r^2 + (z - \zeta)^2} \tag{9.2.46}$$

Also, it is easy to check that

$$T_2(r, z) = \frac{A}{R_2} \tag{9.2.47}$$

where

$$R_2 = \sqrt{r^2 + (z + \zeta)^2} \tag{9.2.48}$$

and

$$A = -\frac{r_0}{4\pi\kappa} \tag{9.2.49}$$

Hence, a solution to the heat conduction problem, by virtue of Equation 9.2.39, takes the form

$$T(r,z) = A\left(\frac{1}{R_1} - \frac{1}{R_2}\right), \quad r > 0, \quad z > 0 \tag{9.2.50}$$

9.2.2.2 Solution to the Thermoelastic Problem

A solution $s = [\mathbf{u}, \mathbf{E}, \mathbf{S}]$ to the thermoelastic problem for the semispace is sought in the form

$$s = s_1 + s_2 \tag{9.2.51}$$

where $s_1 = [\mathbf{u}^{(1)}, \mathbf{E}^{(1)}, \mathbf{S}^{(1)}]$ is a thermoelastic state for an infinite space corresponding to the temperature field given by Equation 9.2.50; and $s_2 = [\mathbf{u}^{(2)}, \mathbf{E}^{(2)}, \mathbf{S}^{(2)}]$ is an elastic state for a semispace produced by the boundary load

$$S_{zz}^{(2)}(r,0) = -S_{zz}^{(1)}(r,0), \quad S_{rz}^{(2)}(r,0) = -S_{rz}^{(1)}(r,0) \quad \text{for } r > 0 \tag{9.2.52}$$

and corresponding to vanishing conditions at infinity.

By using a superposition principle of linear thermoelasticity and the solution of Part A, it is easy to see that

$$s_1 = \hat{s}_1 + \hat{s}_2 \tag{9.2.53}$$

where $\hat{s}_1 = [\hat{\mathbf{u}}^{(1)}, \hat{\mathbf{E}}^{(1)}, \hat{\mathbf{S}}^{(1)}]$ is obtained from Equations 9.2.26 through 9.2.28 in which z is replaced by $z - \zeta$, and $\hat{s}_2 = [\hat{\mathbf{u}}^{(2)}, \hat{\mathbf{E}}^{(2)}, \hat{\mathbf{S}}^{(2)}]$ is obtained from Equations 9.2.26 through 9.2.28 in which r_0 is replaced by $-r_0$, and z by $z + \zeta$. In the expanded form Equation 9.2.53 then reads

$$u_r^{(1)} = u_R^0\, r\left(\frac{1}{R_1} - \frac{1}{R_2}\right), \quad u_\varphi^{(1)} = 0, \quad u_z^{(1)} = u_R^0\left(\frac{z-\zeta}{R_1} - \frac{z+\zeta}{R_2}\right) \tag{9.2.54}$$

$$\begin{gathered}
E_{rr}^{(1)} = u_R^0\left[\frac{1}{R_1} - \frac{1}{R_2} - r^2\left(\frac{1}{R_1^3} - \frac{1}{R_2^3}\right)\right] \\
E_{r\varphi}^{(1)} = 0, \quad E_{rz}^{(1)} = -u_R^0 r\left(\frac{z-\zeta}{R_1^3} - \frac{z+\zeta}{R_2^3}\right) \\
E_{\varphi r}^{(1)} = 0, \quad E_{\varphi\varphi}^{(1)} = u_R^0\left(\frac{1}{R_1} - \frac{1}{R_2}\right), \quad E_{\varphi z}^{(1)} = 0 \\
E_{zr}^{(1)} = -u_R^0 r\left(\frac{z-\zeta}{R_1^3} - \frac{z+\zeta}{R_2^3}\right), \quad E_{z\varphi}^{(1)} = 0, \\
E_{zz}^{(1)} = u_R^0\left[\frac{1}{R_1} - \frac{1}{R_2} - \left(\frac{(z-\zeta)^2}{R_1^3} - \frac{(z+\zeta)^2}{R_2^3}\right)\right]
\end{gathered} \tag{9.2.55}$$

$$
\begin{aligned}
S_{rr}^{(1)} &= -2\mu u_R^0\left[\frac{1}{R_1}-\frac{1}{R_2}+r^2\left(\frac{1}{R_1^3}-\frac{1}{R_2^3}\right)\right]\\
S_{r\varphi}^{(1)} &= 0, \quad S_{rz}^{(1)} = -2\mu u_R^0\, r\left(\frac{z-\zeta}{R_1^3}-\frac{z+\zeta}{R_2^3}\right)\\
S_{\varphi r}^{(1)} &= 0, \quad S_{\varphi\varphi}^{(1)} = -2\mu u_R^0\left(\frac{1}{R_1}-\frac{1}{R_2}\right), \quad S_{\varphi z}^{(1)} = 0,\\
S_{zr}^{(1)} &= -2\mu u_R^0 r\left(\frac{z-\zeta}{R_1^3}-\frac{z+\zeta}{R_2^3}\right), \quad S_{z\varphi}^{(1)} = 0\\
S_{zz}^{(1)} &= -2\mu u_R^0\left[2\left(\frac{1}{R_1}-\frac{1}{R_2}\right)-r^2\left(\frac{1}{R_1^3}-\frac{1}{R_2^3}\right)\right]
\end{aligned}
\tag{9.2.56}
$$

Now, we note that Equations 9.2.56 imply that

$$
S_{zz}^{(1)}(r,0) = 0, \quad S_{zr}^{(1)}(r,0) = 4\mu u_R^0\zeta\,\frac{r}{(r^2+\zeta^2)^{3/2}}, \quad r>0 \tag{9.2.57}
$$

so the boundary conditions 9.2.52 take the form

$$
S_{zz}^{(2)}(r,0) = 0, \quad S_{zr}^{(2)}(r,0) = -4\mu u_R^0\zeta\,\frac{r}{(r^2+\zeta^2)^{3/2}}, \quad r>0 \tag{9.2.58}
$$

To obtain the isothermal elastic state $s_2 = [\mathbf{u}^{(2)}, \mathbf{E}^{(2)}, \mathbf{S}^{(2)}]$ we use Love's solution (see Equations 7.1.93 through 7.1.97 and Example 7.1.8 in which $\theta = \varphi$)

$$
\mathbf{u}^{(2)} = \left[u_r^{(2)},\ 0,\ u_z^{(2)}\right] \tag{9.2.59}
$$

where

$$
\begin{aligned}
u_r^{(2)} &= -\frac{1}{2(1-\nu)}\frac{\partial^2}{\partial r\partial z}\chi\\
u_z^{(2)} &= \frac{1}{2(1-\nu)}\left[2(1-\nu)\nabla^2-\frac{\partial^2}{\partial z^2}\right]\chi
\end{aligned}
\tag{9.2.60}
$$

and

$$
\nabla^2\nabla^2\chi = 0 \tag{9.2.61}
$$

The associated strain and stress tensors are obtained from the formulas

$$
\begin{aligned}
E_{rr}^{(2)} &= -\frac{1}{2(1-\nu)}\frac{\partial^3\chi}{\partial r^2\partial z}, \quad E_{r\varphi}^{(2)} = 0\\
E_{rz}^{(2)} &= \frac{1}{2(1-\nu)}\frac{\partial}{\partial r}\left[(1-\nu)\nabla^2-\frac{\partial^2}{\partial z^2}\right]\chi
\end{aligned}
$$

$$E_{\varphi r}^{(2)} = 0, \quad E_{\varphi\varphi}^{(2)} = -\frac{1}{2(1-\nu)} \frac{1}{r} \frac{\partial^2}{\partial r \partial z} \chi, \quad E_{\varphi z}^{(2)} = 0 \tag{9.2.62}$$

$$E_{zr}^{(2)} = E_{rz}^{(2)}, \quad E_{z\varphi}^{(2)} = 0$$

$$E_{zz}^{(2)} = \frac{1}{2(1-\nu)} \frac{\partial}{\partial z} \left[2(1-\nu)\nabla^2 - \frac{\partial^2}{\partial z^2} \right] \chi$$

and

$$S_{rr}^{(2)} = \frac{\mu}{1-\nu} \frac{\partial}{\partial z} \left(\nu\nabla^2 - \frac{\partial^2}{\partial r^2} \right) \chi, \quad S_{r\varphi}^{(2)} = 0$$

$$S_{rz}^{(2)} = \frac{\mu}{1-\nu} \frac{\partial}{\partial r} \left[(1-\nu)\nabla^2 - \frac{\partial^2}{\partial z^2} \right] \chi$$

$$S_{\varphi r}^{(2)} = 0, \quad S_{\varphi\varphi}^{(2)} = \frac{\mu}{1-\nu} \frac{\partial}{\partial z} \left(\nu\nabla^2 - \frac{1}{r} \frac{\partial}{\partial r} \right) \chi, \quad S_{\varphi z}^{(2)} = 0 \tag{9.2.63}$$

$$S_{zr}^{(2)} = S_{rz}^{(2)}, \quad S_{z\varphi}^{(2)} = 0$$

$$S_{zz}^{(2)} = \frac{\mu}{1-\nu} \frac{\partial}{\partial z} \left[(2-\nu)\nabla^2 - \frac{\partial^2}{\partial z^2} \right] \chi$$

The biharmonic function $\chi = \chi(r, z)$ is to be selected in such a way that the boundary conditions 9.2.58 are satisfied, and

$$s_2 = [\mathbf{u}^{(2)}, \mathbf{E}^{(2)}, \mathbf{S}^{(2)}] \to [\mathbf{0}, \mathbf{0}, \mathbf{0}] \quad \text{as } R = \sqrt{r^2 + z^2} \to \infty$$

A hint to obtain χ comes from the fact that the function $r(r^2 + \zeta^2)^{-3/2}$ occurring in the second of the boundary conditions may be represented by the integral (cf. Equation 9.1.41 in which $z = \zeta$)

$$\frac{r}{(r^2 + \zeta^2)^{3/2}} = -\frac{\partial}{\partial r} \left(\frac{1}{\sqrt{r^2 + \zeta^2}} \right) = -\frac{\partial}{\partial r} \int_0^\infty e^{-\alpha\zeta} J_0(\alpha r) d\alpha \tag{9.2.64}$$

as well as from the observation that the integral

$$\chi(r, z) = \int_0^\infty [A(\alpha) + B(\alpha)\alpha z] \, e^{-\alpha z} J_0(\alpha r) d\alpha \tag{9.2.65}$$

in which $A = A(\alpha)$ and $B = B(\alpha)$ are arbitrary functions, is biharmonic for $r > 0$, $z > 0$, and vanishes as $R = \sqrt{r^2 + z^2} \to \infty$. Substituting Equation 9.2.65 into the equation (cf. Equations 9.2.63)

$$S_{zz}^{(2)} = \frac{\mu}{1-\nu} \frac{\partial}{\partial z} \left[(2-\nu)\nabla^2 - \frac{\partial^2}{\partial z^2} \right] \chi \tag{9.2.66}$$

and using the relation (see Equations 9.1.21 and 9.1.22)

$$\left(\nabla_r^2 + \alpha^2\right) J_0(\alpha r) = 0 \tag{9.2.67}$$

in which

$$\nabla_r^2 = \frac{\partial^2}{\partial r^2} + \frac{1}{r}\frac{\partial}{\partial r} \tag{9.2.68}$$

we arrive at

$$S_{zz}^{(2)}(r,z) = \frac{\mu}{1-\nu}\int_0^\infty \alpha^3\,[A + (1-2\nu)B + \alpha z B]\,e^{-\alpha z} J_0(\alpha r)d\alpha \tag{9.2.69}$$

Hence, the first of boundary conditions 9.2.58 is satisfied provided

$$A(\alpha) = -(1-2\nu)B(\alpha) \tag{9.2.70}$$

Substituting Equation 9.2.70 into Equation 9.2.69 we get

$$S_{zz}^{(2)}(r,z) = \frac{\mu}{1-\nu} z\int_0^\infty \alpha^4 B(\alpha)e^{-\alpha z} J_0(\alpha r)d\alpha \tag{9.2.71}$$

and inserting Equation 9.2.70 into Equation 9.2.65 we obtain

$$\chi(r,z) = -\int_0^\infty [(1-2\nu) - \alpha z]B(\alpha)e^{-\alpha z} J_0(\alpha r)d\alpha \tag{9.2.72}$$

The only unknown function $B = B(\alpha)$ is to be found from the second of boundary conditions 9.2.58. To this end we substitute Equation 9.2.72 into the relation (see Equations 9.2.63)

$$S_{zr}^{(2)} = \frac{\mu}{1-\nu}\frac{\partial}{\partial r}\left[(1-\nu)\nabla^2 - \frac{\partial^2}{\partial z^2}\right]\chi \tag{9.2.73}$$

and get

$$S_{zr}^{(2)}(r,z) = -\frac{\mu}{1-\nu}\int_0^\infty \alpha^3(1-\alpha z)B(\alpha)e^{-\alpha z} J_1(\alpha r)d\alpha \tag{9.2.74}$$

Now, we insert Equation 9.2.64 into the second of boundary conditions 9.2.58 and obtain

$$S_{zr}^{(2)}(r,0) = -4\mu u_R^0 \zeta \int_0^\infty \alpha e^{-\alpha\zeta} J_1(\alpha r)d\alpha \tag{9.2.75}$$

Finally, by equating the RHS of Equation 9.2.74 at $z = 0$ to the RHS of Equation 9.2.75 we have

$$B(\alpha) = 4(1-\nu)u_R^0\zeta\alpha^{-2}e^{-\alpha\zeta} \tag{9.2.76}$$

And, inserting Equation 9.2.76 into Equations 9.2.71 and 9.2.74, we write, respectively

$$S_{zz}^{(2)}(r,z) = 4\mu u_R^0 \zeta z \int_0^\infty \alpha^2 e^{-\alpha(z+\zeta)} J_0(\alpha r) d\alpha \tag{9.2.77}$$

and

$$S_{zr}^{(2)} = -4\mu u_R^0 \zeta \int_0^\infty \alpha(1-\alpha z) e^{-\alpha(z+\zeta)} J_1(\alpha r) d\alpha \tag{9.2.78}$$

The remaining stress components are obtained from the relations (see Equations 9.2.63)

$$S_{rr}^{(2)} = \frac{\mu}{1-\nu} \frac{\partial}{\partial z} \left(\nu \nabla^2 - \frac{\partial^2}{\partial r^2} \right) \chi \tag{9.2.79}$$

$$S_{\varphi\varphi}^{(2)} = \frac{\mu}{1-\nu} \frac{\partial}{\partial z} \left(\nu \nabla^2 - \frac{1}{r}\frac{\partial}{\partial r} \right) \chi \tag{9.2.80}$$

Using the relations

$$\nabla^2 \{[-(1-2\nu) + \alpha z] e^{-\alpha z} J_0(\alpha r)\} = -2\alpha^2 e^{-\alpha z} J_0(\alpha r) \tag{9.2.81}$$

$$\frac{\partial^2}{\partial r^2} J_0(\alpha r) = -\alpha^2 \left[J_0(\alpha r) - \frac{J_1(\alpha r)}{\alpha r} \right] \tag{9.2.82}$$

$$\frac{1}{r}\frac{\partial}{\partial r} J_0(\alpha r) = -\alpha^2 \frac{J_1(\alpha r)}{\alpha r} \tag{9.2.83}$$

and substituting χ from Equation 9.2.72 into Equations 9.2.79 and 9.2.80, we obtain, respectively

$$S_{rr}^{(2)} = \frac{\mu}{1-\nu} \int_0^\infty \left\{ (2-\alpha z) e^{-\alpha z} J_0(\alpha r) - [2(1-\nu) - \alpha z] e^{-\alpha z} \frac{J_1(\alpha r)}{\alpha r} \right\} \alpha^3 B(\alpha) d\alpha \tag{9.2.84}$$

and

$$S_{\varphi\varphi}^{(2)} = \frac{\mu}{1-\nu} \int_0^\infty \left\{ 2\nu e^{-\alpha z} J_0(\alpha r) + [2(1-\nu) - \alpha z] e^{-\alpha z} \frac{J_1(\alpha r)}{\alpha r} \right\} \alpha^3 B(\alpha) d\alpha \tag{9.2.85}$$

Finally, by inserting Equation 9.2.76 into Equations 9.2.84 and 9.2.85, we arrive at, respectively

$$S_{rr}^{(2)}(r,z) = 4\mu u_R^0 \zeta \int_0^\infty \alpha e^{-\alpha(z+\zeta)} \left\{ (2-\alpha z) J_0(\alpha r) - [2(1-\nu) - \alpha z] \frac{J_1(\alpha r)}{\alpha r} \right\} d\alpha \tag{9.2.86}$$

and

$$S_{\varphi\varphi}^{(2)}(r,z) = 4\mu u_R^0 \zeta \int_0^\infty \alpha e^{-\alpha(z+\zeta)} \left\{ 2\nu J_0(\alpha r) + [2(1-\nu) - \alpha z] \frac{J_1(\alpha r)}{\alpha r} \right\} \quad (9.2.87)$$

To find the displacement components associated with the elastic state s_2, we substitute χ from Equation 9.2.72 into Equations 9.2.60$_1$ and 9.2.60$_2$, and obtain, respectively

$$u_r^{(2)}(r,z) = \frac{1}{2(1-\nu)} \int_0^\infty \alpha^2 B(\alpha)(2 - 2\nu - \alpha z) e^{-\alpha z} J_1(\alpha r) d\alpha \quad (9.2.88)$$

and

$$u_z^{(2)}(r,z) = -\frac{1}{2(1-\nu)} \int_0^\infty \alpha^2 B(\alpha)(1 - 2\nu + \alpha z) e^{-\alpha z} J_0(\alpha r) d\alpha \quad (9.2.89)$$

Finally, in view of Equation 9.2.76, we write

$$u_r^{(2)}(r,z) = 2u_R^0 \zeta \int_0^\infty (2 - 2\nu - \alpha z) e^{-\alpha(z+\zeta)} J_1(\alpha r) d\alpha \quad (9.2.90)$$

and

$$u_z^{(2)}(r,z) = -2u_R^0 \zeta \int_0^\infty (1 - 2\nu + \alpha z) e^{-\alpha(z+\zeta)} J_0(\alpha r) d\alpha \quad (9.2.91)$$

Next, we use Equations 9.2.90 and 9.2.91 to find the diagonal components of the strain tensor $\mathbf{E}^{(2)}$ from the formulas (cf. also Equations 9.2.62)

$$E_{rr}^{(2)} = \frac{\partial u_r^{(2)}}{\partial r}, \quad E_{\varphi\varphi}^{(2)} = \frac{u_r^{(2)}}{r}, \quad E_{zz}^{(2)} = \frac{\partial u_z^{(2)}}{\partial z} \quad (9.2.92)$$

Hence, we obtain

$$E_{rr}^{(2)} = 2u_R^0 \zeta \int_0^\infty \alpha (2 - 2\nu - \alpha z) e^{-\alpha(z+\zeta)} \left[J_0(\alpha r) - \frac{J_1(\alpha r)}{\alpha r} \right] d\alpha \quad (9.2.93)$$

$$E_{\varphi\varphi}^{(2)} = 2u_R^0 \frac{\zeta}{r} \int_0^\infty (2 - 2\nu - \alpha z) e^{-\alpha(z+\zeta)} J_1(\alpha r) d\alpha \quad (9.2.94)$$

$$E_{zz}^{(2)} = -2u_R^0 \zeta \int_0^\infty \alpha (2\nu - \alpha z) e^{-\alpha(z+\zeta)} J_0(\alpha r) d\alpha \quad (9.2.95)$$

In addition, the use of the relation

$$E_{rz}^{(2)} = \frac{1}{2\mu} S_{rz}^{(2)} \tag{9.2.96}$$

and Equation 9.2.78, leads to

$$E_{rz}^{(2)} = -2u_R^0 \zeta \int_0^\infty \alpha(1 - \alpha z) e^{-\alpha(z+\zeta)} J_1(\alpha r) d\alpha \tag{9.2.97}$$

It follows from Equations 9.2.93 through 9.2.95 that

$$\operatorname{tr} \mathbf{E}^{(2)} = 4u_R^0 \zeta (1 - 2\nu) \int_0^\infty \alpha e^{-\alpha(z+\zeta)} J_0(\alpha r) d\alpha \tag{9.2.98}$$

Also, from Equations 9.2.77, 9.2.86, and 9.2.87, we have

$$\operatorname{tr} \mathbf{S}^{(2)} = 8u_R^0 \zeta \mu (1 + \nu) \int_0^\infty \alpha e^{-\alpha(z+\zeta)} J_0(\alpha r) d\alpha \tag{9.2.99}$$

Hence, Equations 9.2.98 and 9.2.99 imply the well-known result of isothermal elasticity (cf. Equation 4.1.23)

$$\operatorname{tr} \mathbf{E}^{(2)} = \frac{1}{2\mu} \frac{1 - 2\nu}{1 + \nu} \operatorname{tr} \mathbf{S}^{(2)} \tag{9.2.100}$$

Now, we will show that the integral representation of $s_2 = [\mathbf{u}^{(2)}, \mathbf{E}^{(2)}, \mathbf{S}^{(2)}]$ can be reduced to a closed form involving elementary functions. To this end, we use the results (cf. Equation 9.1.23)

$$\int_0^\infty e^{-\alpha(z+\zeta)} J_0(\alpha r) d\alpha = \frac{1}{R_2} \tag{9.2.101}$$

and

$$\int_0^\infty e^{-\alpha(z+\zeta)} J_1(\alpha r) d\alpha = \frac{r}{R_2} \frac{1}{R_2 + z + \zeta} = \frac{1}{r} \left(1 - \frac{\partial R_2}{\partial z} \right)$$

where

$$R_2 = \sqrt{r^2 + (z + \zeta)^2}, \quad z + \zeta > 0 \quad r > 0 \tag{9.2.102}$$

Note that by differentiation of Equations 9.2.101 and 9.2.102 with respect to z we get, respectively

$$\int_0^\infty \alpha e^{-\alpha(z+\zeta)} J_0(\alpha r) d\alpha = \frac{(z+\zeta)}{R_2^3} \tag{9.2.103}$$

and

$$\int_0^\infty \alpha e^{-\alpha(z+\zeta)} J_1(\alpha r) d\alpha = \frac{r}{R_2^3} \tag{9.2.104}$$

Therefore, by using Equations 9.2.101 through 9.2.104, and 9.2.90 and 9.2.91, the closed-form displacement components are obtained

$$\begin{aligned} u_r^{(2)}(r,z) &= 2u_R^0\zeta \frac{r}{R_2}\left[\frac{2(1-\nu)}{R_2+z+\zeta} - \frac{z}{R_2^2}\right] \\ u_z^{(2)}(r,z) &= -2u_R^0\zeta \frac{1}{R_2}\left[(1-2\nu) + \frac{z(z+\zeta)}{R_2^2}\right] \end{aligned} \tag{9.2.105}$$

To obtain the strain components in a closed form, we find that by Equations 9.2.102 and 9.2.103 we get

$$\int_0^\infty \alpha e^{-\alpha(z+\zeta)} \left[J_0(\alpha r) - \frac{J_1(\alpha r)}{\alpha r}\right] d\alpha = \frac{1}{R_2}\left(\frac{z+\zeta}{R_2^2} - \frac{1}{R_2+z+\zeta}\right) \tag{9.2.106}$$

and the differentiation of Equation 9.2.106 with respect to z leads to

$$\int_0^\infty \alpha^2 e^{-\alpha(z+\zeta)} \left[J_0(\alpha r) - \frac{J_1(\alpha r)}{\alpha r}\right] d\alpha = \frac{1}{R^3}\left(1 - \frac{3r^2}{R_2^2}\right) \tag{9.2.107}$$

From this relation and from Equation 9.2.93 we arrive at

$$E_{rr}^{(2)}(r,z) = 2u_R^0\zeta \left[\frac{2(1-\nu)}{R_2}\left(\frac{z+\zeta}{R_2^2} - \frac{1}{R_2+z+\zeta}\right) - \frac{z}{R_2^3}\left(1 - \frac{3r^2}{R_2^2}\right)\right] \tag{9.2.108}$$

Next, it follows from Equation 9.2.94 that

$$E_{\varphi\varphi}^{(2)}(r,z) = 2u_R^0\zeta \frac{1}{R_2}\left[\frac{2(1-\nu)}{R_2+z+\zeta} - \frac{z}{R_2^2}\right] \tag{9.2.109}$$

To find $E_{zz}^{(2)}(r,z)$ we note that by differentiating Equation 9.2.103 with respect to z we obtain

$$\int_0^\infty \alpha^2 e^{-\alpha(z+\zeta)} J_0(\alpha r) d\alpha = \frac{1}{R_2^3}\left(2 - \frac{3r^2}{R_2^2}\right) \tag{9.2.110}$$

From this relation and from Equations 9.2.103 and 9.2.95 follows

$$E_{zz}^{(2)}(r,z) = -2u_R^0\zeta\left[2\nu\frac{(z+\zeta)}{R_2^3} - \frac{z}{R_2^3}\left(2 - \frac{3r^2}{R_2^2}\right)\right] \tag{9.2.111}$$

Finally, from Equations 9.2.97 and 9.2.104, follows

$$E_{rz}^{(2)}(r,z) = -2u_R^0\zeta r\left[\frac{1}{R_2^3} - \frac{3z(z+\zeta)}{R_2^5}\right] \tag{9.2.112}$$

It is easy to check that (see Equation 9.2.98)

$$E_{rr}^{(2)} + E_{\varphi\varphi}^{(2)} + E_{zz}^{(2)} = 4u_R^0\zeta(1-2\nu)\frac{(z+\zeta)}{R_2^3} \tag{9.2.113}$$

To find the closed-form stress components, note that it follows from Equations 9.2.78 and 9.2.104 that

$$S_{zr}^{(2)}(r,z) = -4\mu u_R^0\zeta r\left[\frac{1}{R_2^3} - \frac{3z(z+\zeta)}{R_2^5}\right] \tag{9.2.114}$$

while Equations 9.2.77 and 9.2.110 lead to

$$S_{zz}^{(2)}(r,z) = 4\mu u_R^0\zeta z\frac{1}{R_2^3}\left(2 - \frac{3r^2}{R_2^2}\right) \tag{9.2.115}$$

Also, from Equations 9.2.86, 9.2.102, 9.2.103, and 9.2.104, we obtain

$$S_{rr}^{(2)}(r,z) = 4\mu u_R^0\zeta\left[\frac{2(z+\zeta)}{R_2^3} - \frac{z}{R_2^3}\left(2 - \frac{3r^2}{R_2^2}\right) - \frac{2(1-\nu)}{R_2}\frac{1}{R_2+z+\zeta} + \frac{z}{R_2^3}\right] \tag{9.2.116}$$

and Equations 9.2.87, 9.2.102 through 9.2.104 yield

$$S_{\varphi\varphi}^{(2)}(r,z) = 4\mu u_R^0\zeta\left[2\nu\frac{(z+\zeta)}{R_2^3} + \frac{2(1-\nu)}{R_2}\frac{1}{R_2+z+\zeta} - \frac{z}{R_2^3}\right] \tag{9.2.117}$$

This completes the derivation of the closed-form stress components of tensor $\mathbf{S}^{(2)}$.

It is easy to check that (see Equation 9.2.99)

$$S^{(2)}_{rr} + S^{(2)}_{\varphi\varphi} + S^{(2)}_{zz} = 8\mu u^0_R \zeta (1+\nu)\frac{(z+\zeta)}{R_2^3} \tag{9.2.118}$$

Finally, using Equations 9.2.54 through 9.2.56, 9.2.105, 9.2.108 and 9.2.109, 9.2.111 and 9.2.112, and 9.2.114 through 9.2.117, the closed form of $s = s_1 + s_2$ is obtained:

The displacement components

$$\begin{aligned} u_r &= u^0_R r\left\{\left(\frac{1}{R_1} - \frac{1}{R_2}\right) + \frac{2\zeta}{R_2}\left[\frac{2(1-\nu)}{R_2+z+\zeta} - \frac{z}{R_2^2}\right]\right\} \\ u_z &= u^0_R\left\{\left(\frac{z-\zeta}{R_1} - \frac{z+\zeta}{R_2}\right) - \frac{2\zeta}{R_2}\left[(1-2\nu) + \frac{z(z+\zeta)}{R_2^2}\right]\right\} \\ u_\varphi &= 0 \end{aligned} \tag{9.2.119}$$

The strain components

$$\begin{aligned} E_{rr} &= u^0_R\left\{\left(\frac{1}{R_1} - \frac{1}{R_2}\right) - r^2\left(\frac{1}{R_1^3} - \frac{1}{R_2^3}\right)\right. \\ &\quad \left. + 2\frac{\zeta}{R_2}\left[2(1-\nu)\left(\frac{z+\zeta}{R_2^2} - \frac{1}{R_2+z+\zeta}\right) - \frac{z}{R_2^2}\left(1 - \frac{3r^2}{R_2^2}\right)\right]\right\} \\ E_{\varphi\varphi} &= u^0_R\left\{\left(\frac{1}{R_1} - \frac{1}{R_2}\right) + 2\frac{\zeta}{R_2}\left[\frac{2(1-\nu)}{R_2+z+\zeta} - \frac{z}{R_2^2}\right]\right\} \\ E_{zz} &= u^0_R\left\{\left(\frac{1}{R_1} - \frac{1}{R_2}\right) - \left[\frac{(z-\zeta)^2}{R_1^3} - \frac{(z+\zeta)^2}{R_2^3}\right]\right. \\ &\quad \left. - 2\frac{\zeta}{R_2^3}\left[2\nu(z+\zeta) - z\left(2 - \frac{3r^2}{R_2^2}\right)\right]\right\} \\ E_{rz} &= -u^0_R r\left\{\left(\frac{z-\zeta}{R_1^3} - \frac{z+\zeta}{R_2^3}\right) + \frac{2\zeta}{R_2^3}\left[1 - \frac{3z(z+\zeta)}{R_2^2}\right]\right\} \\ E_{r\varphi} &= 0, \quad E_{\varphi z} = 0, \end{aligned} \tag{9.2.120}$$

The stress components

$$\begin{aligned} S_{rr} &= -2\mu u^0_R\left\{\left(\frac{1}{R_1} - \frac{1}{R_2}\right) + r^2\left(\frac{1}{R_1^3} - \frac{1}{R_2^3}\right)\right. \\ &\quad \left. - \frac{2\zeta}{R_2}\left[\frac{2(z+\zeta)}{R_2^2} - \frac{z}{R_2^2}\left(2 - \frac{3r^2}{R_2^2}\right) + \frac{z}{R_2^2} - \frac{2(1-\nu)}{R_2+z+\zeta}\right]\right\} \\ S_{\varphi\varphi} &= -2\mu u^0_R\left\{\left(\frac{1}{R_1} - \frac{1}{R_2}\right) - \frac{2\zeta}{R_2}\left[2\nu\frac{(z+\zeta)}{R_2^2} - \frac{z}{R_2^2} + \frac{2(1-\nu)}{R_2+z+\zeta}\right]\right\} \end{aligned}$$

$$S_{zz} = -2\mu u_R^0 \left[2\left(\frac{1}{R_1} - \frac{1}{R_2}\right) - r^2\left(\frac{1}{R_1^3} - \frac{1}{R_2^3}\right) - 2\frac{\zeta z}{R_2^3}\left(2 - \frac{3r^2}{R_2^2}\right)\right]$$

$$S_{rz} = -2\mu u_R^0 r\left\{\left(\frac{z-\zeta}{R_1^3} - \frac{z+\zeta}{R_2^3}\right) + \frac{2\zeta}{R_2^3}\left[1 - \frac{3z(z+\zeta)}{R_2^2}\right]\right\}$$

$$S_{r\varphi} = 0, \quad S_{z\varphi} = 0. \tag{9.2.121}$$

It follows from Equations 9.2.119 that

$$\mathbf{u} \to \mathbf{0} \quad \text{as } r \to \infty \quad \text{and} \quad 0 < z < \infty \tag{9.2.122}$$

and

$$\mathbf{u} \to \mathbf{0} \quad \text{as } z \to \infty \quad \text{and} \quad 0 < r < \infty \tag{9.2.123}$$

while Equations 9.2.120 and 9.2.121 lead to the asymptotic conditions

$$\mathbf{E} \to \mathbf{0}, \quad \mathbf{S} \to \mathbf{0} \quad \text{as } R = \sqrt{r^2 + z^2} \to \infty \tag{9.2.124}$$

Also, Equations 9.2.121 imply that

$$S_{zz}(r,0) = 0, \quad S_{rz}(r,0) = 0 \quad r > 0 \tag{9.2.125}$$

which means that $s = [\mathbf{u}, \mathbf{E}, \mathbf{S}]$ corresponds to a stress free boundary of the semispace with an internal heat source.

9.2.3 Thermoelastic State in a Semispace Caused by a Boundary Heat Exposure

Let a homogeneous isotropic elastic semispace occupy the region described by the inequalities

$$|x_1| < \infty, \quad |x_2| < \infty, \quad x_3 \geq 0 \tag{9.2.126}$$

and let the semispace be subject to a temperature field T that satisfies the field equation

$$\nabla^2 T = 0 \quad \text{for } |x_1| < \infty,\ |x_2| < \infty,\ x_3 > 0 \tag{9.2.127}$$

the boundary condition

$$T(x_1, x_2, 0) = f(x_1, x_2) \quad \text{for } |x_1| < \infty,\ |x_2| < \infty \tag{9.2.128}$$

and the vanishing condition at infinity

$$T \to 0 \quad \text{as } R = \sqrt{x_1^2 + x_2^2 + x_3^2} \to \infty \tag{9.2.129}$$

where $f = f(x_1, x_2)$ is a prescribed function. We are to find a thermoelastic state $s = [\mathbf{u}, \mathbf{E}, \mathbf{S}]$ corresponding to the temperature $T = T(x_1, x_2, x_3)$ and subject to the stress free boundary conditions

$$S_{33}(x_1, x_2, 0) = S_{13}(x_1, x_2, 0) = S_{23}(x_1, x_2, 0) = 0$$

$$\text{for } |x_1| < \infty, \ |x_2| < \infty \tag{9.2.130}$$

and the vanishing condition at infinity

$$s \to [\mathbf{0}, \mathbf{0}, \mathbf{0}] \quad \text{as } R \to \infty \tag{9.2.131}$$

9.2.3.1 Thermoelastic Green's Function for a Semispace with a Boundary Heat Exposure

Let us consider the heat conduction equation in cylindrical coordinates (r, z)

$$\nabla^2 T^* = 0 \quad \text{for } r > 0, \ z > 0 \tag{9.2.132}$$

with the boundary condition

$$T^*(r, 0) = \frac{\delta(r)}{2\pi r} \tag{9.2.133}$$

and the vanishing condition at infinity

$$T^*(r, z) \to 0 \quad \text{as } R = \sqrt{r^2 + z^2} \to \infty \tag{9.2.134}$$

Since (cf. Equations 9.1.1 and 9.1.20)

$$\frac{\delta(r)}{2\pi r} = \frac{1}{2\pi} \int_0^\infty \alpha J_0(\alpha r) d\alpha \tag{9.2.135}$$

and

$$\nabla^2[\mathrm{e}^{-\alpha z} J_0(\alpha r)] = 0 \tag{9.2.136}$$

therefore, a solution to the heat conduction problem described by Equations 9.2.132 through 9.2.134 takes the form

$$T^*(r, z) = \frac{1}{2\pi} \int_0^\infty \alpha e^{-\alpha z} J_0(\alpha r) d\alpha \tag{9.2.137}$$

or, using Equation 9.2.103 with $\zeta = 0$,

$$T^*(r, z) = \frac{1}{2\pi} \frac{z}{R^3} = -\frac{1}{2\pi} \frac{\partial}{\partial z} \left(\frac{1}{R} \right) \tag{9.2.138}$$

Let $s^* = [\mathbf{u}^*, \mathbf{E}^*, \mathbf{S}^*]$ denote a thermoelastic state defined for $r \geq 0$ and $z \geq 0$ that corresponds to the temperature $T^* = T^*(r, z)$ and satisfies the boundary conditions

$$S^*_{zz}(r,0) = 0, \quad S^*_{rz}(r,0) = 0 \quad \text{for } r > 0 \tag{9.2.139}$$

and the vanishing condition at infinity

$$s^* \to [\mathbf{0}, \mathbf{0}, \mathbf{0}] \quad \text{as} \quad R = \sqrt{r^2 + z^2} \to \infty \tag{9.2.140}$$

In the following we are to show that

$$s^* = s_1^* + s_2^* \tag{9.2.141}$$

where s_1^* is the thermoelastic state generated by a thermoelastic displacement potential $\phi^* = \phi^*(r, z)$, and s_2^* is an elastic state derived from Love's function $\chi^* = \chi^*(r, z)$. The state s_1^* is defined by (cf. Equations 9.2.7 through 9.2.9 and 9.2.23 through 9.2.25)

$$s_1^* = [\overline{\mathbf{u}}^*, \overline{\mathbf{E}}^*, \overline{\mathbf{S}}^*] \tag{9.2.142}$$

where

$$\overline{\mathbf{u}}^* = \left[\overline{u}_r^*, 0, \overline{u}_z^*\right] \tag{9.2.143}$$

$$\overline{u}_r^* = \frac{\partial \phi^*}{\partial r}, \quad \overline{u}_z^* = \frac{\partial \phi^*}{\partial z} \tag{9.2.144}$$

$$\overline{\mathbf{E}}^* = \begin{bmatrix} \overline{E}_{rr}^* & 0 & \overline{E}_{rz}^* \\ 0 & \overline{E}_{\varphi\varphi}^* & 0 \\ \overline{E}_{zr}^* & 0 & \overline{E}_{zz}^* \end{bmatrix} \tag{9.2.145}$$

$$\overline{E}_{rr}^* = \frac{\partial^2 \phi^*}{\partial r^2}, \quad \overline{E}_{rz}^* = \frac{\partial^2 \phi^*}{\partial r \partial z}$$

$$\overline{E}_{\varphi\varphi}^* = \frac{1}{r}\frac{\partial \phi^*}{\partial r}, \quad \overline{E}_{zz}^* = \frac{\partial^2 \phi^*}{\partial z^2} \tag{9.2.146}$$

$$\overline{\mathbf{S}}^* = \begin{bmatrix} \overline{S}_{rr}^* & 0 & \overline{S}_{rz}^* \\ 0 & \overline{S}_{\varphi\varphi}^* & 0 \\ \overline{S}_{zr}^* & 0 & \overline{S}_{zz}^* \end{bmatrix} \tag{9.2.147}$$

$$\overline{S}_{rr}^* = -2\mu\left(\frac{1}{r}\frac{\partial \phi^*}{\partial r} + \frac{\partial^2 \phi^*}{\partial z^2}\right), \quad \overline{S}_{rz}^* = 2\mu\frac{\partial^2 \phi^*}{\partial r \partial z}$$

$$\overline{S}_{\varphi\varphi}^* = -2\mu\left(\frac{\partial^2 \phi^*}{\partial r^2} + \frac{\partial^2 \phi^*}{\partial z^2}\right), \quad \overline{S}_{zz}^* = -2\mu\left(\frac{1}{r}\frac{\partial \phi^*}{\partial r} + \frac{\partial^2 \phi^*}{\partial r^2}\right) \tag{9.2.148}$$

and

$$\nabla^2 \phi^* = mT^* \tag{9.2.149}$$

with

$$m = \frac{1+\nu}{1-\nu}\alpha \tag{9.2.150}$$

The state s_2^* is given by (see Equations 9.2.59 through 9.2.63)

$$s_2^* = [\overline{\overline{\mathbf{u}}}^*, \overline{\overline{\mathbf{E}}}^*, \overline{\overline{\mathbf{S}}}^*] \tag{9.2.151}$$

where

$$\overline{\overline{\mathbf{u}}}^* = \left[\overline{\overline{u}}_r^*, 0, \overline{\overline{u}}_z^*\right] \tag{9.2.152}$$

$$\begin{aligned} \overline{\overline{u}}_r^* &= -\frac{1}{2(1-\nu)} \frac{\partial^2}{\partial r \partial z} \chi^* \\ \overline{\overline{u}}_z^* &= \frac{1}{2(1-\nu)} \left[2(1-\nu)\nabla^2 - \frac{\partial^2}{\partial z^2}\right] \chi^* \end{aligned} \tag{9.2.153}$$

$$\overline{\overline{\mathbf{E}}}^* = \begin{bmatrix} \overline{\overline{E}}_{rr}^* & 0 & \overline{\overline{E}}_{rz}^* \\ 0 & \overline{\overline{E}}_{\varphi\varphi}^* & 0 \\ \overline{\overline{E}}_{zr}^* & 0 & \overline{\overline{E}}_{zz}^* \end{bmatrix} \tag{9.2.154}$$

$$\begin{aligned} \overline{\overline{E}}_{rr}^* &= -\frac{1}{2(1-\nu)} \frac{\partial^3 \chi^*}{\partial r^2 \partial z} \\ \overline{\overline{E}}_{rz}^* &= \frac{1}{2(1-\nu)} \frac{\partial}{\partial r} \left[(1-\nu)\nabla^2 - \frac{\partial^2}{\partial z^2}\right] \chi^* \\ \overline{\overline{E}}_{\varphi\varphi}^* &= -\frac{1}{2(1-\nu)} \frac{1}{r} \frac{\partial^2}{\partial r \partial z} \chi^* \\ \overline{\overline{E}}_{zz}^* &= \frac{1}{2(1-\nu)} \frac{\partial}{\partial z} \left[2(1-\nu)\nabla^2 - \frac{\partial^2}{\partial z^2}\right] \chi^* \end{aligned} \tag{9.2.155}$$

$$\overline{\overline{\mathbf{S}}}^* = \begin{bmatrix} \overline{\overline{S}}_{rr}^* & 0 & \overline{\overline{S}}_{rz}^* \\ 0 & \overline{\overline{S}}_{\varphi\varphi}^* & 0 \\ \overline{\overline{S}}_{zr}^* & 0 & \overline{\overline{S}}_{zz}^* \end{bmatrix} \tag{9.2.156}$$

$$\begin{aligned}
\overline{\overline{S}}{}^*_{rr} &= \frac{\mu}{1-\nu}\frac{\partial}{\partial z}\left(\nu\nabla^2 - \frac{\partial^2}{\partial r^2}\right)\chi^* \\
\overline{\overline{S}}{}^*_{rz} &= \frac{\mu}{1-\nu}\frac{\partial}{\partial r}\left[(1-\nu)\nabla^2 - \frac{\partial^2}{\partial z^2}\right]\chi^* \\
\overline{\overline{S}}{}^*_{\varphi\varphi} &= \frac{\mu}{1-\nu}\frac{\partial}{\partial z}\left(\nu\nabla^2 - \frac{1}{r}\frac{\partial}{\partial r}\right)\chi^* \\
\overline{\overline{S}}{}^*_{zz} &= \frac{\mu}{1-\nu}\frac{\partial}{\partial z}\left[(2-\nu)\nabla^2 - \frac{\partial^2}{\partial z^2}\right]\chi^*
\end{aligned} \tag{9.2.157}$$

and Love's function χ^* satisfies the biharmonic equation

$$\nabla^2\nabla^2\chi^* = 0 \tag{9.2.158}$$

To obtain a thermoelastic state s_1^* we note that by applying ∇^2 to Equation 9.2.149 and using Equation 9.2.132 we have

$$\nabla^2\nabla^2\phi^* = 0 \tag{9.2.159}$$

Thus both χ^* and ϕ^* are biharmonic. Let us select ϕ^* in the form

$$\phi^*(r,z) = \int_0^\infty [C(\alpha) + \alpha z D(\alpha)]\mathrm{e}^{-\alpha z} J_0(\alpha r)d\alpha \tag{9.2.160}$$

where $C = C(\alpha)$ and $D = D(\alpha)$ are arbitrary functions on $[0,\infty)$. By using the identities (see Equations 9.2.81 and 9.2.136)

$$\nabla^2[e^{-\alpha z}J_0(\alpha r)] = 0 \tag{9.2.161}$$

and

$$\nabla^2[\alpha z e^{-\alpha z}J_0(\alpha r)] = -2\alpha^2 e^{-\alpha z}J_0(\alpha r) \tag{9.2.162}$$

and substituting ϕ^* from Equation 9.2.160 into Equation 9.2.149 in which T^* is represented by the integral 9.2.137, we obtain

$$-2\int_0^\infty \alpha^2 D(\alpha)e^{-\alpha z}J_0(\alpha r) = \frac{m}{2\pi}\int_0^\infty \alpha e^{-\alpha z}J_0(\alpha r)d\alpha \tag{9.2.163}$$

Hence, we get

$$\alpha D(\alpha) = -\frac{m}{4\pi} \tag{9.2.164}$$

and substituting this into Equation 9.2.160 leads to

$$\phi^*(r,z) = \int_0^\infty \left[C(\alpha) - \frac{mz}{4\pi}\right]e^{-\alpha z}J_0(\alpha r)d\alpha \tag{9.2.165}$$

We note that the function ϕ^*, due to the presence of an arbitrary function $C = C(\alpha)$, generates a whole class of thermoelastic states s_1^* through the formulas 9.2.142 through 9.2.150. In particular, from Equations 9.2.148, we obtain

$$\overline{S}_{zz}^* = -2\mu\nabla_r^2\phi^* = 2\mu\int_0^\infty\left[C(\alpha) - \frac{zm}{4\pi}\right]\alpha^2 e^{-\alpha z}J_0(\alpha r)d\alpha \tag{9.2.166}$$

and

$$\overline{S}_{rz}^* = 2\mu\frac{\partial^2\phi^*}{\partial r\partial z} = 2\mu\int_0^\infty\left\{\left[C(\alpha) - \frac{zm}{4\pi}\right]\alpha + \frac{m}{4\pi}\right\}\alpha e^{-\alpha z}J_1(\alpha r)d\alpha \tag{9.2.167}$$

The thermoelastic state s_1^* may be selected in such a way as to satisfy a zero normal stress boundary condition or a zero shear stress boundary condition. Using results of Sections 9.2.2.1 and 9.2.2.2, we select $C = C(\alpha)$ in such a way that the normal stress vanishes at $z = 0$ (cf. Equations 9.2.58)

$$\overline{S}_{zz}^*(r,0) = 0 \quad \text{for } r > 0 \tag{9.2.168}$$

This condition together with Equation 9.2.166 imply

$$C(\alpha) = 0 \quad \text{for } \alpha > 0 \tag{9.2.169}$$

and, from Equations 9.2.166 and 9.2.167, we get

$$\overline{S}_{zz}^* = -\frac{m\mu}{2\pi}z\int_0^\infty\alpha^2 e^{-\alpha z}J_0(\alpha r)d\alpha \tag{9.2.170}$$

$$\overline{S}_{rz}^* = \frac{m\mu}{2\pi}\int_0^\infty(1-\alpha z)\alpha e^{-\alpha z}J_1(\alpha r)d\alpha \tag{9.2.171}$$

To obtain the remaining stress components as well as the displacement and strain components of the thermoelastic state s_1^* we substitute $C(\alpha) = 0$ into Equation 9.2.165 and receive

$$\phi^*(r,z) = -\frac{mz}{4\pi}\int_0^\infty e^{-\alpha z}J_0(\alpha r)d\alpha \tag{9.2.172}$$

Hence, the substitution of this into Equations 9.2.144, 9.2.146, $9.2.148_1$, and $9.2.148_3$ leads to

$$\overline{u}_r^* = \frac{mz}{4\pi}\int_0^\infty\alpha e^{-\alpha z}J_1(\alpha r)d\alpha \tag{9.2.173}$$

$$\overline{u}_z^* = -\frac{m}{4\pi}\int_0^\infty(1-\alpha z)\mathrm{e}^{-\alpha z}J_0(\alpha r)d\alpha \tag{9.2.174}$$

$$\overline{E}_{rr}^* = \frac{mz}{4\pi}\int_0^\infty\alpha^2 e^{-\alpha z}\left[J_0(\alpha r) - \frac{J_1(\alpha r)}{\alpha r}\right]d\alpha \tag{9.2.175}$$

$$\overline{E}^*_{rz} = \frac{m}{4\pi} \int_0^\infty (1 - \alpha z)\alpha e^{-\alpha z} J_1(\alpha r) d\alpha \tag{9.2.176}$$

$$\overline{E}^*_{\varphi\varphi} = \frac{mz}{4\pi r} \int_0^\infty \alpha e^{-\alpha z} J_1(\alpha r) d\alpha \tag{9.2.177}$$

$$\overline{E}^*_{zz} = \frac{m}{4\pi} \int_0^\infty (2 - z\alpha)\alpha e^{-\alpha z} J_0(\alpha r) d\alpha \tag{9.2.178}$$

$$\overline{S}^*_{rr} = -\frac{m\mu}{2\pi} \int_0^\infty \alpha e^{-\alpha z} \left[(2 - \alpha z) J_0(\alpha r) + \frac{z}{r} J_1(\alpha r)\right] d\alpha \tag{9.2.179}$$

$$\overline{S}^*_{\varphi\varphi} = -\frac{m\mu}{2\pi} \int_0^\infty \alpha e^{-\alpha z} \left[2 J_0(\alpha r) - \frac{z}{r} J_1(\alpha r)\right] d\alpha \tag{9.2.180}$$

The formulas 9.2.170 and 9.2.171, and 9.2.173 through 9.2.180 provide an integral representation of s_1^* with the properties

$$\overline{S}^*_{zz}(r, 0) = 0, \quad \overline{S}^*_{rz}(r, 0) = \frac{m\mu}{2\pi} \int_0^\infty \alpha J_1(\alpha r) d\alpha \tag{9.2.181}$$

Therefore, because of Equations 9.2.139 and 9.2.141, the isothermal elastic state s_2^* defined by Equations 9.2.151 through 9.2.158 should be found in such a way that

$$\overline{\overline{S}}^*_{zz}(r, 0) = 0, \quad \overline{\overline{S}}^*_{rz}(r, 0) = -\frac{m\mu}{2\pi} \int_0^\infty \alpha J_1(\alpha r) d\alpha \tag{9.2.182}$$

With this goal in mind we postulate Love's function χ^* in the form (see Equation 9.2.65)

$$\chi^*(r, z) = \int_0^\infty [A(\alpha) + \alpha z B(\alpha)] e^{-\alpha z} J_0(\alpha r) d\alpha \tag{9.2.183}$$

where $A = A(\alpha)$ and $B = B(\alpha)$ are arbitrary functions.

Since the boundary conditions 9.2.182 are of the type of Equation 9.2.58 in which the shear stress is represented by the integral 9.2.75, a procedure of finding χ^*, similar to that of obtaining Equations 9.2.65 through 9.2.74, leads to the results

$$\chi^*(r, z) = -\int_0^\infty [(1 - 2\nu) - \alpha z] B(\alpha) e^{-\alpha z} J_0(\alpha r) d\alpha \tag{9.2.184}$$

$$\overline{\overline{S}}^*_{zz}(r, z) = \frac{\mu}{1 - \nu} z \int_0^\infty \alpha^4 B(\alpha) e^{-\alpha z} J_0(\alpha r) d\alpha \tag{9.2.185}$$

and

$$\overline{\overline{S}}^*_{rz}(r, z) = -\frac{\mu}{1 - \nu} \int_0^\infty \alpha^3 B(\alpha)(1 - \alpha z) e^{-\alpha z} J_1(\alpha r) d\alpha \tag{9.2.186}$$

Hence, by the second of boundary conditions 9.2.182, we get

$$\alpha^2 B(\alpha) \equiv \hat{B}(\alpha) = \frac{m(1-\nu)}{2\pi} \tag{9.2.187}$$

Substituting Equation 9.2.187 into Equations 9.2.185 and 9.2.186 we get, respectively,

$$\overline{\overline{S}}{}^*_{zz}(r,z) = \frac{m\mu}{2\pi} z \int_0^\infty \alpha^2 e^{-\alpha z} J_0(\alpha r) d\alpha \tag{9.2.188}$$

and

$$\overline{\overline{S}}{}^*_{rz}(r,z) = -\frac{m\mu}{2\pi} \int_0^\infty \alpha(1-\alpha z) e^{-\alpha z} J_1(\alpha r) d\alpha \tag{9.2.189}$$

Also, the substitution of χ^* into Equations $9.2.157_1$ and $9.2.157_3$ leads, respectively, to [see Equations 9.2.84 and 9.2.85 in which $B(\alpha)$ is given by Equation 9.2.187]

$$\overline{\overline{S}}{}^*_{rr}(r,z) = \frac{m\mu}{2\pi} \int_0^\infty \left\{ (2-\alpha z) e^{-\alpha z} J_0(\alpha r) - [2(1-\nu) - \alpha z] e^{-\alpha z} \frac{J_1(\alpha r)}{\alpha r} \right\} \alpha \, d\alpha \tag{9.2.190}$$

and

$$\overline{\overline{S}}{}^*_{\varphi\varphi}(r,z) = \frac{m\mu}{2\pi} \int_0^\infty \left\{ 2\nu e^{-\alpha z} J_0(\alpha r) + [2(1-\nu) - \alpha z] e^{-\alpha z} \frac{J_1(\alpha r)}{\alpha r} \right\} \alpha \, d\alpha \tag{9.2.191}$$

Next, substituting χ^* into Equations $9.2.153_1$ and $9.2.153_2$, we get, respectively (see Equations 9.2.88 and 9.2.89 in which $B(\alpha)$ is given by Equation 9.2.187)

$$\overline{\overline{u}}{}^*_r(r,z) = \frac{m}{4\pi} \int_0^\infty (2 - 2\nu - \alpha z) e^{-\alpha z} J_1(\alpha r) d\alpha \tag{9.2.192}$$

and

$$\overline{\overline{u}}{}^*_z(r,z) = -\frac{m}{4\pi} \int_0^\infty (1 - 2\nu + \alpha z) e^{-\alpha z} J_0(\alpha r) d\alpha \tag{9.2.193}$$

Finally, inserting $\chi*$ from Equation 9.2.184 into Equations 9.2.155 we arrive at (cf. Equations 9.2.93 through 9.2.97)

$$\overline{\overline{E}}{}^*_{rr}(r,z) = \frac{m}{4\pi} \int_0^\infty \alpha(2 - 2\nu - \alpha z) e^{-\alpha z} \left[J_0(\alpha r) - \frac{J_1(\alpha r)}{\alpha r} \right] d\alpha \tag{9.2.194}$$

$$\overline{\overline{E}}^*_{\varphi\varphi}(r,z) = \frac{m}{4\pi r}\int_0^\infty (2-2\nu-\alpha z)e^{-\alpha z}J_1(\alpha r)d\alpha \tag{9.2.195}$$

$$\overline{\overline{E}}^*_{zz}(r,z) = -\frac{m}{4\pi}\int_0^\infty \alpha(2\nu-\alpha z)e^{-\alpha z}J_0(\alpha r)d\alpha \tag{9.2.196}$$

$$\overline{\overline{E}}^*_{rz}(r,z) = -\frac{m}{4\pi}\int_0^\infty \alpha(1-\alpha z)e^{-\alpha z}J_1(\alpha r)d\alpha \tag{9.2.197}$$

The formulas 9.2.188 through 9.2.197 provide an integral representation of s_2^*. Since

$$s^* = s_1^* + s_2^* \tag{9.2.198}$$

where

$$s^* = [\mathbf{u}^*, \mathbf{E}^*, \mathbf{S}^*], \quad s_1^* = [\overline{\mathbf{u}}^*, \overline{\mathbf{E}}^*, \overline{\mathbf{S}}^*], \quad s_2^* = [\overline{\overline{\mathbf{u}}}^*, \overline{\overline{\mathbf{E}}}^*, \overline{\overline{\mathbf{S}}}^*] \tag{9.2.199}$$

therefore, using Equations 9.2.170 and 9.2.171, 9.2.173 through 9.2.180, and 9.2.188 through 9.2.197, we have

$$\begin{aligned} u_r^*(r,z) &= \frac{m(1-\nu)}{2\pi}\int_0^\infty e^{-\alpha z}J_1(\alpha r)d\alpha \\ u_z^*(r,z) &= -\frac{m(1-\nu)}{2\pi}\int_0^\infty e^{-\alpha z}J_0(\alpha r)d\alpha \end{aligned} \tag{9.2.200}$$

$$\begin{aligned} E_{rr}^*(r,z) &= \frac{m(1-\nu)}{2\pi}\int_0^\infty \alpha e^{-\alpha z}\left[J_0(\alpha r) - \frac{J_1(\alpha r)}{\alpha r}\right]d\alpha \\ E_{\varphi\varphi}^*(r,z) &= \frac{m(1-\nu)}{2\pi r}\int_0^\infty e^{-\alpha z}J_1(\alpha r)d\alpha \\ E_{zz}^*(r,z) &= \frac{m(1-\nu)}{2\pi}\int_0^\infty \alpha e^{-\alpha z}J_0(\alpha r)d\alpha \\ E_{rz}^*(r,z) &= 0 \end{aligned} \tag{9.2.201}$$

and

$$\begin{aligned} S_{rr}^*(r,z) &= -\frac{m\mu(1-\nu)}{\pi r}\int_0^\infty e^{-\alpha z}J_1(\alpha r)d\alpha \\ S_{\varphi\varphi}^*(r,z) &= -\frac{m\mu(1-\nu)}{\pi}\int_0^\infty \alpha e^{-\alpha z}\left[J_0(\alpha r) - \frac{J_1(\alpha r)}{\alpha r}\right]d\alpha \\ S_{zz}^*(r,z) &= 0, \quad S_{rz}^*(r,z) = 0 \end{aligned} \tag{9.2.202}$$

It follows from Equations 9.2.201 and 9.2.202 that

$$E^*_{rr} + E^*_{\varphi\varphi} + E^*_{zz} = 2(1 + \nu)\alpha T^* \tag{9.2.203}$$

and

$$S^*_{rr} + S^*_{\varphi\varphi} = -2\mu(1 + \nu)\alpha T^* \tag{9.2.204}$$

Note that Equation 9.2.203, because of Equation 9.2.204, may be reduced to the familiar form (see Equation 4.1.55 in which **E**, **S**, and T are replaced by $\mathbf{E}^*$, $\mathbf{S}^*$, and T^*, respectively, and λ is expressed in terms of μ and ν)

$$\operatorname{tr}\mathbf{E}^* = \frac{1}{2\mu}\frac{1-2\nu}{1+\nu}\operatorname{tr}\mathbf{S}^* + 3\alpha T^* \tag{9.2.205}$$

Finally, by using Equations 9.2.101 through 9.2.103, in which we let $\zeta = 0$, the closed form of s^* is obtained:

The displacement components

$$u^*_r(r,z) = \frac{(1+\nu)\alpha}{2\pi}\frac{r}{R(R+z)}, \quad u^*_z(r,z) = -\frac{(1+\nu)\alpha}{2\pi}\frac{1}{R} \tag{9.2.206}$$

The strain components

$$\begin{aligned} E^*_{rr}(r,z) &= \frac{(1+\nu)\alpha}{2\pi}\frac{1}{R}\left(\frac{z}{R^2} - \frac{1}{R+z}\right) \\ E^*_{\varphi\varphi}(r,z) &= \frac{(1+\nu)\alpha}{2\pi}\frac{1}{R}\frac{1}{R+z} \\ E^*_{zz}(r,z) &= \frac{(1+\nu)\alpha}{2\pi}\frac{z}{R^3} \\ E^*_{rz}(r,z) &= E^*_{r\varphi}(r,z) = 0 \end{aligned} \tag{9.2.207}$$

The stress components

$$\begin{aligned} S^*_{rr}(r,z) &= -\frac{\mu(1+\nu)\alpha}{\pi}\frac{1}{R}\frac{1}{R+z}, \\ S^*_{\varphi\varphi}(r,z) &= -\frac{\mu(1+\nu)\alpha}{\pi R}\left(\frac{z}{R^2} - \frac{1}{R+z}\right) \\ S^*_{zz}(r,z) &= S^*_{rz}(r,z) = S^*_{r\varphi}(r,z) = 0 \end{aligned} \tag{9.2.208}$$

It follows from Equation 9.2.208 that s^* corresponds to a plane state of stress parallel to the boundary $z = 0$ of the semispace. As a result we may formulate the following theorem:

Theorem: Let us consider an elastic semispace on the boundary of which the temperature $T^*(r,0) = \delta(r)/2\pi r$ is prescribed, and assume that there are no heat sources inside the semispace. Let the boundary of the semispace be free of tractions. Then the associated thermoelastic state $s^*(r,z)$ corresponds to a plane state of stress parallel to the boundary $z = 0$. ■

Note that (i) the thermoelastic plane state of stress $s^*(r,z)$ corresponds to the three-dimensional temperature $T^* = T^*(r,z)$, and (ii) $s^*(r,z)$ is not obtained under a hypothesis that a plane state of stress exists on any plane parallel to the boundary $z = 0$.

To define a thermoelastic Green's function for a semispace with a boundary heat exposure, we transform $s^*(r,z)$ to the Cartesian coordinates (x_1, x_2, x_3), by using the formulas (see Equations 9.1.82, and 9.1.84 and 9.1.85 with θ replaced by φ)

$$u_1^* = u_r^* \cos\varphi \quad u_2^* = u_r^* \sin\varphi, \quad u_3^* = u_z^* \tag{9.2.209}$$

and

$$\begin{aligned} S_{11}^* &= S_{rr}^* \cos^2\varphi + S_{\varphi\varphi}^* \sin^2\varphi \\ S_{22}^* &= S_{rr}^* \sin^2\varphi + S_{\varphi\varphi}^* \cos^2\varphi \\ S_{12}^* &= \frac{1}{2}\left(S_{rr}^* - S_{\varphi\varphi}^*\right) \sin 2\varphi \end{aligned} \tag{9.2.210}$$

where

$$\cos\varphi = \frac{x_1}{r}, \qquad \sin\varphi = \frac{x_2}{r} \tag{9.2.211}$$

Substituting u_r^* and u_z^* from Equations 9.2.206 into Equations 9.2.209 we obtain the Cartesian components of the displacement vector $\mathbf{u}^* = \mathbf{u}^*(\mathbf{x})$

$$\begin{aligned} u_1^*(x_1,x_2,x_3) &= \frac{(1+\nu)\alpha}{2\pi} \frac{x_1}{R(R+x_3)} \\ u_2^*(x_1,x_2,x_3) &= \frac{(1+\nu)\alpha}{2\pi} \frac{x_2}{R(R+x_3)} \\ u_3^*(x_1,x_2,x_3) &= -\frac{(1+\nu)\alpha}{2\pi} \frac{1}{R} \end{aligned} \tag{9.2.212}$$

To obtain the Cartesian components of the strain tensor $\mathbf{E}^*$ we use the strain–displacement relation

$$E_{ij}^* = \frac{1}{2}\left(u_{i,j}^* + u_{j,i}^*\right) \quad (i,j = 1,2,3) \tag{9.2.213}$$

and, by Equations 9.2.212, we obtain

$$\begin{aligned} E_{11}^*(x_1,x_2,x_3) &= \frac{(1+\nu)\alpha}{2\pi} \frac{1}{R(R+x_3)} \left[1 - \frac{x_1^2}{R^2}\left(1 + \frac{R}{R+x_3}\right)\right] \\ E_{22}^*(x_1,x_2,x_3) &= \frac{(1+\nu)\alpha}{2\pi} \frac{1}{R(R+x_3)} \left[1 - \frac{x_2^2}{R^2}\left(1 + \frac{R}{R+x_3}\right)\right] \end{aligned}$$

$$E^*_{12}(x_1,x_2,x_3) = -\frac{(1+\nu)\alpha}{2\pi}\frac{x_1x_2}{R^3(R+x_3)}\left(1+\frac{R}{R+x_3}\right)$$
$$E^*_{33}(x_1,x_2,x_3) = \frac{(1+\nu)\alpha}{2\pi}\frac{x_3}{R^3}, \quad E^*_{13}(x_1,x_2,x_3) = E^*_{23}(x_1,x_2,x_3) = 0 \tag{9.2.214}$$

Finally, the substitution S^*_{rr} and $S^*_{\varphi\varphi}$ from Equations 9.2.208 into Equations 9.2.210 leads to

$$S^*_{11}(x_1,x_2,x_3) = \frac{(1+\nu)\alpha\mu}{\pi}\left[\frac{1}{R}\left(\frac{1}{R+x_3}-\frac{x_3}{R^2}\right)-\frac{x_1^2}{R^3(R+x_3)}\left(1+\frac{R}{R+x_3}\right)\right]$$
$$S^*_{22}(x_1,x_2,x_3) = \frac{(1+\nu)\alpha\mu}{\pi}\left[\frac{1}{R}\left(\frac{1}{R+x_3}-\frac{x_3}{R^2}\right)-\frac{x_2^2}{R^3(R+x_3)}\left(1+\frac{R}{R+x_3}\right)\right] \tag{9.2.215}$$

$$S^*_{12}(x_1,x_2,x_3) = -\frac{(1+\nu)\alpha\mu}{\pi}\frac{x_1x_2}{R^3(R+x_3)}\left(1+\frac{R}{R+x_3}\right)$$
$$S^*_{13}(x_1,x_2,x_3) = S^*_{23}(x_1,x_2,x_3) = S^*_{33}(x_1,x_2,x_3) = 0$$

It is easy to check that

$$E^*_{kk}(\mathbf{x}) = \frac{(1+\nu)\alpha}{\pi}\frac{x_3}{R^3} \tag{9.2.216}$$

and

$$S^*_{kk}(\mathbf{x}) = -\frac{(1+\nu)\alpha\mu}{\pi}\frac{x_3}{R^3} \tag{9.2.217}$$

and that Equation 9.2.205 is satisfied.

We note that $s^* = s^*(\mathbf{x})$ satisfies the stress free boundary conditions at $x_3 = 0$ and corresponds to the boundary heat exposure of the form

$$T^*(x_1,x_2,0) = \delta(x_1)\delta(x_2) \tag{9.2.218}$$

where $\delta = \delta(x_1)$ and $\delta = \delta(x_2)$ are the one-dimensional Dirac delta functions on the x_1- and x_2-axis, respectively. Also,

$$s^* \to [\mathbf{0},\mathbf{0},\mathbf{0}] \quad \text{as} \quad R \to \infty \tag{9.2.219}$$

This motivates the following definition:

Definition: A thermoelastic state $\hat{s} = \hat{s}(\mathbf{x},\boldsymbol{\xi})$ is said to be Green's function for a semispace $x_3 \geq 0$ with the stress free boundary conditions and a boundary heat exposure at the point $(\xi_1,\xi_2,0)$ if

$$\hat{s}(\mathbf{x},\boldsymbol{\xi}) = s^*(x_1-\xi_1,x_2-\xi_2,x_3) \tag{9.2.220}$$

where $s^* = s^*(\mathbf{x}) = [\mathbf{u}^*(\mathbf{x}), \mathbf{E}^*(\mathbf{x}), \mathbf{S}^*(\mathbf{x})]$ is the thermoelastic state described by Equations 9.2.212 through 9.2.215. The associated temperature field $\hat{T} = \hat{T}(\mathbf{x}, \boldsymbol{\xi})$ is then given by

$$\hat{T}(\mathbf{x}, \boldsymbol{\xi}) = T^*(x_1 - \xi_1, x_2 - \xi_2, x_3) \qquad (9.2.221)$$

where $T^* = T^*(\mathbf{x})$ is the temperature described by Equation 9.2.138.

9.2.3.2 Thermoelastic State in a Semispace Caused by an Arbitrary Heat Exposure on the Boundary Plane

If a temperature field $T = T(\mathbf{x})$ satisfies the harmonic equation inside the semispace $x_3 \geq 0$ as well as the boundary condition (see Equation 9.2.128)

$$T(x_1, x_2, 0) = f(x_1, x_2) \quad \text{for } |x_1| < \infty,\ |x_2| < \infty \qquad (9.2.222)$$

where $f = f(x_1, x_2)$ is a prescribed function, and if the boundary $x_3 = 0$ is stress free (see Equations 9.2.130), then it follows from Section 9.2.3.1 that a thermoelastic state $s = [\mathbf{u}, \mathbf{E}, \mathbf{S}]$ defined for $|x_1| < \infty, |x_2| < \infty, x_3 \geq 0$ and corresponding to the temperature $T = T(\mathbf{x})$ is represented by the formula

$$s(\mathbf{x}) = \int_{-\infty}^{+\infty}\int_{-\infty}^{+\infty} \hat{s}(\mathbf{x}, \boldsymbol{\xi}) f(\xi_1, \xi_2) d\xi_1 d\xi_2 \qquad (9.2.223)$$

provided the integral on the RHS of this equation exists. If the function $f = f(\xi_1, \xi_2)$ vanishes outside of a finite domain Ω lying in the boundary plane, then Equation 9.2.223 reduces to

$$s(\mathbf{x}) = \int_{\Omega} \hat{s}(\mathbf{x}, \boldsymbol{\xi}) f(\xi_1, \xi_2) d\xi_1 d\xi_2 \qquad (9.2.224)$$

In the case of the temperature exposure $T = T_0 = \text{const}$ in the rectangular region: $|x_1| < a_1, |x_2| < a_2$ in the plane $x_3 = 0$, the formula 9.2.224 takes the form

$$s(\mathbf{x}) = T_0 \int_{-a_1}^{a_1}\int_{-a_2}^{a_2} \hat{s}(\mathbf{x}; \xi_1, \xi_2) d\xi_1 d\xi_2 \qquad (9.2.225)$$

and, in view of Equations 9.2.212, the boundary displacement components are given by the following integrals:

$$\begin{aligned}
u_1(x_1, x_2, 0) &= -\frac{(1+\nu)\alpha T_0}{2\pi} \int_{-a_2}^{a_2} d\xi_2 \int_{-a_1}^{a_1} d\xi_1 \frac{\partial}{\partial \xi_1} \ln\left(\frac{R_0}{L}\right) \\
u_2(x_1, x_2, 0) &= -\frac{(1+\nu)\alpha T_0}{2\pi} \int_{-a_1}^{a_1} d\xi_1 \int_{-a_2}^{a_2} d\xi_2 \frac{\partial}{\partial \xi_2} \ln\left(\frac{R_0}{L}\right) \\
u_3(x_1, x_2, 0) &= -\frac{(1+\nu)\alpha T_0}{2\pi} \int_{-a_1}^{a_1} d\xi_1 \int_{-a_2}^{a_2} d\xi_2 \frac{1}{R_0}
\end{aligned} \qquad (9.2.226)$$

where

$$R_0 = \sqrt{(x_1 - \xi_1)^2 + (x_2 - \xi_2)^2} \tag{9.2.227}$$

and L is a positive constant of the length dimension.

In the following we show that the double integrals in Equations 9.2.226 can be calculated in a closed form. To this end, we note that for a dimensionless parameter $A \neq 0$ and a dimensionless variable u, we have

$$\int \ln\left(u^2 + A^2\right) du = u\left[\ln(u^2 + A^2) - 2\right] + 2A \tan^{-1}\left(\frac{u}{A}\right) \tag{9.2.228}$$

and for a parameter $Y \neq 0$ that has the dimension of a variable v

$$\int \frac{dv}{\sqrt{v^2 + Y^2}} = \ln\left(\frac{v + \sqrt{v^2 + Y^2}}{|Y|}\right) \tag{9.2.229}$$

while for a dimensionless parameter $B \neq 0$ and a dimensionless variable w, we have

$$\begin{aligned} &\int \ln\left(B + \sqrt{B^2 + w^2}\right) dw \\ &= w\left[\ln\left(B + \sqrt{B^2 + w^2}\right) - 1\right] + B \ln\left(\frac{w + \sqrt{B^2 + w^2}}{|B|}\right) \end{aligned} \tag{9.2.230}$$

To prove Equation 9.2.228, we differentiate the RHS of Equation 9.2.228 with respect to u and obtain the integrand of the LHS of Equation 9.2.228. Similarly, to prove Equation 9.2.229, we differentiate the RHS of Equation 9.2.229 with respect to v and arrive at the integrand of the LHS of this equation. Finally, by differentiation of Equation 9.2.230 with respect to w we arrive at an identity that proves Equation 9.2.230.

To find $u_1(x_1, x_2, 0)$ in a closed form, we use Equation $9.2.226_1$ and obtain

$$u_1(x_1, x_2, 0) = -\frac{(1 + \nu)\alpha T_0}{2\pi} \int_{-a_2}^{a_2} d\xi_2 \left[\ln\left(\frac{R_1^-}{L}\right) - \ln\left(\frac{R_1^+}{L}\right)\right] \tag{9.2.231}$$

where

$$\begin{aligned} \frac{R_1^-}{L} &= \left[\left(\frac{x_1 - a_1}{L}\right)^2 + \left(\frac{x_2 - \xi_2}{L}\right)^2\right]^{1/2} \\ \frac{R_1^+}{L} &= \left[\left(\frac{x_1 + a_1}{L}\right)^2 + \left(\frac{x_2 - \xi_2}{L}\right)^2\right]^{1/2} \end{aligned} \tag{9.2.232}$$

By introducing the new variable

$$u = \frac{x_2 - \xi_2}{L}$$

Equation 9.2.231 is rewritten in the form

$$u_1(x_1,x_2,0) = -\frac{(1+\nu)\alpha T_0}{2\pi}L\int\limits_{(x_2-a_2)/L}^{(x_2+a_2)/L}\left\{\ln\left[u^2+\left(\frac{x_1-a_1}{L}\right)^2\right]^{1/2}\right.$$
$$\left.-\ln\left[u^2+\left(\frac{x_1+a_1}{L}\right)^2\right]^{1/2}\right\}du \tag{9.2.233}$$

and, by Equation 9.2.228, we write

$$u_1(x_1,x,0) = -\frac{(1+\nu)\alpha T_0}{2\pi}\left\{\frac{Lu}{2}\ln\frac{\left[u^2+(x_1-a_1)^2/L^2\right]}{\left[u^2+(x_1+a_1)^2/L^2\right]}\right.$$
$$\left.+(x_1-a_1)\tan^{-1}\frac{Lu}{x_1-a_1}-(x_1+a_1)\tan^{-1}\frac{Lu}{x_1+a_1}\right\}_{u=(x_2-a_2)/L}^{u=(x_2+a_2)/L} \tag{9.2.234}$$

Hence, we obtain the closed form of $u_1(x_1,x_2,0)$

$$u_1(x_1,x_2,0) = -\frac{(1+\nu)\alpha T_0}{2\pi}\left[(x_2+a_2)\ln\frac{\sqrt{(x_1-a_1)^2+(x_2+a_2)^2}}{\sqrt{(x_1+a_1)^2+(x_2+a_2)^2}}\right.$$
$$-(x_2-a_2)\ln\frac{\sqrt{(x_1-a_1)^2+(x_2-a_2)^2}}{\sqrt{(x_1+a_1)^2+(x_2-a_2)^2}}$$
$$+(x_1-a_1)\left(\tan^{-1}\frac{x_2+a_2}{x_1-a_1}-\tan^{-1}\frac{x_2-a_2}{x_1-a_1}\right)$$
$$\left.-(x_1+a_1)\left(\tan^{-1}\frac{x_2+a_2}{x_1+a_1}-\tan^{-1}\frac{x_2-a_2}{x_1+a_1}\right)\right] \tag{9.2.235}$$

To calculate $u_2(x_1,x_2,0)$ in a closed form we use Equation 9.2.226$_2$, and obtain

$$u_2(x_1,x_2,0) = -\frac{(1+\nu)\alpha T_0}{2\pi}\int\limits_{-a_1}^{a_1}d\xi_1\left[\ln\left(\frac{R_2^-}{L}\right)-\ln\left(\frac{R_2^+}{L}\right)\right] \tag{9.2.236}$$

where

$$\frac{R_2^-}{L} = \left[\left(\frac{x_1-\xi_1}{L}\right)^2+\left(\frac{x_2-a_2}{L}\right)^2\right]^{1/2}$$
$$\frac{R_2^+}{L} = \left[\left(\frac{x_1-\xi_1}{L}\right)^2+\left(\frac{x_2+a_2}{L}\right)^2\right]^{1/2} \tag{9.2.237}$$

By introducing the new variable

$$v = \frac{x_1-\xi_1}{L}$$

Equation 9.2.236 is reduced to the form

$$u_2(x_1,x_2,0) = -\frac{(1+\nu)\alpha T_0}{2\pi}L\int_{(x_1-a_1)/L}^{(x_1+a_1)/L}\left\{\ln\left[v^2+\left(\frac{x_2-a_2}{L}\right)^2\right]^{1/2} - \ln\left[v^2+\left(\frac{x_2+a_2}{L}\right)^2\right]^{1/2}\right\}dv \tag{9.2.238}$$

and, by Equation 9.2.228, the closed form of $u_2(x_1,x_2,0)$ is obtained

$$\begin{aligned} u_2(x_1,x_2,0) = -\frac{(1+\nu)\alpha T_0}{2\pi}\Bigg[&(x_1+a_1)\ln\frac{\sqrt{(x_1+a_1)^2+(x_2-a_2)^2}}{\sqrt{(x_1+a_1)^2+(x_2+a_2)^2}} \\ &-(x_1-a_1)\ln\frac{\sqrt{(x_1-a_1)^2+(x_2-a_2)^2}}{\sqrt{(x_1-a_1)^2+(x_2+a_2)^2}} \\ &+(x_2-a_2)\left(\tan^{-1}\frac{x_1+a_1}{x_2-a_2}-\tan^{-1}\frac{x_1-a_1}{x_2-a_2}\right) \\ &-(x_2+a_2)\left(\tan^{-1}\frac{x_1+a_1}{x_2+a_2}-\tan^{-1}\frac{x_1-a_1}{x_2+a_2}\right)\Bigg] \end{aligned} \tag{9.2.239}$$

It is easy to observe that Equation 9.2.239 may be obtained from Equation 9.2.235 by replacing the index 1 by the index 2, and the index 2 by the index 1.

Finally, to get $u_3(x_1,x_2,0)$ in a closed form we note that

$$\int_{-a_1}^{a_1}\frac{d\xi_1}{R_0} = \int_{x_1-a_1}^{x_1+a_1}\frac{dv}{\sqrt{v^2+(x_2-\xi_2)^2}} \tag{9.2.240}$$

and, by Equation 9.2.229 with $Y = x_2 - \xi_2$, we write

$$\int_{-a_1}^{a_1}\frac{d\xi_1}{R_0} = \ln\frac{x_1+a_1+\sqrt{(x_1+a_1)^2+(x_2-\xi_2)^2}}{x_1-a_1+\sqrt{(x_1-a_1)^2+(x_2-\xi_2)^2}} \tag{9.2.241}$$

or

$$\int_{-a_1}^{a_1}\frac{d\xi_1}{R_0} = \ln\left[\left(\frac{x_1+a_1}{L}\right)+\sqrt{\left(\frac{x_1+a_1}{L}\right)^2+w^2}\right] - \ln\left[\left(\frac{x_1-a_1}{L}\right)+\sqrt{\left(\frac{x_1-a_1}{L}\right)^2+w^2}\right] \tag{9.2.242}$$

where

$$w = \frac{x_2 - \xi_2}{L} \tag{9.2.243}$$

By integrating Equation 9.2.242 with respect to ξ_2 over the interval $[-a_2, a_2]$, and using Equation 9.2.226$_3$, we get

$$u_3(x_1, x_2, 0) = -\frac{(1+\nu)\alpha T_0}{2\pi} L \int_{(x_2-a_2)/L}^{(x_2+a_2)/L} \left\{ \ln\left[\left(\frac{x_1+a_1}{L}\right) + \sqrt{\left(\frac{x_1+a_1}{L}\right)^2 + w^2} \right] - \ln\left[\left(\frac{x_1-a_1}{L}\right) + \sqrt{\left(\frac{x_1-a_1}{L}\right)^2 + w^2} \right] \right\} dw \tag{9.2.244}$$

Hence, by Equation 9.2.230, we arrive at

$$u_3(x_1, x_2, 0) = -\frac{(1+\nu)\alpha T_0}{2\pi} L \left\{ w \left(\ln\left[\left(\frac{x_1+a_1}{L}\right) + \sqrt{\left(\frac{x_1+a_1}{L}\right)^2 + w^2} \right] - \ln\left[\left(\frac{x_1-a_1}{L}\right) + \sqrt{\left(\frac{x_1-a_1}{L}\right)^2 + w^2} \right] \right) + \left(\frac{x_1+a_1}{L}\right) \ln \frac{\left[w + \sqrt{(x_1+a_1)^2/L^2 + w^2}\right]}{|x_1+a_1|/L} - \left(\frac{x_1-a_1}{L}\right) \ln \frac{\left[w + \sqrt{(x_1-a_1)^2/L^2 + w^2}\right]}{|x_1-a_1|/L} \right\}_{w=(x_2-a_2)/L}^{w=(x_2+a_2)/L} \tag{9.2.245}$$

and the closed form of $u_3(x_1, x_2, 0)$ has the form

$$u_3(x_1, x_2, 0) = -\frac{(1+\nu)\alpha T_0}{2\pi} \left[(x_2+a_2) \ln \frac{(x_1+a_1) + \sqrt{(x_1+a_1)^2 + (x_2+a_2)^2}}{(x_1-a_1) + \sqrt{(x_1-a_1)^2 + (x_2+a_2)^2}} - (x_2-a_2) \ln \frac{(x_1+a_1) + \sqrt{(x_1+a_1)^2 + (x_2-a_2)^2}}{(x_1-a_1) + \sqrt{(x_1-a_1)^2 + (x_2-a_2)^2}} + (x_1+a_1) \ln \frac{(x_2+a_2) + \sqrt{(x_1+a_1)^2 + (x_2+a_2)^2}}{(x_2-a_2) + \sqrt{(x_1+a_1)^2 + (x_2-a_2)^2}} - (x_1-a_1) \ln \frac{(x_2+a_2) + \sqrt{(x_1-a_1)^2 + (x_2+a_2)^2}}{(x_2-a_2) + \sqrt{(x_1-a_1)^2 + (x_2-a_2)^2}} \right] \tag{9.2.246}$$

In particular, if $a_1 = a_2 = 1$ cm, then it follows from Equations 9.2.235, 9.2.239, and 9.2.246 that $u_1(0,0,0) = u_2(0,0,0) = 0$, and

$$u_3(0,0,0) = -\frac{2}{\pi}(1+\nu)\alpha T_0 \ln(3+2\sqrt{2})\ \text{cm} \tag{9.2.247}$$

Therefore, for a temperature exposure on a square region in the stress free plane $x_3 = 0$, the center of the square moves upward by the amount of $|u_3(0,0,0)|$, where $u_3(0,0,0)$ is given by Equation 9.2.247.

In winding up Sections 9.2.3.1 and 9.2.3.2, we note that the formula 9.2.225 may be used to obtain other closed-form fields that define the thermoelastic state s produced in the semispace by a rectangular boundary heat exposure. For example, to find $S_{12} = S_{12}(\mathbf{x})$ in a closed form, we note that (see Equations 9.2.215)

$$S^*_{12}(\mathbf{x}) = \frac{\mu(1+\nu)\alpha T_0}{\pi}\frac{\partial^2}{\partial x_1 \partial x_2}\ln\left(\frac{R+x_3}{L}\right) \tag{9.2.248}$$

where

$$R = \sqrt{x_1^2 + x_2^2 + x_3^2}$$

and, in view of Equation 9.2.225, we have

$$S_{12}(\mathbf{x}) = \frac{\mu(1+\nu)\alpha T_0}{\pi}\int_{-a_1}^{a_1} d\xi_1 \int_{-a_2}^{a_2} d\xi_2 \frac{\partial^2}{\partial \xi_1 \partial \xi_2}\left[\ln\left(\frac{R^*+x_3}{L}\right)\right] \tag{9.2.249}$$

where

$$R^* = \sqrt{(x_1-\xi_1)^2 + (x_2-\xi_2)^2 + x_3^2}$$

Hence, the closed form of $S_{12}(\mathbf{x})$ reads

$$S_{12}(\mathbf{x}) = \frac{\mu(1+\nu)\alpha T_0}{\pi}\ln\left[\frac{x_3+\sqrt{(x_1-a_1)^2+(x_2-a_2)^2+x_3^2}}{x_3+\sqrt{(x_1-a_1)^2+(x_2+a_2)^2+x_3^2}}\right.$$
$$\left.\times\frac{x_3+\sqrt{(x_1+a_1)^2+(x_2+a_2)^2+x_3^2}}{x_3+\sqrt{(x_1+a_1)^2+(x_2-a_2)^2+x_3^2}}\right] \tag{9.2.250}$$

This formula may be found on p. 106 in a book by Witold Nowacki [5].

A number of three-dimensional axially symmetric initial BVP of quasi-static thermoelasticity for a semispace subject to a time-dependent laser induced heat flux on its boundary were solved by the first author (RBH) and his collaborators [6–9].

9.3 TORSION PROBLEM

In Chapter 4, we formulated a mixed problem of elastostatics for a general three-dimensional solid, and this formulation may be applied to an elastic prismatic bar bounded by a cylindrical lateral surface and by a pair of planes normal to the lateral surface, called the bases.

Even when only two opposite couples of concentrated forces tangent to the bases are prescribed, and the lateral surface is stress free, the problem of finding an elastic state $s = [\mathbf{u}, \mathbf{E}, \mathbf{S}]$ in the bar is a rather difficult one. In engineering practice it is sufficient to solve an approximate problem* in which resultant torsion moments are applied at the bases and the lateral surface is stress free. In this section we deal with such an approximate torsion problem.

In Section 9.3.1, we deal with torsion of circular prismatic bars, in Sections 9.3.2 and 9.3.3 we discuss torsion of noncircular bars, in particular, of elliptic and triangular cross sections. In each of these parts an approximate three-dimensional formulation is reduced to a two-dimensional problem for Laplace's or Poisson's equation.

The problems of torsion were analyzed by the fathers of elasticity, such as de Saint-Venant, Prandtl, and Poisson.

9.3.1 Torsion of Circular Prismatic Bars

We consider a circular prismatic bar of length l and radius a, without body forces acting. We take x_3 along the axis of the bar and assume that the bar is fixed at one end, $x_3 = 0$, in the (x_1, x_2) plane, while at the other end, $x_3 = l$, a torsion moment M_3 is applied, see Figure 9.5.

This moment causes the bar to be twisted, and the generators of the circular cylinder deform into helical curves. Because of the symmetry of the cross section, we expect that the cross sections of the bar remain plane, and that they rotate by an angle θ, which is proportional to the distance x_3, see Figure 9.6

$$\theta = \alpha x_3 \tag{9.3.1}$$

with α being the angle of twist per unit length along the x_3-axis.

Experiments show that the radial displacement during torsion of a circular prismatic bar is approximately zero, $u_r \approx 0$, and on account of the fact that the cross sections remain plane, $u_3 = 0$. Since the circumferential displacement of a point P at a distance r from the axis of the bar is proportional to the angle† θ and to the radius r, $u_\theta = \theta\, r = \alpha x_3 r$. Therefore, the displacement components of P in cylindrical coordinates (r, θ, x_3) are

$$u_r = 0, \quad u_\theta = \alpha x_3 r, \quad u_3 = 0 \tag{9.3.2}$$

* In view of the de Saint-Venant principle, a distribution of stresses at a sufficient distance from the ends is not affected by the manner in which the torsion moments are applied at the bases. Also, in practice, at the ends the resultant torsion moments are prescribed, rather than the shear tractions.

† The angle θ does not need to be small. In fact, in long shafts, like those used in oil exploration, θ may achieve values of a number of full turns.

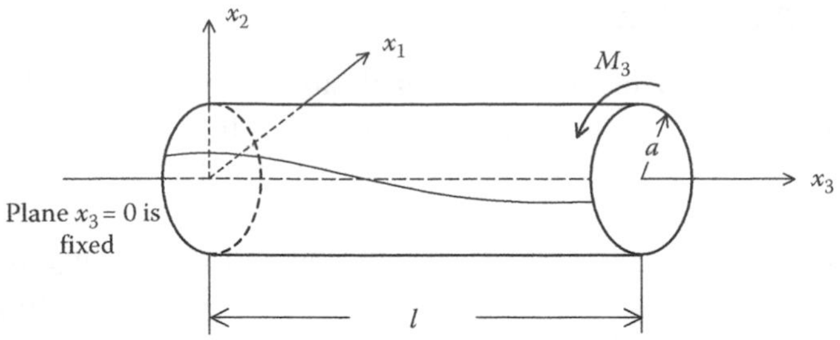

FIGURE 9.5 Torsion of a circular prismatic bar.

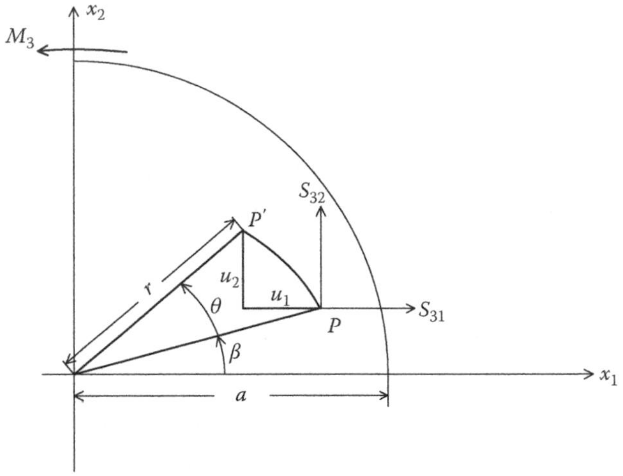

FIGURE 9.6 Rotation by an angle θ and displacement of point P.

Using the strain–displacement relations in cylindrical coordinates [10]

$$E_{rr} = \frac{\partial u_r}{\partial r}, \quad E_{\theta\theta} = \frac{1}{r}\frac{\partial u_\theta}{\partial \theta} + \frac{u_r}{r}, \quad E_{33} = \frac{\partial u_3}{\partial x_3}$$

$$E_{r\theta} = \frac{1}{2}\left(\frac{1}{r}\frac{\partial u_r}{\partial \theta} + \frac{\partial u_\theta}{\partial r} - \frac{u_\theta}{r}\right) \tag{9.3.3}$$

$$E_{r3} = \frac{1}{2}\left(\frac{\partial u_r}{\partial x_3} + \frac{\partial u_3}{\partial r}\right), \quad E_{\theta 3} = \frac{1}{2}\left(\frac{\partial u_\theta}{\partial x_3} + \frac{1}{r}\frac{\partial u_3}{\partial \theta}\right)$$

we obtain all components of the strain tensor to be zero except $E_{\theta 3}$, which is

$$E_{\theta 3} = \frac{\alpha r}{2} \tag{9.3.4}$$

Using the stress–strain relations, we find

$$\begin{aligned} &S_{rr} = S_{\theta\theta} = S_{33} = S_{r\theta} = S_{r3} = 0 \\ &S_{\theta 3} = \mu \alpha r \end{aligned} \tag{9.3.5}$$

Thus, the only nonzero component of the stress tensor is the shear stress $S_{\theta 3}$, which is proportional to r, and its maximum value is at the radius of the bar

$$(S_{\theta 3})_{\max} = (S_{\theta 3})_{r=a} = \mu \alpha a \tag{9.3.6}$$

Integrating over elementary rings of width dr and using Equation 9.3.5, the moment M_3 becomes

$$M_3 = \int_0^a (rS_{\theta 3}) 2\pi r dr = \mu \alpha J \tag{9.3.7}$$

where

$$J = \frac{1}{2}\pi a^4 \tag{9.3.8}$$

is the polar moment of inertia of the cross section about its center. The product μJ by which we multiply α to receive M_3 is called the *torsional rigidity* of the bar.

In case of a hollow circular bar with a_{in} and a_{out} being, respectively, inner and outer radii of the bar, in Equation 9.3.7 the value of J needs to be changed to

$$J = \frac{\pi}{2}\left(a_{\text{out}}^4 - a_{in}^4\right) \tag{9.3.9}$$

The solution presented is associated with the name of C. A. Coulomb.

To present the results in a Cartesian coordinate system, we consider a point $P(x_1, x_2)$ of the cross section that after the application of torsion moment M_3 moves to $P'(x_1+u_1, x_2+u_2)$, see Figure 9.6.

Denoting by β the angle between radius $r = OP$ and the x_1-axis, we have $x_1 = r\cos\beta$ and $x_2 = r\sin\beta$ and

$$\begin{aligned} u_1 &= r\cos(\beta+\theta) - r\cos\beta = x_1(\cos\theta - 1) - x_2\sin\theta \\ u_2 &= r\sin(\beta+\theta) - r\sin\beta = x_1\sin\theta + x_2(\cos\theta - 1) \end{aligned} \tag{9.3.10}$$

If the angle θ is small,* and noting that $\theta = \alpha x_3$, the components of the displacement vector are

$$\begin{aligned} u_1 &\approx -\theta x_2 = -\alpha x_2 x_3 \\ u_2 &\approx \theta x_1 = \alpha x_1 x_3 \end{aligned} \tag{9.3.11}$$

The components of the strain tensor $\mathbf{E}$ are, see Equation 4.1.1,

$$\begin{aligned} &E_{11} = E_{22} = E_{33} = E_{12} = 0 \\ &E_{23} = \frac{1}{2}\alpha x_1, \quad E_{31} = -\frac{1}{2}\alpha x_2 \end{aligned} \tag{9.3.12}$$

* As stated in footnote on page 491, θ may be large for long shafts. However, except in the immediate vicinity of the ends, the strains and stresses in all cross sections are the same. For derivation of the displacement vector u_i $(i = 1, 2, 3)$ we choose a cross section that is rotated by a small angle θ, that is, the one sufficiently close to the $x_3 = 0$ end of the bar.

Using the stress–displacement equations, see Equation 4.1.9,

$$S_{ij} = \lambda u_{k,k}\delta_{ij} + \mu(u_{i,j} + u_{j,i}) \tag{9.3.13}$$

we find the components of the stress tensor **S**

$$\begin{aligned} &S_{11} = S_{22} = S_{33} = S_{12} = 0 \\ &S_{23} = \mu\alpha x_1, \quad S_{31} = -\mu\alpha x_2 \end{aligned} \tag{9.3.14}$$

These components satisfy the equilibrium equation (4.1.34) and the compatibility equation 4.1.35 with $b_i = 0$, and the boundary conditions on the lateral surface of the bar given by

$$S_{ij}n_j = 0 \tag{9.3.15}$$

where vector $\mathbf{n} = (n_1, n_2, 0)$ is a unit outward vector on the lateral surface. Equation 9.3.15, in components, and for the lateral surface of the bar where $n_3 = 0$ reduces to

$$\begin{aligned} S_{11}n_1 + S_{12}n_2 &= 0 \\ S_{21}n_1 + S_{22}n_2 &= 0 \\ S_{31}n_1 + S_{32}n_2 &= 0 \end{aligned} \tag{9.3.16}$$

The torsion moment M_3 is

$$\begin{aligned} M_3 &= \int_A (x_1 S_{23} - x_2 S_{31}) dx_1 dx_2 \\ &= \mu\alpha \int_A \left(x_1^2 + x_2^2\right) dx_1 dx_2 = \mu\alpha J \end{aligned} \tag{9.3.17}$$

where A is the area of the cross section.

The magnitude of shear stress S_t at point P is

$$S_t = \sqrt{S_{23}^2 + S_{31}^2} = \mu\alpha\sqrt{x_1^2 + x_2^2} = \mu\alpha r \tag{9.3.18}$$

9.3.2 Torsion of Noncircular Prismatic Bars

We now consider a noncircular prismatic bar of length l. The bar is fixed at $x_3 = 0$ and the other end, $x_3 = l$, is twisted by a moment M_3. The influence of body forces is neglected.

Navier tried to follow the solution for a circular bar by assuming that the cross sections of a noncircular bar will remain plane after deformation. This assumption was wrong, and this fact may be observed by checking the boundary conditions 9.3.16. Substitution of stresses from Equation 9.3.14 into the third condition of 9.3.16 leads to

$$-\mu\alpha x_2 n_1 + \mu\alpha x_1 n_2 = 0 \tag{9.3.19}$$

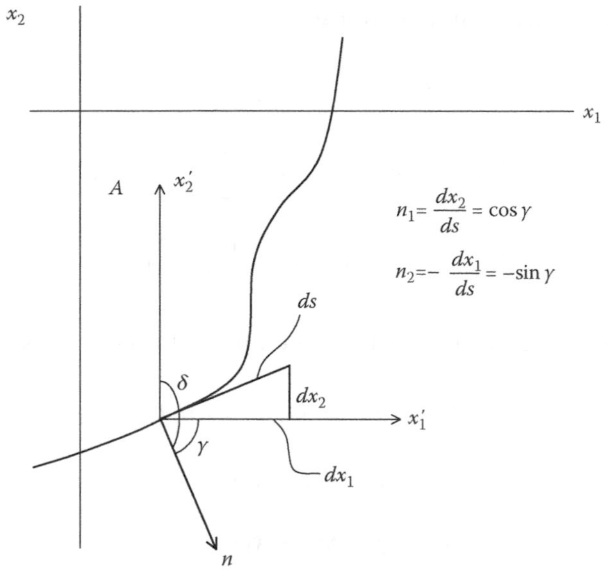

FIGURE 9.7 Segment of the cross section A of a noncircular bar.

where $n_1 = \cos\gamma$ and $n_2 = \cos\delta = -\sin\gamma$, see Figure 9.7. On the other hand, from Figure 9.7 it follows that

$$\frac{dx_1}{ds} = \cos(90 - \gamma) = -\cos\delta = -n_2$$
$$\frac{dx_2}{ds} = \sin(90 - \gamma) = \cos\gamma = n_1 \tag{9.3.20}$$

Substituting these quantities into Equation 9.3.19 and dividing by $\mu\alpha$ we have

$$x_1 dx_1 + x_2 dx_2 = 0 \tag{9.3.21}$$

This equation represents a family of concentric circles

$$x_1^2 + x_2^2 = \text{const} \tag{9.3.22}$$

Therefore, only bars of circular cross section are free from forces applied to the lateral surface if stresses are to be expressed by Equation 9.3.14. In case of a bar of any other cross section there will exist on the boundary a component of stress normal to the boundary, and this violates boundary conditions.

To proceed with the analysis of prismatic bars with noncircular cross section, we part with the assumption that the cross sections remain plane. Instead, we admit warping of a cross section by assuming the displacement vector as

$$\mathbf{u} = [-\alpha x_2 x_3,\ \alpha x_1 x_3,\ \alpha\psi(x_1, x_2)] \tag{9.3.23}$$

where $\psi(x_1, x_2)$ is called a *warping function.* We assume then that all cross sections are subject to the same warping.

Equation 4.1.1 leads to the strain tensor **E** as

$$E_{11} = E_{22} = E_{33} = E_{12} = 0 \tag{9.3.24}$$

$$E_{31} = \frac{1}{2}(u_{1,3} + u_{3,1}) = \frac{\alpha}{2}(\psi_{,1} - x_2) \tag{9.3.25}$$

$$E_{23} = \frac{1}{2}(u_{2,3} + u_{3,2}) = \frac{\alpha}{2}(\psi_{,2} + x_1) \tag{9.3.26}$$

and from the constitutive equation 9.3.13 the components of the stress tensor **S** become

$$S_{11} = S_{22} = S_{33} = S_{12} = 0 \tag{9.3.27}$$

$$S_{31} = \mu\alpha(\psi_{,1} - x_2) \tag{9.3.28}$$

$$S_{23} = \mu\alpha(\psi_{,2} + x_1) \tag{9.3.29}$$

Substituting Equations 9.3.27 through 9.3.29 into the equilibrium equation with zero body forces shows that the warping function $\psi(x_1, x_2)$ is a harmonic function on the cross section A:

$$\nabla^2\psi \equiv \psi_{,11} + \psi_{,22} = 0 \quad \text{on A} \tag{9.3.30}$$

Furthermore, substituting Equations 9.3.28 and 9.3.29 into the third of conditions 9.3.16 results in the equation

$$(\psi_{,1} - x_2)n_1 + (\psi_{,2} + x_1)n_2 = 0 \quad \text{on } \partial A = C \tag{9.3.31}$$

Note that the first two boundary conditions 9.3.16 are satisfied identically.

Equations 9.3.27 through 9.3.29 ensure that on the bases $x_3 = 0, x_3 = l$ of the bar the resultant force $\mathbf{F} = [F_1, F_2]$ is zero and the distribution of shear stresses represents a torsion moment. To prove that the resultant force F_1 in the x_1 direction is zero we write

$$F_1 = \int_A S_{31}dx_1dx_2 = \mu\alpha\int_A(\psi_{,1} - x_2)dx_1dx_2 \tag{9.3.32}$$

The latter expression in view of Equation 9.3.30 is equivalent to

$$F_1 = \mu\alpha\int_A \left\{\left[x_1(\psi_{,1} - x_2)\right]_{,1} + \left[x_1(\psi_{,2} + x_1)\right]_{,2}\right\} dx_1dx_2 \tag{9.3.33}$$

Using now a two-dimensional version of the Divergence Theorem (see Equation 2.3.1), we have

$$F_1 = \mu\alpha\int_C x_1\left(\frac{d\psi}{dn} - x_2n_1 + x_1n_2\right) ds \tag{9.3.34}$$

where C is the boundary curve of the base. Observe that $d\psi/dn$ in the last equation may be expressed by

$$\frac{d\psi}{dn} = \psi_{,1}n_1 + \psi_{,2}n_2 = \psi_{,1}\cos\gamma - \psi_{,2}\sin\gamma \tag{9.3.35}$$

where n_1 and n_2 are directional cosines of the normal to the boundary C with respect to x_1 and x_2, respectively, see Figure 9.7.

Using Equation 9.3.35 we may write the boundary condition 9.3.31 as

$$\frac{d\psi}{dn} = x_2n_1 - x_1n_2 \quad \text{on } C \tag{9.3.36}$$

Because of the condition 9.3.36 it follows from Equation 9.3.34 that $F_1 = 0$. In a similar way it may be proved that the force F_2 in the x_2 direction is zero:

$$F_2 = \int_A S_{23}dx_1dx_2 = 0 \tag{9.3.37}$$

Now, to show that the stresses expressed by Equations 9.3.27 through 9.3.29 lead to a torsion moment, we write

$$M_3 = \int_A (x_1S_{32} - x_2S_{31})dx_1dx_2 \tag{9.3.38}$$

Therefore, by Equations 9.3.28 and 9.3.29

$$M_3 = \mu\alpha\int_A \left(x_1^2 + x_2^2 + x_1\psi_{,2} - x_2\psi_{,1}\right)dx_1dx_2 = \alpha D \tag{9.3.39}$$

Here

$$D = \mu\int_A \left(x_1^2 + x_2^2 + x_1\psi_{,2} - x_2\psi_{,1}\right)dx_1dx_2 \tag{9.3.40}$$

is called the *torsional rigidity of the bar.* Thus, the torsion moment M_3 is proportional to the angle α of twist per unit length and to the rigidity D.

Summing up our discussion, the torsion problem of a noncircular prismatic bar has been solved once a warping function $\psi(x_1, x_2)$ becomes known. Functions $\mathbf{u}, \mathbf{E}$, and $\mathbf{S}$ are determined from Equations 9.3.23, 9.3.24 through 9.3.26, and 9.3.27 through 9.3.29, respectively, while $\psi(x_1, x_2)$ satisfies the equations

$$\nabla^2\psi = 0 \quad \text{on } A \tag{9.3.41}$$

$$\frac{d\psi}{dn} = x_2n_1 - x_1n_2 \quad \text{on } C \tag{9.3.42}$$

In view of Equations 9.3.31 and 9.3.20, we may write the condition 9.3.42 in an equivalent form

$$(\psi_{,1} - x_2)\frac{dx_2}{ds} - (\psi_{,2} + x_1)\frac{dx_1}{ds} = 0 \quad \text{on } C \tag{9.3.43}$$

The problem of finding a warping function $\psi(x_1, x_2)$ is a special case of the Neumann boundary value problem [11] of the potential theory in which one is to find a harmonic function $\psi(x_1, x_2)$ on the region A in such a way that $d\psi/dn$ is prescribed on C. The condition for the existence of a solution $\psi(x_1, x_2)$ to the Neumann problem is that

$$\int_C \frac{d\psi}{dn} ds = 0 \tag{9.3.44}$$

In the case of the torsion problem

$$\int_C \frac{d\psi}{dn} ds = \int_C (x_2 n_1 - x_1 n_2) ds = \int_C (x_2 dx_2 + x_1 dx_1) = 0 \tag{9.3.45}$$

because the integrand $x_2 dx_2 + x_1 dx_1$ is the exact differential of the function

$$\frac{1}{2}\left(x_1^2 + x_2^2\right) + \text{const} \tag{9.3.46}$$

Therefore, the condition 9.3.44 is satisfied.

9.3.2.1 Torsion of an Elliptic Bar

In solving particular torsion problems, we may try various harmonic functions as the warping function $\psi(x_1, x_2)$ and check to which cross section contours C they apply. To solve the torsion problem for an elliptic bar we try

$$\psi = K x_1 x_2 \quad (K = \text{const}) \tag{9.3.47}$$

which is obviously a harmonic function.

Substitution of this function in condition 9.3.43 gives

$$(K-1) x_2 \frac{dx_2}{ds} - (K+1) x_1 \frac{dx_1}{ds} = 0 \tag{9.3.48}$$

This equation may be written as

$$\frac{d}{ds}\left(x_1^2 + \frac{1-K}{1+K} x_2^2\right) = 0 \tag{9.3.49}$$

which after integration with respect to s results in

$$x_1^2 + \frac{1-K}{1+K} x_2^2 = \text{const.} \tag{9.3.50}$$

Observing that the equation of an ellipse with semiaxes a and b and with the center at the origin is

$$x_1^2 + \frac{a^2}{b^2} x_2^2 = a^2 \tag{9.3.51}$$

Equation 9.3.50 becomes identical with Equation 9.3.51 if

$$\frac{1-K}{1+K} = \frac{a^2}{b^2} \tag{9.3.52}$$

or

$$K = \frac{b^2 - a^2}{b^2 + a^2} \tag{9.3.53}$$

Therefore, we have found the warping function for an elliptic bar

$$\psi(x_1, x_2) = \frac{b^2 - a^2}{b^2 + a^2} x_1 x_2 \tag{9.3.54}$$

We now calculate the displacement vector from Equation 9.3.23:

$$\mathbf{u} = \left(-\alpha x_2 x_3, \quad \alpha x_1 x_3, \quad \alpha \frac{b^2 - a^2}{b^2 + a^2} x_1 x_2 \right) \tag{9.3.55}$$

The components of the strain tensor **E** from Equations 9.3.24 through 9.3.26 are

$$E_{11} = E_{22} = E_{33} = E_{12} = 0 \tag{9.3.56}$$

$$E_{31} = -\alpha \frac{a^2}{a^2 + b^2} x_2 \tag{9.3.57}$$

$$E_{23} = \alpha \frac{b^2}{a^2 + b^2} x_1 \tag{9.3.58}$$

The components of the stress tensor from Equations 9.3.27 through 9.3.29 are

$$S_{11} = S_{22} = S_{33} = S_{12} = 0 \tag{9.3.59}$$

$$S_{31} = -2\mu\alpha \frac{a^2}{a^2 + b^2} x_2 \tag{9.3.60}$$

$$S_{23} = 2\mu\alpha \frac{b^2}{a^2 + b^2} x_1 \tag{9.3.61}$$

We observe that the maximum shear stress appears at the ends of the minor axis, at points $(0, b)$ and $(0, -b)$:

$$|S_{31}|_{\max} = 2\mu\alpha \frac{a^2 b}{a^2 + b^2} \tag{9.3.62}$$

The torsion moment is

$$M_3 = \int_A (x_1 S_{23} - x_2 S_{31})\, dx_1 dx_2$$

$$= \frac{2\mu\alpha}{a^2+b^2}\left(b^2 \int_A x_1^2 dx_1 dx_2 + a^2 \int_A x_2^2 dx_1 dx_2\right)$$

$$= \frac{2\mu\alpha}{a^2+b^2}\left(a^2 I_{11} + b^2 I_{22}\right) \tag{9.3.63}$$

where I_{11} and I_{22} are the moments of inertia of an ellipse about the x_1- and x_2-axes, respectively,

$$I_{11} = \frac{1}{4}\pi ab^3, \quad I_{22} = \frac{1}{4}\pi a^3 b \tag{9.3.64}$$

Substitution of I_{11} and I_{22} into Equation 9.3.63 gives the torsion moment as

$$M_3 = \frac{\pi \mu\alpha a^3 b^3}{a^2+b^2} \tag{9.3.65}$$

Therefore, the torsional rigidity of an elliptical bar is

$$D = \frac{\pi \mu a^3 b^3}{a^2+b^2} \tag{9.3.66}$$

It is often important in applications to express stresses in terms of the torsion moment. Using Equation 9.3.65 we find stresses S_{31} and S_{23} in terms of M_3:

$$S_{31} = -\frac{2M_3}{\pi ab^3} x_2$$
$$S_{23} = \frac{2M_3}{\pi a^3 b} x_1 \tag{9.3.67}$$

The resultant shear stress at any point of the elliptical cross section is

$$S_t = \left(S_{31}^2 + S_{23}^2\right)^{1/2} = \frac{2M_3}{\pi ab}\left(\frac{x_1^2}{a^4} + \frac{x_2^2}{b^4}\right)^{1/2} \tag{9.3.68}$$

We return now to the analysis of the warping of an elliptical cross section. The displacement u_3 that is the measure of warping takes the form

$$u_3 = \alpha \frac{b^2-a^2}{b^2+a^2} x_1 x_2 \tag{9.3.69}$$

The equation $u_3 = \text{const}$ represents a family of hyperbolas shown in Figure 9.8 where the curves more distant from the origin correspond to larger u_3. With the torsion moment acting counterclockwise, the solid lines indicate the positive (upward) displacement and the dotted lines indicate the negative (downward) displacement.

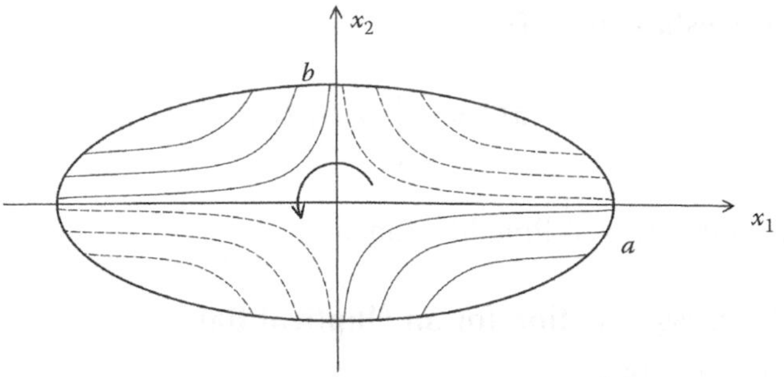

FIGURE 9.8 Warping of the elliptical cross section.

9.3.3 PRANDTL'S STRESS FUNCTION

We will now formulate an alternative approach to the problem of torsion of noncircular bars. To this end, we introduce a function $\phi(x_1, x_2)$, called Prandtl's stress function, defined in terms of the warping function $\psi(x_1, x_2)$ by the following formulas:

$$\begin{aligned} \phi_{,2} &= \mu\alpha(\psi_{,1} - x_2) = S_{31} \\ \phi_{,1} &= -\mu\alpha(\psi_{,2} + x_1) = -S_{23} \end{aligned} \tag{9.3.70}$$

We may see that $\phi(x_1, x_2)$ satisfies Poisson's equation

$$\nabla^2\phi \equiv \phi_{,11} + \phi_{,22} = -2\mu\alpha \tag{9.3.71}$$

The condition 9.3.43 in terms of $\phi(x_1, x_2)$ becomes

$$\phi_{,2}\frac{dx_2}{ds} + \phi_{,1}\frac{dx_1}{ds} = \frac{d\phi}{ds} = 0 \tag{9.3.72}$$

which means that $\phi(x_1, x_2) = \text{const}$ on C. Let us take the arbitrary constant as zero. Thus, the torsion problem is defined by

$$\begin{aligned} \nabla^2\phi &= -2\mu\alpha \quad \text{on } A \\ \phi &= 0 \quad \text{on } C \end{aligned} \tag{9.3.73}$$

The torsion moment is

$$\begin{aligned} M_3 &= \int_A (x_1 S_{23} - x_2 S_{31})\,dx_1 dx_2 = -\int_A (x_1\phi_{,1} + x_2\phi_{,2})\,dx_1 dx_2 \\ &= -\int_A \left[(x_1\phi)_{,1} + (x_2\phi)_{,2}\right] dx_1 dx_2 + 2\int_A \phi\,dx_1 dx_2 \end{aligned} \tag{9.3.74}$$

Now, using the Divergence Theorem (see Equation 2.3.1) we have

$$\int_A \left[(x_1\phi)_{,1} + (x_2\phi)_{,2}\right] dx_1 dx_2 = \int_C \phi(x_1 n_1 + x_2 n_2) ds = 0 \tag{9.3.75}$$

because function ϕ vanishes on C. Thus,

$$M_3 = 2 \int_A \phi \, dx_1 dx_2 \tag{9.3.76}$$

is the torsion moment in terms of Prandtl's function.

9.3.3.1 Prandtl's Stress Function for an Elliptical Bar

We try the function of the form

$$\phi(x_1, x_2) = m\left(\frac{x_1^2}{a^2} + \frac{x_2^2}{b^2} - 1\right) \tag{9.3.77}$$

where m is a constant. This function is zero on the contour C of an ellipse, that is, it satisfies the condition 9.3.73$_2$. Substituting Equation 9.3.77 into Equation 9.3.73$_1$ yields

$$m = -\frac{\mu\alpha}{\dfrac{1}{a^2} + \dfrac{1}{b^2}} \tag{9.3.78}$$

Combining with Equation 9.3.77, we have

$$\phi(x_1, x_2) = -\frac{\mu\alpha}{\dfrac{1}{a^2} + \dfrac{1}{b^2}}\left(\frac{x_1^2}{a^2} + \frac{x_2^2}{b^2} - 1\right) \tag{9.3.79}$$

This function may be used to calculate, for example, the torsion moment M_3 from Equation 9.3.76.

9.3.3.2 Prandtl's Stress Function for a Triangular Bar

Consider a bar with a cross section in the form of an equilateral triangle of altitude $3a$. As in Section 9.3.3.1, we assume Prandtl's stress function in such a way that it is zero on the contour C of the triangle, see Figure 9.9,

$$\phi(x_1, x_2) = m(x_1 - x_2\sqrt{3} + 2a)\left(x_1 + x_2\sqrt{3} + 2a\right)(x_1 - a) \tag{9.3.80}$$

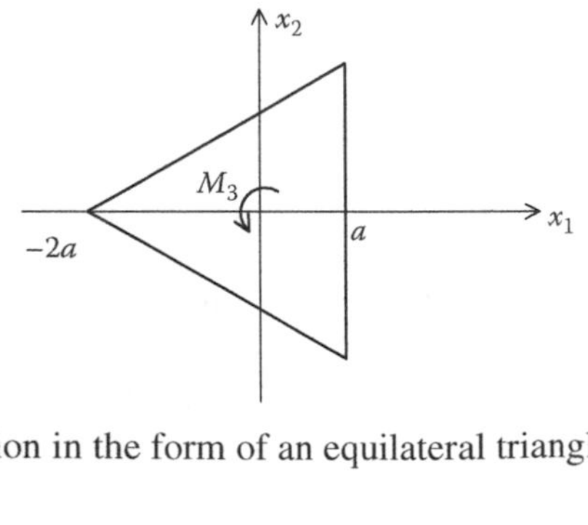

FIGURE 9.9 The cross section in the form of an equilateral triangle.

or

$$\phi(x_1, x_2) = m\left(x_1^3 + 3ax_1^2 - 3x_1x_2^2 + 3ax_2^2 - 4a^3\right) \tag{9.3.81}$$

where m is a constant. Substitution of 9.3.81 into Equation 9.3.73$_1$ yields

$$m = -\frac{\mu\alpha}{6a} \tag{9.3.82}$$

Combining this expression with Equation 9.3.81 yields Prandtl's stress function

$$\phi(x_1, x_2) = -\frac{\mu\alpha}{6a}\left(x_1^3 + 3ax_1^2 - 3x_1x_2^2 + 3ax_2^2 - 4a^3\right) \tag{9.3.83}$$

Using Equations 9.3.70 we find

$$\begin{aligned} S_{31} &= \phi_{,2} = \frac{\mu\alpha}{a}(x_1 - a)x_2 \\ S_{23} &= -\phi_{,1} = \frac{\mu\alpha}{2a}\left(x_1^2 + 2ax_1 - x_2^2\right) \end{aligned} \tag{9.3.84}$$

Along the x_1-axis stress S_{31} is zero, and S_{23} is, see Figure 9.10,

$$S_{23} = \frac{\mu\alpha}{2a}(x_1 + 2a)x_1 \tag{9.3.85}$$

with its maximum value at $x_1 = a$

$$(S_{23})_{\max} = (S_{23})_{x_1=a} = \frac{3}{2}\mu\alpha a \tag{9.3.86}$$

and with its minimum value at $x_1 = -a$

$$(S_{23})_{\min} = (S_{23})_{x_1=-a} = -\frac{\mu\alpha}{2}a \tag{9.3.87}$$

Note that shear stresses are zero at the corners and at the origin, and they are maximum at the midpoints of the sides of the triangle.

The warping function is

$$\psi(x_1, x_2) = \frac{x_2}{6a}\left(3x_1^2 - x_2^2\right) + \text{const} \tag{9.3.88}$$

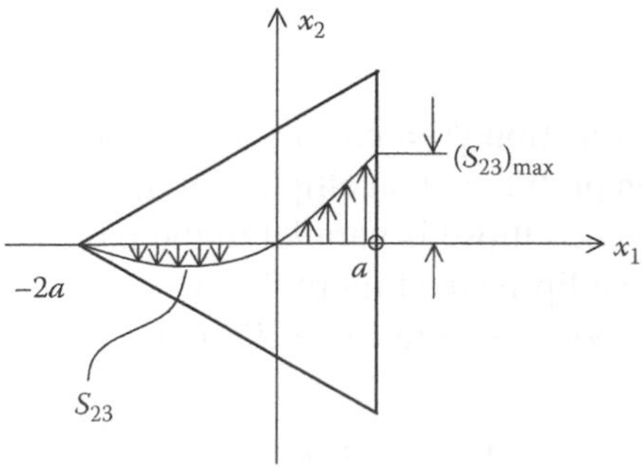

FIGURE 9.10 Distribution of shear stress S_{23} along the x_1-axis.

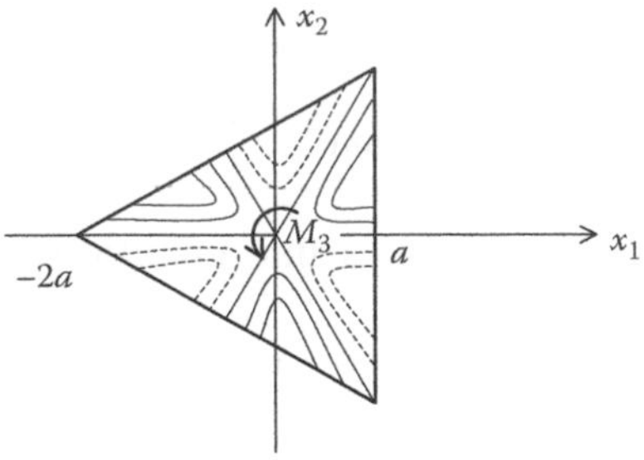

FIGURE 9.11 Lines of $u_3 = \text{const.}$

and we may check that Equations 9.3.70 are satisfied. Therefore, the displacement vector is

$$\mathbf{u} = \left[-\alpha x_2 x_3, \quad \alpha x_1 x_3, \quad \frac{\alpha}{6a} x_2 \left(3x_1^2 - x_2^2\right) \right] \tag{9.3.89}$$

The lines of constant displacement along the x_3 axis, $u_3 = \text{const}$, are shown in Figure 9.11, where solid lines indicate the positive (upward) displacements and dotted lines indicate the negative (downward) displacements.

The torsion moment, found from Equation 9.3.76, is

$$M_3 = \frac{3}{5} \mu \alpha J \tag{9.3.90}$$

where $J = 3\sqrt{3}a^4$ is the polar moment of inertia of the triangle about the origin.

The components of the strain tensor $\mathbf{E}$ are

$$E_{11} = E_{22} = E_{33} = E_{12} = 0 \tag{9.3.91}$$

$$E_{31} = \frac{\alpha x_2}{2a} (x_1 - a) \tag{9.3.92}$$

$$E_{23} = \frac{\alpha}{4a} \left(2ax_1 + x_1^2 - x_2^2\right) \tag{9.3.93}$$

The equations obtained provide a solution to the torsion problem for the bar with a cross section in the form of an equilateral triangle.

PROBLEMS

9.1 Show that the warping function $\psi = \text{const}$ solves the torsion problem of a circular bar.

9.2 Show that in the torsion problem of an elliptic bar, the resultant shear stress S_t at points on a given diameter of the ellipse is parallel to the tangent at the point of intersection of the diameter and the ellipse (see Figure P9.2).

9.3 Show that the torsion moment in terms of Prandtl's stress function $\phi = \phi(x_1, x_2)$ is expressed by

$$M_3 = 2 \int_A \phi(x_1, x_2) dx_1 dx_2$$

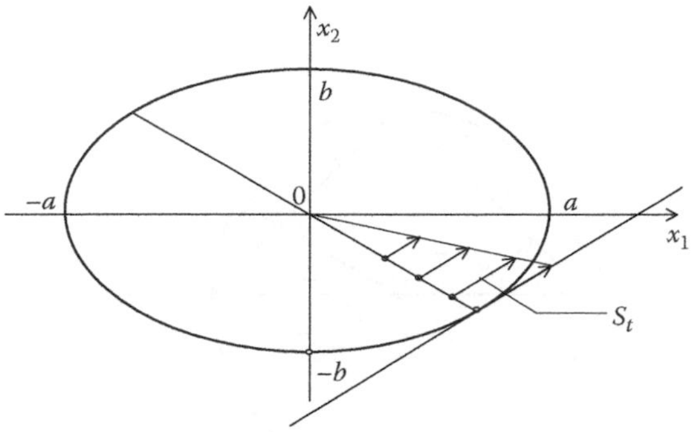

FIGURE P9.2

9.4 Show that Prandtl's stress function $\phi = \phi(x_1, x_2)$ given by

$$\phi(x_1, x_2) = \frac{32\mu\alpha a_1^2}{\pi^3} \sum_{n=1,3,5,\ldots}^{\infty} \frac{\sin\left(\frac{n\pi}{2}\right)}{n^3} \left[1 - \frac{\cosh\left(\frac{n\pi x_2}{2a_1}\right)}{\cosh\left(\frac{n\pi a_2}{2a_1}\right)}\right] \cos\left(\frac{n\pi x_1}{2a_1}\right)$$

solves the torsion problem of a bar with the rectangular cross section: $|x_1| \le a_1$, $|x_2| \le a_2$.

Also, show that in this case the torsion moment

$$M_3 = 2 \int_{-a_1}^{a_1} \int_{-a_2}^{a_2} \phi(x_1, x_2) dx_1 dx_2 = \mu\alpha (2a_1)^3 (2a_2)\, k^*$$

where

$$k^* = \frac{1}{3}\left[1 - \frac{192}{\pi^5}\left(\frac{a_1}{a_2}\right) \sum_{n=1,3,5\ldots}^{\infty} \frac{1}{n^5} \tanh\left(\frac{n\pi a_2}{2a_1}\right)\right]$$

9.5 Show that Prandtl's stress function

$$\phi(r, \theta) = \frac{\mu\alpha}{2}(r^2 - b^2)\left(\frac{2a\cos\theta}{r} - 1\right)$$

defined over the region

$$0 < b \le r \le 2a - b, \quad -\cos^{-1}\left(\frac{b}{2a}\right) \le \theta \le \cos^{-1}\left(\frac{b}{2a}\right)$$

solves the torsion problem of the circular shaft with a circular groove shown in Figure P9.5

In particular, find the stresses S_{13} and S_{23} on the boundary of the shaft.

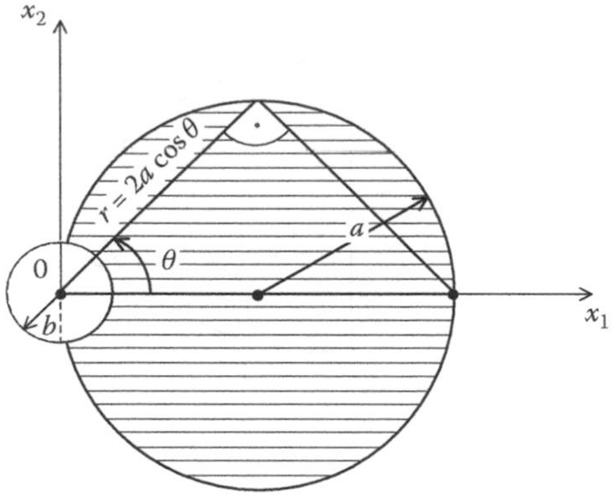

FIGURE P9.5

Hint: Use the polar coordinates

$$x_1 = r\cos\theta, \quad x_2 = r\sin\theta$$

REFERENCES

1. W. Nowacki, *Thermoelasticity*, Pergamon Press and PWN—Polish Scientific Publishers, Warszawa, 1962, p. 319.
2. I. S. Gradshteyn and I. M. Ryzhik, *Tables of Integrals, Sums, and Products*, Gosud. Izdat. Fiz. Mat. Lit., Moscow, 1962, p. 721 (in Russian); English translation: Academic Press, London, U.K., 1980.
3. A. S. Saada, *Elasticity—Theory and Applications*, Pergamon Press, New York, 1974, or Krieger, Malabar, 1987, p. 248.
4. H. S. Carslaw and J. C. Jaeger, *Conduction of Heat in Solids*, 2nd edn., Oxford University Press, Oxford, 1959.
5. W. Nowacki, *Thermoelasticity*, Pergamon Press, PWN—Polish Scientific Publishers Warszawa, 1962.
6. L. G. Hector Jr. and R. B. Hetnarski, Thermal stresses due to a laser pulse: The elastic solution, *J. Appl. Mech.*, 63, March 1996, 38–46.
7. L. G. Hector Jr. and R. B. Hetnarski, Thermal stresses in materials due to laser heating, in *Thermal Stresses IV*, ed., R. B. Hetnarski, Elsevier, Amsterdam, the Netherlands, 1996, Chap. 6, pp. 454–532.
8. W.-S. Kim, L. G. Hector Jr., and R. B. Hetnarski, Thermoelastic stresses in a bounded layer due to repetitively pulsed laser radiation, *Acta Mech.*, 125, 1997, 107–128.
9. P. H. Tehrani, L. G. Hector Jr., R. B. Hetnarski, and M. R. Eslami, Boundary element formulation for thermal stresses during pulsed laser heating, *J. Appl. Mech.*, 68, May 2001, 480–489.
10. A. S. Saada, *Elasticity—Theory and Applications*, Pergamon Press, New York, 1974, or Krieger, Malabar, 1987, pp. 138–139.
11. O. D. Kellogg, *Foundations of Potential Theory*, Springer, Berlin, Germany, Chap. XI, Sec. 12, 1929. Also published by Dover in 1953.

10 Solutions to Particular Two-Dimensional Boundary Value Problems of Elastostatics

In this chapter, the general solutions of Chapter 8 are used to obtain closed-form solutions to typical two-dimensional boundary value problems of isothermal and nonisothermal elastostatics. The solutions include (a) an elastic state corresponding to a concentrated force normal to the boundary of a semispace, (b) an elastic state corresponding to a concentrated force tangent to the boundary of a semispace, (c) the Lamé solution for a thin hollow circular disk subject to uniform pressures on the boundaries, (d) an elastic state in an infinite sheet with a circular hole subject to uniform tension at infinity, (e) a thermoelastic state due to a discontinuous temperature field in an infinite sheet, and (f) a thermoelastic state due to a concentrated heat source in a semi-infinite sheet. A number of end-of-chapter problems are included, and their solutions may be found in the Solutions Manual.

10.1 TWO-DIMENSIONAL SOLUTIONS OF ISOTHERMAL ELASTOSTATICS

In Chapter 8, three types of two-dimensional isothermal elastic states were defined: (i) an elastic state corresponding to a plane strain problem, (ii) an elastic state corresponding to a generalized plane stress problem, and (iii) an elastic state associated with a plane stress problem. The elastic states (i) and (ii) are defined on a two-dimensional domain C_0 of the (x_1, x_2) plane, while the elastic state (iii) is defined, in general, on a three-dimensional domain B. A plane strain elastic state corresponding to given body forces b_α $(\alpha = 1, 2)$ and satisfying suitable boundary conditions on ∂C_0 (∂C_0 is the boundary of C_0) is called a solution to a boundary value problem of elastostatics under plane strain conditions. A similar definition may be introduced for a solution to a boundary value problem of elastostatics related to the generalized plane stress conditions.

Also, a number of general properties of the two-dimensional elastic states were discussed in Chapter 8. For example, it was shown that a stress tensor field $\mathbf{S}$ corresponding to the elastic states (i) and (ii) does not depend on elastic moduli; and a solution to a plane strain boundary value problem generates a solution to a generalized plane stress problem by suitable transformation of elastic constants.

In this section, a number of closed form solutions of isothermal elastostatics corresponding to the plane strain conditions or generalized plane stress conditions are presented. These are: (A) an elastic state corresponding to a concentrated body force in an infinite solid under plane strain conditions, (B) an elastic state corresponding to a concentrated force normal to the boundary of a semi-infinite solid under plane strain conditions, (C) an elastic state

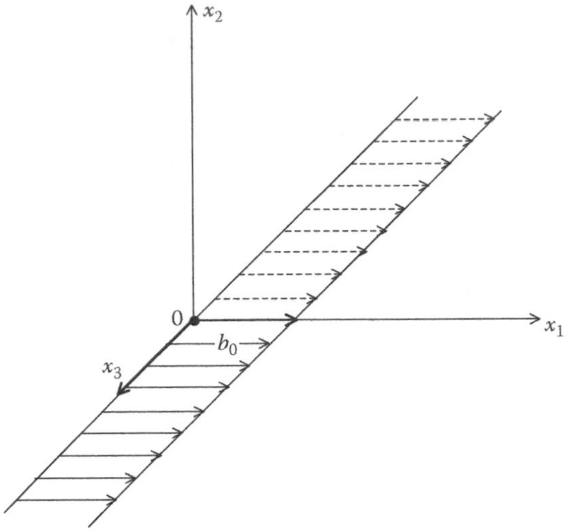

FIGURE 10.1 A body force b_0 acting in an infinite solid under plane strain conditions.

corresponding to a concentrated force tangent to the boundary of a semi-infinite solid under plane strain conditions, (D) the Lamé solution for a thin hollow circular disk subject to uniform pressures on the boundaries, (E) the axially symmetric elastic state in a rotating thin hollow circular disk, and (F) the elastic state in an infinite sheet with a circular hole subject to uniform tension at infinity (Kirsch's problem).

10.1.1 A Concentrated Force in an Infinite Elastic Body under Plane Strain Conditions

Consider a homogeneous isotropic infinite elastic body under plane strain conditions in which a line concentrated force of intensity b_0 acts in the x_1 direction of the Cartesian coordinate system (x_1, x_2, x_3), as shown in Figure 10.1.

To find an elastic state $s = [\mathbf{u}, \mathbf{E}, \mathbf{S}]$ corresponding to the situation shown in Figure 10.1, we recall the fundamental field equations for a plane strain state (see Equations 8.1.6, 8.1.8, 8.1.10, 8.1.12, 8.1.13, and 8.1.17).

The displacement field $\mathbf{u}$ takes the form

$$\mathbf{u} = [u_1\,(x_1, x_2)\,,\ u_2\,(x_1, x_2)\,,\ 0] \tag{10.1.1}$$

The strain–displacement relations read

$$E_{\alpha\beta} = \frac{1}{2}(u_{\alpha,\beta} + u_{\beta,\alpha}) \tag{10.1.2}$$

The equilibrium equations are

$$S_{\alpha\beta,\beta} + b_\alpha = 0 \tag{10.1.3}$$

The constitutive relations take the form

$$S_{\alpha\beta} = 2\mu E_{\alpha\beta} + \lambda E_{\gamma\gamma}\delta_{\alpha\beta} \tag{10.1.4}$$

or

$$E_{\alpha\beta} = \frac{1}{2\mu}\left(S_{\alpha\beta} - \nu S_{\gamma\gamma}\delta_{\alpha\beta}\right) \tag{10.1.5}$$

and

$$S_{33} = \nu S_{\gamma\gamma} \tag{10.1.6}$$

In the problem discussed, Equations 10.1.1 through 10.1.6 are to be satisfied in the infinite plane described by the inequalities

$$|x_1| < \infty, \quad |x_2| < \infty \tag{10.1.7}$$

and the body force b_α in Equation 10.1.3 takes the form

$$b_1 = b_0\delta(x_1)\,\delta(x_2), \quad b_2 = 0 \tag{10.1.8}$$

which means that a concentrated force at the point $(0, 0)$ acts in the x_1 direction. Moreover, we assume that

$$s = [\mathbf{u}, \mathbf{E}, \mathbf{S}] \to [\mathbf{0}, \mathbf{0}, \mathbf{0}] \quad \text{as } r = \sqrt{x_1^2 + x_2^2} \to \infty \tag{10.1.9}$$

To obtain a solution to the problem, we use the two-dimensional version of the Boussinesq–Papkovitch–Neuber solution (see Equations 7.1.2 through 7.1.4)

$$u_\alpha = \psi_\alpha - \frac{1}{4(1-\nu)}\left(x_\beta\psi_\beta + \varphi\right)_{,\alpha} \tag{10.1.10}$$

where

$$\psi_{\alpha,\gamma\gamma} = -\frac{b_\alpha}{\mu} \tag{10.1.11}$$

and

$$\varphi_{,\gamma\gamma} = \frac{1}{\mu}x_\beta b_\beta \tag{10.1.12}$$

The tensor fields $E_{\alpha\beta}$ and $S_{\alpha\beta}$ associated with the field u_α, given by Equations 10.1.10 through 10.1.12, are represented, respectively, by

$$E_{\alpha\beta} = \frac{1}{4(1-\nu)}\left[2(1-2\nu)\psi_{(\alpha,\beta)} - x_\gamma\psi_{\gamma,\alpha\beta} - \varphi_{,\alpha\beta}\right] \tag{10.1.13}$$

and

$$S_{\alpha\beta} = \frac{\mu}{2(1-\nu)}\left[2(1-2\nu)\psi_{(\alpha,\beta)} - x_\gamma\psi_{\gamma,\alpha\beta} + 2\nu\psi_{\gamma,\gamma}\delta_{\alpha\beta} - \varphi_{,\alpha\beta}\right] \tag{10.1.14}$$

Since $b_\alpha = b_0\delta_{\alpha 1}\delta(x_1)\delta(x_2)$, then $x_\beta b_\beta = 0$, and Equation 10.1.12 reduces to the harmonic equation

$$\varphi_{,\gamma\gamma} = 0 \tag{10.1.15}$$

Therefore, letting $\varphi = 0$ in Equations 10.1.10 through 10.1.12 we arrive at the representation

$$u_\alpha = \psi_\alpha - \frac{1}{4(1-\nu)}(x_\beta\psi_\beta)_{,\alpha} \tag{10.1.16}$$

where

$$\psi_{\alpha,\gamma\gamma} = -\frac{b_0}{\mu}\,\delta_{\alpha 1}\delta(x_1)\delta(x_2) \tag{10.1.17}$$

In the following, we are to show that s generated by Equations 10.1.16 and 10.1.17 and Equations 10.1.13 and 10.1.14 with $\varphi = 0$ is a solution to the concentrated force problem provided ψ_α is taken in the form

$$\psi_\alpha = \frac{b_0}{2\pi\mu}\,\delta_{\alpha 1}\ln\frac{L}{r} \tag{10.1.18}$$

where L is a positive constant of the length dimension. To this end, we note that ψ_α given by Equation 10.1.18 satisfies Poisson's equation 10.1.17, and $\psi_\alpha = 0$ for $r = L$. Therefore,

$$\psi_\alpha \to 0 \quad \text{as } r \to L \tag{10.1.19}$$

Next, note that Equation 10.1.16 can be written as

$$u_\alpha = \frac{1}{4(1-\nu)}\left[(3-4\nu)\psi_\alpha - x_\beta\psi_{\beta,\alpha}\right] \tag{10.1.20}$$

Since, because of Equation 10.1.18,

$$\psi_{\alpha,\beta} = -b_0\,\frac{\delta_{\alpha 1}}{2\pi\mu}\,\frac{x_\beta}{r^2}, \qquad \psi_{\gamma,\gamma} = -b_0\,\frac{x_1}{2\pi\mu r^2} \tag{10.1.21}$$

and

$$\psi_{\gamma,\alpha\beta} = -b_0\,\frac{\delta_{\gamma 1}}{2\pi\mu}\,\frac{1}{r^2}\left(\delta_{\alpha\beta} - 2\frac{x_\alpha x_\beta}{r^2}\right) \tag{10.1.22}$$

therefore, Equations 10.1.20 and 10.1.13 and 10.1.14 with $\varphi = 0$ can be written in the form

$$u_\alpha = \frac{b_0}{8\pi\mu(1-\nu)}\left[(3-4\nu)\delta_{\alpha 1}\ln\left(\frac{L}{r}\right) + \frac{x_1 x_\alpha}{r^2}\right] \tag{10.1.23}$$

$$E_{\alpha\beta} = \frac{b_0}{8\pi\mu(1-\nu)}\,\frac{1}{r^2}\left[x_1\left(\delta_{\alpha\beta} - 2\frac{x_\alpha x_\beta}{r^2}\right) - (1-2\nu)\left(\delta_{\alpha 1}x_\beta + \delta_{\beta 1}x_\alpha\right)\right] \tag{10.1.24}$$

and

$$S_{\alpha\beta} = \frac{b_0}{4\pi(1-\nu)}\frac{1}{r^2}\left\{(1-2\nu)\left[x_1\delta_{\alpha\beta} - (\delta_{\alpha 1}x_\beta + \delta_{\beta 1}x_\alpha)\right] - 2\frac{x_1 x_\alpha x_\beta}{r^2}\right\} \tag{10.1.25}$$

Now, we rewrite Equation 10.1.23 in the form

$$u_\alpha = \frac{b_0}{2\pi\mu}\left[\delta_{\alpha 1}\ln\frac{L}{r} - \frac{1}{4(1-\nu)}\frac{\partial}{\partial x_\alpha}\left(x_1\ln\frac{L}{r}\right)\right] \tag{10.1.26}$$

Applying to this equation the operator ∇^2 and using Equations 10.1.17 and 10.1.18, we get

$$\nabla^2 u_\alpha = -\frac{b_0}{\mu}\delta_{\alpha 1}\delta(x_1)\delta(x_2) - \frac{b_0}{2\pi\mu}\frac{1}{4(1-\nu)}\frac{\partial}{\partial x_\alpha}\nabla^2\left(x_1\ln\frac{L}{r}\right) \tag{10.1.27}$$

Next, replacing α by β in Equation 10.1.26 and differentiating with respect to x_β, we obtain

$$u_{\beta,\beta} = \frac{b_0}{2\pi\mu}\left[\frac{\partial}{\partial x_1}\ln\frac{L}{r} - \frac{1}{4(1-\nu)}\nabla^2\left(x_1\ln\frac{L}{r}\right)\right] \tag{10.1.28}$$

Hence, because of Equations 10.1.27 and 10.1.28 we write

$$\begin{aligned}&\nabla^2 u_\alpha + \frac{1}{1-2\nu}u_{\beta,\beta\alpha} + \frac{b_0}{\mu}\delta_{\alpha 1}\delta(x_1)\delta(x_2)\\ &\quad = \frac{b_0}{2\pi\mu}\frac{1}{1-2\nu}\frac{\partial}{\partial x_\alpha}\left[\frac{\partial}{\partial x_1}\ln\frac{L}{r} - \frac{1}{2}\nabla^2\left(x_1\ln\frac{L}{r}\right)\right]\end{aligned} \tag{10.1.29}$$

Since

$$\nabla^2\left(x_1\ln\frac{L}{r}\right) = 2\frac{\partial}{\partial x_1}\ln\frac{L}{r} + x_1\nabla^2\ln\frac{L}{r} \tag{10.1.30}$$

and

$$x_1\nabla^2\ln\frac{L}{r} = -2\pi x_1\delta(x_1)\delta(x_2) = 0 \tag{10.1.31}$$

therefore, it follows from Equation 10.1.29 that

$$u_{\alpha,\beta\beta} + \frac{1}{1-2\nu}u_{\beta,\beta\alpha} + \frac{b_0}{\mu}\delta_{\alpha 1}\delta(x_1)\delta(x_2) = 0 \tag{10.1.32}$$

This completes the proof that Equations 10.1.23 through 10.1.25 represent a solution to the concentrated force problem.

The solution has the properties:

$$\text{(i)}\quad u_\alpha(0, \pm L) = 0, \quad u_2(0, x_2) = 0, \quad x_2 > 0 \tag{10.1.33}$$

$$\text{(ii)}\quad s = [\mathbf{u}, \mathbf{E}, \mathbf{S}] \to [\mathbf{0}, \mathbf{0}, \mathbf{0}] \quad \text{as } x_2 = L \to \infty \text{ and } |x_1| < \infty \tag{10.1.34}$$

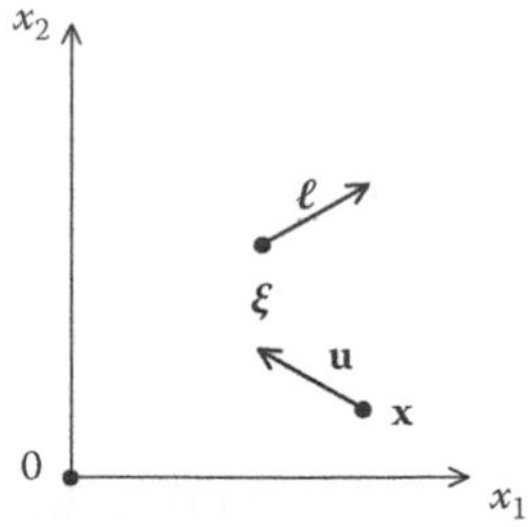

FIGURE 10.2 A concentrated force $\boldsymbol{\ell}$ acting at a point $\boldsymbol{\xi}$ of an infinite solid under plane strain conditions.

Notes:

(1) A displacement field $u_\alpha^{(\beta)}$ due to a body force matrix of the form

$$b_\alpha^{(\beta)} = b_0\, \delta_{\alpha\beta}\delta(x_1)\delta(x_2) \tag{10.1.35}$$

in an infinite plane $|x_1| < \infty$, $|x_2| < \infty$, because of Equation 10.1.23, takes the form

$$u_\alpha^{(\beta)} = \frac{b_0}{8\pi\mu(1-\nu)}\left[(3-4\nu)\delta_{\alpha\beta}\ln\frac{L}{r} + \frac{x_\alpha x_\beta}{r^2}\right] \tag{10.1.36}$$

Hence, a displacement field $u_\alpha = u_\alpha(\mathbf{x}, \boldsymbol{\xi})$ corresponding to an arbitrary force $\boldsymbol{\ell}$ acting at a point $\boldsymbol{\xi} = (\xi_1, \xi_2)$ of an infinite plane, as shown in Figure 10.2, is obtained from the formula

$$u_\alpha(\mathbf{x}, \boldsymbol{\xi}) = U_{\alpha\beta}(\mathbf{x}, \boldsymbol{\xi})\ell_\beta \tag{10.1.37}$$

where

$$U_{\alpha\beta}(\mathbf{x}, \boldsymbol{\xi}) = \frac{1}{8\pi\mu(1-\nu)}\left[(3-4\nu)\delta_{\alpha\beta}\ln\frac{L}{|\mathbf{x}-\boldsymbol{\xi}|} + \frac{(x_\alpha-\xi_\alpha)(x_\beta-\xi_\beta)}{|\mathbf{x}-\boldsymbol{\xi}|^2}\right] \tag{10.1.38}$$

(2) An elastic state $\bar{s} = [\bar{\mathbf{u}}, \bar{\mathbf{E}}, \bar{\mathbf{S}}]$ corresponding to a concentrated force $\bar{b}_\alpha = \bar{b}_0\, \delta_{\alpha 1}\delta(x_1)\delta(x_2)$ in an infinite sheet subject to generalized plane stress conditions is obtained from Equations 10.1.23 through 10.1.25 by replacing b_0 by $\bar{b}_0$ and ν by $\bar{\nu}$, where $\bar{\nu} = \nu/(1+\nu)$. As a result we obtain

$$\bar{u}_\alpha = \frac{\bar{b}_0}{8\pi\mu}\left[(3-\nu)\delta_{\alpha 1}\ln\left(\frac{L}{r}\right) + (1+\nu)\frac{x_1 x_\alpha}{r^2}\right] \tag{10.1.39}$$

$$\bar{E}_{\alpha\beta} = \frac{\bar{b}_0}{8\pi\mu}\frac{1}{r^2}\left[(1+\nu)x_1\left(\delta_{\alpha\beta} - 2\frac{x_\alpha x_\beta}{r^2}\right) - (1-\nu)(\delta_{\alpha 1}x_\beta + \delta_{\beta 1}x_\alpha)\right] \tag{10.1.40}$$

and

$$\bar{S}_{\alpha\beta} = \frac{\bar{b}_0}{4\pi}\frac{1}{r^2}\left\{(1-\nu)\left[x_1\delta_{\alpha\beta} - (\delta_{\alpha 1}x_\beta + \delta_{\beta 1}x_\alpha)\right] - 2(1+\nu)\frac{x_1 x_\alpha x_\beta}{r^2}\right\} \tag{10.1.41}$$

(3) A displacement field $\bar{u}_\alpha^{(\beta)}$ due to a body force matrix of the form

$$\bar{b}_\alpha^{(\beta)} = \bar{b}_0\delta_{\alpha\beta}\delta(x_1)\delta(x_2) \tag{10.1.42}$$

in an infinite sheet under generalized plane stress conditions, in view of Equation 10.1.36, is given by

$$\overline{u}_{\alpha}^{(\beta)} = \frac{\overline{b}_0}{8\pi\mu}\left[(3-\nu)\delta_{\alpha\beta}\ln\left(\frac{L}{r}\right) + (1+\nu)\frac{x_\alpha x_\beta}{r^2}\right] \tag{10.1.43}$$

Hence, a displacement field $\overline{u}_\alpha = \overline{u}_\alpha(\mathbf{x}, \boldsymbol{\xi})$ at a point $\mathbf{x}$ due to a concentrated force $\overline{\boldsymbol{\ell}}$ acting at a point $\boldsymbol{\xi}$ of an infinite sheet is obtained from the formula

$$\overline{u}_\alpha(\mathbf{x}, \boldsymbol{\xi}) = \overline{U}_{\alpha\beta}(\mathbf{x}, \boldsymbol{\xi})\overline{l}_\beta \tag{10.1.44}$$

where

$$\overline{U}_{\alpha\beta} = \frac{1}{8\pi\mu}\left[(3-\nu)\delta_{\alpha\beta}\ln\frac{L}{|\mathbf{x}-\boldsymbol{\xi}|} + (1+\nu)\frac{(x_\alpha-\xi_\alpha)(x_\alpha-\xi_\alpha)}{|\mathbf{x}-\boldsymbol{\xi}|^2}\right] \tag{10.1.45}$$

(4) The elastic states $s = [\mathbf{u}, \mathbf{E}, \mathbf{S}]$ and $\overline{s} = [\overline{\mathbf{u}}, \overline{\mathbf{E}}, \overline{\mathbf{S}}]$ comply with the formulas

$$u_{\alpha,\alpha} = E_{\alpha\alpha} = \frac{1}{2\mu}(1-2\nu)S_{\alpha\alpha} \tag{10.1.46}$$

and

$$\overline{u}_{\alpha,\alpha} = \overline{E}_{\alpha\alpha} = \frac{1-\nu}{E}\overline{S}_{\alpha\alpha} \tag{10.1.47}$$

respectively. Clearly, Equations 10.1.46 and 10.1.47 are identical to Equations 8.1.15 and 8.1.46 with $\alpha = \beta$, respectively.

(5) If $\mathbf{b} = \mathbf{b}(\mathbf{x})$ is a body force prescribed on a two-dimensional domain Ω lying in an infinite solid subject to plane strain conditions then the displacement field $\mathbf{u} = \mathbf{u}(\mathbf{x})$ due to the body force is given by

$$\mathbf{u}(\mathbf{x}) = \int_\Omega \mathbf{U}(\mathbf{x}, \boldsymbol{\xi})\mathbf{b}(\boldsymbol{\xi})d\xi_1 d\xi_2 \tag{10.1.48}$$

A similar formula is valid for a generalized plane stress state.

10.1.2 A Concentrated Force Normal to the Boundary of a Semi-Infinite Solid

Consider a homogeneous isotropic semi-infinite elastic solid under plane strain conditions occupying the region

$$|x_1| < \infty, \quad x_2 \geq 0 \tag{10.1.49}$$

Let a concentrated normal force P_0 be applied to the boundary $x_2 = 0$, as shown in Figure 10.3, and assume that an elastic state $s = [\mathbf{u}, \mathbf{E}, \mathbf{S}]$ corresponding to P_0 is such

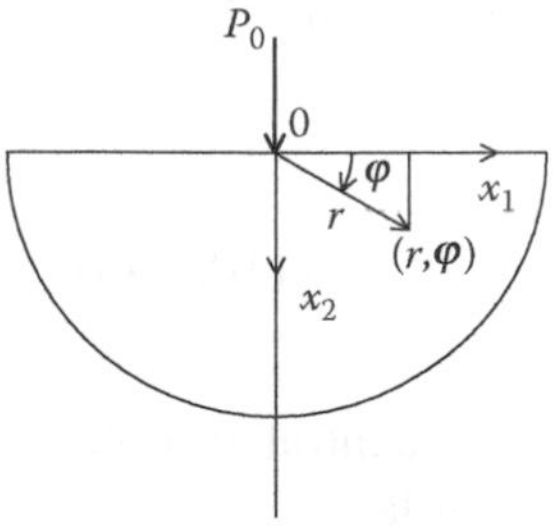

FIGURE 10.3 A concentrated normal force on the boundary of a semispace.

that $|\mathbf{S}| \to 0$ as $r = |\mathbf{x}| \to \infty$. To find s we are to solve the governing equations of two-dimensional elastostatics (see Equations 10.1.1 through 10.1.6) in the region described by the inequalities 10.1.49 subject to the boundary conditions

$$S_{22}(x_1, 0) = -P_0\,\delta(x_1), \quad S_{12}(x_1, 0) = 0 \tag{10.1.50}$$

and the asymptotic condition

$$s \to [\mathbf{u}_0, \mathbf{0}, \mathbf{0}] \quad \text{as } |\mathbf{x}| \to \infty \tag{10.1.51}$$

where $\mathbf{u}_0$ is a constant vector.

A solution to the problem can be obtained by using the Boussinesq–Papkovitch–Neuber representation of a displacement field [see Equations 10.1.10 through 10.1.12 in which $b_\alpha = 0$ and $\psi_\alpha = \psi\,\delta_{\alpha 2}$, $\psi = \psi(x_1, x_2)$]

$$u_1 = -\frac{1}{4(1-\nu)}\frac{\partial \phi}{\partial x_1}, \quad u_2 = \psi - \frac{1}{4(1-\nu)}\frac{\partial \phi}{\partial x_2} \tag{10.1.52}$$

where

$$\nabla^2 \psi = 0, \quad \nabla^2 \varphi = 0 \tag{10.1.53}$$

and

$$\phi = \varphi + x_2 \psi \tag{10.1.54}$$

It follows from Equations 10.1.53 and 10.1.54 that

$$\nabla^2 \phi = 2\frac{\partial \psi}{\partial x_2} \tag{10.1.55}$$

The associated strain tensor components are given by

$$\begin{aligned} E_{11} &= -\frac{1}{4(1-\nu)}\frac{\partial^2 \phi}{\partial x_1^2}, \quad E_{22} = \frac{\partial \psi}{\partial x_2} - \frac{1}{4(1-\nu)}\frac{\partial^2 \phi}{\partial x_2^2} \\ E_{12} &= -\frac{1}{4(1-\nu)}\left[\frac{\partial^2 \phi}{\partial x_1 \partial x_2} - 2(1-\nu)\frac{\partial \psi}{\partial x_1}\right] \end{aligned} \tag{10.1.56}$$

while the stress tensor components take the form

$$S_{11} = -\frac{\mu}{2(1-\nu)}\left(\frac{\partial^2 \phi}{\partial x_1^2} - 2\nu\frac{\partial \psi}{\partial x_2}\right) \tag{10.1.57}$$

$$S_{22} = -\frac{\mu}{2(1-\nu)}\left[\frac{\partial^2 \phi}{\partial x_2^2} - 2(2-\nu)\frac{\partial \psi}{\partial x_2}\right] \tag{10.1.58}$$

$$S_{12} = -\frac{\mu}{2(1-\nu)}\left[\frac{\partial^2 \phi}{\partial x_1 \partial x_2} - 2(1-\nu)\frac{\partial \psi}{\partial x_1}\right] \tag{10.1.59}$$

Note that by Equations 10.1.55 and 10.1.56

$$E_{\gamma\gamma} = \frac{1-2\nu}{2(1-\nu)} \frac{\partial \psi}{\partial x_2} \tag{10.1.60}$$

and by Equations 10.1.55 and 10.1.57 and 10.1.58 we obtain

$$S_{\gamma\gamma} = \frac{\mu}{1-\nu} \frac{\partial \psi}{\partial x_2} \tag{10.1.61}$$

Hence

$$E_{\gamma\gamma} = \frac{1-2\nu}{2\mu} S_{\gamma\gamma} \tag{10.1.62}$$

which coincides with Equation 10.1.46. Now, since, by Equation 10.1.50,

$$S_{22}(x_1, 0) = -P_0\,\delta(x_1) = -\frac{P_0}{\pi} \int_0^\infty \cos \alpha x_1 \, d\alpha \tag{10.1.63}$$

therefore, by Equation 10.1.58, the functions ϕ and ψ may be postulated in the forms

$$\phi = \int_0^\infty [A(\alpha) + B(\alpha)\alpha x_2] e^{-\alpha x_2} \cos \alpha x_1 \, d\alpha \tag{10.1.64}$$

and

$$\psi = \int_0^\infty B(\alpha)\alpha e^{-\alpha x_2} \cos \alpha x_1 \, d\alpha \tag{10.1.65}$$

where $A = A(\alpha)$ and $B = B(\alpha)$ are arbitrary functions on $[0, \infty)$. Clearly, the functions ϕ and ψ are selected in such a way that Equation 10.1.55 is met, ψ is a harmonic function and ϕ is a biharmonic function. The functions $A = A(\alpha)$ and $B = B(\alpha)$ will be selected in such a way that the boundary conditions 10.1.50 are satisfied. By substituting ϕ and ψ from Equations 10.1.64 and 10.1.65, respectively, into Equation 10.1.59, and using the boundary condition

$$S_{12}(x_1, 0) = 0 \tag{10.1.66}$$

we obtain

$$A(\alpha) = -(1-2\nu)B(\alpha) \tag{10.1.67}$$

Hence, by Equations 10.1.64 and 10.1.65 we get

$$\phi = \int_0^\infty [-(1-2\nu) + \alpha x_2] B(\alpha) e^{-\alpha x_2} \cos \alpha x_1 \, d\alpha \tag{10.1.68}$$

and

$$\psi = \int_0^\infty B(\alpha)\alpha e^{-\alpha x_2} \cos \alpha x_1 \, d\alpha \tag{10.1.69}$$

Next, substituting ϕ and ψ from Equations 10.1.68 and 10.1.69 into Equation 10.1.58, and using the boundary condition 10.1.63, we arrive at

$$B(\alpha)\alpha^2 = \frac{2(1-\nu)P_0}{\pi \mu} \tag{10.1.70}$$

To obtain the stress components at any internal point of the semispace, we need to compute the second partial derivatives of ϕ and the first partial derivatives of ψ (see Equations 10.1.57 through 10.1.59). Using Equations 10.1.68 and 10.1.69, we obtain

$$\frac{\partial^2 \phi}{\partial x_1^2} = \int_0^\infty B\alpha^2[(1-2\nu) - \alpha x_2]e^{-\alpha x_2} \cos \alpha x_1 \, d\alpha \tag{10.1.71}$$

$$\frac{\partial^2 \phi}{\partial x_2^2} = -\int_0^\infty B\alpha^2(3 - 2\nu - \alpha x_2)e^{-\alpha x_2} \cos \alpha x_1 \, d\alpha \tag{10.1.72}$$

$$\frac{\partial^2 \phi}{\partial x_1 \partial x_2} = -\int_0^\infty B\alpha^2[2(1-\nu) - \alpha x_2]e^{-\alpha x_2} \sin \alpha x_1 \, d\alpha \tag{10.1.73}$$

and

$$\frac{\partial \psi}{\partial x_1} = -\int_0^\infty B\alpha^2 e^{-\alpha x_2} \sin \alpha x_1 \, d\alpha \tag{10.1.74}$$

$$\frac{\partial \psi}{\partial x_2} = -\int_0^\infty B\alpha^2 e^{-\alpha x_2} \cos \alpha x_1 \, d\alpha \tag{10.1.75}$$

Hence, by substituting Equation 10.1.70 into Equations 10.1.71 through 10.1.75 and using Equations 10.1.57 through 10.1.59, we arrive at the stress components

$$S_{11} = -\frac{P_0}{\pi} \int_0^\infty (1 - \alpha x_2)e^{-\alpha x_2} \cos \alpha x_1 \, d\alpha \tag{10.1.76}$$

$$S_{22} = -\frac{P_0}{\pi} \int_0^\infty (1 + \alpha x_2)e^{-\alpha x_2} \cos \alpha x_1 \, d\alpha \tag{10.1.77}$$

$$S_{12} = -\frac{P_0}{\pi} x_2 \int_0^\infty \alpha e^{-\alpha x_2} \sin \alpha x_1 \, d\alpha \tag{10.1.78}$$

It follows from Equations 10.1.76 through 10.1.78 that the stress tensor $S_{\alpha\beta}$ is independent of the elastic properties of the semi-space [see Note (2) in Section 8.1.1].

Similarly, because of Equations 10.1.56, and 10.1.70 through 10.1.75, we obtain the strain tensor components

$$E_{11} = -\frac{P_0}{2\pi\mu}\int_0^\infty [(1-2\nu) - \alpha x_2]e^{-\alpha x_2}\cos\alpha x_1\, d\alpha \tag{10.1.79}$$

$$E_{22} = -\frac{P_0}{2\pi\mu}\int_0^\infty (1-2\nu+\alpha x_2)^{-\alpha x_2}\cos\alpha x_1\, d\alpha \tag{10.1.80}$$

$$E_{12} = -\frac{P_0}{2\pi\mu}x_2\int_0^\infty \alpha e^{-\alpha x_2}\sin\alpha x_1\, d\alpha \tag{10.1.81}$$

It follows from Equations 10.1.79 through 10.1.81 that the strain tensor $E_{\alpha\beta}$ does depend on the elastic constants μ and ν.

Finally, note that the first partial derivatives of ϕ given by Equation 10.1.68 in which $B=B(\alpha)$ is computed from Equation 10.1.70 are represented by the divergent integrals. Also, ψ given by Equation 10.1.69 with $B=B(\alpha)$ from Equation 10.1.70 is represented by a divergent integral. Therefore, the displacement representation 10.1.52 through 10.1.54 cannot be used to obtain the displacement components u_α. The displacement vector will be obtained from the geometric relations

$$\frac{\partial u_1}{\partial x_1} = E_{11}, \quad \frac{\partial u_2}{\partial x_2} = E_{22} \tag{10.1.82}$$

These relations are equivalent to

$$u_1(x_1,x_2) - u_1(0,x_2) = \int_0^{x_1} E_{11}(u,x_2)du \tag{10.1.83}$$

and

$$u_2(x_1,x_2) - u_2(x_1,0) = \int_0^{x_2} E_{22}(x_1,u)du \tag{10.1.84}$$

Therefore, substituting E_{11} from Equation 10.1.79 into Equation 10.1.83 and performing integration under the integral on $[0,\infty)$, we obtain

$$u_1(x_1,x_2) - u_1(0,x_2) = -\frac{P_0}{2\pi\mu}\int_0^\infty [(1-2\nu) - \alpha x_2]e^{-\alpha x_2}\frac{\sin\alpha x_1}{\alpha}d\alpha \tag{10.1.85}$$

Similarly, substituting E_{22} from Equation 10.1.80 into Equation 10.1.84 and integrating we have

$$u_2(x_1,x_2) - u_2(x_1,0) = -\frac{P_0}{2\pi\mu}\left\{(1-2\nu)\int_0^\infty \frac{1-e^{-\alpha x_2}}{\alpha}\cos\alpha x_1\, d\alpha \right.$$
$$\left. + \int_0^\infty \frac{[1-(1+\alpha x_2)e^{-\alpha x_2}]}{\alpha}\cos\alpha x_1\, d\alpha\right\} \qquad (10.1.86)$$

In the following, we are to show that the integrals in Equations 10.1.76 through 10.1.78, 10.1.79 through 10.1.81, and 10.1.85 and 10.1.86 can be calculated in closed forms. To this end, we recall the useful formulas:

$$\int_0^\infty e^{-\alpha x_2}\cos\alpha x_1\, d\alpha = \frac{x_2}{r^2} = \frac{\partial}{\partial x_2}\ln\frac{r}{L} \qquad (10.1.87)$$

$$\int_0^\infty e^{-\alpha x_2}\sin\alpha x_1\, d\alpha = \frac{x_1}{r^2} = \frac{\partial}{\partial x_1}\ln\frac{r}{L} \qquad (10.1.88)$$

$$\int_0^\infty e^{-\alpha x_2}\frac{\sin\alpha x_1}{\alpha}\, d\alpha = \tan^{-1}\frac{x_1}{x_2} \qquad (10.1.89)$$

where L is a positive constant of the length dimension. An alternative form of Equation 10.1.87 reads

$$\int_0^\infty e^{-\alpha u}\cos\alpha x_1\, d\alpha = \frac{\partial}{\partial u}\ln\left(\frac{R}{L}\right) \qquad (10.1.90)$$

where

$$R = \sqrt{x_1^2 + u^2}, \quad u > 0 \qquad (10.1.91)$$

By integrating Equation 10.1.90 with respect to u over the interval $0 \le u \le x_2$, we obtain

$$\int_0^\infty \frac{1-e^{-\alpha x_2}}{\alpha}\cos\alpha x_1\, d\alpha = \ln\frac{r}{L} - \ln\frac{|x_1|}{L} \qquad (10.1.92)$$

By differentiating Equation 10.1.90 with respect to u and multiplying the resulting equation by u, we get

$$\int_0^\infty \alpha u e^{-\alpha u}\cos\alpha x_1\, d\alpha = -u\frac{\partial^2}{\partial u^2}\ln\left(\frac{R}{L}\right) \qquad (10.1.93)$$

Now, since

$$\int_0^{x_2} u e^{-\alpha u} du = \frac{[1-(1+\alpha x_2)e^{-\alpha x_2}]}{\alpha^2} \qquad (10.1.94)$$

hence, integrating 10.1.93 with respect to u over the interval $[0, x_2]$ we obtain

$$\int_0^\infty \frac{[1-(1+\alpha x_2)e^{-\alpha x_2}]}{\alpha}\cos\alpha x_1\, d\alpha$$

$$= -\int_0^{x_2} u\frac{\partial^2}{\partial u^2}\ln\left(\frac{R}{L}\right)du = -\left\{u\frac{\partial}{\partial u}\left[\ln\frac{R}{L}\right]\right\}_{u=0}^{u=x_2} + \int_0^{x_2}\frac{\partial}{\partial u}\ln\left(\frac{R}{L}\right)du \quad (10.1.95)$$

or

$$\int_0^\infty \frac{[1-(1+\alpha x_2)e^{-\alpha x_2}]}{\alpha}\cos\alpha x_1\, d\alpha = -\frac{x_2^2}{r^2} + \ln\frac{r}{L} - \ln\frac{|x_1|}{L} \quad (10.1.96)$$

The formulas 10.1.87 through 10.1.89, 10.1.92, and 10.1.96 allow us to reduce Equations 10.1.85 and 10.1.86 to the closed forms

$$u_1(x_1,x_2) - u_1(0,x_2) = -\frac{P_0}{2\pi\mu}\left[(1-2\nu)\tan^{-1}\frac{x_1}{x_2} - \frac{x_1 x_2}{r^2}\right] \quad (10.1.97)$$

and

$$u_2(x_1,x_2) - u_2(x_1,0) = -\frac{P_0}{2\pi\mu}\left\{2(1-\nu)\left[\ln\frac{r}{L} - \ln\frac{|x_1|}{L}\right] - \frac{x_2^2}{r^2}\right\} \quad (10.1.98)$$

Hence, if we let

$$u_1(0,x_2) = 0 \quad \text{for } x_2 > 0 \quad (10.1.99)$$

and

$$u_2(x_1,0) = -\frac{P_0(1-\nu)}{\pi\mu}\ln\frac{|x_1|}{L} \quad (10.1.100)$$

which implies that

$$u_2(L,0) = u_2(-L,0) = 0 \quad \text{for } L > 0 \quad (10.1.101)$$

the unique displacement components are obtained

$$u_1 = -\frac{P_0}{2\pi\mu}\left[(1-2\nu)\tan^{-1}\frac{x_1}{x_2} - \frac{x_1 x_2}{r^2}\right] \quad (10.1.102)$$

$$u_2 = -\frac{P_0}{2\pi\mu}\left[2(1-\nu)\ln\frac{r}{L} - \frac{x_2^2}{r^2}\right] \quad (10.1.103)$$

It follows from Equations 10.1.102 and 10.1.103 that

$$u_1 \to 0, \quad u_2 \to u^* \quad \text{as } x_2 = L \to \infty, \ |x_1| < \infty \tag{10.1.104}$$

where

$$u^* = \frac{P_0}{2\pi\mu} \tag{10.1.105}$$

Note that $[P_0] = [\text{Force} \times \text{L}^{-1}]$ and $[\mu] = [\text{Force} \times \text{L}^{-2}]$, therefore $[u^*] = [L]$. As a result, the displacement vector corresponding to the solution of the concentrated force problem for the semispace complies with the asymptotic condition 10.1.51 in which $\mathbf{u}_0 = [0, u^*]$.

To obtain the strain components in a closed form, we note that by Equations 10.1.87 and 10.1.88

$$\int_0^\infty \alpha e^{-\alpha x_2} \cos \alpha x_1 \, d\alpha = -\frac{\partial}{\partial x_2}\left(\frac{x_2}{r^2}\right) = -\frac{1}{r^2}\left(1 - \frac{2x_2^2}{r^2}\right) \tag{10.1.106}$$

and

$$\int_0^\infty \alpha e^{-\alpha x_2} \sin \alpha x_1 \, d\alpha = -\frac{\partial}{\partial x_2}\left(\frac{x_1}{r^2}\right) = 2\frac{x_1 x_2}{r^4} \tag{10.1.107}$$

Hence, by Equations 10.1.79 through 10.1.81, 10.1.87 and 10.1.88, and 10.1.106 and 10.1.107, we obtain

$$E_{11} = -\frac{P_0}{\pi\mu}\frac{x_2}{r^2}\left[(1-\nu) - \frac{x_2^2}{r^2}\right] \tag{10.1.108}$$

$$E_{22} = \frac{P_0}{\pi\mu}\frac{x_2}{r^2}\left(\nu - \frac{x_2^2}{r^2}\right) \tag{10.1.109}$$

$$E_{12} = -\frac{P_0}{\pi\mu}\frac{x_1 x_2^2}{r^4} \tag{10.1.110}$$

Similarly, it follows from Equations 10.1.76 through 10.1.78, 10.1.87 and 10.1.88, and 10.1.106 and 10.1.107 that

$$S_{11} = -\frac{2P_0 x_1^2 x_2}{\pi r^4} \tag{10.1.111}$$

$$S_{22} = -\frac{2P_0 x_2^3}{\pi r^4} \tag{10.1.112}$$

$$S_{12} = -\frac{2P_0 x_1 x_2^2}{\pi r^4} \tag{10.1.113}$$

Also, it follows from Equation 10.1.112 that for every $x_2 > 0$

$$\int_{-\infty}^{+\infty} S_{22}(x_1, x_2)dx_1 = -\frac{2P_0}{\pi}\int_{-\infty}^{+\infty} \frac{x_2^3}{r^4}dx_1$$

$$= \frac{2P_0}{\pi}\frac{x_2^2}{2}\frac{\partial}{\partial x_2}\int_{-\infty}^{+\infty}\frac{dx_1}{r^2} \tag{10.1.114}$$

Since

$$\int_{-\infty}^{+\infty}\frac{dx_1}{r^2} = \frac{\pi}{x_2} \quad \text{for } x_2 > 0 \tag{10.1.115}$$

therefore

$$\int_{-\infty}^{+\infty} S_{22}(x_1, x_2)dx_1 = -P_0 \quad \text{for } x_2 > 0 \tag{10.1.116}$$

An extension of this equation to $x_2 = 0$ leads to the total equilibrium of normal forces on the boundary of the semispace

$$\int_{-\infty}^{+\infty} S_{22}(x_1, 0)dx_1 = -P_0 \tag{10.1.117}$$

while Equation 10.1.113 implies that

$$S_{12}(x_1, 0) = 0 \quad \text{for } |x_1| < \infty \tag{10.1.118}$$

Finally, it follows from Equations 10.1.108 through 10.1.113 that $E_{\alpha\beta} \to 0$ and $S_{\alpha\beta} \to 0$ as $x_2 = L \to \infty$, $|x_1| < \infty$.

This completes the proof that $s = [\mathbf{u}, \mathbf{E}, \mathbf{S}]$ in which $\mathbf{u}$, $\mathbf{E}$, and $\mathbf{S}$ are given by Equations 10.1.102 and 10.1.103, 10.1.108 through 10.1.110, and 10.1.111 through 10.1.113, respectively, represents a solution to the concentrated normal force problem for the semispace under plane strain conditions.

Notes:

(1) By using the transformation formulas from (x_1, x_2) to (r, φ) coordinates

$$\begin{aligned} S_{rr} &= S_{11}\cos^2\varphi + 2S_{12}\sin\varphi\cos\varphi + S_{22}\sin^2\varphi \\ S_{\varphi\varphi} &= S_{11}\sin^2\varphi - 2S_{12}\sin\varphi\cos\varphi + S_{22}\cos^2\varphi \\ S_{r\varphi} &= (S_{22} - S_{11})\sin\varphi\cos\varphi + S_{12}(\cos^2\varphi - \sin^2\varphi) \end{aligned} \tag{10.1.119}$$

in which $S_{\alpha\beta}$ is given by Equations 10.1.111 through 10.1.113 and

$$x_1 = r\cos\varphi, \quad x_2 = r\sin\varphi \tag{10.1.120}$$

we obtain the simple formulas

$$S_{rr} = -\frac{2P_0}{\pi r}\sin\varphi, \quad S_{\varphi\varphi} = S_{r\varphi} = 0 \tag{10.1.121}$$

Therefore, the (r, φ) coordinates coincide with the principal axes of the tensor $S_{\alpha\beta}$.

(2) An elastic state $\overline{s} = [\overline{\mathbf{u}}, \overline{\mathbf{E}}, \overline{\mathbf{S}}]$ corresponding to a concentrated force $\overline{P}_0$ normal to the boundary of a semi-infinite sheet under generalized plane stress conditions is obtained from Equations 10.1.102 and 10.1.103, and 10.1.108 through 10.1.113, in which $[\mathbf{u}, \mathbf{E}, \mathbf{S}]$ is replaced by $[\overline{\mathbf{u}}, \overline{\mathbf{E}}, \overline{\mathbf{S}}]$, P_0 by $\overline{P}_0$, and ν by $\overline{\nu} = \nu/(1+\nu)$. As a result we arrive at

$$\overline{u}_1 = -\frac{\overline{P}_0}{\pi E}\left[(1-\nu)\tan^{-1}\frac{x_1}{x_2} - (1+\nu)\frac{x_1 x_2}{r^2}\right] \tag{10.1.122}$$

$$\overline{u}_2 = -\frac{\overline{P}_0}{\pi E}\left[2\ln\frac{r}{L} - (1+\nu)\frac{x_2^2}{r^2}\right] \tag{10.1.123}$$

$$\overline{E}_{11} = -\frac{2\overline{P}_0}{\pi E}\frac{x_2}{r^2}\left[1 - (1+\nu)\frac{x_2^2}{r^2}\right] \tag{10.1.124}$$

$$\overline{E}_{22} = +\frac{2\overline{P}_0}{\pi E}\frac{x_2}{r^2}\left[\nu - (1+\nu)\frac{x_2^2}{r^2}\right] \tag{10.1.125}$$

$$\overline{E}_{12} = -\frac{\overline{P}_0}{\pi\mu}\frac{x_1 x_2^2}{r^4} \tag{10.1.126}$$

and

$$\overline{S}_{11} = -\frac{2\overline{P}_0}{\pi}\frac{x_1^2 x_2}{r^4} \tag{10.1.127}$$

$$\overline{S}_{22} = -\frac{2\overline{P}_0}{\pi}\frac{x_2^3}{r^4} \tag{10.1.128}$$

$$\overline{S}_{12} = -\frac{2\overline{P}_0}{\pi}\frac{x_1 x_2^2}{r^4} \tag{10.1.129}$$

In Equations 10.1.122 through 10.1.125, $E = 2\mu(1+\nu)$ stands for Young's modulus.

10.1.3 A Concentrated Force Tangent to the Boundary of a Semi-Infinite Solid

Suppose that a concentrated tangent force T_0 is applied to the boundary of a homogeneous isotropic semi-infinite elastic solid under plane strain conditions, as shown in Figure 10.4.

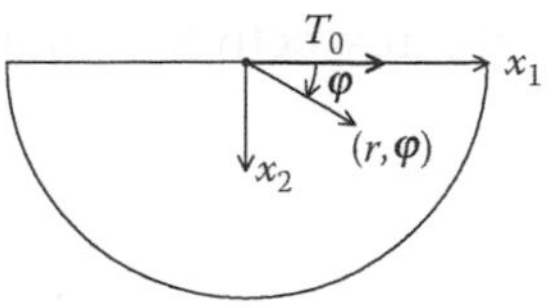

FIGURE 10.4 A concentrated tangent force T_0 on the boundary of a semispace.

An elastic state $s=[\mathbf{u},\mathbf{E},\mathbf{S}]$ corresponding to the situation shown in Figure 10.4 can be found by a method similar to that of Section 10.1.2. The displacement, strain, and stress fields will be found by using the formulas (see Equations 10.1.52 through 10.1.54, 10.1.56, and 10.1.57 through 10.1.59)

$$u_1 = -\frac{1}{4(1-\nu)}\frac{\partial\phi}{\partial x_1}$$
$$u_2 = \psi - \frac{1}{4(1-\nu)}\frac{\partial\phi}{\partial x_2} \quad (10.1.130)$$

$$E_{11} = -\frac{1}{4(1-\nu)}\frac{\partial^2\phi}{\partial x_1^2}$$
$$E_{22} = \frac{\partial\psi}{\partial x_2} - \frac{1}{4(1-\nu)}\frac{\partial^2\phi}{\partial x_2^2} \quad (10.1.131)$$
$$E_{12} = -\frac{1}{4(1-\nu)}\left[\frac{\partial^2\phi}{\partial x_1\partial x_2} - 2(1-\nu)\frac{\partial\psi}{\partial x_1}\right]$$

and

$$S_{11} = -\frac{\mu}{2(1-\nu)}\left(\frac{\partial^2\phi}{\partial x_1^2} - 2\nu\frac{\partial\psi}{\partial x_2}\right) \quad (10.1.132)$$

$$S_{22} = -\frac{\mu}{2(1-\nu)}\left[\frac{\partial^2\phi}{\partial x_2^2} - 2(2-\nu)\frac{\partial\psi}{\partial x_2}\right] \quad (10.1.133)$$

$$S_{12} = -\frac{\mu}{2(1-\nu)}\left[\frac{\partial^2\phi}{\partial x_1\partial x_2} - 2(1-\nu)\frac{\partial\psi}{\partial x_1}\right] \quad (10.1.134)$$

Here, ϕ and ψ are postulated in the forms

$$\phi = \int_0^\infty (A + B\alpha x_2)e^{-\alpha x_2}\sin\alpha x_1\, d\alpha \quad (10.1.135)$$

and

$$\psi = \int_0^\infty B\alpha e^{-\alpha x_2}\sin\alpha x_1\, d\alpha \quad (10.1.136)$$

where the functions $A = A(\alpha)$ and $B = B(\alpha)$ on $[0,\infty)$ are selected in such a way that the boundary conditions are satisfied

$$S_{22}(x_1,0) = 0, \quad S_{12}(x_1,0) = -T_0\,\delta(x_1) = -\frac{T_0}{\pi}\int_0^\infty \cos\alpha x_1\, d\alpha \quad (10.1.137)$$

By substituting ϕ and ψ from Equations 10.1.135 and 10.1.136, respectively, into Equations 10.1.132 through 10.1.134, we have

$$S_{11} = \frac{\mu}{2(1-\nu)} \int_0^\infty [A - (2\nu - \alpha x_2)B]\alpha^2 e^{-\alpha x_2} \sin \alpha x_1 \, d\alpha \tag{10.1.138}$$

$$S_{22} = -\frac{\mu}{2(1-\nu)} \int_0^\infty \{A + [2(1-\nu) + \alpha x_2]B\}\alpha^2 e^{-\alpha x_2} \sin \alpha x_1 \, d\alpha \tag{10.1.139}$$

$$S_{12} = \frac{\mu}{2(1-\nu)} \int_0^\infty [A + (1 - 2\nu + \alpha x_2)B]\alpha^2 e^{-\alpha x_2} \cos \alpha x_1 \, d\alpha \tag{10.1.140}$$

Therefore, it follows from Equations 10.1.139 and 10.1.140 that the boundary conditions 10.1.137 are satisfied provided the functions $A = A(\alpha)$ and $B = B(\alpha)$ satisfy the equations

$$\begin{gathered} A + 2(1-\nu)B = 0 \\ \frac{\mu}{2(1-\nu)}[A + (1-2\nu)B]\alpha^2 = -\frac{T_0}{\pi} \end{gathered} \tag{10.1.141}$$

Hence, we obtain

$$A\alpha^2 = -\frac{4T_0(1-\nu)^2}{\pi\mu}, \quad B\alpha^2 = \frac{2T_0(1-\nu)}{\pi\mu} \tag{10.1.142}$$

and substituting $A\alpha^2$ and $B\alpha^2$ from Equations 10.1.142 into Equations 10.1.138 through 10.1.140 we arrive at

$$S_{11} = -\frac{T_0}{\pi} \int_0^\infty (2 - \alpha x_2)e^{-\alpha x_2} \sin \alpha x_1 \, d\alpha \tag{10.1.143}$$

$$S_{22} = -\frac{T_0}{\pi} x_2 \int_0^\infty \alpha e^{-\alpha x_2} \sin \alpha x_1 \, d\alpha \tag{10.1.144}$$

$$S_{12} = -\frac{T_0}{\pi} \int_0^\infty (1 - \alpha x_2)e^{-\alpha x_2} \cos \alpha x_1 \, d\alpha \tag{10.1.145}$$

To obtain $E_{\alpha\beta}$ $(\alpha, \beta = 1, 2)$ we note that by virtue of Equations 10.1.135 and 10.1.142

$$\phi = \int_0^\infty [-2(1-\nu) + \alpha x_2]Be^{-\alpha x_2} \sin \alpha x_1 \, d\alpha \tag{10.1.146}$$

Substituting then ϕ from Equation 10.1.146 and ψ from Equation 10.1.136 into Equations 10.1.131, and using Equations 10.1.142, we arrive at

$$E_{11} = -\frac{T_0}{2\pi\mu}\int_0^\infty [2(1-\nu) - \alpha x_2]e^{-\alpha x_2}\sin\alpha x_1\, d\alpha \tag{10.1.147}$$

$$E_{22} = \frac{T_0}{2\pi\mu}\int_0^\infty (2\nu - \alpha x_2)e^{-\alpha x_2}\sin\alpha x_1\, d\alpha \tag{10.1.148}$$

$$E_{12} = -\frac{T_0}{2\pi\mu}\int_0^\infty (1 - \alpha x_2)e^{-\alpha x_2}\cos\alpha x_1\, d\alpha \tag{10.1.149}$$

Finally, note that substituting ϕ into the first of Equations 10.1.130 leads to a divergent integral. Therefore, to find u_1 we use Equation 10.1.147 in the form

$$\frac{\partial u_1}{\partial x_1} = -\frac{T_0}{2\pi\mu}\int_0^\infty [2(1-\nu) - \alpha x_2]e^{-\alpha x_2}\sin\alpha x_1\, d\alpha \tag{10.1.150}$$

Integration of this equation with respect x_1 over the interval $[0, x_1]$ yields

$$u_1(x_1, x_2) - u_1(0, x_2) = -\frac{T_0}{2\pi\mu}\int_0^\infty [2(1-\nu) - \alpha x_2]e^{-\alpha x_2}\frac{(1-\cos\alpha x_1)}{\alpha}d\alpha \tag{10.1.151}$$

To find $u_2 = u_2(x_1, x_2)$ we substitute ψ and ϕ from Equations 10.1.136 and 10.1.146, respectively, into the second of Equations 10.1.130, and in view of the second of Equations 10.1.142, we get

$$u_2(x_1, x_2) = \frac{T_0}{2\pi\mu}\int_0^\infty (1 - 2\nu + \alpha x_2)e^{-\alpha x_2}\,\frac{\sin\alpha x_1}{\alpha}\, d\alpha \tag{10.1.152}$$

Now, by integrating Equation 10.1.88 with respect to x_1 over interval $[0, x_1]$ we obtain

$$\int_0^\infty e^{-\alpha x_2}\,\frac{1-\cos\alpha x_1}{\alpha}\, d\alpha = \ln\frac{r}{L} - \ln\frac{|x_2|}{L} \tag{10.1.153}$$

Therefore, by letting $u_1(0, x_2) = 0$ for $x_2 > 0$ in Equation 10.1.151, and by using Equations 10.1.87 through 10.1.89 and 10.1.153, the closed forms of u_1 and u_2 are obtained

$$u_1(x_1, x_2) = \frac{T_0}{2\pi\mu}\left[2(1-\nu)\ln\frac{x_2}{r} + \left(1 - \frac{x_2^2}{r^2}\right)\right] \tag{10.1.154}$$

$$u_2(x_1, x_2) = \frac{T_0}{2\pi\mu}\left[(1-2\nu)\tan^{-1}\frac{x_1}{x_2} + \frac{x_1 x_2}{r^2}\right] \tag{10.1.155}$$

Clearly, $u_1(0, x_2) = 0$, $u_2(0, x_2) = 0$ for $x_2 > 0$, and

$$u_1 \to 0, \quad u_2 \to u_0 \quad \text{as } x_2 = L \to \infty, \ |x_1| < \infty \tag{10.1.156}$$

where

$$u_0 = \frac{T_0(1 - 2\nu)}{4\mu} \tag{10.1.157}$$

Therefore, the displacement field corresponding to a solution of the tangent concentrated force problem for a semispace obeys the asymptotic behavior

$$\mathbf{u} \to [0, u_0] \quad \text{as } x_2 = L \to \infty, \ |x_1| < \infty \tag{10.1.158}$$

in which u_0 depends on both elastic moduli μ and ν, in contrast to a solution of the normal concentrated force problem (see Equations 10.1.104 and 10.1.105) where $\mathbf{u} \to [0, u^*]$ as $x_2 = L \to \infty$, and u^* depends on μ only.

Also, by using Equations 10.1.87 and 10.1.88, the integral representations of $S_{\alpha\beta}$ and $E_{\alpha\beta}$ given by Equations 10.1.143 through 10.1.145 and 10.1.147 through 10.1.149, respectively, may be reduced to the closed forms:

$$S_{11} = -\frac{2T_0}{\pi r^4} x_1^3 \tag{10.1.159}$$

$$S_{22} = -\frac{2T_0}{\pi r^4} x_1 x_2^2 \tag{10.1.160}$$

$$S_{12} = -\frac{2T_0}{\pi r^4} x_1^2 x_2 \tag{10.1.161}$$

and

$$E_{11} = -\frac{T_0 x_1}{\pi \mu r^2}\left(1 - \nu - \frac{x_2^2}{r^2}\right) \tag{10.1.162}$$

$$E_{22} = \frac{T_0 x_1}{\pi \mu r^2}\left(\nu - \frac{x_2^2}{r^2}\right) \tag{10.1.163}$$

$$E_{12} = -\frac{T_0}{\pi \mu r^4} x_1^2 x_2 \tag{10.1.164}$$

It follows from Equations 10.1.159 through 10.1.164 that

$$S_{\alpha\beta} \to 0, \quad E_{\alpha\beta} \to 0 \quad \text{as } x_2 = L \to \infty, \ |x_1| < \infty \tag{10.1.165}$$

This completes the derivation of a closed form solution to the problem of a concentrated force tangent to the boundary of a semi-infinite solid under plane strain conditions.

Notes:

(1) The elastic state $s = [\mathbf{u}, \mathbf{E}, \mathbf{S}]$ obtained in Section 10.1.3 may be easily transformed to the polar coordinates (r, φ). For example, by using the transformation equations 10.1.119 and 10.1.120, Equations 10.1.159 through 10.1.161 lead to

$$S_{rr} = -\frac{2T_0}{\pi r}\cos\varphi, \quad S_{r\varphi} = S_{\varphi\varphi} = 0 \tag{10.1.166}$$

(2) An elastic state $\overline{s} = [\overline{\mathbf{u}}, \overline{\mathbf{E}}, \overline{\mathbf{S}}]$ corresponding to a concentrated force $\overline{T}_0$ tangent to the boundary of a semi-infinite sheet under generalized plane stress conditions is obtained from Equations 10.1.154 and 10.1.155 and 10.1.159 through 10.1.164, in which $[\mathbf{u}, \mathbf{E}, \mathbf{S}]$ is replaced by $[\overline{\mathbf{u}}, \overline{\mathbf{E}}, \overline{\mathbf{S}}]$, T_0 by $\overline{T}_0$, and ν by $\overline{\nu} = \nu/(1+\nu)$. In particular, for the displacement $\overline{\mathbf{u}}$ we have

$$\overline{u}_1(x_1, x_2) = \frac{\overline{T}_0}{\pi E}\left[\ln\left(1 - \frac{x_1^2}{r^2}\right) + (1+\nu)\frac{x_1^2}{r^2}\right] \tag{10.1.167}$$

$$\overline{u}_2(x_1, x_2) = \frac{\overline{T}_0}{\pi E}\left[(1-\nu)\tan^{-1}\frac{x_1}{x_2} + (1+\nu)\frac{x_1 x_2}{r^2}\right] \tag{10.1.168}$$

where $E = 2\mu(1+\nu)$ is Young's modulus.

10.1.4 The Lamé Solution for a Thin Hollow Circular Disk

10.1.4.1 The Field Equations of Two-Dimensional Elastostatics in Polar Coordinates

An elastic state $s = [\overline{\mathbf{u}}, \overline{\mathbf{E}}, \overline{\mathbf{S}}]$ corresponding to the generalized plane stress conditions and zero body forces satisfies the following field equations in the polar coordinates (r, φ) (see Equations 8.1.34, 8.1.41, 8.1.43, and 8.1.46):

The equilibrium equations

$$\frac{\partial}{\partial r}\overline{S}_{rr} + \frac{1}{r}\frac{\partial}{\partial\varphi}\overline{S}_{r\varphi} + \frac{\overline{S}_{rr} - \overline{S}_{\varphi\varphi}}{r} = 0 \tag{10.1.169}$$

$$\frac{1}{r}\frac{\partial}{\partial\varphi}\overline{S}_{\varphi\varphi} + \frac{\partial}{\partial r}\overline{S}_{r\varphi} + \frac{2\overline{S}_{r\varphi}}{r} = 0 \tag{10.1.170}$$

The strain–displacement relations

$$\overline{E}_{rr} = \frac{\partial\,\overline{u}_r}{\partial r}, \quad \overline{E}_{\varphi\varphi} = \frac{\overline{u}_r}{r} + \frac{1}{r}\frac{\partial\,\overline{u}_\varphi}{\partial\varphi} \tag{10.1.171}$$

$$\overline{E}_{r\varphi} = \frac{1}{2}\left(\frac{1}{r}\frac{\partial\,\overline{u}_r}{\partial\varphi} + \frac{\partial\,\overline{u}_\varphi}{\partial r} - \frac{\overline{u}_\varphi}{r}\right) \tag{10.1.172}$$

The constitutive equations

$$\overline{S}_{rr} = \frac{2\mu}{1-\nu}\left(\overline{E}_{rr} + \nu\overline{E}_{\varphi\varphi}\right) \tag{10.1.173}$$

$$\overline{S}_{\varphi\varphi} = \frac{2\mu}{1-\nu}\left(\overline{E}_{\varphi\varphi} + \nu\overline{E}_{rr}\right) \tag{10.1.174}$$

$$\overline{S}_{r\varphi} = 2\mu\overline{E}_{r\varphi} \tag{10.1.175}$$

or

$$\overline{E}_{rr} = \frac{1}{E}\left(\overline{S}_{rr} - \nu\overline{S}_{\varphi\varphi}\right) \tag{10.1.176}$$

$$\overline{E}_{\varphi\varphi} = \frac{1}{E}\left(\overline{S}_{\varphi\varphi} - \nu\overline{S}_{rr}\right) \tag{10.1.177}$$

$$\overline{E}_{r\varphi} = \frac{1+\nu}{E}\overline{S}_{r\varphi} \tag{10.1.178}$$

Here

$$\overline{u} = [\overline{u}_r, \overline{u}_\varphi] \tag{10.1.179}$$

$$\mathbf{E} = \begin{bmatrix} \overline{E}_{rr} & \overline{E}_{r\varphi} & 0 \\ \overline{E}_{\varphi r} & \overline{E}_{\varphi\varphi} & 0 \\ 0 & 0 & \overline{E}_{33} \end{bmatrix} \tag{10.1.180}$$

$$\mathbf{S} = \begin{bmatrix} \overline{S}_{rr} & \overline{S}_{r\varphi} & 0 \\ \overline{S}_{\varphi r} & \overline{S}_{\varphi\varphi} & 0 \\ 0 & 0 & 0 \end{bmatrix} \tag{10.1.181}$$

$$\overline{E}_{33} = -\frac{\nu}{1-\nu}\left(\overline{E}_{rr} + \overline{E}_{\varphi\varphi}\right) \tag{10.1.182}$$

For a general boundary value problem $s = s(r, \varphi)$, which means that $\overline{\mathbf{u}} = \overline{\mathbf{u}}(r, \varphi)$, $\overline{\mathbf{E}} = \overline{\mathbf{E}}(r, \varphi)$, and $\overline{\mathbf{S}} = \overline{\mathbf{S}}(r, \varphi)$. In this case, the equilibrium equations 10.1.169 and 10.1.170 are satisfied identically if $\overline{\mathbf{S}}$ is represented by a stress function $\overline{F} = \overline{F}(r, \varphi)$ in the form

$$\overline{S}_{rr} = \left(\nabla^2 - \frac{\partial^2}{\partial r^2}\right)\overline{F} \tag{10.1.183}$$

$$\overline{S}_{\varphi\varphi} = \frac{\partial^2 \overline{F}}{\partial r^2} \tag{10.1.184}$$

$$\overline{S}_{r\varphi} = -\frac{\partial}{\partial r}\left(\frac{1}{r}\frac{\partial \overline{F}}{\partial \varphi}\right) \tag{10.1.185}$$

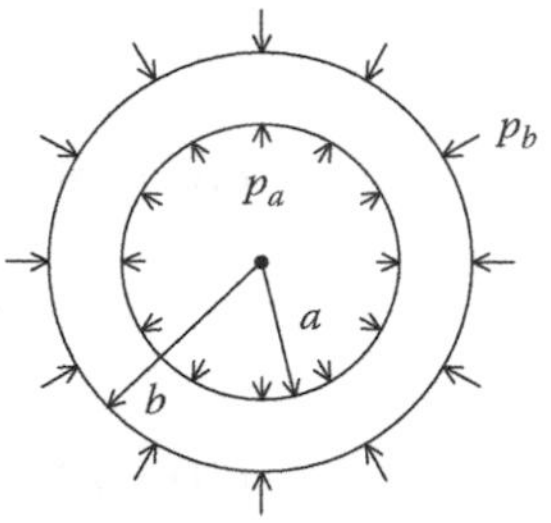

FIGURE 10.5 A thin hollow circular disk with inner radius a and outer radius b and with internal pressure p_a and external pressure p_b.

while the compatibility condition [see (b) in Example 8.1.3]

$$\nabla^2(\overline{S}_{rr} + \overline{S}_{\varphi\varphi}) = 0 \tag{10.1.186}$$

implies that $\overline{F}$ satisfies the biharmonic equation

$$\nabla^2\nabla^2\overline{F} = 0 \tag{10.1.187}$$

Here

$$\nabla^2 = \frac{\partial^2}{\partial r^2} + \frac{1}{r}\frac{\partial}{\partial r} + \frac{1}{r^2}\frac{\partial^2}{\partial \varphi^2} \tag{10.1.188}$$

Therefore, for problems involving boundaries formed by circular arcs, the stress tensor $\overline{\mathbf{S}}$ may be found by using Equations 10.1.183 through 10.1.185 and 10.1.187.

10.1.4.2 The Lamé Problem for a Thin Hollow Circular Disk

We examine a thin hollow circular disk subject to external and internal constant pressures shown in Figure 10.5.

Since p_a and p_b are constants, an elastic state $s = [\overline{\mathbf{u}}, \overline{\mathbf{E}}, \overline{\mathbf{S}}]$ corresponding to the situation shown in Figure 10.5 is axially symmetric, which means that $\overline{\mathbf{u}} = \overline{\mathbf{u}}(r)$, $\overline{\mathbf{E}} = \overline{\mathbf{E}}(r)$, and $\overline{\mathbf{S}} = \overline{\mathbf{S}}(r)$; in addition $\overline{u}_\varphi = 0$. Therefore, s is to satisfy Equations 10.1.171 and 10.1.172, 10.1.176 through 10.1.178, 10.1.183 through 10.1.185, and 10.1.187 in which $\overline{u}_\varphi = 0$ and partial derivatives with respect to φ are omitted. The field equations are now reduced to

The equilibrium equation

$$\frac{\partial}{\partial r}\overline{S}_{rr} + \frac{\overline{S}_{rr} - \overline{S}_{\varphi\varphi}}{r} = 0 \tag{10.1.189}$$

The strain–displacement relations

$$\overline{E}_{rr} = \frac{\partial \overline{u}_r}{\partial r}, \quad \overline{E}_{\varphi\varphi} = \frac{\overline{u}_r}{r}, \quad \overline{E}_{r\varphi} = 0 \tag{10.1.190}$$

The constitutive equations

$$\overline{E}_{rr} = \frac{1}{E}\left(\overline{S}_{rr} - \nu\overline{S}_{\varphi\varphi}\right) \tag{10.1.191}$$

$$\overline{E}_{\varphi\varphi} = \frac{1}{E}\left(\overline{S}_{\varphi\varphi} - \nu\overline{S}_{rr}\right) \tag{10.1.192}$$

The stress components $\overline{S}_{11}$ and $\overline{S}_{\varphi\varphi}$ may be expressed in terms of a function $\overline{F} = \overline{F}(r)$ by the formulas

$$\overline{S}_{rr} = \frac{1}{r}\frac{\partial \overline{F}}{\partial r}, \quad \overline{S}_{\varphi\varphi} = \frac{\partial^2 \overline{F}}{\partial r^2}, \quad \overline{S}_{r\varphi} = 0 \tag{10.1.193}$$

where

$$\left(\frac{\partial^2}{\partial r^2} + \frac{1}{r}\frac{\partial}{\partial r}\right)\left(\frac{\partial^2}{\partial r^2} + \frac{1}{r}\frac{\partial}{\partial r}\right)\overline{F} = 0 \tag{10.1.194}$$

or

$$\left(\frac{1}{r}\frac{\partial}{\partial r}r\frac{\partial}{\partial r}\right)\left(\frac{1}{r}\frac{\partial}{\partial r}r\frac{\partial}{\partial r}\right)\overline{F} = 0 \tag{10.1.195}$$

The Lamé problem for the thin hollow circular disk shown in Figure 10.5 may be now formulated in the following way:

Find an elastic state $s = [\overline{\mathbf{u}}, \overline{\mathbf{E}}, \overline{\mathbf{S}}]$ for $a \leq r \leq b$ that satisfies Equations 10.1.189 through 10.1.192 subject to the boundary conditions

$$\overline{S}_{rr}(a) = -p_a, \quad \overline{S}_{rr}(b) = -p_b \tag{10.1.196}$$

An alternative form of the problem reads: Find a function $\overline{F} = \overline{F}(r)$ that satisfies Equation 10.1.195 for $a < r < b$ and the boundary conditions

$$\left.\frac{\partial \overline{F}}{\partial r}\right|_{r=a} = -a\,p_a, \quad \left.\frac{\partial \overline{F}}{\partial r}\right|_{r=b} = -b\,p_b \tag{10.1.197}$$

To solve the problem involving function $\overline{F}$, we note that a general solution to Equation 10.1.195 takes the form

$$\overline{F} = C_0 + C_1 \ln\frac{r}{L} + C_2 r^2 + C_3 r^2 \ln\frac{r}{L} \tag{10.1.198}$$

where L is a positive constant of the length dimension, and C_0, C_1, C_2, and C_3 are arbitrary constants.

Substitution of $\overline{F}$ from Equation 10.1.198 into Equations 10.1.193 yields

$$\overline{S}_{rr} = C_1 r^{-2} + 2C_2 + C_3\left(2\ln\frac{r}{L} + 1\right) \tag{10.1.199}$$

$$\overline{S}_{\varphi\varphi} = -C_1 r^{-2} + 2C_2 + C_3\left(2\ln\frac{r}{L} + 3\right) \tag{10.1.200}$$

In the following, we show that $C_3 = 0$, and C_1 and C_2 are uniquely determined by the boundary conditions 10.1.197. With this in mind we note that Equations 10.1.190 imply the compatibility condition

$$\overline{E}_{rr} - \frac{\partial}{\partial r}\left(r\overline{E}_{\varphi\varphi}\right) = 0 \tag{10.1.201}$$

Substitution Equations 10.1.191 and 10.1.192 into Equation 10.1.201 yields

$$\left(\overline{S}_{rr} - \overline{S}_{\varphi\varphi}\right)(1+\nu) - r\left(\frac{\partial \overline{S}_{\varphi\varphi}}{\partial r} - \nu\frac{\partial \overline{S}_{rr}}{\partial r}\right) = 0 \tag{10.1.202}$$

Now, because of Equations 10.1.199 and 10.1.200,

$$\frac{\partial \overline{S}_{\varphi\varphi}}{\partial r} = 2C_1 r^{-3} + 2C_3 r^{-1} \tag{10.1.203}$$

$$\frac{\partial \overline{S}_{rr}}{\partial r} = -2C_1 r^{-3} + 2C_3 r^{-1} \tag{10.1.204}$$

Therefore, substituting Equations 10.1.199 and 10.1.200, and 10.1.203 and 10.1.204 into Equation 10.1.202, we obtain

$$C_3 = 0 \tag{10.1.205}$$

Hence, the stress components $\overline{S}_{rr}$ and $\overline{S}_{\varphi\varphi}$, by Equations 10.1.199 and 10.1.200, take the forms

$$\overline{S}_{rr} = C_1 r^{-2} + 2C_2 \tag{10.1.206}$$

$$\overline{S}_{\varphi\varphi} = -C_1 r^{-2} + 2C_2 \tag{10.1.207}$$

Finally, substituting $\overline{S}_{rr}$ given by the Equation 10.1.206 into the boundary conditions 10.1.196, we have

$$C_1 = -\frac{(p_a - p_b)}{(b^2 - a^2)}\, a^2 b^2, \quad C_2 = -\frac{(b^2 p_b - a^2 p_a)}{2(b^2 - a^2)} \tag{10.1.208}$$

As a result, by Equations 10.1.206 through 10.1.208, the stress components $\overline{S}_{rr}$ and $\overline{S}_{\varphi\varphi}$ are obtained

$$\overline{S}_{rr} = -\frac{1}{b^2 - a^2}\left[\frac{a^2 b^2}{r^2}(p_a - p_b) + \left(b^2 p_b - a^2 p_a\right)\right] \tag{10.1.209}$$

$$\overline{S}_{\varphi\varphi} = \frac{1}{b^2 - a^2}\left[\frac{a^2 b^2}{r^2}(p_a - p_b) - \left(b^2 p_b - a^2 p_a\right)\right] \tag{10.1.210}$$

Also, by substituting $\overline{S}_{rr}$ and $\overline{S}_{\varphi\varphi}$ from Equations 10.1.209 and 10.1.210, respectively, into Equations 10.1.191 and 10.1.193, we have

$$\overline{E}_{rr} = -\frac{1}{E(b^2 - a^2)}\left[(1+\nu)(p_a - p_b)\frac{a^2 b^2}{r^2} + (1-\nu)(b^2 p_b - a^2 p_a)\right] \tag{10.1.211}$$

$$\overline{E}_{\varphi\varphi} = \frac{1}{E(b^2 - a^2)}\left[(1+\nu)(p_a - p_b)\frac{a^2 b^2}{r^2} - (1-\nu)(b^2 p_b - a^2 p_a)\right] \tag{10.1.212}$$

The last equation together with the second equation of 10.1.190 imply the radial displacement formula

$$\overline{u}_r = \frac{r}{E(b^2 - a^2)}\left[(1+\nu)(p_a - p_b)\frac{a^2 b^2}{r^2} - (1-\nu)(b^2 p_b - a^2 p_a)\right] \tag{10.1.213}$$

This completes the derivation of a solution to the Lamé problem.

Notes:

(1) If $p_a \to p$ and $p_b \to p$, where p is a constant pressure, then $s \to s_0 = [\overline{\mathbf{u}}^{(0)}, \overline{\mathbf{E}}^{(0)}, \overline{\mathbf{S}}^{(0)}]$, where

$$\overline{\mathbf{u}}^{(0)} = [\overline{u}_r^{(0)}, 0] \tag{10.1.214}$$

$$\overline{u}_r^{(0)} = -\frac{r(1-\nu)}{E} p \tag{10.1.215}$$

$$\overline{E}_{rr}^{(0)} = -\frac{1-\nu}{E} p, \quad \overline{E}_{\varphi\varphi}^{(0)} = -\frac{1-\nu}{E} p \tag{10.1.216}$$

$$\overline{S}_{rr}^{(0)} = -p, \quad \overline{S}_{\varphi\varphi}^{(0)} = -p \tag{10.1.217}$$

Hence, if a hollow circular disk is subject to the external and internal pressures of the same magnitude p, the associated elastic state is independent of the disk radii, and there is a homogeneous state of strains and stresses in the disk.

(2) If $p_b = 0$ then

$$\overline{S}_{rr} = \frac{a^2 p_a}{b^2 - a^2}\left(1 - \frac{b^2}{r^2}\right) \tag{10.1.218}$$

$$\overline{S}_{\varphi\varphi} = \frac{a^2 p_a}{b^2 - a^2}\left(1 + \frac{b^2}{r^2}\right) \tag{10.1.219}$$

Hence, since $a \le r \le b$, we write

$$\overline{S}_{rr} \le 0, \quad \overline{S}_{\varphi\varphi} > 0, \quad a \le r \le b \tag{10.1.220}$$

and

$$\overline{S}_{\varphi\varphi}(b) \le \overline{S}_{\varphi\varphi}(r) \le \overline{S}_{\varphi\varphi}(a), \quad a \le r \le b \tag{10.1.221}$$

In addition, if $b - a$ is a small number then it follows from Equation 9.2.219 that

$$\overline{S}_{\varphi\varphi} \approx \overline{S}_{\varphi\varphi}^{*} = \frac{a}{b-a} p_a \tag{10.1.222}$$

Finally, by letting $b \to \infty$ in Equations 10.1.218 and 10.1.219, we arrive at the stresses corresponding to a two-dimensional static representation of an explosion:

$$\overline{S}_{rr} = -p_a \frac{a^2}{r^2} \tag{10.1.223}$$

$$\overline{S}_{\varphi\varphi} = p_a \frac{a^2}{r^2} \tag{10.1.224}$$

(3) A plane strain elastic state $s = [\mathbf{u}, \mathbf{E}, \mathbf{S}]$ in a thick-walled cylinder subject to external and internal pressures shown in Figure 10.5 may be obtained from Equations 10.1.209 through 10.1.213 by replacing ν by $\nu/(1-\nu)$ and $[\overline{\mathbf{u}}, \overline{\mathbf{E}}, \overline{\mathbf{S}}]$ by $[\mathbf{u}, \mathbf{E}, \mathbf{S}]$. After such a transformation we arrive at

$$S_{rr} = \overline{S}_{rr}, \quad S_{\varphi\varphi} = \overline{S}_{\varphi\varphi} \tag{10.1.225}$$

$$E_{rr} = -\frac{1}{2\mu(b^2 - a^2)}\left[(p_a - p_b)\frac{a^2b^2}{r^2} + (1-2\nu)(b^2p_b - a^2p_a)\right] \tag{10.1.226}$$

$$E_{\varphi\varphi} = \frac{1}{2\mu(b^2 - a^2)}\left[(p_a - p_b)\frac{a^2b^2}{r^2} - (1-2\nu)(b^2p_b - a^2p_a)\right] \tag{10.1.227}$$

and

$$u_r = \frac{r}{2\mu(b^2 - a^2)}\left[(p_a - p_b)\frac{a^2b^2}{r^2} - (1-2\nu)(b^2p_b - a^2p_a)\right] \tag{10.1.228}$$

It follows from Equations 10.1.225 through 10.1.227 that for $p_a = p_b = p$, where p is a constant pressure, there is a homogeneous state of strains and stresses in the hollow cylinder.

10.1.5 Axially Symmetric Elastic State in a Rotating Thin Hollow Circular Disk [1]

Consider a rotating hollow circular disk as shown in Figure 10.6.

The disk is made of a homogeneous isotropic elastic material, and its boundaries $r = a$ and $r = b$ are assumed to be stress free. A generalized plane stress state $s = [\overline{\mathbf{u}}, \overline{\mathbf{E}}, \overline{\mathbf{S}}]$ in the disk is then axially symmetric, and it obeys the equations (see Equations 10.1.189 through 10.1.192, and Equation 10.1.202):

The equilibrium equation

$$\frac{d}{dr}\overline{S}_{rr} + \frac{\overline{S}_{rr} - \overline{S}_{\varphi\varphi}}{r} + \rho\omega^2 r = 0 \tag{10.1.229}$$

The strain–displacement relations

$$\overline{E}_{rr} = \frac{d\overline{u}_r}{dr}, \quad \overline{E}_{\varphi\varphi} = \frac{\overline{u}_r}{r} \tag{10.1.230}$$

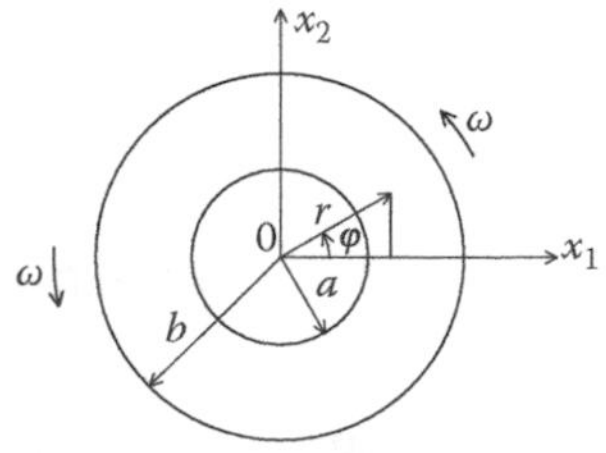

FIGURE 10.6 A thin circular disk rotating with an angular velocity ω.

The constitutive equations

$$\overline{E}_{rr} = \frac{1}{E}\left(\overline{S}_{rr} - \nu\overline{S}_{\varphi\varphi}\right) \tag{10.1.231}$$

$$\overline{E}_{\varphi\varphi} = \frac{1}{E}\left(\overline{S}_{\varphi\varphi} - \nu\overline{S}_{rr}\right) \tag{10.1.232}$$

The stress compatibility condition

$$(1+\nu)\,\frac{\overline{S}_{rr}-\overline{S}_{\varphi\varphi}}{r}-\frac{d}{dr}\left(\overline{S}_{\varphi\varphi}-\nu\overline{S}_{rr}\right)=0 \tag{10.1.233}$$

The boundary conditions

$$\overline{S}_{rr}(a)=\overline{S}_{rr}(b)=0 \tag{10.1.234}$$

In Equation 10.1.229, $\rho\omega^2 r=\overline{b}_r$ represents a centrifugal force in the radial direction; ρ is a density of the disk, and ω is a constant angular velocity; hence $[\overline{b}_r]=[\mathrm{ML}^{-2}\mathrm{T}^{-2}]$, where M is a mass unit, L is a length unit, and T stands for a time unit. To find s we rewrite the equilibrium equation 10.1.229 in the form

$$\frac{\overline{S}_{rr}-\overline{S}_{\varphi\varphi}}{r}=-\frac{d\overline{S}_{rr}}{dr}-\rho\omega^2 r \tag{10.1.235}$$

By substituting Equation 10.1.235 into Equation 10.1.233, we have

$$-\frac{d\overline{S}_{\varphi\varphi}}{dr}=\frac{d\overline{S}_{rr}}{dr}+(1+\nu)\rho\omega^2 r \tag{10.1.236}$$

Next, multiplying Equation 10.1.235 by r, we get

$$-\overline{S}_{\varphi\varphi}=-\overline{S}_{rr}-\left[r\frac{d\overline{S}_{rr}}{dr}+\rho\omega^2 r^2\right] \tag{10.1.237}$$

Finally, differentiating Equation 10.1.237 with respect to r, by virtue of Equation 10.1.236, we arrive at the single equation for $\overline{S}_{rr}=\overline{S}_{rr}(r)$:

$$\frac{d}{dr}\left(r\frac{d\overline{S}_{rr}}{dr}\right)+2\frac{d\overline{S}_{rr}}{dr}+(3+\nu)\rho\omega^2 r=0 \tag{10.1.238}$$

Now, if we let

$$\frac{d\overline{S}_{rr}}{dr}=f(r) \tag{10.1.239}$$

we reduce Equation 10.1.238 to the first-order differential equation for f:

$$r\frac{df}{dr}+3f=-(3+\nu)\rho\omega^2 r \tag{10.1.240}$$

It is easy to check that a general solution of Equation 10.1.240 takes the form

$$f=f_0 r^{-3}-\frac{1}{4}(3+\nu)\rho\omega^2 r \tag{10.1.241}$$

where f_0 is an arbitrary constant. Hence, by Equation 10.1.239, a function $\overline{S}_{rr} = \overline{S}_{rr}(r)$ that satisfies the boundary condition $\overline{S}_{rr}(a) = 0$ is given by

$$\overline{S}_{rr}(r) = f_0 \int_a^r u^{-3} du - \frac{1}{4}(3+\nu)\rho\omega^2 \int_a^r u\, du \tag{10.1.242}$$

By substituting $\overline{S}_{rr}(r)$ from this equation into the second of boundary conditions 10.1.234, we obtain

$$f_0 = \frac{1}{4}(3+\nu)a^2b^2\rho\omega^2 \tag{10.1.243}$$

This equation together with Equation 10.1.242 imply that

$$\overline{S}_{rr} = \frac{3+\nu}{8}\,\frac{\rho\omega^2}{r^2}(r^2 - a^2)(b^2 - r^2) \tag{10.1.244}$$

Substituting $\overline{S}_{rr}$ from Equation 10.1.245 into Equation 10.1.237, we receive

$$\overline{S}_{\varphi\varphi} = \frac{3+\nu}{8}\,\frac{\rho\omega^2}{r^2}\left[\left(a^2 + b^2 - \frac{1+3\nu}{3+\nu}r^2\right)r^2 + a^2b^2\right] \tag{10.1.245}$$

Next, if $\overline{S}_{rr}$ and $\overline{S}_{\varphi\varphi}$ from Equations 10.1.244 and 10.1.245, respectively, are inserted into Equations 10.1.231 and 10.1.232, the strain components $\overline{E}_{rr}$ and $\overline{E}_{\varphi\varphi}$ are obtained

$$\begin{aligned} \overline{E}_{rr} &= \frac{3+\nu}{16\mu}\rho\omega^2\left[\frac{1-\nu}{1+\nu}(a^2+b^2) - \frac{a^2b^2}{r^2} - \frac{3(1-\nu)}{3+\nu}r^2\right] \\ \overline{E}_{\varphi\varphi} &= \frac{3+\nu}{16\mu}\rho\omega^2\left[\frac{1-\nu}{1+\nu}(a^2+b^2) + \frac{a^2b^2}{r^2} - \frac{(1-\nu)}{3+\nu}r^2\right] \end{aligned} \tag{10.1.246}$$

Finally, it follows from Equations 10.1.230 and 10.1.246 that

$$\overline{u}_r = \frac{3+\nu}{16\mu}\,\rho\omega^2 r\left[\frac{1-\nu}{1+\nu}(a^2+b^2) + \frac{a^2b^2}{r^2} - \frac{1-\nu}{3+\nu}r^2\right] \tag{10.1.247}$$

Hence, a solution to the rotating disk problem is represented by an elastic state $s = [\overline{\mathbf{u}}, \overline{\mathbf{E}}, \overline{\mathbf{S}}]$ in which $\overline{\mathbf{u}}$, $\overline{\mathbf{E}}$, and $\overline{\mathbf{S}}$ are given by Equations 10.1.247, 10.1.246, and 10.1.244 and 10.1.245, respectively.

For a solid rotating disk, $a = 0$, so letting $a \to 0$ in Equations 10.1.244 through 10.1.247, we have

$$\overline{u}_r^{\,0} = \frac{(3+\nu)(1-\nu)}{16\mu}\rho\omega^2 r\left(\frac{1}{1+\nu}\,b^2 - \frac{1}{3+\nu}\,r^2\right) \tag{10.1.248}$$

$$\overline{E}_{rr}^{\,0} = \frac{(3+\nu)(1-\nu)}{16\mu}\rho\omega^2\left(\frac{1}{1+\nu}\,b^2 - \frac{3}{3+\nu}\,r^2\right) \tag{10.1.249}$$

$$\overline{E}_{\varphi\varphi}^{\,0} = \frac{(3+\nu)(1-\nu)}{16\mu}\rho\omega^2\left(\frac{1}{1+\nu}\,b^2 - \frac{1}{3+\nu}\,r^2\right) \tag{10.1.250}$$

and

$$\overline{S}_{rr}^{0} = \frac{3+\nu}{8}\rho\omega^2(b^2 - r^2) \tag{10.1.251}$$

$$\overline{S}_{\varphi\varphi}^{0} = \frac{3+\nu}{8}\rho\omega^2\left(b^2 - \frac{1+3\nu}{3+\nu}r^2\right) \tag{10.1.252}$$

In particular, it follows from Equations 10.1.251 and 10.1.252 that

$$\overline{S}_{rr}(0) = \overline{S}_{\varphi\varphi}^{0}(0) = \frac{(3+\nu)\rho\omega^2 b^2}{8} \tag{10.1.253}$$

Also, it follows from Equations 10.1.244 and 10.1.245 that $\overline{S}_{rr}$ and $\overline{S}_{\varphi\varphi}$ have the properties:

$$\text{(i)} \quad \max_{a\le r\le b} \overline{S}_{rr}(r) = \overline{S}_{rr}\left(\sqrt{ab}\right) = \frac{3+\nu}{8}\rho\omega^2(b-a)^2 \tag{10.1.254}$$

$$\text{(ii)} \quad \max_{a\le r\le b} \overline{S}_{\varphi\varphi}(r) = \overline{S}_{\varphi\varphi}(a) = \frac{3+\nu}{8}\rho\omega^2\left[\frac{2(1-\nu)}{3+\nu}a^2 + 2b^2\right] \tag{10.1.255}$$

$$\text{(iii)} \quad \max_{a\le r\le b} \overline{S}_{rr}(r) \le \max_{a\le r\le b} \overline{S}_{\varphi\varphi}(r) \tag{10.1.256}$$

$$\text{(iv)} \quad \overline{S}_{rr}(r) \ge 0, \quad \overline{S}_{\varphi\varphi}(r) > 0 \quad \text{for } 0 < a \le r \le b \tag{10.1.257}$$

$$\text{(v)} \quad \overline{S}_{\varphi\varphi}(a) \to 2\overline{S}_{\varphi\varphi}^{0}(0) \quad \text{as } a \to 0 \tag{10.1.258}$$

The property (v) implies that by making a small circular hole at the center of a rotating disk, we double the maximum stress, i.e., a stress concentration occurs in a neighborhood of the small hole. The limit

$$k^* = \lim_{a\to 0} \frac{\overline{S}_{\varphi\varphi}(a)}{\overline{S}_{\varphi\varphi}^{0}} = 2 \tag{10.1.259}$$

is called a *stress concentration factor* in the problem of a rotating disk with a small circular hole.

Notes:

(1) A plane strain elastic state $s = [\mathbf{u}, \mathbf{E}, \mathbf{S}]$ in a rotating thick-walled cylinder may be obtained from Equations 10.1.244 through 10.1.247 by replacing there ν by $\nu/(1-\nu)$ and $[\overline{\mathbf{u}}, \overline{\mathbf{E}}, \overline{\mathbf{S}}]$ by $[\mathbf{u}, \mathbf{E}, \mathbf{S}]$.

(2) A plane strain elastic state $s_0 = [\mathbf{u}_0, \mathbf{E}_0, \mathbf{S}_0]$ in a rotating solid cylinder with the radius $r = b$ may be obtained from Equations 10.1.248 through 10.1.252, by replacing ν by $\nu/(1-\nu)$ and $[\overline{\mathbf{u}}^0, \overline{\mathbf{E}}^0, \overline{\mathbf{S}}^0]$ by $[\mathbf{u}_0, \mathbf{E}_0, \mathbf{S}_0]$.

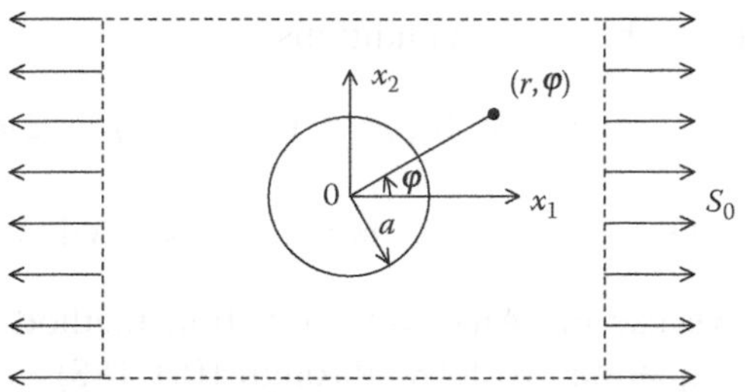

FIGURE 10.7 A thin elastic sheet with a circular hole of radius a under tension.

10.1.6 A Sheet with a Circular Hole Subject to Uniform Tension at Infinity (Kirsch's Problem) [2]

Consider a thin homogeneous isotropic infinite elastic sheet with a circular hole of radius a under uniform tension S_0 in the x_1 direction, as shown in Figure 10.7.

The sheet is supposed to be in a generalized plane stress state with regard to the (x_1, x_2) plane, and the boundary of the hole is to be stress free.

The problem of finding an elastic state $s = [\overline{\mathbf{u}}, \overline{\mathbf{E}}, \overline{\mathbf{S}}]$ corresponding to the situation shown in Figure 10.7 is called Kirsch's problem. The problem is a genuine two-dimensional problem of elastostatics, and s may be treated as a function of x_1 and x_2, or as a function of r and φ, where

$$0 < a \le r < \infty, \quad 0 \le \varphi \le 2\pi \tag{10.1.260}$$

Therefore, a solution to the problem is to obey the equations (see Equations 10.1.169 through 10.1.178):

The equilibrium equations

$$\frac{\partial}{\partial r}\overline{S}_{rr} + \frac{1}{r}\frac{\partial}{\partial \varphi}\overline{S}_{r\varphi} + \frac{\overline{S}_{rr} - \overline{S}_{\varphi\varphi}}{r} = 0 \tag{10.1.261}$$

$$\frac{1}{r}\frac{\partial}{\partial \varphi}\overline{S}_{\varphi\varphi} + \frac{\partial}{\partial r}\overline{S}_{r\varphi} + \frac{2\overline{S}_{r\varphi}}{r} = 0 \tag{10.1.262}$$

The strain–displacement relations

$$\begin{aligned} &\overline{E}_{rr} = \frac{\partial \overline{u}_r}{\partial r}, \quad \overline{E}_{\varphi\varphi} = \frac{\overline{u}_r}{r} + \frac{1}{r}\frac{\partial \overline{u}_\varphi}{\partial \varphi}, \\ &\overline{E}_{r\varphi} = \frac{1}{2}\left(\frac{1}{r}\frac{\partial \overline{u}_r}{\partial \varphi} + \frac{\partial \overline{u}_\varphi}{\partial r} - \frac{\overline{u}_\varphi}{r}\right) \end{aligned} \tag{10.1.263}$$

The constitutive relations

$$\overline{E}_{rr} = \frac{1}{E}\left(\overline{S}_{rr} - \nu\overline{S}_{\varphi r}\right) \tag{10.1.264}$$

$$\overline{E}_{\varphi\varphi} = \frac{1}{E}\left(\overline{S}_{\varphi\varphi} - \nu\overline{S}_{rr}\right) \tag{10.1.265}$$

$$\overline{E}_{r\varphi} = \frac{1+\nu}{E}\overline{S}_{r\varphi} \tag{10.1.266}$$

In addition, s will satisfy the boundary conditions

$$\overline{S}_{rr}(a,\varphi) = \overline{S}_{r\varphi}(a,\varphi) = 0, \quad 0 \le \varphi \le 2\pi \tag{10.1.267}$$

$$\overline{S}_{11}(x_1,x_2) \to \pm S_0 \quad \text{as } x_1 \to \pm\infty,\ |x_2| < \infty \tag{10.1.268}$$

To solve the problem, we use the Airy stress function method (see Example 8.1.4). By using the formulas (see Equations 10.1.183 through 10.1.188)

$$\overline{S}_{rr} = \left(\nabla^2 - \frac{\partial^2}{\partial r^2}\right)\overline{F} \tag{10.1.269}$$

$$\overline{S}_{\varphi\varphi} = \frac{\partial^2 \overline{F}}{\partial r^2} \tag{10.1.270}$$

$$\overline{S}_{r\varphi} = -\frac{\partial}{\partial r}\left(\frac{1}{r}\frac{\partial \overline{F}}{\partial \varphi}\right) \tag{10.1.271}$$

where $\overline{F} = \overline{F}(r,\varphi)$ is the Airy stress function that satisfies the biharmonic equation

$$\nabla^2\nabla^2\overline{F} = 0 \tag{10.1.272}$$

and

$$\nabla^2 = \frac{\partial^2}{\partial r^2} + \frac{1}{r}\frac{\partial}{\partial r} + \frac{1}{r^2}\frac{\partial^2}{\partial \varphi^2} \tag{10.1.273}$$

we find that the equilibrium equations 10.1.261 and 10.1.262 are satisfied identically.

Hence, $\overline{F}$ is to satisfy Equation 10.1.272 as well the boundary conditions 10.1.267 and 10.1.268.

To this end, we look for s in the form

$$s = s_1 + s_2 \tag{10.1.274}$$

where

$$s_1 = \left[\overline{\mathbf{u}}^{(1)}, \overline{\mathbf{E}}^{(1)}, \overline{\mathbf{S}}^{(1)}\right], \quad s_2 = \left[\overline{\mathbf{u}}^{(2)}, \overline{\mathbf{E}}^{(2)}, \overline{\mathbf{S}}^{(2)}\right] \tag{10.1.275}$$

and s_1 is an elastic state in an infinite sheet without hole under tension S_0 in the x_1 direction while s_2 is an elastic state in an infinite sheet with a circular hole of radius a subject to such a boundary load on $r = a$ that s is a solution to the problem.

First, we observe that s_1 may be defined in Cartesian coordinates by the formulas

$$\overline{\mathbf{u}}^{(1)} = \left[\frac{1}{E}x_1 S_0,\ -\frac{\nu}{E}x_2 S_0\right] \tag{10.1.276}$$

$$\overline{E}_{11}^{(1)} = \frac{1}{E}S_0, \quad \overline{E}_{22}^{(1)} = -\frac{\nu}{E}S_0, \quad \overline{E}_{12}^{(1)} = 0 \tag{10.1.277}$$

$$\overline{S}_{11}^{(1)} = S_0, \quad \overline{S}_{22}^{(1)} = 0, \quad \overline{S}_{12}^{(1)} = 0 \tag{10.1.278}$$

Next, note that s_1 may be generated by the stress function

$$\overline{F}^{(1)} = \frac{1}{2} S_0 \, x_2^2 \tag{10.1.279}$$

In polar coordinates (r, φ), that are related to Cartesian coordinates (x_1, x_2) by the formulas

$$x_1 = r\cos\varphi, \quad x_2 = r\sin\varphi \tag{10.1.280}$$

the function $F^{(1)}$ takes the form

$$\overline{F}^{(1)} = \frac{1}{2} S_0 \, r^2 \sin^2\varphi = \frac{1}{4} S_0 \, r^2 (1 - \cos 2\varphi) \tag{10.1.281}$$

Substituting $\overline{F}^{(1)}$ from Equation 10.1.281 into Equations 10.1.269 through 10.1.271, we get the components of $\overline{\mathbf{S}}^{(1)}$ in polar coordinates

$$\overline{S}_{rr}^{(1)} = \frac{S_0}{2}(1 + \cos 2\varphi) \tag{10.1.282}$$

$$\overline{S}_{\varphi\varphi}^{(1)} = \frac{S_0}{2}(1 - \cos 2\varphi) \tag{10.1.283}$$

$$\overline{S}_{r\varphi}^{(1)} = -\frac{S_0}{2} \sin 2\varphi \tag{10.1.284}$$

The associated strain components, because of Equations 10.1.264 through 10.1.266 and Equations 10.1.282 through 10.1.284, take the form

$$\overline{E}_{rr}^{(1)} = \frac{S_0}{2E}[(1 - \nu) + (1 + \nu)\cos 2\varphi] \tag{10.1.285}$$

$$\overline{E}_{\varphi\varphi}^{(1)} = \frac{S_0}{2E}[(1 - \nu) - (1 + \nu)\cos 2\varphi] \tag{10.1.286}$$

$$\overline{E}_{r\varphi}^{(1)} = -\frac{S_0}{2E}(1 + \nu)\sin 2\varphi \tag{10.1.287}$$

Finally, by using the transformation formulas

$$\begin{aligned} \overline{u}_r^{(1)} &= \overline{u}_1^{(1)} \cos\varphi + \overline{u}_2^{(1)} \sin\varphi \\ \overline{u}_\varphi^{(1)} &= -\overline{u}_1^{(1)} \sin\varphi + \overline{u}_2^{(1)} \cos\varphi \end{aligned} \tag{10.1.288}$$

the polar components of $\overline{\mathbf{u}}^{(1)}$ are obtained

$$\overline{u}_r^{(1)} = \frac{S_0 r}{2E}[(1 - \nu) + (1 + \nu)\cos 2\varphi] \tag{10.1.289}$$

$$\overline{u}_\varphi^{(1)} = -\frac{S_0 r}{2E}(1 + \nu)\sin 2\varphi \tag{10.1.290}$$

Hence, the elastic state s_1 may be described either in Cartesian coordinates by using Equations 10.1.276 through 10.1.278 or in polar coordinates by using Equations 10.1.282 through 10.1.284, 10.1.285 through 10.1.287, and 10.1.289 and 10.1.290.

To find s_2 we postulate a stress function $\overline{F}^{(2)} = \overline{F}^{(2)}(r, \varphi)$ in the form

$$\overline{F}^{(2)} = A \ln \frac{r}{L} + (B + Cr^{-2}) \cos 2\varphi \tag{10.1.291}$$

where A, B, and C are arbitrary constants, and L is a positive constant of the length dimension. Then, it is easy to show that

$$\nabla^2 \nabla^2 \overline{F}^{(2)} = 0 \quad \text{for } a < r < \infty,\ 0 \le \varphi \le 2\pi \tag{10.1.292}$$

and the stress components $\overline{S}_{rr}^{(2)}$, $\overline{S}_{\varphi\varphi}^{(2)}$, and $\overline{S}_{r\varphi}^{(2)}$ are given by (see Equations 10.1.269 through 10.1.271)

$$\overline{S}_{rr}^{(2)} = Ar^{-2} - 2(2Br^{-2} + 3Cr^{-4}) \cos 2\varphi \tag{10.1.293}$$

$$\overline{S}_{\varphi\varphi}^{(2)} = -Ar^{-2} + 6Cr^{-4} \cos 2\varphi \tag{10.1.294}$$

$$\overline{S}_{r\varphi}^{(2)} = -2(Br^{-2} + 3Cr^{-4}) \sin 2\varphi \tag{10.1.295}$$

To determine A, B, and C we rewrite the boundary conditions 10.1.267 in the form

$$\overline{S}_{rr}^{(1)}(a) + \overline{S}_{rr}^{(2)}(a) = 0 \tag{10.1.296}$$

$$\overline{S}_{r\varphi}^{(1)}(a) + \overline{S}_{r\varphi}^{(2)}(a) = 0 \tag{10.1.297}$$

By substituting $\overline{S}_{rr}^{(1)}$ from Equation 10.1.282 and $\overline{S}_{rr}^{(2)}$ from Equation 10.1.293 into Equation 10.1.296 and $\overline{S}_{r\varphi}^{(1)}$ from Equation 10.1.284 and $\overline{S}_{r\varphi}^{(2)}$ from Equation 10.1.295 into Equation 10.1.297, we write

$$\begin{aligned} Aa^{-2} - 2\left(2Ba^{-2} + 3Ca^{-4}\right) \cos 2\varphi + \frac{S_0}{2}(1 + \cos 2\varphi) &= 0 \\ -2\left(Ba^{-2} + 3Ca^{-4}\right) \sin 2\varphi - \frac{S_0}{2} \sin 2\varphi &= 0 \end{aligned} \tag{10.1.298}$$

Since Equations 10.1.298 are satisfied for any $\varphi \in [0, 2\pi]$, the following algebraic equations should be met

$$A = -\frac{S_0}{2} a^2 \tag{10.1.299}$$

and

$$\begin{aligned} 2Ba^{-2} + 3Ca^{-4} &= \frac{S_0}{4} \\ Ba^{-2} + 3Ca^{-4} &= -\frac{S_0}{4} \end{aligned} \tag{10.1.300}$$

Hence, we obtain

$$B = \frac{S_0}{2}a^2, \quad C = -\frac{S_0}{4}a^4 \tag{10.1.301}$$

As a result, by Equations 10.1.293 through 10.1.295, 10.1.299, and 10.1.301, we arrive at the stress components $\overline{S}_{rr}^{(2)}$, $\overline{S}_{\varphi\varphi}^{(2)}$, and $\overline{S}_{r\varphi}^{(2)}$

$$\overline{S}_{rr}^{(2)} = -\frac{S_0}{2}\frac{a^2}{r^2}\left[1 + \left(4 - 3\frac{a^2}{r^2}\right)\cos 2\varphi\right] \tag{10.1.302}$$

$$\overline{S}_{\varphi\varphi}^{(2)} = \frac{S_0}{2}\frac{a^2}{r^2}\left(1 - 3\frac{a^2}{r^2}\right)\cos 2\varphi \tag{10.1.303}$$

$$\overline{S}_{r\varphi}^{(2)} = -\frac{S_0}{2}\frac{a^2}{r^2}\left(2 - 3\frac{a^2}{r^2}\right)\sin 2\varphi \tag{10.1.304}$$

The strain components $\overline{E}_{rr}^{(2)}$, $\overline{E}_{\varphi\varphi}^{(2)}$, and $\overline{E}_{r\varphi}^{(2)}$, and the displacement components $\overline{u}_r^{(2)}$ and $\overline{u}_\varphi^{(2)}$ may be obtained by using the formulas

$$\overline{\mathbf{E}}^{(2)} = \overline{\mathbf{E}} - \overline{\mathbf{E}}^{(1)}, \quad \overline{\mathbf{u}}^{(2)} = \overline{\mathbf{u}} - \overline{\mathbf{u}}^{(1)} \tag{10.1.305}$$

provided the fields $\overline{\mathbf{E}}$ and $\overline{\mathbf{u}}$ are known.

In the following, we show that $\overline{\mathbf{E}}$ and $\overline{\mathbf{u}}$ may be uniquely generated from $\overline{\mathbf{S}}$, where

$$\overline{\mathbf{S}} = \overline{\mathbf{S}}^{(1)} + \overline{\mathbf{S}}^{(2)} \tag{10.1.306}$$

and $\overline{\mathbf{S}}^{(1)}$ and $\overline{\mathbf{S}}^{(2)}$ are represented by Equations 10.1.282 through 10.1.284 and Equations 10.1.302 through 10.1.304, respectively. To this end note that, in view of Equations 10.1.282 through 10.1.284 and 10.1.302 through 10.1.304, we write Equation 10.1.306 in components

$$\overline{S}_{rr} = \frac{S_0}{2}\left[1 - \frac{a^2}{r^2} + \left(1 - 4\frac{a^2}{r^2} + 3\frac{a^4}{r^4}\right)\cos 2\varphi\right] \tag{10.1.307}$$

$$\overline{S}_{\varphi\varphi} = \frac{S_0}{2}\left[1 + \frac{a^2}{r^2} - \left(1 + 3\frac{a^4}{r^4}\right)\cos 2\varphi\right] \tag{10.1.308}$$

$$\overline{S}_{r\varphi} = -\frac{S_0}{2}\left(1 + 2\frac{a^2}{r^2} - 3\frac{a^4}{r^4}\right)\sin 2\varphi \tag{10.1.309}$$

Also note that an alternative form of the constitutive relations 10.1.264 through 10.1.266 reads

$$\overline{E}_{rr} = \frac{1}{2\mu}\left(\overline{S}_{rr} - \overline{\nu}\,\mathrm{tr}\,\overline{\mathbf{S}}\right) \tag{10.1.310}$$

$$\overline{E}_{\varphi\varphi} = \frac{1}{2\mu}\left(\overline{S}_{\varphi\varphi} - \overline{\nu}\,\mathrm{tr}\,\overline{\mathbf{S}}\right) \tag{10.1.311}$$

$$\overline{E}_{r\varphi} = \frac{1}{2\mu}\overline{S}_{r\varphi} \tag{10.1.312}$$

where

$$\overline{\nu} = \frac{\nu}{1+\nu}, \qquad \operatorname{tr}\overline{\mathbf{S}} = \overline{S}_{rr} + \overline{S}_{\varphi\varphi} \tag{10.1.313}$$

Now, it follows from Equations 10.1.307 and 10.1.308 that

$$\operatorname{tr}\overline{\mathbf{S}} = S_0\left(1 - 2\,\frac{a^2}{r^2}\cos 2\varphi\right) \tag{10.1.314}$$

Hence, by using Equations 10.1.307 through 10.1.314, we arrive at the strain components

$$\overline{E}_{rr} = \frac{S_0}{4\mu}\left\{1 - 2\overline{\nu} - \frac{a^2}{r^2} + \left[1 - 4(1-\overline{\nu})\frac{a^2}{r^2} + 3\frac{a^4}{r^4}\right]\cos 2\varphi\right\} \tag{10.1.315}$$

$$\overline{E}_{\varphi\varphi} = \frac{S_0}{4\mu}\left[1 - 2\overline{\nu} + \frac{a^2}{r^2} - \left(1 - 4\overline{\nu}\,\frac{a^2}{r^2} + 3\frac{a^4}{r^4}\right)\cos 2\varphi\right] \tag{10.1.316}$$

$$\overline{E}_{r\varphi} = -\frac{S_0}{4\mu}\left(1 + 2\frac{a^2}{r^2} - 3\frac{a^4}{r^4}\right)\sin 2\varphi \tag{10.1.317}$$

To find $\overline{u}_r$ and $\overline{u}_\varphi$ we proceed in the following way. By virtue of the first of Equations 10.1.263 and 10.1.315 we rewrite Equation 10.1.315 in the form

$$\frac{\partial \overline{u}_r}{\partial r} = \frac{S_0}{4\mu}\left\{1 - 2\overline{\nu} - \frac{a^2}{r^2} + \left[1 - 4(1-\overline{\nu})\frac{a^2}{r^2} + 3\frac{a^4}{r^4}\right]\cos 2\varphi\right\} \tag{10.1.318}$$

By integrating this equation with respect to r over the interval $[a, r]$, we arrive at

$$\begin{aligned}\overline{u}_r(r,\varphi) = \overline{u}_r(a,\varphi) + \frac{S_0 r}{4\mu}\left(1 - \frac{a}{r}\right)\bigg\{1 - 2\overline{\nu} - \frac{a}{r}\\ + \left[1 - (3-4\overline{\nu})\frac{a}{r} + \frac{a^2}{r^2} + \frac{a^3}{r^3}\right]\cos 2\varphi\bigg\}\end{aligned} \tag{10.1.319}$$

Next, it follows from the second of Equations 10.1.263 that

$$\frac{\partial \overline{u}_\varphi}{\partial \varphi} = r\overline{E}_{\varphi\varphi} - \overline{u}_r \tag{10.1.320}$$

Therefore, substituting $\overline{E}_{\varphi\varphi}$ and $\overline{u}_r$ from Equations 10.1.316 and 10.1.319, respectively, into Equation 10.1.320 leads to

$$\frac{\partial \overline{u}_\varphi}{\partial \varphi} = \frac{S_0 r}{2\mu}\left\{(1-\overline{\nu})\frac{a}{r} - \left[1 - 2(1-\overline{\nu})\frac{a}{r} + 2(1-2\overline{\nu})\frac{a^2}{r^2} + \frac{a^4}{r^4}\right]\cos 2\varphi\right\} - \overline{u}_r(a,\varphi) \tag{10.1.321}$$

By integrating this equation with respect to φ over the interval $[0, \varphi]$, we obtain

$$\overline{u}_\varphi(r,\varphi) = \overline{u}_\varphi(0,r) - \int_0^\varphi \overline{u}_r(a,\theta)d\theta + \frac{S_0 r}{4\mu}\left\{2(1-\overline{\nu})\frac{a}{r}\varphi - \left[1 - 2(1-\overline{\nu})\frac{a}{r} + 2(1-\overline{\nu})\frac{a^2}{r^2} + \frac{a^4}{r^4}\right]\sin 2\varphi\right\} \tag{10.1.322}$$

Also, it follows from the third of Equations 10.1.263 and from Equation 10.1.312 that

$$\overline{S}_{r\varphi} = \mu\left(\frac{1}{r}\frac{\partial \overline{u}_r}{\partial \varphi} + \frac{\partial \overline{u}_\varphi}{\partial r} - \frac{\overline{u}_\varphi}{r}\right) \tag{10.1.323}$$

Hence, substituting $\overline{S}_{r\varphi}$ from Equation 10.1.309, $\overline{u}_r$ from Equation 10.1.319, and $\overline{u}_\varphi$ from Equation 10.1.322 into Equation 10.1.323, we get

$$\begin{aligned}
&-\frac{S_0}{2}\left(1 + 2\frac{a^2}{r^2} - 3\frac{a^4}{r^4}\right)\sin 2\varphi \\
&= -\frac{S_0}{2}\left\{(1-\overline{\nu})\frac{a}{r}\varphi + \left[1 - 3(1-\overline{\nu})\frac{a}{r} + 2\frac{a^2}{r^2} - 3\frac{a^4}{r^4}\right]\sin 2\varphi\right\} \\
&+ \mu\left[\frac{\partial \overline{u}_\varphi}{\partial r}(r,0) - \frac{\overline{u}_\varphi}{r}(r,0) + \frac{1}{r}\frac{\partial \overline{u}_r}{\partial \varphi}(a,\varphi) + \frac{1}{r}\int_0^\varphi \overline{u}_r(a,\theta)d\theta\right]
\end{aligned} \tag{10.1.324}$$

Since φ is an arbitrary angle, $\varphi \in [0, 2\pi]$, Equation 10.1.324 is satisfied if and only if

$$\begin{aligned}
&r\frac{\partial \overline{u}_\varphi}{\partial r}(r,0) - \overline{u}_\varphi(r,0) + \frac{\partial \overline{u}_r}{\partial \varphi}(a,\varphi) + \int_0^\varphi \overline{u}_r(a,\theta)d\theta \\
&= \frac{S_0 a}{2\mu}(1-\overline{\nu})(\varphi - 3\sin 2\varphi)
\end{aligned} \tag{10.1.325}$$

Now, if we let

$$\overline{u}_\varphi(r,0) = 0 \quad \text{for } a < r < \infty \tag{10.1.326}$$

Equation 10.1.325 reduces to the integro-differential equation for $\overline{u}_r(a,\varphi) \equiv u(\varphi)$:

$$\frac{du}{d\varphi} + \int_0^\varphi u(\theta)\,d\theta = \frac{S_0 a}{2\mu}(1-\overline{\nu})(\varphi - 3\sin 2\varphi) \tag{10.1.327}$$

In addition, if we let

$$\overline{u}_r(a,0) = 0 \tag{10.1.328}$$

then it follows from Equation 10.1.327 that $u = u(\varphi)$ satisfies the differential equation

$$u''(\varphi) + u(\varphi) = \frac{S_0(1-\overline{\nu})}{2\mu} a(1 - 6\cos 2\varphi) \tag{10.1.329}$$

subject to the initial conditions

$$u(0) = u'(0) = 0 \tag{10.1.330}$$

Here prime indicates derivatives with respect to φ.

It is easy to check that the only solution to Equation 10.1.329 that satisfies the conditions 10.1.330 is

$$u(\varphi) = \frac{S_0(1-\overline{\nu})}{2\mu} a(1 - 3\cos\varphi + 2\cos 2\varphi) \tag{10.1.331}$$

Therefore, by substituting $u(\varphi) \equiv \overline{u}_r(a,\varphi)$ from Equation 10.1.331 into Equations 10.1.319 and 10.1.322, and taking into account the condition 10.1.326, we get the single-valued components $\overline{u}_r$ and $\overline{u}_\varphi$:

$$\overline{u}_r = \frac{S_0 r}{4\mu}\left\{1 - 2\overline{\nu} + \frac{a^2}{r^2} - 6(1-\overline{\nu})\frac{a}{r}\cos\varphi \right.$$

$$\left. + \left[1 + 4(1-\overline{\nu})\frac{a^2}{r^2} - \frac{a^4}{r^4}\right]\cos 2\varphi\right\} \tag{10.1.332}$$

$$\overline{u}_\varphi = \frac{S_0 r}{4\mu}\left\{6(1-\overline{\nu})\frac{a}{r}\sin\varphi - \left[1 + 2(1-2\overline{\nu})\frac{a^2}{r^2} + \frac{a^4}{r^4}\right]\sin 2\varphi\right\} \tag{10.1.333}$$

Hence, a solution to Kirsch's problem is represented by an elastic state $s = [\overline{\mathbf{u}}, \overline{\mathbf{E}}, \overline{\mathbf{S}}]$ in which $\overline{\mathbf{u}}$, $\overline{\mathbf{E}}$, and $\overline{\mathbf{S}}$ are given by Equations 10.1.332 and 10.1.333, 10.1.315 through 10.1.317, and 10.1.307 through 10.1.309, respectively.

The closed-form solution to the problem allows us to analyze s for a large range of geometric and physical parameters. Most interesting is to consider extremum properties of the stress field $\overline{\mathbf{S}}$. It follows from Equations 10.1.307 through 10.1.309 that the maximum stress that occurs at $r = a$

$$\overline{S}_{\varphi\varphi}(a,\varphi) = S_0(1 - 2\cos 2\varphi) \tag{10.1.334}$$

is largest when $\varphi = \pi/2$ or $3\pi/2$, that is, at the sides of the hole where the tangent vector is parallel to the direction of the applied stress S_0,

$$\max_{\varphi\in[0,\pi/2]} \overline{S}_{\varphi\varphi}(a,\varphi) = \overline{S}_{\varphi\varphi}\left(a, \frac{\pi}{2}\right) = 3S_0 \tag{10.1.335}$$

Also, if follows from Equation 10.1.334 that

$$\min_{\varphi\in[0,\pi/2]} \overline{S}_{\varphi\varphi}(a,\varphi) = \overline{S}_{\varphi\varphi}(a,0) = -S_0 \tag{10.1.336}$$

The ratio

$$K = \frac{\overline{S}_{\varphi\varphi}(a, \pi/2)}{\overline{S}^{(1)}_{\varphi\varphi}(a, \pi/2)} = \frac{3S_0}{S_0} = 3 \tag{10.1.337}$$

is called the *stress concentration factor* for the problem. Clearly, K is independent of the radius of the hole, so the solution is a general one. Stress concentration problems based on elasticity theory are significant in a study of structural fatigue, which is a part of fracture mechanics.

Notes:

(1) The ratio

$$K_r = \frac{\overline{u}_r(a, \pi/2)}{\overline{u}^{(1)}_r(a, \pi/2)} = \frac{1}{\nu} \quad \left(0 < \nu < \frac{1}{2}\right) \tag{10.1.338}$$

may be called the *displacement concentration factor* for the problem. It follows from Equation 10.1.338 that $K_r \to +\infty$ as $\nu \to 0 + 0$.

(2) The method of finding solution to Kirsch's problem may be applied to determine an elastic state in the infinite sheet as shown in Figure 10.8.

The stress tensor $\overline{\mathbf{S}}$ corresponding to the situation shown in Figure 10.8 is given by

$$\overline{\mathbf{S}} = \overline{\mathbf{S}}^{*} + \overline{\mathbf{S}}^{**} \tag{10.1.339}$$

where $\overline{\mathbf{S}}^{*}$ is represented by Equations 10.1.307 through 10.1.309 in which S_0 is replaced by S_1 and $\overline{\mathbf{S}}^{**}$ is represented by Equations 10.1.307 through 10.1.309 in which S_0 is replaced by S_2 and φ by $\varphi + \pi/2$. Equation 10.1.339 in components takes the form

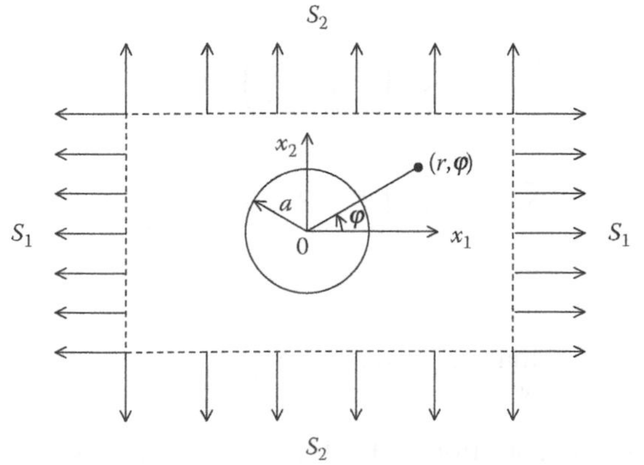

FIGURE 10.8 An infinite elastic sheet with a circular hole under tension in the two perpendicular directions.

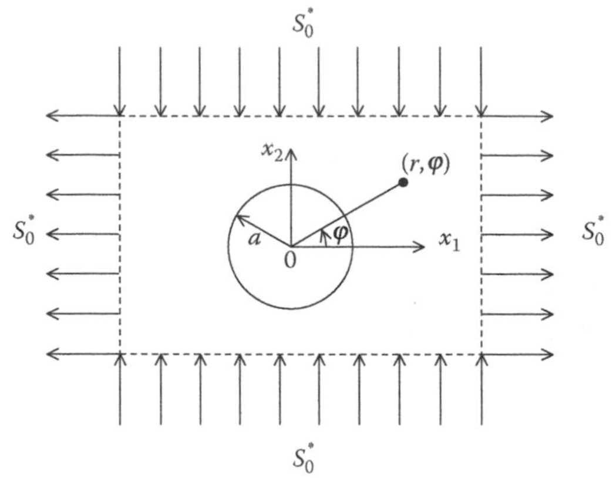

FIGURE 10.9 An infinite sheet with a circular hole under an equal tension in one direction and compression in the perpendicular direction.

$$\overline{S}_{rr} = \frac{S_1 + S_2}{2}\left(1 - \frac{a^2}{r^2}\right) + \frac{S_1 - S_2}{2}\left(1 - 4\frac{a^2}{r^2} + 3\frac{a^4}{r^4}\right)\cos 2\varphi \quad (10.1.340)$$

$$\overline{S}_{\varphi\varphi} = \frac{S_1 + S_2}{2}\left(1 + \frac{a^2}{r^2}\right) - \frac{S_1 - S_2}{2}\left(1 + 3\frac{a^4}{r^4}\right)\cos 2\varphi \quad (10.1.341)$$

$$\overline{S}_{r\varphi} = -\frac{S_1 - S_2}{2}\left(1 + 2\frac{a^2}{r^2} - 3\frac{a^4}{r^4}\right)\sin 2\varphi \quad (10.1.342)$$

The associated strain tensor $\overline{\mathbf{E}}$ and displacement vector $\overline{\mathbf{u}}$ may be obtained in a similar way.

(3) If an infinite sheet with a circular hole is subject to a uniform tension $S_1 = S_0^* > 0$ in the x_1 direction and a uniform pressure $S_2 = -S_0^* < 0$ in the x_2 direction, as shown in Figure 10.9 then, letting $S_1 = S_0^*$ and $S_2 = -S_0^*$ in Equations 10.1.340 through 10.1.342, we have

$$\overline{S}_{rr} = S_0^*\left(1 - 4\frac{a^2}{r^2} + 3\frac{a^4}{r^4}\right)\cos 2\varphi \quad (10.1.343)$$

$$\overline{S}_{\varphi\varphi} = -S_0^*\left(1 + 3\frac{a^4}{r^4}\right)\cos 2\varphi \quad (10.1.344)$$

$$\overline{S}_{r\varphi} = -S_0^*\left(1 + 2\frac{a^2}{r^2} - 3\frac{a^4}{r^4}\right)\sin 2\varphi \quad (10.1.345)$$

and

$$\max_{\varphi \in [0, \pi/2]} \overline{S}_{\varphi\varphi}(a, \varphi) = \overline{S}_{\varphi\varphi}(a, \pi/2) = 4S_0^* \quad (10.1.346)$$

Hence, the stress concentration factor for the situation shown in Figure 10.9 is

$$K^* = 4 \quad (10.1.347)$$

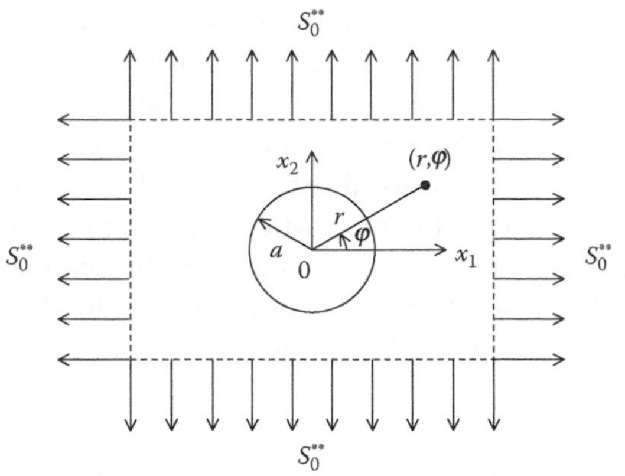

FIGURE 10.10 An infinite sheet with a circular hole under a uniform biaxial tension.

(4) For an infinite sheet as shown in Figure 10.10 by letting $S_1 = S_0^{**}$ and $S_2 = S_0^{**}$ in Equations 10.1.340 through 10.1.342, we have

$$\overline{S}_{rr} = S_0^{**}\left(1 - \frac{a^2}{r^2}\right) \tag{10.1.348}$$

$$\overline{S}_{\varphi\varphi} = S_0^{**}\left(1 + \frac{a^2}{r^2}\right) \tag{10.1.349}$$

$$\overline{S}_{r\varphi} = 0 \tag{10.1.350}$$

and the associated stress concentration factor

$$K^{**} = 2 \tag{10.1.351}$$

Also, it follows from Equations 10.1.348 and 10.1.349 and from the inequality $S_0^{**} > 0$ that

$$\overline{S}_{rr} \geq 0, \quad \overline{S}_{\varphi\varphi} > 0 \quad \text{for } r \geq a \tag{10.1.352}$$

If we let $S_0^{**} = -\hat{S}_0$ $(\hat{S}_0 > 0)$ in Equations 10.1.348 and 10.1.349, we obtain the stress in an infinite sheet with a circular hole under a uniform biaxial pressure. In this case the stress concentration factor is also equal to 2.

(5) It follows from Equations 10.1.307 through 10.1.309, 10.1.340 through 10.1.342, 10.1.343 through 10.1.345, and 10.1.348 and 10.1.349 that the stress concentration in a sheet with a circular hole is quite localized and decays fast with an increasing distance from the hole.

10.2 TWO-DIMENSIONAL SOLUTIONS OF NONISOTHERMAL ELASTOSTATICS

In Section 8.1.2, two-dimensional boundary value problems of nonisothermal elastostatics were formulated, and general properties of a solution of the theory were discussed. In particular, it was shown how stress field $S_{\alpha\beta}$ corresponding to a temperature T and to a plane strain state generated a stress field $\overline{S}_{\alpha\beta}$ corresponding to a temperature $\overline{T}$ and to a generalized plane stress state. In this section, we outline typical methods of solving the problems of nonisothermal elastostatics corresponding to a generalized plane stress state, and present a number of closed-form solutions for a body with stress free boundary. In Section 10.2.1, two methods of finding a solution to a boundary value problem for a thin elastic sheet subject to a two-dimensional temperature $\overline{T}$ are presented. The closed-form solutions are the following: (A) a thermoelastic state due to a discontinuous temperature field in an infinite sheet, (B) thermal stresses due to a discontinuous temperature field in a semi-infinite thermoelastic sheet, (C) a thermoelastic state due to a heat source at the center of a thin circular disk, and (D) thermal stresses due to a concentrated heat source in a semi-infinite elastic sheet.

10.2.1 Two Methods of Finding a Solution to a Boundary Value Problem for a Thin Thermoelastic Sheet

A thermoelastic state $s = [\overline{\mathbf{u}}, \overline{\mathbf{E}}, \overline{\mathbf{S}}]$ corresponding to a temperature $\overline{T} = \overline{T}(\mathbf{x})$ and generalized plane stress conditions with regard to (x_1, x_2) plane, obey the field equations (see Equations 8.1.71, 8.1.72, 8.1.79, 8.1.84, and 8.1.85)

The strain–displacement relations

$$\overline{E}_{\alpha\beta} = \frac{1}{2}\left(\overline{u}_{\alpha,\beta} + \overline{u}_{\beta,\alpha}\right) \quad \text{on } C_0 \tag{10.2.1}$$

The equilibrium equations

$$\overline{S}_{\alpha\beta,\beta} = 0 \quad \text{on } C_0 \tag{10.2.2}$$

The constitutive relations

$$\overline{S}_{\alpha\beta} = 2\mu\left(\overline{E}_{\alpha\beta} + \frac{\overline{\nu}}{1-2\overline{\nu}}\overline{E}_{\gamma\gamma}\delta_{\alpha\beta} - \frac{1}{1-2\overline{\nu}}\alpha\overline{T}\delta_{\alpha\beta}\right) \quad \text{on } C_0 \tag{10.2.3}$$

or

$$\overline{E}_{\alpha\beta} = \frac{1}{2\mu}\left(\overline{S}_{\alpha\beta} - \overline{\nu}\overline{S}_{\gamma\gamma}\delta_{\alpha\beta}\right) + \alpha\overline{T}\delta_{\alpha\beta} \quad \text{on } C_0 \tag{10.2.4}$$

where

$$\overline{\nu} = \frac{\nu}{1+\nu} \tag{10.2.5}$$

In Equations 10.2.3 and 10.2.4, $\overline{T} = \overline{T}(x_1, x_2)$ is a prescribed temperature field on C_0, where C_0 is a two-dimensional domain of (x_1, x_2) plane. In general, $\overline{T}$ satisfies Poisson's equation (see Equation 8.1.96)

$$\nabla^2\overline{T} = -\frac{Q}{\kappa} \quad \text{on } C_0 \tag{10.2.6}$$

Note that $Q/\kappa = W/k$, where Q is the prescribed heat supply field, W stands for the internal heat generated per unit volume per unit time, k denotes the thermal conductivity, and κ

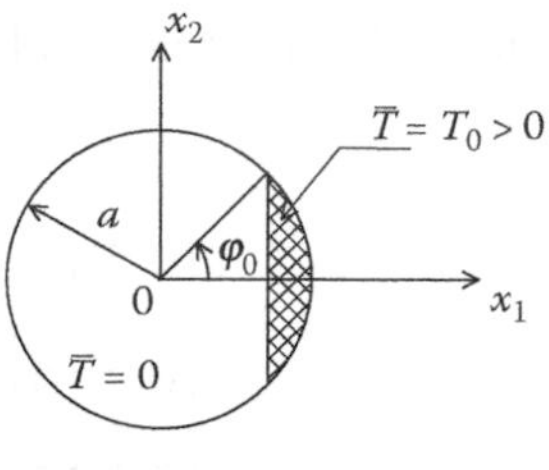

FIGURE 10.11 A thin circular disk of radius a under a discontinuous temperature field: $\overline{T}(r,\varphi) = T_0 H(r\cos\varphi - a\cos\varphi_0)$, where (r,φ) are polar coordinates and $H = H(x)$ is the Heaviside function.

means the thermal diffusivity. We recall that $\kappa = k/\rho c$, where ρ is the density of the material and c is the specific heat. When $Q = 0$ and $\overline{T}$ is piecewise constant on C_0, Equation 10.2.6 is satisfied identically on C_0 except for the discontinuity lines (see Figure 10.11).

A thermoelastic state $s = [\overline{\mathbf{u}}, \overline{\mathbf{E}}, \overline{\mathbf{S}}]$ corresponding to a temperature $\overline{T} = \overline{T}(\mathbf{x})$ on C_0 may be obtained by using the formula

$$s(\mathbf{x}) = \int_{C_0} s^*(\mathbf{x}, \boldsymbol{\xi})\overline{T}(\boldsymbol{\xi})d\xi_1 d\xi_2 \qquad (10.2.7)$$

where $s^* = [\overline{\mathbf{u}}^*, \overline{\mathbf{E}}^*, \overline{\mathbf{S}}^*]$ is a thermoelastic state corresponding to a temperature $\overline{T}^* = \overline{T}^*(\mathbf{x}, \boldsymbol{\xi})$ of the form

$$\overline{T}^*(\mathbf{x}, \boldsymbol{\xi}) = \delta(x_1 - \xi_1)\,\delta(x_2 - \xi_2) \quad \mathbf{x} \in C_0, \quad \boldsymbol{\xi} \in C_0 \qquad (10.2.8)$$

which represents a *nucleus of thermoelastic strain* located at a point (ξ_1, ξ_2).

If s^* satisfies the stress-free boundary condition

$$\overline{\mathbf{S}}^*(\mathbf{x}, \boldsymbol{\xi})\mathbf{n}(\mathbf{x}) = \mathbf{0} \quad \text{for } \mathbf{x} \in \partial C_0, \quad \boldsymbol{\xi} \in C_0 \qquad (10.2.9)$$

then s^* is a solution to a boundary value problem of two-dimensional nonisothermal elastostatics corresponding to the temperature field $\overline{T}^*$ and the stress free boundary ∂C_0, while s is a solution to the problem described by Equations 10.2.1 through 10.2.3 and the boundary condition

$$\overline{\mathbf{S}}(\mathbf{x})\mathbf{n}(\mathbf{x}) = \mathbf{0} \quad \text{for } \mathbf{x} \in \partial C_0 \qquad (10.2.10)$$

In the following, we confine ourselves to the boundary value problems with a stress free boundary only. Since the method of finding a thermoelastic state s^* that corresponds to the temperature $\overline{T}^*$ and the boundary condition 10.2.9 is similar to that of finding s corresponding to the temperature $\overline{T}$ and the boundary condition 10.2.10, first we outline *a thermoelastic displacement potential method* of finding s. To this end we let

$$s = s_1 + s_2 \qquad (10.2.11)$$

where $s_1 = [\overline{\mathbf{u}}^{(1)}, \overline{\mathbf{E}}^{(1)}, \overline{\mathbf{S}}^{(1)}]$ is generated from a potential $\overline{\phi} = \overline{\phi}(\mathbf{x})$ by the formula

$$\overline{u}_\alpha^{(1)} = \overline{\phi}_{,\alpha} \qquad (10.2.12)$$

and $s_2 = [\overline{\mathbf{u}}^{(2)}, \overline{\mathbf{E}}^{(2)}, \overline{\mathbf{S}}^{(2)}]$ is a solution to a two-dimensional isothermal boundary value problem on C_0 selected in such a way that s satisfies Equations 10.2.1 through 10.2.3 and 10.2.10.

Substituting $\overline{u}_\alpha^{(1)}$ given by Equation 10.2.12 into Equations 10.2.1 through 10.2.3 we find that these equations are satisfied provided

$$\overline{E}_{\alpha\beta}^{(1)} = \overline{\phi}_{,\alpha\beta} \tag{10.2.13}$$

$$\overline{S}_{\alpha\beta}^{(1)} = 2\mu\left(\overline{\phi}_{,\alpha\beta} - \nabla^2\overline{\phi}\,\delta_{\alpha\beta}\right) \tag{10.2.14}$$

$$\nabla^2\overline{\phi} = m_0\overline{T} \tag{10.2.15}$$

where

$$m_0 = (1+\nu)\alpha \tag{10.2.16}$$

Therefore, s_1 may be generated by Equations 10.2.12 through 10.2.14 in which the thermoelastic displacement potential $\overline{\phi}$ satisfies Poisson's equation 10.2.15.

To obtain $s_2 = [\overline{\mathbf{u}}^{(2)}, \overline{\mathbf{E}}^{(2)}, \overline{\mathbf{S}}^{(2)}]$ we note that s_2 satisfies Equations 10.2.1 through 10.2.3 with $\overline{T} = 0$ on C_0 subject to the boundary condition

$$\overline{\mathbf{S}}^{(2)}\mathbf{n} = -\overline{\mathbf{S}}^{(1)}\mathbf{n} \quad \text{on } \partial C_0 \tag{10.2.17}$$

It follows from Example 8.1.4 that s_2 may be generated by an Airy stress function $\overline{F} = \overline{F}(x_1, x_2)$ that satisfies the biharmonic equation

$$\nabla^2\nabla^2\overline{F} = 0 \quad \text{on } C_0 \tag{10.2.18}$$

subject to the boundary condition

$$(\delta_{\alpha\beta}\nabla^2\overline{F} - \overline{F}_{,\alpha\beta})n_\beta = -\overline{S}_{\alpha\beta}^{(1)}n_\beta \quad \text{on } \partial C_0 \tag{10.2.19}$$

Since

$$\overline{S}_{\alpha\beta}^{(2)} = \delta_{\alpha\beta}\nabla^2\overline{F} - \overline{F}_{,\alpha\beta} \quad \text{on } C_0 \tag{10.2.20}$$

therefore, the stress tensor $\overline{S}_{\alpha\beta}$ corresponding to the thermoelastic state s is expressed in terms of $\overline{\phi}$ and $\overline{F}$ in the form

$$\overline{S}_{\alpha\beta} = \left(\delta_{\alpha\beta}\nabla^2 - \frac{\partial^2}{\partial x_\alpha \partial x_\beta}\right)(\overline{F} - 2\mu\overline{\phi}) \tag{10.2.21}$$

Once $\overline{S}_{\alpha\beta}$ has been found, $\overline{E}_{\alpha\beta}$ is computed from Equations 10.2.4, and $\overline{u}_\alpha$ is obtained by integration of Equation 10.2.1 in which $\overline{E}_{\alpha\beta}$ are known functions.

An alternative method of finding s that satisfies Equations 10.2.1 through 10.2.3 and 10.2.10 is based on solving the following boundary value problem (see Example 8.1.10).

Find a function $\overline{\chi} = \overline{\chi}(x_1, x_2)$ on C_0 that satisfies the equation

$$\nabla^2\nabla^2\overline{\chi} = -E\alpha\nabla^2\overline{T} \quad \text{on } C_0 \tag{10.2.22}$$

subject to the boundary condition

$$(\nabla^2\overline{\chi}\,\delta_{\alpha\beta} - \overline{\chi}_{,\alpha\beta})n_\beta = 0 \quad \text{on } \partial C_0 \tag{10.2.23}$$

If a solution to the problem 10.2.22 and 10.2.23 is found, the stress tensor $\overline{S}_{\alpha\beta}$ corresponding to the thermoelastic state s is obtained from the equation

$$\overline{S}_{\alpha\beta} = \nabla^2\overline{\chi}\,\delta_{\alpha\beta} - \overline{\chi}_{,\alpha\beta} \tag{10.2.24}$$

The associated strain components $\overline{E}_{\alpha\beta}$ are then computed from Equations 10.2.4, and the displacement components $\overline{u}_\alpha$ are obtained by integration of Equations 10.2.1 in which $\overline{E}_{\alpha\beta}$ are known functions.

It may be shown (see Equations 8.1.105 and 8.1.106) that if the temperature $\overline{T}$ satisfies Poisson's equation 10.2.6, then the problem described by Equations 10.2.22 and 10.2.23 is equivalent to finding a function $\overline{\chi} = \overline{\chi}(x_1, x_2)$ that satisfies the equation

$$\nabla^2\nabla^2\overline{\chi} = \frac{E\alpha Q}{\kappa} \quad \text{on } C_0 \tag{10.2.25}$$

subject to the boundary conditions

$$\overline{\chi} = \frac{\partial\overline{\chi}}{\partial n} = 0 \quad \text{on } \partial C_0 \tag{10.2.26}$$

Hence, to find the stress tensor $\overline{S}_{\alpha\beta}$ corresponding to s one may use either Equation 10.2.21 in which $\overline{\phi}$ and $\overline{F}$ are suitably selected, or Equation 10.2.24 in which $\overline{\chi}$ is a solution to the boundary value problem 10.2.25 and 10.2.26.

10.2.2 A Thermoelastic State due to a Discontinuous Temperature Field in an Infinite Sheet

Consider a homogeneous isotropic infinite thermoelastic sheet described by the inequalities

$$|x_1| < \infty, \quad |x_2| < \infty \tag{10.2.27}$$

and subject to the temperature field (see Figure 10.12).

$$\overline{T}(x_1, x_2) = T_0[H(x_1 + a_1) - H(x_1 - a_1)] \cdot [H(x_2 + a_2) - H(x_2 - a_2)] \tag{10.2.28}$$

where T_0 is a constant temperature and $H = H(x)$ is the Heaviside function.

To obtain a thermoelastic state produced by the temperature $\overline{T}$ shown in Figure 10.12 we use the formula (see Equation 10.2.7)

$$s(\mathbf{x}) = T_0 \int_{-a_1}^{a_1} \int_{-a_2}^{a_2} s^*(\mathbf{x}, \boldsymbol{\xi})\, d\xi_1 d\xi_2 \tag{10.2.29}$$

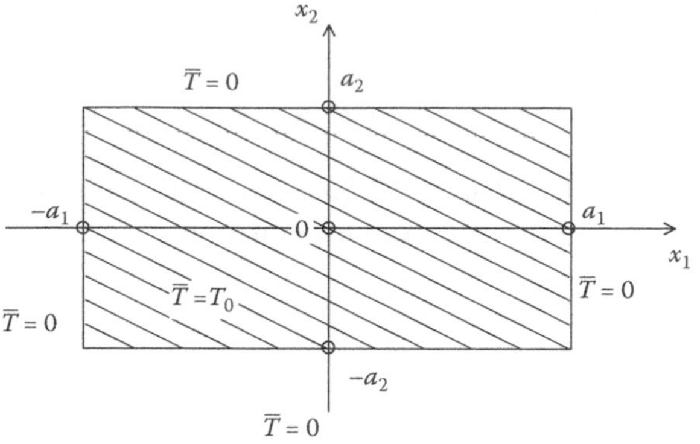

FIGURE 10.12 The rectangular distribution of a temperature field $\overline{T}$ in an infinite thermoelastic sheet.

where $s^* = s^*(\mathbf{x}, \boldsymbol{\xi})$ is a thermoelastic state corresponding to the temperature $\overline{T}^*(\mathbf{x}, \boldsymbol{\xi}) = \delta(x_1 - \xi_1)\delta(x_2 - \xi_2)$ in an infinite sheet. Since s^* is defined on the whole (x_1, x_2) plane, therefore, the domain C_0 of the previous section is the whole plane and s^* may be found by the thermoelastic displacement potential method in which $(\overline{\phi}, \overline{F}) = (\overline{\phi}^*, 0)$, and $\overline{\phi}^*$ satisfies Poisson's equation

$$\nabla^2 \overline{\phi}^* = m_0 \delta(x_1 - \xi_1)\, \delta(x_2 - \xi_2) \tag{10.2.30}$$

Here

$$\nabla^2 = \frac{\partial^2}{\partial x_1^2} + \frac{\partial^2}{\partial x_2^2} \tag{10.2.31}$$

$\mathbf{x} = (x_1, x_2) \in E^2$, and $\boldsymbol{\xi} = (\xi_1, \xi_2)$ is a fixed point of E^2.

If we note that a solution to Equation 10.2.30 is given by (see Equation 10.1.18)

$$\overline{\phi}^*(\mathbf{x}, \boldsymbol{\xi}) = -\frac{m_0}{2\pi} \ln \frac{L}{\overline{r}} \tag{10.2.32}$$

where

$$\overline{r} = \sqrt{(x_1 - \xi_1^2) + (x_2 - \xi_2)^2} \tag{10.2.33}$$

and L is a positive constant of the length dimension, then the thermoelastic state $s^* = s^*(\mathbf{x}, \boldsymbol{\xi})$ is obtained in the form (see Equations 10.2.12 through 10.2.14 in which $\overline{\phi}$ is replaced by $\overline{\phi}^*$)

$$\overline{u}_\alpha^*(\mathbf{x}, \boldsymbol{\xi}) = -\frac{m_0}{2\pi} \frac{\partial}{\partial x_\alpha} \ln \frac{L}{\overline{r}} \tag{10.2.34}$$

$$\overline{E}_{\alpha\beta}^*(\mathbf{x}, \boldsymbol{\xi}) = -\frac{m_0}{2\pi} \frac{\partial^2}{\partial x_\alpha \partial x_\beta} \ln \frac{L}{\overline{r}} \tag{10.2.35}$$

$$\overline{S}_{\alpha\beta}(\mathbf{x}, \boldsymbol{\xi}) = 2\mu \left(-\frac{m_0}{2\pi}\right) \left(\frac{\partial^2}{\partial x_\alpha \partial x_\beta} - \delta_{\alpha\beta} \nabla^2\right) \ln \frac{L}{\overline{r}} \tag{10.2.36}$$

Next, using the relation

$$\frac{\partial}{\partial x_\alpha}\overline{r} = -\frac{\partial}{\partial \xi_\alpha}\overline{r} \tag{10.2.37}$$

we reduce Equations 10.2.34 through 10.2.36 to the form useful when computing the integral 10.2.29:

$$\overline{u}_\alpha^*(\mathbf{x},\boldsymbol{\xi}) = -\frac{m_0}{2\pi}\frac{\partial}{\partial \xi_\alpha}\ln\left(\frac{\overline{r}}{L}\right) \tag{10.2.38}$$

$$\overline{E}_{\alpha\beta}^*(\mathbf{x},\boldsymbol{\xi}) = \frac{m_0}{2\pi}\frac{\partial^2}{\partial \xi_\alpha \partial \xi_\beta}\ln\left(\frac{\overline{r}}{L}\right) \tag{10.2.39}$$

$$\overline{S}_{\alpha\beta}^*(\mathbf{x},\boldsymbol{\xi}) = \frac{m_0\mu}{\pi}\left(\frac{\partial^2}{\partial \xi_\alpha \partial \xi_\beta} - \delta_{\alpha\beta}\nabla^2\right)\ln\left(\frac{\overline{r}}{L}\right) \tag{10.2.40}$$

where

$$\nabla^2 = \frac{\partial^2}{\partial \xi_1^2} + \frac{\partial^2}{\partial \xi_2^2} \tag{10.2.41}$$

To obtain the displacement field $\overline{u}_\alpha = \overline{u}_\alpha(\mathbf{x})$ corresponding to s, we let $\alpha = 1$ and from Equations 10.2.29 and 10.2.38 we write

$$\begin{aligned}\overline{u}_1(\mathbf{x}) &= T_0\int_{-a_2}^{a_2} d\xi_2 \int_{-a_1}^{a_1} d\xi_1 \overline{u}_1^*(\mathbf{x},\boldsymbol{\xi}) = -\frac{m_0T_0}{2\pi}\int_{-a_2}^{a_2} d\xi_2 \int_{-a_1}^{a_1} d\xi_1 \frac{\partial}{\partial \xi_1}\ln\left(\frac{\overline{r}}{L}\right)\\ &= -\frac{m_0T_0}{2\pi}\left\{\int_{-a_2}^{a_2}\left[\ln\sqrt{\left(\frac{x_1-a_1}{L}\right)^2 + \left(\frac{x_2-\xi_2}{L}\right)^2}\right.\right.\\ &\left.\left. -\ln\sqrt{\left(\frac{x_1+a_1}{L}\right)^2 + \left(\frac{x_2-\xi_2}{L}\right)^2}\right]\right\} d\xi_2\end{aligned} \tag{10.2.42}$$

Now

$$\int_{-a_2}^{a_2}\ln\sqrt{\left(\frac{x_1-a_1}{L}\right)^2 + \left(\frac{x_2-\xi_2}{L}\right)^2}\, d\xi_2 = L\int_{(x_2-a_2)/L}^{(x_2+a_2)/L}\ln\sqrt{u^2+\beta^2}\, du \tag{10.2.43}$$

where

$$\beta = \frac{x_1-a_1}{L} \tag{10.2.44}$$

Since

$$\int \ln\sqrt{u^2+\beta^2}\,du = u(\ln\sqrt{u^2+\beta^2}-1)+\beta\tan^{-1}\frac{u}{\beta} \tag{10.2.45}$$

therefore, the integral 10.2.43 becomes

$$\begin{aligned}
&\int_{-a_2}^{a_2} \ln\sqrt{\left(\frac{x_1-a_1}{L}\right)^2+\left(\frac{x_2-\xi_2}{L}\right)^2}\,d\xi_2 \\
&= (x_2+a_2)\left[\ln\sqrt{\left(\frac{x_2+a_2}{L}\right)^2+\left(\frac{x_1-a_1}{L}\right)^2}-1\right]+(x_1-a_1)\tan^{-1}\frac{x_2+a_2}{x_1-a_1} \\
&\quad -(x_2-a_2)\left[\ln\sqrt{\left(\frac{x_2-a_2}{L}\right)^2+\left(\frac{x_1-a_1}{L}\right)^2}-1\right]-(x_1-a_1)\tan^{-1}\frac{x_2-a_2}{x_1-a_1}
\end{aligned} \tag{10.2.46}$$

By replacing a_1 by $-a_1$ in this equation we get

$$\begin{aligned}
&\int_{-a_2}^{a_2} \ln\sqrt{\left(\frac{x_1+a_1}{L}\right)^2+\left(\frac{x_2-\xi_2}{L}\right)^2}\,d\xi_2 \\
&= (x_2+a_2)\left[\ln\sqrt{\left(\frac{x_1+a_1}{L}\right)^2+\left(\frac{x_2+a_2}{L}\right)^2}-1\right]+(x_1+a_1)\tan^{-1}\frac{x_2+a_2}{x_1+a_1} \\
&\quad -(x_2-a_2)\left[\ln\sqrt{\left(\frac{x_1+a_1}{L}\right)^2+\left(\frac{x_2-a_2}{L}\right)^2}-1\right]-(x_1+a_1)\tan^{-1}\frac{x_2-a_2}{x_1+a_1}
\end{aligned} \tag{10.2.47}$$

Hence, Equations 10.2.42, and 10.2.46 and 10.2.47 imply the closed-form formula for $\overline{u}_1(\mathbf{x})$ in which L is absent

$$\begin{aligned}
\overline{u}_1(\mathbf{x}) = -\frac{m_0T_0}{2\pi}\Bigg[&(x_2+a_2)\ln\frac{\sqrt{(x_1-a_1)^2+(x_2+a_2)^2}}{\sqrt{(x_1+a_1)^2+(x_2+a_2)^2}} \\
&-(x_2-a_2)\ln\frac{\sqrt{(x_1-a_1)^2+(x_2-a_2)^2}}{\sqrt{(x_1+a_1)^2+(x_2-a_2)^2}} \\
&+(x_1-a_1)\left(\tan^{-1}\frac{x_2+a_2}{x_1-a_1}-\tan^{-1}\frac{x_2-a_2}{x_1-a_1}\right) \\
&-(x_1+a_1)\left(\tan^{-1}\frac{x_2+a_2}{x_1+a_1}-\tan^{-1}\frac{x_2-a_2}{x_1+a_1}\right)\Bigg]
\end{aligned} \tag{10.2.48}$$

By replacing the index 1 by 2 and the index 2 by 1 in Equation 10.2.48 we arrive at the closed-form formula for $\overline{u}_2(\mathbf{x})$:

$$\overline{u}_2(\mathbf{x}) = -\frac{m_0 T_0}{2\pi}\left[(x_1+a_1)\ln\frac{\sqrt{(x_1+a_1)^2+(x_2-a_2)^2}}{\sqrt{(x_1+a_1)^2+(x_2+a_2)^2}}\right.$$
$$-\ (x_1-a_1)\ln\frac{\sqrt{(x_1-a_1)^2+(x_2-a_2)^2}}{\sqrt{(x_1-a_1)^2+(x_2+a_2)^2}}$$
$$+\ (x_2-a_2)\left(\tan^{-1}\frac{x_1+a_1}{x_2-a_2}-\tan^{-1}\frac{x_1-a_1}{x_2-a_2}\right)$$
$$\left.-\ (x_2+a_2)\left(\tan^{-1}\frac{x_1+a_1}{x_2+a_2}-\tan^{-1}\frac{x_1-a_1}{x_2+a_2}\right)\right] \tag{10.2.49}$$

Note that Equation 10.2.49 may also be obtained by using Equations 10.2.29 and 10.2.38 with $\alpha = 2$, and by taking the steps similar to those leading to Equation 10.2.48.

To compute the strain components $\overline{E}_{\alpha\beta}$ we note that by Equations 10.2.39 and 10.2.40

$$\overline{E}^*_{11}(\mathbf{x},\boldsymbol{\xi}) = -\frac{1}{2\mu}\overline{S}^*_{22}(\mathbf{x},\boldsymbol{\xi}) \tag{10.2.50}$$

$$\overline{E}^*_{22}(\mathbf{x},\boldsymbol{\xi}) = -\frac{1}{2\mu}\overline{S}^*_{11}(\mathbf{x},\boldsymbol{\xi}) \tag{10.2.51}$$

$$\overline{E}^*_{12}(\mathbf{x},\boldsymbol{\xi}) = \frac{1}{2\mu}\overline{S}^*_{12}(\mathbf{x},\boldsymbol{\xi}) \tag{10.2.52}$$

Hence, because of Equation 10.2.29, we have

$$\overline{E}_{11}(\mathbf{x}) = -\frac{1}{2\mu}\overline{S}_{22}(\mathbf{x}) \tag{10.2.53}$$

$$\overline{E}_{22}(\mathbf{x}) = -\frac{1}{2\mu}\overline{S}_{11}(\mathbf{x}) \tag{10.2.54}$$

$$\overline{E}_{12}(\mathbf{x}) = \frac{1}{2\mu}\overline{S}_{12}(\mathbf{x}) \tag{10.2.55}$$

Therefore, $\overline{E}_{\alpha\beta}$ may be computed from Equations 10.2.53 through 10.2.55 if $\overline{S}_{\alpha\beta}$ is known.

To obtain $\overline{S}_{\alpha\beta}$ we begin with $\overline{S}_{11}$. From Equation 10.2.40 with $\alpha = \beta = 1$ follows

$$\overline{S}^*_{11}(\mathbf{x},\boldsymbol{\xi}) = -\frac{m_0\mu}{\pi}\frac{\partial^2}{\partial\xi_2^2}\ln\left(\frac{\overline{r}}{L}\right) \tag{10.2.56}$$

Hence and by Equation 10.2.29, we have

$$\overline{S}_{11}(\mathbf{x}) = T_0\int_{-a_1}^{a_1} d\xi_1\int_{-a_2}^{a_2} d\xi_2\,\overline{S}^*_{11}(\mathbf{x},\boldsymbol{\xi})$$
$$= -\frac{m_0 T_0\mu}{\pi}\int_{-a_1}^{a_1} d\xi_1\int_{-a_2}^{a_2} d\xi_2\,\frac{\partial^2}{\partial\xi_2^2}\ln\left(\frac{\overline{r}}{L}\right) \tag{10.2.57}$$

Since

$$\int_{-a_2}^{a_2} d\xi_2 \frac{\partial^2}{\partial \xi_2^2} \ln\left(\frac{\bar{r}}{L}\right) = \left[\frac{\partial}{\partial \xi_2} \ln\left(\frac{\bar{r}}{L}\right)\right]_{\xi_2=-a_2}^{\xi_2=a_2} = -\left[\frac{(x_2-\xi_2)}{\bar{r}^2}\right]_{\xi_2=-a_2}^{\xi_2=a_2}$$

$$= -\left[\frac{(x_2-a_2)}{(x_1-\xi_1)^2+(x_2-a_2)^2} - \frac{(x_2+a_2)}{(x_1-\xi_1)^2+(x_2+a_2)^2}\right] \tag{10.2.58}$$

therefore, it follows from Equation 10.2.57 that

$$\overline{S}_{11}(\mathbf{x}) = \frac{m_0 T_0 \mu}{\pi}\left[(x_2-a_2)\int_{-a_1}^{a_1} \frac{d\xi_1}{(x_1-\xi_1)^2+(x_2-a_2)^2}\right.$$

$$\left. - (x_2+a_2)\int_{-a_1}^{a_1} \frac{d\xi_1}{(x_1-\xi_1)^2+(x_2+a_2)^2}\right] \tag{10.2.59}$$

Since

$$\int_{-a_1}^{a_1} \frac{d\xi_1}{(x_1-\xi_1)^2+(x_2-a_2)^2} = \int_{x_1-a_1}^{x_1+a_1} \frac{du}{u^2+(x_2-a_2)^2}$$

$$= \frac{1}{(x_2-a_2)}\left(\tan^{-1}\frac{x_1+a_1}{x_2-a_2} - \tan^{-1}\frac{x_1-a_1}{x_2-a_2}\right) \tag{10.2.60}$$

and

$$\int_{-a_1}^{a_1} \frac{d\xi_1}{(x_1-\xi_1)^2+(x_2+a_2)^2} = \int_{x_1-a_1}^{x_1+a_1} \frac{du}{u^2+(x_2+a_2)^2}$$

$$= \frac{1}{(x_2+a_2)}\left(\tan^{-1}\frac{x_1+a_1}{x_2+a_2} - \tan^{-1}\frac{x_1-a_1}{x_2+a_2}\right) \tag{10.2.61}$$

therefore, Equation 10.2.59 yields

$$\overline{S}_{11}(\mathbf{x}) = \frac{\mu m_0 T_0}{\pi}\left(\tan^{-1}\frac{x_1+a_1}{x_2-a_2} - \tan^{-1}\frac{x_1-a_1}{x_2-a_2}\right.$$

$$\left. -\tan^{-1}\frac{x_1+a_1}{x_2+a_2} - \tan^{-1}\frac{x_1-a_1}{x_2+a_2}\right) \tag{10.2.62}$$

In a similar way, we obtain

$$\overline{S}_{22}(\mathbf{x}) = \frac{\mu m_0 T_0}{\pi}\left(\tan^{-1}\frac{x_2+a_2}{x_1-a_1} - \tan^{-1}\frac{x_2-a_2}{x_1-a_1}\right.$$
$$\left. - \tan^{-1}\frac{x_2+a_2}{x_1+a_1} - \tan^{-1}\frac{x_2-a_2}{x_1+a_1}\right) \quad (10.2.63)$$

Note that Equation 10.2.63 is obtained from 10.2.62 in which the index 1 is replaced by 2, and the index 2 by 1.

Finally, Equation 10.2.40 with $\alpha = 1, \beta = 2$ leads to

$$\overline{S}^*_{12}(\mathbf{x}, \boldsymbol{\xi}) = \frac{\mu m_0 T_0}{\pi}\frac{\partial^2}{\partial \xi_1 \partial \xi_2}\ln\left(\frac{\overline{r}}{L}\right) \quad (10.2.64)$$

and, by Equation 10.2.29, we have

$$\overline{S}_{12}(\mathbf{x}) = \frac{\mu m_0 T_0}{\pi}\int_{-a_1}^{a_1} d\xi_1 \frac{\partial}{\partial \xi_1}\int_{-a_2}^{a_2} d\xi_2 \frac{\partial}{\partial \xi_2}\ln\left(\frac{\overline{r}}{L}\right) \quad (10.2.65)$$

or

$$\overline{S}_{12}(\mathbf{x}) = \frac{\mu m_0 T_0}{\pi}\ln\left[\frac{\sqrt{(x_1-a_1)^2+(x_2-a_2)^2}}{\sqrt{(x_1-a_1)^2+(x_2+a_2)^2}}\cdot\frac{\sqrt{(x_1+a_1)^2+(x_2+a_2)^2}}{\sqrt{(x_1+a_1)^2+(x_2-a_2)^2}}\right] \quad (10.2.66)$$

The thermoelastic state $s = s(\mathbf{x})$ corresponding to the rectangular distribution of temperature shown in Figure 10.12 is now represented by

$$s(\mathbf{x}) = [\overline{\mathbf{u}}(\mathbf{x}), \overline{\mathbf{E}}(\mathbf{x}), \overline{\mathbf{S}}(\mathbf{x})] \quad (10.2.67)$$

where

$$\overline{\mathbf{u}}(\mathbf{x}) = [\overline{u}_1(\mathbf{x}), \overline{u}_2(\mathbf{x})] \quad (10.2.68)$$

and $\overline{u}_1$ and $\overline{u}_2$ are given by Equations 10.2.48 and 10.2.49, respectively,

$$\overline{\mathbf{E}}(\mathbf{x}) = \begin{bmatrix} \overline{E}_{11}(\mathbf{x}) & \overline{E}_{12}(\mathbf{x}) \\ \overline{E}_{12}(\mathbf{x}) & \overline{E}_{22}(\mathbf{x}) \end{bmatrix} \quad (10.2.69)$$

where $\overline{E}_{\alpha\beta}(\mathbf{x})$, $(\alpha, \beta = 1, 2)$ are given by Equations 10.2.53 through 10.2.55, and

$$\overline{S}(\mathbf{x}) = \begin{bmatrix} \overline{S}_{11}(\mathbf{x}) & \overline{S}_{12}(\mathbf{x}) \\ \overline{S}_{12}(\mathbf{x}) & \overline{S}_{22}(\mathbf{x}) \end{bmatrix} \quad (10.2.70)$$

where $\overline{S}_{11}$, $\overline{S}_{22}$, and $\overline{S}_{12}$ are given by Equations 10.2.62, 10.2.63, and 10.2.66, respectively.

It follows from Equations 10.2.62 and 10.2.63 and from the identity

$$\tan^{-1}(z) + \tan^{-1}\left(\frac{1}{z}\right) = \frac{\pi}{2} \quad \text{for } z > 0 \tag{10.2.71}$$

that

$$\overline{S}_{11}(\mathbf{x}) + \overline{S}_{22}(\mathbf{x}) = -2\mu m_0 \overline{T}(\mathbf{x}) \tag{10.2.72}$$

where

$$\overline{T}(\mathbf{x}) = T_0[H(x_1 + a_1) - H(x_1 - a_1)][H(x_2 + a_2) - H(x_2 - a_2)] \tag{10.2.73}$$

This result complies with the thermoelastic displacement potential method in which $\overline{u}_\alpha = \overline{\phi}_{,\alpha}$, and

$$\overline{S}_{\gamma\gamma} = -2\mu\nabla^2\overline{\phi} = -2\mu m_0 \overline{T} \tag{10.2.74}$$

Also, it follows from Equations 10.2.62 and 10.2.63 that a hoop stress exhibits a jump across a temperature discontinuity line. To prove this let us consider the rectangular region of (x_1, x_2) plane as shown in Figure 10.13 and concentrate on the temperature discontinuity line $x_1 = a_1$, $|x_2| < a_2$. For such a line the hoop stress is $\overline{S}_{22}(\mathbf{x})$ and the normal to the line is $\mathbf{n} = [1, 0]$. To prove that $\overline{S}_{22}$ suffers a jump at a particular point of the line, we confine ourselves to the point $(a_1, 0)$, and introduce the notations

$$\overline{S}_{22}^{+} = \overline{S}_{22}(a_1 + 0, 0), \quad \overline{S}_{22}^{-} = \overline{S}_{22}(a_1 - 0, 0) \tag{10.2.75}$$

A jump of $\overline{S}_{22}$ at $(a_1, 0)$ is then defined by

$$[\![\overline{S}_{22}]\!](a_1, 0) = \overline{S}_{22}^{+} - \overline{S}_{22}^{-} \tag{10.2.76}$$

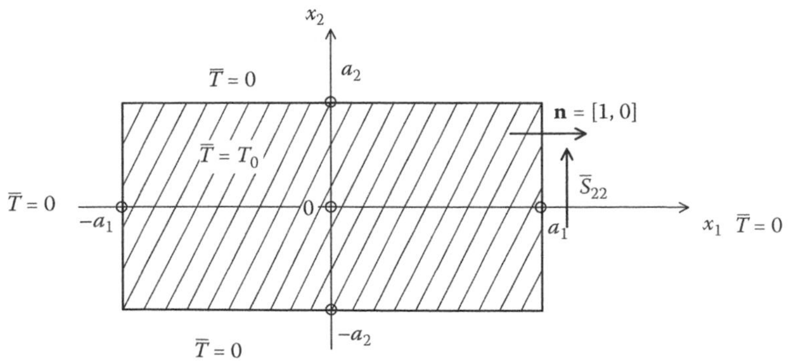

FIGURE 10.13 The temperature discontinuity line $x_1 = a_1$, $|x_2| < a_2$ as a discontinuity line for $\overline{S}_{22}$.

To compute $\overline{S}_{22}^{+}$ we let $(x_1, x_2) = (a_1 + 0, 0)$ in Equation 10.2.63 and obtain

$$\overline{S}_{22}^{+} = \frac{\mu m_0 T_0}{\pi}\left[\tan^{-1}(\infty) + \tan^{-1}(\infty) - \tan^{-1}\left(\frac{a_2}{2a_1}\right) - \tan^{-1}\left(\frac{a_2}{2a_1}\right)\right]$$
$$= \frac{\mu m_0 T_0}{\pi}\left[\pi - 2\tan^{-1}\left(\frac{a_2}{2a_1}\right)\right] \quad (10.2.77)$$

Similarly, by letting $(x_1, x_2) = (a_1 - 0, 0)$ in Equation 10.2.63 we obtain

$$\overline{S}_{22}^{-} = \frac{\mu m_0 T_0}{\pi}\left[-\tan^{-1}(\infty) - \tan^{-1}(\infty) - \tan^{-1}\left(\frac{a_2}{2a_1}\right) - \tan^{-1}\left(\frac{a_2}{2a_1}\right)\right]$$
$$= \frac{\mu m_0 T_0}{\pi}\left[-\pi - 2\tan^{-1}\left(\frac{a_2}{2a_1}\right)\right] \quad (10.2.78)$$

Hence, Equations 10.2.76 through 10.2.78 lead to

$$[\![\overline{S}_{22}]\!](a_1, 0) = 2\mu m_0 T_0 \quad (10.2.79)$$

This completes the proof that the line $x_1 = a_1$, $|x_2| < a_2$ is a discontinuity line for $\overline{S}_{22}$.

An inspection of $\overline{S}_{11}$ given by Equation 10.2.62 indicates that $\overline{S}_{11}(a_1 - 0, 0) = \overline{S}_{11}(a_1 + 0, 0)$, that is, $\overline{S}_{11}(\mathbf{x})$ is a continuous function at $(a_1, 0)$. Hence, taking the jump of Equation 10.2.72 at the point $(a_1, 0)$ and using Equation 10.2.79 yields

$$[\![\overline{S}_{\gamma\gamma}]\!](a_1, 0) = 2\mu m_0 T_0 \quad (10.2.80)$$

Also, it follows from Equations 10.2.53 and 10.2.79 that

$$[\![\overline{E}_{11}]\!](a_1, 0) = -m_0 T_0 \quad (10.2.81)$$

Finally, it follows from Equations 10.2.62 and 10.2.63 and 10.2.66 that if $a_1 = a_2 = a > 0$, then the following behavior of $\overline{S}_{\alpha\beta}$ in a neighborhood of the corner $(x_1, x_2) = (a, a)$ is observed:

$$\overline{S}_{11}(\mathbf{x}) \to 0, \quad \overline{S}_{22}(\mathbf{x}) \to 0, \quad \overline{S}_{12}(\mathbf{x}) \to -\infty \quad \text{as } (x_1, x_2) \to (a + 0, a + 0) \quad (10.2.82)$$

10.2.3 Thermal Stresses due to a Discontinuous Temperature Field in a Semi-Infinite Thermoelastic Sheet

10.2.3.1 Thermal Stresses due to a Nucleus of Thermoelastic Strain in a Semi-Infinite Sheet

Consider a homogeneous isotropic semi-infinite sheet: $|x_1| < \infty$, $x_2 \geq 0$ with a stress free boundary $x_2 = 0$ in which a nucleus of thermoelastic strain is acting at the point $(0, \xi_2)$ $(\xi_2 > 0)$ (see Figure 10.14)

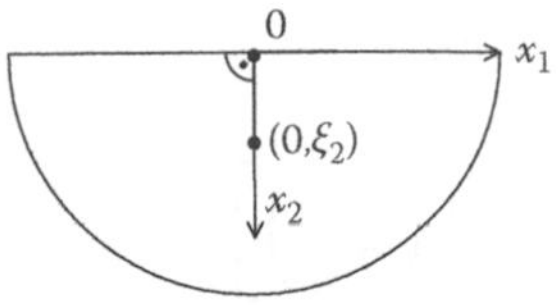

FIGURE 10.14 A nucleus of thermoelastic strain at a point $(0, \xi_2)$.

Therefore, the temperature $\overline{T}^*$ of the semi-infinite solid takes the form

$$\overline{T}^*(\mathbf{x}, \boldsymbol{\xi}) = \delta(x_1)\delta(x_2 - \xi_2) \tag{10.2.83}$$

and the boundary conditions read

$$\overline{S}^*_{22}(x_1, 0; \boldsymbol{\xi}) = \overline{S}^*_{12}(x_1, 0; \boldsymbol{\xi}) = 0 \quad \text{for every} \quad |x_1| < \infty \quad \text{and } \boldsymbol{\xi} = (0, \xi_2) \tag{10.2.84}$$

Here, $\overline{\mathbf{S}}^*(\mathbf{x}, \boldsymbol{\xi})$ is a stress field corresponding to the thermoelastic state $s^*(\mathbf{x}, \boldsymbol{\xi}) = [\overline{\mathbf{u}}^*(\mathbf{x}, \boldsymbol{\xi}), \overline{\mathbf{E}}^*(\mathbf{x}, \boldsymbol{\xi}), \overline{\mathbf{S}}^*(\mathbf{x}, \boldsymbol{\xi})]$ produced by the temperature $\overline{T}^* = \overline{T}^*(\mathbf{x}, \boldsymbol{\xi})$ in the semi-infinite sheet shown in Figure 10.14.

To obtain the stress components $\overline{S}^*_{\alpha\beta}$ $(\alpha, \beta = 1, 2)$ we use the thermoelastic displacement potential method outlined in Section 10.2.1. According to the method $\overline{S}^*_{\alpha\beta}$ may be computed from the formula (see Equation 10.2.21)

$$\overline{S}^*_{\alpha\beta} = \left(\delta_{\alpha\beta}\nabla^2 - \frac{\partial^2}{\partial x_\alpha \partial x_\beta}\right)\left(\tilde{F}^* - 2\mu\tilde{\phi}^*\right) \tag{10.2.85}$$

where $\tilde{\phi}^* = \tilde{\phi}^*(\mathbf{x}, \boldsymbol{\xi})$ is a solution to Poisson's equation

$$\nabla^2\tilde{\phi}^* = m_0\overline{T}^* \quad \text{for } |x_1| < \infty, \quad x_2 > 0 \tag{10.2.86}$$

and $\tilde{F}^* = \tilde{F}^*(\mathbf{x}, \boldsymbol{\xi})$ is a biharmonic function; and these functions are selected in such a way that the stress free boundary conditions 10.2.84 are satisfied, and in addition

$$\overline{S}^*_{\alpha\beta} \to 0 \quad \text{as } |\mathbf{x}| \to \infty \tag{10.2.87}$$

To obtain a solution to Equation 10.2.86 we note that Poisson's equation (10.2.86) may be extended to the whole (x_1, x_2) plane in the form

$$\nabla^2\tilde{\phi}^* = m_0\,\delta(x_1)[\delta(x_2 - \xi_2) - \delta(x_2 + \xi_2)] \quad \text{for } |x_1| < \infty, \quad |x_2| < \infty \tag{10.2.88}$$

and a restriction of Equation 10.2.88 to the semi-infinite space shown in Figure 10.14 reduces to Poisson's equation 10.2.86. The extended Equation 10.2.88 amounts to a situation in which there are two thermoelastic nuclei of the intensities $+m_0$ and $-m_0$ acting at the points $(0, \xi_2)$ and $(0, -\xi_2)$, respectively, of an infinite plane (x_1, x_2).

Proceeding in a way similar to that of solving Equation 10.2.30, we find that a solution to Equation 10.2.88 may be taken in the form

$$\tilde{\phi}^*(\mathbf{x}, \boldsymbol{\xi}) = -\frac{m_0}{2\pi}\left(\ln\frac{L}{\overline{r}_1} - \ln\frac{L}{\overline{r}_2}\right) \tag{10.2.89}$$

where

$$\overline{r}_1 = \sqrt{x_1^2 + (x_2 - \xi_2)^2}, \quad \overline{r}_2 = \sqrt{x_1^2 + (x_2 + \xi_2)^2} \tag{10.2.90}$$

and L is a positive constant of the length dimension.

The associated displacement $\tilde{u}^*_\alpha$, strain $\tilde{E}^*_{\alpha\beta}$, and stress $\tilde{S}^*_{\alpha\beta}$ are given by

$$\tilde{u}^*_\alpha = \frac{m_0}{2\pi}\frac{\partial}{\partial x_\alpha}\left(\ln\frac{\overline{r}_1}{L} - \ln\frac{\overline{r}_2}{L}\right) \tag{10.2.91}$$

$$\tilde{E}^*_{\alpha\beta} = \frac{m_0}{2\pi}\frac{\partial^2}{\partial x_\alpha \partial x_\beta}\left(\ln\frac{\overline{r}_1}{L} - \ln\frac{\overline{r}_2}{L}\right) \tag{10.2.92}$$

$$\tilde{S}^*_{\alpha\beta} = \frac{m_0\mu}{\pi}\left(\frac{\partial^2}{\partial x_\alpha x_\beta} - \delta_{\alpha\beta}\nabla^2\right)\left(\ln\frac{\overline{r}_1}{L} - \ln\frac{\overline{r}_2}{L}\right) \tag{10.2.93}$$

In particular, for the stress components $\tilde{S}^*_{22}$ and $\tilde{S}^*_{12}$ we obtain

$$\tilde{S}^*_{22} = -\frac{m_0\mu}{\pi}\frac{\partial^2}{\partial x_1^2}\left(\ln\frac{\overline{r}_1}{L} - \ln\frac{\overline{r}_2}{L}\right) \tag{10.2.94}$$

$$\tilde{S}^*_{12} = \frac{m_0\mu}{\pi}\frac{\partial^2}{\partial x_1 \partial x_2}\left(\ln\frac{\overline{r}_1}{L} - \ln\frac{\overline{r}_2}{L}\right) \tag{10.2.95}$$

Since

$$\frac{\partial}{\partial x_2}\ln\frac{\overline{r}_1}{L} = \frac{x_2 - \xi_2}{\overline{r}_1^2}, \qquad \frac{\partial}{\partial x_2}\ln\frac{\overline{r}_2}{L} = \frac{x_2 + \xi_2}{\overline{r}_2^2} \tag{10.2.96}$$

therefore

$$\frac{\partial}{\partial x_2}\left(\ln\frac{\overline{r}_1}{L} - \ln\frac{\overline{r}_2}{L}\right) = \frac{x_2 - \xi_2}{\overline{r}_1^2} - \frac{x_2 + \xi_2}{\overline{r}_2^2} \tag{10.2.97}$$

and it follows from Equations 10.2.94 and 10.2.95 and 10.2.97 that

$$\tilde{S}^*_{22}(x_1, 0; \boldsymbol{\xi}) = 0, \quad |x_1| < \infty \tag{10.2.98}$$

and

$$\tilde{S}^*_{12}(x_1, 0; \boldsymbol{\xi}) = -\frac{2m_0\mu\xi_2}{\pi}\frac{\partial}{\partial x_1}\left(\frac{1}{\overline{r}_0^2}\right) \quad |x_1| < \infty \tag{10.2.99}$$

where

$$\overline{r}_0 = \sqrt{x_1^2 + \xi_2^2} \tag{10.2.100}$$

Also note that

$$\int_0^\infty e^{-\alpha\xi_2}\sin\alpha x_1\, d\alpha = \frac{x_1}{\overline{r}_0^2} \tag{10.2.101}$$

Differentiating Equation 10.2.101 with respect to ξ_2 we obtain

$$\int_0^\infty e^{-\alpha\xi_2}\alpha \sin\alpha x_1 \, d\alpha = \frac{2x_1\xi_2}{\overline{r}_0^4} \tag{10.2.102}$$

Since, by virtue of Equation 10.2.99,

$$\tilde{S}_{12}^*(x_1, 0; \boldsymbol{\xi}) = -\frac{2m_0\mu}{\pi}\left(-\frac{2x_1\xi_2}{\overline{r}_0^4}\right) \tag{10.2.103}$$

therefore, it follows from Equations 10.2.102 and 10.2.103 that the boundary condition 10.2.103 may be represented in the integral form

$$\tilde{S}_{12}^*(x_1, 0; \boldsymbol{\xi}) = \frac{2m_0\mu}{\pi}\int_0^\infty e^{-\alpha\xi_2}\alpha \sin\alpha x_1 \, d\alpha \tag{10.2.104}$$

Also, it follows from Equation 10.2.85 that

$$\overline{S}_{\alpha\beta}^* = \tilde{S}_{\alpha\beta}^* + \tilde{S}_{\alpha\beta}^{**} \tag{10.2.105}$$

where

$$\tilde{S}_{11}^{**} = \tilde{F}_{,22}^*, \quad \tilde{S}_{22}^{**} = \tilde{F}_{,11}^*, \quad \tilde{S}_{12}^{**} = -\tilde{F}_{,12}^* \tag{10.2.106}$$

Therefore, to satisfy the stress free boundary conditions (see Equations 10.2.84)

$$\overline{S}_{22}^*(x_1, 0; \boldsymbol{\xi}) = \overline{S}_{12}^*(x_1, 0; \boldsymbol{\xi}) = 0 \quad \text{for } |x_1| < \infty \tag{10.2.107}$$

the biharmonic function $\tilde{F}^* = \tilde{F}^*(\mathbf{x}, \boldsymbol{\xi})$ is taken in the form

$$\tilde{F}^*(\mathbf{x}, \boldsymbol{\xi}) = \int_0^\infty [A(\alpha) + \alpha B(\alpha)x_2]e^{-\alpha x_2} \cos\alpha x_1 \, d\alpha \tag{10.2.108}$$

where the functions $A = A(\alpha)$ and $B = B(\alpha)$ are determined from the boundary conditions 10.2.107.

Substituting $\tilde{F}^*$ from Equation 10.2.108 into Equations 10.2.106 yields

$$\tilde{S}_{11}^{**} = \int_0^\infty e^{-\alpha x_2}\alpha^2[A - B(2 - \alpha x_2)] \cos\alpha x_1 \, d\alpha \tag{10.2.109}$$

$$\tilde{S}_{22}^{**} = -\int_0^\infty e^{-\alpha x_2}\alpha^2(A + B\alpha x_2) \cos\alpha x_1 \, d\alpha \tag{10.2.110}$$

$$\tilde{S}_{12}^{*} = -\int_0^\infty e^{-\alpha x_2}\alpha^2[A - B(1 - \alpha x_2)] \sin\alpha x_1 \, d\alpha \tag{10.2.111}$$

Now, because of Equation 10.2.98

$$\tilde{S}_{22}^{**}(x_1, 0; \boldsymbol{\xi}) = 0 \quad |x_1| < \infty \tag{10.2.112}$$

it follows from Equation 10.2.110 that

$$A(\alpha) = 0 \quad \forall\, \alpha > 0 \tag{10.2.113}$$

and Equations 10.2.109 through 10.2.111 reduce to

$$\tilde{S}_{11}^{**} = -\int_0^\infty e^{-\alpha x_2}\alpha^2(2 - \alpha x_2)B(\alpha)\cos\alpha x_1\, d\alpha \tag{10.2.114}$$

$$\tilde{S}_{22}^{**} = -x_2\int_0^\infty e^{-\alpha x_2}\alpha^3 B(\alpha)\cos\alpha x_1\, d\alpha \tag{10.2.115}$$

$$\tilde{S}_{12}^{**} = \int_0^\infty e^{-\alpha x_2}\alpha^2(1 - \alpha x_2)B(\alpha)\sin\alpha x_1\, d\alpha \tag{10.2.116}$$

Finally, substituting Equations 10.2.104 and 10.2.116 into the boundary condition

$$\tilde{S}_{12}^{*}(x_1, 0; \xi_2) + \tilde{S}_{12}^{**}(x_1, 0; \xi_2) = 0 \quad \text{for } |x_1| < \infty \tag{10.2.117}$$

we find that

$$\alpha^2 B(\alpha) = -\frac{2m_0\mu}{\pi}\alpha e^{-\alpha\xi_2} \tag{10.2.118}$$

Hence, the integral representations of $\tilde{S}_{\alpha\beta}^{**}$ are obtained

$$\tilde{S}_{11}^{**} = \frac{2m_0\mu}{\pi}\int_0^\infty e^{-\alpha(x_2+\xi_2)}\alpha(2 - \alpha x_2)\cos\alpha x_1\, d\alpha \tag{10.2.119}$$

$$\tilde{S}_{22}^{**} = \frac{2m_0\mu}{\pi}x_2\int_0^\infty e^{-\alpha(x_2+\xi_2)}\alpha^2\cos\alpha x_1\, d\alpha \tag{10.2.120}$$

$$\tilde{S}_{12}^{**} = -\frac{2m_0\mu}{\pi}\int_0^\infty e^{-\alpha(x_2+\xi_2)}\alpha(1 - \alpha x_2)\sin\alpha x_1\, d\alpha \tag{10.2.121}$$

Also, it follows from Equations 10.2.108, 10.2.113, and 10.2.118 that

$$\tilde{F}^{*} = -\frac{2m_0\mu}{\pi}x_2\int_0^\infty e^{-\alpha(x_2+\xi_2)}\cos\alpha x_1\, d\alpha \tag{10.2.122}$$

Hence, if we use the formula

$$\int_0^\infty e^{-\alpha(x_2+\xi_2)}\cos\alpha x_1\, d\alpha = \frac{x_2+\xi_2}{\overline{r}_2^2} = \frac{\partial}{\partial x_2}\ln\left(\frac{\overline{r}_2}{L}\right) \tag{10.2.123}$$

on account of Equation 10.2.122, the closed form of $\tilde{F}^*$ is obtained

$$\tilde{F}^* = -\frac{2m_0\mu}{\pi} x_2 \frac{\partial}{\partial x_2} \ln\left(\frac{\overline{r}_2}{L}\right) \tag{10.2.124}$$

Now, if we introduce the function $\tilde{\chi}^*$ by the formula

$$\tilde{\chi}^* = \tilde{F}^* - 2\mu\tilde{\phi}^* \tag{10.2.125}$$

where $\tilde{\phi}^*$ is given by Equation 10.2.89, we find that

$$\tilde{\chi}^* = -\frac{2m_0\mu}{\pi}\left\{x_2 \frac{\partial}{\partial x_2} \ln\left(\frac{\overline{r}_2}{L}\right) + \frac{1}{2}\left[\ln\left(\frac{\overline{r}_1}{L}\right) - \ln\left(\frac{\overline{r}_2}{L}\right)\right]\right\} \tag{10.2.126}$$

The associated thermal stresses $\overline{S}^*_{\alpha\beta}$ are then computed from the formulas

$$\overline{S}^*_{11} = \tilde{\chi}^*_{,22}, \quad \overline{S}^*_{22} = \tilde{\chi}^*_{,11}, \quad \overline{S}^*_{12} = -\tilde{\chi}^*_{,12} \tag{10.2.127}$$

Substituting $\tilde{\chi}^*$ from Equation 10.2.126 into Equations 10.2.127, the thermal stresses due to a nucleus of thermoelastic strain in a semi-infinite sheet shown in Figure 10.14 become

$$\begin{aligned}\overline{S}^*_{11} = &- A\left\{\frac{1}{2}\left[\frac{x_1^2 - (x_2 - \xi_2)^2}{\overline{r}_1^4} + 3\frac{x_1^2 - (x_2 + \xi_2)^2}{\overline{r}_2^4}\right]\right.\\ &\left. - 2\frac{x_2(x_2 + \xi_2)}{\overline{r}_2^6}\left[3x_1^2 - (x_2 + \xi_2)^2\right]\right\}\end{aligned} \tag{10.2.128}$$

$$\begin{aligned}\overline{S}^*_{22} = &- A\left\{\frac{1}{2}\left[\frac{(x_2 - \xi_2)^2 - x_1^2}{\overline{r}_1^4} - \frac{(x_2 + \xi_2)^2 - x_1^2}{\overline{r}_2^4}\right]\right.\\ &\left. + 2\frac{x_2(x_2 + \xi_2)}{\overline{r}_2^6}\left[3x_1^2 - (x_2 + \xi_2)^2\right]\right\}\end{aligned} \tag{10.2.129}$$

$$\overline{S}^*_{12} = - Ax_1\left\{\frac{x_2 - \xi_2}{\overline{r}_1^4} + \frac{x_2 + \xi_2}{\overline{r}_2^4} + 2\frac{x_2}{\overline{r}_2^6}\left[x_1^2 - 3(x_2 + \xi_2)^2\right]\right\} \tag{10.2.130}$$

where

$$A = \frac{2m_0\mu}{\pi} \tag{10.2.131}$$

It follows from Equations 10.2.129 and 10.2.130 that

$$\overline{S}^*_{22}(x_1, 0; 0, \xi_2) = \overline{S}^*_{12}(x_1, 0; 0, \xi_2) = 0 \quad \text{for every } |x_1| < \infty,\ \xi_2 > 0 \tag{10.2.132}$$

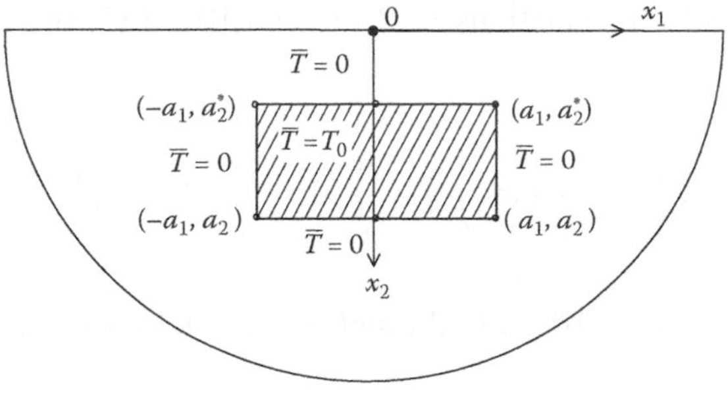

FIGURE 10.15 A discontinuous rectangular distribution of temperature in a semi-infinite sheet.

and

$$\overline{S}^*_{\alpha\alpha} = -2A \frac{\partial^2}{\partial x_2^2} \ln\left(\frac{\overline{r}_2}{L}\right) \tag{10.2.133}$$

The results obtained will be used in Section 10.2.3.2 to calculate the thermal stresses due to a rectangular distribution of temperature in a semi-infinite thermoelastic sheet.

10.2.3.2 Thermal Stresses due to a Rectangular Distribution of Temperature in a Semi-Infinite Solid

Let a homogeneous isotropic semi-infinite thermoelastic sheet: $|x_1| < \infty, x_2 \geq 0$ with a stress free boundary $x_2 = 0$ be subject to a discontinuous rectangular distribution of temperature as shown in Figure 10.15.

Let $S^*_{\alpha\beta}(\mathbf{x}, \boldsymbol{\xi})$ denote the thermal stresses at a point $\mathbf{x} = (x_1, x_2)$ due to a nucleus of thermoelastic strain at a point $\boldsymbol{\xi} = (\xi_1, \xi_2)$ of a semi-infinite sheet $|x_1| < \infty, x_2 \geq 0$ with the stress free boundary $x_2 = 0$. Then it follows from Section 10.2.3.1 (see Equations 10.2.126 and 10.2.127) that

$$S^*_{11} = \chi^*_{,22}, \quad S^*_{22} = \chi^*_{,11}, \quad S^*_{12} = -\chi^*_{,12} \tag{10.2.134}$$

where

$$\chi^* = -A\left[x_2 \frac{\partial}{\partial x_2}\left(\ln \frac{r_2}{L}\right) + \frac{1}{2}\left(\ln \frac{r_1}{L} - \ln \frac{r_2}{L}\right)\right] \tag{10.2.135}$$

and

$$r_1 = \sqrt{(x_1 - \xi_1)^2 + (x_2 - \xi_2)^2}, \quad r_2 = \sqrt{(x_1 - \xi_1)^2 + (x_2 + \xi_2)^2} \tag{10.2.136}$$

In the following, we take advantage of the identities

$$\frac{\partial r_2}{\partial x_2} = \frac{\partial r_2}{\partial \xi_2}, \quad \frac{\partial r_2}{\partial x_1} = -\frac{\partial r_2}{\partial \xi_1} \tag{10.2.137}$$

$$\frac{\partial r_1}{\partial x_1} = -\frac{\partial r_1}{\partial \xi_1}, \quad \frac{\partial r_1}{\partial x_2} = -\frac{\partial r_1}{\partial \xi_2} \tag{10.2.138}$$

In particular, it follows from Equations 10.2.135 and 10.2.137$_1$ that an alternative form of χ^* is

$$\chi^* = -A\left\{x_2 \frac{\partial}{\partial \xi_2}\left(\ln \frac{r_2}{L}\right) + \frac{1}{2}\left(\ln \frac{r_1}{L} - \ln \frac{r_2}{L}\right)\right\} \tag{10.2.139}$$

Also, on account of Equations 10.2.137 through 10.2.139, for the second partial derivatives of χ^*, we obtain

$$\frac{\partial^2 \chi^*}{\partial x_1^2} = -\frac{A}{2}\frac{\partial}{\partial \xi_1}\left[\frac{\xi_1 - x_1}{r_1^2} - \frac{\xi_1 - x_1}{r_2^2} + 2x_2 \frac{\partial}{\partial \xi_2}\left(\frac{\xi_1 - x_1}{r_2^2}\right)\right] \tag{10.2.140}$$

$$\frac{\partial^2 \chi^*}{\partial x_2^2} = -\frac{A}{2}\frac{\partial}{\partial \xi_2}\left[\frac{\xi_2 - x_2}{r_1^2} + 3\frac{\xi_2 + x_2}{r_2^2} + 2x_2 \frac{(x_1 - \xi_1)^2 - (x_2 + \xi_2)^2}{r_2^4}\right] \tag{10.2.141}$$

$$\frac{\partial^2 \chi^*}{\partial x_1 \partial x_2} = -\frac{A}{2}\frac{\partial^2}{\partial \xi_1 \partial \xi_2}\left(\ln \frac{r_1}{L} - \ln \frac{r_2}{L} - 2x_2 \frac{\xi_2 + x_2}{r_2^2}\right) \tag{10.2.142}$$

The thermal stresses $\overline{S}_{\alpha\beta}(\mathbf{x})$ due to a rectangular distribution of temperature in a semi-infinite sheet shown in Figure 10.15 are then obtained by the formula

$$\overline{S}_{\alpha\beta}(\mathbf{x}) = T_0 \int_{-a_1}^{a_1} \int_{a_2^*}^{a_2} S^*_{\alpha\beta}(\mathbf{x}, \boldsymbol{\xi})\, d\xi_1 d\xi_2 \tag{10.2.143}$$

where S^*_{22}, S^*_{11}, and S^*_{12} are determined by Equations 10.2.134 and 10.2.140 through 10.2.142.

In components, we obtain

$$\overline{S}_{11}(\mathbf{x}) = -\frac{AT_0}{2} \int_{-a_1}^{a_1} \int_{a_2^*}^{a_2} \frac{\partial}{\partial \xi_2}\left[\frac{\xi_2 - x_2}{r_1^2} + 3\frac{\xi_2 + x_2}{r_2^2} \right.$$

$$\left. + 2x_2 \frac{(x_1 - \xi_1)^2 - (x_2 + \xi_2)^2}{r_2^4}\right] d\xi_1 d\xi_2 \tag{10.2.144}$$

$$\overline{S}_{22}(\mathbf{x}) = -\frac{AT_0}{2} \int_{-a_1}^{a_1} \int_{a_2^*}^{a_2} \frac{\partial}{\partial \xi_1}\left[\frac{\xi_1 - x_1}{r_1^2} - \frac{\xi_1 - x_1}{r_2^2} \right.$$

$$\left. + 2x_2 \frac{\partial}{\partial \xi_2}\left(\frac{\xi_1 - x_1}{r_2^2}\right)\right] d\xi_1 d\xi_2 \tag{10.2.145}$$

$$\overline{S}_{12}(\mathbf{x}) = \frac{AT_0}{2} \int_{-a_1}^{a_1} \int_{a_2^*}^{a_2} \frac{\partial^2}{\partial \xi_1 \partial \xi_2}\left[\ln \frac{r_1}{L} - \ln \frac{r_2}{L} - 2x_2 \frac{\xi_2 + x_2}{r_2^2}\right] d\xi_1 d\xi_2 \tag{10.2.146}$$

Hence, it follows from Equation 10.2.144 that

$$\overline{S}_{11}(\mathbf{x}) = -\frac{AT_0}{2}\int_{-a_1}^{a_1}\left\{\left[\frac{(a_2 - x_2)}{(x_1-\xi_1)^2+(x_2-a_2)^2} - \frac{\left(a_2^* - x_2\right)}{(x_1-\xi_1)^2+\left(x_2-a_2^*\right)^2}\right]\right.$$

$$+3\left[\frac{(a_2 + x_2)}{(x_1-\xi_1)^2+(x_2+a_2)^2} - \frac{\left(a_2^* + x_2\right)}{(x_1-\xi_1)^2+\left(x_2+a_2^*\right)^2}\right]$$

$$+2x_2\left[\frac{1}{(x_1-\xi_1)^2+(x_2+a_2)^2} - \frac{1}{(x_1-\xi_1)^2+\left(x_2+a_2^*\right)^2}\right.$$

$$\left.\left.-2\left(\frac{(x_2+a_2)^2}{[(x_1-\xi_1)^2+(x_2+a_2)^2]^2} - \frac{\left(x_2+a_2^*\right)^2}{[(x_1-\xi_1)^2+\left(x_2+a_2^*\right)2]^2}\right)\right]\right\}d\xi_1 \tag{10.2.147}$$

Now, if we note that for any positive β

$$\int \frac{du}{u^2+\beta^2} = \frac{1}{\beta}\tan^{-1}\frac{u}{\beta} \tag{10.2.148}$$

and

$$\int \frac{du}{(u^2+\beta^2)^2} = \frac{1}{2\beta^2}\left(\frac{1}{\beta}\tan^{-1}\frac{u}{\beta} + \frac{u}{u^2+\beta^2}\right) \tag{10.2.149}$$

we reduce Equation 10.2.147 to the form

$$\overline{S}_{11}(\mathbf{x}) = \frac{AT_0}{2}\left\{\tan^{-1}\frac{x_1+a_1}{x_2-a_2} - \tan^{-1}\frac{x_1-a_1}{x_2-a_2} - \tan^{-1}\frac{x_1+a_1}{x_2-a_2^*}\right.$$

$$+\tan^{-1}\frac{x_1-a_1}{x_2-a_2^*} - 3\left[\tan^{-1}\frac{x_1+a_1}{x_2+a_2} - \tan^{-1}\frac{x_1-a_1}{x_2+a_2} - \tan^{-1}\frac{x_1+a_1}{x_2+a_2^*}\right.$$

$$\left.+\tan^{-1}\frac{x_1-a_1}{x_2+a_2^*}\right] - 2x_2\left[\frac{(x_1-a_1)}{(x_1-a_1)^2+(x_2+a_2)^2} - \frac{(x_1-a_1)}{(x_1-a_1)^2+\left(x_2+a_2^*\right)^2}\right.$$

$$\left.\left.-\frac{(x_1+a_1)}{(x_1+a_1)^2+(x_2+a_2)^2} + \frac{(x_1+a_1)}{(x_1+a_1)^2+(x_2+a_2^*)^2}\right]\right\} \tag{10.2.150}$$

Similarly, it follows from Equation 10.2.145 that

$$\overline{S}_{22}(\mathbf{x}) = \frac{AT_0}{2}\left\{\tan^{-1}\frac{x_2 - a_2^*}{x_1 - a_1} - \tan^{-1}\frac{x_2 - a_2}{x_1 - a_1} - \tan^{-1}\frac{x_2 + a_2}{x_1 - a_1}\right.$$

$$+ \tan^{-1}\frac{x_2 + a_2^*}{x_1 - a_1} - \tan^{-1}\frac{x_2 - a_2^*}{x_1 + a_1} + \tan^{-1}\frac{x_2 - a_2}{x_1 + a_1} + \tan^{-1}\frac{x_2 + a_2}{x_1 + a_1}$$

$$- \tan^{-1}\frac{x_2 + a_2^*}{x_1 + a_1} + 2x_2\left[\frac{(x_1 - a_1)}{(x_1 - a_1)^2 + (x_2 + a_2)^2} - \frac{(x_1 - a_1)}{(x_1 - a_1)^2 + \left(x_2 + a_2^*\right)^2}\right.$$

$$\left.\left. - \frac{(x_1 + a_1)}{(x_1 + a_1)^2 + (x_2 + a_2)^2} + \frac{(x_1 + a_1)}{(x_1 + a_1)^2 + (x_2 + a_2^*)^2}\right]\right\} \tag{10.2.151}$$

Finally, computing the double integral in Equation 10.2.146, we have

$$\overline{S}_{12}(\mathbf{x}) = \frac{AT_0}{2}\left\{\ln\left[\frac{\sqrt{(x_1 - a_1)^2 + (x_2 - a_2)^2}}{\sqrt{(x_1 - a_1)^2 + \left(x_2 - a_2^*\right)^2}}\,\frac{\sqrt{(x_1 + a_1)^2 + \left(x_2 - a_2^*\right)^2}}{\sqrt{(x_1 + a_1)^2 + (x_2 - a_2)^2}}\right.\right.$$

$$\left.\times\,\frac{\sqrt{(x_1 + a_1)^2 + (x_2 + a_2)^2}}{\sqrt{(x_1 + a_1)^2 + \left(x_2 + a_2^*\right)^2}}\,\frac{\sqrt{(x_1 - a_1)^2 + \left(x_2 + a_2^*\right)^2}}{\sqrt{(x_1 - a_1)^2 + (x_2 + a_2)^2}}\right]$$

$$- 2x_2\left[\frac{(x_2 + a_2)}{(x_1 - a_1)^2 + (x_2 + a_2)^2} - \frac{(x_2 + a_2)}{(x_1 + a_1)^2 + (x_2 + a_2)^2}\right.$$

$$\left.\left. - \frac{\left(x_2 + a_2^*\right)}{(x_1 - a_1)^2 + \left(x_2 + a_2^*\right)^2} + \frac{\left(x_2 + a_2^*\right)}{(x_1 + a_1)^2 + \left(x_2 + a_2^*\right)^2}\right]\right\} \tag{10.2.152}$$

By letting $x_2 = 0$ in Equations 10.2.151 and 10.2.152 we obtain

$$\overline{S}_{22}(x_1, 0) = \overline{S}_{12}(x_1, 0) = 0 \quad \text{for } |x_1| < \infty \tag{10.2.153}$$

Also, it follows from the identity

$$\tan^{-1}z + \tan^{-1}\left(\frac{1}{z}\right) = \frac{\pi}{2} \quad \text{for } z > 0 \tag{10.2.154}$$

and from Equations 10.2.150 and 10.2.151 that

$$\overline{S}_{\gamma\gamma}(\mathbf{x}) = 2AT_0\left(\tan^{-1}\frac{x_1 - a_1}{x_2 + a_2} - \tan^{-1}\frac{x_1 - a_1}{x_2 + a_2^*}\right.$$

$$\left.- \tan^{-1}\frac{x_1 + a_1}{x_2 + a_2} + \tan^{-1}\frac{x_1 + a_1}{x_2 + a_2^*}\right) \tag{10.2.155}$$

Finally, if we introduce the notations

$$\overline{S}_{11}(0, a_2 + 0) = \overline{S}_{11}^{+} \tag{10.2.156}$$

$$\overline{S}_{11}(0, a_2 - 0) = \overline{S}_{11}^{-} \tag{10.2.157}$$

$$[\![\overline{S}_{11}]\!](0, a_2) = \overline{S}_{11}^{+} - \overline{S}_{11}^{-} \tag{10.2.158}$$

then it follows from Equation 10.2.150 that

$$[\![\overline{S}_{11}]\!](0, a_2) = 2m_0\mu T_0 \tag{10.2.159}$$

This means that the hoop stress $\overline{S}_{11}$ at the point $(0, a_2)$ of the temperature discontinuity line $|x_1| < a_1$, $x_2 = a_2$ (see Figure 10.15) exhibits a jump of the magnitude $2m_0\mu T_0$. A similar result was obtained for an infinite sheet subject to a rectangular distribution of temperature (see Equation 10.2.79).

10.2.4 A Thermoelastic State due to a Heat Source at the Center of a Thin Circular Disk

It was shown in Section 8.1.2 (see Equations 8.1.101 through 8.1.108) that in the case of a generalized plane stress state corresponding to a harmonic temperature change $\overline{T}$ on C_0 and a stress free boundary ∂C_0, the thermal stresses vanish throughout the body. In this section, we discuss an axially symmetric thermoelastic state in a thin circular disk with a stress-free boundary corresponding to a concentrated heat source at the center of the disk, and show that in this case the stress free disk theorem (see Equation 8.1.104) is not true. The discussion is confined to a situation shown in Figure 10.16 in which a circular disk of radius a is subject to a concentrated heat source of intensity Q_0 at the center $r = 0$ of the disk.

It is assumed that the temperature change $\overline{T} = \overline{T}(r)$ satisfies Poisson's equation (see Equation 10.2.6)

$$\nabla_r^2 \overline{T} = -\frac{Q_0}{\kappa}\frac{\delta(r)}{2\pi r} \quad 0 < r < a \tag{10.2.160}$$

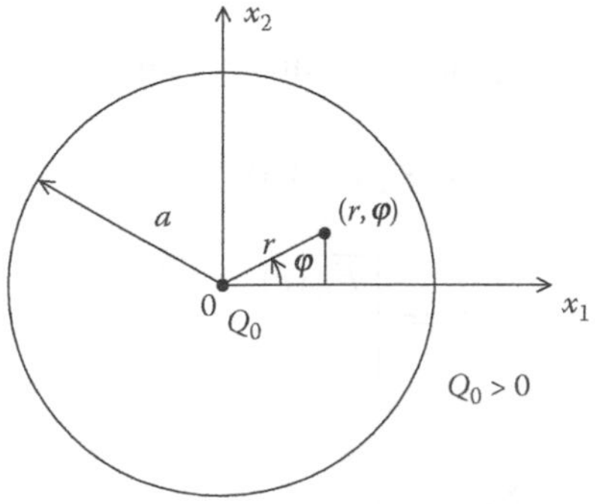

FIGURE 10.16 A circular disk of radius a subject to a heat source at its center.

subject to the boundary condition

$$\overline{T} = 0 \quad \text{for } r = a \tag{10.2.161}$$

In Equation 10.2.160

$$\nabla_r^2 = \frac{1}{r}\frac{d}{dr}r\frac{d}{dr} \tag{10.2.162}$$

$\delta = \delta(r)$ is the Dirac delta function and r is a polar coordinate as shown in Figure 10.16

$$x_1 = r\cos\varphi, \quad x_2 = r\sin\varphi \quad 0 \le r < \infty, \quad 0 \le \varphi \le 2\pi \tag{10.2.163}$$

Note that a unique solution to the boundary value problem 10.2.160 and 10.2.161 takes the form

$$\overline{T}(r) = \frac{Q_0}{2\pi\kappa}\ \ln\frac{a}{r} \tag{10.2.164}$$

and it follows from Equation 10.2.164 that

$$\overline{T}(0) = +\infty, \quad \overline{T}(a) = 0 \tag{10.2.165}$$

and

$$-\kappa\frac{\partial\overline{T}}{\partial r}(a) = \frac{Q_0}{2\pi a} \tag{10.2.166}$$

Therefore, an action of the concentrated heat source corresponds to a constant heat flux over the boundary of the disk.

A thermoelastic state $s = [\overline{\mathbf{u}}, \overline{\mathbf{E}}, \overline{\mathbf{S}}]$ in the disk corresponding to the temperature $\overline{T} = \overline{T}(r)$ depends on r only, and complies with the stress free boundary conditions

$$\overline{S}_{rr}(a) = 0, \quad \overline{S}_{r\varphi}(a) = 0 \tag{10.2.167}$$

The fields $\overline{\mathbf{u}}$, $\overline{\mathbf{E}}$, and $\overline{\mathbf{S}}$, written in polar coordinates, take the forms

$$\overline{\mathbf{u}}(r) = [\overline{u}_r(r), 0] \tag{10.2.168}$$

$$\overline{\mathbf{E}}(r) = \begin{bmatrix} \overline{E}_{rr}(r) & 0 \\ 0 & \overline{E}_{\varphi\varphi}(r) \end{bmatrix} \tag{10.2.169}$$

$$\overline{\mathbf{S}}(r) = \begin{bmatrix} \overline{S}_{rr}(r) & 0 \\ 0 & \overline{S}_{\varphi\varphi}(r) \end{bmatrix} \tag{10.2.170}$$

and it follows from Section 10.2.1 (see Equations 10.2.25 and 10.2.26 reduced to polar coordinates in the case of axial symmetry) that these fields may be generated by a function $\chi = \chi(r)$ that satisfies the equation

$$\nabla_r^2 \nabla_r^2 \chi = \frac{E\alpha Q_0}{\kappa} \frac{\delta(r)}{2\pi r} \tag{10.2.171}$$

subject to the boundary conditions

$$\chi(a) = \chi'(a) = 0 \tag{10.2.172}$$

and the finiteness conditions at $r = 0$

$$\chi(0) < \infty, \quad \chi'(0) < \infty \tag{10.2.173}$$

where $(') = d/dr$. It will be shown later that the conditions 10.2.173 lead to a physically meaningful solution to the problem.

The stress components $\overline{S}_{rr}$ and $\overline{S}_{\varphi\varphi}$ associated with the function χ are obtained from the formulas [see Equation 10.2.24 written in polar coordinates (r, φ); also see Equations 10.1.183 through 10.1.185 for an isothermal case]

$$\overline{S}_{rr} = \frac{1}{r}\frac{\partial}{\partial r}\chi, \quad \overline{S}_{\varphi\varphi} = \frac{\partial^2}{\partial r^2}\chi \tag{10.2.174}$$

while the strain components $\overline{E}_{rr}$ and $\overline{E}_{\varphi\varphi}$ are obtained from the constitutive relations [see Equation 10.2.4 written in polar coordinates (r, φ); see also Equations 10.1.176 through 10.1.178 for an isothermal case]

$$\overline{E}_{rr} = \frac{1}{E}\left(\overline{S}_{rr} - \nu \overline{S}_{\varphi\varphi}\right) + \alpha \overline{T} \tag{10.2.175}$$

$$\overline{E}_{\varphi\varphi} = \frac{1}{E}\left(\overline{S}_{\varphi\varphi} - \nu \overline{S}_{rr}\right) + \alpha \overline{T} \tag{10.2.176}$$

Finally, the radial displacement $\overline{u}_r$ is obtained from the relation

$$\overline{u}_r(r) = \int_0^r \overline{E}_{rr}(u)du \tag{10.2.177}$$

or from the equation

$$\overline{u}_r(r) = r\overline{E}_{\varphi\varphi}(r) \tag{10.2.178}$$

Note that Equation 10.2.177 complies with a radial symmetry of the problem in which $\overline{u}_r(0) = 0$.

Therefore, to find the thermoelastic state $s = s(r)$ in the disk, a suitable solution to the problem 10.2.171 through 10.2.173 is needed. To this end, we observe that the nonhomogeneous biharmonic equation 10.2.171 represents Poisson's equation for an unknown function $\nabla_r^2 \chi$; integrating this Poisson's equation in a way similar to that of the heat conduction equation 10.2.160, we have

$$\nabla_r^2 \chi = M \ln \frac{r}{a} + C_0 \tag{10.2.179}$$

where

$$M = \frac{E\alpha Q_0}{2\pi \kappa} \tag{10.2.180}$$

and C_0 is an arbitrary constant. To find a solution to Poisson's equation 10.2.179, note that from Equations 10.2.162 follows that Equation 10.2.179 is equivalent to

$$\frac{d}{dr}\left(r \frac{d\chi}{dr} \right) = Mr \ln \frac{r}{a} + C_0 r \tag{10.2.181}$$

Hence, integrating this equation over the interval (r, a), and using the second of boundary conditions 10.2.172 yields

$$-r \frac{d\chi}{dr} = -\frac{1}{4} M \left[2r^2 \ln \frac{r}{a} + (a^2 - r^2) \right] + \frac{C_0}{2}(a^2 - r^2) \tag{10.2.182}$$

Next, it follows from the second of finiteness conditions 10.2.173 $[\chi'(0) < \infty]$ and from Equation 10.2.182 that C_0 in Equation 10.2.182 should be selected in such a way that

$$\left(\frac{C_0}{2} - \frac{M}{4} \right)(a^2 - r^2) = 0 \tag{10.2.183}$$

Hence, we get

$$C_0 = \frac{M}{2} \tag{10.2.184}$$

and Equation 10.2.182 is reduced to the form

$$\frac{d\chi}{dr} = \frac{M}{2} r \ln \frac{r}{a} \tag{10.2.185}$$

Finally, integrating Equation 10.2.185 over the interval (r, a), and using the first of boundary conditions 10.2.172 $[\chi(a) = 0]$ we obtain

$$\chi(r) = \frac{M}{8}\left(2r^2 \ln \frac{r}{a} + a^2 - r^2 \right) \tag{10.2.186}$$

Also, note that the function χ given by Equation 10.2.186 satisfies the first of finiteness conditions 10.2.173 $[\chi(0) < \infty]$ as

$$r^2 \ln \frac{r}{a} \to 0 \quad \text{as } r \to 0 \tag{10.2.187}$$

Therefore, it follows from Equations 10.2.174 through 10.2.178 and 10.2.186 that the thermoelastic state $s = s(r)$ in the disk is represented by

$$s(r) = [\overline{\mathbf{u}}(r), \overline{\mathbf{E}}(r), \overline{\mathbf{S}}(r)] \tag{10.2.188}$$

where $\overline{\mathbf{u}}$, $\overline{\mathbf{E}}$, and $\overline{\mathbf{S}}$ are given by Equations 10.2.168, 10.2.169, and 10.2.170, respectively, in which

$$\overline{u}_r(r) = \frac{Q_0\alpha}{4\pi\kappa}\, r\left[(1+\nu)\ln\frac{a}{r} + 1\right] \tag{10.2.189}$$

$$\overline{E}_{rr}(r) = \frac{Q_0\alpha}{4\pi\kappa}\left[(1+\nu)\ln\frac{a}{r} - \nu\right] \tag{10.2.190}$$

$$\overline{E}_{\varphi\varphi}(r) = \frac{Q_0\alpha}{4\pi\kappa}\left[(1+\nu)\ln\frac{a}{r} + 1\right] \tag{10.2.191}$$

$$\overline{S}_{rr}(r) = -\frac{Q_0\alpha E}{4\pi\kappa}\ln\frac{a}{r} \tag{10.2.192}$$

$$\overline{S}_{\varphi\varphi}(r) = -\frac{Q_0\alpha E}{4\pi\kappa}\left(\ln\frac{a}{r} - 1\right) \tag{10.2.193}$$

Now, if we introduce the dimensionless fields

$$u_r^*(\xi) = \frac{4\pi\kappa}{Q_0\alpha}\,\frac{\overline{u}_r}{a} \tag{10.2.194}$$

$$E_{rr}^*(\xi) = \frac{4\pi\kappa}{Q_0\alpha}\,\overline{E}_{rr}(r), \quad E_{\varphi\varphi}^*(\xi) = \frac{4\pi\kappa}{Q_0\alpha}\,\overline{E}_{\varphi\varphi}(r) \tag{10.2.195}$$

$$S_{rr}^*(\xi) = \frac{4\pi\kappa}{Q_0\alpha E}\,\overline{S}_{rr}(r), \quad S_{\varphi\varphi}^*(\xi) = \frac{4\pi\kappa}{Q_0\alpha E}\,\overline{S}_{\varphi\varphi}(r) \tag{10.2.196}$$

where

$$\xi = \frac{a}{r}, \quad 1 \le \xi \le \infty \tag{10.2.197}$$

the formulas 10.2.189 through 10.2.193 are reduced to the dimensionless form

$$u_r^*(\xi) = (1+\nu)\frac{\ln\xi}{\xi} + \frac{1}{\xi} \tag{10.2.198}$$

$$E_{rr}^*(\xi) = (1+\nu)\ln\xi - \nu \tag{10.2.199}$$

$$E_{\varphi\varphi}^*(\xi) = (1+\nu)\ln\xi + 1 \tag{10.2.200}$$

$$S_{rr}^*(\xi) = -\ln\xi \tag{10.2.201}$$

$$S_{\varphi\varphi}^*(\xi) = -(\ln\xi - 1) \tag{10.2.202}$$

Note that the points $\xi = 1$ and $\xi = \infty$ correspond to $r = a$ and $r = 0$, respectively.

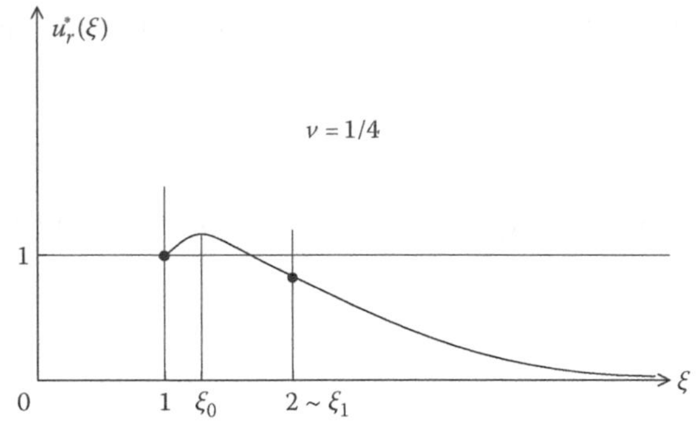

FIGURE 10.17 The function $u_r^*(\xi)$ over the interval $1 \leq \xi \leq \infty$.

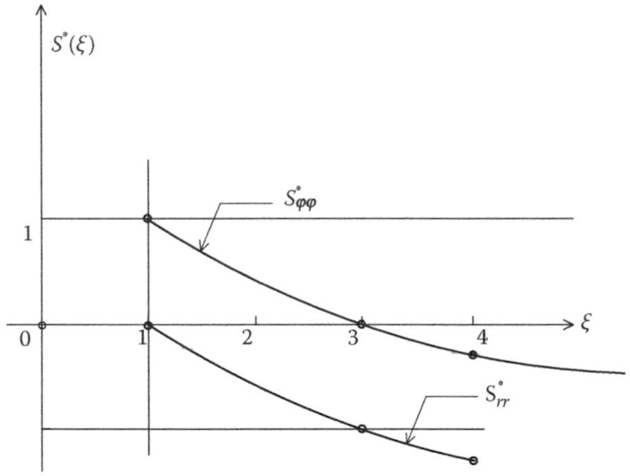

FIGURE 10.18 The stresses $S^*_{rr}(\xi)$ and $S^*_{\varphi\varphi}(\xi)$ over the interval $1 \leq \xi \leq \infty$.

The dimensionless displacement $u_r^*(\xi)$ calculated at $\nu = 1/4$ is shown in Figure 10.17 while the dimensionless stress components $S^*_{rr}(\xi)$ and $S^*_{\varphi\varphi}(\varphi)$ are plotted in Figure 10.18.

It follows from Figure 10.17 that u_r^* attains a maximum slightly greater than 1 at $\xi_0 \sim 1.22$, and $\xi_1 \sim 2$ is an inflection point for u_r^*; also, $u_r^* \to 0$ as $\xi \to \infty$. Figure 10.18 shows that S^*_{rr} and $S^*_{\varphi\varphi}$ are represented by the decreasing functions of $\xi \in [1, \infty]$, and $S^*_{rr} \to -\infty$ as $\xi \to \infty$ and $S^*_{\varphi\varphi} \to -\infty$ as $\xi \to \infty$. The unboundedness of the stress components at the center of the disk $\xi = \infty$ $(r = 0)$ corresponds to the concentrated heat source at $\xi = \infty$.

In concluding this section, we note that the thermoelastic state in the disk is well defined if the radius of the disk is finite. If $a \to \infty$, the functions 10.2.189 through 10.2.193 become unbounded for any $0 < r < \infty$. This means that a thermoelastic problem for an infinite sheet subject to a concentrated heat source is not well posed.

10.2.5 Thermal Stresses due to a Concentrated Heat Source in a Semi-Infinite Sheet

Suppose that at the point $(0, \xi_2)$ $(\xi_2 > 0)$ of a homogeneous isotropic semi-infinite elastic sheet $|x_1| < \infty$, $x_2 \geq 0$, there acts a concentrated heat source of intensity Q_0 (see Figure 10.19).

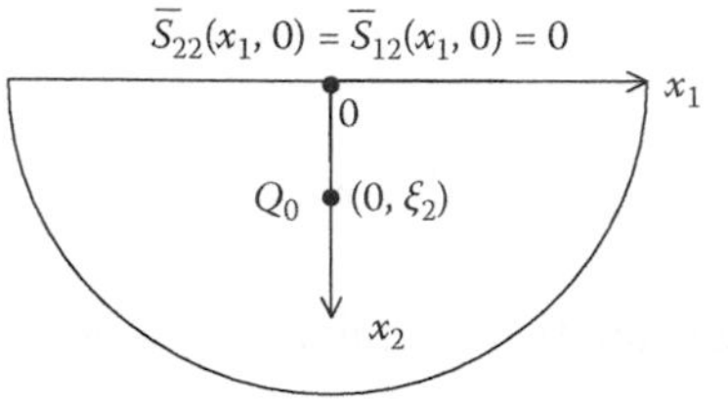

FIGURE 10.19 A semi-infinite thermoelastic sheet subject to a concentrated heat source at $(0, \xi_2)$.

Also, suppose that the boundary $x_2 = 0$ of the semi-infinite body be stress free, that is,

$$\overline{S}_{22}(x_1, 0) = \overline{S}_{12}(x_1, 0) = 0 \quad \text{for } |x_1| < \infty \tag{10.2.203}$$

and a stress field $\overline{\mathbf{S}} = \overline{\mathbf{S}}(\mathbf{x})$ due to the heat source be vanishing as $|\mathbf{x}| \to \infty$.

A problem of finding the stress field $\overline{\mathbf{S}}(\mathbf{x})$ for every point of the solid is the one that belongs to the class of problems described by Equations 10.2.25 and 10.2.26. For the situation shown in Figure 10.19, $\overline{\mathbf{S}} = \overline{\mathbf{S}}(\mathbf{x})$ is obtained from the formula

$$\overline{S}_{\alpha\beta} = \nabla^2 \overline{\chi}\, \delta_{\alpha\beta} - \overline{\chi}_{,\alpha\beta} \tag{10.2.204}$$

where the function $\overline{\chi} = \overline{\chi}(\mathbf{x})$ satisfies the nonhomogeneous biharmonic equation

$$\nabla^2 \nabla^2 \overline{\chi} = \frac{E\alpha Q_0}{\kappa}\, \delta(x_1)\, \delta(x_2 - \xi_2) \quad \text{for } |x_1| < \infty,\ x_2 > 0 \tag{10.2.205}$$

the homogeneous boundary conditions

$$\overline{\chi}_{,11}(x_1, 0) = 0, \quad \overline{\chi}_{,12}(x_1, 0) = 0 \quad \text{for } |x_1| < \infty \tag{10.2.206}$$

and the vanishing conditions at infinity

$$\overline{\chi}_{,22} \to 0, \quad \overline{\chi}_{,11} \to 0, \quad \overline{\chi}_{,12} \to 0 \quad \text{as } |\mathbf{x}| \to \infty \tag{10.2.207}$$

We look for $\overline{\chi}$ in the form

$$\overline{\chi} = \chi^* + \chi^{**} \tag{10.2.208}$$

where $\chi^* = \chi^*(\mathbf{x})$ is a solution to the equation

$$\nabla^2 \nabla^2 \chi^* = \frac{E\alpha Q_0}{\kappa}\, \delta(x_1)\, [\delta(x_2 - \xi_2) - \delta(x_2 + \xi_2)]$$

$$\text{for } |x_1| < \infty,\ |x_2| < \infty \tag{10.2.209}$$

such that

$$\chi^* \to 0 \quad \text{as } |\mathbf{x}| \to \infty \tag{10.2.210}$$

and the function $\chi^{**} = \chi^{**}(\mathbf{x})$ satisfies the biharmonic equation

$$\nabla^2\nabla^2\chi^{**} = 0 \quad \text{for } |x_1| < \infty, \quad x_2 > 0 \tag{10.2.211}$$

the boundary conditions

$$\chi^{**}_{,11} = -\chi^{*}_{,11}, \quad \chi^{**}_{,12} = -\chi^{*}_{,12} \quad \text{for } |x_1| < \infty \tag{10.2.212}$$

and suitable vanishing conditions at infinity.

Let

$$A = \frac{E\alpha Q_0}{\kappa} \tag{10.2.213}$$

Since

$$\delta(x_1) = \frac{1}{\pi}\int_0^\infty \cos \alpha x_1 \, d\alpha \tag{10.2.214}$$

therefore, Equation 10.2.209 may be written as

$$\nabla^2\nabla^2\chi^* = \frac{A}{\pi^2}\int_0^\infty \cos \alpha x_1 \, d\alpha \int_0^\infty [\cos \beta(x_2 - \xi_2) - \cos \beta(x_2 + \xi_2)] \, d\beta \tag{10.2.215}$$

and a solution to this equation takes the form

$$\chi^* = \frac{A}{\pi^2}\int_0^\infty d\alpha \int_0^\infty d\beta \cos \alpha x_1 \frac{[\cos \beta(x_2 - \xi_2) - \cos \beta(x_2 + \xi_2)]}{(\alpha^2 + \beta^2)^2} \tag{10.2.216}$$

Next, differentiating the identity

$$\int_0^\infty \frac{\cos \beta(x_2 - \xi_2)}{\alpha^2 + \beta^2} d\beta = \frac{\pi}{2\alpha} e^{-\alpha|x_2 - \xi_2|} \tag{10.2.217}$$

with respect to α yields

$$\int_0^\infty \frac{\cos \beta(x_2 - \xi_2)}{(\alpha^2 + \beta^2)^2} d\beta = \frac{\pi}{4} \frac{(1 + \alpha|x_2 - \xi_2|)}{\alpha^3} e^{-\alpha|x_2 - \xi_2|} \tag{10.2.218}$$

Therefore, it follows from Equations 10.2.216 and 10.2.218 that χ^* is represented by the single integral

$$\chi^*(\mathbf{x}) = \frac{A}{4\pi}\int_0^\infty \frac{\cos\alpha x_1}{\alpha^3}\{[1+\alpha|x_2-\xi_2|]e^{-\alpha|x_2-\xi_2|} - [1+\alpha(x_2+\xi_2)]e^{-\alpha(x_2+\xi_2)}\}\,d\alpha \tag{10.2.219}$$

An alternative form of Equation 10.2.219 reads

$$\chi^*(\mathbf{x}) = \begin{cases} \chi^{(1)}(\mathbf{x}) & \text{for } 0 \le x_2 < \xi_2,\ |x_1| < \infty \\ \chi^{(2)}(\mathbf{x}) & \text{for } \xi_2 < x_2 < \infty,\ |x_1| < \infty \end{cases} \tag{10.2.220}$$

where

$$\chi^{(1)}(\mathbf{x}) = \frac{A}{4\pi}\int_0^\infty \frac{\cos\alpha x_1}{\alpha^3}\{[1-\alpha(x_2-\xi_2)]\,e^{\alpha(x_2-\xi_2)} - [1+\alpha(x_2+\xi_2)]\,e^{-\alpha(x_2+\xi_2)}\}\,d\alpha \tag{10.2.221}$$

and

$$\chi^{(2)}(\mathbf{x}) = \frac{A}{4\pi}\int_0^\infty \frac{\cos\alpha x_1}{\alpha^3}\{[1+\alpha(x_2-\xi_2)]e^{-\alpha(x_2-\xi_2)} - [1+\alpha(x_2+\xi_2)]e^{-\alpha(x_2+\xi_2)}\}\,d\alpha \tag{10.2.222}$$

The stress components $S^*_{\alpha\beta}$ $(\alpha,\beta = 1,2)$ generated by χ^* are given by

$$S^*_{11} = \chi^*_{,22},\quad S^*_{22} = \chi^*_{,11},\quad S^*_{12} = -\chi^*_{,12} \tag{10.2.223}$$

Hence, it follows from Equations 10.2.219 through 10.2.221 that

$$S^*_{11} = \chi^{(1)}_{,22},\quad S^*_{22} = \chi^{(1)}_{,11},\quad S^*_{12} = -\chi^{(1)}_{,12}\quad \text{for } |x_1| < \infty,\ 0 \le x_2 < \xi_2 \tag{10.2.224}$$

where $\chi^{(1)} = \chi^{(1)}(\mathbf{x})$ is given by Equation 10.2.221. Substituting $\chi^{(1)}$ from Equation 10.2.221 into Equations 10.2.223 we obtain for $0 \le x_2 < \xi_2$

$$S^*_{11} = -\frac{A}{4\pi}\left\{\int_0^\infty \frac{[e^{-\alpha(\xi_2-x_2)} - e^{-\alpha(x_2+x_2)}]}{\alpha}\cos\alpha x_1\,d\alpha + (x_2-\xi_2)\int_0^\infty e^{-\alpha(\xi_2-x_2)}\cos\alpha x_1\,d\alpha + (x_2+\xi_2)\int_0^\infty e^{-\alpha(\xi_2+x_2)}\cos\alpha x_1\,d\alpha\right\} \tag{10.2.225}$$

$$S_{22}^{*} = -\frac{A}{4\pi}\left\{\int_0^\infty \frac{[e^{-\alpha(\xi_2-x_2)} - e^{-\alpha(\xi_2+x_2)}]}{\alpha}\cos\alpha x_1\, d\alpha \right.$$
$$- (x_2-\xi_2)\int_0^\infty e^{-\alpha(\xi_2-x_2)}\cos\alpha x_1\, d\alpha$$
$$\left. - (x_2+\xi_2)\int_0^\infty e^{-\alpha(\xi_2+x_2)}\cos\alpha x_1\, d\alpha\right\} \tag{10.2.226}$$

$$S_{12}^{*} = -\frac{A}{4\pi}\left[(x_2-\xi_2)\int_0^\infty e^{-\alpha(\xi_2-x_2)}\sin\alpha x_1\, d\alpha \right.$$
$$\left. - (x_2+\xi_2)\int_0^\infty e^{-\alpha(\xi_2+x_2)}\sin\alpha x_1\, d\alpha\right] \tag{10.2.227}$$

By letting $x_2 = 0$ in Equations 10.2.226 and 10.2.227 we receive

$$S_{22}^{*}(x_1,0) = 0, \quad S_{12}^{*}(x_1,0) = \frac{A\xi_2}{2\pi}\int_0^\infty e^{-\alpha\xi_2}\sin\alpha x_1\, d\alpha \tag{10.2.228}$$

These boundary values of $S_{\alpha\beta}^{*}$ together with the boundary conditions 10.2.212 imply the integral representation of the function χ^{**}

$$\chi^{**}(\mathbf{x}) = \int_0^\infty [A(\alpha) + B(\alpha)\alpha x_2]e^{-\alpha x_2}\cos\alpha x_1\, d\alpha \tag{10.2.229}$$

The stresses $S_{\alpha\beta}^{**}$ associated with χ^{**} are given by

$$S_{11}^{**} = \chi_{,22}^{**}, \quad S_{22}^{**} = \chi_{,11}^{**}, \quad S_{12}^{**} = -\chi_{,12}^{**} \tag{10.2.230}$$

In particular, it follows from Equations 10.2.229 and 10.2.230 that

$$S_{22}^{**} = -\int_0^\infty [A(\alpha) + B(\alpha)\alpha x_2]\alpha^2 e^{-\alpha x_2}\cos\alpha x_1\, d\alpha \tag{10.2.231}$$

Hence, the first of boundary conditions 10.2.212 in the form

$$S_{22}^{*}(x_1,0) + S_{22}^{**}(x_1,0) = 0 \quad |x_1| < \infty \tag{10.2.232}$$

together with the first of Equations 10.2.228 imply that

$$A(\alpha) = 0 \quad \text{for } \alpha > 0 \tag{10.2.233}$$

and Equation 10.2.229 reduces to the form

$$\chi^{**}(\mathbf{x}) = x_2\int_0^\infty \alpha B(\alpha)e^{-\alpha x_2}\cos\alpha x_1\, d\alpha \tag{10.2.234}$$

As a result, the stresses $S^{**}_{\alpha\beta}$ associated with χ^{**} are represented by the integrals

$$S^{**}_{11} = -\int_0^\infty B(\alpha)(2-\alpha x_2)\alpha^2 e^{-\alpha x_2}\cos\alpha x_1\, d\alpha \tag{10.2.235}$$

$$S^{**}_{22} = -x_2\int_0^\infty B(\alpha)\alpha^3 e^{-\alpha x_2}\cos\alpha x_1\, d\alpha \tag{10.2.236}$$

$$S^{**}_{12} = \int_0^\infty B(\alpha)(1-\alpha x_2)\alpha^2 e^{-\alpha x_2}\sin\alpha x_1\, d\alpha \tag{10.2.237}$$

Finally, using the second of boundary conditions 10.2.212 in the form

$$S^{*}_{12}(x_1,0) + S^{**}_{12}(x_1,0) = 0 \quad |x_1| < \infty \tag{10.2.238}$$

by virtue of the second of Equations 10.2.228 and 10.2.237, we obtain

$$B(\alpha)\alpha^2 = -\frac{A\xi_2}{2\pi}\, e^{-\alpha\xi_2} \tag{10.2.239}$$

Hence, Equations 10.2.235 through 10.2.237 take the forms

$$S^{**}_{11} = \frac{A\xi_2}{2\pi}\int_0^\infty (2-\alpha x_2)e^{-\alpha(x_2+\xi_2)}\cos\alpha x_1\, d\alpha \tag{10.2.240}$$

$$S^{**}_{22} = \frac{A\xi_2}{2\pi}x_2\int_0^\infty \alpha e^{-\alpha(x_2+\xi_2)}\cos\alpha x_1\, d\alpha \tag{10.2.241}$$

$$S^{**}_{12} = -\frac{A\xi_2}{2\pi}\int_0^\infty (1-\alpha x_2)e^{-\alpha(x_2+\xi_2)}\sin\alpha x_1\, d\alpha \tag{10.2.242}$$

In the following, we show that the integral representations of $S^{*}_{\alpha\beta}$, given by Equations 10.2.225 through 10.2.227, and $S^{**}_{\alpha\beta}$, given by Equations 10.2.240 through 10.2.242, can be expressed in terms of elementary functions. To this end we note that

$$\int_0^\infty e^{-\alpha y}\cos\alpha x\, d\alpha = \frac{y}{x^2+y^2} \quad \text{for } y>0 \tag{10.2.243}$$

$$\int_0^\infty e^{-\alpha y}\sin\alpha x\, d\alpha = \frac{x}{x^2+y^2} \quad \text{for } y>0 \tag{10.2.244}$$

$$\int_0^\infty \frac{\cos\alpha x}{\alpha}\left(e^{-\alpha y_1} - e^{-\alpha y_2}\right) d\alpha = -\ln\frac{\sqrt{x^2+y_1^2}}{\sqrt{x^2+y_2^2}} \quad \text{for } y_1>0,\ y_2>0 \tag{10.2.245}$$

Therefore, taking advantage of the formulas 10.2.243 through 10.2.245, the closed forms of $S^*_{\alpha\beta}$ are obtained

$$S^*_{11} = \frac{A}{4\pi}\left\{\ln\frac{\sqrt{x_1^2+(x_2-\xi_2)^2}}{\sqrt{x_1^2+(x_2+\xi_2)^2}} + \left[\frac{(x_2-\xi_2)^2}{x_1^2+(x_2-\xi_2)^2} - \frac{(x_2+\xi_2)^2}{x_1^2+(x_2+\xi_2)^2}\right]\right\} \tag{10.2.246}$$

$$S^*_{22} = \frac{A}{4\pi}\left\{\ln\frac{\sqrt{x_1^2+(x_2-\xi_2)^2}}{\sqrt{x_1^2+(x_2+\xi_2)^2}} - \left[\frac{(x_2-\xi_2)^2}{x_1^2+(x_2-\xi_2)^2} - \frac{(x_2+\xi_2)^2}{x_1^2+(x_2+\xi_2)^2}\right]\right\} \tag{10.2.247}$$

$$S^*_{12} = -\frac{Ax_1}{4\pi}\left[\frac{(x_2-\xi_2)}{x_1^2+(x_2-\xi_2)^2} - \frac{(x_2+\xi_2)}{x_1^2+(x_2+\xi_2)^2}\right] \tag{10.2.248}$$

Note that the closed-form formulas 10.2.246 through 10.2.248 have been obtained by using Equations 10.2.225 through 10.2.227 valid for $|x_1| < \infty, 0 \le x_2 < \xi_2$. It is easy to show that if $\chi^{(1)}$ is replaced by $\chi^{(2)}$ in Equations 10.2.223, then $S^*_{\alpha\beta}$ generated by $\chi^{(2)}$ may also be reduced to Equations 10.2.246 through 10.2.248. Therefore, Equations 10.2.246 through 10.2.248 hold true for any point of the semi-infinite solid.

To show that the integrals 10.2.240 through 10.2.242 can be expressed in terms of elementary functions, let us observe that the differentiation of Equation 10.2.243 with respect to x and y, respectively, yields

$$\int_0^\infty \alpha e^{-\alpha y}\sin\alpha x\, d\alpha = \frac{2xy}{(x^2+y^2)^2} \tag{10.2.249}$$

and

$$\int_0^\infty \alpha e^{-\alpha y}\cos\alpha x\, d\alpha = -\frac{(x^2-y^2)}{(x^2+y^2)^2} \tag{10.2.250}$$

Hence, using Equations 10.2.243 through 10.2.245, and 10.2.249 and 10.2.250, the stress components given by Equations 10.2.235 through 10.2.237 are obtained in the forms

$$S^{**}_{11} = \frac{A\xi_2}{2\pi}\left\{2\frac{(x_2+\xi_2)}{x_1^2+(x_2+\xi_2)^2} - x_2\frac{[(x_2+\xi_2)^2-x_1^2]}{[x_1^2+(x_2+\xi_2)^2]^2}\right\} \tag{10.2.251}$$

$$S^{**}_{22} = \frac{A\xi_2 x_2}{2\pi}\left\{\frac{(x_2+\xi_2)^2-x_1^2}{[x_1^2+(x_2+\xi_2)^2]^2}\right\} \tag{10.2.252}$$

$$S^{**}_{12} = -\frac{A\xi_2 x_1}{2\pi}\left\{\frac{1}{x_1^2+(x_2+\xi_2)^2} - 2\frac{x_2(x_2+\xi_2)}{[x_1^2+(x_2+\xi_2)^2]^2}\right\} \tag{10.2.253}$$

If we introduce the notations

$$r_1 = \sqrt{x_1^2+(x_2-\xi_2)^2}, \quad r_2 = \sqrt{x_1^2+(x_2+\xi_2)^2} \tag{10.2.254}$$

Equations 10.2.246 through 10.2.248, and 10.2.251 through 10.2.253 are, respectively, reduced to

$$S_{11}^{*} = -\frac{A}{4\pi}\left[\ln\frac{r_2}{r_1} + x_1^2\left(\frac{1}{r_1^2} - \frac{1}{r_2^2}\right)\right] \tag{10.2.255}$$

$$S_{22}^{*} = -\frac{A}{4\pi}\left[\ln\frac{r_2}{r_1} - x_1^2\left(\frac{1}{r_1^2} - \frac{1}{r_2^2}\right)\right] \tag{10.2.256}$$

$$S_{12}^{*} = -\frac{Ax_1}{4\pi}\left(\frac{x_2 - \xi_2}{r_1^2} - \frac{x_2 + \xi_2}{r_2^2}\right) \tag{10.2.257}$$

and

$$S_{11}^{**} = -\frac{A}{4\pi}2\xi_2\left[x_2\frac{(x_2 + \xi_2)^2 - x_1^2}{r_2^4} - 2\frac{(x_2 + \xi_2)}{r_2^2}\right] \tag{10.2.258}$$

$$S_{22}^{**} = -\frac{A}{4\pi}(-2)\xi_2 x_2\left[\frac{(x_2 + \xi_2)^2 - x_1^2}{r_2^4}\right] \tag{10.2.259}$$

$$S_{12}^{**} = -\frac{A}{4\pi}(2)\xi_2 x_1\left[\frac{1}{r_2^2} - \frac{2x_2(x_2 + \xi_2)}{r_2^4}\right] \tag{10.2.260}$$

Therefore, the thermal stresses $\overline{S}_{\alpha\beta}$ due to the concentrated heat source in the semi-infinite solid shown in Figure 10.19 are represented by the formula

$$\overline{S}_{\alpha\beta} = S_{\alpha\beta}^{*} + S_{\alpha\beta}^{**} \tag{10.2.261}$$

where $S_{\alpha\beta}^{*}$ and $S_{\alpha\beta}^{**}$ are given by Equations 10.2.255 through 10.2.257 and 10.2.258 through 10.2.260, respectively. In components, Equation 10.2.261 takes the form

$$\overline{S}_{11} = -\frac{A}{4\pi}\left\{\ln\frac{r_2}{r_1} + x_1^2\left(\frac{1}{r_1^2} - \frac{1}{r_2^2}\right) - \frac{2\xi_2}{r_2^4}\left[(x_2 + 2\xi_2)r_2^2 + 2x_1^2 x_2\right]\right\} \tag{10.2.262}$$

$$\overline{S}_{22} = -\frac{A}{4\pi}\left\{\ln\frac{r_2}{r_1} - x_1^2\left(\frac{1}{r_1^2} - \frac{1}{r_2^2}\right) - \frac{2\xi_2 x_2}{r_2^4}\left[(x_2 + \xi_2)^2 - x_1^2\right]\right\} \tag{10.2.263}$$

$$\overline{S}_{12} = -\frac{Ax_1}{4\pi}\left[\frac{x_2 - \xi_2}{r_1^2} - \frac{x_2 + \xi_2}{r_2^2} + \frac{2\xi_2}{r_2^4}\left(x_1^2 + \xi_2^2 - x_2^2\right)\right] \tag{10.2.264}$$

It follows from Equations 10.2.262 through 10.2.264 that

$$\overline{S}_{22}(x_1, 0) = \overline{S}_{12}(x_1, 0) = 0 \quad \text{for } |x_1| < \infty \tag{10.2.265}$$

and

$$\overline{S}_{11}(x_1, 0) = \frac{A}{\pi}\frac{\xi_2^2}{x_1^2 + \xi_2^2} \tag{10.2.266}$$

If we introduce the dimensionless variables

$$\xi = x_1/\xi_2, \quad \overline{S}_1(\xi) = \frac{\pi}{A}\overline{S}_{11}(x_1, 0) \tag{10.2.267}$$

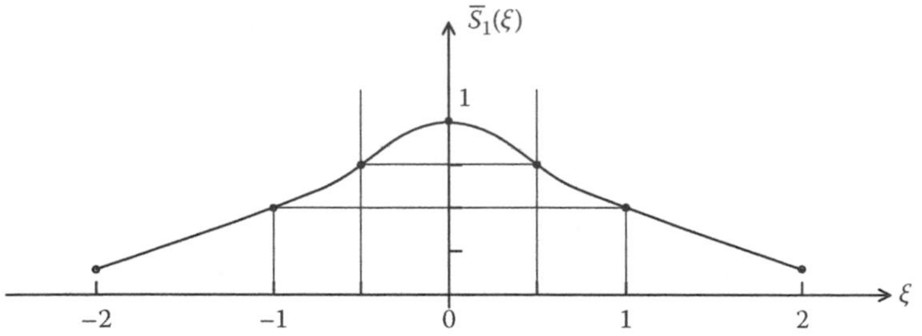

FIGURE 10.20 A dimensionless hoop stress on the boundary of the semi-infinite sheet subject to an internal concentrated heat source.

then

$$\overline{S}_1(\xi) = \frac{1}{1+\xi^2}, \quad |\xi| < \infty \tag{10.2.268}$$

The function $\overline{S}_1 = \overline{S}_1(\xi)$ represents a dimensionless hoop stress on the boundary of the semi-infinite sheet subject to the concentrated heat source beneath the stress free boundary. A graph of $\overline{S}_1(\xi)$ is shown in Figure 10.20.

In concluding this section, we note that a temperature field $\overline{T} = \overline{T}(\mathbf{x})$ produced by the concentrated heat source in the semi-infinite solid $|x_1| < \infty, 0 \leq x_2 < \infty$ in which the stress free boundary $x_2 = 0$ is kept at a zero temperature, takes the form

$$\overline{T}(\mathbf{x}) = \frac{Q_0}{2\pi\kappa} \ln \frac{r_2}{r_1} \tag{10.2.269}$$

where r_1 and r_2 are given by Equation 10.2.254.

PROBLEMS

10.1 Find an elastic state $s = [\mathbf{u}, \mathbf{E}, \mathbf{S}]$ corresponding to a concentrated body force in an interior of a homogeneous and isotropic semispace $|x_1| < \infty$, $x_2 > 0$, under plane strain conditions, when the boundary of semispace is stress free and the elastic state satisfies suitable asymptotic conditions at infinity.

10.2 Find an elastic state $s = [\mathbf{u}, \mathbf{E}, \mathbf{S}]$ corresponding to a concentrated body force in an interior of a homogeneous and isotropic semispace $|x_1| < \infty$, $x_2 > 0$, under plane strain conditions, when the boundary of semispace is clamped and the elastic state vanishes at infinity.

10.3 Suppose that a homogeneous isotropic infinite elastic wedge, subject to generalized plane stress conditions, is loaded in its plane by a concentrated force $\boldsymbol{\ell}$ applied at its tip (see Figure P10.3)

$$x_1 = r\cos\varphi \quad \theta = \frac{\pi}{2} - \varphi$$

$$x_2 = r\sin\varphi \quad |\theta| \leq \alpha$$

Show that the stress components $\overline{S}_{rr}$, $\overline{S}_{r\varphi}$, and $\overline{S}_{\varphi\varphi}$ corresponding to the force $\boldsymbol{\ell}$ and vanishing at infinity take the form

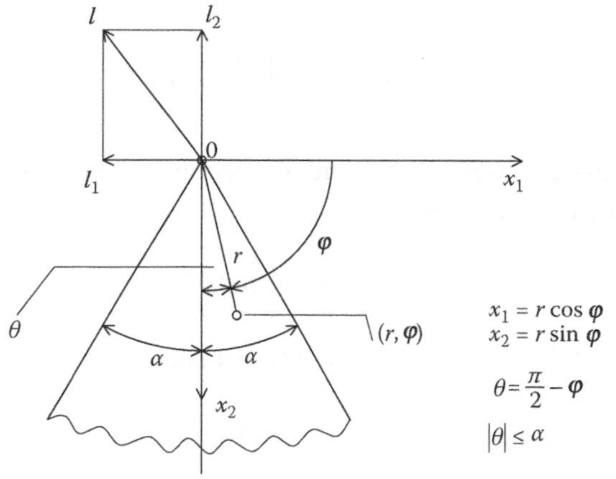

FIGURE P10.3

$$\overline{S}_{rr}(r,\varphi) = \frac{2l_1\cos\varphi}{r(2\alpha - \sin 2\alpha)} + \frac{2l_2\sin\varphi}{r(2\alpha + \sin 2\alpha)}$$

$$\overline{S}_{r\varphi}(r,\varphi) = \overline{S}_{\varphi\varphi}(r,\varphi) = 0$$

for every $0 < r < \infty$, $\frac{\pi}{2} - \alpha \le \varphi \le \frac{\pi}{2} + \alpha$. Note that $l_1 < 0$ and $l_2 < 0$ and $\overline{S}_{rr}$ is infinite for $\alpha \to 0$ and $r > 0$.

10.4 Show that for a homogeneous isotropic infinite elastic wedge under generalized plane stress conditions loaded by a concentrated moment M at its tip (see Figure P10.4)

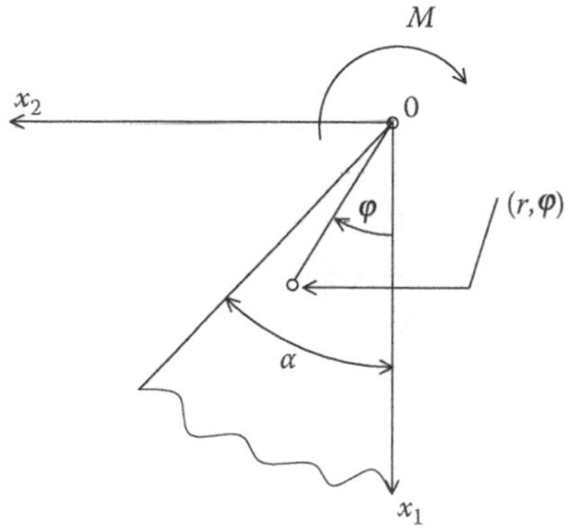

FIGURE P10.4

the stress components $\overline{S}_{rr}$, $\overline{S}_{r\varphi}$, and $\overline{S}_{\varphi\varphi}$ vanishing at infinity take the form

$$\overline{S}_{rr}(r,\varphi) = \frac{2M}{r^2}\frac{\sin(2\varphi - \alpha)}{\sin\alpha - \alpha\cos\alpha}$$

$$\overline{S}_{r\varphi}(r,\varphi) = -\frac{M}{r^2}\frac{\cos(2\varphi - \alpha) - \cos\alpha}{\sin\alpha - \alpha\cos\alpha}$$

$$\overline{S}_{\varphi\varphi}(r,\varphi) = 0 \quad \text{for every } r > 0,\ 0 < \varphi < \alpha$$

where

$$M = -r \int_0^{\alpha} (\overline{S}_{r\varphi} r)\, d\varphi$$

Note that the stress components $\overline{S}_{rr}$ and $\overline{S}_{r\varphi}$ become unbounded for $\alpha = \alpha^*$, where α^* is the only root of the equation

$$\sin\alpha^* - \alpha^* \cos\alpha^* = 0$$

that is, for $\alpha^* = 257.4°$. Hence, the solution makes sense for an elastic wedge that obeys the condition $0 < \alpha < \alpha^*$.

10.5 Consider a homogeneous isotropic infinite elastic strip under generalized plane stress conditions:

$|x_1| \leq 1$, $|x_2| < \infty$ subject to the temperature field of the form

$$\overline{T}(x_1, x_2) = T_0[1 - H(x_2)] \tag{a}$$

where T_0 is a constant temperature and $H = H(x)$ is the Heaviside function

$$H(x) = \begin{cases} 1 & \text{for} \quad x > 0 \\ \frac{1}{2} & \text{for} \quad x = 0 \\ 0 & \text{for} \quad x < 0 \end{cases} \tag{b}$$

Note that in this case, we complemented the definition of the Heaviside function by specifying its value at $x = 0$.

(see Figure P10.5)

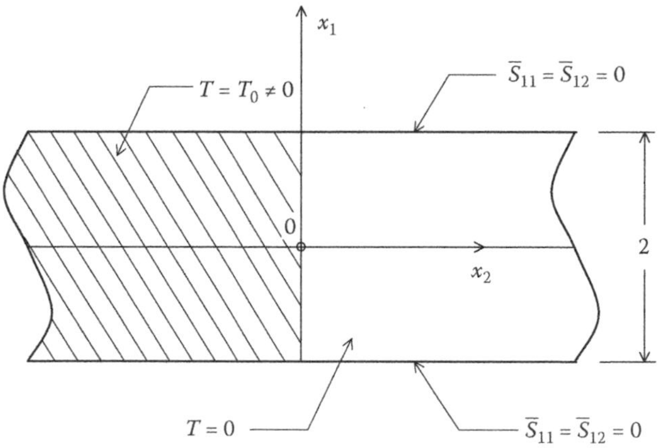

FIGURE P10.5

Show that the stress tensor field $\overline{\mathbf{S}} = \overline{\mathbf{S}}(x_1, x_2)$ corresponding to the discontinuous temperature (a) is represented by the sum

$$\overline{\mathbf{S}} = \overline{\mathbf{S}}^{(1)} + \overline{\mathbf{S}}^{(2)} \tag{c}$$

where

$$\overline{S}_{11}^{(1)} = -E\alpha T_0[1 - H(x_2)], \quad \overline{S}_{22}^{(1)} = \overline{S}_{12}^{(1)} = 0 \tag{d}$$

and

$$\overline{S}_{11}^{(2)} = F_{,22}, \quad \overline{S}_{22}^{(2)} = F_{,11}, \quad \overline{S}_{12}^{(2)} = -F_{,12} \tag{e}$$

where the biharmonic function $F = F(x_1, x_2)$ is given by

$$F(x_1, x_2) = E\alpha T_0 \left[\frac{x_2^2}{4} + \frac{2}{\pi} \int_0^\infty (A \cosh \beta x_1 + B\beta x_1 \sinh \beta x_1) \frac{\sin \beta x_2}{\beta^3} d\beta \right] \tag{f}$$

$$A = \frac{\sinh \beta + \beta \cosh \beta}{\sinh 2\beta + 2\beta}, \quad B = -\frac{\sinh \beta}{\sinh 2\beta + 2\beta} \tag{g}$$

Hint: Note that

$$\begin{aligned} \overline{S}_{11}^{(2)}(\pm 1, x_2) &= -\overline{S}_{11}^{(1)}(\pm 1, x_2) \\ &= \frac{E\alpha T_0}{2} \left(1 - \frac{2}{\pi} \int_0^\infty \frac{\sin \beta x_2}{\beta} d\beta \right) \quad \text{for } |x_2| < \infty \end{aligned}$$

REFERENCES

1. S. Timoshenko and J. N. Goodier, *Theory of Elasticity*, 3rd edn., McGraw-Hill, New York, 1970, pp. 80–82.
2. G. Kirsch, Die Theorie der Elastizität und die Bedurfnisse der Festigkeitslehre, *Z. VDI*, 42, 1898, 797.

11 Solutions to Particular Three-Dimensional Initial–Boundary Value Problems of Elastodynamics

In this chapter, both the displacement and stress languages are used to present a number of closed-form solutions of three-dimensional isothermal and nonisothermal elastodynamics. The solutions of isothermal elastodynamics include (a) the plane progressive elastic waves in E^3, (b) the elastic waves produced by a moving point force in E^3, (c) the stress waves due to the initial stress and stress-rate fields in E^3, and (d) the elastic waves generated by a pressurization of a spherical cavity in E^3. A decomposition formula for the stress energy density of a progressive wave [see (a)] is obtained here for the first time. Also, the solutions (b) and (c) are obtained here for the first time. The solutions of nonisothermal elastodynamics include (1) a radially symmetric thermoelastic process corresponding to an instantaneous concentrated heat source in an infinite body, and (2) a radially symmetric thermoelastic process corresponding to an instantaneous spherical temperature inclusion in an infinite body. In addition, Saint-Venant's principle of isothermal elastodynamics for a semi-infinite inhomogeneous anisotropic elastic cylinder, in terms of stresses, is formulated here for the first time. Problems on radial symmetry with solutions are presented, and their solutions are provided in the Solutions Manual.

11.1 THREE-DIMENSIONAL SOLUTIONS OF ISOTHERMAL ELASTODYNAMICS

In Section 4.2.4, the formulations of an initial–boundary value problem (IBVP) of isothermal elastodynamics were discussed. In particular, a mixed problem in terms of displacements and a mixed problem in terms of stresses were analyzed. Also, it was shown to what extent a displacement formulation could be embedded in a stress formulation, and how a stress formulation could be used to describe a propagation of elastic waves in a body with defects. In this section, we offer a number of closed-form solutions of isothermal elastodynamics for a three-dimensional unbounded body. The following problems are dealt with:

(A) The plane progressive elastic waves in E^3
(B) The elastic waves due to application of an instantaneous concentrated body force in E^3
(C) The elastic waves produced by a moving point force in E^3
(D) The stress waves due to the initial stress and stress-rate fields in E^3
(E) The elastic waves generated by a pressurization of a spherical cavity in an infinite body

The elastic waves of Part (A) are described in both pure displacement and pure stress languages. A natural pure stress description of incompatible elastodynamics is used to study the stress waves in Part (D); the elastic waves of Parts (B), (C), and (E) are described using the displacement language only.

11.1.1 The Plane Progressive Waves in a Homogeneous Anisotropic Elastic Unbounded Body

11.1.1.1 The Displacement Progressive Waves in E^3

Consider a homogeneous anisotropic infinite elastic solid, characterized by a constant density $\varrho > 0$ and a fourth-order elasticity tensor **C**. A *displacement progressive wave* in such a solid is defined as a solution to the displacement equation of motion (see Equation 4.2.2)

$$\operatorname{div} \mathbf{C}[\nabla \mathbf{u}] - \varrho \ddot{\mathbf{u}} = \mathbf{0} \quad \text{on } E^3 \times (-\infty, +\infty) \tag{11.1.1}$$

of the form

$$\mathbf{u}(\mathbf{x}, t) = \mathbf{a} f(\mathbf{x} \cdot \mathbf{m} - ct) \quad \text{on } E^3 \times (-\infty, +\infty) \tag{11.1.2}$$

where

$f = f(s)$ is a real-valued function of class C^2 on $(-\infty, +\infty)$ with $f''(s) \not\equiv 0$
a and **m** are unit vectors
c is *a* positive constant

The vectors **a** and **m** are called the *direction of motion* and the *direction of propagation*, respectively, while c is called the *velocity of propagation*.

It follows from Equation 11.1.2 that

$$\mathbf{u}(\mathbf{x}, 0) = \mathbf{a} f(\mathbf{x} \cdot \mathbf{m}), \quad \dot{\mathbf{u}}(\mathbf{x}, 0) = -c\mathbf{a} f'(\mathbf{x} \cdot \mathbf{m}) \quad \text{on } E^3 \tag{11.1.3}$$

Therefore, the displacement progressive wave is a solution to Equation 11.1.1 corresponding to the initial conditions 11.1.3, provided the function $f = f(s)$, the parameter $c > 0$, and the vectors **a** and **m** are prescribed.

The name of a progressive wave for the function $\mathbf{u} = \mathbf{u}(\mathbf{x}, t)$ comes from the fact that **u** takes a constant value vector on any plane defined by

$$P_t = \{\mathbf{x} : \mathbf{x} \cdot \mathbf{m} - ct = s_0\} \tag{11.1.4}$$

where $s_0 \in (-\infty, \infty)$, $t \geq 0$, and the plane, treated as a function of t, is moving with the velocity c in the direction **m**.

In the following we are to show that **u** of the form 11.1.2 represents a progressive wave if and only if **a** satisfies the *Fresnel–Hadamard propagation condition*:

$$[\mathbf{A}(\mathbf{m}) - c^2 \mathbf{1}]\mathbf{a} = \mathbf{0} \tag{11.1.5}$$

where, for a given unit vector **m**, **A(m)** represents the *acoustic tensor for the direction* **m**, defined by

$$\mathbf{A}(\mathbf{m})\mathbf{a} = \varrho^{-1} \mathbf{C}[\mathbf{a} \otimes \mathbf{m}]\mathbf{m} \quad \text{for every vector } \mathbf{a} \tag{11.1.6}$$

In components, Equation 11.1.5 takes the form

$$[A_{ik}(\mathbf{m}) - c^2\delta_{ik}]a_k = 0 \tag{11.1.7}$$

where

$$A_{ik}(\mathbf{m}) = \varrho^{-1}C_{ijkl}m_j m_l \tag{11.1.8}$$

To prove that **u** given by Equation 11.1.2 represents a progressive wave if and only if the condition 11.1.5 is satisfied, we rewrite Equations 11.1.1 and 11.1.2 in components; they are, respectively,

$$C_{ijkl}u_{k,lj} - \varrho\ddot{u}_i = 0 \tag{11.1.9}$$

and

$$u_i = a_i f(x_a m_a - ct) \tag{11.1.10}$$

The differentiation of Equation 11.1.10 leads to

$$\ddot{u}_i = c^2 a_i f''(x_a m_a - ct) \tag{11.1.11}$$

$$u_{k,lj} = a_k m_l m_j f''(x_a m_a - ct) \tag{11.1.12}$$

Substituting Equations 11.1.11 and 11.1.12 into Equation 11.1.9, and using the condition $f''(s) \not\equiv 0$ on $(-\infty, +\infty)$, we have

$$(\varrho^{-1}C_{ijkl}m_j m_l - c^2\delta_{ik})a_k = 0 \tag{11.1.13}$$

Therefore, if u_i of the form 11.1.10 is a progressive wave, then the condition 10.1.5 is satisfied. Conversely, if a_k is a nonvanishing solution of Equation 11.1.13, then multiplying Equation 11.1.13 by $f''(x_a m_a - ct) \not\equiv 0$ we find that u_i, given by Equation 11.1.10, satisfies Equation 11.1.9, which means that u_i represents a displacement progressive wave. This completes the proof.

Hence, a necessary and sufficient condition that **u** defined by Equation 11.1.2 be a progressive wave is that **a** is an eigenvector of the acoustic tensor **A**(**m**) corresponding to an eigenvalue c^2.

Since **u** is defined over the moving plane P_t (see Equation 11.1.4), it may be split into orthogonal parts that are normal and tangential to the plane, by using the formula

$$\mathbf{u} = \mathbf{u}^{\perp} + \mathbf{u}^{\parallel} \tag{11.1.14}$$

where

$$\mathbf{u}^{\perp} = (\mathbf{m} \cdot \mathbf{u})\mathbf{m} \tag{11.1.15}$$

and

$$\mathbf{u}^{\parallel} = (\mathbf{1} - \mathbf{m} \otimes \mathbf{m})\mathbf{u} \tag{11.1.16}$$

Note that

$$\mathbf{u}^{\perp} \cdot \mathbf{u}^{\parallel} = 0 \tag{11.1.17}$$

$$\mathbf{u} \cdot \mathbf{m} = \mathbf{u}^{\perp} \cdot \mathbf{m} \tag{11.1.18}$$

$$\mathbf{u}^{\parallel} \cdot \mathbf{m} = 0 \tag{11.1.19}$$

The decomposition formula 11.1.14, when related to the progressive wave 11.1.2, leads to the following definitions. A progressive wave is

$$\text{longitudinal} \Leftrightarrow \mathbf{u}^{\parallel} = \mathbf{0} \tag{11.1.20}$$

$$\text{transverse} \Leftrightarrow \mathbf{u}^{\perp} = \mathbf{0} \tag{11.1.21}$$

or, a progressive wave is

$$\text{longitudinal} \Leftrightarrow \mathbf{a} \times \mathbf{m} = \mathbf{0} \tag{11.1.22}$$

$$\text{transverse} \Leftrightarrow \mathbf{a} \cdot \mathbf{m} = 0 \tag{11.1.23}$$

From Equation 11.1.10 follows

$$\operatorname{div} \mathbf{u} = \mathbf{a} \cdot \mathbf{m} f'(\mathbf{x} \cdot \mathbf{m} - ct) \tag{11.1.24}$$

$$\operatorname{curl} \mathbf{u} = \mathbf{m} \times \mathbf{a} f'(\mathbf{x} \cdot \mathbf{m} - ct) \tag{11.1.25}$$

therefore, a progressive wave is

$$\text{longitudinal} \Leftrightarrow \operatorname{curl} \mathbf{u} = \mathbf{0} \tag{11.1.26}$$

$$\text{transverse} \Leftrightarrow \operatorname{div} \mathbf{u} = 0 \tag{11.1.27}$$

In other words, a progressive wave is longitudinal if the direction of motion and the direction of propagation coincide, and a progressive wave is transverse if the direction of motion is orthogonal to the direction of propagation.

The strain field **E** associated with a progressive wave **u** may also be split into orthogonal parts which are normal and tangential to the plane P_t, by using the formula [1]

$$\mathbf{E} = \mathbf{E}^{\perp} + \mathbf{E}^{\parallel} \tag{11.1.28}$$

where

$$\mathbf{E}^{\perp} = 2\,\mathrm{sym}(\mathbf{m} \otimes \mathbf{E}\mathbf{m}) - (\mathbf{m} \cdot \mathbf{E}\mathbf{m})\mathbf{m} \otimes \mathbf{m} \tag{11.1.29}$$

and

$$\mathbf{E}^{\parallel} = (\mathbf{1} - \mathbf{m} \otimes \mathbf{m})\mathbf{E}(\mathbf{1} - \mathbf{m} \otimes \mathbf{m}) \tag{11.1.30}$$

and the following relations hold true

$$\mathbf{E}^{\perp} \cdot \mathbf{E}^{\parallel} = 0 \tag{11.1.31}$$

$$\mathbf{E}\mathbf{m} = \mathbf{E}^{\perp}\mathbf{m} \tag{11.1.32}$$

$$\mathbf{E}^{\parallel}\mathbf{m} = \mathbf{0} \tag{11.1.33}$$

The function $\mathcal{E}_E(\mathbf{E})$ defined by

$$\mathcal{E}_E(\mathbf{E}) = \frac{1}{2}\mathbf{E} \cdot \mathbf{C}[\mathbf{E}] \tag{11.1.34}$$

represents the *strain energy density* of the progressive wave 11.1.2, while the functions $\mathcal{E}_E(\mathbf{E}^{\perp})$ and $\mathcal{E}_E(\mathbf{E}^{\parallel})$ represent the "*normal*" and "*tangential*" strain energy densities, respectively, associated with the wave 11.1.2. Substituting $\mathbf{u}$ from Equation 11.1.2 into the strain–displacement relation

$$\mathbf{E} = \hat{\nabla}\mathbf{u} \tag{11.1.35}$$

we have

$$\mathbf{E} = \text{sym}(\mathbf{a} \otimes \mathbf{m}) f'(\mathbf{x} \cdot \mathbf{m} - ct) \tag{11.1.36}$$

Hence, and from Equations 11.1.29 and 11.1.30, we obtain

$$\mathbf{E}^{\perp} = \text{sym}\,(\mathbf{a} \otimes \mathbf{m}) f'(s_0) \tag{11.1.37}$$

and

$$\mathbf{E}^{\parallel} = \mathbf{0} \tag{11.1.38}$$

Therefore, it follows from Equation 11.1.34 that

$$\mathcal{E}_E(\mathbf{E})|_{P_t} = \mathcal{E}_E(\mathbf{E}^{\perp}) = \frac{1}{2}[f'(s_0)]^2(\mathbf{m} \otimes \mathbf{a}) \cdot \mathbf{C}[\mathbf{m} \otimes \mathbf{a}] \tag{11.1.39}$$

Now, since $|\mathbf{a}| = 1$, therefore, multiplying Equation 11.1.5 by $\mathbf{a}$ in the dot sense, we get

$$\mathbf{a} \cdot (\mathbf{C}[\mathbf{m} \otimes \mathbf{a}]\mathbf{m}) = \varrho c^2 \tag{11.1.40}$$

or, by using the identity

$$\mathbf{a} \cdot (\mathbf{C}[\mathbf{m} \otimes \mathbf{a}]\mathbf{m}) = (\mathbf{m} \otimes \mathbf{a}) \cdot \mathbf{C}[\mathbf{m} \otimes \mathbf{a}] \tag{11.1.41}$$

we find

$$(\mathbf{m} \otimes \mathbf{a}) \cdot \mathbf{C}[\mathbf{m} \otimes \mathbf{a}] = \varrho c^2 \tag{11.1.42}$$

Hence, Equations 11.1.39 and 11.1.42 imply the familiar form of the energy

$$E = Mc^2 \tag{11.1.43}$$

where

$$E = \mathcal{E}_E(\mathbf{E}^{\perp})L^3 \tag{11.1.44}$$

$$M = \frac{1}{2}L^3[f'(s_0)]^2\varrho \tag{11.1.45}$$

and L stands for a parameter of the length dimension. Note that

$$[\mathbf{a}] = [\mathbf{m}] = [f'(s_0)] = [1] \tag{11.1.46}$$

therefore, from Equations 11.1.44 and 11.1.45 we get

$$[E] = [\text{Force} \times \text{Length}], \quad [M] = [\text{Mass}] \tag{11.1.47}$$

The problem of existence of a progressive wave, that is, the problem of the existence of a solution to the eigenproblem for Equation 11.1.5 is settled by the following theorem [2]

Theorem 1: If C is symmetric and strongly elliptic (see Equations 3.3.98 and 3.3.99), there exist, for every direction $\mathbf{m}$, three orthogonal directions of motion $\mathbf{a}_1$, $\mathbf{a}_2$, $\mathbf{a}_3$ and three associated velocities of propagation c_1, c_2, c_3 for progressive waves. ■

Proof: If C is symmetric, then $\mathbf{A}(\mathbf{m})$ is symmetric. To prove this we note that by Equation 11.1.8

$$A_{ik}(\mathbf{m}) = \varrho^{-1}C_{ijkl}m_jm_l \tag{11.1.48}$$

Since $C_{ijkl} = C_{klij}$, therefore

$$C_{ijkl}m_jm_l = C_{klij}m_jm_l = C_{kjil}m_lm_j \tag{11.1.49}$$

and Equations 11.1.48 and 11.1.49 imply that

$$A_{ik}(\mathbf{m}) = A_{ki}(\mathbf{m}) \quad \text{for every } \mathbf{m} \tag{11.1.50}$$

The symmetry of $\mathbf{A}(\mathbf{m})$ implies that $\mathbf{A}(\mathbf{m})$ has at least three orthogonal principal directions $\mathbf{a}_1$, $\mathbf{a}_2$, $\mathbf{a}_3$ and three associated principal values c_1^2, c_2^2, c_3^2 for every $\mathbf{m}$.

Also, the strong ellipticity of C implies that

$$(\mathbf{m} \otimes \mathbf{a}) \cdot \mathsf{C}[\mathbf{m} \otimes \mathbf{a}] > 0 \quad \text{for every } \mathbf{a} \text{ and } \mathbf{m} \tag{11.1.51}$$

Therefore, by Equation 11.1.42, c_1, c_2, and c_3 are real, and this completes the proof of Theorem 1.

Theorem 1 holds true for an arbitrary homogeneous anisotropic elastic solid. In a particular case of a homogeneous isotropic solid, Theorem 1 implies the following result:

Theorem 2: For an isotropic material, there are but two types of progressive waves: longitudinal and transverse; a longitudinal wave propagates with the velocity $c = c_1 = \sqrt{(\lambda + 2\mu)/\varrho}$, while a transverse wave propagates with the velocity $c = c_2 = \sqrt{\mu/\varrho}$. ■

Proof: For an isotropic material with Lamé moduli λ and μ we have

$$\mathbf{C}\,[\mathbf{E}] = 2\mu\mathbf{E} + \lambda(\operatorname{tr}\mathbf{E})\mathbf{1} \quad \text{for } \mathbf{E} = \mathbf{E}^T \tag{11.1.52}$$

Hence,

$$\mathbf{C}[\operatorname{sym}(\mathbf{a} \otimes \mathbf{m})] = \mathbf{C}[\mathbf{a} \otimes \mathbf{m}] = 2\mu \operatorname{sym}(\mathbf{a} \otimes \mathbf{m}) + (\mathbf{a} \cdot \mathbf{m})\mathbf{1} \tag{11.1.53}$$

and

$$(\mathbf{C}[\mathbf{a} \otimes \mathbf{m}])\mathbf{m} = [\mu(\mathbf{1} - \mathbf{m} \otimes \mathbf{m}) + (\lambda + 2\mu)\mathbf{m} \otimes \mathbf{m}]\mathbf{a} \tag{11.1.54}$$

Dividing this equation by $\varrho > 0$, and using the definition of $\mathbf{A}(\mathbf{m})$ (see Equation 11.1.6), we obtain

$$\mathbf{A}(\mathbf{m}) = c_1^2\mathbf{m} \otimes \mathbf{m} + c_2^2(\mathbf{1} - \mathbf{m} \otimes \mathbf{m}) \tag{11.1.55}$$

where

$$c_1^2 = \frac{\lambda + 2\mu}{\varrho}, \quad c_2^2 = \frac{\mu}{\varrho} \tag{11.1.56}$$

Therefore, the line spanned by $\mathbf{m}$ is the characteristic space for c_1^2, while the plane perpendicular to $\mathbf{m}$ is the characteristic space for c_2^2. This completes the proof of Theorem 2.

11.1.1.2 The Stress Progressive Waves in E^3

In this section, pure stress equations of motion of linear homogeneous anisotropic elastodynamics are used to present a *tensorial classification of elastic waves* associated with the vectorial progressive waves of Section 11.1.1.1.

We represent the stress progressive wave $\mathbf{S}$ on $E^3 \times (-\infty, +\infty)$ in the form

$$\mathbf{S}(\mathbf{x}, t) = \overset{\circ}{\mathbf{S}}\, \psi(\mathbf{x} \cdot \mathbf{m} - ct) \tag{11.1.57}$$

where ψ is a real-valued scalar function of class C^2 on $(-\infty, +\infty)$ with $\psi''(s) \not\equiv 0$, $\overset{\circ}{\mathbf{S}}$ is a unit second-order symmetric tensor, $\mathbf{m}$ is a unit vector, and c is a positive constant. Equation 11.1.57 represents a plane stress progressive wave propagating in the direction $\mathbf{m}$ with velocity $c > 0$.

Similarly to the decomposition formula for the strain tensor **E** (see Equations 11.1.28 through 11.1.33) we can write **S** as the sum of the "normal" and "tangential" parts with respect to the plane $P_t = \{\mathbf{x} : \mathbf{x} \cdot \mathbf{m} - ct = s_0\}$, ($s_0 = $ const.), viz.,

$$\mathbf{S} = \mathbf{S}^{\perp} + \mathbf{S}^{\parallel} \tag{11.1.58}$$

where the normal component is

$$\mathbf{S}^{\perp} = 2\,\mathrm{sym}\,(\mathbf{m} \otimes \mathbf{Sm}) - (\mathbf{m} \cdot \mathbf{Sm})\mathbf{m} \otimes \mathbf{m} \tag{11.1.59}$$

and the tangential component is

$$\mathbf{S}^{\parallel} = (\mathbf{1} - \mathbf{m} \otimes \mathbf{m})\mathbf{S}(\mathbf{1} - \mathbf{m} \otimes \mathbf{m}) \tag{11.1.60}$$

Note that (see Equations 11.1.31 through 11.1.33)

$$\mathbf{S}^{\perp} \cdot \mathbf{S}^{\parallel} = 0, \quad \mathbf{S}^{\perp}\mathbf{m} = \mathbf{Sm}, \quad \mathbf{S}^{\parallel}\mathbf{m} = \mathbf{0} \tag{11.1.61}$$

In a homogeneous anisotropic elastic medium with zero body force, the tensor **S** must satisfy the stress equation of motion (see Equation 4.2.9 with $\mathbf{B} = \mathbf{0}$ and $\varrho = $ const.)

$$\hat{\nabla}(\mathrm{div}\,\mathbf{S}) - \varrho\mathsf{K}[\ddot{\mathbf{S}}] = \mathbf{0} \tag{11.1.62}$$

where $\hat{\nabla}$ is the symmetric gradient operator, K is the compliance tensor, and ϱ is the density of the medium.

From Equation 11.1.57,

$$\hat{\nabla}(\mathrm{div}\,\mathbf{S}) = \mathrm{sym}\,(\mathbf{m} \otimes \overset{\circ}{\mathbf{S}}\,\mathbf{m})\psi'' \tag{11.1.63}$$

and

$$\mathrm{curl\,curl}\,\mathbf{S} = \mathbf{H}[\overset{\circ}{\mathbf{S}}]\psi'' \tag{11.1.64}$$

where (in Cartesian components)

$$H_{ijkl} = \epsilon_{ipk}\epsilon_{jql}m_p m_q \tag{11.1.65}$$

$$\psi'' = \psi''(s), \quad s = \mathbf{x} \cdot \mathbf{m} - ct \tag{11.1.66}$$

and ϵ_{ijk} is the three-dimensional alternator, therefore,

$$\mathrm{curl\,curl}\,\mathbf{S} = \mathbf{0} \Leftrightarrow \mathbf{S}^{\parallel} = \mathbf{0} \tag{11.1.67}$$

and

$$\hat{\nabla}(\mathrm{div}\mathbf{S}) = \mathbf{0} \Leftrightarrow \mathbf{S}^{\perp} = \mathbf{0} \tag{11.1.68}$$

Hence, a stress progressive wave is

$$\text{“normal''} \Leftrightarrow \mathbf{H}[\overset{\circ}{\mathbf{S}}] = \mathbf{0} \qquad (11.1.69)$$

$$\text{“tangential''} \Leftrightarrow \overset{\circ}{\mathbf{S}}\,\mathbf{m} = \mathbf{0} \qquad (11.1.70)$$

Also, it follows from Equation 11.1.57 that

$$\ddot{\mathbf{S}} = c^2\,\overset{\circ}{\mathbf{S}}\,\psi'' \qquad (11.1.71)$$

Therefore, substituting Equations 11.1.63 and 11.1.71 into Equation 11.1.62, we arrive at the propagation condition for a stress progressive wave

$$\text{sym}(\mathbf{m}\otimes \overset{\circ}{\mathbf{S}}\,\mathbf{m}) - \varrho c^2 \mathsf{K}[\overset{\circ}{\mathbf{S}}] = \mathbf{0} \qquad (11.1.72)$$

Since $\mathsf{K} = \mathsf{C}^{-1}$, where C is the elasticity tensor, an equivalent form of Equation 11.1.72 reads

$$\mathsf{C}\,[\mathbf{m}\otimes \overset{\circ}{\mathbf{S}}\,\mathbf{m}] - \varrho c^2\,\overset{\circ}{\mathbf{S}} = \mathbf{0} \qquad (11.1.73)$$

By applying the operator $\mathbf{H}[\cdot]$ to Equation 11.1.72, and using the identity

$$\mathbf{H}[\text{sym}(\mathbf{m}\otimes \overset{\circ}{\mathbf{S}}\,\mathbf{m})] = \mathbf{0} \qquad (11.1.74)$$

we find that a solution $\overset{\circ}{\mathbf{S}}$ to Equation 11.1.72 must satisfy the compatibility condition

$$\mathbf{H}[\mathsf{K}[\overset{\circ}{\mathbf{S}}]] = \mathbf{0} \qquad (11.1.75)$$

Note that this condition is satisfied identically if $\overset{\circ}{\mathbf{S}}$ corresponds to a displacement progressive wave of the form 11.1.2, that is, if we set

$$\overset{\circ}{\mathbf{S}} = \mathsf{C}\,[\text{sym}\,(\mathbf{m}\otimes\mathbf{a})] = \mathsf{C}\,[\mathbf{m}\otimes\mathbf{a}] \qquad (11.1.76)$$

where $\mathbf{a}$ is an arbitrary vector. Moreover, if $\mathbf{a}$ satisfies the Fresnel–Hadamard propagation condition (see Equation 11.1.5), then $\overset{\circ}{\mathbf{S}}$ given by Equation 11.1.76 represents a solution to Equation 11.1.72.

Taking the inner product of Equation 11.1.72 with $\overset{\circ}{\mathbf{S}}$, we write

$$c(\overset{\circ}{\mathbf{S}}) = \frac{|\overset{\circ}{\mathbf{S}}\,\mathbf{m}|}{\left\{\varrho\,\overset{\circ}{\mathbf{S}}\cdot\mathsf{K}[\overset{\circ}{\mathbf{S}}]\right\}^{1/2}} \qquad (11.1.77)$$

where $c(\overset{\circ}{\mathbf{S}})$ is the velocity of a tensorial wave. Note that $c(\overset{\circ}{\mathbf{S}})$ represents the velocity of a normal wave if and only if $\mathbf{H}[\overset{\circ}{\mathbf{S}}] = \mathbf{0}$, and one can put $c(\overset{\circ}{\mathbf{S}}^{\|}) = 0$ if $\overset{\circ}{\mathbf{S}}^{\|}\cdot\mathsf{K}[\overset{\circ}{\mathbf{S}}^{\|}] = 0$.

An alternative form of the velocity formula 11.1.77 is obtained if we note that the function

$$\mathcal{E}_S(\mathbf{S}) = \frac{1}{2}\mathbf{S}\cdot\mathsf{K}[\mathbf{S}] \tag{11.1.78}$$

represents the *stress energy density* of the progressive wave 11.1.57, while the functions

$$\mathcal{E}_S(\mathbf{S}^{\perp}) = \frac{1}{2}\mathbf{S}^{\perp}\cdot\mathsf{K}[\mathbf{S}^{\perp}] \tag{11.1.79}$$

and

$$\mathcal{E}_S(\mathbf{S}^{\parallel}) = \frac{1}{2}\mathbf{S}^{\parallel}\cdot\mathsf{K}[\mathbf{S}^{\parallel}] \tag{11.1.80}$$

represent the "normal" *and* "tangential" *stress energy densities*, respectively, associated with the wave 11.1.57.

Taking the inner product of Equation 11.1.72 with $\overset{\circ}{\mathbf{S}}{}^{\parallel}$ and using the third of Equations 11.1.61 as well as symmetry of K, we arrive at $(c>0)$

$$\overset{\circ}{\mathbf{S}}\cdot\mathsf{K}[\overset{\circ}{\mathbf{S}}{}^{\parallel}] = 0 \tag{11.1.81}$$

Multiplying this by $\psi^2 > 0$ we have

$$\mathbf{S}\cdot\mathsf{K}[\mathbf{S}^{\parallel}] = 0 \tag{11.1.82}$$

This formula shows that the *work done by a stress progressive wave on the strain associated with the* "tangential" *part of the wave is always zero.*

Next, from Equation 11.1.81 and the relation

$$\overset{\circ}{\mathbf{S}}\cdot\mathsf{K}[\mathbf{S}^{\circ}] = \overset{\circ}{\mathbf{S}}{}^{\perp}\cdot\mathsf{K}[\overset{\circ}{\mathbf{S}}{}^{\perp}] + \overset{\circ}{\mathbf{S}}{}^{\parallel}\cdot\mathsf{K}[\overset{\circ}{\mathbf{S}}{}^{\parallel}] + 2\,\overset{\circ}{\mathbf{S}}{}^{\perp}\cdot\mathsf{K}[\overset{\circ}{\mathbf{S}}{}^{\parallel}] \tag{11.1.83}$$

we obtain

$$\overset{\circ}{\mathbf{S}}\cdot\mathsf{K}[\overset{\circ}{\mathbf{S}}] = \overset{\circ}{\mathbf{S}}{}^{\perp}\cdot\mathsf{K}[\overset{\circ}{\mathbf{S}}{}^{\perp}] - \overset{\circ}{\mathbf{S}}{}^{\parallel}\cdot\mathsf{K}[\overset{\circ}{\mathbf{S}}{}^{\parallel}] \tag{11.1.84}$$

This together with the definition of $\mathcal{E}(\mathbf{S})$ (see Equation 11.1.78) implies that

$$\mathcal{E}_S(\mathbf{S}) = \mathcal{E}_S(\mathbf{S}^{\perp}) - \mathcal{E}_S(\mathbf{S}^{\parallel}) \tag{11.1.85}$$

Therefore, the *stress energy density of a progressive wave* 11.1.57 *is the difference between the* "normal" *and* "tangential" *stress energy densities.*

Since $\mathcal{E}_S(\mathbf{S}) > 0$ for every $\mathbf{S} \neq \mathbf{0}$, Equation 11.1.85 implies that

$$\mathcal{E}_S(\mathbf{S}^{\perp}) > \mathcal{E}_S(\mathbf{S}^{\parallel}) \tag{11.1.86}$$

that is, the "normal" *stress energy density is greater than the* "tangential" *stress energy density*.

Finally, it follows from Equations 11.1.77 and 11.1.84 and 11.1.85 that an alternative formula for the wave velocity reads

$$c(\overset{\circ}{\mathbf{S}}) = c(\overset{\circ\perp}{\mathbf{S}}) \left[1 - \frac{\mathcal{E}_S(\overset{\circ\parallel}{\mathbf{S}})}{\mathcal{E}_S(\overset{\circ\perp}{\mathbf{S}})} \right]^{-1/2} \tag{11.1.87}$$

where

$$c(\overset{\circ\perp}{\mathbf{S}}) = \frac{|\overset{\circ\perp}{\mathbf{S}}\,\mathbf{m}|}{\left\{ \varrho\,\overset{\circ\perp}{\mathbf{S}} \cdot \mathsf{K}[\overset{\circ\perp}{\mathbf{S}}] \right\}^{1/2}} \tag{11.1.88}$$

The formula 11.1.87 allows us to compute the velocity of a plane progressive wave in any constant density homogeneous anisotropic elastic body if $c(\overset{\circ\perp}{\mathbf{S}})$ and the ratio $\mathcal{E}_S(\overset{\circ\parallel}{\mathbf{S}})/\mathcal{E}_S(\overset{\circ\perp}{\mathbf{S}})$ are available. The formula 10.1.87 also implies that

$$c(\overset{\circ}{\mathbf{S}}) \geq c(\overset{\circ\perp}{\mathbf{S}}) \tag{11.1.89}$$

Hence, in any homogeneous anisotropic linear elastic body the *velocity of a stress progressive wave is bounded from below by the velocity of the normal part of the wave*. Also, in the general case, because of Equation 11.1.85 and since $\mathcal{E}_S(\mathbf{S}) > 0$, *a pure* "tangential" *stress progressive wave cannot propagate in the body*. These are features brought out explicitly in the stress formulation, but have not been obtained in the displacement formulation.

In the following we investigate the velocity formula 11.1.77 for admissible solutions of the form 11.1.76 for different orientations of vector **a** with a view to characterizing stress waves and their velocities in a homogeneous isotropic body.

Case 1: Vector a is parallel to the wave normal m, that is, $\mathbf{a} = \alpha\mathbf{m}$ ($\alpha \neq 0$). In this case

$$\mathbf{m} \otimes \mathbf{a} = \alpha\mathbf{m} \otimes \mathbf{m} \tag{11.1.90}$$

and by Equation 11.1.76 we obtain

$$\overset{\circ}{\mathbf{S}} = \alpha(2\mu\mathbf{m} \otimes \mathbf{m} + \lambda\mathbf{1}) \tag{11.1.91}$$

where λ and μ are the Lamé moduli. Hence

$$\overset{\circ}{\mathbf{S}}\,\mathbf{m} = \alpha(\lambda + 2\mu)\mathbf{m} \tag{11.1.92}$$

and

$$|\overset{\circ}{\mathbf{S}}\,\mathbf{m}| = |\alpha|(\lambda + 2\mu) \tag{11.1.93}$$

Also,

$$\mathsf{K}[\overset{\circ}{\mathbf{S}}] = \frac{1}{2\mu}\left[\overset{\circ}{\mathbf{S}} - \frac{\lambda}{3\lambda + 2\mu}(\mathrm{tr}\ \overset{\circ}{\mathbf{S}})\mathbf{1}\right] \tag{11.1.94}$$

Hence

$$\overset{\circ}{\mathbf{S}} \cdot \mathsf{K}[\overset{\circ}{\mathbf{S}}] = \frac{1}{2\mu}\left[\overset{\circ}{\mathbf{S}} \cdot \overset{\circ}{\mathbf{S}} - \frac{\lambda}{3\lambda + 2\mu}(\mathrm{tr}\ \overset{\circ}{\mathbf{S}})^2\right] \tag{11.1.95}$$

Substitution of $\overset{\circ}{\mathbf{S}}$ from Equation 11.1.91 into Equation 11.1.95 yields

$$\overset{\circ}{\mathbf{S}} \cdot \mathsf{K}[\overset{\circ}{\mathbf{S}}] = \alpha^2(\lambda + 2\mu) \tag{11.1.96}$$

Finally, Equations 11.1.77, 11.1.93, and 11.1.96 yield

$$c(\overset{\circ}{\mathbf{S}}) = \frac{|\overset{\circ}{\mathbf{S}}\mathbf{m}|}{\left\{\varrho\,\overset{\circ}{\mathbf{S}} \cdot \mathsf{K}[\overset{\circ}{\mathbf{S}}]\right\}^{1/2}} = \left(\frac{\lambda + 2\mu}{\varrho}\right)^{1/2} \tag{11.1.97}$$

which recovers the familiar wave velocity associated with the longitudinal waves in a displacement formulation of homogeneous isotropic elastodynamics.

Taking the "normal" and "tangential" parts of $\overset{\circ}{\mathbf{S}}$ given by Equation 11.1.91 we write

$$\overset{\circ}{\mathbf{S}}{}^{\perp} = \alpha(\lambda + 2\mu)\mathbf{m} \otimes \mathbf{m} \tag{11.1.98}$$

and

$$\overset{\circ}{\mathbf{S}}{}^{\parallel} = \alpha\lambda(\mathbf{1} - \mathbf{m} \otimes \mathbf{m}) \tag{11.1.99}$$

Therefore, the "normal" and "tangential" stress energy densities are given, respectively, by (see Equations 11.1.78 through 11.1.80)

$$\mathcal{E}_S(\mathbf{S}^{\perp}) = \frac{1}{2}\psi^2\,\overset{\circ}{\mathbf{S}}{}^{\perp} \cdot \mathsf{K}[\overset{\circ}{\mathbf{S}}{}^{\perp}] = \frac{(\lambda + 2\mu)^2}{2\mu}\,\frac{(\lambda + \mu)}{3\lambda + 2\mu}\,\alpha^2\psi^2 \tag{11.1.100}$$

and

$$\mathcal{E}_S(\mathbf{S}^{\parallel}) = \frac{1}{2}\psi^2\,\overset{\circ}{\mathbf{S}}{}^{\parallel} \cdot \mathsf{K}[\overset{\circ}{\mathbf{S}}{}^{\parallel}] = \frac{\lambda^2}{2\mu}\,\frac{\lambda + 2\mu}{3\lambda + 2\mu}\,\alpha^2\psi^2 \tag{11.1.101}$$

The total stress energy density is obtained from the equation (see Equation 11.1.85)

$$\mathcal{E}_S(\mathbf{S}) = \mathcal{E}_S(\mathbf{S}^{\perp}) - \mathcal{E}_S(\mathbf{S}^{\parallel}) = \frac{1}{2}(\lambda + 2\mu)\alpha^2\psi^2 \tag{11.1.102}$$

This case reveals a characteristic feature of a stress progressive wave 11.1.57 in homogeneous isotropic bodies: for a stress wave propagating with the velocity $c_1 = \sqrt{(\lambda + 2\mu)/\varrho}$ the total stress energy density is measured by the positive difference between the "normal" and "tangential" stress energy densities.

If we introduce the ratio

$$\varphi = \frac{\mathcal{E}_S(\mathbf{S}^{\parallel})}{\mathcal{E}_S(\mathbf{S}^{\perp})} \tag{11.1.103}$$

we obtain

$$\varphi = \varphi(\nu) = \frac{2\nu^2}{1 - \nu} \tag{11.1.104}$$

where ν is Poisson's ratio. Therefore, since

$$-1 < \nu < \frac{1}{2} \tag{11.1.105}$$

we have

$$0 \leq \varphi(\nu) < 1 \tag{11.1.106}$$

For a typical value $\nu = 1/3$, we get $\varphi = 1/3$, whereas for rubber-like materials with $\nu \to \frac{1}{2} - 0$, we get $\varphi \to 1$.

Case 2: Vector a is perpendicular to the wave normal m. In this case $\mathbf{a} \cdot \mathbf{m} = 0$ and hence $\text{tr}\,(\mathbf{a} \otimes \mathbf{m}) = 0$. Therefore, by Equation 11.1.76, we get

$$\overset{\circ}{\mathbf{S}} = 2\mu \,\text{sym}\,(\mathbf{m} \otimes \mathbf{a}) \tag{11.1.107}$$

Hence

$$\overset{\circ}{\mathbf{S}}\,\mathbf{m} = \mu\mathbf{a}, \quad \overset{\circ}{\mathbf{S}} \cdot \overset{\circ}{\mathbf{S}} = 2\mu^2 \mathbf{a} \cdot \mathbf{a}, \quad \text{tr}\,\overset{\circ}{\mathbf{S}} = 0 \tag{11.1.108}$$

Following the steps in Case 1, the formula 11.1.77 yields

$$c(\overset{\circ}{\mathbf{S}}) = \frac{\mu|\mathbf{a}|}{(\varrho\mu|\mathbf{a}|^2)^{1/2}} = \left(\frac{\mu}{\varrho}\right)^{1/2} \tag{11.1.109}$$

This is the familiar shear-wave velocity of homogeneous isotropic elastodynamics (see Equations 11.1.56).

From Equations 11.1.59 and 11.1.60 and 11.1.107, we have

$$\overset{\circ}{\mathbf{S}}{}^{\perp} = 2\mu \text{sym}(\mathbf{m} \otimes \mathbf{a}), \quad \overset{\circ}{\mathbf{S}}{}^{\parallel} = \mathbf{0} \tag{11.1.110}$$

The stress progressive wave is now purely normal, and the associated "normal" stress energy density is given by

$$\mathcal{E}_S(\mathbf{S}) = \mathcal{E}_S(\mathbf{S}^{\perp}) = \frac{1}{2}\mu \mathbf{a}^2 \psi^2 \tag{11.1.111}$$

Hence, for a stress progressive wave corresponding to the transverse displacement waves the total stress energy density is associated entirely with the normal component of the wave.

Conclusions

1. A purely tensorial treatment of homogeneous anisotropic elastodynamics including a classification of stress wave types and their velocities is possible.
2. When a stress plane progressive wave is decomposed into "normal" and "tangential" parts with respect to the wave front, both parts can propagate with the same velocity in a homogeneous anisotropic elastic body.
3. The total stress energy density of the wave in a homogeneous anisotropic elastic body is represented by the positive difference between "normal" and "tangential" stress energy densities; this implies that a pure "tangential" stress wave cannot propagate in the body.
4. For a homogeneous isotropic elastic body, the velocities of stress progressive waves associated with the longitudinal and transverse displacement waves can be recovered using a general stress-velocity formula. It turns out that a stress wave propagating with the velocity $c_1 = (\lambda + 2\mu)^{1/2}/\varrho^{1/2}$ can be decomposed into "normal" and "tangential" parts in such a way that both the "normal" and "tangential" stress energy densities are strictly positive; for a stress wave propagating with the velocity $c_2 = \mu^{1/2}/\varrho^{1/2}$ only "normal" stress energy is carried by the wave front.

Similar conclusions hold true regarding the decomposition of a stress progressive wave propagating in a transversely isotropic homogeneous elastic body with particular orientations of the axis of symmetry [3].

11.1.2 Elastic Waves in a Homogeneous Isotropic Infinite Solid due to the Application of an Instantaneous Concentrated Force

Suppose that a homogeneous isotropic infinite elastic solid, referred to a Cartesian coordinate system $\{x_i\}$, is initially at rest, and an instantaneous concentrated force is applied to the origin $\mathbf{x} = \mathbf{0}$ of the system at time $t = 0 + 0$. To find an elastic process $p = [\mathbf{u}, \mathbf{E}, \mathbf{S}]$ (see Section 4.2.3) corresponding to such a force we may use either the displacement or the stress formulation of the problem. In the following, we employ the displacement formulation while the use of the stress formulation for other problems will be considered later. Therefore, we wish to find a solution $\mathbf{u} = \mathbf{u}(\mathbf{x}, t)$ to the displacement equation of motion (see Equations 7.2.1 and 7.2.2)

$$\Box_2^2 \mathbf{u} + \left[\left(\frac{c_1}{c_2} \right)^2 - 1 \right] \nabla (\operatorname{div} \mathbf{u}) + \frac{\mathbf{b}}{\mu} = \mathbf{0} \quad \text{on } E^3 \times (0, \infty) \tag{11.1.112}$$

subject to the homogeneous initial conditions

$$\mathbf{u}(\mathbf{x},0)=\mathbf{0}, \quad \dot{\mathbf{u}}(\mathbf{x},0)=\mathbf{0}, \quad \mathbf{x}\in E^3 \tag{11.1.113}$$

and suitable vanishing conditions as $|\mathbf{x}|\to\infty$ for every $t>0$. In Equation 11.1.112

$$\Box_2^2=\nabla^2-\frac{1}{c_2^2}\frac{\partial^2}{\partial t^2}, \quad \frac{1}{c_1^2}=\frac{\varrho}{\lambda+2\mu}, \quad \frac{1}{c_2^2}=\frac{\varrho}{\mu} \tag{11.1.114}$$

and

$$\mathbf{b}=\mathbf{b}(\mathbf{x},t)=\boldsymbol{\ell}\delta(\mathbf{x})\delta(t) \tag{11.1.115}$$

where

$$\delta(\mathbf{x})=\delta(x_1)\delta(x_2)\delta(x_3) \tag{11.1.116}$$

is a three-dimensional Dirac delta function and $\boldsymbol{\ell}$ is a constant vector. Since $[\mathbf{b}]=[\text{Force}\times \text{L}^{-3}]$, where L is a constant of the length dimension, and $[\delta(x_1)]=[\delta(x_2)]=[\delta(x_3)]=[\text{L}^{-1}]$, and $[\delta(t)]=[\text{T}^{-1}]$, therefore $[\boldsymbol{\ell}]=[\text{Force}\times\text{T}]$; here T is a constant of the time dimension.

To obtain a solution to the problem we use the Cauchy–Kovalevski–Somigliana representation of $\mathbf{u}$ (see Equations 7.2.40 and 7.2.41)

$$\mathbf{u}=\Box_1^2\mathbf{g}+\left(\frac{c_2^2}{c_1^2}-\mathbf{1}\right)\nabla(\operatorname{div}\mathbf{g}) \tag{11.1.117}$$

where $\mathbf{g}=\mathbf{g}(\mathbf{x},t)$ is a solution to the biwave equation

$$\Box_1^2\Box_2^2\mathbf{g}=-\frac{\mathbf{b}}{\mu} \tag{11.1.118}$$

and

$$\Box_1^2=\nabla^2-\frac{1}{c_1^2}\frac{\partial^2}{\partial t^2} \tag{11.1.119}$$

It follows from Equation 11.1.117 that $\mathbf{u}$ satisfies the homogeneous initial conditions 11.1.113 provided $\mathbf{g}$ complies with the conditions

$$\frac{\partial^k}{\partial t^k}\mathbf{g}(\mathbf{x},0)=\mathbf{0}, \quad k=0,1,2,3, \quad \text{for } \mathbf{x}\in E^3 \tag{11.1.120}$$

Therefore, from Equation 11.1.118 in which $\mathbf{b}$ is replaced by the RHS of Equation 11.1.115, and by Equations 11.1.120, the vector field $\mathbf{g}=\mathbf{g}(\mathbf{x},t)$ may be taken in the form

$$\overline{\mathbf{g}}=\mathbf{G}\boldsymbol{\ell} \tag{11.1.121}$$

where $\mathbf{G} = \mathbf{G}(\mathbf{x}, t)$ is a second-order symmetric tensor field on $E^3 \times [0, \infty)$ that satisfies the biwave equation

$$\Box_1^2 \Box_2^2 \mathbf{G} = -\frac{\mathbf{1}}{\mu}\delta(\mathbf{x})\delta(t) \quad \text{on } E^3 \times (0, \infty) \tag{11.1.122}$$

the homogeneous initial conditions

$$\frac{\partial^k}{\partial t^k}\mathbf{G}(\mathbf{x}, 0) = \mathbf{0}, \quad k = 0, 1, 2, 3 \tag{11.1.123}$$

and suitable vanishing conditions at infinity.

In the following, a solution $\mathbf{G}$ to the problem 11.1.122 and 11.1.123 will be found by a Laplace transform technique. Let $f = f(\mathbf{x}, t)$ be a function on $E^3 \times [0, \infty)$, and let $\bar{f} = \bar{f}(\mathbf{x}, p)$ denote the Laplace transform of f:

$$Lf \equiv \bar{f}(\mathbf{x}, p) = \int_0^\infty e^{-pt} f(\mathbf{x}, t)\, dt \tag{11.1.124}$$

where p is the Laplace transform parameter.

Applying the operator L to Equation 11.1.122 and using the homogeneous initial conditions 11.1.123 we obtain

$$\left(\nabla^2 - \frac{p^2}{c_1^2}\right)\left(\nabla^2 - \frac{p^2}{c_2^2}\right)\overline{\mathbf{G}} = -\frac{\mathbf{1}}{\mu}\delta(\mathbf{x}) \tag{11.1.125}$$

Since

$$\left[\left(\nabla^2 - \frac{p^2}{c_1^2}\right)\left(\nabla^2 - \frac{p^2}{c_2^2}\right)\right]^{-1} = \left(\frac{p^2}{c_1^2} - \frac{p^2}{c_2^2}\right)^{-1}\left[\left(\nabla^2 - \frac{p^2}{c_1^2}\right)^{-1} - \left(\nabla^2 - \frac{p^2}{c_2^2}\right)^{-1}\right] \tag{11.1.126}$$

and

$$\frac{1}{\nabla^2 - k^2}\delta(\mathbf{x}) = -\frac{1}{4\pi}\frac{e^{-k|\mathbf{x}|}}{|\mathbf{x}|}, \quad k > 0, \tag{11.1.127}$$

therefore, a solution to Equation 10.1.125 that goes to zero as $|\mathbf{x}| \to \infty$ takes the form

$$\overline{\mathbf{G}}(\mathbf{x}, p) = -A_0 \mathbf{1}\frac{1}{p^2}\left(\frac{e^{-\frac{p}{c_1}R} - e^{-\frac{p}{c_2}R}}{R}\right) \tag{11.1.128}$$

where $R = |\mathbf{x}|$, and

$$A_0 = \frac{1}{4\pi\varrho}\left[1 - \left(\frac{c_2}{c_1}\right)^2\right]^{-1} > 0 \tag{11.1.129}$$

Since

$$L^{-1}\left\{\frac{e^{-\alpha p}}{p^2}\right\} = H(t-\alpha)(t-\alpha), \quad \alpha > 0 \tag{11.1.130}$$

where $H = H(t)$ is the Heaviside function

$$H(t) = \begin{Bmatrix} 1 & \text{for } t > 0 \\ 0 & \text{for } t < 0 \end{Bmatrix} \tag{11.1.131}$$

therefore, applying the operator L^{-1} to Equation 11.1.128 we arrive at the simple formula

$$\mathbf{G}(\mathbf{x}, t) = -A_0 \frac{\mathbf{1}}{R}\left[\left(t-\frac{R}{c_1}\right)H\left(t-\frac{R}{c_1}\right) - \left(t-\frac{R}{c_2}\right)H\left(t-\frac{R}{c_2}\right)\right] \tag{11.1.132}$$

As a result, a solution to the problem 11.1.112 and 11.1.113 in which $\mathbf{b}$ is given in the form 11.1.115, takes the closed form

$$\mathbf{u} = \square_1^2(\mathbf{G}\boldsymbol{\ell}) + \left(\frac{c_2^2}{c_1^2} - 1\right)\nabla[\text{div}\,(\mathbf{G}\boldsymbol{\ell})] \tag{11.1.133}$$

where

$$\mathbf{G}\boldsymbol{\ell} = -A_0 \frac{\boldsymbol{\ell}}{R}\left[\left(t-\frac{R}{c_1}\right)H\left(t-\frac{R}{c_1}\right) - \left(t-\frac{R}{c_2}\right)H\left(t-\frac{R}{c_2}\right)\right] \tag{11.1.134}$$

To derive an alternative form of $\mathbf{u}$ in which the vector fields $\square_1^2(\mathbf{G}\boldsymbol{\ell})$ and $\nabla[\text{div}(\mathbf{G}\boldsymbol{\ell})]$ are computed explicitly we apply the Laplace operator L to Equations 11.1.133 and 11.1.134, respectively, and have

$$\bar{\mathbf{u}} = \left(\nabla^2 - \frac{p^2}{c_1^2}\right)\bar{\mathbf{g}} + \left(\frac{c_2^2}{c_1^2} - 1\right)\nabla(\text{div}\,\bar{\mathbf{g}}) \tag{11.1.135}$$

where

$$\bar{\mathbf{g}} = -\frac{\boldsymbol{\ell}^0}{p^2}\left(\frac{e^{-\frac{p}{c_1}R} - e^{-\frac{p}{c_2}R}}{R}\right), \quad \boldsymbol{\ell}^0 = A_0\boldsymbol{\ell} \tag{11.1.136}$$

In components, Equations 11.1.135 and 11.1.136, respectively, take the forms

$$\bar{u}_i = \left(\nabla^2 - \frac{p^2}{c_1^2}\right)\bar{g}_i + \left(\frac{c_2^2}{c_1^2} - 1\right)\bar{g}_{k,ki} \tag{11.1.137}$$

and

$$\bar{g}_i = -l_i^0 \frac{1}{p^2}\left(\frac{e^{-\frac{p}{c_1}R}}{R} - \frac{e^{-\frac{p}{c_2}R}}{R}\right) \tag{11.1.138}$$

By computing the first and second partial derivatives of $\overline{g}_i$, we obtain

$$\overline{g}_{i,k} = l_i^0 \frac{1}{p^2} \frac{x_k}{R^3} \left[\left(1 + R\frac{p}{c_1}\right) e^{-R\frac{p}{c_1}} - \left(1 + R\frac{p}{c_2}\right) e^{-R\frac{p}{c_2}} \right] \tag{11.1.139}$$

and

$$\overline{g}_{i,kl} = -\frac{l_i^0}{p^2} \left\{ \left[\frac{x_k x_l}{R^5} \left(3 + 3R\frac{p}{c_1} + R^2\frac{p^2}{c_1^2}\right) - \frac{\delta_{kl}}{R^3}\left(1 + R\frac{p}{c_1}\right) \right] e^{-R\frac{p}{c_1}} - \left[\frac{x_k x_l}{R^5} \left(3 + 3R\frac{p}{c_2} + R^2\frac{p^2}{c_2^2}\right) - \frac{\delta_{kl}}{R^3}\left(1 + R\frac{p}{c_2}\right) \right] e^{-R\frac{p}{c_2}} \right\} \quad (i,k,l = 1,2,3) \tag{11.1.140}$$

Hence, using Equations 11.1.139 and 11.1.140 we get

$$\overline{g}_{k,k} = \frac{l_k^0}{p^2} \frac{x_k}{R^3} \left\{ \left(1 + R\frac{p}{c_1}\right) e^{-R\frac{p}{c_1}} - \left(1 + R\frac{p}{c_2}\right) e^{-R\frac{p}{c_2}} \right\} \tag{11.1.141}$$

$$\overline{g}_{k,ki} = \frac{l_k^0}{p^2} \left\{ \frac{\delta_{ik}}{R^3} \left[\left(1 + R\frac{p}{c_1}\right) e^{-R\frac{p}{c_1}} - \left(1 + R\frac{p}{c_2}\right) e^{-R\frac{p}{c_2}} \right] - \frac{x_i x_k}{R^5} \left[\left(3 + 3R\frac{p}{c_1} + R^2\frac{p^2}{c_1^2}\right) e^{-R\frac{p}{c_1}} - \left(3 + 3R\frac{p}{c_2} + R^2\frac{p^2}{c_2^2}\right) e^{-R\frac{p}{c_2}} \right] \right\} \tag{11.1.142}$$

and

$$\overline{g}_{i,kk} = -l_i^0 \left(\frac{1}{c_1^2} \frac{e^{-R\frac{p}{c_1}}}{R} - \frac{1}{c_2^2} \frac{e^{-R\frac{p}{c_2}}}{R} \right) \tag{11.1.143}$$

Note that from Equations 11.1.138 and 11.1.143

$$\left(\nabla^2 - \frac{p^2}{c_1^2}\right) \overline{g}_i = l_i^0 \left(\frac{1}{c_2^2} - \frac{1}{c_1^2} \right) \frac{e^{-R\frac{p}{c_2}}}{R} \tag{11.1.144}$$

or, using the definition of l_i^0 (see Equations 11.1.129 and 11.1.136),

$$\left(\nabla^2 - \frac{p^2}{c_1^2}\right) \overline{g}_i = \frac{l_i}{4\pi\mu} \frac{e^{-R\frac{p}{c_2}}}{R} \tag{11.1.145}$$

Substituting Equations 11.1.142 and 11.1.145 into Equation 11.1.137, we obtain $\overline{u}_i$ in an explicit form.

By applying the operator L^{-1} to Equations 11.1.142 and 11.1.145, we have, respectively

$$
\begin{aligned}
g_{k,ki} = l_k^0 \Bigg\{ & \frac{\delta_{ik}}{R^3} \Bigg[\left(t - \frac{R}{c_1}\right) H\left(t - \frac{R}{c_1}\right) - \left(t - \frac{R}{c_2}\right) H\left(t - \frac{R}{c_2}\right) \\
& + \frac{R}{c_1} H\left(t - \frac{R}{c_1}\right) - \frac{R}{c_2} H\left(t - \frac{R}{c_2}\right) \Bigg] \\
& - \frac{x_i x_k}{R^5} \Bigg[3\left(t - \frac{R}{c_1}\right) H\left(t - \frac{R}{c_1}\right) - 3\left(t - \frac{R}{c_2}\right) H\left(t - \frac{R}{c_2}\right) \\
& + 3\frac{R}{c_1} H\left(t - \frac{R}{c_1}\right) - 3\frac{R}{c_2} H\left(t - \frac{R}{c_2}\right) \\
& + \frac{R^2}{c_1^2} \delta\left(t - \frac{R}{c_1}\right) - \frac{R^2}{c_2^2} \delta\left(t - \frac{R}{c_2}\right) \Bigg] \Bigg\} \\
= -l_k^0 \Bigg\{ & \left(\frac{3x_i x_k}{R^2} - \delta_{ik}\right) \frac{t}{R^3} \left[H\left(t - \frac{R}{c_1}\right) - H\left(t - \frac{R}{c_2}\right) \right] \\
& + \frac{x_i x_k}{R^3} \left[\frac{1}{c_1^2} \delta\left(t - \frac{R}{c_1}\right) - \frac{1}{c_2^2} \delta\left(t - \frac{R}{c_2}\right) \right] \Bigg\}
\end{aligned}
\tag{11.1.146}
$$

and

$$
\square_1^2 g_i = \frac{l_i}{4\pi\mu} \frac{1}{R} \delta\left(t - \frac{R}{c_2}\right) \tag{11.1.147}
$$

Finally, taking into account the relation

$$
-l_k^0 \left(\frac{c_2^2}{c_1^2} - 1\right) = \frac{l_k}{4\pi\varrho} \tag{11.1.148}
$$

and applying the inverse Laplace transform operator L^{-1} to Equation 11.1.137, and using Equations 11.1.146 and 11.1.147, gives

$$
u_i = U_{ij} l_j \tag{11.1.149}
$$

where [4]

$$
\begin{aligned}
U_{ij} = \frac{1}{4\pi\varrho} \Bigg\{ & \frac{\delta_{ij}}{c_2^2} \frac{1}{R} \delta\left(t - \frac{R}{c_2}\right) + t\left(\frac{3x_i x_j}{R^5} - \frac{\delta_{ij}}{R^3}\right) \left[H\left(t - \frac{R}{c_1}\right) - H\left(t - \frac{R}{c_2}\right) \right] \\
& + \frac{x_i x_j}{R^3} \left[\frac{1}{c_1^2} \delta\left(t - \frac{R}{c_1}\right) - \frac{1}{c_2^2} \delta\left(t - \frac{R}{c_2}\right) \right] \Bigg\}
\end{aligned}
\tag{11.1.150}
$$

If we note that for every $c > 0$

$$
\frac{R}{c} \delta\left(t - \frac{R}{c}\right) = t\,\delta\left(t - \frac{R}{c}\right) \tag{11.1.151}
$$

and

$$\frac{1}{c^2}\delta\left(t-\frac{R}{c}\right)=\frac{1}{Rc}\frac{R}{c}\delta\left(t-\frac{R}{c}\right)=\frac{t}{Rc}\delta\left(t-\frac{R}{c}\right) \tag{11.1.152}$$

we reduce Equation 11.1.150 to an equivalent form

$$\begin{aligned}U_{ij}=\frac{t}{4\pi\varrho R^2}&\left\{\frac{\delta_{ij}}{c_2}\delta\left(t-\frac{R}{c_2}\right)+\left(\frac{3x_ix_j}{R^3}-\frac{\delta_{ij}}{R}\right)\left[H\left(t-\frac{R}{c_1}\right)-H\left(t-\frac{R}{c_2}\right)\right]\right.\\&\left.+\frac{x_ix_j}{R^2}\left[\frac{1}{c_1}\delta\left(t-\frac{R}{c_1}\right)-\frac{1}{c_2}\delta\left(t-\frac{R}{c_2}\right)\right]\right\}\end{aligned} \tag{11.1.153}$$

Note that

$$[\mathbf{U}]=\left[T\varrho^{-1}L^{-3}\right] \tag{11.1.154}$$

and

$$[\boldsymbol{\ell}]=[\text{Force}\times\text{T}]=[\varrho L^4\text{T}^{-1}] \tag{11.1.155}$$

Therefore,

$$[\mathbf{u}]=[\mathbf{U}\boldsymbol{\ell}]=[L] \tag{11.1.156}$$

To find the strain tensor **E** associated with **u** we take the symmetric gradient of Equation 11.1.149, and receive

$$\mathbf{E}=\hat{\nabla}\mathbf{u}=\hat{\nabla}(\mathbf{U}\boldsymbol{\ell}) \tag{11.1.157}$$

In components this equation takes the form

$$E_{ij}=u_{(i,j)}=V_{ijk}l_k \tag{11.1.158}$$

where the third-order tensor V_{ijk} is given by

$$\begin{aligned}V_{ijk}=-\frac{1}{4\pi\varrho}&\left\{\frac{x_{(i}\delta_{kj)}}{c_2^2R^3}\left[\delta\left(t-\frac{R}{c_2}\right)+\frac{R}{c_2}\dot{\delta}\left(t-\frac{R}{c_2}\right)\right]\right.\\&+\frac{3t}{R^5}\left(\frac{5x_ix_jx_k}{R^2}-x_k\delta_{ij}-2x_{(i}\delta_{kj)}\right)\left[H\left(t-\frac{R}{c_1}\right)-H\left(t-\frac{R}{c_2}\right)\right]\\&+\frac{1}{R^4}\left(\frac{3x_ix_jx_k}{R^2}-x_{(i}\delta_{kj)}\right)\left[\frac{1}{c_1}H\left(t-\frac{R}{c_1}\right)-\frac{1}{c_2}H\left(t-\frac{R}{c_2}\right)\right]\\&+\frac{1}{R^3}\left(\frac{3x_ix_jx_k}{R^2}-x_k\delta_{ij}-x_{(i}\delta_{kj)}\right)\left[\frac{1}{c_1^2}\delta\left(t-\frac{R}{c_1}\right)-\frac{1}{c_2^2}\delta\left(t-\frac{R}{c_2}\right)\right]\\&\left.+\frac{x_ix_jx_k}{R^4}\left[\frac{1}{c_1^3}\dot{\delta}\left(t-\frac{R}{c_1}\right)-\frac{1}{c_2^2}\dot{\delta}\left(t-\frac{R}{c_2}\right)\right]\right\}\end{aligned} \tag{11.1.159}$$

Note that

$$[V_{ijk}] = [(\text{Force})^{-1}\text{T}^{-1}] \tag{11.1.160}$$

therefore,

$$[V_{ijk}l_k] = [1] \tag{11.1.161}$$

By taking the trace of V_{ijk} with respect to the indices ij, we find

$$\begin{aligned} V_{aak} = -\frac{1}{4\pi\varrho}\frac{x_k}{R^2}\Bigg\{&\left[\frac{1}{c_1^3}\dot{\delta}\left(t-\frac{R}{c_1}\right)-\frac{1}{c_2^3}\dot{\delta}\left(t-\frac{R}{c_2}\right)\right]\\ &-\frac{1}{R}\left[\frac{1}{c_1^2}\delta\left(t-\frac{R}{c_1}\right)-\frac{1}{c_2^2}\delta\left(t-\frac{R}{c_2}\right)\right]+\frac{2}{R^2}\left[\frac{1}{c_1}H\left(t-\frac{R}{c_1}\right)-\frac{1}{c_2}H\left(t-\frac{R}{c_2}\right)\right]\\ &+\frac{1}{c_2^2 R}\left[\delta\left(t-\frac{R}{c_2}\right)+\frac{R}{c_2}\dot{\delta}\left(t-\frac{R}{c_2}\right)\right]\Bigg\} \end{aligned} \tag{11.1.162}$$

Hence, using the formula

$$\mathbf{S} = 2\mu\hat{\nabla}\mathbf{u} + \lambda\,\text{tr}\,(\hat{\nabla}\mathbf{u})\mathbf{1} \tag{11.1.163}$$

the stress components S_{ij} corresponding to u_i are obtained in the form

$$S_{ij} = T_{ijk}l_k \tag{11.1.164}$$

where

$$T_{ijk} = 2\mu\left[V_{ijk} - \left(1 - \frac{1}{2}\frac{c_1^2}{c_2^2}\right)V_{aak}\delta_{ij}\right] \tag{11.1.165}$$

and V_{ijk} and V_{aak} are given by Equations 11.1.159 and 11.1.162, respectively. This completes the task of finding the elastic process $p = [\mathbf{u}, \mathbf{E}, \mathbf{S}]$ corresponding to the action of an instantaneous concentrated force in a homogeneous isotropic infinite elastic body.

If an initially undisturbed homogeneous isotropic infinite elastic body is subject to a body force $\mathbf{b} = \mathbf{b}(\mathbf{x}, t)$ on $\Omega \times [0, \infty)$, where Ω is a bounded domain of E^3, then a displacement vector $\hat{\mathbf{u}} = \hat{\mathbf{u}}(\mathbf{x}, t)$ produced by such a load is obtained from the formula

$$\hat{\mathbf{u}}(\mathbf{x}, t) = \int_0^t \int_\Omega \mathbf{U}(\mathbf{x} - \boldsymbol{\xi}, t - \tau)\,\mathbf{b}(\boldsymbol{\xi}, \tau)\,dv(\boldsymbol{\xi})\,d\tau \tag{11.1.166}$$

where the second-order tensor field $\mathbf{U} = \mathbf{U}(\mathbf{x}, t)$ is defined by Equation 11.1.153.

11.1.3 The Elastic Waves Produced by a Moving Point Force in an Infinite Solid

Assume that a point force of magnitude l acting in the x_3 direction of the Cartesian coordinate system $\{x_i\}$ is applied suddenly at $\mathbf{x} = \mathbf{0}$ and $t = 0 + 0$ in a homogeneous isotropic infinite elastic body, and is then maintained at a position that moves with a constant velocity v

along the positive x_3-axis. A displacement field $\mathbf{u}=\mathbf{u}(\mathbf{x},t)$ produced by such a load then satisfies the displacement equation of motion 11.1.112 in which

$$b_i(\mathbf{x},t) = l\delta_{i3}\delta(x_1)\delta(x_2)\delta(x_3 - vt)H(t) \quad (i = 1,2,3) \tag{11.1.167}$$

subject to the homogeneous initial conditions 11.1.113 and suitable vanishing conditions at $|\mathbf{x}| = \infty$ for every $t > 0$. Following the steps in Section 11.1.2, and substituting b_i given by Equation 11.1.167 into the formula 11.1.166 in which Ω stands for an infinite domain, we find

$$u_i(\mathbf{x},t) = l\int_{-\infty}^{+\infty} d\xi_1 \int_{-\infty}^{+\infty} d\xi_2 \int_{-\infty}^{+\infty} d\xi_3 \int_0^t d\tau\, \delta(\xi_1)\,\delta(\xi_2)\,\delta(\xi_3 - v\tau)$$
$$\times\, U_{i3}(x_1 - \xi_1, x_2 - \xi_2, x_3 - \xi_3; t - \tau) \quad (i = 1,2,3) \tag{11.1.168}$$

or, by using a filtering property of the Dirac delta function, [see Equation (g) in Example 4.1.15]

$$u_i(\mathbf{x},t) = l\int_0^t U_{i3}(x_1, x_2, x_3 - v\tau; t - \tau)\, d\tau \tag{11.1.169}$$

Here U_{i3} is obtained from Equations 11.1.153 in the form

$$U_{\alpha 3}(x_1,x_2,x_3 - v\tau; t-\tau) = \frac{x_\alpha(x_3 - v\tau)(t-\tau)}{4\pi\varrho R_\tau^4} \times \left\{ \frac{3}{R_\tau}\left[H\left(t - \tau - \frac{R_\tau}{c_1}\right) - H\left(t - \tau - \frac{R_\tau}{c_2}\right)\right] + \left[\frac{1}{c_1}\delta\left(t - \tau - \frac{R_\tau}{c_1}\right) - \frac{1}{c_2}\delta\left(t - \tau - \frac{R_\tau}{c_2}\right)\right]\right\} \quad \alpha = 1,2 \tag{11.1.170}$$

and

$$U_{33}(x_1,x_2,x_3 - v\tau; t-\tau) = \frac{(t-\tau)}{4\pi\varrho R_\tau^2}\left\{\frac{1}{c_2}\delta\left(t-\tau-\frac{R_\tau}{c_2}\right) + \frac{1}{R_\tau}\left[3\frac{(x_3 - v\tau)^2}{R_\tau^2} - 1\right]\left[H\left(t-\tau-\frac{R_\tau}{c_1}\right) - H\left(t-\tau-\frac{R_\tau}{c_2}\right)\right] + \frac{(x_3 - v\tau)^2}{R_\tau^2}\left[\frac{1}{c_1}\delta\left(t-\tau-\frac{R_\tau}{c_1}\right) - \frac{1}{c_2}\delta\left(t-\tau-\frac{R_\tau}{c_2}\right)\right]\right\} \tag{11.1.171}$$

where

$$R_\tau^2 = x_1^2 + x_2^2 + (x_3 - v\tau)^2 \tag{11.1.172}$$

It follows from Equations 11.1.169 through 11.1.172 that $\mathbf{u} = \mathbf{u}(\mathbf{x}, t)$ possesses axial symmetry with respect to the x_3-axis. Therefore, if we introduce cylindrical coordinates by the formulas

$$x_1 = r\cos\varphi, \quad x_2 = r\sin\varphi, \quad x_3 = z \tag{11.1.173}$$

and note that

$$u_r = u_1\cos\varphi + u_2\sin\varphi, \quad u_3 = u_z, \tag{11.1.174}$$

we reduce Equations 11.1.169 through 11.1.172 to the form

$$u_r(r, x_3; t) = l\int_0^t U_r(r, x_3 - v\tau; t - \tau)\, d\tau \tag{11.1.175}$$

$$u_3(r, x_3; t) = l\int_0^t U_3(r, x_3 - v\tau; t - \tau)\, d\tau \tag{11.1.176}$$

where

$$\begin{aligned} U_r(r, x_3 - v\tau; t - \tau) = {} & \frac{r(x_3 - v\tau)(t - \tau)}{4\pi\varrho R_\tau^4} \\ & \times \left\{\frac{3}{R_\tau}\left[H\left(t - \tau - \frac{R_\tau}{c_1}\right) - H\left(t - \tau - \frac{R_\tau}{c_2}\right)\right]\right. \\ & \left. + \left[\frac{1}{c_1}\delta\left(t - \tau - \frac{R_\tau}{c_1}\right) - \frac{1}{c_2}\delta\left(t - \tau - \frac{R_\tau}{c_2}\right)\right]\right\} \end{aligned} \tag{11.1.177}$$

$$\begin{aligned} U_3(r, x_3 - v\tau; t - \tau) = {} & \frac{(t - \tau)}{4\pi\varrho R_\tau^2}\left\{\frac{1}{c_2}\delta\left(t - \tau - \frac{R_\tau}{c_2}\right)\right. \\ & + \frac{1}{R_\tau}\left[3\frac{(x_3 - v\tau)^2}{R_\tau^2} - 1\right]\left[H\left(t - \tau - \frac{R_\tau}{c_1}\right) - H\left(t - \tau - \frac{R_\tau}{c_2}\right)\right] \\ & \left. + \frac{(x_3 - v\tau)^2}{R_\tau^2}\left[\frac{1}{c_1}\delta\left(t - \tau - \frac{R_\tau}{c_1}\right) - \frac{1}{c_2}\delta\left(t - \tau - \frac{R_\tau}{c_2}\right)\right]\right\} \end{aligned} \tag{11.1.178}$$

and

$$R_\tau^2 = r^2 + (x_3 - v\tau)^2 \tag{11.1.179}$$

By letting $r = 0$ in Equations 11.1.177 through 11.1.179 we obtain

$$U_r(0, x_3 - v\tau; t - \tau) = 0 \quad \text{for } t \geq \tau \text{ and } |x_3 - v\tau| > 0 \tag{11.1.180}$$

and

$$U_3\left(0, x_3 - v\tau; t - \tau\right) = \frac{1}{4\pi\varrho v^3}\frac{(t-\tau)}{(\tau - x_3/v)^2}\left\{M_1\delta\left(t - \tau - M_1\left|\tau - x_3/v\right|\right)\right.$$

$$\left. + \frac{2}{|\tau - x_3/v|}\left[H\left(t - \tau - M_1\left|\tau - x_3/v\right|\right) - H\left(t - \tau - M_2\left|\tau - x_3/v\right|\right)\right]\right\}$$

$$\text{for } t \geq \tau \text{ and } |x_3 - v\tau| > 0 \tag{11.1.181}$$

where

$$M_i = \frac{v}{c_i} \quad (i = 1, 2) \tag{11.1.182}$$

Here, M_1 and M_2 are the Mach numbers of the moving force relative to the longitudinal and transverse waves, respectively. In the following we analyze waves generated by the displacement

$$u_3(x_3, t) = l\int_0^t U_3(0, x_3 - v\tau;\, t - \tau)\, d\tau \tag{11.1.183}$$

in the *subsonic case* ($M_1 < M_2 < 1$); similar analysis can be carried out in the *transonic* ($M_1 < 1, M_2 > 1$) and *supersonic* ($M_2 > M_1 > 1$) cases. Also, we restrict the analysis to $x_3 > 0$.

To this end we represent $U_3(0, x_3 - v\tau; t - \tau)$ by a sum

$$U_3(0, x_3 - v\tau; t - \tau) = U_3^{\delta}(0, x_3 - v\tau, t - \tau) + U_3^{H}(0, x_3 - v\tau, t - \tau) \tag{11.1.184}$$

where

$$U_3^{\delta}(0, x_3 - v\tau, t - \tau) = \frac{M_1}{4\pi\varrho v^3}\frac{(t-\tau)}{(\tau - x_3/v)^2}\delta[\varphi_1(t, \tau)] \tag{11.1.185}$$

$$U_3^{H}(0, x_3 - v\tau, t - \tau) = \frac{2}{4\pi\varrho v^3}\frac{t-\tau}{(\tau - x_3/v)^2}\frac{1}{|\tau - x_3/v|}$$

$$\times \{H[\varphi_1(t, \tau)] - H[\varphi_2(t, \tau)]\} \tag{11.1.186}$$

and

$$\varphi_i(t, \tau) = t - \tau - M_i\,|\tau - x_3/v| \quad (i = 1, 2) \tag{11.1.187}$$

Next, we introduce the following partition of the semi-infinite time interval

$$[0, \infty) = [0, M_1 x_3/v] \cup (M_1 x_3/v, M_2 x_3/v] \cup (M_2 x_3/v, x_3/v] \cup (x_3/v, \infty) \tag{11.1.188}$$

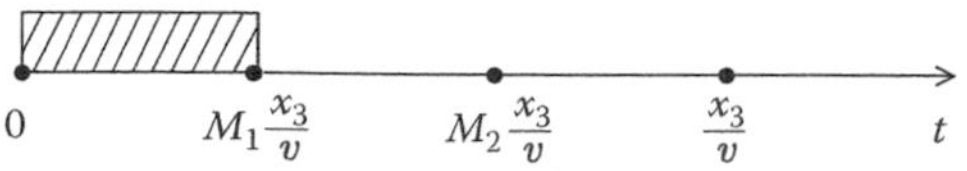

FIGURE 11.1 A range of time (shaded) for Case (a).

and calculate the integral 11.1.183 in four cases (a), (b), (c), and (d) corresponding to the four intervals on the RHS of Equation 11.1.188.

Case (a): $0 \le t < M_1 x_3/v, \quad M_1 < 1 - \sqrt{1 - M_2}$

In this case time ranges over the shaded interval are shown in Figure 11.1.

Since $M_1 < M_2 < 1$ and $0 \le t < \tau$, hence $\tau < x_3/v$, and $|\tau - x_3/v| = -(\tau - x_3/v)$, and Equations 11.1.185 and 11.1.186, respectively, take the forms

$$U_3^{\delta}(0, x_3 - v\tau; t - \tau) = \frac{M_1}{4\pi\varrho v^3} \frac{(t-\tau)}{(\tau - x_3/v)^2} \delta[\varphi_1(t,\tau)] \tag{11.1.189}$$

and

$$U_3^{H}(0, x_3 - v\tau, t - \tau) = -\frac{2}{4\pi\varrho v^3} \frac{(t-\tau)}{(\tau - x_3/v)^3} \{H[\varphi_1(t,\tau)] - H[\varphi_2(t,\tau)]\} \tag{11.1.190}$$

where

$$\varphi_1(t,\tau) = (1 - M_1)(\tau_1 - \tau) \tag{11.1.191}$$

$$\varphi_2(t,\tau) = (1 - M_2)(\tau_2 - \tau) \tag{11.1.192}$$

and

$$\tau_1 = \frac{t - M_1 x_3/v}{1 - M_1}, \quad \tau_2 = \frac{t - M_2 x_3/v}{1 - M_2} \tag{11.1.193}$$

Since $\tau_1 < 0$ and $\tau_2 < 0$, therefore $\varphi_1(t,\tau) < 0$ and $\varphi_2(t,\tau) < 0$, and Equations 11.1.189 and 11.1.190 imply that

$$U_3^{\delta}(0, x_3 - v\tau; t - \tau) = U_3^{H}(0, x_3 - v\tau; t - \tau) = 0 \tag{11.1.194}$$

and it follows from Equations 11.1.183 and 11.1.184 that

$$u_3(x_3, t) = 0 \quad \text{for } 0 \le t < M_1 x_3/v \tag{11.1.195}$$

Case (b): $M_1 x_3/v < t < M_2 x_3/v, \quad M_1 < 1 - \sqrt{1 - M_2}$

A range of time for this case is shown in Figure 11.2. Also, in this case $\tau < x_3/v$; therefore, it follows from Equations 11.1.183, 11.1.184, and 11.1.189 through 11.1.193 that

$$u_3(x_3, t) = u_3^{\delta}(x_3, t) + u_3^{H}(x_3, t) \tag{11.1.196}$$

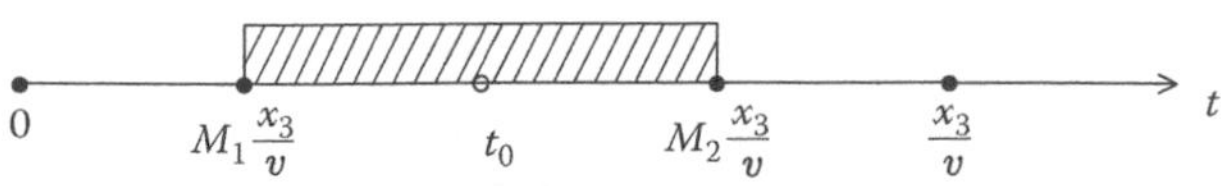

FIGURE 11.2 A range of time (shaded) for Case (b).

where

$$u_3^\delta(x_3,t) = \frac{lM_1}{4\pi\varrho v^3}\int_0^t \frac{(t-\tau)}{(\tau - x_3/v)^2}\,\delta[\varphi_1(t,\tau)]\,d\tau \tag{11.1.197}$$

$$u_3^H(x_3,t) = -\frac{2l}{4\pi\varrho v^3}\int_0^t \frac{(t-\tau)}{(\tau - x_3/v)^3}\,\{H[\varphi_1(t,\tau)] - H[\varphi_2(t,\tau)]\}\,d\tau \tag{11.1.198}$$

while $\varphi_1(t,\tau)$ and $\varphi_2(t,\tau)$ are given by Equations 11.1.191 and 11.1.192, respectively. Since

$$\varphi_1(t,\tau) = -(1-M_1)(\tau - \tau_1) \tag{11.1.199}$$

and $1 - M_1 > 0$, therefore

$$\delta[\varphi_1(t,\tau)] = \frac{1}{1-M_1}\delta(\tau - \tau_1) \tag{11.1.200}$$

and from Equation 11.1.197 we find

$$u_3^\delta(x_3,t) = \frac{l}{4\pi\varrho v^3}\,\frac{M_1}{1-M_1}\int_0^t \frac{(t-\tau)}{(\tau - x_3/v)^2}\delta(\tau-\tau_1)\,d\tau \tag{11.1.201}$$

Also, since

$$\varphi_2(t,\tau) = -(1-M_2)(\tau - \tau_2) \tag{11.1.202}$$

and $1 - M_2 > 0$, from Equation 11.1.198 we get

$$u_3^H(x_3,t) = -\frac{2l}{4\pi\varrho v^3}\int_0^t \frac{(t-\tau)}{(\tau - x_3/v)^3}\,[H(\tau_1 - \tau) - H(\tau_2 - \tau)]\,d\tau \tag{11.1.203}$$

Since $M_1\,x_3/v < t < M_2\,x_3/v$, therefore, the integral in Equation 11.1.201 can be split into two integrals, and then

$$u_3^\delta(x_3,t) = A\left[\int_0^{M_1x_3/v} \frac{t-\tau}{(\tau - x_3/v)^2}\delta(\tau-\tau_1)\,d\tau + \int_{M_1x_3/v}^{t} \frac{t-\tau}{(\tau - x_3/v)^2}\delta(\tau-\tau_1)\,d\tau\right] \tag{11.1.204}$$

where

$$A = \frac{l}{4\pi\varrho v^3}\,\frac{M_1}{1-M_1} \tag{11.1.205}$$

Similarly, Equation 11.1.203 can be written in the form

$$u_3^H(x_3,t) = -B\left[\int_0^{M_1x_3/v} \frac{t-\tau}{(\tau - x_3/v)^3}[H(\tau_1-\tau) - H(\tau_2-\tau)]\,d\tau \right.$$
$$\left. + \int_{M_1x_3/v}^{t} \frac{t-\tau}{(\tau - x_3/v)^3}[H(\tau_1-\tau) - H(\tau_2-\tau)]\,d\tau\right] \quad (11.1.206)$$

where

$$B = \frac{2l}{4\pi\varrho v^3} \quad (11.1.207)$$

Now we consider the two subcases of case (b).

Case (b.1): $0 < M_1\,x_3/v < t < M_1\,x_3/v + M_1(1-M_1)\,x_3/v$

Case (b.2): $M_1\,x_3/v + M_1(1-M_1)\,x_3/v < t < M_2\,x_3/v$

Note that splitting case (b) into (b.1) and (b.2) cases makes sense since the inequality $M_1 < 1 - \sqrt{1-M_2}$ guarantees that $t_0 \equiv M_1\,x_3/v + M_1(1-M_1)x_3/v = M_1(2-M_1)\,x_3/v$ belongs to the interval $(M_1\,x_3/v, M_2\,x_3/v)$ shown in Figure 11.2.

Case (b.1): It follows from the definition of τ_1 and τ_2 (see Equations 11.1.193) that in this case

$$0 < \tau_1 < M_1\,x_3/v, \quad \tau_2 < 0 \quad (11.1.208)$$

and Equations 11.1.204 and 11.1.206, respectively, reduce to

$$u_3^\delta(x_3,t) = A\,\frac{t-\tau_1}{(\tau_1 - x_3/v)^2} \quad (11.1.209)$$

and

$$u_3^H(x_3,t) = -B\int_0^{\tau_1} \frac{(t-\tau)}{(\tau - x_3/v)^3}\,d\tau \quad (11.1.210)$$

Since

$$\int \frac{(t-\tau)}{(\tau-a)^3}\,d\tau = (\tau-a)^{-1} - \frac{1}{2}(t-a)(\tau-a)^{-2} \quad \text{for } a \neq \tau \quad (11.1.211)$$

therefore,

$$u_3^H(x_3,t) = -B\left[\frac{1}{\tau_1 - x_3/v} - \frac{1}{2}\frac{(t-x_3/v)}{(\tau_1 - x_3/v)^2} + \frac{(t+x_3/v)}{2\,(x_3/v)^2}\right] \quad (11.1.212)$$

Next, substituting τ_1 from the first of Equations 11.1.193 into Equations 11.1.209 and 11.1.212 gives, respectively,

$$u_3^\delta(x_3, t) = -A\,\frac{M_1(1 - M_1)}{(t - x_3/v)} \tag{11.1.213}$$

and

$$u_3^H(x_3, t) = -B\left[\frac{1 - M_1^2}{2\,(t - x_3/v)} + \frac{t + x_3/v}{2\,(x_3/v)^2}\right] \tag{11.1.214}$$

Finally, using the definition of A and B (see Equations 11.1.205 and 11.1.207) we have

$$u_3(x_3, t) = u_3^\delta(x_3, t) + u_3^H(x_3, t) = -\frac{l}{4\pi\rho v^3}\,\frac{t^2}{(x_3/v)^2\,(t - x_3/v)} \tag{11.1.215}$$

Case (b.2): In this case $\tau < x_3/v$, $\tau_2 < 0$, and

$$M_1\,x_3/v < \frac{t - M_1\,x_3/v}{1 - M_1} = \tau_1 < t \tag{11.1.216}$$

and from Equation 11.1.201 we get

$$u_3^\delta(x_3, t) = A\int\limits_{M_1 x_3/v}^{t}\frac{(t - \tau)}{(\tau - x_3/v)^2}\,\delta(\tau - \tau_1)\,d\tau = A\frac{(t - \tau_1)}{(\tau_1 - x_3/v)^2} = -\frac{M_1^2}{4\pi\rho v^3}\,\frac{l}{(t - x_3/v)} \tag{11.1.217}$$

Also, from Equation 11.1.203 we have

$$u_3^H(x_3, t) = -B\int\limits_0^{\tau_1}\frac{(t - \tau)}{(\tau - x_3/v)^3}\,d\tau \tag{11.1.218}$$

Therefore, Equations 11.1.217 and 11.1.218 are identical to Equations 11.1.209 and 11.1.210, respectively. Hence,

$$u_3(x_3, t) = -\frac{l}{4\pi\rho v^3}\,\frac{t^2}{(x_3/v)^2\,(t - x_3/v)} \tag{11.1.219}$$

By combining the results obtained in cases (b.1) and (b.2) we find that $u_3(x_3, t)$ is given by Equation 11.1.215 (or Equation 11.1.219) for every $M_1\,x_3/v < t < M_2\,x_3/v$.

Case (c): $M_2\,x_3/v < t < x_3/v$, $M_1 < 1 - \sqrt{1 - M_2}$; a range of time for this case is shown in Figure 11.3.

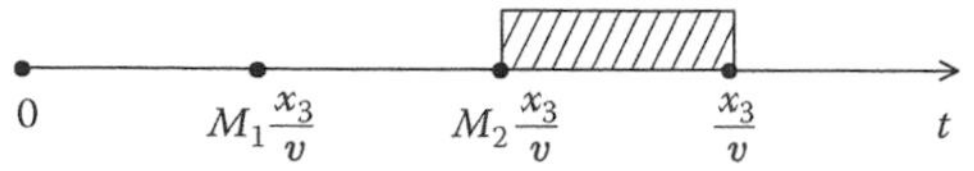

FIGURE 11.3 A range of time (shaded) for Case (c).

Since $\tau < t < x_3/v$ then $\tau < x_3/v$. Also, the inequalities $M_2\, x_3/v < t < x_3/v$ and $0 < M_1\, x_3/v < M_2\, x_3/v$ imply that

$$0 < M_1\, x_3/v < M_2\, x_3/v < t < x_3/v \tag{11.1.220}$$

Hence,

$$0 < \tau_1 < t, \quad 0 < \tau_2 < t \tag{11.1.221}$$

and from Equations 11.1.201 and 11.1.203 we get, respectively,

$$u_3^{\delta}(x_3, t) = A \frac{t - \tau_1}{(\tau_1 - x_3/v)^2} = -\frac{M_1^2}{4\pi \rho v^3} \frac{l}{(t - x_3/v)} \tag{11.1.222}$$

and

$$\begin{aligned} u_3^H(x_3, t) &= -B \left[\int_0^{\tau_1} \frac{(t - \tau)}{(\tau - x_3/v)^3} d\tau - \int_0^{\tau_2} \frac{(t - \tau)}{(\tau - x_3/v)^3} d\tau \right] \\ &= -\frac{l}{4\pi \rho v^3} \left[\frac{t^2 - M_1^2\, (x_3/v)^2}{(x_3/v)^2\, (t - x_3/v)} - \frac{t^2 - M_2^2\, (x_3/v)^2}{(x_3/v)^2\, (t - x_3/v)} \right] = -\frac{\left(M_2^2 - M_1^2\right)}{4\pi \rho v^3} \frac{l}{(t - x_3/v)} \end{aligned} \tag{11.1.223}$$

As a result, we obtain

$$u_3(x_3, t) = u_3^{\delta}(x_3, t) + u_3^H(x_3, t) = -\frac{lM_2^2}{4\pi \rho v^3} \frac{1}{(t - x_3/v)} \quad \text{for } M_2\, x_3/v < t < x_3/v \tag{11.1.224}$$

Case (d): $t > x_3/v$ (see Figure 11.4).

In this case, the integral in Equation 11.1.183 is split into two integrals

$$u_3(x_3, t) = l \left[\int_0^{x_3/V} U_3(0, x_3 - v\tau; t - \tau)\, d\tau + \int_{x_3/V}^{t} U_3(0, x_3 - v\tau;\; t - \tau)\, d\tau \right] \tag{11.1.225}$$

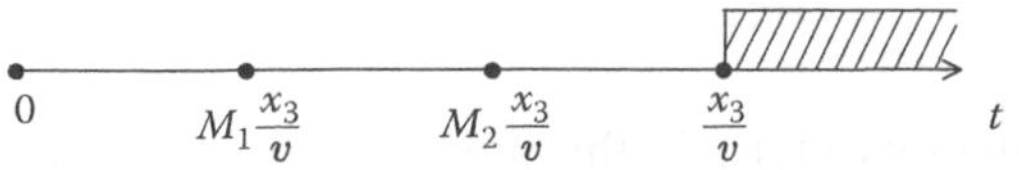

FIGURE 11.4 A range of time (shaded) for Case (d).

where

$$U_3(0, x_3 - v\tau; t - \tau) = U_3^\delta(0, x_3 - v\tau; t - \tau) + U_3^H(0, x_3 - v\tau; t - \tau) \tag{11.1.226}$$

$$U_3^\delta(0, x_3 - v\tau, t - \tau) = \frac{1}{4\pi\rho v^3} \frac{M_1}{1 - M_1} \frac{t - \tau}{(\tau - x_3/v)^2} \delta(\tau - \tau_1) \tag{11.1.227}$$

$$U_3^H(0, x_3 - v\tau; t - \tau) = -\frac{2}{4\pi\rho v^3} \frac{(t - \tau)}{(\tau - x_3/v)^3} [H(\tau_1 - \tau) - H(\tau_2 - \tau)]$$

$$\text{for } 0 < \tau < x_3/v \tag{11.1.228}$$

and

$$U_3^\delta(0, x_3 - v\tau; t - \tau) = U_3^\delta(0, x_3 - v\tau; t - \tau) + U_3^H(0, x_3 - v\tau; t - \tau) \tag{11.1.229}$$

$$U_3^\delta(0, x_3 - v\tau; t - \tau) = \frac{1}{4\pi\rho v^3} \frac{M_1}{1 + M_1} \frac{t - \tau}{(\tau - x_3/v)^2} \delta\left(\tau - \tau_1^0\right) \tag{11.1.230}$$

$$U_3^H(0, x_3 - v\tau; t - \tau) = \frac{2}{4\pi\rho v^3} \cdot \frac{t - \tau}{(\tau - x_3/v)^3} \left\{H\left(\tau_1^0 - \tau\right) - H\left(\tau_2^0 - \tau\right)\right\}$$

$$\text{for } x_3/v < \tau < t \tag{11.1.231}$$

Here

$$\tau_i^0 = \frac{t + M_i x_3/v}{1 + M_i} \quad (i = 1, 2) \tag{11.1.232}$$

Now, we show that the first integral on the RHS of Equation 11.1.225 vanishes. To this end we note that the inequality $0 < \tau < x_3/v < t$ implies that

$$\tau_1 - \tau > \tau_1 - x_3/v = \frac{t - x_3/v}{1 - M_1} > 0 \tag{11.1.233}$$

and

$$\tau_2 - \tau > \tau_2 - x_3/v = \frac{t - x_3/v}{1 - M_2} > 0 \tag{11.1.234}$$

Hence, we obtain

$$\delta(\tau - \tau_1) = \delta(\tau_1 - \tau) = 0 \tag{11.1.235}$$

$$H(\tau_1 - \tau) - H(\tau_2 - \tau) = 0 \tag{11.1.236}$$

and it follows from Equations 11.1.226 through 11.1.228 that the first integral on the RHS of Equation 11.1.225 vanishes.

As a result, we have

$$u_3(x_3,t) = u_3^{\delta}(x_3,t) + u_3^{H}(x_3,t) \tag{11.1.237}$$

where

$$u_3^{\delta}(x_3,t) = \frac{l}{4\pi\rho v^3}\frac{M_1}{1+M_1}\int\limits_{x_3/v}^{t}\frac{t-\tau}{(\tau - x_3/v)^2}\,\delta\left(\tau-\tau_1^0\right)d\tau \tag{11.1.238}$$

and

$$u_3^{H}(x_3,t) = \frac{2l}{4\pi\rho v^3}\int\limits_{x_3/v}^{t}\frac{t-\tau}{(\tau - x_3/v)^3}\left[H\left(\tau_1^0-\tau\right)-H\left(\tau_2^0-\tau\right)\right]d\tau \tag{11.1.239}$$

To compute the integrals in Equations 11.1.238 and 11.1.239 note that

$$x_3/v < \tau_1^0 < t \tag{11.1.240}$$

and

$$x_3/v < \tau_2^0 < t \tag{11.1.241}$$

Hence, it follows from Equation 11.1.238 that

$$u_3^{\delta}(x_3,t) = \frac{l}{4\pi\rho v^3}\frac{M_1}{1+M_1}\frac{t-\tau_1^0}{\left(\tau_1^0 - x_3/v\right)^2} = \frac{M_1^2}{4\pi\rho v^3}\frac{l}{(t-x_3/v)} \tag{11.1.242}$$

Also, it follows from Equation 11.1.239 that

$$u_3^{H}(x_3,t) = \frac{2l}{4\pi\rho v^3}\left[\int\limits_{x_3/v}^{\tau_1^0}\frac{(t-\tau)}{(\tau - x_3/v)^3}\,d\tau - \int\limits_{x_3/v}^{\tau_2^0}\frac{(t-\tau)}{(\tau - x_3/v)^3}\,d\tau\right] \tag{11.1.243}$$

Note that the integrals on the RHS of this equation are singular as both have the same integrand that tends to infinity as $\tau \to x_3/v + 0$. If they are computed in the sense of a principal value, the result is

$$\int\limits_{x_3/v}^{\tau_1^0}\frac{(t-\tau)}{(\tau - x_3/v)^3}\,d\tau = \lim_{\epsilon\to 0}\left[F\left(\tau_1^0\right) - F\left(x_3/v+\epsilon\right)\right] \tag{11.1.244}$$

$$\int\limits_{x_3/v}^{\tau_2^0}\frac{(t-\tau)}{(\tau - x_3/v)^3}\,d\tau = \lim_{\epsilon\to 0}\left[F\left(\tau_2^0\right) - F\left(x_3/v+\epsilon\right)\right] \tag{11.1.245}$$

where (see Equation 11.1.211)

$$F(\tau) = (\tau - x_3/v)^{-1} - \frac{1}{2}(t - x_3/v)(\tau - x_3/v)^{-2} \tag{11.1.246}$$

By subtracting Equation 11.1.245 from Equation 11.1.244 gives

$$\int_{x_3/v}^{\tau_1^0} \frac{(t-\tau)}{(\tau - x_3/v)^3} d\tau - \int_{x_3/v}^{\tau_2^0} \frac{(t-\tau)}{(\tau - x_3/v)^3} d\tau = F\left(\tau_1^0\right) - F\left(\tau_2^0\right) \tag{11.1.247}$$

Substituting Equation 11.1.247 into the RHS of Equation 11.1.243, and using the definitions of τ_1^0 and τ_2^0 (see Equations 11.1.232) we get

$$u_3^H(x_3, t) = \frac{M_2^2 - M_1^2}{4\pi\rho v^3} \frac{l}{(t - x_3/v)} \tag{11.1.248}$$

Finally, it follows from Equations 11.1.237, 11.1.242, and 11.1.248 that

$$u_3(x_3, t) = \frac{M_2^2}{4\pi\rho v^3} \frac{l}{(t - x_3/v)} \quad \text{for } t > x_3/v \tag{11.1.249}$$

This completes the discussion of Cases (a)–(d).

To complete the analysis of the displacement wave $u_3 = u_3(x_3, t)$ for $t \geq 0$ and $x_3 > 0$, we explore the behavior of this function at the points $t_1 = M_1 x_3/v$, $t_2 = M_2 x_3/v$, $t_3 = x_3/v$, and $t = \infty$ for a fixed $x_3 > 0$.

It follows from Equations 11.1.195 and 11.1.219 that

$$u_3\left(x_3, M_1\, x_3/v - 0\right) = 0 \tag{11.1.250}$$

and

$$u_3\left(x_3, M_1\, x_3/v + 0\right) = \frac{l}{4\pi\rho v^3} \frac{M_1^2}{1 - M_1} \frac{1}{(x_3/v)} \tag{11.1.251}$$

Hence

$$[\![u_3]\!](t_1) = \frac{l}{4\pi\rho v^3} \frac{M_1^2}{1 - M_1} \frac{1}{(x_3/v)} \tag{11.1.252}$$

where

$$[\![u_3]\!](t_1) = u_3(x_3, t_1 + 0) - u_3(x_3, t_1 - 0) \tag{11.1.253}$$

represents a jump of u_3 at $t = t_1$.

Also, it follows from Equations 11.1.219 and 11.1.224 that

$$u_3\,(x_3, M_2\,x_3/v - 0) = \frac{l}{4\pi\rho v^3}\,\frac{M_2^2}{1 - M_2}\,\frac{1}{(x_3/v)} \tag{11.1.254}$$

$$u_3\,(x_3, M_2\,x_3/v + 0) = \frac{l}{4\pi\rho v^3}\,\frac{M_2^2}{1 - M_2}\,\frac{1}{(x_3/v)} \tag{11.1.255}$$

Hence, we find

$$[\![u_3]\!](t_2) = 0 \tag{11.1.256}$$

Finally, using Equations 11.1.224 and 11.1.249 we find, respectively,

$$u_3\,(x_3, x_3/v - 0) = +\infty \tag{11.1.257}$$

and

$$u_3\,(x_3, x_3/v + 0) = +\infty \tag{11.1.258}$$

Therefore,

$$u_3(x_3, t) \to +\infty \quad \text{as } t \to x_3/v \tag{11.1.259}$$

The results are summarized in the following theorem:

Theorem 1: An elastic wave produced by the suddenly applied concentrated force l that moves with a constant velocity v in the positive direction of the x_3-axis of the cylindrical coordinates (r, φ, x_3) in a homogeneous isotropic infinite solid, and restricted to points of the positive x_3-axis, is represented by the vector field

$$\mathbf{u}(x_3, t) = [0, 0, u_3(x_3, t)] \tag{11.1.260}$$

where

$$u_3(x_3, t) = \begin{cases} 0 & \text{for } 0 \le t < M_1 x_3/v \\ -\dfrac{l}{4\pi\rho v^3}\,\dfrac{t^2}{(x_3/v)^2(t - x_3/v)} & \text{for } M_1 x_3/v < t < M_2 x_3/v \\ -\dfrac{lM_2^2}{4\pi\rho v^3}\,\dfrac{1}{(t - x/v)} & \text{for } M_2 x_3/v < t < x_3/v \\ \dfrac{lM_2^2}{4\pi\rho v^3}\,\dfrac{1}{(t - x_3/v)} & \text{for } t > x_3/v \end{cases} \tag{11.1.261}$$

In Equation 11.1.261, M_1 and M_2 are the Mach numbers of the moving force relative to the longitudinal and transverse waves, respectively, restricted by the inequalities

$$M_1 < 1 - \sqrt{1 - M_2}, \quad 0 < M_1 < M_2 < 1 \tag{11.1.262}$$

The profile of the wave, based on Equation 11.1.260, is shown in Figure 11.5. ■

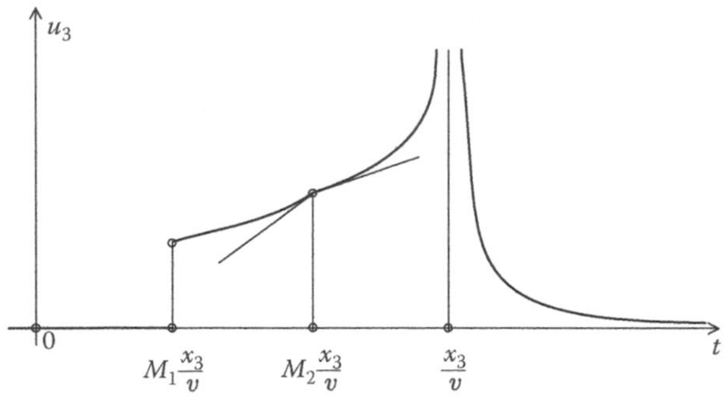

FIGURE 11.5 Profile of the wave produced by a moving concentrated force.

Note that the point $t_2 = M_2\, x_3/v$ is a point of discontinuity of the velocity field $\dot{u}_3(x_3, t)$, and this is indicated by two different tangent lines to u_3 at $t = t_2$. Also note that $u_3 \to +\infty$ at the point of application of the moving force, and $u_3 \to 0$ as $t \to \infty$, which means that the wave has a transient character.

Note:

The hypothesis $M_1 < 1 - \sqrt{1 - M_2}$ in Theorem 1 is satisfied, for example, for steel in which $M_1 > 0.2$. If $M_1 > 1 - \sqrt{1 - M_2}$, a similar theorem involving a closed-form wave profile may be formulated.

11.1.4 The Stress Waves due to the Initial Stress and Stress-Rate Fields in an Unbounded Solid

11.1.4.1 A Solution to the Homogeneous Scalar Wave Equation Subject to Initial Conditions in an Infinite Space

We now prove a theorem that will be widely used in Section 11.1.4.2.

Theorem 1: A solution of the scalar wave equation

$$\left(\nabla^2 - \frac{1}{c^2}\frac{\partial^2}{\partial t^2}\right)\phi(\mathbf{x}, t) = 0 \quad (c > 0) \quad (\mathbf{x}, t) \in E^3 \times [0, \infty) \tag{11.1.263}$$

subject to the initial conditions

$$\phi(\mathbf{x}, 0) = \phi_0(\mathbf{x}), \quad \dot{\phi}(\mathbf{x}, 0) = \dot{\phi}_0(\mathbf{x}) \quad \mathbf{x} \in \Omega \subset E^3 \tag{11.1.264}$$

admits the representation

$$\phi(\mathbf{x}, t) = t\mathcal{M}_{\mathbf{x},ct}(\dot{\phi}_0) + \frac{\partial}{\partial t}[t\mathcal{M}_{\mathbf{x},ct}(\phi_0)] \tag{11.1.265}$$

where, for any function $g = g(\mathbf{x})$ on Ω, the symbol $\mathcal{M}_{\mathbf{x},ct}(g)$ stands for the *mean value of g over the surface of a ball with its center at* $\mathbf{x}$ *and with the radius ct*:

$$\mathcal{M}_{\mathbf{x},ct}(g) = \frac{1}{4\pi}\int_0^{2\pi}\int_0^{\pi} g(\mathbf{x} + \mathbf{n}ct)\sin\theta\, d\theta\, d\varphi \tag{11.1.266}$$

In Equation 11.1.266 $\mathbf{n}$ is the unit outward normal vector on the spherical surface

$$\mathbf{n} = [\sin\theta\cos\varphi,\ \sin\theta\sin\varphi,\ \cos\theta] \tag{11.1.267}$$

■

Proof: Let $\overline{\phi} = \overline{\phi}(\mathbf{x}, p)$ denote the Laplace transform of $\phi = \phi(\mathbf{x}, t)$, that is,

$$L\{\phi(\mathbf{x}, t)\} \equiv \overline{\phi}(\mathbf{x}, p) = \int_0^\infty e^{-pt}\phi(\mathbf{x}, t)\, dt \tag{11.1.268}$$

where p is the transform parameter. Applying operator L to Equation 11.1.263 and using the initial conditions 11.1.264 we obtain

$$\left(\nabla^2 - \frac{p^2}{c^2}\right)\overline{\phi}(\mathbf{x}, p) = -\frac{1}{c^2}[\dot{\phi}_0(\mathbf{x}) + p\phi_0(\mathbf{x})] \quad \mathbf{x} \in \Omega \tag{11.1.269}$$

Since for a fixed point $\boldsymbol{\xi} \in \Omega$

$$\left(\nabla^2 - \frac{p^2}{c^2}\right)\frac{e^{-\frac{p}{c}|\mathbf{x}-\boldsymbol{\xi}|}}{|\mathbf{x}-\boldsymbol{\xi}|} = -4\pi\,\delta(\mathbf{x}-\boldsymbol{\xi}) \tag{11.1.270}$$

where $\delta(\cdot)$ stands for the Dirac delta function, then, for any function $g = g(\boldsymbol{\xi})$ on Ω, multiplying Equation 11.1.270 by $g(\boldsymbol{\xi})$ and integrating over Ω, we find

$$\left(\nabla^2 - \frac{p^2}{c^2}\right)\frac{1}{4\pi}\int_\Omega \frac{e^{-\frac{p}{c}|\mathbf{x}-\boldsymbol{\xi}|}}{|\mathbf{x}-\boldsymbol{\xi}|} g(\boldsymbol{\xi})\, dv(\boldsymbol{\xi}) = -g(\mathbf{x}) \tag{11.1.271}$$

or, in an alternative form,

$$-\frac{1}{c^2}\left(\nabla^2 - \frac{p^2}{c^2}\right)^{-1} g(\mathbf{x}) = \frac{1}{4\pi c^2}\int_\Omega \frac{e^{-\frac{p}{c}|\mathbf{x}-\boldsymbol{\xi}|}}{|\mathbf{x}-\boldsymbol{\xi}|} g(\boldsymbol{\xi})\, dv(\boldsymbol{\xi}) \tag{11.1.272}$$

Now, applying the operator L^{-1} to this equation we arrive at

$$L^{-1}\left\{-\frac{1}{c^2}\left(\nabla^2 - \frac{p^2}{c^2}\right)^{-1} g(\mathbf{x})\right\} = \frac{1}{4\pi c^2}\int_\Omega \frac{g(\boldsymbol{\xi})}{|\mathbf{x}-\boldsymbol{\xi}|}\delta\left(t - \frac{|\mathbf{x}-\boldsymbol{\xi}|}{c}\right) dv(\boldsymbol{\xi}) \tag{11.1.273}$$

In the following we show that the volume integral on the RHS of Equation 11.1.273 may be expressed in terms of the mean value of the function g over the surface of a ball. Let $|\mathbf{x}-\boldsymbol{\xi}| = R$.

Since

$$\frac{1}{c^2R}\delta\left(t - \frac{R}{c}\right) = \frac{1}{c^2R}\delta\left(\frac{R}{c} - t\right) = \frac{1}{c^2R}\delta\left[\frac{1}{c}(R - ct)\right]$$

$$= \frac{1}{cR}\delta(R - ct) = \frac{R}{ct}\frac{t}{R^2}\delta(R - ct) = \frac{t}{R^2}\delta(R - ct) \tag{11.1.274}$$

hence

$$g^*(\mathbf{x},t) \stackrel{\text{df}}{=} \frac{1}{4\pi c^2}\int_\Omega \frac{g(\boldsymbol{\xi})}{|\mathbf{x}-\boldsymbol{\xi}|}\delta\left(t-\frac{|\mathbf{x}-\boldsymbol{\xi}|}{c}\right)dv(\boldsymbol{\xi}) = \frac{t}{4\pi}\int_\Omega \frac{g(\boldsymbol{\xi})}{|\mathbf{x}-\boldsymbol{\xi}|^2}\delta\left(|\mathbf{x}-\boldsymbol{\xi}|-ct\right)dv(\boldsymbol{\xi}) \tag{11.1.275}$$

Introduce the spherical coordinates with the origin at point $\mathbf{x}$:

$$\boldsymbol{\xi}-\mathbf{x} = R\,\mathbf{n} \tag{11.1.276}$$

where

$$\mathbf{n} = [\cos\varphi\sin\theta,\ \sin\varphi\sin\theta,\ \cos\theta] \tag{11.1.277}$$

Then Equation 11.1.275 is reduced to the form

$$g^*(\mathbf{x},t) = \frac{t}{4\pi}\int_0^R\int_0^{2\pi}\int_0^{\pi}\delta(R-ct)g(\mathbf{x}+R\mathbf{n})\sin\theta\,d\theta\,d\varphi dR \tag{11.1.278}$$

or, by using the filtering property of the Dirac delta function, to

$$g^*(\mathbf{x},t) = t\mathcal{M}_{\mathbf{x},ct}(g) \tag{11.1.279}$$

where $\mathcal{M}_{\mathbf{x},ct}(g)$ is the mean value of g over the surface of a ball with center at $\mathbf{x}$ and with radius ct (see Equation 11.1.266). Finally, using Equations 11.1.273, 11.1.275, and 11.1.279, we arrive at

$$L^{-1}\left\{-\frac{1}{c^2}\left(\nabla^2-\frac{p^2}{c^2}\right)^{-1} g(\mathbf{x})\right\} = t\mathcal{M}_{\mathbf{x},ct}(g) \tag{11.1.280}$$

Now, it follows from Equation 11.1.269 that

$$\overline{\phi}(\mathbf{x},p) = -\frac{1}{c_1^2}\left(\nabla^2-\frac{p^2}{c^2}\right)^{-1}[\dot{\phi}_0(\mathbf{x})+p\phi_0(\mathbf{x})] \tag{11.1.281}$$

Therefore, applying the operator L^{-1} to Equation 11.1.281 and using the relation 11.1.280 we arrive at the representation 11.1.265. This completes the proof of Theorem 1. □

Theorem 1 implies that for $\phi_0(\mathbf{x}) = 0$ on Ω the solution $\phi = \phi(\mathbf{x},t)$ vanishes outside of an interval $0 < t_1 < t < t_2$ that determines a spherical shell with the center at $\mathbf{x}$ bounded by the radii $R_1 = ct_1$ and $R_2 = ct_2$, and containing Ω (see Figure 11.6).

If $t_1 < t < t_2$ then $\mathcal{M}_{\mathbf{x},ct}(\dot{\phi}_0) \neq 0$ over an intersection Σ of the support of $\dot{\phi}_0$ and the spherical surface of radius ct.

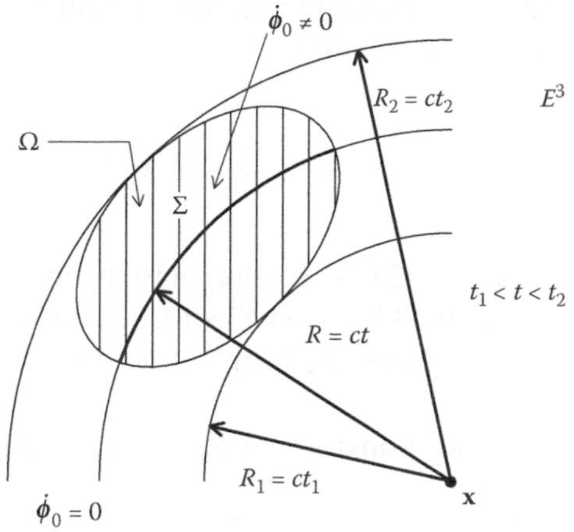

FIGURE 11.6 Intersection Σ of a support of $\dot{\phi}_0$ with a spherical surface of radius ct.

11.1.4.2 The Stress Waves Produced by the Initial Data in a Homogeneous Isotropic Infinite Elastic Solid

The stress waves produced by the initial stress and stress-rate fields of compact support in a homogeneous isotropic infinite elastic solid are described by a solution to the following pure stress initial-value problem. Find a stress field $\mathbf{S} = \mathbf{S}(\mathbf{x}, t)$ on $E^3 \times [0, \infty)$ that satisfies the equation (see Equation 4.2.14 with $\mathbf{b} = \mathbf{0}$)

$$\hat{\nabla}(\operatorname{div}\mathbf{S}) - \frac{\rho}{2\mu}\left[\ddot{\mathbf{S}} - \frac{\lambda}{3\lambda + 2\mu}(\operatorname{tr}\ddot{\mathbf{S}})\mathbf{1}\right] = \mathbf{0} \quad \text{on } E^3 \times (0, \infty) \tag{11.1.282}$$

subject to the initial conditions

$$\mathbf{S}(\mathbf{x}, 0) = \mathbf{S}^0(\mathbf{x}), \quad \dot{\mathbf{S}}(\mathbf{x}, 0) = \dot{\mathbf{S}}^0(\mathbf{x}), \quad \mathbf{x} \in \Omega \subset E^3 \tag{11.1.283}$$

where $\mathbf{S}^0(\mathbf{x})$ and $\dot{\mathbf{S}}^0(\mathbf{x})$ are prescribed second-order symmetric tensor fields on Ω. In components, Equations 11.1.282 and 11.1.283, respectively, take the forms

$$S_{(ik,kj)} - \frac{\rho}{2\mu}\left(\ddot{S}_{ij} - \frac{\lambda}{3\lambda + 2\mu}\ddot{S}_{kk}\,\delta_{ij}\right) = 0 \quad \text{on } E^3 \times [0, \infty) \tag{11.1.284}$$

and

$$S_{ij}(\mathbf{x}, 0) = S_{ij}^0(\mathbf{x}), \quad \dot{S}_{ij}(\mathbf{x}, 0) = \dot{S}_{ij}^0(\mathbf{x}), \quad \mathbf{x} \subset \Omega \subset E^3 \tag{11.1.285}$$

By letting $i = j = k$ in Equation 11.1.284, we have

$$\ddot{S}_{kk} = \frac{3\lambda + 2\mu}{\rho} S_{ab,ab} \tag{11.1.286}$$

Therefore, if we substitute Equation 11.1.286 into Equation 11.1.284 and integrate the result twice with respect to time, because of Equation 11.1.285 we arrive at an alternative formulation of the initial-value problem: Find a tensor field $S_{ij} = S_{ij}(\mathbf{x}, t)$ on $E^3 \times [0, \infty)$ that satisfies the integro-differential equation

$$S_{ij} = S_{ij}^0 + t\dot{S}_{ij}^0 + \rho^{-1} \int_0^t (t - \tau)[2\mu S_{(ik,kj)} + \lambda S_{ab,ab}\delta_{ij}]\, d\tau \quad \text{on } E^3 \times [0, \infty) \tag{11.1.287}$$

Note that S_{ij} given by Equation 11.1.287 is a solution to the initial value problem 11.1.284 and 11.1.285 provided the tensor field $S_{(ik,kj)}$ on $E^3 \times [0, \infty)$ is available. In the following we outline a method of finding $S_{(ik,kj)}$ using the formulation described by Equations 11.1.284 and 11.1.285.

Let $\overline{S}_{ij} = \overline{S}_{ij}(\mathbf{x}, p)$ be the Laplace transform of $S_{ij} = S_{ij}(\mathbf{x}, t)$ with respect to time, that is,

$$L\{S_{ij}\} \equiv \overline{S}_{ij}(\mathbf{x}, p) = \int_0^\infty e^{-pt} S_{ij}(\mathbf{x}, t)\, dt \tag{11.1.288}$$

where p is the transform parameter. By applying the operator L to Equation 11.1.284 and using the initial conditions 11.1.285, we write

$$\overline{S}_{(ik,kj)} - \frac{\rho p^2}{2\mu}\left(\overline{S}_{ij} - \frac{\lambda}{3\lambda + 2\mu}\overline{S}_{kk}\,\delta_{ij}\right) = -\rho p^2 \overline{B}_{ij} \tag{11.1.289}$$

where

$$p^2 \overline{B}_{ij} = \frac{1}{2\mu}\left(\dot{S}_{ij}^0 - \frac{\lambda}{3\lambda + 2\mu}\dot{S}_{kk}^0 \delta_{ij}\right) + \frac{p}{2\mu}\left(S_{ij}^0 - \frac{\lambda}{3\lambda + 2\mu} S_{kk}^0\,\delta_{ij}\right) \tag{11.1.290}$$

Next, by letting $j = a$ and getting rid of the index parentheses in Equation 11.1.289 we find

$$\overline{S}_{ik,ka} + \overline{S}_{ak,ki} - \frac{\rho p^2}{\mu}\left(\overline{S}_{ia} - \frac{\lambda}{3\lambda + 2\mu}\overline{S}_{kk}\,\delta_{ia}\right) = -2\rho p^2 \overline{B}_{ia} \tag{11.1.291}$$

Now, applying the operator $\partial^2/\partial x_a \partial x_j$ to Equation 11.1.291 and taking the symmetric part of the result with respect to the indexes i and j, gives

$$\left(\nabla^2 - \frac{p^2}{c_2^2}\right)\overline{S}_{(ia,aj)} = -\left(\rho\frac{p^2}{\mu}\frac{\lambda}{3\lambda + 2\mu}\overline{S}_{kk} + \overline{S}_{ab,ab}\right)_{,ij} - 2\rho p^2 \overline{B}_{(ia,aj)} \tag{11.1.292}$$

where

$$c_2^2 = \mu/\rho \tag{11.1.293}$$

Finally, the application of the operator $\left(\nabla^2 - p^2/c_1^2\right)$ to Equation 11.1.292, where $c_1^2 = (\lambda + 2\mu)/\rho$, results in

$$\left(\nabla^2 - \frac{p^2}{c_1^2}\right)\left(\nabla^2 - \frac{p^2}{c_2^2}\right)\overline{S}_{(ia,aj)} = -\overline{F}_{ij} \tag{11.1.294}$$

where

$$\overline{F}_{ij} = \left(\nabla^2 - \frac{p^2}{c_1^2}\right)\left(\overline{S}_{ab,ab} + \frac{p^2}{c_2^2}\frac{\lambda}{3\lambda+2\mu}\overline{S}_{kk}\right)_{,ij} + \left(\nabla^2 - \frac{p^2}{c_1^2}\right)\left[2\rho p^2 \overline{B}_{(ia,aj)}\right] \quad (11.1.295)$$

Clearly, a solution of Equation 11.1.294 in E^3 is available provided $\overline{F}_{ij}$ is a known tensor field on E^3. In the following, we show that $\overline{F}_{ij}$ may be expressed entirely in terms of the stress and stress-rate initial fields, and of the transform parameter p.

First, it follows from the definition of $\overline{B}_{ij}$ (see Equation 11.1.290) that the last term on the RHS of Equation 11.1.295 may be expressed in terms of S^0_{ij}, $\dot{S}^0_{ij}$, and p. To show that the term $\left(\nabla^2 - p^2/c_1^2\right)(\cdot)$ on the RHS of 11.1.295 may be expressed in terms of S^0_{ij}, $\dot{S}^0_{ij}$, and p, we proceed in the following way.

By applying the operator $\left(\delta_{ij}\nabla^2 - \partial^2/\partial x_i \partial x_j\right)$ to Equation 11.1.289 and using the identity

$$\left(\delta_{ij}\nabla^2 - \frac{\partial^2}{\partial x_i \partial x_j}\right)\overline{S}_{(ik,kj)} = 0 \quad (11.1.296)$$

we obtain

$$\left(\delta_{ij}\nabla^2 - \frac{\partial^2}{\partial x_i \partial x_j}\right)\left(\overline{S}_{ij} - \frac{\lambda}{3\lambda+2\mu}\overline{S}_{kk}\delta_{ij}\right) = 2\mu\left(\nabla^2 \overline{B}_{aa} - \overline{B}_{ab,ab}\right) \quad (11.1.297)$$

or

$$\frac{\lambda+2\mu}{3\lambda+2\mu}\nabla^2 \overline{S}_{aa} - \overline{S}_{ab,ab} = 2\mu(\nabla^2 \overline{B}_{aa} - \overline{B}_{ab,ab}) \quad (11.1.298)$$

Also, by letting $i = j = a$ in Equation 11.1.289 we get

$$-\frac{\rho p^2}{3\lambda+2\mu}\overline{S}_{aa} + \overline{S}_{ab,ab} = -\rho p^2 \overline{B}_{aa} \quad (11.1.299)$$

Now, because of Equation 11.1.290, we get

$$2\mu(\nabla^2 \overline{B}_{aa} - \overline{B}_{ab,ab}) = -\frac{1}{p^2}\left[\left(\dot{S}^0_{ab,ab} - \frac{\lambda+2\mu}{3\lambda+2\mu}\dot{S}^0_{aa,bb}\right) + p\left(S^0_{ab,ab} - \frac{\lambda+2\mu}{3\lambda+2\mu}S^0_{aa,bb}\right)\right] \quad (11.1.300)$$

and

$$\overline{B}_{aa} = \frac{1}{p^2}\frac{1}{3\lambda+2\mu}\left(\dot{S}^0_{aa} + pS^0_{aa}\right) \quad (11.1.301)$$

Hence, substituting Equations 11.1.300 and 11.1.301 into Equations 11.1.298 and 11.1.299, respectively, and eliminating $\overline{S}_{ab,ab}$, we arrive at the single equation for $\overline{S}_{aa}$:

$$\left(\nabla^2 - \frac{p^2}{c_1^2}\right)\overline{S}_{aa} = -\frac{1}{c_1^2}\overline{f} \quad (11.1.302)$$

where

$$\bar{f} = \dot{S}^0_{aa} + pS^0_{aa} + \frac{(3\lambda + 2\mu)}{\rho}\left[p^{-1}\left(S^0_{ab,ab} - \frac{\lambda + 2\mu}{3\lambda + 2\mu}S^0_{aa,bb}\right)\right.$$

$$\left. + p^{-2}\left(\dot{S}^0_{ab,ab} - \frac{\lambda + 2\mu}{3\lambda + 2\mu}\dot{S}^0_{aa,bb}\right)\right] \tag{11.1.303}$$

Also, by substituting Equations 11.1.300 and 11.1.301 into Equations 11.1.298 and 11.1.299, respectively, and eliminating $\overline{S}_{aa}$, we arrive at the simple equation for $\overline{S}_{ab,ab}$:

$$\left(\nabla^2 - \frac{p^2}{c_1^2}\right)\overline{S}_{ab,ab} = -\frac{1}{c_1^2}\bar{g} \tag{11.1.304}$$

where

$$\bar{g} = \dot{S}^0_{ab,ab} + pS^0_{ab,ab} \tag{11.1.305}$$

Next, if we introduce the notation

$$\overline{H}_{ij} = 2\rho c_2^2 p^2 \overline{B}_{(ia,aj)} \tag{11.1.306}$$

then it follows from Equation 11.1.290 that

$$\overline{H}_{ij} = \left(\dot{S}^0_{(ia,aj)} - \frac{\lambda}{3\lambda + 2\mu}\dot{S}^0_{aa,ij}\right) + p\left(S^0_{(ia,aj)} - \frac{\lambda}{3\lambda + 2\mu}S^0_{aa,ij}\right) \tag{11.1.307}$$

Finally, substituting Equations 11.1.302, 11.1.304, and 11.1.306 into Equation 11.1.295, we arrive at

$$\overline{F}_{ij} = -\frac{1}{c_1^2}\left(\bar{g}_{,ij} + \frac{p^2}{c_2^2}\frac{\lambda}{3\lambda + 2\mu}\bar{f}_{,ij}\right) + \frac{1}{c_2^2}\left(\nabla^2 - \frac{p^2}{c_1^2}\right)\overline{H}_{ij} \tag{11.1.308}$$

Hence, on account of Equations 11.1.303, 11.1.305, and 11.1.307, $\overline{F}_{ij}$ is expressible in terms of S^0_{ij}, $\dot{S}^0_{ij}$, and p.

Now, to obtain S_{aa}, $S_{ab,ab}$, and $S_{(ia,aj)}$ in the space–time domain, we use the results of Theorem 1 of Section 11.1.4.1.

First, to obtain S_{aa} we rewrite Equation 11.1.302 in the form

$$\overline{S}_{aa} = -\frac{1}{c_1^2}\left(\nabla^2 - \frac{p^2}{c_1^2}\right)^{-1}\bar{f} \tag{11.1.309}$$

By applying the operator L^{-1} to this equation and using the formula (see Equation 11.1.280)

$$L^{-1}\left\{-\frac{1}{c^2}\left(\nabla^2 - \frac{p^2}{c^2}\right)^{-1}g(\mathbf{x})\right\} = t\mathcal{M}_{\mathbf{x},ct}(g) \tag{11.1.310}$$

that is valid for any function $g = g(\mathbf{x})$ on Ω, and Equation 11.1.303, we find

$$S_{aa}(\mathbf{x},t) = t\mathcal{M}_{\mathbf{x},c_1 t}\left(\dot{S}^0_{aa}\right) + \frac{\partial}{\partial t}\left[t\mathcal{M}_{\mathbf{x},c_1 t}\left(S^0_{aa}\right)\right]$$

$$+ \frac{3\lambda + 2\mu}{\rho}\int_0^t \tau\left[\mathcal{M}_{\mathbf{x},c_1\tau}\left(S^0_{ab,ab} - \frac{\lambda + 2\mu}{3\lambda + 2\mu}S^0_{aa,bb}\right)\right.$$

$$\left. + (t-\tau)\mathcal{M}_{\mathbf{x},c_1\tau}\left(\dot{S}^0_{ab,ab} - \frac{\lambda + 2\mu}{3\lambda + 2\mu}\dot{S}^0_{aa,bb}\right)\right] d\tau \tag{11.1.311}$$

Next, to find $S_{ab,ab}$ we rewrite Equation 11.1.304 in the form

$$\overline{S}_{ab,ab} = -\frac{1}{c_1^2}\left(\nabla^2 - \frac{p^2}{c_1^2}\right)^{-1}\overline{g} \tag{11.1.312}$$

By applying the operator L^{-1} to this equation and using Equations 11.1.305 and 11.1.310, we arrive at the simple result

$$S_{ab,ab}(\mathbf{x},t) = t\mathcal{M}_{\mathbf{x},c_1 t}\left(\dot{S}^0_{ab,ab}\right) + \frac{\partial}{\partial t}\left[t\mathcal{M}_{\mathbf{x},c_1 t}\left(S^0_{ab,ab}\right)\right] \tag{11.1.313}$$

Finally, to get $S_{(ia,aj)}$, we reduce Equation 11.1.294 to the form

$$\overline{S}_{(ia,aj)} = \frac{1}{c_1^2}\left[\left(\nabla^2 - \frac{p^2}{c_1^2}\right)\left(\nabla^2 - \frac{p^2}{c_2^2}\right)\right]^{-1}\left(\frac{p^2}{c_2^2}\frac{\lambda}{3\lambda + 2\mu}\overline{f}_{,ij} + \overline{g}_{,ij}\right) - \frac{1}{c_2^2}\left(\nabla^2 - \frac{p^2}{c_2^2}\right)^{-1}\overline{H}_{ij} \tag{11.1.314}$$

Since, for any function $h = h(\mathbf{x})$ on Ω,

$$\left[\left(\nabla^2 - \frac{p^2}{c_1^2}\right)\left(\nabla^2 - \frac{p^2}{c_2^2}\right)\right]^{-1} h(\mathbf{x}) = \left(\frac{p^2}{c_1^2} - \frac{p^2}{c_2^2}\right)^{-1}$$

$$\times \left[\left(\nabla^2 - \frac{p^2}{c_1^2}\right)^{-1} h(\mathbf{x}) - \left(\nabla^2 - \frac{p^2}{c_2^2}\right)^{-1} h(\mathbf{x})\right] \tag{11.1.315}$$

therefore, by applying the operator L^{-1} to Equation 11.1.315 and using the formula 11.1.310 we get

$$L^{-1}\left\{\left[\left(\nabla^2 - \frac{p^2}{c_1^2}\right)\left(\nabla^2 - \frac{p^2}{c_2^2}\right)\right]^{-1} h(\mathbf{x})\right\}$$

$$= \frac{c_1^2 c_2^2}{c_1^2 - c_2^2}\int_0^t (t-\tau)\tau\left\{c_1^2\mathcal{M}_{\mathbf{x},c_1\tau}(h) - c_2^2\mathcal{M}_{\mathbf{x},c_2\tau}(h)\right\} d\tau \tag{11.1.316}$$

Note that an alternative form of Equation 11.1.316 reads

$$L^{-1}\left\{\left[\left(\nabla^2 - \frac{p^2}{c_1^2}\right)\left(\nabla^2 - \frac{p^2}{c_2^2}\right)\right]^{-1} p^2 h(\mathbf{x})\right\} = \frac{c_1^2 c_2^2}{c_1^2 - c_2^2}t\left[c_1^2\mathcal{M}_{\mathbf{x},c_1 t}(h) - c_2^2\mathcal{M}_{\mathbf{x},c_2 t}(h)\right] \tag{11.1.317}$$

Hence, by applying the operator L^{-1} to Equation 11.1.314 and using Equations 11.1.303, 11.1.305, and 11.1.307, as well as the inversion formulas 11.1.310, 11.1.316, and 11.1.317, we arrive at

$$\begin{aligned}
S_{(ia,aj)}(\mathbf{x},t) = & \int_0^t (t-\tau)\tau \left[c_1^2 \mathcal{M}_{\mathbf{x},c_1\tau}\left(\dot{F}_{ij}^0\right) - c_2^2 \mathcal{M}_{\mathbf{x},c_2\tau}\left(\dot{F}_{ij}^0\right)\right] d\tau \\
& + \int_0^t \tau \left[c_1^2 \mathcal{M}_{\mathbf{x},c_1\tau}\left(F_{ij}^0\right) - c_2^2 \mathcal{M}_{\mathbf{x},c_2\tau}\left(F_{ij}^0\right)\right] d\tau \\
& + t\left[c_1^2 \mathcal{M}_{\mathbf{x},c_1t}\left(\dot{G}_{ij}^0\right) - c_2^2 \mathcal{M}_{\mathbf{x},c_2t}\left(\dot{G}_{ij}^0\right)\right] \\
& + \frac{\partial}{\partial t}\left\{t\left[c_1^2 \mathcal{M}_{\mathbf{x},c_1t}\left(G_{ij}^0\right) - c_2^2 \mathcal{M}_{\mathbf{x},c_2t}\left(G_{ij}^0\right)\right]\right\} \\
& + t\mathcal{M}_{\mathbf{x},c_2t}\left(\dot{I}_{ij}^0\right) + \frac{\partial}{\partial t}\left[t\mathcal{M}_{\mathbf{x},c_2t}\left(I_{ij}^0\right)\right]
\end{aligned} \tag{11.1.318}$$

where

$$\dot{F}_{ij}^0 = \left(\dot{S}_{ab,ab}^0 - \frac{c_1^2}{c_1^2 - c_2^2}\frac{\lambda}{3\lambda+2\mu}\dot{S}_{aa,bb}^0\right)_{,ij} \tag{11.1.319}$$

$$F_{ij}^0 = \left(S_{ab,ab}^0 - \frac{c_1^2}{c_1^2 - c_2^2}\frac{\lambda}{3\lambda+2\mu}S_{aa,bb}^0\right)_{,ij} \tag{11.1.320}$$

$$\dot{G}_{ij}^0 = \frac{1}{c_1^2 - c_2^2}\frac{\lambda}{3\lambda+2\mu}\dot{S}_{aa,ij}^0 \tag{11.1.321}$$

$$G_{ij}^0 = \frac{1}{c_1^2 - c_2^2}\frac{\lambda}{3\lambda+2\mu}S_{aa,ij}^0 \tag{11.1.322}$$

$$\dot{I}_{ij}^0 = \dot{S}_{(ia,aj)}^0 - \frac{\lambda}{3\lambda+2\mu}\dot{S}_{aa,ij}^0 \tag{11.1.323}$$

and

$$I_{ij}^0 = S_{(ia,aj)}^0 - \frac{\lambda}{3\lambda+2\mu}S_{aa,ij}^0 \tag{11.1.324}$$

As a result we formulate the following theorem:

Theorem 2: A solution to the stress initial value problem 11.1.284 and 11.1.285 takes the form 11.1.287 in which $S_{ab,ab}$ and $S_{(ia,aj)}$ are given by Equations 11.1.313 and 11.1.318, respectively. ■

Note:

The problem described by Equations 11.1.284 and 11.1.285 cover both a conventional formulation of elastodynamics in which the initial strain and strain-rate fields, associated with the initial stress and stress-rate fields through Hooke's law, satisfy the strain and strain-rate compatibility conditions at $t = 0$, and a formulation of elastodynamics with defects in which the initial stress and stress-rate fields do not satisfy compatibility conditions. Hence, a solution in the form 11.1.287 is valid in both the conventional and nonconventional elastodynamics.

11.1.4.3 The Volume Change Waves Produced by an Initial Spherical Pressure in a Homogeneous Isotropic Infinite Elastic Solid

Assume that the initial stress and stress-rate fields in the problem 11.1.284 and 11.1.285 take the forms

$$S^0_{ij}(\mathbf{x}) = -p\,\delta_{ij}H(a - |\mathbf{x}|) \tag{11.1.325}$$

$$\dot{S}^0_{ij}(\mathbf{x}) = 0 \tag{11.1.326}$$

where $p > 0$ and $a > 0$ are prescribed constants, and $H = H(x)$ is the Heaviside function. This means that at the time $t = 0$ a uniform pressure of magnitude p is applied inside the sphere with center at $\mathbf{x} = \mathbf{0}$ and with radius a, while the exterior domain of the sphere is kept at a zero stress level.

The discontinuity of S^0_{ij} across the spherical surface $|\mathbf{x}| = a$ implies a discontinuity of the associated initial strain field

$$E^0_{ij}(\mathbf{x}) = -\frac{p\delta_{ij}}{3\lambda + 2\mu} H(a - |\mathbf{x}|) \tag{11.1.327}$$

and this amounts to the existence of an initial defect in the infinite body.

The stress waves produced by the initial spherical pressure in the body may be studied using the stress representation formula 11.1.287 in which $S_{(ia,aj)}$ is restricted to the initial conditions 11.1.325 and 11.1.326. In the following, we discuss the associated volume change waves only, that is, the waves described by the field $E_{aa}(\mathbf{x}, t)$. To this end we take trace of Equation 11.1.287 and obtain

$$S_{aa}(\mathbf{x}, t) = S^0_{aa}(\mathbf{x}) + t\dot{S}^0_{aa}(\mathbf{x}) + \frac{3\lambda + 2\mu}{\rho} \int_0^t (t - \tau) S_{ab,ab}(\mathbf{x}, \tau)\, d\tau \tag{11.1.328}$$

Substituting $S_{ab,ab}(\mathbf{x}, t)$ from Equation 11.1.313 into Equation 11.1.328, and using the volume change formula

$$E_{aa}(\mathbf{x}, t) = \frac{1}{3\lambda + 2\mu} S_{aa}(\mathbf{x}, t) \tag{11.1.329}$$

leads to

$$E_{aa}(\mathbf{x}, t) = \frac{1}{3\lambda + 2\mu}\left[S^0_{aa}(\mathbf{x}) + t\dot{S}^0_{aa}(\mathbf{x})\right]$$
$$+ \frac{1}{\rho}\int_0^t (t - \tau)\left\{\tau \mathcal{M}_{\mathbf{x},c_1\tau}\left(\dot{S}^0_{ab,ab}\right) + \frac{\partial}{\partial \tau}\left[\tau \mathcal{M}_{\mathbf{x},c_1\tau}\left(S^0_{ab,ab}\right)\right]\right\} d\tau \tag{11.1.330}$$

Note that Equation 11.1.330 holds true for the arbitrary initial stress and stress-rate fields. By restricting this equation to the initial conditions 11.1.325 and 11.1.326 the simple volume change formula is obtained

$$\hat{E}_{aa}(\mathbf{x},t) = \frac{1}{3\lambda+2\mu} S^0_{aa}(\mathbf{x}) + \frac{1}{\rho}\int_0^t \tau \mathcal{M}_{\mathbf{x},c_1\tau}\left(S^0_{ab,ab}\right) d\tau \tag{11.1.331}$$

Hence, an analysis of the volume change waves produced by the initial spherical pressure in the infinite body has been reduced to that for the integral on the RHS of Equation 11.1.331. To discuss the integral we rewrite Equation 11.1.325 in the form

$$S^0_{ab}(\mathbf{x}) = -p\delta_{ab} H(a-R) \tag{11.1.332}$$

where

$$R = |\mathbf{x}| \tag{11.1.333}$$

Hence, we have

$$S^0_{aa}(\mathbf{x}) = -3pH(a-R) \tag{11.1.334}$$

$$S^0_{ab,b}(\mathbf{x}) = -p\delta_{ab}(-R_{,b})\delta(a-R) = p\frac{x_a}{R}\delta(R-a) = \frac{p}{a}x_a\,\delta(R-a) \tag{11.1.335}$$

and

$$S^0_{ab,ab}(\mathbf{x}) = \frac{3p}{a}\left[\delta(R-a) + \frac{R}{3}\delta'(R-a)\right] \tag{11.1.336}$$

and, by using the definition of the mean value of a function over the surface of a ball with center at $\mathbf{x}$ and with radius $c_1\tau$, we get

$$\mathcal{M}_{\mathbf{x},c_1\tau}\left(S^0_{ab,ab}\right) = \frac{3p}{4\pi a}\int_0^{2\pi}\int_0^{\pi}\left[\delta(|\mathbf{x}+\mathbf{n}c_1\tau|-a) + \frac{1}{3}|\mathbf{x}+\mathbf{n}c_1\tau|\delta'(|\mathbf{x}+\mathbf{n}c_1\tau|-a)\right]\sin\theta\, d\theta\, d\varphi \tag{11.1.337}$$

where

$$\mathbf{n} = [\sin\theta\cos\varphi, \sin\theta\sin\varphi, \cos\theta] \tag{11.1.338}$$

In the following, we compute the function $\hat{E}_{aa}(\mathbf{x},t)$ at $\mathbf{x}=\mathbf{x}_1=\mathbf{0}$, and $\mathbf{x}=\mathbf{x}_2=(0,0,-h)$ $(h>a)$ for every $t\geq 0$.

11.1.4.3.1 Case 1: Analysis of the Volume Change Waves at the Center of the Spherical Pressure Inclusion

By letting $\mathbf{x}=\mathbf{0}$ in Equation 11.1.337 we find

$$\begin{aligned}\mathcal{M}_{\mathbf{0},c_1\tau}\left(S^0_{ab,ab}\right) &= \frac{3p}{4\pi a}\int_0^{2\pi}\int_0^{\pi}\left[\delta(c_1\tau-a) + \frac{1}{3}c_1\tau\delta'(c_1\tau-a)\right]\sin\theta\, d\theta\, d\varphi \\ &= \frac{3p}{a}\left[\delta(c_1\tau-a) + \frac{1}{3}c_1\tau\delta'(c_1\tau-a)\right]\end{aligned} \tag{11.1.339}$$

Hence

$$\int_0^t \tau \mathcal{M}_{\mathbf{0},c_1\tau}\left(S^0_{ab,ab}\right) d\tau = \frac{3p}{a}\int_0^t \tau\left[\delta(c_1\tau - a) + \frac{1}{3}c_1\tau\delta'(c_1\tau - a)\right] d\tau \tag{11.1.340}$$

Since

$$\int_0^t \tau\delta(c_1\tau - a)\, d\tau = \frac{1}{c_1^2}\int_{-a}^{c_1 t - a}(u+a)\delta(u)du$$

$$= \frac{1}{c_1^2}H\left(t - \frac{a}{c_1}\right)\int_{-a}^{c_1 t - a}(u+a)\delta(u)\, du = \frac{a}{c_1^2}H\left(t - \frac{a}{c_1}\right) \tag{11.1.341}$$

and

$$\int_0^t \tau^2\delta'(c_1\tau - a)\, d\tau = \frac{1}{c_1^3}\int_{-a}^{c_1 t - a}(u+a)^2\delta'(u)\, du$$

$$= \frac{1}{c_1^3}\left[(u+a)^2\delta(u)\Big|_{u=-a}^{u=c_1 t - a} - 2\int_{-a}^{c_1 t - a}(u+a)\delta(u)\, du\right]$$

$$= \frac{1}{c_1^3}\left[c_1^2 t^2\delta(c_1 t - a) - 2aH\left(t - \frac{a}{c_1}\right)\right]$$

$$= \frac{1}{c_1^3}\left[c_1 t^2\delta\left(t - \frac{a}{c_1}\right) - 2aH\left(t - \frac{a}{c_1}\right)\right]$$

$$= \frac{1}{c_1^3}\left[\frac{a^2}{c_1}\delta\left(t - \frac{a}{c_1}\right) - 2aH\left(t - \frac{a}{c_1}\right)\right] \tag{11.1.342}$$

therefore, Equation 11.1.340 takes the form

$$\int_0^t \tau \mathcal{M}_{\mathbf{0},c_1\tau}\left(S^0_{ab,ab}\right) d\tau = \frac{p}{c_1^2}\left[H\left(t - \frac{a}{c_1}\right) + \frac{a}{c_1}\delta\left(t - \frac{a}{c_1}\right)\right] \tag{11.1.343}$$

Substituting Equation 11.1.343 into Equation 11.1.331 taken at $\mathbf{x} = \mathbf{0}$, and using Equation 11.1.334, we get

$$\hat{E}_{aa}(\mathbf{0}, t) = -E^*_{aa}\left\{1 - \frac{3\lambda + 2\mu}{3(\lambda + 2\mu)}\left[H\left(t - \frac{a}{c_1}\right) + \frac{a}{c_1}\delta\left(t - \frac{a}{c_1}\right)\right]\right\} \tag{11.1.344}$$

where

$$E^*_{aa} = \frac{3p}{3\lambda + 2\mu} \tag{11.1.345}$$

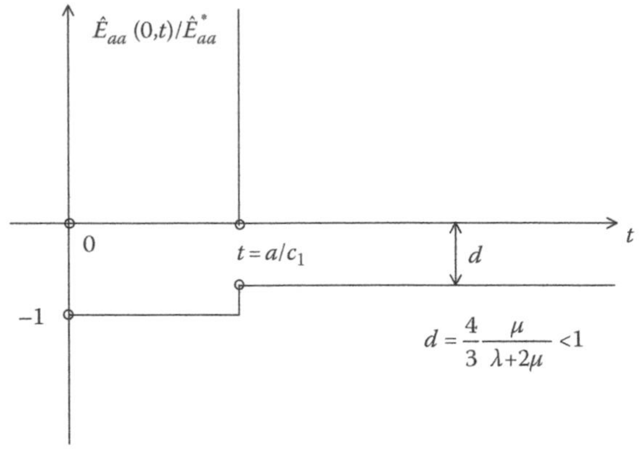

FIGURE 11.7 The volume change at the center of spherical pressure inclusion as a function of time.

The function $\left[\hat{E}_{aa}(\mathbf{0}, t)/E^*_{aa}\right]$ for $t \geq 0$ is shown in Figure 11.7. It follows from Figure 11.7 that the volume change $\hat{E}_{aa}(\mathbf{0}, t)$ is negative and equal to the initial value $\left(-E^*_{aa}\right)$ over the interval $0 \leq t < a/c_1$; it jumps to $+\infty$ at $t = a/c_1$; and attains a negative value greater than the initial value over the interval $t > a/c_1$. The value of the function $\hat{E}_{aa}(\mathbf{0}, t)/E^*_{aa}$ at $t = a/c_1$ is shown in Figure 11.7 by a vertical semi-infinite line.

11.1.4.3.2 Case 2: Analysis of the Volume Change Waves at a Point Lying outside of the Spherical Pressure Inclusion

By letting $\mathbf{x} = \mathbf{x}_2 = (0, 0, -h)$ $(h > a)$ in Equation 11.1.331 we find

$$\hat{E}_{aa}(\mathbf{x}_2, t) = \frac{1}{\rho}\int_0^t \tau \mathcal{M}_{\mathbf{x}_2, c_1\tau}\left(S^0_{ab,ab}\right) d\tau \tag{11.1.346}$$

To compute the integral 11.1.346 we need to find the length of the vector $\mathbf{y} = \mathbf{x}_2 + \mathbf{n}c_1\tau$ (see Equations 11.1.337 and 11.1.338). To this end consider the situation shown in Figure 11.8.

It follows from the definition of $\mathbf{n}$ (see Equation 11.1.338) and from Figure 11.8 that

$$|\mathbf{y}| = \sqrt{h^2 - 2c_1\tau h\cos\theta + c_1^2\tau^2} \tag{11.1.347}$$

where θ has the same meaning as in the integral 11.1.337 in which $\mathbf{x} = \mathbf{x}_2$. Therefore, the integrand in the integral 11.1.337 at $\mathbf{x} = \mathbf{x}_2$ is independent of φ, and we have

$$\mathcal{M}_{\mathbf{x}_2, c_1\tau}\left(S^0_{ab,ab}\right) = \frac{3p}{2a}\int_0^\pi \left[\delta(|\mathbf{y}| - a) + \frac{1}{3}|\mathbf{y}|\delta'(|\mathbf{y}| - a)\right] \sin\theta\, d\theta \tag{11.1.348}$$

where $|\mathbf{y}| = |\mathbf{y}|(\theta, \tau)$ is given by Equation 11.1.347.

To compute the integral 11.1.348 we let

$$u(\theta, \tau) = |\mathbf{y}| - a \tag{11.1.349}$$

and, because of Equation 11.1.347,

$$du = \frac{c_1\tau h}{|\mathbf{y}|}\sin\theta\, d\theta \tag{11.1.350}$$

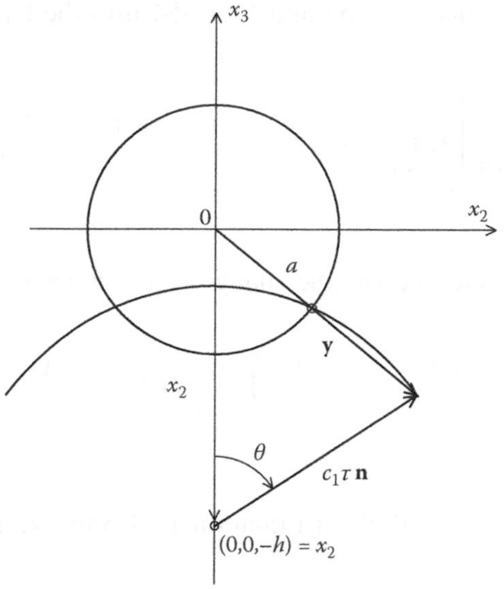

FIGURE 11.8 The sphere at center $\mathbf{x}_2$ and of radius $c_1\tau$ intersecting the spherical support of the initial stress at center $\mathbf{x} = \mathbf{0}$ and of radius a for $(h-a)/c_1 < \tau < (h+a)/c_1$.

or

$$\sin\theta\, d\theta = \frac{|\mathbf{y}|}{c_1\tau h}du = \frac{u+a}{c_1\tau h}du \tag{11.1.351}$$

Hence, the integral 11.1.348 can be reduced to the form

$$\mathcal{M}_{\mathbf{x}_2,c_1\tau}\left(S^0_{ab,ab}\right) = \frac{3p}{2ac_1\tau h}\int\limits_{|h-c_1\tau|-a}^{|h+c_1\tau|-a}\left[(u+a)\delta(u) + \frac{1}{3}(u+a)^2\delta'(u)\right]du \tag{11.1.352}$$

Since $h > a$, therefore

$$\int\limits_{|h-c_1\tau|-a}^{|h+c_1\tau|-a}(u+a)\delta(u)\,du = aH\left(\frac{a}{c_1} - \left|\tau - \frac{h}{c_1}\right|\right) \tag{11.1.353}$$

and

$$\begin{aligned}
&\frac{1}{3}\int\limits_{|h-c_1\tau|-a}^{|h+c_1\tau|-a}(u+a)^2\delta'(u)\,du \\
&= \frac{1}{3}\left[(u+a)^2\delta(u)\Big|_{u=|h-c_1\tau|-a}^{u=h+c_1\tau-a} - 2\int\limits_{|h-c_1\tau|-a}^{|h+c_1\tau|-a}(u+a)\delta(u)\,du\right] \\
&= -\frac{a}{3}\left[\frac{c_1}{a}\left(\tau - \frac{h}{c_1}\right)^2\delta\left(\left|\tau - \frac{h}{c_1}\right| - \frac{a}{c_1}\right) + 2H\left(\frac{a}{c_1} - \left|\tau - \frac{h}{c_1}\right|\right)\right]
\end{aligned} \tag{11.1.354}$$

Hence, substituting Equations 11.1.353 and 11.1.354 into the RHS of Equation 11.1.352, we find

$$\tau \mathcal{M}_{\mathbf{x}_2, c_1\tau}\left(S^0_{ab,ab}\right) = \frac{p}{2c_1 h}\left[H\left(\frac{a}{c_1} - \left|\tau - \frac{h}{c_1}\right|\right) - \frac{c_1}{a}\left(\tau - \frac{h}{c_1}\right)^2 \delta\left(\left|\tau - \frac{h}{c_1}\right| - \frac{a}{c_1}\right)\right] \tag{11.1.355}$$

and, using Equation 11.1.346, the volume change formula has the form

$$\hat{E}_{aa}(\mathbf{x}_2, t) = \frac{p}{2\rho c_1 h}\int_0^t \left[H\left(\frac{a}{c_1} - \left|\tau - \frac{h}{c_1}\right|\right) - \frac{c_1}{a}\left(\tau - \frac{h}{c_1}\right)^2 \delta\left(\left|\tau - \frac{h}{c_1}\right| - \frac{a}{c_1}\right)\right] d\tau \tag{11.1.356}$$

To calculate the integral on the RHS of Equation 11.1.356, we introduce the partition of time interval

$$[0, \infty) = \left[0, \frac{h-a}{c_1}\right) \cup \left[\frac{h-a}{c_1}, \frac{h+a}{c_1}\right) \cup \left[\frac{h+a}{c_1}, \infty\right) \tag{11.1.357}$$

and consider the three cases corresponding to the three subintervals in Equation 11.1.357.

Case (i): $0 \le t < \dfrac{h-a}{c_1}$

In this case

$$0 < \tau < t < \frac{h-a}{c_1} < \frac{h}{c_1} \tag{11.1.358}$$

Hence

$$\tau - \frac{h}{c_1} < 0 \tag{11.1.359}$$

and

$$\frac{a}{c_1} - \left|\tau - \frac{h}{c_1}\right| = \tau - \frac{h-a}{c_1} < 0 \tag{11.1.360}$$

$$H\left(\frac{a}{c_1} - \left|\tau - \frac{h}{c_1}\right|\right) = H\left(\tau - \frac{h-a}{c_1}\right) = 0 \tag{11.1.361}$$

$$\delta\left(\left|\tau - \frac{h}{c_1}\right| - \frac{a}{c_1}\right) = \delta\left(\tau - \frac{h-a}{c_1}\right) = 0 \tag{11.1.362}$$

and, from Equation 11.1.356 we find

$$\hat{E}_{aa}(\mathbf{x}_2, t) = 0 \quad \text{for } 0 \le t < \frac{h-a}{c_1} \tag{11.1.363}$$

Case (ii): $\dfrac{h-a}{c_1} < t < \dfrac{h+a}{c_1}$

This case will be resolved into the two subcases, (ii 1) and (ii 2).

Subcase (ii 1): $\dfrac{h-a}{c_1} < t < \dfrac{h}{c_1}$

In this case

$$0 \leq \tau < t < \frac{h}{c_1} \tag{11.1.364}$$

hence

$$\tau - \frac{h}{c_1} < 0 \tag{11.1.365}$$

and

$$\int_0^t \{\cdot\}\, d\tau = \int_0^{(h-a)/c_1} \{\cdot\}\, d\tau + \int_{(h-a)/c_1}^t \{\cdot\}\, d\tau \tag{11.1.366}$$

where $\{\cdot\}$ stands for the integrand in Equation 11.1.356. Now, the first integral on the RHS of Equation 11.1.366 vanishes because of the relations 11.1.359 through 11.1.362. The second integral in Equation 11.1.366 takes the form

$$\begin{aligned}\int_{(h-a)/c_1}^t \{\cdot\}\, d\tau &= \int_{(h-a)/c_1}^t \left[H\left(\tau - \frac{h-a}{c_1}\right) - \frac{c_1}{a}\left(\tau - \frac{h}{c_1}\right)^2 \delta\left(\tau - \frac{h-a}{c_1}\right)\right] d\tau \\ &= \int_0^{t-(h-a)/c_1} \left[H(u) - \frac{a}{c_1}\delta(u)\right] du = H\left(t - \frac{h}{c_1}\right)\left(t - \frac{h}{c_1}\right)\end{aligned} \tag{11.1.367}$$

Hence

$$\hat{E}_{aa}(\mathbf{x}_2, t) = \frac{p}{2\rho c_1 h}\left(t - \frac{h}{c_1}\right) \quad \text{for } \frac{h-a}{c_1} < t < \frac{h}{c_1} \tag{11.1.368}$$

Subcase (ii 2): $\dfrac{h}{c_1} < t < \dfrac{h+a}{c_1}$

In this case

$$\int_0^t \{\cdot\}\, d\tau = \int_0^{(h-a)/c_1} \{\cdot\}\, d\tau + \int_{(h-a)/c_1}^{h/c_1} \{\cdot\}\, d\tau + \int_{h/c_1}^t \{\cdot\}\, d\tau \tag{11.1.369}$$

The first integral on the RHS of this equation coincides with the first integral on the RHS of Equation 11.1.366; hence it is equal to zero. The second integral takes the form

$$\int_{(h-a)/c_1}^{h/c_1} \{\cdot\}\, d\tau = \int_0^{a/c_1} \left[H(u) - \frac{a}{c_1}\,\delta(u)\right] du = 0 \tag{11.1.370}$$

Finally, the third integral on the RHS of Equation 11.1.369 takes the form

$$\int_{h/c_1}^{t} \{\cdot\}\, d\tau = \int_{h/c_1}^{t} \left[H\left(\frac{h+a}{c_1} - \tau\right) - \frac{c_1}{a}\left(\tau - \frac{h}{c_1}\right)^2 \delta\left(\tau - \frac{h+a}{c_1}\right) \right] d\tau$$

$$= \int_{(h+a)/c_1 - t}^{a/c_1} H(u)du - \frac{c_1}{a} \int_{-a/c_1}^{t-(h+a)/c_1} \left(v + \frac{a}{c_1}\right)^2 \delta(v)\, dv$$

$$= \frac{a}{c_1} - \left(\frac{h+a}{c_1} - t\right) = t - \frac{h}{c_1} \tag{11.1.371}$$

Hence, it follows from Equations 11.1.369 through 11.1.371 that

$$\hat{E}_{aa}(\mathbf{x}_2, t) = \frac{p}{2\rho c_1 h}\left(t - \frac{h}{c_1}\right) \quad \text{for } \frac{h}{c_1} < t < \frac{h+a}{c_1} \tag{11.1.372}$$

and, from Equations 11.1.368 and 11.1.372 we find

$$\hat{E}_{aa}(\mathbf{x}_2, t) = \frac{p}{2\rho c_1 h}\left(t - \frac{h}{c_1}\right) \quad \text{for } \frac{h-a}{c_1} < t < \frac{h+a}{c_1} \tag{11.1.373}$$

Case (iii): $t > \dfrac{h+a}{c_1}$

In this case we obtain

$$\int_{0}^{t} \{\cdot\}\, d\tau = \int_{(h/c)_1}^{t} \left[H\left(\frac{h+a}{c_1} - \tau\right) - \frac{c_1}{a}\left(\tau - \frac{h}{c_1}\right)^2 \delta\left(\tau - \frac{h+a}{c_1}\right) \right] d\tau$$

$$= \int_{0}^{a/c_1} H(u)du - \frac{c_1}{a} \int_{-a/c_1}^{t-(h+a)/c_1} \left(v + \frac{a}{c_1}\right)^2 \delta(v)\, dv = \frac{a}{c_1} - \frac{c_1}{a}\left(\frac{a}{c_1}\right)^2 = 0 \tag{11.1.374}$$

As a result we have

$$\hat{E}_{aa}(\mathbf{x}_2, t) = \begin{cases} 0 & \text{for } 0 \le t < \dfrac{h-a}{c_1} \\ \dfrac{p}{2\rho c_1 h}\left(t - \dfrac{h}{c_1}\right) & \text{for } \dfrac{h-a}{c_1} \le t \le \dfrac{h+a}{c_1} \\ 0 & \text{for } t > \dfrac{h+a}{c_1} \end{cases} \tag{11.1.375}$$

The function $f(t) \equiv \hat{E}_{aa}(\mathbf{x}_2, t)/(p/2\rho c_1 h)$ is shown in Figure 11.9. It follows from Figure 11.9 that except for $t = h/c_1$ at which the volume change reverses its sign from negative to positive values, the interval $\left(\dfrac{h-a}{c_1}, \dfrac{h+a}{c_1}\right)$ represents a support of the volume

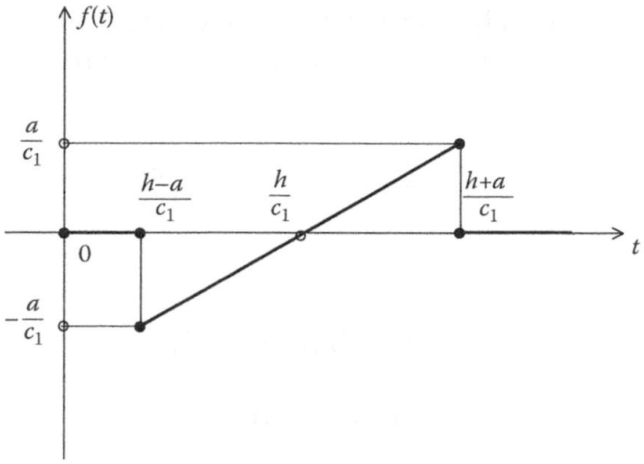

FIGURE 11.9 The volume change at a point outside of the spherical pressure inclusion as a function of time.

change at $\mathbf{x}=\mathbf{x}_2$. This means that an intersection of the spherical surface at the center at $\mathbf{x}_2$ and of radius $c_1 t$ with the initial stress spherical support is not empty for $\dfrac{h-a}{c_1}<t<\dfrac{h+a}{c_1}$ (see Figure 11.8).

Finally, we note that the analysis of Cases (1) and (2) is representative for a discussion of the volume change waves, since the pure stress initial value problem, described by Equations 11.1.284 and 11.1.285 in which the initial stress and stress-rate fields are given by Equations 11.1.325 and 11.1.326, referred to the spherical coordinates with the center at $\mathbf{x}=\mathbf{0}$, has a polar symmetry. The case $0<h<a$, when a point of observation of the waves lies inside of the spherical pressure inclusion, may be treated in a way similar to Case (2).

11.1.5 The Elastic Waves Generated by a Pressurization of a Spherical Cavity in an Infinite Body

11.1.5.1 The Elastic Waves Generated by Suddenly Applied Uniform Pressure on a Spherical Cavity in an Infinite Medium

Consider an initial–boundary value problem of homogeneous isotropic elastodynamics in which we are to find an elastic process corresponding to zero initial conditions and a suddenly applied uniform pressure $p_0>0$ on the surface of a spherical cavity of radius $r=a$ in an infinite solid (see Figure 11.10).

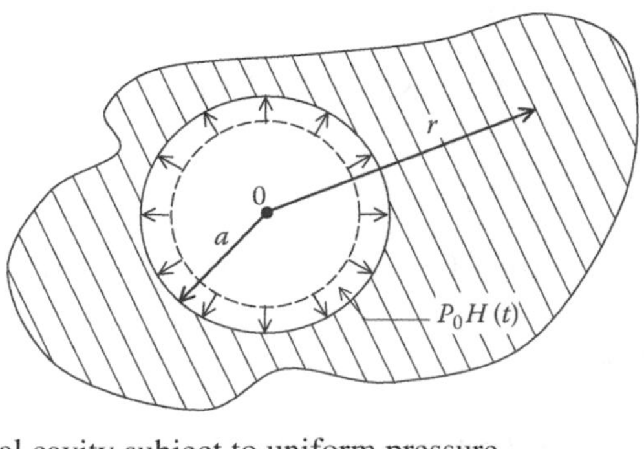

FIGURE 11.10 Spherical cavity subject to uniform pressure.

Since the pressure is uniformly distributed over the cavity surface, the elastic process, referred to the spherical coordinates (r, φ, θ) with the center at the cavity center, has a polar symmetry, which means that

$$s = s(r,t) = [\mathbf{u}(r,t),\ \mathbf{E}(r,t),\ \mathbf{S}(r,t)] \tag{11.1.376}$$

where

$$\mathbf{u}(r,t) = [u_r(r,t), 0, 0] \tag{11.1.377}$$

$$\mathbf{E}(r,t) = \begin{bmatrix} E_{rr}(r,t) & 0 & 0 \\ 0 & E_{\varphi\varphi}(r,t) & 0 \\ 0 & 0 & E_{\theta\theta}(r,t) \end{bmatrix} \tag{11.1.378}$$

and

$$\mathbf{S}(r,t) = \begin{bmatrix} S_{rr}(r,t) & 0 & 0 \\ 0 & S_{\varphi\varphi}(r,t) & 0 \\ 0 & 0 & S_{\theta\theta}(r,t) \end{bmatrix} \tag{11.1.379}$$

As a result, a solution to the problem satisfies the following set of equations: The strain–displacement relations

$$\begin{gathered} E_{rr} = \frac{\partial u_r}{\partial r}, \quad E_{\varphi\varphi} = \frac{u_r}{r}, \quad E_{\theta\theta} = \frac{u_r}{r} \\ E_{r\varphi} = E_{r\theta} = E_{\varphi\theta} = 0 \quad \text{for } r > a,\ t > 0 \end{gathered} \tag{11.1.380}$$

the equation of motion

$$\frac{\partial S_{rr}}{\partial r} + 2\frac{(S_{rr} - S_{\theta\theta})}{r} = \rho \frac{\partial^2 u_r}{\partial t^2} \quad \text{for } r > a,\ t > 0 \tag{11.1.381}$$

the constitutive relations

$$\begin{gathered} S_{rr} = 2\mu E_{rr} + \lambda \operatorname{tr} \mathbf{E}, \quad S_{\varphi\varphi} = 2\mu E_{\varphi\varphi} + \lambda \operatorname{tr} \mathbf{E} \\ S_{\theta\theta} = S_{\varphi\varphi}, \quad S_{r\varphi} = S_{r\theta} = S_{\varphi\theta} = 0 \quad \text{for } r > a,\ t > 0 \end{gathered} \tag{11.1.382}$$

the initial conditions

$$u_r(r,0) = 0, \quad \dot{u}_r(r,0) = 0 \quad \text{for } r > a \tag{11.1.383}$$

and the boundary condition

$$S_{rr}(a,t) = -p_0 H(t) \quad \text{for } t > 0 \tag{11.1.384}$$

where $H = H(t)$ is the Heaviside function. In addition, the solution is to vanish as $r \to \infty$ for every $t > 0$.

To find the solution we let

$$u_r(r,t) = \frac{\partial \phi}{\partial r} \tag{11.1.385}$$

where $\phi = \phi(r,t)$ is a scalar potential defined for $r \geq a,\ t \geq 0$. Substituting of Equation 11.1.385 into Equations 11.1.380 yields

$$E_{rr} = \frac{\partial^2 \phi}{\partial r^2}, \quad E_{\varphi\varphi} = \frac{1}{r}\frac{\partial \phi}{\partial r}, \quad E_{\theta\theta} = \frac{1}{r}\frac{\partial \phi}{\partial r} \tag{11.1.386}$$

Hence, we write

$$\operatorname{tr}\mathbf{E} = \nabla^2 \phi \tag{11.1.387}$$

where

$$\nabla^2 = \frac{\partial^2}{\partial r^2} + \frac{2}{r}\frac{\partial}{\partial r} \tag{11.1.388}$$

and, using Equations 11.1.382, 11.1.386, and 11.1.387, we get

$$S_{rr} = 2\mu \frac{\partial^2 \phi}{\partial r^2} + \lambda \nabla^2 \phi \tag{11.1.389}$$

$$S_{\theta\theta} = 2\mu \frac{1}{r}\frac{\partial \phi}{\partial r} + \lambda \nabla^2 \phi \tag{11.1.390}$$

Also, substituting u_r from Equation 11.1.385, and S_{rr} and $S_{\theta\theta}$ from Equations 11.1.389 and 11.1.390, respectively, into Equation 11.1.381 we arrive at the equation of motion in the form

$$(\lambda + 2\mu)\frac{\partial}{\partial r}(\nabla^2 \phi) = \rho \frac{\partial}{\partial r}\left(\frac{\partial^2 \phi}{\partial t^2}\right) \tag{11.1.391}$$

Clearly, this equation is satisfied if ϕ is a solution of the wave equation

$$\left(\nabla^2 - \frac{1}{c_1^2}\frac{\partial^2}{\partial t^2}\right)\phi = 0 \tag{11.1.392}$$

where

$$\frac{1}{c_1^2} = \frac{\rho}{\lambda + 2\mu} \tag{11.1.393}$$

Since an equivalent form of Equation 11.1.392 is

$$\frac{\partial^2}{\partial r^2}(r\phi) - \frac{1}{c_1^2}\frac{\partial^2}{\partial t^2}(r\phi) = 0 \tag{11.1.394}$$

then a solution to the problem is obtained if ϕ satisfies the equation

$$\frac{\partial^2}{\partial r^2}(r\phi) - \frac{1}{c_1^2}\frac{\partial^2}{\partial t^2}(r\phi) = 0 \quad \text{for } r > a,\ t > 0 \tag{11.1.395}$$

the initial conditions

$$\phi(r,0) = 0 \quad \text{for } r > a \tag{11.1.396}$$

$$\dot{\phi}(r,0) = 0 \quad \text{for } r > a \tag{11.1.397}$$

and the boundary condition

$$\left[2\mu\frac{\partial^2\phi}{\partial r^2} + \frac{\lambda}{c_1^2}\frac{\partial^2\phi}{\partial t^2}\right]_{r=a} = -p_0 H(t) \quad \text{for } t > 0 \tag{11.1.398}$$

In addition, $\phi = \phi(r,t)$ and its partial derivatives of a finite order should vanish as $r \to \infty$ for every $t > 0$.

To solve the problem 11.1.395 through 11.1.398 we use the Laplace transform technique. Let $\overline{\phi} = \overline{\phi}(r,p)$ be the Laplace transform of $\phi = \phi(r,t)$ with respect to time t, that is,

$$L\phi \equiv \overline{\phi}(r,p) = \int_0^\infty e^{-pt}\phi(r,t)dt \tag{11.1.399}$$

where p is the transform parameter. Applying the Laplace transform to Equations 11.1.395 and 11.1.398, respectively, and using the homogeneous initial conditions 11.1.396 and 11.1.397, we find

$$\left(\frac{d^2}{dr^2} - \frac{p^2}{c_1^2}\right)\left[r\overline{\phi}(r,p)\right] = 0 \quad \text{for } r > a \tag{11.1.400}$$

and

$$\left(2\mu\,\frac{\partial^2\overline{\phi}}{\partial r^2} + \frac{\lambda}{c_1^2}p^2\,\overline{\phi}\right)_{r=a} = -\frac{p_0}{p} \tag{11.1.401}$$

Hence, a solution of Equation 11.1.400 that vanishes as $r \to \infty$ takes the form

$$\overline{\phi}(r,p) = A\frac{e^{-\frac{p}{c_1}r}}{r} \tag{11.1.402}$$

where A is a constant. Next, substituting $\overline{\phi}$ from Equation 11.1.402 into the boundary condition 11.1.401, we obtain

$$A = -\frac{a}{\rho}\,\frac{p_0}{p}\,\frac{e^{\frac{p}{c_1}a}}{[(p+\alpha)^2 + \beta^2]} \tag{11.1.403}$$

where

$$\alpha = \frac{2\mu}{\lambda + 2\mu}\frac{c_1}{a}, \quad \beta = \frac{2}{a}\sqrt{\frac{\mu}{\rho}\frac{\lambda + \mu}{\lambda + 2\mu}} \tag{11.1.404}$$

or

$$\alpha = \frac{1 - 2\nu}{1 - \nu}\frac{c_1}{a}, \quad \beta = \frac{\sqrt{1 - 2\nu}}{1 - \nu}\frac{c_1}{a} \tag{11.1.405}$$

Here ν stands for Poisson's ratio.

It then follows from Equations 11.1.402 and 11.1.403 that

$$\overline{\phi}(r, p) = -\frac{p_0}{\rho}\left(\frac{a}{r}\right)\frac{1}{p}\frac{e^{-\frac{p}{c_1}(r-a)}}{[(p + \alpha)^2 + \beta^2]}, \quad r \geq a \tag{11.1.406}$$

By applying the inverse Laplace transform to this equation gives

$$L^{-1}\{\overline{\phi}\} = \phi = -\frac{p_0}{\rho}\left(\frac{a}{r}\right)L^{-1}\left\{\frac{1}{p}\frac{e^{-\frac{p}{c_1}(r-a)}}{[(p + \alpha)^2 + \beta^2]}\right\} \tag{11.1.407}$$

Since

$$L^{-1}\left\{\frac{1}{(p + \alpha)^2 + \beta^2}\right\} = e^{-\alpha t}\frac{\sin \beta t}{\beta} \tag{11.1.408}$$

and

$$L^{-1}\left\{\frac{e^{-\frac{p}{c_1}(r-a)}}{p}\right\} = H\left(t - \frac{r - a}{c_1}\right) \tag{11.1.409}$$

where $H(.)$ is the Heaviside function, then, by the convolution theorem, we find

$$\phi(r, t) = -\frac{p_0}{\rho}\frac{a}{r}\frac{1}{\beta}H\left(t - \frac{r - a}{c_1}\right)\int_0^{t-\frac{r-a}{c_1}} e^{-\alpha\tau}\sin \beta\tau \, d\tau \tag{11.1.410}$$

By using the formula

$$\int e^{-\alpha\tau}\sin \beta\tau \, d\tau = -\frac{e^{-\alpha\tau}}{\alpha^2 + \beta^2}(\alpha \sin \beta\tau + \beta \cos \beta\tau) \tag{11.1.411}$$

Equation 11.1.407 is reduced to

$$\phi(r, t) = -\frac{p_0}{\rho}\frac{a}{r}\frac{1}{\alpha^2 + \beta^2}H\left(t - \frac{r - a}{c_1}\right)\left\{1 - \left[\cos \beta\left(t - \frac{r - a}{c_1}\right)\right.\right.$$
$$\left.\left. + \frac{\alpha}{\beta}\sin \beta\left(t - \frac{r - a}{c_1}\right)\right]\exp\left[-\alpha\left(t - \frac{r - a}{c_1}\right)\right]\right\} \tag{11.1.412}$$

Now, since

$$\frac{1}{\alpha^2+\beta^2}=\frac{\rho a^2}{4\mu},\quad \frac{\alpha}{\beta}=\sqrt{1-2\nu} \tag{11.1.413}$$

therefore,

$$\phi(r,t)=-\frac{a^3p_0}{4\mu r}H(\hat{t})\left[1-\left(\cos\beta\hat{t}+\sqrt{1-2\nu}\sin\beta\hat{t}\right)e^{-\alpha\hat{t}}\right] \tag{11.1.414}$$

where

$$\hat{t}=t-\frac{r-a}{c_1} \tag{11.1.415}$$

By substituting $\phi=\phi(r,t)$ from Equation 11.1.414 into Equations 11.1.385, 11.1.389, and 11.1.390 we find, respectively,

$$u_r(r,t)=\frac{a^3p_0}{4\mu r^2}H(\hat{t})\left[1-(\cos\beta\hat{t}+\sqrt{1-2\nu}\sin\beta\hat{t})e^{-\alpha\hat{t}}+2\left(\frac{r}{a}\right)\sqrt{1-2\nu}\sin\beta\hat{t}e^{-\alpha\hat{t}}\right] \tag{11.1.416}$$

$$\begin{aligned}S_{rr}(r,t)=-p_0\left(\frac{a}{r}\right)^3H(\hat{t})\Bigg[&1-(\cos\beta\hat{t}+\sqrt{1-2\nu}\sin\beta\hat{t})e^{-\alpha\hat{t}}\\&+2\left(\frac{r}{a}\right)\sqrt{1-2\nu}\sin\beta\hat{t}e^{-\alpha\hat{t}}+\left(\frac{r}{a}\right)^2(\cos\beta\hat{t}-\sqrt{1-2\nu}\sin\beta\hat{t})e^{-\alpha\hat{t}}\Bigg]\end{aligned} \tag{11.1.417}$$

and

$$\begin{aligned}S_{\theta\theta}(r,t)=\frac{p_0}{2}\left(\frac{a}{r}\right)^3H(\hat{t})\Bigg[&1-(\cos\beta\hat{t}+\sqrt{1-2\nu}\sin\beta\hat{t})e^{-\alpha\hat{t}}\\&+2\left(\frac{r}{a}\right)\sqrt{1-2\nu}\sin\beta\hat{t}e^{-\alpha\hat{t}}-\frac{2\nu}{1-\nu}\left(\frac{r}{a}\right)^2(\cos\beta\hat{t}-\sqrt{1-2\nu}\sin\beta\hat{t})e^{-\alpha\hat{t}}\Bigg]\end{aligned} \tag{11.1.418}$$

Hence, we have

$$\begin{aligned}\operatorname{tr}\mathbf{S}=S_{rr}+2S_{\theta\theta}&=(3\lambda+2\mu)\operatorname{tr}\mathbf{E}\\&=-p_0\frac{1+\nu}{1-\nu}\left(\frac{a}{r}\right)H(\hat{t})(\cos\beta\hat{t}-\sqrt{1-2\nu}\sin\beta\hat{t})e^{-\alpha\hat{t}}\end{aligned} \tag{11.1.419}$$

$$S_{rr}(a,t)=-p_0H(t) \tag{11.1.420}$$

and

$$S_{\theta\theta}(a,t)=\frac{p_0}{2}H(t)\left[1-\frac{1+\nu}{1-\nu}(\cos\beta t-\sqrt{1-2\nu}\sin\beta t)e^{-\alpha t}\right] \tag{11.1.421}$$

To obtain the strain tensor components, note that by substituting $\phi = \phi(r,t)$ from Equation 11.1.414 into Equations 11.1.386 we arrive at

$$E_{rr}(r,t) = -\frac{p_0}{2\mu}\left(\frac{a}{r}\right)^3 H(\hat{t})\Bigg[1-(\cos\beta\hat{t}+\sqrt{1-2\nu}\sin\beta\hat{t})e^{-\alpha\hat{t}}$$
$$+2\left(\frac{r}{a}\right)\sqrt{1-2\nu}\sin\beta\hat{t}e^{-\alpha\hat{t}}+\frac{1-2\nu}{1-\nu}\left(\frac{r}{a}\right)^2(\cos\beta\hat{t}-\sqrt{1-2\nu}\sin\beta\hat{t})e^{-\alpha\hat{t}}\Bigg] \tag{11.1.422}$$

and

$$E_{\theta\theta}(r,t) = E_{\varphi\varphi}(r,t) = \frac{p_0}{4\mu}\left(\frac{a}{r}\right)^3 H(\hat{t})\Bigg[1-(\cos\beta\hat{t}+\sqrt{1-2\nu}\sin\beta\hat{t})e^{-\alpha\hat{t}}$$
$$+2\left(\frac{r}{a}\right)\sqrt{1-2\nu}\sin\beta\hat{t}e^{-\alpha\hat{t}}\Bigg] \tag{11.1.423}$$

Thus, the elastic waves produced by a suddenly applied uniform pressure on the spherical cavity in an infinite medium are represented by the process (see Equations 11.1.376 through 11.1.379)

$$s(r,t) = [\mathbf{u}(r,t),\ \mathbf{E}(r,t),\ \mathbf{S}(r,t)] \tag{11.1.424}$$

where **u**, **E**, and **S** are given by Equations 11.1.377, 11.1.378, and 11.1.379, respectively; in which $u_r = u_r(r,t)$ is obtained from Equation 11.1.416; $E_{rr} = E_{rr}(r,t)$ and $E_{\theta\theta} = E_{\theta\theta}(r,t)$ are given by Equations 11.1.422 and 11.1.423, respectively; and $S_{rr} = S_{rr}(r,t)$ and $S_{\theta\theta} = S_{\theta\theta}(r,t)$ are obtained from Equations 11.1.417 and 11.1.418, respectively.
It follows from these formulas that

$$s(r,t) \to \hat{s}(r) \quad \text{as } t \to \infty \tag{11.1.425}$$

where

$$\hat{s}(r) = [\hat{\mathbf{u}}(r);\ \hat{\mathbf{E}}(r),\ \hat{\mathbf{S}}(r)] \tag{11.1.426}$$

$$\hat{\mathbf{u}}(r) = a\frac{p_0}{4\mu}\left(\frac{a}{r}\right)^2[1,0,0] \tag{11.1.427}$$

$$\hat{\mathbf{E}}(r) = \frac{p_0}{2\mu}\left(\frac{a}{r}\right)^3\begin{bmatrix}-1 & 0 & 0\\ 0 & 1/2 & 0\\ 0 & 0 & 1/2\end{bmatrix} \tag{11.1.428}$$

and

$$\hat{\mathbf{S}}(r) = -p_0\left(\frac{a}{r}\right)^3\begin{bmatrix}1 & 0 & 0\\ 0 & -1/2 & 0\\ 0 & 0 & -1/2\end{bmatrix} \tag{11.1.429}$$

Then, the dynamic process $s = s(r,t)$ approaches an isochoric elastic state $\hat{s} = \hat{s}(r)$ as t tends to infinity.

11.1.5.2 The Periodic Vibrations of an Infinite Elastic Solid with a Spherical Cavity

Suppose that a uniform time-periodic normal pressure is applied to the boundary of a spherical cavity in a homogeneous isotropic infinite elastic solid (see Figure 11.11).

Let the period of the load be $T = 2\pi/\omega_0$, where ω_0 is a prescribed frequency ($\omega_0 > 0$). We look for a time-periodic elastic process with polar symmetry of the form

$$s(r,t) = \hat{s}(r,\omega_0)e^{-i\omega_0 t} \tag{11.1.430}$$

where

$$\hat{s}(r,\omega_0) = [\hat{\mathbf{u}}(r,\omega_0),\ \hat{\mathbf{E}}(r,\omega_0),\ \hat{\mathbf{S}}(r,\omega_0)] \tag{11.1.431}$$

$$\hat{\mathbf{u}}(r,\omega_0) = [\hat{u}_r(r,\omega_0), 0, 0] \tag{11.1.432}$$

$$\hat{\mathbf{E}}(r,\omega_0) = \begin{bmatrix} \hat{E}_{rr}(r,\omega_0) & 0 & 0 \\ 0 & \hat{E}_{\varphi\varphi}(r,\omega_0) & 0 \\ 0 & 0 & \hat{E}_{\theta\theta}(r,\omega_0) \end{bmatrix} \tag{11.1.433}$$

and

$$\hat{\mathbf{S}}(r,\omega_0) = \begin{bmatrix} \hat{S}_{rr}(r,\omega_0) & 0 & 0 \\ 0 & \hat{S}_{\varphi\varphi}(r,\omega_0) & 0 \\ 0 & 0 & \hat{S}_{\theta\theta}(r,\omega_0) \end{bmatrix} \tag{11.1.434}$$

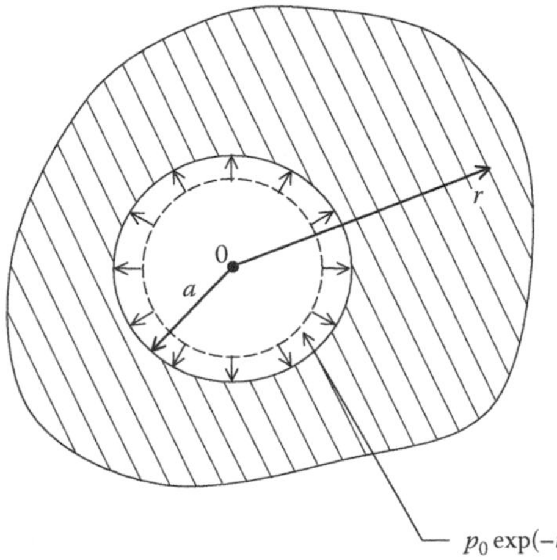

FIGURE 11.11 A uniform pressure p_0 on the surface of a spherical cavity $r = a$ oscillating with a frequency ω_0.

Note that the actual time-periodic elastic process that has a physical meaning may be characterized by either the real or imaginary part of the right-hand side of Equation 11.1.430.

It follows from Equations 11.1.380 through 11.1.382, and 11.1.385 through 11.1.394 that such a process is generated by a displacement potential

$$\phi(r,t) = \hat{\phi}(r,\omega_0)e^{-i\omega_0 t} \tag{11.1.435}$$

where $\hat{\phi} = \hat{\phi}(r,\omega_0)$ satisfies the field equation

$$\left(\frac{d^2}{dr^2} + k_1^2\right)(r\hat{\phi}) = 0 \quad \text{for } r > a \tag{11.1.436}$$

subject to the boundary condition

$$\left[2\mu\frac{d^2\hat{\phi}}{dr^2} - \lambda k_1^2\hat{\phi}\right]_{r=a} = -p_0 \tag{11.1.437}$$

and suitable vanishing conditions at infinity. Here,

$$k_1 = \frac{\omega_0}{c_1} \tag{11.1.438}$$

A general solution to Equation 11.1.436 takes the form

$$\hat{\phi}(r,\omega_0) = \hat{A}\,\frac{e^{ik_1 r}}{r} + \hat{B}\,\frac{e^{-ik_1 r}}{r} \tag{11.1.439}$$

where $\hat{A}$ and $\hat{B}$ are arbitrary constants. Since the function

$$\hat{\phi}_+(r,\omega_0) = e^{-i\omega_0 t}\,\frac{e^{-ik_1 r}}{r} = \frac{e^{-i(k_1 r+\omega_0 t)}}{r} \tag{11.1.440}$$

represents an *incoming wave*, that is, a spherical wave propagating from $r = \infty$ to $r = a$, while the function

$$\hat{\phi}_-(r,\omega_0) = e^{-i\omega_0 t}\frac{e^{ik_1 r}}{r} = \frac{e^{i(k_1 r-\omega_0 t)}}{r} \tag{11.1.441}$$

stands for an *outgoing wave*, that is, a wave propagating from $r=a$ to $r=\infty$, and no spherical waves are sent from infinity, we let $\hat{B}=0$ in Equation 11.1.439, and obtain

$$\hat{\phi}(r,\omega_0) = \hat{A}\,\frac{e^{ik_1 r}}{r} \tag{11.1.442}$$

A comparison between the functions $\overline{\phi}(r,p)$ and $\hat{\phi}(r,\omega_0)$, given by Equations 11.1.402 and 11.1.442, respectively, as well as between the boundary conditions 11.1.401 and 11.1.437, respectively, leads to a conclusion that

$$\hat{\phi}(r,\omega_0) = \left[p\overline{\phi}(r,p)\right]_{p=-i\omega_0} \tag{11.1.443}$$

Thus, in view of Equation 11.1.406, we find

$$\hat{\phi}(r,\omega_0) = -\frac{p_0}{\rho}\left(\frac{a}{r}\right)\frac{e^{ik_1(r-a)}}{\left[(\alpha - i\omega_0)^2 + \beta^2\right]} \tag{11.1.444}$$

where α and β are given by (see Equation 11.1.405)

$$\alpha = \frac{1-2\nu}{1-\nu}\frac{c_1}{a}, \quad \beta = \frac{\sqrt{1-2\nu}}{1-\nu}\frac{c_1}{a} \tag{11.1.445}$$

An alternative form of $\hat{\phi} = \hat{\phi}(r,\omega_0)$ is given by Equation 11.1.442 in which

$$\hat{A} = -\frac{p_0}{\rho}a\frac{e^{-ik_1 a}}{[(\alpha - i\omega_0)^2 + \beta^2]} \tag{11.1.446}$$

By differentiating Equation 11.1.442 we obtain

$$\frac{\partial\hat{\phi}}{\partial r} = -\frac{\hat{A}}{r^2}(1 - ik_1 r)e^{ik_1 r} \tag{11.1.447}$$

$$\frac{\partial^2\hat{\phi}}{\partial r^2} = \frac{\hat{A}}{r^3}\left(2 - 2ik_1 r - k_1^2 r^2\right)e^{ik_1 r} \tag{11.1.448}$$

Hence,

$$\nabla^2\hat{\phi} = -k_1^2\hat{\phi} \tag{11.1.449}$$

and

$$\hat{u}_r(r,\omega_0) = \frac{\partial\hat{\phi}}{\partial r} = \frac{a}{\rho}\frac{p_0}{r^2}\frac{(1 - ik_1 r)}{[(\alpha - i\omega_0)^2 + \beta^2]}e^{ik_1(r-a)} \tag{11.1.450}$$

$$\hat{E}_{rr}(r,\omega_0) = \frac{\partial^2\hat{\phi}}{\partial r^2} = -\frac{a}{\rho}\frac{p_0}{r^3}\frac{\left(2 - 2ik_1 r - k_1^2 r^2\right)}{[(\alpha - i\omega_0)^2 + \beta^2]}e^{-ik_1(r-a)} \tag{11.1.451}$$

$$\hat{E}_{\varphi\varphi}(r,\omega_0) = \hat{E}_{\theta\theta}(r,\omega_0) = \frac{1}{r}\frac{\partial\hat{\phi}}{\partial r} = \frac{a}{\rho}\frac{p_0}{r^3}\frac{(1 - ik_1 r)}{[(\alpha - i\omega_0)^2 + \beta^2]}e^{ik_1(r-a)} \tag{11.1.452}$$

The stress components are

$$\begin{aligned}\hat{S}_{rr}(r,\omega_0) &= 2\mu\hat{E}_{rr}(r,\omega_0) + \lambda\,\mathrm{tr}\,\hat{\mathbf{E}}(r,\omega_0)\\ &= -p_0\left(\frac{a}{r}\right)^3\frac{\left[\alpha^2 + \beta^2 - \omega_0^2(r/a)^2 - 2i\omega_0\alpha\,(r/a)\right]}{[(\alpha - i\omega_0)^2 + \beta^2]}e^{ik_1(r-a)}\end{aligned} \tag{11.1.453}$$

and

$$\hat{S}_{\theta\theta}(r,\omega_0) = 2\mu\hat{E}_{\theta\theta}(r,\omega_0) + \lambda\,\mathrm{tr}\,\hat{\mathbf{E}}(r,\omega_0)$$
$$= \frac{p_0}{2}\left(\frac{a}{r}\right)^3 \frac{\{\alpha^2+\beta^2+[2\nu/(1-\nu)]\omega_0^2\,(r/a)^2 - 2i\omega_0\alpha\,(r/a)\}}{[(\alpha-i\omega_0)^2+\beta^2]}\,e^{ik_1(r-a)} \tag{11.1.454}$$

The real-valued time-periodic vibrations of an infinite elastic solid with a spherical cavity may now be derived from the formula

$$s^*(r,t) \equiv Re[\hat{s}(r,\omega_0)e^{-i\omega_0 t}] \tag{11.1.455}$$

where $\hat{s} = \hat{s}(r,\omega_0)$ is given by Equations 11.1.431 through 11.1.434, and 11.1.450 through 11.1.454.

In particular, we find

$$u_r^*(r,t) = a\,\frac{p_0}{4\mu}\left(\frac{a}{r}\right)^2 \frac{(\alpha^2+\beta^2)}{\left[\left(\alpha^2+\beta^2-\omega_0^2\right)^2+4\omega_0^2\alpha^2\right]}$$
$$\times\left\{\left[\left(\alpha^2+\beta^2-\omega_0^2\right)+2\omega_0^2\alpha\left(\frac{r}{c_1}\right)\right]\cos\omega_0\left(t-\frac{r-a}{c_1}\right)\right.$$
$$\left. -\,\omega_0\left[\left(\alpha^2+\beta^2-\omega_0^2\right)\left(\frac{r}{c_1}\right)-2\alpha\right]\sin\omega_0\left(t-\frac{r-a}{c_1}\right)\right\} \tag{11.1.456}$$

and

$$S_{rr}^*(r,t) = -p_0\left(\frac{a}{r}\right)^3 \frac{1}{\left[\left(\alpha^2+\beta^2-\omega_0^2\right)^2+4\omega_0^2\alpha^2\right]}$$
$$\times\left\{\left[\left(\alpha^2+\beta^2-\omega_0^2\right)\left(\alpha^2+\beta^2-\omega_0^2\frac{r^2}{a^2}\right)+4\omega_0^2\alpha^2\frac{r}{a}\right]\cos\omega_0\left(t-\frac{r-a}{c_1}\right)\right.$$
$$\left. -2\omega_0\alpha\left[\left(\alpha^2+\beta^2-\omega_0^2\right)\frac{r}{a}-\left(\alpha^2+\beta^2-\omega_0^2\frac{r^2}{a^2}\right)\right]\sin\omega_0\left(t-\frac{r-a}{c_1}\right)\right\} \tag{11.1.457}$$

$$S_{\varphi\varphi}^*(r,t) = \frac{p_0}{2}\left(\frac{a}{r}\right)^3 \frac{1}{\left[\left(\alpha^2+\beta^2-\omega_0^2\right)^2+4\omega_0^2\alpha^2\right]}$$
$$\times\left\{\left[\left(\alpha^2+\beta^2-\omega_0^2\right)\left(\alpha^2+\beta^2+\frac{2\nu}{1-\nu}\omega_0^2\frac{r^2}{a^2}\right)\right.\right.$$
$$\left.+\,4\omega_0^2\alpha^2\frac{r}{a}\right]\cos\omega_0\left(t-\frac{r-a}{c_1}\right)-2\omega_0\alpha\left[\left(\alpha^2+\beta^2-\omega_0^2\right)\frac{r}{a}\right.$$
$$\left.\left.-\left(\alpha^2+\beta^2+\frac{2\nu}{1-\nu}\omega_0^2\frac{r^2}{a^2}\right)\right]\sin\omega_0\left(t-\frac{r-a}{c_1}\right)\right\} \tag{11.1.458}$$

By letting $\omega_0 \to 0$ in Equations 11.1.456, 11.1.457, and 11.1.458, respectively, we have

$$u_r^*(r,t) \to \hat{u}_r(r) \equiv a\frac{p_0}{4\mu}\left(\frac{a}{r}\right)^2 \quad \text{as } \omega_0 \to 0 \tag{11.1.459}$$

$$S_{rr}^*(r,t) \to \hat{S}_{rr}(r) \equiv -p_0\left(\frac{a}{r}\right)^3 \quad \text{as } \omega_0 \to 0 \tag{11.1.460}$$

and

$$S_{\varphi\varphi}^*(r,t) \to \hat{S}_{\varphi\varphi}(r) \equiv \frac{p_0}{2}\left(\frac{a}{r}\right)^3 \quad \text{as } \omega_0 \to 0 \tag{11.1.461}$$

Hence, the harmonic process $s^*(r,t)$ approaches an isochoric elastic state $\hat{s}(r)$ if the period of vibrations $T = 2\pi/\omega_0$ goes to infinity (see also Equations 11.1.425 through 11.1.429).

Also, the formulas 11.1.456 through 11.1.458 take a simple form if $\omega_0 = \omega^*$, where

$$\omega^* = \sqrt{\alpha^2 + \beta^2} = 2\left(\frac{c_2}{a}\right) = \sqrt{\frac{2(1-2\nu)}{(1-\nu)}}\left(\frac{c_1}{a}\right) \tag{11.1.462}$$

In this case, the result is

$$\tilde{u}_r^*(r,t) = a\frac{p_0}{8\mu}\left(\frac{a}{r}\right)^2\sqrt{\frac{2-2\nu}{1-2\nu}}\sqrt{1+\left(\frac{\omega^* r}{c_1}\right)^2}\sin\left[\omega^*\left(t-\frac{r-a}{c_1}\right)+\tan^{-1}\left(\frac{\omega^* r}{c_1}\right)\right] \tag{11.1.463}$$

where $\tilde{u}_r^* = \tilde{u}_r^*(r,t)$ stands for $u_r^* = u_r^*(r,t)$ restricted to $\omega_0 = \omega^*$. By letting $p_0 = \mu$ and $r = a$ in Equation 11.1.463 we have

$$\tilde{u}_r^*(a,t) = \frac{a\sqrt{2}}{8}\sqrt{\frac{3-5\nu}{1-2\nu}}\sin\left(\omega^* t + \tan^{-1}\sqrt{\frac{2(1-\nu)}{1-\nu}}\right) \tag{11.1.464}$$

This implies the useful estimate

$$\left|\tilde{u}_r^*(a,t)\right| \le \frac{a\sqrt{2}}{8}\sqrt{\frac{3-5\nu}{1-2\nu}} \quad \text{for } t \ge 0 \tag{11.1.465}$$

In particular, for $\nu = 0.25$ we get

$$\left|\tilde{u}_r^*(a,t)\right| \le 0.33a \quad \text{for } t \ge 0 \tag{11.1.466}$$

Thus, in a problem of periodic vibrations of a homogeneous isotropic infinite elastic body with a spherical cavity of radius $r = a$, in which a uniform pressure $p_0 = \mu$ oscillating with the frequency ω^* is applied to the surface $r = a$, a maximum amplitude of the radial displacement on $r = a$ does not exceed $0.33a$.

11.2 THREE-DIMENSIONAL SOLUTIONS OF NONISOTHERMAL ELASTODYNAMICS

In Section 4.2.2, the field equations of a dynamic theory of thermal stresses were discussed. In particular, it was shown that for a homogeneous isotropic body an elastic process $s = s(\mathbf{x}, t) = [\mathbf{u}(\mathbf{x}, t), \mathbf{E}(\mathbf{x}, t), \mathbf{S}(\mathbf{x}, t)]$, corresponding to zero body forces and a temperature change $T = T(\mathbf{x}, t)$, satisfies the following system of field equations (see Equations 4.2.53 through 4.2.56):

the strain-displacement relations

$$\mathbf{E} = \frac{1}{2}\left(\nabla\mathbf{u} + \nabla\mathbf{u}^T\right) \tag{11.2.1}$$

the equation of motion

$$\operatorname{div}\mathbf{S} = \rho\ddot{\mathbf{u}} \tag{11.2.2}$$

the stress–strain–temperature relation

$$\mathbf{S} = 2\mu\mathbf{E} + \lambda(\operatorname{tr}\mathbf{E})\mathbf{1} - (3\lambda + 2\mu)\alpha T\mathbf{1} \tag{11.2.3}$$

Here, $\mathbf{u}$, $\mathbf{E}$, and $\mathbf{S}$ represent the displacement, strain, and stress fields, respectively, and they depend on both $\mathbf{x}$ and t, with $\mathbf{x} \in B$ and $t \geq 0$. The temperature field $T = T(\mathbf{x}, t)$ satisfies the parabolic heat conduction equation

$$\nabla^2 T - \frac{1}{\kappa}\dot{T} = -\frac{Q}{\kappa} \tag{11.2.4}$$

We recall that $Q/\kappa = W/k$, where Q is a prescribed heat supply field, W stands for the internal heat generated per unit volume per unit time, k denotes the thermal conductivity, and κ means the thermal diffusivity. We note the dimensions: $[Q] = \text{K/s}$ and $[W] = \text{N}/(\text{m}^2\,\text{s})$.

It was shown in Example 4.2.4 that a particular elastic process of the theory is generated by a scalar potential $\phi = \phi(\mathbf{x}, t)$ through the relations

$$\mathbf{u} = \nabla\phi \tag{11.2.5}$$

$$\mathbf{E} = \nabla\nabla\phi \tag{11.2.6}$$

$$\mathbf{S} = 2\mu\left(\nabla\nabla\phi - \nabla^2\phi\mathbf{1}\right) + \rho\ddot{\phi}\mathbf{1} \tag{11.2.7}$$

where ϕ satisfies the nonhomogeneous wave equation

$$\Box_1^2\phi = mT \tag{11.2.8}$$

with

$$\Box_1^2 = \nabla^2 - \frac{1}{c_1^2}\frac{\partial^2}{\partial t^2}, \quad m = \frac{3\lambda + 2\mu}{\lambda + 2\mu}\alpha, \quad \frac{1}{c_1^2} = \frac{\lambda + 2\mu}{\rho} \tag{11.2.9}$$

In this section, we use the potential process described by Equations 11.2.5 through 11.2.9 to obtain a number of solutions to particular initial–boundary value problems of the dynamical theory of thermal stresses for an infinite body. The solutions include (A) dynamic thermal stresses produced by an instantaneous concentrated source of heat in an infinite body, (B) harmonic vibrations of an infinite elastic body produced by a concentrated source of heat, and (C) dynamic thermal stresses produced by an instantaneous spherical temperature inclusion in an infinite body. The solutions (A)–(C) are presented in closed forms involving elementary and error functions.

11.2.1 Dynamic Thermal Stresses Produced by an Instantaneous Concentrated Source of Heat in an Infinite Body

Suppose that a homogeneous isotropic infinite elastic solid, referred to a spherical coordinate system (r, φ, θ), is initially at rest, and an instantaneous concentrated heat source is applied to the origin of the system at time $t=0+0$. The temperature field T produced by such a heat source is then spherically symmetric with respect to the origin $r=0$, that is, $T=T(r,t)$, and satisfies the heat conduction equation (see Equation 11.2.4)

$$\nabla^2 T - \frac{1}{\kappa}\dot{T} = -\frac{Q_0}{\kappa}\frac{\delta(r)}{4\pi r^2}\delta(t) \quad \text{for } r \geq 0, \quad t \geq 0 \tag{11.2.10}$$

the initial condition

$$T(r,0) = 0 \quad \text{for } r \geq 0 \tag{11.2.11}$$

and suitable vanishing conditions at infinity. In Equation 11.2.10 $\delta(r)$ and $\delta(t)$ are, respectively, the Dirac delta functions of radius r and of time t, and κ means the diffusivity, and Q_0 denotes the heat source intensity, and

$$\frac{\delta(r)}{4\pi r^2} = \delta(x_1)\,\delta(x_2)\,\delta(x_3) \tag{11.2.12}$$

$$x_1 = r\cos\varphi\sin\theta, \quad x_2 = r\sin\varphi\sin\theta, \quad x_3 = \cos\theta \tag{11.2.13}$$

$$0 \leq \varphi \leq 2\pi, \quad 0 \leq \theta \leq \pi, \quad 0 \leq r \leq \infty \tag{11.2.14}$$

An elastic process corresponding to the temperature $T = T(r,t)$ is described by a scalar potential $\phi = \phi(r,t)$ through the relations [see Equations 11.2.5 through 11.2.9 written in spherical coordinates $(r.\varphi, \theta)$]

$$s(r,t) = [\mathbf{u}(r,t),\ \mathbf{E}(r,t),\ \mathbf{S}(r,t)] \tag{11.2.15}$$

where

$$\mathbf{u}(r,t) = [u_r(r,t), 0, 0] \tag{11.2.16}$$

$$\mathbf{E}(r,t) = \begin{bmatrix} E_{rr}(r,t) & 0 & 0 \\ 0 & E_{\varphi\varphi}(r,t) & 0 \\ 0 & 0 & E_{\theta\theta}(r,t) \end{bmatrix} \tag{11.2.17}$$

$$\mathbf{S}(r,t) = \begin{bmatrix} S_{rr}(r,t) & 0 & 0 \\ 0 & S_{\varphi\varphi}(r,t) & 0 \\ 0 & 0 & S_{\theta\theta}(r,t) \end{bmatrix} \tag{11.2.18}$$

and

$$u_r(r,t) = \frac{\partial \phi}{\partial r} \tag{11.2.19}$$

$$E_{rr}(r,t) = \frac{\partial^2 \phi}{\partial r^2}, \quad E_{\varphi\varphi}(r,t) = E_{\theta\theta}(r,t) = \frac{1}{r}\frac{\partial \phi}{\partial r} \tag{11.2.20}$$

$$\begin{aligned} S_{rr}(r,t) &= -4\mu\,\frac{1}{r}\frac{\partial \phi}{\partial r} + \rho\ddot{\phi} \\ S_{\varphi\varphi}(r,t) = S_{\theta\theta}(r,t) &= -2\mu\left(\frac{\partial^2 \phi}{\partial r^2} + \frac{1}{r}\frac{\partial \phi}{\partial r}\right) + \rho\ddot{\phi} \end{aligned} \tag{11.2.21}$$

The function $\phi = \phi(r,t)$ satisfies the equation

$$\left(\nabla^2 - \frac{1}{c_1^2}\frac{\partial^2}{\partial t^2}\right)\phi = mT \quad \text{for } r \geq 0,\ t \geq 0 \tag{11.2.22}$$

the initial conditions

$$\phi(r,0) = \dot{\phi}(r,0) = 0 \quad \text{for } r \geq 0 \tag{11.2.23}$$

and suitable vanishing conditions as $r \to \infty$ for every $t > 0$.

Note that the homogeneous initial conditions 11.2.11 and 11.2.23 imply that at $t=0$ the body is kept at a constant temperature $\theta_0 > 0$ ($T = \theta - \theta_0 = 0$ at $t=0$), and $u_r(r,0) = 0$ and $\dot{u}_r(r,0) = 0$ for every $r \geq 0$.

To obtain a pair of functions (T, ϕ) that generates the process $s = s(r,t)$, we transform Equations 11.2.10 and 11.2.11, and 11.2.15 through 11.2.23 to a dimensionless form in the following way. Let x_0, t_0, Q_0, ϕ_0, and S_0, in this order, denote the length, time, heat source, potential, and stress units, defined by

$$\begin{gathered} x_0 = \frac{\kappa}{c_1}, \quad t_0 = \frac{\kappa}{c_1^2} \\ Q_0 = \theta_0 x_0^3, \quad \phi_0 = \frac{Q_0 m}{x_0}, \quad S_0 = \frac{\rho \phi_0}{t_0^2} \end{gathered} \tag{11.2.24}$$

Introduce the dimensionless independent variables

$$R = \frac{r}{x_0}, \quad \tau = \frac{t}{t_0} \tag{11.2.25}$$

and the dimensionless fields

$$\hat{T} = \hat{T}(R,\tau) = \frac{T}{\theta_0}, \quad \hat{\phi} = \hat{\phi}(R,\tau) = \frac{\phi}{\phi_0}, \quad \hat{u}_R(R,\tau) = \frac{u_r}{u_0},$$

$$\hat{\mathbf{E}}(R,\tau) = \frac{1}{E_0}\mathbf{E}, \quad \hat{\mathbf{S}}(R,\tau) = \frac{1}{S_0}\mathbf{S} \tag{11.2.26}$$

where

$$u_0 = \frac{\phi_0}{x_0}, \quad E_0 = \frac{\phi_0}{x_0^2} \tag{11.2.27}$$

Since

$$\delta(r) = \delta(x_0 R) = \frac{1}{x_0}\delta(R) \tag{11.2.28}$$

and

$$\delta(t) = \delta(\tau t_0) = \frac{1}{t_0}\delta(\tau) \tag{11.2.29}$$

therefore, the temperature problem 11.2.10 and 11.2.11 may be transformed to the dimensionless form. Find a function $\hat{T} = \hat{T}(R,\tau)$ for $r \geq 0$, $\tau \geq 0$ that satisfies the equation

$$\nabla^2\hat{T} - \frac{\partial}{\partial\tau}\hat{T} = -\frac{\delta(R)}{4\pi R^2}\delta(\tau) \quad \text{for } R > 0, \quad \tau > 0 \tag{11.2.30}$$

the initial condition

$$\hat{T}(R,0) = 0 \quad \text{for } R > 0 \tag{11.2.31}$$

and the vanishing condition

$$\hat{T}(R,\tau) \to 0 \quad \text{as } R \to +\infty \quad \text{for } \tau > 0 \tag{11.2.32}$$

In Equation 11.2.30

$$\nabla^2 = \frac{\partial^2}{\partial R^2} + \frac{2}{R}\frac{\partial}{\partial R} \tag{11.2.33}$$

Similarly, Equations 11.2.15 through 11.2.21 are transformed to the dimensionless equations

$$\hat{s}(R,\tau) = [\hat{\mathbf{u}}(R,\tau), \hat{\mathbf{E}}(R,\tau), \hat{\mathbf{S}}(R,\tau)] \tag{11.2.34}$$

$$\hat{\mathbf{u}}(R,\tau) = [\hat{u}_R(R,\tau), 0, 0] \tag{11.2.35}$$

$$\hat{\mathbf{E}}(R,\tau) = \begin{bmatrix} \hat{E}_{RR}(R,\tau) & 0 & 0 \\ 0 & \hat{E}_{\varphi\varphi}(R,\tau) & 0 \\ 0 & 0 & \hat{E}_{\theta\theta}(R,\tau) \end{bmatrix} \tag{11.2.36}$$

$$\hat{\mathbf{S}}(R,\tau) = \begin{bmatrix} \hat{S}_{RR}(R,\tau) & 0 & 0 \\ 0 & \hat{S}_{\varphi\varphi}(R,\tau) & 0 \\ 0 & 0 & \hat{S}_{\theta\theta}(R,\tau) \end{bmatrix} \tag{11.2.37}$$

where

$$\hat{u}_R(R,\tau) = \frac{\partial \hat{\phi}}{\partial R} \tag{11.2.38}$$

$$\hat{E}_{RR}(R,\tau) = \frac{\partial^2 \hat{\phi}}{\partial R^2}, \quad \hat{E}_{\varphi\varphi}(R,\tau) = \hat{E}_{\theta\theta}(R,\tau) = \frac{1}{R}\frac{\partial \hat{\phi}}{\partial R} \tag{11.2.39}$$

and

$$\hat{S}_{RR}(R,\tau) = \frac{\partial^2 \hat{\phi}}{\partial \tau^2} - 2\frac{1-2\nu}{1-\nu}\frac{1}{R}\frac{\partial \hat{\phi}}{\partial R} \tag{11.2.40}$$

$$\hat{S}_{\varphi\varphi}(R,\tau) = \hat{S}_{\theta\theta}(R,\tau) = \frac{\partial^2 \hat{\phi}}{\partial \tau^2} - \frac{1-2\nu}{1-\nu}\left(\frac{\partial^2 \hat{\phi}}{\partial R^2} + \frac{1}{R}\frac{\partial \hat{\phi}}{\partial R}\right) \tag{11.2.41}$$

Here ν stands for Poisson's ratio. Finally, Equations 11.2.22 and 11.2.23 are reduced to the dimensionless form

$$\left(\nabla^2 - \frac{\partial^2}{\partial \tau^2}\right)\hat{\phi} = \hat{T} \quad \text{for } R > 0,\ \tau > 0 \tag{11.2.42}$$

$$\hat{\phi}(R,0) = \dot{\hat{\phi}}(R,0) = 0 \quad \text{for } R > 0 \tag{11.2.43}$$

$$\hat{\phi}(R,\tau) \to 0 \quad \text{as } R \to \infty \quad \text{and } \tau > 0 \tag{11.2.44}$$

Since, in view of Equation 11.2.42,

$$\frac{\partial^2 \hat{\phi}}{\partial R^2} = \hat{T} + \left(\frac{\partial^2}{\partial \tau^2} - \frac{2}{R}\frac{\partial}{\partial R}\right)\hat{\phi} \tag{11.2.45}$$

then, substituting $\partial^2\hat{\phi}/\partial R^2$ from Equation 11.2.45 into the RHS of Equation 11.2.41, an alternative form of Equations 11.2.40 and 11.2.41 is established:

$$\begin{aligned} \hat{S}_{RR} &= \frac{\partial^2 \hat{\phi}}{\partial \tau^2} - \frac{2(1-2\nu)}{1-\nu}\frac{1}{R}\hat{u}_R \\ \hat{S}_{\varphi\varphi} = \hat{S}_{\theta\theta} &= \frac{\nu}{1-\nu}\frac{\partial^2 \hat{\phi}}{\partial \tau^2} + \frac{1-2\nu}{1-\nu}\left(\frac{1}{R}\hat{u}_R - \hat{T}\right) \end{aligned} \tag{11.2.46}$$

Thus, to find the stress components, it is sufficient to have the fields $\ddot{\hat{\phi}}$, $\hat{u}_R$, and $\hat{T}$. In the following, a pair $(\hat{T}, \hat{\phi})$ will be found by the Laplace transform technique.

Let $\bar{f} = \bar{f}(R,p)$ be the Laplace transform of a function $f = f(R,\tau)$:

$$L\{f\} \equiv \bar{f}(R,p) = \int_0^\infty e^{-p\tau} f(R,\tau)\, d\tau \tag{11.2.47}$$

where p is the dimensionless transform parameter. By applying the operator L to Equation 11.2.30 and using the homogeneous initial condition 11.2.31 we obtain

$$(\nabla^2 - p)\bar{\hat{T}} = -\frac{\delta(R)}{4\pi R^2}, \quad R \geq 0 \tag{11.2.48}$$

Similarly, by applying the operator L to Equation 11.2.42 and using the homogeneous initial conditions 11.2.43 we write

$$(\nabla^2 - p^2)\bar{\hat{\phi}} = \bar{\hat{T}}, \quad R \geq 0 \tag{11.2.49}$$

Hence, eliminating $\bar{\hat{T}}$ from Equations 11.2.48 and 11.2.49 leads to

$$(\nabla^2 - p^2)(\nabla^2 - p)\bar{\hat{\phi}} = -\frac{\delta(R)}{4\pi R^2}, \quad R \geq 0 \tag{11.2.50}$$

Now, note that for a positive parameter k, a solution of the equation

$$(\nabla^2 - k^2)\bar{f}(R,p) = -\frac{\delta(R)}{4\pi R^2} \quad \text{for } R \geq 0 \tag{11.2.51}$$

that vanishes as $R \to \infty$ takes the form

$$\bar{f}(R,p) = -(\nabla^2 - k^2)^{-1}\frac{\delta(R)}{4\pi R^2} = \frac{e^{-kR}}{4\pi R} \tag{11.2.52}$$

Therefore, letting p be a real valued positive parameter, and applying the operator $(\nabla^2 - p)^{-1}$ to Equation 11.2.48 we get

$$\bar{\hat{T}}(R,p) = -(\nabla^2 - p)^{-1}\frac{\delta(R)}{4\pi R^2} = \frac{e^{-R\sqrt{p}}}{4\pi R} \tag{11.2.53}$$

To derive the function $\bar{\hat{\phi}} = \bar{\hat{\phi}}(R,p)$, we apply the operator

$$[(\nabla^2 - p^2)(\nabla^2 - p)]^{-1}$$

to Equation 11.2.50 and arrive at

$$\begin{aligned} \bar{\hat{\phi}}(R,p) &= -\left[(\nabla^2 - p^2)\left(\nabla^2 - p\right)\right]^{-1}\frac{\delta(R)}{4\pi R^2} \\ &= -\frac{1}{p(p-1)}\left\{\left(\nabla^2 - p^2\right)^{-1} - \left(\nabla^2 - p\right)^{-1}\right\}\frac{\delta(R)}{4\pi R^2} \end{aligned} \tag{11.2.54}$$

Then, using Equation 11.2.52 we get

$$\bar{\hat{\phi}}(R,p) = \frac{1}{4\pi}\frac{1}{p(p-1)}\frac{e^{-Rp} - e^{-R\sqrt{p}}}{R} \quad (11.2.55)$$

Now, we note that [5,6]

$$L^{-1}\left\{e^{-R\sqrt{p}}\right\} = \frac{R}{\sqrt{4\pi\tau^3}}e^{-\frac{R^2}{4\tau}} \quad (11.2.56)$$

$$L^{-1}\left\{\frac{e^{-R\sqrt{p}}}{p-1}\right\} = \frac{1}{2}\left[e^{\tau-R}\,\text{erfc}\left(\frac{R}{2\sqrt{\tau}} - \sqrt{\tau}\right) + e^{\tau+R}\,\text{erfc}\left(\frac{R}{2\sqrt{\tau}} + \sqrt{\tau}\right)\right] \quad (11.2.57)$$

$$L^{-1}\left\{\frac{e^{-R\sqrt{p}}}{p}\right\} = \text{erfc}\left(\frac{R}{2\sqrt{\tau}}\right) \quad (11.2.58)$$

and

$$L^{-1}\left\{\bar{f}(R,p)e^{-Rp}\right\} = f(R,\tau - R)H(\tau - R) \quad (11.2.59)$$

where, for any real valued x,

$$\text{erfc}\,(x) = 1 - \text{erf}\,(x) \quad (11.2.60)$$

$$\text{erf}\,(x) = \frac{2}{\sqrt{\pi}}\int_0^x e^{-u^2}\,du \quad (11.2.61)$$

$$\text{erfc}\,(x) = \frac{2}{\sqrt{\pi}}\int_x^\infty e^{-u^2}\,du \quad (11.2.62)$$

Here, erf (x) stands for the error function, erfc (x) is the complementary error function, and H (.) means the Heaviside function. Hence, by applying the operator L^{-1} to Equation 11.2.53 and using Equation 11.2.56 we get

$$\hat{T}(R,\tau) = \frac{1}{4\pi}\frac{1}{\sqrt{4\pi\tau^3}}e^{-\frac{R^2}{4\tau}} \quad (11.2.63)$$

Also, by using the decomposition

$$\frac{1}{p(p-1)} = \frac{1}{p-1} - \frac{1}{p} \quad (11.2.64)$$

and the formulas

$$L^{-1}\left\{\frac{1}{p-1}\right\} = e^{\tau} \quad (11.2.65)$$

$$L^{-1}\left\{\frac{1}{p}\right\} = 1 \quad (11.2.66)$$

as well as Equations 11.2.55 through 11.2.59, we arrive at

$$\hat{\phi}(R,\tau) = \frac{1}{4\pi R}\left[\left(e^{\tau-R} - 1\right)H(\tau - R) + \operatorname{erfc}\left(\frac{R}{2\sqrt{\tau}}\right) - U(R,\tau)\right] \tag{11.2.67}$$

where

$$U(R,\tau) = \frac{1}{2}\left[e^{\tau-R}\operatorname{erfc}\left(\frac{R}{2\sqrt{\tau}} - \sqrt{\tau}\right) + e^{\tau+R}\operatorname{erfc}\left(\frac{R}{2\sqrt{\tau}} + \sqrt{\tau}\right)\right] \tag{11.2.68}$$

To calculate the function $\ddot{\hat{\phi}}(R,\tau)$ we may differentiate Equation 11.2.67 twice with respect to τ, or use the formula (see Equation 11.2.55 multiplied by p^2)

$$p^2\overline{\hat{\phi}}(R,p) = \frac{1}{4\pi R}\left[\left(e^{-Rp} - e^{-R\sqrt{p}}\right) + \frac{1}{p-1}\left(e^{-Rp} - e^{-R\sqrt{p}}\right)\right] \tag{11.2.69}$$

Since

$$L^{-1}\left\{e^{-Rp}\right\} = \delta(\tau - R) \tag{11.2.70}$$

therefore, application of the operator L^{-1} to Equation 11.2.69 and using Equations 11.2.56 through 11.2.59 gives

$$\ddot{\hat{\phi}}(R,\tau) = \frac{1}{4\pi R}\left[\delta(\tau - R) + e^{\tau-R}H(\tau - R) - \frac{R}{\sqrt{4\pi\tau^3}}e^{-\frac{R^2}{4\tau}} - U(R,\tau)\right] \tag{11.2.71}$$

Finally, by differentiating Equation 11.2.67 with respect to R, we arrive at

$$\begin{aligned}\frac{\partial\hat{\phi}}{\partial R}(R,\tau) = -\frac{1}{4\pi R^2}\Big\{&\left[(1+R)e^{\tau-R} - 1\right]H(\tau - R)\\ &+ \operatorname{erfc}\left(\frac{R}{2\sqrt{\tau}}\right) - U(R,\tau) - RV(R,\tau)\Big\}\end{aligned} \tag{11.2.72}$$

and

$$\begin{aligned}\frac{\partial^2\hat{\phi}}{\partial R^2}(R,\tau) = &\frac{1}{4\pi R}\left[\delta(\tau - R) + e^{\tau-R}H(\tau - R) - U(R,\tau)\right]\\ &+ \frac{2}{4\pi R^3}\Big\{\left[(1+R)e^{\tau-R} - 1\right]H(\tau - R)\\ &+ \operatorname{erfc}\left(\frac{R}{2\sqrt{\tau}}\right) - U(R,\tau) - RV(R,\tau)\Big\}\end{aligned} \tag{11.2.73}$$

Here, the function $V = V(R,\tau)$ is defined by

$$V(R,\tau) = \frac{1}{2}\left[e^{\tau-R}\operatorname{erfc}\left(\frac{R}{2\sqrt{\tau}} - \sqrt{\tau}\right) - e^{\tau+R}\operatorname{erfc}\left(\frac{R}{2\sqrt{\tau}} + \sqrt{\tau}\right)\right] \tag{11.2.74}$$

Note that the pair (U, V) satisfies the relations

$$U(R,\tau) = -\frac{\partial}{\partial R} V(R,\tau)$$
$$V(R,\tau) = -\frac{\partial}{\partial R} U(R,\tau) - \frac{1}{\sqrt{\pi\tau}} e^{-\frac{R^2}{4\tau}} \tag{11.2.75}$$

By substituting the functions $\hat{T}$, $\ddot{\hat{\phi}}$, and $\partial\hat{\phi}/\partial R$ from Equations 11.2.63, 11.2.71, and 11.2.72, respectively, into Equations 11.2.46 we arrive at the stress components

$$\hat{S}_{RR}(R,\tau) = \frac{1}{4\pi R}\left[\delta(\tau - R) + e^{\tau - R}H(\tau - R) - U(R,\tau) - \frac{R}{\sqrt{4\pi\tau^3}} e^{-\frac{R^2}{4\tau}}\right] + \frac{2(1-2\nu)}{4\pi(1-\nu)}\frac{1}{R^3}\left\{[(1+R)\,e^{\tau-R} - 1]H(\tau - R) + \operatorname{erfc}\left(\frac{R}{2\sqrt{\tau}}\right) - U(R,\tau) - RV(R,\tau)\right\} \tag{11.2.76}$$

$$\hat{S}_{\theta\theta}(R,\tau) = \frac{1}{4\pi R}\frac{\nu}{1-\nu}\left[\delta(\tau - R) + e^{\tau - R}H(\tau - R) - U(R,\tau) - \frac{R}{\sqrt{4\pi\tau^3}} e^{-\frac{R^2}{4\tau}}\right] - \frac{(1-2\nu)}{4\pi(1-\nu)}\frac{1}{R^3}\left\{\left[(1+R)e^{\tau-R} - 1\right]H(\tau - R) + \operatorname{erfc}\left(\frac{R}{2\sqrt{\tau}}\right) - U(R,\tau) - RV(R,\tau)\right\} - \frac{(1-2\nu)}{4\pi(1-\nu)}\frac{1}{\sqrt{4\pi\tau^3}} e^{-\frac{R^2}{4\tau}} \tag{11.2.77}$$

Also, by substituting $\hat{T}$ and $\ddot{\hat{\phi}}$ from Equations 11.2.63 and 11.2.71, respectively, into the volume change formula

$$\operatorname{tr}\hat{\mathbf{E}} = \ddot{\hat{\phi}} + \hat{T} \tag{11.2.78}$$

we find

$$\operatorname{tr}\hat{\mathbf{E}} = \frac{1}{4\pi R}\left[\delta(\tau - R) + e^{\tau - R}H(\tau - R) - U(R,\tau)\right] \tag{11.2.79}$$

11.2.1.1 Analysis of the Solution

It follows from Equation 11.2.63 that the temperature change $\hat{T} = \hat{T}(R,\tau)$ due to the instantaneous concentrated heat source in an infinite body has a diffusive character, which means that $\hat{T}$ "propagates" with an infinite speed throughout the body. The infinite speed paradox is temperated by the fact that $\hat{T}(R,\tau) \to 0$ as $R \to \infty$ for every $\tau > 0$.

It follows from Equations 11.2.34 through 11.2.41, 11.2.46, and 11.2.71 through 11.2.79 that the thermoelastic process $\hat{s} = \hat{s}(R,\tau)$ corresponding to the temperature $\hat{T} = \hat{T}(R,\tau)$ takes the form

$$\hat{s}(R,\tau) = \hat{s}_d(R,\tau) + \hat{s}_w(R,\tau) \tag{11.2.80}$$

where $\hat{s}_d = [\hat{\mathbf{u}}^{(d)}, \hat{\mathbf{E}}^{(d)}, \hat{\mathbf{S}}^{(d)}]$ and $\hat{s}_w = [\hat{\mathbf{u}}^{(w)}, \hat{\mathbf{E}}^{(w)}, \hat{\mathbf{S}}^{(w)}]$ represent the diffusive and wave processes, respectively. For example, using Equations 11.2.72, 11.2.76 and 11.2.77 yields

$$\hat{u}_R^{(w)}(R,\tau) = -\frac{1}{4\pi R^2}\left[(1+R)e^{\tau-R} - 1\right]H(\tau - R) \tag{11.2.81}$$

$$\hat{S}_{RR}^{(w)}(R,\tau) = \frac{1}{4\pi R}\,\delta(\tau - R) + \frac{1}{4\pi R}\left\{e^{\tau-R} + \frac{2(1-2\nu)}{(1-\nu)} \times \frac{1}{R^2}[(1+R)e^{\tau-R} - 1]\right\}H(\tau - R) \tag{11.2.82}$$

$$\hat{S}_{\theta\theta}^{(w)}(R,\tau) = \frac{1}{4\pi R}\frac{\nu}{1-\nu}\,\delta(\tau - R) + \frac{1}{4\pi R}\left\{\frac{\nu}{1-\nu}\,e^{\tau-R} - \frac{(1-2\nu)}{(1-\nu)}\frac{1}{R^2}[(1+R)e^{\tau-R} - 1]\right\}H(\tau - R) \tag{11.2.83}$$

Therefore, $\hat{s}_w$ represents a spherical wave in which $\tau = R$ is a front propagating with a unit velocity from $R = 0$ to $R = \infty$. Let $[\![f]\!](R)$ denote the jump of a function $f = f(R,\tau)$ across the front $\tau = R$

$$[\![f]\!](R) = f(R, R-0) - f(R, R+0) \tag{11.2.84}$$

Then it follows from Equation 11.2.81 that

$$[\![\hat{u}_R^{(w)}]\!](R) = \frac{1}{4\pi R} \tag{11.2.85}$$

Also, it follows from Equations 11.2.82 and 11.2.83 that

$$\left[\!\!\left[\hat{S}_{RR}^{(w)}\right]\!\!\right](R) = -\frac{1}{4\pi R}\left[1 + \frac{2(1-2\nu)}{1-\nu}\frac{1}{R}\right] \tag{11.2.86}$$

$$\left[\!\!\left[\hat{S}_{\theta\theta}^{(w)}\right]\!\!\right](R) = -\frac{1}{4\pi R}\left(\frac{\nu}{1-\nu} - \frac{1-2\nu}{1-\nu}\frac{1}{R}\right) \tag{11.2.87}$$

Since

$$\left[\!\!\left[\hat{u}_R^{(d)}\right]\!\!\right](R) = 0 \tag{11.2.88}$$

therefore

$$\left[\!\!\left[\hat{u}_R\right]\!\!\right](R) = \frac{1}{4\pi R} \tag{11.2.89}$$

Hence, the radial displacement $\hat{u}_R$ exhibits a finite jump across the front $\tau = R$.

Also, since

$$\left[\!\!\left[\hat{S}_{RR}^{(d)}\right]\!\!\right](R) = \left[\!\!\left[\hat{S}_{\theta\theta}^{(d)}\right]\!\!\right](R) = 0 \tag{11.2.90}$$

therefore

$$\begin{aligned} \left[\!\left[\hat{S}_{RR}\right]\!\right](R) &= \left[\!\left[\hat{S}_{RR}^{(w)}\right]\!\right](R) \\ \left[\!\left[\hat{S}_{\theta\theta}\right]\!\right](R) &= \left[\!\left[\hat{S}_{\theta\theta}^{(W)}\right]\!\right](R) \end{aligned} \tag{11.2.91}$$

Note that a delta profile in a neighborhood of the front $\tau = R$ should be imposed on the discontinuities of $\hat{S}_{RR}$ and $\hat{S}_{\theta\theta}$ across $\tau = R$ to describe completely the behavior of the stresses in a neighborhood of the front $\tau = R$. For a time behavior of the radial displacement $\hat{u}_R(R, \tau)$ as $R \to 0$ we derive the formula

$$\hat{u}_R(R, \tau) \to \frac{1}{8\pi}\delta(\tau) \quad \text{as } R \to 0 \tag{11.2.92}$$

Finally, the thermoelastic process $\hat{s} = \hat{s}(R, \tau)$ is transient in time, that is, for $R > 0$

$$\hat{s}(R, \tau) = [\hat{\mathbf{u}}(R, \tau),\ \hat{\mathbf{E}}(R, \tau),\ \hat{\mathbf{S}}(R, \tau)] \to [\mathbf{0}, \mathbf{0}, \mathbf{0}] \quad \text{as } \tau \to \infty \tag{11.2.93}$$

If the Dirac delta functions in Equations 11.2.76 and 11.2.77 are ignored, we arrive at the dynamic thermal stress components obtained to the authors' knowledge for the first time by Witold Nowacki [7].

11.2.2 Harmonic Vibrations of an Infinite Elastic Body Produced by a Concentrated Source of Heat

Assume that at the origin of the spherical coordinates (R, φ, θ), located in a homogeneous isotropic infinite elastic solid, there acts a concentrated heat source, the intensity of which varies harmonically with time; and we are to find a harmonic temperature and associated thermoelastic process in the solid. By introducing the dimensionless independent variables R and τ as well as the dimensionless temperature $\hat{T} = \hat{T}(R, \tau)$ and potential $\hat{\phi} = \hat{\phi}(R, \tau)$ (see Equations 11.2.25 and 11.2.26) and proceeding in a way similar to that of Section 11.2.1, we reduce the problem to that of finding a pair $(\hat{T}, \hat{\phi})$ that satisfies the field equations

$$\left(\nabla^2 - \frac{\partial}{\partial\tau}\right)\hat{T} = -\frac{\delta(R)}{4\pi R^2}\, e^{-i\Omega\tau}, \quad R \geq 0 \tag{11.2.94}$$

$$\left(\nabla^2 - \frac{\partial^2}{\partial\tau^2}\right)\hat{\phi} = \hat{T}, \quad R \geq 0 \tag{11.2.95}$$

subject to suitable vanishing conditions at infinity. Here, Ω is a dimensionless frequency ($\Omega > 0$), and $i = \sqrt{-1}$.

By looking for the pair $(\hat{T}, \hat{\phi})$ in the form

$$\hat{T}(R, \tau) = e^{-i\Omega\tau}\, T_0(R, \Omega) \tag{11.2.96}$$

$$\hat{\phi}(R, \tau) = e^{-i\Omega\tau}\, \phi_0(R, \Omega) \tag{11.2.97}$$

we find that $(\hat{T}, \hat{\phi})$ satisfies Equations 11.2.94 and 11.2.95 provided (T_0, ϕ_0) satisfies the equations

$$(\nabla^2 + i\Omega)T_0 = -\frac{\delta(R)}{4\pi R^2} \tag{11.2.98}$$

$$(\nabla^2 + \Omega^2)\phi_0 = T_0, \quad R \geq 0 \tag{11.2.99}$$

The associated displacement $\hat{u}_R = \hat{u}_R(R, \tau)$ and the stress components $\hat{S}_{RR} = \hat{S}_{RR}(R, \tau)$ and $\hat{S}_{\theta\theta} = \hat{S}_{\theta\theta}(R, \tau)$ take the forms

$$\hat{u}_R(R, \tau) = e^{-i\Omega\tau}\, u_R^{(0)}(R, \Omega) \tag{11.2.100}$$

$$\hat{S}_{RR}(R, \tau) = e^{-i\Omega\tau}\, S_{RR}^{(0)}(R, \Omega) \tag{11.2.101}$$

$$\hat{S}_{\theta\theta}(R, \tau) = e^{-i\Omega\tau}\, S_{\theta\theta}^{(0)}(R, \Omega) \tag{11.2.102}$$

where (see Equations 11.2.38 and 11.2.46)

$$u_R^{(0)} = \frac{\partial \phi_0}{\partial R} \tag{11.2.103}$$

$$S_{RR}^{(0)} = -\Omega^2\phi_0 - \frac{2(1-2\nu)}{1-\nu}\frac{1}{R}\, u_R^{(0)} \tag{11.2.104}$$

$$S_{\theta\theta}^{(0)} = -\frac{\nu}{1-\nu}\Omega^2\phi_0 + \frac{1-2\nu}{1-\nu}\left(\frac{1}{R}u_R^{(0)} - T_0\right) \tag{11.2.105}$$

By proceeding in a way similar to that of deriving the pair $(\overline{\hat{T}}, \overline{\hat{\phi}})$ of Section 11.2.1, we find that a pair (T_0, ϕ_0) that represents an outgoing wave from $R=0$ to $R=\infty$ takes the form (see Equations 11.2.53 and 11.2.55 in which we let $p = -i\Omega$)

$$T_0(R, \Omega) = \frac{e^{-R\sqrt{-i\Omega}}}{4\pi R} \tag{11.2.106}$$

$$\phi_0(R, \Omega) = \frac{1}{4\pi R}\frac{e^{-R(-i\Omega)} - e^{-R\sqrt{-i\Omega}}}{(-i\Omega)(-i\Omega - 1)} \tag{11.2.107}$$

where

$$\sqrt{-i\Omega} = \sqrt{\frac{\Omega}{2}}(1-i) \tag{11.2.108}$$

Therefore, it follows from Equations 11.2.96 and 11.2.97 that

$$\hat{T}(R, \tau) = \frac{e^{-R\sqrt{\frac{\Omega}{2}}}}{4\pi R} e^{-i\Omega\left(\tau - \frac{R}{\sqrt{2\Omega}}\right)} \tag{11.2.109}$$

$$\hat{\phi}(R, \tau) = \frac{1}{4\pi R}\frac{e^{-i\Omega(\tau - R)} - e^{-R\sqrt{\frac{\Omega}{2}}} \cdot e^{-i\Omega\left(\tau - \frac{R}{\sqrt{2\Omega}}\right)}}{\Omega(i - \Omega)} \tag{11.2.110}$$

Now, if we introduce the notation

$$f^*(R,\tau) \equiv Re[\hat{f}(R,\tau)] = Re[e^{-i\Omega\tau} f_0(R,\Omega)] \tag{11.2.111}$$

where $\hat{f}(R,\tau) = e^{-i\Omega\tau} f_0(R,\Omega)$ is an arbitrary function and $Re[\hat{f}]$ stands for the real part of $\hat{f}$, then by taking the real parts of Equations 11.2.109 and 11.2.110, respectively, we get

$$T^*(R,\tau) = \frac{1}{4\pi R} e^{-R\sqrt{\frac{\Omega}{2}}} \cos\Omega\left(\tau - \frac{R}{\sqrt{2\Omega}}\right) \tag{11.2.112}$$

and

$$\phi^*(R,\tau) = -\frac{1}{4\pi R}\frac{1}{1+\Omega^2}\left\{\cos\Omega(\tau - R) - e^{-R\sqrt{\frac{\Omega}{2}}}\cos\Omega\left(\tau - \frac{R}{\sqrt{2\Omega}}\right)\right.$$

$$\left. + \frac{1}{\Omega}\left[\sin\Omega(\tau - R) - e^{-R\sqrt{\frac{\Omega}{2}}}\sin\Omega\left(\tau - \frac{R}{\sqrt{2\Omega}}\right)\right]\right\} \tag{11.2.113}$$

It follows from Equations 11.2.112 and 11.2.113 that

$$(T^*, \phi^*) \to (0,0) \quad \text{as } R \to \infty \text{ and } \Omega > 0 \tag{11.2.114}$$

and

$$T^* \to \frac{1}{4\pi R}, \quad \phi^* \to -\infty \quad \text{as } \Omega \to 0, R > 0 \tag{11.2.115}$$

Therefore, for a long period of vibrations ($2\pi/\Omega \to \infty$) the temperature T^* approaches a steady-state field (see Equation 9.2.5) while the potential ϕ^* reveals a pathological behavior. This is related to the fact that in a steady concentrated heat source problem of thermoelasticity for an infinite body, a potential generating the solution is unbounded as $R \to \infty$ (see Equation 9.2.14).

In the following we show that a pair (T^*, ϕ^*) defined by Equations 11.2.112 and 11.2.113 generates the displacement u_R^* and the stress components S_{RR}^* and $S_{\theta\theta}^*$ in such a way that $u_R^* \to u_R^\circ$, $S_{RR}^* \to S_{RR}^\circ$, $S_{\theta\theta}^* \to S_{\theta\theta}^\circ$ as $\Omega \to 0$ for $R > 0$, where u_R°, S_{RR}°, and $S_{\theta\theta}^\circ$ are the displacement and stress components of the steady-state problem. To this end we differentiate Equation 11.2.113 with respect to R and obtain

$$u_R^*(R,\tau) = \frac{\partial \phi^*}{\partial R} = \frac{1}{4\pi R^2}\frac{1}{1+\Omega^2}\left\{\cos\Omega(\tau - R) - R\Omega\sin\Omega(\tau - R)\right.$$

$$- e^{-R\sqrt{\frac{\Omega}{2}}}\left[\left(1 + R\sqrt{\frac{\Omega}{2}}\right)\cos\Omega\left(\tau - \frac{R}{\sqrt{2\Omega}}\right) - R\sqrt{\frac{\Omega}{2}}\sin\Omega\left(\tau - \frac{R}{\sqrt{2\Omega}}\right)\right]$$

$$+ \frac{1}{\Omega}[\sin\Omega(\tau - R) + R\Omega\cos\Omega(\tau - R)] - \frac{1}{\Omega}e^{-R\sqrt{\frac{\Omega}{2}}}$$

$$\left.\left[\left(1 + R\sqrt{\frac{\Omega}{2}}\right)\sin\Omega\left(\tau - \frac{R}{\sqrt{2\Omega}}\right) + R\sqrt{\frac{\Omega}{2}}\cos\Omega\left(\tau - \frac{R}{\sqrt{2\Omega}}\right)\right]\right\}$$

$$\tag{11.2.116}$$

The stress components are obtained from the formulas (see Equations 11.2.104 and 11.2.105)

$$S_{RR}^*(R,\tau) = -\Omega^2\phi^*(R,\tau) - \frac{2(1-2\nu)}{1-\nu}\frac{u_R^*(R,\tau)}{R} \tag{11.2.117}$$

$$S_{\theta\theta}^*(R,\tau) = -\frac{\nu}{1-\nu}\Omega^2\phi^*(R,\tau) + \frac{1-2\nu}{1-\nu}\left[\frac{u_R^*(R,\tau)}{R} - T^*(R,\tau)\right] \tag{11.2.118}$$

where the functions $T^* = T^*(R,\tau)$, $\phi^* = \phi^*(R,\tau)$, and $u_R^* = u_R^*(R,\tau)$ are given by Equations 11.2.112, 11.2.113, and 11.2.116, respectively.

By letting $\Omega \to 0$ in Equation 11.2.116 we get

$$u_R^*(R,\tau) \to u_R^\circ \equiv \frac{1}{8\pi} \quad \text{as } \Omega \to 0 \tag{11.2.119}$$

Also, it follows from Equation 11.2.113 that

$$\Omega^2\phi^*(R,\tau) \to 0 \quad \text{as } \Omega \to 0 \tag{11.2.120}$$

Hence, by letting $\Omega \to 0$ in Equations 11.2.117 and 11.2.118 and using Equations 11.2.119 and 11.2.120 we have

$$S_{RR}^*(R,\tau) \to S_{RR}^\circ \equiv -\frac{(1-2\nu)}{4\pi(1-\nu)}\frac{1}{R} \quad \text{as } \Omega \to 0 \tag{11.2.121}$$

and

$$S_{\theta\theta}^*(R,\tau) \to S_{\theta\theta}^\circ \equiv -\frac{(1-2\nu)}{8\pi(1-\nu)}\frac{1}{R} \quad \text{as } \Omega \to 0 \tag{11.2.122}$$

The displacement u_R° and stresses S_{RR}° and $S_{\theta\theta}^\circ$ represent a solution to the steady-state concentrated heat source problem for an infinite body (see Equations 9.2.15 and 9.2.18). In particular, S_{RR}° and $S_{\theta\theta}^\circ$ satisfy the equilibrium equation

$$\frac{\partial}{\partial R}S_{RR}^\circ + \frac{2}{R}\left(S_{RR}^\circ - S_{\theta\theta}^\circ\right) = 0 \tag{11.2.123}$$

Finally, by letting $R \to 0$ in Equation 11.2.116 we arrive at

$$u_R^*(R,\tau) \to \frac{\cos\Omega\tau}{8\pi} \quad \text{as } R \to 0 \tag{11.2.124}$$

This result reduces to that of Equation 11.2.119 if $\Omega = 0$.

11.2.3 Dynamic Thermal Stresses Produced by an Instantaneous Spherical Temperature Inclusion in an Infinite Body

Assume that a homogeneous isotropic infinite elastic body, referred to a spherical coordinate system (R, φ, θ), and initially at rest, be subject to an instantaneous spherical temperature inclusion of the form

$$\hat{T}(R,\tau) = T_0 H(a-R)\delta(\tau) \tag{11.2.125}$$

and a spherically symmetric thermoelastic process

$$\hat{s}(R,\tau) = [\hat{\mathbf{u}}(R,\tau), \hat{\mathbf{E}}(R,\tau), \hat{\mathbf{S}}(R,\tau)] \tag{11.2.126}$$

produced in the body by the discontinuous temperature change $\hat{T}$ is to be found. Here, $\hat{T}$, $\hat{\mathbf{u}}$, $\hat{\mathbf{E}}$, and $\hat{\mathbf{S}}$ are the dimensionless temperature, displacement, strain, and stress fields, respectively, defined by Equations 11.2.26; and $H = H(a - R)$ is the Heaviside function, while $\delta(\tau)$ is the Dirac delta function

$$\delta(\tau) = \frac{dH(\tau)}{d\tau} \tag{11.2.127}$$

The parameters T_0 and a represent a dimensionless constant temperature and a dimensionless radius of the temperature inclusion, respectively. The nonvanishing components of $\hat{\mathbf{u}}$ and $\hat{\mathbf{S}}$ are given by the formulas (see Equations 11.2.38 and 11.2.46)

$$\hat{u}_R(R,\tau) = \frac{\partial\hat{\phi}}{\partial R} \tag{11.2.128}$$

and

$$\hat{S}_{RR}(R,\tau) = \frac{\partial^2\hat{\phi}}{\partial\tau^2} - \frac{2(1-2\nu)}{1-\nu}\frac{\hat{u}_R}{R} \tag{11.2.129}$$

$$\hat{S}_{\varphi\varphi}(R,\tau) = \hat{S}_{\theta\theta}(R,\tau) = \frac{\nu}{1-\nu}\frac{\partial^2\hat{\phi}}{\partial\tau^2} + \frac{1-2\nu}{1-\nu}\left(\frac{\hat{u}_R}{R} - \hat{T}\right) \tag{11.2.130}$$

where the potential $\hat{\phi} = \hat{\phi}(R,\tau)$ satisfies the nonhomogeneous wave equation

$$\left(\nabla^2 - \frac{\partial^2}{\partial\tau^2}\right)\hat{\phi} = \hat{T} \quad \text{for } R > 0, \quad \tau > 0 \tag{11.2.131}$$

subject to the homogeneous initial conditions

$$\hat{\phi}(R,0) = \dot{\hat{\phi}}(R,0) = 0 \tag{11.2.132}$$

and suitable vanishing conditions at infinity.

Note that in the problem under consideration the temperature $\hat{T} = \hat{T}(R,\tau)$ does not satisfy a parabolic heat conduction equation in the classic sense since $\hat{T}$ is a product of the spatially discontinuous Heaviside function and the time-dependent Dirac delta function. Then, $\hat{T}$ is a singular function in the space–time domain: $R \geq 0$, $\tau \geq 0$ with a singularity at the point $(R = a, \tau = 0)$.

To find a solution $\hat{\phi} = \hat{\phi}(R,\tau)$ to the initial–boundary value problem 11.2.131 and 11.2.132 we use the Laplace transform technique. Let $\bar{f} = \bar{f}(R,p)$ be the Laplace transform of a function $f = f(R,\tau)$

$$L\{f\} \equiv \bar{f}(R,p) = \int_0^\infty e^{-p\tau} f(R,\tau)\, d\tau \tag{11.2.133}$$

where p is the transform parameter. By applying the operator L to Equation 11.2.131 and using the homogeneous initial conditions 11.2.132 we write

$$(\nabla^2 - p^2)\bar{\hat{\phi}} = \bar{\hat{T}} \tag{11.2.134}$$

A solution to this equation may be represented by the integral

$$\bar{\hat{\phi}}(R,p) = \int_0^\infty \bar{\hat{\phi}}^*(R, R_0; p)\bar{\hat{T}}(R_0, p)\, dR_0 \tag{11.2.135}$$

where $\bar{\hat{\phi}}^*(R, R_0; p)$ satisfies the equation

$$(\nabla^2 - p^2)\bar{\hat{\phi}}^*(R, R_0; p) = \delta(R - R_0) \tag{11.2.136}$$

subject to the vanishing condition

$$\bar{\hat{\phi}}^*(R, R_0; p) \to 0 \quad \text{as } R \to \infty \tag{11.2.137}$$

To find $\bar{\hat{\phi}}^*$ we multiply Equation 11.2.136 by R and have

$$\left(\frac{d^2}{dR^2} - p^2\right)\left(R\bar{\hat{\phi}}^*\right) = R\delta(R - R_0) = R_0\delta(R - R_0) \tag{11.2.138}$$

Since

$$R_0\delta(R - R_0) = \frac{2}{\pi}R_0 \int_0^\infty \sin\alpha R_0 \sin\alpha R\, d\alpha \tag{11.2.139}$$

therefore, a solution of Equation 11.2.138 may be represented by the integral

$$R\bar{\hat{\phi}}^*(R, R_0; p) = -\frac{2}{\pi}R_0 \int_0^\infty \frac{\sin\alpha R_0 \sin\alpha R}{\alpha^2 + p^2}\, d\alpha \tag{11.2.140}$$

or, using the identity

$$2\sin x \sin y = \cos(x - y) - \cos(x + y) \tag{11.2.141}$$

by the formula

$$R\bar{\hat{\phi}}^*(R, R_0; p) = -\frac{1}{\pi}R_0 \left[\int_0^\infty \frac{\cos\alpha(R - R_0)}{\alpha^2 + p^2}\, d\alpha - \int_0^\infty \frac{\cos\alpha(R + R_0)}{\alpha^2 + p^2}\, d\alpha\right] \tag{11.2.142}$$

Since

$$\int_0^\infty \frac{\cos\alpha x}{\alpha^2 + p^2}\, d\alpha = \frac{\pi}{2}\frac{e^{-p|x|}}{p} \quad \text{for } p > 0 \tag{11.2.143}$$

hence Equation 11.2.142 reduces to the form

$$R\overline{\hat{\phi}}^{*}(R, R_0; p) = -\frac{R_0}{2p}\left[e^{-p|R-R_0|} - e^{-p(R+R_0)}\right] \tag{11.2.144}$$

This completes the derivation of a solution to Equation 11.2.136.

By applying the operator L^{-1} to Equation 11.2.144 we get

$$\hat{\phi}^{*}(R, R_0; \tau) = -\frac{R_0}{2R}\{H[\tau - |R - R_0|] - H[\tau - (R + R_0)]\} \tag{11.2.145}$$

Finally, by applying the operator L^{-1} to Equation 11.2.135 we arrive at the integral representation of $\hat{\phi} = \hat{\phi}(R, \tau)$

$$\hat{\phi}(R, \tau) = \int_0^{\tau}\int_0^{\infty} \hat{\phi}^{*}(R, R_0; \tau - u)\hat{T}(R_0, u)\, dR_0\, du \tag{11.2.146}$$

where $\hat{T}$ and $\hat{\phi}^{*}$ are given by Equations 11.2.125 and 11.2.145, respectively.* Therefore, substituting $\hat{T}$ and $\hat{\phi}^{*}$ from Equations 11.2.125 and 11.2.145, respectively, into Equation 11.2.146 we obtain

$$\hat{\phi}(R, \tau) = -\frac{T_0}{2R}\int_0^{a} R_0\{H[\tau - |R - R_0|] - H[\tau - (R + R_0)]\}\, dR_0 \tag{11.2.147}$$

Calculating the integral on the RHS of Equation 11.2.147 we get

$$\begin{aligned}\hat{\phi}(R, \tau) = \frac{T_0}{2R}\Bigg\{&\Bigg\langle\frac{1}{2}[\tau^2 - (a - R)^2]H[\tau - (a - R)] \\ &- \frac{1}{2}[\tau^2 - (a + R)^2]H[\tau - (a + R)]\Bigg\rangle H(a - R) \\ &+ \Bigg\langle\frac{1}{2}[\tau^2 - (R - a)^2]H[\tau - (R - a)] \\ &- \frac{1}{2}[\tau^2 - (R + a)^2]H[\tau - (R + a)]\Bigg\rangle H(R - a)\Bigg\}\end{aligned} \tag{11.2.148}$$

* The integral representation 11.2.146 involving an arbitrary distribution of temperature $\hat{T} = \hat{T}(R, \tau)$ was obtained apparently for the first time [8].

Hence, by differentiating Equation 11.2.148 with respect to R we find

$$\hat{u}_R(R,\tau) = \frac{\partial \hat{\phi}}{\partial R} = -\frac{T_0}{4R^2}(\tau^2 + R^2 - a^2) \times \left\{ \Big\langle H[\tau - (a-R)] - H[\tau - (a+R)] \Big\rangle H(a-R) + \Big\langle H[\tau - (R-a)] - H[\tau - (a+R)] \Big\rangle H(R-a) \right\} \tag{11.2.149}$$

Also, by differentiating Equation 11.2.148 twice with respect to τ we have

$$\ddot{\hat{\phi}}(R,\tau) = \frac{T_0}{2R} \left\{ \Big\langle H[\tau - (a-R)] - H[\tau - (a+R)] + \tau\delta[\tau - (a-R)] - \tau\delta[\tau - (a+R)] \Big\rangle H(a-R) + \Big\langle H[\tau - (R-a)] - H[\tau - (R+a)] + \tau\delta[\tau - (R-a)] - \tau\delta[\tau - (R+a)] \Big\rangle H(R-a) \right\} \tag{11.2.150}$$

Substituting $\hat{u}_R$ and $\ddot{\hat{\phi}}$ from Equations 11.2.149 and 11.2.150, respectively, into Equation 11.2.129 we find

$$\hat{S}_{RR}(R,\tau) = \frac{T_0}{2R} \left\{ \left\langle \left[1 + \frac{1-2\nu}{1-\nu} \frac{\tau^2 + R^2 - a^2}{R^2} \right] \times [H(\tau - (a-R)) - H(\tau - (a+R))] + \tau[\delta(\tau - (a-R)) - \delta(\tau - (a+R))] \right\rangle H(a-R) + \left\langle \left[1 + \frac{1-2\nu}{1-\nu} \frac{\tau^2 + R^2 - a^2}{R^2} \right] \times [H(\tau - (R-a)) - H(\tau - (R+a))] + \tau[\delta(\tau - (R-a)) - \delta(\tau - (R+a))] \right\rangle H(R-a) \right\} \tag{11.2.151}$$

Also, by substituting $\hat{T}$, $\hat{u}_R$, and $\ddot{\hat{\phi}}$ from Equations 11.2.125, 11.2.149, and 11.2.150, respectively, into Equation 11.2.130 we get

$$\hat{S}_{\varphi\varphi}(R,\tau) = \hat{S}_{\theta\theta}(R,\tau) = \frac{T_0}{2R}\left\{\left\langle\left[\frac{\nu}{1-\nu} - \frac{1-2\nu}{2(1-\nu)}\frac{\tau^2+R^2-a^2}{R^2}\right]\right.\right.$$
$$\times\,[H(\tau-(a-R)) - H(\tau-(a+R))]$$
$$\left.+\frac{\nu}{1-\nu}\tau[\delta(\tau-(a-R)) - \delta(\tau-(a+R))]\right\rangle H(a-R)$$
$$+\left\langle\left[\frac{\nu}{1-\nu} - \frac{1-2\nu}{2(1-\nu)}\frac{\tau^2+R^2-a^2}{R^2}\right]\right.$$
$$\times\,[H(\tau-(R-a)) - H(\tau-(R+a))]$$
$$\left.+\frac{\nu}{1-\nu}\tau[\delta(\tau-(R-a)) - \delta(\tau-(R+a))]\right\rangle H(R-a)$$
$$\left.-\frac{2(1-2\nu)}{1-\nu}R\delta(\tau)H(a-R)\right\} \qquad (11.2.152)$$

Finally, using Equations 11.2.125 and 11.2.150, we arrive at the volume change formula

$$\mathrm{tr}\,\hat{\mathbf{E}} = \nabla^2\hat{\phi} = \hat{T} + \ddot{\hat{\phi}} = T_0\left\{\delta(\tau) + \tau\frac{\delta[\tau-(a-R)] - \delta[\tau-(a+R)]}{2R}\right.$$
$$\left.+\frac{H[\tau-(a-R)] - H[\tau-(a+R)]}{2R}\right\}H(a-R)$$
$$+T_0\left\{\tau\frac{\delta[\tau-(R-a)] - \delta[\tau-(R+a)]}{2R}\right.$$
$$\left.+\frac{H[\tau-(R-a)] - H[\tau-(R+a)]}{2R}\right\}H(R-a) \qquad (11.2.153)$$

11.2.3.1 Analysis of the Solution

It follows from Equations 11.2.149 through 11.2.153 that the thermoelastic displacement potential $\hat{\phi} = \hat{\phi}(R,\tau)$ as well as the displacement $\hat{u}_R = \hat{u}_R(R,\tau)$ and the stress components $\hat{S}_{RR} = \hat{S}_{RR}(R,\tau)$ and $\hat{S}_{\theta\theta} = \hat{S}_{\theta\theta}(R,\tau)$ represent spherical waves with the wave fronts:

$$\tau-(a-R)=0 \quad \text{and } \tau-(a+R)=0 \quad \text{if } a>R$$

$$\tau-(R-a)=0 \quad \text{and } \tau-(R+a)=0 \quad \text{if } R>a$$

If we define the jump of a function $f = f(R,\tau)$ across a front $\tau - \tau(R) = 0$ by the formula (see Equation 11.2.84)

$$[\![f]\!](\tau(R)) = f(R,\tau(R)-0) - f(R,\tau(R)+0) \qquad (11.2.154)$$

where $\tau = \tau(R)$ is a prescribed positive function, then by taking the jump operation on Equation 11.2.148, we find

$$[\![\hat{\phi}]\!](a - R) = [\![\hat{\phi}]\!](a + R) = 0 \quad \text{for } a > R \tag{11.2.155}$$

and

$$[\![\hat{\phi}]\!](R - a) = [\![\hat{\phi}]\!](R + a) = 0 \quad \text{for } R > a \tag{11.2.156}$$

Therefore, $\hat{\phi} = \hat{\phi}(R, \tau)$ is a continuous function on the wave fronts; since $\hat{\phi}$ is continuous otherwise, then $\hat{\phi} = \hat{\phi}(R, \tau)$ is a continuous function for every $R \geq 0$ and $\tau \geq 0$.

To discuss the radial displacement $\hat{u}_R$ we consider two cases: (i) $a > R$ and (ii) $R > a$.

Case (i): $a > R$. In this case

$$\hat{u}_R(R, \tau) = 0 \quad \text{for } 0 < \tau < a - R \tag{11.2.157}$$

$$\hat{u}_R(R, \tau) = -\frac{T_0}{4R^2}(\tau^2 + R^2 - a^2) = -\frac{T_0}{4R^2}\left(\tau - \sqrt{a^2 - R^2}\right)\left(\tau + \sqrt{a^2 - R^2}\right)$$

$$\text{for } 0 < a - R < \tau < a + R \tag{11.2.158}$$

and

$$\hat{u}_R(R, \tau) = 0 \quad \text{for } \tau > a + R \tag{11.2.159}$$

Hence,

$$\hat{u}_R(R, a - R - 0) = 0 \tag{11.2.160}$$

$$\hat{u}_R(R, a - R + 0) = \frac{T_0}{2R}(a - R) \tag{11.2.161}$$

$$\hat{u}_R(R, a + R - 0) = -\frac{T_0}{2R}(a + R) \tag{11.2.162}$$

$$\hat{u}_R(R, a + R + 0) = 0 \tag{11.2.163}$$

and $\hat{u}_R = \hat{u}_R(R, \tau)$ exhibits the jumps on the wave fronts as shown in Figure 11.12.

Although $\hat{u}_R$ shown in Figure 11.12 is displayed for a particular set of input parameters ($T_0 = 4, a = 2, R = 1$), the graph reveals a general pattern of the displacement wave inside the spherical inclusion ($a > R$).

Case (ii): $R > a$. If follows from Equation 11.2.149 that in this case

$$\hat{u}_R(R, \tau) = 0 \quad \text{for } 0 < \tau < R - a \tag{11.2.164}$$

$$\hat{u}_R(R, \tau) = -\frac{T_0}{4R^2}(\tau^2 + R^2 - a^2) < 0 \quad \text{for } R - a < \tau < R + a \tag{11.2.165}$$

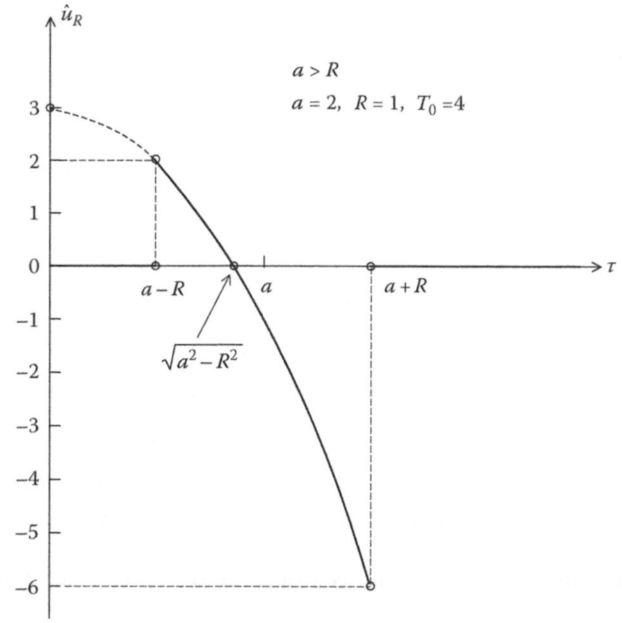

FIGURE 11.12 The radial displacement $\hat{u}_R$ as a function of τ for a fixed R inside the spherical temperature inclusion.

and

$$\hat{u}_R(R, \tau) = 0 \quad \text{for } \tau > R + a \tag{11.2.166}$$

In addition,

$$\hat{u}_R(R, R - a + 0) = -\frac{T_0}{2R}(R - a) \tag{11.2.167}$$

and

$$\hat{u}_R(R, R + a - 0) = -\frac{T_0}{2R}(R + a) \tag{11.2.168}$$

By letting $T_0 = 4$, $a = 2$; $R = 3$ in Equation 11.2.165 we write

$$\hat{u}_R(3, \tau) = -\frac{1}{9}(\tau^2 + 5) \quad \text{for } 1 < \tau < 5 \tag{11.2.169}$$

Plot of $\hat{u}_R(3, \tau)$ for $\tau \geq 0$ is shown in Figure 11.13. Note that in contrast to the interior of the spherical region ($a < R$) in which the displacement $\hat{u}_R$ changes sign from positive to negative over a finite time interval, the displacement $\hat{u}_R$ is strictly negative and decreasing over a finite time interval outside the spherical region ($R > a$).

To plot the stress component $\hat{S}_{RR} = \hat{S}_{RR}(R, \tau)$ as a function of time τ, consider cases (i) and (ii) separately.

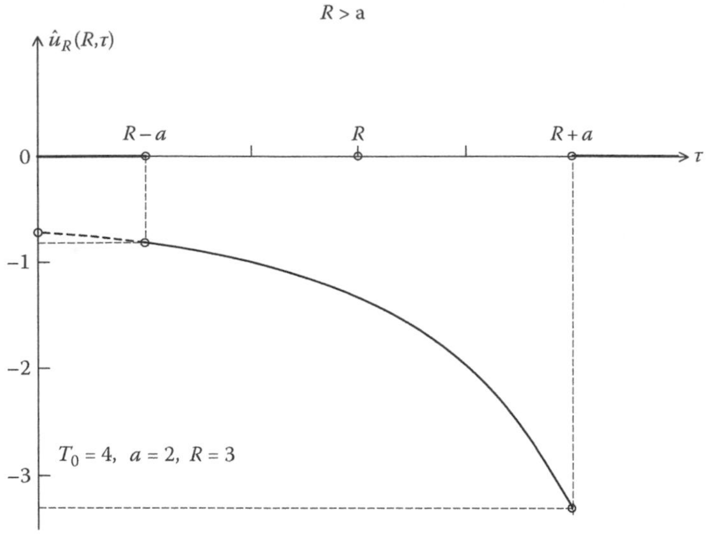

FIGURE 11.13 The radial displacement $\hat{u}_R$ as a function of τ for a fixed R outside the spherical inclusion ($R > a$).

Case (i): $a > R$. In this case Equation 11.2.151 implies that

$$\hat{S}_{RR}(R, \tau) = 0 \quad \text{for } 0 < \tau < a - R \tag{11.2.170}$$

$$\hat{S}_{RR}(R, \tau) = \frac{T_0}{2R}\left(1 + \frac{1-2\nu}{1-\nu}\frac{\tau^2 + R^2 - a^2}{R^2}\right)$$

$$\text{for } a - R < \tau < a + R \tag{11.2.171}$$

$$\hat{S}_{RR}(R, \tau) = 0 \quad \text{for } \tau > R + a \tag{11.2.172}$$

Note that the Dirac delta functions with supports on the wave fronts $\tau = a - R$ and $\tau = a + R$ are not included in Equations 11.2.170 through 11.2.172; and the finite interval $a - R < \tau < a + R$ represents a support of $\hat{S}_{RR}$ on the semi-infinite time interval $\tau \geq 0$. By letting $T_0 = 4$, $a = 2$, and $R = 1$, and $\nu = 1/4$ in Equation 11.2.171 we get

$$\hat{S}_{RR}(1, \tau) = \frac{4}{3}\left(\tau^2 - \frac{3}{2}\right) \quad \text{for } 1 < \tau < 3 \tag{11.2.173}$$

A plot of the function $\hat{S}_{RR}(1, \tau)$ for $\tau \geq 0$ is shown in Figure 11.14

Figure 11.14 shows that for a particular set of input parameters the stress wave $\hat{S}_{RR}(R, \tau)$ has a support over the interval $a - R < \tau < a + R$, it becomes compressive for $a - R < \tau < \tau_0 < a + R$ and expansive for $\tau_0 < \tau < a + R$. It follows from Equations 11.2.170 through 11.2.172 that this type of behavior of the stress wave remains valid provided

$$\frac{a^2}{R^2} > 1 + \frac{1-\nu}{1-2\nu} \tag{11.2.174}$$

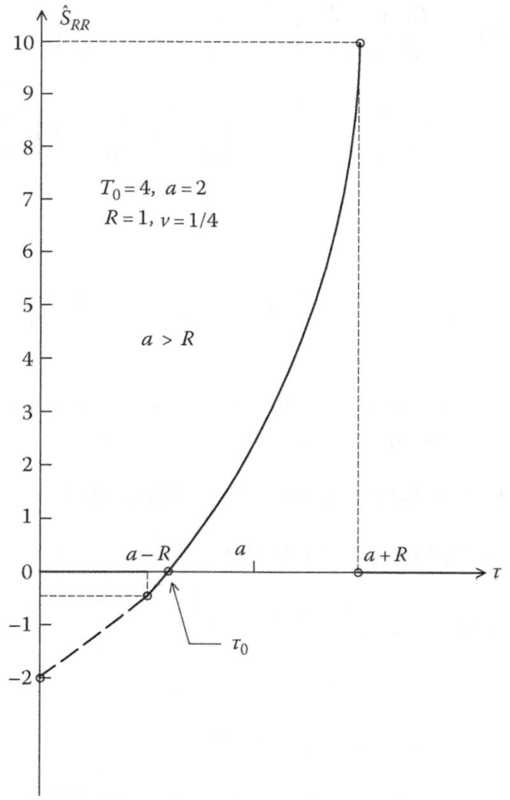

FIGURE 11.14 The radial stress $\hat{S}_{RR}(R, \tau)$ on τ-axis for a fixed R inside the spherical inclusion $(a > R)$. The Dirac delta contributions are not shown in the figure.

In this case τ_0 defined by

$$\tau_0 = \sqrt{a^2 - R^2 - R^2 \frac{1-\nu}{1-2\nu}} \tag{11.2.175}$$

is positive and

$$\hat{S}_{RR}(R, \tau_0) = 0 \tag{11.2.176}$$

If the ratio a/R is restricted by the inequality

$$1 < \frac{a}{R} < \sqrt{1 + \frac{1-\nu}{1-2\nu}} \tag{11.2.177}$$

the radial stress $\hat{S}_R(R, \tau)$ represents an expansive spherical wave over its whole support interval $a - R < \tau < a + R$. In this case

$$\hat{S}_{RR}(R, a-R+0) = \frac{T_0}{2R}\frac{1-2\nu}{1-\nu}\left(\frac{1-\nu}{1-2\nu}+2-2\frac{a}{R}\right)$$
$$> \frac{T_0}{2R}\frac{1-2\nu}{1-\nu}\left[1+\left(\sqrt{\frac{1-\nu}{1-2\nu}}-1\right)^2\right] > 0 \qquad (11.2.178)$$

and

$$\hat{S}_{RR}(R, a+R-0) = \frac{T_0}{2R}\frac{1-2\nu}{1-\nu}\left(\frac{1-\nu}{1-2\nu}+2\frac{R+a}{R}\right) > 0 \qquad (11.2.179)$$

Also, if the inequality 11.2.177 holds true, the radial stress wave is represented by a concave parabolic curve restricted to the interval $a-R<\tau<a+R$.

Case (ii): $R>a$. In this case Equation 11.2.151 implies that

$$\hat{S}_{RR}(R,\tau) = 0 \quad \text{for } 0<\tau<R-a \qquad (11.2.180)$$

$$\hat{S}_{RR}(R,\tau) = \frac{T_0}{2R}\left(1+\frac{1-2\nu}{1-\nu}\frac{\tau^2+R^2-a^2}{R^2}\right)$$
$$\text{for } R-a<\tau<R+a \qquad (11.2.181)$$

$$\hat{S}_{RR}(R,\tau) = 0 \quad \text{for } \tau>R+a \qquad (11.2.182)$$

Letting $T_0=4$, $a=2$, $R=3$, and $\nu=1/4$ in Equation 11.2.181 yields

$$\hat{S}_{RR}(3,\tau) = \frac{2}{3}\left(1+\frac{2}{3}\frac{\tau^2+5}{9}\right) \quad \text{for } 1<\tau<5 \qquad (11.2.183)$$

A plot of $\hat{S}_{RR}(3,\tau)$ for $\tau\geq 0$ is shown in Figure 11.15.

Figure 11.15 shows that $\hat{S}_{RR}(R,\tau)$ for $R>a$ is switched on at $\tau=R-a+0$ to a compressive wave, the amplitude of which increases until $\tau=R+a-0$, and jumps to a zero level for $\tau>R+a$.

A discussion on the hoop stress wave $\hat{S}_{\theta\theta}=\hat{S}_{\theta\theta}(R,\tau)$ is similar to that of the radial stress wave, so it is omitted here.

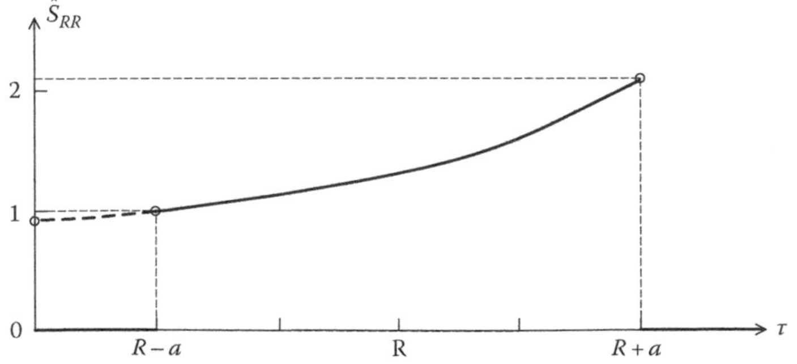

FIGURE 11.15 The radial stress $\hat{S}_{RR}(R,\tau)$ on τ-axis for a fixed R outside the spherical inclusion ($R>a$). The effects due to the Dirac delta function are not shown in the figure.

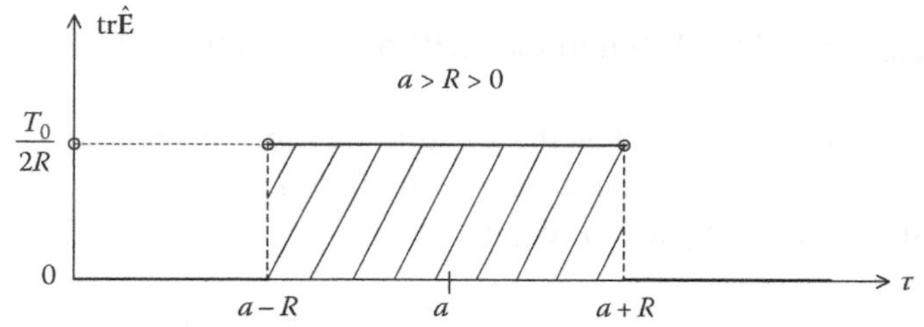

FIGURE 11.16 The volume change $\text{tr}\,\hat{\mathbf{E}}$ as a function of time τ for a fixed R inside the temperature inclusion.

To conclude this section, we discuss briefly the volume change of the body subject to the dynamic spherical temperature inclusion. The volume change $\text{tr}\,\hat{\mathbf{E}}$ restricted to the internal points of the temperature inclusion ($a > R$) is plotted in Figure 11.16 in which effects due to the Dirac delta function are not shown (see Equation 11.2.153).

The volume change shown in Figure 11.16 holds true if $R > 0$. For $R = 0$ the simple formula has the form

$$\text{tr}\,\hat{\mathbf{E}}(0,\tau) = T_0[\delta(\tau) + a\dot{\delta}(\tau - a)] \tag{11.2.184}$$

To prove this equation we note that Equation 11.2.153 restricted to $0 < R < a$ yields

$$\text{tr}\,\hat{\mathbf{E}}(R,\tau) = T_0\left[\delta(\tau) + \tau\frac{\delta[\tau-(a-R)] - \delta[\tau-(a+R)]}{2R} + \frac{H[\tau-(a-R)] - H[\tau-(a+R)]}{2R}\right] \tag{11.2.185}$$

Since

$$\frac{\delta[\tau-(a-R)] - \delta[\tau-(a+R)]}{2R} \to \dot{\delta}(\tau-a) \quad \text{as } R \to 0 \tag{11.2.186}$$

and

$$\frac{H[\tau-(a-R)] - H[\tau-(a+R)]}{2R} \to \delta(\tau-a) \quad \text{as } R \to 0 \tag{11.2.187}$$

therefore, by letting $R \to 0$ in Equation 11.2.185 we get

$$\text{tr}\,\hat{\mathbf{E}}(R,\tau) \to T_0[\delta(\tau) + \tau\dot{\delta}(\tau-a) + \delta(\tau-a)] \quad \text{as } R \to 0 \tag{11.2.188}$$

Since

$$(\tau - a)\delta(\tau - a) = 0 \tag{11.2.189}$$

hence, by differentiating Equation 11.2.189 with respect to τ we have

$$\tau\dot{\delta}(\tau-a) + \delta(\tau-a) = a\dot{\delta}(\tau-a) \tag{11.2.190}$$

By substituting Equation 11.2.190 into the RHS of the relation 11.2.188 we arrive at the result 11.2.184.

The formula 11.2.184 implies that the volume change of an infinite body due to an instantaneous spherical temperature inclusion, calculated at the center of the inclusion, is a sum of the diffusive and wave parts, as expected.

11.3 SAINT-VENANT'S PRINCIPLE OF ELASTODYNAMICS IN TERMS OF STRESSES

In this section, a pure stress initial–boundary value problem for a nonhomogeneous anisotropic semi-infinite elastic cylinder loaded on its end face is formulated, and a decay estimate for the stress energy $U = U(l, t)$ of a portion of the cylinder beyond a distance $l > 0$ from the load region stored over the time interval $[0, t]$ is obtained. The result represents a dynamic counterpart of the classic static Saint-Venant's principle for a semi-infinite cylinder due to Toupin [9] for $t > l/c$, where c is a velocity expressed in terms of the lower bounds on the density and compliance fields of the cylinder. The stress energy $U(l, t)$ is shown to vanish for $t < l/c$.

11.3.1 SAINT-VENANT'S PRINCIPLE OF ELASTODYNAMICS FOR A SEMI-INFINITE CYLINDER

Let B denote a semi-infinite nonhomogeneous anisotropic elastic cylinder with end face C_0 as shown in Figure 11.17 and let C_l be the intersection with B of a plane perpendicular to an axis $\mathbf{k}$ of the cylinder and a distance l from C_0. Suppose that we are looking for a solution to the following pure stress initial–boundary value problem. Find a stress field $\mathbf{S} = \mathbf{S}(\mathbf{x}, t)$ on $\overline{B} \times [0, \infty)$ that satisfies the field equation (see Equation 4.2.9 with $\mathbf{B} = \mathbf{0}$)

$$\hat{\nabla}\left(\rho^{-1}\operatorname{div}\mathbf{S}\right) - \mathsf{K}[\ddot{\mathbf{S}}] = \mathbf{0} \quad \text{on } B \times [0, \infty) \tag{11.3.1}$$

the initial conditions

$$\mathbf{S}(\mathbf{x}, 0) = \mathbf{0}, \quad \dot{\mathbf{S}}(\mathbf{x}, 0) = \mathbf{0} \quad \text{for } \mathbf{x} \in B \tag{11.3.2}$$

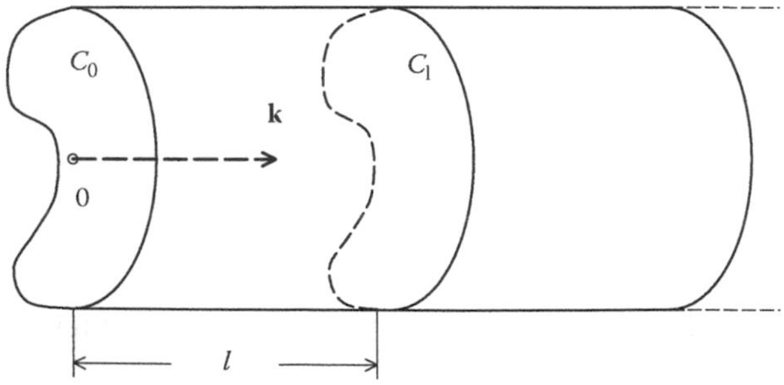

FIGURE 11.17 Configuration of a semi-infinite cylinder.

the boundary conditions

$$\mathbf{Sn} = \mathbf{0} \quad \text{on } (\partial B - C_0) \times [0, \infty) \tag{11.3.3}$$

$$\mathbf{Sn} = \mathbf{p} \neq \mathbf{0} \quad \text{on } C_0 \times [0, \infty) \tag{11.3.4}$$

and the asymptotic conditions

$$\int_{C_l} \mathbf{Sn}\, da \to \mathbf{0} \quad \text{as } l \to \infty \quad \text{for } t \geq 0 \tag{11.3.5}$$

$$\int_{C_l} \mathbf{x} \times (\mathbf{Sn})\, da \to \mathbf{0} \quad \text{as } l \to \infty \quad \text{for } t \geq 0 \tag{11.3.6}$$

$$\int_{C_l} \rho^{-1}(\operatorname{div} \mathbf{S}) \cdot (\dot{\mathbf{S}}\mathbf{n})\, da \to 0 \quad \text{as } l \to \infty \quad \text{for } t \geq 0 \tag{11.3.7}$$

Here, $\rho = \rho(\mathbf{x})$ and $\mathsf{K} = \mathsf{K}(\mathbf{x})$ are the density and compliance fourth-order tensor fields, respectively; $\mathbf{n}$ stands for the unit outer vector normal to a surface, and $\mathbf{p} = \mathbf{p}(\mathbf{x}, t)$ is a prescribed load. The functions $\rho = \rho(\mathbf{x})$ and $\mathsf{K} = \mathsf{K}(\mathbf{x})$ satisfy the inequalities

$$0 < \rho_m \leq \rho(\mathbf{x}) \leq \rho_M < \infty \quad \text{on } \overline{B} \tag{11.3.8}$$

and

$$0 < k_m |\mathbf{A}|^2 \leq \mathbf{A} \cdot \mathsf{K}[\mathbf{A}] \leq k_M |\mathbf{A}|^2 < \infty \quad \text{on } \overline{B}$$

$$\text{and for every} \quad \mathbf{A} = \mathbf{A}^T \neq \mathbf{0} \tag{11.3.9}$$

where ρ_m and ρ_M denote the minimum and maximum of $\rho = \rho(\mathbf{x})$ on $\overline{B}$, respectively; while k_m and k_M denote the minimum and maximum of $\mathsf{K} = \mathsf{K}(\mathbf{x})$ on $\overline{B}$, respectively.

Let $B(l)$ denote the semi-infinite cylinder defined by

$$B(l) = \{\mathbf{x} \in \overline{B} : \mathbf{x} \cdot \mathbf{k} > l\} \quad l \geq 0 \tag{11.3.10}$$

i.e., $B(l)$ is a semi-infinite cylinder with the end face C_l, see Figure 11.17. Let $\mathbf{u} = \mathbf{u}(\mathbf{x}, t)$ be the displacement field satisfying the equation of motion (see Equation 4.2.2 with $\mathbf{b} = \mathbf{0}$)

$$\operatorname{div} \mathbf{S} = \rho \ddot{\mathbf{u}} \quad \text{on } \overline{B} \times [0, \infty) \tag{11.3.11}$$

By integrating Equation 11.3.11 over the finite cylinder $B - B(l)$, using the divergence theorem, and the boundary conditions 11.3.3 and 11.3.4, we get

$$\int_{C_0} \mathbf{p}\, da + \int_{C_l} \mathbf{Sn}\, da = \int_{B-B(l)} \rho \ddot{\mathbf{u}}\, dv \tag{11.3.12}$$

Next, by letting $l \to \infty$ in Equation 11.3.12, and taking into account the asymptotic condition 11.3.5 as well as the fact that $B - B(l) \to B$ as $l \to \infty$, we obtain

$$\int_{C_0} \mathbf{p}\, da = \int_{B} \rho \ddot{\mathbf{u}}\, dv \quad \text{for } t \geq 0 \tag{11.3.13}$$

Similarly, by multiplying Equation 11.3.11 by $\mathbf{x}$ in the sense of vector product, integrating the result over $B - B(l)$, using the divergence theorem, the symmetry of $\mathbf{S}$, and letting $l \to \infty$, on account of Equations 11.3.3 and 11.3.4 and 11.3.6, we get

$$\int_{C_0} \mathbf{x} \times \mathbf{p}\, da = \int_{B} \rho \mathbf{x} \times \ddot{\mathbf{u}}\, dv \quad \text{for } t \geq 0 \tag{11.3.14}$$

Therefore, Equations 11.3.13 and 11.3.14 are the necessary conditions for the existence of a solution to the problem 11.3.1 through 11.3.7. They assert that the resultant force and resultant moment applied to the end face C_0 are equilibrated by the total acceleration force and total acceleration moment, respectively, for every $t \geq 0$. Note that these global dynamic equilibrium conditions are obtained under the asymptotic hypotheses 11.3.5 and 11.3.6.

We are now in a position to formulate and prove the dynamic version of Saint-Venant's principle for a semi-infinite cylinder.

Theorem: Let $\mathbf{S} = \mathbf{S}(\mathbf{x}, t)$ be a solution to the problem described by Equations 11.3.1 through 11.3.7. Define the total stress energy associated with the solution and stored in the semi-infinite cylinder $B(l)$ over the time interval $[0, t]$ by

$$U(l, t) = \frac{1}{2} \int_0^t \int_{B(l)} \{\rho^{-1} (\operatorname{div} \mathbf{S}) \cdot (\operatorname{div} \mathbf{S}) + \dot{\mathbf{S}} \cdot \mathsf{K}[\dot{\mathbf{S}}]\}\, dv\, d\tau \tag{11.3.15}$$

Then, there exists a positive constant c of the velocity dimension such that for any $l > 0$ the following estimates hold true:

$$U(l, t) = 0 \quad \text{for } 0 \leq t < \frac{l}{c} \tag{11.3.16}$$

and

$$U(l, t) \leq U(0, t) \exp\left(-\frac{l}{ct}\right) \quad \text{for } t > \frac{l}{c} \geq 0 \tag{11.3.17}$$

■

Proof: First, we show that if $\mathbf{S}$ is a solution to the problem, then the function $U = U(l, t)$ defined by Equation 11.3.15 may be represented in the form

$$U(l, t) = \int_0^t \int_{C_l} \left[(t - \tau) \rho^{-1} (\operatorname{div} \mathbf{S}) \cdot (\dot{\mathbf{S}}\mathbf{n})\right] da\, d\tau \tag{11.3.18}$$

To this end we rewrite Equation 11.3.1 in components and have

$$(\rho^{-1} S_{(ik,k)})_{,j)} - K_{ijkl} \ddot{S}_{kl} = 0 \tag{11.3.19}$$

By multiplying Equation 11.3.19 by $\dot{S}_{ij}$ and using the symmetry of S_{ij} and K_{ijkl}, we get

$$(\rho^{-1}S_{ik,k}\dot{S}_{ij})_{,j} - \rho^{-1}\dot{S}_{ij,j}S_{ik,k} - \frac{1}{2}\frac{\partial}{\partial t}(\dot{S}_{ij}K_{ijkl}\dot{S}_{kl}) = 0 \tag{11.3.20}$$

Let $h > 0$. Integrating Equation 11.3.20 over the finite cylinder $B(l) - B(l+h)$, using the divergence theorem, and taking into account the fact that the lateral surface of the cylinder is stress free, we arrive at

$$\int_{C_l} \rho^{-1}(\operatorname{div}\mathbf{S}) \cdot (\dot{\mathbf{S}}\mathbf{n})\, da + \int_{C_{l+h}} \rho^{-1}(\operatorname{div}\mathbf{S}) \cdot (\dot{\mathbf{S}}\mathbf{n})\, da = \ddot{U}(l,t) - \ddot{U}(l+h,t) \tag{11.3.21}$$

where $U = U(l,t)$ is defined by Equation 11.3.15.

Next, by letting $h \to \infty$ in Equation 11.3.21, using the asymptotic condition 11.3.7 as well as the fact that $B(l) - B(l+h) \to B(l)$ as $h \to \infty$, we get

$$\ddot{U}(l,t) = \int_{C_l} \rho^{-1}(\operatorname{div}\mathbf{S}) \cdot (\dot{\mathbf{S}}\mathbf{n})\, da \tag{11.3.22}$$

Finally, integrating Equation 11.3.22 twice with respect to time, and using the homogeneous initial conditions 11.3.2, we arrive at Equation 11.3.18.

Also, note that the function $U = U(l,t)$ given by Equation 11.3.15 may be written as

$$U(l,t) = \frac{1}{2}\int_0^t \int_l^\infty \int_{C_l} \{\rho^{-1}(\operatorname{div}\mathbf{S}) \cdot (\operatorname{div}\mathbf{S}) + \dot{\mathbf{S}} \cdot \mathsf{K}[\dot{\mathbf{S}}]\}\, da\, dx_3\, d\tau \tag{11.3.23}$$

where x_3 is the integration variable on the axis of a cylinder spanned by the unit vector $\mathbf{k}$ (see Figure 11.17); and, by differentiating Equation 11.3.23 with respect to l, we obtain

$$\frac{\partial}{\partial l}U(l,t) = -\frac{1}{2}\int_0^t \int_{C_l} \{\rho^{-1}(\operatorname{div}\mathbf{S}) \cdot (\operatorname{div}\mathbf{S}) + \dot{\mathbf{S}} \cdot \mathsf{K}[\dot{\mathbf{S}}]\}\, da\, d\tau \tag{11.3.24}$$

Next, we note that for any two functions $f = f(\mathbf{x},t)$ and $g = g(\mathbf{x},t)$ defined on the Cartesian product $C_l \times [0,t]$ and integrable on $C_l \times [0,t]$ the Schwartz inequality is satisfied

$$\int_0^t \int_{C_l} f \cdot g\, da\, d\tau \le \left(\int_0^t \int_{C_l} f^2\, da\, d\tau\right)^{1/2} \left(\int_0^t \int_{C_l} g^2\, da\, d\tau\right)^{1/2} \tag{11.3.25}$$

By letting $f = \rho^{-\frac{1}{2}}(\operatorname{div}\mathbf{S})$ and $g = \rho^{-\frac{1}{2}}(\dot{\mathbf{S}}\mathbf{n})$ in the inequality 11.3.25 we get

$$\int_0^t \int_{C_l} \rho^{-1}(\operatorname{div}\mathbf{S}) \cdot (\dot{\mathbf{S}}\mathbf{n})\, da\, d\tau \le \left(\int_0^t \int_{C_l} \rho^{-1}(\operatorname{div}\mathbf{S})^2\, da\, d\tau\right)^{1/2}$$

$$\times \left(\int_0^t \int_{C_l} \rho^{-1}(\dot{\mathbf{S}}\mathbf{n})^2\, da\, d\tau\right)^{1/2} \tag{11.3.26}$$

Hence, it follows from Equation 11.3.18 and from the inequality $t - \tau \leq t$ that

$$U(l,t) \leq t \left(\int_0^t \int_{C_l} \rho^{-1} (\operatorname{div} \mathbf{S})^2 \, da \, d\tau \right)^{1/2} \left(\int_0^t \int_{C_l} \rho^{-1} (\dot{\mathbf{S}}\mathbf{n})^2 \, da \, d\tau \right)^{1/2} \tag{11.3.27}$$

Applying the inequality

$$(AB)^{1/2} \leq \frac{1}{2}(\epsilon A + \epsilon^{-1} B) \tag{11.3.28}$$

where A, B, and ϵ are nonnegative scalars with $\epsilon > 0$, to the inequality 11.3.27, we arrive at

$$U(l,t) \leq \frac{t}{2} \left[\epsilon^{-1} \int_0^t \int_{C_l} \rho^{-1} (\operatorname{div} \mathbf{S})^2 \, da \, d\tau + \epsilon \int_0^t \int_{C_l} \rho^{-1} (\dot{\mathbf{S}}\mathbf{n})^2 \, da \, d\tau \right] \tag{11.3.29}$$

Now, by the hypothesis 11.3.8 and 11.3.9, we arrive at the estimates

$$\frac{1}{\rho} \leq \frac{1}{\rho_m} \quad \text{on } C_l \tag{11.3.30}$$

and

$$|\dot{\mathbf{S}}\mathbf{n}|^2 \leq |\dot{\mathbf{S}}|^2 \leq \frac{1}{k_m} \, \dot{\mathbf{S}} \cdot \mathbf{K}[\dot{\mathbf{S}}] \quad \text{on } C_l \times [0,t] \tag{11.3.31}$$

Hence, from Equation 11.3.29 follows

$$U(l,t) \leq \frac{t}{2} \left\{ \frac{\epsilon}{\rho_m k_m} \int_0^t \int_{C_l} \dot{\mathbf{S}} \cdot \mathbf{K}[\dot{\mathbf{S}}] \, da \, d\tau + \epsilon^{-1} \int_0^t \int_{C_l} \rho^{-1} (\operatorname{div} \mathbf{S})^2 \, da \, d\tau \right\}$$

$$\leq \frac{t}{2} \gamma(\epsilon) \int_0^t \int_{C_l} \{ \rho^{-1} (\operatorname{div} \mathbf{S})^2 + \dot{\mathbf{S}} \cdot \mathbf{K}[\dot{\mathbf{S}}] \} \, da \, d\tau \tag{11.3.32}$$

where

$$\gamma(\epsilon) = \epsilon^{-1} + \epsilon k_m^{-1} \rho_m^{-1} \tag{11.3.33}$$

The function $\gamma = \gamma(\epsilon)$, treated as a function of ϵ on $(0, \infty)$, attains a minimum at $\epsilon = \epsilon_0 = (\rho_m k_m)^{1/2}$. Therefore, letting $\epsilon = \epsilon_0$ in Equation 11.3.32, and introducing the notation

$$c = \gamma(\epsilon_0) = \frac{2}{\sqrt{\rho_m k_m}} \tag{11.3.34}$$

we arrive at the estimate

$$U(l,t) \leq \frac{1}{2} ct \int_0^t \int_{C_l} \{\rho^{-1}(\operatorname{div}\mathbf{S})^2 + \dot{\mathbf{S}} \cdot \mathbf{K}[\dot{\mathbf{S}}]\}\, da\, d\tau \tag{11.3.35}$$

It follows from the definition of c (see Equation 11.3.34) that c has dimension of a velocity. Also, differentiating Equation 11.3.18 with respect to t yields

$$\frac{\partial}{\partial t} U(l,t) = \int_0^t \int_{C_l} \rho^{-1}(\operatorname{div}\mathbf{S}) \cdot (\dot{\mathbf{S}}\mathbf{n})\, da\, d\tau \tag{11.3.36}$$

and, by virtue of the inequality 11.3.26, we get

$$\frac{\partial}{\partial t} U(l,t) \leq \left[\int_0^t \int_{C_l} \rho^{-1}(\operatorname{div}\mathbf{S})^2\, da\, d\tau \right]^{1/2} \left[\int_0^t \int_{C_l} \rho^{-1}(\dot{\mathbf{S}}\mathbf{n})^2\, da\, d\tau \right]^{1/2} \tag{11.3.37}$$

By estimating the RHS of the inequality 11.3.37 in a way similar to that leading from Equation 11.3.27 to Equation 11.3.35, we arrive at

$$\frac{\partial}{\partial t} U(l,t) \leq \frac{1}{2} c \int_0^t \int_{C_l} \{\rho^{-1}(\operatorname{div}\mathbf{S})^2 + \dot{\mathbf{S}} \cdot \mathbf{K}[\dot{\mathbf{S}}]\}\, da\, d\tau \tag{11.3.38}$$

Hence, applying Equation 11.3.24 leads to

$$\frac{1}{c} \frac{\partial}{\partial t} U(l,t) + \frac{\partial}{\partial l} U(l,t) \leq 0 \tag{11.3.39}$$

Also, because of Equations 11.3.24 and 11.3.35, we may write

$$\frac{1}{ct} U(l,t) + \frac{\partial}{\partial l} U(l,t) \leq 0 \tag{11.3.40}$$

Next, by integrating the inequality 11.3.39 along the line $l = ct$ in the plane (l,t) from $(0,0)$ to (ct,t) we find that

$$U(ct;t) \leq U(0,0) \tag{11.3.41}$$

On account of the homogeneous initial conditions, $U(0,0) = 0$. Moreover, it follows from Equation 11.3.24 that

$$\frac{\partial}{\partial l} U(l,t) \leq 0 \quad \text{for } t \geq 0 \tag{11.3.42}$$

Hence, $U(l,t)$ is a nonincreasing function of l, that is,

$$U(l,t) \leq U(ct,t) \quad \text{for } l > ct \geq 0 \tag{11.3.43}$$

Since $U(0,0) = 0$, the inequalities 11.3.41 and 11.3.43 imply that

$$0 \leq U(l,t) \leq 0 \quad \text{for } l > ct \geq 0 \tag{11.3.44}$$

This completes the proof of the first part of the theorem (see Equation 11.3.16) that is a kind of a domain of influence result for the pure stress initial–boundary value problem [10].

To prove the spatial decay estimate 11.3.17 we restrict the time interval to the range $t > l/c \geq 0$, and multiply the inequality 11.3.40 by exp (l/ct), to obtain

$$\frac{\partial}{\partial l}\left[\exp\left(\frac{l}{ct}\right) U(l,t)\right] \leq 0 \quad \text{for } t > \frac{l}{c} \geq 0 \tag{11.3.45}$$

Integrating the inequality 11.3.45 with respect to l over the interval $[0, l]$, we arrive at

$$U(l,t) \leq U(0,t)\exp\left(-\frac{l}{ct}\right) \quad \text{for } t > \frac{l}{c} \geq 0 \tag{11.3.46}$$

This completes the proof of the theorem.

11.3.2 Concluding Remarks

- The dynamic version of Saint-Venant's principle for a nonhomogeneous anisotropic semi-infinite elastic cylinder asserts that a total stress energy $U = U(l,t)$ of a portion of the cylinder beyond a distance l from the loaded region, stored over the time interval $[0,t]$, vanishes for $0 \leq t < l/c$, and decays exponentially with increasing l for $t > l/c \geq 0$. Here c is a velocity expressed in terms of the lower bounds on the density and compliance tensor fields of the cylinder. The result is consistent with a domain of influence theorem for a pure stress initial–boundary value problem of linear elastodynamics, and for $t > l/c$ it is similar to the classic static Saint-Venant's principle for a semi-infinite cylinder in which an exponential decay of the elastic energy away from the end load is observed.
- A characteristic feature of the dynamic version of Saint-Venant's principle is that the time-dependent end load complies with the global dynamic equilibrium conditions: the resultant force and the resultant moment of the end tractions are equilibrated by the total acceleration force and the total acceleration moment of the cylinder, respectively, for every $t > 0$; in the static case the end load is time-independent and self-equilibrated.
- The dynamic version of Saint-Venant's principle is related to a pure stress initial–boundary value problem for a semi-infinite cylinder. A dynamic version of Saint-Venant's principle for a semi-infinite cylinder using the displacement language of elastodynamics may be formulated by the relations [11]

$$\hat{U}(l,t) = 0 \quad \text{for } t < \frac{l}{\hat{c}} \tag{11.3.47}$$

and

$$\hat{U}(l,t) \leq \hat{U}(0,t)\left(1 - \frac{l}{\hat{c}t}\right) \quad \text{for } t > \frac{l}{\hat{c}} \tag{11.3.48}$$

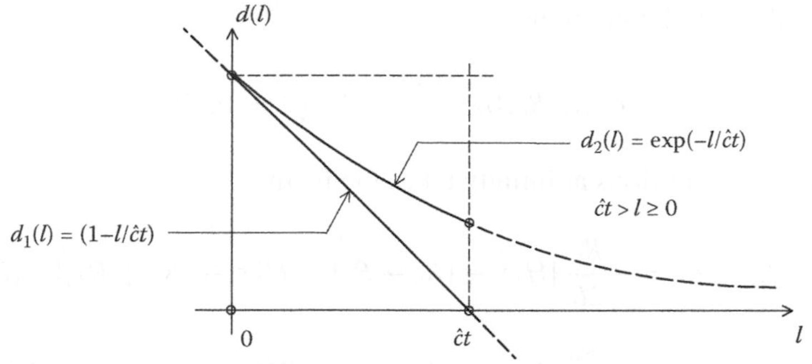

FIGURE 11.18 The decay functions $d_1(l)$ and $d_2(l)$ corresponding to the energy estimates 11.3.48 and 11.3.51, respectively.

Here, $\hat{U} = \hat{U}(l, t)$ is a total energy of $B(l)$ stored over the time interval $[0, t]$, expressed in terms of the displacement vector $\mathbf{u} = \mathbf{u}(\mathbf{x}, t)$; and $\hat{c}$ is a positive constant of the velocity dimension. Since

$$1 - \frac{l}{\hat{c}t} \leq \exp\left(-\frac{l}{\hat{c}t}\right) \quad \text{for } t > \frac{l}{\hat{c}} \geq 0 \tag{11.3.49}$$

and

$$\hat{U}(0, t) \geq 0 \quad \text{for } t > \frac{l}{\hat{c}} \geq 0 \tag{11.3.50}$$

therefore, the inequality 11.3.48 implies that $\hat{U} = \hat{U}(l, t)$ also satisfies the exponential decay estimate

$$\hat{U}(l, t) \leq \hat{U}(0, t) \exp\left(-\frac{l}{\hat{c}t}\right) \quad \text{for } t > \frac{l}{\hat{c}} \geq 0 \tag{11.3.51}$$

The decay functions for the estimates 11.3.48 and 11.3.51 are shown in Figure 11.18.

PROBLEMS

11.1 A dimensionless temperature field $\hat{T}^* = \hat{T}^*(R, R_0; \tau)$ is assumed in the form

$$\hat{T}^*(R, R_0; \tau) = \delta(R - R_0)\delta(\tau) \tag{a}$$

for any point of an infinite body referred to a spherical system of coordinates (R, φ, θ) $(R \geq 0, R_0 > 0, \tau \geq 0)$. Show that a solution $\hat{\phi}^* = \hat{\phi}^*(R, R_0; \tau)$ to the nonhomogeneous wave equation

$$\left(\nabla^2 - \frac{\partial^2}{\partial \tau^2}\right)\hat{\phi}^* = \hat{T}^* \quad \text{for } R \geq 0, \tau \geq 0 \tag{b}$$

subject to the initial conditions

$$\hat{\phi}^*(R, R_0; 0) = \dot{\hat{\phi}}^*(R, R_0; 0) = 0, \ R \geq 0 \tag{c}$$

and vanishing conditions at infinity takes the form

$$\hat{\phi}^*(R, R_0; \tau) = -\frac{R_0}{2R}\{H[\tau - (R_0 - R)] - H[\tau - (R_0 + R)]\}H(R_0 - R)$$
$$-\frac{R_0}{2R}\{H[\tau - (R - R_0)] - H[\tau - (R + R_0)]\}H(R - R_0) \tag{d}$$

for every $R > 0$, $R_0 > 0$, $\tau > 0$.

11.2 Compute the radial displacement $\hat{u}_R^* = \hat{u}_R^*(R, R_0; \tau)$, and the stress components $\hat{S}_{RR}^* = \hat{S}_{RR}^*(R, R_0; \tau)$ and $\hat{S}_{\theta\theta}^* = \hat{S}_{\theta\theta}^*(R, R_0; \tau)$ generated by the potential $\hat{\phi}^* = \hat{\phi}^*(R, R_0; \tau)$ of Problem 11.1.
Hint: Use Equations 11.2.128 through 11.2.130.

11.3 Show that if $\hat{T} = \hat{T}(R, \tau)$ satisfies a parabolic heat conduction equation for $R \geq 0$, $\tau > 0$, then the radial displacement $\hat{u}_R = \hat{u}_R(R, \tau)$ and the stresses $\hat{S}_{RR} = \hat{S}_{RR}(R, \tau)$ and $\hat{S}_{\theta\theta} = \hat{S}_{\theta\theta}(R, \tau)$ produced by the temperature $\hat{T}$ in an infinite body $R \geq 0$ and for $\tau \geq 0$ are generated by the potential $\hat{\phi} = \hat{\phi}(R, \tau)$ given by the double integral

$$\hat{\phi}(R, \tau) = \int_0^{\tau}\int_0^{\infty} \hat{\phi}^*(R, R_0; \tau - u)\hat{T}(R_0, u)\, dR_0\, du \tag{a}$$

where $\hat{\phi}^* = \hat{\phi}^*(R, R_0; \tau)$ is the thermoelastic displacement potential of Problem 11.1.

11.4 The Laplace transform of the temperature $\hat{T} = \hat{T}(R, \tau)$ due to an instantaneous concentrated source of heat in an infinite body $R \geq 0$ takes the form (see Equation 11.2.53)

$$\bar{\hat{T}}(R, p) = \frac{e^{-R\sqrt{p}}}{4\pi R} \tag{a}$$

where p is the Laplace transform parameter. By applying the Laplace transform to Equation (a) of Problem 11.3, show that

$$\bar{\hat{\phi}}(R, p) = \frac{1}{4\pi p(p-1)}\,\frac{e^{-Rp} - e^{-R\sqrt{p}}}{R} \tag{b}$$

where $\bar{\hat{\phi}} = \bar{\hat{\phi}}(R, p)$ is the Laplace transform of $\hat{\phi} = \hat{\phi}(R, \tau)$ associated with $\hat{T} = \hat{T}(R, \tau)$.

11.5 Find a thermoelastic displacement potential $\hat{\phi}^{**} = \hat{\phi}^{**}(R, R_0; \tau)$ for an infinite body with a stress free spherical cavity of radius $R = a$ corresponding to the temperature field

$$\hat{T}^{**}(R; R_0; \tau) = \delta(R - R_0)\delta(\tau) \quad \text{for } R \geq a,\ R_0 > a,\ \tau > 0 \tag{a}$$

11.6 The temperature field $\hat{T} = \hat{T}(R, \tau)$ produced by a sudden heating of a spherical cavity of radius $R = a$ in an infinite body $R \geq a$ to a constant temperature T_0 is given by the formula

$$\hat{T}(R,\tau) = \frac{T_0 a}{R} \operatorname{erfc}\left(\frac{R-a}{\sqrt{4\tau}}\right) \quad \text{for } R \geq a \quad \text{and } \tau \geq 0 \tag{a}$$

Show that the associated potential $\hat{\phi} = \hat{\phi}(R, \tau)$ that generates the displacement $\hat{u}_R = \hat{u}_R(R, \tau)$, and the stress components $\hat{S}_{RR} = \hat{S}_{RR}(R, \tau)$ and $\hat{S}_{\theta\theta} = \hat{S}_{\theta\theta}(R, \tau)$ in an infinite body $R \geq a$ subject to the stress free boundary condition

$$\hat{S}_{RR}(a, \tau) = 0 \quad \text{for } \tau > 0 \tag{b}$$

and zero stresses at infinity, admits the integral representation

$$\hat{\phi}(R,\tau) = T_0 a \int_a^{\infty} \int_0^{\tau} \operatorname{erfc}\left(\frac{R_0 - a}{\sqrt{4u}}\right) \frac{1}{R_0} \hat{\phi}^{**}(R, R_0; \tau - u) du\, dR_0 \tag{c}$$

where $\hat{\phi}^{**}$ is the potential obtained in Problem 11.5.

REFERENCES

1. J. Ignaczak and C. R. A. Rao, A tensorial classification of elastic waves, *J. Acoust. Soc. Am.*, 93(1), 1993, 17–21, Eqs. (2)–(5).
2. M. E. Gurtin, *The Linear Theory of Elasticity*, in Encyclopedia of Physics, chief editor: S. Flügge, vol. VIa/2, editor: C. Truesdell, Springer, Berlin, Germany, 1972, p. 247.
3. J. Ignaczak and C. R. A. Rao, A tensorial classification of elastic waves, *J. Acoust. Soc. Am.*, 93(1), 1993, 17–21.
4. A. C. Eringen and E. S. Suhubi, *Elastodynamics*, vol. II, *Linear Theory*, Academic Press, New York, 1975, p. 400, Eq. (5.10.30).
5. H. S. Carslaw and J. C. Jaeger, *Operational Methods in Applied Mathematics*, Dover Publications, New York, 1963, pp. 354–355.
6. N. Noda, R. B. Hetnarski and Y. Tanigawa, *Thermal Stresses*, 2nd edn., Taylor & Francis, New York, 2003, p. 104.
7. W. Nowacki, A dynamical problem of thermoelasticity, *Archiwum Mechaniki Stosowanej*, 9(3), 1957, 325–334.
8. W. Piechocki and J. Ignaczak, Certain problems of dynamic thermal distortion in thermoelasticity, *Archiwum Mechaniki Stosowanej*, 1960, 12(2), 1960, 259–278.
9. R. A. Toupin, Saint-Venant's principle, *Arch. Rat. Mech. Anal.*, 18, 1965, 83–96.
10. J. Ignaczak, Domain of influence theorem for stress equations of motion of linear elastodynamics, *Bull. Acad. Polon., Sci. Ser. Tech.*, 22, 1974, 465–470.
11. S. Chirita and R. Quintanilla, On Saint-Venant's principle in linear elastodynamics, *J. Elasticity*, 42, 1996, 201–215.

11 (a) The temperature field $T = T(R, t)$ produced by a sudden heating of a spherical cavity of radius R_0 in an infinite body $R \geq R_0$ to a constant temperature T_0 is given by the formula

[illegible]

[illegible]

12 Solutions to Particular Two-Dimensional Initial–Boundary Value Problems of Elastodynamics

In this chapter, both the displacement and stress languages of two-dimensional elastodynamics are used to consider a number of typical isothermal and nonisothermal elastic waves. The waves of isothermal elastodynamics include (a) two-dimensional plane progressive waves in a homogeneous isotropic infinite elastic body, (b) the surface waves propagating in a homogeneous isotropic elastic semispace, and (c) the stress waves produced by the initial stress and stress-rate fields in an infinite elastic body; while those of nonisothermal elastodynamics include (1) dynamic thermal stresses in an infinite elastic sheet subject to a discontinuous rectangular temperature field, and (2) dynamic thermal stresses produced by an instantaneous concentrated heat source in an infinite elastic sheet. The two-dimensional stress progressive waves, the stress waves due to the initial stress and stress-rate fields, as well as the dynamic thermal stresses due to an instantaneous concentrated heat source, are discussed here for the first time. A number of problems, with solutions provided in the Solutions Manual, conclude the chapter.

12.1 TWO-DIMENSIONAL SOLUTIONS OF ISOTHERMAL ELASTODYNAMICS

In Section 8.2.1, the plane strain and generalized plane stress problems of homogeneous isotropic and isothermal elastodynamics were formulated. In this section, closed-form solutions to particular plane strain and so-called antiplane problems of the isothermal elastodynamics are obtained. The problems include

(A) Two-dimensional plane progressive waves in a homogeneous isotropic infinite elastic body
(B) The surface waves propagating in a homogeneous isotropic elastic semispace
(C) The propagation of SH waves in a semispace overlaid by a solid layer
(D) The stress waves produced by the initial stress and stress-rate fields in E^2

In Part (A), both the displacement and stress characterizations of a two-dimensional progressive wave are presented. In Part (B), a natural pure stress language is used to recover a classic solution of the problem of surface waves in a semispace as well as to provide a new stress energy partition formula for such waves. In Part (C), the dispersion of elastic waves in a composite solid is discussed, while in Part (D), the stress waves produced by the initial stress and stress-rate fields that may or may not satisfy the compatibility conditions in E^2 are analyzed.

12.1.1 Two-Dimensional Plane Progressive Waves in a Homogeneous Isotropic Infinite Elastic Body

12.1.1.1 The Plane Displacement Progressive Waves

Consider a homogeneous isotropic infinite elastic solid referred to the rectangular coordinates (x_1, x_2, x_3). *A plane strain displacement progressive wave* referred to the (x_1, x_2) plane is defined as a solution to the displacement equation of motion [see Equation (a) with $b_\alpha = 0$ in Example 8.2.1]

$$\mu u_{\alpha,\gamma\gamma} + (\lambda + \mu) u_{\gamma,\gamma\alpha} = \rho \ddot{u}_\alpha \quad \text{on } E^2 \times (-\infty, +\infty) \tag{12.1.1}$$

of the form

$$u_\alpha(x_\gamma, t) = a_\alpha f(x_\gamma n_\gamma - ct) \quad \text{on } E^2 \times (-\infty, +\infty) \tag{12.1.2}$$

where
$f = f(s)$ is a real-valued function of class C^2 on $(-\infty, +\infty)$ with $f''(s) \neq 0$
a_α and n_α are unit vectors
c is a positive constant

The vectors **a** and **n** define the *direction of motion* and the *direction of propagation*, respectively, while c is called the *velocity of propagation*.

Note that (i) the vector field $u_\alpha = u_\alpha(x_\gamma, t)$ takes a constant value on any straight line of the plane (x_1, x_2) defined by

$$P_t = \{\mathbf{x} : x_\gamma n_\gamma - ct = s_0\} \tag{12.1.3}$$

where $s_0 = \text{const}$, and (ii) the line is moving with the velocity c in the direction **n** as t ranges over a time interval (t_0, t_1); $(0 \leq t_0 < t_1 < \infty)$.

By substituting u_α from Equation 12.1.2 into Equation 12.1.1 and dividing the result by $f''(s) \neq 0$, $[s \equiv x_\gamma n_\gamma - ct]$, we find

$$\mu a_\alpha + (\lambda + \mu) a_\gamma n_\gamma n_\alpha - c^2 a_\alpha = 0 \tag{12.1.4}$$

or

$$\left[\left(c^2 - c_2^2\right)\delta_{\alpha\beta} - \left(c_1^2 - c_2^2\right) n_\alpha n_\beta\right] a_\beta = 0 \tag{12.1.5}$$

where

$$c_1^2 = \frac{\lambda + 2\mu}{\rho}, \quad c_2^2 = \frac{\mu}{\rho} \tag{12.1.6}$$

A nontrivial solution a_α to Equation 12.1.5 exists if and only if c^2 is a real-valued solution to the algebraic equation

$$\det\left[\left(c^2 - c_2^2\right)\delta_{\alpha\beta} - \left(c_1^2 - c_2^2\right)n_\alpha n_\beta\right] = 0 \tag{12.1.7}$$

or

$$\left(c^2 - c_2^2\right)\left(c^2 - c_1^2\right) = 0 \tag{12.1.8}$$

Therefore, the plane strain displacement progressive waves can propagate in a homogeneous isotropic infinite elastic medium with two velocities $c = c_1$ and $c = c_2$. If $c = c_1$, Equation 12.1.5 implies that

$$a_\alpha = (n_\beta a_\beta)n_\alpha \tag{12.1.9}$$

Hence, the displacement vector associated with a plane strain progressive wave propagating with the velocity c_1 is in the direction of propagation, which means that such a vector represents a *longitudinal wave*. If $c = c_2$, Equation 12.1.5 implies that

$$a_\beta n_\beta = 0 \tag{12.1.10}$$

Thus, the displacement vector associated with a plane strain progressive wave propagating with the velocity c_2 is in a direction perpendicular to the direction of propagation, that is, it represents a *transverse* or *shear wave*.

Let $u_i = u_i(x_\alpha, t)$, $i = 1, 2, 3$, be a vector field defined by

$$u_\alpha = u_\alpha(x_\gamma, t), \quad u_3 = u_3(x_\alpha, t) = 0 \tag{12.1.11}$$

where u_α is a plane strain displacement progressive wave of the form 12.1.2. Then the vector field

$$\hat{u}_i(x_\alpha, t) = \epsilon_{ijk}u_{k,j}, \quad i, j, k = 1, 2, 3 \tag{12.1.12}$$

in components takes the form

$$\hat{u}_1 = 0, \quad \hat{u}_2 = 0, \quad \hat{u}_3 = u_{2,1} - u_{1,2} \tag{12.1.13}$$

and for a longitudinal wave, on account of 12.1.9, we obtain

$$\hat{u}_3 = (a_2 n_1 - a_1 n_2)f'(x_\gamma n_\gamma - ct) = [(n_\gamma a_\gamma)n_2 n_1 - (n_\gamma a_\gamma)n_1 n_2]f'(x_\gamma n_\gamma - ct) = 0 \tag{12.1.14}$$

Therefore, *the longitudinal wave associated with a plane strain progressive wave is irrotational or dilatational*.

Also, since for a transverse wave, because of Equations 12.1.2 and 12.1.10,

$$u_{\alpha,\alpha} = 0 \tag{12.1.15}$$

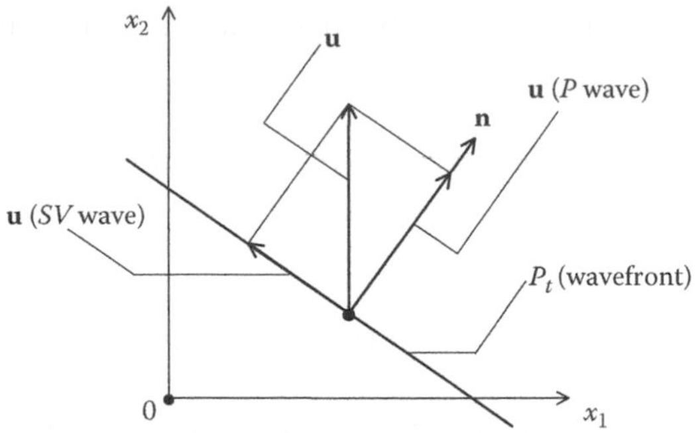

FIGURE 12.1 Two-dimensional plane waves: *SV* and *P* waves.

the transverse wave associated with a plane strain progressive wave is equivoluminal or distortional.

Let the (x_1, x_2) plane be a vertical plane as shown in Figure 12.1. The displacement progressive wave **u** can be decomposed into the longitudinal and transverse components with respect to the wavefront P_t.

The longitudinal component of the wave is called a "P wave" (*primary wave*), while the transverse component is called "SV-wave" (*secondary or vertically polarized shear wave*). These definitions come from geophysics: since $c_1 > c_2$, a longitudinal wave arrives first and a transverse wave second in an earthquake.

The strain tensor $E_{\alpha\beta}$ related to the displacement progressive wave 12.1.2 is obtained from the formula

$$E_{\alpha\beta} = u_{(\alpha,\beta)} = a_{(\alpha}n_{\beta)}f'(x_\gamma n_\gamma - ct) \tag{12.1.16}$$

while the stress tensor $S_{\alpha\beta}$ is given by

$$S_{\alpha\beta} = \left[2\mu a_{(\alpha}n_{\beta)} + \lambda\delta_{\alpha\beta}a_\gamma n_\gamma\right]f'(x_\alpha n_\alpha - ct) \tag{12.1.17}$$

Hence, the elastic energy density of a plane strain progressive wave 12.1.2 is computed from the formula

$$\mathcal{E}(x_\alpha, t) = \frac{1}{2}E_{\alpha\beta}S_{\alpha\beta} = \left[\mu a_\alpha a_\alpha + \frac{\lambda}{2}(a_\gamma n_\gamma)^2\right]\left[f'(x_\alpha n_\alpha - ct)\right]^2 \tag{12.1.18}$$

12.1.1.2 The Antiplane Displacement Progressive Waves

Substituting the displacement field $\mathbf{u} = \mathbf{u}(x_\alpha, t)$ of the form

$$\mathbf{u}(x_\alpha, t) = [0, 0, u_3(x_\alpha, t)] \tag{12.1.19}$$

into the displacement equation of motion 4.2.13 with $\mathbf{b} = \mathbf{0}$, we obtain

$$\mu u_{3,\alpha\alpha} = \rho\ddot{u}_3 \quad \text{on } E^2 \times (-\infty, +\infty) \tag{12.1.20}$$

An *antiplane displacement progressive wave* is defined as a solution to Equation 12.1.20 of the form

$$u_3(x_\alpha, t) = g(x_\gamma n_\gamma - ct) \tag{12.1.21}$$

where
  $g = g(s)$ is a real-valued function of class C^2 on $(-\infty, +\infty)$ with $g''(s) \neq 0$
  $\mathbf{n}$ is a unit vector
  c is a positive constant

Substituting 12.1.21 in 12.1.20 we find that

$$c = c_2 = \sqrt{\frac{\mu}{\rho}} \tag{12.1.22}$$

Hence, a function

$$u_3(x_\gamma, t) = g(x_\gamma n_\gamma - c_2 t) \tag{12.1.23}$$

represents an antiplane progressive wave propagating with the velocity c_2, and for a vertical (x_1, x_2) plane u_3 is called an "SH wave" (*horizontally polarized shear wave*), see Figure 12.2.

The strain components E_{31} and E_{32}, and the stress components S_{31} and S_{32}, associated with the wave 12.1.23 are given by the formulas

$$E_{31} = \frac{1}{2} u_{3,1} = \frac{1}{2} n_1 g'(x_\alpha n_\alpha - c_2 t) \tag{12.1.24}$$

$$E_{32} = \frac{1}{2} u_{3,2} = \frac{1}{2} n_2 g'(x_\alpha n_\alpha - c_2 t) \tag{12.1.25}$$

and

$$S_{31} = 2\mu E_{31} = \mu n_1 g'(x_\alpha n_\alpha - c_2 t) \tag{12.1.26}$$

$$S_{32} = 2\mu E_{32} = \mu n_2 g'(x_\alpha n_\alpha - c_2 t) \tag{12.1.27}$$

Hence, the elastic energy density of an antiplane progressive wave is given by the simple formula

$$\mathcal{E}(x_\alpha, t) = E_{13} S_{13} + E_{23} S_{23} = \frac{1}{2} \mu [g'(x_\alpha n_\alpha - c_2 t)]^2 \tag{12.1.28}$$

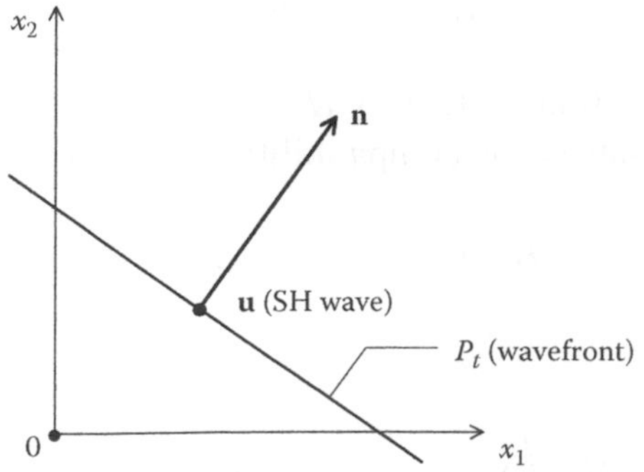

FIGURE 12.2 Two-dimensional SH wave.

12.1.1.3 The Stress Progressive Waves Corresponding to a Plane Strain State

To discuss a two-dimensional stress progressive wave, we note that the stress equations of motion for a homogeneous isotropic infinite elastic solid subject to plane strain conditions take the form [see Example 8.2.3, Equation (i) with $b_\alpha = 0$]

$$S_{(\alpha\gamma,\gamma\beta)} - \frac{\rho}{2\mu}\left(\ddot{S}_{\alpha\beta} - \nu \ddot{S}_{\gamma\gamma}\delta_{\alpha\beta}\right) = 0 \quad \text{on } E^2 \times (-\infty, +\infty) \tag{12.1.29}$$

Representing a stress progressive wave $S_{\alpha\beta}$ on $E^2 \times (-\infty, +\infty)$ in the form

$$S_{\alpha\beta}(x_\gamma, t) = \overset{\circ}{S}_{\alpha\beta}\ \psi(x_\gamma n_\gamma - ct) \tag{12.1.30}$$

where $\psi = \psi(s)$ is a real-valued scalar function of class C^2 on $(-\infty, +\infty)$ with $\psi''(s) \neq 0$, $\overset{\circ}{S}_{\alpha\beta}$ is a unit second-order symmetric tensor, n_γ is a unit vector, and c is a positive constant; and substituting Equation 12.1.30 into Equation 12.1.29, and dividing the result by $\psi''(s) \neq 0$, we find

$$\overset{\circ}{S}_{(\alpha\gamma}\ n_\gamma n_{\beta)} - \frac{\rho c^2}{2\mu}\left(\overset{\circ}{S}_{\alpha\beta} - \nu\ \overset{\circ}{S}_{\gamma\gamma}\ \delta_{\alpha\beta}\right) = 0 \tag{12.1.31}$$

Let $H_{\alpha\beta}$ be the second-order symmetric tensor defined by

$$H_{\alpha\beta} = \epsilon_{\alpha\gamma 3}\epsilon_{\beta\delta 3} n_\gamma n_\delta \tag{12.1.32}$$

where $\epsilon_{\alpha\beta 3}$ is the three-dimensional alternator ϵ_{ijk} restricted to $i = \alpha$, $j = \beta$, $k = 3$; $\alpha, \beta = 1, 2$. Then, we have

$$\begin{aligned} H_{\alpha\beta}\ \overset{\circ}{S}_{(\alpha\gamma}\ n_\gamma n_{\beta)} &= H_{\alpha\beta}\ \overset{\circ}{S}_{\alpha\beta}\ n_\gamma n_\beta = \epsilon_{\alpha\gamma 3}\epsilon_{\beta\delta 3} n_\gamma n_\delta\ \overset{\circ}{S}_{\alpha\epsilon}\ n_\epsilon n_\beta \\ &= \epsilon_{\alpha\gamma 3} n_\gamma\ \overset{\circ}{S}_{\alpha\epsilon}\ n_\epsilon \epsilon_{\beta\delta 3} n_\beta n_\delta \end{aligned} \tag{12.1.33}$$

Since

$$\epsilon_{\beta\delta 3} n_\beta n_\delta = 0 \tag{12.1.34}$$

therefore,

$$H_{\alpha\beta}\ \overset{\circ}{S}_{(\alpha\gamma}\ n_\gamma n_{\beta)} = 0 \tag{12.1.35}$$

Hence, by multiplying Equation 12.1.31 by $H_{\alpha\beta}$, we find that a tensorial wave $S_{\alpha\beta}$ defined in Equation 12.1.30 must satisfy the compatibility condition

$$H_{\alpha\beta}\left(\overset{\circ}{S}_{\alpha\beta} - \nu\ \overset{\circ}{S}_{\gamma\gamma}\ \delta_{\alpha\beta}\right) = 0 \tag{12.1.36}$$

or, since $\nu = \lambda/2(\lambda + \mu)$,

$$H_{\alpha\beta}\left(\overset{\circ}{S}_{\alpha\beta} - \frac{\lambda}{2\lambda + 2\mu}\ \overset{\circ}{S}_{\gamma\gamma}\ \delta_{\alpha\beta}\right) = 0 \tag{12.1.37}$$

This condition is satisfied identically if we set (see Equation 12.1.17)

$$\overset{\circ}{S}_{\alpha\beta} = 2\mu a_{(\alpha} n_{\beta)} + \lambda \delta_{\alpha\beta} a_\gamma n_\gamma \tag{12.1.38}$$

where a_α is an arbitrary vector. To prove that $\overset{\circ}{S}_{\alpha\beta}$ given by Equation 12.1.38 satisfies Equation 12.1.37 note that, by virtue of Equation 12.1.38,

$$\overset{\circ}{S}_{\gamma\gamma} = (2\lambda + 2\mu) a_\gamma n_\gamma \tag{12.1.39}$$

Hence

$$\overset{\circ}{S}_{\alpha\beta} - \frac{\lambda}{2\lambda + 2\mu} \overset{\circ}{S}_{\gamma\gamma}\, \delta_{\alpha\beta} = 2\mu a_{(\alpha} n_{\beta)} \tag{12.1.40}$$

Substituting Equation 12.1.40 into Equation 12.1.37 and dividing the result by $2\mu \neq 0$, we arrive at

$$H_{\alpha\beta} a_{(\alpha} n_{\beta)} = H_{\alpha\beta} a_\alpha n_\beta = \epsilon_{\alpha\gamma 3} \epsilon_{\beta\delta 3} n_\gamma n_\delta a_\alpha n_\beta = 0 \tag{12.1.41}$$

since $\epsilon_{\beta\delta 3} n_\delta n_\beta = 0$. This completes the proof that $\overset{\circ}{S}_{\alpha\beta}$ given by Equation 12.1.38 satisfies Equation 12.1.37.

Moreover, if a_α satisfies Equation 12.1.5, the tensor $\overset{\circ}{S}_{\alpha\beta}$ given by Equation 12.1.38 represents a solution to Equation 12.1.31.

Let $\overset{\circ}{S}_{\alpha\beta}$ be a solution to Equation 12.1.31. By taking the inner product of the tensorial equation 12.1.31 with the tensor $\overset{\circ}{S}_{\alpha\beta}$, we obtain

$$c(\mathbf{n}, \overset{\circ}{\mathbf{S}}) = \frac{|\overset{\circ}{\mathbf{S}}\,\mathbf{n}|}{\left\{(\rho/2\mu)\left[\overset{\circ}{\mathbf{S}} \cdot \overset{\circ}{\mathbf{S}} - \nu(\mathrm{tr}\, \overset{\circ}{\mathbf{S}})^2\right]\right\}^{1/2}} \tag{12.1.42}$$

Hence, a velocity of the stress progressive wave may be treated as a function of a pair $(\mathbf{n}, \overset{\circ}{\mathbf{S}})$, in which $\mathbf{n}$ and $\overset{\circ}{\mathbf{S}}$ represent a unit vector and a unit tensor, respectively, and, in addition, $\overset{\circ}{\mathbf{S}}$ satisfies the compatibility condition 12.1.36.

To describe other properties of the stress progressive wave in terms of the pair $(\mathbf{n}, \overset{\circ}{\mathbf{S}})$, we decompose $\overset{\circ}{S}_{\alpha\beta}$ into the normal and tangential parts with respect to the front P_t (see (12.1.3)) as

$$\overset{\circ}{S}_{\alpha\beta} = \overset{\circ\perp}{S}_{\alpha\beta} + \overset{\circ\parallel}{S}_{\alpha\beta} \tag{12.1.43}$$

where

$$\overset{\circ\perp}{S}_{\alpha\beta} = (\delta_{\alpha\mu} n_\nu n_\beta + \delta_{\beta\mu} n_\nu n_\alpha - n_\alpha n_\beta n_\mu n_\nu)\, \overset{\circ}{S}_{\mu\nu} \tag{12.1.44}$$

and

$$\overset{\circ\parallel}{S}_{\alpha\beta} = (\delta_{\alpha\mu} - n_\alpha n_\mu)(\delta_{\beta\nu} - n_\beta n_\nu)\, \overset{\circ}{S}_{\mu\nu} \tag{12.1.45}$$

Then

$$\overset{\circ\perp}{S}_{\alpha\beta}\overset{\circ\parallel}{S}_{\alpha\beta} = 0, \quad \overset{\circ\parallel}{S}_{\alpha\beta}\, n_\beta = 0, \quad \overset{\circ}{S}_{\alpha\beta}\, n_\beta = \overset{\circ\perp}{S}_{\alpha\beta}\, n_\beta \tag{12.1.46}$$

Note that $\overset{\circ\perp}{S}_{\alpha\beta}$ and $\overset{\circ\parallel}{S}_{\alpha\beta}$ represent the normal and tangential parts of $\overset{\circ}{S}_{\alpha\beta}$ with respect to P_t, and, because of Equations 12.1.44 and 12.1.46, we receive

$$\overset{\circ\perp}{S}_{\alpha\beta} = \overset{\circ\perp}{S}_{\alpha\gamma}\, n_\gamma n_\beta + \overset{\circ\perp}{S}_{\beta\gamma}\, n_\gamma n_\alpha - \overset{\circ\perp}{S}_{\mu\nu}\, n_\mu n_\nu n_\alpha n_\beta \tag{12.1.47}$$

and

$$\overset{\circ\perp}{S}_{\alpha\beta}\overset{\circ\perp}{S}_{\alpha\beta} = 2\, \overset{\circ\perp}{S}_{\alpha\gamma}\, n_\gamma\, \overset{\circ\perp}{S}_{\alpha\delta}\, n_\delta - \left(\overset{\circ\perp}{S}_{\mu\nu} n_\nu n_\mu\right)^2 \tag{12.1.48}$$

The decomposition formula 12.1.43 allows us to obtain estimates for the stress energy of a progressive wave in terms of "normal" *and* "tangential" *stress energies*. The function $\mathcal{E}(\mathbf{S})$ defined by

$$\mathcal{E}(\mathbf{S}) = \frac{1}{4\mu}\left[\mathbf{S}\cdot\mathbf{S} - \nu(\operatorname{tr}\mathbf{S})^2\right] \tag{12.1.49}$$

represents the *stress energy of the progressive wave* 12.1.30. The functions $\mathcal{E}(\mathbf{S})^\perp$ and $\mathcal{E}(\mathbf{S}^\parallel)$ are defined as the "normal" *and* "tangential" stress *energies* of the wave 12.1.30. Since

$$\overset{\circ\parallel}{S}_{\alpha\beta}\, n_\beta = 0 \tag{12.1.50}$$

therefore,

$$\overset{\circ\parallel}{S}_{\alpha\beta} \cdot \overset{\circ}{S}_{(\alpha\gamma}\, n_\gamma n_{\beta)} = \overset{\circ\parallel}{S}_{\alpha\beta}\overset{\circ}{S}_{\alpha\gamma}\, n_\gamma n_\beta = 0 \tag{12.1.51}$$

and multiplying Equation 12.1.31 by $\overset{\circ\parallel}{S}_{\alpha\beta}\, \psi^2$, we get

$$S^\parallel_{\alpha\beta}\left(S_{\alpha\beta} - \nu S_{\gamma\gamma}\delta_{\alpha\beta}\right) = 0 \tag{12.1.52}$$

By substituting Equation 12.1.43 into Equation 12.1.49 and using Equation $12.1.46_1$, we obtain

$$\begin{aligned}\mathcal{E}(\mathbf{S}) &= \frac{1}{4\mu}\left[\left(\mathbf{S}^\perp + \mathbf{S}^\parallel\right)\cdot\left(\mathbf{S}^\perp + \mathbf{S}^\parallel\right) - \nu\left(\operatorname{tr}\mathbf{S}^\perp + \operatorname{tr}\mathbf{S}^\parallel\right)^2\right] \\ &= \mathcal{E}\left(\mathbf{S}^\perp\right) + \mathcal{E}\left(\mathbf{S}^\parallel\right) - \frac{\nu}{2\mu}\left(\operatorname{tr}\mathbf{S}^\perp\right)\left(\operatorname{tr}\mathbf{S}^\parallel\right)\end{aligned} \tag{12.1.53}$$

Also, by substituting Equation 12.1.43 into Equation 12.1.52 and using Equation $12.1.46_1$, we receive

$$\mathbf{S}^{\parallel} \cdot \mathbf{S}^{\parallel} - \nu \left(\operatorname{tr} \mathbf{S}^{\perp} + \operatorname{tr} \mathbf{S}^{\parallel} \right) \operatorname{tr} \mathbf{S}^{\parallel} = 0 \tag{12.1.54}$$

or

$$2\mathcal{E}\left(\mathbf{S}^{\parallel}\right) = \frac{\nu}{2\mu} \left(\operatorname{tr} \mathbf{S}^{\perp} \right) \left(\operatorname{tr} \mathbf{S}^{\parallel} \right) \tag{12.1.55}$$

Hence, it follows from Equations 12.1.53 and 12.1.55 that

$$\mathcal{E}(\mathbf{S}) = \mathcal{E}\left(\mathbf{S}^{\perp}\right) - \mathcal{E}\left(\mathbf{S}^{\parallel}\right) \tag{12.1.56}$$

Therefore, *the stress energy of a progressive wave is the difference between the* "normal" *and* "tangential" *stress energies.*

Since

$$\mathcal{E}(\mathbf{S}) > 0 \quad \text{for every } \mathbf{S} \neq \mathbf{0} \tag{12.1.57}$$

then, by virtue of the relation 12.1.56,

$$\mathcal{E}\left(\mathbf{S}^{\perp}\right) > \mathcal{E}\left(\mathbf{S}^{\parallel}\right) \tag{12.1.58}$$

that is, *the* "normal" *stress energy of a progressive wave is greater than the* "tangential" *stress energy of the wave.* Also, it follows from Equation 12.1.56 and the inequality 12.1.57 that *a pure tangential stress progressive wave cannot propagate in the body.*

Finally, it follows from Equations 12.1.42 and 12.1.56 that

$$c(\mathbf{n}, \overset{\circ}{\mathbf{S}}) = c\left(\mathbf{n}, \overset{\circ}{\mathbf{S}}{}^{\perp}\right) \left[1 - \frac{\mathcal{E}\left(\overset{\circ}{\mathbf{S}}{}^{\parallel}\right)}{\mathcal{E}\left(\overset{\circ}{\mathbf{S}}{}^{\perp}\right)} \right]^{-1/2} \tag{12.1.59}$$

which implies that

$$c\left(\mathbf{n}, \overset{\circ}{\mathbf{S}}\right) \geq c\left(\mathbf{n}, \overset{\circ}{\mathbf{S}}{}^{\perp}\right) \tag{12.1.60}$$

that is, *the velocity of a stress progressive wave is bounded from below by the velocity of normal part of the wave.*

Note that the properties of a stress progressive wave described by Equations 12.1.56 and 12.1.58 through 12.1.60 are complementary to those describing a plane strain displacement progressive wave of Section 12.1.1.1, and they are brought out explicitly only in the stress formulation. In this sense, the pure stress language of elastodynamics makes a positive difference in a study of the plane strain progressive waves.

To wind up this section, we take advantage of formula 12.1.42 to recover the familiar wave velocities c_1 and c_2, and verify the difference stress energy formula 12.1.56 for both these velocities. To this end we consider the following two cases.

12.1.1.3.1 Case 1: Vector a in Equation 12.1.38 Is Parallel to the Wave Normal n, That Is, a = kn, k ≠ 0

In this case, Equation 12.1.38 takes the form

$$\overset{\circ}{S}_{\alpha\beta} = k\left(2\mu n_\alpha n_\beta + \lambda\delta_{\alpha\beta}\right) \tag{12.1.61}$$

Hence, we receive

$$\overset{\circ}{S}_{\alpha\beta}\, n_\beta = (\lambda + 2\mu)kn_\alpha, \quad \overset{\circ}{S}_{\gamma\gamma} = 2(\lambda + \mu)k \tag{12.1.62}$$

$$\overset{\circ}{S}_{\alpha\beta}\overset{\circ}{S}_{\alpha\beta} = 2k^2\left[(\lambda + \mu)^2 + \mu^2\right] \tag{12.1.63}$$

$$\frac{1}{2\mu}\left[\overset{\circ}{S}_{\alpha\beta}\overset{\circ}{S}_{\alpha\beta} - \frac{\lambda}{2\lambda + 2\mu}\left(\overset{\circ}{S}_{\gamma\gamma}\right)^2\right] = k^2(\lambda + 2\mu) \tag{12.1.64}$$

Substituting $12.1.62_1$ and 12.1.64 into the velocity formula 12.1.42 we obtain

$$c(\mathbf{n}, \overset{\circ}{\mathbf{S}}) = c_1 = \sqrt{\frac{\lambda + 2\mu}{\rho}} \tag{12.1.65}$$

which recovers the familiar wave speed associated with the longitudinal waves in a displacement formulation of homogeneous isotropic elastodynamics.

Taking the "normal" and "tangential" parts of $\overset{\circ}{\mathbf{S}}$ given by Equation 12.1.61, we obtain, respectively,

$$\overset{\circ\perp}{S}_{\alpha\beta} = k(\lambda + 2\mu)n_\alpha n_\beta \tag{12.1.66}$$

and

$$\overset{\circ\parallel}{S}_{\alpha\beta} = -k\lambda(n_\alpha n_\beta - \delta_{\alpha\beta}) \tag{12.1.67}$$

Therefore, the "normal" and "tangential" stress energies are given by (see Equation 12.1.49)

$$\mathcal{E}(\mathbf{S}^\perp) = \frac{1}{4\mu}\left[\mathbf{S}^\perp \cdot \mathbf{S}^\perp - \frac{\lambda}{2\lambda + 2\mu}\left(\operatorname{tr}\mathbf{S}^\perp\right)^2\right] = \frac{k^2\psi^2}{8}\frac{(\lambda + 2\mu)^3}{\mu(\lambda + \mu)} \tag{12.1.68}$$

and

$$\mathcal{E}(\mathbf{S}^\parallel) = \frac{1}{4\mu}\left[\mathbf{S}^\parallel \cdot \mathbf{S}^\parallel - \frac{\lambda}{2\lambda + 2\mu}\left(\operatorname{tr}\mathbf{S}^\parallel\right)^2\right] = \frac{k^2\psi^2}{8}\frac{\lambda^2(\lambda + 2\mu)}{\mu(\lambda + \mu)} \tag{12.1.69}$$

respectively. The total stress energy is obtained from the equation (see Equation 12.1.56)

$$\mathcal{E}(\mathbf{S}) = \mathcal{E}\left(\mathbf{S}^\perp\right) - \mathcal{E}\left(\mathbf{S}^\parallel\right) = \frac{1}{2}k^2\psi^2(\lambda + 2\mu) \tag{12.1.70}$$

If we introduce the ratio $\varphi = \mathcal{E}(\mathbf{S}^{\|})/\mathcal{E}(\mathbf{S}^{\perp})$, we get $\varphi = \varphi(\nu) = \nu^2/(1-\nu)^2$, where ν is Poisson's ratio. Since $0 < \nu < 1/2$, $\varphi(\nu) < 1$, as expected (see the inequality 12.1.58). Also note that the parameter k in Equations 12.1.61 through 12.1.64, and 12.1.66 through 12.1.70 is defined by the normalization condition $\overset{\circ}{\mathbf{S}} \cdot \overset{\circ}{\mathbf{S}} = 1$, and, on account of Equation 12.1.63, may be taken in the form

$$k = \frac{1}{\sqrt{2}\left[(\lambda+\mu)^2+\mu^2\right]^{1/2}} \tag{12.1.71}$$

12.1.1.3.2 Case 2: Vector a in Equation 12.1.38 Is Perpendicular to the Wave Normal n

In this case, Equation 12.1.38 takes the form

$$\overset{\circ}{S}_{\alpha\beta} = 2\mu a_{(\alpha} n_{\beta)} \tag{12.1.72}$$

and we obtain

$$\overset{\circ}{S}_{\gamma\gamma} = 0, \quad \overset{\circ}{S}_{\alpha\beta}\, n_\beta = \mu a_\alpha, \quad \overset{\circ}{S}_{\alpha\beta}\overset{\circ}{S}_{\alpha\beta} = 2\mu^2 a_\alpha a_\alpha \tag{12.1.73}$$

Substituting Equations 12.1.73 into the velocity formula 12.1.42

$$c\left(\mathbf{n}, \overset{\circ}{\mathbf{S}}\right) = c_2 = \sqrt{\frac{\mu}{\rho}} \tag{12.1.74}$$

This is the familiar shear wave velocity of homogeneous isotropic elastodynamics. From Equations 12.1.44 through 12.1.45 and 12.1.72 we have

$$\overset{\circ\perp}{S}_{\alpha\beta} = 2\mu a_{(\alpha} n_{\beta)}, \quad \overset{\circ\|}{S}_{\alpha\beta} = 0 \tag{12.1.75}$$

Hence, the stress progressive wave 12.1.30 is now purely normal, and the associated "normal" stress energy is given by

$$\mathcal{E}(\mathbf{S}) = \mathcal{E}\left(\mathbf{S}^{\perp}\right) = \frac{1}{2}\mu \mathbf{a}^2 \psi^2 \tag{12.1.76}$$

Since $\overset{\circ}{S}_{\alpha\beta}\overset{\circ}{S}_{\alpha\beta} = 1$, therefore, $\mathbf{a}^2 = 1/2\mu^2$, and Equation 12.1.76 reduces to

$$\mathcal{E}(\mathbf{S}) = \frac{1}{4\mu}\psi^2 \tag{12.1.77}$$

Hence, for a stress progressive wave corresponding to the transverse displacement waves the total stress energy is associated entirely with the normal component of the wave.

12.1.2 The Surface Waves Propagating in a Homogeneous Isotropic Elastic Semispace: Rayleigh Waves

The surface waves propagating in a semispace $|x_1| < \infty$, $x_2 \geq 0$ can be defined as the plane strain waves confined to a thin layer adjacent to a stress free boundary $x_2 = 0$. They can be described using the displacement or stress language of two-dimensional dynamics. In the following, a pure stress formulation of the problem will be used to study the surface waves in a semispace [1].

Consider a homogeneous isotropic elastic solid that occupies the semispace $|x_1| < \infty$, $x_2 \geq 0$, subject to the dynamic plane strain conditions. The stress equation of motion for the solid takes the form (see Equation 12.1.29)

$$S_{(\alpha\gamma,\gamma\beta)} - \frac{\rho}{2\mu}\left(\ddot{S}_{\alpha\beta} - \nu\ddot{S}_{\gamma\gamma}\delta_{\alpha\beta}\right) = 0 \quad \text{for } |x_1| < \infty, x_2 > 0 \tag{12.1.78}$$

where

$S_{\alpha\beta} = S_{\alpha\beta}(x_\gamma, t)$ is the stress tensor
ρ is the density
μ and ν are the shear modulus and Poisson's ratio, respectively

A solution $S_{\alpha\beta}$ of Equation 12.1.78 is said to be a *surface wave* or a *Rayleigh wave* [2,3], propagating in a direction parallel to the stress-free boundary $x_2 = 0$ of the semispace if

$$S_{\alpha\beta}(x_1, x_2; t) = \hat{S}_{\alpha\beta}(x_2)e^{i(sx_1 - pt)} \tag{12.1.79}$$

where s and p are positive constants, $i^2 = -1$,

$$\hat{S}_{22}(0) = \hat{S}_{12}(0) = 0 \tag{12.1.80}$$

and

$$\hat{S}_{\alpha\beta}(x_2) \to 0 \quad \text{as } x_2 \to \infty \tag{12.1.81}$$

If such a solution exists, the tensor field 12.1.79 represents a surface wave propagating with the velocity $c_R = p/s$ in the x_1 direction. The period T and the length L of the wave are then given by $T = 2\pi/p$ and $L = 2\pi/s$, respectively. In the following, we show that there are a tensor field $\hat{S}_{\alpha\beta} = \hat{S}_{\alpha\beta}(x_2)$ on $[0, \infty)$ and a velocity $c_R = p/s$, such that $S_{\alpha\beta} = S_{\alpha\beta}(x_\gamma, t)$ given by Equation 12.1.79 represents a stress surface wave. To this end we rewrite Equation 12.1.78 in components

$$\begin{aligned} &S_{1\gamma,\gamma 1} - \frac{1}{2c_2^2}\left[(1-\nu)\ddot{S}_{11} - \nu\ddot{S}_{22}\right] = 0 \\ &S_{2\gamma,\gamma 2} - \frac{1}{2c_2^2}\left[(1-\nu)\ddot{S}_{22} - \nu\ddot{S}_{11}\right] = 0 \\ &S_{1\gamma,\gamma 2} + S_{2\gamma,\gamma 1} - \frac{1}{c_2^2}\ddot{S}_{12} = 0 \end{aligned} \tag{12.1.82}$$

where

$$\frac{1}{c_2^2} = \frac{\rho}{\mu} \tag{12.1.83}$$

Substituting Equations 12.1.79 into Equations 12.1.82 and omitting the factor $\exp[\,i(sx_1 - pt)]$ we find that the stress components $\hat{S}_{11}$, $\hat{S}_{22}$, and $\hat{S}_{12}$ must satisfy the equations

$$\begin{aligned}
&-s^2\left[1-(1-\nu)\frac{\Omega}{2}\right]\hat{S}_{11} - s^2\nu\frac{\Omega}{2}\hat{S}_{22} + isD\hat{S}_{12} = 0\\
&-s^2\nu\frac{\Omega}{2}\hat{S}_{11} + \left[D^2 + s^2(1-\nu)\frac{\Omega}{2}\right]\hat{S}_{22} + isD\hat{S}_{12} = 0\\
&isD(\hat{S}_{11} + \hat{S}_{22}) + [D^2 - s^2(1-\Omega)]\hat{S}_{12} = 0
\end{aligned} \tag{12.1.84}$$

where

$$D = \frac{d}{dx_2} \quad \text{and } \Omega = \left(\frac{c_R}{c_2}\right)^2 \tag{12.1.85}$$

By eliminating first $\hat{S}_{11}$ and then $\hat{S}_{12}$ from Equations 12.1.84 we obtain the equivalent system of equations

$$\hat{S}_{11} = -\frac{1}{s^2(1-\Omega/2)}\left(D^2 + s^2\Omega/2\right)\hat{S}_{22} \tag{12.1.86}$$

$$\hat{S}_{12} = \frac{1}{2is}\frac{1}{s^2(1-\Omega)}D\left\{\frac{1}{1-\kappa}\frac{\Omega/2}{1-\Omega/2}\left[D^2 - s^2(1-\kappa\Omega)\right] - 2s^2\frac{1-\Omega}{1-\Omega/2}\right\}\hat{S}_{22} \tag{12.1.87}$$

$$\left[D^2 - s^2(1-\Omega)\right]\left[D^2 - s^2(1-\kappa\Omega)\right]\hat{S}_{22} = 0 \tag{12.1.88}$$

where

$$\kappa = \frac{1-2\nu}{2-2\nu} \tag{12.1.89}$$

Note that because of Equation 12.1.86

$$\hat{S}_{11} + \hat{S}_{22} = -\frac{1}{s^2(1-\Omega/2)}\left[D^2 - s^2(1-\Omega)\right]\hat{S}_{22} \tag{12.1.90}$$

Hence, applying the operator $[D^2 - s^2(1 - \kappa\Omega)]$ to Equation 12.1.90 and using Equation 12.1.88 we arrive at the formula

$$\left[D^2 - s^2(1-\kappa\Omega)\right]\hat{S}_{\gamma\gamma} = 0 \tag{12.1.91}$$

Also, note that because of Equation 12.1.89

$$\nu = \frac{1-2\kappa}{2-2\kappa} \tag{12.1.92}$$

Therefore, since $-1 < \nu < 1/2$, we obtain

$$0 < \kappa < \frac{3}{4} \tag{12.1.93}$$

and

$$1 - \kappa\Omega > 0 \quad \text{for } \Omega < 1 \tag{12.1.94}$$

Hence, if $\Omega < 1$ the stress component $\hat{S}_{22} = \hat{S}_{22}(x_2)$ that satisfies Equation 12.1.88 on $[0, \infty)$ and the boundary conditions

$$\hat{S}_{22}(0) = 0, \quad \hat{S}_{22}(\infty) = 0 \tag{12.1.95}$$

takes the form

$$\hat{S}_{22}(x_2) = S_0 \psi(x_2) \tag{12.1.96}$$

where

$$\psi(x_2) = e^{-x_2 h_1} - e^{-x_2 h_2} \tag{12.1.97}$$

$$h_1 = s\sqrt{1-\kappa\Omega}, \quad h_2 = s\sqrt{1-\Omega} \tag{12.1.98}$$

and S_0 is a real-valued constant of the stress dimension.

Since $h_1 > h_2$, therefore

$$\hat{S}_{22}(x_2) \geq 0 \quad \text{for } S_0 < 0 \quad \text{and } x_2 \geq 0 \tag{12.1.99}$$

Now, substituting Equation 12.1.96 into Equation 12.1.87, we find

$$\hat{S}_{12} = S_0 \frac{1}{2ish_2^2(1-\Omega/2)(1-\kappa)} D \left\{ \left[\frac{\Omega\left(h_1^2 - h_2^2\right)}{2 + 2h_2^2(1-\kappa)} \right] e^{-h_2 x_2} - 2h_2^2(1-\kappa) e^{-x_2 h_1} \right\} \tag{12.1.100}$$

and using the boundary condition

$$\hat{S}_{12}(0) = 0 \tag{12.1.101}$$

we arrive at the relation

$$-h_2 \left[\frac{\Omega\left(h_1^2 - h_2^2\right)}{2 + 2h_2^2(1-\kappa)} \right] + 2h_2^2 h_1 (1-\kappa) = 0 \tag{12.1.102}$$

On account of Equation 12.1.98, the relation 12.1.102 is equivalent to the algebraic equation

$$(1-\Omega)^2 - 4\sqrt{1-\kappa\Omega}\,\sqrt{1-\Omega} = 0 \tag{12.1.103}$$

Also, because of Equation 12.1.102, Equation 12.1.100 may be reduced to the simple one

$$\hat{S}_{12}(x_2) = -iS_0\,\frac{\sqrt{1-\kappa\Omega}}{1-\Omega/2}\,\psi(x_2) \tag{12.1.104}$$

where $\psi = \psi(x_2)$ is given by Equation 12.1.97. Finally, substituting $\hat{S}_{22}$ from Equation 12.1.96 into Equation 12.1.90 we receive

$$\hat{S}_{11}(x_2) = -S_0\left[\psi(x_2) + \frac{(1-\kappa)\Omega}{1-\Omega/2}e^{-x_2h_1}\right] \tag{12.1.105}$$

Hence, the tensor field $S_{\alpha\beta} = S_{\alpha\beta}(x_\gamma, t)$ given by Equation 12.1.79 represents a surface wave in the semispace if $\hat{S}_{22}$, $\hat{S}_{12}$, and $\hat{S}_{11}$ are computed from Equations 12.1.96, 12.1.104, and 12.1.105, respectively, and Ω is a real-valued root of Equation 12.1.103 from the interval $(0, 1)$. It can be shown [4] that there is only one root of Equation 12.1.103 for $0 \le \nu < 1/2$ in $(0, 1)$. For example, for $\nu = 0$ we have

$$\Omega = 3 - \sqrt{5} \tag{12.1.106}$$

and for $\nu = 1/4$ we find

$$\Omega = 2 - 2/\sqrt{3} \tag{12.1.107}$$

Hence, using Equation $12.1.85_2$, for $\nu = 0$ and $\nu = 1/4$ we arrive at $c_R = 0.8740c_2$ and $c_R = 0.9194c_2$, respectively.

Note that $S_{\alpha\beta} = S_{\alpha\beta}(x_\gamma, t)$ given by Equation 12.1.79 represents a complex-valued surface wave. If $\Sigma_{\alpha\beta} = \Sigma_{\alpha\beta}(x_\gamma, t)$ denotes the real part of $S_{\alpha\beta} = S_{\alpha\beta}(x_\gamma, t)$, then, the use of Equations 12.1.79, 12.1.96, 12.1.104, and 12.1.105 gives

$$\Sigma_{11} = -S_0\left[\psi(x_2) + \frac{(1-\kappa)\Omega}{1-\Omega/2}e^{-x_2h_1}\right]\cos\varphi \tag{12.1.108}$$

$$\Sigma_{22} = S_0\psi(x_2)\cos\varphi \tag{12.1.109}$$

$$\Sigma_{12} = S_0\frac{\sqrt{1-\kappa\Omega}}{1-\Omega/2}\,\psi(x_2)\sin\varphi \tag{12.1.110}$$

where

$$\varphi = s(x_1 - c_R t) \tag{12.1.111}$$

Also, it follows from Equations 12.1.109 and 12.1.110 that

$$\frac{\Sigma_{12}^2}{a^2} + \frac{\Sigma_{22}^2}{b^2} = 1 \tag{12.1.112}$$

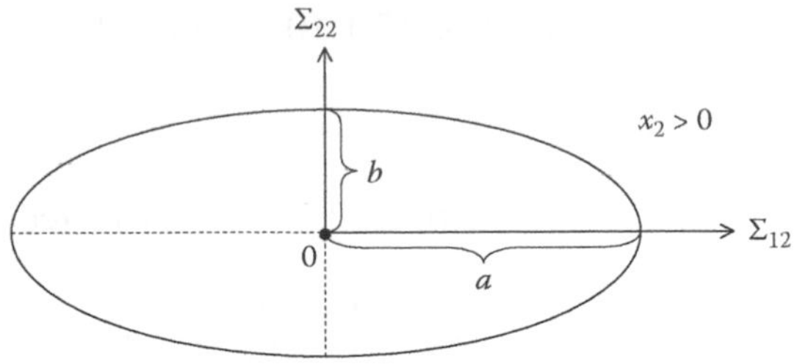

FIGURE 12.3 The stress elliptical orbit associated with a surface wave ($\nu = 0$).

where

$$a = S_0 \frac{\sqrt{1-\kappa\Omega}}{1-\Omega/2}\psi(x_2), \quad b = S_0\psi(x_2) \quad \text{for } x_2 > 0 \tag{12.1.113}$$

Hence, the stress vector $\mathbf{s} = (\Sigma_{12}, \Sigma_{22})$ associated with the surface wave is represented by an elliptical orbit in the $(\Sigma_{12}, \Sigma_{22})$ plane (see Figure 12.3).

Finally, it follows from Equation 12.1.97 that the function $\psi = \psi(x_2)$ has the properties

(i) $\psi(x_2) \leq 0 \quad \text{for } x_2 \geq 0$

(ii) $\psi'(x_0) = 0 \quad \text{for } x_0 = \dfrac{1}{h_1 - h_2}\ln\dfrac{h_1}{h_2}$

(iii) $\psi''(x_0) > 0$

(iv) $\psi'(0) < 0$

Hence, if $S_0 < 0$ and $\varphi = \varphi_0 \in (0, \pi/2)$, then Σ_{22} represents a nonnegative function of x_2 that increases over the interval $(0, x_0)$ and after reaching a maximum at x_0 decays asymptotically to zero. A similar behavior is revealed by Σ_{12} treated as a function of x_2. A plot of Σ_{22} for a fixed $S_0 < 0$ and $\varphi_0 \in (0, \pi/2)$ is shown in Figure 12.4.

Using the formula

$$x_0 = \frac{1}{h_1 - h_2}\ln\frac{h_1}{h_2} = \frac{2L}{\pi}\frac{\sqrt{1-\Omega}}{\Omega^2}\ln\frac{(1-\Omega/2)^2}{1-\Omega} \tag{12.1.114}$$

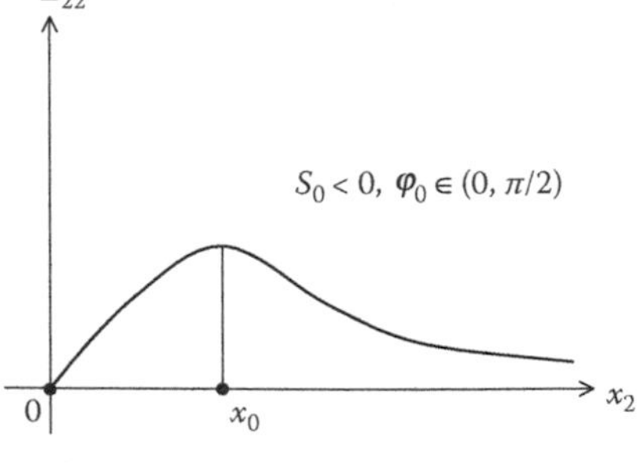

FIGURE 12.4 Amplitude Σ_{22} of a surface wave versus depth of the semispace.

where L is the surface wave length, we find that for $\nu = 1/4$ when $\Omega = 2 - 2/\sqrt{3}$

$$x_0 = 0.2689L \tag{12.1.115}$$

Note:

For a fixed depth of the semispace, say $\hat{x}_2 > 0$, $\Sigma_{\alpha\beta}$ represents a plane progressive wave propagating with the velocity c_R along the stress free boundary $x_2 = 0$. The front of the wave is described by the moving plane

$$\varphi \equiv sx_1 - pt = \varphi_0 = \text{const} \tag{12.1.116}$$

If $\Sigma_{\alpha\beta}$ is written as the sum of the "normal" and "tangential" parts with respect to the front, viz.:

$$\Sigma_{\alpha\beta} = \Sigma^{\perp}_{\alpha\beta} + \Sigma^{\parallel}_{\alpha\beta} \tag{12.1.117}$$

where

$$\Sigma^{\perp}_{\alpha\beta} = \begin{bmatrix} \Sigma_{11} & \Sigma_{12} \\ \Sigma_{12} & 0 \end{bmatrix} \tag{12.1.118}$$

and

$$\Sigma^{\parallel}_{\alpha\beta} = \begin{bmatrix} 0 & 0 \\ 0 & \Sigma_{22} \end{bmatrix} \tag{12.1.119}$$

then substituting Equation 12.1.117 into the stress energy density formula

$$\mathcal{E}(\mathbf{S}) = \frac{1}{4\mu}\left[\mathbf{\Sigma}\cdot\mathbf{\Sigma} - \nu(\text{tr}\,\mathbf{\Sigma})^2\right] \tag{12.1.120}$$

we obtain

$$\mathcal{E}(\mathbf{\Sigma}) = \mathcal{E}\left(\mathbf{\Sigma}^{\perp}\right) + M\left(\hat{x}_2, \nu\right)\mathcal{E}\left(\mathbf{\Sigma}^{\parallel}\right) \tag{12.1.121}$$

where $\mathcal{E}(\mathbf{\Sigma}^{\perp})$ and $\mathcal{E}(\mathbf{\Sigma}^{\parallel})$ represent the "normal" and "tangential" stress energy densities, respectively, and

$$M(\hat{x}_2, \nu) = \frac{1+\nu}{1-\nu}\left\{1 - \lambda^*(\nu)\left[e^{\hat{x}_2(h_1 - h_2)} - 1\right]^{-1}\right\} \tag{12.1.122}$$

$$\lambda^*(\nu) = 2\beta\frac{\nu}{1-\nu}, \quad \beta = \frac{(1-\kappa)\Omega}{1-\Omega/2} \tag{12.1.123}$$

It follows from Equations 12.1.122 and 12.1.123 that

$$M\left(\hat{x}_2, 0\right) = 1 \quad \text{for}\ \hat{x}_2 > 0 \tag{12.1.124}$$

and

$$M\left(x_2^*, \nu\right) = 0 \quad \text{for } 0 < \nu < 1/2 \tag{12.1.125}$$

where

$$x_2^* = \frac{2L}{\pi} \frac{\sqrt{1-\Omega}}{\Omega^2} \ln\left[1 + \lambda^*(\nu)\right] > 0 \tag{12.1.126}$$

Therefore,

$$\mathcal{E}(\mathbf{\Sigma}) = \mathcal{E}\left(\mathbf{\Sigma}^{\perp}\right) + \mathcal{E}\left(\mathbf{\Sigma}^{\parallel}\right) \quad \text{for } \hat{x}_2 > 0 \quad \text{and } \nu = 0 \tag{12.1.127}$$

and

$$\mathcal{E}(\mathbf{\Sigma}) = \mathcal{E}\left(\mathbf{\Sigma}^{\perp}\right) \quad \text{for } \hat{x}_2 = x_2^* \quad \text{and } 0 < \nu < 1/2 \tag{12.1.128}$$

The stress energy density decomposition formula 12.1.121 was obtained apparently for the first time by Jakubowska [5]. Also, note that the pure stress treatment of surface waves in a homogeneous isotropic elastic semispace is a restriction of the one for a nonhomogeneous isotropic elastic semispace proposed by Ignaczak [1]; see also [6–8].

12.1.3 The Propagation of SH Waves in a Semispace Overlaid by a Solid Layer: Love Waves

An SH wave related to (x_1, x_2) plane is associated with a displacement vector $\mathbf{u} = \mathbf{u}(x_\alpha, t)$ of the form

$$\mathbf{u}(x_\alpha, t) = [0, 0, u_3(x_\alpha, t)] \tag{12.1.129}$$

Substituting $\mathbf{u}$ given by Equation 12.1.129 into the displacement equation of homogeneous isotropic three-dimensional elastodynamics with zero body forces (see Equation 4.2.13 with $\mathbf{b} = \mathbf{0}$) we find

$$\mu u_{3,\alpha\alpha} = \rho \ddot{u}_3 \tag{12.1.130}$$

The associated stress tensor $\mathbf{S} = \mathbf{S}(x_\alpha, t)$ takes the form

$$\mathbf{S} = \begin{bmatrix} 0 & 0 & S_{13} \\ 0 & 0 & S_{23} \\ S_{13} & S_{23} & 0 \end{bmatrix} \tag{12.1.131}$$

where

$$S_{13} = S_{31} = \mu u_{3,1} \tag{12.1.132}$$

$$S_{23} = S_{32} = \mu u_{3,2} \tag{12.1.133}$$

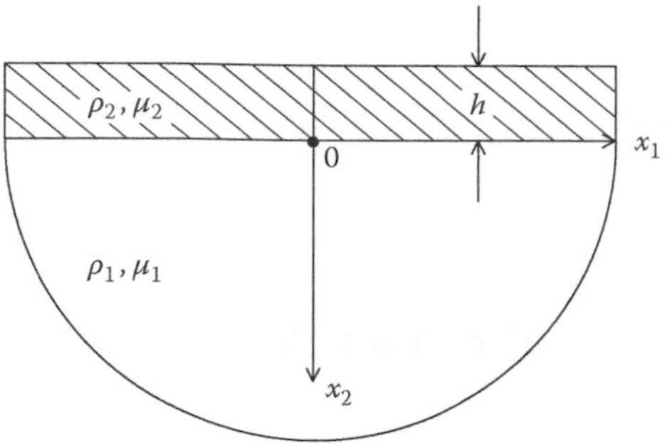

FIGURE 12.5 A layer of thickness h over the semispace.

Consider a homogeneous isotropic semispace $|x_1| < \infty$, $x_2 > 0$ with density ρ_1 and shear modulus μ_1, overlaid by a homogeneous isotropic layer $|x_1| < \infty$, $-h < x_2 < 0$ with density ρ_2 and shear modulus μ_2 as shown in Figure 12.5.

Let $s_1 = [\mathbf{u}^{(1)}, \mathbf{E}^{(1)}, \mathbf{S}^{(1)}]$ and $s_2 = [\mathbf{u}^{(2)}, \mathbf{E}^{(2)}, \mathbf{S}^{(2)}]$ denote SH waves in the semispace $|x_1| < \infty, x_2 > 0$ and the layer $|x_1| < \infty$, $-h < x_2 < 0$, respectively. A pair (s_1, s_2) is said to represent a *Love wave* in the composite shown in Figure 12.5 if it satisfies the conditions:

$$\mu_2 u^{(2)}_{3,\alpha\alpha} = \rho_2 \ddot{u}^{(2)}_3 \quad \text{for } |x_1| < \infty,\ -h < x_2 < 0,\ t > 0 \tag{12.1.134}$$

$$\mu_1 u^{(1)}_{3,\alpha\alpha} = \rho_1 \ddot{u}^{(1)}_3 \quad \text{for } |x_1| < \infty,\ x_2 > 0,\ t > 0 \tag{12.1.135}$$

$$S^{(2)}_{23}(x_1, -h, t) = 0 \quad \text{for } |x_1| < \infty,\ t > 0 \tag{12.1.136}$$

$$u^{(2)}_3(x_1, 0, t) = u^{(1)}_3(x_1, 0, t) \quad \text{for } |x_1| < \infty,\ t > 0 \tag{12.1.137}$$

$$S^{(2)}_{23}(x_1, 0, t) = S^{(1)}_{23}(x_1, 0, t) \quad \text{for } |x_1| < \infty,\ t > 0 \tag{12.1.138}$$

and

$$s_1 \to [\mathbf{0}, \mathbf{0}, \mathbf{0}] \quad \text{as } x_2 \to \infty,\ |x_1| < \infty,\ t > 0 \tag{12.1.139}$$

Note that Equations 12.1.134 and 12.1.135 are the field equations to be satisfied by $u_3^{(2)}$ and $u_3^{(1)}$, respectively; Equation 12.1.136 means that the boundary $x_2 = -h$ is stress free, while Equations 12.1.137 and 12.1.138 guarantee the continuity of the displacement and stress vectors at $x_2 = 0$, respectively. Finally, Equation 12.1.139 implies that the wave is confined to a thin layer adjacent to the interface $x_2 = 0$. Also note that Equations 12.1.134 and 12.1.135 may be written as

$$\left[\nabla^2 - \frac{1}{\left(c_2^{(2)}\right)^2} \frac{\partial^2}{\partial t^2}\right] u_3^{(2)} = 0 \quad \text{for } |x_1| < \infty,\ -h < x_2 < 0,\ t > 0 \tag{12.1.140}$$

$$\left[\nabla^2 - \frac{1}{\left(c_2^{(1)}\right)^2} \frac{\partial^2}{\partial t^2}\right] u_3^{(1)} = 0 \quad \text{for } |x_1| < \infty,\ x_2 > 0,\ t > 0 \tag{12.1.141}$$

where

$$c_2^{(1)} = \sqrt{\frac{\mu_1}{\rho_1}}, \quad c_2^{(2)} = \sqrt{\frac{\mu_2}{\rho_2}} \tag{12.1.142}$$

To find s_1 and s_2 we let

$$u_3^{(2)}(x_\alpha, t) = U^{(2)}(x_2)e^{is(x_1 - ct)} \tag{12.1.143}$$

and

$$u_3^{(1)}(x_\alpha, t) = U^{(1)}(x_2)e^{is(x_1 - ct)} \tag{12.1.144}$$

where $U^{(1)} = U^{(1)}(x_2)$ and $U^{(2)} = U^{(2)}(x_2)$ are unknown functions, s is a wave number, and c is an unknown velocity of the wave ($i^2 = -1$).

Substituting $u_3^{(2)}$ and $u_3^{(1)}$ into Equations 12.1.140 and 12.1.141, respectively, and omitting the exponential factor, we obtain

$$\left(\frac{d^2}{dx_2^2} + \beta_2^2\right) U^{(2)} = 0 \quad \text{for } -h < x_2 < 0 \tag{12.1.145}$$

and

$$\left(\frac{d^2}{dx_2^2} - \beta_1^2\right) U^{(1)} = 0 \quad \text{for } x_2 > 0 \tag{12.1.146}$$

where

$$\beta_2^2 = s^2\left[\frac{c^2}{\left(c_2^{(2)}\right)^2} - 1\right], \quad \beta_1^2 = s^2\left[1 - \frac{c^2}{\left(c_2^{(1)}\right)^2}\right] \tag{12.1.147}$$

If c satisfies the inequality

$$c_2^{(2)} < c < c_2^{(1)} \tag{12.1.148}$$

and if β_1 and β_2 are selected to be the positive numbers

$$\beta_1 = s\sqrt{1 - \frac{c^2}{\left(c_2^{(1)}\right)^2}}, \quad \beta_2 = s\sqrt{\frac{c^2}{\left(c_2^{(2)}\right)^2} - 1} \tag{12.1.149}$$

the solutions $U^{(2)}$ and $U^{(1)}$ to Equations 12.1.145 and 12.1.146, respectively, may be taken in the forms

$$U^{(2)}(x_2) = Ae^{i\beta_2 x_2} + Be^{-i\beta_2 x_2} \tag{12.1.150}$$

and

$$U^{(1)}(x_2) = Ce^{-\beta_1 x_2} \tag{12.1.151}$$

where A, B, and C are arbitrary constants. In this case, Equations 12.1.143 and 12.1.144 reduce to

$$u_3^{(2)} = Ae^{i(sx_1-\omega t+\beta_2 x_2)} + Be^{i(sx_1-\omega t-\beta_2 x_2)} \tag{12.1.152}$$

and

$$u_3^{(1)} = Ce^{-\beta_1 x_2}e^{i(sx_1-\omega t)} \tag{12.1.153}$$

respectively, where $\omega = sc$ is a frequency. Clearly, $u_3^{(2)}$ represents a plane wave propagating back and forth within the layer $|x_1| < \infty$, $-h < x_2 < 0$; while $u_3^{(1)}$ represents a harmonic wave confined to a neighborhood of the interface $x_2 = 0$, that is, $u_3^{(1)} \to 0$ as $x_2 \to \infty$. This implies that s_1 generated by $u_3^{(1)}$ satisfies the asymptotic condition 12.1.139. Therefore, to show that $u_3^{(2)}$ and $u_3^{(1)}$ represent a Love wave in the composite, it is sufficient to prove that there is a velocity c as well as the nonvanishing constants A, B, and C such that the boundary conditions 12.1.136 through 12.1.138 are satisfied. To this end we note that (see Equation 12.1.133)

$$S_{23}^{(2)} = \mu_2 u_{3,2}^{(2)} = i\mu_2\beta_2\left[Ae^{i(sx_1-\omega t-\beta_2 x_2)} - Be^{-(sx_1-\omega t-\beta_2 x_2)}\right] \tag{12.1.154}$$

and

$$S_{23}^{(1)} = \mu_1 u_{3,2}^{(1)} = -\mu_1\beta_1 Ce^{-\beta_1 x_2+i(sx_1-\omega t)} \tag{12.1.155}$$

and reduce the boundary condition 12.1.136 through 12.1.138 to the form

$$Ae^{-i\beta_2 h} - Be^{i\beta_2 h} = 0 \tag{12.1.156}$$

$$A + B - C = 0 \tag{12.1.157}$$

$$i\mu_2\beta_2(A-B) + \mu_1\beta_1 C = 0 \tag{12.1.158}$$

or

$$Ae^{-i\beta_2 h} - Be^{i\beta_2 h} = 0 \tag{12.1.159}$$

$$A\,(i\mu_2\beta_2 + \mu_1\beta_1) - B\,(i\mu_2\beta_2 - \mu_1\beta_1) = 0 \tag{12.1.160}$$

Equating to zero the determinant of the system 12.1.159 and 12.1.160 we write

$$\tan\beta_2 h = \frac{\mu_1\beta_1}{\mu_2\beta_2} \tag{12.1.161}$$

or, using Equations 12.1.149,

$$\tan\left\{sh\left[\frac{c^2}{\left(c_2^{(2)}\right)^2} - 1\right]^{1/2}\right\} = \frac{\mu_1\left[1 - c^2/\left(c_2^{(1)}\right)^2\right]^{1/2}}{\mu_2\left[c^2/\left(c_2^{(2)}\right)^2 - 1\right]^{1/2}} \tag{12.1.162}$$

Hence, $u_3^{(2)}$ and $u_3^{(1)}$ represent a Love wave if c is a positive root of Equation 12.1.162. It follows from the inequality 12.1.148 that a solution c to Equation 12.1.162 can exist if

$$m = \frac{c_2^{(2)}}{c_2^{(1)}} < 1 \tag{12.1.163}$$

and an equivalent form of Equation 12.1.162 reads

$$\tan h\beta_2 = m\frac{\mu_1}{\mu_2}\left[\left(\frac{\beta_0}{\beta_2}\right)^2 - 1\right]^{1/2} \equiv \varphi(\beta_2) \tag{12.1.164}$$

where

$$\beta_0 = s\left(\frac{1}{m^2} - 1\right)^{1/2} \tag{12.1.165}$$

Equation 12.1.164 treated as a transcendental equation for a root β_2 will have positive roots if $\beta_2 < \beta_0$, and the roots can be obtained by a graphic method, see Figure 12.6.

It follows from Figure 12.6 that the smallest root β_2^* of Equation 12.1.164 lies in the interval

$$0 < \beta_2^* < \frac{\pi}{2h} \tag{12.1.166}$$

To the smallest root β_2^* there corresponds the first mode of oscillation of the composite. To other real roots of Equation 12.1.164 there correspond higher modes of oscillation of the body. The value $\beta_2 = \beta_0$ in Figure 12.6 corresponds to $c = c_2^{(1)}$. In this case

$$h\beta_2 = h\beta_0 = (n-1)\pi, \quad n = 2, 3, 4, \ldots \tag{12.1.167}$$

or

$$\lambda = \frac{2h(m^{-2} - 1)^{1/2}}{n-1}, \quad n > 1 \tag{12.1.168}$$

where $\lambda = 2\pi/s$ is the wavelength. Hence, the nth mode ($n > 1$) of the Love wave propagating with the velocity $c = c_2^{(1)}$ has the wavelength given by Equation 12.1.168. Also,

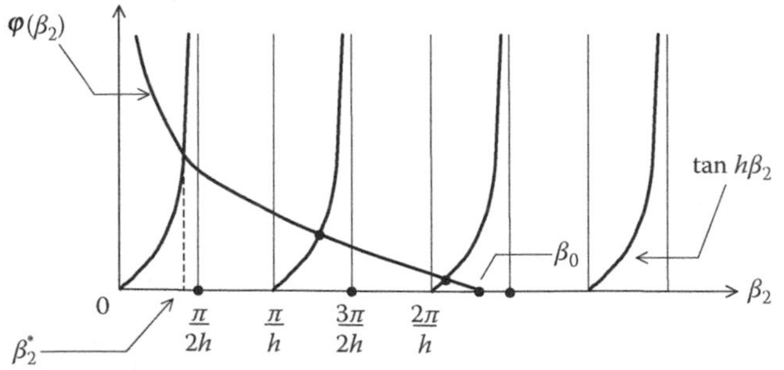

FIGURE 12.6 Solution of Equation 12.1.164.

it follows from Equation 12.1.162 that a Love wave may propagate with the velocity $c = c_2^{(2)} + 0$ provided

$$\frac{2\pi h}{\lambda} = \frac{\pi}{2}(2n-1)\left[\frac{c^2}{\left(c_2^{(2)}\right)^2} - 1\right]^{-1/2} \to \infty \qquad n = 1, 2, 3, \ldots \tag{12.1.169}$$

which means that the Love wave propagates with the velocity $c_2^{(2)} + 0$ provided its length is small in comparison to the layer thickness h.

Finally, note that any positive root of Equation 12.1.164 depends on the wave number s. Also, the velocity c corresponding to the root depends on s. In this sense a Love wave is *dispersive*, that is, it changes its shape as it travels.

12.1.4 The Stress Waves Produced by the Initial Stress and Stress-Rate Fields in E^2

12.1.4.1 A Solution to the Homogeneous Wave Equation Subject to Initial Conditions in E^2

In this section, we prove a two-dimensional counterpart of Theorem 1 from Section 11.1.4 (see Equations 11.1.263 through 11.1.265).

Theorem 1: A solution of the scalar wave equation

$$\left(\nabla^2 - \frac{1}{c^2}\frac{\partial^2}{\partial t^2}\right)\phi(\mathbf{x}, t) = 0, \quad c > 0, \quad (\mathbf{x}, t) \in E^2 \times [0, \infty) \tag{12.1.170}$$

subject to the initial conditions

$$\phi(\mathbf{x}, 0) = \phi_0(\mathbf{x}), \quad \dot{\phi}(\mathbf{x}, 0) = \dot{\phi}_0(\mathbf{x}), \quad \mathbf{x} \in \Omega \subset E^2 \tag{12.1.171}$$

admits the representation

$$\phi(\mathbf{x}, t) = t\,\mathcal{N}_{\mathbf{x},ct}(\dot{\phi}_0) + \frac{\partial}{\partial t}\left[t\,\mathcal{N}_{\mathbf{x},ct}(\phi_0)\right] \tag{12.1.172}$$

Here, for any function $f = f(\mathbf{x})$ on Ω, the symbol $\mathcal{N}_{\mathbf{x},ct}(f)$ is defined by

$$\mathcal{N}_{\mathbf{x},ct}(f) = \frac{1}{2\pi ct}\int_{\Sigma(\mathbf{x},ct)} \frac{f(\xi_1, \xi_2)\, d\xi_1\, d\xi_2}{\sqrt{c^2t^2 - (x_1 - \xi_1)^2 - (x_2 - \xi_2)^2}} \tag{12.1.173}$$

where $\Sigma(\mathbf{x}, ct)$ stands for the circular region with its center at $\mathbf{x}$ and with the radius ct, that is,

$$\Sigma(\mathbf{x}, ct) = \{\boldsymbol{\xi} : |\boldsymbol{\xi} - \mathbf{x}| < ct\} \tag{12.1.174}$$

■

Proof: Let $\overline{\phi} = \overline{\phi}(\mathbf{x}, p)$ denote the Laplace transform of $\phi = \phi(\mathbf{x}, t)$ with respect to t, that is,

$$L\phi \equiv \overline{\phi}(\mathbf{x}, p) = \int_0^\infty e^{-pt}\phi(\mathbf{x}, t)\, dt \tag{12.1.175}$$

where p is the transform parameter. Applying the operator L to Equation 12.1.170 and using the initial conditions 12.1.171 we obtain

$$\left(\nabla^2 - \frac{p^2}{c^2}\right)\overline{\phi}(\mathbf{x}, p) = -\frac{1}{c^2}\left[\dot{\phi}_0(\mathbf{x}) + p\phi_0(\mathbf{x})\right], \quad \mathbf{x} \in \Omega \subset E^2 \tag{12.1.176}$$

Since

$$\left(\nabla_3^2 - \frac{p^2}{c^2}\right)\frac{e^{-\frac{p}{c}|\mathbf{x}-\boldsymbol{\xi}|}}{|\mathbf{x}-\boldsymbol{\xi}|} = -4\pi\,\delta(\mathbf{x}-\boldsymbol{\xi}) \quad \text{on } E^3 \tag{12.1.177}$$

where

$$\nabla_3^2 = \nabla^2 + \frac{\partial^2}{\partial x_3^2}, \quad \mathbf{x}, \boldsymbol{\xi} \in E^3 \tag{12.1.178}$$

therefore, integrating Equation 12.1.177 over x_3 from $x_3 = -\infty$ to $x_3 = +\infty$, and using the result [9]

$$\int_{-\infty}^{+\infty}\frac{e^{-\frac{p}{c}|\mathbf{x}-\boldsymbol{\xi}|}}{|\mathbf{x}-\boldsymbol{\xi}|}\, dx_3 = 2\int_0^\infty \frac{e^{-\frac{p}{c}\sqrt{\rho^2+x_3^2}}}{\sqrt{\rho^2+x_3^2}}\, dx_3 = 2K_0\left(\frac{p}{c}\rho\right) \tag{12.1.179}$$

where $K_0 = K_0(x)$ is the modified Bessel function of the second kind and order zero, and $\rho = \rho(\mathbf{x}, \boldsymbol{\xi})$ is given by

$$\rho(\mathbf{x}, \boldsymbol{\xi}) = \sqrt{(x_1-\xi_1)^2 + (x_2-\xi_2)^2} \tag{12.1.180}$$

we find

$$\left(\nabla^2 - \frac{p^2}{c^2}\right)K_0\left(\frac{p}{c}\rho\right) = -2\pi\,\delta(x_1-\xi_1)\delta(x_2-\xi_2) \tag{12.1.181}$$

Let $g = g(\mathbf{x})$ be a function on $\Omega \subset E^2$. Multiplying Equation 12.1.181 by $g = g(\boldsymbol{\xi})$ and integrating over Ω we have

$$\left(\nabla^2 - \frac{p^2}{c^2}\right)\int_\Omega K_0\left(\frac{p}{c}\rho\right)g(\boldsymbol{\xi})\, d\boldsymbol{\xi} = -2\pi\, g(\mathbf{x}), \quad \mathbf{x} \in \Omega \subset E^2 \tag{12.1.182}$$

or

$$-\frac{1}{c^2}\left(\nabla^2 - \frac{p^2}{c^2}\right)^{-1} g(\mathbf{x}) = \frac{1}{2\pi c^2}\int_\Omega K_0\left(\frac{p}{c}\rho\right)g(\boldsymbol{\xi})\, d\boldsymbol{\xi} \tag{12.1.183}$$

Since [10]

$$L^{-1}\left\{K_0\left(\frac{p}{c}\rho\right)\right\} = H\left(t-\frac{\rho}{c}\right)\left(t^2-\frac{\rho^2}{c^2}\right)^{-1/2} \tag{12.1.184}$$

where $H = H(x)$ is the Heaviside function, therefore, applying the operator L^{-1} to Equation 12.1.183 we arrive at

$$L^{-1}\left\{-\frac{1}{c^2}\left(\nabla^2-\frac{p^2}{c^2}\right)^{-1} g(\mathbf{x})\right\} = t\,\mathcal{N}_{\mathbf{x},ct}(g) \tag{12.1.185}$$

where the symbol $\mathcal{N}_{\mathbf{x},ct}(g)$ is defined in Equation 12.1.173. Hence, applying the operator $L^{-1}\{(\nabla^2 - p^2/c^2)^{-1}\}$ to Equation 12.1.176 and using Equation 12.1.185, we arrive at the representation 12.1.172. This completes the proof of Theorem 1. □

Notes:

(1) By introducing the polar coordinates (r,φ) with the origin at $\mathbf{x}$

$$\xi_1 - x_1 = r\cos\varphi, \quad \xi_2 - x_2 = r\sin\varphi \tag{12.1.186}$$

Equation 12.1.173 may be reduced to the form

$$\mathcal{N}_{\mathbf{x},ct}(f) = \frac{1}{2\pi ct}\int_0^{ct} dr \int_0^{2\pi} d\varphi \frac{r}{\sqrt{c^2t^2-r^2}} f(x_1 + r\cos\varphi, x_2 + r\sin\varphi) \tag{12.1.187}$$

(2) As opposed to the wave produced by a localized initial disturbance in E^3 for which the wave amplitude vanishes outside of a finite time interval [see Remarks after Theorem 1 of Section 11.1], the wave produced by a localized initial disturbance in E^2 has a "tail" over the semi-infinite time interval $t > 0$. To prove this consider the situation shown in Figure 12.7 in which $\phi_0 = 0$ and $\dot{\phi}_0 \neq 0$ on Ω. Let Ω be contained in a circular ring region $ct_1 < |\mathbf{x} - \boldsymbol{\xi}| < ct_2$, and let $\partial\Omega$ be tangent to the circles $|\mathbf{x} - \boldsymbol{\xi}| = ct_1$ and $|\mathbf{x} - \boldsymbol{\xi}| = ct_2$ at the points $\mathbf{x}_1$ and $\mathbf{x}_2$, respectively (see Figure 12.7). If $ct < |\mathbf{x} - \mathbf{x}_1|$ then $\mathcal{N}_{\mathbf{x},ct}(f) = 0$. If $|\mathbf{x} - \mathbf{x}_1| < ct < |\mathbf{x} - \mathbf{x}_2|$ then $\mathcal{N}_{\mathbf{x},ct}(f) \neq 0$. Since the integration in Equation 12.1.187 is over the circular region with the origin at $\mathbf{x}$ and with the radius ct for every $t > 0$, therefore $\mathcal{N}_{\mathbf{x},ct}(f) \neq 0$ for $|\mathbf{x} - \mathbf{x}_1| < ct < \infty$, which means that the wave represented by ϕ is not localized for $t > 0$, that is, it has a "tail."

(3) A "tail" may be also observed for the two-dimensional point source solution 12.1.184. The situation is shown in Figure 12.8 in which

$$\varphi(\rho,t) = H(t-\rho/c)(t^2-\rho^2/c^2)^{-1/2}$$

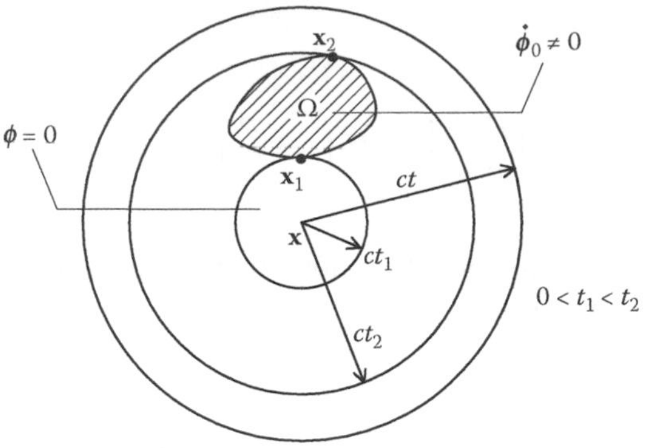

FIGURE 12.7 A support of the initial disturbance $\dot{\phi}_0 \equiv f$ in a ring region.

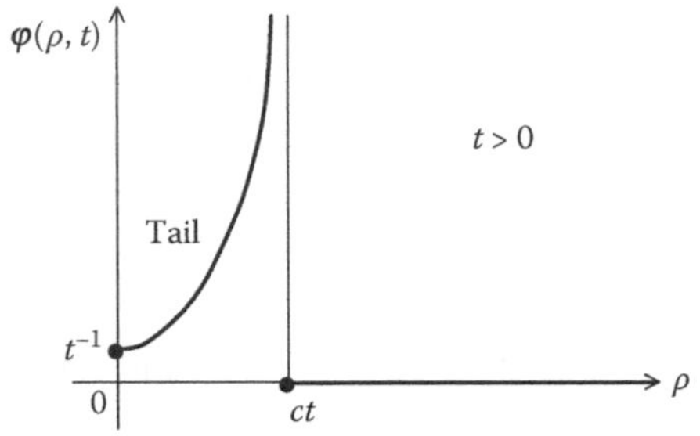

FIGURE 12.8 Two-dimensional wave function for an impulsive source at a fixed time $t > 0$.

12.1.4.2 The Stress Waves Produced by the Initial Data in E^2

The stress waves propagating in a homogeneous isotropic infinite elastic solid subject to the initial stress and stress-rate fields under plane strain conditions are described by a solution to the following pure stress initial value problem. Find a stress field $S_{\alpha\beta} = S_{\alpha\beta}(\mathbf{x}, t)$ on $E^2 \times [0, \infty)$ that satisfies the field equation (see Equation 12.1.29)

$$S_{(\alpha\gamma,\gamma\beta)} - \frac{\rho}{2\mu}\left[\ddot{S}_{\alpha\beta} - \frac{\lambda}{2(\lambda+\mu)}\delta_{\alpha\beta}\ddot{S}_{\gamma\gamma}\right] = 0 \quad \text{on } E^2 \times [0, \infty) \tag{12.1.188}$$

subject to the initial conditions

$$S_{\alpha\beta}(\mathbf{x}, 0) = S^{\circ}_{\alpha\beta}(\mathbf{x}), \quad \dot{S}_{\alpha\beta}(\mathbf{x}, 0) = \dot{S}^{\circ}_{\alpha\beta}(\mathbf{x}), \quad \mathbf{x} \in \Omega \subset E^2 \tag{12.1.189}$$

where $S^{\circ}_{\alpha\beta}(\mathbf{x})$ and $\dot{S}^{\circ}_{\alpha\beta}(\mathbf{x})$ are prescribed second-order symmetric tensor fields on Ω.

By letting $\alpha = \beta$ in Equation 12.1.188 we find

$$S_{\alpha\beta,\alpha\beta} - \frac{\rho}{2\lambda + 2\mu}\ddot{S}_{\gamma\gamma} = 0 \tag{12.1.190}$$

Hence, by eliminating $\ddot{S}_{\gamma\gamma}$ from Equations 12.1.188 and 12.1.190 we obtain

$$\rho\ddot{S}_{\alpha\beta} = 2\mu S_{(\alpha\beta,\gamma\beta)} + \lambda S_{\delta\epsilon,\delta\epsilon}\delta_{\alpha\beta} \tag{12.1.191}$$

Integrating Equation 12.1.191 twice with respect to t and using the initial conditions 12.1.189 we get

$$S_{\alpha\beta}(\mathbf{x},t) = S^{\circ}_{\alpha\beta}(\mathbf{x}) + t\dot{S}^{\circ}_{\alpha\beta}(\mathbf{x}) + \rho^{-1}\int_0^t (t-\tau)\left[2\mu S_{(\alpha\gamma,\gamma\beta)}(\mathbf{x},\tau) + \lambda S_{\delta\epsilon,\delta\epsilon}(\mathbf{x},\tau)\delta_{\alpha\beta}\right] d\tau \tag{12.1.192}$$

Therefore, an alternative formulation of the problem reads: Find a tensor field $S_{\alpha\beta} = S_{\alpha\beta}(\mathbf{x},t)$ on $E^2 \times [0,\infty)$ that satisfies the integro-differential equation 12.1.192. Note that $S_{\alpha\beta}$ given by Equation 12.1.192 is a solution to the problem provided the tensor field $S_{(\alpha\gamma,\gamma\beta)}$ on $E^2 \times [0,\infty)$ is available. In the following, we outline a Laplace transform technique of finding this field.

Let $\overline{S}_{\alpha\beta} = \overline{S}_{\alpha\beta}(\mathbf{x},p)$ be the Laplace transform of $S_{\alpha\beta} = S_{\alpha\beta}(\mathbf{x},t)$ with respect to time, that is,

$$L\{S_{\alpha\beta}\} = \overline{S}_{\alpha\beta}(\mathbf{x},p) = \int_0^\infty e^{-pt} S_{\alpha\beta}(\mathbf{x},t)\, dt \tag{12.1.193}$$

where p is the transform parameter. By applying the operator L to Equation 12.1.188, and using the initial conditions 12.1.189, we obtain

$$\overline{S}_{(\alpha\gamma,\gamma\beta)} - \frac{\rho p^2}{2\mu}\left(\overline{S}_{\alpha\beta} - \frac{\lambda}{2\lambda+2\mu}\overline{S}_{\gamma\gamma}\delta_{\alpha\beta}\right) = -\rho p^2 \overline{B}_{\alpha\beta} \tag{12.1.194}$$

where

$$p^2\overline{B}_{\alpha\beta} = \frac{1}{2\mu}\left(\dot{S}^{\circ}_{\alpha\beta} - \frac{\lambda}{2\lambda+2\mu}\dot{S}^{\circ}_{\gamma\gamma}\delta_{\alpha\beta}\right) + \frac{p}{2\mu}\left(S^{\circ}_{\alpha\beta} - \frac{\lambda}{2\lambda+2\mu}S^{\circ}_{\gamma\gamma}\delta_{\alpha\beta}\right) \tag{12.1.195}$$

Next, letting $\beta = \delta$ and getting rid of the index parentheses in Equation 12.1.194, we have

$$\overline{S}_{\alpha\gamma,\gamma\delta} + \overline{S}_{\delta\gamma,\gamma\alpha} - \frac{\rho p^2}{\mu}\left(\overline{S}_{\alpha\delta} - \frac{\lambda}{2\lambda+2\mu}\overline{S}_{\gamma\gamma}\delta_{\alpha\delta}\right) = -2\rho p^2 \overline{B}_{\alpha\delta} \tag{12.1.196}$$

Now, by applying the operator $\partial^2/\partial x_\delta \partial x_\beta$ to Equation 12.1.196 and taking the symmetric part of the result with respect to α and β, we obtain

$$\left(\nabla^2 - \frac{p^2}{c_2^2}\right)\overline{S}_{(\alpha\gamma,\gamma\beta)} = -\frac{\partial^2}{\partial x_\alpha \partial x_\beta}\left(\overline{S}_{\gamma\delta,\gamma\delta} + \frac{\rho p^2}{\mu}\frac{\lambda}{2\lambda+2\mu}\overline{S}_{\gamma\gamma}\right) - 2\rho p^2 \overline{B}_{(\alpha\delta,\delta\beta)} \tag{12.1.197}$$

where

$$c_2^2 = \frac{\mu}{\rho} \tag{12.1.198}$$

Finally, by applying the operator $(\nabla^2 - p^2/c_1^2)$ to Equation 12.1.197, where $c_1^2 = (\lambda + 2\mu)/\rho$, we arrive at

$$\left(\nabla^2 - \frac{p^2}{c_1^2}\right)\left(\nabla^2 - \frac{p^2}{c_2^2}\right)\overline{S}_{(\alpha\gamma,\gamma\beta)} = -\overline{F}_{\alpha\beta} \tag{12.1.199}$$

where

$$\overline{F}_{\alpha\beta} = \left(\nabla^2 - \frac{p^2}{c_1^2}\right)\frac{\partial^2}{\partial x_\alpha \partial x_\beta}\left(\overline{S}_{\gamma\delta,\gamma\delta} + \frac{\rho p^2}{\mu}\frac{\lambda}{2\lambda + 2\mu}\overline{S}_{\gamma\gamma}\right) + 2\rho p^2\left(\nabla^2 - \frac{p^2}{c_1^2}\right)\overline{B}_{(\alpha\delta,\delta\beta)} \tag{12.1.200}$$

In the following, we show that $\overline{F}_{\alpha\beta}$ can be expressed entirely in terms of $S^\circ_{\alpha\beta}$ and $\dot{S}^\circ_{\alpha\beta}$, and p. First, it follows from the definition of $\overline{B}_{\alpha\beta}$ (see Equation 12.1.195) that the last term on the RHS of Equation 12.1.200 may be expressed in terms of $S^\circ_{\alpha\beta}$, $\dot{S}^\circ_{\alpha\beta}$, and p.

Next, by applying the operator $(\delta_{\alpha\beta}\nabla^2 - \partial^2/\partial x_\alpha \partial x_\beta)$ to Equation 12.1.194 and using the identity

$$\left(\delta_{\alpha\beta}\nabla^2 - \frac{\partial^2}{\partial x_\alpha \partial x_\beta}\right)\overline{S}_{(\alpha\gamma,\gamma\beta)} = 0 \tag{12.1.201}$$

we write

$$\left(\delta_{\alpha\beta}\nabla^2 - \frac{\partial^2}{\partial x_\alpha \partial x_\beta}\right)\left(\overline{S}_{\alpha\beta} - \frac{\lambda}{2\lambda + 2\mu}\overline{S}_{\gamma\gamma}\delta_{\alpha\beta}\right) = 2\mu\left(\delta_{\alpha\beta}\nabla^2 - \frac{\partial^2}{\partial x_\alpha \partial x_\beta}\right)\overline{B}_{\alpha\beta} \tag{12.1.202}$$

or

$$\frac{\lambda + 2\mu}{2\lambda + 2\mu}\nabla^2\overline{S}_{\alpha\alpha} - \overline{S}_{\alpha\beta,\alpha\beta} = 2\mu\left(\overline{B}_{\alpha\alpha,\beta\beta} - \overline{B}_{\alpha\beta,\alpha\beta}\right) \tag{12.1.203}$$

Also, by letting $\alpha = \delta$ in Equation 12.1.196, we write

$$\overline{S}_{\alpha\gamma,\alpha\gamma} - \frac{\rho p^2}{2\lambda + 2\mu}\overline{S}_{\gamma\gamma} = -\rho p^2\overline{B}_{\alpha\alpha} \tag{12.1.204}$$

Hence, by eliminating $\overline{S}_{\alpha\beta,\alpha\beta}$ from Equations 12.1.203 and 12.1.204 we get

$$\left(\nabla^2 - \frac{p^2}{c_1^2}\right)\overline{S}_{\alpha\alpha} = -\frac{1}{c_1^2}\overline{f} \tag{12.1.205}$$

where

$$\overline{f} = (2\lambda + 2\mu)\left[p^2\overline{B}_{\alpha\alpha} - 2\frac{\mu}{\rho}\left(\overline{B}_{\alpha\alpha,\beta\beta} - \overline{B}_{\alpha\beta,\alpha\beta}\right)\right] \tag{12.1.206}$$

or, using of Equation 12.1.195,

$$\bar{f} = \dot{S}^{\circ}_{\alpha\alpha} + pS^{\circ}_{\alpha\alpha} + \frac{2\lambda + 2\mu}{\rho}\left\{p^{-1}\left(S^{\circ}_{\alpha\beta,\alpha\beta} - \frac{\lambda + 2\mu}{2\lambda + 2\mu}S^{\circ}_{\alpha\alpha,\beta\beta}\right)\right.$$
$$\left. + p^{-2}\left(\dot{S}^{\circ}_{\alpha\beta,\alpha\beta} - \frac{\lambda + 2\mu}{2\lambda + 2\mu}\dot{S}^{\circ}_{\alpha\alpha,\beta\beta}\right)\right\} \tag{12.1.207}$$

Now, by applying $(\nabla^2 - p^2/c_1^2)$ to Equation 12.1.204 and using Equation 12.1.205, we have

$$\left(\nabla^2 - \frac{p^2}{c_1^2}\right)\bar{S}_{\alpha\gamma,\gamma\alpha} = -\frac{1}{c_1^2}\,\bar{g} \tag{12.1.208}$$

where

$$\bar{g} = \dot{S}^{\circ}_{\alpha\beta,\alpha\beta} + pS^{\circ}_{\alpha\beta,\alpha\beta} \tag{12.1.209}$$

Next, we introduce the notation

$$\bar{P}_{\alpha\beta} = 2\rho c_2^2 p^2 \bar{B}_{(\alpha\gamma,\gamma\beta)} \tag{12.1.210}$$

and, using of Equation 12.1.195, we obtain

$$\bar{P}_{\alpha\beta} = \dot{S}^{\circ}_{(\alpha\gamma,\gamma\beta)} - \frac{\lambda}{2\lambda + 2\mu}\dot{S}^{\circ}_{\gamma\gamma,\alpha\beta} + p\left(S^{\circ}_{(\alpha\gamma,\gamma\beta)} - \frac{\lambda}{2\lambda + 2\mu}\,S^{\circ}_{\gamma\gamma,\alpha\beta}\right) \tag{12.1.211}$$

Finally, using Equations 12.1.200, 12.1.205, 12.1.208, and 12.1.211, we get

$$\bar{F}_{\alpha\beta} = -\frac{1}{c_1^2}\,\frac{\partial^2}{\partial x_\alpha \partial x_\beta}\left(\bar{g} + \frac{\rho p^2}{\mu}\,\frac{\lambda}{2\lambda + 2\mu}\bar{f}\right) + \frac{1}{c_2^2}\left(\nabla^2 - \frac{p^2}{c_1^2}\right)\bar{P}_{\alpha\beta} \tag{12.1.212}$$

It follows from Equations 12.1.207, 12.1.209, and 12.1.211 that RHS of Equation 12.1.212 is expressed entirely in terms of $S^{\circ}_{\alpha\beta}$, $\dot{S}^{\circ}_{\alpha\beta}$, and p; and this completes the proof that $\bar{F}_{\alpha\beta}$ is expressed in terms of $S^{\circ}_{\alpha\beta}$, $\dot{S}^{\circ}_{\alpha\beta}$, and p.

Now, we note that a unique solution of Equation 12.1.205 in E^2 takes the form

$$\bar{S}_{\alpha\alpha} = -\frac{1}{c_1^2}\left(\nabla^2 - \frac{p^2}{c_1^2}\right)^{-1}\bar{f} \tag{12.1.213}$$

Similarly, a unique solution of Equation 12.1.208 takes the form

$$\bar{S}_{\alpha\gamma,\alpha\gamma} = -\frac{1}{c_1^2}\left(\nabla^2 - \frac{p^2}{c_1^2}\right)^{-1}\bar{g} \tag{12.1.214}$$

Hence, applying the operator L^{-1} to Equation 12.1.213, and using the inversion formula 12.1.185 together with Equation 12.1.207, we find

$$S_{\alpha\alpha}(\mathbf{x},t) = t\mathcal{N}_{\mathbf{x},c_1 t}(\dot{S}^{\circ}_{\alpha\alpha}) + \frac{\partial}{\partial t}\left[t\mathcal{N}_{\mathbf{x},c_1 t}(S^{\circ}_{\alpha\alpha})\right]$$

$$+\frac{2\lambda+2\mu}{\rho}\int_0^t \tau\left\{\mathcal{N}_{\mathbf{x},c_1\tau}\left(S^{\circ}_{\alpha\beta,\alpha\beta} - \frac{\lambda+2\mu}{2\lambda+2\mu}S^{\circ}_{\alpha\alpha,\beta\beta}\right)\right.$$

$$\left.+(t-\tau)\mathcal{N}_{\mathbf{x},c_1\tau}\left(\dot{S}^{\circ}_{\alpha\beta,\alpha\beta} - \frac{\lambda+2\mu}{2\lambda+2\mu}\dot{S}^{\circ}_{\alpha\alpha,\beta\beta}\right)\right\}d\tau \tag{12.1.215}$$

where the operator $\mathcal{N}_{\mathbf{x},ct}(\cdot)$ is defined by Equation 12.1.173. Similarly, applying the operator L^{-1} to Equation 12.1.214 and using Equations 12.1.185 and 12.1.209, we arrive at

$$S_{\alpha\beta,\alpha\beta}(\mathbf{x},t) = t\mathcal{N}_{\mathbf{x},c_1 t}(\dot{S}^{\circ}_{\alpha\beta,\alpha\beta}) + \frac{\partial}{\partial t}\left[t\mathcal{N}_{\mathbf{x},c_1 t}(S^{\circ}_{\alpha\beta,\alpha\beta})\right] \tag{12.1.216}$$

To obtain $\overline{S}_{(\alpha\gamma,\gamma\beta)}$ we apply the operator $\left[\left(\nabla^2 - p^2/c_1^2\right)\left(\nabla^2 - p^2/c_2^2\right)\right]^{-1}$ to Equation 12.1.199, and using Equation 12.1.212, we obtain

$$\overline{S}_{(\alpha\gamma,\gamma\beta)} = \frac{1}{c_1^2}\left[\left(\nabla^2 - \frac{p^2}{c_1^2}\right)\left(\nabla^2 - \frac{p^2}{c_2^2}\right)\right]^{-1}$$

$$\times\left(\frac{p^2}{c_2^2}\frac{\lambda}{2\lambda+2\mu}\overline{f}_{,\alpha\beta} + \overline{g}_{,\alpha\beta}\right) - \frac{1}{c_2^2}\left(\nabla^2 - \frac{p^2}{c_2^2}\right)^{-1}\overline{P}_{\alpha\beta} \tag{12.1.217}$$

Since for any function $h = h(\mathbf{x})$ on $\Omega \subset E^2$

$$\left[\left(\nabla^2 - \frac{p^2}{c_1^2}\right)\left(\nabla^2 - \frac{p^2}{c_2^2}\right)\right]^{-1} h(\mathbf{x}) = \left(\frac{p^2}{c_1^2} - \frac{p^2}{c_2^2}\right)^{-1}$$

$$\times\left[\left(\nabla^2 - \frac{p^2}{c_1^2}\right)^{-1} h(\mathbf{x}) - \left(\nabla^2 - \frac{p^2}{c_2^2}\right)^{-1} h(\mathbf{x})\right] \tag{12.1.218}$$

therefore, applying L^{-1} to Equation 12.1.218 and using the inversion formula (see Equation 12.1.185)

$$L^{-1}\left\{-\frac{1}{c^2}\left(\nabla^2 - \frac{p^2}{c^2}\right)^{-1} h(\mathbf{x})\right\} = t\mathcal{N}_{\mathbf{x},ct}(h) \tag{12.1.219}$$

we have

$$L^{-1}\left\{\left[\left(\nabla^2 - \frac{p^2}{c_1^2}\right)\left(\nabla^2 - \frac{p^2}{c_2^2}\right)\right]^{-1} h(\mathbf{x})\right\}$$

$$= \frac{c_1^2 c_2^2}{c_1^2 - c_2^2}\int_0^t (t-\tau)\tau\left[c_1^2\mathcal{N}_{\mathbf{x},c_1\tau}(h) - c_2^2\mathcal{N}_{\mathbf{x},c_2\tau}(h)\right]d\tau \tag{12.1.220}$$

Also, multiplying Equation 12.1.218 by p^2 and applying the operator L^{-1} to the result, by virtue of Equation 12.1.219, we arrive at

$$L^{-1}\left\{\left[\left(\nabla^2 - \frac{p^2}{c_1^2}\right)\left(\nabla^2 - \frac{p^2}{c_2^2}\right)\right]^{-1} p^2 h(\mathbf{x})\right\} = \frac{c_1^2 c_2^2}{c_1^2 - c_2^2}\left[c_1^2 t \mathcal{N}_{\mathbf{x},c_1 t}(h) - c_2^2 t \mathcal{N}_{\mathbf{x},c_2 t}(h)\right] \tag{12.1.221}$$

Hence, applying the operator L^{-1} to Equation 12.1.217 and using Equations 12.1.207, 12.1.209, and 12.1.211 we obtain

$$\begin{aligned} S_{(\alpha\gamma,\gamma\beta)}(\mathbf{x},t) &= \int_0^t (t-\tau)\left[c_1^2 \tau\, \mathcal{N}_{\mathbf{x},c_1\tau}\left(\dot{M}^\circ_{\alpha\beta}\right) - c_2^2 \tau\, \mathcal{N}_{\mathbf{x},c_2\tau}\left(\dot{M}^\circ_{\alpha\beta}\right)\right] d\tau \\ &+ \int_0^t \left[c_1^2 \tau \mathcal{N}_{\mathbf{x},c_1\tau}\left(M^\circ_{\alpha\beta}\right) - c_2^2 \tau \mathcal{N}_{\mathbf{x},c_2\tau}\left(M^\circ_{\alpha\beta}\right)\right] d\tau \\ &+ \left[c_1^2 t \mathcal{N}_{\mathbf{x},c_1 t}\left(\dot{N}^\circ_{\alpha\beta}\right) - c_2^2 t \mathcal{N}_{\mathbf{x},c_2 t}\left(\dot{N}^\circ_{\alpha\beta}\right)\right] \\ &+ \frac{\partial}{\partial t}\left[c_1^2 t \mathcal{N}_{\mathbf{x},c_1 t}\left(N^\circ_{\alpha\beta}\right) - c_2^2 t \mathcal{N}_{\mathbf{x},c_2 t}\left(N^\circ_{\alpha\beta}\right)\right] \\ &+ t \mathcal{N}_{\mathbf{x},c_2 t}\left(\dot{L}^\circ_{\alpha\beta}\right) + \frac{\partial}{\partial t}\left[t \mathcal{N}_{\mathbf{x},c_2 t}\left(L^\circ_{\alpha\beta}\right)\right] \end{aligned} \tag{12.1.222}$$

where the tensor fields $\dot{M}^\circ_{\alpha\beta}$, $M^\circ_{\alpha\beta}$, $\dot{N}^\circ_{\alpha\beta}$, $N^\circ_{\alpha\beta}$, $\dot{L}^\circ_{\alpha\beta}$, and $L^\circ_{\alpha\beta}$ are defined on $\Omega \subset E^2$ through the formulas

$$\dot{M}^\circ_{\alpha\beta} = \frac{\partial^2}{\partial x_\alpha \partial x_\beta}\left(\dot{S}^\circ_{\mu\nu,\mu\nu} - \frac{c_1^2}{c_1^2 - c_2^2}\frac{\lambda}{2\lambda + 2\mu}\dot{S}^\circ_{\mu\mu,\nu\nu}\right) \tag{12.1.223}$$

$$M^\circ_{\alpha\beta} = \frac{\partial^2}{\partial x_\alpha \partial x_\beta}\left(S^\circ_{\mu\nu,\mu\nu} - \frac{c_1^2}{c_1^2 - c_2^2}\frac{\lambda}{2\lambda + 2\mu}S^\circ_{\mu\mu,\nu\nu}\right) \tag{12.1.224}$$

$$\dot{N}^\circ_{\alpha\beta} = \frac{1}{c_1^2 - c_2^2}\frac{\lambda}{2\lambda + 2\mu}\dot{S}^\circ_{\mu\mu,\alpha\beta} \tag{12.1.225}$$

$$N^\circ_{\alpha\beta} = \frac{1}{c_1^2 - c_2^2}\frac{\lambda}{2\lambda + 2\mu}S^\circ_{\mu\mu,\alpha\beta} \tag{12.1.226}$$

$$\dot{L}^\circ_{\alpha\beta} = \dot{S}^\circ_{(\alpha\gamma,\gamma\beta)} - \frac{\lambda}{2\lambda + 2\mu}\dot{S}^\circ_{\mu\mu,\alpha\beta} \tag{12.1.227}$$

and

$$L^\circ_{\alpha\beta} = S^\circ_{(\alpha\gamma,\gamma\beta)} - \frac{\lambda}{2\lambda + 2\mu}S^\circ_{\mu\mu,\alpha\beta} \tag{12.1.228}$$

Hence, we arrive at the following theorem.

Theorem 2: A solution to the stress initial value problem described by Equations 12.1.188 and 12.1.189 takes the form 12.1.192 in which the fields $S_{\delta\epsilon,\delta\epsilon}$ and $S_{(\alpha\gamma,\gamma\beta)}$ are given by Equations 12.1.216 and 12.1.222, respectively. ■

Notes:

(1) The solution 12.1.192 represents stress waves of classic two-dimensional elastodynamics if $S^{\circ}_{\alpha\beta}$ and $\dot{S}^{\circ}_{\alpha\beta}$ are given by

$$S^{\circ}_{\alpha\beta} = 2\mu u^{\circ}_{(\alpha,\beta)} + \lambda u^{\circ}_{\gamma,\gamma}\delta_{\alpha\beta} \tag{12.1.229}$$

and

$$\dot{S}^{\circ}_{\alpha\beta} = 2\mu \dot{u}^{\circ}_{(\alpha,\beta)} + \lambda \dot{u}^{\circ}_{\gamma,\gamma}\delta_{\alpha\beta} \tag{12.1.230}$$

respectively, where $u^{\circ}_{\alpha} = u^{\circ}_{\alpha}(\mathbf{x})$ and $\dot{u}^{\circ}_{\alpha} = \dot{u}^{\circ}_{\alpha}(\mathbf{x})$ are prescribed vector fields on Ω. Otherwise, the solution describes stress waves propagating in an *elastic solid with initial defects*.

(2) If $S^{\circ}_{\alpha\beta}$ and $\dot{S}^{\circ}_{\alpha\beta}$ satisfy the equilibrium equations

$$S^{\circ}_{\alpha\beta,\beta} = 0, \quad \dot{S}^{\circ}_{\alpha\beta,\beta} = 0 \quad \text{on } \Omega \subset E^2 \tag{12.1.231}$$

they may be represented by

$$S^{\circ}_{\alpha\beta} = \left(\delta_{\alpha\beta}\nabla^2 - \frac{\partial^2}{\partial x_\alpha \partial x_\beta}\right) F^{\circ} \quad \text{on } \Omega \tag{12.1.232}$$

and

$$\dot{S}^{\circ}_{\alpha\beta} = \left(\delta_{\alpha\beta}\nabla^2 - \frac{\partial^2}{\partial x_\alpha \partial x_\beta}\right) \dot{F}^{\circ} \quad \text{on } \Omega \tag{12.1.233}$$

where the functions $F^{\circ} = F^{\circ}(\mathbf{x})$ and $\dot{F}^{\circ} = \dot{F}^{\circ}(\mathbf{x})$ are the Airy stress functions. If, in addition, F° and $\dot{F}^{\circ}$ are biharmonic functions on Ω, then $S^{\circ}_{\alpha\beta}$ and $\dot{S}^{\circ}_{\alpha\beta}$ satisfy the stress and stress-rate compatibility conditions

$$S^{\circ}_{\mu\mu,\nu\nu} = 0, \quad \dot{S}^{\circ}_{\mu\mu,\nu\nu} = 0 \quad \text{on } \Omega \tag{12.1.234}$$

In this case, a solution to the problem takes the form

$$S_{\alpha\beta}(\mathbf{x}, t) = S^{\circ}_{\alpha\beta}(\mathbf{x}) + t\dot{S}^{\circ}_{\alpha\beta}(\mathbf{x}) + 2\frac{\mu}{\rho}\int_0^t (t-\tau) S_{(\alpha\gamma,\gamma\beta)}(\mathbf{x}, \tau)\, d\tau \tag{12.1.235}$$

where

$$\begin{aligned} S_{(\alpha\gamma,\gamma\beta)}(\mathbf{x}, t) = {} & \frac{1}{c_1^2 - c_2^2}\frac{\lambda}{2\lambda + 2\mu}\Big\{\left[c_1^2 t \mathcal{N}_{\mathbf{x},c_1 t}\left(\dot{S}^{\circ}_{\mu\mu,\alpha\beta}\right) - c_2^2 t \mathcal{N}_{\mathbf{x},c_2 t}\left(\dot{S}^{\circ}_{\mu\mu,\alpha\beta}\right)\right] \\ & + \frac{\partial}{\partial t}\left[c_1^2 t \mathcal{N}_{\mathbf{x},c_1 t}\left(S^{\circ}_{\mu\mu,\alpha\beta}\right) - c_2^2 t \mathcal{N}_{\mathbf{x},c_2 t}\left(S^{\circ}_{\mu\mu,\alpha\beta}\right)\right]\Big\} \\ & - \frac{\lambda}{2\lambda + 2\mu}\left\{t\mathcal{N}_{\mathbf{x},c_2 t}\left(\dot{S}^{\circ}_{\mu\mu,\alpha\beta}\right) + \frac{\partial}{\partial t}\left[t\mathcal{N}_{\mathbf{x},c_2 t}\left(S^{\circ}_{\mu\mu,\alpha\beta}\right)\right]\right\} \end{aligned} \tag{12.1.236}$$

In particular, we obtain

$$S_{\alpha\alpha}(\mathbf{x},t) = S^{\circ}_{\alpha\alpha}(\mathbf{x}) + t\dot{S}^{\circ}_{\alpha\alpha}(\mathbf{x}) \tag{12.1.237}$$

(3) The tensor fields $S^{\circ}_{\alpha\beta}$ and $\dot{S}^{\circ}_{\alpha\beta}$ may be obtained by solving suitable traction boundary value problems of two-dimensional elastostatics for the domain Ω when the stress vector $t_\alpha = S^{\circ}_{\alpha\beta}n_\beta$ and the stress-rate vector $\dot{t}_\alpha = \dot{S}^{\circ}_{\alpha\beta}n_\beta$, respectively, are prescribed on $\partial\Omega$. They may be also identified with solutions to suitable mixed boundary value problems of two-dimensional elastostatics, or with the internal stress and stress-rate fields of a continuum theory of dislocations [11].

12.2 TWO-DIMENSIONAL SOLUTIONS OF NONISOTHERMAL ELASTODYNAMICS

In this section, the two closed-form solutions of two-dimensional dynamic theory of thermal stresses introduced in Section 8.2.2 are presented: (A) dynamic thermal stresses in an infinite elastic sheet subject to a discontinuous temperature field, and (B) dynamic thermal stresses produced by an instantaneous concentrated heat source in an infinite elastic sheet. In case (A) the solution is generated by a thermoelastic displacement wave potential, while in case (B) the solution is a sum of wave and diffusive parts.

12.2.1 Dynamic Thermal Stresses in an Infinite Elastic Sheet Subject to a Discontinuous Temperature Field

Suppose that a homogeneous isotropic infinite elastic sheet obeying generalized plane stress conditions, described by the inequalities: $|x_1| < \infty$, $|x_2| < \infty$, and initially at rest, be subject to a temperature field $\overline{T} = \overline{T}(x_1, x_2; t)$ of the form

$$\overline{T}(x_1, x_2; t) = T_0[H(x_1 + a_1) - H(x_1 - a_1)][H(x_2 + a_2) - H(x_2 - a_2)]\delta(t) \tag{12.2.1}$$

where T_0 is a constant temperature, a_1 and a_2 are positive parameters of the length dimension; while $H = H(x)$ and $\delta = \delta(t)$ represent the Heaviside and Dirac functions, respectively. Hence, the sheet is heated instantaneously at $t = 0+0$ to a constant temperature T_0 over the rectangular region $|x_1| < a_1$, $|x_2| < a_2$. A thermoelastic process $\overline{p} = [\overline{\mathbf{u}}, \overline{\mathbf{E}}, \overline{\mathbf{S}}]$ corresponding to the temperature $\overline{T} = \overline{T}(\mathbf{x}, t)$ is described by a potential $\overline{\phi} = \overline{\phi}(\mathbf{x}, t)$ through the formulas (see Part (B) in Example 8.2.13)

$$\overline{\mathbf{u}} = \nabla\overline{\phi} \quad \text{on } E^2 \times [0, \infty) \tag{12.2.2}$$

$$\overline{\mathbf{E}} = \nabla\nabla\overline{\phi} \quad \text{on } E^2 \times [0, \infty) \tag{12.2.3}$$

and

$$\overline{\mathbf{S}} = 2\mu\left(\nabla\nabla\overline{\phi} - \nabla^2\overline{\phi}\mathbf{1}\right) + \rho\ddot{\overline{\phi}}\mathbf{1} \quad \text{on } E^2 \times [0, \infty) \tag{12.2.4}$$

where $\overline{\phi} = \overline{\phi}(\mathbf{x}, t)$ satisfies the nonhomogeneous wave equation

$$\overline{\Box}^2_1\,\overline{\phi} = \overline{m}\,\overline{T} \quad \text{on } E^2 \times [0, \infty) \tag{12.2.5}$$

Here

$$\overline{\square}_1^2 = \nabla^2 - \frac{1}{\overline{c}_1^2}\frac{\partial^2}{\partial t^2} \tag{12.2.6}$$

$$\frac{1}{\overline{c}_1^2} = \frac{\rho}{\overline{\lambda} + 2\mu} = \frac{1-\nu}{2}\frac{\rho}{\mu}, \quad \overline{m} = (1+\nu)\alpha \tag{12.2.7}$$

In the following, we let $\overline{c}_1 = c$. Since the infinite sheet is initially at rest, the process $\overline{p} = \overline{p}(\mathbf{x}, t)$ satisfies the homogeneous initial conditions

$$\overline{p}(\mathbf{x}, 0) = \dot{\overline{p}}(\mathbf{x}, 0) = 0 \quad \text{for } |x_1| < \infty, \quad |x_2| < \infty \tag{12.2.8}$$

Note that the conditions 12.2.8 are satisfied provided

$$\overline{\phi}(\mathbf{x}, 0) = \dot{\overline{\phi}}(\mathbf{x}, 0) = 0 \quad \text{for } |x_1| < \infty, \quad |x_2| < \infty \tag{12.2.9}$$

Let $\overline{\phi}^* = \overline{\phi}^*(\mathbf{x}, \boldsymbol{\xi}; t)$ be a solution of the wave equation

$$\overline{\square}_1^2 \overline{\phi}^* = \delta(x_1 - \xi_1)\delta(x_2 - \xi_2)\delta(t) \tag{12.2.10}$$

subject to the initial conditions

$$\overline{\phi}^*(\mathbf{x}, \boldsymbol{\xi}; 0) = \dot{\overline{\phi}}^*(\mathbf{x}, \boldsymbol{\xi}; 0) = 0 \tag{12.2.11}$$

where $\mathbf{x}, \boldsymbol{\xi} \in E^2$. Then, it follows from Equations 12.1.183 and 12.1.184 in which $g(\mathbf{x}) = \delta(\mathbf{x} - \boldsymbol{\xi})$ that

$$\overline{\phi}^*(\mathbf{x}, \boldsymbol{\xi}; t) = -\frac{1}{2\pi} H\left(t - \frac{|\mathbf{x} - \boldsymbol{\xi}|}{c}\right)\left[t^2 - \frac{|\mathbf{x} - \boldsymbol{\xi}|^2}{c^2}\right]^{-1/2} \tag{12.2.12}$$

and a solution $\overline{\phi} = \overline{\phi}(\mathbf{x}, t)$ to Equation 12.2.5 subject to the initial conditions 12.2.9 takes the form

$$\overline{\phi}(\mathbf{x}, t) = -\frac{\overline{m}}{2\pi}\int_{-a_1}^{a_1} d\xi_1 \int_{-a_2}^{a_2} d\xi_2 \, \frac{H(t - |\mathbf{x} - \boldsymbol{\xi}|/c)}{[t^2 - |\mathbf{x} - \boldsymbol{\xi}|^2/c^2]^{1/2}} \tag{12.2.13}$$

Since

$$\frac{\partial}{\partial x_\alpha}|\mathbf{x} - \boldsymbol{\xi}| = -\frac{\partial}{\partial \xi_\alpha}|\mathbf{x} - \boldsymbol{\xi}| \quad \text{for } \alpha = 1, 2 \tag{12.2.14}$$

therefore, differentiating Equation 12.2.13 with respect to x_1 and using Equation 12.2.2, we obtain

$$\bar{u}_1(\mathbf{x},t) = \frac{\overline{m}}{2\pi}\int_{-a_2}^{a_2} d\xi_2 \int_{-a_1}^{a_1} d\xi_1 \frac{\partial}{\partial \xi_1}\left\{\frac{H(t-|\mathbf{x}-\boldsymbol{\xi}|/c)}{[t^2-|\mathbf{x}-\boldsymbol{\xi}|^2/c^2]^{1/2}}\right\}$$

$$= \frac{\overline{m}c}{2\pi}\int_{-a_2}^{a_2} d\xi_2 \left\{\frac{H\left[ct-\sqrt{(x_1-a_1)^2+(x_2-\xi_2)^2}\right]}{\sqrt{c^2t^2-(x_1-a_1)^2-(x_2-\xi_2)^2}} - \frac{H\left[ct-\sqrt{(x_1+a_1)^2+(x_2-\xi_2)^2}\right]}{\sqrt{c^2t^2-(x_1+a_1)^2-(x_2-\xi_2)^2}}\right\} \tag{12.2.15}$$

Hence, taking into account the formula

$$\int \frac{du}{\sqrt{a^2-u^2}} = \sin^{-1}\frac{u}{a} \quad \text{for } |a|>|u| \tag{12.2.16}$$

we reduce Equation 12.2.15 to the form:

$$\begin{aligned}\bar{u}_1(x_1,x_2;t) = \frac{\overline{m}c}{2\pi}\Bigg\{ & H\left[ct-\sqrt{(x_1-a_1)^2+(x_2+a_2)^2}\right]\sin^{-1}\frac{(x_2+a_2)}{\sqrt{c^2t^2-(x_1-a_1)^2}} \\ & - H\left[ct-\sqrt{(x_1-a_1)^2+(x_2-a_2)^2}\right]\sin^{-1}\frac{(x_2-a_2)}{\sqrt{c^2t^2-(x_1-a_1)^2}} \\ & - H\left[ct-\sqrt{(x_1+a_1)^2+(x_2+a_2)^2}\right]\sin^{-1}\frac{(x_2+a_2)}{\sqrt{c^2t^2-(x_1+a_1)^2}} \\ & + H\left[ct-\sqrt{(x_1+a_1)^2+(x_2-a_2)^2}\right]\sin^{-1}\frac{(x_2-a_2)}{\sqrt{c^2t^2-(x_1+a_1)^2}}\Bigg\}\end{aligned} \tag{12.2.17}$$

In a similar way, by differentiating Equation 12.2.13 with respect to x_2, and using Equations 12.2.14, we write

$$\begin{aligned}\bar{u}_2(x_1,x_2;t) = \frac{\overline{m}c}{2\pi}\Bigg\{ & H\left[ct-\sqrt{(x_1+a_1)^2+(x_2-a_2)^2}\right]\sin^{-1}\frac{(x_1+a_1)}{\sqrt{c^2t^2-(x_2-a_2)^2}} \\ & - H\left[ct-\sqrt{(x_1-a_1)^2+(x_2-a_2)^2}\right]\sin^{-1}\frac{(x_1-a_1)}{\sqrt{c^2t^2-(x_2-a_2)^2}} \\ & - H\left[ct-\sqrt{(x_1+a_1)^2+(x_2+a_2)^2}\right]\sin^{-1}\frac{(x_1+a_1)}{\sqrt{c^2t^2-(x_2+a_2)^2}} \\ & + H\left[ct-\sqrt{(x_1-a_1)^2+(x_2+a_2)^2}\right]\sin^{-1}\frac{(x_1-a_1)}{\sqrt{c^2t^2-(x_2+a_2)^2}}\Bigg\}\end{aligned} \tag{12.2.18}$$

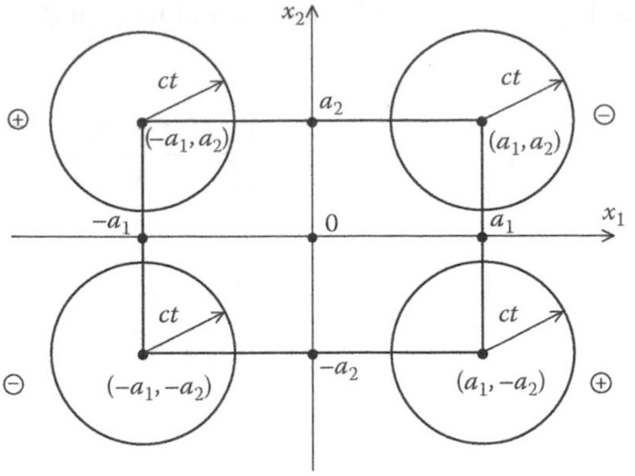

FIGURE 12.9 The circular wavefronts around four corners of the rectangular region. The sign $+(-)$ represents the positive (negative) amplitude of a circular wave.

It follows from Equation 12.2.17 that the displacement $\overline{u}_1$ is a sum of the four circular waves with the centers at four corners of the rectangular region: (a_1, a_2), $(a_1, -a_2)$, $(-a_1, -a_2)$, and $(-a_1, a_2)$, and with the same radius ct. For sufficiently small time, the displacement $\overline{u}_1$ is localized around the four corners as shown in Figure 12.9. As time progresses, the displacement waves expand to contain the whole rectangular region together with its neighborhood of a finite diameter, and as $t \to \infty$, $\overline{u}_1 \to 0$ for any point of the sheet that is located at a finite distance from the origin $\mathbf{x} = \mathbf{0}$. Also, it follows from Equation 12.2.17 that $\overline{u}_1(0, 0; t) = 0$ for every $t > 0$.

Since the displacement $\overline{u}_2 = \overline{u}_2(x_1, x_2; t)$, given by Equation 12.2.18, has a form similar to that of $\overline{u}_1 = \overline{u}_1(x_1, x_2; t)$, the displacement waves described by $\overline{u}_2 = \overline{u}_2(x_1, x_2; t)$ reveal a behavior that is similar to that of $\overline{u}_1 = \overline{u}_1(x_1, x_2; t)$. In particular, it follows from Equation 12.2.18 that $\overline{u}_2(0, 0; t) = 0$ for $t > 0$.

The strain tensor $\overline{E}_{\alpha\beta}$ corresponding to the displacement vector $\overline{u}_\alpha$ may be obtained from the relation

$$\overline{E}_{\alpha\beta} = \frac{1}{2}(\overline{u}_{\alpha,\beta} + \overline{u}_{\beta,\alpha}) \tag{12.2.19}$$

while the associated stress tensor $\overline{S}_{\alpha\beta}$ may be found from the constitutive equation (see Equation 8.2.28)

$$\overline{S}_{\alpha\beta} = 2\mu\left(\overline{E}_{\alpha\beta} + \frac{\nu}{1-\nu}\overline{E}_{\gamma\gamma}\delta_{\alpha\beta} - \frac{1+\nu}{1-\nu}\alpha\overline{T}\delta_{\alpha\beta}\right) \tag{12.2.20}$$

In particular, the shear stress $\overline{S}_{12}$ is obtained in the form

$$\overline{S}_{12}(x_1, x_2; t) = -\frac{\overline{m}\mu c}{\pi}\left\{\frac{H\left[ct - \sqrt{(x_1-a_1)^2 + (x_2-a_2)^2}\right]}{\sqrt{c^2t^2 - (x_1-a_1)^2 - (x_2-a_2)^2}}\right.$$

$$- \frac{H\left[ct - \sqrt{(x_1+a_1)^2 + (x_2-a_2)^2}\right]}{\sqrt{c^2t^2 - (x_1+a_1)^2 - (x_2-a_2)^2}}$$

$$- \frac{H\left[ct - \sqrt{(x_1 - a_1)^2 + (x_2 + a_2)^2}\right]}{\sqrt{c^2t^2 - (x_1 - a_1)^2 - (x_2 + a_2)^2}}$$

$$\left. + \frac{H\left[ct - \sqrt{(x_1 + a_1)^2 + (x_2 + a_2)^2}\right]}{\sqrt{c^2t^2 - (x_1 + a_1)^2 - (x_2 + a_2)^2}} \right\} \tag{12.2.21}$$

It follows from Equation 12.2.21 that $\overline{S}_{12}$ is represented by a sum of four circular waves with the centers at four corners of the rectangular region shown in Figure 12.9. Each of the waves behaves as the two-dimensional wave produced by an impulsive source located at a corner of the rectangular region (see Figure 12.9).

12.2.2 Dynamic Thermal Stresses Produced by an Instantaneous Concentrated Source of Heat in an Infinite Elastic Sheet

Suppose that a homogeneous isotropic infinite elastic sheet referred to the polar coordinates (r, φ), $0 \leq r \leq \infty$, $0 \leq \varphi \leq 2\pi$; and initially at rest be subject to an instantaneous concentrated heat source of intensity Q_0 at $r=0$. A temperature field $\overline{T}$ corresponding to such a heat source is then independent of φ, that is, $\overline{T}=\overline{T}(r,t)$, and satisfies the parabolic heat conduction equation (see Equation 8.2.29)

$$\left(\nabla_r^2 - \frac{1}{\kappa}\frac{\partial}{\partial t}\right)\overline{T} = -\frac{Q_0}{\kappa}\frac{\delta(r)}{2\pi r}\delta(t) \quad \text{for } r > 0, \quad t > 0 \tag{12.2.22}$$

subject to the homogeneous initial condition

$$\overline{T}(r, 0) = 0 \quad \text{for } r > 0 \tag{12.2.23}$$

and the vanishing condition at infinity

$$\overline{T}(r, t) \to 0 \quad \text{as } r \to \infty \quad \text{and } t > 0 \tag{12.2.24}$$

In Equation 12.2.22 ∇_r^2 stands for the Laplacian

$$\nabla_r^2 = \frac{\partial^2}{\partial r^2} + \frac{1}{r}\frac{\partial}{\partial r} \tag{12.2.25}$$

where
$\delta = \delta(r)$ is the Dirac delta function
κ is the thermal diffusivity of the sheet

Since the temperature field $\overline{T}$ depends on r and t only, the heat flux that is normal to the sheet plane vanishes for every $t > 0$ and for each point of the sheet except for $r=0$. A thermoelastic process $\overline{p}=\overline{p}(r,t)=[\overline{\mathbf{u}}, \overline{\mathbf{E}}, \overline{\mathbf{S}}]$ associated with the temperature $\overline{T}=\overline{T}(r,t)$ can be described in terms of a thermoelastic displacement potential $\overline{\phi}=\overline{\phi}(r,t)$ by the relations (see Equation 12.2.2 through 12.2.4 restricted to polar coordinates)

$$\overline{\mathbf{u}} = [\overline{u}_r(r,t), 0] \tag{12.2.26}$$

where

$$\overline{u}_r(r,t) = \frac{\partial \overline{\phi}}{\partial r} \tag{12.2.27}$$

$$\overline{\mathbf{E}} = \begin{bmatrix} \overline{E}_{rr} & 0 \\ 0 & \overline{E}_{\varphi\varphi} \end{bmatrix} \tag{12.2.28}$$

where

$$\overline{E}_{rr} = \frac{\partial^2 \overline{\phi}}{\partial r^2}, \quad \overline{E}_{\varphi\varphi} = \frac{1}{r}\frac{\partial \overline{\phi}}{\partial r} \tag{12.2.29}$$

and

$$\overline{\mathbf{S}} = \begin{bmatrix} \overline{S}_{rr} & 0 \\ 0 & \overline{S}_{\varphi\varphi} \end{bmatrix} \tag{12.2.30}$$

where

$$\overline{S}_{rr} = -2\mu \frac{1}{r}\frac{\partial \overline{\phi}}{\partial r} + \rho \ddot{\overline{\phi}} \tag{12.2.31}$$

$$\overline{S}_{\varphi\varphi} = -2\mu \frac{\partial^2 \overline{\phi}}{\partial r^2} + \rho \ddot{\overline{\phi}} \tag{12.2.32}$$

The function $\overline{\phi} = \overline{\phi}(r,t)$ satisfies the wave equation

$$\left(\nabla_r^2 - \frac{1}{c^2}\frac{\partial^2}{\partial t^2}\right)\overline{\phi} = \overline{m}\overline{T} \quad \text{for } r > 0,\ t > 0 \tag{12.2.33}$$

subject to the initial conditions

$$\overline{\phi}(r,0) = 0, \quad \dot{\overline{\phi}}(r,0) = 0 \quad \text{for } r > 0 \tag{12.2.34}$$

and suitable vanishing conditions as $r \to \infty$ for every finite $t > 0$.

By applying the operator $(\nabla_r^2 - \kappa^{-1}\partial/\partial t)$ to Equation 12.2.33 and using Equation 12.2.22 we find

$$\left(\nabla_r^2 - \frac{1}{\kappa}\frac{\partial}{\partial t}\right)\left(\nabla_r^2 - \frac{1}{c^2}\frac{\partial^2}{\partial t^2}\right)\overline{\phi} = -\frac{\overline{m}Q_0}{\kappa}\frac{\delta(r)}{2\pi r}\delta(t) \tag{12.2.35}$$

Let $\tilde{f} = \tilde{f}(r,p)$ be the Laplace transform of a function $f = f(r,t)$:

$$L\{f\} \equiv \tilde{f}(r,p) = \int_0^\infty e^{-pt} f(r,t)\, dt \tag{12.2.36}$$

where p is the transform parameter.

By applying the operator L to Equation 12.2.22 and using the homogeneous initial condition 12.2.23 we get

$$\left(\nabla_r^2 - \frac{p}{\kappa}\right)\tilde{\bar{T}} = -\frac{Q_0}{\kappa}\frac{\delta(r)}{2\pi r} \tag{12.2.37}$$

Since for any $k > 0$ (see Equations 9.2.37 and 12.1.181)

$$-\left(\nabla_r^2 - k^2\right)^{-1}\frac{\delta(r)}{r} = K_0(kr) \tag{12.2.38}$$

where $K_0 = K_0(kr)$ is the modified Bessel function of second kind and of zero order, therefore, a solution to Equation 12.2.37 that vanishes as $r \to \infty$ takes the form

$$\tilde{\bar{T}}(r,p) = -\frac{Q_0}{2\pi\kappa}\left(\nabla_r^2 - \frac{p}{\kappa}\right)^{-1}\frac{\delta(r)}{r} = \frac{Q_0}{2\pi\kappa}K_0\left(r\sqrt{\frac{p}{\kappa}}\right) \tag{12.2.39}$$

To obtain a solution to Equation 12.2.35 in the Laplace transform domain, note that by letting $t = 0$ in Equation 12.2.33 and using Equations 12.2.23 and 12.2.34 we arrive at the three homogeneous initial conditions for $\overline{\phi} = \overline{\phi}(r,t)$:

$$\overline{\phi}(r,0) = 0, \quad \frac{\partial\overline{\phi}}{\partial t}(r,0) = 0, \quad \frac{\partial^2\overline{\phi}}{\partial t^2}(r,0) = 0 \quad \text{for } r > 0 \tag{12.2.40}$$

Hence, by applying the operator L to Equation 12.2.35 and using the conditions 12.2.40 we obtain

$$\left(\nabla_r^2 - \frac{p}{\kappa}\right)\left(\nabla_r^2 - \frac{p^2}{c^2}\right)\tilde{\overline{\phi}} = -\frac{\overline{m}Q_0}{\kappa}\frac{\delta(r)}{2\pi r} \tag{12.2.41}$$

Since

$$\left(\nabla_r^2 - \frac{p}{\kappa}\right)^{-1}\left(\nabla_r^2 - \frac{p^2}{c^2}\right)^{-1} = \left(\frac{p^2}{c^2} - \frac{p}{\kappa}\right)^{-1}\left[\left(\nabla_r^2 - \frac{p^2}{c^2}\right)^{-1} - \left(\nabla_r^2 - \frac{p}{\kappa}\right)^{-1}\right] \tag{12.2.42}$$

therefore, applying the operator 12.2.42 to Equation 12.2.41 and using the formula 12.2.38 we find that a solution to Equation 12.2.41 that vanishes as $r \to \infty$ takes the form

$$\tilde{\overline{\phi}}(r,p) = \frac{\overline{m}Q_0}{2\pi\kappa}\left(\frac{p^2}{c^2} - \frac{p}{\kappa}\right)^{-1}\left[K_0\left(r\frac{p}{c}\right) - K_0\left(r\sqrt{\frac{p}{\kappa}}\right)\right] \tag{12.2.43}$$

Now, we take advantage of the formulas [10]

$$L^{-1}\{K_0(a\sqrt{p})\} = \frac{1}{2t}e^{-a^2/4t}, \quad (a > 0) \tag{12.2.44}$$

and

$$L^{-1}\{K_0(bp)\} = H(t-b)(t^2-b^2)^{-1/2}, \quad (b>0) \tag{12.2.45}$$

Also, note that Equation 12.2.43 can be written as

$$\tilde{\overline{\phi}}(r,p) = A\left(\frac{1}{p-c^2/\kappa} - \frac{1}{p}\right)\left[K_0\left(\frac{r}{c}p\right) - K_0\left(\frac{r}{\sqrt{\kappa}}\sqrt{p}\right)\right] \tag{12.2.46}$$

where

$$A = \frac{\overline{m}Q_0}{2\pi} \tag{12.2.47}$$

To obtain the temperature $\overline{T} = \overline{T}(r,t)$ in the space–time domain, we apply the operator L^{-1} to Equation 12.2.39 and from Equation 12.2.44, we find

$$\overline{T}(r,t) = \frac{Q_0}{4\pi\kappa}\frac{1}{t}e^{-\frac{r^2}{4\kappa t}} \tag{12.2.48}$$

and the dimensions of Q_0 are $[Q_0] = \text{K m}^2$. Therefore, for any fixed $t>0$ the temperature $\overline{T}$ is represented by a Gaussian surface on the (r,φ) plane. Since $\overline{T}(r,t)>0$ for any $r>0$ and $t>0$, a "signal" represented by $\overline{T}$ propagates from $r=0$ to $r=\infty$ with an infinite velocity, and in this sense the function $\overline{T}$ has a diffusive character.

To find the potential $\overline{\phi} = \overline{\phi}(r,t)$ we apply the operator L^{-1} to Equation 12.2.46, use the formulas 12.2.44 and 12.2.45 as well as the convolution relation

$$L^{-1}\left\{\tilde{f}(r,p)\tilde{g}(r,p)\right\} = \int_0^t f(r,t-\tau)g(r,\tau)\,d\tau \tag{12.2.49}$$

valid for any two Laplace tranformable functions $f = f(r,t)$ and $g = g(r,t)$; and write

$$\overline{\phi}(r,t) = \overline{\phi}_w(r,t) + \overline{\phi}_d(r,t) \tag{12.2.50}$$

where

$$\overline{\phi}_w(r,t) = AH\left(t-\frac{r}{c}\right)\int_{r/c}^{t}\frac{\left[e^{(c^2/\kappa)(t-\tau)}-1\right]}{\sqrt{\tau^2-r^2/c^2}}\,d\tau \tag{12.2.51}$$

and

$$\overline{\phi}_d(r,t) = -\frac{1}{2}A\int_0^t\frac{\left[e^{(c^2/\kappa)(t-\tau)}-1\right]}{\tau}\,e^{-r^2/4\kappa\tau}\,d\tau \tag{12.2.52}$$

The functions $\overline{\phi}_w$ and $\overline{\phi}_d$ represent the wave and diffusive parts of $\overline{\phi}$, respectively, in the following sense. Since $\overline{\phi}_w = 0$ for $t<r/c$ and $\overline{\phi}_w>0$ for $t>r/c$, the equation $t=r/c$

represents a circular wavefront for $\overline{\phi}_w$: a signal represented by $\overline{\phi}_w$ and traveling with the velocity c from $r = 0$ to $r = \infty$ is to reach a circle $r = r_0 > 0$ in the sheet plane at time $t_0 = r_0/c + 0$. Hence, $\overline{\phi}_w$ represents a circular wave propagating with a finite velocity. A diffusive character of $\overline{\phi}_d$ is revealed by the inequality $\overline{\phi}_d < 0$ for any $r > 0$ and $t > 0$, that is, a "signal" represented by $\overline{\phi}_d$ travels from $r = 0$ to $r = \infty$ with an infinite velocity.

To discuss a process $\overline{p} = \overline{p}(r,t)$ generated by $\overline{\phi} = \overline{\phi}(r,t)$, we study behavior of the functions $\overline{\phi}_w = \overline{\phi}_w(r,t)$ and $\overline{\phi}_d = \overline{\phi}_d(r,t)$ on the (r,t) plane. First, we show that both these functions are finite throughout the plane. To this end we note that

$$\int_{r/c}^{t} \frac{\left[e^{(c^2/\kappa)(t-\tau)} - 1\right]}{\sqrt{\tau^2 - r^2/c^2}} d\tau \leq e^{c^2 t/\kappa} \int_{r/c}^{t} \frac{e^{-(c^2/\kappa)\tau}}{\sqrt{\tau^2 - r^2/c^2}} d\tau + \int_{r/c}^{t} \frac{d\tau}{\sqrt{\tau^2 - r^2/c^2}}$$

$$\leq \left(1 + e^{c^2 t/\kappa}\right) \int_{r/c}^{t} \frac{d\tau}{\sqrt{\tau^2 - r^2/c^2}} \tag{12.2.53}$$

Since for any dimensionless x and a such that $x > |a| > 0$

$$\int \frac{dx}{\sqrt{x^2 - a^2}} = \ln(x + \sqrt{x^2 - a^2}) \tag{12.2.54}$$

hence, the inequality 12.2.53 can be reduced to

$$\int_{r/c}^{t} \frac{\left[e^{(c^2/\kappa)(t-\tau)} - 1\right]}{\sqrt{\tau^2 - r^2/c^2}} d\tau \leq \left(1 + e^{c^2 t/\kappa}\right) \ln\left(\frac{t + \sqrt{t^2 - r^2/c^2}}{r/c}\right) \tag{12.2.55}$$

This implies that $\overline{\phi}_w(r,t)$ is finite for every $r > 0$ and $t > r/c$.

Since, by Equation 12.2.51, $\overline{\phi}_w(r,t) = 0$ for $0 < t < r/c$, therefore $\overline{\phi}_w(r,t)$ is finite for every $r > 0$ and $t > 0$. To show that $\overline{\phi}_d(r,t)$ is finite throughout the plane (r,t) except for $r = 0$ and $t = 0$, we note that

$$\int_{0}^{t} \frac{\left[e^{(c^2/\kappa)(t-\tau)} - 1\right]}{\tau} e^{-r^2/4\kappa\tau} d\tau \leq \left(1 + e^{c^2 t/\kappa}\right) \int_{0}^{t} \frac{e^{-r^2/4\kappa\tau}}{\tau} d\tau = -\left(1 + e^{c^2 t/\kappa}\right) Ei\left(-\frac{r^2}{4\kappa t}\right) \tag{12.2.56}$$

where

$$Ei(-x) = -\int_{x}^{\infty} \frac{e^{-v}}{v} dv, \quad x > 0 \tag{12.2.57}$$

Since the exponential integral $Ei(-x)$ in Equation 12.2.57 is finite for every $x > 0$, therefore, by Equation 12.2.52 and the inequality 12.2.56, the function $\overline{\phi}_d(r,t)$ is finite for every $r > 0$ and $t > 0$.

To show that the radial displacement $\overline{u}_r = \overline{u}_r(r,t) = \partial\overline{\phi}/\partial r$ is finite for every $r > 0$ and $t > 0$, we proceed in the following way. Using the relation

$$K_0'(x) = -K_1(x), \quad x > 0 \tag{12.2.58}$$

and differentiating Equation 12.2.46 with respect to r, we find

$$\frac{\partial\tilde{\overline{\phi}}}{\partial r}(r,p) = -A\,\frac{c^2}{\kappa}\,\frac{1}{p - c^2/\kappa}\left[\frac{1}{c}K_1\left(\frac{r}{c}p\right) - \frac{1}{\sqrt{\kappa}}\,\frac{1}{\sqrt{p}}K_1\left(\frac{r}{\sqrt{\kappa}}\sqrt{p}\right)\right] \tag{12.2.59}$$

Now, from Equations 12.2.44 and 12.2.45 we obtain

$$\int_0^\infty e^{-pt}\frac{1}{2t}\,e^{-\frac{a^2}{4t}}\,dt = K_0(a\sqrt{p}), \quad a > 0 \tag{12.2.60}$$

and

$$\int_0^\infty e^{-pt}\,\frac{H(t-b)}{\sqrt{t^2-b^2}}\,dt = K_0(bp), \quad b > 0 \tag{12.2.61}$$

Hence, the differentiation of Equations 12.2.60 and 12.2.61 with respect to p, and the use of the relation 12.2.58 yields

$$\int_0^\infty e^{-pt}\,\frac{e^{-a^2/4t}}{a}\,dt = \frac{1}{\sqrt{p}}\,K_1(a\sqrt{p}), \quad a > 0 \tag{12.2.62}$$

and

$$\int_0^\infty e^{-pt}\,\frac{t}{b}\,\frac{H(t-b)}{\sqrt{t^2-b^2}}\,dt = K_1(bp), \quad b > 0 \tag{12.2.63}$$

respectively, or

$$L^{-1}\left\{\frac{1}{\sqrt{p}}K_1(a\sqrt{p})\right\} = \frac{1}{a}e^{-a^2/4t}, \quad a > 0 \tag{12.2.64}$$

and

$$L^{-1}\left\{K_1(bp)\right\} = \frac{t}{b}\,\frac{H(t-b)}{\sqrt{t^2-b^2}}, \quad b > 0 \tag{12.2.65}$$

Finally, applying the operator L^{-1} to Equation 12.2.59, and using the convolution formula 12.2.49 together with Equations 12.2.64 and 12.2.65, we obtain

$$\frac{\partial\overline{\phi}}{\partial r}(r,t) = \frac{\partial\overline{\phi}_w}{\partial r}(r,t) + \frac{\partial\overline{\phi}_d}{\partial r}(r,t) \tag{12.2.66}$$

where

$$\frac{\partial \overline{\phi}_w}{\partial r} = -\frac{A}{r}\frac{c^2}{\kappa}H\left(t-\frac{r}{c}\right)\int_{r/c}^{t} e^{(c^2/\kappa)(t-\tau)}\frac{\tau}{\sqrt{\tau^2-r^2/c^2}}\,d\tau \tag{12.2.67}$$

and

$$\frac{\partial \overline{\phi}_d}{\partial r} = \frac{A}{r}\frac{c^2}{\kappa}\int_{0}^{t} e^{(c^2/\kappa)(t-\tau)}e^{-r^2/4\kappa\tau}\,d\tau \tag{12.2.68}$$

Since

$$\int_{r/c}^{t} e^{-c^2\tau/\kappa}\frac{\tau}{\sqrt{\tau^2-r^2/c^2}}\,d\tau = e^{-c^2t/\kappa}\sqrt{t^2-\frac{r^2}{c^2}} + \frac{c^2}{\kappa}\int_{r/c}^{t} e^{-c^2\tau/\kappa}\sqrt{\tau^2-\frac{r^2}{c^2}}\,d\tau \tag{12.2.69}$$

therefore, an alternative form of Equations 12.2.66 through 12.2.68 reads

$$\overline{u}_r(r,t) = \overline{u}_r^{(w)}(r,t) + \overline{u}_r^{(d)}(r,t) \tag{12.2.70}$$

where

$$\overline{u}_r^{(w)} \equiv \frac{\partial \overline{\phi}_w}{\partial r} = -\frac{A}{r}\frac{c^2}{\kappa}H\left(t-\frac{r}{c}\right) \times \left[\sqrt{t^2-\frac{r^2}{c^2}} + \frac{c^2}{\kappa}\int_{r/c}^{t} e^{(c^2/\kappa)(t-\tau)}\sqrt{\tau^2-\frac{r^2}{c^2}}\,d\tau\right] \tag{12.2.71}$$

and

$$\overline{u}_r^{(d)} \equiv \frac{\partial \overline{\phi}_d}{\partial r} = \frac{A}{r}\frac{c^2}{\kappa}\int_{0}^{t} e^{(c^2/\kappa)(t-\tau)} \times e^{-r^2/4\kappa\tau}\,d\tau \tag{12.2.72}$$

The functions $\overline{u}_r^{(w)}$ and $\overline{u}_r^{(d)}$ represent the wave and diffusive parts of the radial displacement $\overline{u}_r$, respectively, and both these parts are finite for every $r > 0$ and $t > 0$.

To find the stress component $\overline{S}_{rr} = \overline{S}_{rr}(r,t)$ we use the formula 12.2.31 in which $\partial\overline{\phi}/\partial r = \overline{u}_r$ is given by Equations 12.2.70 through 12.2.72, and $\ddot{\overline{\phi}}$ is computed in the following way. By multiplying Equation 12.2.43 by p^2 and using the decomposition

$$\frac{p^2}{p^2/c^2 - p/\kappa} = c^2\left(1 + \frac{c^2/\kappa}{p - c^2/\kappa}\right) \tag{12.2.73}$$

we obtain

$$p^2\tilde{\bar{\phi}}(r,p) = A\,\frac{c^2}{\kappa}\left(1+\frac{c^2/\kappa}{p-c^2/\kappa}\right)\left[K_0\left(r\frac{p}{c}\right)-K_0\left(r\sqrt{\frac{p}{\kappa}}\right)\right] \tag{12.2.74}$$

Now, if the operator L^{-1} is applied to Equation 12.2.74, and the relations 12.2.44 and 12.2.45 are taken into account, we obtain

$$\ddot{\bar{\phi}}(r,t) = \ddot{\bar{\phi}}_w(r,t) + \ddot{\bar{\phi}}_d(r,t) \tag{12.2.75}$$

where

$$\ddot{\bar{\phi}}_w(r,t) = A\frac{c^2}{\kappa}H\left(t-\frac{r}{c}\right)\left[\frac{1}{\sqrt{t^2-r^2/c^2}}+\frac{c^2}{\kappa}\int_{r/c}^{t}\frac{e^{(c^2/\kappa)(t-\tau)}}{\sqrt{\tau^2-r^2/c^2}}\,d\tau\right] \tag{12.2.76}$$

and

$$\ddot{\bar{\phi}}_d(r,t) = -\frac{Ac^2}{2\kappa}\left[\frac{e^{-r^2/4\kappa t}}{t}+\frac{c^2}{\kappa}\int_0^t e^{(c^2/\kappa)(t-\tau)}\times\frac{e^{-r^2/4\kappa\tau}}{\tau}\,d\tau\right] \tag{12.2.77}$$

Note that the function $\ddot{\bar{\phi}} = \ddot{\bar{\phi}}(r,t)$ is finite for every $r > 0$ and $t > 0$, except for the wavefront $t = r/c$ at which it becomes unbounded, that is,

$$\ddot{\bar{\phi}}(r,t) \to +\infty \quad \text{as } t \to r/c + 0 \tag{12.2.78}$$

The radial stress component is now computed from the formula

$$\bar{S}_{rr}(r,t) = -\frac{2\mu}{r}\frac{\partial\bar{\phi}}{\partial r} + \rho\ddot{\bar{\phi}} = \bar{S}_{rr}^{(w)}(r,t) + \bar{S}_{rr}^{(d)}(r,t) \tag{12.2.79}$$

where

$$\begin{aligned}\bar{S}_{rr}^{(w)}(r,t) = A\,\rho\,\frac{c^2}{\kappa}H\left(t-\frac{r}{c}\right)&\left[\frac{1}{\sqrt{t^2-r^2/c^2}}+\frac{2}{r^2}\frac{\mu}{\rho}\sqrt{t^2-r^2/c^2}\right.\\ &\left.+\frac{c^2}{\kappa}\int_{r/c}^{t}e^{(c^2/\kappa)(t-\tau)}\left(\frac{1}{\sqrt{\tau^2-r^2/c^2}}+\frac{2}{r^2}\frac{\mu}{\rho}\sqrt{\tau^2-r^2/c^2}\right)d\tau\right]\end{aligned} \tag{12.2.80}$$

and

$$\bar{S}_{rr}^{(d)}(r,t) = -\frac{A\rho}{2}\frac{c^2}{\kappa}\left[\frac{e^{-r^2/4\kappa t}}{t}+\frac{c^2}{\kappa}\int_0^t e^{(c^2/\kappa)(t-\tau)-r^2/4\kappa\tau}\left(\frac{1}{\tau}+\frac{4}{r^2}\frac{\mu}{\rho}\frac{\kappa}{c^2}\right)d\tau\right] \tag{12.2.81}$$

To compute the hoop stress $\overline{S}_{\varphi\varphi}(r,t)$ we differentiate Equation 12.2.59 with respect to r, and obtain

$$\frac{\partial^2\tilde{\overline{\phi}}}{\partial r^2}(r,p) = -A\,\frac{c^2}{\kappa}\,\frac{1}{p - c^2/\kappa}\left[\frac{p}{c^2}\,K_1'\left(\frac{r}{c}p\right) - \frac{1}{\kappa}\,K_1'\left(\frac{r}{\sqrt{\kappa}}\sqrt{p}\right)\right] \quad (12.2.82)$$

Now, by differentiating Equation 12.2.62 with respect to a, we find

$$\int_0^\infty e^{-pt}\,\frac{d}{da}\left(\frac{e^{-a^2/4t}}{a}\right)dt = K_1'(a\sqrt{p}) \quad (12.2.83)$$

Hence

$$L^{-1}\left\{K_1'(a\sqrt{p})\right\} = -\left(\frac{1}{a^2} + \frac{1}{2t}\right)e^{-a^2/4t} \quad (12.2.84)$$

Similarly, by differentiating Equation 12.2.63 with respect to p, we obtain

$$\int_0^\infty e^{-pt}\,\frac{t^2}{b^2}\,\frac{H(t-b)}{\sqrt{t^2-b^2}}\,dt = -K_1'(bp) \quad (12.2.85)$$

which is equivalent to

$$L^{-1}\left\{K_1'(bp)\right\} = -\frac{t^2}{b^2}\,\frac{H(t-b)}{\sqrt{t^2-b^2}} \quad (12.2.86)$$

Therefore, applying the operator L^{-1} to Equation 12.2.82 and using the convolution relation 12.2.49, we have

$$\frac{\partial^2\overline{\phi}}{\partial r^2}(r,t) = \frac{\partial^2\overline{\phi}_w}{\partial r^2}(r,t) + \frac{\partial^2\overline{\phi}_d}{\partial r^2}(r,t) \quad (12.2.87)$$

where

$$\frac{\partial^2\overline{\phi}_w}{\partial r^2} = \frac{A}{r^2}\,\frac{c^2}{\kappa}H\left(t - \frac{r}{c}\right)\left[\frac{t^2}{\sqrt{t^2 - r^2/c^2}} + \frac{c^2}{\kappa}\int_{r/c}^{t} e^{(c^2/\kappa)(t-\tau)}\,\frac{\tau^2}{\sqrt{\tau^2 - r^2/c^2}}\,d\tau\right] \quad (12.2.88)$$

and

$$\frac{\partial^2\overline{\phi}_d}{\partial r^2} = -A\,\frac{c^2}{\kappa}\int_0^t\left(\frac{1}{r^2} + \frac{1}{2\kappa\tau}\right)e^{(c^2/\kappa)(t-\tau) - r^2/4\kappa\tau}\,d\tau \quad (12.2.89)$$

Finally, substituting $\ddot{\overline{\phi}}$ from Equation 12.2.75 and $\partial^2\overline{\phi}/\partial r^2$ from Equation 12.2.87 into Equation 12.2.32 we obtain

$$\overline{S}_{\varphi\varphi}(r,t) = \overline{S}_{\varphi\varphi}^{(w)}(r,t) + \overline{S}_{\varphi\varphi}^{(d)}(r,t) \quad (12.2.90)$$

where

$$\overline{S}_{\varphi\varphi}^{(w)}(r,t) = A\rho\frac{c^2}{\kappa}H\left(t-\frac{r}{c}\right)\left\{\frac{1}{\sqrt{t^2-r^2/c^2}} - \frac{2}{r^2}\frac{\mu}{\rho}\frac{t^2}{\sqrt{t^2-r^2/c^2}} + \frac{c^2}{\kappa}\int\limits_{r/c}^{t} e^{(c^2/\kappa)(t-\tau)}\left[\frac{1}{\sqrt{\tau^2-r^2/c^2}} - \frac{2}{r^2}\frac{\mu}{\rho}\frac{\tau^2}{\sqrt{\tau^2-r^2/c^2}}\right]d\tau\right\} \tag{12.2.91}$$

and

$$\overline{S}_{\varphi\varphi}^{(d)}(r,t) = -\frac{A\rho}{2}\frac{c^2}{\kappa}\left\{\frac{e^{-r^2/4\kappa t}}{t} + \frac{c^2}{\kappa}\int\limits_{0}^{t} e^{(c^2/\kappa)(t-\tau)-r^2/4\kappa\tau}\left[\frac{1}{\tau} - \frac{4\mu}{\rho}\frac{\kappa}{c^2}\left(\frac{1}{r^2}+\frac{1}{2\kappa\tau}\right)\right]d\tau\right\} \tag{12.2.92}$$

Since

$$1 - \frac{2}{r^2}\frac{\mu}{\rho}t^2 \to \nu \quad \text{as } t \to \frac{r}{c} + 0 \tag{12.2.93}$$

therefore, it follows from Equations 12.2.79 through 12.2.81 and 12.2.90 through 12.2.92 that

$$\overline{S}_{rr}(r,t) \to +\infty \quad \text{and } \overline{S}_{\varphi\varphi}(r,t) \to +\infty \quad \text{as } t \to \frac{r}{c} + 0 \tag{12.2.94}$$

It is easy to show that the pair $(\overline{T}, \overline{\phi})$ given by Equations 12.2.48 and 12.2.50 satisfies Equation 12.2.33. Also, it is easy to prove that the stress components $\overline{S}_{rr}$ and $\overline{S}_{\varphi\varphi}$ given by Equations 12.2.79 and 12.2.90, respectively, comply with the condition

$$\overline{S}_{rr} + \overline{S}_{\varphi\varphi} = -2\mu\overline{m}\,\overline{T} + (1+\nu)\rho\ddot{\overline{\phi}} \tag{12.2.95}$$

Equation 12.2.95 is a direct consequence of Equations 12.2.31 through 12.2.33.

PROBLEMS

12.1 Find the dynamic thermal stresses in an infinite elastic sheet with a quiescent past subject to the temperature $\overline{T}^*$ of the form

$$\overline{T}^*(\mathbf{x},t) = T_0\delta(x_1)\delta(x_2)\delta(t)$$

where

T_0 is a constant temperature
$\delta = \delta(x)$ is the Dirac delta function

12.2 Find the dynamic thermal stresses in an infinite elastic sheet with a quiescent past subject to the temperature $\hat{T}^*$ of the form

$$\hat{T}^*(\mathbf{x}, t) = T_0 \delta(x_1)\delta(x_2)H(t)$$

where

T_0 is a constant temperature

$\delta = \delta(x)$ and $H = H(t)$ represent the Dirac delta and Heaviside functions, respectively

12.3 Find the dynamic shear stress $\overline{S}_{12}$ in an infinite elastic sheet with a quiescent past subject to a temperature $\overline{T}$ of the form

$$\overline{T}(\mathbf{x}, t) = T_0[H(x_1 + a_1) - H(x_1 - a_1)] \cdot [H(x_2 + a_2) - H(x_2 - a_2)]H(t)$$

where

T_0 is a constant temperature

$H = H(t)$ is the Heaviside function

a_1 and a_2 are positive parameters of the length dimension

12.4 An infinite elastic body described by the inequalities

$$0 \le r \le \infty, \quad 0 \le \varphi \le 2\pi, \quad |x_3| \le \infty \tag{a}$$

is subject to a line heat source of the form

$$Q(r, t) = \frac{Q_0 H(t)\delta(r)}{2\pi r} \tag{b}$$

Use a method similar to that of Section 12.2.2 to find the temperature $T = T(r, t)$ and associated thermal stress components $S_{rr} = S_{rr}(r, t)$ and $S_{\varphi\varphi} = S_{\varphi\varphi}(r, t)$ produced by the line heat source. Assume that the body is initially at rest, which means that $T(r, 0) = 0$, $u_r(r, 0) = 0$, and $\dot{u}_r(r, 0) = 0$, where $u_r = u_r(r, t)$ is the radial displacement corresponding to the heat source.

12.5 An infinite elastic body described by the inequalities

$$0 \le r \le \infty, \quad 0 \le \varphi \le 2\pi, \quad |x_3| < \infty \tag{a}$$

is subject to a time-periodic line heat source of the form

$$Q(r, t) = Q_0 e^{i\omega t}\delta(r)/2\pi r, \quad i = \sqrt{-1} \tag{b}$$

where $\omega > 0$ is the frequency. Show that the temperature $T = T(r, t)$ and the thermoelastic displacement potential $\phi = \phi(r, t)$ corresponding to the heat source take the form

$$T(r, t) = \frac{Q_0 e^{i\omega t}}{2\pi\kappa} K_0(r\sqrt{i\omega/\kappa}) \tag{c}$$

and

$$\phi(r,t) = \frac{Q_0 m e^{i\omega t}}{2\pi\kappa\left(i\omega/\kappa + \omega^2/c_1^2\right)}\left[K_0\left(r\sqrt{i\omega/\kappa}\right) - K_0(ir\omega/c_1)\right] \quad \text{(d)}$$

where $K_0 = K_0(z)$ is the modified Bessel function of the second kind and zero order; κ stands for the thermal diffusivity, c_1 is the longitudinal velocity, and

$$m = \frac{1+\nu}{1-\nu}\alpha \quad \text{(e)}$$

Here, ν and α are Poisson's ratio and the coefficient of linear thermal expansion, respectively.

REFERENCES

1. J. Ignaczak, Rayleigh waves in a nonhomogeneous isotropic elastic semispace, *Archiwum Mechaniki Stosowanej*, 15(3), 1963, 341–345.
2. Lord Rayleigh (John William Strutt), On waves propagated along the plane surface of an elastic solid, *Proc. London Math. Soc.*, 17, 1885, 4–11.
3. Y. C. Fung, *Foundations of Solid Mechanics*, Prentice-Hall, Englewood Cliffs, NJ, 1965, pp. 178–181.
4. A. C. Eringen and E. S. Suhubi, *Elastodynamics*, vol. II, *Linear Theory*, Academic Press, New York, 1975, pp. 518–524.
5. M. Jakubowska, Partition of energy of the surface waves in a homogeneous isotropic elastic semispace, *Bull. Military Univ. Technol.*, XLV(12), 1996, 15–24.
6. C. R. A. Rao, Surface waves in an elastic semispace with a stable nonhomogeneity, *Pure Appl. Geophys.*, 75, 1969, 31–41.
7. T. Klecha, Surface stress waves in a nonhomogeneous elastic half-space, Parts I and II, *Arch. Mech.*, 48(3), 1996, 493–539.
8. T. Rożnowski, Surface waves in an isotropic elastic semispace with small nonhomogeneity, *Bull. Acad. Polon. Sci., Ser. Tech.*, 25(2), 1977, 67–77.
9. I. S. Gradshteyn and I. M. Ryzhik, *Tables of Integrals, Sums, Series, and Products*, Gosud. Izdat. Fiz. Mat. Lit., Moscow, 1962, Eq. (9) with $\nu = 0$, p. 973, (in Russian); English translation: Academic Press, London, U.K., 1980.
10. H. S. Carslaw and J. C. Jaeger, *Operational Methods in Applied Mathematics,* Dover Publications, New York, 1963, p. 355.
11. E. Kröner, Continuum theory of defects, in *Physics of Defects*, Eds. R. Balian, M. Kleman, and J. P. Poirier, North-Holland, Amsterdam, the Netherlands, 1981, pp. 255–262.

13 One-Dimensional Solutions of Elastodynamics

When the data of an initial–boundary value problem of elastodynamics depend on a single space variable and on time t, a solution to the problem satisfies the one-dimensional field equations subject to suitable initial and boundary conditions. In this chapter, a number of typical one-dimensional solutions of homogeneous isotropic isothermal and nonisothermal elastodynamics are obtained in a closed-form using the Laplace transform technique. The isothermal solutions include (a) one-dimensional stress waves in an infinite or semi-infinite elastic solid subject to the initial stress and stress-rate fields, and (b) one-dimensional stress waves in a semispace subject to a uniform dynamic boundary pressure. The non-isothermal solutions cover (1) dynamic thermal stresses in an infinite or semi-infinite elastic solid subject to a plane instantaneous heat source, and (2) dynamic thermal stresses in a semispace subject to a sudden heating of the boundary plane. Also, the solution (1) is applied to obtain the integral representation of a dynamic thermoelastic response of a semispace to a laser pulse. The chapter ends with problems, and their solutions are provided in the Solutions Manual.

13.1 ONE-DIMENSIONAL SOLUTIONS OF ISOTHERMAL ELASTODYNAMICS

The one-dimensional field equations of isothermal elastodynamics are obtained by letting an elastic process $p = [\mathbf{u}, \mathbf{E}, \mathbf{S}]$ and the body force vector $\mathbf{b}$ depend on a single space variable $x \equiv x_1$ and on time t. In addition, if we assume that an elastic body is homogeneous and isotropic, and the displacement and body force vector fields take the forms

$$\mathbf{u}(\mathbf{x}, t) = [u(x, t), 0, 0] \tag{13.1.1}$$

and

$$\mathbf{b}(\mathbf{x}, t) = [b(x, t), 0, 0] \tag{13.1.2}$$

respectively, then the strain and stress tensor fields are given by the matrices:

$$\mathbf{E}(\mathbf{x}, t) = \begin{bmatrix} E_{11}(x, t) & 0 & 0 \\ 0 & 0 & 0 \\ 0 & 0 & 0 \end{bmatrix} \tag{13.1.3}$$

where

$$E_{11}(x,t) = \frac{\partial u}{\partial x} \tag{13.1.4}$$

and

$$\mathbf{S}(\mathbf{x},t) = \begin{bmatrix} S_{11}(x,t) & 0 & 0 \\ 0 & S_{22}(x,t) & 0 \\ 0 & 0 & S_{33}(x,t) \end{bmatrix} \tag{13.1.5}$$

where

$$S_{11}(x,t) \equiv S(x,t) = (\lambda + 2\mu)\frac{\partial u}{\partial x} \tag{13.1.6}$$

$$S_{22}(x,t) = S_{33}(x,t) = \lambda\frac{\partial u}{\partial x} \tag{13.1.7}$$

Finally, the equation of motion takes the form

$$\frac{\partial S}{\partial x} + b = \rho\frac{\partial^2 u}{\partial t^2} \tag{13.1.8}$$

(see Equations 4.2.1 through 4.2.3 restricted to a one-dimensional homogeneous isotropic case).

By substituting Equation 13.1.6 into Equation 13.1.8 we arrive at the one-dimensional displacement equation of motion

$$\left(\frac{\partial^2}{\partial x^2} - \frac{1}{c^2}\frac{\partial^2}{\partial t^2}\right)u = -\frac{b}{\lambda + 2\mu} \tag{13.1.9}$$

where

$$c \equiv c_1 = \sqrt{\frac{\lambda + 2\mu}{\rho}} \tag{13.1.10}$$

represents the longitudinal wave velocity.

Also, by applying the operator $\partial/\partial x$ to Equation 13.1.8 and using Equation 13.1.6 we obtain the one-dimensional stress equation of motion

$$\left(\frac{\partial^2}{\partial x^2} - \frac{1}{c^2}\frac{\partial^2}{\partial t^2}\right)S = -\frac{\partial b}{\partial x} \tag{13.1.11}$$

In the following, a number of the initial–boundary value problems for Equation 13.1.9 or Equation 13.1.11 will be formulated, and closed-form solutions to these problems will be presented. The discussion includes

(A) One-dimensional stress waves in an infinite elastic solid subject to the initial stress and stress-rate fields
(B) One-dimensional stress waves in an infinite elastic solid subject to a body force field
(C) One-dimensional stress waves in a half-space with free boundary subject to the initial stress and stress-rate fields
(D) One-dimensional stress waves in a half-space with free boundary subject to a body force field
(E) One-dimensional stress waves in a half-space subject to a uniform dynamical boundary pressure
(F) One-dimensional stress waves in a finite strip with free boundaries subject to the initial stress and stress-rate fields

13.1.1 One-Dimensional Stress Waves in an Infinite Elastic Solid Subject to the Initial Stress and Stress-Rate Fields

13.1.1.1 One-Dimensional Green's Function for an Infinite Space Corresponding to the Homogeneous Initial Conditions

In this section, we are to solve the following initial–boundary value problem. Find a function $G = G(x, \xi; t)$ for $|x| < \infty$, $|\xi| < \infty$, $t > 0$, that satisfies the field equation

$$\left(\frac{\partial^2}{\partial x^2} - \frac{1}{c^2}\frac{\partial^2}{\partial t^2}\right) G = -\delta(x-\xi)\delta(t)$$

$$\text{for } |x| < \infty,\ |\xi| < \infty,\ t > 0 \tag{13.1.12}$$

the initial conditions

$$G(x, \xi; 0) = 0, \quad \frac{\partial}{\partial t} G(x, \xi; 0) = 0$$

$$\text{for } |x| < \infty,\ |\xi| < \infty \tag{13.1.13}$$

and the vanishing condition at infinity

$$G(x, \xi; t) \to 0 \quad \text{as } |x| \to +\infty$$

$$\text{for every } \xi \in (-\infty, +\infty) \quad \text{and } t > 0 \tag{13.1.14}$$

Let $\bar{f} = \bar{f}(x, p)$ be the Laplace transform of a function $f = f(x, t)$ with respect to time t

$$L\{f(x,t)\} \equiv \bar{f}(x, p) = \int_0^\infty e^{-pt} f(x, t)dt \tag{13.1.15}$$

where p is the transform parameter. By applying the operator L to Equation 3.1.12 and using the homogeneous initial conditions 13.1.13 we reduce the problem 13.1.12 through 13.1.14 to the following one. Find a function $\overline{G} = \overline{G}(x, \xi; p)$ that satisfies the equation

$$\left(\frac{d^2}{dx^2} - \frac{p^2}{c^2}\right)\overline{G} = -\delta(x-\xi) \quad \text{for } |x| < \infty,\ |\xi| < \infty,\ p > 0 \tag{13.1.16}$$

and the vanishing condition

$$\overline{G}(x,\xi;p) \to 0 \quad \text{as } |x| \to \infty, \quad |\xi| < \infty,\ p > 0 \tag{13.1.17}$$

To solve the problem 13.1.16 and 13.1.17 we use the integral representation of the Dirac delta function [1]

$$\delta(x-\xi) = \frac{1}{\pi}\int_0^\infty \cos\alpha(x-\xi)d\alpha \tag{13.1.18}$$

By substituting Equation 13.1.18 into the RHS of Equation 13.1.16 and looking for $\overline{G}$ in the form of a cosine integral, we obtain

$$\overline{G}(x,\xi;p) = \frac{1}{\pi}\int_0^\infty \frac{\cos\alpha(x-\xi)}{\alpha^2 + (p/c)^2}d\alpha \tag{13.1.19}$$

or

$$\overline{G}(x,\xi;p) = \frac{1}{\pi}\frac{\pi}{2}\frac{e^{-(p/c)|x-\xi|}}{(p/c)} = \frac{c}{2}\frac{e^{-(p/c)|x-\xi|}}{p} \tag{13.1.20}$$

Hence, by applying the operator L^{-1} to Equation 13.1.20, we have

$$G(x,\xi;t) = \frac{c}{2}H\left(t - \frac{|x-\xi|}{c}\right) \tag{13.1.21}$$

where $H = H(t)$ is the Heaviside function. The function $G = G(x,\xi;t)$ is Green's function for an infinite space corresponding to the homogeneous initial conditions in the following sense. For a given function $f = f(x,t)$ on $(-\infty,+\infty)\times[0,\infty)$, a solution $F = F(x,t)$ of the equation

$$\left(\frac{\partial^2}{\partial x^2} - \frac{1}{c^2}\frac{\partial^2}{\partial t^2}\right)F = -f(x,t) \quad \text{for } |x| < \infty,\ t > 0 \tag{13.1.22}$$

subject to the homogeneous initial conditions

$$F(x,0) = 0 \quad \text{for } |x| < \infty \tag{13.1.23}$$

and

$$\frac{\partial}{\partial t}F(x,0) = 0 \quad \text{for } |x| < \infty \tag{13.1.24}$$

and the vanishing condition at infinity

$$F(x,t) \to 0 \quad \text{as } |x| \to \infty \quad \text{and } t > 0 \tag{13.1.25}$$

takes the form

$$F(x,t) = \int_0^t \int_{-\infty}^{+\infty} G(x,\xi;t-\tau) f(\xi,\tau) d\xi\, d\tau \qquad (13.1.26)$$

By substituting $G = G(x,\xi;t)$ from Equation 13.1.21 into Equation 13.1.26 an alternative form of Equation 13.1.26 is found

$$F(x,t) = \frac{c}{2} \int_0^t \int_{x-c(t-\tau)}^{x+c(t-\tau)} f(\xi,\tau) d\xi\, d\tau \qquad (13.1.27)$$

In particular, a displacement wave that satisfies the equation (see Equation 13.1.9)

$$\left(\frac{\partial^2}{\partial x^2} - \frac{1}{c^2} \frac{\partial^2}{\partial t^2} \right) u = -\frac{1}{\lambda + 2\mu} b(x,t) \quad \text{for } |x| < \infty,\ t > 0 \qquad (13.1.28)$$

subject to the conditions

$$u(x,0) = 0, \quad \frac{\partial}{\partial t} u(x,0) = 0 \quad \text{for } |x| < \infty \qquad (13.1.29)$$

and

$$u(x,t) \to 0 \quad \text{as } |x| \to \infty \quad \text{and } t > 0 \qquad (13.1.30)$$

is represented by the integral

$$u(x,t) = \frac{c}{2} \frac{1}{\lambda + 2\mu} \int_0^t \int_{x-c(t-\tau)}^{x+c(t-\tau)} b(\xi,\tau) d\xi\, d\tau \quad \text{for } |x| < \infty,\ t \geq 0 \qquad (13.1.31)$$

Note that the displacement wave 13.1.31 is well defined for any $b = b(x,t)$ integrable over a finite rectangular region: $[-x_0, x_0] \times [0,t]$, where $x_0 < \infty$ and $t < \infty$. In particular, the formula 13.1.31 makes sense if $b(x,t)$ is piecewise continuous over the rectangular region.

13.1.1.2 One-Dimensional Stress Waves in an Infinite Solid Subject to the Initial Stress and Stress-Rate Fields: D'Alembert Solution

Assume that a homogeneous isotropic infinite elastic solid $|x| < \infty$ is subject to initial stress and stress-rate fields only. Then a stress wave produced by such fields satisfies the equation (see Equation 13.1.11 with $b = 0$)

$$\left(\frac{\partial^2}{\partial x^2} - \frac{1}{c^2} \frac{\partial^2}{\partial t^2} \right) S = 0 \quad \text{for } |x| < \infty,\ t > 0 \qquad (13.1.32)$$

and the initial conditions

$$S(x,0) = S_0(x), \quad \frac{\partial}{\partial t} S(x,0) = \dot{S}_0(x) \quad \text{for } |x| < \infty \qquad (13.1.33)$$

where $S_0 = S_0(x)$ and $\dot{S}_0 = \dot{S}_0(x)$ are prescribed functions on $(-\infty, +\infty)$. To obtain a solution $S = S(x,t)$ to the initial value problem 13.1.32 and 13.1.33 we use the Laplace transform technique. By applying the operator L (see Equation 13.1.15) to Equation 13.1.32 and using the initial conditions 13.1.33 we have

$$\left(\frac{d^2}{dx^2} - \frac{p^2}{c^2}\right)\overline{S} = -\frac{1}{c^2}\left[pS_0(x) + \dot{S}_0(x)\right] \quad \text{for } |x| < \infty,\ p > 0 \qquad (13.1.34)$$

Now, it follows from Equations 13.1.16 and 13.1.20 that

$$\left(\frac{d^2}{dx^2} - \frac{p^2}{c^2}\right)\left[\frac{c}{2}\frac{e^{-(p/c)|x-\xi|}}{p}\right] = -\delta(x-\xi) \quad \text{for } |x| < \infty,\ |\xi| < \infty,\ p > 0 \qquad (13.1.35)$$

Hence, multiplying Equation 13.1.35 by $[pS_0(\xi) + \dot{S}_0(\xi)]/c^2$ and integrating the result over $\xi \in (-\infty, +\infty)$ we find the integral representation of $\overline{S} = \overline{S}(x,p)$:

$$\overline{S}(x,p) = \frac{1}{2c}\left[\int_{-\infty}^{+\infty} S_0(\xi)e^{-(p/c)|x-\xi|}\,d\xi + \frac{1}{p}\int_{-\infty}^{+\infty} \dot{S}_0(\xi)e^{-(p/c)|x-\xi|}\,d\xi\right] \quad \text{for } |x| < \infty,\ p > 0 \qquad (13.1.36)$$

Finally, by applying the operator L^{-1} to Equation 13.1.36 we arrive at

$$S(x,t) = \frac{1}{2c}\left[\int_{-\infty}^{+\infty} S_0(\xi)\delta\left(t - \frac{|x-\xi|}{c}\right)d\xi + \int_{-\infty}^{+\infty} \dot{S}_0(\xi)H\left(t - \frac{|x-\xi|}{c}\right)d\xi\right] \quad \text{for } |x| < \infty,\ t \geq 0 \qquad (13.1.37)$$

where $\delta = \delta(t)$ and $H = H(t)$ are the Dirac delta and Heaviside functions, respectively. Since

$$\int_{-\infty}^{+\infty} S_0(\xi)\delta\left(t - \frac{|x-\xi|}{c}\right)d\xi = \int_{-\infty}^{x} S_0(\xi)\delta\left(t - \frac{x-\xi}{c}\right)d\xi + \int_{x}^{\infty} S_0(\xi)\delta\left(t - \frac{\xi - x}{c}\right)d\xi = c\left[\int_{-\infty}^{t} S_0(x - ct + cu)\delta(u)du + \int_{-\infty}^{t} S_0(x + ct - cv)\delta(v)dv\right] = c[S_0(x-ct) + S_0(x+ct)] \qquad (13.1.38)$$

and

$$\int_{-\infty}^{+\infty} \dot{S}_0(\xi)H\left(t - \frac{|x-\xi|}{c}\right)d\xi = c\left[\int_{-\infty}^{t} \dot{S}_0(x - ct + cu)H(u)du + \int_{-\infty}^{t} \dot{S}_0(x + ct - cv)H(v)dv\right] = \int_{x-ct}^{x+ct} \dot{S}_0(\xi)d\xi \qquad (13.1.39)$$

therefore, an alternative form of Equation 13.1.37 reads

$$S(x,t) = \frac{1}{2}[S_0(x-ct) + S_0(x+ct)] + \frac{1}{2c}\int_{x-ct}^{x+ct} \dot{S}_0(\xi)d\xi \quad \text{for } |x| < \infty,\ t \geq 0 \quad (13.1.40)$$

Solution 13.1.40 is known as a *solution of the d'Alembert type.*

If $\dot{S}_0(x) = 0$ for $|x| < \infty$ and $S_0 = S_0(x)$ is represented by the Dirac delta function

$$S_0(x) = S_0\delta(x) \quad \text{for } |x| < \infty \quad (13.1.41)$$

where S_0 is a constant of the dimension [Stress × Length], then

$$S(x,t) = \frac{1}{2}S_0\,[\delta(x-ct) + \delta(x+ct)] \quad (13.1.42)$$

Hence, a unit initial stress delta profile produces a sum of the two moving delta waves: the one of magnitude 1/2 moving with velocity c in the x direction, and the other of the same magnitude moving with velocity c in the opposite direction. If the initial stress 13.1.41 complies with the relation

$$S_0 = S_0(x) = (\lambda + 2\mu)E_0(x) = (\lambda + 2\mu)\frac{\partial}{\partial x}u_0(x) \quad (13.1.43)$$

where $u_0 = u_0(x)$ is a displacement in the x direction, then the function $S = S(x,t)$ given by Equation 13.1.42 represents the stress wave of a one-dimensional incompatible elastodynamics in which the displacement $u_0 = u_0(x)$ undergoes a finite jump across the plane $x = 0$.

Finally, note that the one-dimensional displacement initial value problem of classic elastodynamics in which the initial displacement $u_0 = u_0(x)$ and the initial velocity $\dot{u}_0 = \dot{u}_0(x)$ are prescribed on $(-\infty, +\infty)$ (see Equation 13.1.9 with $b = 0$)

$$\left(\frac{\partial^2}{\partial x^2} - \frac{1}{c^2}\frac{\partial^2}{\partial t^2}\right)u = 0 \quad \text{for } |x| < \infty,\ t > 0 \quad (13.1.44)$$

$$u(x,0) = u_0(x), \qquad \frac{\partial}{\partial t}u(x,0) = \dot{u}_0(x) \quad \text{for } |x| < \infty \quad (13.1.45)$$

can be solved in a way similar to that of the stress problem 13.1.32 and 13.1.33, and we arrive at the *d'Alembert solution*

$$u(x,t) = \frac{1}{2}\{u_0(x-ct) + u_0(x+ct)\} + \frac{1}{2c}\int_{x-ct}^{x+ct} \dot{u}_0(\xi)d\xi \quad (13.1.46)$$

If the initial data of the stress problem are suitably related to the initial data of the displacement problem, the stress wave $S = S(x,t)$ given by Equation 13.1.40 corresponds to the displacement wave $u = u(x,t)$ given by Equation 13.1.46, which means that $p = [u, E, S]$ is a solution to a one-dimensional displacement initial value problem of classic elastodynamics for an infinite space.

13.1.2 One-Dimensional Stress Waves in an Infinite Elastic Solid Subject to a Body Force Field

In this case the stress waves are represented by a solution to the field equation (see Equation 13.1.10)

$$\left(\frac{\partial^2}{\partial x^2} - \frac{1}{c^2}\frac{\partial^2}{\partial t^2}\right) S = -\frac{\partial b}{\partial x} \quad \text{for } |x| < \infty,\ t > 0 \tag{13.1.47}$$

subject to the homogeneous initial conditions

$$S(x, 0) = 0, \quad \frac{\partial}{\partial t} S(x, 0) = 0 \quad \text{for } |x| < \infty \tag{13.1.48}$$

Since this problem is a part of the one described by Equations 13.1.23 through 13.1.25, a solution to Equations 13.1.47 and 13.1.48 takes the form (see Equation 13.1.27)

$$S(x, t) = \frac{c}{2}\int_0^t \int_{x-c(t-\tau)}^{x+c(t-\tau)} \frac{\partial b}{\partial \xi}(\xi, \tau)\,d\xi\,d\tau \tag{13.1.49}$$

or

$$S(x, t) = \frac{c}{2}\int_0^t \{b[x + c(t - \tau), \tau] - b[x - c(t - \tau), \tau]\}\,d\tau \tag{13.1.50}$$

For an impulsive rectangular body force

$$b(x, t) = b_0[H(x + a) - H(x - a)]\delta(t) \tag{13.1.51}$$

where $a > 0$, and b_0 is a constant of the dimension [Stress × Velocity^{-1}], we have

$$S(x, t) = \frac{c}{2} b_0 \{H(x + a + ct) - H(x + a - ct) + H(x - a - ct) - H(x - a + ct)\} \tag{13.1.52}$$

Figure 13.1 shows a graph of $S = S(x, t)$ for $x = -a$, $x = 0$, and $x = a$ as a function of time t.

It follows from Figure 13.1 that for a small ratio a/c the impulsive rectangular body force produces the two self-equilibrated constant stresses of short duration: a tension at $x = -a$ and a compression at $x = a$, while the middle cross section $x = 0$ remains unstressed for every $t > 0$.

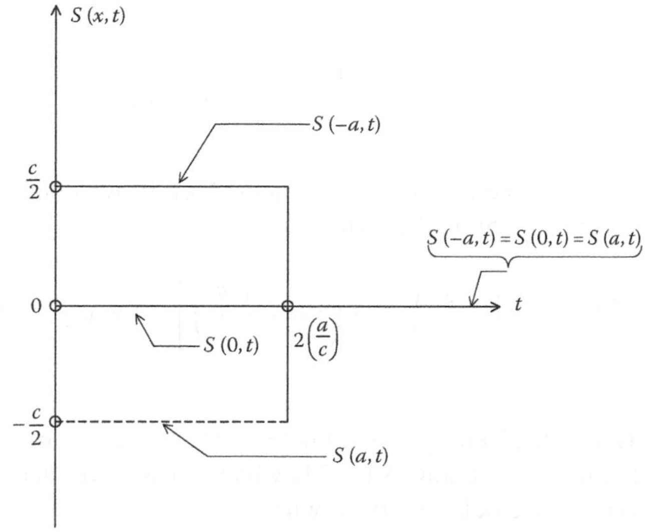

FIGURE 13.1 Stress $S(x, t)$ due to an impulsive rectangular body force as a function of time t for $x = -a$, $x = 0$, $x = a$, and $b_0 = 1$.

13.1.3 ONE-DIMENSIONAL STRESS WAVES IN A HALF-SPACE WITH FREE BOUNDARY SUBJECT TO THE INITIAL STRESS AND STRESS-RATE FIELDS

13.1.3.1 One-Dimensional Green's Function for a Half-Space Corresponding to Homogeneous Boundary and Initial Conditions

We will now solve the following initial–boundary value problem. Find a function $G = G(x, \xi; t)$ for $x \geq 0$, $\xi > 0$, $t \geq 0$, that satisfies the field equation

$$\left(\frac{\partial^2}{\partial x^2} - \frac{1}{c^2}\frac{\partial^2}{\partial t^2}\right) G = -[\delta(x-\xi) - \delta(x+\xi)]\delta(t) \quad \text{for } x > 0,\ \xi > 0,\ t > 0 \quad (13.1.53)$$

the initial conditions

$$G(x, \xi; 0) = 0, \quad \frac{\partial}{\partial t} G(x, \xi; 0) = 0 \quad \text{for } x > 0,\ \xi > 0 \quad (13.1.54)$$

the boundary condition

$$G(0, \xi; t) = 0 \quad \text{for } \xi > 0,\ t > 0 \quad (13.1.55)$$

and the vanishing condition

$$G(x, \xi; t) \to 0 \quad \text{as } x \to +\infty \quad \text{fo every } \xi > 0 \quad \text{and } t > 0 \quad (13.1.56)$$

To solve the problem 13.1.53 through 13.1.56 we note that using the Laplace transform technique similar to that of Section 13.1.1.1 (see Equations 13.1.12 through 13.1.22) we write

$$\overline{G}(x, \xi; p) = -\left(\frac{d^2}{dx^2} - \frac{p^2}{c^2}\right)^{-1} [\delta(x-\xi) - \delta(x+\xi)] \quad (13.1.57)$$

or

$$\overline{G}(x, \xi; p) = \frac{c}{2}\left[\frac{e^{-(p/c)|x-\xi|}}{p} - \frac{e^{-(p/c)(x+\xi)}}{p}\right] \quad \text{for every } x \geq 0,\ \xi > 0,\ p > 0 \tag{13.1.58}$$

Therefore, by applying the operator L^{-1} to Equation 13.1.58 we obtain a solution to the problem 13.1.53 through 13.1.56 in the form

$$G(x, \xi; t) = \frac{c}{2}\left[H\left(t - \frac{|x-\xi|}{c}\right) - H\left(t - \frac{x+\xi}{c}\right)\right] \quad \text{for } x \geq 0,\ \xi > 0,\ t \geq 0 \tag{13.1.59}$$

The function $G = G(x, \xi; t)$ given by Equation 13.1.59 represents Green's function for a half-space that will be used in Section 13.1.3.2 in which stress waves in a half-space subject to initial stress and stress-rate fields are dealt with.

13.1.3.2 One-Dimensional Stress Waves in a Half-Space with Free Boundary Subject to Initial Stress and Stress-Rate Fields

We will now study stress waves characterized by the initial–boundary value problem. Find a stress $S = S(x, t)$ for $x \geq 0$, $t \geq 0$, that satisfies the field equation

$$\left(\frac{\partial^2}{\partial x^2} - \frac{1}{c^2}\frac{\partial^2}{\partial t^2}\right) S = 0 \quad \text{for } x > 0,\ t > 0 \tag{13.1.60}$$

the initial conditions

$$S(x, 0) = S_0(x), \quad \frac{\partial}{\partial t} S(x, 0) = \dot{S}_0(x) \quad \text{for } x \geq 0 \tag{13.1.61}$$

the boundary condition

$$S(0, t) = 0 \quad \text{for } t \geq 0 \tag{13.1.62}$$

and the vanishing condition

$$S(x, t) \to 0 \quad \text{as } x \to \infty \quad \text{and } t > 0 \tag{13.1.63}$$

In addition, to make the initial and boundary conditions compatible at $x = 0$ and $t = 0$ we assume that the initial fields $S_0(x)$ and $\dot{S}_0(x)$ can be extended to the whole space $(-\infty, +\infty)$ in an even manner:

$$S_0(x) = S_0(-x), \quad S_0(0) = 0 \quad \text{for } |x| < \infty \tag{13.1.64}$$

and

$$\dot{S}_0(x) = \dot{S}_0(-x), \quad \dot{S}_0(0) = 0 \quad \text{for } |x| < \infty \tag{13.1.65}$$

By applying the Laplace transform operator L to Equation 13.1.60 and using the initial conditions 13.1.61 we find

$$\left(\frac{\partial^2}{\partial x^2}-\frac{p^2}{c^2}\right)\overline{S}=-\frac{1}{c^2}\left[pS_0(x)+\dot{S}_0(x)\right]\quad \text{for } x>0,\ p>0 \tag{13.1.66}$$

The conditions 13.1.62 and 13.1.63 in the Laplace transform domain take the forms

$$\overline{S}(0,p)=0\quad \text{for } p>0 \tag{13.1.67}$$

and

$$\overline{S}(x,p)\to 0\quad \text{as } x\to\infty\quad \text{and } p>0 \tag{13.1.68}$$

respectively. It follows from the properties of Green's function of Section 13.1.3.1 that a solution $\overline{S}=\overline{S}(x,p)$ to the problem 13.1.66 through 13.1.68 can be represented by the integral

$$\overline{S}(x,p)=\frac{1}{c^2}\int_0^\infty \overline{G}(x,\xi;p)\left[pS_0(\xi)+\dot{S}_0(\xi)\right]d\xi\quad \text{for } x\geq 0,\ p>0 \tag{13.1.69}$$

where $\overline{G}=\overline{G}(x,\xi;p)$ is given by Equation 13.1.58. Hence, substituting $\overline{G}$ from Equation 13.1.58 into Equation 13.1.69, and applying the operator L^{-1} to the result we get

$$\begin{aligned}S(x,t)=\frac{1}{2c}&\left\{\int_0^\infty S_0(\xi)\left[\delta\left(t-\frac{|x-\xi|}{c}\right)-\delta\left(t-\frac{x+\xi}{c}\right)\right]d\xi\right.\\&\left.+\int_0^\infty \dot{S}_0(\xi)\left[H\left(t-\frac{|x-\xi|}{c}\right)-H\left(t-\frac{x+\xi}{c}\right)\right]d\xi\right\}\quad \text{for } x\geq 0,\ t\geq 0\end{aligned} \tag{13.1.70}$$

In the following, an alternative form of Equation 13.1.70 will be obtained. To this end, note that

$$\begin{aligned}\int_0^\infty S_0(\xi)\delta\left(t-\frac{|x-\xi|}{c}\right)d\xi&=\int_0^x S_0(\xi)\delta\left(t-\frac{x-\xi}{c}\right)d\xi\\&+\int_x^\infty S_0(\xi)\delta\left(t+\frac{x-\xi}{c}\right)d\xi=c\left[H(x-ct)S_0(x-ct)+S_0(x+ct)\right]\end{aligned} \tag{13.1.71}$$

$$\int_0^\infty S_0(\xi)\delta\left(t-\frac{x+\xi}{c}\right)d\xi=cH(ct-x)S_0(ct-x) \tag{13.1.72}$$

$$\int_0^{\infty} \dot{S}_0(\xi) H\left(t - \frac{|x-\xi|}{c}\right) d\xi = \int_0^{x} \dot{S}_0(\xi) H\left(t - \frac{x-\xi}{c}\right) d\xi$$

$$+ \int_x^{\infty} \dot{S}_0(\xi) H\left(t + \frac{x-\xi}{c}\right) d\xi = H(x-ct) \int_{x-ct}^{x} \dot{S}_0(\xi) d\xi$$

$$+ \int_x^{x+ct} \dot{S}_0(\xi) d\xi + H(ct-x) \int_0^{x} \dot{S}_0(\xi) d\xi \qquad (13.1.73)$$

and

$$\int_0^{\infty} \dot{S}_0(\xi) H\left(t - \frac{x+\xi}{c}\right) d\xi = H(ct-x) \int_0^{ct-x} \dot{S}_0(\xi) d\xi \qquad (13.1.74)$$

Therefore, because of Equations 13.1.71 through 13.1.74, an alternative form of Equation 13.1.70 reads

$$S(x,t) = \frac{1}{2}[H(x-ct)S_0(x-ct) + S_0(x+ct) - H(ct-x)S_0(ct-x)]$$

$$+ \frac{1}{2c}\left[H(x-ct) \int_{x-ct}^{x} \dot{S}_0(\xi) d\xi + \int_x^{x+ct} \dot{S}_0(\xi) d\xi\right.$$

$$\left. + H(ct-x) \int_0^{x} \dot{S}_0(\xi) d\xi - H(ct-x) \int_0^{ct-x} \dot{S}_0(\xi) d\xi\right]$$

$$\text{for every } x \geq 0 \quad \text{and } t \geq 0 \qquad (13.1.75)$$

By differentiating Equation 13.1.75 with respect to t and using the condition $S_0(0) = 0$ (see Equations 13.1.64) we find

$$\frac{\partial}{\partial t} S(x,t) = -\frac{c}{2}\left[H(x-ct)S_0'(x-ct)\right.$$

$$\left. - S_0'(x+ct) + H(ct-x)S_0'(ct-x)\right] + \frac{1}{2}\left[H(x-ct)\dot{S}_0(x-ct)\right.$$

$$\left. + \dot{S}_0(x+ct) - H(ct-x)\dot{S}_0(ct-x) + \delta(x-ct) \int_0^{x} \dot{S}_0(\xi) d\xi\right] \qquad (13.1.76)$$

where prime on S_0 stands for the derivative of $S_0 = S_0(x)$. Using Equations 13.1.75 and 13.1.76, it is easy to show that $S = S(x,t)$ given by Equation 13.1.75 does satisfy Equation 13.1.60 subject to the conditions 13.1.61 through 13.1.63. Also, one can show that if $S_0 = S_0(x)$ and $\dot{S}_0 = \dot{S}_0(x)$ do not satisfy the one-dimensional stress–displacement and stress rate–displacement rate relations, respectively, then $S = S(x,t)$ represents a stress wave in a free boundary half-space subject to initially distributed defects.

13.1.4 One-Dimensional Stress Waves in a Half-Space with Free Boundary Subject to a Body Force Field

The stress waves in a half-space with free boundary subject to a body force $b = b(x, t)$ are described by a solution to the problem (see Equation 13.1.11). Find a function $S = S(x, t)$ on $[0, \infty) \times [0, \infty)$ that satisfies the field equation

$$\left(\frac{\partial^2}{\partial x^2} - \frac{1}{c^2}\frac{\partial^2}{\partial t^2}\right) S = -\frac{\partial b}{\partial x} \quad \text{for } x > 0,\ t > 0 \tag{13.1.77}$$

the initial conditions

$$S(x, 0) = 0, \quad \frac{\partial}{\partial t} S(x, 0) = 0 \quad \text{for } x > 0 \tag{13.1.78}$$

the boundary condition

$$S(0, t) = 0 \quad \text{for } t \geq 0 \tag{13.1.79}$$

and the vanishing condition at infinity

$$S(x, t) \to 0 \quad \text{as } x \to \infty \quad \text{and } t > 0 \tag{13.1.80}$$

By using the Laplace transform technique similar to that of Section 13.1.3 we find that (see Equation 13.1.69)

$$\overline{S}(x, p) = \int_0^\infty \overline{G}(x, \xi; p) \frac{\partial \overline{b}}{\partial \xi}(\xi, p) d\xi \tag{13.1.81}$$

where

$$\overline{G}(x, \xi; p) = \frac{c}{2}\left[\frac{e^{-(p/c)|x-\xi|}}{p} - \frac{e^{-(p/c)(x+\xi)}}{p}\right] \tag{13.1.82}$$

Hence, by applying the operator L^{-1} to Equation 13.1.81 we have

$$S(x, t) = \frac{c}{2}\int_0^t\int_0^\infty \left[H\left(t - \tau - \frac{|x-\xi|}{c}\right) - H\left(t - \tau - \frac{x+\xi}{c}\right)\right]\frac{\partial b}{\partial \xi}(\xi, \tau) d\xi\, d\tau \tag{13.1.83}$$

or (see Equations 13.1.73 and 13.1.74)

$$S(x, t) = \frac{c}{2}\int_0^t \left\{H[x - c(t-\tau)] \int_{x-c(t-\tau)}^{x} \frac{\partial b}{\partial \xi}(\xi, \tau) d\xi + \int_x^{x+c(t-\tau)} \frac{\partial b}{\partial \xi}(\xi, \tau) d\xi \right.$$
$$\left. - H[c(t-\tau) - x] \int_0^{c(t-\tau)-x} \frac{\partial b}{\partial \xi}(\xi, \tau) d\xi \right\} d\tau \tag{13.1.84}$$

Finally, by computing the integrals over ξ we obtain

$$S(x,t) = \frac{c}{2}\int_0^t \{[b(x,\tau) - b[x - c(t-\tau),\tau]]$$

$$\times H[x - c(t-\tau)] + b[x + c(t-\tau),\tau] - b(x,\tau)$$

$$- [b[c(t-\tau) - x,\tau] - b(0,\tau)]H[c(t-\tau) - x]\}\, d\tau \tag{13.1.85}$$

For an impulsive unit body force at $x = x_0 > 0$

$$b(x,t) = b^*\delta(x - x_0)\delta(t) \tag{13.1.86}$$

where b^* is a constant of the dimension [Stress × Time] and the stress wave is represented by the simple formula

$$S(x,t) = \frac{c}{2}\, b^* \{[\delta(x - x_0) - \delta(x - ct - x_0)]H(x - ct) + \delta(x + ct - x_0)$$

$$- \delta(x - x_0) - \delta(ct - x - x_0)H(ct - x)\} \tag{13.1.87}$$

It follows from Equation 13.1.87 that the homogeneous initial conditions 13.1.78, the boundary condition 13.1.79, as well as the vanishing condition 13.1.80 are satisfied identically by the impulsive body force stress wave propagating in a half-space with free boundary.

13.1.5 One-Dimensional Stress Waves in a Half-Space Subject to a Uniform Dynamic Boundary Pressure

In this case stress waves are described by a solution to the following problem. Find a function $S = S(x,t)$ on $[0,\infty) \times [0,\infty)$ that satisfies the field equation

$$\left(\frac{\partial^2}{\partial x^2} - \frac{1}{c^2}\frac{\partial^2}{\partial t^2}\right) S = 0 \quad \text{for } x > 0,\ t > 0 \tag{13.1.88}$$

the homogeneous initial conditions

$$S(x,0) = 0, \quad \frac{\partial}{\partial t}S(x,0) = 0 \quad \text{for } x > 0 \tag{13.1.89}$$

the boundary condition

$$S(0,t) = -s(t) \quad \text{for } t > 0 \tag{13.1.90}$$

and the vanishing condition at infinity

$$S(x,t) \to 0 \quad \text{as } x \to +\infty \quad \text{and } t > 0 \tag{13.1.91}$$

Here, $s = s(t)$ is a prescribed function on $[0,\infty)$ such that $s(0) = 0$.

By using the Laplace transform with respect to time t we reduce the problem to the one in the Laplace transform domain:

$$\left(\frac{\partial^2}{\partial x^2} - \frac{p^2}{c^2}\right)\overline{S} = 0 \quad \text{for } x > 0,\ p > 0 \tag{13.1.92}$$

$$\overline{S}(0,p) = -\overline{s}(p) \quad \text{for } p > 0 \tag{13.1.93}$$

and

$$\overline{S}(x,p) \to 0 \quad \text{as } x \to +\infty,\ p > 0 \tag{13.1.94}$$

Next, we note that a solution to the problem 13.1.92 through 13.1.94 takes the form

$$\overline{S}(x,p) = -\overline{s}(p)e^{-(p/c)x} \quad \text{for } x > 0,\ p > 0 \tag{13.1.95}$$

Hence, by applying the operator L^{-1} to Equation 13.1.95 we find the simple formula

$$S(x,t) = -H\left(t - \frac{x}{c}\right) s\left(t - \frac{x}{c}\right) \quad \text{for } x \geq 0,\ t \geq 0 \tag{13.1.96}$$

This formula shows that the stress at a point $x > 0$ at time $t > 0$ is the pressure applied on the boundary at a previous time $t - x/c$. Therefore, each pressure signal propagates toward the points $x > 0$ with a velocity c. Also, note that by differentiating Equation 13.1.96 with respect to x and using the condition $s(0) = 0$, we obtain

$$\begin{aligned}\frac{\partial}{\partial x}S(x,t) &= \frac{1}{c}\delta\left(t - \frac{x}{c}\right)s\left(t - \frac{x}{c}\right) + \frac{1}{c}H\left(t - \frac{x}{c}\right)\frac{d}{dt}s\left(t - \frac{x}{c}\right)\\ &= \frac{1}{c}H\left(t - \frac{x}{c}\right)\frac{d}{dt}s\left(t - \frac{x}{c}\right)\end{aligned} \tag{13.1.97}$$

Let $u = u(x,t)$ be the displacement corresponding to $S = S(x,t)$ and satisfying the homogeneous initial conditions. Then, by integrating the equation of motion

$$\frac{\partial^2 u}{\partial t^2} = \frac{1}{\rho}\frac{\partial S}{\partial x} \tag{13.1.98}$$

with respect to time we have

$$\frac{\partial}{\partial t}u(x,t) = \frac{1}{\rho}\int_0^t \frac{\partial S}{\partial x}(x,\tau)d\tau \tag{13.1.99}$$

Next, by substituting Equation 13.1.97 to Equation 13.1.99, using the condition $s(0) = 0$, and Equation 13.1.96, we get

$$\frac{\partial}{\partial t}u(x,t) = -\frac{1}{\rho c}S(x,t) \tag{13.1.100}$$

Equation 13.1.100 relates the velocity of a material point at x to the stress at x and, combined with Equation 13.1.96, implies that when the surface traction is compressive, the particle velocity is in the same direction as the wave velocity, while, for a tensile force, the particle velocity is in the opposite direction [2].

13.1.6 One-Dimensional Stress Waves in a Finite Strip with Free Boundaries Subject to the Initial Stress and Stress-Rate Fields

The problem is now formulated in the following way. Find a solution $S = S(x,t)$ on $[0,l] \times [0,\infty)$ $(l > 0)$ to the field equation

$$\left(\frac{\partial^2}{\partial x^2} - \frac{1}{c^2}\frac{\partial^2}{\partial t^2}\right) S = 0 \quad \text{for } 0 < x < l,\ t > 0 \tag{13.1.101}$$

subject to the initial conditions

$$S(x,0) = S_0(x), \qquad \frac{\partial}{\partial t} S(x,0) = \dot{S}_0(x) \quad \text{for } 0 < x < l \tag{13.1.102}$$

and the boundary conditions

$$S(0,t) = S(l,t) = 0, \quad \text{for } t > 0 \tag{13.1.103}$$

where $S_0 = S_0(x)$ and $\dot{S}_0 = \dot{S}_0(x)$ are prescribed functions. To solve this problem we apply the operator L (see Equation 13.1.15) to Equations 13.1.101 and 13.1.103, and using Equations 13.1.102 we find

$$\left(\frac{\partial^2}{\partial x^2} - \frac{p^2}{c^2}\right) \overline{S} = -\frac{1}{c^2}\left[pS_0(x) + \dot{S}_0(x)\right] \quad \text{for } 0 < x < l,\ p > 0 \tag{13.1.104}$$

and

$$\overline{S}(0,p) = 0, \quad \overline{S}(l,p) = 0 \quad \text{for } p > 0 \tag{13.1.105}$$

respectively. In the following, a solution S will be obtained under the hypothesis that the functions $S_0 = S_0(x)$ and $\dot{S}_0 = \dot{S}_0(x)$ satisfy the conditions

$$S_0(0) = S_0(l) = 0, \quad \dot{S}_0(0) = \dot{S}_0(l) = 0 \tag{13.1.106}$$

which means that the initial data are compatible with the homogeneous boundary conditions at $(x,t) = (0,0)$ and $(x,t) = (l,0)$. Then, the functions $S_0 = S_0(x)$ and $\dot{S}_0 = \dot{S}_0(x)$ can be represented by the sine series [3]

$$S_0(x) = \sum_{n=1}^{\infty} S_{0n}\psi_n(x) \quad 0 \le x \le l \tag{13.1.107}$$

and

$$\dot{S}_0(x) = \sum_{n=1}^{\infty} \dot{S}_{0n}\psi_n(x) \quad 0 \leq x \leq l \tag{13.1.108}$$

where

$$\psi_n(x) = \left(\frac{2}{l}\right)^{1/2} \sin\frac{n\pi x}{l}, \quad n = 1, 2, 3, ... \tag{13.1.109}$$

$$S_{0n} = \int_0^l S_0(\xi)\psi_n(\xi)d\xi, \quad \dot{S}_{0n} = \int_0^l \dot{S}_0(\xi)\psi(\xi)d\xi \tag{13.1.110}$$

As a result, the field equation 13.1.104 is reduced to the form

$$\left(\frac{\partial^2}{\partial x^2} - \frac{p^2}{c^2}\right)\overline{S} = -\frac{1}{c^2}\left[p\sum_{n=1}^{\infty} S_{0n}\psi_n(x) + \sum_{n=1}^{\infty} \dot{S}_{on}\psi_n(x)\right] \tag{13.1.111}$$

Note that

$$\frac{d^2\psi_n}{dx^2} = -\left(\frac{n\pi}{l}\right)^2 \psi_n \quad \text{and } \psi_n(0) = \psi_n(l) = 0 \ \text{ for } n = 1, 2, 3, ... \tag{13.1.112}$$

Hence, a solution to the problem 13.1.104 and 13.1.105 takes the form

$$\overline{S}(x, p) = p\sum_{n=1}^{\infty} \frac{S_{0n}\psi_n(x)}{p^2 + (n\pi c/l)^2} + \sum_{n=1}^{\infty} \frac{\dot{S}_{0n}\psi_n(x)}{p^2 + (n\pi c/l)^2} \tag{13.1.113}$$

Finally, by applying the operator L^{-1} to Equation 13.1.113 we arrive at

$$S(x, t) = \sum_{n=1}^{\infty}\left[S_{0n}\cos\left(\frac{n\pi ct}{l}\right) + \dot{S}_{0n}\frac{l}{n\pi c}\sin\left(\frac{n\pi ct}{l}\right)\right]\psi_n(x) \tag{13.1.114}$$

Therefore, the stress $S = S(x, t)$ is a sum of functions periodic in time with period $T_n = 2\pi/\omega_n$ ($\omega_n = n\pi c/l$) and frequency $\nu_n = T_n^{-1}$. To each n there corresponds one term in the series 13.1.114, and this term represents the nth mode of oscillation with frequency ν_n.

For an initial stress of intensity S_0 concentrated in the middle of the strip ($x = l/2$):

$$S_0(x) = S_0\delta\left(x - \frac{l}{2}\right), \quad \dot{S}_0(x) = 0 \tag{13.1.115}$$

we find

$$S_{0n} = S_0\int_0^l \delta\left(x - \frac{l}{2}\right)\psi_n(x)dx = S_0\psi_n\left(\frac{l}{2}\right) = S_0\left(\frac{2}{l}\right)^{1/2}\sin\frac{n\pi}{2}$$

$$= S_0\left(\frac{2}{l}\right)^{1/2}(-1)^{n+1} \quad \text{for } n = 1, 3, 5, ... \tag{13.1.116}$$

and

$$S(x,t) = S_0 \frac{2}{l} \sum_{k=1}^{\infty} \sin \frac{(2k-1)\pi x}{l} \cos \frac{(2k-1)\pi ct}{l} \tag{13.1.117}$$

Here S_0 is a constant of the dimension [Stress × Length], see Equation 3.1.41.

This is a series representation of the dynamic response of the strip subject to an initial concentrated stress.

13.2 ONE-DIMENSIONAL SOLUTIONS OF NONISOTHERMAL ELASTODYNAMICS

One-dimensional initial–boundary value problems of nonisothermal homogeneous isotropic elastodynamics for an infinite or semi-infinite space may be described in terms of a pair (ϕ, T), where $\phi = \phi(x,t)$ represents a thermoelastic displacement potential and $T = T(x,t)$ is a temperature field; the space variable x ranges over the interval $(-\infty, +\infty)$ or $[0, \infty)$; and the time t belongs to the interval $[0, \infty)$. A one-dimensional thermoleastic process $p(x,t) = [\mathbf{u}(x,t), \mathbf{E}(x,t), \mathbf{S}(x,t)]$ is then associated with the pair (ϕ, T) through the relations [see Equation 4.2.56 and Equations (a) through (e) in Example 4.2.4 restricted to a one-dimensional case in which T and ϕ depend on $x = x_1$ and t only]

$$\mathbf{u}(x,t) = \left[\frac{\partial \phi}{\partial x}, 0, 0 \right] \tag{13.2.1}$$

$$\mathbf{E}(x,t) = \begin{bmatrix} \dfrac{\partial^2 \phi}{\partial x^2} & 0 & 0 \\ 0 & 0 & 0 \\ 0 & 0 & 0 \end{bmatrix} \tag{13.2.2}$$

and

$$\mathbf{S}(x,t) = \begin{bmatrix} \rho \dfrac{\partial^2 \phi}{\partial t^2} & 0 & 0 \\ 0 & \rho \dfrac{\partial^2 \phi}{\partial t^2} - 2\mu \dfrac{\partial^2 \phi}{\partial x^2} & 0 \\ 0 & 0 & \rho \dfrac{\partial^2 \phi}{\partial t^2} - 2\mu \dfrac{\partial^2 \phi}{\partial x^2} \end{bmatrix} \tag{13.2.3}$$

where $\phi = \phi(x,t)$ satisfies the one-dimensional nonhomogeneous wave equation

$$\left(\frac{\partial^2}{\partial x^2} - \frac{1}{c_1^2} \frac{\partial^2}{\partial t^2} \right) \phi = mT \tag{13.2.4}$$

In Equation 13.2.4, c_1 and m are given by the relations

$$c_1 = \sqrt{\frac{\lambda + 2\mu}{\rho}}, \quad m = \frac{3\lambda + 2\mu}{\lambda + 2\mu} \alpha_t \tag{13.2.5}$$

where
ρ is the density
α_t is the coefficient of linear thermal expansion
λ and μ are the Lamé moduli; moreover, $T = T(x,t)$ is a solution of the one-dimensional parabolic heat conduction equation

$$\left(\frac{\partial^2}{\partial x^2} - \frac{1}{\kappa}\frac{\partial}{\partial t}\right) T = -\frac{1}{\kappa} Q(x,t) \tag{13.2.6}$$

Note that $Q(x,t)/\kappa = W(x,t)/k$, where $Q(x,t)$ is a prescribed heat supply field, $W(x,t)$ stands for the internal heat generated per unit volume per unit time, k denotes the thermal conductivity, and κ means the thermal diffusivity.

If we introduce the dimensionless space and time variables

$$\hat{x} = \frac{x}{x_0}, \quad \hat{t} = \frac{t}{t_0} \tag{13.2.7}$$

where

$$x_0 = \frac{\kappa}{c_1}, \quad t_0 = \frac{\kappa}{c_1^2} \tag{13.2.8}$$

and the dimensionless fields

$$\begin{aligned} \hat{u} &= \frac{u}{u_0}, \quad \hat{\mathbf{E}} = \frac{\mathbf{E}}{E_0}, \quad \hat{\mathbf{S}} = \frac{\mathbf{S}}{S_0} \\ \hat{T} &= \frac{T}{T_0}, \quad \hat{\phi} = \frac{\phi}{\phi_0}, \quad \hat{Q} = \frac{Q}{Q_0} \end{aligned} \tag{13.2.9}$$

where

$$\begin{aligned} u_0 &= \frac{\phi_0}{x_0}, \quad E_0 = \frac{\phi_0}{x_0^2}, \quad S_0 = \frac{\rho\phi_0}{t_0^2} \\ \phi_0 &= T_0 x_0^2 m, \quad Q_0 = \frac{T_0\kappa}{x_0^2} \end{aligned} \tag{13.2.10}$$

and $T_0 > 0$ is a reference temperature; and if we denote the dimensionless quantities by the same symbols as dimensional ones, then Equations 13.2.1 through 13.2.4 and Equation 13.2.6 take the dimensionless form

$$\mathbf{u}(x,t) = \left[\frac{\partial \phi}{\partial x}, 0, 0\right] \tag{13.2.11}$$

$$\mathbf{E}(x,t) = \begin{bmatrix} \dfrac{\partial^2 \phi}{\partial x^2} & 0 & 0 \\ 0 & 0 & 0 \\ 0 & 0 & 0 \end{bmatrix} \tag{13.2.12}$$

$$\mathbf{S}(x,t) = \begin{bmatrix} \dfrac{\partial^2\phi}{\partial t^2} & 0 & 0 \\ 0 & \dfrac{\partial^2\phi}{\partial t^2} - \dfrac{1-2\nu}{1-\nu}\dfrac{\partial^2\phi}{\partial x^2} & 0 \\ 0 & 0 & \dfrac{\partial^2\phi}{\partial t^2} - \dfrac{1-2\nu}{1-\nu}\dfrac{\partial^2\phi}{\partial x^2} \end{bmatrix} \tag{13.2.13}$$

$$\left(\frac{\partial^2}{\partial x^2} - \frac{\partial^2}{\partial t^2}\right)\phi = T \tag{13.2.14}$$

and

$$\left(\frac{\partial^2}{\partial x^2} - \frac{\partial}{\partial t}\right)T = -Q \tag{13.2.15}$$

respectively. In Equation 13.2.13 ν stands for Poisson's ratio.

By introducing the notation

$$S_{11}(x,t) \equiv S(x,t) \tag{13.2.16}$$

and applying the operator $\partial^2/\partial t^2$ to Equation 13.2.14 we arrive at the one-dimensional dimensionless stress–temperature equation of nonisothermal elastodynamics

$$\left(\frac{\partial^2}{\partial x^2} - \frac{\partial^2}{\partial t^2}\right)S = \frac{\partial^2 T}{\partial t^2} \tag{13.2.17}$$

Finally, by applying the operator $(\partial^2/\partial x^2 - \partial/\partial t)$ to Equation 13.2.17 and using Equation 13.2.15 we obtain

$$\left(\frac{\partial^2}{\partial x^2} - \frac{\partial}{\partial t}\right)\left(\frac{\partial^2}{\partial x^2} - \frac{\partial^2}{\partial t^2}\right)S = -\frac{\partial^2 Q}{\partial t^2} \tag{13.2.18}$$

In the following, the three initial–boundary value problems, related to the field equations 13.2.15, 13.2.17, and 13.2.18, will be discussed:

(A) Dynamic thermal stresses in an infinite elastic solid subject to a plane instantaneous heat source
(B) Dynamic thermal stresses in a half-space with free boundary subject to a plane internal instantaneous heat source
(C) Dynamic thermal stresses in a half-space subject to a sudden heating of the boundary plane—the Danilovskaya problem. In all cases closed-form solutions are obtained, and an analysis of the solution is presented.

13.2.1 Dynamic Thermal Stresses in an Infinite Elastic Solid Subject to a Plane Instantaneous Heat Source

We look for a solution to the following initial–boundary value problem. Find a stress $S = S(x,t)$ on $(-\infty, +\infty) \times [0, \infty)$ that satisfies the field equation [see Equation 13.2.18

with $Q = \delta(x)\delta(t)$]

$$\left(\frac{\partial^2}{\partial x^2} - \frac{\partial}{\partial t}\right)\left(\frac{\partial^2}{\partial x^2} - \frac{\partial^2}{\partial t^2}\right) S = -\frac{\partial^2}{\partial t^2}[\delta(t)\delta(x)] \quad \text{on } (-\infty, +\infty) \times (0, \infty) \tag{13.2.19}$$

subject to the initial conditions

$$S(x,0) = \frac{\partial}{\partial t} S(x,0) = \frac{\partial^2}{\partial t^2} S(x,0) = 0 \quad \text{for } |x| < \infty \tag{13.2.20}$$

and the vanishing condition

$$S(x,t) \to 0 \quad \text{as } |x| \to \infty, \ t > 0 \tag{13.2.21}$$

where $\delta = \delta(x)$ is the Dirac delta function. Hence, we look for the dynamic thermal stress $S = S(x,t)$ due to a unit instantaneous heat source applied on the plane $x = 0$ of a homogeneous isotropic infinite elastic space $|x| < \infty$. The conditions $13.2.20_1$ and $13.2.20_2$ assert that the initial stress and stress-rate fields vanish throughout the solid, while the condition $13.2.20_3$ is implied by the field equations 13.2.15, 13.2.17, the conditions $13.2.20_1$ through $13.2.20_2$, and the hypotheses

$$T(x,0) = Q(x,0) = \frac{\partial}{\partial t} Q(x,0) = 0 \quad \text{for } |x| < \infty \tag{13.2.22}$$

In the following, we assume that the functions T and Q do satisfy the conditions 13.2.22. As a result, the temperature $T = T(x,t)$ satisfies the field equation [see Equation 13.2.15 with $Q = \delta(x)\delta(t)$]

$$\left(\frac{\partial^2}{\partial x^2} - \frac{\partial}{\partial t}\right) T = -\delta(x)\delta(t) \quad \text{for } |x| < \infty, \ t > 0 \tag{13.2.23}$$

subject to the conditions

$$T(x,0) = 0 \quad \text{for } |x| < \infty \tag{13.2.24}$$

$$T(x,t) \to 0 \quad \text{as } |x| \to \infty, \ t > 0 \tag{13.2.25}$$

Let $\bar{f} = \bar{f}(x,p)$ denote the Laplace transform of a function $f = f(x,t)$

$$L\{f\} \equiv \bar{f}(x,p) = \int_0^\infty e^{-pt} f(x,t)dt \tag{13.2.26}$$

where p is the Laplace transform parameter. Applying the operator L to Equation 13.2.23, and taking into account the conditions 13.2.24 through 13.2.25 we find

$$\left(\frac{\partial^2}{\partial x^2} - p\right)\bar{T} = -\delta(x) \quad \text{for } |x| < \infty, \ p > 0 \tag{13.2.27}$$

and

$$\overline{T}(x,p) \to 0 \quad \text{as } |x| \to \infty,\ p > 0 \tag{13.2.28}$$

Next, by using the Dirac delta representation

$$\delta(x) = \frac{1}{\pi}\int_0^\infty \cos\alpha x\, d\alpha \tag{13.2.29}$$

we find that a solution to Equation 13.2.27 that satisfies the condition 13.2.28 takes the form

$$\overline{T}(x,p) = \frac{1}{\pi}\int_0^\infty \frac{\cos\alpha x}{\alpha^2 + p} d\alpha \tag{13.2.30}$$

or

$$\overline{T}(x,p) = \frac{1}{2}\frac{e^{-|x|\sqrt{p}}}{\sqrt{p}} \tag{13.2.31}$$

Finally, by applying the operator L^{-1} to Equation 13.2.31 we obtain [4]

$$T(x,t) = \frac{1}{2}\frac{1}{\sqrt{\pi t}}\, e^{-x^2/4t} \tag{13.2.32}$$

It follows from this formula that for a fixed time t, the function $T = T(x,t)$ is represented by a Gaussian profile that attains a maximum at $x = 0$ and vanishes at $|x| = \infty$.

To obtain the thermal stress $S = S(x,t)$ corresponding to the temperature $T = T(x,t)$, we apply the operator L to Equation 13.2.19 and using the conditions 13.2.20 and 13.2.21 we arrive at

$$\left(\frac{\partial^2}{\partial x^2} - p\right)\left(\frac{\partial^2}{\partial x^2} - p^2\right)\overline{S} = -p^2\delta(x) \quad \text{for } |x| < \infty,\ p > 0 \tag{13.2.33}$$

and

$$\overline{S}(x,p) \to 0 \quad \text{as } |x| \to \infty,\ p > 0 \tag{13.2.34}$$

Next, substituting the integral representation of $\delta = \delta(x)$ into Equation 13.2.33 (see Equation 13.2.29) we find that a solution to Equation 13.2.33 subject to the condition 13.2.34 takes the form

$$\overline{S}(x,p) = -\frac{p^2}{\pi}\int_0^\infty \frac{\cos\alpha x}{(\alpha^2 + p)(\alpha^2 + p^2)} d\alpha \tag{13.2.35}$$

or

$$\overline{S}(x,p) = \frac{1}{\pi}\frac{p}{p-1}\left(\int_0^\infty \frac{\cos\alpha x}{\alpha^2 + p^2} d\alpha - \int_0^\infty \frac{\cos\alpha x}{\alpha^2 + p} d\alpha\right) \tag{13.2.36}$$

or

$$\overline{S}(x,p) = \frac{1}{2}\frac{p}{p-1}\left(\frac{e^{-|x|p}}{p} - \frac{e^{-|x|\sqrt{p}}}{\sqrt{p}}\right) \tag{13.2.37}$$

An alternative form of this equation reads

$$\overline{S}(x,p) = \frac{1}{2}\left(\frac{e^{-|x|p}}{p-1} + \frac{\partial}{\partial|x|}\frac{e^{-|x|\sqrt{p}}}{p-1}\right) \tag{13.2.38}$$

Now [5]

$$L^{-1}\left\{\frac{e^{-|x|p}}{p-1}\right\} = e^{t-|x|}H(t-|x|) \tag{13.2.39}$$

and

$$L^{-1}\left\{\frac{e^{-|x|\sqrt{p}}}{p-1}\right\} = U(|x|,t) \tag{13.2.40}$$

where

$$U(|x|,t) = \frac{1}{2}e^{t}\left[e^{-|x|}\,\text{erfc}\left(\frac{|x|}{2\sqrt{t}} - \sqrt{t}\right) + e^{|x|}\text{erfc}\left(\frac{|x|}{2\sqrt{t}} + \sqrt{t}\right)\right] \tag{13.2.41}$$

In Equation 13.2.41, the function erfc (z) is the complementary error function defined as

$$\text{erfc}\,(z) = 1 - \text{erf}(z) = \frac{2}{\sqrt{\pi}}\int_{z}^{\infty} e^{-\xi^2}d\xi \tag{13.2.42}$$

Hence, by applying the operator L^{-1} to Equation 13.2.38 we obtain

$$\begin{aligned} S(x,t) &= \frac{1}{2}\left[e^{t-|x|}H(t-|x|) + \frac{\partial}{\partial|x|}U(|x|,t)\right] \\ &= \frac{1}{2}\left[e^{t-|x|}H(t-|x|) - \frac{1}{\sqrt{\pi t}}e^{-x^2/4t} - V(|x|,t)\right] \end{aligned} \tag{13.2.43}$$

where

$$V(|x|,t) = \frac{1}{2}e^{t}\left[e^{-|x|}\text{erfc}\left(\frac{|x|}{2\sqrt{t}} - \sqrt{t}\right) - e^{|x|}\text{erfc}\left(\frac{|x|}{2\sqrt{t}} + \sqrt{t}\right)\right] \tag{13.2.44}$$

It follows from Equation 13.2.43 that

$$S(x,t) = S_w(x,t) + S_d(x,t) \tag{13.2.45}$$

where $S_w(x,t)$ and $S_d(x,t)$ represent the wave and diffusive parts of $S(x,t)$, respectively, defined by

$$S_w(x,t) = \frac{1}{2}e^{t-|x|}H(t-|x|) \tag{13.2.46}$$

and

$$S_d(x,t) = -\frac{1}{2}\left[\frac{e^{-x^2/4t}}{\sqrt{\pi t}} + V(|x|,t)\right] \tag{13.2.47}$$

The wave part has two plane fronts propagating with a unit velocity in the opposite directions: $t+x=0$ and $t-x=0$ for every $|x|<\infty$ and $t>0$, which means that $S_w>0$ for $|x|<t$, and $S_w=0$ for $|x|>t$. Moreover,

$$S_w(x,|x|+0) - S_w(x,|x|-0) = \frac{1}{2} \tag{13.2.48}$$

$$S_d(x,|x|+0) - S_d(x,|x|-0) = 0 \tag{13.2.49}$$

Therefore, for a fixed value of x the stress $S(x,t)$ undergoes the jump across the plane $|x|=t$

$$S(x,|x|+0) - S(x,|x|-0) = \frac{1}{2} \quad \text{for } |x| < \infty \tag{13.2.50}$$

On the other hand, the diffusive part $S_d = S_d(x,t)$ is felt instantaneously at any distance $|x|$ from the heat source plane $x=0$, which means that $S_d(x,t)<0$ for every $|x|<\infty$ and $t>0$.

Note that if the instantaneous heat source plane is located at a cross section $x=\xi$ of the infinite solid, then a pair (T^*, S^*) corresponding to such a heat source is obtained from the formulas (see Equations 13.2.32 and 13.2.43)

$$T^*(x,\xi;t) = \frac{1}{2}\frac{1}{\sqrt{\pi t}}\,e^{-(x-\xi)^2/4t} \tag{13.2.51}$$

and

$$S^*(x,\xi;t) = \frac{1}{2}\left[e^{t-|x-\xi|}H(t-|x-\xi|) - \frac{e^{-(x-\xi)^2/4t}}{\sqrt{\pi t}} - V(|x-\xi|,t)\right] \tag{13.2.52}$$

Finally, for an arbitrary heat source function $Q=Q(x,t)$ on $(-\infty,+\infty)\times[0,\infty)$ the associated temperature $T=T(x,t)$ and stress $S=S(x,t)$ are obtained from the formulas

$$T(x,t) = \int_0^t \int_{-\infty}^{+\infty} T^*(x,\xi;t-\tau)Q(\xi,\tau)d\xi\, d\tau \tag{13.2.53}$$

and

$$S(x,t) = \int_0^t \int_{-\infty}^{+\infty} S^*(x,\xi;t-\tau)Q(\xi,\tau)d\xi\, d\tau \tag{13.2.54}$$

where T^* and S^* are given by Equations 13.2.51 and 13.2.52, respectively.

13.2.2 Dynamic Thermal Stresses in a Half-Space with Free Boundary Subject to an Instantaneous Internal Plane Heat Source

We consider now a homogeneous isotropic semi-infinite thermoelastic solid $x \geq 0$, initially at rest, and subject to the action of an instantaneous heat source distributed on the plane $x = x_0 > 0$. The bounding plane is stress free and maintained at zero temperature. The vanishing of the heat source-induced temperature $T = T(x, x_0; t)$ and stress $S = S(x, x_0; t)$ at infinity is also assumed. Therefore, the function $T = T(x, x_0; t)$ is to satisfy the field equation [see Equation 13.2.15 with $Q = \delta(x - x_0)\delta(t)$]

$$\left(\frac{\partial^2}{\partial x^2} - \frac{\partial}{\partial t}\right) T = -\delta(x - x_0)\delta(t) \quad \text{for } x > 0,\ x_0 > 0,\ t > 0 \tag{13.2.55}$$

subject to the initial condition

$$T(x, x_0; 0) = 0 \quad \text{for } x > 0,\ x_0 > 0 \tag{13.2.56}$$

the boundary condition

$$T(0, x_0; t) = 0 \quad \text{for } x_0 > 0,\ t > 0 \tag{13.2.57}$$

and the vanishing condition at infinity

$$T(x, x_0; t) \to 0 \quad \text{as } x \to \infty,\ x_0 > 0,\ t > 0 \tag{13.2.58}$$

while the function $S = S(x, x_0; t)$ satisfies the field equation [see Equation 13.2.19 with $Q = \delta(x - x_0)\delta(t)$]

$$\left(\frac{\partial^2}{\partial x^2} - \frac{\partial}{\partial t}\right)\left(\frac{\partial^2}{\partial x^2} - \frac{\partial^2}{\partial t^2}\right) S = -\frac{\partial^2}{\partial t^2}[\delta(x - x_0)\delta(t)] \quad \text{for } x > 0,\ t > 0,\ x_0 > 0 \tag{13.2.59}$$

the initial conditions

$$S(x, x_0; 0) = \frac{\partial}{\partial t} S(x, x_0; 0) = \frac{\partial^2}{\partial t^2} S(x, x_0; 0) = 0 \quad \text{for } x > 0,\ x_0 > 0 \tag{13.2.60}$$

the boundary condition

$$S(0, x_0; t) = 0 \quad \text{for } t > 0,\ x_0 > 0 \tag{13.2.61}$$

and the vanishing condition at infinity

$$S(x, x_0; t) \to 0 \quad \text{as } x \to \infty,\ x_0 > 0,\ t > 0 \tag{13.2.62}$$

To find a solution to the initial–boundary value problem 13.2.55 and 13.2.56, we use the Laplace transform technique. By applying the operator L (see Equation 13.2.26) to Equations 13.2.55 and 13.2.57, and using the conditions 13.2.56 and 13.2.58, we find

$$\left(\frac{\partial^2}{\partial x^2} - p\right)\overline{T} = -\delta(x - x_0) \quad \text{for } x > 0,\ x_0 > 0,\ p > 0 \tag{13.2.63}$$

$$\overline{T}(0, x_0; p) = 0 \quad \text{for } x_0 > 0,\ p > 0 \tag{13.2.64}$$

and

$$\overline{T}(x, x_0; p) \to 0 \quad \text{as } x \to \infty,\ x_0 > 0,\ p > 0 \tag{13.2.65}$$

Since (see Equations 13.2.27 and 13.2.31)

$$\left(\frac{\partial^2}{\partial x^2} - p\right)\frac{1}{2}\frac{e^{-|x-x_0|\sqrt{p}}}{\sqrt{p}} = -\delta(x - x_0) \tag{13.2.66}$$

we look for a solution $\overline{T}$ to Equations 13.2.63 through 13.2.65 in the form

$$\overline{T}(x, x_0; p) = Ae^{-x\sqrt{p}} + \frac{1}{2}\frac{e^{-|x-x_0|\sqrt{p}}}{\sqrt{p}} \tag{13.2.67}$$

where $A = A(x_0, p)$ is to be found from the boundary condition 13.2.64. By letting $x = 0$ in Equation 13.2.67 and substituting into Equation 13.2.64 we obtain

$$A(x_0, p) = -\frac{1}{2}\frac{e^{-x_0\sqrt{p}}}{\sqrt{p}} \tag{13.2.68}$$

Hence,

$$\overline{T}(x, x_0; p) = \frac{1}{2}\left[\frac{e^{-|x-x_0|\sqrt{p}} - e^{-(x+x_0)\sqrt{p}}}{\sqrt{p}}\right] \tag{13.2.69}$$

and by applying the operator L^{-1} to this equation we obtain (see Equations 13.2.31 and 13.2.32)

$$T(x, x_0; t) = \frac{1}{2\sqrt{\pi t}}\left[e^{-(x-x_0)^2/4t} - e^{-(x+x_0)^2/4t}\right] \quad \text{for } x \geq 0,\ x_0 > 0,\ t \geq 0 \tag{13.2.70}$$

In a similar way, by applying the operator L to Equations 13.2.59 and 13.2.61, and taking into account the conditions 13.2.60 and 13.2.62, we arrive at the boundary value problem for $\overline{S}$:

$$\left(\frac{\partial^2}{\partial x^2} - p\right)\left(\frac{\partial^2}{\partial x^2} - p^2\right)\overline{S} = -p^2\delta(x - x_0) \quad \text{for } x > 0,\ x_0 > 0,\ p > 0 \tag{13.2.71}$$

$$\overline{S}(0, x_0; p) = 0 \quad \text{for } x_0 > 0,\ p > 0 \tag{13.2.72}$$

and

$$\overline{S}(x, x_0; p) \to 0 \quad \text{as } x \to \infty,\ x_0 > 0, p > 0 \tag{13.2.73}$$

Since (see Equations 13.2.33 and 13.2.37)

$$\left(\frac{\partial^2}{\partial x^2} - p\right)\left(\frac{\partial^2}{\partial x^2} - p^2\right)\left[\frac{1}{2}\frac{p}{p-1}\left(\frac{e^{-|x-x_0|p}}{p} - \frac{e^{-|x-x_0|\sqrt{p}}}{\sqrt{p}}\right)\right]$$
$$= -p^2\delta(x - x_0) \tag{13.2.74}$$

therefore, a solution to the problem 13.2.71 through 13.2.73 takes the form

$$\bar{S}(x, x_0; p) = \frac{1}{2}\frac{p}{p-1}\left\{\left[\frac{e^{-|x-x_0|p} - e^{-(x+x_0)p}}{p}\right]\right.$$
$$\left. - \left[\frac{e^{-|x-x_0|\sqrt{p}} - e^{-(x+x_0)\sqrt{p}}}{\sqrt{p}}\right]\right\} \quad \text{for } x \geq 0,\ x_0 > 0,\ p > 0 \tag{13.2.75}$$

Finally, by applying the operator L^{-1} to Equation 13.2.75 and proceeding in a way similar to that of obtaining Equation 13.2.45, we find

$$S(x, x_0; t) = S_w(x, x_0; t) + S_d(x, x_0; t) \quad \text{for } x \geq 0,\ x_0 > 0,\ t > 0 \tag{13.2.76}$$

where

$$S_w(x, x_0; t) = \frac{1}{2}\left\{e^{t-|x-x_0|}H[t - |x - x_0|] - e^{t-(x+x_0)}H[t - (x + x_0)]\right\} \tag{13.2.77}$$

$$S_d(x, x_0; t) = -\frac{1}{2}\left\{\frac{e^{-(x-x_0)^2/4t} - e^{-(x+x_0)^2/4t}}{\sqrt{\pi t}}\right.$$
$$\left. + V[|x - x_0|; t] - V[(x + x_0); t]\right\} \tag{13.2.78}$$

and the function $V = V(|x|; t)$ is given by (see Equation 13.2.44)

$$V(|x|; t) = \frac{1}{2}e^t\left[e^{-|x|}\text{erfc}\left(\frac{|x|}{2\sqrt{t}} - \sqrt{t}\right) - e^{|x|}\text{erfc}\left(\frac{|x|}{2\sqrt{t}} + \sqrt{t}\right)\right] \tag{13.2.79}$$

It follows from Equations 13.2.76 through 13.2.79 that the functions $S_w = S_w(x, x_0; t)$ and $S_d = S_d(x, x_0; t)$ represent the wave and diffusive parts of S, respectively. This means that $S_d \neq 0$ for every $0 < x < \infty$ and $t > 0$; while $S_w \neq 0$ for $t > x + x_0$ or $|x - x_0| < t < \mathbf{x} + x_0$, and $S_w = 0$ otherwise.

Since

$$S_w(x, x_0; t) \to 0 \quad \text{as } t \to |x - x_0| - 0 \tag{13.2.80}$$

$$S_w(x, x_0; t) \to \frac{1}{2} \quad \text{as } t \to |x - x_0| + 0 \tag{13.2.81}$$

and

$$S_w(x, x; t) \rightarrow \frac{1}{2} e^{x+x_0-|x-x_0|} \quad \text{as } t \rightarrow x + x_0 - 0 \tag{13.2.82}$$

$$S_w(x, x; t) \rightarrow \frac{1}{2}\left[e^{x+x_0-|x-x_0|} - e^{x+x_0-(x+x_0)}\right] \quad \text{as } t \rightarrow x + x_0 + 0 \tag{13.2.83}$$

and, since $S_d = S_d(x, x_0; t)$ is a continuous function on the wavefronts $t - |x - x_0| = 0$ and $t - (x + x_0) = 0$, we write

$$S(x, x_0; |x - x_0| + 0) - S(x, x_0; |x - x_0| - 0) = \frac{1}{2} \tag{13.2.84}$$

and

$$S[x, x_0; (x + x_0) + 0] - S[x, x_0; (x + x_0) - 0] = -\frac{1}{2} \tag{13.2.85}$$

Hence, the stress $S = S(x, x_0; t)$ undergoes a jump of magnitude $1/2$ across the moving front $t - |x - x_0| = 0$, and a jump of the same magnitude but of opposite sign across the moving front $t - (x + x_0) = 0$.

A pair (T, S) in which $T = T(x, x_0; t)$ and $S = S(x, x_0; t)$ are given by Equations 13.2.70 and 13.2.76, respectively, represents Green's function for a semispace $x \geq 0$ subject to an instantaneous heat source distributed over the plane $x = x_0 > 0$. If we look for a solution $(\hat{T}, \hat{S})$ of a semispace problem in which a heat source function takes the form

$$Q(x, t) = Y(t)e^{-ax} \quad \text{for } x \geq 0,\ t \geq 0 \tag{13.2.86}$$

where

$$Y(t) = Y_0 t^n \exp(-bt^m) \tag{13.2.87}$$

and Y_0, a, b, m, and n are positive constants, then the functions $\hat{T} = \hat{T}(x, t)$ and $\hat{S} = \hat{S}(x, t)$ are computed from the formulas

$$\hat{T}(x, t) = \int_0^t Y(\tau)\left[\int_0^\infty e^{-ax_0} T(x, x_0; t - \tau) dx_0\right] d\tau \tag{13.2.88}$$

and

$$\hat{S}(x, t) = \int_0^t Y(\tau)\left[\int_0^\infty e^{-ax_0} S(x, x_0; t - \tau) dx_0\right] d\tau \tag{13.2.89}$$

where $T = T(x, x_0; t)$ and $S = S(x, x_0; t)$ are given by Equations 13.2.70 and 13.2.76, respectively. The pair $(\hat{T}, \hat{S})$ represents then a thermoelastic response of the semispace to a laser pulse.* The function $Q = Q(x, t)$ given by Equations 13.2.86 and 13.2.87 represents a

* For a similar problem, see [6].

laser-induced heat of the semispace; the function $Y = Y(t)$ describes the "skewed" Gaussian temporal profile of the laser; and a^{-1} stands for an absorption coefficient. The laser pulse is of a short duration if the number $\xi = t_f^n \exp\left(-bt_f^m\right)$, where t_f is a pulse activation time, is small (e.g., $\xi = 0.001$). For such a pulse, its rising time t_r is computed from the formula $t_r = (n/mb)^{1/m}$, and for $b = 10$, $n = 4$, and $m = 2$ we obtain $t_r = 0.4472$. Recall that the problem is discussed in a dimensionless setting. For aluminum, the space and time units are given by $x_0 = 1.36 \times 10^{-6}$ cm and $t_0 = 0.21 \times 10^{-11}$ s (see Equations 13.2.8). Hence, a real rising time of the laser is of order 0.09×10^{-11} s.

Note:

In the laser problem for a half-space in which a laser beam action is converted instantaneously into a heat supply of the form

$$Q(x,t) = Q_0[H(t) - H(t - t_0)]\exp(-ax)$$

where $H = H(t)$ is the Heaviside function, $Q_0 > 0$, and t_0 is a switch-off time, the following results hold true: (i) the laser-induced heat makes the semispace expand proportionally to the temperature. At a fixed cross section of the semispace, the temperature attains a maximum at a time greater than the switch-off time, and the maximum decreases along the semispace depth; as a result, the material located far from the boundary expands less than that of a boundary layer, (ii) initially, at a given cross section far from the boundary the stress wave is compressive with the magnitude of compression decreasing with the depth, while near the boundary the stress wave becomes tensile, and (iii) for times greater than the switch-off time, the stress magnitude increases to its maximum first and then decays to zero as time t goes to infinity [7–9].

13.2.3 Dynamic Thermal Stresses in a Half-Space Subject to a Sudden Heating of the Boundary Plane: Danilovskaya Problem

Let us consider the dynamic thermal stresses in a homogeneous isotropic elastic semispace $x \geq 0$ due to a sudden heating of the plane $x = 0$. A temperature $T = T(x,t)$ is to satisfy the field equation (see Equation 13.2.15 with $Q = 0$)

$$\left(\frac{\partial^2}{\partial x^2} - \frac{\partial}{\partial t}\right) T = 0 \quad \text{for } x > 0,\ t > 0 \qquad (13.2.90)$$

the initial condition

$$T(x,0) = 0 \quad \text{for } x > 0 \qquad (13.2.91)$$

the boundary condition

$$T(0,t) = H(t) \quad \text{for } t > 0 \qquad (13.2.92)$$

and the vanishing condition at infinity

$$T(x,t) \to 0 \quad \text{as } x \to +\infty,\ t > 0 \qquad (13.2.93)$$

where $H(t)$ is the Heaviside function. The associated stress $S = S(x,t)$ is to satisfy the field equation (see Equation 13.2.17)

$$\left(\frac{\partial^2}{\partial x^2} - \frac{\partial^2}{\partial t^2}\right) S = \frac{\partial^2 T}{\partial t^2} \quad \text{for } x > 0,\ t > 0 \tag{13.2.94}$$

subject to the conditions

$$S(x,0) = 0, \ \frac{\partial}{\partial t} S(x,0) = 0 \quad \text{for } x > 0 \tag{13.2.95}$$

$$S(0,t) = 0 \quad \text{for } t \geq 0 \tag{13.2.96}$$

and

$$S(x,t) \to 0 \quad \text{as } x \to \infty,\ t > 0 \tag{13.2.97}$$

Hence, the boundary $x = 0$ is free of tractions for $t \geq 0$, and both T and S are to vanish as $x \to +\infty$ for every time $t > 0$. Also, it follows from Equation 13.2.90 taken at $t = 0$ and the homogeneous initial condition 13.2.91 that

$$T(x,0) = 0, \quad \frac{\partial}{\partial t} T(x,0) = 0 \quad \text{for } x > 0 \tag{13.2.98}$$

The problem of finding a pair (T,S) that satisfies Equations 13.2.90 through 13.2.98 is called a *Danilovskaya problem* [10].

To solve the problem, we use the Laplace transform technique with respect to time t. By applying the operator L (see Equation 13.2.26) to Equations 13.2.90 and 13.2.92, and using the conditions 13.2.91 and 13.2.93, we have

$$\left(\frac{\partial^2}{\partial x^2} - p\right) \overline{T} = 0 \quad \text{for } x > 0,\ p > 0 \tag{13.2.99}$$

$$\overline{T}(0,p) = \frac{1}{p} \quad \text{for } p > 0 \tag{13.2.100}$$

and

$$\overline{T}(x,p) \to 0 \quad \text{as } x \to \infty,\ p > 0 \tag{13.2.101}$$

Hence,

$$\overline{T}(x,p) = \frac{1}{p} e^{-x\sqrt{p}} \quad \text{for } x > 0,\ p > 0 \tag{13.2.102}$$

and by applying the operator L^{-1} to Equation 13.2.103 we obtain [4]

$$T(x,t) = \text{erfc}\left(\frac{x}{2\sqrt{t}}\right) \quad \text{for } x \geq 0,\ t \geq 0 \tag{13.2.103}$$

Similarly, by applying the operator L to Equations 13.2.94 and 13.2.96 and using the conditions 13.2.95, 13.2.97, and 13.2.98 we arrive at the equations

$$\left(\frac{\partial^2}{\partial x^2} - p^2\right)\bar{S} = p^2\bar{T} \quad \text{for } x > 0,\ p > 0 \tag{13.2.104}$$

$$\bar{S}(0,p) = 0 \quad \text{for } p > 0 \tag{13.2.105}$$

and

$$\bar{S}(x,p) \to 0 \quad \text{as } x \to \infty,\ p > 0 \tag{13.2.106}$$

Since $\bar{T}$ is given by Equation 13.2.102, an alternative form of Equations 13.2.104 through 13.2.106 reads

$$\left(\frac{\partial^2}{\partial x^2} - p^2\right)\bar{S} = pe^{-x\sqrt{p}} \quad \text{for } x > 0,\ p > 0 \tag{13.2.107}$$

$$\bar{S}(0,p) = 0 \quad \text{for } p > 0 \tag{13.2.108}$$

and

$$\bar{S}(x,p) \to 0 \quad \text{as } x \to \infty,\ p > 0 \tag{13.2.109}$$

It is easy to check that a solution $\bar{S}$ to Equations 13.2.107 through 13.2.109 takes the form

$$\bar{S}(x,p) = \frac{1}{p-1}\left(e^{-xp} - e^{-x\sqrt{p}}\right) \quad \text{for } x \geq 0,\ p > 0 \tag{13.2.110}$$

Therefore, applying the operator L^{-1} to Equation 13.2.110 and using the formulas 13.2.39 and 13.2.40 we arrive at

$$S(x,t) = \left\{e^{t-x}H(t-x) - U(x,t)\right\} \tag{13.2.111}$$

where

$$U(x,t) = \frac{1}{2}e^t\left[e^{-x}\text{erfc}\left(\frac{x}{2\sqrt{t}} - \sqrt{t}\right) + e^{x}\text{erfc}\left(\frac{x}{2\sqrt{t}} + \sqrt{t}\right)\right] \tag{13.2.112}$$

It follows from Equations 13.2.111 and 13.2.112 that

$$S(x,t) = S_w(x,t) + S_d(x,t) \tag{13.2.113}$$

where S_w and S_d represent the wave and diffusive parts of S, respectively; and

$$S_w(x,t) = e^{t-x}H(t-x) \tag{13.2.114}$$

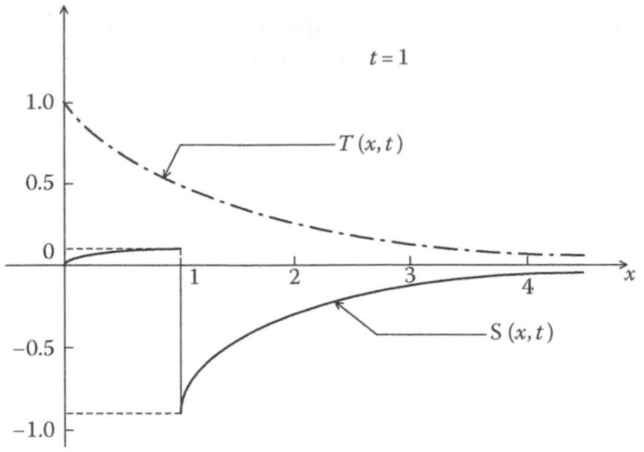

FIGURE 13.2 Temperature $T = T(x, t)$ and stress $S = S(x, t)$ as functions of x for $t = 1$. Note a stress jump of magnitude 1 at $x = 1$.

and

$$S_d(x, t) = -U(x, t) \tag{13.2.115}$$

Hence $S_d < 0$ for every $x > 0$ and $t > 0$, while $S_w > 0$ for $t > x$ and $S_w = 0$ for $t < x$. In addition

$$S_w(x, x+0) - S_w(x, x-0) = S(x, x+0) - S(x, x-0) = 1 \tag{13.2.116}$$

In Figure 13.2 graphs of $T = T(x, t)$ and $S = S(x, t)$ are shown for $t = 1$ and $0 < x < \infty$ (see also the graphs by T. Mura) [11].

The graph of $S(x, t)$ from Figure 13.2 has been used as a vignette on the cover page of the *Journal of Thermal Stresses* since its release in 1978 [12].

PROBLEMS

13.1 Using the displacement characterization of one-dimensional homogeneous isotropic elastodynamics, find the displacement $u = u(x, t)$ in a half-space $x \geq 0$ that is initially at rest and subject to a uniform dynamic pressure $s = s(t)$ on the boundary $x = 0$. Also, show that the particle velocity $\dot{u} = \dot{u}(x, t)$ is related to the stress $S = S(x, t)$ by the equation $S(x, t) = -\rho c \dot{u}(x, t)$, where ρ and c are the density and the longitudinal velocity of the body, respectively.

13.2 Solve a one-dimensional initial–boundary value problem for a half-space with fixed boundary subject to the initial disturbances $u(x, 0) = u_0(x)$, $\dot{u}(x, 0) = \dot{u}_0(x)$ for $x \geq 0$, where $u_0(x)$ and $\dot{u}_0(x)$ are prescribed functions on $[0, \infty)$.

13.3 Find the displacement $u = u(x, t)$ in a homogeneous isotropic elastic half-space $x \geq 0$ with free boundary subject to the initial disturbances as in Problem 13.2.

13.4 Find the displacement $u = u(x, t)$ in a homogeneous isotropic elastic finite strip $0 \leq x \leq l$ fixed at $x = 0$ and subject to an impulsive unit traction at $x = l$ [i.e., $S(l, t) = \delta(t)$, where $S = S(x, t)$ is the stress associated with $u = u(x, t)$, and $\delta = \delta(t)$

is the delta function]. The initial conditions are assumed to be homogeneous, i.e., $u(x,0)=0$ and $\dot{u}(x,0)=0$ for $0\le x\le l$.

13.5 Let $x=0$ be an interface between two homogeneous isotropic elastic half-spaces of different material properties, and assume that (ρ_-,λ_-,μ_-) and (ρ_+,λ_+,μ_+) denote the material properties of the $(-\infty,0)$ and $(0,+\infty)$ half-spaces, respectively. Also, assume that an incident stress wave $S^{(i)}=S^{(i)}(t-x/c_-)$, where $S^{(i)}(s)=0$ for $s<0$ and $c_-=\sqrt{(\lambda_-+2\mu_-)/\rho_-}$, strikes the interface $x=0$ in such a way that it is completely reflected. Let $S^{(r)}=S^{(r)}(x,t)$ be the reflected stress wave. Show that the total stress $S=S(x,t)$ on $|x|<\infty$ and $t>0$ is represented by

$$S(x,t)=\begin{cases} S^{(i)}(t-x/c_-)+S^{(r)}(x,t) & \text{for } x\le 0\\ 0 & \text{for } x\ge 0\end{cases}$$

where

$$S^{(r)}(x,t)=-S^{(i)}(t+x/c_-)$$

This case occurs if the half-space $[0,\infty)$ is a vacuum.

13.6 Assume that an incident stress wave $S^{(i)}=S^{(i)}(t-x/c_-)$ strikes the interface $x=0$ between the two half-spaces introduced in Problem 13.5 in such a way that a part of $S^{(i)}$ is reflected and a part is transmitted across the interface. This case occurs when the total stress and the particle velocity are continuous at $x=0$ for every $t>0$. Show that the total stress wave propagating in the two half-space solid is represented by the formula

$$S(x,t)=\begin{cases} S^{(i)}\,(t-x/c_-)+S^{(r)}\,(t+x/c_-) & \text{for } x\le 0\\ S^{(t)}\,(t-x/c_+) & \text{for } x\ge 0\end{cases}$$

where $S^{(r)}=S^{(r)}(t+x/c_-)$ and $S^{(t)}=S^{(t)}(t-x/c_+)$ are the reflected and transmitted stress waves, respectively, given by

$$S^{(r)}(t+x/c_-)=c^{(r)}S^{(i)}(t+x/c_-)$$

$$S^{(t)}(t-x/c_+)=c^{(t)}S^{(i)}(t-x/c_+)$$

and

$$c^{(r)}=-\frac{1-(\rho_+c_+)/(\rho_-c_-)}{1+(\rho_+c_+)/(\rho_-c_-)}$$

$$c^{(t)}=2\frac{(\rho_+c_+)/(\rho_-c_-)}{1+(\rho_+c_+)/(\rho_-c_-)}$$

$$c_+=\sqrt{\frac{\lambda_++2\mu_+}{\rho_+}},\qquad c_-=\sqrt{\frac{\lambda_-+2\mu_-}{\rho_-}}$$

The dimensionless constants $c^{(r)}$ and $c^{(t)}$ are called the *reflection and transmission coefficients* for the one dimensional stress wave motion in the nonhomogeneous two half-space medium.

Hint: Use the relation $S(x,t) = -\rho c\dot{u}(x,t)$ from Problem 13.1 for a stress wave propagating in the positive direction of x with a velocity c to satisfy the particle velocity continuity condition at the interface $x=0$ [13].

13.7 Find the dynamic thermal stresses produced by a plane source of heat that varies harmonically with time in an infinite elastic solid.
Hint: Assume the heat source function as $Q(x,t) = e^{i\omega t}\delta(x)$, where $\omega > 0$ is a prescribed frequency, and use Equations 13.2.15 and 13.2.18.

13.8 Find the dynamic thermal stresses produced by a plane heat source of the form $Q(x,t) = H(t)\delta(x)$ in an infinite elastic solid which is initially at rest. Here $H = H(t)$ and $\delta = \delta(x)$ are the Heaviside and Dirac delta functions, respectively.

13.9 Find the dynamic thermal stresses in a half-space $x \geq 0$ subject to the boundary heating $T(0,t) = T_0 t^2 \exp(-at)$ $(a > 0, T_0 > 0)$ when the boundary is stress free, the body is initially at rest, and both the temperature $T = T(x,t)$ and the induced stress $S = S(x,t)$ vanish as $x \to \infty$ for $t > 0$.

13.10 An instantaneous nucleus of thermoelastic strain distributed over the plane $x = 0$ in an infinite solid can be identified with the temperature

$$T^*(x,t) = \delta(x)\,\delta(t) \quad \text{for } |x| < \infty,\ t > 0$$

Find the dynamic thermal stresses $S^* = S^*(x,t)$ produced by the nucleus provided the infinite body is initially at rest, that is, $S^*(x,0) = 0$, $\dot{S}^*(x,0) = 0$ for $|x| < \infty$.

13.11 Find the dynamic thermal stresses in a semispace $x \geq 0$ due to the action of an instantaneous nucleus of thermoelastic strain distributed over the plane $x = x_0 > 0$, when the boundary $x = 0$ is stress free, and the semispace is initially at rest.

13.12 Let $S = S(x, x_0; t)$ be the stress field obtained in Problem 13.11. Show that the dynamic thermal stress $S = S(x,t)$ produced in a semispace with free boundary by an arbitrary temperature field $T = T(x,t)$ on $[0,\infty) \times [0,\infty)$ takes the form

$$S(x,t) = \int_0^t \int_0^\infty S(x, x_0; t-\tau)\,T(x_0,\tau)\,dx_0\,d\tau$$

Also, use the formula to get the closed-form solution to Problem 13.9.

REFERENCES

1. I. N. Sneddon, *Fourier Transforms,* McGraw-Hill, New York, 1955, p. 32.
2. A. C. Eringen and E. S. Suhubi, *Elastodynamics*, vol. II, *Linear Theory*, Academic Press, New York, 1975, p. 464.
3. A. C. Eringen and E. S. Suhubi, *Elastodynamics*, vol. II, *Linear Theory*, Academic Press, New York, 1975, pp. 467–469.
4. H. S. Carslaw and J. C. Jaeger, *Operational Methods in Applied Mathematics*, Dover Publications, New York, 1963, p. 354.
5. H. S. Carslaw and J. C. Jaeger, *Operational Methods in Applied Mathematics*, Dover Publications, New York, 1963, p. 355.
6. R. B. Hetnarski and J. Ignaczak, Generalized thermoelasticity: Response of semi-space to a short laser pulse, *J. Thermal Stresses*, 17, 1994, 337–396.

7. V. I. Danilovskaya, The dynamic thermal stresses produced in an elastic semispace by a flux of radiation energy incident on the boundary of semispace, (in Russian), *Izv. Akad. Nauk SSSR*, 9, 1959, 129–132.
8. A. Gałka and R. Wojnar, One-dimensional dynamic thermal stresses generated in an elastic half-space by laser pulses, *J. Thermal Stresses*, 18, 1995, 113–140.
9. L. G. Hector Jr. and R. B. Hetnarski, Thermal stresses due to a laser pulse: The elastic solution, *J. Appl. Mech. (ASME)*, 63, March 1996, 38–46.
10. V. I. Danilovskaya, Thermal stresses in an elastic semispace due to a sudden heating of its boundary, *Prik. Mat. Mekh.*, 3, 1950, 14 (in Russian).
11. T. Mura, Thermal strains and stresses in a transient state, *Proc. Soc. Jap. Nat. Congress Appl. Mech.*, 2, 1952, 9.
12. J. Ignaczak, Recollections on Richard B. Hetnarski, *J. Thermal Stresses*, 21, 1998, 181–183.
13. J. D. Achenbach, *Wave propagation in Elastic Solids*, North Holland, Amsterdam, the Netherlands, 1990, pp. 26–29.

VERBA VOLANT, SCRIPTA MANENT

Appendix A: Coupled and Generalized Thermoelasticity

A.1 INTRODUCTION

In the classic uncoupled thermoelasticity, the field of the temperature is governed by the partial differential equation, called the Fourier equation of heat conduction, which does not contain any elastic terms. Moreover, since this equation is of parabolic type, the temperature effects described by the solutions are felt instantly up to infinity. This theory, the theory of uncoupled thermal stresses, is considered in Chapter 4. In Section 4.2.2, the field equations of dynamic theory of thermal stresses, in which there is no coupling between the mechanical and thermal fields, were introduced. In such a theory, the transient thermal stresses are produced by a time-dependent temperature field that satisfies a parabolic heat conduction equation which is separate from the dynamic displacement–temperature field equations (see Equations 4.2.56 and 4.2.58). This theory does not explain why, for example, a steel specimen as a result of a standard tension test gets hot.

As early as 1837, Duhamel [1] proposed a set of equations that would remedy this situation. However, the more extensive research in this area started a few decades ago. It was in 1956 that the theory of coupled thermoelasticity was introduced. Although this theory provides the coupling between the temperature field and the stress field, thus removing the paradox that elastic changes have no effect on the temperature, it uses the displacement–temperature equations of hyperbolic–parabolic type. These equations reveal the infinite speed of propagation of the effects, and thus, they are not consistent with experiments. Maxwell [2] was the first to note the hyperbolic nature of thermal disturbances in gases, which became known as the *second sound*. Beginning in 1967, theories of so-called *generalized thermoelasticity* started to appear.* In these theories, the governing equations of classic coupled thermoelasticity are modified in such a way that the resulting displacement–temperature field equations become of hyperbolic type and, therefore, predict finite speeds of propagation of thermo-elastic waves. In this Appendix, the basic field equations and some observations on these theories will be shown:

(i) "Coupled thermoelasticity," initiated in 1956 by Biot [5].
(ii) "Generalized thermoelasticity," proposed in 1967 by Lord and Shulman (the L–S theory) [6].
(iii) "Temperature rate dependent thermoelasticity," introduced in 1972 by Green and Lindsay (the G–L theory) [7].
(iv) "Thermoelasticity without energy dissipation," formulated in 1993 by Green and Naghdi (the G–N theory) [8].

* Description of some of these theories is contained in [3]. The theories of generalized thermoelasticity are extensively treated in a monograph dedicated entirely to this subject [4].

Since the theories (i)–(iv) are restricted to small interactions between thermal and mechanical fields, they describe, for example, a dynamics of a thermoelastic rod subject to a tensile test with thermal insulation within a pre-fracture range.

A.2 COUPLED THERMOELASTICITY

The governing equations of coupled thermoelasticity for a linear homogeneous isotropic material have the form:

The equation of motion

$$\operatorname{div}\mathbf{S} + \mathbf{b} = \rho\,\ddot{\mathbf{u}}, \quad \mathbf{S} = \mathbf{S}^T \tag{A.2.1}$$

The strain–displacement relation

$$\mathbf{E} = \frac{1}{2}\left(\nabla\mathbf{u} + \nabla\mathbf{u}^T\right) \tag{A.2.2}$$

Hooke's law

$$\mathbf{S} = 2\mu\mathbf{E} + \lambda(\operatorname{tr}\mathbf{E})\mathbf{1} - \gamma\, T\,\mathbf{1}, \quad T = \theta - \theta_0 \tag{A.2.3}$$

The energy equation

$$\nabla^2 T - \frac{1}{\kappa}\dot{T} - \frac{\gamma\,\theta_0}{k}\operatorname{tr}\dot{\mathbf{E}} = -\frac{Q}{\kappa} \tag{A.2.4}$$

Combining the first three equations leads to the displacement–temperature equation of motion

$$\mu\,\nabla^2\mathbf{u} + (\lambda + \mu)\nabla(\operatorname{div}\mathbf{u}) - \gamma\nabla T + \mathbf{b} = \rho\ddot{\mathbf{u}} \tag{A.2.5}$$

Equations A.2.4 and A.2.5 are the displacement–temperature equations of coupled thermoelasticity for a solid elastic body. For the full description of the problem, appropriate initial and boundary conditions for thermal and mechanical loads should be stated.

A.2.1 COUPLED PROBLEM FOR AN INFINITE BODY

As an example, it will be shown now an approach to a solution of a coupled problem of spherical stress and temperature waves in an infinite elastic space with a quiescent past [9].* It is assumed that there acts an instantaneous point source of heat $Q = Q_0\delta(t)\,\delta(R)$, where R is the radial distance from the source to a representative point in the space. Neglecting the body force $\mathbf{b}$ in Equation A.2.5 and writing the energy equation A.2.4 in slightly different form, the fundamental equations are

$$\mu\nabla^2\mathbf{u} + (\lambda + \mu)\nabla(\operatorname{div}\mathbf{u}) - \gamma\nabla T = \rho\ddot{\mathbf{u}} \tag{A.2.6}$$

* The contents of the paper are the topic of Section 3.3 of [10]. The section contains a more extensive presentation of the paper than that in the Appendix, as it covers the entire solution, the figures, and the analysis of the results.

and

$$\nabla^2 T - \frac{1}{\kappa}\dot{T} - \frac{\gamma\,\theta_0}{k}\,\mathrm{div}\,\dot{\mathbf{u}} = -\frac{Q(P,t)}{\kappa} \tag{A.2.7}$$

where P denotes the position of a representative point. Expressing the displacement vector $\mathbf{u}$ as a sum of two parts, an irrotational part and a solenoidal part [11],

$$\mathbf{u} = \nabla\phi + \nabla\times\boldsymbol{\Psi} \tag{A.2.8}$$

where
  ϕ is a scalar potential
  $\boldsymbol{\Psi}$ is a vector potential

and substituting Equation A.2.8 into Equations A.2.6 and A.2.7 we obtain

$$\begin{aligned} \nabla^2\phi - \frac{1}{c_1^2}\ddot{\phi} &= \frac{\gamma}{\lambda+2\mu}T \\ \nabla^2\Psi_i - \frac{1}{c_2^2}\ddot{\Psi}_i &= 0 \quad i=1,2,3 \\ \nabla^2 T - \frac{1}{\kappa}\dot{T} - \frac{\gamma\,\theta_0}{k}\nabla^2\dot{\phi} &= -\frac{Q(P,t)}{\kappa} \end{aligned} \tag{A.2.9}$$

where c_1 and c_2 are the speed of propagation of the elastic longitudinal wave and the speed of the shear wave, respectively.

Elimination of T between the first and the third of Equations A.2.9 results in a single differential equation for ϕ

$$\left(\nabla^2 - \frac{1}{\kappa}\frac{\partial}{\partial t}\right)\left(\nabla^2 - \frac{1}{c_1^2}\frac{\partial^2}{\partial t^2}\right)\phi - \frac{\gamma^2\,\theta_0}{(\lambda+2\mu)k}\nabla^2\dot{\phi} = -\frac{\gamma}{\lambda+2\mu}\frac{Q(P,t)}{\kappa} \tag{A.2.10}$$

The plan is to find the scalar potential ϕ from Equation A.2.10 and from it both the stresses and the temperature may be determined.

Substituting the expression for $Q(P,t)$, introducing a nondimensional *coupling parameter* ϵ defined as

$$\epsilon = \frac{\gamma^2\,\theta_0}{c\,\rho\,(\lambda+2\mu)}$$

and applying the Laplace transform to Equation A.2.10 subject to homogeneous initial conditions results in

$$\left[\left(\nabla^2 - \frac{p}{\kappa}\right)\left(\nabla^2 - \frac{p^2}{c_1^2}\right) - \frac{\epsilon}{\kappa}p\,\nabla^2\right]\bar{\phi} = -\frac{\gamma}{\lambda+2\mu}\frac{Q_0\,\delta(R)}{\kappa} \tag{A.2.11}$$

Treating Laplacians as numbers, and using the notation $h_1 = p/c_1$, where $c_1 = \sqrt{(\lambda+2\mu)/\rho}$, and $h_2 = \sqrt{p/\kappa}$, where p is the Laplace transform parameter, we find

$$\overline{\phi} = -\frac{\gamma}{\lambda+2\mu}\frac{Q_0}{\kappa}\frac{1}{\left(\nabla^2-h_1^2\right)\left(\nabla^2-h_2^2\right)-\epsilon h_2^2\nabla^2}\delta(R)$$

$$= -\frac{\gamma}{\lambda+2\mu}\frac{Q_0}{\kappa}\frac{1}{\left(\nabla^2-h_1^2\right)\left(\nabla^2-h_2^2\right)}\left[\frac{1}{1-h_2^2\epsilon\frac{\nabla^2}{\left(\nabla^2-h_1^2\right)\left(\nabla^2-h_2^2\right)}}\right]\delta(R) \qquad \text{(A.2.12)}$$

Now, the expression in square brackets is expanded in the power series in ϵ:

$$\overline{\phi} = -\frac{\gamma}{\lambda+2\mu}\frac{Q_0}{\kappa}\frac{1}{\left(\nabla^2-h_1^2\right)\left(\nabla^2-h_2^2\right)}\left\{1+h_2^2\epsilon\frac{\nabla^2}{\left(\nabla^2-h_1^2\right)\left(\nabla^2-h_2^2\right)}\right.$$

$$\left.+\left[h_2^2\epsilon\frac{\nabla^2}{\left(\nabla^2-h_1^2\right)\left(\nabla^2-h_2^2\right)}\right]^2+\cdots\right\}\delta(R) \qquad \text{(A.2.13)}$$

or

$$\overline{\phi} = A\left[\sum_{n=0}^{\infty}\frac{\left(h_2^2\epsilon\nabla^2\right)^n}{\left(\nabla^2-h_1^2\right)^{n+1}\left(\nabla^2-h_2^2\right)^{n+1}}\right][-4\pi\delta(R)] \qquad \text{(A.2.14)}$$

where

$$A = \frac{\gamma}{4\pi(\lambda+2\mu)}\frac{Q_0}{\kappa} \qquad \text{(A.2.15)}$$

Since for an infinite space [12]

$$\left(\nabla^2-h_1^2\right)\frac{\exp(-h_1R)}{R} = -4\pi\delta(R)$$
$$\left(\nabla^2-h_2^2\right)\frac{\exp(-h_2R)}{R} = -4\pi\delta(R) \qquad \text{(A.2.16)}$$

by combining these two equations we obtain

$$\frac{1}{\left(\nabla^2-h_1^2\right)\left(\nabla^2-h_2^2\right)}[-4\pi\delta(R)] = \frac{\exp(-h_1R)-\exp(-h_2R)}{\left(h_1^2-h_2^2\right)R} \qquad \text{(A.2.17)}$$

Next, replacing h_1^2 by $\omega_1 h_1^2$ and h_2^2 by $\omega_2 h_2^2$, where ω_1 and ω_2 are nondimensional parameters, we get

$$\frac{1}{\left(\nabla^2-\omega_1h_1^2\right)\left(\nabla^2-\omega_2h_2^2\right)}[-4\pi\delta(R)]$$

$$= \frac{\exp(-\sqrt{\omega_1}h_1R)-\exp(-\sqrt{\omega_2}h_2R)}{\left(\omega_1h_1^2-\omega_2h_2^2\right)R} \qquad \text{(A.2.18)}$$

The consecutive derivatives of $\left(\nabla^2 - \omega_1 h_1^2\right)^{-1}$ with respect to ω_1 are:

$$\begin{aligned}
\frac{\partial}{\partial \omega_1}\left(\nabla^2 - \omega_1 h_1^2\right)^{-1} &= 1 \cdot h_1^2 \left(\nabla^2 - \omega_1 h_1^2\right)^{-2} \\
\frac{\partial^2}{\partial \omega_1^2}\left(\nabla^2 - \omega_1 h_1^2\right)^{-1} &= 1 \cdot 2 \cdot \left(h_1^2\right)^2 \left(\nabla^2 - \omega_1 h_1^2\right)^{-3} \\
&\vdots \\
\frac{\partial^n}{\partial \omega_1^n}\left(\nabla^2 - \omega_1 h_1^2\right)^{-1} &= n!\, \left(h_1^2\right)^n \left(\nabla^2 - \omega_1 h_1^2\right)^{-n-1}
\end{aligned} \tag{A.2.19}$$

Similarly

$$\frac{\partial^n}{\partial \omega_2^n}\left(\nabla^2 - \omega_2 h_2^2\right)^{-1} = n!\, \left(h_2^2\right)^n \left(\nabla^2 - \omega_2 h_2^2\right)^{-n-1} \tag{A.2.20}$$

Substituting expressions from Equations A.2.19 and A.2.20 with $\omega_1 = \omega_2 = 1$ in Equation A.2.14 yields

$$\bar{\phi} = A\left.\left\{\sum_{n=0}^{\infty} \frac{\epsilon^n(\nabla^2)}{(n!)^2}\, \frac{1}{\left(h_1^2\right)^n}\, \frac{\partial^{2n}}{\partial \omega_1^n\, \partial \omega_2^n}\, \frac{1}{\left(\nabla^2 - \omega_1 h_1^2\right)\left(\nabla^2 - \omega_2 h_2^2\right)} [-4\pi\,\delta(R)]\right\}\right|_{\omega_1=\omega_2=1} \tag{A.2.21}$$

Substituting Equation A.2.18 into Equation A.2.21 and using $L^{-1}\{\bar{f}(p)\}$ for denoting the inverse Laplace transform, we obtain

$$\begin{aligned}
\phi(R,t) = L^{-1}\{\bar{\phi}(R,p)\} = A &\left\{\sum_{n=0}^{\infty} \frac{\epsilon^n(\nabla^2)}{(n!)^2}\, \frac{1}{R}\right. \\
&\left.\left. \times \frac{\partial^{2n}}{\partial \omega_1^n\, \partial \omega_2^n} L^{-1}\left[\frac{\exp\left(-\sqrt{\omega_1} h_1 R\right) - \exp\left(-\sqrt{\omega_2} h_2 R\right)}{\left(h_1^2\right)^n \left(\omega_1 h_1^2 - \omega_2 h_2^2\right) R}\right]\right\}\right|_{\omega_1=\omega_2=1}
\end{aligned} \tag{A.2.22}$$

To make Equation A.2.22 more suitable for inverse Laplace transformation, the following relations are used:

$$(\nabla^2)^n \frac{\exp\left(-\sqrt{\omega_1} h_1 R\right)}{R} = \omega_1^n h_1^{2n} \frac{\exp\left(-\sqrt{\omega_1} h_1 R\right)}{R} \tag{A.2.23}$$

and

$$(\nabla^2)^n \frac{\exp\left(-\sqrt{\omega_2} h_2 R\right)}{R} = \omega_2^n h_2^{2n} \frac{\exp\left(-\sqrt{\omega_2} h_2 R\right)}{R} \tag{A.2.24}$$

Substitution of Equations A.2.23 and A.2.24 into Equation A.2.22 yields

$$\phi(R,t) = L^{-1}\{\overline{\phi}(R,p)\} = A\left\{\sum_{n=0}^{\infty} \frac{\epsilon^n(\nabla^2)}{(n!)^2}\frac{1}{R}\right.$$

$$\left.\times\frac{\partial^{2n}}{\partial\omega_1^n\,\partial\omega_2^n}L^{-1}\left[\frac{\omega_1^n h_1^{2n}\exp\left(-\sqrt{\omega_1}h_1 R\right) - \omega_2^n h_2^{2n}\exp\left(-\sqrt{\omega_2}h_2 R\right)}{\left(h_1^2\right)^n\left(\omega_1 h_1^2 - \omega_2 h_2^2\right)R}\right]\right\} \quad \text{(A.2.25)}$$

with the expression in braces taken at $\omega_1 = \omega_2 = 1$. This represents a power series with respect to the coupling parameter ϵ. The first term, for $n = 0$, yields the solution to the classic (uncoupled) problem, and the following terms provide the effects of the coupling. In a nondimensional setting in which the units of Section 11.2.1 are used, the first term of Equation A.2.25 coincides with a potential of uncoupled dynamic thermoelasticity given by Equation 11.2.71.

We recall that the Laplace transform parameter p appears in h_1 and h_2. After performing the indicated operations, both ω_1 and ω_2 are made equal one. For inversion of Laplace transforms, some of the formulas of Laplace transform (Part C) provided in the front matter are needed. Once $\phi(R,t)$ is found, the stresses and the temperature are calculated from the formulas

$$S_{RR} = -\frac{4\mu}{R}\frac{\partial\phi}{\partial R} + \rho\frac{\partial^2\phi}{\partial t^2} \quad \text{(A.2.26)}$$

$$S_{\varphi\varphi} = S_{\vartheta\vartheta} = -2\mu\left(\frac{1}{R}\frac{\partial\phi}{\partial R} + \frac{\partial^2\phi}{\partial R^2}\right) + \rho\frac{\partial^2\phi}{\partial t^2} \quad \text{(A.2.27)}$$

$$S_{R\varphi} = S_{\varphi\vartheta} = S_{R\vartheta} = 0 \quad \text{(A.2.28)}$$

$$T = \frac{\lambda + 2\mu}{\gamma}\left(\frac{\partial^2\phi}{\partial R^2} + \frac{2}{R}\frac{\partial\phi}{\partial R} - \frac{1}{c_1^2}\frac{\partial^2\phi}{\partial t^2}\right) \quad \text{(A.2.29)}$$

This approach to one of basic problems of coupled and generalized thermoelasticity illustrates the fact that tackling the problems formulated by these theories is not a simple matter, and only few such problems have been solved in a closed form so far.

A.3 GENERALIZED THERMOELASTICITY WITH ONE RELAXATION TIME

A.3.1 The Lord–Shulman Theory, or the L–S Theory

An inhomogeneous anisotropic thermoelastic solid with one relaxation time is described by the following field equations:

The geometric relations

$$\mathbf{E} = \frac{1}{2}\left(\nabla\mathbf{u} + \nabla\mathbf{u}^T\right) \quad \text{(A.3.1)}$$

The equation of motion

$$\operatorname{div}\mathbf{S} + \mathbf{b} = \rho\ddot{\mathbf{u}}, \quad \mathbf{S} = \mathbf{S}^T \quad \text{(A.3.2)}$$

The law of conservation of energy

$$\theta_0 \dot{\eta} = -\operatorname{div} \mathbf{q} + r \tag{A.3.3}$$

where r stands for the heat supply.

The constitutive relations

$$\mathbf{S} = \mathbf{C}[\nabla \mathbf{u}] + \mathbf{M}\theta \tag{A.3.4}$$

$$\theta_0 \, \eta = -\theta_0 \mathbf{M} \cdot \mathbf{E} + c\, T \tag{A.3.5}$$

$$\mathbf{q} + t_0 \dot{\mathbf{q}} = -\mathbf{K}\nabla T \quad \text{where } T = \theta - \theta_0, \;\; \theta_0 > 0 \tag{A.3.6}$$

where **C**, **M**, **K**, η, and c stand for the elasticity fourth-order tensor, the stress–temperature second-order tensor, the heat conductivity second-order tensor, the entropy field, and the specific heat scalar field, respectively, while t_0, ($t_0 > 0$), denotes the relaxation time in the L–S theory. Equation A.3.6 is also called the Maxwell–Cattaneo law, and the theory described by Equation A.3.1 through A.3.6 is sometimes referred to as the thermoelasticity with one relaxation time. It is shown [4] that the L–S theory obeys the governing equations of a linear extended thermoelasticity in which the first and the second laws of thermodynamics are satisfied identically. It follows from Equations A.3.1 through A.3.6 that by letting $t_0 = 0$ in Equation A.3.6 and by ignoring the first term on the RHS of Equation A.3.5, we obtain the governing equations of the dynamic theory of thermal stresses for an inhomogeneous anisotropic thermoelastic solid with no coupling between the mechanical and thermal fields. Also, by letting $t_0 = 0$ in Equations A.3.6 the system A.3.1 through A.3.6 represents the field equations of classic dynamic coupled thermoelasticity for an inhomogeneous anisotropic solid.

We are to recall some results from which it follows that an ordered array of functions $[\mathbf{u},\mathbf{E},\mathbf{S},T,\eta,\mathbf{q}]$ that comply with Equations A.3.1 through A.3.6 describes a thermoelastic wave propagating with a finite speed. The functions are defined on a Cartesian product on $\overline{B} \times [0, \infty)$, where $\overline{B}$ is a domain occupied by the L–S model, and $[0, \infty)$ is the time interval.

By elimination of four functions from the six that define a thermoelastic wave, one can obtain the field equations of the L–S theory in terms of various pairs of mechanical and thermal variables, such as $(\mathbf{u}, T)$, $(\mathbf{u}, \mathbf{q})$, $(\mathbf{S},T)$, and $(\mathbf{S}, \mathbf{q})$; and a pair of thermomechanical variables (,), formed from the variables that define a thermoelastic wave, corresponds to the wave if the remaining variables of the wave can be restored from the chosen pair. For example, a pair $(\mathbf{u},T)$ that satisfies the displacement–temperature field equations of the L–S theory, subject to suitable initial and boundary conditions, is a pair corresponding to a thermoelastic wave because it generates the fields $\mathbf{E}, \mathbf{S}, \eta$, and $\mathbf{q}$ in such a way that the ordered array of functions $[\mathbf{u}, \mathbf{E}, \mathbf{S},T,\eta, \mathbf{q}]$ represents a thermoelastic wave corresponding to an external thermomechanical load applied to the body $\overline{B}$ over the time interval $[0, \infty)$. An initial–boundary value problem for the pair $(\mathbf{u},T)$ in which the initial conditions are imposed on the displacement $\mathbf{u}$, velocity $\dot{\mathbf{u}}$, temperature T and temperature rate $\dot{T}$, is called the displacement–temperature characterization of a thermoelastic wave of the L–S theory. Similarly, a pair $(\mathbf{S}, \mathbf{q})$ that satisfies the stress–heat flux field equation of the L–S theory subject to suitable stress–heat flux initial and boundary conditions is a pair corresponding to a thermoelastic wave because it generates the fields $\mathbf{u}, \mathbf{E}, T$, and η in such a way that the array of functions $[\mathbf{u}, \mathbf{E}, \mathbf{S}, T, \eta, \mathbf{q}]$ represents a thermoelastic wave of the L–S theory [13]. In particular, a mixed displacement–temperature initial–boundary value problem for an inhomogeneous anisotropic body of the L–S theory, obtained by eliminating the fields $\mathbf{E}$, $\mathbf{S}$, η, and $\mathbf{q}$ from Equations A.3.1 through A.3.6 reads:

Find a pair $(\mathbf{u}, T)$ that satisfies the field equations

$$\begin{aligned} &\operatorname{div} \mathbf{C}[\nabla \mathbf{u}] - \rho \ddot{\mathbf{u}} + \operatorname{div}(T\mathbf{M}) = -\mathbf{b} \\ &\operatorname{div}(\mathbf{K}\nabla T) - c\dot{\widehat{T}} + \theta_0 \mathbf{M} \cdot \nabla \dot{\widehat{\mathbf{u}}} = -\widehat{r} \quad \text{on } B \times [0, \infty) \end{aligned} \tag{A.3.7}$$

the initial conditions

$$\begin{aligned} &\mathbf{u}(\cdot, 0) = \mathbf{u}_0, \quad \dot{u}(\cdot, 0) = \dot{\mathbf{u}}_0 \\ &T(\cdot, 0) = \vartheta_0, \quad \dot{T}(\cdot, 0) = \dot{\vartheta}_0 \quad \text{on } B \end{aligned} \tag{A.3.8}$$

and the boundary conditions

$$\begin{aligned} \mathbf{u} = \mathbf{u}' &\quad \text{on } \partial B_1 \times (0, \infty) \\ (\mathbf{C}[\nabla \mathbf{u}] + \mathbf{M}T)\mathbf{n} = \mathbf{s}' &\quad \text{on } \partial B_2 \times (0, \infty) \\ T = T' &\quad \text{on } \partial B_3 \times (0, \infty) \\ -(\mathbf{K}\nabla T) \cdot \mathbf{n} = q' &\quad \text{on } \partial B_4 \times (0, \infty) \end{aligned} \tag{A.3.9}$$

Here, the operator $\widehat{\ }$ is defined by

$$\widehat{f} = f + t_0 \dot{f} \tag{A.3.10}$$

for any function

$$f = f(\mathbf{x}, t) \quad \text{on } \overline{B} \times [0, \infty) \tag{A.3.11}$$

In Equation A.3.9 $(\partial B_1, \partial B_2)$ and $(\partial B_3, \partial B_4)$ are two partitions of the boundary ∂B of B such that

$$\begin{aligned} &\partial B = \partial B_1 \cup \partial B_2 = \partial B_3 \cup \partial B_4 \\ &\partial B_1 \cap \partial B_2 = \partial B_3 \cap \partial B_4 = \emptyset \end{aligned} \tag{A.3.12}$$

and $\mathbf{n} = \mathbf{n}(\mathbf{x})$ is the unit outward vector normal to ∂B at $\mathbf{x}$. Finally, $(\mathbf{u}_0, \dot{\mathbf{u}}_0, \vartheta_0, \dot{\vartheta}_0)$ in Equation A.3.8 and $(\mathbf{u}', \mathbf{s}', T', q')$ in Equation A.3.9 are prescribed systems of functions that determine the initial and boundary thermomechanical loads, respectively.

A displacement–temperature wave corresponding to the problem A.3.7 through A.3.9 is produced by an external thermomechanical load that is represented by the system of functions

$$(\mathbf{b}, r;\ \mathbf{u}_0, \dot{\mathbf{u}}_0, \vartheta_0, \dot{\vartheta}_0;\ \mathbf{u}', \mathbf{s}', T', \mathrm{q}') \tag{A.3.13}$$

Let $B(t)$ denote a support of the thermomechanical load for a fixed time t, that is, the set of points of $\overline{B}$ on which the load does not vanish over the time interval $[0, t]$. Let C, $(C > 0)$, be a constant of the velocity dimension that satisfies the inequality

$$C \geq \max(C_1, C_2) \tag{A.3.14}$$

where

$$\begin{aligned} C_1 &= \sup_{B,|\mathbf{m}|=1} \left\{ \frac{1}{2}\left(\frac{\theta_0}{\rho c}\right)^{1/2} |\mathbf{M}| + \left\{ |\mathbf{A}| + \left[\frac{1}{2}\left(\frac{\theta_0}{\rho c}\right)^{1/2} |\mathbf{M}|\right]^2 \right\}^{1/2} \right\} \\ C_2 &= \sup_{B} \left\{ \frac{1}{2}\left(\frac{\theta_0}{\rho c}\right)^{1/2} |\mathbf{M}| + \left\{ \frac{|\mathbf{K}|}{t_0 c} + \left[\frac{1}{2}\left(\frac{\theta_0}{\rho c}\right)^{1/2} |\mathbf{M}|\right]^2 \right\}^{1/2} \right\} \end{aligned} \tag{A.3.15}$$

Here, $\sup_B\{f(\mathbf{x})\}$ denotes a maximum of a function $f = f(x)$ on $\overline{B}$, and $\mathbf{A}(\mathbf{x}, \mathbf{m})$ is the second-order *acoustic tensor in the propagation direction* $\mathbf{m}$ which is defined for any unit vector $\mathbf{m}$ and any vector $\mathbf{a}$ by the relation (see Equation 11.1.6)

$$\mathbf{A}(\mathbf{x}, \mathbf{m})\mathbf{a} = \rho^{-1}(\mathbf{x})\mathbf{C}[\mathbf{a} \otimes \mathbf{m}]\mathbf{m} \tag{A.3.16}$$

We note that Equation A.3.16 is an extension of Equation 11.1.6 to include an inhomogeneous anisotropic thermoelastic solid.

Let $S(\mathbf{x}, Ct)$ denote an open ball in E^3 with radius Ct and center at $\mathbf{x}$. We shall call the *domain of influence of the thermomechanical load at the instant t for the mixed problem* A.3.7 through A3.9 the set

$$B^*(t) = \{\mathbf{x} \in \overline{B} : B(t) \cap \overline{S(\mathbf{x}, Ct)} \neq \emptyset\} \tag{A.3.17}$$

where C is defined by Equations A.3.14 and A.3.15. The following theorem shows that on $[0, t]$ the thermomechanical load of the mixed problem has no influence on points outside of $B^*(t)$.

Theorem 1: (Domain of Influence Theorem for Mixed Displacement–Temperature Problems of the L–S Theory.) [14]. Let $(\mathbf{u}, T)$ be a solution to the problem A.3.7 through A.3.9. Then

$$\mathbf{u} = \mathbf{0}, \quad T = 0 \quad \text{on } \{\overline{B} - B^*(t)\} \times [0, t] \tag{A.3.18}$$

■

Theorem 1 implies that for a finite time t and for a bounded support of the thermomechanical load $B(t)$, the thermoelastic disturbance generated by the pair $(\mathbf{u}, T)$ satisfying A.3.7 through A.3.9 vanishes outside of a bounded domain $B^*(t)$ that depends on the load support, the bounds for the thermomechanical constitutive fields, and the relaxation time t_0. This theorem also shows that the thermoelastic disturbance propagates as a wave from the domain $B(t)$ with a finite speed equal to or less than the speed C defined by the relations A.3.14 and A.3.15. It follows from the definition of C that $C \to \infty$ as $t_0 \to 0$. Therefore, if the relaxation time tends to zero, the thermoelastic disturbance described by $(\mathbf{u}, T)$ gains an infinite speed, as should be expected, since in this case the L–S theory reduces to the coupled hyperbolic–parabolic thermoelasticity. The definition of velocity C also implies that for a particular inhomogeneous and anisotropic thermoelastic solid of the L–S theory in which the acoustic and conductivity tensor fields are relatively small, that is, when

$$|\mathbf{A}| \ll \frac{1}{4}\left(\frac{\theta_0}{\rho c}\right)|\mathbf{M}|^2, \quad |\mathbf{K}| \ll \frac{1}{4}\left(\frac{t_0\theta_0}{\rho}\right)|\mathbf{M}|^2 \tag{A.3.19}$$

the maximum speed of a thermoelastic wave is

$$C_0 = \sup_B \left\{\left(\frac{\theta_0}{\rho c}\right)^{1/2}|\mathbf{M}|\right\} \tag{A.3.20}$$

This formula shows that for an inhomogeneous anisotropic thermoelastic body in which the acoustic and heat conductivity tensor fields are relatively small, the maximum speed of a thermoelastic wave in the L–S theory is dominated by a suitably scaled stress–temperature tensor field. Also, note that if $|\mathbf{M}|$ is relatively small, the formula for C_1 reduces to that of a domain of influence theorem of classic isothermal elastodynamics [15,16], while the formula for C_2 reduces to that of a domain of influence theorem for an inhomogeneous anisotropic rigid heat conductor. Finally, for a finite value of $|\mathbf{M}|$, the velocities C_1 and C_2 represent upper bounds for the velocities of a quasi-mechanical and of a quasi-thermal wave, respectively, propagating in the inhomogeneous anisotropic L–S model.

A.4 TEMPERATURE RATE-DEPENDENT THERMOELASTICITY

A.4.1 THE GREEN–LINDSAY THEORY, OR THE G–L THEORY

The governing equation of the G–L theory for an inhomogeneous anisotropic thermoelastic solid are of the form:

The geometric relations

$$\mathbf{E} = \frac{1}{2}\left(\nabla\mathbf{u} + \nabla\mathbf{u}^T\right) \tag{A.4.1}$$

The equation of motion

$$\operatorname{div}\mathbf{S} + \mathbf{b} = \rho\ddot{\mathbf{u}}, \quad \mathbf{S} = \mathbf{S}^T \tag{A.4.2}$$

The law of conservation of energy

$$\theta_0\dot{\eta} = -\operatorname{div}\mathbf{q} + r \tag{A.4.3}$$

The constitutive relations

$$\mathbf{S} = \mathsf{C}[\nabla \mathbf{u}] + \mathbf{M}\left(T + t_1 \dot{T}\right) \tag{A.4.4}$$

$$\theta_0\, \eta = -\theta_0\, \mathbf{M} \cdot \mathbf{E} + c\left(T + t_0 \dot{T}\right) \tag{A.4.5}$$

$$\mathbf{q} = -\mathbf{K}\nabla T \tag{A.4.6}$$

Here, t_0 and t_1 are the relaxation times, and other symbols were defined earlier. We note that t_0 here is different from that of Equation A.3.6. The appearance of two relaxation times, t_0 and t_1, in Equations A.4.1 through A.4.6 constitutes a difference between the L–S theory and the G–L theory. The G–L theory is based on a generalized temperature rate-dependent dissipation inequality that is identically satisfied provided $t_1 \geq t_0 > 0$ and it follows from A.4.6 that the G–L model obeys also the classic dissipation inequality

$$\mathbf{q} \cdot \nabla T \leq 0 \tag{A.4.7}$$

Finally, note that the G–L model reduces to that of classic dynamical coupled thermoelasticity if $t_1 = t_0 = 0$. A mixed displacement-temperature initial–boundary value problem for an inhomogeneous anisotropic body of the G–L theory, obtained by eliminating the fields **E**, **S**, η, and **q** from Equations A.4.1 through A.4.6, reads as follows:

Find a pair $(\mathbf{u}, T)$ that satisfies the field equations

$$\begin{aligned} &\operatorname{div} \mathsf{C}[\nabla \mathbf{u}] - \rho \ddot{\mathbf{u}} + \operatorname{div}\left[\mathbf{M}\left(T + t_1 \dot{T}\right)\right] = -\mathbf{b} \\ &\operatorname{div}(\mathbf{K}\nabla T) - c\left(\dot{T} + t_0 \ddot{T}\right) + \theta_0 \mathbf{M} \cdot \nabla \dot{\mathbf{u}} = -r \quad \text{on } B \times [0, \infty) \end{aligned} \tag{A.4.8}$$

the initial conditions

$$\begin{aligned} &\mathbf{u}(\cdot, 0) = \mathbf{u}_0, \quad \dot{u}(\cdot, 0) = \dot{\mathbf{u}}_0 \\ &T(\cdot, 0) = \vartheta_0, \quad \dot{T}(\cdot, 0) = \dot{\vartheta}_0 \quad \text{on } B \end{aligned} \tag{A.4.9}$$

and the boundary conditions

$$\begin{aligned} \mathbf{u} &= \mathbf{u}' \quad \text{on } \partial B_1 \times (0, \infty) \\ \left[\mathsf{C}[\nabla \mathbf{u}] + \mathbf{M}(T + t_1 \dot{T})\right]\mathbf{n} &= \mathbf{s}' \quad \text{on } \partial B_2 \times (0, \infty) \\ T &= T' \quad \text{on } \partial B_3 \times (0, \infty) \\ -(\mathbf{K}\nabla T) \cdot \mathbf{n} &= q' \quad \text{on } \partial B_4 \times (0, \infty) \end{aligned} \tag{A.4.10}$$

Clearly, the set $B(t)$ from Section A.3 is also a support of the thermomechanical load at an instant t for the problem A.4.8 through A.4.10. A domain of influence of the thermomechanical load at an instant t for the mixed problem A.4.8 through A.4.10 is defined as

$$B^*(t) = \{\mathbf{x} \in \overline{B} : B(t) \cap \overline{S(\mathbf{x}, Ct)} \neq \emptyset\} \tag{A.4.11}$$

where C is a constant of the velocity dimension that satisfies the inequality

$$C \geq \max\left(C_1', C_2'\right) \tag{A.4.12}$$

in which

$$C_1' = \sup_{B,|\mathbf{m}|=1} \left\{ \frac{1}{2}\left(\frac{\theta_0}{\rho c}\right)^{1/2} |\mathbf{M}| + \left\{ |\mathbf{A}| + \left[\frac{1}{2}\left(\frac{\theta_0}{\rho c}\right)^{1/2} |\mathbf{M}|\right]^2 \right\}^{1/2} \right\}$$

$$C_2' = \sup_{B} \left\{ \frac{1}{2}\frac{t_1}{t_0}\left(\frac{\theta_0}{\rho c}\right)^{1/2} |\mathbf{M}| + \left\{ \frac{|\mathbf{K}|}{t_0\, c} + \left[\frac{1}{2}\frac{t_1}{t_0}\left(\frac{\theta_0}{\rho c}\right)^{1/2} |\mathbf{M}|\right]^2 \right\}^{1/2} \right\} \tag{A.4.13}$$

With regard to the mixed displacement–temperature problems characterized by Equations A.4.8 through A.4.10, the following theorem holds true:

Theorem 2: (Domain of Influence Theorem for Mixed Displacement–Temperature Problems of the G–L Theory.) Let $(\mathbf{u}, T)$ be a solution to the system A.4.8 through A.4.10. Then

$$\mathbf{u} = \mathbf{0}, \quad T = 0 \quad \text{on } \{\overline{B} - B^*(t)\} \times [0, t] \tag{A.4.14}$$

where $B^*(t)$ is given by the relations A.4.11 through A.4.13.* A physical interpretation of Theorem 2 is similar to that of Theorem 1. Moreover, the definition of C (see A.4.12 and A.4.13) implies that the velocities C_1' and C_2' correspond, respectively, to the maximum speed of a quasi-mechanical and of a quasi-thermal wave propagating in the G–L model; and for $|\mathbf{M}| = 0$ they reduce to the maximum speeds of a pure mechanical and a pure thermal wave, respectively. Also, for a particular inhomogeneous anisotropic thermoelastic solid in which the acoustic and heat conductivity tensor fields are relatively small, that is, when

$$|\mathbf{A}| \ll \frac{1}{4}\left(\frac{\theta_0}{\rho\, c}\right)|\mathbf{M}|^2, \quad |\mathbf{K}| \ll \frac{1}{4}\left(\frac{t_1}{t_0}\right)^2\left(\frac{t_0\theta_0}{\rho}\right)|\mathbf{M}|^2 \tag{A.4.15}$$

the maximum speed of a thermoelastic wave is

$$C_0' = \sup_{B} \left\{ \frac{t_1}{t_0}\left(\frac{\theta_0}{\rho\, c}\right)^{1/2} |\mathbf{M}| \right\} \tag{A.4.16}$$

In addition, if (C_1, C_2) stands for a pair of velocities in the L–S theory (see A.3.15), and if the thermomechanical constitutive fields of the L–S and the G–L models, such as $\mathbf{K}$ and $\mathbf{M}$, are identical, we have the results for $t_1 \geq t_0 > 0$

$$C_1' = C_1 \quad \text{and} \quad C_2' \geq C_2 \tag{A.4.17}$$

---

* The proof of this theorem may be found in [17].

and for $t_1 = t_0 > 0$

$$C_2' = C_2 \tag{A.4.18}$$

Therefore, the following observations are in order. If the supports of the thermomechanical load in a mixed displacement–temperature problem of the L–S and G–L theories are the same, then (i) the domain of influence of the G–L theory is not smaller than that of the L–S theory, and (ii) the domain of influence of the G–L theory coincides with that of the L–S theory if $t_1 = t_0$. ■

We note that Theorem 2 covers a conventional mixed displacement–temperature problems of the G–L theory in which the initial conditions are imposed on the pair $(\mathbf{u}, T)$. For a domain of influence theorem associated with a pure stress–temperature initial–boundary value problem of the G–L theory that admits initially distributed thermoelastic defects, the reader is referred to a survey article on the domain of influence results in generalized thermoelasticity [18].

Finally, we note that particular domain of influence results in the L–S and G–L theories have been obtained in a number of papers devoted to the potential temperature waves propagating in a homogeneous isotropic thermoelastic body. This type of waves is generated by a pair $(\mathbf{u}, T)$ in which $\mathbf{u} = \nabla\phi$, where ϕ is a scalar potential on $\overline{B} \times [0, \infty)$, and for a one-dimensional domain B it covers all transient plane thermoelastic waves in the L–S and G–L theories. In the one-dimensional initial–boundary value problem for a semispace $x \geq 0$ subject to a thermomechanical load on the boundary $x = 0$, a domain of influence is identified with the boundary layer $0 \leq x \leq v_2 t$, where v_2 is the greater speed of a decomposition theorem of the G–L theory.

A theory of one-dimensional thermoelastic waves produced by an instantaneous plane source of heat in homogeneous isotropic infinite and semi-infinite models of the G–L type was presented by the authors [19], and the application of this theory to the analysis of the response of a semi-infinite G–L model to short laser pulses was presented by the authors [20]. The mathematical theory of a potential-temperature initial–boundary value problem that accommodates asymptotic behavior of the waves analyzed by the authors [19] for $t \to \infty$ was presented by the late Professor Gaetano Fichera in two papers [21,22]. Also, a comparison between the wave forms observed experimentally in an aluminum plate subject to a high-power Nd-YAG laser pulse and the thermoelastic waves propagating in a G–L plate loaded by a laser-induced heat supply, was presented [23,24].

A.5 THERMOELASTICITY WITHOUT ENERGY DISSIPATION

A.5.1 THE GREEN–NAGHDI THEORY, OR THE G–N THEORY

The G–N model, proposed by Green and Naghdi [8], is described by the system of partial differential equations in which, in comparison to the classic thermoelastic system, the Fourier law of heat conduction is replaced by a heat flux rate–temperature gradient relation:

The geometric relations

$$\mathbf{E} = \frac{1}{2}\left(\nabla\mathbf{u} + \nabla\mathbf{u}^T\right) \tag{A.5.1}$$

The equation of motion

$$\operatorname{div}\mathbf{S} + \mathbf{b} = \rho\ddot{\mathbf{u}}, \quad \mathbf{S} = \mathbf{S}^T \tag{A.5.2}$$

The law of conservation of energy

$$\theta_0\dot{\eta} = -\operatorname{div}\mathbf{q} + r \tag{A.5.3}$$

The constitutive relations

$$\mathbf{S} = \mathsf{C}[\nabla\mathbf{u}] + \mathbf{M}\,T \tag{A.5.4}$$

$$\theta_0\,\eta = -\theta_0\,\mathbf{M}\cdot\mathbf{E} + c\,T \tag{A.5.5}$$

$$\dot{\mathbf{q}} = -\mathbf{K}^*\nabla T \tag{A.5.6}$$

Here, $\mathbf{K}^*$ is a symmetric positive definite second-order tensor field of the dimension

$$[\mathbf{K}^*] = [\mathbf{K}\,\mathrm{s}^{-1}] \tag{A.5.7}$$

where
 K is the heat conductivity second-order tensor field
 s stands for second

By eliminating the fields **E**, **S**, η, and **q** from Equations A.5.1 through A.5.6 the following mixed displacement–temperature initial–boundary value problem for a homogeneous anisotropic body of the G–N theory can be formulated:

Find a pair $(\mathbf{u}, T)$ that satisfies the field equations

$$\begin{aligned}&\operatorname{div}\mathsf{C}[\nabla\mathbf{u}] - \rho\ddot{\mathbf{u}} + \operatorname{div}(T\,\mathbf{M}) = -\mathbf{b}\\&\operatorname{div}(\mathbf{K}^*\nabla T) - c\ddot{T} + \theta_0\mathbf{M}\cdot\nabla\ddot{\mathbf{u}} = -\dot{r} \quad \text{on } B\times[0,\infty)\end{aligned} \tag{A.5.8}$$

the initial conditions

$$\begin{aligned}&\mathbf{u}(\cdot,0) = \mathbf{u}_0, \quad \dot{u}(\cdot,0) = \dot{\mathbf{u}}_0\\&T(\cdot,0) = \vartheta_0, \quad \dot{T}(\cdot,0) = \dot{\vartheta}_0 \quad \text{on } B\end{aligned} \tag{A.5.9}$$

and the boundary conditions

$$\begin{aligned}\mathbf{u} &= \mathbf{u}' \quad \text{on } \partial B_1\times(0,\infty)\\(\mathsf{C}[\nabla\mathbf{u}] + \mathbf{M}T)\mathbf{n} &= \mathbf{s}' \quad \text{on } \partial B_2\times(0,\infty)\\T &= T' \quad \text{on } \partial B_3\times(0,\infty)\\-(\mathbf{K}^*\nabla T)\cdot\mathbf{n} &= \dot{q}' \quad \text{on } \partial B_4\times(0,\infty)\end{aligned} \tag{A.5.10}$$

A displacement–temperature wave corresponding to the problem A.5.8 through A.5.10 is produced by the external thermomechanical load A.3.13 in which q' is replaced by a prescribed scalar field $\dot{q}'$. Also, since the displacement–temperature energy equation A.5.8$_2$ does not contain the temperature rate $\dot{T}$, a solution $(\mathbf{u}, T)$ to the problem represents an undamped thermoelastic wave, and this motivates the name of the G–N theory as thermoelasticity without energy dissipation. For the L–S model the displacement–temperature energy equation A.3.7$_2$ does contain the temperature rate $\dot{T}$. Similarly, $\dot{T}$ is included in the displacement–temperature energy equation in Equation A.4.8$_2$ of the G–L theory. This is a reason why the L–S and the G–L models represent materials transmitting damped thermoelastic waves. The existence of damped thermoelastic waves in the L–S and the G–L theories has been presented in a number of papers devoted to theoretical aspects [25,26] and application aspects [13,19,20] of these theories.

A domain of influence theorem for the displacement–temperature problem A.5.8 through A.5.10, similar to Theorems 1 and 2, has not been obtained so far, except for a homogeneous isotropic thermoelastic body. In the following, we recall a domain of influence result for the homogeneous isotropic case [27]. To this end, we replace the fields $\mathbf{C}, \mathbf{M}$, and $\mathbf{K}^*$ in Equations A.5.1 through A.5.10 by their isotropic counterparts, and note that physical properties of the model are represented by the material constants

$$[\rho,\ c;\ \lambda,\ \mu;\ k^*,\ \theta_0;\ \alpha] \tag{A.5.11}$$

subject to the inequalities

$$\begin{aligned} &\rho > 0,\ c > 0,\ \theta_0 > 0, \quad k^* > 0 \\ &\mu > 0,\ 3\lambda + 2\mu > 0, \quad \alpha \neq 0 \end{aligned} \tag{A.5.12}$$

We recall that ρ, c; λ, μ; θ_0; and α stand for the density, the specific heat, the Lamé constants; the reference temperature, and the coefficient of thermal expansion, respectively. The quantity k^* denotes a material constant characteristic for the theory. We note that $\mathbf{K}^* = k^*\,\mathbf{1}$. Also note that the set $B(t)$ of Section A.3, with q' replaced by $\dot{q}'$, is a support of the thermomechanical load at an instant t for the problem A.5.8 through A.5.10, restricted to the isotropic solid.

Consider next the set $B_r(t)$ of all points $\mathbf{x}$ of $\overline{B}$ whose distance to $B(t)$ is equal to or less than r, $(r > 0)$, that is,

$$B_r(t) = \left\{\mathbf{x} \in \overline{B} : B(t) \cap \overline{S(\mathbf{x}, r)} \neq \emptyset\right\} \tag{A.5.13}$$

where $\overline{S(\mathbf{x}, r)}$ is the closure of the ball of the radius r and the center at point $\mathbf{x}$. Since $B(t)$ is nonempty set and $\overline{B}$ is bounded, it follows that $B_r(t)$ is well defined for any $r \in [0, L(t)]$, where $L(t)$ is the maximum distance of the points of $\overline{B}$ to $B(t)$. Also, introduce a constant C of the velocity dimension by

$$C = C^*\sqrt{1+\sigma} \quad \text{where } C^* = \sqrt{\frac{\mu^*}{\rho}} \tag{A.5.14}$$

with

$$\sigma = \left\{a + \frac{1}{4}[1 - a(1+b)]^2\right\}^{1/2} - \frac{1}{2}[1 - a(1+b)] \tag{A.5.15}$$

and

$$\mu^* = \max[2\mu,\ 3\lambda + 2\mu] \tag{A.5.16}$$

with

$$a = \frac{3\gamma^2\theta_0}{c\,\mu^*} \quad \text{and } b = \frac{k^*\rho}{3\,\gamma^2\theta_0} \quad \text{with } \gamma = (3\lambda + 2\mu)\alpha \tag{A.5.17}$$

Then the following theorem holds true:

Theorem 3: (Domain of Influence Theorem for a Homogeneous Isotropic G–N Model.)* Let ($\mathbf{u}$, T) be a solution to the problem A.5.8 through A.5.10 restricted to a homogeneous isotropic solid. Then

$$\mathbf{u} = \mathbf{0},\ T = 0 \quad \text{on } \left\{\overline{B} - B_r(t)\right\} \times [0, t] \quad \text{for r} \geq C\,t \tag{A.5.18}$$

Theorem 3 is similar to that of the classic isothermal elastodynamics [28]. The analogy could be expected as both the isothermal elastic waves and the thermoelastic waves of the G–N theory propagate without energy dissipation. A uniqueness theorem for a natural stress–entropy flux initial–boundary value problem of the G–N exists [29]. The continuous dependence of a solution to a displacement–temperature initial–boundary value problem on the thermomechanical load in this theory was established [30] and the undamped character of one-dimensional thermoelastic waves in the G–N theory was also shown [31]. ■

A.6 CLOSING REMARKS

The four theories presented in Sections A.2 through A.5, that is, the coupled theory and the three theories of generalized thermoelasticity, are not the only ones that have been proposed during the recent decades in an attempt to create the theory with fields of stresses and the temperature coupled, and then to describe the thermoelastic waves in a solid. These theories are, however, in the authors' opinion, representative in discussing the subject.

It should be mentioned that the three theories of the generalized thermoelasticity discussed in the Appendix, that is, the L–S, the G–L and the G–N theory, have not been verified by a lab experiment up to date. To study particular thermoelastic waves described by a solution to an initial–boundary value problem in these theories, the approximate values of material functions are usually used. For example, to numerically discuss an analytical solution to a problem in the L–S theory, the value of the relaxation time is taken from experimental results for a rigid heat conductor, while the remaining material parameters are identified with those of an elastic body under isothermal or adiabatic conditions. In particular, finding materials that would comply with any of the three theories of the generalized thermoelasticity remains a challenge for experimental researchers in the field of thermoelastic waves. In the authors' opinion, the velocity formulas A.3.15 and A.4.13 should prove useful in a laboratory to find the genuine thermoelastic properties of models of L–S and G–L type, respectively.

---

* The proof of this theorem is given in the cited paper by Nappa [27].

As for the analytical description of the L–S, G–L, and G–N theories, there exists a unified presentation of these three theories that utilizes two unifier terms [32,33].*

REFERENCES

1. J.-M.-C. Duhamel, Second mémoire, sur les phenomenes thermo-mechaniques, *J. de L'École Polytechnique*, 15(25), 1837, 1–57.
2. J. C. Maxwell, On the dynamical theory of gases, *Philos. Trans. Royal Soc. London*, 157, 1867, 49–88.
3. R. B. Hetnarski and M. R. Eslami, *Thermal Stresses—Advanced Theory and Applications*, Springer, Berlin, Germany, 2009, Chaps. 2 and 8.
4. J. Ignaczak and M. Ostoja-Starzewski, *Thermoelasticity with Finite Wave Speeds*, Oxford University Press, Oxford, U.K., 2010.
5. M. Biot, Thermoelasticity and irreversible thermodynamics, *J. Appl. Phy.*, 27, 1956, 240–253.
6. H. Lord and Y. Shulman, Generalized dynamic theory of thermoelasticity, *J. Mech. Phys. Solids*, 15, 1967, 299–309.
7. A. E. Green and K. A. Lindsay, Thermoelasticity, *J. Elasticity*, 2, 1971, 1–7.
8. A. E. Green and P. M. Naghdi, Thermoelasticity without energy dissipation, *J. Elasticity*, 31, 1993, 189–208.
9. R. B. Hetnarski, Solution of the coupled problem of thermoelasticity in the form of series of functions, *Arch. Mech. Stos.*, 4, 1964, 919–941.
10. W. Nowacki, *Dynamic Problems of Thermoelasticity*, eds. P. H. Francis and R. B. Hetnarski, Nordhoff International Publishing, Leyden, and PWN—Polish Scientific Publishers, Warszawa, 1975, pp. 256–269.
11. W. Nowacki, *Thermoelasticity*, 2nd edn., Pergamon Press, Oxford, U.K., and PWN—Polish Scientific Publishers, Warszawa, 1986, p. 52.
12. P. M. Morse and H. Feshbach, *Methods of Theoretical Physics,* vol. 1, McGraw-Hill, New York, 1953.
13. J. Ignaczak, Generalized thermoelasticity and its applications, in: *Thermal Stresses III*, Ed. R. B. Hetnarski, North Holland, Amsterdam, the Netherlands, 1989, pp. 279–354.
14. J. Ignaczak, B. Carbonaro, and R. Russo, Domain of influence theorem in thermoelasticity with one relaxation time, *J. Thermal Stresses*, 9, 1986, 79–91.
15. M. E. Gurtin, *The Linear Theory of Elasticity*, Encyclopedia of Physics, chief editor: S. Flügge, vol. VIa/2, editor: C. Truesdell, Springer, Berlin, Germany, 1972.
16. A. C. Eringen and E. S. Suhubi, *Elastodynamics,* vol. 2, Academic Press, New York, 1975.
17. B. Carbonaro and J. Ignaczak, Some theorems in temperature-rate dependent thermoelasticity for unbounded domains, *J. Thermal Stresses*, 10, 1987, 193–220.
18. J. Ignaczak, Domain of influence results in generalized thermoelasticity—A survey, *Appl. Mech. Rev.*, 44, 1991, 375–382.
19. R. B. Hetnarski and J. Ignaczak, Generalized thermoelasticity: Closed-form solutions, *J. Thermal Stresses*, 16, 1993, 473–498.
20. R. B. Hetnarski and J. Ignaczak, Generalized thermoelasticity: Response of semi-space to a short laser pulse, *J. Thermal Stresses*, 17, 1994, 377–396.
21. G. Fichera, A boundary-value problem connected with response of semi-space to a short laser pulse, *Rend. Mat. Acc. Lincei*, 9(8), 1997, 197–228.
22. G. Fichera, On an initial-boundary value problem describing response of semi-space to a short laser pulse. *Uspehi Mat. Nauk*, 52(2), 1997, 191–192 (in Russian).
23. C. S. Suh and C. P. Burger, Thermoelastic modeling of laser-induced stress waves in plates, *J. Thermal Stresses*, 21, 1998, 829–847.

* This unified presentation is described in [34].

24. C. S. Suh and C. P. Burger, Effects of thermomechanical coupling and relaxation times on wave spectrum in dynamic theory of generalized thermoelasticity, *J. Appl. Mech.*, 65, 1998, 605–613.
25. J. D. Achenbach, The influence of heat conduction on propagating stress jumps, *J. Mech. Phys. Solids*, 16, 1968, 273–282.
26. J. Ignaczak, Decomposition theorem for thermoelasticity with finite wave speeds. *J. Thermal Stresses*, 1, 1978, 41–52.
27. L. Nappa, Spatial decay estimates for the evolution equations of linear thermoelasticity without energy dissipation, *J. Thermal Stresses*, 21, 1998, 581–592.
28. S. Chirita and R. Quintanilla, On Saint-Venant's principle in linear elastodynamics, *J. Elasticity*, 42, 1996, 201–215.
29. D. S. Chandrasekharaiah, A uniqueness theorem in the theory of thermoelasticity without energy dissipation, *J. Thermal Stresses*, 19, 1996, 267–272.
30. D. Iesan, On the theory of thermoelasticity without energy dissipation, *J. Thermal Stresses*, 21, 1998, 295–307.
31. D. S. Chandrasekharaiah, One-dimensional wave propagation in the linear theory of thermoelasticity without energy dissipation, *J. Thermal Stresses*, 19, 1996, 695–710.
32. A. Bagri and M. R. Eslami, A unified generalized thermoelasticity formulation; application to thick functionally graded cylinders, *J. Thermal Stresses*, 30, 2007, 911–930.
33. A. Bagri and M. R. Eslami, A unified generalized thermoelasticity: Solution for cylinders and spheres, *Int. J. Mech. Sci.*, 49, 2007, 1325–1335.
34. R. B. Hetnarski and M. R. Eslami, *Thermal Stresses—Advanced Theory and Applications*, Springer, Berlin, Germany, 2009, pp. 75–79.

Name Index

A

B

C

D

E

F

G

H

I

J

K

L

M

N

O

P

Q

R

S

T

V

W

Y

Subject Index

A

B

C

D

E

R

S

T

U

V

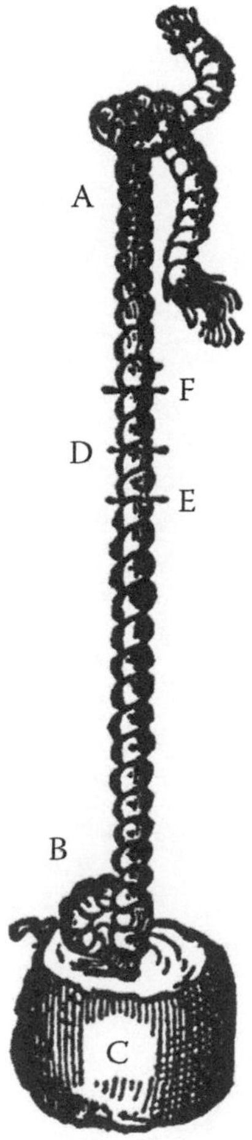
A
F
D
E
B
C

Printed and bound by CPI Group (UK) Ltd, Croydon, CR0 4YY
21/10/2024
01777100-0007